Peter Widmayer Francisco Triguero
Rafael Morales Matthew Hennessy
Stephan Eidenbenz Ricardo Conejo (Eds.)

Automata, Languages and Programming

29th International Colloquium, ICALP 2002
Málaga, Spain, July 8-13, 2002
Proceedings

 Springer

Series Editors

Gerhard Goos, Karlsruhe University, Germany
Juris Hartmanis, Cornell University, NY, USA
Jan van Leeuwen, Utrecht University, The Netherlands

Volume Editors

Peter Widmayer
Stephan Eidenbenz
ETH Zürich, Institute of Theoretical Computer Science, ETH Zentrum
8092 Zürich, Switzerland
E-mail: {widmayer, eidenben}@inf.ethz.ch

Francisco Triguero
Rafael Morales
Ricardo Conejo
University of Málaga, Department of Languages and Sciences of the Computation
E.T.S. de Ingeniería Informática, Campus de Teatinos, 29071 Málaga, Spain
E-mail: {triguero, morales, conejo}@lcc.uma.es

Matthew Hennessy
University of Sussex, School of Cognitive and Computing Sciences
Falmer, Brighton, BN1 9QN, United Kingdom
E-mail: matthewh@cogs.susx.ac.uk

Cataloging-in-Publication Data applied for

Die Deutsche Bibliothek - CIP-Einheitsaufnahme

Automata, languages and programming : 29th international colloquium ;
proceedings / ICALP 2002, Málaga, Spain, July 8 - 13, 2002. Peter Widmayer
... (ed.). - Berlin ; Heidelberg ; New York ; Barcelona ; Hong Kong ; London
; Milan ; Paris ; Tokyo : Springer, 2002
   (Lecture notes in computer science ; Vol. 2380)
   ISBN 3-540-43864-5

CR Subject Classification (1998):F, D, C.2-3, G.1-2, I.3, E.1-2

ISSN 0302-9743
ISBN 3-540-43864-5 Springer-Verlag Berlin Heidelberg New York

Springer-Verlag Berlin Heidelberg New York
a member of BertelsmannSpringer Science+Business Media GmbH

http://www.springer.de

© Springer-Verlag Berlin Heidelberg 2002
Printed in Germany

Typesetting: Camera-ready by author, data conversion by PTP-Berlin, Stefan Sossna e.K.
Printed on acid-free paper   SPIN 10870465   06/3142   5 4 3 2 1 0

Lecture Notes in Computer Science 2380

Edited by G. Goos, J. Hartmanis, and J. van Leeuwen

Springer

Berlin
Heidelberg
New York
Barcelona
Hong Kong
London
Milan
Paris
Tokyo

Preface

This volume contains the contributions of the 29th International Colloquium on Automata, Languages, and Programming (ICALP) held July 8–13, 2002 in Málaga, Spain. The diversity and the quality of the scientific program of ICALP 2002 serve as evidence that theoretical computer science with all its aspects continues to be a vibrant field and is increasingly turning into a valuable and abundant source of ideas and concepts for other research disciplines and applications.

ICALP has been the main annual scientific event of the European Association for Theoretical Computer Science (EATCS) for almost thirty years, making ICALP one of the best-established conferences in theoretical computer science and EATCS one of the world's largest and oldest theoretical computer science organizations.

To reflect the two parts of the EATCS journal Theoretical Computer Science (TCS), ICALP's technical program is split into two parts: while track A is devoted to algorithms, automata, complexity, and games, track B focuses on logic, semantics, and theory of programming.

In response to the call for papers, the program committee received 269 submissions from 34 countries: 204 for track A and 65 for track B. Each paper was reviewed by at least three program committee members who were often assisted by subreferees. The committee met on March 16, 2002 in Málaga and selected 83 papers (64 from track A and 19 from track B) for presentation. The program committee was overwhelmed by the quality and the quantity of submissions and had to reject many good submissions because of the limited number of available time-slots. We wish to thank all authors who submitted extended abstracts and all referees who helped in the thorough evaluation process.

The EATCS best paper award was given to Lars Engebretsen, Jonas Holmerin, and Alexander Russell for their track A contribution *Inapproximability Results for Equations over Finite Groups*. Seth Pettie's paper entitled *A Faster All-pairs Shortest Path Algorithm for Real-Weighted Sparse Graphs* was given the best student paper award for track A. Thomas Colombet received the best student paper award for his track B contribution *On Families of Graph Having a Decidable First Order Theory with Reachability*.

In addition to the accepted contributions, this volume contains the award-winning lectures by Maurice Nivat who received the EATCS award and by Géraud Sénizergues who was awarded the Gödel prize, as well as the invited lectures by Manuel Hermenegildo, Heikki Mannila, Madhav Marathe, Andrew Pitts, and John Reif.

The following workshops were held as satellite events of ICALP 2002 with Inmaculada Fortes and Llanos Mora as coordinators: Computability and Complexity in Analysis (CCA 2002), Algorithmic Methods and Models for Optimization of Railways (ATMOS 2002), 7th International Workshop on Formal Methods for Industrial Critical Systems, Foundations of Wide Area Network Computing, Formal Methods and Component Interaction, and Unification in Non-classical Logics. Another significant event in conjunction with ICALP 2002 was held on

July 11: the presentations of the projects funded by the "FET-Global Computing proactive initiative" (http://www.cordis.lu/ist/fetgc.htm) were given in the
Global Computing Event (organized by the IST of the EC).

We sincerely thank the organizers of ICALP 2002 for all their work (especially
José Luis Triviño for his work in the preparation of these proceedings), and the
Escuela Técnica Superior de Ingeniería Informática (Computer Science School)
of Málaga University (Spain) for their support in the organization.

We also thank the sponsors of ICALP 2002, namely Unicaja (the first savings bank of Andalucía), the University of Málaga, the Science and Technology
Minister (under contract TIC2000-3086-E), and Málaga City Council.

July 2002

Peter Widmayer
Francisco Triguero
Rafael Morales
Matthew Hennessy
Stephan Eidenbenz
Ricardo Conejo

Program Committee

Track A

Ricardo Baeza-Yates (U. Chile)
Volker Diekert (U. Stuttgart)
Paolo Ferragina (U. Pisa)
Catherine Greenhill (U. Melbourne)
Torben Hagerup (U. Frankfurt)
Johan Håstad (KTH, Stockholm)
Gabriel Istrate (Los Alamos)
Claire Kenyon (U. Paris XI)
Der-Tsai Lee (Acad. Sinica, Taipei)
Heikki Mannila (Nokia, Helsinki)
Elvira Mayordomo (U. Zaragoza)
Helmut Prodinger (U. Witwatersrand)
Jan van Leeuwen (U. Utrecht)
Paul Vitányi (CWI, Amsterdam)
Peter Widmayer (ETH Zürich) (Chair)
Gerhard Woeginger (T.U. Graz)
Christos Zaroliagis (U. Patras)

Track B

Martín Abadi (U. California, Santa Cruz)
Roberto Amadio (U. Provence)
Gilles Barthe (INRIA-SophiaAntipolis)
Manfred Droste (University of Technology Dresden)
Cédric Fournet (Microsoft Cambridge)
Matthew Hennessy (U. Sussex) (Chair)
Furio Honsell (U. Udine)
Peter O'Hearn (Queen Mary & Westfield C. London)
Fernando Orejas (U.P. Catalunya)
Ernesto Pimentel (U. Málaga)
David Sands (Chalmers University of Technology and Göteborg University)
Dave Schmidt (U. Kansas)
Gheorghe Stefanescu (U. Bucharest)
Vasco Vasconcelos (U. Lisbon)
Thomas Wilke (U. Kiel)

Organizing Committee

Buenaventura Clares (U. Granada)
Ricardo Conejo (U. Málaga)
Inmaculada Fortes (U. Málaga)
Llanos Mora (U. Málaga)
Rafael Morales (Co-chair, U. Málaga)
Marlon Núñez (U. Málaga)
José Luis Pérez de la Cruz (U. Málaga)
Gonzalo Ramos (U. Málaga)
Francisco Triguero (Co-chair, U. Málaga)
José Luis Triviño (U. Málaga).

List of Reviewers

Udo Adamy
Pankaj Agarwal
Susanne Albers
Jürgen Albert
Eric Allender
Helmut Alt
Ernst Althaus
Carme Alvarez
Klaus Ambos-Spies
Christoph Ambühl
Luzi Anderegg
Lars Arge
Stefan Arnborg
Vikraman Arvind
Tetsuo Asano
James Aspnes
Albert Atserias
Peter Auer
Vincenzo Auletta
Franz Aurenhammer
Holger Austinat
Laszlo Babai
Paolo Baldan
Cristina Bazgan
Paul Beame
Nick Benton
Martin Berger
Marco Bernardo
Valérie Berthé
Armin Biere
Avrim Blum
Luc Boasson
Hans Bodlaender
Bernard Boigelot
Ahmed Bouajjani
Julian Bradfield
Andreas Brandstädt
Hajo Broersma
Manuel Bronstein
David Bryant
Peter Bühlman

Harry Buhrman
Joshua Buresh-Oppenheim
Peter Bürgisser
Luis Caires
Carlos Caleiro
Cecile Capponi
Luca Cardelli
J. Orestes Cerdeira
Ioannis Chatzigiannakis
Otfried Cheong
Christian Choffrut
Marek Chrobak
Dave Clarke
Andrea Clementi
Bruno Codenotti
Ed Coffman jr.
Paolo Coppola
Andrea Corradini
Jordi Cortadella
José Félix Costa
Nadia Creignou
Pierluigi Crescenzi
Janos Csirik
Karel Culik
Arthur Czumaj
Silvano Dal Zilio
Victor Dalmau
Mark de Berg
Philippe de Groote
Robert de Simone
Sven de Vries
Erik Demaine
Luca Di Gaspero
Pietro Di Gianantonio
Gabriele di Stefano
Josep Diaz
Irit Dinur
Christoph Duerr
Francisco Durán
Jérôme Durand-Lose
Christoph Dürr

Abbas Edalat

Thomas Ehrhard

Stephan Eidenbenz

Susan Eisenbach

Lars Engebretsen

Leah Epstein

Thomas Erlebach

Juan Luis Esteban

Kousha Etessami

Uri Feige

Joan Feigenbaum

Stefan Felsner

Antonio Fernandez

Afonso Ferreira

Lisa Fleischer

Patrik Floreen

Mario Florido

Riccardo Focadi

Steven Fortune

Dimitris Fotakis

Pierre Fraigniaud

Antonio Frangioni

Kimmo Fredriksson

Zoltan Fueloep

Nicola Galesi

M. Mar Gallardo

Naveen Garg

Leszek Gasieniec

Paul Gastin

Ricard Gavalda

Raffaele Giancarlo

Christian Glasser

Gillem Godoy

Ashish Goel

Michel Goemans

Andreas Goerdt

Mikael Goldmann

Gaston Gonnet

Pavel Goralcik

Peter Grabner

Erich Grädel

Dima Grigoriev

Roberto Grossi

Sudipto Guha

Jörgen Gustavsson

Michel Habib

Reiner Haehnle

Gustav Hast

Lane. A Hemaspaandra

Hugo Herbelin

Montserrat Hermo

lrich Hertrampf

Alejandro Hevia

Daniel Hirschkoff

Tony Hoare

Frank Hoffmann

Jonas Holmerin

Kohei Honda

Hendrik Jan Hoogeboom

Han Hoogeveen

Juraj Hromkovic

Tsan-sheng Hsu

Michael Huth

Hans Huttel

Lucian Ilie

Guiseppe Italiano

Gabor Ivanyos

Radha Jagadeesan

David Janin

Alan Jeffrey

Mark Jerrum

David Johnson

David Juedes

Jan Juerjens

Marcin Jurdzinski

Christos Kaklamanis

Erich Kaltofen

Juhani Karhumäki

Marek Karpinski

Shmuel Katz

Steven Kautz

Dimitris Kavvadias

Rainer Kemp

Andrew Kennedy

Lefteris Kirousis

Daniel Kirsten

Marcos Kiwi

Arnold Knopfmacher

Ming-Tat Ko

Johannes Köbler

Mike Paterson
Gheorghe Paun
Andrzej Pelc
Ricardo Peña
Paolo Penna
José L. Pérez de la Cruz
Giuseppe Persiano
Anotine Petit
I. Petre
Sylvain Peyronnet
Cynthia A. Phillips
Andrea Pietracaprina
Nick Pippenger
Nadia Pisanti
Toniann Pitassi
Loic Pottier
Athanasios Poulakidas
Christophe Prieur
Kirk Pruhs
Geppino Pucci
Pavel Pudlak
David Pym
Elisa Quintarelli
Yuval Rabani
Mathieu Raffinot
Rajeev Raman
Ivan Rapaport
A. Ravara
S.S. Ravi
Ran Raz
Anna Redz
Mireille Regnier
Klaus Reinhardt
E. Remila
Arend Rensink
Antonio Restivo
Jürgen Richter-Gebert
James Riely
Eike Ritter
Hein Roehrig
Simona Ronchi della Rocca
Pierre Rosenstiehl
Tim Roughgarden
Claudio Russo
Tobias Ryden

Marie-France Sagot
Peter Sanders
José Carlos E. Santo
Guido Schaefer
Rene Schott
Rainer Schuler
Robert Sedgewick
Helmut Seidl
Géraud Sénizergues
Maria Serna
Jiri Sgall
Jeffrey Shallit
Ron Shamir
Mark Sheilds
Abhi Shelat
Alexander Shen
Sott Shenker
David Shmoys
M. Amin Shokrollahi
Klaas Sikkel
Riccardo Silvestri
Spiros Sioutas
Martin Skutella
Michiel Smid
Roberto Solis-Oba
Joel Spencer
Paul Spirakis
Christoph Sprenger
Renzo Sprugnoli
Anand Srivastav
Ludwig Staiger
Yannis Stamatiou
Robert Stärk
Frank Stephan
Colin Stirling
Jens Stoye
Ting-Yi Sung
Wojciech Szpankowski
Li Tan
Eva Tardos
Gerard Tel
Sebastiaan Terwijn
Denis Thérien
Mayur Thakur
Dimitrios Thilikos

Table of Contents

Molecular Assembly and Computation: From Theory to Experimental Demonstrations

John H. Reif*

Department of Computer Science, Duke University Box 90129,
Durham, NC 27708-0129.
reif@cs.duke.edu.

Abstract. While the topic of Molecular Computation would have appeared even a half dozen years ago to be purely conjectural, it now is an emerging subfield of computer science with the development of its theoretical basis and a number of moderate to large-scale experimental demonstrations. This paper focuses on a subarea of Molecular Computation known as *DNA self-assembly*. Self-assembly is the spontaneous self-ordering of substructures into superstructures driven by the selective affinity of the substructures. DNA provides a molecular scale material for effecting this programmable self-assembly, using the selective affinity of pairs of DNA strands to form DNA nanostructures. DNA self-assembly is the most advanced and versatile system known for programmable construction of patterned systems on the molecular scale. The methodology of DNA self-assembly begins with the synthesis of single-strand DNA molecules that self-assemble into macromolecular building blocks called DNA tiles. These tiles have sticky ends that match the sticky ends of other DNA tiles, facilitating further assembly into large structures known as DNA tiling lattices. In principal you can make the DNA tiling assemblies form any computable two- or three-dimensional pattern, however complex, with the appropriate choice of the tile's component DNA. This paper overviews the evolution of DNA self-assembly techniques from pure theory to experimental practice. We describe how some theoretical developments have made a major impact on the design of self-assembly experiments, as well as a number of theoretical challenges remaining in the area of DNA self-assembly. We descuss algorithms and software for the design, simulation and optimization of DNA tiling assemblies. We also describe the first experimental demonstrations of DNA self-assemblies that execute molecular computations and the assembly of patterned objects at the molecular scale. Recent experimental results indicate that this technique is scalable. Molecular imaging devices such as atomic force microscopes and transmission electron microscopes allow visualization of self-assembled two-dimensional DNA tiling lattices composed of hundreds of thousands of tiles. These assemblies can be used as scaffolding on which to position molecular electronics and robotics components with precision and specificity. The programmability lets this scaffolding have the patterning required for fabricating complex devices made of these components.

* John Reif is supported by DARPA/AFSOR Contract F30602-01-2-0561, NSF ITR Grant EIA-0086015, DARPA/NSF Grant CCR-9725021.

P. Widmayer et al. (Eds.): ICALP 2002, LNCS 2380, pp. 1–21, 2002.

1   Introduction

There is a long history of theoretical ideas in computer science that have led to major practical advances in experimental and applied computer science: for example, formal language and automata theory led to practical programming language design and parsing techniques, and number theoretic algorithms led to the development of public key cryptography forming the basis of many of the cryptographic protocols currently used on the internet.

This paper describes the development of the theory of self-assembly starting with a theory of tiling in the 1960's (that first established that tilings can form any computable 2D pattern), and its on-going evolution, including theoretical works on its parallel complexity as well as sophisticated stochastic and kinetic theories of self-assembly. We also describe our experimental demonstrations of DNA nanostructure self-assembly to execute computation and to form 2D lattices with regular patterns. See [Reif, et al 01] for a more detailed survey of current experimental work in self-assembled DNA nanostructures. Also, see [Reif, 98 and 02] for comprehensive surveys of the larger field of DNA computation (also known as biomolecular computation).

Motivation: The Need to Form Complex Patterned Objects at the Molecular Scale. As a motivating application of self-assembled nanostructures, we first briefly describe a major challenge to be faced by the field of computer science in the immediate future, namely the scale limitations of known fabrication techniques for microelectronics and MEMS(microelectrocal mechanical systems). We describe how bottom-up techniques based on self-assembly may be used to address and overcome this challenge.

Top-Down Techniques. The microelectronics industry currently uses optical lithography for constructing microelectronic and MEMS devices on silicon chips. Because of wavelength resolution limits, it is unlikely that optical lithography will scale much below a few nanometers. It has been projected that in approximately 15 years, this top-down method for patterning microelectronics will reach its ultimate limits and rate of progress in miniaturization of microelectronics will either halt or be very much reduced. It should be noticed that the use of lithography in microelectronics and MEMS is but one example of how our current engineering technology is based entirely on top-down manufacturing methods – whereas engineering progress at the molecular scale will require replacement of the current engineering technology by bottom-up self-assembly methods. Other top-down approaches to assembling nanoscale objects (such as microelectronics or MEMS) use ebeam lithography or a macroscale instrument, such as a scanning probe microscope, that can move and probe at molecular-size scales. Major obstacles to using such instruments to construct complex devices such as microelectronics at the molecular scale include the sequential nature of these technologies and their controllability and scalability. Although certain ebeam lithography systems and scanning probe microscopes can make use of a small amount of parallelism (e.g., a number of simultaneous probes), those numbers are dwarfed by the vast number of molecules that need to be manipulated.

To overcome this key problem of patterning structures below a few nanometers and into the molecular scale, we need new approaches. Known patterning methods used for manufacture at the microscale can be categorized as either top-down or bottom-up.

Bottom-Up methods for Self-assembly at the Molecular Scale. All known bottom-up approaches for patterned assembly rely on self-assembly. Self-assembly is the spontaneous self-ordering of substructures into superstructures driven by the selective affinity of the substructures. Self-assembly processes are well studied in biological systems such as the cell, in chemistry, and protein engineering. How can we program them to design new structures or to execute computations at the molecular scale?

(i) Cellular Self-Assembly. One can get inspiration from biological cells which operate on the nanoscale and use bottom-up methods for assembly. Cells perform a multiplicity of self-assembly tasks, including the self-assembly of cell walls (via lipids), of microtubules, etc. Many of these biological self-assembly processes utilize the specificity of ligand affinities to direct the self-assembly. However it is not at all easy to reprogram these biological cells for specialized assemblies at the nanoscale since the machinery available in biological cells is exceptionally difficult to predict and control, and we are only now beginning to understand the complexity of its control systems.

(ii) Chemical Self-Assembly. Self-assembly methods are well known in chemistry and have long been used for the self-assembly of lipid or polymer layers, but they generally result in structures that have limited complexity and are not readily programmable.

(iii) Protein Self-Assembly. Protein engineering is a bottom-up approach that offers great promise for molecular devices, but we do not yet fully understand the folding rules for protein assembly. The main difficulty here is that when you attempt to engineer proteins, you have at this time only a very limited degree of predictability in the resulting protein conformations.

(iv) DNA self-assembly. Finally, DNA self-assembly is a bottom-up approach which entails the spontaneous self-assembly of DNA strands into nanostructures driven by selective hybridization among DNA strands. As we shall see, in contrast to the other methods for self-assembly just mentioned, DNA self-assembly is readily programmable and has already been experimentally demonstrated.

Goals and Organization of this Paper. The goal of this paper is describe techniques for self-assembly of DNA tiling arrays and applications of this technology, including DNA computation. *Section 2* overviews the emerging theory of self-assembly starting from the theory of Domino Tiling Problems developed in the 1960s to some kinetic and stochastic theoretical models of tiling self-assembly processes. Turing-universality, NP completeness, and program-size complexity results self-assembly processes are cited. Also, we describe the parallel depth complexity of self-assembled tilings and in particular, linear self-assemblies which have been used in practice. *Section 3* describes the experimental demonstration of self-assembled DNA tiling lattices. This section introduces various classes of DNA nanostructures known as DNA tiles, and describe some the software developed for the design of the DNA strands composing these tiles. We also describe some 2D DNA tiling lattice assemblies and their visualization by atomic force and electron microscopes. *Section 4* describes our experimental demonstration of DNA tiling computations, which used linear DNA tiling assemblies. *Section 5* concludes the paper.

2   The Theory of Self-Assembly

This Section overviews the emerging theory of self-assembly. We note a number of techniques that have been used in the experimental demonstrations of DNA tiling assemblies, as described in the following Section 3.

Domino Tiling Problems. The theoretical basis for self-assembly has its roots in *Domino Tiling Problems*(also known as Wang tilings) as defined by Wang [Wang61] (Also see the comprehensive text [Grunbaum, et al, 87]). The input is a finite set of unit size square tiles, each of whose sides are labeled with symbols over a finite alphabet. Additional restrictions may include the initial placement of a subset of these tiles, and the dimensions of the region where tiles must be placed. Assuming an arbitrarily large supply of each tile, the problem is to place the tiles, without rotation (a criterion that cannot apply to physical tiles), to completely fill the given region so that each pair of abutting tiles have identical symbols on their contacting sides. (See Figure 1 for 'Smart Bricks' tiling assembly generalized to polygons.)

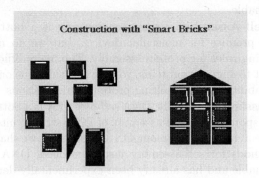

Fig. 1. A tiling assembly using 'Smart Bricks' with affinity between colored pads.

Theoretical Models of Tiling Self-assembly Processes. Domino tiling problems do not presume or require a specific process for tiling. Winfree [Winfree95] is responsible for invention of a key idea: self-assembly processes can be used for computation via the construction of DNA tiling lattices. (These will be further discussed in Section 3.) The sides of the tiles are assumed to have some methodology for selective affinity, which we call *pads*. Pads function as programmable binding domains, which hold together the tiles. Each pair of pads have specified binding strengths The self-assembly process is initiated by a singleton tile (the *seed tile*) and proceeds by tiles binding together at their pads to form aggregates known as *tiling assemblies*. The preferential matching of tile pads facilitates the further assembly into tiling assemblies.

Using the kinetic modeling techniques of physical chemistry, [Winfree98] developed a kinetic model for the self-assembly of DNA tiles. Following the classical literature of models for crystal assembly processes, [Winfree95] considers assembly processes where the tiling assembly is only augmented by singleton tiles (known in crystallography as *monomers*) which bind to the assembly at their tile pads. The likelihood of a particular tile binding at (or disassociating from) a particular site of the assembly is assumed to be a fixed probability dependent on that tile's concentration, the respective pad's binding affinity, and a temperature parameter. Winfree [W98] developed software for discrete time simulation of the tiling assembly processes, using approximate probabilities for the insertion or removal individual tiles from the assembly. These simulations gave an approximation to the kinetics of self-assembly chemistry and provided some validation of the feasibility of tiling self-assembly processes. Using this software as a basis, Yuan [Guangwei00] at Duke developed improved (sped up by use of an improved method for computing on/of likelihood suggested by Winfree) simulation software with a Java interface (http://www.cs.duke.edu/~yuangw/project/test.html) for a number of example tilings, such as string tilings for integer addition and XOR computations. In spite of an extensive literature on the kinetics of the assembly of regular crystalline lattices, the fundamental thermodynamic and kinetic aspects of self-assembly of tiling assemblies are still not yet well understood. For example, the affect of distinct tile concentrations and different relative numbers of tiles is not yet known; for this there is the possible application of Le Chatelier's principle.

[Adleman, et al 00] developed stochastic differential equation models for self-assembly of tiling assemblies and determined equilibrium probability distributions convergence rates for some 1-dimensional self-assemblies. His model allowed for binding between subassemblies and assumed a fixed probability for tile binding events independent of the size of tile assemblies. Since the movement of tile assemblies may depend on their size (and thus mass), this model might in the future be refined to make the probability for tile binding events dependent of the size of tile assemblies.

Meso-Scale Physical Simulation of Tile Assemblies. Pad binding mechanisms for the preferential matching of tile sides can be provided by various methods: (i) *molecular affinity*, using for example hydrogen bonding of complementary DNA or RNA bases, (ii) *magnetic attraction*, e.g., pads with magnetic orientations constructed by curing the polymer/ferrite composites in the presence of strong magnet fields, and also pads with patterned strips of magnetic orientations, (iii) *capillary force*, using hydrophobic/hydrophilic (capillary) effects at surface boundaries that generate lateral forces, (iv) *shape complementarity* (or conformational affinity), using the shape of the tile sides to hold them together. There is a variety of distinct materials for tiles, at a variety of scales:

(a) Molecular-Scale Tiling Assemblies have tiles of size up to a few hundred Angstroms. Specifically, DNA tiles will be the focus of our discussions in the following sections.

(a) Meso-Scale Tiling Assemblies have tiles of size a few millimeters up to a few centimeters. Whitesides at Harvard University has developed and tested multiple technologies [Zhao, et al, 98] [Xia et al, 98a,98b], [Bowden,et al 98], [Harder,et al 00] for meso-scale self-assembly, using capillary forces, shape complementarity, and magnetic forces (see http://www-chem.harvard.edu/GeorgeWhitesides.html). [Rothemund, 2000] also gave some meso-scale tiling assemblies using polymer tiles on fluid boundaries with pads that use hydrophobic/hydrophilic forces. A materials science group at the U. of Wisconsin also tested meso-scale self-assembly using magnetic tiles (http://mrsec.wisc.edu/edetc/selfassembly). These meso-scale tiling assemblies were demonstrated by a number of methods, including placement of tiles on a liquid surface interface (e.g., at the interface between two liquids of distinct density or on the surface of an air/liquid interface). These meso-scale tiling experiments have used mechanical agitation with shakers to provide a temperature setting for the assembly kinetics (that is, a temperature setting is made by fixing the rate and intensity of shaker agitation). These meso-scale tilings also have potential to illustrate fundamental thermodynamic and kinetic aspects of self-assembly chemistry.

Optimization of Tiling Assembly Processes. There are various techniques that may promote assembly processes in practice. One important technique is the tuning of the parameters (tile concentration, temperature, etc.) governing the kinetics of the process. [Adleman,et al 02] considers the problem of determining tile concentrations for given assemblies and conjectures this problem is $\sharp P$ complete. Nevertheless, the above tiling simulation software may be useful in the determination of heuristic estimates for parameters values such as tile concentration.

Various other techniques may improve convergence rates to the intended assembly. A blockage of tiling assembly process can occur if an incorrect tile binds in a unintended location of the assembly. While such a tile may be dislodged by the kinetics of subsequent time steps, it still may slow down the convergence rate of the tiling assembly process to the intended final assembly. To reduce the possibility of blockages of tiling assembly processes, [Reif97] proposed the use of distinct tile pads for distinct time steps during the assembly. [Reif97] also described the use of self-assembled tiling *nano-frames* to constrain the region of the tiling assemblies.

Turing-universal and NP Complete Self-assemblies. Domino tiling problems over an infinite domain with only a constant number of tiles were first proved by [Berger66] to be undecidable. This and subsequent proofs [Berger66, Robinson71, Wang75] rely on constructions where tiling patterns simulate single-tape Turing Machines or cellular arrays[Winfree95]. They require only a constant number of distinct tiles. These undecidability proof techniques allow [Winfree95] to show (he used the first and last layers of the assembly for input and output of computations, respectively) that computation by self-assembly is Turing-universal and so tiling self-assemblies can theoretically provide arbitrarily complex assemblies even with a constant number of distinct tile types. [Winfree,96] also demonstrated various families of assemblies which can be viewed as computing languages from families(e.g., regular, CFL, etc.) of the Chomsky hierarchy.

[LewisPapa81, Moore00] proved the NP-completeness of Domino tiling problems over polynomial-size regions. Subsequently, [Winfree96], [Jonoska97, Jonoska98] and [Lagoudakis and LaBean,99] proposed the use of self-assembly processes (in the context of DNA tiling and nanostructures) to solve NP complete combinatorial search problems such as SAT and graph coloring. However, the practical scale of these tiling methods to solve NP complete problems is limited to only moderate size problems at best.

Program-size Complexity of Tiling Self-assemblies. The programming of tiling assemblies is determined simply by the set of tiles, their pads, and the choice of the initial seed tile. A basic issue is the number of distinct tile types required to produce a specified tile assembly. The *program size complexity* of a specified tiling is the number of distinct tiles (with replacement) to produce it. [Rothemund and Winfree, 00b] show that the assembly of an $n \times n$ size square can be done using $\theta(\log n / \log \log n)$ distinct tiles. They also show that largest square uniquely produced by a tiling of a given number of distinct tiles grows faster than any computable function. [Adleman,et al 02] recently gave program size complexity bounds for tree shaped assemblies.

Massively Parallel Computation by Tiling. In computation by self-assembly, parallelism reveals itself in many ways. Each superstructure may contain information representing a different calculation (*global parallelism*). Due to the extremely small size of DNA strands, as many as 10^{18} DNA tiling assemblies may be made simultaneously in a small test tube. Growth on each individual superstructure may also occur at many locations simultaneously via *local parallelism*. The *depth* of a tiling superstructure is the maximum number of self-assembly reactions experienced by any substructure (the depth of the graph of reaction events), and the *size* of a superstructure is the number of tiles it contains. Likewise for the number of layers. For example, a superstructure consisting of an array of $n \times m$ tiles, where $n > m$ has depth m. Tiling systems with low depth, small size, and few layers are considered desirable, motivating the search for efficient computations performed by such systems. [Reif97] was the first to consider the parallel depth complexity of tiling assemblies and gave DNA self-assemblies of linear size and logarithmic depth for a number of fundamental problems (e.g., prefix computation, permutation, integer addition and subtraction, finite state automata simulation, and string fingerprinting) that form the basis for the design of many parallel algorithms. Furthermore, [Reif97] showed these elementary operations can be combined to perform more complex computations, such as bitonic sorting and general circuit evaluation with polylog depth assemblies.

Linear Self-Assemblies. Tiling systems that produce only superstructures with k layers, for some constant k, are said to use *linear self-assembly*. [Reif97] gave some simple linear tiling self-assemblies for integer addition as well as related operations (e.g., prefix XOR summing of n Boolean bits). These were the basis of the DNA tiling experiments of [Mao,00] that demonstrated the first example of DNA computation using DNA assembly, as described in Section 3.

These linear tilings were refined by [Winfree, Eng, and Rozenberg,00] to a class of String Tilings that have been the basis for further ongoing DNA tiling experiments of [LaBean, et al 00] of integer addition described in Section 3.

3   The Practice: Self-Assembly of DNA Tiling Lattices

DNA Hybridization. Single-strand DNA is a polymer that consists of a sequence of four types of bases grouped into two disjoint pairs known as Watson-Crick complementary pairs that can bind together through hydrogen bonding in an operation known as hybridization. DNA enjoys a unique advantage for a nanostructure construction material because two single strands of DNA can be designed and constructed by the experimental scientist to be selectively sticky and bind together to form doubly stranded DNA. (see Figure 2.) Hybridization is much more likely to occur if the DNA base sequences are complementary that is, if the component bases are Watson-Crick pairs and the temperature and salinity are set appropriately. The resulting doubly stranded DNA is relatively rigid and forms the well-known double-helix geometry. If the sticky single-strand segments that hybridize abut doubly stranded segments of DNA, you can use an enzymic reaction known as ligation to concatenate these segments.

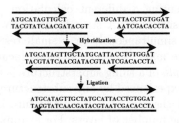

Fig. 2. Hybridization of sticky single-strand DNA segments.

DNA Nanostructures. [Seeman, 82] first pioneered DNA structure nanofabrication in the 1980s by assembling a multitude of DNA nanostructures (such as rings, cubes, and octahedrons) using DNA branched junctions and remains a leader in this area [Seeman, 98, Seeman et al 94, 98, 99]. However, these early DNA nanostructures weren't very rigid. To increase the rigidity of DNA nanostructures, Seeman made use of a DNA nanostructure known as a DNA crossover [Seeman, 82 and Seeman et al 89] (also known as a *branched Holiday junction*) which consists of two doubly stranded DNA, each having a single strand that crosses over to the other. Pairs of crossovers, known as double crossovers, provide a significant increase in rigidity of a DNA nanostructure. Also, certain crossovers (known as antiparallel crossovers) cause a reversal in the direction of strand propagation following the exchange of the strand to a new helix.

DNA Tiles. These are quite rigid and stable DNA nanostructures that are formed from multiple DNA antiparallel crossovers. DNA tiles typically have a

roughly rectangular geometry. These tiles come in multiple varieties that differ
from one another in the geometry of strand exchange and the topology of the
strand paths through the tile. The first DNA tiles developed [Winfree, et al
86,98] were known as double-crossover (DX) tiles and composed of two DNA
double helices with two crossovers. LaBean, Reif, and Seeman [LaBean, et al 00]
have developed some novel DNA tiles known as triple-crossover(TX) tiles that
are composed of three DNA double helices with four crossovers. These TX tiles
have properties that can facilitate one and two dimensional tiling assemblies and
computations. (See Figure 3.)

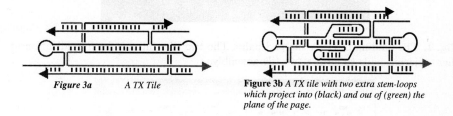

Figure 3a   A TX Tile
Figure 3b *A TX tile with two extra stem-loops*
which project into (black) and out of (green) the
plane of the page.

Fig. 3. (a) A triple-crossover tile and (b) a triple-crossover tile that has two extra
stem-loops that project into (black) and out of (green) the plane of the page.

Each DNA tile is designed to match the ends of certain other DNA tiles, a
process that facilitates the assembly into tiling lattices. In particular, DNA tiles
are designed to contain several short sections of unpaired, single-strand DNA
(ssDNA) extending from the ends of selected helices (often called 'sticky ends')
that function as programmable binding domains, which are the *tile pads*. Both
double- and triple-crossover tiles are useful for doing tiling assemblies. The DX
tiles provide up to four pads for encoding associations with neighboring tiles,
whereas the TX tiles provide up to six pads that are designed to function as
binding domains with other DNA tiles. Use of pads with complementary base
sequences provides control the neighbor relations of tiles in the final assembly.
In particular, the tile pads hybridize to the pads of other chosen DNA tiles.
Individual tiles interact by binding with other specific tiles through hybridization
of their pads to self-assemble into desired superstructures. (See Figure 4.)

Software for the optimized design of DNA tiles was first developed in Mathlab
by Winfree. This software used a greedy search method to optimize the choice
of DNA strands comprising the DNA tiles. The software was improved at Duke
Univ. by Bo [Bo01] to allow for a more sophisticated optimization heuristic
(providing improved sequence specificity of the DNA words used for tile pads,
minimizing the likelihood of incorrect hybridizations from non-matching pads),
to include more realistic models of DNA hybridization, and to provide a Java
interface. (see Figure 5.)

DNA Tiling Lattices. These are superstructures built up from smaller component structures(DNA tiles). Individual DNA tiles interact by annealing with
other specific tiles via their ssDNA pads to self-assemble into desired super-

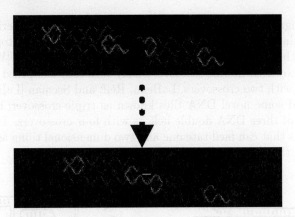

Fig. 4. The binding of DNA tile pad pairs. The two tiles interact by hybridization at their adjacent pads to form a two-tile assembly.

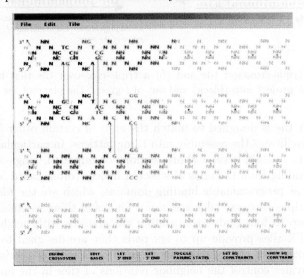

Fig. 5. An illustration of our Java software for design of a TX tile.

structures. These lattices can be either: **(a)** *non-computational,* containing a fairly small number of distinct tile types in a repetitive, periodic pattern; or **(b)** *computational,* containing a larger number of tile types with more complicated association rules which perform a computation during lattice assembly. The direct assembly of DNA lattices from component ssDNA has been demonstrated for non-computational DNA lattices described below.

Winfree and Seeman [Winfree, et al 98 and Mao, 99] demonstrated the self-assembly of two-dimensional periodic lattices consisting of at hundreds of thousands of double-crossover tiles, which is strong evidence of this approach's scalability. In addition, LaBean, Reif, and Seeman [LaBean, et al 00] have constructed DNA TX molecules which produced tiling lattices of even larger numbers of

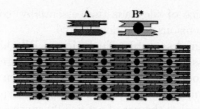

Fig. 6. AB* Array. *Lattice formed from two varieties of DNA tiles, including one (B*) containing an extra loop of DNA projecting out of the lattice plane, faciliting atomic force microscope imaging of the lattice.*

tiles. Both classes of self-assembled DNA lattices were observed through atomic force microscopy (AFM), a mechanical scanning process that provides images of molecular structures on a two-dimensional plate, as well as by use of transmission electron microscopy (TEM). Distinguishing surface features can be designed into individual tiles by slightly modifying the DNA strands composing the tiles. These modified DNA strands form short loops that protrude above the tile. (See Figure 6.)

To enhance definition, we have also affixed metallic balls to these DNA loops using known methods for affixing gold balls to DNA. Surface features – such as two-dimensional banding patterns – have been programmed into these DNA lattices by using DNA tiles that assemble into regular repetitive patterns.

These topographical features were observed on the DNA tiling lattices with atomic force and transmission electron microscopy imaging devices [Winfree et al., 98; Liu et al., 99; Mao, et al 99]. (See Figure 7.)

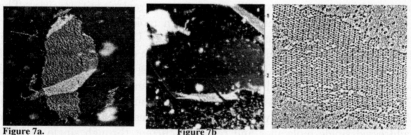

Figure 7a.
Figures 7a and 7b give AFM images of DNA lattices with TX tiles 3-4 microns on a side.

Figure 7b

Figure7c. TEM image of platinum rotary shadowed TX lattice.

Fig. 7. DNA tiling lattices.

These tiling assemblies had no fixed limit on their size. Recall that [Reif97] introduced the concept of a *nano-frame*, which is a self-assembled nanostructure that constrains the subsequent timing assembly (e.g., to a fixed size rectangle). A tiling assembly might be designed to be *self-delimitating* (growing to only a fixed size) by the choice of tile pads that essentially 'count' to their intended boundaries in the dimensions to be delimitated.

Directed Nucleation Assembly Techniques. We have recently developed another method for assembly of complex patterns, where an input DNA strand is synthesized that encodes the required pattern, and then specified tiles assemble around blocks of this input DNA strand, forming the required 1D or 2D pattern of tiles. This method makes the use of artificially synthesized DNA strands that specify the pattern and around which 2D DNA tiles assemble into the specified pattern; in this method, the permanent features of the 2D pattern are generated uniquely for each case. (See Figure 8.)

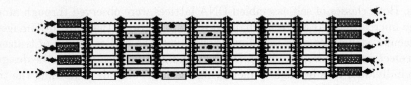

Fig. 8. In red is an input pattern strand of DNA. It encodes a 2D pattern in modified row major order (each odd row traversing alternately left to right, and each even row traversing right to left). Specific DNA tiles self-assemble around each segment of this input 'pattern' strand of DNA. Then the tiles self-assemble into a 2D tiling lattice with a pattern determined by the pattern strand. A small instance of this method has been successfully executed where up to 10 TX tiles assembled around a preformed scaffold DNA strand.

Application to Layout of Molecular-Scale Circuit Components. Molecular-scale circuits have the potential of replacing the traditional microelectronics with densities up to millions of times current circuit densities. There have been a number of recent efforts to design molecular circuit components ([Petty et al 95] [Aviram,Ratner98]). Tour at Rice Univ. in collaboration with Reed at Yale have designed and demonstrated [Chen et al 99] organic molecules that act as conducting wires [Reed et al.97],[Zhou99] and also rectifying diodes (showing negative differential resistance (NDR), and as well as [CRR+,99], [RCR+,00], and have the potential to provide dynamic random access memory (DRAM) cells. These generally use $\sim 1,000$ molecules per device, but they have also addressed single molecules and recorded current through single molecules [BAC+96], [RZM+97]. These molecular electronic components make conformational changes when they do electrical switching. One key open problem in molecular electronics is to develop molecular electronic components that exhibit restoration of a signal to binary values; one possible approach may be to make use of multi-component assemblies that exhibit cooperative thresholding. A key problem is to develop methods for assembling these molecular electronic components into a molecular scale circuit. Progress in the molecular circuit assembly problem could have revolutionary impact on the electronic industry, since it is one of the key problems delaying the development of molecular-scale circuits. As explained in the Introduction, the usual approach of laying out circuits by top-down techniques (e.g., lithography) is not practical at the molecular scale; instead bottom-up approaches (e.g., self-assembly) need to be used. Hence this may be a key area of application of DNA tiling assemblies. There are a number of possible methods for the selective attachment of the molecular electronic components to particular tiles of the DNA tiling array, using annealing: (i) using DNA as a selective assembly glue for linking chemistry between and molecular electronics [Tour and Bunz, 00], and (ii) the use of gold beads with attached DNA strands that can hybridize at selected locations of a self-assembled DNA arrays, so the molecular electronics components may self-assemble between the gold breads. Also, DNA lattices may be useful as a foundation upon which to grow nano-scale gold wires. This might be done by depositions of gold from colloid onto nano-spheres immobilized on DNA tiling lattices. Molecular probe devices may be used to test the electrical properties of the resulting molecular circuit attached to the DNA tiling array. *Computational* lattices (as opposed to regular, non-computational lattices), may also be employed to provide for the layout of highly complex circuits, e.g., the layout of the electronic components of an arithmetic unit. (For a discussion of possible schemes for encorporating molecular motors into tiling assemblies, see [Reif, et al 00].)

4 Computation by DNA Self-Assembly

Programming Self-Assembly of DNA Tilings. Programming DNA self-assembly of tilings amounts to the design of the pads of the DNA tiles (recall these are sticky ends of ssDNA that function as programmable binding domains, and that individual tiles interact by annealing with other specific tiles via their ssDNA pads to self-assemble into desired superstructures). The use of pads with complementary base sequences allows the neighbor relations of tiles in the final assembly to be intimately controlled; thus the only large-scale superstructures formed during assembly are those that encode valid mappings of input to output. The self-assembly approach for computation only uses four laboratory steps: (i) mixing the input oligonucleotides to form the DNA tiles, (ii) allowing the tiles to self-assemble into superstructures, (iii) ligating strands that have been co-localized, and (iv) then performing a single separation to identify the correct output.

The Speed of Computing via DNA Tiling Assemblies (compared with silicon-based computing.) The speed of DNA tiling assemblies is limited by the annealing time, which can be many minutes, and can be 10^{10} slower than a conventional computer. A DNA computation via self-assembly must take into account the fact that the time to execute an assembly can range from a few minutes up to hours. Therefore, a reasonable assessment of the power of DNA computation must take into account both the speed of operation as well as the degree of massive parallelism. Nevertheless, the massive parallelism (both within assemblies and also via the parallel construction of distinct assemblies) possibly ranging up to 10^{18} provides a potential that may be advantageous for classes of computational problems that can be parallelized.

String-Tiles: A Mechanism for Small-Depth Tiling. An approach for small-depth computations is to compress several tile layers into single tiles, so that the simplest form of linear self-assembly suffices. Linear self-assembly schemes for integer addition were first described by [Reif97]; in this scheme each tile performed essentially the operation of a single carry-bit logic step. This linear self-assembly approach works particularly well when the topology and routing of the strands in the DNA tiles is carefully considered, leading to the notion of string tiles. The concept of string tile assemblies derives from Eng's observation that allowing neighboring tiles in an assembly to associate by two sticky ends on each side, he could increase the computational complexity of languages generated by linear self-assembly. [Winfree99a] showed that by allowing contiguous strings of DNA to trace through individual tiles and the entire assembly multiple times, surprisingly sophisticated calculations can be performed with 1-layer linear assemblies of string tiles. The TAE tiles recently developed by [LaBean, et al 99] are particularly useful as string tiles.

Input/Output to Tiling Assemblies Using Scaffold and Reporter Strands. Recall that the TX tiles are constructed of three double-helices linked by strand exchange. The TX tiles have an interesting property, namely that certain distinguished single stranded DNA (to be called scaffold and reporter strands, respectively) wind through all the tiles of a tiling assembly. This property provides a more sophisticated method for input and output of DNA computations in string tiling assemblies. In particular, there are two types of ; the TAE tile contains an Even (and the TAO tiles contains an Odd) number of helical half-turns between crossover points. Even spacing of crossovers of the TAE tile allows reporter strands to stretch straight through each helix from one side of the tile to the other. These reporter segments are used for building up a long strand which records inputs and outputs for the entire assembly computations.

(a) Input via Scaffold Strands: We take as input the scaffold strands and which encode the data input to the assembly computation. They are long DNA strands capable of serving as nucleation points for assembly. Preformed, multimeric scaffold strands are added to the hybridization/annealing mixture in place of the monomeric oligo corresponding to the tile's reporter segment. The remaining portion of the component ssDNA comprising the tiles are also added. In the resulting annealing process, tiles assemble around the scaffold strand, automatically forming a chain of connected tiles which can subsequently be used as the input layer in a computational assembly.

(b) Output via Reporter Strands: After ligation of the tiling assembly (this joins together each tile's segments of the reporter strands), the reporter strand provides an encoding of the output of the tiling assembly computation (and typically also the inputs). Note this input/output can occur in parallel for multiple distinct tiling assemblies. Finally, the tiling assembly is disassembled by denaturing (e.g., via heating) and the resulting ssDNA Reporter Strands provide the result (these may be used as scaffold strands for later cycles of assembly computation, or the readout may be by PCR, restriction cutting, sequencing, or DNA expression chips).

One Dimensional DNA Tiling Computations for Parallel Arithmetic.
We now outline (See Figure 9.) procedures for using the string tiles described above that self-assemble into linear tiling assemblies to perform massively parallel arithmetic. [LaBean, et al 99] describes string tile systems that compute binary number addition (where the binary numbers are encoded by strands of DNA) by using two distinct sets of sticky-ends between adjacent tiles in the assembly to effectively communicate the values of the carry-bits. (They can also be used for computation of bit-wise XOR of Boolean vectors encoded by strands of DNA.) The assemblies result in the appending of these strands to the addition sums. For computations on specific inputs, these procedures make use of the scaffold strands mentioned above. The inputs are self-assembled strands of DNA composed of sequences DNA words encoding the pairs of binary numbers to be summed. Otherwise, the input tiles can be (using known techniques uses for the assembly of combinatorial libraries of DNA strands) randomly assembled and thereby generate a molecular look-up table in which each reporter strand encodes the random inputs and resultant outputs of a single calculation. After denaturing the assemblies back to individual strands, one may sample the resulting reporter strands to verify the outputs are correctly computed. A sufficient number of DNA tile molecules provide full coverage of all possible n-bit input strings. Such look-up tables may be useful as input for further computations as they represent a unique library of sequences with a complex structural theme. An experimental demonstration of an XOR tiling computation based on TAO tiles is reported in [Mao, LaBean, Reif, and Seeman, 00].

Two Dimensional DNA Tiling Computations. In the immediate future, it may be possible to extend the one dimensional DNA tiling assembly methods to two dimensional tilings, and to demonstrate these methods experimentally. One interesting goal is integer multiplication. The most direct and relatively straightforward way is to multiply via repeated additions and bit shifts, applying known VLSI systolic array architecture designs for integer multiplication. This would require a two dimensional $n \times n$ tiling assembly, with some increased complexity over the linear assembly for integer addition. Two dimensional computational tilings may also be used to do logical processing. [Lagoudakis and LaBean,99] proposed a 2D DNA self-assembly for Boolean variable satisfiability, which uses parallel construction of multiple self-assembling 2D DNA lattices to solve the problem. Such methods for solving combinatorial search problems do not scale well with the input size (the number of parallel tiling assemblies grows exponentially with the number of Boolean variables of the formula). However, similar constructions may be used for evaluating Boolean queries and circuits in massively parallel fashion, for multiple input settings of the input Boolean variable, and in this context it may be appropriate to consider the Boolean formulas a to be of fixed size.

Three Dimensional DNA Tiling Computations. There is a number of possible methods for executing computations experimentally on 3D DNA lattices, providing computations with (implicit) data movement in three dimensions. Matrix inner product might be executed by a three dimensional computational tiling by applying known VLSI systolic array architecture designs for matrix

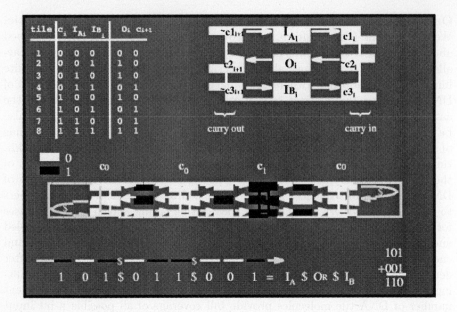

Fig. 9. String Tile Addition with TAE Building Blocks. *Upper left shows the truth table for addition; one tile type will be required for each row in the table. Upper right shows a schematic of a tile including the sequence positions for encoding values for input bits (IAi and IBi), the output bit (Oi), and the carry bit values on the tile's sticky-ends. The center schematic shows a five tile complex carrying out the addition of two 3-bit numbers. Arrows indicate the trace of the reporter strand as it winds through the entire assembly three times. The left and right extreme tiles act to reroute the reporter strand back through the lattice. The two corner tiles have been successfully built and shown to properly associate with one another.*

inner product. Another possible three dimensional computational tiling is that of the time-evolution (time is the third dimension of the tiling) of a 2D cellular automata (e.g., 2D cellular automata simulation of fluid flow).

5   Conclusion

The self-assembly of DNA tilings is a promising emerging method for molecular scale constructions and computation. We have overviewed the theory of DNA tiling self-assemblies and noted a number of open problems. We have discussed the potential advantages of self-assembly techniques for DNA computation; particularly the decreased number of laboratory steps required. We also discussed the potential broader technological impacts of DNA tiling lattices and identified some technological impacts of non-computational DNA assemblies: including their use as substrates for surface chemistry and particularly molecular electronics, robotics. Many of these applications are dependent on the further development of the appropriate attachment chemistry between DNA and the molecules attached to the arrays.

Error Control in DNA Tiling Assemblies. A chief challenge in DNA tiling self-assemblies is the control of assembly errors. As stated above, two dimensional self-assembled *non-computational* tilings have been demonstrated (and imaged via atom force microscopy) that involve up to a hundred thousand tiles. Certain of these appear to suffer from relatively low defect rates, perhaps in the order of less than a fraction of a percentage or less. But even such low error rates should be reduced. The factors influencing these defect rates are not yet well understood and there are no known estimates on the error-rates for self-assembled *computation* tilings, since such tilings have been achieved only very recently and have only been done on a very small scale(error rates appear to be less than 5% [Mao et al 00]). There is reason (see the construction of a potential assembly blockage described in [Reif, 98]) to believe that in computational tilings, defect errors may be more prevalent; and moreover, they can have catastrophic effects on the computation. Experiments need to be done to determine the error rates of the various types of self-assembly reactions, both computational and non-computational. There are a number of possible methods to decrease errors in DNA tilings: *(a) Annealing Temperature Optimization* is a well known technique used in hybridization and also crystallization experiments. It can be used decrease in defect rates at the expense in increased overall annealing time duration. In the context of DNA tiling lattices, the parameters for the temperature variation that minimize defects have not yet been determined. *(b) Error Control by Redundancy.* There are a number of ways to introduce redundancy into a computational tiling assembly. One simple method that can be developed for linear tiling assemblies, is to replace each tile with a stack of three tiles executing the same function, and then add additional tiles that essentially 'vote' on the pad associations associated with these redundant tiles. This results in a tiling of increased complexity but still linear size. This error resistant design can easily be applied to the integer addition linear tiling described above, and similar redundancy methods may be applied to higher dimension tilings. *(c) Error Control by step-wise assembly:* [Reif, 98] suggested the use of serial self-assembly to decrease errors in self-assembly. It is as yet uncertain which of these methods will turn out to be effective and it is likely that a combination of at least a few of the methods will prove most effective.

References

1. Adleman, L., Molecular Computation of Solution to Combinatorial Problems, Science, 266, 1021, (1994).
2. Adleman, L., Toward a mathematical theory of self-assembly, USC Tech Report, January 2000.
3. Adleman,L., Q. Cheng, A. Goel, M. Huang, D. Kempe, P. M. de Moisset and P. W. K. Rothemund, Combinatorial Optimization Problems in Self-Assembly, Thirty-Fourth Annual ACM Symposium on Theory of Computing, Montréal, Québec, Canada, May 19–21, 2002. http://www-scf.usc.edu/~pwkr/papers.html
4. Adleman,L., Qi Cheng, Ashish Goel, Ming-Deh Huang and Hal Wasserman: Linear Self-Assemblies: equilibria, entropy, and convergence rates. Abstract Draft April 2000.

5. Adleman,L., Q. Cheng, A. Goel, M. Huang, D. Kempe, P. M. de Moisset and P. W. K. Rothemund, Combinatorial Optimization Problems in Self-Assembly, *Thirty-Fourth Annual ACM Symposium on Theory of Computing*, Montréal, Québec, Canada, May 19–21, 2002.

6. Alivisatos, A.P., K.P. Johnsson, X. Peng, T.E. Wilson, C.J. Loweth, M. P. Bruchez Jr., P. G. Schultz, *Organization of 'nanocrystal molecules' using DNA*, Nature, bf 382, 609–611, August 1996.

7. Aviram and M. Ratner (Eds), Molecular Electronics: Science and Technology, Annals of the New York Academy of Sciences, New York, Vol. 852 (1998).

8. Berger, R. The Undecidability of the Domino Problem, Memoirs of the American Mathematical Society, 66 (1966).

9. Bo, Guo Master Thesis Computing by DNA Self-Assembly. Oct, 2001.

10. Bowden, N.; Brittain, S.; Evans, A. G., Hutchinson, J. W. and Whitesides, G. M. Spontaneous formation of ordered structures in thin films of metals supported on an elastomeric polymer, Nature 1998, 393, 146–149.

11. J. Chen, M. A. Reed, A. M. Rawlett, and J. M. Tour, Observation of a Large On-Off Ratio and Negative Differential Resistance in an Molecular Electronic Device, Science, Vol 286, 19, Nov. 1999, p 1550–1552.

12. E. Coven, N. Jonoska: DNA Hybridization, Shifts of Finite type and Tiling of the Integers, to appear in: Words, Sequences, Languages: where computer science and linguistics meet, published by Kluwer, edited by Carlos Martin-Vide (2001).

13. Du, S.M., S. Zhang and N.C. Seeman, DNA Junctions, Antijunctions and Meso-junctions, Biochem., 31, 10955-7pt963, (1992).

14. Eng, T., Linear DNA self-assembly with hairpins generates the equivalent of linear context-free grammars, 3rd DIMACS Meeting on DNA Based Computers, Univ. of Penn. (June, 1997).

15. Fu, T.-J., and N.C. Seeman, DNA Double Crossover Structures, Biochemistry, 32, 3211–3220, (1993).

16. Gehani, A., T. H. LaBean, and J.H. Reif, DNA-based Cryptography, 5th DIMACS Workshop on DNA Based Computers, MIT, June, 1999. DNA Based Computers, V, DIMACS Series in Discrete Mathematics and Theoretical CS (ed. E. Winfree), American Mathematical Society,
2000. www.cs.duke.edu/~reif/paper/DNAcrypt/crypt.ps

17. Grunbaum, S., Branko, and G.C. Shepard, Tilings and Patterns, H Freeman and Company, Chapter 11, (1987).

18. Guangwei,Y. "Simulation of DNA Self-assembly", MS Thesis, Duke University, 2000.

19. Harder, P.; Grunze, M.; Dahint, R.; Whitesides, G. M. and Laibinis, P. E., Molecular Conformation in Oligo(ethylene glycol)-Terminated Self-Assembled Monolayers on Gold and Silver Surfaces Determines Their Ability To Resist Protein Adsorption, J. Phys. Chem. B 1998,102, 426–436.

20. Jonoska, N., S. Karl, M. Saito: Creating 3-Dimensional Graph Structures With DNA in DNA based computer III (Editors: H. Rubin, D. Wood) DIMACS series in Discrete Math. and Theoretical Comp. Sci. vol 48 (1999) 123–136.

21. Jonoska, N., S. Karl, M. Saito: Three dimensional DNA structures in computing (to appear in BioSystems, 2001).

22. Jonoska, N., 3D DNA patterns and Computing (to appear) collection of papers on Patterns in Biology edited by M. Gromov and A. Carbone (to be published by World Scientific), 2001.

23. Jonoska, N., S. Karl: Ligation Experiments in Computing with DNA, Proceedings of 1997 IEEE International Conference on Evolutionary Computation (ICEC'97), April 13-16, (1997) 261–265.

24. Jonoska, N., S. Karl: A molecular computation of the road coloring problem, in DNA based computer II (Editors: L. Landwaber, E. Baum) DIMACS series in Discrete Math. and Theoretical Comp. Sci. vol 44 (1999) 87–96,

25. Jonoska, N., S. Karl, M. Saito: Graph structures in DNA computing in Computing with Bio-Molecules, theory and experiments, (editor Gh. Paun) Springer-Verlag (1998), 93–110.

26. LaBean, T. H., E. Winfree, and J.H. Reif, Experimental Progress in Computation by Self-Assembly of DNA Tilings, Proc. DNA Based Computers V: June 14-16, 1999. DIMACS Series in Discrete Mathematics and Theoretical Computer Science, E. Winfree and D. K. Gifford, editors, American Mathematical Society, Providence, RI, vol. 54, 2000, pp. 123–140. http://www.cs.duke.edu/~thl/tilings/labean.ps

27. LaBean, T. H., Yan, H., Kopatsch, J., Liu, F., Winfree, E., Reif, J.H. and Seeman, N.C., The construction, analysis, ligation and self-assembly of DNA triple crossover complexes, J. Am. Chem. Soc. 122, 1848–1860 (2000).
www.cs.duke.edu/~reif/paper/DNAtiling/tilings/JACS.pdf

28. Lagoudakis, M. G., T. H. LaBean, 2D DNA Self-Assembly for Satisfiability, 5th International Meeting on DNA Based Computers(DNA5), MIT, Cambridge, MA, (June, 1999). DIMACS Series in Discrete Mathematics and Theoretical Computer Science, vol.44, American Mathematical Society, ed. E. Winfree, (1999).

29. Lewis, H.R., and C.H. Papadimitriou, Elements of the Theory of Computation, Prentice-Hall, pages 296–300 and 345-348 (1981).

30. Li, X.J., X.P. Yang, J. Qi, and N.C. Seeman, Antiparallel DNA Double Crossover Molecules as Components for Nanoconstruction, J. Am. Chem. Soc., 118, 6131–6140, (1996).

31. Liu, F., H. Wang and N.C. Seeman, Short Extensions to Sticky Ends for DNA Nanotechnology and DNA-Based Computation, Nanobiology 4, 257–262 (1999).

32. Liu, F., R. Sha and N.C. Seeman, Modifying the Surface Features of Two-Dimensional DNA Crystals, J. Am. Chem. Soc. 121, 917–922 (1999).

33. Liu, F., M.F. Bruist and N.C. Seeman, Parallel Helical Domains in DNA Branched Junctions Containing 5', 5' and 3', 3' Linkages, Biochemistry 38, 2832–2841 (1999).

34. Mao, C., W. Sun and N.C. Seeman, Designed Two-Dimensional DNA Holliday Junction Arrays Visualized by Atomic Force Microscopy, J. Am. Chem. Soc. 121, 5437–5443 (1999).

35. Mao, C., LaBean, T.H. Reif, J.H., Seeman, Logical Computation Using Algorithmic Self-Assembly of DNA Triple-Crossover Molecules, Nature, vol. 407, Sept. 2000, pp. 493–495; C. Erratum: Nature 408, 750-750 (2000)
www.cs.duke.edu/~reif/paper/SELFASSEMBLE/AlgorithmicAssembly.pdf

36. Petty, M. C., M. R. Bryce and D. Bloor (Eds.), An Introduction to Molecular Electronics, Oxford University Press, New York (1995).

37. Reed, M. A., C. Zhou, C. J. Muller, T. P. Burgin and J. M. Tour, Conductance of a molecular junction, Science, Vol. 278, pages 252–254, October 10, 1997.

38. Reed, M. A. and J. M. Tour Computing with Molecules, Scientific American, June 2000. http://www.scientificamerican.com/2000/0600issue/0600reed.html

39. Reif, J.H., Local Parallel Biomolecular Computation, Third Annual DIMACS Workshop on DNA Based Computers, University of Pennsylvania, June 23–26, 1997. Published in DNA Based Computers, III, DIMACS Series in Discrete Mathematics and Theoretical Computer Science, Vol 48 (ed. H. Rubin), American Mathematical Society, 1999, p. 217–254.
http://www.cs.duke.edu/~reif/paper/Assembly.ps Assembly.fig.ps

40. Reif, J.H., Paradigms for Biomolecular Computation, First International Conference on Unconventional Models of Computation, Auckland, New Zealand, January 1998. Published in Unconventional Models of Computation, edited by C.S. Calude, J. Casti, and M.J. Dinneen, Springer Publishers, 1998, pp 72–93.
http://www.cs.duke.edu/~reif/paper/paradigm.ps

41. Reif, J.H., T.H. LaBean, and N.C. Seeman, Challenges and Applications for Self-Assembled DNA Nanostructures, Proc. Sixth International Workshop on DNA-Based Computers, Leiden, The Netherlands, June, 2000. DIMACS Series in Discrete Mathematics and Theoretical Computer Science, Edited by A. Condon and G. Rozenberg. Lecture Notes in Computer Science, Springer-Verlag, Berlin Heidelberg, vol. 2054, 2001, pp. 173–198:
www.cs.duke.edu/~reif/paper/SELFASSEMBLE/selfassemble.pdf

42. Reif, J.H., DNA Lattices: A Programmable Method for Molecular Scale Patterning and Computation, Special issue on Bio-Computation, Computer and Scientific Engineering Journal of the Computer Society. 2002.

43. Reif, J.H., The Emergence of the Discipline of Biomolecular Computation, invited paper to the special issue on Biomolecular Computing, New Generation Computing, edited by Masami Hagiya, Masayuki Yamamura, and Tom Head, 2002. [PostScript][PDF]
http://www.cs.duke.edu/~reif/paper/NGCsurvey/NGCsurvey.pdf

44. Robinson, R.M. Undecidablility and Nonperiodicity for Tilings of the Plane, Inventiones Mathematicae, 12, 177–209, (1971).

45. Rothemund, P.W.K., Using lateral capillary forces to compute by self-assembly, Proc. Nat. Acad. Sci. (USA) 97, 984–989 (2000).

46. Rothemund, P.W.K., and E. Winfree, The Program-Size Complexity of Self-Assembled Squares, Symposium on Theory of Computing Portland, OR, (May, 2000).

47. Seeman, N.C. Nucleic Acid Junctions and Lattices, J. Theor. Biol., 99, 237–247, (1982).

48. Seeman, N.C., DNA Engineering and its Application to Nanotechnology, Trends in Biotech. 17, 437–443 (1999).

49. Seeman, N.C., Nucleic Acid Nanostructures and Topology. Angewandte Chemie. 110, 3408–3428 (1998); Angewandte Chemie International Edition 37, 3220–3238 (1998).

50. Seeman, N. C., J.-H. Chen, N.R. Kallenbach, Gel electrophoretic analysis of DNA branched junctions, Electrophoresis, 10, 345–354, (1989).

51. Seeman, N. C., F. Liu, C. Mao, X. Yang, L.A. Wenzler, R. Sha, W. Sun, Z. Shen, X. Li, J. Qi, Y. Zhang, T. Fu, J.-H. Chen, and E. Winfree, Two Dimensions and Two States in DNA Nanotechnology, Journal of Biomolecular Structure and Dynamics, ISSN 0739–1102, Conversion 11, Issue 1, June, 1999.

52. Seeman, N. C., H. Wang, X. Yang, F. Liu, C. Mao, W. Sun, L.A. Wenzler, Z. Shen, R. Sha, H. Yan, M.H. Wong, P. Sa-Ardyen, B. Lui, H. Qiu, X. Li, J. Qi, S.M. Du, Y. Zhang, J.E. Mueller, T.-J. Fu, Y. Wang, amd J. Chen, New Motifs in DNA nanotechnology, Nanotechnology 9, p 257–273 (1998).

53. Seeman, N. C., Y. Zhang, and J. Chen, DNA nanoconstructions, J. Vac. Sci. Technol., 12:4, 1895–1905, (1994).

54. Wang, H., In Proc. Symp. Math. Theory of Automata, 23–26 (Polytechnic Press, New York, 1963).

55. Winfree, E. Simulations of Computing by Self-Assembly. In Proceedings of the Fourth Annual Meeting on DNA Based Computers, held at the University of Pennsylvania, June 16–19, 1998.

56. Winfree, E., X. Yang, N.C. Seeman, Universal Computation via Self-assembly of DNA: Some Theory and Experiments, 2nd Annual DIMACS Meeting on DNA Based Computers, Princeton, June, 1996.

57. Winfree, E., T. Eng, and G. Rozenberg, String Tile Models for DNA Computing by Self-Assembly, Proc. Sixth International Workshop on DNA-Based Computers, Leiden, The Netherlands, June, 2000. DIMACS Series in Discrete Mathematics and Theoretical Computer Science, Edited by A. Condon and G. Rozenberg. Lecture Notes in Computer Science, Springer-Verlag, Berlin Heidelberg, vol. 2054, 2001, pp. 63–88.

58. Winfree, E., Furong Liu, Lisa A. Wenzler, Nadrian C. Seeman (1998) Design and Self-Assembly of Two Dimensional DNA Crystals. Nature 394: 539–544, 1998.

59. Winfree, E., X. Yang, and N.C. Seeman, Universal Computation via Self-assembly of DNA: Some Theory and Experiments, Proc. DNA Based Computers II: June 10-12, 1996, DIMACS Series in Discrete Mathematics and Theoretical Computer Science, vol. 44, L. F. Landweber and E. B. Baum, editors, American Mathematics Society, Providence, RI, 1998, pp. 191–213.

60. Winfree, E., On the Computational Power of DNA Annealing and Ligation, Proc. DNA-Based Computers: April 4, 1995, In DIMACS Series in Discrete Mathematics and Theoretical Computer Science, Richard Lipton and E. B. Baum, editors, American Mathematical Society, Providence, RI, vol. 27, 1996, pp. 199–211.

61. Winfree, E., F. Liu, L. A. Wenzler, and N.C. Seeman, Design and Self-Assembly of Two-Dimensional DNA Crystals, Nature 394, 539–544 (1998).

62. Xia, Y. and Whitesides, G. M., Soft Lithography, Annu. Rev. Mater. Sci. 1998, 28, 153–184. Yan, L.; Zhao, X.-M. and Whitesides, G. M. Patterning a Preformed, Reactive SAM Using Microcontact Printing, J. Am. Chem. Soc. 1998,120, 6179-6180.

63. Yan,H., X. Zhang, Z. Shen and N.C. Seeman, A robust DNA mechanical device controlled by hybridization topology, Nature, 415, 62–65, 2002.

64. Yang, X., L.A. Wenzler, J. Qi, X. Li and N.C. Seeman, Ligation of DNA Triangles Containing Double Crossover Molecules, J. Am. Chem. Soc. 120, 9779–9786 (1998).

65. Zhao, X.; Votruba, P.G.; Strother, T.C.; Ellison, M.D.; Smith, L.M. and Hamers, R.J. Formation of Organic Films on Si(111) Surfaces via Photochemical Reaction (in preparation), 1999.

66. Zhou, C., et al., Appl. Phys. Lett. 71, 611 (1997).

67. Zhou, C., Atomic and Molecular Wires, Ph.D. thesis, Yale University (1999).

Towards a Predictive Computational Complexity Theory

Madhav V. Marathe*

Basic and Applied Simulation Science (D-2), P. O. Box 1663, MS M997
Los Alamos National Laboratory, Los Alamos NM 87545. marathe@lanl.gov

Over the last three decades, language recognition models of computation and associated resource bounded reductions have played a central role in characterizing the computational complexity of combinatorial problems. However, due to their generality, these concepts have inherent limitations – they typically ignore the underlying structure and semantics of the problem instances. Thus they are generally not "robust" in terms of simultaneously classifying variants of the original problem.

For example consider the two well studied satisfiability problems 3SAT and 1-3SAT (One-in-Three SAT). Highly efficient resource-bounded reductions (quasi-linear time and linear space) exist from 3SAT to 1-3SAT showing that 1-3SAT is **NP**-hard. But such efficient resource-bounded reductions cannot generally be used to characterize simultaneously, the complexity of the variants such as MAX-1-3SAT (optimization version) or #1-3SAT (counting version) or Q-1-3SAT (quantified version), or H-1-3SAT (hierarchically specified version) or even PL-1-3SAT (planar version).

In the past, the complexities of these variants have been characterized by devising "new" reductions for each variant. In this talk, I will outline research efforts aimed at developing a *predictive computational complexity theory*. The goal here is to characterize simultaneously (and in a predictive manner) the relative complexity and approximability of a large class of combinatorial problems.

The theory outlined here is based on an algebraic model of problems called *generalized constraint satisfaction problems* and four *simple yet natural concepts/techniques* outlined below. As will be discussed, a large number of results in the literature can be understood in terms of the these concepts. Several of these concepts have been used in the literature at least at an intuitive level for a number of years.

Algebraic Model: A natural model for representing combinatorial problems is the generalized constraint satisfaction model. The model can be seen as en extension of the model first proposed in Schaefer [64]. Informally speaking, an instance of a problem represented using this model consists of a set of constraints applied to a collection of variables. More formally, let D be an arbitrary (*not*

---

* Work supported by the Department of Energy under Contract W-7405-ENG-36.

P. Widmayer et al. (Eds.): ICALP 2002, LNCS 2380, pp. 22–31, 2002.

necessarily finite) nonempty set; C a finite set of constant symbols denoting elements of D; and **S** and **T** are arbitrary finite sets of finite-arity relations on D. An **S**-*clause* (a constant free **S**-*clause*) is a relation in **S** applied to variables and constants (to variables only). An **S**-*formula* (a constant free **S**-*formula*) is a finite nonempty conjunction of **S**-*clauses* (constant free **S**-*clauses*). Viewed algebraically, this yields a system of simultaneous equations over a given algebraic structure. A large class of combinatorial problems can be expressed as generalized constraint satisfaction problems by changing any subset of the following parameters:

(i)   the type of constraints (e.g. monotone, XOR, etc.),
(ii) the size of domain and the associated algebraic structure (e.g. the domain is $\{0,1\}$ and the underlying structure being the Boolean algebra; the domain is **Z** and the algebraic structure being a field, etc.),
(iii) the quantifiers associated with the variables (existential, universal, stochastic, etc.),
(iv) the underlying specification used to specify the problem instances (natural "flat representations", hierarchical or periodic specifications, etc),
(v)   the graph theoretic structure representing the interaction/dependence between variables and constraints (planar, treewidth, geometric intersection, etc.), and
(vi) the objective function (decision, optimization, counting, unique, ambiguous, etc).

For example, the problem 3SAT has four types of constraints over Boolean variables [64,17] that are all existentially quantified. The objective is to decide if the formula is satisfiable. The problem PL-3SAT is a restriction of the problem 3SAT in which the clause-variable bipartite graph is planar. Q-3SAT is a problem like 3SAT except that the variables can be universally or existentially quantified. In general, we denote the problem of determining the satisfiability of finite conjunctions of relations in **S** applied to variables (applied to variables and constant symbols in C) by SAT(**S**) (by $\text{SAT}_C(\mathbf{S})$.) More generally we use Π-SAT(**S**) and Π-$\text{SAT}_C(\mathbf{S})$ to be any one of the variant satisfiability problems: MAX-SAT(**S**), #-SAT(**S**), UNIQUE-SAT(**S**), etc. From a computational standpoint, one advantage of representing problems in this algebraic model is that problem instances now have a combinatorial (graph) structure determined by which variables appear in which clauses. This allows us to develop a theory that preserves the structure of individual instances as we transform one problem instance into another problem instance. A second advantage is that many classical combinatorial problems especially graph problems can be transformed into appropriate generalized constraint satisfaction problems in a way that preserves (or nearly preserves) the underlying graph structure.

Characterizing the computational complexity of such generalized constraint satisfaction problems has been a theme of several articles. See [6], [15], [16], [17], [18], [23], [28], [33], [37], [42], [43] and the references therein. This research is motivated by a number of considerations, including the following:

A. Versions/variants of these problems are widely applicable in modeling both real-life and abstract mathematical problems.

B.   These problems, have played fundamental role in the development of discrete complexity theory, providing prototypical complete problems for various complexity classes.

Underlying Techniques: The basic tools for characterizing the relative complexities of various generalized constraint satisfaction problems are the following simple as well as natural concepts:

1. *Relational/Algebraic Definability:* Let **S** and **T** be sets of relations/algebraic constraints on a common domain D. Relational definability formalizes the intuitive concept that the relations in **S** are *expressible* (or, extending the terminology from [64] are *representable*) by finite conjunctions of the relations in **T**. This concept was formalized by Schaefer [64] to study the decision complexity of generalized satisfiability problems (both unquantified and quantified) and in [14], [15], [16], [17], [18], [35], [36], [37], [39], [43] for counting and optimization problems. In [43], the concept is termed as "implementations". This central concept forms the basis of most of the results on the complexity of generalized constraint satisfaction problems including complete classifications of such problems. See [17] for a recent survey article.

2. *Local Replacements and Simultaneous Reductions:* Reductions by local replacement have been used extensively in the literature (e.g., see [28]). The first step in formalizing this concept is to separate the concept of replacement from that of reduction. Local replacement constructs target instances from source instances by replacing each object (e.g. clause/variable in a formula) by a collection of objects (e.g. conjunction of clauses) in the target instance. Local replacements have a number of desirable properties. They usually preserve the graph structure as well as the specification structure used to describe the problem, are extremely efficient in terms of the resources used, and preserve power and polynomial indices [67]. When a replacement preserves the property (semantics) we are interested in, we call it a reduction. This part usually requires one to use some form of relational definability. For example, reductions that preserve decision complexity are simply called "reductions" in the literature. **L**-reductions are an example of approximation scheme preserving reductions and parsimonious reductions are those reductions that preserve the number of solutions. A natural question to ask in this context is: Can we design a single transformation that is decision preserving, approximation preserving, number preserving, etc. ? We have found that for a large class of algebraic problems, it is indeed possible to devise such transformations; we call them *simultaneous reductions.* (see [37,39] and related results in [17,43]). For example, a (**L** + *parsimonious* + **A**)-reduction is a reduction that is simultaneously an **L**-reduction, a *parsimonious*-reduction and an **A**-reduction. The existence of multi-purpose reductions, for algebraic problems, may not be surprising. But what we find surprising is the existence

of simultaneous reductions for a wide class of natural algebraic problems. Simultaneous reductions obtained via local replacement based transformations combine the best of both these concepts: *they simultaneously preserve a variety of semantics and also preserve structure of instances.* By structure of instances we usually mean the variable-clause interaction graph structure, the structure of the specification used to specify the problem, the structure of proof trees in case of quantified problems, the structure of the phase space when talking about dynamical systems, etc.

For example, we can show that there is a local replacement based transformation from the problem 3SAT to the problem 1-3SAT that is simultaneously a (*decision* + **L** + **A** + *parsimonious*)-reduction. As a result using the known results on complexity of 3SAT and its variants, we **simultaneously** obtain the following: 1-3SAT is **NP**-hard, #-1-3SAT is **#P**-complete, MAX-1-3SAT is **APX**-complete, MAX-DONES-1-3SAT is **MAX-Π_1**-complete, Q-1-3SAT is **PSPACE**-hard, Q-MAX-1-3SAT is hard to approximate beyond a certain $\epsilon > 0$ unless **P** = **PSPACE** (quantified versions of the problem), H-1-3SAT and PL-H-1-3SAT (hierarchically specified versions of the problem) are **PSPACE**-hard, H-MAX-1-3SAT is hard to approximate beyond a certain $\epsilon > 0$ unless **P** = **PSPACE**, etc. Note that once an appropriate local replacement based simultaneous reduction is constructed between 3SAT and 1-3SAT, the relative complexities of their variants followed directly from the properties of such reductions. It is in this sense that we use the term predictive complexity theory.

3. *Level Treewidth Decomposition and Partial Expansion*: This is a generalization of an elegant technique due to Baker [7], and Hochbaum and Maass [32] for obtaining approximation schemes for certain graph-theoretic and geometric problems. Consider a problem Π which can be solved by a divide-and-conquer approach with a performance guarantee of ρ. The technique in [7,32] allows us to bound the error of the simple divide-and-conquer approach by applying it iteratively and choosing the best solution among these iterations as the solution to Π. The two important properties needed to obtain such algorithms are as follows:

   a) The ability to decompose the given formula into variable-disjoint subformulas such that an optimal (or near optimal) solution to each subformula can be obtained in polynomial time.

   b) The ability to obtain a near-optimal solution for the whole formula by merging the optimal solution obtained for each sub-formula.

A central concept used in this context is the level-treewidth property and is also referred to as locally tree-decomposable property [30]. It extends the earlier concept called diameter-treewidth property proposed by Eppstein [20]. These ideas are further extended in [5,16,22,29,30,41,44]. Partial expansion is an extension of this concept that can be applied to succinctly specified objects [40,54]. Once again, we briefly discuss an example of a general result. The result serves to illustrate how one can obtain easiness results using the above discussed concepts.

A (δ, g)-*almost planar graph* is a graph with vertex set V together with a genus g layout with at most $\delta \cdot |V|$ crossovers. Let $\mathbf{S}$ be a fixed finite set of finite arity relations $\{R_1, \ldots, R_q\}$. The optimization problem MAX-RELATION($\mathbf{S}$) is the following: Given a set of terms $\{t_1, t_2, \ldots, t_m\}$, where each term t_i is of the form $f(x_{i_1}, x_{i_2}, \ldots, x_{i_r})$ for some $f \in \mathbf{S}$, assign values to each x_i, $1 \leq i \leq n$, so as to maximize the number of satisfied terms. First, by using the level-treewidth decomposition concepts, we show that for each fixed finite set $\mathbf{S}$ and fixed δ, $g \geq 0$, there is a **PTAS** for the problem MAX-RELATION($\mathbf{S}$) when restricted to instances whose bipartite graphs are (δ, g)-almost planar. In the next step, we show that a number of important classes of combinatorial and graph problems (e.g. maximum independent set) when restricted to (δ, g)-almost planar instances can be reduced to appropriate problems MAX-RELATION($\mathbf{S}$). The reductions are based on local transformations and have two important properties: (i) they can be carried out in **NC** and (ii) if a problem instance is (δ, g)-almost planar, then the instance of MAX-RELATION($\mathbf{S}$) obtained as a result of the reduction is (δ', g')-almost planar, where δ' and g' are functions of δ and g (independent of n). Thus, each of these problems has an **NCAS** when restricted to (δ, g)-almost planar instances. We refer to such reductions as *structure preserving* **NC-L**-reductions. The reductions are in fact reductions by local replacement. As a result by our discussion about properties of simultaneous reductions by local transformation, we get that the level-restricted hierarchical versions of these problems have a **PTAS** [54]. Note the decision versions of these hierarchical problems are **PSPACE**-hard. The results provide a syntactic (algebraic) class of problems, namely, $(0, 0)$-almost planar MAX-RELATION($\mathbf{S}$), whose closure under **L**-reductions defines one characterization for problems that have a **PTAS**. They illustrate the *positive* use of **L**-reductions: use of **L**-reductions in devising **PTAS** rather than in proving non-approximability results. See [44,69] for similar results.

4. *Algebraic/Predicate Decomposability*: Predicate decomposability is a method for replacing a large unbounded arity predicate by a conjunction of fixed arity predicates. The concept is a special case of relational definability that is typically useful in obtaining easiness results. In such cases we also require that the transformation preserve the structure of the source instance. For decision problem this is exactly the concept used to reduce SAT to 3SAT. Predicate decomposability is the semantic analogue of the level-treewidth concept: the former is used to decompose predicates while the latter is used to decompose the graphs (formulas). The concept is most useful in situations when we want to obtain easiness results for generalized constraint satisfaction problems when the arity of clauses is not fixed. In such a case we decompose the larger predicates in a way that preserves the graph structure [41]. In many cases, the concept of predicate decomposability can be combined with the level-treewidth concept and the positive use of reductions to yield exact and approximate algorithms

for many algebraic problem in which the arity of relations in **S** grows polynomially with instance size.

Applications: We will highlight the concepts by discussing a few representative applications. They include:

(A) [Upper bounds] Efficient approximation algorithms and approximation schemes with provable performance guarantees for large classes of natural **NP-**, **PSPACE-** and **NEXPTIME-**hard optimization problems.

(B) [Lower bounds] General results characterizing simultaneously, the decision, counting, optimization and approximation complexities of generalized constraint satisfaction problems and applications of these results to combinatorial problems arising in games, logic, graph theory and discrete dynamical systems.

(C) Further insights on recently proposed ideas on periodically and hierarchically specified problems and its implications to understanding the relationship between various computational complexity classes (e.g Freedman's ideas on approaching the **P** vs **NP** question).

Acknowledgements. I thank Anne Condon, Joan Feigenbaum, Harry Hunt III, Gabriel Istrate, Anil Kumar, Rajeev Motwani, S.S. Ravi, Daniel Rosenkrantz, Venkatesh Radhakrishnan, Richard Stearns, Madhu Sudan and Egon Wanke, for their collaboration on related topics, fruitful discussions and pointers to related literature. Some of the ideas developed here were motivated by the questions and discussions in Condon et.al. [12,13], Garey & Johnson [28], Papadimitriou [60] Lengauer and Wagner [47] and Orlin [57].

References

1. S. Agarwal and A. Condon. On approximation algorithms for hierarchical MAX-SAT. *J. of Algorithms*, 26, 1998, pp. 141-165.
2. E. Amaldi and V. Kann. The complexity and approximability of finding maximum feasible subsystems of linear relations. *Theoretical Computer Science (TCS)* 147, pp. 181-210, 1995.
3. S. Arnborg, B. Courcelle, A. Proskurowski and D. Seese. An algebraic theory of graph reduction. *J. of the ACM (JACM)*, 40, 1993, pp. 1134-1164.
4. S. Arora, C. Lund, R. Motwani, M. Sudan and M. Szegedy. Proof verification and hardness of approximation problems. *J.of the ACM (JACM)*, 45, 1998, pp. 501-555.
5. S. Arora. Polynomial time approximation scheme for Euclidean TSP and other geometric problems. *J.of the ACM (JACM)*, 45(5), pp. 753-682, 1998.
6. G. Ausiello, P. Crescenzi, G. Gambosi, V. Kann, A. Marchetti-Spaccamela and M. Protasi. *Complexity and approximation: Combinatorial optimization problems and their approximability properties*, Springer Verlag, 1999.

7. B. S. Baker. "Approximation algorithms for NP-complete problems on planar graphs", *J. of the ACM (JACM)*, Vol. 41, No. 1, Jan. 1994, pp. 153–180.
8. R. Beigel, W. Gasarch. " On the complexity of finding the chromatic number of recursive graphs," Parts I and II, *Annals of Pure and Applied Logic*, 45, 1989, pp. 1-38 and 227-247.
9. J.L. Bentley, T. Ottmann, and P. Widmayer. The complexity of manipulating hierarchically defined sets of intervals. *Advances in Computing Research* 1, F.P. Preparata(ed.), 1983, pp. 127-158.
10. U. Bertele and F. Brioschi. *Nonserial Dynamic Programming*, Academic Press, NY, 1972.
11. H. Bodlaender Dynamic programming on graphs of bounded treewidth. *Proc. 15th International Colloquium on Automata Languages and Programming (ICALP)*, LNCS Vol. 317, 1988, pp. 105-118.
12. A. Condon, J. Feigenbaum, C. Lund and P. Shor. Probabilistically checkable debate systems and approximation algorithms for PSPACE-Hard Functions. *Chicago Journal of Theoretical Computer Science,* Vol. 1995, No. 4. http://www.cs.uchicago.edu/publications/cjtcs/articles/1995/4/ contents.html.
13. A. Condon, J. Feigenbaum, C. Lund and P. Shor. Random debaters and the hardness of approximating stochastic functions. *SIAM J. Computing*, 26, 1997, pp. 369-400.
14. N. Creignou. A dichotomy theorem for maximum generalized satisfiability problems. *J. of Computer and System Sciences (JCSS)*, 51, 1995, pp. 511-522.
15. N. Creignou and M. Hermann. Complexity of generalized satisfiability counting problems. *Information and Computation* 125(1), pp. 1-12, 1996.
16. N. Creignou, S. Khanna, and M. Sudan Complexity classifications of Boolean constraint satisfaction problems SIAM Monographs on Discrete Mathematics and Applications 7, 2001.
17. N. Creignou, S. Khanna, and M. Sudan. Complexity classifications of Boolean constraint satisfaction problems. *SIGACT News*, Volume 32, 4(121), Complexity Theory Column 34, pp. 24-33, November 2001.
18. V. Dalmau. Computational complexity of problems over generalized formulas. Ph.D. thesis, Dept. of Computer Science, U. Politecnica De Catalunya.
19. R. Downey and M. Fellows. *Parameterized Complexity*, Springer Verlag, 1998.
20. D. Eppstein "Subgraph Isomorphism in planar graphs and related problems," *6th ACM-SIAM Symposium on Discrete Algorithms (SODA)*, 1995, pp. 632-640.
21. A. Ehrenfeucht, J. Engelfriet and G. Rosenberg. Finite Languages for Representation of Finite Graphs. *J. Computer and System Sciences (JCSS)*, (52), pp. 170-184, 1996.
22. T. Erlebach, K. Jansen and E. Seidel. Polynomial time approximation schemes for geometric Graphs. *Proc. 12th Annual ACM-SIAM Symposium on Discrete Algorithms (SODA)*, 2001, pp. 671-679.
23. T. Feder and M. Vardi. The computational structure of monotone monadic SNP and constraint satisfaction: A study through datalog and group Theory. *SIAM J. Computing*, 28(1): 57-104 (1998).
24. J. Feigenbaum. "Games, complexity classes and approximation algorithms," invited talk at the *International Congress on Mathematics*, Berlin, 1998.
25. A. Fraenkel and Y. Yesha. Complexity of problems in games, graphs, and algebraic equations. *Discrete Mathematics,* 1, 1979, pp. 15-30.
26. M. Freedman. *k*-SAT on groups and Undecidability. *Proc. 30th ACM Annual Symposium on Theory of Computing (STOC)*. 1998, pp. 572-576.

27. M. Freedman. Limits, logic and computation. *Proc. National Academy of Sciences, USA,* 95, 1998, pp. 95-97.

28. M. Garey and D. Johnson. *Computers and Intractability: A Guide to the Theory of NP-Completeness.* W. H. Freeman, San Francisco, 1979.

29. M. Grohe. Local tree-width, excluded minors, and approximation algorithms. To appear in *Combinatorica,* 2002.

30. M. Grohe and M. Frick. Deciding first-order properties of locally tree-decomposable structures. To appear in *J. of the ACM (JACM).* Conference version appeared in *Proc. 26th International Colloquium on Automata, Languages, and Programming,* Lecture Notes in Computer Science 1644, Springer-Verlag, 1999.

31. T. Hirst and D. Harel. "Taking it to the Limit: On infinite variants of NP-complete problems," *Proc. 8th IEEE Annual Conference on Structure in Complexity Theory,* 1993, pp. 292-304.

32. D. S. Hochbaum and W. Maass. Approximation schemes for covering and packing problems in image processing and VLSI. *J. of the ACM (JACM),* 32(1), pp. 130-136, 1985.

33. D. Hochbaum(Ed.). *Approximation Algorithms for NP-Hard Problems.* PWS Publishing Company, Boston, MA, 1997.

34. F. Höfting, T. Lengauer, and E. Wanke. Processing of hierarchically defined graphs and graph families. *Data Structures and Efficient Algorithms* (Final Report on the DFG Special Joint Initiative), Springer-Verlag, LNCS 594, 1992, pp. 44-69.

35. H. Hunt III, R. Stearns and M. Marathe. Generalized CNF satisfiability problems and non-efficient approximability. *Proc. 9th Annual IEEE Conf. on Structure in Complexity Theory,* 1994, pp.356-366.

36. H. Hunt III, M. Marathe. V. Radhakrishnan, S. Ravi, D. Rosenkrantz and R. Stearns. Parallel approximation schemes for a class of planar and near planar combinatorial problems. *Information and Computation,* 173(1), Feb. 2002, pp. 40–63.

37. H. Hunt III, R. Stearns and M. Marathe. Relational representability, local reductions and the complexity of generalized satisfiability problem. submitted. Technical Report No. LA-UR-00-6108, Los Alamos National Laboratory.

38. H. Hunt III, M. Marathe, V. Radhakrishnan and R. Stearns. The complexity of planar counting problems. *SIAM J. Computing,* 27, 1998, pp.1142-1167.

39. H. Hunt III, R. Stearns and M. Marathe. Strongly local reductions and the complexity/efficient approximability of algebra and optimization on abstract algebraic structures. *International Conference on Symbolic and Algebraic Computations (ISAAC),* July 2001.

40. H. Hunt III, M. Marathe, V. Radhakrishnan, S. Ravi, D. Rosenkrantz and R. Stearns. NC-approximation schemes for NP- and PSPACE-hard problems for geometric graphs. *J. Algorithms,* 26, 1998, pp. 238-274.

41. H. Hunt III, R. Jacob, M.V. Marathe, D. Rosenkrantz and R.E. Stearns. Towards syntactic characterizations of approximation schemes via predicate and graph decompositions. Technical Report No. LA-UR-97-479, Los Alamos National Laboratory, January 1997.

42. P. Jeavons, D. Cohen and M. Gyssens. Closure properties of constraints. *J. of the ACM, (JACM)* 44(4):527-548, July 1997.

43. S. Khanna, M. Sudan L. Trevisan and D. Williamson. The approximability of constraint satisfaction problems. *SIAM J. on Computing,* 30(6), pp. 1863–1920, March 2001.

44. S. Khanna and R. Motwani "Towards a syntactic characterization of PTAS" *Proc. 28th Annual ACM Symposium on Theory of Computing, (STOC)*, pp. 329-337, Philadelphia, PA May 1996.

45. M. Krentel. The Complexity of optimization problems. *J. Computer and System Sciences (JCSS)* 36, pp. 490-509, 1988.

46. R. Ladner. Polynomial space counting problems. *SIAM J. Computing*, 18(6), pp. 1087-1097, 1989.

47. T. Lengauer and K. Wagner. The correlation between the complexities of non-hierarchical and hierarchical versions of graph problems. *J. Computer and System Sciences (JCSS)*, Vol. 44, 1992, pp. 63-93.

48. D. Lichtenstein. Planar formulae and their uses. *SIAM J. Computing*, Vol 11, No. 2, May 1982 , pp. 329-343.

49. P.D. Lincoln, J.C. Mitchell and A. Scederov. Optimization complexity of linear logic proof games. *Theoretical Computer Science*, 227 (1999) pp. 299-331.

50. R. Lipton and R. Tarjan. Applications of a planar separator theorem. *SIAM J. Computing*, 9, 1980, pp.615-627.

51. M. Littman, S. Majercik, and T. Pitassi. Stochastic Boolean satisfiability. *Journal of Artificial Intelligence*, Kluwer Publications, 2000.

52. S. MacLane and G. Birkhoff, *Algebra*, Macmillan, NY 1967.

53. M. Marathe, H. Hunt III, D. Rosenkrantz and R. Stearns. Theory of periodically specified problems:complexity and approximability. *Proc. 13th IEEE Conf. on Computational Complexity*, 1998.

54. M. Marathe, H. Hunt III, R. Stearns and V. Radhakrishnan. Approximation algorithms for PSPACE-hard hierarchically and periodically specified problems. *SIAM J. Computing*, 27(5), pp. 1237-1261, Oct. 1998.

55. M. Marathe, H. Hunt III, R. Stearns and V. Radhakrishnan. Complexity of hierarchically and 1-dimensional periodically specified problems. *AMS-DIMACS Volume Series on Discrete Mathematics and Theoretical Computer Science: Workshop on Satisfiability Problem: Theory and Application*, 35, November 1997.

56. G. L. Miller, S. H. Teng, W. Thurston and S. A. Vavasis. Separators for sphere packings and nearest neighbor graphs. *J. of the ACM (JACM)* , 44(1), pp. 1-29, Jan. 1997.

57. J. Orlin. The Complexity of dynamic/periodic languages and optimization Problems. Sloan W.P. No. 1679-86 July 1985, Working paper, Alfred P. Sloan School of Management, MIT, Cambridge, MA 02139. A Preliminary version of the paper appears in *Proc. 13th ACM Annual Symposium on Theory of Computing (STOC)*, 1978, pp. 218-227.

58. A. Panconesi and D. Ranjan. Quantifiers and approximations. *Theoretical Computer Science (TCS)*, 107, 1993, pp. 145-163.

59. C. Papadimitriou. Games against nature. *J. Computer and System Sciences (JCSS)* , 31, 1985, pp. 288-301.

60. C. Papadimitriou. *Complexity Theory*. Addison-Wesley, Reading, MA, 1994.

61. C. Papadimitriou and M. Yannakakis. Optimization, approximation, and complexity classes. *J. Computer and System Sciences (JCSS)*, 43, 1991, pp. 425-440.

62. C. Robinson. *Dynamical Systems Stability, Symbolic Dynamics, and Chaos* 2nd Edition. CRC Press, Boca Raton, Florida, 1999.

63. J. Saxe. Two papers on graph embedding problems. *Technical Report, Dept of Comp. Science,* Carnegie Mellon University, Pittsburg, CMU-CS-80-102, 1980.

64. T. Schaefer. The complexity of satisfiability problems. *Proc. 10th Annual ACM Symposium on Theory of Computing (STOC)*, 1978, pp. 216-226.

65. C. Schnorr. Satisfiability is quasilinear complete in NQL. *J. of the ACM (JACM)*, 25(1):136-145, January 1978.
66. S.K. Shukla, H.B. Hunt III, D.J. Rosenkrantz, and R.E. Stearns. On the complexity of relational problems for finite state processes. *Proc. International Colloquium on Automata, Programming, and Languages (ICALP)* ,1996, pp. 466-477.
67. R. Stearns and H. Hunt III. Power indices and easier hard problems. *Math. Systems Theory* 23, pp. 209-225, 1990.
68. R. Stearns and H. Hunt III. An Algebraic model for combinatorial problems. *SIAM J. Computing*, Vol. 25, April 1996, pp. 448–476.
69. L. Trevisan. *Reductions and (Non-)Approximability.* Ph.D. Thesis, Dipartimento di Scienze dell'Informazione, University of Rome, "La Sapienza", Italy, 1997.
70. L. Valiant. The complexity of enumeration and reliability problems. *SIAM J. Computing*, 8(3), August 1979 , pp. 410-421.
71. D. Zuckerman. On unapproximable versions of NP-complete problems. *SIAM J. Computing*, 25(6), pp. 1293-1304, 1996.

Equivariant Syntax and Semantics
(Abstract of Invited Talk)*

Andrew M. Pitts

University of Cambridge Computer Laboratory, Cambridge CB3 0FD, UK
Andrew.Pitts@cl.cam.ac.uk

Abstract. The notion of *symmetry* in mathematical structures is a powerful tool in many branches of mathematics. The talk presents an application of this notion to programming language theory.

Algebraic Syntax and Semantics

Since the 1970s at least, universal algebra has played an important role in programming language theory (see [17] for a survey). For example, from the point of view of 'algebraic semantics', a programming language is specified by a signature of sorts and typed function symbols. Then its abstract syntax, i.e. the set of well-formed parse trees of the language, is given by the *initial algebra* for this signature; and a denotational semantics for the language is given by the unique homomorphism of algebras from the initial algebra to some algebra of meanings. This algebraic viewpoint has useful computational consequences that one sees directly in the development of term-rewriting systems [1] and in the notion of user-declared datatype occurring in functional programming languages such as ML [19] or Haskell [22]. The initial algebra property also has useful logical consequences, since it gives rise to principles of structural recursion and induction that are fundamental for proving properties of the programming language. The fact that these principles can be automatically generated from the language's signature has facilitated incorporating them into general-purpose systems for machine-assisted reasoning (cf. the inductive sets package of the HOL system [12], or Coq's inductive types [4]).

In fact if one dips into the 1985 volume edited by Nivat and Reynolds [20] representing the state of the art of algebraic semantics when it graduated from the novel to the routine, one sees that this attractively simple framework is only adequate for rather simple languages. For example, to see denotational semantics for languages containing recursive features as just algebra homomorphisms, one must mix universal algebra with *domain theory* [13] and consider the (interesting!) complications of 'continuous algebras'. And it is only recently, with the work of Plotkin and Turi reported in [26], that (certain kinds of) *structural operational semantics* [25] have been fitted convincingly into the algebraic framework.

---

* Research funded by UK EPSRC grant GR/R07615 and Microsoft Research Ltd.

P. Widmayer et al. (Eds.): ICALP 2002, LNCS 2380, pp. 32–36, 2002.

Getting in a Bind

In my opinion, the biggest defect of the algebraic approach to syntax and semantics in its traditional form is that it cannot cope convincingly with programming language constructs that involve *binding*. By the latter I mean constructs where the names of certain entities (such as program variables, or function parameters) can be changed, consistently and subject to freshness constraints within a certain textual scope, without changing the meaning of the program. Such 'statically scoped' binding constructs are very common. Any semantics of them has to identify programs that only differ up to bound names—*α-equivalent* programs, as one says. Furthermore, many transformations on the syntax of such programs (such as ones involving capture-avoiding substitution) are only really meaningful up to α-equivalence. Therefore one would really prefer to work with a representation of syntax that abstracts away from differences caused by named variants. In the conventional algebraic approach one has to define a suitable notion of α-equivalence language-by-language, and then quotient the algebra of parse trees by it. In doing so, one looses the initial algebra property and hence looses automatically generated principles of structural recursion/induction that apply directly to parse trees modulo α-equivalence.

Because issues to do with binders and α-equivalence are so prevalent, tiresome and easy to get wrong, it is highly desirable to have a mathematical foundation for syntax involving binders that enables one to generalise the initial algebra view of syntax from parse trees to parse trees modulo α-equivalence. Ever since Church [3], formal systems for *total functions*, i.e. various forms of typed λ-calculus, have been used to model variable-binding operations; see [18] for an excellent recent survey of this so-called *higher-order abstract syntax* and related notions. However, this approach does not provide the simple generalisation of the initial algebra viewpoint one would like. True, it is possible to develop structural induction/recursion principles for higher-order abstract syntax (see [6], [14] and [15] for example), but the logical niceties are rather subtle, involving one or more layers of carefully constrained metalanguage in which syntax representation, and reasoning about it, have to take place. It reminds me a bit of the situation for non-standard analysis, whose logical subtleties hindered widespread take-up by 'users'.

A Fresh Approach

In fact it is possible to have a mathematical foundation for syntax involving binders which directly generalises the classical initial algebra view of syntax from parse trees to parse trees modulo α-equivalence. The papers by Fiore, Plotkin and Turi [7] and Gabbay and myself [10,11] present two different solutions within the general framework of *(pre)sheaf toposes* [16] (although the connection with sheaf theory is not to the fore in the second work, which chooses to present a set-theoretical version of its model). The work by Fiore et al provides a very nice, syntax-independent and algebraic treatment of the notion of *nameless*, or *de Bruijn terms* [5]; whereas that by Gabbay and myself hinges upon a syntax-independent characterisation of *freshness* of bound names in terms of permutative renaming. Both approaches can characterise parse trees modulo α-equivalence as initial algebras. But as for the work on higher-order abstract syntax mentioned above, workers in 'applied semantics' will ask to what extent the logical (and in this case category- and set-theoretic) subtleties hinder the take-up of these ideas in, say, systems for machine-assisted reasoning about syntax and semantics, or in logic- and functional-programming languages for metaprogramming?

In the talk I will try to give a positive answer to this question. I will show that the model of abstract syntax with binders in [10,11] can be formulated as quite a simple and (I hope) appealing generalisation of the classical initial algebra treatment of syntax without binders; and that this generalisation has good logical and computational properties. The presentation is based on the notion of *nominal set* from [23]. The whole approach stems from the somewhat surprising observation that all of the concepts we need (α-equivalence, freshness, name-binding, ...) can be—and, I would claim, should be—defined purely in terms of the operation of *swapping* pairs of names. Thus the mathematical framework takes into account certain symmetries on objects induced by name-swapping; and the the notion *equivariance*, i.e. of invariance for these symmetries, plays a prominent role in the theory.

The talk will introduce nominal sets and their use for modelling abstract syntax with binders. I will survey some of the logical and computational consequences of this 'equivariant syntax and semantics' that have been produced so far: Gabbay's packages for FM-set theory and higher order logic [8,9] in Isabelle [21]; a functional programming language incorporating our notion of binding and freshness [24] currently being implemented by Shinwell; a 'nominal' version of first-order equational logic, unification and term-rewriting (work in progress by Gabbay, Urban and myself); domain theory in nominal sets (work in progress by Shinwell); the use of name swapping and quantification over fresh names in a spatial logic for concurrent processes by Caires and Cardelli [2].

References

[1] F. Baader and T. Nipkow. *Term Rewriting and All That.* Cambridge University Press, 1998.

[2] L. Caires and L. Cardelli. A spatial logic for concurrency (part I). In N. Kobayashi and B. C. Pierce, editors, *Theoretical Aspects of Computer Software, 4th International Symposium, TACS 2001, Sendai, Japan, October 29-31, 2001, Proceedings*, volume 2215 of *Lecture Notes in Computer Science*, pages 1–38. Springer-Verlag, Berlin, 2001.

[3] A. Church. A formulation of the simple theory of types. *Journal of Symbolic Logic*, 5:56–68, 1940.

[4] The Coq proof assistant. Institut National de Recherche en Informatique et en Automatique, France, ⟨coq.inria.fr⟩.

[5] N. G. de Bruijn. Lambda calculus notation with nameless dummies, a tool for automatic formula manipulation, with application to the Church-Rosser theorem. *Indag. Math.*, 34:381–392, 1972.

[6] J. Despeyroux, F. Pfenning, and C. Schürmann. Primitive recursion for higher-order abstract syntax. In *Typed Lambda Calculus and Applications, 3rd International Conference*, volume 1210 of *Lecture Notes in Computer Science*, pages 147–163. Springer-Verlag, Berlin, 1997.

[7] M. P. Fiore, G. D. Plotkin, and D. Turi. Abstract syntax and variable binding. In *14th Annual Symposium on Logic in Computer Science*, pages 193–202. IEEE Computer Society Press, Washington, 1999.

[8] M. J. Gabbay. *A Theory of Inductive Definitions with α-Equivalence: Semantics, Implementation, Programming Language*. PhD thesis, University of Cambridge, 2000.

[9] M. J. Gabbay. FM-HOL, a higher-order theory of names. In *Thirty Five years of Automath, Heriot-Watt University, Edinburgh*. Informal proceedings, 2002.

[10] M. J. Gabbay and A. M. Pitts. A new approach to abstract syntax involving binders. In *14th Annual Symposium on Logic in Computer Science*, pages 214–224. IEEE Computer Society Press, Washington, 1999.

[11] M. J. Gabbay and A. M. Pitts. A new approach to abstract syntax with variable binding. *Formal Aspects of Computing*. Special issue in honour of Rod Burstall. To appear.

[12] M. J. C. Gordon and T. F. Melham. *Introduction to HOL. A theorem proving environment for higher order logic*. Cambridge University Press, 1993.

[13] C. A. Gunter and D. S. Scott. Semantic domains. In J. van Leeuwen, editor, *Handbook of Theoretical Computer Science*, volume B, pages 633–674. North-Holland, 1990.

[14] M. Hofmann. Semantical analysis of higher-order abstract syntax. In *14th Annual Symposium on Logic in Computer Science*, pages 204–213. IEEE Computer Society Press, Washington, 1999.

[15] F. Honsell, M. Miculan, and I. Scagnetto. An Axiomatic Approach to Meta-reasoning on Nominal Algebras in HOAS. In *28th International Colloquium on Automata, Languages and Programming, ICALP 2001, Crete, Greece, July 2001, Proceedings*, volume 2076 of *Lecture Notes in Computer Science*, pages 963–978. Springer-Verlag, Heidelberg, 2001.

[16] S. MacLane and I. Moerdijk. *Sheaves in Geometry and Logic. A First Introduction to Topos Theory*. Springer-Verlag, New York, 1992.

[17] K. Meinke and J. V. Tucker. Universal algebra. In S. Abramsky, D. M. Gabbay, and T. S. E. Maibaum, editors, *Handbook of Logic in Computer Science*, volume 1, pages 189–411. Oxford University Press, 1992.

[18] D. Miller. Abstract syntax for variable binders: An overview. In John Lloyd et al, editors, *Computational Logic - CL 2000 First International Conference London, UK, July 24-28, 2000 Proceedings*, volume 1861 of *Lecture Notes in Artificial Intelligence*, pages 239–253. Springer-Verlag, 2000.

[19] R. Milner, M. Tofte, R. Harper, and D. MacQueen. *The Definition of Standard ML (Revised)*. MIT Press, 1997.

[20] M. Nivat and J. C. Reynolds, editors. *Algebraic Methods in Semantics*. Cambridge University Press, 1985.

[21] L. C. Paulson. *Isabelle: A Generic Theorem Prover*, volume 828 of *Lecture Notes in Computer Science*. Springer-Verlag, Berlin, 1994.

[22] S. L. Peyton Jones et al, editors. *Report on the Programming Language Haskell 98. A Non-strict Purely Functional Language*. February 1999. Available from <www.haskell.org>.

[23] A. M. Pitts. Nominal logic, a first order theory of names and binding. Submitted for publication (a preliminary version appeared in the *Proceedings of the 4th International Symposium on Theoretical Aspects of Computer Software* (TACS 2001), LNCS 2215, Springer-Verlag, 2001, pp 219–242), March 2002.

[24] A. M. Pitts and M. J. Gabbay. A metalanguage for programming with bound names modulo renaming. In R. Backhouse and J. N. Oliveira, editors, *Mathematics of Program Construction. 5th International Conference, MPC2000, Ponte de Lima, Portugal, July 2000. Proceedings*, volume 1837 of *Lecture Notes in Computer Science*, pages 230–255. Springer-Verlag, Heidelberg, 2000.

[25] G. D. Plotkin. A structural approach to operational semantics. Technical Report DAIMI FN-19, Aarhus University, 1981.

[26] D. Turi and G. D. Plotkin. Towards a mathematical operational semantics. In *12th Annual Symposium on Logic in Computer Science*, pages 280–291. IEEE Computer Society Press, Washington, 1997.

L(A) = L(B)? Decidability Results from Complete Formal Systems

Géraud Sénizergues
(Gödel Prize 2002)

University of Bordeaux I Nouvelle, Bordeaux, France
ges@labri.u-bordeaux.fr, http://dept-info.labri.u-bordeaux.fr/~ges/

(Laudatio)

In his paper (Theoretical Computer Science, Vol 251 (2001), pages 1–166), the author gives a positive solution to the decidability of the equivalence problem for deterministic pushdown automata: Given two languages L_1 and L_2 accepted by deterministic pushdown automata decide whether $L_1=L_2$. The problem was posed by S. Ginsburg and S. Greibach in 1966 and various subcases were shown to be decidable over years. However, the full question remained elusive until it was finally settled by the awarded. He showed the problem to be decidable. The paper not only settles the equivalence problem for deterministic context-free languages, but also develops an entire machinery of new techniques which are likely to be useful in other contexts. They have already found useful in semantics of programming languages.

P. Widmayer et al. (Eds.): ICALP 2002, LNCS 2380, p. 37, 2002.
© Springer-Verlag Berlin Heidelberg 2002

Discrete Tomography: Reconstruction under Periodicity Constraints

Alberto Del Lungo[1], Andrea Frosini[1], Maurice Nivat[2], and Laurent Vuillon[2]

[1] Dipartimento di Matematica, Università di Siena, Via del Capitano 15, 53100, Siena, Italy {dellungo,frosini}@unisi.it
[2] Laboratoire d'Informatique, Algorithmique, Fondements et Applications (LIAFA) Université Denis Diderot 2, place Jussieu 75251 Paris Cedex 05, France {Maurice.Nivat,Laurent.Vuillon}@liafa.jussieu.fr

Abstract. This paper studies the problem of reconstructing binary matrices that are only accessible through few evaluations of their discrete X-rays. Such question is prominently motivated by the demand in material science for developing a tool for the reconstruction of crystalline structures from their images obtained by high-resolution transmission electron microscopy. Various approaches have been suggested for solving the general problem of reconstructing binary matrices that are given by their discrete X-rays in a number of directions, but more work have to be done to handle the ill-posedness of the problem. We can tackle this ill-posedness by limiting the set of possible solutions, by using appropriate a priori information, to only those which are reasonably typical of the class of matrices which contains the unknown matrix that we wish to reconstruct. Mathematically, this information is modelled in terms of a class of binary matrices to which the solution must belong. Several papers study the problem on classes of binary matrices on which some connectivity and convexity constraints are imposed.
We study the reconstruction problem on some new classes consisting of binary matrices with periodicity properties, and we propose a polynomial-time algorithm for reconstructing these binary matrices from their orthogonal discrete X-rays.

Keywords: Combinatorial problem, discrete tomography, binary matrix, polyomino, periodic constraint, discrete X-rays.

1 Introduction

The present paper studies the possibility of determining the geometrical aspects of a discrete physical structure whose interior is accessible only through a small number of measurements of the atoms lying along a fixed set of directions. This is the central theme of *Discrete Tomography* and the principal motivation of this study is in the attempt to reconstruct three-dimensional crystals from two-dimensional images taken by a transmission electron microscope. The quantitative analysis of these images can be used to determine the number of atoms in atomic lines in certain directions [22,27]. The question is to deduce the local

P. Widmayer et al. (Eds.): ICALP 2002, LNCS 2380, pp. 38–56, 2002.
© Springer-Verlag Berlin Heidelberg 2002

atomic structure of the crystal from the atomic line count data. The goal is to use the reconstruction technique for quality control in VLSI (Very Large Scale Integration) technology. Before showing the results of this paper, we give a brief survey of the relevant contributions in Discrete Tomography.

Clearly, the best known and most important part of the general area of tomography is *Computerized Tomography*, an invaluable tool in medical diagnosis and many other areas including biology, chemistry and material science. Computerized Tomography is the process of obtaining the density distribution within a physical structure from multiple X-rays. More formally, we attempts to reconstruct a density function $f(x)$ for x in $\mathbb{R}^2$ or $\mathbb{R}^3$ from knowledge of its line integral $X_f(L) = \int_L f(x)dx$ for lines L through the space. This line integral is the *X-ray* of $f(x)$ along L. The mapping $f \to X_f$ is known as the *Radon transform*. The mathematics of Computerized Tomography is quite well understood. Appropriate quadratures [28,29] of the Radon inversion formula are used, with concepts from calculus and continuous mathematics playing the main role.

Usually, the physical structure has a very big variety of density values, and so a large number of X-rays are necessary to ensure the accurate reconstruction of their distribution. In some cases the structure that we want to reconstruct has only a small number of possible values. For example, a large number of objects encountered in industrial computerized tomography (for the purpose of non-destructive testing or reverse engineering) [9] are made of a single homogenous material. In many of these applications there are strong technical reasons why only a few X-rays of the structure can be physically determined. *Discrete Tomography* is the area of Computerized Tomography in which these special cases are studied. The name Discrete Tomography is due to Larry Shepp, who organized the first meeting devoted to the topics in 1994.

An example of such a case is the above-mentioned problem of determining local atomic structure of a crystal from the atomic line count data. In a simple but highly relevant model suggested by Peter Schwander and Larry Shepp the possible atom locations in a unit cell of a crystal are defined on the integer lattice $\mathbb{Z}^3$, while the electron beams are modeled as lines parallel to a given direction. The presence of an atom at a specific location corresponds to a pixel value one at the location; the absence of an atom corresponds to a pixel value zero. The number of atoms along certain lines through the crystal (i.e., the sum of pixel values along those lines) define the *discrete X-rays* of the atomic structure with respect to the family of lines involved. Since in practice, one degree of freedom for moving the imaging device is used to control the position of the crystal, the view directions for which data are provided lie all in the same plane. This means that the 3D-problem leads itself to a 2D-slice-by-slice reconstruction. Therefore, the crystal is represented by a binary matrix and its discrete X-ray along a direction u is an integral vector giving the sum of its elements on each line parallel to u (see Fig. 1).

Measurements are usually only available along two, three or four directions, which is much less than what is typical used in Computerized Tomography (a few hundred). In fact, the electron microscope makes measurements at the atomic

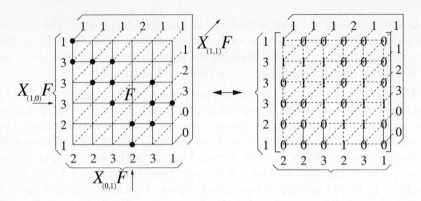

Fig. 1. A subset F of $\mathbb{Z}^2$ with the corresponding binary matrix. $X_{(1,0)}F, X_{(0,1)}F$ and $X_{(1,1)}F$ are the discrete X-rays in the directions $(1,0)$, $(0,1)$ and $(1,1)$.

level and uses high energy (and so deeply penetrating) rays which can corrupt the crystal itself. So, it can take only two, three or four images of the crystal before the energy of the radiations destroys it, or at least changes permanently its atomical configuration so that the subsequent radiations will see something different from the original one.

Now, the problem is to invert the *discrete Radon transform*, i.e., to reconstruct the binary matrix from this small set of discrete X-rays. More precisely, the basic question is to determine, given a set of directions $u_1, \ldots, u_k$ and a set of integral vectors $X_1, \ldots X_k$, whether there exists a binary matrix F whose discrete X-rays along $u_1, \ldots u_k$ are $X_1, \ldots, X_k$. The general methods of Computerized Tomography cannot be used effectively if the number of X-rays is so small, and they seems unlikely to work in practice.

Discrete Tomography has its own mathematical theory mostly based on discrete mathematics. It has some strong connection with combinatorics and geometry. We wish to point out that the mathematical techniques developed in Discrete Tomography have applications in other fields such as: image processing [30], statistical data security [21], biplane angiography [25], graph theory [2] and so on. As a survey of the state of the art of Discrete Tomography we can suggest the book [20].

Interestingly, mathematicians have been concerned with abstract formulations of these problems before the emergence of the practical applications. Many problems of Discrete Tomography were first discussed as combinatorial problems during the late 1950s and early 1960s. In 1957 Ryser [26] and Gale [13] gave a necessary and sufficient condition for a pair of vectors being the discrete X-rays of a binary matrix along horizontal and vertical directions. The discrete X-rays in horizontal and vertical directions are equal to *row* and *column sums* of the matrix. They gave an exact combinatorial characterization of the row and column sums that correspond to a binary matrix, and they derived a fast $O(nm)$ time algorithm for reconstructing a matrix, where n and m denote its sizes. We refer the reader to an excellent survey on the binary matrices with given row and column sums by Brualdi [8].

The space of solutions of the reconstruction problem, however, is really huge and in general quite impossible to control. A good idea may seem to start increasing the number of X-rays one by one in order to decrease the number of solutions. Unfortunately, the reconstruction problem becomes intractable when the number of X-rays is greater than two, as proved in [14]. This means that (unless $P = NP$) exact reconstructions require, in general, an exponential amount of time. In polynomial time only approximate solutions can be expected. In this context, an approximate solution is close to the optimal one if its discrete X-rays in the set of prescribed directions are close to those of the original set. Various approaches have been suggested for solving the problem [12,31,32]. Recently, an interesting method [18] for finding an approximate solutions has been proposed. Even though the reconstruction problem is intractable, some simple algorithms proposed in [18] have good worst-case bounds and they perform even better in computational practice.

Unluckly, this is not still enough. During the last meeting devoted to Discrete Tomography, Gabor T. Herman [19] and Peter Gritzmann [17] stress the fact that various approaches have been suggested for solving the general problem of reconstructing binary matrices that are given by their discrete X-rays in a small number of directions, but more work has to be done to handle the ill-posedness of the problem. In fact, the relevant measure for the quality of a solution of the problem would be its deviation from the original matrix. Hence in order to establish this deviation we would have to know the real binary matrix. However, the goal is to find this unknown original binary matrix so we can only consider measures for the quality of a solution based on the given input discrete X-rays. We have a good solution in this sense if its discrete X-rays in the prescribed directions are close to those of the original matrix. Unfortunately, if the input data do not uniquely determine the matrix even a solution having the given discrete X-rays may be very different from the unknown original matrix. It is shown in [14] that extremely small changes in the data may lead to entirely different solutions. Consequently, the problem is ill-posed, and in a strict mathematical setting we are not able to solve this problem and get the correct solution.

In most practical application we have some a priori information about the images that have to be reconstructed. So, we can tackle the algorithmic challenges induced by the ill-posedness by limiting the class of possible solutions using appropriate prior information. The reconstruction algorithms can take advantage of this further information to reconstruct the binary images.

A first approach is given in [24], where it is posed the hypothesis that the binary matrix is a typical member of a class of binary matrices having a certain Gibbs distribution. Then, by using this information we can limit the class of possible solutions to only those which are close to the given unknown binary matrix. A modified Metropolis algorithm based on the known Gibbs prior provides a good tool to move the reconstruction process toward the correct solution when the discrete X-rays by themselves are not sufficient to find such solution.

A second approach modelled a priori information in terms of a subclass of binary images to which the solution must belong. Several papers study the problem on classes of binary matrices having convexity or connectivity properties. By using these geometric properties we reduce the class of possible solutions. For instance, there is a uniqueness result [15] for the subclass of *convex binary matrices* (i.e., finite subsets F of $\mathbb{Z}^n$ such that $F = \mathbb{Z}^n \cap conv(F)$). It is proved that a convex binary matrix is uniquely determined by its discrete X-rays in certain prescribed sets of four directions or in any seven non-parallel coplanar directions. Moreover, there are efficient algorithms for reconstructing binary matrices of these subclasses defined by convexity or connectivity properties. For example, there are polynomial time algorithms to reconstruct *hv-convex polyominoes* [4, 11,5] (i.e., two-dimensional binary matrices which are 4-connected and convex in the horizontal and vertical directions) and convex binary matrices [6,7] from their discrete X-rays. At the moment, several researchers are studying the following stability question: given a binary matrix having some connectivity and convexity properties and its discrete X-rays along three or four directions, is it possible that small changes in the data lead to "dramatic" change in the binary matrix ?

In this paper, we take the second approach into consideration, and we propose some new subclasses consisting of binary matrices with periodicity properties. The periodicity is a natural constraint and it has not yet been studied in Discrete Tomography. We provide a polynomial-time algorithm for reconstructing $(p, 1)$-periodical binary matrices from their discrete X-rays in the horizontal and vertical directions (i.e., row and column sums). The basic idea of the algorithm is to determine a polynomial transformation of our reconstruction problem to 2-Satisfiability problem which can be solved in linear time [3]. A similar approach has been described in [4,10].

We wish to point out that this paper is only an initial approach to the problem of reconstructing binary matrices having periodicity properties from a small number of discrete X-rays. There are many open problems on these classes of binary matrices of interest to researchers in Discrete Tomography and related fields: the problem of uniqueness, the problem of reconstruction from three or more X-rays, the problem of reconstructing binary matrices having convexity and periodicity properties, and so on.

2   Definitions and Preliminaries

Notations. Let $A^{m \times n}$ be a binary matrix, $r_i = \sum_{j=1}^{n} a_{i,j}$ and $c_j = \sum_{i=1}^{m} a_{i,j}$, for each $1 \le i \le m$ and $1 \le j \le n$. We define $R = (r_1, \ldots, r_m)$ and $C = (c_1, \ldots, c_n)$ as the vectors of *row* and *column sums* of A, respectively. The enumeration of the rows and columns of A starts with row 1 and column 1 which intersect in the upper left position of A. A realization of (R, C) is a matrix B whose row and column sums are R and C.

A binary matrix $A^{m \times n}$ is said to be (p, q)-*periodical* if $a_{i,j} = 1$ implies that

$a_{i+p, j+q} = 1$ if $1 \le i + p \le m$ and $1 \le j + q \le n$,

$a_{i-p, j-q} = 1$ if $1 \le i - p \le m$ and $1 \le j - q \le n$.

Such a matrix is said to have *period* (p, q).

For any given couple (x, y) such that $a_{x,y} = 1$ we define the set P of propagation of the value in position (x, y) in direction (p, q), $P = \{(x + kp, y + kq) | 1 \le x + kp \le m, 1 \le y + kq \le n, k \in \mathbb{Z}\}$. Such set is called a *line*. Each line has a *starting point*, which is its leftmost point, and an *ending point*, which is its rightmost point. We say that a line starts on column j and ends on column j' when its starting and ending points are on column j and j', respectively.

The notion of *box* is a crucial part for our work. Let A be a (p, q)-periodical matrix. From the periodicity it follows that if there exists an index $1 \le i \le m$ such that

$r_i = r_{i+p} + k$, then the positions on row i, from column $n - q + 1$ to column n, contain at least k elements equal to 1. Such positions form a box at the end of row i and will be addressed to as *right box* (rt);

$r_i + k = r_{i+p}$, then on row $i+p$, from column 1 to column q we have k elements equal to 1. Such positions form a box at the beginning of the row i and will be addressed to as *left box* (lt);

We define the upper and lower boxes (up and lw respectively) on columns in the same way (see Fig. 2), if there exists an index $1 \le j \le n$ such that

$c_j = c_{j+q} + k$ then the positions on column j, from row $m - p + 1$ to row m, contain at least k elements equal to 1. Such positions form a box at the end of column j and will be addressed to as lower box (lw);

$c_j + k = c_{j+q}$ then the positions on column $j + q$, from row 1 to row p, contain at least k elements equal to 1. Such positions form a box at the beginning of column j and will be addressed to as upper box (up);

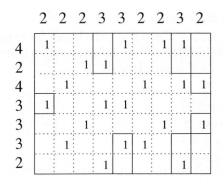

Fig. 2. A $(2, 1)$-periodical binary matrix with upper (up), lower (lw), right (rt) and left (lt) boxes.

Definitions of polyominoes. A *polyomino* P is a finite union of elementary cells of the lattice $\mathbb{Z} \times \mathbb{Z}$ whose interior is connected. This means that, for any pair of cells of P there exists a lattice path in P connecting them (see Fig. 3(a)). A lattice path is a path made up of horizontal and vertical unitary steps. These sets are well-known combinatorial objects [16] and are called *digital 4-connected sets* in discrete geometry and computer vision. We point out that a polyomino can be easily represented by a binary matrix.

A polyomino is said to be *v-convex* [*h-convex*], when its intersection with any vertical [horizontal] line is convex. A polyomino is *hv-convex* or *simply convex* when it is both horizontal and vertical convex. A *parallelogram polyomino* is a polyomino whose boundary consists of two non intersecting paths (except at their origin and extremity) having only north or west steps. Fig. 3 shows polyominoes having the above-mentioned geometric properties.

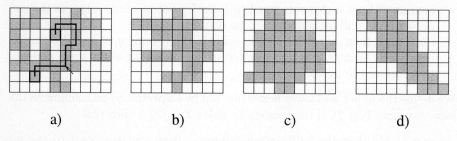

a)   b)   c)   d)

Fig. 3. a) A polyomino. b) A *h*-convex polyomino. c) A *hv*-convex polyomino. d) A parallelogram polyomino.

3   Periodicity $(1, 1)$

Let A be a (1,1)-periodical matrix.

By definition of boxes for $p = 1$ and $q = 1$ the boxes are reduced to only one cell and the integer k of the definition takes only the values 0 or 1. If there exists an index $1 \leq i \leq m$ or $1 \leq j \leq n$ such that

$r_i = r_{i+1} + 1$ then $a_{i,n} = 1$.
$r_i + 1 = r_{i+1}$ then $a_{i+1,1} = 1$.
$c_j = c_{j+1} + 1$ then $a_{m,j} = 1$.
$c_j + 1 = c_{j+1}$ then $a_{1,j+1} = 1$.

A preprocessing part uses the previous box properties to extract the fixed part (called F) of the reconstruction matrix. The following algorithm performed on a given pair of vectors (R, C) gives, if a solution exists, the fixed part (namely the matrix F) and a pair of vectors (R', C') of the mobile part.

Algorithm 1

Input: A pair of integral vectors (R, C);
Output: If $PB = 0$, then it gives Matrix F and couple of integral vector (R', C');
      or If $PB = 1$, then Failure in the reconstruction;
Initialization: $\ell := 0, PB := 0, R^{(\ell)} := R, C^{(\ell)} := C, F = 0^{m \times n}$;
while $R^{(\ell)}$ and $C^{(\ell)}$ are positive non homogeneous vector and $PB = 0$ **do**
    while $R^{(\ell)}$ is positive non homogeneous vector and $PB = 0$ **do**
        $R^{(\ell+1)} := R^{(\ell)}, C^{(\ell+1)} := C^{(\ell)}$;
        determine first index s.t. $r_i^{(\ell)} \neq r_{i+1}^{(\ell)}$;
        if $r_i^{(\ell)} = r_{i+1}^{(\ell)} + 1$ **then** $x := i, y := n$, **Propagation**(x, y, F, ℓ);
        else if $r_i^{(\ell)} + 1 = r_{i+1}^{(\ell)}$ **then** $x := i + 1, y := 1$, **Propagation**(x, y, F, ℓ)
            else $PB := 1$;
        $\ell := \ell + 1$;
    end while;
    while $C^{(\ell)}$ is positive non homogeneous vector and $PB = 0$ **do**
        $R^{(\ell+1)} := R^{(\ell)}, C^{(\ell+1)} := C^{(\ell)}$;
        determine first index s.t. $c_j^{(\ell)} \neq c_{j+1}^{(\ell)}$;
        if $c_j^{(\ell)} = c_{j+1}^{(\ell)} + 1$ **then** $x := m, y := j$, **Propagation**(x, y, F, ℓ)
        else if $c_j^{(\ell)} + 1 = c_{j+1}^{(\ell)}$ **then** $x := 1, y := j + 1$, **Propagation**(x, y, F, ℓ)
            else $PB := 1$;
        $\ell := \ell + 1$;
    end while;
end while;

Propagation(x, y, F, ℓ)

$P = \{(x + k, y + k) | 1 \leq x + k \leq m, 1 \leq y + k \leq n, k \in \mathbb{Z}\}$;
For all $(i, j) \in P$ **do** $F_{i,j} = 1, r_i^{(\ell+1)} = r_i^{(\ell)} - 1, c_j^{(\ell+1)} = c_j^{(\ell)} - 1$;

The main program finds the fixed 1's by considering the differences between the values of the pair of vectors. For each fixed 1, the procedure Propagation fills by periodicity $(1, 1)$ the matrix F and decreases the row and column sum of the current matrix.

At the end of the preprocessing part both row and column vectors are homogeneous ($r_i' = \rho$ for all $1 \leq i \leq m$ and $c_j' = \gamma$ for all $1 \leq j \leq n$).

Now, either (R', C') are zero vectors and then the solution is unique and equal to F, or we perform a reconstruction from homogeneous X-rays with periodicity $(1, 1)$.

Since the vectors are homogeneous we can extend the periodicity on a torus. Indeed, suppose that $r_i' = r_{i+1}'$ with $r_i' = \sum_{j=1}^n a_{i,j}'$ and $r_{i+1}' = \sum_{j=1}^n a_{i+1,j}'$. By periodicity $a_{i,j}' = a_{i+1,j+1}'$, for $j = 1, \ldots, n - 1$, implies that $a_{i,n}' = a_{i+1,1}'$, for $i = 1, \ldots, m - 1$. In other terms the values of the matrix A' are mapped on a cylinder. The same argument in column proves that the values of the matrix A' are mapped on a torus. That is if $a_{i,j}' = 1$ then $a_{i+1 \mod m, j+1 \mod n}' = 1$ and $a_{i-1 \mod m, j-1 \mod n}' = 1$.

So, a solution is formed by loops, namely a beginning in (i, j) with $a'_{i,j} = 1$ and a propagation by periodicity $(1, 1)$ until the position (i, j). All the loops have the same length. As the vectors are homogeneous, we can compute the number of loops. Using this strong condition and the algorithm of Ryser [26] in the first row in order to place the loops, we can reconstruct easily a solution in $O(mn)$ time.

Another remark is the arithmetical nature of the stability of the solution. We can prove that if m and n are relatively prime then there is only one solution. Indeed in this case, to perform a reconstruction of a binary matrix with homogenous vector, we have only one loop felling the whole matrix and then either the matrix A' is full of 1's or full of 0's and nothing between because of the toric conditions.

Proposition 1. *Let $R \in \mathbb{N}^m$ and $C \in \mathbb{N}^n$. If $\gcd(n, m) = 1$, then there is at most a $(1, 1)$-periodical matrix having row and column sums equal to (R, C).*

For example, if we perform a reconstruction with a matrix with m rows and $m + 1$ columns and periodicity $(1, 1)$ then the solution is unique if it exists.

Example 1. For $R = (2, 2, 1, 2), C = (2, 1, 2, 2)$, the algorithm gives the matrix

$$F = \begin{matrix} 0\ 0\ 1\ 0 \\ 0\ 0\ 0\ 1 \\ 0\ 0\ 0\ 0 \\ 1\ 0\ 0\ 0 \end{matrix}$$

and the vectors $R = (1, 1, 1, 1), C = (1, 1, 1, 1)$.
We can reconstruct two solutions for $R = (2, 2, 1, 2), C = (2, 1, 2, 2)$:

$$A'_1 = \begin{matrix} 1\ 0\ 1\ 0 \\ 0\ 1\ 0\ 1 \\ 0\ 0\ 1\ 0 \\ 1\ 0\ 0\ 1 \end{matrix}$$

and

$$A'_2 = \begin{matrix} 0\ 0\ 1\ 1 \\ 1\ 0\ 0\ 1 \\ 0\ 1\ 0\ 0 \\ 1\ 0\ 1\ 0 \end{matrix}.$$

For $R = (1, 2, 2, 2), C = (2, 1, 2, 1, 1)$, the algorithm gives a fixed part and $R = (0, 0, 0, 0), C = (0, 0, 0, 0, 0)$, then the solution is unique.

$$F = \begin{matrix} 0\ 0\ 1\ 0\ 0 \\ 1\ 0\ 0\ 1\ 0 \\ 0\ 1\ 0\ 0\ 1 \\ 1\ 0\ 1\ 0\ 0 \end{matrix}.$$

4   Periodicity $(p, 1)$ with $1 < p < m$

Let A be a matrix with periodicity $(p, 1)$.
The preprocessing part uses only the row sums in order to find the fixed part of the reconstruction. In fact, by definition of boxes for $q = 1$ the horizontal boxes

are reduced to only one cell and the integer k of the definition takes only the values 0 or 1. If there exists an index $1 \leq i \leq m$ such that

$r_i = r_{i+p} + 1$ then $a_{i,n} = 1$.
$r_i + 1 = r_{i+1}$ then $a_{i+p,1} = 1$.

A preprocessing part uses the previous box properties to extract the fixed part (called F) of the reconstruction matrix. The following algorithm performed on a given pair of vectors (R, C) gives, if a solution exists, the fixed part (namely the matrix F) and a pair of (R', C') of the mobile part.

Algorithm 2

Input: A pair of integral vectors (R, C);
Output: If $PB = 0$, then it gives Matrix F and couple of integral vector (R', C');
         or If $PB = 1$ Failure in the reconstruction;
Initialisation: $\ell := 0, PB := 0, R^{(\ell)} := R, C^{(\ell)} := C, F = 0^{m \times n}$;
while $R^{(\ell)}$ is positive non homogeneous vector and $PB = 0$ **do**
    $R^{(\ell+1)} := R^{(\ell)}, C^{(\ell+1)} := C^{(\ell)}$;
    determine first index s.t. $r_i^{(\ell)} \neq r_{i+p}^{(\ell)}$;
    if $r_i^{(\ell)} = r_{i+p}^{(\ell)} + 1$ **then** $x := i, y := n$, **Propagation**(x, y, F, ℓ);
    else if $r_i^{(\ell)} + 1 = r_{i+p}^{(\ell)}$ **then** $x := i + p, y := 1$, **Propagation**(x, y, F, ℓ)
        else $PB := 1$;
    $\ell := \ell + 1$;
end while;

Propagation(x, y, F, ℓ)

$P = \{(x + kp, y + k) | 1 \leq x + kp \leq m, 1 \leq y + k \leq n, k \in \mathbb{Z}\}$;
For all $(i, j) \in P$ **do** $F_{i,j} = 1, r_i^{(\ell+1)} = r_i^{(\ell)} - 1, c_j^{(\ell+1)} = c_j^{(\ell)} - 1$;

The main program finds the fixed 1's by considering the differences between the values of the row sums. For each fixed 1, the procedure Propagation fills by periodicity $(p, 1)$ the matrix F and decreases the row and column sum of the current matrix.

At the end of the preprocessing part the row vector sum R' of A' has the same value on indices in arithmetical progression of rank p: $r_i' = r_{i+p}' = r_{i+2p}' \cdots = r_{i+(\ell-1)p}'$ where $\ell = L$ or $L + 1$. This set of element of the row sums R' of A' is called $line$ of R'. The minimum length of each line of R' is $L = \lfloor \frac{m}{p} \rfloor$. The number of lines of length $L + 1$ and L of R' is $n_{L+1} = m \mod p$ and $n_L = p - n_{L+1}$, respectively.

Example 2. If $(p, 1) = (2, 1)$, and $R = (2, 3, 2, 4, 3, 4, 2), C = (3, 4, 3, 3, 4, 2, 1)$, the algorithm gives the matrix

$$F = \begin{matrix} 0\ 0\ 0\ 0\ 1\ 0\ 0 \\ 0\ 0\ 0\ 0\ 0\ 0\ 0 \\ 0\ 0\ 0\ 0\ 0\ 1\ 0 \\ 1\ 0\ 0\ 0\ 0\ 0\ 0 \\ 1\ 0\ 0\ 0\ 0\ 0\ 1 \\ 0\ 1\ 0\ 0\ 0\ 0\ 0 \\ 0\ 1\ 0\ 0\ 0\ 0\ 0 \end{matrix}$$

and the new vectors are $R' = (1, 3, 1, 3, 1, 3, 1), C' = (1, 2, 3, 3, 3, 1, 0)$. Since $m = 7$ and $p = 2$, we have that $L = 3, n_L = 1, n_{L+1} = 1$. So, $R' = (1, 3, 1, 3, 1, 3, 1)$ contains a line of length $L = 3$ and a line of length $L + 1 = 4$. Since the lines of length $L = 3$ and $L + 1$ are $(*, 3, *, 3, *, 3, *)$ and $(1, *, 1, *, 1, *, 1)$, respectively, we have that matrix A' contains a line of length three lines of length $L = 3$ starting from the second row and $L + 1 = 4$ starting from the first row.

Now, we prove now the values of A' are mapped on a cylinder. We have $r'_i = r'_{i+p}$ with $r'_i = \sum_{j=1}^{n} a'_{i,j}$ and $r'_{i+p} = \sum_{j=1}^{n} a'_{i+p,j}$. From the periodicity $a'_{i,j} = a'_{i+p,j+1}$, for $j = 1, \cdots, n - 1$, it follows that $a'_{i+p,1} = a'_{i,n}$, for $i = 1, \cdots m - p$. In other terms the values of the matrix A' are mapped on a cylinder.

Thus a 1 on the first p rows (at position $(x, y), 1 \leq x \leq p, 1 \leq y \leq n$) can be extended by periodicity on the matrix A' by 1's in positions $(x + kp, y + k \mod n)$ with $k = 0, \cdots, \ell - 1$ where $\ell = L$ or $L + 1$. The matrix A' is in particular composed on a cylinder of lines in direction $(p, 1)$ of length L or $L + 1$. In addition to that the number of lines of length $L + 1$ is exactly $n_{L+1} = m \mod p$ and the number of lines of length L is $n_L = p - n_{L+1}$.

4.1   A Reduction to the Problem of Reconstructing a Special Class of H-Convex Binary Matrices Lying on a Cylinder

Let A' be a solution for a given (R', C') with $r'_i = r'_{i+p} = r'_{i+2p} \cdots = r'_{i+(\ell-1)}$ where $\ell = L$ or $L + 1$. We now perform a reduction of reconstruction of $(p, 1)$-periodical matrix A' on a cylinder to a reconstruction of a special h-convex matrix A'' on a cylinder.

By the previous construction, matrix A' is formed by lines $(x + kp, j + k \mod n)$ with $k = 0, \ldots, \ell - 1$ where $\ell = L$ or $L + 1$. The starting points of the lines is the set of position $S = \{(x, y) | 1 \leq x \leq p, 1 \leq y \leq n, a'_{x,y} = 1\}$. S is ordered by: $(x, y) \leq (x', y')$ if and only if $x \geq x'$ and $y \leq y'$ (i.e., we proceed from bottom to up and from left to right).

Let S' be the set S with an extra index of the rank in the previous order. Each element of S' is a triple (x, y, o), where (x, y) is an element of S and o the rank in the order. Now, we can describe the reduction.

Reduction. Let $(x, y, o) \in S'$. The point (x, y) is the starting point of a line of matrix A' having length L or $L + 1$. This line gives a horizontal bar of 1's begins in position $(o, y + r \mod n)$ with $r = 1, \ldots, \ell$ where $\ell = L$ if $x > n_{L+1}$ or $\ell = L + 1$ if $x \leq n_{L+1}$. The set of these horizontal bars gives the h-convex matrix A''.

Notice that, this transformation makes the column sum of A'' equal to the column sum of A'. In Section 4.3, we will show the inverse reduction that provides A' from A''.

Example 3. Let us take the following matrix A' with periodicity $(2,1)$ and $R' = (3,2,3,2,3)$ and $C' = (3,3,2,1,2,2)$ into consideration.

$$A' = \begin{array}{l} 1\ 0\ 1\ 0\ 0\ 1 \\ 1\ 0\ 0\ 0\ 1\ 0 \\ 1\ 1\ 0\ 1\ 0\ 0 \\ 0\ 1\ 0\ 0\ 0\ 1 \\ 0\ 1\ 1\ 0\ 1\ 0 \end{array}$$

Since $m = 5$ and $p = 2$, we have that $L = 2$, $n_{L+1} = 1$. The matrix A' is composed of three lines of length $L + 1 = 3$ and two lines of length $L = 2$. The starting points in the first two rows (the two first indices are the position in the matrix A' and the last index is the rank in the order) are:

$$S' = \{(2,1,1),(1,1,2),(1,3,3),(2,5,4),(1,6,5)\}.$$

The transformation gives the following h-convex matrix A'' mapped on a cylinder with three bars of length three and two bars of length two.

$$A'' = \begin{array}{l} 1\ 1\ 0\ 0\ 0\ 0 \\ 1\ 1\ 1\ 0\ 0\ 0 \\ 0\ 0\ 1\ 1\ 1\ 0 \\ 0\ 0\ 0\ 0\ 1\ 1 \\ 1\ 1\ 0\ 0\ 0\ 1 \end{array}$$

The column sums of A'' are equal to $C' = (3,3,2,1,2,2)$.

We point out that the order on the starting points adds the following constraints:
Condition 1. On each column of A'' can start at most n_L bars of length L and at most n_{L+1} bars of length $L + 1$. Moreover, by proceeding from up to down on the column at first we find the bars of the length L and then the bars of the length $L + 1$ (see Fig. 4).
We denote the class of h-convex binary matrices lying on a cylinder and satisfying condition 1 by $\mathcal{HC}(n_L, n_{L+1})$.
By this property, the matrix A'' consists of four disjoint zones B, C, E and P whose boundaries are three paths having only north or west steps, and $a''(i,j) = 1$ for $(i,j) \in C \cup P$ and $a''(i,j) = 0$ for $(i,j) \in B \cup E$ (see Fig. 5 and the matrix A'' of the previous example). Notice that, the matrices of the class $\mathcal{HC}(n_L, n_{L+1})$ are set of parallelogram polyominoes lying on a cylinder.

From the reduction it follows that the problem of reconstructing a $(p,1)$-periodical binary matrix A' having row and column sums (R', C') (output of Algorithm 1) is equivalent to the problem of reconstructing a binary matrix A'' of $\mathcal{HP}(n_L, n_{L+1})$ having column sums C', m rows of length L and $L + 1$. We denote this reconstruction problem on the cylinder by **RHC** problem.

In the following subsection, we determine a polynomial transformation of **RHC** problem to 2-Satisfiability problem (2-SAT).

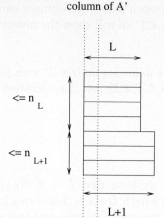

column of A'

Fig. 4. A column of a binary matrix satisfying condition 1.

4.2   A Reduction to the 2-SAT Problem

Given an instance I of **RHC** problem, we want to build a 2-SAT formula Ω (a formula in conjunctive normal form, where each clause has at most two literals) whose satisfiability is linked to the existence of a solution for I in such a way: if Ω is satisfiable, then we are able to reconstruct a solution for I in P-time and, vice versa, each solution of I gives an evaluation of the variables satisfying Ω in P-time. We will do not show the proofs of the lemmas of this section for brevity's sake. Let I be an instance of **RHC** problem; that is:

- two integers n_L, n_{L+1};
- a couple (L, C), where L and $L+1$ are the only possible values of the row sums of a binary matrix A'' of $\mathcal{HC}(n_L, n_{L+1})$, solution of I, and $C = (c_1, \ldots, c_n)$ is its column sums.
- an integer m which denotes the number of rows of A''.

The formula Ω that we want to construct is the conjunction of three 2-SAT formulas: Ω_1 which encodes the geometrical constraints of A'', Ω_2 which gives the consistency of A'' with the couple (L, C) and, finally, Ω_3 which imposes the constraints of condition 1 on each column of A''. The variables of the formula Ω belong to the union of the four disjoint sets of variables:

$$\mathcal{B} = \{b(i,j) : 1 \leq i \leq m, 1 \leq j \leq n\}, \qquad \mathcal{C} = \{c(i,j) : 1 \leq i \leq m, 1 \leq j \leq n\},$$

$$\mathcal{P} = \{p(i,j) : 1 \leq i \leq m, 1 \leq j \leq n\}, \qquad \mathcal{E} = \{e(i,j) : 1 \leq i \leq m, 1 \leq j \leq n\}.$$

We use the variables of the set $\mathcal{X}_i$, with $\mathcal{X}_i \in \{\mathcal{B}, \mathcal{C}, \mathcal{P}, \mathcal{E}\}$, to represent the four disjoint zones B, C, P, E inside A.

Coding in Ω_1 the geometrical constraints of A''. Formula Ω_1 is the conjunction of the following sets of clauses:

$$Corners = \bigwedge_{i,j} \begin{cases} (x(i,j) \Rightarrow x(i-1,j)) \wedge (x(i,j) \Rightarrow x(i,j+1)) \text{ for } x \in \mathcal{C} \bigcup \mathcal{E} \\ \\ (x(i,j) \Rightarrow x(i+1,j)) \wedge (x(i,j) \Rightarrow x(i,j-1)) \text{ for } x \in \mathcal{B} \bigcup \mathcal{P} \end{cases}$$

$$Disj \quad = \bigwedge_{i,j} \left\{ (b(i,j) \Rightarrow \overline{c}(i,j)) \wedge (p(i,j) \Rightarrow b(i,j)) \wedge (e(i,j) \Rightarrow c(i,j)) \right\}$$

$$Compl \quad = \bigwedge_{i,j} \{ \overline{b}(i,j) \Rightarrow c(i,j) \}$$

$$Anch \quad = \{ \overline{e}(1,L) \wedge \overline{p}(m,L+1) \wedge p(r,1) \}$$

with $1 < r \leq m$.

Definition 1. *Let V_1 be an evaluation of the variables in $\mathcal{B}, \mathcal{C}, \mathcal{P}, \mathcal{E}$ which satisfies Ω_1. We define the binary matrix A'' of size $m \times n$ as follows:*

$$(c(i,j) = 1 \ \wedge \ e(i,j) = 0) \Rightarrow a''(i,j) = 1 \ , \qquad p(i,j) = 1 \Rightarrow a''(i,j) = 1 \ ,$$
$$(b(i,j) = 1 \ \wedge \ p(i,j) = 0) \Rightarrow a''(i,j) = 0 \ , \qquad e(i,j) = 0 \Rightarrow a''(i,j) = 0.$$

It is immediate to check that A'' is well defined.

The matrix A'' contains the four zones B, C, E and P of A such that: $(i,j) \in X$, with $X \in \{B, C, E, P\}$ if and only if $x(i,j) = 1$ (see Fig. 5). From $Corners$, $Disj, Compl$ and $Anch$ we deduce the following properties these four zones:

Lemma 1. *i)* $\{B, C\}$ *is a partition of A'', $P \subseteq B$ and $E \subseteq C$;*

ii) the boundary of zones B, C, E and P is made up of three paths having only north or west steps;

iii) there does not exist a column of A'' containing both points of P and points of E.

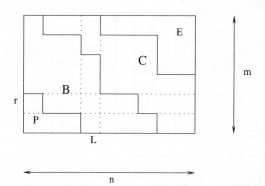

Fig. 5. The matrix A'' and the four zones B, C, E and P of A defined by the formulas *Corners, Disj, Compl* and *Anch*.

Coding in Ω_2 the bound of the row and column sums of A''. The formula Ω_2 is the conjunction of the following sets of clauses:

$$LBC = \bigwedge_{i,j} \left\{ \begin{array}{l} \text{if } j > L,\, e(i,j) \Rightarrow \overline{b}(i + c_j, j) \\[2ex] \text{if } j \leq L,\, b(i,j) \Rightarrow p(i + m - c_j, j) \end{array} \right\}$$

$$UBC = \bigwedge_{i,j} \left\{ \begin{array}{l} \text{if } j > L,\, \overline{e}(i,j) \Rightarrow b(i + c_j, j) \\[2ex] \text{if } j \leq L,\, \overline{b}(i,j) \Rightarrow \overline{p}(i + m - c_j, j) \end{array} \right\}$$

$$UBR = \bigwedge_{i,j} \left\{ \begin{array}{l} \text{if } i < r,\, \overline{b}(i,j) \Rightarrow e(i, j + L + 1) \\[2ex] \text{if } i \geq r,\, p(i,j) \Rightarrow \overline{c}(i, j + n - L - 1) \end{array} \right\}$$

$$LBR = \bigwedge_{i,j} \left\{ \begin{array}{l} \text{if } i < r,\, b(i,j) \Rightarrow \overline{e}(i, j + L) \\[2ex] \text{if } i \geq r,\, \overline{p}(i,j) \Rightarrow c(i, j + n - L) \end{array} \right\}.$$

The formulas LBC and UBC give a lower and an upper bound for the column sums of A. The formula LBR express that the row sums are greater than L. Finally, the formula UBR express that the row sums are smaller than $L + 1$. More precisely,

Lemma 2. *Let A'' be the binary matrix defined by means of the valuation V_2 which satisfies $\Omega_1 \wedge \Omega_2$ as in Definition 1. We have that:*

i)   the column sums of A'' are equal to $C = (c_1, \ldots, c_n)$;
ii) each row sum of A'' has value L or $L + 1$ (see Figure 5).

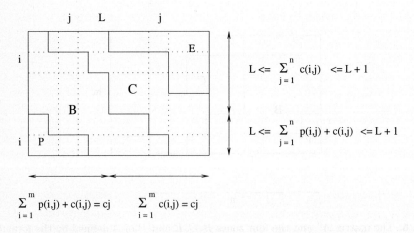

Fig. 6. The matrix A'' with the bounds on row and column sums.

Coding in Ω_3 the maximum number of bars of length L and $L + 1$ starting on each column of A''. The formula Ω_3 is the conjunction of the following sets of clauses:

$$
BB_L \;=\; \bigwedge_{i,j}
\begin{cases}
\text{if } i \le m - c_n, & e(i,j) \Rightarrow \overline{b}(i - n_{L+1}, j - L - 1) \\[4pt]
\text{if } m - c_n < i < r, & r - (m - c_n) - 1 > n_{L+1} \Rightarrow \overline{b}(m - c_n + 1, n - L) \\[4pt]
\text{if } (i \ge r \wedge j \le L), & \overline{p}(i,j) \Rightarrow \overline{b}(i - n_{L+1}, j + n - L - 1)
\end{cases}
$$

$$
BB_{L+1} = \bigwedge_{i,j}
\begin{cases}
\text{if } i \le n - c_n, & c(i,j) \Rightarrow e(i - n_L, j + L) \\[4pt]
\text{if } n - c_n < i < r, & r - (m - c_n) - 1 > z_1 \Rightarrow \overline{c}(m - c_n + n_L + 1, n - L) \\[4pt]
\text{if } i \ge r, & c(i,j) \Rightarrow \overline{p}(i - n_L, j + L - n)
\end{cases}
$$

Lemma 3. *Let A'' be the binary matrix defined by means of the valuation V_3 which satisfies $\Omega_1 \wedge \Omega_2 \wedge \Omega_3$ as in Definition 1. We have that:*

i)   on each column of A'' can start at most n_L bars of length L,

ii) on each column of A'' can start at most n_{L+1} bars of length $L + 1$.

By Lemmas 1 and 3, matrix $A'' \in \mathcal{HP}(n_L, n_{L+1})$. By Lemma 2, matrix A'' satisfies the tomographic constraints. Therefore:

Theorem 1. *$\Omega_1 \wedge \Omega_2 \wedge \Omega_3$ is satisfiable if and only if there is a binary matrix A'' of $\mathcal{HP}(n_L, n_{L+1})$ having column sum C', m rows of length L and $L + 1$.*

Since $\Omega_1 \wedge \Omega_2 \wedge \Omega_3$ is a boolean formula in conjunctive normal form with at most two literals in each clause, by Theorem 1 we have a polynomial time transformation of **HRC** problem to 2-SAT problem which can be solved in linear time [3].

4.3   Final Step

By performing the previous reduction and an algorithm for solving 2-SAT problem [1], we obtain a matrix A'' of $\mathcal{HP}(n_L, n_{L+1})$ having column sums C', m rows of length L and $L + 1$, where $n_{L+1} = m \mod p$ and $n_L = p - n_{L+1}$. Now, for determining a $(p, 1)$-periodical matrix A' having row and column sums equal to (R', C'), we have to perform the inverse of the reduction defined in Section 4.1. We point out that (R', C') is the output of Algorithm 2. This inverse reduction should provides A' from A''. The following algorithm describes this inverse reduction.

Algorithm 3

Input: the matrix A'' whose column sums is vector C', and vector R'
  which is homogeneous with respect to p (R' is the output of Algorithm 2);
Output: the matrix A' having row and column sums (R', C');
Step 1: determine the two vectors $C^{(1)} = (c_1^{(1)}, \ldots, c_n^{(1)})$, $C^{(2)} = (c_1^{(2)}, \ldots, c_n^{(2)})$
  such that $c_i^{(1)}$ and $c_i^{(2)}$ are the number of bars starting from the i-th
  column of A'' having length L and $L + 1$, respectively, ;
Step 2: construct the matrix A' of size $m \times n$ in such a way:
  if $1 \leq i \leq (m - p)$ and $1 \leq j \leq n$, we set $a'(i, j) = 0$;
  if $(m - p + 1) \leq i \leq (m - p + n_{L+1})$ and $1 \leq j \leq n$, **then** perform Ryser's
  reconstruction algorithm on the row and column vectors:
  $(r_{m-p+1}, \ldots, r_{m-p+n_{L+1}})$ and $C^{(1)}$;
  if $(m - p + n_{L+1} + 1) \leq i \leq m$ and $1 \leq j \leq n$, **then** perform Ryser's
  reconstruction algorithm on the row and column vectors:
  $(r_{m-p+n_{L+1}+1}, \ldots, r_m)$ and $C^{(2)}$;
Step 3: for all $m - p + 1 \leq i \leq m$ and $1 \leq j \leq n$,
  if $a'(i, j) = 1$ **then** $a'(i - pk, (j + k) \mod n) = 1$ with $1 \leq i - pk \leq m - p$.

Proposition 2. *The matrix A' which is the output of the Algorithm 3 is $(p, 1)$-periodical and it has row and column sums equal to (R', C').*

We do not show the proof of this Proposition for brevity's sake.
By performing Algorithm 1, the reduction of Section 4.1, the reduction to 2-SAT problem, a linear-time algorithm for solving 2-SAT problem [1], and Algorithm 3, we obtain a $(p, 1)$-periodical binary matrix A having row and column sums equal to (R, C). Since each step can be performed in polynomial time, we have that:

Theorem 2. *The problem of reconstructing $(p, 1)$-periodical binary matrices from their row and column sums can be solved in polynomial time.*

5   Conclusions

Our main purpose has been to introduce periodicity properties in terms relevant for Discrete Tomography. The periodicity is a natural constraint and it has not yet been studied in this field. The motivation of this study is in the attempt to tackle the ill-posedness of the reconstruction problem by limiting the class of possible solutions using appropriate prior information. This means that, we modelled a priori information in terms of a subclass of binary images to which the solution must belong.

By using the periodicity properties we reduce the class of possible solutions. For instance, we proved a uniqueness result for the class of binary matrices having period $(1, 1)$. We have shown a simple greedy algorithm for reconstructing this class of matrices from their row and column sums. This reconstruction problem becomes more difficult for the binary matrices having period $(p, 1)$ or $(1, q)$. We have described a polynomial-time algorithm for solving this problem which use a reduction to 2-Satisfiability problem. We stress the fact that an interesting property of this approach is that it can be used for reconstructing parallegram polyominoes lying a cylinder from row and column sums.

The future challenges concern the reconstruction of binary matrices with a generical period (p, q). We wish to point out that this paper is only an initial approach to the problem of reconstructing binary matrices having periodicity properties from a small number of discrete X-rays. Lot of work should be done to understand such environment: we only challenge the reconstruction problem from two X-rays in some special cases, but MANY consistency, reconstruction and uniqueness problems can be reformulated imposing periodical constraints.

References

1. A. Alpers, P. Gritzmann and L. Thorens, Stability and Instability in Discrete Tomography, *Lecture Notes in Computer Sciece 2243, Digital and Image Geometry*, G. Bertrand, A. Imiya, R. Klette (eds.), (2002) 175-186.
2. R. P. Anstee, Invariant sets of arcs in network flow problems, *Discrete Applied Mathematics*, 13 1-7 (1986).
3. B. Aspvall, M.F. Plass and R.E. Tarjan, A linear-time algorithm for testing the truth of certain quantified Boolean formulas, *Information Processing Letters*, 8 (1979) 121-123.
4. E. Barcucci, A. Del Lungo, M. Nivat and R. Pinzani, Reconstructing convex polyominoes from their horizontal and vertical projections, *Theoretical Computer Science*, 155 (1996) 321-347.
5. E. Barcucci, S. Brunetti, A. Del Lungo and M. Nivat, Reconstruction of lattice sets from their horizontal, vertical and diagonal X-rays, *Discrete Mathematics* 241 (2001) 65-78.
6. S. Brunetti, A. Daurat, Reconstruction of Discrete Sets From Two or More Projections in Any Direction, to appear in *Theoretical Computer Science*.
7. S. Brunetti, A. Daurat and A. Del Lungo, Approximate X-rays reconstruction of special lattice sets, *Pure Mathematics and Applications*, vol. 11, No. 3, (2000) 409-425.
8. R. A. Brualdi, Matrices of zeros and ones with fixed row and column sum vectors, *Lin. Algebra and Its Applications* **33**, 159-231 (1980).
9. J. A. Browne, M. Koshy and J. H. Stanley, On the application of discrete tomography to CT-assisted engineering and design, *International Journal Imaging Systems and Technology*, vol. 9 n.2/3 (1998) 85-98.
10. M. Chrobak, C. Dürr, Reconstructing hv-Convex Polyominoes from Orthogonal Projections, *Information Processing Letters*, 69 (1999) 283-289.
11. A. Del Lungo and M. Nivat, Reconstruction of connected sets from two projections, in *Discrete Tomography: Foundations, Algorithms and Applications*, G.T. Herman and A. Kuba (eds.), Birkhauser, Boston, MA, USA, (1999) 85-113.
12. P. Fishburn, P. Schwander, L. Shepp and R. J. Vanerbei, The discrete Radon tranform and its approximate inversion via linear programming, *Discrete Applied Mathematics*, 75 (1997) 39-61.
13. D. Gale, A theorem on flows in networks, *Pacif. J. Math.*, 7 (1957) 1073-1082.

14. R. J. Gardner, P. Gritzmann and D. Prangenberg, On the computational complexity of reconstructing lattice sets from their X-rays, *Discrete Mathematics*, 202 (1999) 45-71.

15. R.J. Gardner and P. Gritzmann, Discrete tomography: determination of finite sets by X-rays, *Trans. Amer. Math. Soc.* 349 2271-2295 (1997).

16. S. W. Golomb, *Polyominoes*, Revised and Expanded Edition, (Princeton University Press), 1994.

17. P. Gritzmann, Title of the talk: *Recent results, open problems and future challenges*, Workshop Discrete Tomography: Algoritms and Applications, Siena, Italy, 2000 (see http://www.cs.gc.cuny.edu/ gherman/index2new.html).

18. P. Gritzmann, S. de Vries and M. Wingelmann, Approximating binary images from discrete X-rays, *SIAM J. Optimizat.*, 11, No.2, (2000) 522-546.

19. G.T. Herman, Title of the talk: *Bayesian binary tomography with Gibbs distributions as priors*, Workshop Discrete Tomography: Algoritms and Applications; Siena, Italy, 2000 (see http://www.cs.gc.cuny.edu/ gherman/index2new.html).

20. G. T. Herman and A. Kuba (eds.), *Discrete Tomography: Foundations, Algorithms and Applications*, Birkhauser Boston, Cambridge, MA (1999).

21. R. W. Irving and M. R. Jerrum, Three-dimensional statistical data security problems, *SIAM Journal of Computing*, 23 (1994) 170-184.

22. C. Kiesielolowski, P. Schwander, F. H. Baumann, M. Seibt, Y. Kim and A. Ourmazd, An approach to quantitative hight-resolution transmission electron microscopy of crystalline materials, *Ultramicroscopy*, 58 (1995) 131-155.

23. D. Kölzov, A. Kuba and A. Volcic, An algorithm for reconstructing convex bodies from their projections, *Disc. & Comp. Geom.* 4 205-237 (1989).

24. S. Matej, A. Vardi, G. T. Hermann and E. Vardi, Binary tomography using Gibbs priors, in *discrete tomography: foundations, algorithms and applications*, G.T. Herman and A. Kuba (eds.), Birkhauser, Boston, MA, USA, (1999) 191-212.

25. G. P. M. Prause and D. G. W. Onnasch, Binary reconstruction of the heart chambers from biplane angiographic image sequence, *IEEE Transactions Medical Imaging*, 15 (1996) 532-559.

26. H. Ryser, Combinatorial Mathematics, The Carus Mathematical Monographs, Vol. 14, *Math. Assoc. America*, 1963.

27. P. Schwander, C. Kiesielolowski, M. Seibt, F. H. Baumann, Y. Kim and A. Ourmazd, Mapping projected potential, interfacial roughness, and composition in general crystalline solids by quantitative transmission electron microscopy, *Physical Review Letters*, 71 (1993) 4150-4153.

28. L. A. Shepp and B. F. Logan, The Fourier reconstruction of a head, *IEEE Trans. Nucl. Sci.*, NS-21 (1974) 21-43.

29. L. A. Shepp and J. B. Kruskal, Computerized tomography: the new medical X-ray technology, *Amer. Math. Montly*, 85 (1978) 420-439.

30. A. R. Shliferstein and Y. T. Chien, Switching components and the ambiguity problem in the reconstruction of pictures from their projections, *Pattern Recognition*, 10 (1978) 327-340.

31. R. Tijdeman and L. Hajdu, An algorithm for discrete tomography, *Linear Algebra and Its Applications*, 339 (2001) 147-169.

32. Y. Vardi and D. Lee, Discrete Radon transform and its approximate inversion via the EM algorithm, *Int. J. Imaging Sci. Tech.*, 9 (1998) 155-173.

Local and Global Methods in Data Mining: Basic Techniques and Open Problems

Heikki Mannila[1,2]

[1] HIIT Basic Research Unit, University of Helsinki, Department of Computer Science, PO Box 26, FIN-00014 University of Helsinki, Finland
Heikki.Mannila@cs.helsinki.fi, http://www.cs.helsinki.fi/u/mannila
[2] Laboratory of Computer and Information Science, Helsinki University of Technology, PO Box 5400, FIN-02015 HUT, Finland

Abstract. Data mining has in recent years emerged as an interesting area in the boundary between algorithms, probabilistic modeling, statistics, and databases. Data mining research can be divided into global approaches, which try to model the whole data, and local methods, which try to find useful patterns occurring in the data. We discuss briefly some simple local and global techniques, review two attempts at combining the approaches, and list open problems with an algorithmic flavor.

1 Introduction

Data mining has in recent years emerged as an interesting area in the boundary between algorithms, probabilistic modeling, statistics, and databases. A working definition of the area is the following [27]:

> Data mining is the analysis of (often large) observational data sets to find unsuspected relationships and to summarize the data in novel ways that are both understandable and useful to the data analyst.

The question of finding interesting relationships from large masses of data has, of course, been the main occupation of statisticians for a long time. Is there anything new in data mining? While many data mining methods and software packages rely heavily on statistical techniques, there are several new ideas in the area that can be traced to computer science rather than to traditional statistics. Examples of these include flexible predictive modeling methods, some hidden variable techniques, pattern discovery algorithms, scalability issues, and the analysis of heterogenous information sources (such as the Web). One of the driving forces behind these developments is the emphasis in data mining on the analysis of observational data: the data set has not been collected to answer a particular question. For a longer discussion on the differences between data mining and statistics, see [27,26,42].

The data mining literature contains examples of at least two research traditions. The first tradition, probabilistic modeling, views data mining as the task of *approximating the joint distribution*. Once we understand the joint distribution of the data, we can compute all sorts of statistics from it. In this tradition,

P. Widmayer et al. (Eds.): ICALP 2002, LNCS 2380, pp. 57–68, 2002.

the idea is to develop modeling and description methods that incorporate an understanding of the generative process producing the data; thus the approach is global in nature.

The other tradition can be summarized in the slogan: *data mining is the technology of fast counting*. The idea is that one starts with certain types of aggregate information that can be computed efficiently, and then uses those aggregate quantities to describe the data. The most prominent example of this type of work is are association rules [2]. This tradition typically aims at discovering frequently occurring patterns. Each pattern and its frequency indicate only a local property of the data, and a pattern can be understood without having information about the rest of the data. While the patterns by themselves are local, the collection of patterns gives global information about the data set; we will discuss some techniques related to this in Section 4. The interaction between local and global approaches is one of the key issues in data mining.

In this paper we give a brief overview of pattern discovery and probabilistic modeling, outlining the basics of two techniques. We present some recent work that aims at combining the two traditions, and list some open problems. The paper is not a survey of data mining: several important and interesting areas, such as support vector machines in data mining or large-scale decision tree induction, are not treated at all. Systems issues are also not considered.

The rest of this paper is organized as follows. We start in Section 2 by describing some methods for pattern discovery. In Section 3 we review a powerful and general probabilistic modeling technique, mixture modeling. In Section 4 we consider how pattern discovery and approximation of distributions could be connected together, i.e., how the two research traditions could be combined. Section 5 gives a selection of algorithmically flavored open problems in the area.

As an example in the paper we use large-dimensional discrete (or even 0-1) data. Such data occurs, e.g., in the analysis of large document collections. Using the term vector representation such a data set can be considered to be a collection of pairs (d, w), where d is the identification of the document and w is the word occurring in the document. Formulated in another way, the data is a large matrix $A(d, w)$, where the columns correspond to words and the rows correspond to documents. The entry $A(d, w)$ indicates whether word w occurred in the document d, or how many times w occurred in d. The task is to find something interesting from this data set (presumably, to solve some underlying question such as efficient retrieval of relevant documents).

Same type of data occurs in various other applications, too. The log of a web site can be abstractly viewed as a set of triples (h, p, t), where h is the identification of the host from which the access came, p is the identification of the page accessed, and t is the time of the access. If we overlook the time of the access, we get a similar form as above. As another example, consider the case of market basket data, i.e., data containing information about what products were bought in transactions at a supermarket. The rows correspond to transactions, and the columns correspond to products.

Such data sets can be large: they can have millions or hundreds of millions of rows, and the number of columns can be tens of thousands. The matrices are sparse, however: most of the entries are 0.

2   Pattern Discovery

The goal in pattern discovery is to find "interesting" patterns that hold in the data. Such a pattern can for example be a boolean expression (for row-oriented data) or a regular expression (for sequential data).

The simplest case of pattern discovery is finding *association rules* [2]. Consider a relation schema (table header) $R = \{A_1, \ldots, A_p\}$, where the domain of all attributes is the set $\{0, 1\}$. Let d be a data set over R, i.e., a set of binary vectors of length p. Let $\sigma > 0$ be a threshold. A subset $X \subseteq R$ is σ-*frequent* or just frequent, if at least a fraction of σ of all the rows in d have a 1 in all columns of X. That is, $X = \{B_1, \ldots, B_k\}$ is frequent if the result of the query $B_1 = 1 \land B_2 = 1 \land \cdots \land B_k = 1$ contains at least a fraction of σ of all the rows of d. We denote this fraction by $f(X, d)$ or just by $f(X)$, if d is clear from the context.

An *association rule* is an expression $X \Rightarrow Y$, where X and Y are subsets of R. The *accuracy* of the rule is simply the conditional probability of having all variables of Y equal to 1, given that all variables of X are equal to 1:

$$P(Y \mid X) = \frac{f(X \cup Y)}{f(X)}.$$

There exists lots of algorithms for finding all σ-frequent sets in a data set. From these, it is straightforward to find all association rules $X \Rightarrow Y$ such that $f(X) \geq \sigma$. The algorithms for finding frequent sets work in time $O(pnC)$, where the n is the number of rows in the data, p is the number of observations, and C is the number of candidate sets encountered in the search. A subset $Y \subseteq R$ is a *candidate*, if all proper subsets of Y are frequent. The algorithms typically work by first finding the frequent sets of size 1, then building candidates of size 2, finding out which of those are actually frequent, then building candidates of size 3, etc. (see [3,4,33,13]). The basic idea has been varied in many ways, and there are some quite nice data structure issues involved in the efficient implementation of the algorithms. See [26] for a good summary.

The basic association rule and frequent set formalism handles only binary data and positive connections between attributes; there is no real negation nor disjunction (although they can be added, see, e.g., [31]). The output contains lots of rules, so additional tools are needed for helping the user find the interesting rules.

The data sets for which the frequent sets can be applied can have a large number of attributes (hundreds) and lots of rows (millions). The drawback of the method is that the matrix has to be sparse, or at least the number of attributes in the largest frequent set has to be small (since otherwise we would be flooded with frequent sets, as all subsets of a frequent set are frequent).

The algorithm(s) for finding frequent sets actually work for any conjunctive class of patterns. That is, given some atomic formulae β_i whose truth for a row of the data can be determined quickly, we can search for all the conjunctions $\beta_{i_1} \wedge \cdots \wedge \beta_{i_k}$ such that at least a fraction of σ of the rows of the database satisfy the conjunction. Again, we have to assume that the number of such conjunctions is not prohibitively large.

The algorithm actually searches in the lattice of subsets defined by conjunctions of attributes (or basic concepts) [32]. The monotonicity property (a subset of a frequent set is frequent) makes the search easier: we know when to stop. Adding, e.g., disjunctions destroys monotonicity, and thus the search is more difficult. Finding significant rules etc. tends to be much harder than finding frequent patterns, although even there interesting progress has been made (see, e.g., [8]). See [19] for a fascinating approach to assigning interestingness to frequent sets.

As noted above, finding frequent sets (or frequently occurring patterns from some class of patterns) is a local approach to data mining: we search for simple descriptions that are true of a reasonably large fraction of the data set. Each pattern can be understood in isolation, and there is no direct attempt at a global description of the data. However, as the algorithms are complete, i.e., the find all patterns from a class of patterns that satisfy certain conditions, they provide also some global information.

3   Mixture Modeling

Now we turn to the other tradition of data mining, global modeling of the data. There exists a large number of different techniques for modeling, and we will only touch on one general approach.

The basic goal in modeling can be described as follows. Given a set of variables $S = \{B_1, \ldots, B_p\}$, which we for simplicity assume to have finite domains, find a representation F of the underlying distribution such that for each potential observation $x = (x_1, \ldots, x_p)$ the quantity $F(x)$ gives the probability of seeing x.

Suppose now that we have a way of specifying the probability of an observation x by using a function G and some parameters θ, i.e., we could write $G(x) = G(x; \theta)$. Assume further that we feel that a single set of parameters is not sufficient to capture the variability present in the data. Then we can select K different functions $G_1, \ldots, G_K$ and parameter combinations $\theta_1, \ldots, \theta_K$ and model the data using a weighted average of these:

$$G(x) = \sum_{k=1}^{K} \pi_k G_k(x; \theta_k). \tag{1}$$

(In most cases, the form of the component function G_k does not vary with k.)

A typical application of this *mixture modeling* approach is in clustering. Given a set of points in an Euclidean space we can write $G_i(x; \theta_i) = \exp(-\|x - \theta_i\|)$

to mimic points being generated from K different sources with Gaussian error. Or, in the case of binary vectors over the set $R = \{A_1, \ldots, A_p\}$ of attributes, a cluster center can be specified as a vector $c_j = (c_{j1}, \ldots, c_{jp}) \in [0,1]^p$. The likelihood of an observation $t = (t_1, \ldots, t_p)$ can be defined as

$$G_k(t; c_j) = g(t; c_j) = \prod_{k=1}^{p} c_{jk}^{t_k} (1 - c_{jk})^{1-t_k}. \tag{2}$$

The total likelihood of the data would be as in (1).

The mixture modeling framework is quite strong. A somewhat surprising feature is that the mixture coefficients π_k and the individual parameters θ_k can be found using a (conceptually) simple method, the EM-algorithm (expectation maximization). Roughly speaking, what is needed is the ability to fit one-component models: if that can be done, the K-component models can be fitted. For general material on mixture models, see, e.g., [22,35], and for a nice example on how to apply the concepts in clustering of complex objects, see [11],

4   Patterns and Distributions

The previous sections have described some relatively general methods for finding patterns and for modeling data. In this section we describe some work that aims at combining these two approaches. The basic question we pose is the following. If we know the frequencies of certain patterns, what do we actually know about the underlying joint distribution of the data?

Consider the case of frequent sets. Suppose we know the frequencies $f(X)$ for all subsets X such that $f(X) \geq \sigma$. What does this tell us about the full distribution? A possible approach is to use inclusion-exclusion [31,38]. For $\sigma = 0$ the frequencies of frequent set determine the distribution uniquely; however, $\sigma = 0$ means $2^{|R|}$ frequent sets, and therefore the approach is infeasible. For $\sigma > 0$, we can use an truncated version of the inclusion-exclusion method. As an example, we can compute the number $g(\alpha)$ of rows that satisfy the disjunction $\alpha \equiv A_1 = 1 \vee \cdots \vee A_k = 1$ by

$$g(\alpha) = g(A_1 = 1 \vee \cdots \vee A_k = 1) = \sum_{Y \subseteq \{A_1, \ldots, A_k\}} (-1)^{|Y|+1} f(Y). \tag{3}$$

If only the frequencies of sets in $\mathcal{D}$ are known, we have the truncated sum

$$g(\mathcal{D})(\alpha) = g_{\mathcal{D}}(A_1 = 1 \vee \cdots \vee A_k = 1) = \sum_{Y \subseteq \{A_1, \ldots, A_k\}, Y \in \mathcal{D}} (-1)^{|Y|+1} f(Y). \tag{4}$$

An open question is how large the difference $|g(\alpha) - g_{\mathcal{D}}(\alpha)|$ can be (see Section 5). The inclusion-exclusion approach does not try to build a model of the distribution; in some respects, it can be considered to be a worst-case approach.

Given the frequencies of the frequent sets, could we somehow form a distribution? There are several possible distributions satisfying the constraints of the frequent sets; thus we need additional assumptions to obtain a unique distribution. As an example, suppose we have three 0-1 valued attributes A, B, and C, and assume we know that in dataset d we have $f(A) = 0.5$, $f(B) = 0.4$, and $f(C) = 0.2$. (We write $f(A)$ for $f(\{A\})$ and $f(AB)$ for $f(\{A, B\})$.) A natural assumption in this case is the independence of attributes. To estimate the frequency $f(ABC)$, i.e., the probability of the event $A = 1 \wedge B = 1 \wedge C = 1$, we obtain $f(ABC) = f(A)f(B)f(C) = 0.04$.

If we have information only about individual attributes, then independence is a natural assumption. (See [24] for a possible explanation for the usefulness of the independence assumption.) What if we have more data? For example, suppose somebody would tell us, in addition to the frequencies of the three attributes, also that $f(AB) = 0.3$ and $f(BC) = 0.2$. The *maximum entropy approach* (see, e.g., [18]) states that one selects the distribution that satisfies the constraints and, among the distributions with this property, has the highest entropy. It can be shown that under fairly general conditions such a distribution (the maximum entropy distribution) exists and that it can be found using a simple algorithm, the iterative scaling method [18]. In the above example, for the maximum entropy distribution satisfying the 5 constraints (on A, B, C, AB, and BC) we have $f(ABC) = 0.15$.

The maximum entropy method can be applied to information produced by pattern discovery algorithms. Suppose we have computed the frequent sets of a threshold σ. That is, for certain sets $X_i \subseteq R$ we know the frequencies $f(X_i)$ in the data. These can be viewed as constraints on the underlying (unknown) distribution g. Given a set $Y \subseteq R$ of attributes we can compute the maximum entropy distribution for Y as follows. We take all sets X_i such that $X_i \subseteq Y$, and find the maximum entropy distribution that satisfies these constraints. The complexity of this is dominated by the number of states in the distribution, i.e., $2^{|Y|}$. Thus the method cannot be applied to very large sets of attributes. Note that R can be arbitrarily large; the restriction is on the size of the set Y of variables for which we want to find the joint distribution. The maximum entropy method produces very good approximations to the joint density [38,37]. Overall, understanding what pattern data tells us about distributions is an interesting problem.

5   Open Problems and Research Themes

This section lists some algorithmically oriented problems in data mining. The list is biased, and the problems are fairly different in nature. Some are explicit algorithmic questions, while others are just vague topics.

5.1   Pattern Discovery, Mixture Modeling, and Combinations

Most of the pattern discovery algorithms have focused on reasonably simple patterns such as association rules. It is not clear how far the existing methods can be generalized.

Problem 1 *Consider 0-1 data and a class $\mathcal{P}$ of patterns defined as boolean expressions. Under what conditions for $\mathcal{P}$ is there an efficient algorithm for finding those patterns $\theta \in \mathcal{P}$ that are true for at least a given fraction of rows of the data set?*

Many results exist already; see, e.g., [9,40,39].

In pattern discovery one often has to analyze different but similar data sets: transactions from last month and from this month etc.

Problem 2 *Given a class $\mathcal{P}$ of patterns and two data sets d and d', find a good way of representing the patterns of $\mathcal{P}$ that are true in d', assuming the user knows the patterns that are true in d. That is, explain the differences between d and d' (from the point of view of the pattern class $\mathcal{P}$).*

See [12,23] for an interesting approaches, and [14] for finding distances between subtables.

For example in biological and telecommunications applications finding patterns on sequences would be desirable. There are algorithms for finding sequential patterns or episodes in sequences [5,34,41], but they typically provide very partial descriptions of sequences. For more global descriptions the work on learning finite-state automata and on hidden Markov models is obviously relevant.

Problem 3 *Find a good concept class and algorithm for learning simple global descriptions of sequences of events.*

Mixture modeling is a powerful technique, and the results are quite general, showing convergence under weak assumptions. The results, however, do not tell much about the computational complexity of the problem.

Problem 4 *Can anything be said about computational complexity of the mixture modeling problem?*

Problem 5 *Maximum entropy is a way of inducing distributions from frequent sets. Are there other approaches for the same task?*

The preceding discussion has concentrated on row-oriented data. Similar questions can also be asked about string data.

Problem 6 *Given a long string s and a threshold σ. For a string w, denote by $h(w)$ the number of times w occurs as a substring in s. Suppose we know $h(w)$ for all w such that $h(w) \geq \sigma$. What do we know about the string s?*

Frequent sets are perhaps the simplest form of patterns, and using them in approximations leads to interesting issues.

Problem 7 *(Not very important.) Given a collection $\mathcal{L}$ of subsets of R such that for all $X \in \mathcal{L}$ we have $|X| = s$. The upper shadow [10] $\partial \mathcal{L}$ of $\mathcal{L}$ is the collection of sets of size $s+1$ whose all subsets are in $\mathcal{L}$. What is the complexity of computing the upper shadow?*

It is easy to see the following.

Proposition 8 *For any boolean query α there exists sets X_i and coefficients u_i such that for any dataset d we have*

$$g(\alpha, d) = \sum_i u_i f(X_i, d)$$

As an example, for $\alpha \equiv A_3 = 1 \wedge (A_1 = 0 \vee A_2 = 1)$ we can write $g(\alpha, d) = f(A_3, d) - f(A_3 A_1, d) + f(A_3 A_1 A_2, d)$. The following is probably known, but I have not been able to find any references.

Problem 9 *(Not very important.) How large do the coefficients u_i have to be in the expression of Proposition 8?*

The following is more interesting.

Problem 10 *Given a Boolean formula α, consider the inclusion-exclusion description of $g(\alpha, d)$ given in Proposition 8. Let $g_D(, \alpha, d)$ be the sum when only frequent sets from D are considered. How large can the the approximation error $e(\alpha, D) = |g(\alpha) - g_D(\alpha)|$ be as a function of structural properties of D and α.*

One approach to this problem leads to the following combinatorial conjecture. Given a downwards closed collection D of subsets of R, define the positive border $Bd^+(D)$ to be the maximal elements of D, and the negative border $Bd(D)$ to be the minimal elements of $P(R) \setminus D$.

Conjecture 11 *For $\alpha = \bigvee_{A \in X} A$ we have*

$$e(\alpha, D) \leq \sum_{Y \in Bd^+(D)} f(Y) + \sum_{Y \in Bd^-(D)} f(Y).$$

This bound is in most cases quite weak, but proving it seems difficult. The conjecture would follow from the following.

Conjecture 12 *For downwards closed collection D we have*

$$\sum_{\emptyset \neq Y \in D} (-1)^{|Y|+1} \leq |Bd^+(D)| + |Bd^-(D)|$$

The notions of positive and negative border are connected to the question of finding transversals of hypergraphs. Given a hypergraph, a transversal is a minimal set of vertices that intersects all edges. The following problem has lots of interesting applications also outside data mining.

Problem 13 *Find a polynomial time algorithm (in the size of the input and output) for enumerating all transversals.*

See [21] for recent results and [20] for a list of equivalent problems. In data mining the problem arises also in connection with finding functional dependencies.

5.2   Other Topics

Dimensionality reduction. Many of the discrete datasets encountered in data mining have very high dimensionality: document databases can have hundreds of thousands of dimensions. Of course, the data is very sparse, so the true dimensionality is less. Still, there is a clear need for dimensionality reduction techniques.

Problem 14 *Why does LSI (latent semantic indexing) [17] work for discrete data?*

The seminal paper [36] provided part of the answer, and [7] gives additional results. Still, there seems to be lots of open issues.

Problem 15 *Randomized rounding [1] seems to work very well for computing low-rank approximations. What are the boundaries of this approach?*

Optimal summary statistics. We have discussed the use of frequent sets as summary statistics for answering queries, or, more generally, for describing the data set.

Frequent sets happen to be simple to compute. It is not at all clear that they are the optimal summary statistics to compute from the data.

Problem 16 *Given a relation schema R and a class of boolean queries Q for datasets over R and an integer k. Find k queries $\alpha_1, \ldots, \alpha_k$ such that for all datasets d over R we can answer any query $\beta \in Q$ on d using the answers of α_i on d.*

Similarity of attributes in discrete data sets.

Problem 17 *Given a set of attributes $S = \{B_1, \ldots, B_p\}$, where each B_i can obtain values from a finite domain D_i, find a good way of defining the similarity of values in each D_i.*

Some attempts have been made [25,16,15]. Especially the algorithm in [25] is intriguing: why does it work? What does it do?

Framework for data mining.

Problem 18 *What is a good theoretical framework for data mining?*

See [30] for a relatively recent discussion on the alternatives. The approach in the paper [28] deserves very careful study.

Privacy. When applied to personal data any data analysis method can lead to problems. Data mining is not an exception. There are powerful new developments on the area of finding ways of guaranteeing privacy in data analysis [6,29].

Problem 19 *Explore the boundaries of provably privacy-preserving data mining.*

References

1. D. Achlioptas and F. McSherry. Fast computation of low-rank approximations. In *STOC 01*, pages 611–618, 2001.
2. R. Agrawal, T. Imielinski, and A. Swami. Mining association rules between sets of items in large databases. In P. Buneman and S. Jajodia, editors, *Proceedings of ACM SIGMOD Conference on Management of Data (SIGMOD'93)*, pages 207 – 216, Washington, D.C., USA, May 1993. ACM.
3. R. Agrawal, H. Mannila, R. Srikant, H. Toivonen, and A. I. Verkamo. Fast discovery of association rules. In U. M. Fayyad, G. Piatetsky-Shapiro, P. Smyth, and R. Uthurusamy, editors, *Advances in Knowledge Discovery and Data Mining*, pages 307 – 328. AAAI Press, Menlo Park, CA, 1996.
4. R. Agrawal and R. Srikant. Fast algorithms for mining association rules in large databases. In *Proceedings of the Twentieth International Conference on Very Large Data Bases (VLDB'94)*, pages 487 – 499, Sept. 1994.
5. R. Agrawal and R. Srikant. Mining sequential patterns. In *Proceedings of the Eleventh International Conference on Data Engineering (ICDE'95)*, pages 3 – 14, Taipei, Taiwan, Mar. 1995.
6. R. Agrawal and R. Srikant. Privacy-preserving data mining. In *ACM SIGMOD*, 2000.
7. Y. Azar, A. Fiat, A. R. Karlin, F. McSherry, and J. Saia. Spectral analysis of data. In *ACM Symposium on Theory of Computing*, 2000.
8. R. J. Bayardo Jr. and R. Agrawal. Mining the most interesting rules. In *Proc. of the Fifth ACM SIGKDD Int'l Conf. on Knowledge Discovery and Data Mining*, pages 145–154, 1999.
9. R. J. Bayardo Jr., R. Agrawal, and D. Gunopulos. Constraint-based rule mining in large, dense databases. *Data Mining and Knowledge Discovery*, 4(2/3):217–240, 2000.
10. B. Bollobás. *Combinatorics*. Cambridge University Press, Cambridge, 1986.
11. I. V. Cadez, S. Gaffney, and P. Smyth. A general probabilistic framework for clustering individuals and objects. In *KDD 2000*, pages 140–149, 2000.
12. S. Chakrabarti, S. Sarawagi, and B. Dom. Mining surprising patterns using temporal description length. In A. Gupta, O. Shmueli, and J. Widom, editors, *VLDB'98*, pages 606–617. Morgan Kaufmann, 1998.
13. E. Cohen, M. Datar, S. Fujiwara, A. Gionis, P. Indyk, R. Motwani, J. D. Ullman, and C. Yang. Finding interesting associations without support pruning. *IEEE Transactions on Knowledge and Data Engineering*, 13(1):64–78, 2001.
14. G. Cormode, P. Indyk, N. Koudas, and S. Muthukrishnan. Fast mining of massive tabular data via approximate distance computations. In *ICDE2002*, 2002.
15. G. Das and H. Mannila. Context-based similarity measures for categorical databases. In *PKDD 2000*, 2000.
16. G. Das, H. Mannila, and P. Ronkainen. Similarity of attributes by external probes. In R. Agrawal, P. Stolorz, and G. Piatetsky-Shapiro, editors, *Proceedings of the Fourth International Conference on Knowledge Discovery and Data Mining (KDD'98)*, pages 23 – 29, New York, NY, USA, Aug. 1998. AAAI Press.
17. S. C. Deerwester, S. T. Dumais, T. K. Landauer, G. W. Furnas, and R. A. Harshman. Indexing by latent semantic analysis. *Journal of the American Society of Information Science*, 41(6):391–407, 1990.
18. S. Della Pietra, V. J. Della Pietra, and J. D. Lafferty. Inducing features of random fields. *IEEE Transactions on Pattern Analysis and Machine Intelligence*, 19(4):380–393, 1997.

19. W. DuMouchel and D. Pregibon. Empirical bayes screening for multi-item associations. In *KDD-2001*, pages 67–76, 2001.
20. T. Eiter and G. Gottlob. Identifying the minimal transversals of a hypergraph and related problems. *SIAM Journal on Computing*, 24(6):1278 – 1304, Dec. 1995.
21. T. Eiter, G. Gottlob, and K. Makino. New results on monotone dualization and generating hypergraph transversals. In *STOC'02*, 2002.
22. B. Everitt and D. Hand. *Finite Mixture Distributions*. Chapman and Hall, London, 1981.
23. V. Ganti, J. Gehrke, and R. Ramakrishnan. Demon: Mining and monitoring evolving data. *IEEE Transactions on Knowledge and Data Engineering*, 13(1):50–63, 2001.
24. A. Garg and D. Roth. Understanding probabilistic classifiers. In *ECML 2001*, pages 179–191, 2001.
25. D. Gibson, J. M. Kleinberg, and P. Raghavan. Clustering categorical data: An approach based on dynamical systems. In A. Gupta, O. Shmueli, and J. Widom, editors, *VLDB'98*, pages 311–322. Morgan Kaufmann, 1998.
26. J. Han and M. Kamber. *Data Mining: Concepts and Techniques*. Morgan Kaufmann, 2000.
27. D. Hand, H. Mannila, and P. Smyth. *Principles of Data Mining*. MIT Press, 2001.
28. J. M. Kleinberg, C. H. Papadimitriou, and P. Raghavan. A microeconomic view of data mining. *Data Mining and Knowledge Discovery*, 2(4):311–324, 1998.
29. Y. Lindell and B. Pinkas. Privacy preserving data mining. In *Crypto 2000*, pages 36–54. Springer-Verlag, 2000.
30. H. Mannila. Theoretical frameworks for data mining. *SIGKDD Explorations*, 1(2):30–32, 2000.
31. H. Mannila and H. Toivonen. Multiple uses of frequent sets and condensed representations. In E. Simoudis, J. Han, and U. Fayyad, editors, *Proceedings of the Second International Conference on Knowledge Discovery and Data Mining (KDD'96)*, pages 189 – 194, Portland, Oregon, Aug. 1996. AAAI Press.
32. H. Mannila and H. Toivonen. Levelwise search and borders of theories in knowledge discovery. *Data Mining and Knowledge Discovery*, 1(3):241 – 258, Nov. 1997.
33. H. Mannila, H. Toivonen, and A. I. Verkamo. Efficient algorithms for discovering association rules. In U. M. Fayyad and R. Uthurusamy, editors, *Knowledge Discovery in Databases, Papers from the 1994 AAAI Workshop (KDD'94)*, pages 181 – 192, Seattle, Washington, USA, July 1994. AAAI Press.
34. H. Mannila, H. Toivonen, and A. I. Verkamo. Discovery of frequent episodes in event sequences. *Data Mining and Knowledge Discovery*, 1(3):259 – 289, Nov. 1997.
35. G. McLachlan and D. Peel. *Finite Mixture Distributions*. Wiley, New York, 2000.
36. C. H. Papadimitriou, P. Raghavan, H. Tamaki, and S. Vempala. Latent semantic indexing: A probabilistic analysis. In *Proc. 17th ACM Symposium on the Principles of Database Systems (PODS'98)*, pages 159–168, Seattle, WA, June 1998.
37. D. Pavlov, H. Mannila, and P. Smyth. Probabilistic models for query approximation with large sparse binary data sets. In *Sixteenth Conference on Uncertainty in Artificial Intelligence (UAI-00)*, 2000.
38. D. Pavlov, H. Mannila, and P. Smyth. Beyond independence: Probabilistic models for query approximation on binary transaction data. Technical Report Technical Report UCI-ICS TR-01-09, Information and Computer Science Department, UC Irvine, 2001.
39. J. Pei, J. Han, and L. V. S. Lakshmanan. Mining frequent item sets with convertible constraints. In *ICDE 2001*, pages 433–442, 2001.

40. J. Pei, J. Han, H. Lu, S. Nishio, S. Tang, and D. Yang. H-mine: Hyper-structure mining of frequent patterns in large databases. In *Proc. 2001 Int. Conf. on Data Mining (ICDM'01)*, 2001.
41. R. Ramakrishnan and J. Gehrke. *Database Management Systems (2nd ed.)*. McGraw-Hill, 2001.
42. P. Smyth. Data mining at the interface of computer science and statistics. In *Data Mining for Scientific and Engineering Applications*, 2002. To appear.

Program Debugging and Validation Using Semantic Approximations and Partial Specifications

M. Hermenegildo, G. Puebla, F. Bueno, and P. López-García

School of Computer Science
Technical University of Madrid, Spain
herme@fi.upm.es, http://www.clip.dia.fi.upm.es/~herme

(Extended Abstract)

The technique of Abstract Interpretation [11] has allowed the development of sophisticated program analyses which are provably correct and practical. The semantic approximations produced by such analyses have been traditionally applied to *optimization* during program compilation. However, recently, novel and promising applications of semantic approximations have been proposed in the more general context of program *validation* and *debugging* [3,9,7].

We study the case of (Constraint) Logic Programs (CLP), motivated by the comparatively large body of approximation domains, inference techniques, and tools for abstract interpretation-based semantic analysis which have been developed to a powerful and mature level for this programming paradigm (see, e.g., [23,8,18,6,12] and their references). These systems can approximate at compile-time a wide range of properties, from directional types to determinacy or termination, always safely, and with a significant degree of precision. Thus, our approach is to take advantage of these advances in program analysis tools within the context of program validation and debugging, rather than using traditional proof-based methods (e.g., [1,2,13,17,28]), developing new tools and procedures, such as specific concrete [4,15,16] or abstract [9,10] diagnosers and declarative debuggers, or limiting error detection to run-time checking [28].

In this talk we discuss these issues and present a framework for combined static/dynamic validation and debugging using semantic approximations [7,26, 21] which is meant to be a part of an advanced program development environment comprising a variety of co-existing tools [14]. Program validation and detection of errors is first performed at compile-time by inferring properties of the program via abstract interpretation-based static analysis and comparing this information against (partial) specifications written in terms of assertions. Such assertions are linguistic constructions which allow expressing properties of programs.

Classical examples of assertions are type declarations (e.g., in the context of (C)LP those used by [22,27,5]). However, herein we are interested in supporting a more general setting in which assertions can be of a much more general nature, stating additionally other properties, some of which cannot always be determined statically for all programs. These properties may include properties defined by means of user programs and extend beyond the predefined set which may be

P. Widmayer et al. (Eds.): ICALP 2002, LNCS 2380, pp. 69–72, 2002.
© Springer-Verlag Berlin Heidelberg 2002

natively understandable by the available static analyzers. Also, in the proposed framework only a small number of (even zero) assertions may be present in the program, i.e., the assertions are *optional*. In general, we do not wish to limit the programming language or the language of assertions unnecessarily in order to make the validity of the assertions statically decidable (and, consequently, the proposed framework needs to deal throughout with *approximations*). We present a concrete language of assertions which allows writing this kind of (partial) specifications for (C)LP [25].

The assertion language is also used by the compiler to express both the information inferred by the analysis and the results of the comparisons performed against the specifications.[1] These comparisons can result in proving statically (i.e., at compile-time) that the assertions hold (i.e., they are validated) or that they are violated, and thus bugs have been detected. User-provided assertions (or *parts* of assertions) which cannot be statically proved nor disproved are optionally translated into run-time tests. Both the static and the dynamic checking are provably *safe* in the sense that all errors flagged are definite violations of the specifications.

We illustrate the practical usefulness of the framework by demonstrating what is arguably the first and most complete implementation of these ideas: CiaoPP, the Ciao system preprocessor [24,20]. Ciao is a public-domain, next-generation (constraint) logic programming system, which supports ISO-Prolog, but also, selectively for each module, pure logic programming, functions, constraints, objects, or higher-order. Ciao is specifically designed to a) be highly extendible and b) support modular program analysis, debugging, and optimization. The latter tasks are performed in an integrated fashion by CiaoPP.

CiaoPP, which incorporates analyses developed by several groups in the community, uses abstract interpretation to infer properties of program predicates and literals, including types, modes and other variable instantiation properties, non-failure, determinacy, bounds on computational cost, bounds on sizes of terms in the program, etc. It processes modules separately, performing incremental analysis. CiaoPP can find errors at compile-time (or perform partial verification), by checking how programs call system libraries and also by checking assertions present in the program or in other modules used by the program. This allows detecting errors in user programs even if they contain no assertions. In addition, CiaoPP also performs program transformations and optimizations such as multiple abstract specialization, parallelization (including granularity control), and inclusion of run-time tests for assertions which cannot be checked completely at compile-time.

The implementation of the preprocessor is generic in that it can be easily customized to different CLP systems and dialects and in that it is designed to

---

[1] Interestingly, the assertions are also quite useful for generating documentation automatically using [19]

allow the integration of additional analyses in a simple way (for example, it has been adapted for use with the CHIP CLP(fd) system).

More info: For more information, full versions of selected papers and technical reports, and/or to download Ciao and other related systems please access http://www.clip.dia.fi.upm.es/.

Keywords: Global Analysis, Debugging, Verification, Parallelization, Optimization, Abstract Interpretation.

References

[1] K. R. Apt and E. Marchiori. Reasoning about Prolog programs: from modes through types to assertions. *Formal Aspects of Computing*, 6(6):743–765, 1994.

[2] K. R. Apt and D. Pedreschi. Reasoning about termination of pure PROLOG programs. *Information and Computation*, 1(106):109–157, 1993.

[3] F. Bourdoncle. Abstract debugging of higher-order imperative languages. In *Programming Languages Design and Implementation'93*, pages 46–55, 1993.

[4] J. Boye, W. Drabent, and J. Małuszyński. Declarative diagnosis of constraint programs: an assertion-based approach. In *Proc. of the 3rd. Int'l Workshop on Automated Debugging–AADEBUG'97*, pages 123–141, Linköping, Sweden, May 1997. U. of Linköping Press.

[5] F. Bueno, D. Cabeza, M. Carro, M. Hermenegildo, P. López-García, and G. Puebla. The Ciao Prolog System. Reference Manual. TR CLIP3/97.1, School of Computer Science, Technical University of Madrid (UPM), August 1997. System and on-line version of the manual available at http://clip.dia.fi.upm.es/Software/Ciao/.

[6] F. Bueno, D. Cabeza, M. Hermenegildo, and G. Puebla. Global Analysis of Standard Prolog Programs. In *European Symposium on Programming*, number 1058 in LNCS, pages 108–124, Sweden, April 1996. Springer-Verlag.

[7] F. Bueno, P. Deransart, W. Drabent, G. Ferrand, M. Hermenegildo, J. Maluszynski, and G. Puebla. On the Role of Semantic Approximations in Validation and Diagnosis of Constraint Logic Programs. In *Proc. of the 3rd. Int'l Workshop on Automated Debugging–AADEBUG'97*, pages 155–170, Linköping, Sweden, May 1997. U. of Linköping Press.

[8] B. Le Charlier and P. Van Hentenryck. Experimental Evaluation of a Generic Abstract Interpretation Algorithm for Prolog. *ACM Transactions on Programming Languages and Systems*, 16(1):35–101, 1994.

[9] M. Comini, G. Levi, M. C. Meo, and G. Vitiello. Proving properties of logic programs by abstract diagnosis. In M. Dams, editor, *Analysis and Verification of Multiple-Agent Languages, 5th LOMAPS Workshop*, number 1192 in Lecture Notes in Computer Science, pages 22–50. Springer-Verlag, 1996.

[10] M. Comini, G. Levi, M. C. Meo, and G. Vitiello. Abstract diagnosis. *Journal of Logic Programming*, 39(1–3):43–93, 1999.

[11] P. Cousot and R. Cousot. Abstract Interpretation: a Unified Lattice Model for Static Analysis of Programs by Construction or Approximation of Fixpoints. In *4th. ACM Symp. on Principles of Programming Languages*, pages 238–252, 1977.

[12] M. García de la Banda, M. Hermenegildo, M. Bruynooghe, V. Dumortier, G. Janssens, and W. Simoens. Global Analysis of Constraint Logic Programs. *ACM Transactions on Programming Languages and Systems*, 18(5):564–615, September 1996.

[13] P. Deransart. Proof methods of declarative properties of definite programs. *Theoretical Computer Science*, 118:99–166, 1993.

[14] P. Deransart, M. Hermenegildo, and J. Maluszynski. *Analysis and Visualization Tools for Constraint Programming*. Number 1870 in LNCS. Springer-Verlag, September 2000.

[15] W. Drabent, S. Nadjm-Tehrani, and J. Małuszyński. The Use of Assertions in Algorithmic Debugging. In *Proceedings of the Intl. Conf. on Fifth Generation Computer Systems*, pages 573–581, 1988.

[16] W. Drabent, S. Nadjm-Tehrani, and J. Maluszynski. Algorithmic debugging with assertions. In H. Abramson and M.H.Rogers, editors, *Meta-programming in Logic Programming*, pages 501–522. MIT Press, 1989.

[17] G. Ferrand. Error diagnosis in logic programming. *J. Logic Programming*, 4:177–198, 1987.

[18] J.P. Gallagher and D.A. de Waal. Fast and precise regular approximations of logic programs. In Pascal Van Hentenryck, editor, *Proc. of the 11th International Conference on Logic Programming*, pages 599–613. MIT Press, 1994.

[19] M. Hermenegildo. A Documentation Generator for (C)LP Systems. In *International Conference on Computational Logic, CL2000*, number 1861 in LNAI, pages 1345–1361. Springer-Verlag, July 2000.

[20] M. Hermenegildo, F. Bueno, G. Puebla, and P. López-García. Program Analysis, Debugging and Optimization Using the Ciao System Preprocessor. In *1999 International Conference on Logic Programming*, pages 52–66, Cambridge, MA, November 1999. MIT Press.

[21] M. Hermenegildo, G. Puebla, and F. Bueno. Using Global Analysis, Partial Specifications, and an Extensible Assertion Language for Program Validation and Debugging. In K. R. Apt, V. Marek, M. Truszczynski, and D. S. Warren, editors, *The Logic Programming Paradigm: a 25–Year Perspective*, pages 161–192. Springer-Verlag, July 1999.

[22] P. Hill and J. Lloyd. *The Goedel Programming Language*. MIT Press, Cambridge MA, 1994.

[23] K. Muthukumar and M. Hermenegildo. Compile-time Derivation of Variable Dependency Using Abstract Interpretation. *Journal of Logic Programming*, 13(2/3):315–347, July 1992.

[24] G. Puebla, F. Bueno, and M. Hermenegildo. A Generic Preprocessor for Program Validation and Debugging. In P. Deransart, M. Hermenegildo, and J. Maluszynski, editors, *Analysis and Visualization Tools for Constraint Programming*, number 1870 in LNCS, pages 63–107. Springer-Verlag, September 2000.

[25] G. Puebla, F. Bueno, and M. Hermenegildo. An Assertion Language for Constraint Logic Programs. In P. Deransart, M. Hermenegildo, and J. Maluszynski, editors, *Analysis and Visualization Tools for Constraint Programming*, number 1870 in LNCS, pages 23–61. Springer-Verlag, September 2000.

[26] G. Puebla, F. Bueno, and M. Hermenegildo. Combined Static and Dynamic Assertion-Based Debugging of Constraint Logic Programs. In *Logic-based Program Synthesis and Transformation (LOPSTR'99)*, number 1817 in LNCS, pages 273–292. Springer-Verlag, 2000.

[27] Z. Somogyi, F. Henderson, and T. Conway. The execution algorithm of Mercury: an efficient purely declarative logic programming language. *JLP*, 29(1–3), October 1996.

[28] E. Vetillard. *Utilisation de Declarations en Programmation Logique avec Constraintes*. PhD thesis, U. of Aix-Marseilles II, 1994.

Inapproximability Results for Equations over Finite Groups

Lars Engebretsen[1], Jonas Holmerin[1], and Alexander Russell[2]

[1] Department of Numerical Analysis and Computer Science, Royal Institute of Technology, SE-100 44 Stockholm, Sweden. {enge,joho}@kth.se,
[2] Department of Computer Science and Engineering, University of Connecticut, Storrs, CT 06269. acr@cse.uconn.edu

Abstract. An *equation* over a finite group G is an expression of form $w_1 w_2 \ldots w_k = 1_G$, where each w_i is a variable, an inverted variable, or a constant from G; such an equation is *satisfiable* if there is a setting of the variables to values in G so that the equality is realized. We study the problem of simultaneously satisfying a family of equations over a finite group G and show that it is **NP**-hard to approximate the number of simultaneously satisfiable equations to within $|G| - \epsilon$ for any $\epsilon > 0$. This generalizes results of Håstad, who established similar bounds under the added condition that the group G is Abelian.

1 Introduction

Many fundamental computational problems can be naturally posed as questions concerning the simultaneous solvability of families of equations over finite groups. This connection has been exploited to achieve a variety of strong inapproximability results for problems such as Max Cut, Max Di-Cut, Exact Satisfiability, and Vertex Cover (see, e.g., [11,19]). A chief technical ingredient in these hardness results is a tight lower bound on the approximability of the problem of simultaneously satisfying equations over a finite *Abelian* group G. In this article we extend these results to cover all finite groups.

An *equation* in variables $x_1, \ldots, x_n$ over a group G is an expression of form $w_1 \ldots w_k = 1_G$, where each w_i is either a variable, an inverted variable, or a group constant and 1_G denotes the identity element. A *solution* is an assignment of the variables to values in G that realizes the equality. A collection of equations $\mathcal{E}$ over the same variables induces a natural optimization problem, the problem of determining the maximum number of simultaneously satisfiable equations in $\mathcal{E}$. We let EQ_G denote this optimization problem. The special case where a variable may only appear *once* in each equation is denoted EQ_G^1; when each equation has single occurrences of exactly k variables, the problem is denoted $\mathrm{EQ}_G^1[k]$. Our main theorem asserts that $\mathrm{EQ}_G^1[3]$ (and hence EQ_G^1 and EQ_G) cannot be approximated to within $|G| - \epsilon$ for any $\epsilon > 0$; this is tight.

As mentioned above, EQ_G is tightly related to a variety of familiar optimization problems. When $G = \mathbf{Z}_2$, for example, instances of EQ_G where exactly two

P. Widmayer et al. (Eds.): ICALP 2002, LNCS 2380, pp. 73–84, 2002.

variables occur in each clause (i.e., $EQ^1_{Z_2}[2]$) correspond precisely to the familiar optimization problem Max Cut, the problem of determining the largest number of edges which cross some bipartition of an undirected graph. If, for example, $G = S_3$, the (non-Abelian) symmetric group on three letters, then the problem of maximizing the number of bichromatic edges in a three coloring of a given graph can be reduced to EQ_G [10]. See [11,19] for other examples. The general problem has also been studied due to applications to the fine structure of $\mathbf{NC^1}$ [3,10] (specializing the framework of Barrington, et al. [4]). Finally, the problem also naturally gives rise to a variety of well-studied combinatorial enumeration problems: see, e.g., [5,9,17] and [16, pp. 110ff.].

If G is Abelian and $\mathcal{E}$ is a collection of equations over G, the randomized approximation algorithm which independently assigns each variable to a uniformly selected value in G satisfies an expected $1/|G|$ fraction of the equations[1]. (This can be efficiently derandomized by the method of conditional expectation.) This same approximation algorithm applies to EQ^1_G for any finite group G. In 1997, Håstad [11] showed that if $\mathbf{P} \neq \mathbf{NP}$ and G is Abelian, then no polynomial time approximation algorithm can approximate EQ_G to within $|G| - \epsilon$ for any $\epsilon > 0$. The main theorem of this paper shows that this same lower bound holds for all finite groups.

Theorem 1. *For any finite group G and any constant $\epsilon > 0$, it is $\mathbf{NP}$-hard to approximate $EQ^1_G[3]$ to within $|G| - \epsilon$.*

The paper is organized as follows: After an overview of our contribution in Sec. 2 we briefly describe the representation theory of finite groups and the generalization of the so called *long code* to non-Abelian groups in Sections 3 and 4. The main theorem then appears in Section 5. Most proofs have been omitted from this extended abstract for lack of space.

2   Overview of Our Results

A burst of activity focusing on the power of various types of interactive proof systems in the 80s and early 90s culminated in the PCP theorem [1], which asserts the startling fact that $\mathbf{NP}$-languages can be probabilistically checked by a verifier that uses logarithmic randomness, always accepts a correct proof, rejects an incorrect proof with probability at least $1/2$, and examines but a *constant* number of bits of the proof.

There is an approximation-preserving reduction from conjunctive normal form Boolean formulas containing exactly 3 literals per clause (E3-Sat) to E3-Sat formulas where each variable occurs in exactly five clauses [7,12]. Coupling the PCP theorem and this reduction shows that for every language L in $\mathbf{NP}$, an arbitrary instance x can be transformed into an E3-Sat formula $\phi_{x,L}$ with the following property: if $x \in L$, then $\phi_{x,L}$ is satisfiable, and if $x \notin L$ then at most a fraction $\mu < 1$ of the clauses can be satisfied. (Here μ is independent of the language and the instance.)

---

[1] This is under the assumption that each equation is independently satisfiable.

In his seminal paper [11], Håstad introduced a methodology for proving lower bounds for constraint satisfaction problems. On a high level, the method can be viewed as a simulation of the well-known two-prover one-round (2P1R) protocol for E3-Sat where the verifier sends a variable to one prover and a clause containing that variable to the other prover, accepting if the returned assignments are consistent and satisfy the clause. It follows from Raz's parallel repetition theorem [13] that if the 2P1R protocol is repeated u times in parallel and applied to the formula $\phi_{x,L}$ above then the verifier always accepts an unsatisfiable formula with probability at most c^u where $c < 1$ is independent of u.

To prove his approximation hardness result for equations over finite Abelian groups, Håstad constructed a proof system where the verifier tests a given assignment of variables $x_1, \ldots, x_n$ to group values to determine if it satisfies an equation selected at random from a family of equations $\mathcal{E}$. As each equation involves a constant number of variables, this can be tested with a constant number of oracle queries. He then, in essence, reduced the problem of finding a strategy for the 2P1R protocol for E3-Sat to the problem of finding an assignment which satisfies many of the group equations by showing that if the verifier (that checks satisfiability of the group equation) accepts with high probability, there is a strategy for the provers in the 2P1R protocol that makes the verifier of that protocol accept with high probability. The hardness result follows since it is known that the verifier in the latter protocol cannot accept an unsatisfiable instance of E3-Sat with high probability.

To adapt Håstad's method to equations over arbitrary finite groups we need to overcome a couple of technical difficulties. To establish the connection between the proof system that tests a group equation and the 2P1R protocol for E3-Sat, the acceptance probability of the former protocol is "arithmetized." For the case of an Abelian finite group G, this is straightforward: The acceptance probability can be written as a sum of $|G|$ terms. If the acceptance probability is large, there has to be at least one large term in the sum. Håstad then proceeds by expanding this allegedly large term in its Fourier expansion and then uses the Fourier coefficients to devise a strategy for the provers in the 2P1R game for E3-Sat. Specifically, the probability distribution induced by the Fourier coefficients is used to construct a probabilistic strategy for the provers in the 2P1R game. Roughly speaking, the acceptance probability of the verifier in the 2P1R game is large because some pair of related Fourier coefficients is large.

For non-Abelian groups, the way to arithmetize the test turns out to require a sum of the traces of products of certain matrices given by the representation theory of the group in question. As in Håstad's case, we find that if the acceptance probability of the linear test is large, there has to be one product of matrices with a large trace. Our next step is to expand this matrix product in its Fourier series. Unfortunately, the Fourier expansion of each entry in those matrices contains matrices that could be very large; consequently, the Fourier expansion of the entire trace contains a product of matrices with potentially huge dimension. Thus, the fact that this trace is large does not necessarily mean that the individual entries in the matrices are large and directly using the entries in the matrices to construct the probabilistic strategy for the provers in the 2P1R game does not appear to work. Instead, and this is our first main technical contribution, we prove that the terms in the Fourier expansion corresponding to matrices with large dimension cannot contribute much to the value of the trace. Having done that, we know that the terms corresponding to matrices with reasonably small dimension actually sum up to a significantly large value and we use those terms to construct a strategy for the provers in the 2P1R game; this is our second main technical contribution.

3   Representation Theory and the Fourier Transform

In this section, we give a short account of the representation theory needed to state and prove our results. For more details, we refer the reader to the excellent accounts by Serre [14] and Terras [18].

The traditional Fourier transform, as appearing in, say, signal processing [6], algorithm design [15], or PCPs [11], focuses on decomposing functions $f : G \to C$ defined over an Abelian group G. This "decomposition" proceeds by writing f as a linear combination of *characters* of the group G. Unfortunately, this same procedure cannot work over a non-Abelian group since in this case there are not enough characters to span the space of all functions from G into C; the theory of group representations fills this gap, being the natural framework for Fourier analysis over non-Abelian groups and shall be the primary tool utilized in the analysis the "non-Abelian PCPs" introduced in Section 4.

Group representation theory studies realizations of groups as collections of matrices: specifically, a *representation* of a group G associates a matrix with each group element so that the group multiplication operation corresponds to normal matrix multiplication. Such an association gives an embedding of the group into $\mathrm{GL}(V)$, the group of invertible linear operators on a finite dimensional C-vector space V. (Note that if the dimension of V is 1, then this is exactly the familiar notion of character used in the Fourier analysis over Abelian groups.)

Definition 1. *Let G be a finite group. A* representation *of G is a homomorphism $\gamma : G \to \mathrm{GL}(V)$; the dimension of V is denoted by d_γ and called the dimension of the representation.*

Two representations are immediate: the *trivial representation* has dimension 1 and maps everything to 1. The permutation action of a group on itself gives rise to the *left regular representation*. Concretely, let V be a $|G|$-dimensional vector space with an orthogonal basis $B = \{e_g : g \in G\}$ indexed by elements of G. Then the *left regular representation* $\mathrm{reg} : G \to \mathrm{GL}(V)$ is given by $\mathrm{reg}(g) : e_h \mapsto e_{gh}$; the matrix associated with $\mathrm{reg}(g)$ is simply the permutation matrix given by mapping each element h of G to gh.

If γ is a representation, then for each group element g, $\gamma(g)$ is a linear operator and, as mentioned above, can be identified with a matrix. We denote by $(\gamma(g)_{i,j})$ the matrix corresponding to $\gamma(g)$. Two representations γ and θ of G

are *equivalent* if they have the same dimension and there is a change of basis U so that $U\gamma(g)U^{-1} = \theta(g)$ for all g.

If $\gamma\colon G \to \mathrm{GL}(V)$ is a representation and $W \subseteq V$ is a subspace of V, we say that W is *invariant* if $\gamma(g)(W) \subset W$ for all g. If the only invariant subspaces are $\{0\}$ and V, we say that γ is *irreducible*. Otherwise, γ does have a non-trivial invariant subspace W_0 and notice that by restricting each $\gamma(g)$ to W_0 we obtain a new representation. When this happens, it turns out that there is always another invariant subspace W_1 so that $V = W_0 \oplus W_1$ and in this case we write $\gamma = \gamma_0 \oplus \gamma_1$, where γ_0 and γ_1 are the representations obtained by restricting to W_0 and W_1. This is equivalent to the existence of a basis in which the $\gamma(g)$ are all block diagonal, where the matrix of $\gamma(g)$ consists of $\gamma_0(g)$ on the first block and $\gamma_1(g)$ on the second block. In this way, any representation can be decomposed into a sum of irreducible representations.

For a finite group G, there are only a finite number of irreducible representations up to isomorphism; we let $\hat{G}$ denote the set of distinct irreducible representations of G. It is not hard to show that any irreducible representation is equivalent to a representation where each $\gamma(g)$ is unitary, and we will always work under this assumption.

There is also a natural product of representations, the *tensor product*. For our purposes, the best way to think of the tensor product $A \otimes B$ of two matrices $A = (a_{ij})$ and $B = (b_{ij})$ is to let it be a matrix indexed by pairs so that $(A \otimes B)_{(i,k);(j,\ell)} = a_{ij}b_{k\ell}$. If γ and θ are representations of G and H, respectively, we define $\gamma \otimes \theta$ to be the representation of $G \times H$ given by $(\gamma \otimes \theta)(g,h) = \gamma(g) \otimes \theta(h)$.

Proposition 1. *Let G and H be finite groups. Then the irreducible representations of $G \times H$ are precisely $\{\gamma \otimes \theta \mid \gamma \in \hat{G}, \theta \in \hat{H}\}$. Furthermore, each of these representations is distinct.*

We also need a way to denote the so called *inner trace* of a tensor product. For a matrix M indexed by pairs $(i,k);(j,\ell)$ the inner trace is denoted by $\mathrm{Tr}\, M$ and it is defined by $(\mathrm{Tr}\, M)_{ij} = \sum_k M_{(i,k),(j,k)}$. We let tr denote the normal trace.

For a representation γ, the function $g \mapsto \mathrm{tr}\,\gamma(g)$ is called the *character corresponding to γ* and is denoted by χ_γ. Note that χ_γ takes values in C even if γ has dimension larger than 1. Our main use of the character comes from the following fact:

Proposition 2. *Let g be an element of G. Then $\sum_{\gamma\in\hat{G}} d_\gamma\chi_\gamma(g) = |G|$ if $g = 1_G$ and $\sum_{\gamma\in\hat{G}} d_\gamma\chi_\gamma(g) = 0$ otherwise.*

We now proceed to describe the Fourier transform of functions from an arbitrary finite group G to C. Let f be a function from G to C and γ be an irreducible representation of G. Then

$$\hat{f}_\gamma = \frac{1}{|G|} \sum_{g\in G} f(g)\gamma(g) \tag{1}$$

is the *Fourier coefficient of f at γ*. Moreover, f can be written as a Fourier series

$$f(g) = \sum_{\gamma \in \hat{G}} d_\gamma \operatorname{tr}(\hat{f}_\gamma(\gamma(g))^*) = \sum_{\gamma \in \hat{G}} \sum_{1 \le i \le d_\gamma} \sum_{1 \le j \le d_\gamma} d_\gamma \langle f \mid \gamma_{ij} \rangle \gamma_{ij}(g). \qquad (2)$$

where $(\cdot)^*$ denotes conjugate transpose and $\langle \cdot \mid \cdot \rangle$ denotes the conventional inner product $\langle f \mid h \rangle = \frac{1}{|G|} \sum_{g \in G} f(g) h(g)^*$ on the space of functions from G to C. (If we wish to make explicit the group over which the inner product is taken, we write $\langle f \mid h \rangle_G$). As in the Abelian case, the Fourier coefficients satisfy Plancherel's equality:

$$\langle f \mid f \rangle = \sum_{\gamma \in \hat{G}} \sum_{1 \le i \le d_\gamma} \sum_{1 \le j \le d_\gamma} d_\gamma |\langle f \mid \gamma_{ij} \rangle|^2. \qquad (3)$$

We also need to use the Fourier transform on functions $f \colon G \to \operatorname{End}(V)$, where $\operatorname{End}(V)$ is the set of linear maps from the vector space V to itself; we here identify $\operatorname{End}(V)$ with the space of all $\dim V \times \dim V$ matrices over C. Now, for a representation γ of G, we define

$$\hat{f}_\gamma = \frac{1}{|G|} \sum_{g \in G} f(g) \otimes \gamma(g). \qquad (4)$$

Treating the $f(g)$ as matrices, this is nothing more than the component-wise Fourier transform of the function f. The reason for grouping them together into these tensor products is the following: Let $f, h \colon G \to \operatorname{End}(V)$ be two such functions, and define their convolution as

$$(f * h)(g) = \frac{1}{|G|} \sum_{t \in G} f(t) h(t^{-1} g).$$

Then it turns out that $(\widehat{f * h})_\gamma = \hat{f}_\gamma \hat{h}_\gamma$. In this case, the Fourier series is

$$f(g) = \sum_\gamma d_\gamma \operatorname{Tr}\left(\hat{f}_\gamma (\boldsymbol{I} \otimes \gamma(g^{-1}))\right) \qquad (5)$$

where $\operatorname{Tr} M$ is the inner trace.

4   The Non-Abelian Long Code and Its Fourier Transform

Let U be a set of variables of size u. We denote by $\{0, 1\}^U$ all possible assignments to variables in U. For a finite group G, denote by $\mathcal{F}_G(U)$ the set of all functions $f \colon \{0, 1\}^U \to G$.

Definition 2. *Let $x \in \{0, 1\}^U$. Define the long G-code of x to be the function $\mathcal{LC}_G(x) \colon \mathcal{F}_G(U) \to G$ defined as $\mathcal{LC}_G(x)(f) = f(x)$.*

In our setting, a verifier will have access to a purported long code $A \colon \mathcal{F}_G(U) \to G$. We will be interested in the Fourier transform of such functions composed with a representation of G. To this end, first note that $\mathcal{F}_G(U)$ is a group in a natural way, with the group operation being pointwise multiplication of functions. Furthermore, this group can be identified with G^{2^u}. Let $\gamma \in \hat{G}$. Then $\gamma \circ A$ is a function $G^{2^u} \to \mathrm{End}(V)$. To be able to reason about the Fourier transform of such a function, we need to know something about the irreducible representations of G^k.

It follows from Proposition 1 that all the irreducible representations of G^k are constructed as $\rho = \rho_1 \otimes \rho_2 \ldots \otimes \rho_k$, where each ρ_i is an irreducible representation of G. In the case when $G^k = G^{2^u} = \mathcal{F}_G(U)$, we will write

$$\rho = \bigotimes_{x \in \{0,1\}^U} \rho_x, \quad \text{and} \quad \rho(f) = \bigotimes_{x \in \{0,1\}^U} \rho_x(f(x)).$$

Let $\mathbf{1}$ be the trivial representation of G, i.e, $\mathbf{1}(g) = (1)$ for all $g \in G$. We define the *size* $|\rho|$ of an irreducible representation ρ of $\mathcal{F}_G(U)$ to be the number of x such that $\rho_x \neq \mathbf{1}$.

For our proofs to go through, we need to assume that the Fourier coefficients have a certain form. By employing certain access conventions — so called *folding* and *conditioning* — in the verifier, we can ensure that the Fourier coefficients indeed have the form that we need.

Definition 3. *Partition $\mathcal{F}_G(U)$ into equivalence classes by the relation $\sim$, where $f \sim h$ if there is $g \in G$ such that $f = gh$ (that is, $\forall w, f(w) = gh(w)$). Write $\tilde{f}$ for the equivalence class of f. Define A_G, A left-folded over G by choosing a representative for each equivalence class and defining $A_G(h) = gA(f)$, if $h = gf$ and f is the chosen representative for the equivalence class $\tilde{h}$.*

Definition 4. *Define A_{inv}, A folded over inverse, by, for each pair f, f^{-1} such that $f \neq f^{-1}$, choosing one of f, f^{-1} (say f) and setting $A_{\mathrm{inv}}(f) = A(f)$ and $A_{\mathrm{inv}}(f^{-1}) = A(f)^{-1}$. If $f = f^{-1}$ and $A(f)^2 \neq 1_G$, set $A_{\mathrm{inv}}(f) = 1_G$.*

Note that A_{inv} has the property that, for all f, $A_{\mathrm{inv}}(f^{-1}) = A_{\mathrm{inv}}(f)^{-1}$.

Definition 5. *For a Boolean function $h \colon \{0,1\}^U \to \{0,1\}$ we define A_h, A conditioned upon h by $A_h(f) = A(f \wedge h)$, where*

$$f \wedge h(w) = \begin{cases} 1_G & \text{if } h(w) = 0, \\ f(w) & \text{otherwise.} \end{cases}$$

Note that we may condition upon h and fold over inverse at the same time; since $(f \wedge h)^{-1} = f^{-1} \wedge h$, $A_{\mathrm{inv},h} = (A_{\mathrm{inv}})_h$ has both the property that $A_{\mathrm{inv},h}(f^{-1}) = A_{\mathrm{inv},h}(f)^{-1}$ and $A_{\mathrm{inv},h}(f) = A_{\mathrm{inv},h}(f \wedge h)$.

While we do not discuss the specific effects of these conventions in this extended abstract, we do mention that if A is a table folded over G and B is a table folded over inverse, $(\widehat{\gamma \circ A})_\mathbf{1} = 0$ for any $\gamma \in \hat{G} \setminus \{\mathbf{1}\}$ and the matrix $(\widehat{\gamma \circ B})_\rho$ is Hermitian if we pick unitary bases for γ and ρ.

5   The Main Result

The starting point for our PCPs will be the standard two-prover one-round protocol for **NP** which we will now describe. Recall that it is possible to reduce any problem in **NP** to the problem of distinguishing between the two cases that an E3-Sat formula ϕ is satisfiable, or that at most a fraction μ of the clauses are satisfiable [1,7,12]. The formula has the property that each variable occurs in exactly five clauses. This gives rise to a natural two-prover one-round protocol, described next. Given an E3-Sat formula ϕ, the verifier picks u clauses $(C_1, \ldots, C_u)$ each uniformly at random from the instance. For each C_i, it also picks a variable x_i from C_i uniformly at random. The verifier then sends $(x_1, \ldots, x_u)$ to P_1 and the clauses $(C_1, \ldots, C_u)$ to P_2. It receives an assignment to $(x_1, \ldots, x_u)$ from P_1 and an assignment to the variables in $(C_1, \ldots, C_u)$ from P_2, and accepts if these assignments are consistent and satisfy $C_1 \wedge \cdots \wedge C_u$. The completeness of this proof system is 1, and it follows by a general result by Raz [13] that the soundness is at most c_μ^u, where $c_\mu < 1$ is some constant depending on μ but not on u or the size of the instance.

Definition 6. *A* Standard Written G-Proof *consists of a table* $A_U : \mathcal{F}_G(U) \to G$ *for each set* U *of u variables and a table* $A_W : \mathcal{F}_G(W) \to G$ *for each set* W *of variables induced by some u clauses.*

For each U and W, the tables A_U and A_W are supposed to be encodings of answers from the provers P_1 and P_2 on query U and W respectively.

Definition 7. *A Standard Written G-Proof is a* correct *proof for formula ϕ if x is a satisfying assignment to ϕ and* $A_U = \mathcal{LC}^G(x|_U)$ *and* $A_W = \mathcal{LC}^G(x|_W)$.

5.1   The Protocol

The protocol is similar to that used by Håstad [11] to prove inapproximability of equations over Abelian group; the only difference is in the coding of the proof. The tables corresponding to sets U are left-folded over G and the tables corresponding to sets W are conditioned upon $\phi_W = C_{i_1} \wedge \cdots \wedge C_{i_k}$ and folded over inverse. The verifier is given in Figure 1. The completeness is straightforward:

Lemma 1. *The verifier in Figure 1 has completeness at least* $1 - (1 - |G|^{-1})\epsilon$.

5.2   Analysis of the Soundness

The analysis follows the now standard approach. The ultimate goal is to prove that if the verifier accepts some proof π with probability $|G|^{-1} + \delta_1$, then it is possible to extract strategy from π for the 2P1R game with success probability that does not depend on u, the number of parallel repetitions in the 2P1R game. If ϕ is not satisfiable we then arrive at a contradiction since the soundness of the 2P1R game can be made arbitrarily small by increasing u.

    To this end, we first apply Proposition 2 to get an expression for the acceptance probability. Since $|G|^{-1} \sum_{\gamma \in \hat{G}} d_\gamma \chi_\gamma (A_{U,G}(f) A_{W,\phi_W}(h) A_{W,\phi_W}((fh)^{-1}e))$

Input: A formula ϕ and oracle access to a Standard Written G-Proof with parameter u.

1. Select uniformly at random a set $W = \{C_{i_1}, \ldots, C_{i_u}\}$ of u clauses. Let $\phi_W = \bigwedge_{j=1}^{k} C_{i_j}$.
2. Construct U by choosing a variable uniformly at random from each C_{i_k}.
3. Select uniformly at random $f \in \mathcal{F}_G(U)$.
4. Select uniformly at random $h \in \mathcal{F}_G(W)$.
5. Choose e, such that, independently for all $y \in \{-1, 1\}^W$,
   a) With probability $1 - \epsilon$, $e(y) = 1_G$.
   b) With probability ϵ, $e(y) = g$, where $g \in G$ is chosen uniformly at random.
6. If $A_{U,G}(f)A_{W,\mathrm{inv},\phi_W}(h)A_{W,\mathrm{inv},\phi_W}((fh)^{-1}e) = 1_G$ then accept, else reject.

Fig. 1. The verifier.

is an indicator of the event that the verifier accepts, the acceptance probability can be written as:

$$|G|^{-1} \sum_{\gamma \in \hat{G}} d_\gamma \, \mathrm{E}_{f,h,U,W}\left[\chi_\gamma\left(A_{U,G}(f)A_{W,\phi_W}(h)A_{W,\phi_W}\left((fh)^{-1}e\right) \right) \right] =$$

$$|G|^{-1} + |G|^{-1} \sum_{\gamma \in \hat{G}\setminus\{\mathbf{1}\}} d_\gamma \, \mathrm{E}_{f,h,U,W}\left[\chi_\gamma\left(A_{U,G}(f)A_{W,\phi_W}(h)A_{W,\phi_W}\left((fh)^{-1}e\right) \right) \right].$$

Hence if the verifier accepts with probability $|G|^{-1} + \delta_1$, there must be some non-trivial irreducible representation γ of G such that

$$\left| \mathrm{E}_{f,h,U,W}\left[\chi_\gamma\left(A_{U,G}(f)A_{W,\phi_W}(h)A_{W,\phi_W}\left((fh)^{-1}e\right) \right) \right] \right| > d_\gamma \delta_1.$$

We then proceed by applying Fourier-inversion to $A_{U,G}$ and B_{W,ϕ_W}. More precisely, we first apply Fourier-inversion to B_{W,ϕ_W}, resulting in:

Lemma 2. *Suppose that the verifier in Figure 1 accepts with probability $|G|^{-1} + \delta_1$. Then there exists a non-trivial representation γ of G such that*

$$\left| \mathrm{E}_{f,U,W}\left[\mathrm{tr}\left(A_\gamma(f) \sum_{\rho \in \hat{H}} d_\rho (1-\epsilon)^{|\rho|} \, \mathrm{Tr}\left(\hat{B}_{\gamma,\rho}^2 \left(I_{d_\gamma} \otimes \rho(f) \right) \right) \right) \right] \right| > d_\gamma \delta_1.$$

where $A_\gamma = \gamma \circ A_{U,G}$, $H = \mathcal{F}_G(W)$, and $\hat{B}_{\gamma,\rho} = \frac{1}{|H|} \sum_{h \in H} \gamma(A_{W,\phi_W}(h)) \otimes \rho(h)$.

We then prove that, for every fixed f, many terms in the resulting sum are very small. Specifically, we prove that there is a constant c such that the contribution of terms which correspond to ρ with $|\rho| > c$ contribute at most $d_\gamma \delta_1 / 2$.

Lemma 3. *For any positive integer* c,

$$\left| \mathrm{tr}\left(A_\gamma(f) \sum_{\substack{\rho \in \hat{H} \\ |\rho| > c}} d_\rho (1 - \epsilon)^{|\rho|} \mathrm{Tr}\left(\hat{B}_\rho^2 (\boldsymbol{I}_{d_\gamma} \otimes \rho(f)) \right) \right) \right| \leq d_\gamma (1 - \epsilon)^c \qquad (6)$$

where $A_\gamma = \gamma \circ A_{U,G}$, $H = \mathcal{F}_G(W)$, *and* $\hat{B}_\rho = \frac{1}{|H|} \sum_{h \in H} \gamma(A_{W,\phi_W}(h)) \otimes \rho(h)$.

Corollary 1. *Suppose that* $c = \lceil (\log \delta_1 - 1) / \log(1 - \epsilon) \rceil$. *Then*

$$\left| \mathrm{E}_{f,U,W}\left[\mathrm{tr}\left(A_\gamma(f) \sum_{\substack{\rho \in \hat{H} \\ |\rho| \geq c}} d_\rho (1 - \epsilon)^{|\rho|} \mathrm{Tr}\left(\hat{B}_\rho^2 (\boldsymbol{I}_{d_\gamma} \otimes \rho(f)) \right) \right) \right] \right| < \frac{d_\gamma \delta_1}{2}$$

where $A_\gamma = \gamma \circ A_{U,G}$, $H = \mathcal{F}_G(W)$, *and* $\hat{B}_\rho = \frac{1}{|H|} \sum_{h \in H} \gamma(A_{W,\phi_W}(h)) \otimes \rho(h)$.

Having done that, we use the remaining terms to extract a strategy for the 2P1R game. This involves using Fourier-inversion on $A_{U,G}$.

Lemma 4. *If*

$$\left| \mathrm{E}_{f,U,W}\left[\mathrm{tr}\left(A_\gamma(f) \sum_{\substack{\rho \in \hat{H} \\ |\rho| < c}} d_\rho (1 - \epsilon)^{|\rho|} \mathrm{Tr}\left(\hat{B}_\rho^2 (\boldsymbol{I}_{d_\gamma} \otimes \rho(f)) \right) \right) \right] \right| \geq \delta \qquad (7)$$

then there is a strategy for the provers in the 2P1R protocol with success proba-bility at least $\delta^2 / c |G|^{2c} d_\gamma^6$.

Finally, we put together these two parts and establish the soundness of the verifier.

Lemma 5. *For any* $\delta > 0$, *and for any choice of the parameter* ϵ *of the verifier, there is a choice of the parameter* u *of the 2P1R game such that the soundness is at most* $|G|^{-1} + \delta$

Proof. Suppose that ϕ is not satisfiable and there is a proof which the verifier accepts with probability $|G|^{-1} + \delta_1$. By Lemma 2, for this proof, there is a non-trivial irreducible representation γ of G such that

$$\left| \mathrm{E}_{f,U,W}\left[\mathrm{tr}\left(A_\gamma(f) \sum_{\rho \in \hat{H}} d_\rho (1 - \epsilon)^{|\rho|} \mathrm{Tr}\left(\hat{B}_{\gamma,\rho}^2 (\boldsymbol{I}_{d_\gamma} \otimes \rho(f)) \right) \right) \right] \right| > d_\gamma \delta_1.$$

Combining this with Corollary 1,

$$\left| \mathrm{E}_{f,U,W}\left[\mathrm{tr}\left(A_\gamma(f) \sum_{\substack{\rho \in \hat{H} \\ |\rho| < c}} d_\rho (1 - \epsilon)^{|\rho|} \mathrm{Tr}\left(\hat{B}_\rho^2 (\boldsymbol{I}_{d_\gamma} \otimes \rho(f)) \right) \right) \right] \right| \geq \frac{d_\gamma \delta_1}{2}$$

provided that $c = \lceil (\log \delta_1 - 1) / \log(1 - \epsilon) \rceil$. Thus the conditions of Lemma 4 holds with $\delta = d_\gamma \delta_1 / 2$, and there is a strategy for the 2P1R game with success proba-bility at least $\delta_1^2 / 4c |G|^{2c} d_\gamma^4$. Choosing the parameter u such that the soundness of the 2P1R game is smaller than this expression, we get a contradiction.

5.3   Hardness of Approximating $\mathrm{EQ}_G^1[3]$

Proof (of Theorem 1). By Lemma 1 and Lemma 5 for any constants $\epsilon > 0$ and $\delta > 0$ it is possible to choose the parameters of the verifier in Figure 1 such that it is **NP**-hard to distinguish between the case that there is a proof which the verifier accepts with probability $1 - \epsilon$, and the case that there is no proof which is accepted with probability more than $|G|^{-1} + \delta$.

Now we create a system of equations in the obvious way; the variables corresponds to the positions in the proofs, and we add an equation for each random string corresponding to the test made for this random string. One may think that these would always be on the form $xyz = 1_G$, but this is not the case due to folding over G and over inverse, and in general an equation is of the form $gx y^a z^b = 1_G$, where g is a group constant and $a, b \in \{1, -1\}$. There is also a technicality in that there is a small probability that $g = (fh)^{-1}e$ in the protocol, and thus a variable occurs more than once. But this happens with probability $o(1)$, and therefore such equations may be omitted from the instance. Hence we may construct instances of $\mathrm{EQ}_G^1[3]$ in which it is hard to distinguish between the case that $1 - \epsilon$ of all equations are satisfiable, and the case that at most $|G|^{-1} + \epsilon$ of all equations are satisfiable, and we are done.

6   Open Questions

An interesting question is that of *satisfiable instances*. Some problems, such as E3-Sat retain their inapproximability properties even when restricted to satisfiable instances. This is not the case for $\mathrm{EQ}_G^1[k]$ when G is a finite Abelian group, since if such a system is satisfiable a solution may be found essentially by Gaussian elimination. However, when G is non-Abelian, deciding whether a system of equations over G is satisfiable is **NP**-complete [10], so it seems reasonable that the problem over non-Abelian groups retains some hardness of approximation for satisfiable instances. However, the following simple argument shows that we can not hope, even for the non-Abelian groups, for a lower bound of $|G|^{-1} + \delta$: Given an instance σ of $EQ_G^1[k]$ over some non-Abelian group G, we construct an instance σ' over $EQ_H^1[k]$, where $H = G/G'$ and G' is the commutator subgroup of G, i.e., the subgroup generated by the elements $\{g^{-1}h^{-1}gh : g, h \in G\}$. The instance σ' is the same as σ, except that all group constants are replaced by their equivalence class in G/G'. Now since H is an Abelian group, we can solve over H. The solution is an assignment of cosets to the variables. We then construct a random solution of x by for each variable choosing a random element in the corresponding coset. Now the value of left hand side of each equation will be uniformly distributed in the coset of the right hand side, and thus we will satisfy an expected number of $|G'|^{-1}$ of all equations.

Acknowledgements. We would like to thank Johan Håstad for useful discussions.

References

1. Sanjeev Arora, Carsten Lund, Rajeev Motwani, Madhu Sudan, and Márió Szegedy. Proof verification and the hardness of approximation problems. *Journal of the ACM*, 45(3):501–555, May 1998.
2. Sanjeev Arora and Shmuel Safra. Probabilistic checking of proofs: A new characterization of NP. *Journal of the ACM*, 45(1):70–122, January 1998.
3. David Mix Barrington, Pierre McKenzie, Cristopher Moore, Pascal Tesson, and Denis Thérien. Equation satisfiability and program satisfiability for finite monoids. In *Proceedings of the 25th Annual Symposium on Mathematical Foundations of Computer Science*, 2000.
4. David A. Mix Barrington, Howard Straubing, Denis Thérien. Non-uniform automata over groups. *Information and Computation*, 89(2): 109-132, 1990
5. François Bédard and Alain Goupil. The poset of conjugacy classes and decomposition of products in the symmetric group. *Canad. Math. Bull.*, 35(2):152–160, 1992.
6. Harry Dym and Henry P. McKean. *Fourier Series and Integrals*, volume 14 of *Probability and Mathematical Statistics*. Academic Press, 1972.
7. Uriel Feige. A threshold of $\ln n$ for approximating set cover. *Journal of the ACM*, 45(4):634–652, July 1998.
8. Uriel Feige, Shafi Goldwasser, László Lovász, Shmuel Safra, and Márió Szegedy. Interactive proofs and the hardness of approximating cliques. *Journal of the ACM*, 43(2):268–292, March 1996.
9. Harold Finkelstein and K. I. Mandelberg. On solutions of "equations in symmetric groups". *Jornal of Combinatorial Theory, Series A*, 25(2):142–152, 1978.
10. Mikael Goldmann and Alexander Russell. The computational complexity of solving systems of equations over finite groups. In *Proceedings of the Fourteenth Annual IEEE Conference on Computational Complexity*, Atlanta, GA, May 1999.
11. Johan Håstad. Some optimal inapproximability results. *Journal of the ACM*, 48(4):798–859, July 2001.
12. Christos H. Papadimitriou and Mihalis Yannakakis. Optimization, approximation, and complexity classes. *Journal of Computer and System Sciences*, 43(3):425–440, December 1991.
13. Ran Raz. A parallel repetition theorem. *SIAM Journal on Computing*, 27(3):763–803, June 1998.
14. Jean-Pierre Serre. *Linear Representations of Finite Groups*, volume 42 of *Graduate Texts in Mathematics*. Springer-Verlag, New York, 1977.
15. Arnold Schönhage and Volker Strassen. Schnelle Multiplikation großer Zahlen. *Computing*, 7:281–292, 1971.
16. Richard P. Stanley. *Enumerative Combinatorics, Volume 2*, volume 62 of *Cambridge Studies in Advances Mathematics*. Cambridge University Press, 1999.
17. Sergej P. Strunkov. On the theory of equations on finite groups. *Izvestija Rossijskoj Akademii Nauk, Serija Matematičeskaja*, 59(6):171–180, 1995.
18. Audrey Terras. *Fourier Analysis on Finite Groups and Applications*, volume 43 of *London Mathematical Society student texts*. Cambridge University Press, Cambridge, 1999.
19. Uri Zwick. Approximation algorithms for constraint satisfaction problems involving at most three variables per constraint. In *Proceedings of the Ninth Annual ACM-SIAM Symposium on Discrete Algorithms*, pages 201–210, San Francisco, California, 25–27 January 1998.

A Faster All-Pairs Shortest Path Algorithm for Real-Weighted Sparse Graphs*

Seth Pettie

Department of Computer Sciences
The University of Texas at Austin
Austin, TX 78712
seth@cs.utexas.edu

Abstract. We present a faster all-pairs shortest paths algorithm for arbitrary real-weighted directed graphs. The algorithm works in the fundamental *comparison-addition model* and runs in $O(mn+n^2 \log \log n)$ time, where m and n are the number of edges & vertices, respectively. This is strictly faster than Johnson's algorithm (for arbitrary edge-weights) and Dijkstra's algorithm (for positive edge-weights) when $m = o(n \log n)$ and matches the running time of Hagerup's APSP algorithm, which assumes *integer* edge-weights and a more powerful model of computation.

1 Introduction

Nearly all known shortest path algorithms can be easily separated into two groups: those which assume *real*-weighted graphs, where reals are manipulated only by *comparison* and *addition* operations, and those which assume *integer*-weighted graphs and a suite of RAM-type operations to act on the edge-weights. The standard algorithms, established early on by Dijkstra [Dij59], Bellman-Ford, Floyd-Warshall and others (see [CLR90]), all work in the comparison-addition model with real edge-weights. Since then most progress on shortest paths problems has come by assuming integral edge-weights. Techniques based on scaling, integer matrix multiplication and fast integer sorting only work with integer edge-weights, and until recently [PR02] it appeared as though the *component hierarchy* approach used in [Tho99,Hag00] also required integers. We refer the reader to a recent survey paper [Z01] for more background and references.

The state of the art in APSP for real-weighted, sparse directed graphs is (surprisingly) a combination of two standard textbook algorithms. Johnson [J77] showed that an arbitrarily-weighted graph is reducible to a positively-weighted graph such that shortest paths remain unchanged (assuming no negative cycles and hence 'shortest path' is well-defined.) The reduction makes one call to the Bellman-Ford algorithm and takes $O(mn)$ time. For positively-weighted graphs, Dijkstra's algorithm [Dij59,FT87] solves the single-source shortest path problem (SSSP) in $O(m + n \log n)$ time, implying an APSP algorithm for *arbitrarily*-weighted graphs running in $O(mn + n^2 \log n)$ time. This is the best bound to

---

* This work was supported by Texas Advanced Research Program Grant 003658-0029-1999, NSF Grant CCR-9988160, and an MCD Graduate Fellowship.

P. Widmayer et al. (Eds.): ICALP 2002, LNCS 2380, pp. 85–97, 2002.
© Springer-Verlag Berlin Heidelberg 2002

date for directed graphs, however there is a faster algorithm [PR02] for *undirected* APSP running in $O(mn\alpha(m,n))$ time.

The recent component hierarchy (CH) based algorithms [Tho99,Hag00,PR02] either reduce or eliminate the sorting bottleneck in Dijkstra's algorithm. Thorup, who first described the CH approach, showed that *undirected* SSSP on non-negative, integer-weighted graphs can be solved in $O(m)$ time, assuming edge-weights are subject to typical RAM operations. This immediately implies an $O(mn)$ time APSP algorithm. Hagerup [Hag00] assumed the same non-negative integer/RAM model and showed SSSP on *directed* graphs can be solved in $O(m\log\log C + n\log\log n)$ time, and APSP in $O(mn + n^2\log\log n)$ time. Here C is the largest integer edge weight. Recently Pettie & Ramachandran [PR02] adapted Thorup's algorithm to *real*-weighted graphs and the pointer machine model [Tar79], which is weaker than the RAM, yielding an undirected APSP algorithm running in $O(mn\alpha(m,n))$ time, where α is the inverse-Ackermann function. Pettie [Pet02b] gave a CH-based APSP algorithm for directed graphs that performs $O(mn\log\alpha(m,n))$ comparison/addition operations, however there is no known implementation of this algorithm with a similar asymptotic overhead. The CH approach also turns out to be practical; an experimental study of Pettie et al. [PRS02] of a simplified version of [PR02] shows it to be decisively faster than Dijkstra's algorithm, if the one-time cost of constructing the CH is offset by a sufficient number of SSSP computations.

In this paper we adapt the component hierarchy approach to real-weighted directed graphs, giving an APSP algorithm running in $O(mn + n^2\log\log n)$ time. Our algorithm differs from previous CH-based algorithms in its overall structure. In [Tho99,Hag00,PR02], a component hierarchy is constructed in one phase, and in the next phase SSSP can be computed from any s sources by s non-interacting processes. In Section 4.3 we show that if we adhere to the basic CH framework, solving s SSSP computations by non-interacting processes requires $\Omega(s(m + n\log n))$ comparisons/additions. Thus, a qualitatively new approach is required. In our algorithm we have a three-phase structure. After the CH is constructed (phase 1), we take the time to gather a slew of *approximate* shortest path-related statistics (phase 2) which will allow us to compute APSP faster in phase 3. For a parameter k, if we spend $O(mn\lceil\frac{\log n}{k}\rceil)$ time in phase 2, the s-sources shortest paths problem is solved in phase 3 in $O(s\cdot(m + n\log k + n\log\log n))$ time. Setting $k = \log n$, $s = n$ gives us the claimed APSP result, though improvements over Dijkstra's algorithm [Dij59,FT87] can still be had for $s = \omega(m/\log n)$.

2   Preliminaries

The input is a weighted, directed graph $G = (V, E, \ell)$ where $|V| = n, |E| = m$, and $\ell : E \to \mathbb{R}$ assigns a real *length* to every edge. The length of a path is defined to be the sum of its constituent edge lengths. We let $d(u, v)$ denote the length of the shortest path from u to v, or ∞ if none exists. The single-source shortest paths problem is to compute $d(s, v)$ for some *source* vertex s and every vertex v while the all-pairs shortest path problem is to compute $d(u, v)$ for all u, v. Generalizing the d notation, let $d(u, H)$ (resp. $d(H, u)$) be the shortest distance from u to any vertex in the subgraph H (from any vertex in H to u). H may also be an object that is associated with a subgraph, not necessarily the subgraph itself. It was mentioned in the introduction that the APSP problem is reducible in $O(mn)$ time to one of the same size but having only positive edge lengths. We therefore assume that $\ell : E \rightarrow \mathbb{R}^+$ assigns only positive lengths.

2.1   The Comparison-Addition Model

In the comparison-addition model real numbers are only subject to *comparisons* and *additions*. Comparisons determine the larger of two given reals, and addition of existing reals is the only means for generating new reals. A comparison-addition based algorithm, which is modeled as a decision tree with additions, chooses which operations to make based on the outcomes of previous comparisons.

This model cannot distinguish between integers and arbitrary reals, and cannot produce a specific integer in a real variable. Therefore, when we say some variable or quantity is an *integer*, we mean that it is kept in an integer variable. The only additional property assumed of integers is that they may be used to index an array. We will only produce polynomially-bounded integers, whereas reals are assumed to take on arbitrary values.

3   Dijkstra's Algorithm

Dijkstra's SSSP algorithm visits vertices in order of increasing distance from the source s. It maintains a set S of visited vertices, initially empty, and a *tentative* distance $D(v)$ for all $v \in V$ satisfying the following invariant.

Invariant 0 *For $v \in S$ $D(v) = d(s, v)$ and for $v \notin S$ $D(v)$ is the shortest distance from s to v using only intermediate vertices from S.*

Dijkstra's method for growing the set S while maintaining Invariant 0 is to visit vertices *greedily*. In each step, Dijkstra's algorithm identifies the vertex $v \notin S$ with minimum tentative distance, sets $S := S \cup \{v\}$, and updates tentative distances. This involves *relaxing* each outgoing edge (v, w), setting $D(w) := \min\{D(w), D(v) + \ell(v, w)\}$. The algorithm halts when $S = V$, and therefore $D(v) = d(s, v)$ — the tentative distances equal the shortest distances.

Component hierarchy-based algorithms also maintain Invariant 0, though in a non-greedy fashion. Throughout the paper D, S, s mean the same thing as in Dijkstra's algorithm, and the terms "visit" and "relax" are essentially the same.

4   The Component Hierarchy

In this paper we do not assume a familiarity with the component hierarchy approach to shortest paths; nevertheless, [Tho99,Hag00,PR02] are highly recommended reading.

Hagerup [Hag00] constructed a CH for directed graphs and positive integer edge lengths in $O(m \log \log C)$ time, where C is the largest edge weight. For real edge lengths his method can be adapted, using techniques akin to those in [PR02], to run in $O(m \log \log r + \log r)$ time, where r is the ratio of the maximum-to-minimum edge length. Below we define a component hierarchy for real-weighted directed graphs; it can be constructed in $O(m \log n)$ time using a combination of the techniques from [Hag00,PR02]. We omit the proof.

4.1   The CH for Real-Weighted Directed Graphs

Assume w.l.o.g. that G is strongly connected. This can be enforced without altering the finite distances by adding an n-cycle with very long edges. As in [PR02] we first produce the edge lengths in sorted order: $\ell_1, \ldots, \ell_m$. We then find a set of "normalizing" edge lengths $\{\ell_1\} \cup \{\ell_j : \ell_j > n \cdot \ell_{j-1}\}$. Let r_k be the kth smallest normalizing edge. For each edge j between r_k and $r_{k+1} - 1$ we determine the i s.t. $2^i \ell_{r_k} \le \ell_j < 2^{i+1} \ell_{r_k}$. In other words, we find a factor 2 approximation of every edge length divided by its normalizing edge length.[1]

The CH is composed of layered *strata*, where stratum k, level i nodes correspond to the strongly connected components (SCCs) of the graph restricted to edges with length less than $\ell_{r_k} \cdot 2^i$. If x is a stratum k, level i node we let $norm(x) = \ell_{r_k} \cdot 2^{i-1}$; Most quantities relating to x will be measured in units of $norm(x)$. Let C_x denote the SCC associated with a CH node x, and let $diam(C_x)$ (the diameter) be the longest shortest path between two distinct vertices in C_x. The children of x, $\{x_1, \ldots, x_\nu\}$ are those stratum k, level $i-1$ CH nodes whose SCCs $\{C_{x_1}, \ldots, C_{x_\nu}\}$ are subgraphs of C_x. (If $i = 0$, i.e. x is at the "bottom" of its stratum, then its children are the stratum $k-1$ nodes of maximum level, $\{x_j\}$, s.t. the $\{C_{x_j}\}$ are subgraphs of C_x.) Let C_x^c be derived from C_x by contracting the SCCs $\{C_{x_j}\}$, and C_x^{DAG} be derived from C_x^c by removing edges with length at least $norm(x)$. That is, there is a correspondence between vertices in C_x^c and the children of x in the component hierarchy; *we will frequently use the same notation to refer to both.* It is convenient to think of single-child nodes in the CH being spliced out, hence the children of a node are not necessarily all on the same stratum/level, but the CH is linear in size.

The following lemma, variants of which were used in [Tho99,Hag00,PR02], is useful for associating the running time of our algorithm with certain CH statistics. Refer to [Pet02] for a proof.

Lemma 1. $(i) \sum_{x \in CH} |V(C_x^c)| \le 2n$ $(iii) \left| \{x \in CH : \frac{diam(C_x)}{norm(x)} > k\} \right| \le \frac{8n}{k}$

$(ii) \sum_{x \in CH} \frac{diam(C_x)}{norm(x)} \le 8n$

4.2   Computing SSSP

Component hierarchy-based algorithms also maintain Dijkstra's Invariant 0. However, they do not necessarily visit vertices in increasing distance from the

---

[1] We would be content to let ℓ_1 be the only normalizing edge length. The problem is that it could take an arbitrary amount of time to find the 2-approximations defined above. We use many normalizing edge lengths in order to bound the time to find 2-approximations in terms of m and n.

source. Recall that the D-value of a vertex was its tentative distance from the source s. We extend the D notation to CH nodes as follows:

$$D(x) \stackrel{\text{def}}{=} \min_{v \in C_x}\{D(v)\} \qquad \text{where } x \text{ is any CH node}$$

The Visit procedure, given below, takes a CH node x and some interval $[a, b)$ and visits all vertices $v \in C_x$ whose $d(s, v)$-values lie in $[a, b)$. If C_x is a single vertex and $D(x) \in [a, b)$, we mark C_x as visited and relax all its outgoing edges. Otherwise we delegate the responsibility of visiting vertices in $[a, b)$ to the children of x. SSSP are computed from s by setting $S = \emptyset$, $D(s) = 0$, $D(v) = \infty$ for $v \neq s$ and calling $\text{Visit}(root(CH), [0, \infty))$. One may refer to [Tho99,Hag00,PR02] for more detailed descriptions of the basic component hierarchy algorithm or proofs of its correctness. We will call a node *active* if it has been visited at least once, and *inactive* otherwise.

    $\text{Visit}(x, [a, b))$
        If C_x is a single vertex and $D(x) \in [a, b)$ then
            Visit C_x:
            Let $S := S \cup \{C_x\}$
(1)   Relax C_x's outgoing edges
            Return.
        If $\text{Visit}(x, \cdot)$ is being called for the first time, then
(2)   *Initialize x's bucket array:*
            Create $\lceil diam(C_x)/norm(x)\rceil + 1$ buckets
            Let the first bucket start at t_0, a real number s.t. $a \leq t_0 \leq D(x)$.
            Label bucket j with its associated interval:
                $[t_0 + j \cdot norm(x), t_0 + (j + 1) \cdot norm(x))$.
(3)   Bucket x's children by their D-values.
        t refers to the start of the current bucket's interval (Initially $t = t_0$.)
        While $S \cap C_x \neq C_x$ and $t < b$
            While bucket $[t, t + norm(x))$ is not empty
(4)   Choose a suitable node y from bucket $[t, t + norm(x))$
                $\text{Visit}(y, [t, t + norm(x)))$
                If $S \cap C_y \neq C_y$, put y in bucket $[t + norm(x), t + 2 \cdot norm(x))$
        $t := t + norm(x)$

Some lines which need elaboration are marked by a number.

1. Visiting vertices and relaxing edges is done just as in Dijkstra's algorithm. Relaxing an edge (u, v) may cause an inactive ancestor of v in the CH to be bucketed or *re-bucketed* if relaxing (u, v) caused its D-value (tentative distance) to decrease.
2. Buckets in the bucket array represent consecutive intervals of width $norm(x)$, which together form an interval that contains $d(s, v)$ for all $v \in C_x$. We will refer to buckets by their place in the bucket array (e.g. the first bucket) or by the endpoints of the interval they represent (e.g. bucket $[t, t + norm(x))$).

There is some subtlety to choosing the starting point t_0 of the first bucket. The concern is that we may have a *fractional* interval left over[2] if b, the end of the given interval, is not aligned with $t_0 + q \cdot norm(x)$ for some q. As in [PR02], we choose the initial t_0 as follows: if $D(x) + diam(C_x) < b$ then we will not reach b anyway and the alignment problem does not arise; set $t_0 = D(x)$. Otherwise, count back from b in units of $norm(x)$; find the minimum q such that $t_0 = b - q \cdot norm(x) \le D(x)$. One can also show that, because of the wide separation in edge-lengths between strata, the fractional interval problem does not arise when Visit makes inter-stratum recursive calls. Indeed, this motivated our definition of strata.

3. Logically speaking, a child y of x appears in bucket $\lfloor \frac{D(y)-t_0}{norm(x)} \rfloor$. (In our algorithm this invariant is relaxed somewhat — see Section 5.1.) We use Gabow's [G85] split-findmin data structure to maintain the D-values of CH nodes.

4. Hagerup noted that Invariant 0 is not maintained if nodes from the same bucket are visited in any order; this is in contrast to the [Tho99,PR02] algorithms for undirected graphs, where nodes may be visited in arbitrary order. In [Hag00] it is shown that Invariant 0 can be maintained if nodes from the same bucket are visited in an order consistent with a topological ordering of C_x^{DAG}. Hagerup first assigns numbers in $\{1, 2, \ldots, |V(C_x^c)|\}$ to the vertices in C_x^c consistent with such an ordering, then uses a van Emde Boas heap [vEKZ77] to prioritize nodes within the same bucket. The overhead for managing the van Emde Boas structure is $O(n \log \log n)$ in total.

A CH node y is bucketed on two occasions: when its parent node x is first visited (item 2) or when some edge $(\cdot, v)$, $v \in C_y$ is relaxed (item 1). We will actually think of the first kind of bucketing operation as an edge relaxation too. When x is first visited, $D(y)$ corresponds to a path P_{sy} from s to C_y, hence bucketing y according to its D-value is tantamount to *re-relaxing* the last edge in P_{sy}. We are concerned with both kinds of edge relaxations, of which there are no more than $m + 2n = O(m)$.

4.3   A Lower Bound on Component Hierarchy-Type Algorithms

It is not difficult to show that Dijkstra's algorithm is "just as hard as sorting", that is, producing the vertices in order of their d-values is just as hard as sorting n numbers. This implies an $\Omega(m + n \log n)$ lower bound on the complexity of Dijkstra's algorithm *in the comparison-addition model*, and tells us that we must alter our approach or strengthen the model in order to obtain faster shortest path algorithms. In this section we give a similar lower bound on Hagerup's algorithm [Hag00], or indeed any algorithm of the component hierarchy type. We show that, even given the graph's CH, such an algorithm requires $\Omega(m + n \log n)$ comparisons/additions.

---

[2] Having fractional intervals left over is not a problem in terms of correctness, but it does complicate the analysis.

Let $sep(u, v)^3$ be maximal such that any cycle containing both u and v must have an edge with length at least $sep(u, v)$. All CH-based algorithms satisfy the following Property:

Property 1. If $d(s, u) + sep(u, v) \leq d(s, v)$, then u must be visited before v.

A permutation of the vertices is *compatible* with a certain edge-length function if visiting the vertices in that order does not violate Property 1. We show that there is a directed graph and a large family of distinct edge-length functions, no two of which share a compatible permutation. We should point out that our lower bound does not extend to undirected graphs [PR02], though similar arguments do give significantly weaker lower bounds for undirected graphs.

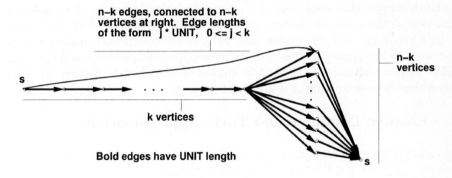

Fig. 1. The "broom" graph.

The graph depicted in Figure 1 is organized a little like a broom. It has a "broom stick" of k vertices, whose head is the source s and whose tail connects to the remaining $n - k$ vertices (the "bush"), each of which is connected back to s by an edge (s appears twice to simplify the figure). All these edges have equal length *UNIT*. Additionally, there are $n - k$ edges directed from s to each of the vertices in the bush, having lengths of the form $j \cdot UNIT$ for j a natural less than k. One may easily confirm two facts: $sep(u, v) = UNIT$ regardless of u, v, and there are exactly k^{n-k} possible edge length functions. Let $\mathcal{L}$ be the subset of edge length functions which, for edges from s to the "bush", assign an equal number to length $j \cdot UNIT$, for $0 \leq j < k$. Assuming w.l.o.g. that k divides n, $|\mathcal{L}| = (n - k)!/(\frac{n}{k} - 1)!^k$. One can also see that for any two length functions $\ell^1, \ell^2 \in \mathcal{L}$, there is always a pair of vertices u, v in the broom's bush such that $d(s, u) < d(s, v)$ w.r.t. ℓ^1, but $d(s, v) < d(s, u)$ w.r.t. ℓ^2. Since $d(s, u) < d(s, v)$ implies $d(s, u) \leq d(s, v) + sep(u, v)$, no permutation of the vertices can be compatible with both ℓ^1 and ℓ^2, and at least $\log |\mathcal{L}|$ comparisons must be made to choose a compatible permutation. For $k \leq n/2$, $\log |\mathcal{L}| = \Omega(n \log k)$. One can

---

[3] One can show that the CH implicitly gives a 2-approximation to the *sep* function.

repeat this lower bound argument for any source in the broom's bush. Theorem 1 is then easily proved.

Theorem 1. *Suppose $sep(u,v)$ is already known, for all vertices u,v. Any SSSP algorithm obeying Property 1 must make $\Omega(m + \min\{n \log r, n \log n\})$ comparison/addition operations, where the source can be any of $n - o(n)$ vertices and r bounds the ratio of any two edge-lengths.*

4.4   Our Approach

The lower bound on computing SSSP with Hagerup's algorithm does not imply a proportionate lower bound on solving SSSP multiple times on the same graph. As we shall see in later sections, there is actually a high degree of *dependence* between shortest paths in the same graph, however, identifying the dependencies is much simpler than exploiting them algorithmically. In Section 5 we define a function Δ that, were it known, could be used to calculate SSSP in $O(m + n \log \log n)$ time (cost of constructing the CH excluded). Our approach is to calculate a close approximation $\hat{\Delta}$ to the more important parts of the Δ function. The SSSP algorithm, assuming $\hat{\Delta}$ is known, is described in Section 5.1. Our algorithm for constructing $\hat{\Delta}$ is given in Section 6.

5   Relative Distances and Their Approximations

Recall that for $z \in CH$, $d(u,z) = d(u, C_z)$ is the shortest distance from u to any vertex in C_z. For $u \in V(G), y \in V(C_x^c)$ we define Δ_x as

$$\Delta_x(u,y) \stackrel{\text{def}}{=} d(u,y) - d(u,x)$$

and for ϵ_x chosen later, we define $\hat{\Delta}_x$ as satisfying

$$|\hat{\Delta}_x(u,y) - \Delta_x(u,y)| < \epsilon_x, \quad \text{and} \quad \frac{\hat{\Delta}_x(u,y)}{\epsilon_x} \text{ is an integer.}$$

In other words, we represent $\hat{\Delta}$ as an *integer* multiple of ϵ_x, *not* as a real.

5.1   The Algorithm

We present our SSSP algorithm assuming the component hierarchy is constructed and $\hat{\Delta}_x(u,v)$ is known for all CH nodes x such that $diam(C_x)/norm(x) \geq k$. We address the computation of $\hat{\Delta}$-values in Section 6 The algorithm runs in time $O(m + n \log k + n \log \log n)$ time.

Consider a particular CH node x. As in Hagerup's algorithm, x has an associated bucket array of at least $\lceil diam(C_x)/norm(x) \rceil + 1$ buckets, each representing a real interval of width $norm(x)$. When x becomes active we *label* each bucket with the real endpoints of its associated interval in $O(diam(C_x)/norm(x))$ time. If $diam(C_x)/norm(x) < k$ we use the lazy bucketing scheme (for real numbers!)

used in [PR02]. First bucketing a child of x takes $O(\log k)$ amortized time and all decrease-keys take $O(1)$ amortized time. Hence the bucketing costs associates with all nodes x s.t. $diam(C_x) < k \cdot norm(x)$ is $O(m + n \log k)$. For nodes with larger diameters we use the known $\hat{\Delta}$-values to handle their bucketing tasks in $O(m)$ time as described below.

Hagerup's algorithm maintains the invariant that for any active CH node x, the children are bucketed according to their D-values (or appear in no bucket if they have infinite D-values). We relax Hagerup's invariant as follows.

Invariant 1 *Let x be an active CH node, y an inactive child of x. If $\hat{\Delta}_x(\cdot, \cdot)$ is known, then y appears in the correct bucket or it is known that $D(y)$ will decrease in the future, in which case y appears in no bucket.*

As a matter of correctness this is clearly just as good as always bucketing CH nodes correctly. The question is, what is a certificate that some node's D-value will decrease in the future? This is where the $\hat{\Delta}_x$ function comes into play.

Suppose we are relaxing an edge (u, v), $v \in C_y$, and $y \in C_x^c$ is an inactive child of an active CH node x. Note that unless v has already been visited, there is always such a pair y, x. If relaxing (u, v) decreases $D(y)$ then we attempt to bucket y by its D-value. Let $f \in C_x$ be such that $d(s, f) = d(s, x)$ (i.e. f is the earliest vertex visited in C_x). We know $d(s, f)$ lies within x's first bucket's associated interval. Furthermore, by definition of Δ_x, $d(s, y) = d(s, f) + \Delta_x(s, y)$; therefore *if $D(v) = d(s, y)$, then* it must be the case that y belongs in bucket $\lfloor \Delta_x(s, y)/norm(x) \rfloor$ or the following bucket. By Invariant 1 we must bucket y only if it is the case that $d(s, y) = D(v)$.

We actually cannot narrow the number of eligible buckets down to two, because we are working with $\hat{\Delta}_x$, a discrete approximation to Δ_x. According to its definition, $|\hat{\Delta}_x - \Delta_x| < \epsilon_x$, and therefore, if $D(v) = d(s, y)$, y must be bucketed in one of the $\lceil \frac{norm(x) + 2\epsilon_x}{norm(x)} \rceil + 1$ contiguous buckets beginning with bucket $\lfloor \frac{\hat{\Delta}_x(s, y) - \epsilon_x}{norm(x)} \rfloor$. We fix $\epsilon_x = norm(x)/2$, hence the number of eligible buckets is at most 3 and a bucketing operation requires only 2 comparisons and $O(1)$ time.

The other costs of our algorithm are the same as Hagerup's: $O(n \log \log n)$ for managing the van Emde Boas structure and $O(m\alpha(m, n)) = O(m + n \log \log n)$ time for updating and managing D-values. Together with Lemma 2, proved later, and the fact that APSP is solved with n SSSP computations, Theorem 2 follows.

Theorem 2. *The all-pairs shortest path problem on arbitrarily-weighted, directed graphs can be solved in $O(mn + n^2 \log \log n)$ time, where the only operations used on edge-lengths are comparisons & additions.*

6   The Computation of $\hat{\Delta}$

We show in this section that for any CH node x, all $\hat{\Delta}_x(\cdot, \cdot)$-values can be computed in $O(m \log n + m|V(C_x^c)| + n \frac{diam(C_x)}{norm(x)})$ time. Lemma 2, stated below, then follows directly from the bounds established in Lemma 1.

Lemma 2. *For $k = \Omega(\log n)$, $\hat{\Delta}_x(u,v)$ can be computed for all CH nodes x such that $diam(C_x)/norm(x) \geq k$, $u \in V(G)$, $v \in V(C_x^c)$, in $O(mn)$ time.*

The computation of $\hat{\Delta}_x(\cdot, \cdot)$ begins by computing $d(v, x)$ for all $v \in V(G)$. This is really just a single source shortest path problem in disguise. Let G^T be the graph derived by reversing the direction of all edges in G. Computing all $d(\cdot, x)$ is equivalent to computing single source shortest paths in the graph derived from G^T by contracting C_x, from the source vertex C_x.

Next we perform several SSSP computations, for each vertex in C_x^c, in a graph with the same topology as G^T but having a new, non-negative *integral* edge-length function. We guarantee that the distance to any vertex is bounded by $2n \cdot \lceil diam(C_x)/norm(x) \rceil$, and therefore a simple, bucket-based implementation of Dijkstra's algorithm can be used. We perform the $|V(C_x^c)|$ SSSP computations simultaneously, in order to bound the number of empty buckets scanned. The following lemma is straightforward.

Lemma 3. *If, in a positive integer-weighted graph, $d(u,v) \leq N$ for all $u \in S$ and vertices v, SSSP can be computed from sources in S in $O(|S| \cdot m + N)$ time.*

Let $G^f = (V(G), E(G), f)$ be the graph G, substituting a new length function f, and let d^f be the distance function for G^f. Define $\epsilon'_x = \epsilon_x/n = norm(x)/2n$ and define $\delta_x : E(G) \to \mathbb{R}$, $\hat{\delta}_x : E(G) \to \mathbb{N}$ as

$$\delta_x(u_1, u_2) \overset{\text{def}}{=} \frac{d(u_2, x) + \ell(u_1, u_2) - d(u_1, x)}{\epsilon'_x}$$

$$\hat{\delta}_x(u_1, u_2) \overset{\text{def}}{=} \lfloor \delta_x(u_1, u_2) \rfloor \quad \text{or} \quad \infty \quad \text{if} \quad \delta_x(u_1, u_2) > diam(C_x)/\epsilon'_x$$

Lemma 4. *$G^{\hat{\delta}_x}$ is computable in $O(m \log n)$ time.*

Proof. We cannot compute $\hat{\delta}_x(u_1, u_2) = \lfloor \frac{d(u_2,x) + \ell(u_1,u_2) - d(u_1,x)}{\epsilon'_x} \rfloor$ in the obvious way: ϵ'_x is a not-necessarily-computable real number, and division, subtraction and floor operations are not available for reals. However, if $\hat{\delta}_x(u_1, u_2) \neq \infty$,

$$\hat{\delta}_x(u_1, u_2) = \max\{j \; : \; 2n \cdot d(u_1, x) + j \cdot norm(x) \leq 2n \cdot (d(u_2, x) + \ell(u_1, u_2))\}$$

Note also that $norm(x)$ is already computed and that $2n \cdot d(u_1, x)$ and $2n \cdot (d(u_2, x) + \ell(u_1, u_2))$ are computable in $O(\log n)$ time. We compute $\hat{\delta}_x(u_1, u_2)$ in $O(\log \frac{diam(C_x)}{\epsilon'_x}) = O(\log n)$ time by first generating the values

$$\{norm(x), 2\,norm(x), 4\,norm(x), \ldots, 2^{\lceil \log \frac{diam(C_x)}{\epsilon'_x} \rceil} norm(x)\}$$

using simple doubling, then using these values to perform a binary search to find the maximal j satisfying the inequality above. ♠

Lemma 5. *For $u \in V(G), y \in V(C_x^c)$, $d^{\delta_x}(u, y) \leq 2n \frac{diam(C_x)}{norm(x)} \leq 4n^2$*

Proof. Let P_{uy} be a shortest path from u to C_y in G^{δ_x}. Then $d^{\delta_x}(u, y) = \sum_{e \in P_{uy}} \delta_x(e) = (\ell(P_{uy}) - d(u, x))/\epsilon'_x$. Together with the inequality $\ell(P_{uy}) \leq d(u, x) + diam(C_x)$, we have $d^{\delta_x}(u, y) \leq 2n \frac{diam(C_x)}{norm(x)}$, and by the construction of the component hierarchy, $\frac{diam(C_x)}{norm(x)} \leq 2n$. ♠

Compute-$\hat{\Delta}_x$:

1. Generate $G^{\hat{\delta}_x}$

2. $\forall u \in V(G), y \in V(C_x^c)$, compute $d^{\hat{\delta}_x}(u, y)$.

3. Set $\hat{\Delta}_x(u, y) = \epsilon_x \cdot \lceil \frac{\epsilon'_x \cdot d^{\hat{\delta}_x}(u,y)}{\epsilon_x} \rceil$

By Lemmas 3, 4 & 5, Steps 1 & 2 take $O(m(|V(C_x^c)| + \log n) + n \cdot \frac{diam(C_x)}{norm(x)})$ time. Step 3 is not (and cannot!) be executed as stated because ϵ_x is not necessarily computable. Recall from Section 5 that we represent $\hat{\Delta}_x$ as an integer multiple of ϵ_x, hence Step 3 involves no real-number operations at all. The correctness of this scheme is established in the following lemmas.

Lemma 6. $d^{\delta_x}(u, v) - d^{\hat{\delta}_x}(u, v) \in [0, n - 1)$.

Proof. The lower bound, $d^{\delta_x}(u, v) \geq d^{\hat{\delta}_x}(u, v)$, follows from the fact that all relevant edge lengths in $G^{\hat{\delta}_x}$ are shorter than in G^{δ_x}. For the upper bound, observe that for any shortest path P_{uv} in $G^{\hat{\delta}_x}$, $\delta_x(P_{uv}) \leq \hat{\delta}_x(P_{uv}) + |P_{uv}|$, with equality only if $|P_{uv}| = 0$, i.e. $u \in C_v$. The upper bound follows. ♠

Lemma 7. $\Delta_x(u, y) = \epsilon'_x \cdot d^{\delta_x}(u, y)$ and $\Delta_x(u, y) - \epsilon'_x \cdot d^{\hat{\delta}_x}(u, y) \in [0, \epsilon_x)$

Proof. Denote by $\langle u_1, u_2, \ldots, u_k \rangle$ a path from u_1 to u_k. Then

$$d^{\delta_x}(u, y) = \min_{j, \langle u=u_1, \ldots, u_j \in C_y \rangle} \left\{ \sum_{i=1}^{j-1} \delta_x(u_i, u_{i+1}) \right\}$$

$$= \min_{j, \langle u=u_1, \ldots, u_j \in C_y \rangle} \left\{ \frac{\ell(\langle u_1, \ldots, u_j \rangle) + d(u_j, x) - d(u_1, x)}{\epsilon'_x} \right\}$$

$$= \frac{d(u, y) - d(u, x)}{\epsilon'_x} = \frac{\Delta_x(u, y)}{\epsilon'_x}$$

The second part, $\Delta_x(u, y) - \epsilon'_x \cdot d^{\hat{\delta}_x}(u, y) \in [0, \epsilon_x)$, then follows from Lemma 6 and the definition of $\epsilon'_x = \epsilon_x/n$. ♠

Lemma 8. *Step 3 sets $\hat{\Delta}_x$ correctly, i.e.*

$$|\hat{\Delta}_x(u, y) - \Delta_x(u, y)| < \epsilon_x \quad \text{and} \quad \frac{\hat{\Delta}_x(u, y)}{\epsilon_x} \text{ is an integer.}$$

Proof. The second requirement, that $\hat{\Delta}_x(u,y)/\epsilon_x$ be an integer, is obvious. From the definition of the ceiling function,

$$\epsilon'_x \cdot d^{\hat{\delta}_x}(u,y) \le \epsilon_x \cdot \lceil \frac{\epsilon'_x \cdot d^{\hat{\delta}_x}(u,y)}{\epsilon_x} \rceil < \epsilon'_x \cdot d^{\hat{\delta}_x}(u,y) + \epsilon_x$$

From Lemma 7 we know that

$$\epsilon'_x \cdot d^{\hat{\delta}_x}(u,y) \le \Delta_x(u,y) = \epsilon'_x \cdot d^{\delta_x}(u,y) < \epsilon'_x \cdot d^{\hat{\delta}_x}(u,y) + \epsilon_x$$

Hence $|\Delta_x(u,y) - \epsilon_x \cdot \lceil \frac{\epsilon'_x \cdot d^{\delta_x}(u,y)}{\epsilon_x} \rceil| < \epsilon_x$. ♠

6.1 Proof of Lemma 2

Let $T(m,n,k)$ be the time to compute $\hat{\Delta}_x$ for all CH nodes x such that $\frac{diam(C_x)}{norm(x)} \ge k$. Using the bounds proved in Section 6, we have

$$T(m,n,k) = \sum_{x\,:\,\frac{diam(C_x)}{norm(x)} \ge k} O(m \log n + m|V(C_x^c)| + n\frac{diam(C_x)}{norm(x)})$$

$$= O(8mn \log n/k + 2mn + 8n^2) \quad \{\text{Lemma 1(i),(ii) \& (iii)}\}$$

$$= O(mn \lceil \frac{\log n}{k} \rceil) \quad \{\text{W.l.o.g. } m \ge n\}$$

and

$$T(m,n,k \ge \log n) = O(mn)$$

References

[CLR90]   T. Cormen, C. Leiserson, R. Rivest. *Intro. to Algorithms.* MIT Press, 1990.

[Dij59]   E. W. Dijkstra. A note on two problems in connexion with graphs. In *Numer. Math.*, 1 (1959), 269-271.

[FT87]   M. L. Fredman, R. E. Tarjan. Fibonacci heaps and their uses in improved network optimization algorithms. In *JACM* 34 (1987), 596–615.

[G85]   H. N. Gabow. A scaling algorithm for weighted matching on general graphs. In *Proc. FOCS 1985*, 90–99.

[Hag00]   T. Hagerup. Improved shortest paths on the word RAM. In *Proc. ICALP 2000*, LNCS volume 1853, 61–72.

[J77]   D. B. Johnson. Efficient algorithms for shortest paths in sparse networks. *J. Assoc. Comput. Mach.* 24 (1977), 1–13.

[Pet02]   S. Pettie. A faster all-pairs shortest path algorithm for real-weighted sparse graphs. UTCS Technical Report CS-TR-02-13, February, 2002.

[Pet02b]   S. Pettie. On the comparison-addition complexity of all-pairs shortest paths. UTCS Technical Report CS-TR-02-21, April, 2002.

[PRS02]   S. Pettie, V. Ramachandran, S. Sridhar. Experimental evaluation of a new shortest path algorithm. *Proceedings of ALENEX 2002.*

[PR02]   S. Pettie, V. Ramachandran. Computing shortest paths with comparisons and additions. *Proceedings of SODA 2002*, 267–276.

[Tar79]   R. E. Tarjan. A class of algorithms which require nonlinear time to maintain disjoint sets. *J. Comput. Syst. Sci.* 18 (1979), no. 2, 110–127.

[Tho99]   M. Thorup. Undirected single source shortest paths with positive integer weights in linear time. *J. Assoc. Comput. Mach.* 46 (1999), no. 3, 362–394.

[vEKZ77]   P. van Emde Boas, R. Kaas, E. Zijlstra. Design and implementation of an efficient priority queue. *Math. Syst. Theory* 10 (1977), 99–127.

[Z01]   U. Zwick. Exact and approximate distances in graphs – A survey. Updated version at `http://www.cs.tau.ac.il/ zwick/`, *Proc. of 9th ESA* (2001), 33–48.

On Families of Graphs Having a Decidable First Order Theory with Reachability

Thomas Colcombet

Irisa, Campus de Beaulieu, 35042, Rennes, France
Thomas.Colcombet@irisa.fr

Abstract. We consider a new class of infinite graphs defined as the smallest solution of equational systems with vertex replacement operators and unsynchronised product. We show that those graphs have an equivalent internal representation as graphs of recognizable ground term rewriting systems. Furthermore, we show that, when restricted to bounded tree-width, those graphs are isomorphic to hyperedge replacement equational graphs. Finally, we prove that on a wider family of graphs — interpretations of trees having a decidable monadic theory — the first order theory with reachability is decidable.

1 Introduction

Automatic verification of properties on programs is one of the challenging problems tackled by modern theoretical computer-science. An approach to this kind of problems is to translate the program in a graph, the property in a logic formula and to use a generic algorithm which automatically solves the satisfaction of the formula over the graph. The use of potentially unbounded data structures such as integers, or stacks in programs leads to infinite graphs. Thus, algorithms dealing with infinite graphs are needed. Many families of infinite graphs have been recently described. For the simplest one, it is possible to verify automatically powerful formulas. For the most complex families, nearly nothing can be said. We are interested here into three logics. The less expressive is the first order logic. The first order logic with reachability extends it with a reachability predicate. The most expressive is the monadic (second order) logic.

The reader may find a survey on infinite graphs in [16]. The study of infinite graphs started with pushdown graphs [11]. Vertices are words and edges correspond to the application of a finite set of prefix rewriting rules. The first extension is the family of HR-equational graphs [5] defined as the smallest solutions of equational systems with hyperedge replacement (HR) operators. The more general family of prefix recognizable graphs [3] is defined internally as systems of prefix rewriting by recognizable sets of rules (instead of finite sets for pushdown graphs). VR-equational graphs are defined as the smallest solutions of equational systems with vertex replacement (VR) operators. VR-equational graphs are isomorphic to prefix recognizable graphs [1]. All those families of graphs share a decidable monadic theory (those results are, in some sense, extensions of the

P. Widmayer et al. (Eds.): ICALP 2002, LNCS 2380, pp. 98–109, 2002.

famous decidability result of Rabin [12]). Some more general families have also been introduced. Automatic graphs are defined by synchronized transducers on words [14,2]. Only the first order theory remains decidable and the reachability problem cannot be handled anymore. The class of rational graphs is defined by general transductions [10]. The first order theory is not decidable anymore. The common point of all those families is that they are (explicitly or not) defined as rewriting systems of words. This is not true anymore with the ground term rewriting systems [9]. Vertices are now terms and transitions are described by a finite set of ground rewriting rules. A practical interest of this family is that it has a decidable first order theory with reachability (but not a decidable monadic theory) [7,4]. Studies have also been pursued on external properties of graphs. VR-equational graphs of bounded tree-width (this notion is known for long in the theory of finite graphs, see [13] for a survey) are HR-equational graphs [1]. Pushdown graphs are VR-equational graphs of finite degree [3]. Ground term rewriting systems of bounded tree-width are also pushdown graphs [9].

In this paper, we define a new family of infinite graphs, namely, the VRP-equational graphs. It is a natural extension of the VR-equational graphs (solution of equational systems with vertex replacement operators) with an unsynchronised product operator (VRP stands for vertex replacement with product). VRP-equational graphs are formally defined as interpretations of regular infinite trees. The first result of this paper gives an equivalent internal representation to VRP-equational graphs — the recognizable ground term rewriting systems. Secondly, we study the VRP-equational graphs of bounded tree-width and prove that those graphs are isomorphic to the HR-equational graphs. Finally, we show the decidability of the first order theory with reachability on a more general family of graphs, the graphs obtained by VRP-interpretation of infinite trees having a decidable monadic theory. Recent results tend to prove that important families of trees have a decidable monadic theory (algebraic trees [6] and higher order trees with a safety constraint [8]).

The remaining of this article is organized as follows. Section 2 gives the basic definitions. Section 3 describes VRP-equational graphs. The following three sections are independent. Section 4 introduces recognizable ground term rewriting systems and states the isomorphism with VRP-equational graphs. In Section 5 we study the VRP-equational graphs of bounded tree-width. In Section 6 we study the first order logic with reachability of VRP-interpretation of trees.

2   Definitions

We design by $\mathbb{N}$ the set of integers. The notation $[n]$ stands for $\{0, \ldots, n-1\}$. Let S be a set of symbols, S^* is the set of words with letters in S. The empty word is written ε. The length of a word w is written $|w|$.

Let Θ be a finite set of base types. A *typed alphabet* $\mathcal{F}$ (over Θ) is a family $(\mathcal{F}_\tau)_{\tau \in \Theta^* \times \Theta}$ where for all τ, $\mathcal{F}_\tau$ is the set of symbols of type τ. We assume that alphabets are finite. We will use notations such as $\mathcal{F} = \{f : \tau \mid f \in \mathcal{F}_\tau\}$ instead of describing the $\mathcal{F}_\tau$ sets separately. The types of the form (ε, θ) are simply written θ. For any symbol f in $\mathcal{F}_{(\theta_0 \ldots \theta_{i-1}, \theta)}$, the base type θ is the *codomain* of f, the integer i is the *arity* and the base type θ_k is the type of the $k + 1$-th argument of f.

A *tree* t (over the typed alphabet $\mathcal{F}$) is a function from $\mathbb{N}^*$ into $\mathcal{F}$ with a non-empty prefix closed domain D_t. The elements of D_t are called *nodes*, and the node ε is the *root* of the tree. A tree t is *well typed* if for all node v and all integer k, vk is a node if and only if k is smaller than the arity of $t(v)$, and, in this case, the codomain of $t(vk)$ is the type of the $k + 1$-th argument of $t(v)$. The *type* of a well typed tree is the codomain of its root symbol. Let t be a well typed tree and v one of its nodes, the *subtree* of t rooted at v is the well typed tree t^v defined for all node vu by $t^v(u) = t(vu)$. The set of subtrees of t is $sub(t)$. A tree t is *regular* if $sub(t)$ is finite.

We call *terms* the trees of finite domain. We denote $\mathcal{T}_\theta^\infty(\mathcal{F})$ the set of well typed trees over $\mathcal{F}$ of type θ, and $\mathcal{T}_\theta(\mathcal{F})$ the set of well typed terms over $\mathcal{F}$ of type θ. $\mathcal{T}^\infty(\mathcal{F})$ is the set of all well typed trees of any type (resp. $\mathcal{T}(\mathcal{F})$ for terms).

Here, we only consider infinite trees over ranked alphabets. A *ranked alphabet* is an alphabet typed over only one base type. Types are uniquely identified by their arity. We write n instead of (θ^n, θ) (where θ is the only base type). We want to describe such infinite trees as limits of sequences of 'growing' terms. We slightly extend the alphabet and equip the corresponding trees with a structure of *complete partial order* (cpo) in which sequences of growing terms are chains, and limits are least upper bounds. Formaly, let us define the new typed alphabet $\mathcal{F}_\perp = \mathcal{F} \cup \{\perp : 0\}$ where $\perp$ is a new symbol. We define a binary relation $\sqsubseteq$ over $\mathcal{T}^\infty(\mathcal{F}_\perp)$. Let t_1 and t_2 be trees, then $t_1 \sqsubseteq t_2$ states if $D_{t_1} \subseteq D_{t_2}$ and for all nodes v of t_1, $t_1(v) = t_2(v)$ or $t_1(v) = \perp$. This relation is a cpo — it has a smallest element, the tree with symbol $\perp$ at root, and the least upper bound of a chain of trees $(t_i)_{i \in \mathbb{N}}$ is $\sqcup t$ with $(\sqcup t)(v) = f$ where f is the symbol such that $v \in D_{t_j}$ and $t_j(v) = f$ for all $j \geq k$ for some k. Let t be a tree of $\mathcal{T}^\infty(\mathcal{F}_\perp)$, the *cut at depth* n of t, written $t \downarrow_n$, is the term defined by $t \downarrow_n (v) = t(v)$ for all $v \in D_t$ such that $|v| < n$, and $t \downarrow_n (v) = \perp$ for all $v \in D_t$ such that $|v| = n$. The sequence $(t \downarrow_n)_{n \in \mathbb{N}}$ is a chain of terms of least upper bound t.

3   VRP-Graphs

In this section, we describe the cpo of colored graphs (or simply graphs) and the VRP operators working on them. The interpretation of those operators over infinite regular trees defines the VRP-equational graphs.

From now on, we fix a finite set A of *labels*. *Colored graphs* (or simply graphs) are triple (V, E, η), where V is a countable set of *vertices*, $E \subseteq V \times A \times V$ is the set of edges and η is a mapping from V into a finite set (of colors). If G is a graph, we write V_G its set of vertices, E_G its set of edges and η_G its color mapping. For simplicity, we will assume that there exists an integer N such that the range of η is $[N]$. The set of graphs with coloring functions ranging in $[N]$ is $\mathcal{G}_N$.

We define the relation $\subseteq$ over graphs of $\mathcal{G}_N$ by $G \subseteq G'$ if $V_G \subseteq V_{G'}$, $E_G \subseteq E_{G'}$ and $\eta_G = \eta_{G'}|_{V_G}$ (η_G is the restriction of $\eta_{G'}$ over V_G). This relation is a cpo: the smallest element is the empty graph and the least upper bound is written $\sqcup$ (it corresponds to the union of vertices, the union of edges and the 'union' of coloring functions).

Given a graph G and given a one-to-one function Φ with domain containing V_G, $\Phi(G) = (V', E', \eta')$ is defined by:

$$V' = \Phi(V_G)$$
$$E' = \{(\Phi(v), e, \Phi(v')) \mid (v, e, v') \in E_G\}$$
$$\eta'(\Phi(v)) = \eta_G(v) \ .$$

If G and G' are two graphs such that there is an injective mapping Φ verifying $\Phi(G) = G'$ then G and G' are said *isomorphic*, written $G \sim G'$.

We define now the five basic operations on graphs used in VRP-equational graphs. The four first are the classical VR operations. The fifth is new.

Single vertex constant: for $n \in [N]$, $\dot{n} = (\{0\}, \emptyset, \{0 \mapsto n\})$.

Recoloring: for ϕ mapping from $[N]$ into $[N]$, $[\phi](V, E, \eta) = (V, E, \phi \circ \eta)$.

Edges adding: for $n, n' \in [N]$, $e \in A$, $[n \overset{e}{\bowtie} n']G = (V_G, E', \eta_G)$
   with $E' = E_G \cup \{(v, e, v') \mid \eta_G(v) = n, \ \eta_G(v') = n'\}$.

Disjoint union: $(V_0, E_0, \eta_0) \oplus (V_1, E_1, \eta_1) = (V, E, \eta)$
   with $\begin{cases} V = \{0\} \times V_0 \ \cup \ \{1\} \times V_1 \\ E = \{((\alpha, v), e, (\alpha, v')) \mid \alpha \in [2], (v, e, v') \in E_\alpha\} \\ \eta(\alpha, v) = \eta_\alpha(v) \qquad \text{for } \alpha \in [2] \ . \end{cases}$

Unsynchronised product: $(V_0, E_0, \eta_0) \otimes (V_1, E_1, \eta_1) = (V', E', \eta')$
   with $\begin{cases} V' = V_0 \times V_1 \\ E' = \{((v_0, v_1), e, (v'_0, v_1)) \mid (v_0, e, v'_0) \in E_0, \ v_1 \in V_1\} \\ \qquad \cup \ \{((v_0, v_1), e, (v_0, v'_1)) \mid v_0 \in V_0, \ (v_1, e, v'_1) \in E_1\} \\ \eta'(v_0, v_1) = \eta_0(v_0) + \eta_1(v_1) \bmod N \ . \end{cases}$

Let us remark that — up to isomorphism — $\oplus$ and $\otimes$ are commutative and associative and $\otimes$ is distributive over $\oplus$. The empty graph is an absorbing element for $\otimes$ and the neutral element for $\oplus$ and the graph $\dot{0}$ is neutral for $\otimes$.

Let us now define the ranked alphabets $\mathcal{F}_N^{\mathrm{VR}}$ and $\mathcal{F}_N^{\mathrm{VRP}}$:

$$\begin{aligned} \mathcal{F}_N^{\mathrm{VR}} &= \{\dot{n}_N : 0 \mid n \in [N]\} \\ &\cup \{[\phi]_N : 1 \mid \phi \text{ mapping from } [N] \text{ to } [N]\} \\ &\cup \{[n \overset{e}{\bowtie} n']_N : 1 \mid n, n' \in [N], \ e \in A\} \\ &\cup \{\oplus_N : 2\} \end{aligned}$$

$$\mathcal{F}_N^{\mathrm{VRP}} = \mathcal{F}_N^{\mathrm{VR}} \ \cup \ \{\otimes_N : 2\}$$

The trees of $\mathcal{T}^\infty(\mathcal{F}_N^{\mathrm{VR}})$ are called *VR-trees*, and the trees of $\mathcal{T}^\infty(\mathcal{F}_N^{\mathrm{VRP}})$ are *VRP-trees*. We define the *VRP-interpretation* $[\![\ ]\!]_N$ as the natural interpretation of terms of $\mathcal{T}(\mathcal{F}_N^{\mathrm{VRP}})$ over $\mathcal{G}_N$.

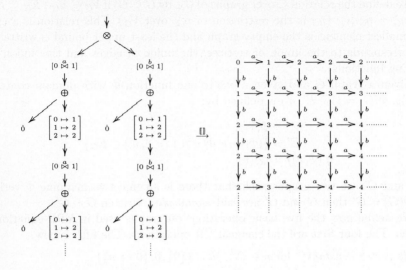

Fig. 1. A VRP-tree producing the infinite grid — assume $N = 5$

$$[\![\dot{c}_N]\!]_N = \dot{c}$$
$$[\![[\phi]_N(t)]\!]_N = [\phi][\![t]\!]_N$$
$$[\![[c \overset{e}{\bowtie} c']_N(t)]\!]_N = [c \overset{e}{\bowtie} c'][\![t]\!]_N$$

$$[\![\oplus_N(s,t)]\!]_N = [\![s]\!]_N \oplus [\![t]\!]_N$$
$$[\![\otimes_N(s,t)]\!]_N = [\![s]\!]_N \otimes [\![t]\!]_N$$

VRP-interpretation is extended by continuity to any tree — let $[\![\bot]\!]_N$ be the empty graph, the interpretation of a tree $t \in \mathcal{T}^\infty(\mathcal{F}_N^{\mathrm{VRP}})$ is $[\![t]\!]_N = \cup_k [\![t \downarrow_k]\!]_N$. This definition makes sense since all operators are continuous with respect to the $\subseteq$ order.

Definition 1. *The VR-equational graphs (resp. VRP-equational graphs) are the VRP-interpretations of the regular VR-trees (resp. VRP-trees).*

Figure 1 gives the example of the infinite grid described by a VRP-tree.

Remark 1. Given two VRP-equational graphs, we cannot yet combine them easily because the number of colors used may be different in the two underlying VRP-trees, and VRP-interpretation depends of this value. The problem comes from the color mapping used in the definition of $\otimes$ — it depends of N. In order to get rid of this drawback, we use VRP-trees without overflows: a VRP-tree t has *no overflow* if for all node u such that $t^u = \otimes_N(s, s')$, the sum of maximum colors appearing in $[\![s]\!]_N$ and $[\![s']\!]_N$ is strictly smaller than N. Under this constraint, the color mapping $\eta'(v_0, v_1) = \eta_0(v_0) + \eta_1(v_1) \bmod N$ is a simple sum and does not depend of N anymore. Increasing the value of N everywhere in a VRP-tree without overflow (and accordingly extending the recoloring operators) do not change the VRP-interpretation of the tree.

Furthermore if a VRP-tree has overflows, it is easy to obtain an equivalent VRP-tree without overflow by doubling the value of N and replacing everywhere $\otimes_N(s, s')$ by $[\mathrm{mod}_N]_{2N}(\otimes_{2N}(s, s'))$.

For now and on, we make the assumption that all VRP-trees have no overflow and we omit everywhere the N indices. We also allow ourself to increase the value of N whenever it is needed.

4   Internal Representation

In this part, we give to VRP-equational graphs an equivalent internal representation (Theorem 1). Instead of describing those graphs as the interpretation of regular trees, we describe them by explicitly giving the set of vertices and explicitly defining the edge relation.

We use here deterministic bottom-up tree automata without final states. Let $\mathcal{F}$ be a typed alphabet over Θ. A *deterministic tree automaton* is a tuple $(\mathcal{F}, Q, \delta)$, where $(Q_\theta)_{\theta \in \Theta}$ is a finite alphabet of states (of arity 0), and δ is a function which maps for all type $\tau = (\theta_0 \ldots \theta_{n-1}, \theta)$ tuples of $\mathcal{F}_\tau \times Q_{\theta_0} \times \cdots \times Q_{\theta_{n-1}}$ into Q_θ. The function δ is naturally extended by induction into a function from $\mathcal{T}_\theta(\mathcal{F} \cup Q)$ into Q_θ (for any type θ).

A set $T \subseteq \mathcal{T}(\mathcal{F})$ is *recognizable* if there exists a deterministic automaton $\mathcal{A} = (\mathcal{F}, Q, \delta)$ and a set $F \subseteq Q$ (of final states) verifying that $t \in T$ iff $\delta(t) \in F$. Similarly, a relation $R \subseteq \mathcal{T}(\mathcal{F}) \times A \times \mathcal{T}(\mathcal{F})$ is *recognizable* if there exists a deterministic tree automaton $\mathcal{A} = (\mathcal{F}, Q, \delta)$ and a set $F \subseteq Q \times A \times Q$ verifying that $(t, e, t') \in R$ iff $(\delta(t), e, \delta(t')) \in F$. Such a relation R *preserves* types if for all $(t, e, t') \in R$, t and t' have the same type.

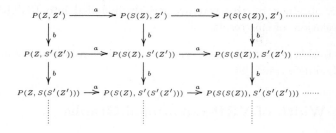

Fig. 2. The infinite grid as a recognizable ground term rewriting system

Let $\mathcal{F}$ be a typed alphabet over the set of types Θ, $\theta_r \in \Theta$ be a (root) type, and R a recognizable subset of $\mathcal{T}(\mathcal{F}) \times A \times \mathcal{T}(\mathcal{F})$ preserving types. The *(recognizable) ground term rewriting system* (RGTRS) are not colored graphs defined by a recognizable relation R. The vertices are terms of $\mathcal{T}_{\theta_r}(\mathcal{F})$, and there is an edge labelled by e between two terms t_1 and t_2 if one can obtain t_2 by replacing a subterm t_1' by t_2' in t_1, with $(t_1', e, t_2') \in R$. Formally, the graph of ground term rewriting by R is the graph:

$$GTR_{\theta_r}(R) = (\mathcal{T}_{\theta_r}(\mathcal{F}), \{(t_1, e, t_2) \mid \exists u \in D_{t_1} \cap D_{t_2}, \\ (t_1^u, e, t_2^u) \in R, \quad t_1|_{(D_{t_1} - u\mathbb{N}^*)} = t_2|_{(D_{t_2} - u\mathbb{N}^*)}\})$$

Figure 2 gives an example of the infinite grid as a RGTRS. The types are $\Theta = \{\theta_r, \theta, \theta'\}$ where θ_r is the root type. The set of symbols is $\mathcal{F} = \{P : (\theta\theta', \theta_r), Z : \theta, Z' : \theta', S : (\theta, \theta), S' : (\theta', \theta')\}$. The recognizable relation used is $R = \{(Z, a, S(Z)), (Z', b, S'(Z'))\}$.

Theorem 1. *The VRP-equational graphs are exactly the RGTRS up to isomorphism and color removal.*

The idea of the translation is almost straightforward. One identifies subtrees of regular VRP-trees with types, and colors with states of the automaton. The same construction can also be applied for non regular VRP-trees. It leads to an infinite set of types.

Remark 2. In the RGTRS, the typing policy is used as a technique for restricting the set of vertices. The prefix recognizable graphs are defined as prefix rewriting systems of words (or equivalently linear terms), but the vertices are restricted by a general recognizable (rational) language. Restricting vertices of RGTRS with a recognizable relation increase the power of the system and leads to graphs which do not have a decidable reachability problem (whereas RGTRS have, see Theorem 3). However the RGTRS restricted to linear terms gives exactly the prefix recognizable graphs up to isomorphism[1].

Remark 3. Another technique for restricting vertices is to use a root vertex and keep only vertices reachable from the root. This approach is used by Löding [9]. The GTRS obtained by this way are strictly included into RGTRS of finite rewriting relation. of this result.

According to this remark, the next section can be seen as an extension to infinite degree of Löding's result [9].

5   Tree-Width of VRP-Equational Graphs

In this section we study the VRP-equational graphs of bounded tree-width. We show that those graphs are exactly the HR-equational graphs (Theorem 2).

Definition 2. *A tree decomposition of a graph G is an undirected tree (a connected undirected graph without cycle) $T = (V_T, E_T)$ with subsets of V_G as vertices such that:*

---

[1] In fact, one can restrict RGTRS by a top-down deterministic tree automaton without changing the power of the system. In general, top-down deterministic tree automaton are less expressive than recognizability. On linear terms, both are equivalent.

1. *for any edge $(v, e, v') \in E_G$, there is a vertex $N \in V_T$ such that $v, v' \in N$.*
2. *for any vertex $v \in V_G$, the set of vertices containing v, $\{N \in V_t \mid v \in N\}$ is a connected subpart of T.*

The tree-width of a graph G is the smallest $k \in \mathbb{N} \cup \{\infty\}$ such that G admits a tree decomposition with vertices of cardinality smaller than or equal to $k + 1$. If k is finite, the graph is of bounded tree-width.

Intuitively, a graph is of tree-width k if it can be obtained by 'pumping up at width k' a tree. An equivalent definition is that a graph is of tree-width k if it can be obtained by a (non-deterministic) graph grammar with at most $k + 1$ vertices per rule. This notion does not rely on orientation nor on labelling of edges. It is defined up to isomorphism and is continuous. Let $\leq$ be the binary relation defined by $G \leq G'$ if G is isomorphic to a subgraph of G' up to relabelling and edge reversal. It has the important property that if $G \leq G'$ then the tree-width of G is smaller or equal to the tree-width of G'.

Given a regular VRP-tree t, we aim at eliminating the $\otimes$ operators of the tree under the constraint of bounded tree-width. We then obtain a regular VR-tree of isomorphic interpretation. To this purpose, we assume that t is *normalized*: none of its subtrees has the empty graph as interpretation (normalization is possible if $[\![t]\!]$ is not empty). Under this constraint, for any subtree s of t, we have $[\![s]\!] \leq [\![t]\!]$ (this would not be true anymore if t was not normalized because the empty graph is an absorbing element of $\otimes$). The first consequence of normalization is that all subtrees of t have an interpretation of bounded tree-width.

First step: The first step of the transformation eliminates subtrees of the form $s = \otimes(s_0, s_1)$ appearing infinitely in a branch of the tree. Using normalization, we obtain that for all n, either $[\![s_0]\!]^n \leq [\![s]\!]$, either $[\![s_1]\!]^n \leq [\![s]\!]$ (with $G^0 = \dot{0}$ and $G^{n+1} = G \otimes G^n$). Lemma 1 handles such cases.

Lemma 1. *Let G be a graph such that for all n, $G^n \leq G'$ and G' is of bounded tree-width, then all edges in G are loops[2] (of the form (x, e, x)).*

In fact, the unsynchronised product of a graph G with a graph with only loops does not behaves like a real product. It leads to the graph G duplicated a finite or an infinite number of times, with colors changed and possibly loops added. The same result can be obtained without product, and furthermore, this transformation can be performed simultaneously everywhere in a tree. The first step of the transformation amounts to apply this technique until there is no branch of the tree with an infinite number of product operators.

Second step: After the first step of the transformation, among the subtrees with $\otimes$ at root, at least one does not contain any other $\otimes$ operator. Let $\otimes(s_0, s_1)$ be this subtree. Its interpretation is of bounded tree-width, and s_0 and s_1 are VR-trees. Lemmas 2 and 3 then hold.

---

[2] If G has a non-looping edge, then G^n has the hypercube of dimension n as subgraph. The hypercubes do not have a bounded tree-width, thus G' is not of bounded tree-width.

Lemma 2. *If the unsynchronised product of two graphs is of bounded tree-width then one of the graphs has all its connected components of bounded size.*

Lemma 3. *VR-equational graphs are closed by unsynchronised product with graphs having all their connected components of bounded size[3].*

It follows that $[\![\otimes(s_0, s_1)]\!]$ is a VR-equational graph. It is then sufficient to replace in the tree all occurrences of $\otimes(s_0, s_1)$ by a VR-tree of isomorphic interpretation. After iteration of this process, there is no more product operator left, and thus the original graph is VR-equational up to isomorphism.

The HR-equational graphs are originally described as the smallest solution of equational systems with hyperedge replacement operators. Barthelmann has proved that those graphs were exactly the VR-equational graphs of bounded tree-width (see [1]). Theorem 2 concludes.

Theorem 2. *The VRP-equational graphs of bounded tree-width are exactly the HR-equational graphs up to isomorphism.*

It is interesting as the HR-equational graphs have a decidable monadic theory whereas VRP-equational graphs do not (the infinite grid is the classical counter-example). In Section 6 we extend the known result that the first order theory with reachability remains decidable for VRP-equational graphs.

6   Decidability of Logic

In this section, we prove that for all trees t which have a decidable monadic theory, $[\![t]\!]$ has a decidable first order theory with reachability predicate (Theorem 3).

6.1   Monadic Second Order Logic

During this part, t will be a (not necessarily regular) VRP-tree. We use here the monadic second order logic on VRP-trees. We do not define precisely what this logic is (the reader may find a description of it in [15]). Informally, we are allowed in monadic formula to use the classical logic connectives (e.g $\vee, \wedge, \neg$), existential ($\exists$) and universal ($\forall$) quantifiers over variables of first and monadic second order. Variables of first order are interpreted as nodes of the tree t and are written in small letters. Monadic second order variables are interpreted as sets of nodes and are written in capital letters. If x is a first order variable then $x0$ represents its left child, and $x1$ the right one (this notation is compatible with the definition of nodes as words). The predicates allowed are equality of first order variables (e.g $x = y0$) and membership (e.g $x \in X$). The first order constant ε is the root of the tree. The symbols of t are described by a finite set of second order constants of same name. For instance, expressing that the symbol at node x is $\otimes$ is written $x \in \otimes$.

---

[3] Notice that it changes the number of colors used.

We also use for simplicity classical operations on sets (e.g. $\cup, \subseteq, \ldots$) and quantification over known finite sets (colors, labels and functions from colors to colors). We also write $x\mathbb{N}^*$ the set of nodes under x. Those extensions can be encoded into the monadic formulas without any difficulty.

If Ψ is a closed monadic formula, we write $t \vDash \Psi$ to express the satisfaction of Ψ by t.

As an example, we define the predicate $\text{finitetree}(r, X)$ which is satisfied if X is a finite connected subset of D_t 'rooted' at r.

$$\begin{aligned} \text{childof}(x, y) &\equiv x = y0 \wedge (\forall \, n, \; y \notin \dot{n}) \\ &\vee x = y1 \wedge (y \in \oplus \; \vee \; y \in \otimes) \end{aligned}$$

$$\begin{aligned} \text{finitetree}(r, X) &\equiv r \in X \; \wedge \; \forall x \in X, \; x = r \text{ xor } \exists y \in X, \text{childof}(x, y) \\ &\wedge \neg(\exists W \subseteq X, W \neq \emptyset \wedge \; \forall x \in W, \exists y \in W, \text{childof}(y, x)) \end{aligned}$$

The predicate $\text{childof}(x, y)$ means that x appears just under y in the tree t. Then, we define a finite connected subset X 'rooted' at r as a set such that every element, except the root r, has a father in the set and which do not contain an infinite branch W.

6.2   First Order Theory with Reachability of VRP-Interpretations

Our first goal is to encode the VRP-interpretation of t into monadic formulas. To obtain this encoding we first remark that each vertex of the resulting graph can be uniquely identified with a special kind of finite subset of D_t. Let t be a VRP-tree and r one of its nodes, it amounts to associate to each vertex of $[\![t^r]\!]$ a non-empty finite connected subpart of D_t rooted at r. This finite rooted subset can be seen as the equivalent term of a RGTRS system.

If $t(r) = \dot{n}$ then $\{r\}$ encodes the only vertex of $[\![t^r]\!]$.

If $t(r) = [\phi]$ or $t(r) = [n \overset{e}{\bowtie} n']$ the vertices of $[\![t^r]\!]$ are exactly the vertices of $[\![t^{r0}]\!]$. Let X be the encoding of a vertex of $[\![t^{r0}]\!]$, then $\{r\} \cup X$ encodes the same vertex in $[\![t^r]\!]$.

If $t(r) = \oplus$, then a vertex of $[\![t^r]\!]$ has its origin either in $[\![t^{r0}]\!]$, either in $[\![t^{r1}]\!]$. Let v be a vertex of $[\![t^{r\alpha}]\!]$ (for some $\alpha \in \{0, 1\}$) and X its encoding, then $\{r\} \cup X$ uniquely encodes the vertex in $[\![t^r]\!]$.

Finally, if $t(r) = \otimes$, then each vertex of $[\![t^r]\!]$ originates from both a vertex of $[\![t^{r0}]\!]$ and a vertex of $[\![t^{r1}]\!]$. Let X_0 be the encoding of the origin vertex in $[\![t^{r0}]\!]$ and X_1 be the encoding of the origin vertex in $[\![t^{r1}]\!]$, then $\{r\} \cup X_0 \cup X_1$ uniquely encodes the vertex in $[\![t^r]\!]$.

We can translate this description into a monadic predicate $\text{vertex}(r, X)$ which is satisfied by t if X encodes a vertex of $[\![t^r]\!]$. We can also describe a monadic predicate $\text{color}_c(r, X)$ which is satisfied by t if the vertex encoded by X has color c in $[\![t^r]\!]$, and a predicate $\text{edge}_e(r, X, X')$ which is satisfied by t if there is an edge of label e between the vertex encoded by X and the vertex encoded by X' in $[\![t^r]\!]$. The validity of the approach is formalized in Lemma 4.

Lemma 4. *For all r, the graph $[\![t^r]\!]$ is isomorphic to $G_r = (V_r, E_r, \eta_r)$ with:*

$$V_r = \{X \subseteq D_t \mid t \vDash \text{vertex}(r, X)\}$$
$$E_r = \{(X, e, X') \mid t \vDash \text{edge}_e(r, X, X')\}$$
$$\eta_r(X) = n \quad \text{such that} \quad t \vDash \text{color}_n(r, X) .$$

We define also the monadic predicate $\text{path}_{A'}(X, X')$ which satisfies the following lemma.

Lemma 5. *Let $A' \subseteq A$ be a set of labels and let X, X' be two vertices of G_ε. There is a path with labels in A' between X and X' in G_ε iff $t \vDash \text{path}_{A'}(X, X')$.*

The proof of this result is technical : it involves the encoding of simpler problems such as "*is there a path between a node of color n and a node of color n' ?*".

Using the two previous lemmas, it is easy to translate a first order formula with reachability predicate over $[\![t]\!]$ into a monadic formula over t. The main theorem is then straightforward.

Theorem 3. *The VRP-interpretation of a tree having a decidable monadic second order theory has a decidable first order theory with reachability predicate.*

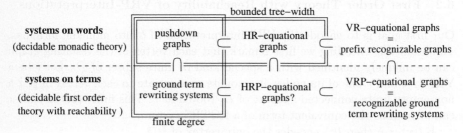

Fig. 3. A partial classification of families of graphs

7   Conclusion

System on words where already well known. This paper is a step toward a corresponding classification for systems on terms. The hierarchy obtained is depicted in Figure 3. The ground term rewriting systems must be understood recognizable ground term rewriting systems of finite rewriting relations (see Remark 3). Notice that there is probably a natural family of graphs — HRP-equational graphs — between ground term rewriting systems and VRP-equational graphs.

An open question is the nature of the family of graphs obtained by ε-closure of VRP-equational graphs (VR-equational graphs are closed by this operation). It also has a first order theory with reachability predicate decidable (ε-closure preserves it). We assume that this family of graphs strictly contains the family of VRP-equational graphs.

Lastly, we have studied the more general family of graphs obtained by VRP-interpretation of infinite trees with a decidable monadic second order theory. Those graphs have a first order theory with reachability predicate decidable. In fact, it is probable that weak monadic second order theory is sufficient.

Acknowledgments. Many thanks to Didier Caucal who has introduced me to this topic and gave helpful remarks. Thanks to Thierry Cachat, Emmanuelle Garel and Tanguy Urvoy for reading previous versions of this work. Thanks to Marion Fassy for her support.

References

1. K. Barthelmann. When can equational simple graphs be generated by hyperedge replacement? Technical report, University of Mainz, 1998.
2. A. Blumensath and E. Grädel. Automatic Structures. In *Proceedings of 15th IEEE Symposium on Logic in Computer Science LICS 2000*, pages 51–62, 2000.
3. D. Caucal. On infinite transition graphs having a decidable monadic theory. In Icalp 96, volume 1099 of *LNCS*, pages 194–205, 1996.
4. H. Comon, M. Dauchet, R. Gilleron, F. Jacquemard, D. Lugiez, S. Tison, and M. Tommasi. Tree automata techniques and applications. Available on: http://www.grappa.univ-lille3.fr/tata, 1997.
5. B. Courcelle. *Handbook of Theoretical Computer Science*, chapter Graph rewriting: an algebraic and logic approach. Elsevier, 1990.
6. B. Courcelle. The monadic second order logic of graphs ix: Machines and their behaviours. In *Theoretical Computer Science*, volume 151, pages 125–162, 1995.
7. M. Dauchet and S. Tison. The theory of ground rewrite systems is decidable. In *Fifth Annual IEEE Symposium on Logic in Computer Science*, pages 242–248. IEEE Computer Society Press, June 1990.
8. T. Knapik, D. Niwinski, and P. Urzyczyn. Higher-order pushdown trees are easy. In M. Nielsen, editor, *FOSSACS'2002*, 2002.
9. C. Löding. Ground tree rewriting graphs of bounded tree width. In *STACS 02*, 2002.
10. C. Morvan. On rational graphs. In J. Tiuryn, editor, *FOSSACS'2000*, volume 1784 of *LNCS*, pages 252–266, 2000.
11. D. Muller and P. Schupp. The theory of ends, pushdown automata, and second-order logic. *Theoretical Computer Science*, 37:51–75, 1985.
12. M.O. Rabin. Decidability of second-order theories and automata on infinite trees. *Trans. Amer. Math. soc.*, 141:1–35, 1969.
13. D. Seese. The structure of models of decidable monadic theories of graphs. *Annals of Pure and Applied Logic*, 53(2):169–195, 1991.
14. G. Sénizergues. Definability in weak monadic second-order logic of some infinite graphs. In *Dagstuhl seminar on Automata theory: Infinite computations, Warden, Germany*, volume 28, page 16, 1992.
15. W. Thomas. Languages, automata, and logic. *Handbook of Formal Language Theory*, 3:389–455, 1997.
16. W. Thomas. A short introduction to infinite automata. In W. Kuich, editor, *DLT'2001*, 2001.

Heuristically Optimized Trade-Offs: A New Paradigm for Power Laws in the Internet

Alex Fabrikant[1], Elias Koutsoupias[2*], and Christos H. Papadimitriou[1**]

[1] Computer Science Division
University of California at Berkeley
Soda Hall 689
Berkeley CA 94720, U.S.A.
alexf@csua.berkeley.edu, christos@cs.berkeley.edu
[2] University of Athens and UCLA
Computer Science Department
3731 Boelter Hall
Los Angeles, CA 90095, U.S.A.
elias@cs.ucla.edu

Abstract. We propose a plausible explanation of the power law distributions of degrees observed in the graphs arising in the Internet topology [Faloutsos, Faloutsos, and Faloutsos, SIGCOMM 1999] based on a toy model of Internet growth in which two objectives are optimized simultaneously: "last mile" connection costs, and transmission delays measured in hops. We also point out a similar phenomenon, anticipated in [Carlson and Doyle, Physics Review E 1999], in the distribution of file sizes. Our results seem to suggest that power laws tend to arise as a result of complex, multi-objective optimization.

1 Introduction

It was observed in [5] that the degrees of the Internet graph (both the graph of routers and that of autonomous systems) obey a sharp *power law*. This means that the distribution of the degrees is such that the probability that a degree is larger than D is about $cD^{-\beta}$ for some constant c and $\beta > 0$ (they observe βs between 2.15 and 2.48 for various graphs and years). They go on to observe similar distributions in Internet-related quantities such as the number of hops per message, and, even more mysteriously, the largest eigenvalues of the Internet graph. This observation has led to a revision of the graph generation models used in the networking community [14], among other important implications. To date, there has been no theoretical model of Internet growth that predicts this phenomenon. Notice that such distributions are incompatible with random graphs in the $G_{n,p}$ model and the law of large numbers, which yield exponential distributions.

* Research supported in part by NSF and the IST Program.
** Research supported in part by an NSF ITR grant, and an IBM faculty development award.

P. Widmayer et al. (Eds.): ICALP 2002, LNCS 2380, pp. 110–122, 2002.
© Springer-Verlag Berlin Heidelberg 2002

Power laws have been observed over the past century in income distribution [13], city populations [15,6], word frequencies [10], and literally hundreds of other domains including, most notably, the degrees of the world-wide web graph [9]; they have been termed "the signature of human activity" (even though they do occasionally arise in nature)[1]. There have been several attempts to explain power laws by so-called generative models (see [12] for a technical survey). The vast majority of such models fall into one large category (with important differences and considerable technical difficulties, of course) that can be termed *scale-free growth* or *preferential attachment* (or, more playfully, *"the rich get richer"*). That is, if the growth of individuals in a population follows a stochastic process that is independent of the individual's size (so that larger individuals attract more growth), then a power law will result (see, from among dozens of examples, [6] for an elegant argument in the domain of city populations, and [8] for a twist involving copying in the world-wide web, crucial for explaining some additional peculiarities of the web graph, such as the abundance of small bipartite graphs).

Highly optimized tolerance (HOT, [2]) is perhaps the other major class of models predicting power laws. In HOT models, power laws are thought to be the result of optimal yet reliable design in the presence of a certain hazard. In a typical example, the power law observed in the distribution of the size of forest fires is attributed to the firebreaks, cleverly distributed and optimized over time so as to minimize the risk of uncontrolled spread of fire. The authors of [2] refer briefly to the Internet topology and usage, and opine that the power law phenomena there are also due to the survivability built in the Internet and its protocols (whereas it is well known that this aspect of the Internet has not had a significant influence on its development beyond the very beginning [3]).

In this paper we propose a simple and primitive model of Internet growth, and prove that, under very general assumptions and parameter values, it results in power-law-distributed degrees. By "power law" we mean here that the probability that a degree is larger than d is *at least* $d^{-\beta}$ for some $\beta > 0$. In other words, we do not pursue here sharp convergence results *à la* [8,4,1], content to bound the distribution away from exponential ones. Extensive experiments suggest that much stronger results actually hold.

In our model a tree is built as nodes arrive uniformly at random in the unit square (the shape is, as usual, inconsequential). When the i-th node arrives, it attaches itself on one of the previous nodes. But which one? One intuitive objective to minimize is the Euclidean distance between the two nodes. But a newly arrived node plausibly also wants to connect to a node in the tree that is "centrally located", that is, its hop distance (graph theoretic distance with edge lengths ignored) to the other nodes is as small as possible. The former objective captures the "last mile" costs, the latter the operation costs due to communication delays. Node i attaches itself to the node j that minimizes the weighted sum of the two objectives:

$$min_{j<i}\alpha \cdot d_{ij} + h_j,$$

---

[1] They are certainly the product of one particular kind of human activity: looking for power laws...

where d_{ij} is the Euclidean distance, and h_j is some measure of the "centrality" of node j, such as (a) the average number of hops from other nodes; (b) the maximum number of hops from another node; (c) the number of hops from a fixed center of the tree; our experiments show that all three measures result in similar power laws, even though we only prove it for (c). α is a parameter, best thought as a function of the final number n of points, gauging the relative importance of the two objectives.

We are not claiming, of course, that this process is an accurate model of the way the Internet grows. But we believe it is interesting that a simple and primitive model of this form leads to power law phenomena. Our model attempts to capture in a simple way the *trade-offs* that are inherent in networking, but also in all complex human activity (arguably, such trade-offs are key manifestations of the aforementioned complexity).

The behavior of the model depends crucially on the value of α, and our main result (Theorem 1) fathoms this dependency: If α is less than a particular constant depending on the shape of the region, then Euclidean distances are not important, and the resulting network is easily seen to be a *star*—the ultimate in degree concentration, and, depending on how you look at it, the exact opposite, or absurd extreme, of a power law. If α grows at least as fast as $\sqrt{n}$, where n is the final number of points, then Euclidean distance becomes *too* important, and the resulting graph is a dynamic version of the Euclidean minimum spanning tree, in which high degrees do occur, but with exponentially vanishing probability (our proof of this case is a geometric argument). Again, no power law. If, however, α is *anywhere in between* — is larger than a certain constant, but grows slower than $\sqrt{n}$ if at all — then, almost certainly, the degrees obey a power law. This part is proved by a combinatorial-geometric argument, in which we show that for any value for the desired degree, there are likely to be enough nodes with large enough "regions of influence," disjoint from one another, such that any future node falling into this region is certain to have an edge to the given node. Our technique proves a lower bound of $\beta = \frac{1}{6}$ for $\alpha = o(n^{1/3})$ (and smaller bounds for $\alpha = o(\sqrt{n})$), while our experiments (see Section 2.2) suggest that the true value is around 0.6-0.9.

In Section 3 we prove a result in a different but not unrelated domain: we present a simple (naïve is more accurate) model of file creation, inspired by [2], and prove that it predicts a power law in the distribution of file sizes under very broad assumptions. The model is this: We have a set of n data items, all of the same size, that we must partition into files. The i-th item has *popularity p_i* — say, the expected number of times it will be retrieved for Internet transmission each day. We want to partition the items into files so that the following two objectives are minimized (a) total transmission costs (the sum over all partitions of the product of the partition size times the total partition popularity), and (b) the total number of files. It is easy to see that the optimum partition will include items in sorted order of popularity, and can be found by dynamic programming. We show that, when the popularities are i.i.d. from any one of a large class of distributions (encompassing the uniform, exponential, Gaussian, power law, etc. distributions), then the optimum file sizes are, almost surely, power-law distributed. The technical requirement on the distribution from which the p_i's are drawn is essentially that the cumulative distribution Φ do not start exponentially slow at zero (see Theorem 2). The authors of [2] propose a similar model, and make an observation in the same direction: They consider a few examples of distributions of the p_i's (*not* distributions from which they are drawn, as in our model, but distributions of the drawn samples) and for these they point out that the file sizes obey a power law.

Our results seem to suggest that power laws are perhaps the manifestation of *trade-offs*, complicated optimization problems with multiple and conflicting objectives — arguably one of the hallmarks of advanced technology, society, and life. Our framework generalizes the HOT class of models proposed in [2], in the sense that HOT models are the trade-offs in which reliable design is one of the objectives being optimized.

As it turns out, within our proposed conceptual framework also lies a classical and beautiful model by Mandelbrot [10]. Suppose that you want to design the optimum language, that is, the optimum set of frequencies $f_1 \leq f_2 \leq \cdots \leq f_n$ assigned to n words. The length of the i-th word is presumably $\log i$. You want to maximize the information transmitted, which is the entropy of the f_i's $(-\sum f_i \log f_i)$, divided by the expected transmission cost, $\sum f_i \log i$. The frequencies that achieve the optimum: a power law! Mandelbrot's multi-objective optimization differs from our two examples in that of his two objectives one is minimized and the other maximized; hence he considers their ratio, instead of their weighted sum (the two are obviously related by Lagrange multipliers).

2   A Model of Internet Growth

2.1   The Main Result

Consider a sequence of points $p_0, p_1, \ldots, p_n$ in the unit square, distributed uniformly at random. We shall define a sequence of undirected trees $T_0, T_1, \ldots$ on these points, with T_0 the tree consisting of p_0. Define h_i to be the number of hops from p_i to p_0 in T_i, and d_{ij} the Euclidean distance between points i^2 and j. Let α be a fixed number (we allow it though to be a function of n). Then T_i is defined as T_{i-1} with the point i and the edge $[i,j]$ added, where $j < i$ minimizes $f_i(j) = \alpha d_{ij} + h_j$. Let $T = T_n$. We will denote by $N_k(i)$ the neighborhood $\{j | [i,j] \in T_k\}$ of i in T_k; similarly, $N(i)$ will denote the neighborhood of i in T.

Theorem 1. *If T is generated as above, then:*

(1) If $\alpha < 1/\sqrt{2}$, then T is a star with p_0 as its center.
(2) If $\alpha = \Omega(\sqrt{n})$, then the degree distribution of T is exponential, that is, the expected number of nodes that have degree at least D is at most $n^2 \exp(-cD)$ for some constant c: $E\left[|\{i : \text{degree of } i \geq D\}|\right] < n^2 \exp(-cD)$.

---

[2] p_i and "point i" are used interchangeably throughout

(3) *If $\alpha \geq 4$ and $\alpha = o(\sqrt{n})$, then the degree distribution of T is a power law; specifically, the expected number of nodes with degree at least D is greater than $c \cdot (D/n)^{-\beta}$ for some constants c and β (that may depend on α): $E\left[|\{i: \text{ degree of } i \geq D\}|\right] > c(D/n)^{-\beta}$. Specifically, for $\alpha = o(\sqrt[3]{n^{1-\epsilon}})$ the constants are: $\beta \geq 1/6$ and $c = O(\alpha^{-1/2})$.*

Proof. We prove each case separately. The proof of the third case is the more involved.

(1) The first case follows immediately from the objective function; since $d_{ij} < \sqrt{2}$ for all i, j, and $h_j \geq 1$ for all $j \neq 0$, $f_i(0) < 1 \leq f_i(j)$ for $j \neq 0$, so every node p_i will link to p_0, creating a star.

(2) To obtain the exponential bound for (2), we consider the degree of any point p_i as consisting of 2 components — one due to geometrically "short" links, $S(i) = |\{j \in N(i) \,|\, d_{ij} \leq \frac{4}{\alpha}\}|$, and one due to "long" links, $L(i) = |\{j \in N(i) \,|\, d_{ij} > \frac{4}{\alpha}\}|$. By the union bound, $\Pr[\text{degree}_i \geq D] \leq \Pr[S(i) \geq D/2] + \Pr[L(i) \geq D/2]$.

For a fixed i and $\alpha \geq c_0\sqrt{n}$, any points contributing to $S(i)$ must fall into a circle of area $16\pi\alpha^{-2} \leq \pi c_0^{-2}n^{-1}$. Thus, $S(i)$ is (bounded by) a sum of Bernoulli trials, with $E[S(i)] = 16\pi\alpha^{-2}n < c$ for a constant c depending only on c_0. By the Chernoff-Hoeffding bound, for $D > 3c$, $\Pr[S(i) > D/2] \leq \exp(-\frac{(D-2c)^2}{D+4c}) \leq \exp(-D/21)$.

For the other component, define $L_x(i) = |\{j \in N(i) \,|\, d_{ij} \in [x, \frac{3}{2}x]\}|$ (the number of points attached to i in a distance between x and $\frac{3}{2}x$ from point i). We will first show that $L_x(i) < 14$ for any $x \geq \frac{4}{\alpha}$. Indeed a geometric argument shows if points p_j and $p_{j'}$, $j < j'$, are both between x and $\frac{3}{2}x$ away from p_i, then $p_{j'}$ would prefer p_j over p_i whenever $|\angle p_j p_i p_{j'}| < c = \cos^{-1}(43/48)$ (see Figure 1); the bound on the angle would force $\alpha d_{ij'} > \alpha d_{jj'} + 1$ while $|h_j - h_i| \leq 1$. Since $c > \frac{2\pi}{14}$, $L_x(i) < 14$. We now bound $L(i)$ as follows: $L(i) = \sum_{k=1}^{-\log_{\frac{3}{2}}\delta_i} L_{(\frac{3}{2})^{-k}}(i) \leq -14\log_{\frac{3}{2}}\delta_i$ where δ_i is defined as $\max\{\frac{4}{\alpha}, \min_j d_{ij}\}$. Since points are distributed uniformly at random, $\Pr[\delta_i \leq y] \leq 1 - (1 - \pi y^2)^{(n-1)} \leq \pi(n-1)y^2$. Therefore, $\Pr[L(i) \geq D/2] \leq \Pr[-14\log_{\frac{3}{2}}\delta_i \geq D/2] \leq \pi(n-1)(\frac{3}{2})^{-D/14}$, completing the proof of (2).

It is worth noting that the only property used for this bound is $|h_j - h_i| \leq 1$ for $j \in N(i)$; this holds for a broad class of hop functions, including all 3 listed in the introduction.

(3) To derive the power law in (3), we concentrate on points close to p_0. While experimental evidence suggests the presence of high-degree points contributing to the power law throughout the area, the proof is rendered more tractable by considering only points $j \in N(0)$, with $d_{0j} \leq 2/\alpha$. Without loss of generality, we assume that p_0 is located at least $2/\alpha$ from the area boundary; the argument carries over to cases where it is near border with only a slight change in C.

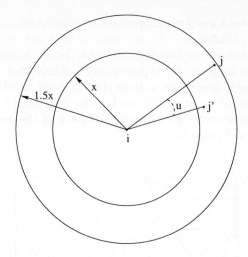

Fig. 1. If $u < cos^{-1}(43/48)$ then point j' prefers point j over point i because $\alpha d_{ij'} > \alpha d_{jj'} + 1$ for $x \geq 4/\alpha$.

First, we prove 2 lemmas for a point i which arrives so that $i \in N(0)$ and $\frac{3}{2\alpha} > d_{i0} > \frac{1}{\alpha}$. Let $r(i) = d_{i0} - \frac{1}{\alpha}$.

Lemma 1. *Every point arriving after i inside the circle of radius $\frac{1}{4}r(i)$ around i will link to i.*

Proof. Since i was linked to 0 on arrival, there was no $j \in N(0)$ prior to its arrival such that $\alpha d_{ij} + 1 < \alpha d_{i0}$, i.e. within distance $r(i)$ from it. Also, if a point j arrives after i so that $d_{ij} < \frac{1}{2}r(i)$, it will not link to 0, since by the triangle inequality $d_{j0} > 1/\alpha + \frac{1}{2}r(i)$. Now, if a point j' arrives after i so that $d_{ij'} < \frac{1}{4}r(i)$, it can't be linked to 0; for all $j \in N(0)\setminus\{i\}$, $d_{jj'} > \frac{1}{4}r(i)$, so j' would rather link to i; and for all other j, $h_j \geq 2$, so $f_{j'}(j) \geq 2$, while $f_{j'}(i) \leq \alpha\frac{1}{4}r(i) + 1 \leq \frac{9}{8}$. Thus, any such point j' will definitely link to i.

Lemma 2. *No point j will link to i unless $|\angle p_j p_0 p_i| \leq \sqrt{2.5\alpha r(i)}$ and $d_{j0} \geq \frac{1}{2}r(i) + 1/\alpha$.*

Proof. Note that if $f_j(0) < f_j(i)$, j will not link to i since i is not the optimal choice. That constraint is equivalent to $d_{j0} < d_{ij} + 1/\alpha$, which defines a region outside the cusp around p_i of a hyperbola with foci at p_0 and p_i and major axis length $1/\alpha$. The asymptotes for this hyperbola each make an angle of $\arctan \sqrt{\alpha^2 r(i)^2 + 2\alpha r(i)} \leq \arctan \sqrt{2.5\alpha r(i)} \leq \sqrt{2.5\alpha r(i)}$ with the segment $\overline{p_0 p_i}$, intersecting it at the midpoint m of $\overline{p_0 p_i}$. Since $|\angle p_i m x| \geq |\angle p_i p_0 x|$ for any point x, this guarantees that any point p_j inside the cusp around p_i satisfies $|\angle p_j p_0 p_i| \leq \sqrt{2.5\alpha r(i)}$. The triangle inequality also guarantees that any such p_j will also satisfy $d_{j0} \geq \frac{1}{2}r(i) + 1/\alpha$. Thus, any point not satisfying both of these will not link to i.

Lemma 1 provides a way for an appropriately positioned $i \in N(0)$ to establish a circular "region of influence" around itself so that any points landing there afterward will contribute to the degree of that point. Since point placement is independent and uniform, its exponentially unlikely that the circle will be populated with much fewer points than the expected number, thus giving a stochastic lower bound on $\deg i$. We use Lemma 2 to lower-bound the number of $i \in N(0)$ with sufficiently large $r(i)$. For a sketch of the geometrical features of the proof, refer to Figure 2.

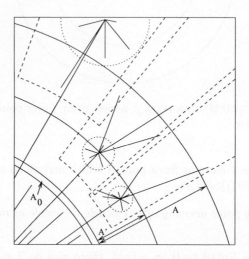

Fig. 2. A hypothetical small window about $1/\alpha$ from the root, of size on the order of $\rho^{2/3}$; figure not entirely to scale. Dotted contours indicate "regions of influence" provided by Lemma 1; dashed contours indicate the "sector" areas defined by Lemma 2. Note that the latter may overlap, while the former cannot.

Here, we only treat the case for $\alpha = o(\sqrt[3]{n^{1-\epsilon}})$, which yields $\beta = 1/6$; the case for $\alpha = o(\sqrt{n})$ can be analyzed similarly, but with β approaching 0 as α approaches $\Theta(\sqrt{n})$.

For any $\epsilon > 0$ and sufficiently high n, suppose $\alpha = o(\sqrt[3]{n^{1-2\epsilon}})$, and let $D \leq \frac{n^{1-\epsilon}}{256\alpha^3}$. Set, with foresight, $\rho = 4\sqrt{D/n}$ and $m = \lceil \frac{1}{2\rho} \rceil$, and consider T_m. Specifically, consider the sets of points in T_m falling within annuli A_0, A, and A' around p_0, with radius ranges $[1/\alpha, 1/\alpha + \rho]$, $(1/\alpha + \rho, 1/\alpha + \rho^{2/3}]$, and $(1/\alpha + \rho, 1/\alpha + 0.5\rho^{2/3}]$, respectively. By our choice of ρ, any $j \in N_m(0)$ in A' will have a region of influence of area at least $\pi D/n$, thus expected to contain $\pi D(n - m)/n > \pi D/2$ points that link to j in T. By our choice of m, the number of points expected to arrive in A_0 is $1/\alpha + \rho/2 < \frac{1}{2}$.

Consider any point i arriving to A'. It cannot link to any j with $h_j \geq 2$, since $f_i(0) < 2 \leq f_i(j)$. By Lemma 2, it cannot link to any $j \in N(0)$ outside the outer radius of A. By triangle inequality, it cannot link to any $j \in N(0)$ such that $d_{j0} < 1/\alpha$ (since $d_{i0} - d_{ij} < 1$). Thus, it can only link to 0 or any $j \in N(0)$ that lies in $A_0 \cup A$.

But, to link to a $j \in N(0)$ which is in $A_0 \cup A$, i must, by Lemma 2 obey $|\angle p_i p_0 p_j| < \sqrt{2.5\alpha r(j)} < \sqrt{2.5\alpha\rho^{1/3}}$. Thus, each time a new point arrives in $A_0 \cup A$ and links to $N(0)$, it "claims" a sector of A' of angle no larger than $\sqrt{10\alpha\rho^{1/3}}$; i.e. no point arriving outside that sector can link to j (note that we disregard the constraint on d_{j0}; it is not needed here). The number of points in T_m expected to be in A' is $m(\rho^{2/3}/\alpha - 2\rho/\alpha + 0.25\rho^{4/3} - \rho^2) > 1/(4\rho^{1/3})$. This can be cast as an occupancy problem by partitioning annulus $A_0 \cup A$ into $N = 1/(8\sqrt{\alpha}\rho^{1/3})$ congruent sectors[3] of angle $16\pi\sqrt{\alpha}\rho^{1/3} > \sqrt{10\alpha\rho^{1/3}}$. Each partition is considered occupied if a point has landed in it or either of its adjacent partitions and linked to p_0, and, by the above argument, a point landing in the intersection of an unoccupied partition and A' will link to p_0, so the number of partitions occupied after m points arrive is at most $3|N(0) \cap (A \cup A_0)|$. By the Chernoff bound, with probability at least $p_1 = 1 - \exp(-N/8)$, at least $1/(8\rho^{1/3})$ points in T_m land in A'. Note that if a point lands in the intersection of a partition and A', that partition is definitely occupied, so:

$$p_2 = \Pr[N/2 \text{ partitions occupied by points in } T_m]$$

$$= 1 - \sum_{k=N/2}^{N} \binom{n}{k} \Pr[k \text{ partitions unoccupied}]$$

$$\geq p_1 \left(1 - \sum_{k=N/2}^{N} \binom{n}{k} \left(1 - \frac{k}{N} \right)^{\sqrt{\alpha} N} \right)$$

$$\geq p_1 \left(1 - \sum_{k=N/2}^{N} \binom{n}{k} \frac{1}{2^{2N}} \right)$$

$$\geq p_1(1 - 2^{-N})$$

Hence, with probability $p_2 \geq 1 - 2\exp(-N/8)$, there are at least $N/6$ points in $N(0) \cap (A \cup A_0)$. By the Chernoff bound again, we find that the probability that more than $N/12$ of these are in A_0 is $\exp(-\frac{2}{m}(N/12 - 1/2)^2) < \exp(-N/8)$, so, with probability at least $1 - 3\exp(-N/8)$, there are $N/12$ points in $N(0) \cap A$, each with expected degree at least $\pi D/2$ in T.

Lastly, by another application of the Chernoff bound, the probability that the degree of any such point is less than D is $\exp(-\frac{1}{2}N(1 - 2/\pi)^2) < \exp(-N/20)$. Thus, with exponentially high probability $1 - (3 + N/24)\exp(-N/20)$, for $C = \frac{1}{2^{16/3}3}n^{-5/6}\alpha^{-1/2}$, and any D in the above specified range, the probability that a randomly chosen point in T has degree at least D is at least $N/24n = CD^{-1/6}$. That is, the degree distribution is lower-bounded by a power law at least up to a constant power of n.

---

[3] Note that $N = \Theta(D^{-1/6}n^{1/6}\alpha^{-1/2}) = \Omega(n^{\epsilon/6})$. Also, since the analysis here assumes $N \geq 1$, it only applies for $D \leq n/(2048\alpha^3)$.

Allowing p_0 to be placed closer than $2/\alpha$ from the border causes only a fraction of the annuli to be within the region for some values, but since at least a quarter of each annulus will always be within the region, N changes only by a small constant factor. To extend the above argument to $\alpha = o(\sqrt{n})$, the outer radii of A and A' would have to be reduced, making the ring thinner and allowing us to partition it into asymptotically more sectors. However, much fewer of them will be occupied, leading to a decrease in β.

2.2   Experiments

An implementation of both this model and several natural variations on it has shown that the cumulative density function (c.d.f.) of the degree distribution produced indeed appears to be a power law, as verified by a good linear fit of the logarithm of the c.d.f. with respect to the logarithm of the degree for all but the highest observed degrees. Using $n = 100,000$ and $\alpha \leq 100$, the β values observed from the slope of the linear fit ranged approximately between 0.6 and 0.9. When we used higher values of α, the range of D where the c.d.f. exhibited linear behavior shrunk too much to allow good estimation of β.

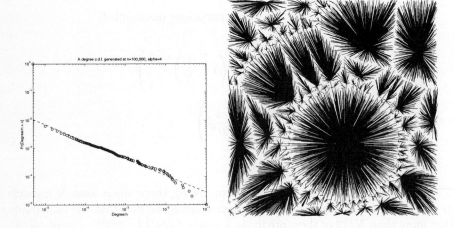

Fig. 3. c.d.f. and the associated tree generated for $\alpha = 4$ and $n = 100,000$ (only the first $10,000$ nodes are shown)

The c.d.f. generated for 2 tests are shown in the Appendix. Specifically, Figures 3 and 4 show the c.d.f. for $n = 100,000$ and $\alpha = 4$ and $\alpha = 20$ (the straight line shown is not actually a linear fit of the entire data set, but visually approximates a large part of the data). We also show the associated trees. (Actually due to enormous (postscript) file sizes, these are only the partially trees $T_{10,000}$.) The full trees (which do not differ substantially) and a more complete set of experimental results can be found in
http://research.csua.berkeley.edu/~alexf/hot/.

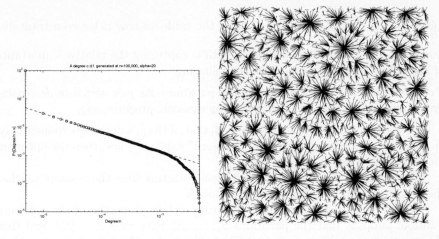

Fig. 4. c.d.f. and the associated tree generated for $\alpha = 20$ and $n = 100,000$ (only the first 10,000 nodes are shown)

Tests with varying values of n and fixed α produced consistent values for β throughout, as expected.

Furthermore, replacing the "rooted" h_j centrality measure with other alternatives mentioned in the introduction also produced power-law behavior in the c.d.f., with very similar values for β. Also, the "maximum hops" centrality measure, which is identical to the "rooted" measure provided that the first node remains the global minimum of h_j, while being more decentralized (and thus better mirroring a real network), was indeed observed to retain p_0 as the "most central" node in more than 75% of the cases.

Clearly the experiments suggest sharper power laws than our theoretical results and they also seem to occur for much wider range of the parameters. Our proofs could in fact be improved somewhat to give tighter results but we opted for simplicity.

It is straightforward to extend the proof of the main theorem (Theorem 1) to higher dimensions and other metrics —with different constants of course. The experiments indeed verify that the power law behavior also holds when the 2-dimensional square was replaced by a circle, as well as by hypercubes or hyperspheres in $\mathbb{R}^d$ for various $d \geq 2$ with the L_i (for various $i \geq 1$) or L_∞ metrics. Interestingly, one-dimensional domains do not seem to give rise to power laws.

3   A Model of File Creation

Suppose that we are given n positive real numbers $p_1, \ldots, p_n$, intuitively capturing the "popularity" of n data items, the expected number of times each will be retrieved. We wish to find a partition Π of the items into sets so as to minimize

$\min_{\Pi} \left[\sum_{S \in \Pi} |S| \cdot \sum_{i \in S} p_i \right] + \alpha \cdot |\Pi|$. That is, the trade-off now is between transmission costs and file creation overhead, with α capturing the relative importance of the latter. It is very easy to see the following:

Proposition 1. *The optimum solution partitions the p_i's sorted in decreasing order, and can be found in $O(n^2)$ time by dynamic programming.*

It is shown by a rough calculation in [2] that, if the p_i's decrease exponentially, or polynomially, or according to a "Gaussian" $\exp(-ci^2)$ law, then the optimum file sizes obey a power law.

In this section we prove a different result, starting from the assumption that the p_i's are drawn i.i.d. from a distribution f.

We start with two inequalities capturing certain properties of the optimum solution. Suppose that the partitions $S_1, \ldots, S_k$ have sizes $s_i = |S_i|$, and that the average item in S_i is a_i. It is not hard to see that the following is true:

Lemma 3. $s_i + s_{i+1} \geq \sqrt{\alpha/a_i}$, and $s_i \leq \sqrt{2\alpha/a_i}$.

The proof uses that, by optimality, it is not advantageous to merge two sets, or to split one in the middle.

Consider now the cumulative distribution Φ of f, and its inverse Ψ. That is, $\Psi(x)$ is the least y for which $\Pr[z \leq y] \geq x$. It is useful to think of $\Psi(y/n)$ as the expected number of elements with popularity smaller than y. Let $g = \Psi(\log n/n)$.

Lemma 4. *Almost certainly in the optimum solution there are at least $y/2\sqrt{2\alpha/g}$ sets of size at least $\sqrt{\alpha/2\Psi(2y/n)}$.*

Sketch of proof: With high probability, the popularity of the smallest element is no bigger than g, and, for large enough $y \leq n$, there are at least y elements with popularities smaller than $\Psi(2y/n)$. By the previous lemma, the sets that contain these elements have sizes that satisfy $s_i + s_{i+1} \geq \sqrt{\alpha/\Psi(2y/n)}$ and $s_i \leq \sqrt{2\alpha/g}$. Thus, these elements are divided into at least $y/\sqrt{2\alpha/g}$ sets (by the second inequality), half of them of size at least $\frac{1}{2}\sqrt{\alpha/\Psi(2y/n)}$ (by the first inequality). □

From this lemma we conclude the following for the distribution of file sizes:

$$\Pr[\text{size of a file} \geq \frac{1}{2}\sqrt{\alpha/\Psi(2y/n)}] \geq y/2n\sqrt{2\alpha/g}.$$

Now set $x = \frac{1}{2}\sqrt{\alpha/\Psi(2y/n)}$ or $\Psi(2y/n) = \alpha/4x^2$ or $2y/n = \Phi(\alpha/4x^2)$ or $y = n\Phi(\alpha/4x^2)/2$. Therefore we have

Theorem 2. *In the distribution of file sizes induced by the optimum solution,*

$$\Pr[\text{size of a file} \geq x] \geq \Phi(\alpha/4x^2)\sqrt{g/32\alpha}.$$

It follows that, with high probability, the file sizes are power-law distributed whenever $\lim_{z \to 0} \Phi(z)/z^c > 0$ for some $c > 0$; this condition does not hold only for distributions f that are extremely slow to start at zero. For example, any continuous distribution f that has $f(0) > 0$ (such as the exponential, normal, uniform, etc.) gives a power law.

4   Discussion and Open Problems

We have observed that power laws can result from trade-offs between various aspects of performance or fitness that must be optimized simultaneously (deployment vs. operational costs in one example, file creation overhead and transmission costs in another). In contrast, when the trade-off is weak (in that one of the criteria is overwhelmingly important), exponentially concentrated distributions result. This seems to be a new genre of a rigorous argument predicting power laws. Needless to say, our network model does exhibit a behavior of the form "the rich get richer," just like many previous models of such phenomena:Nodes that arrived early are more likely to both have high degree and small hop cost, and thus to attract new nodes. The point is, in our model this is not a primitive (and hard to defend) assumption, but a rather sophisticated consequence of assumptions that are quite simple, local, and "behavioral."

By extending our tree generation algorithm to non-tree graphs (attach each arriving node to the few most advantageous nodes, where the number of new edges is appropriately distributed to produce graphs with the correct average degree > 1) we obtain an interesting network generation model, about which we can prove next to nothing. However, it was suggested to us by Ramesh Govindan [7] that the graphs produced by this more elaborate model are passably Internet-like, in that they seem to satisfy several other observed properties of the Internet graph, besides the degree distribution.

It would be very interesting to extend our results to other definitions of the "hop" cost, and to strengthen them by proving stronger power laws (bigger exponents than we can prove by our simple techniques, hopefully closer to the ones observed in our experiments). It would be wonderful to identify more situations in which multi-criterion optimization leads to power laws, or, even more ambitiously, of generic situations in which multi-criterion optimization can be proved sufficient to create solutions with power-law-distributed features.

Finally, [5] observed another intriguing power law, besides the one for the degrees analyzed here: The largest *eigenvalues* of the Internet graph appear to also fit a power law. In joint work with Milena Mihail [11], we have discovered a way of explaining the power law distribution of the eigenvalues. Briefly, it is a corollary of the degree distribution: If a graph consists of a few nodes of very degrees $d_1 \geq d_2 \geq \cdots \geq d_k$ plus a few other edges, then with high probability the k largest eigenvalues will be about $\sqrt{d_1}, \sqrt{d_2}, \ldots, \sqrt{d_k}$. Indeed, in the power law observed in [5] the exponent of the eigenvalues are roughly half those of the degrees. The small difference can be explained by the fact that [5] examines only the 20 highest eigenvalues; these correspond to extreme degrees that do not fit exactly the eigenvalue power law (as is quite evident in the figures in [5]).

References

1. Béla Bollobás, Oliver Riordan, Joel Spencer, and Gábor Tusnády. The degree sequence of a scale-free random graph process. *Random Structures and Algorithms*, 18(3):279–290, 2001.

2. J. M. Carlson and J. Doyle. Highly optimized tolerance: a mechanism for power laws in designed systems. *Physics Review E*, 60(2):1412–1427, 1999.
3. Manuel Castells. *The Internet Galaxy: Re ections on the Internet, Business, and Society*. Oxford, 2001. http://www.oup.co.uk/isbn/0-19-924153-8.
4. Colin Cooper and Alan M. Frieze. A general model of undirected web graphs. In *ESA*, pages 500–511, 2001.
5. C. Faloutsos, M. Faloutsos, and P. Faloutsos. On power-law relationships of the internet topology. In *Proc. SIGCOMM*, 1999.
6. X. Gabaix. Zipf's law for cities: an explanation. *Quarterly Journal of Economics*, 114:739–767, 1999.
7. R. Govindan. Private communication.
8. R. Kumar, P. Raghavan, S. Rajagopalan, D. Sivakumar, A. Tomkins, and E. Upfal. Stochastic models for the web graph. In *Proceedings of the 41st Annual Symposium on Foundations of Computer Science*, pages 57–65, 2000.
9. R. Kumar, P. Raghavan, S. Rajagopalan, and A. Tomkins. Extracting large scale knowledge bases from the Web. In *Proceedings of the 25th VLDB Conference*, 1999.
10. B. Mandlebrot. An informational theory of the statistical structure of languages. In W. Jackson, editor, *Communication theory*, pages 486–502. Betterworth, 1953.
11. M. Mihail and C. Papadimitriou. Power law distribution of the Internet eigenvalues: An explanation. Unpublished.
12. M. Mitzenmacher. A brief history of generative models for power law and lognormal distributions. Manuscript.
13. V. Pareto. *Cours d'Economie Politique*. Dronz, Geneva Switzerland, 1896.
14. H. Tangmunarunkit, R. Govindan, S. Jamin, S. Shenker, and W. Willinger. Network topologies, power laws, and hierarchy. Technical Report 01-746, Computer Science Department, University of Southern California, 2001. To appear in Computer Communication Review.
15. G. Zipf. *Human behavior and the principle of least effort*. Addison-Wesley, Cambridge MA, 1949.

The Structure and Complexity of Nash Equilibria for a Selfish Routing Game[*]

Dimitris Fotakis[1], Spyros Kontogiannis[1], Elias Koutsoupias[2],
Marios Mavronicolas[3], and Paul Spirakis[1,4]

[1] Max Planck Institut für Informatik, 66123 Saarbrücken, Germany,
`fotakis@mpi-sb.mpg.de, spyros@mpi-sb.mpg.de`
[2] Department of Computer Science, University of California at Los Angeles, &
Department of Informatics, University of Athens, Greece, `elias@cs.ucla.edu`.
[3] Department of Computer Science, University of Cyprus, PO Box 20537, Nicosia
CY-1678, Cyprus, `mavronic@ucy.ac.cy`.
[4] Computer Engineering and Informatics Dept., University of Patras, & Computer
Technology Institute, PO Box 1122, 261 10 Patras, Greece, `spirakis@cti.gr`.

Abstract. In this work, we study the combinatorial structure and the computational complexity of Nash equilibria for a certain game that models *selfish routing* over a network consisting of m parallel *links*. We assume a collection of n *users*, each employing a *mixed strategy*, which is a probability distribution over links, to control the routing of its own assigned *traffic*. In a *Nash equilibrium*, each user selfishly routes its traffic on those links that minimize its *expected latency cost*, given the network congestion caused by the other users. The *social cost* of a Nash equilibrium is the expectation, over all random choices of the users, of the maximum, over all links, *latency* through a link.

We embark on a systematic study of several algorithmic problems related to the computation of Nash equilibria for the selfish routing game we consider. In a nutshell, these problems relate to deciding the existence of a Nash equilibrium, constructing a Nash equilibrium with given support characteristics, constructing the *worst* Nash equilibrium (the one with *maximum* social cost), constructing the *best* Nash equilibrium (the one with *minimum* social cost), or computing the social cost of a (given) Nash equilibrium. Our work provides a comprehensive collection of efficient algorithms, hardness results (both as $\mathcal{NP}$-hardness and $\#\mathcal{P}$-completeness results), and structural results for these algorithmic problems. Our results span and contrast a wide range of assumptions on the syntax of the Nash equilibria and on the parameters of the system.

1 Introduction

Nash equilibrium [14] is arguably the most important solution concept in Game Theory [15]. It may be viewed to represent a steady state of the play of a strategic

[*] This work has been partially supported by the IST Program of the European Union under contract numbers IST-1999-14186 (**ALCOM-FT**) and IST-2001-33116 (**FLAGS**), and by funds from the Joint Program of Scientific and Technological Collaboration between Greece and Cyprus.

game in which each player holds an accurate opinion about the (expected) behavior of other players and acts rationally. Despite the apparent simplicity of the concept, computation of Nash equilibria in finite games has been long observed to be difficult (cf. [10,19]); in fact, it is arguably one of the few, most important algorithmic problems for which no polynomial-time algorithms are known. Indeed, Papadimitriou [18, p. 1] actively advocates the problem of computing Nash equilibria as one of the most significant open problems in Theoretical Computer Science today.

In this work, we embark on a systematic study of the computational complexity of Nash equilibria in the context of a simple *selfish routing* game, originally introduced by Koutsoupias and Papadimitriou [7], that we describe here. We assume a collection of n users, each employing a *mixed strategy*, which is a probability distribution over m parallel *links*, to control the shipping of its own assigned *traffic*. For each link, a *capacity* specifies the rate at which the link processes traffic. In a Nash equilibrium, each user selfishly routes its traffic on those links that minimize its *expected latency cost*, given the network congestion caused by the other users. A user's *support* is the set of those links on which it may ship its traffic with non-zero probability. The *social cost* of a Nash equilibrium is the expectation. over all random choices of the users, of the maximum, over all links, *latency* through a link.

We are interested in algorithmic problems related to the computation of Nash equilibria for the selfish routing game we consider. More specifically, we aim at determining the computational complexity of the following prototype problems, assuming that users' traffics and links' capacities are given: Given users' supports, decide whether there exists a Nash equilibrium; if so, determine the corresponding users' (mixed) strategies (this is an existence and computation problem). Decide whether there exists a Nash equilibrium; if so, determine the corresponding users' supports and (mixed) strategies (this is an existence and computation problem). Determine the supports of the *worst* (or the *best*) Nash equilibrium (these are optimization problems). Given a Nash equilibrium, determine its social cost (this turns out to be a hard counting problem.

Our study distinguishes between *pure* Nash equilibria, where each user chooses exactly one link (with probability one), and *mixed* Nash equilibria, where the choices of each user are modeled by a probability distribution over links. We also distinguish in some cases between models of *uniform capacities*, where all link capacities are equal, and of *arbitrary capacities*; also, we do so between models of *identical traffics*, where all user traffics are equal, and of *arbitrary traffics*.

Contribution. We start with pure Nash equilibria. By the linearity of the expected latency cost functions we consider, a *mixed*, but not necessarily pure, Nash equilibrium always exists. The first result (Theorem 1), remarked by Kurt Mehlhorn, establishes that a pure Nash equilibrium always exists. To this end, we continue to present an efficient, yet simple algorithm (Theorem 2) that computes a pure Nash equilibrium.

We proceed to consider the related problems BEST NASH EQUILIBRIUM SUPPORTS and WORST NASH EQUILIBRIUM SUPPORTS of determining either the *best* or the *worst* pure Nash equilibrium (with respect to social cost), respectively. Not surprisingly, we show that both are $\mathcal{NP}$-hard (Theorems 3 and 4).

We now turn to mixed Nash equilibria. Our first major result here is an efficient and elegant algorithm for computing a mixed Nash equilibrium (Theorem 5). More specifically, the algorithm computes a *generalized fully mixed* Nash equilibrium; this is a generalization of *fully mixed* Nash equilibria [9].

We continue to establish that for the model of uniform capacities, and assuming that there are only *two* users, the worst mixed Nash equilibrium (with respect to social cost) is the fully mixed Nash equilibrium (Theorem 6). In close relation, we have attempted to obtain an analog of this result for the model of arbitrary capacities. We establish that *any* mixed Nash equilibrium, in particular the worst one, incurs a social cost that does not exceed 49.02 times the social cost of the fully mixed Nash equilibrium (Theorem 7). Theorems 6 and 7 provide together substantial evidence about the "completeness" of the fully mixed Nash equilibrium: it appears that it suffices, in general, to focus on bounding the social cost of the fully mixed Nash equilibrium and then use reduction results (such as Theorems 6 and 7) to obtain bounds for the general case.

We then shift gears to study the computational complexity of NASH EQUILIBRIUM SOCIAL COST. We have obtained both negative and positive results here. We first show that the problem is $\#\mathcal{P}$-complete (see, e.g., [16]) in general for the case of mixed Nash equilibria (Theorem 8). On the positive side, we get around the established hardness of computing *exacly* the social cost of any mixed Nash equilibrium by presenting a fully polynomial, randomized approximation scheme for computing the social cost of any given mixed Nash equilibrium to any required degree of approximation (Theorem 9).

We point out that the polynomial algorithms we have presented for the computation of pure and mixed Nash equilibria (Theorems 2 and 5, respectively) are the *first* known polynomial algorithms for the problem (for either the general case of a strategic game with a finite number of strategies, or even for a specific game). On the other hand, the hardness results we have obtained (Theorems 3, 4, and 8) indicate that optimization and counting problems in Computational Game Theory may be hard even when restricted to *specific,* simple games such as the selfish routing game considered in our work.

Related Work. The selfish routing game considered in this paper was first introduced by Koutsoupias and Papadimitriou [7] as a vehicle for the study of the price of selfishness for routing over non-cooperative networks, like the Internet. This game was subsequently studied in the work of Mavronicolas and Spirakis [9], where fully mixed Nash equilibria were introduced and analyzed. In both works, the aim had been to quantify the amount of performance loss in routing due to selfish behavior of the users. (Later studies of the selfish routing game from the same point of view, that of performance, include the works by Koutsoupias *et al.* [6], and by Czumaj and Vöcking [1].) Unlike these previous papers, our work considers the selfish routing game from the point of view of computational complexity and attempts to classify certain algorithmic problems related to the computation of Nash equilibria of the game with respect to their computational complexity.

Extensive surveys of algorithms and techniques from the literature of Game Theory for the computation of Nash equilibria of general bimatrix games in either strategic or extensive form appear in [10,19]. All known such algorithms incur exponential running time, with the seminal algorithm of Lemke and Howson [8] being the prime example. Issues of computational complexity for the computation of Nash equilibria in general games have been raised by Megiddo [11], Megiddo and Papadimitriou [12], and Papadimitriou [17]. The $\mathcal{NP}$-hardness of computing a Nash equilibrium of a *general* bimatrix game with maximum payoff has been established by Gilboa and Zemel [3]. Similar in motivation and spirit to our paper is the very recent paper by Deng *et al.* [2], which proves complexity, approximability and inapproximability results for the problem of computing an exchange equilibrium in markets with indivisible goods.

2   Framework

Most of our definitions are patterned after those in [7, Sections 1 & 2] and [9, Section 2].

We consider a *network* consisting of a set of m parallel *links* $1, 2, \ldots, m$ from a *source* node to a *destination* node. Each of n *network users* $1, 2, \ldots, n$, or *users* for short, wishes to route a particular amount of traffic along a (non-fixed) link from source to destination. (Throughout, we will be using subscripts for users and superscripts for links.) Denote w_i the *traffic* of user $i \in [n]$. Define the $n \times 1$ *traffic vector* $\mathbf{w}$ in the natural way. Assume throughout that $m > 1$ and $n > 1$.

A *pure strategy* for user $i \in [n]$ is some specific link. a *mixed strategy* for user $i \in [n]$ is a probability distribution over pure strategies; thus, a mixed strategy is a probability distribution over the set of links. The *support* of the mixed strategy for user $i \in [n]$, denoted *support(i)*, is the set of those pure strategies (links) to which i assigns positive probability. A *pure strategy profile* is represented by an n-tuple $\langle \ell_1, \ell_2, \ldots, \ell_n \rangle \in [m]^n$; a *mixed strategy profile* is represented by an $n \times m$ *probability matrix* $\mathbf{P}$ of nm probabilities p_i^j, $i \in [n]$ and $j \in [m]$, where p_i^j is the probability that user i chooses link j.

For a probability matrix $\mathbf{P}$, define *indicator variables* $I_i^\ell \in \{0,1\}$, $i \in [n]$ and $\ell \in [m]$, such that $I_i^\ell = 1$ if and only if $p_i^\ell > 0$. Thus, the support of the mixed strategy for user $i \in [n]$ is the set $\{\ell \in [m] \mid I_i^\ell = 1\}$. For each link $\ell \in [m]$, define the *view* of link ℓ, denoted *view(ℓ)*, as the set of users $i \in [n]$ that potentially assign their traffics to link ℓ; so, $view(\ell) = \{i \in [n] \mid I_i^\ell = 1\}$. A link $\ell \in [m]$ is *solo* [9] if $|view(\ell)| = 1$; thus, there is exactly one user, denoted $s(\ell)$, that considers a solo link ℓ.

Syntactic Classes of Mixed Strategies. By a *syntactic class* of mixed strategies, we mean a class of mixed strategies with common support characteristics. A mixed strategy profile $\mathbf{P}$ is *fully mixed* [9] if for all users $i \in [n]$ and links $j \in [m]$,

$I_i^j = 1$. Throughout, we will be considering a pure strategy profile as a special case of a mixed strategy profile. in which all (mixed) strategies are pure. We proceed to define two new variations of fully mixed strategy profiles. A mixed strategy profile $\mathbf{P}$ is *generalized fully mixed* if there exists a subset Links $\subseteq [m]$ such that for each pair of a user $i \in [n]$, and a link $j \in [m]$, $I_i^j = 1$ if $j \in$ Links and 0 if $j \notin$ Links. Thus, the fully mixed strategy profile is the special case of generalized fully mixed strategy profiles where Links $= [m]$.

Cost Measures. Denote $c^\ell > 0$ the *capacity* of link $\ell \in [m]$, representing the rate at which the link processes traffic. So, the *latency* for traffic w through link ℓ equals w/c^ℓ. In the model of *uniform capacities*, all link capacities are equal to c, for some constant $c > 0$; link capacities may vary arbitrarily in the model of *arbitrary capacities*. For a pure strategy profile $\langle \ell_1, \ell_2, \ldots, \ell_n \rangle$, the *latency cost for user* $i \in [n]$, denoted λ_i, is $(\sum_{k:\ell_k=\ell_i} w_k)/c^{\ell_i}$; that is, the latency cost for user i is the latency of the link it chooses. For a mixed strategy profile $\mathbf{P}$, denote W^ℓ the *expected traffic* on link $\ell \in [m]$; clearly, $W^\ell = \sum_{i=1}^n p_i^\ell w_i$. Given $\mathbf{P}$, define the $m \times 1$ *expected traffic vector* $\mathbf{W}$ induced by $\mathbf{P}$ in the natural way. Given $\mathbf{P}$, denote Λ^ℓ the *expected latency* on link $\ell \in [m]$; clearly, $\Lambda^\ell = \frac{W^\ell}{c^\ell}$. Define the $m \times 1$ *expected latency vector* Λ in the natural way. For a mixed strategy profile $\mathbf{P}$, the *expected latency cost* for user $i \in [n]$ on link $\ell \in [m]$, denoted λ_i^ℓ, is the expectation, over all random choices of the remaining users, of the latency cost for user i had its traffic been assigned to link ℓ; thus, $\lambda_i^\ell = \frac{w_i + \sum_{k=1, k\neq i} p_k^\ell w_k}{c^\ell} = \frac{(1-p_i^\ell)w_i + W^\ell}{c^\ell}$. For each user $i \in [n]$, the *minimum expected latency cost,* denoted λ_i, is the minimum, over all links $\ell \in [m]$, of the expected latency cost for user i on link ℓ; thus, $\lambda_i = \min_{\ell \in [m]} \lambda_i^\ell$. For a probability matrix $\mathbf{P}$, define the $n \times 1$ *minimum expected latency cost vector* λ induced by $\mathbf{P}$ in the natural way.

Associated with a traffic vector $\mathbf{w}$ and a mixed strategy profile $\mathbf{P}$ is the *social cost* [7, Section 2], denoted $\mathsf{SC}(\mathbf{w}, \mathbf{P})$, which is the expectation, over all random choices of the users, of the maximum (over all links) latency of traffic through a link; thus, $\mathsf{SC}(\mathbf{w}, \mathbf{P}) = \sum_{\langle \ell_1, \ell_2, \ldots, \ell_n \rangle \in [m]^n} \left(\prod_{k=1}^n p_k^{\ell_k} \cdot \max_{\ell \in [m]} \frac{\sum_{k:\ell_k=\ell} w_k}{c^\ell} \right)$. Note that $\mathsf{SC}(\mathbf{w}, \mathbf{P})$ reduces to the maximum latency through a link in the case of pure strategies. On the other hand, the *social optimum* [7, Section 2] associated with a traffic vector $\mathbf{w}$, denoted $\mathsf{OPT}(\mathbf{w})$, is the *least possible* maximum (over all links) latency of traffic through a link; thus, $\mathsf{OPT}(\mathbf{w}) = \min_{\langle \ell_1, \ell_2, \ldots, \ell_n \rangle \in [m]^n} \max_{\ell \in [m]} \frac{\sum_{k:\ell_k=\ell} w_k}{c^\ell}$. Note that while $\mathsf{SC}(\mathbf{w}, \mathbf{P})$ is defined in relation to a mixed strategy profile $\mathbf{P}$, $\mathsf{OPT}(\mathbf{w})$ refers to the *optimum* pure strategy profile.

Nash Equilibria. We are interested in a special class of mixed strategies called Nash equilibria [14] that we describe below. Formally, the probability matrix $\mathbf{P}$ is a *Nash equilibrium* [7, Section 2] if for all users $i \in [n]$ and links $\ell \in [m]$, $\lambda_i^\ell = \lambda_i$ if $I_i^\ell = 1$, and $\lambda_i^\ell > \lambda_i$ if $I_i^\ell = 0$. Thus, each user assigns its traffic with positive probability only on links (possibly more than one of them) for which its expected latency cost is minimized; this implies that there is no incentive for a user to unilaterally deviate from its mixed strategy in order to avoid links on which its expected latency cost is higher than necessary.

For each link $\ell \in [m]$, denote $\widetilde{c^\ell} = c^\ell/(\sum_{j=1}^{n} c^j)$, the *normalized capacity* of link ℓ. The following result due to Mavronicolas and Spirakis [9, Theorem 14] provides necessary and sufficient conditions for the existence (and uniqueness) of Nash equilibria in the case of fully mixed strategies, assuming that all traffics are identical.

Lemma 1 (Mavronicolas and Spirakis [9]). *Consider the case of fully mixed strategy profiles, under the model of arbitrary capacities. Assume that all traffics are identical. Then, for all links $\ell \in [m]$, $\widetilde{c^\ell} \in \left(\frac{1}{m+n-1} , \frac{n}{m+n-1} \right)$ if and only if there exists a Nash equilibrium, which must be unique.*

We remark that although, apparently, Lemma 1 determines a collection of $2m$ necessary and sufficient conditions (m pairs with two conditions per pair) for a fully mixed Nash equilibrium, the fact that all normalized capacities sum to 1 implies that each pair reduces to one condition (say the one establishing the lower bound for c^ℓ, $\ell \in [m]$. Furthermore, all m conditions hold if (and only if) the one for $\min_{\ell \in [m]} c^\ell$ holds. Thus, Lemma 1 establishes that existence of a fully mixed Nash equilibrium can be decided in $\Theta(m)$ time by finding the minimum capacity c^{ℓ_0} and checking whether or not the corresponding normalized capacity $\widetilde{c^{\ell_0}}$ satisfies $\widetilde{c^{\ell_0}} > \frac{1}{m+n-1}$. (This observation is due to B. Monien [13].)

Algorithmic Problems. We now formally define several algorithmic problems related to Nash equilibria. A typical instance is defined by: a number n of users; a number m of links; for each user i, a rational number $w_i > 0$, called the *traffic* of user i; for each link j, a rational number $c^j > 0$, called the *capacity* of link j.

In NASH EQUILIBRIUM SUPPORTS, we want to compute indicator variables $I_i^j \in \{0, 1\}$, where $1 \leq i \leq n$ and $1 \leq j \leq m$, that support a Nash equilibrium for the system of the users and the links.

In BEST NASH EQUILIBRIUM SUPPORTS, we seek the user supports corresponding to the Nash equilibrium with the *minimum* social cost for the given system of users and links.

In WORST NASH EQUILIBRIUM SUPPORTS, we seek the user supports defining the Nash equilibrium with the *maximum* social cost for the given system of users and links.

NASH EQUILIBRIUM SOCIAL COST is a problem of a somehow counting nature. In addition to the user traffics and the link capacities, an instance is defined by a *Nash equilibrium* **P** for the system of the users and the links, and we want to compute the social cost of the Nash equilibrium **P**.

3   Pure Nash Equilibria

We start with a preliminary result remarked by Kurt Mehlhorn.

Theorem 1. *There exists at least one pure Nash equilibrium.*

Proof sketch. Consider the universe of pure strategy profiles. Each such profile induces a *sorted* expected latency vector $\mathbf{\Lambda} = \langle \Lambda^1, \Lambda^2, \ldots, \Lambda^m \rangle$, such that $\Lambda^1 \geq \Lambda^2 \geq \ldots \geq \Lambda^m$, in the natural way. (Rearrangement of links may be necessary to guarantee that the expected latency vector is sorted.) Consider the lexicographically minimum expected latency vector $\mathbf{\Lambda}_0$ and assume that it corresponds to a pure strategy profile $\mathbf{P}_0$. We will argue that $\mathbf{P}_0$ is a (pure) Nash equilibrium. Indeed, assume, by way of contradiction, that $\mathbf{P}_0$ is *not* a Nash equilibrium. By definition of Nash equilibrium, there exists a user $i \in [n]$ assigned by $\mathbf{P}_0$ to link $j \in [m]$, and a link $\kappa \in [m]$ such that $\Lambda^j > \Lambda^\kappa + \frac{w_i}{c^\kappa}$. Construct now from $\mathbf{P}_0$ a pure strategy profile $\widehat{\mathbf{P}_0}$ which is identical to $\mathbf{P}_0$ except that user i is now assigned to link κ. Denote $\widehat{\Lambda_0} = \langle \widehat{\Lambda^1}, \widehat{\Lambda^2}, \ldots, \widehat{\Lambda^m} \rangle$ the traffic vector induced by $\widehat{\mathbf{P}_0}$. By construction, $\widehat{\Lambda^j} = \Lambda^j - \frac{w_i}{c^j} < \Lambda^j$, while by construction and assumption, $\widehat{\Lambda^\kappa} = \Lambda^\kappa + \frac{w_i}{c^\kappa} < \Lambda^j$. Since $\mathbf{\Lambda}_0$ is sorted in non-increasing order and $\Lambda^\kappa + \frac{w_i}{c^\kappa} < \Lambda^j$, Λ^j precedes Λ^κ in $\mathbf{\Lambda}_0$. Clearly, all entries preceding Λ^j in $\mathbf{\Lambda}_0$ remain unchanged in $\widehat{\Lambda_0}$. Consider now the j-th entry of $\widehat{\Lambda_0}$. There are three possibilities. The j-th entry of $\widehat{\Lambda_0}$ is either $\widehat{\Lambda^j}$, or $\widehat{\Lambda^\kappa}$, or some entry of $\mathbf{\Lambda}_0$ that followed Λ^j in $\mathbf{\Lambda}_0$ and remained unchanged in $\widehat{\Lambda_0}$. We obtain a contradiction in all possible cases. □

We remark that the proof of Theorem 1 establishes that the lexicographically minimum expected traffic vector represents a (pure) Nash equilibrium. Since there are exponentially many pure strategy profiles and that many expected traffic vectors, Theorem 1 only provides an *inefficient* proof of existence of pure Nash equilibria (cf. Papadimitriou [17]).

Computing a Pure Nash Equilibrium. We show:

Theorem 2. NASH EQUILIBRIUM SUPPORTS *is in* $\mathcal{P}$ *when restricted to pure equilibria.*

Proof sketch. We present a polynomial-time algorithm A_{pure} that computes the supports of a pure Nash equilibrium. Roughly speaking, the algorithm A_{pure} works in a greedy fashion; it considers each of the user traffics in non-increasing order and assigns it to the link that minimizes (among all links) the latency cost of the user had its traffic been assigned to that link. Clearly, the supports computed by A_{pure} represent a pure strategy profile. We will show that this profile is a Nash equilibrium. We argue inductively on the number of i iterations, $1 \leq i \leq n$, of the main loop of A_{pure}. We prove that the system of users and links is in Nash equilibrium after each such iteration. □

(This nice observation is due to B. Monien [13].) We remark that A_{pure} can be viewed as a variant of Graham's Longest Processing Time (LPT [4]) algorithm for assigning tasks to identical machines. Nevertheless, since in our case the links may have different capacities, our algorithm instead of choosing the link that will first become idle, it actually chooses the link that minimizes the completion time of the specific task (i.e., the load of a machine prior to the assignment of the task under consideration, plus the overhead of this task). Clearly, this greedy algorithm leads to an assignment which is, as we establish, a Nash equilibrium.

Computing the Supports of the Best or Worst Pure Nash Equilibria. We show:

Theorem 3. BEST NASH EQUILIBRIUM SUPPORTS *is $\mathcal{NP}$-hard.*

Proof sketch. Reduction from BIN PACKING (see, e.g., [16]).   □

Theorem 4. WORST NASH EQUILIBRIUM SUPPORTS *is $\mathcal{NP}$-hard when restricted to pure equilibria.*

Proof sketch. Reduction from BIN PACKING (see, e.g., [16]).   □

4   Mixed Nash Equilibria

We present a polynomial upper bound on the complexity of computing a mixed Nash equilibrium for the case where all traffics are identical. We show:

Theorem 5. *Assume that all traffics are identical. Then,* NASH EQUILIBRIUM SUPPORTS *is in $\mathcal{P}$ when it asks for the supports of a mixed equilibrium.*

Proof sketch. We present a polynomial-time algorithm A_{gfm} that computes the supports of a generalized fully mixed Nash equilibrium. We start with an informal description of A_{gfm}. In a preprocessing step, A_{gfm} sorts all capacities and computes all normalized capacities. Roughly speaking, A_{gfm} considers all subsets of fast links, starting with the set of all links; for each subset, it checks whether there exists a Nash equilibrium for the system of all users and the links in the subset, by using Lemma 1 (and the discussion following it). The algorithm A_{gfm} stops when it finds one; else, it drops the slowest link in the subset and continues recursively. Assume wlog that $c^1 \geq c^2 \geq \ldots \geq c^m$. For any integer m', where $1 \leq m' \leq m$, call a set of links $\{\ell_1, \ldots, \ell_{m'}\}$ a *fast link set*. So, we observe that A_{gfm} examines all generalized fully mixed strategy profiles for a system of all users and a fast link set. Hence, to establish correctness for A_{gfm}, we need to show that at least one of the generalized fully mixed strategy profiles for a system of all users and a fast link set is a Nash equilibrium. We show this by induction on m.   □

We note that the preprocessing step of A_{gfm} takes $\Theta(m \lg m) + \Theta(m) = \Theta(m \lg m)$ time. Next, the initial step of A_{gfm} (which considers all links) checks the validity of a single condition (by the discussion following Lemma 1). After this, the loop is executed at most $m - 1$ times. For $1 \leq m' \leq m - 1$, the m'-th execution checks the validity of a single condition (by the discussion following Lemma 1) and the validity of an additional condition (from the definition of Nash equilibrium). Thus, the time complexity of A_{gfm} is at most $\Theta(m \lg m) + \sum_{1 \leq m' \leq m-1} 2 = \Theta(m \lg m)$.

A Characterization of the Worst Mixed Nash Equilibrium. We first prove a structural property of mixed Nash equilibria, which we then use to provide a syntactic characterization of the worst mixed Nash equilibrium under the model of uniform capacities.

For the following proposition, recall the concepts of solo link and view. In addition, let us say that a user *crosses* another user if their supports cross each other, i.e. their supports are neither disjoint nor the same.

Proposition 1. *In any Nash equilibrium* **P** *under the model of uniform capacities,* **P** *induces no solo link considered by a user that crosses another user.*

Proof. Assume that **P** induces a solo link ℓ considered by a user $s(\ell)$ that crosses another user; thus, there exists another link $\ell_0 \in support(s(\ell))$ and a user $i_0 \in view(\ell_0)$, so that $p_{i_0}^{\ell_0} > 0$. Therefore, $\lambda_{s(\ell)}^{\ell_0} = w_{s(\ell)} + p_{i_0}^{\ell_0} w_{i_0} > w_{s(\ell)} = \lambda_{s(\ell)}^{\ell}$, which contradicts the hypothesis that **P** is a Nash equilibrium. □

Theorem 6. *Consider the model of uniform capacities and assume that $n = 2$. Then, the worst Nash equilibrium is the fully mixed Nash equilibrium.*

Proof sketch. Assume wlog that $w_1 \geq w_2$, and consider any Nash equilibrium **P**. If **P** is pure, we observe that it is not possible for both users to have the same pure strategy. This implies that the social cost of any pure equilibrium is $\max\{w_1, w_2\} = w_1$. If **P** is a mixed equilibrium, Proposition 1 implies that there are only two cases to consider: either $support(1) \cap support(2) = \emptyset$ or $support(1) = support(2)$. In the former case, the social cost is w_1. In the latter case, the only possible Nash equilibrium is the fully mixed one having social cost $\mathsf{SC}(\mathbf{w}, \mathbf{P}) = w_1 + w_2 \sum_{\ell \in [m]} p_1^\ell p_2^\ell = w_1 + w_2 \cdot \frac{1}{m}$. □

Worst Versus Fully Mixed Nash Equilibria. We show:

Theorem 7. *Consider the model of identical traffics. Then, the social cost of the worst mixed Nash equilibrium is at most* 49.02 *times the social cost of any generalized fully mixed Nash equilibrium.*

Proof sketch. Assume that $c^1 \geq c^2 \geq \ldots \geq c^m$. Let $C_{\text{tot}} = \sum_{\ell=1}^m c^\ell$. Wlog, we can assume that the minimum link capacity is 1 (i.e., $c^m = 1$) and that all users have a unit amount of traffic (i.e. for all $i \in [n]$, $w_i = w = 1$). We use **1** to denote the corresponding traffic vector. It can be easily verified that it suffices to show the following:

Lemma 2. *Let $c^m = 1$, $c^1 \geq 2$, and $n \geq \min\{3 \ln m, C_{\text{tot}} - m + 2\}$. In addition, let $\overline{\mathbf{P}}$ be the generalized fully mixed strategy profile computed by the algorithm A_{gfm}, and let $\mathbf{P}$ be a strategy profile corresponding to an arbitrary Nash equilibrium. Then,* $49.02\,\mathsf{SC}(\mathbf{1}, \overline{\mathbf{P}}) \geq \mathsf{SC}(\mathbf{1}, \mathbf{P})$.

Proof sketch. To distinguish between the expected latencies of the generalized fully mixed Nash equilibrium defined by $\overline{\mathbf{P}}$ and the expected latencies of the Nash equilibrium defined by $\mathbf{P}$, throughout our proof, we use $\overline{\Lambda^1}, \ldots, \overline{\Lambda^m}$ to denote the expected link latencies in the generalized fully mixed equilibrium computed by the algorithm A_{gfm}, and $\Lambda^1, \ldots, \Lambda^m$ to denote the expected link latencies in $\mathbf{P}$. In addition, we use $\overline{\Lambda_{\max}}$ and $\Lambda_{\max}$ to denote the maximum link latency in the $\overline{\mathbf{P}}$ and $\mathbf{P}$, respectively. We first show some bounds on the expected link latencies for any Nash equilibrium. The first bound states that, for any non-solo link $\ell \in [m]$, the expected latency of ℓ in any Nash equilibrium is bounded from above by a small constant factor times the expected latency of the same link in the generalized fully mixed equilibrium computed by A_{gfm}, i.e. $\Lambda^\ell \leq 2\left(1 + \frac{1}{|view(\ell)|-1}\right) \overline{\Lambda^\ell}$, From now on, the analysis focuses on mixed strategy profiles $\mathbf{P}$. Then, we upper bound the probability that the maximum latency of the generalized fully mixed equilibrium does not exceed a given number μ by proving that for any $\mu \in \left[\overline{\Lambda^1}, \frac{n}{c^1} \right]$, $\Pr[\overline{\Lambda_{\max}} < \mu] \leq 4 \exp\left(-\sum_{j=1}^{m} \left(\frac{\overline{\Lambda^j}}{2\mu} \right)^{\mu c^j} \right)$. We also bound from above the probability that the maximum latency of the equilibrium $\mathbf{P}$ is greater than or equal to a given number μ. In particular, we prove that for any $\mu > \frac{n}{C_{\text{tot}}} + \frac{m-1}{C_{\text{tot}}}$, $\Pr[\Lambda_{\max} \geq \mu] \leq \sum_{j=1}^{m} \left(\frac{e\Lambda^j}{\mu} \right)^{\mu c^j}$. So, we combine these to show that there exists a number $\mu^* \geq \overline{\Lambda^1}$ such that $\mathsf{SC}(1, \overline{\mathbf{P}}) \geq \frac{\mu^*}{3}$ and $6e(1 + 0.0018625)\mu^* \geq \mathsf{SC}(1, \mathbf{P})$.   □

The proof is now complete.   □

Computing the Social Cost of a Mixed Nash Equilibrium. We show:

Theorem 8. NASH EQUILIBRIUM SOCIAL COST *is $\#\mathcal{P}$-complete when restricted to mixed equilibria.*

Proof sketch. First of all, we remark that given a set of integer weights $J = \{w_1, \ldots, w_n\}$ and an integer $C \geq \frac{w_1 + \cdots + w_n}{2}$, it is $\#\mathcal{P}$-complete to count the number of subsets of J with total weight at most C, since this corresponds to counting the number of feasible solutions of a KNAPSACK instance (see, e.g., [16]). Therefore, given n Bernoulli random variables Y_i, each taking an integer value w_i with probability $\frac{1}{2}$ and 0 otherwise, and an integer C as above, it is $\#\mathcal{P}$-complete to compute the probability that $Y = \sum_{i=1}^{n} Y_i$ exceeds C, i.e. $\Pr(Y \leq C)$. We show that two calls to a (hypothetical) oracle computing the social cost of a given mixed Nash equilibrium suffice to compute the above probability.

Given the random variables Y_i, we consider three identical capacity links (denoted as links 0, 1, 2 respectively) and $n + 1$ users, where the user 0 has traffic C and the user i, $i \in [n]$, has traffic w_i. Since $C \geq \frac{w_1 + \cdots + w_n}{2}$, if user 0 chooses link 0 with certainty (i.e. $p_0^0 = 1$) and each of the remaining users chooses link 1 or 2 with probability $\frac{1}{2}$ (i.e. $p_i^1 = p_i^2 = \frac{1}{2}$), this mixed strategy profile corresponds to a Nash equilibrium. In addition, since w_i's are integers, the social cost equals $\mathsf{SC}_1 = C + 2\sum_{B=C+1}^{\infty} \Pr(Y \geq B)$. If we increase the traffic of user 0 to $C + 1$, the social cost becomes $\mathsf{SC}_2 = C + 1 + 2\sum_{B=C+2}^{\infty} \Pr(Y \geq B)$. Therefore, $2\Pr(Y \geq C + 1) = 1 + \mathsf{SC}_1 - \mathsf{SC}_2$, and since C and w_i's are integers, $\Pr(Y \leq C) = 1 - \Pr(Y \geq C + 1) = \frac{1 - \mathsf{SC}_1 + \mathsf{SC}_2}{2}$.   □

Approximating the Social Cost of a Mixed Nash Equilibrium. We show:

Theorem 9. *Consider the model of uniform capacities. Then, there exists a fully polynomial, randomized approximation scheme for* NASH EQUILIBRIUM SOCIAL COST.

Proof sketch. The idea of the scheme is to define an efficiently samplable random variable Λ which accurately estimates the social cost of the given Nash equilibrium $\mathbf{P}$ on a (given) traffic vector $\mathbf{w}$. For this, we design the following experiment, where N is a fixed parameter that will be specified later: "Repeat N times the random experiment of assigning each user to a link in its support according to the (given) Nash probabilities. Define Λ to be the random variable representing the maximum latency (over all links); for each experiment E_i, $1 \leq i \leq N$, denote Λ_i the measured value for Λ. Output the mean $\frac{\sum_{r=1}^{N} \Lambda_r}{N}$ of the measured values." By the Strong Law of Large Numbers (see, e.g., [5, Section 7.5]), it follows that $\left| \frac{\sum_{r=1}^{N} \Lambda_r}{N} - \mathsf{SC}(\mathbf{w}, \mathbf{P}) \right| \leq \varepsilon \, \mathsf{SC}(\mathbf{w}, \mathbf{P})$, for any constant $\varepsilon > 0$ provided that $N \geq \frac{1}{\varepsilon} \mathsf{SC}(\mathbf{w}, \mathbf{P})$. By the results in [1,6], $\mathsf{SC}(\mathbf{w}, \mathbf{P}) = O\left(\frac{\lg n}{\lg \lg n} \right) \cdot \mathsf{OPT}(\mathbf{w})$. Since $\mathsf{OPT}(\mathbf{w}) \leq \sum_{i=1}^{n} w_i$, it follows that $\mathsf{SC}(\mathbf{w}, \mathbf{P}) = O\left(\frac{\lg n}{\lg \lg n} \right) \cdot \sum_{i=1}^{n} w_i$. It suffices to take N to be $\frac{1}{\varepsilon}$ times this upper bound on $\mathsf{SC}(\mathbf{w}, \mathbf{P})$. $\square$

5   Open Problems

Our work leaves open numerous interesting questions that are directly related to our results. We list a few of them here: What is the time complexity of computing the supports of a pure Nash equilibrium? Theorem 2 shows that it is $O(n \lg n + nm) = O(n \max\{\lg n, m\})$. Can this be further improved? Consider the *specific* pure Nash equilibria that are computed by the algorithm that is implicit in the proof of Theorem 1 and the algorithm A_{pure} in the proof of Theorem 2. It would be interesting to study how well these specific pure Nash equilibria approximate the worst one (in terms of social cost). What is the complexity of computing the supports of a generalized fully mixed Nash equilibrium? Theorem 5 shows that it is $O(m \lg m)$ in the case where all traffics are identical. Can this be further improved? Nothing is known about the general case, where traffics are not necessarily identical.

It is tempting to conjecture that Theorem 6 holds for all values of $n \geq 2$. In addition, we conjecture that the generalized fully mixed strategy is actually the worst-case Nash equilibrium for identical traffics and capacitated links (Theorem 7 proves that it is already within constant factor from the worst case social cost).

Besides these directly related open problems, we feel that the most significant extension of our work would be to study other specific games and classify their instances according to the computational complexity of computing the Nash equilibria of the game. We hope that our work provides an initial solid ground for such studies.

Some additional results on the combinatorial structure and the computational complexity of Nash equilibria for the selfish routing game considered in this paper were obtained recently in a follow-up work by Burkhard Monien [13].

Acknowledgments. We thank Kurt Mehlhorn for contributing Theorem 1 to our work. We thank Burkhard Monien for some extremely helpful discussions on our work, and for contributing to our work an improvement to the time complexity (from $\Theta(m^2)$ to $\Theta(m \lg m)$) of Algorithm A_{gfm} (Theorem 5). We thank Berthold Vöcking for a fruitful discussion right after a seminar talk on

Nash equilibria for selfish routing given by Paul Spirakis at Max-Planck-Institut für Informatik. That discussion contributed some initial ideas for the design of Algorithm A_{pure} (Theorem 2). We also thank Rainer Feldman for helpful comments on an earlier version of this paper. Finally, we thank Costas Busch for bringing to our attention the allegoric movie *A Beautiful Mind.*

References

1. A. Czumaj and B. Vöcking, "Tight Bounds for Worst-Case Equilibria", *Proceedings of the 13th Annual ACM Symposium on Discrete Algorithms,* January 2002.
2. X. Deng, C. Papadimitriou and S. Safra, "On the Complexity of Equilibria", *Proceedings of the 34th Annual ACM Symposium on Theory of Computing,* May 2002.
3. I. Gilboa and E. Zemel, "Nash and Correlated Equilibria: Some Complexity Considerations", *Games and Economic Behavior,* Vol. 1, pp. 80–93, 1989.
4. R. L. Graham, "Bounds on Multiprocessing Timing Anomalies," *SIAM Journal on Applied Mathematics,* Vol. 17, pp. 416–429, 1969.
5. G. R. Grimmett and D. R. Stirzaker, *Probability and Random Processes,* Oxford Science Publications, Second Edition, 1992.
6. E. Koutsoupias, M. Mavronicolas and P. Spirakis, "Approximate Equilibria and Ball Fusion," *Proceedings of the 9th International Colloquium on Structural Information and Communication Complexity,* June 2002.
7. E. Koutsoupias and C. H. Papadimitriou, "Worst-case Equilibria," *Proceedings of the 16th Symposium on Theoretical Aspects of Computer Science,* LNCS 1563, pp. 404–413, 1999.
8. C. E. Lemke and J. T. Howson, "Equilibrium Points of Bimatrix Games," *Journal of the Society for Industrial and Applied Mathematics,* Vol. 12, pp. 413–423, 1964.
9. M. Mavronicolas and P. Spirakis, "The Price of Selfish Routing," *Proceedings of the 33rd Annual ACM Symposium on Theory of Computing,* pp. 510–519, 2001.
10. R. D. McKelvey and A. McLennan, "Computation of Equilibria in Finite Games," in *Handbook of Computational Economics,* H. Amman, D. Kendrick and J. Rust eds., pp. 87–142, 1996.
11. N. Megiddo, "A Note on the Complexity of P-Matrix LCP and Computing an Equilibrium," Research Report RJ6439, IBM Almaden Research Center, San Jose, CA95120, 1988.
12. N. Megiddo and C. H. Papadimitriou, "On Total Functions, Existence Theorems, and Computational Complexity," *Theoretical Computer Science,* Vol. 81, No. 2, pp. 317–324, 1991.
13. B. Monien, Personal Communication, April 2002.
14. J. F. Nash, "Non-cooperative Games," *Annals of Mathematics,* Vol. 54, No. 2, pp. 286–295, 1951.
15. M. J. Osborne and A. Rubinstein, *A Course in Game Theory,* MIT Press, 1994.
16. C. H. Papadimitriou, *Computational Complexity,* Addison-Wesley, 1994.
17. C. H. Papadimitriou, "On the Complexity of the Parity Argument and Other Inefficient Proofs of Existence," *Journal of Computer and System Sciences,* Vol. 48, No. 3, pp. 498–532, June 1994.
18. C. H. Papadimitriou, "Algorithms, Games and the Internet," *Proceedings of the 28th International Colloquium on Automata, Languages and Programming,* LNCS 2076, pp. 1–3, 2001.
19. B. von Stengel, "Computing Equlibria for Two-Person Games," in *Handbook of Game Theory,* Vol. 3, R. J. Aumann and S. Hart eds., North Holland, 1998.

Control Message Aggregation in Group Communication Protocols

Sanjeev Khanna[1], Joseph (Seffi) Naor[2], and Dan Raz[2]

[1] Dept. of Computer & Information Science, University of Pennsylvania,
Philadelphia, PA 19104
`sanjeev@cis.upenn.edu`,
[2] Computer Science Dept., Technion, Haifa 32000, Israel
`{naor,danny}@cs.technion.ac.il`

Abstract. Reliable data transmission protocols between a sender and a receiver often use feedback from receiver to sender to acknowledge correct data delivery. Such feedback is typically sent as control messages by receiver nodes. Since sending of control messages involves communication overhead, many protocols rely on aggregating a number of control messages and sending them together as a single packet over the network. On the other hand, the delays in the transmission of control messages may reduce the rate of data transmission from the sender. Thus, there is a basic tradeoff between the communication cost of control messages and the effect of delaying them.

We develop a rigorous framework to study the aggregation of control packets for multicast and other hierarchical network protocols. We define the multicast aggregation problem and design efficient online algorithms for it, both centralized and distributed.

1 Introduction

Reliable transmission of data across an unreliable packet network (such as IP) requires end to end protocols (such as TCP) that use feedback. Such protocols use control messages to set up connections, to acknowledge the correct reception of (part of) the data, and possibly to control the transmission rate. This introduces a communication overhead, that may result in decreasing the overall effective throughput of the network. Since control messages are typically small, one can aggregate several control messages into a single packet in order to reduce the communication overhead. However, such an approach introduces extra delay since a control message may need to wait at the aggregation points for additional control messages. In many cases (e.g. TCP) delayed control messages may result in a reduction of the overall transmission protocol throughput. Thus, there is a tradeoff between the amount of reduction in the communication cost one can get from aggregating control messages, and the affect of the possible delay caused by it. This paper studies this tradeoff in the context of multicast and hierarchical protocols.

We use the term *multicast protocols* to describe a transmission protocol that delivers data from a single sender to a (possibly large) set of receivers. Reliable

P. Widmayer et al. (Eds.): ICALP 2002, LNCS 2380, pp. 135–146, 2002.

multicast protocols provide a mechanism to ensure that every receiver receives all the data. Since the underlying network (IP) is unreliable (i.e. packets may be lost), such protocols must detect errors (i.e. lost packets) in order to request their retransmission. One approach is to have the sender detect that an error has occurred, this is done by making each receiver acknowledge receiving each packet. A common practice is to send a special control packet called ACK (positive acknowledgment) for each received packet. However, this may result in an implosion of control packets being sent to the sender. An alternative is to make each receiver independently responsible for detecting its errors, e.g., a gap in the sequence numbers may signal a lost packet. In such a case the common practice is to send a NAK (negative acknowledgment). In this case, an implosion of control packets may occur as well when the same packet is lost for many receivers.

Several reliable multicast protocols were developed (see for example [7]) which use different techniques to overcome the ACK/NAK implosion problem. One commonly used technique is to use control message aggregation where several control messages to the root are aggregated into a single control message, thus saving on the communication cost at the expense of delaying individual messages to achieve aggregation. Most protocols use ad hoc methods based on user-defined timers to perform such an aggregation. Consider, for example, the Local Group based Multicast Protocol (LGMP) [4]. This protocol uses the local group concept (LGC) in order to overcome the ACK-implosion problem. Local groups are organized in an hierarchical structure. Each receiver sends control message to its local group controller, which aggregates the data, tries to recover locally, and when needed reports to its local group controller (which belongs to a higher level in the hierarchy tree). LGMP uses timer timeout in order to decide when to send a control message up the tree (see Section 2.4 of [4]).

Using hierarchical structures is not restricted to multicast algorithms. Many network protocols use hierarchical structures to address the scalability problem. Examples include the PNNI [8] routing algorithm in ATM and the RSVP [9] protocols. In many of these protocols there is a need to report status to the higher level in the hierarchy. This leads to exactly the same tradeoff between the amount of control information and the delay in the reported information.

A very recent example is the Gathercast mechanism proposed by Badrinath and Sudame [1]. Gathercast proposes to aggregate small control packets, such as TCP acknowledge messages, sent to the same host (or to hosts in the same (sub)-network). The authors show that such an aggregation significantly improves performances under certain conditions. In the proposed scheme there is a time bound on the amount of delay a packet can suffer at the gathering point, and once this timer has expired the packet is sent.

1.1 Our Model and Results

This paper offers a theoretical framework to study the aggregation of control messages in such situations. Specifically, we investigate the global optimization problem of data aggregation in multicast trees. We are given a multicast tree with a *communication cost* associated with each link. When a packet arrives at a receiver node (leaf or internal node), a control message (typically an ACK control message) has to be sent to the root. Nodes can delay control messages in order to aggregate them, and save on the communication cost. The cost paid by a control message to traverse a tree link is independent of the number of aggregated control messages that it represents. Also, note that control messages that originate at the same receiver node as well as the ones originating at different nodes are allowed to be aggregated. However, the delay of each original control message contributes to a *delay cost*. The total delay cost is defined to be the *sum* of the delays of all the messages. Thus, our optimization problem can be stated as follows: Given a multicast tree and a sequence of packet arrivals at the receiver nodes, determine a schedule of minimum cost (communication cost plus the delay cost) to send back the control messages to the root. We refer to this problem as the *multicast aggregation* problem. This is an online optimization problem in which decisions must be made based upon current state without knowing future events.

We present both centralized and distributed online algorithms for the multicast aggregation problem. Our centralized online algorithm assumes a global information model, where a central entity determines how control messages are sent. However, future arrivals of messages are not known to the algorithm. Our distributed online algorithm is a "local" algorithm that makes decisions in the nodes of the tree based on the packets waiting there. Clearly, in practice, multicast aggregation algorithms are distributed, thus making the centralized model to be largely of theoretical interest. However, we believe that studying the centralized online model provides important insight into the combinatorial structure of the problem. Indeed, it turns out that the multicast aggregation problem is highly non-trivial even in the centralized model. Both of our algorithms are based on a natural strategy that tries to balance the communication costs and the delay costs incurred by the control messages.

For the centralized online case, we give an $O(\log \alpha)$-competitive algorithm, where α is the total communication cost associated with the multicast tree. We also show that our analysis of the competitive factor of the algorithm is tight. For the distributed online case, we give an $O(h \cdot \log \alpha)$-competitive algorithm, where h is the height of the multicast tree. We show a lower bound of $\Omega(\sqrt{h})$ on the competitive ratio of *any* distributed online algorithm which is *oblivious*, i.e., uses only local information. This notion will be defined more precisely later.

In order to study the performance of our distributed algorithms in practice, we conducted a performance study using simulations. We compared our algorithm to two commonly used methods: one that reports all events immediately, and another that works with timers. It turns out that in most scenarios our distributed online algorithm outperforms the other heuristics, and in some relevant cases it performs significantly (up to 40%) better. One of the most notable characteristics of our online algorithm is its robustness, i.e., it performs well across a broad spectrum of scenarios. It follows from our simulations that in a sense, our online algorithm works like a *self adjusting timer*, since on one hand it aggregates messages, but on the other hand it does not wait too long before sending them. Due to lack of space the detailed description of the simulation results are omitted from this version of the paper.

1.2   Related Work

Dooly, Goldman and Scott [3] study the problem of aggregating TCP ACK packets over a single link. They observed that the off-line case of the single link version can be optimally solved by using dynamic programming. For the online case of this problem they designed a 2-competitive algorithm in the spirit of rent-to-buy algorithms. Bortnikov and Cohen [2] give several online heuristics for a local scheduling problem for the more general hierarchical case.

A model similar to ours was introduced by Papadimitriou and Servan-Schreiber [6] in the context of organization theory (see also [5]). They model an organization as a tree where the messages arrive at the leaves and must be sent to the root as soon as possible. Messages model pieces of information about the "world" outside the organization, and the "boss" needs to have an up-to-date view of the world as soon as possible. These messages are allowed to be aggregated and the objective function is the sum of the communication cost and the delay cost. However, their work is primarily focused on the case where message arrivals are modeled by a Poisson process.

2   The Model

In this section we formally define our model and assumptions. We are given a rooted tree T that we refer to as the multicast tree. This tree may be a real multicast tree, or a tree describing a hierarchical structure of a protocol, where each link actually represents a path in the real network. We view the tree T as being directed from the leaves towards the root r so that each arc is uniquely defined by its tail. Thus, for each node $v \in T$ (except for the root) we denote by $e(v)$ the arc leaving it. Each arc (a tree edge) has a communication cost (or just cost) denoted by $c(\cdot)$. We assume that the communication cost of each arc is at least 1.

Packets arrive at the tree nodes over time which is slotted. An arrival of a packet at a tree node generates a control message (ACK) that needs to be delivered to the root. Our goal is to minimize communication cost of control messages by aggregating them as they make their way up the tree. We denote by τ the time at which the last packet arrives. In general, packets can arrive at any node, internal node or leaf. However, we can assume without loss of generality that packets arrive only at leaves.

For a given packet p, let $t_a(p)$ denote the time at which it arrives at a leaf node v. As mentioned before, for each packet we need to send a control message to the root. Control messages waiting at a node accumulate delay and the delay accumulated until time t by the control message for a packet p is denoted by $d_t(p)$. We assume that each control message must wait at least one unit of time at a leaf before being sent out. Thus, the delay accumulated by a control message is at least 1. From here on, we will identify each packet with its control message, and simply use the word packet and control messages interchangeably. Our delay metric is the sum of the delays of the packets, where the delay of each packet is linear in the time it waits at a node, i.e. $d_t(p) = \beta(t - t_a(p))$ for some constant β.

In general, nodes can aggregate as many packets as needed and send them up the tree to their parent. We make the simplifying assumption of no propagation delay along the links (this avoids having to deal with synchronization issues). In the distributed model, we assume that at each time step t, each node v may aggregate awaiting packets and send the aggregated packet to its parent. The cost of sending the message is $c(e(v))$, which is independent of the number of aggregated packets. In the centralized model we assume that at each time step t, an online algorithm broadcasts a subtree T_t (possibly empty) that collects all packets present at the leaves of T_t and sends them to the root. These packets are aggregated together in a bottom-up manner so that each link of the tree T_t is traversed exactly once during this process. We refer to such an action as broadcasting the subtree T_t, and the cost of this broadcast is given by $\mathsf{cost}(T_t) = \sum_{v \in T_t} c(e(v))$.

To summarize, our total cost is the sum of the delay costs of all the packets (*total delay cost*) together with the total communication cost. In the centralized model, the total communication cost is the sum of the costs of the subtrees that are broadcast. In the distributed model, the communication cost is the sum of the costs of the tree edges that are used by the (aggregated) messages.

3   The Centralized Online Algorithm

In this section we present log-competitive centralized online algorithms for our problem. We assume a global information model, where a central entity determines how control messages are sent. However, future arrivals of messages are not known to the algorithm. Our online algorithm is based on the following natural strategy: at any time t, we broadcast a maximal subtree (wrt containment) such that the cost of the subtree is roughly equal to the accumulated delay of the packets at its leaves.

Let $d_t(p)$ denote the delay accumulated by a packet p until time t. Our algorithm broadcasts at each time t, a maximal subtree $T' \subseteq T$ that satisfies

$$\sum_{p \in T'} d_t(p) \geq \mathsf{cost}(T'),$$

where $p \in T'$ ranges over all packets waiting at any leaf node in T'. It is easily seen that at each broadcast of a tree T' by the online algorithm, we must also have $\sum_{p \in T'} d_t(p) \leq 2 \cdot \mathsf{cost}(T')$.

Fix an optimal solution OPT and let $\mathcal{T}^* = \{T_1, ..., T_\tau\}$ denote the trees (possibly empty) broadcast by OPT. Let P denote the set of all packets received during the algorithm's execution. For any packet $p \in$ P, let $\mathsf{delay}(p)$ and $\mathsf{delay}^*(p)$ denote the delay incurred by a packet p in the online solution and the optimal solution, respectively. Clearly, the cost C^* incurred by OPT is given by

$$C^* = \sum_{T_i \in \mathcal{T}^*} \text{cost}(T_i) + \sum_{p \in P} \text{delay}^*(p).$$

Define $L = \{p \mid \text{delay}(p) > \text{delay}^*(p)\}$ to be the set of packets that the online algorithm broadcasts later than OPT (late packets), and let $E = \{p \mid \text{delay}(p) \leq \text{delay}^*(p)\}$ be the set of packets that are broadcast no later than OPT (early packets). The key fact that we will use in our analysis is the following lemma that relates the delay incurred by the late packets in the online algorithm to one in the optimal solution.

Lemma 1.

$$\sum_{p \in L} \text{delay}(p) \leq \sum_{p \in L} \text{delay}^*(p) + 4 \left(\sum_{T_t \in \mathcal{T}^*} \text{cost}(T_t) \right) \cdot \log \alpha$$

Proof. Consider the set $L_t \subseteq L$ of packets that OPT sends at time t in a tree $T_t \in \mathcal{T}^*$. Let L_t denote this subset of packets and let $\ell = |L_t|$. Define $t_i = t + (1/\beta) \cdot 2^i (\text{cost}(T_t)/|L_t|)$. We claim that the number of packets in L_t that are still alive at time t_i is at most $|L_t|/2^i$. Suppose not, then the total accumulated delay of these packets exceeds $\text{cost}(T_t)$. Since they have not yet been broadcast, this contradicts the online broadcasting rule. Thus at time $t_{1+\lceil \log(\text{cost}(T_t)) \rceil}$, no packets from the set L_t remain. The total delay incurred by packets in L_t in the online algorithm is thus bounded as below:

$$\sum_{p \in L_t} \text{delay}(p) \leq \sum_{p \in L_t} \text{delay}^*(p) + \sum_{i=1}^{1+\lceil \log(\text{cost}(T_t)) \rceil} \beta \cdot \left(\frac{|L_t|}{2^{i-1}} \right) \left(\frac{1}{\beta} \right) \left(\frac{2^i \text{cost}(T_t)}{|L_t|} \right)$$

$$= \sum_{p \in L_t} \text{delay}^*(p) + 2 \log \left(\text{cost}(T_t) \right) \cdot \text{cost}(T_t) + 4\text{cost}(T_t)$$

$$\leq \sum_{p \in L_t} \text{delay}^*(p) + 4 \log \left(\text{cost}(T_t) \right) \cdot \text{cost}(T_t)$$

Since $\text{cost}(T_t) \leq \alpha$, the lemma now follows by summing over all sets $L_1, ..., L_\tau$.

Recall that in the centralized algorithm, a subtree $T' \subseteq T$ is broadcast if it satisfies $\sum_{p \in T'} d_t(p) \geq \text{cost}(T')$. Therefore, the communication cost of the algorithm is no more than the delay cost, and the cost C incurred by the online centralized algorithm is given by:

$$C \leq 2 \sum_{p \in P} \text{delay}(p) \leq 2 \left(\sum_{p \in E} \text{delay}^*(p) + \sum_{p \in L} \text{delay}(p) \right)$$

$$\leq 2 \left(\sum_{p \in E} \text{delay}^*(p) + \sum_{p \in L} \text{delay}^*(p) \right) + 2 \left(4 \left(\sum_{T_i \in \mathcal{T}^*} \text{cost}(T_i) \right) \cdot \log \alpha \right)$$

$$= 2 \left(\sum_{p \in P} \text{delay}^*(p) + 4 \left(\sum_{T_i \in \mathcal{T}^*} \text{cost}(T_i) \right) \log \alpha \right) \leq 8C^* \cdot \log \alpha$$

We note that we are not trying to optimize constants. Thus we have the following result.

Theorem 1. *There is an $O(\log \alpha)$-competitive centralized algorithm for the multicast aggregation problem.*

3.1 A Lower Bound

We now show that our analysis above is essentially tight. Consider a two-level tree T as shown in Figure 1. Assume $\beta = 1$ for clarity of exposition. The tree T has a single edge coming into root r from u with a cost of k^{2k+1} for some integer k. The node u has k children $v_0, ..., v_k$ where the cost of each edge (v_i, u) is k^{2k}. The total cost α of the tree T is thus $2k^{k+1}$. We will now describe a packet arrival sequence such that our online algorithm pays $O(\log \alpha / \log \log \alpha)$ times the optimal cost on this sequence. The arrival sequence consists of a sequence of blocks. The jth block comprises of arrival of packets at each leaf at time $t_j = (2j)k^{2k+1}$ where $j \geq 0$. The leaf node v_i receives $k^{2(k-i)}$ packets in each block. It is easily seen that the optimal solution is to immediately broadcast the entire tree at time $t_j + 1$.

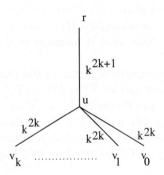

Fig. 1. A lower bound for our online algorithm

We now analyze the behavior of our online algorithm on this sequence. Let T_i denote the subtree of T that only contains the nodes r, u and v_i. First observe that all requests that arrive in any block j are broadcast before the arrival of the requests in the next block; this follows from the fact that the starting time of two consecutive blocks are at least α time units apart. So it suffices to analyze the cost paid by the online algorithm in each block. In each block j, the first subtree is broadcast at time $t_j + (k+1)$; node v_0 accumulates a delay that is equal to $\text{cost}(T_0)$ and we broadcast tree T_0. Observe that by our choice of edge weights, no other leaf v_i is able to connect to this subtree at this time. The next subtree is now broadcast at time $t_j + (k^3 + k^2)$ and is simply the subtree T_1. In general, at time $t_j + (k^{2i+1} + k^{2i})$ we broadcast subtree T_i. Thus the total cost paid by our algorithm in each block is $O(k \cdot k^{2k+1})$ which is $O(\log \alpha / \log \log \alpha)$ times the cost paid by the optimal solution.

4   The Distributed Online Model

We now present a distributed version of the centralized online algorithm. Consider node v at time t, and denote by $P(v, t)$ the set of packets that are waiting at v at time t. The set $P(v, t)$ is aggregated into a single message, $m(v, t)$, waiting to be sent from v. We denote by $\text{delay}(m(v, t))$ the delay associated with $m(v, t)$. At each unit of time, $\text{delay}(m(v, t))$ is increased by $\beta |P(v, t)|$, i.e., each message in the set $P(v, t)$ contributes towards the delay accumulated by $m(v, t)$. We note that if node v receives a message m' at time t', then it aggregates m' with $m(v, t')$, and the delay associated with the aggregated message is $\text{delay}(m') + \text{delay}(m(v, t'))$. We assume here that each message must wait at least one unit of time at a node before being sent out. Thus, the delay accumulated by a message is at least β times the length of the path from the leaf (where it originated) to the root.

The algorithm sends message $m(v, t)$ at time t if $\text{delay}(m(v, t)) \geq c(e(v))$. Suppose node u is the head of arc $e(v)$. Then, the delay associated with message $m(v, t)$ upon arrival at u is $\text{delay}(m(v, t)) - c(e(v))$, i.e., message $m(v, t)$ has "paid" for crossing $e(v)$. Note that if message $m(v, t)$ is sent at time t, then $\text{delay}(m(v, t)) \leq 2c(e(v))$.

We now analyze the competitive factor of the distributed online algorithm. Fix an optimal solution OPT and let $\mathcal{T}^* = \{T_1, ..., T_\tau\}$ denote the trees (possibly empty) broadcast by OPT. Let P denote the set of all packets received during the algorithm's execution. For any packet $p \in$ P, let $\text{delay}(p)$ and $\text{delay}^*(p)$ denote the delay incurred by a packet p in the online solution and the optimal solution, respectively. Clearly, the cost C^* incurred by OPT is given by

$$C^* = \sum_{T_i \in \mathcal{T}^*} \text{cost}(T_i) + \sum_{p \in P} \text{delay}^*(p).$$

As in the previous section, define $L = \{p \mid \text{delay}(p) > \text{delay}^*(p)\}$ to be the set of packets that reached the root in the online algorithm later than the time they reached the root in OPT, and let $E = \{p \mid \text{delay}(p) \leq \text{delay}^*(p)\}$ be the set of packets that reached the root in the online algorithm no later than the time they reached the root in OPT. The key fact that we will use in our analysis is the following lemma that relates the delay incurred by the late packets in the online algorithm to one in the optimal solution.

Lemma 2.

$$\sum_{p \in L} \text{delay}(p) \leq \sum_{p \in L} \text{delay}^*(p) + \left(8 \sum_{T_t \in \mathcal{T}^*} \text{cost}(T_t) + 4 \sum_{p \in P} \text{delay}^*(p) \right) \cdot h \cdot \log \alpha$$

Proof. Consider a $T_t \in \mathcal{T}^*$. Denote by W the set of late packets that are broadcast in tree T_t. Define the following sequences, $\{t_i\}$ and $\{W_i\}$, where $W_i \subseteq W$ denotes the packets that have not reached the root by time t_i. Define $t_0 = t$ and $W_0 = W$; t_i $(i \geq 1)$ is defined to be the first time since t_{i-1} by which the packets belonging to W_i have accumulated delay of at least $2 \cdot \text{cost}(T_t)$. Since t_i is the earliest time at which this event occurs, the accumulated delay cannot exceed $2 \cdot \text{cost}(T_t) + \sum_{p \in T_t} \text{delay}^*(p)$. (Since the packets in OPT also had to wait one unit of time.)

We now claim that for all $i \geq 1$, $|W_i| \leq |W_{i-1}|/2$. Suppose that this is not the case. Then, the delay accumulated by the packets that are alive at time t_i is:

$$\sum_{v \in T_t} \sum_{p \in v} \text{delay}(p) > \text{cost}(T_t)$$

Also,

$$\text{cost}(T_t) = \sum_{v \in T_t} \text{cost}(e(v))$$

Hence, there exists a node v such that the delay accumulated by the packets alive at v at time t_i is strictly greater than $\text{cost}(e(v))$. This is a contradiction, since according to the algorithm node v should have sent its packets before time t_i.

We now define the potential at time t_i, Φ_i, to be the sum of the distances of the packets belonging to W_i from the root of the tree T. The distance of a node v from the root is defined to be the number of links in the path from v to the root. Clearly, $\Phi_i \leq |W_i| h$, and since $|W_i| \leq |W_{i-1}|/2$, we get

$$\Phi_i \leq \Phi_{i-1} - \frac{|W_{i-1}|}{2} \leq \Phi_{i-1} \cdot \left(1 - \frac{1}{2h} \right)$$

Hence, by time t_f, where $f = 4h \log(h|W|)$, we get that $\Phi_f = 0$. The total delay incurred by packets in W in the distributed online algorithm is thus bounded as below:

$$\sum_{p \in W} \text{delay}(p) \leq \sum_{p \in W} \text{delay}^*(p) + \left(8\text{cost}(T_t) + 4 \sum_{p \in W} \text{delay}^*(p) \right) h \log(h|W|)$$

We now claim that $|W|$ can be bounded by $|T_t| \cdot \alpha$. To see this, observe that any node that contains more than α packets will never be part of the late set. Thus, $|W| \leq |T_t| \cdot \alpha$, and $\log(h|W|)$ is $O(\log \alpha)$, since we assume that the cost of each tree arc is at least 1.

The lemma now follows by summing over all sets of late packets in the trees $\{T_1, ..., T_\tau\}$.

Recall that in the distributed algorithm, if message $m(v,t)$ is sent at time t, then $\mathsf{delay}(m(v,t)) \geq c(e(v))$. Therefore, the communication cost of the algorithm is no more than the delay cost, and the cost C incurred by the online distributed algorithm is given by:

$$C \leq 2 \sum_{p \in P} \mathsf{delay}(p)$$

$$\leq 2 \left(\sum_{p \in E} \mathsf{delay}^*(p) + \sum_{p \in L} \mathsf{delay}(p) \right)$$

$$\leq 2 \left(\sum_{p \in E} \mathsf{delay}^*(p) + \sum_{p \in L} \mathsf{delay}^*(p) \right)$$

$$+2 \left(\left(8 \sum_{T_i \in T^*} \mathsf{cost}(T_i) + 4 \sum_{p \in P} \mathsf{delay}^*(p) \right) \cdot h \cdot \log \alpha \right)$$

$$= 2 \sum_{p \in P} \mathsf{delay}^*(p)$$

$$+2 \left(\left(8 \sum_{T_i \in \mathcal{T}^*} \mathsf{cost}(T_i) + 4 \sum_{p \in P} \mathsf{delay}^*(p) \right) \cdot h \cdot \log \alpha \right)$$

$$\leq 16 C^* \cdot h \cdot \log \alpha$$

Thus we have the following result.

Theorem 2. *There is an $O(h \cdot \log \alpha)$-competitive distributed algorithm for the multicast aggregation problem.*

4.1   A Lower Bound

We now provide evidence that it is inherently difficult to obtain significantly better bounds on the competitive ratio.

A distributed online algorithm is called *oblivious* if decisions at each node are based solely upon the static local information available at the node. In particular, in such algorithms the wait time of a packet is independent of the location of the node in the tree. We note that all the distributed algorithms considered in this paper are oblivious. In general, being oblivious is a desirable property of distributed algorithms, since non-oblivious algorithms require a dynamic updating mechanism at each node that informs it about changes in the hierarchical structure. Such a mechanism can be both expensive and complicated to implement.

Consider a path of length h, as shown in Figure 2, where the vertices on the path are $r = v_1, \ldots, v_{h+1} = z$, such that r is the root and z is the (only) receiver. The cost of each link in the path is h. We assume that $\beta = 1$. We will now describe a packet arrival sequence such that any oblivious online algorithm pays $O(\sqrt{h})$ times the optimal cost on this sequence.

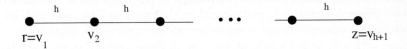

Fig. 2. A lower bound for oblivious online algorithms

Denote by ALG the online algorithm. The packets arrive one-by-one. At time 0, packet p_1 arrives at z; for $i \geq 2$, packet p_i arrives at z when packet p_{i-1} leaves node z. Denote by $wait(h)$ the time a single packet waits in z before being sent out towards r. Since the algorithm is oblivious, and since the local static information at all the nodes is the same (one out-going link with a cost of h), the same waiting time applies to all nodes on the way from z to r. Thus, the total waiting time of each packet paid by the online algorithm is $h \cdot wait(h)$ and the communication cost of each packet is h^2. Then, the cost of each packet in the online algorithm is

$$\mathsf{cost}(\text{ALG}) = h \cdot wait(h) + h^2$$

We now derive an upper bound on the cost of OPT, the optimal algorithm. We partition the packets into blocks of size $O(\sqrt{h})$. The packets in each block will be broadcast together to the root r. Clearly, the communication cost of each block in OPT is h^2, and thus the communication cost per packet is $h \cdot \sqrt{h}$. Next we bound the average delay cost of a packet in OPT. Since each block has $\sqrt{h}$ packets, a packet waits $O(\sqrt{h} \cdot wait(h))$ at node z, and then one unit at each node on its way to r. Adding the communication and the delay costs, we get that the average total cost of a packet in OPT is

$$O(\sqrt{h} \cdot wait(h) + h + h \cdot \sqrt{h})$$

Now if $h \leq wait(h)$, then OPT is $O(\sqrt{h} \cdot wait(h))$ while the online algorithm pays $\Omega(h \cdot wait(h))$, and if $h > wait(h)$, then OPT is $O(\sqrt{h} \cdot h)$ while the online algorithm pays $\Omega(h^2)$, yielding the desired lower bound. We conclude with the following theorem.

Theorem 3. *The competitive ratio of any oblivious distributed online algorithm for the multicast aggregation problem is at least $\Omega(\sqrt{h})$.*

Notice that the lower bound that we proved for the centralized online algorithm also applies to our distributed online algorithm, yielding a lower bound of $\Omega(\sqrt{h} + \log \alpha / \log \log \alpha)$.

5 Concluding Remarks

Many important questions remain open. One direction would be to study the performance of non-oblivious algorithms. It would be very useful to understand how much can be gained from the knowledge of the hierarchical tree structure. Another research avenue is to find algorithms that work well for trees with special properties. For example, many group communication protocols might have a very flat tree, with a bounded height of, say, two or three levels. Can they use better aggregation protocols? Another direction which is worth looking into is a model where events in the input sequence are not independent. For example, consider the case where the same tree is used for both multicasting and for collecting Acks (NAKs). In this case, packet loss will trigger an event in all the leaves belonging to its subtree. This requires a different model for the input sequence. Another interesting problem is whether there exists a centralized online algorithm with a constant competitive factor. In this model we are only able to show a constant factor lower bound on the competitive ratio of any algorithm.

Acknowledgments. We thank Guy Even, Baruch Schieber, Bruce Shepherd, and Francis Zane for many stimulating discussions. A special thanks to Yair Bartal for his insightful comments on many aspects of this work.

References

1. B. R. Badrinath and P. Sudame. Gathercast: The design and implementation of a programmable agregation mechanism for the internet. Submitted for publication, 1999.
2. E. Bortnikov and R. Cohen. Schemes for scheduling of control messages by hierarchical protocols. In *IEEE INFOCOM'98*, March 1998.
3. D. R. Dooly, S.A. Goldman, and S. D. Scott. On-line analysis of the TCP acknowledgement delay problem. *Journal of the ACM*, 48:243–273, 2001.
4. M.Hofmann. A generic concept for large-scale multicast. In B. Plattner, editor, *International Zurich Seminar on Digital Communication*, number 1044, pages 95 – 106. Springer Verlag, February 1996.
5. C. Papadimitriou. Computational aspects of organization theory. In *ESA '96*, 1996.
6. C. Papadimitriou and E. Servan-Schreiber. The origins of the deadline: optimizing communication in organization. Workshop on Complexity in Economic Games, Aix-en-Provence, 1999. To appear, Handbook on the Economics of Information.
7. Reliable multicast protocols.
   http://www.tascnets.com/newtascnets/Projects/Mist/Documents/
   RelatedLinks.html.
8. The ATM Forum technical committee. Private network-network interface specification version 1. 0 (PNNI). Technical report, March 1996. af-pnni-0055.000.
9. Lixia Zhang, Stephen Deering, Deborah Estrin, Scott Shenker, and Daniel Zappala. RSVP: a new resource ReSerVation protocol. *IEEE Network Magazine*, 7(5):8 – 18, September 1993.

Church-Rosser Languages vs. UCFL[*]

Tomasz Jurdziński[1,2] and Krzysztof Loryś[2]

[1] Institute of Computer Science, Chemnitz University of Technology
[2] Institute of Computer Science, Wrocław University

Abstract. The class of growing context-sensitive languages (GCSL) was proposed as a naturally defined subclass of context-sensitive languages whose membership problem is solvable in polynomial time. In this paper we concentrate on the class of Church Rosser Languages (CRL), the "deterministic counterpart" of GCSL. We prove the conjecture (stated in [9]) that the set of palindromes is not in CRL. This implies that CFL∩co-CFL as well as UCFL∩co-UCFL are not included in CRL, where UCFL denotes the class of unambiguous context-free languages. Our proof uses a novel pumping technique, which is of independent interest.

1 Introduction and Related Work

Families of languages that lay between context-free languages (CFL) and context-sensitive languages (CSL) have been a field of intensive studies for many years. The motivation was to find a family which possesses both an acceptable computational complexity and sufficient expressibility power. Neither CSL nor CFL fulfil these demands. For the first class the membership problem is PSPACE-complete what makes it in its full generality too hard for some practical applications. On the other hand context-free grammars are e.g. not powerful enough to express all syntactical aspects of programming languages.

In the past many modifications extending context-free grammars or reducing the power of context-sensitive grammars have been studied. One of the most interesting proposals was presented by Dahlhaus and Warmuth in [6]. They considered grammars with strictly growing rules, i.e. rules which righthand side is longer than the lefthand one. The resulting class, the class of growing context-sensitive languages (GCSL), in a natural way complements the Chomsky hierarchy and possesses several very nice properties. First of all, immediately from the definition follows a linear bound on the length of derivations, so GCSL is contained in nondeterministic linear-time. As far as deterministic time is considered, Dahlhaus and Warmuth showed a rather astonishing result that GCSL is contained in polynomial-time. Later, Buntrock and Loryś [3,4] showed that this class is closed under many operations and forms an abstract family of languages. They also showed that it may be characterized by less restricted grammars. Finally, Buntrock and Otto [5] (see also [2]) gave a characterization of GCSL by nondeterministic machine model, *shrinking two-pushdown automata* (sTPDA). They also initiated study on the deterministic version of these automata.

---

[*] Partially supported by DFG grant GO 493/1-1 and Komitet Badań Naukowych, grant 8 T11C 04419.

Unexpectedly it turned out that the class of languages recognized by deterministic sTPDA is equal to the class of Church-Rosser languages (CRL), that was introduced by McNaughton et al. [9] as languages defined by finite, length-reducing and confluent string-rewriting systems. First, Buntrock and Otto [5] showed that deterministic TPDA recognize exactly the class gCRL of generalized Church-Rosser languages. Then, the final step was done by Niemann and Otto [14], who showed that the classes CRL and gCRL coincide.

From practical point of view CRL has a nice property that the membership problem has linear time complexity and contains many important languages that are not context-free. However, on the other hand its weakness seems to be evident. Already McNaughton et al. [9] conjectured that CRL does not include all context-free languages. They stated also the conjecture that even a very simple context-free language, namely the set of palindromes over two-letter alphabet, is not in CRL. The incomparability of CRL and CFL was finally proved by Buntrock and Otto [5]. The result was obtained as a consequence of the fact that CRL is closed on complementation (whereas CFL is not) and there are context-free languages whose complements are not even in GCSL. The question about palindromes was restated in [14] and in a stronger form by Beaudry et al. [1] who posed the question whether the set of unambiguous context-free languages, UCFL, is included in CRL. Observe that the complement of palindromes is in CFL too, so the techniques used hitherto seem to be far insufficient for proving the conjecture that palindromes are not in CRL.

2 New Result

We show that the set of palindromes is not a Church-Rosser language, answering the open question stated in [9]. By w^R we denote the reversal of the word w, i.e. if $w = w(1)\dots w(m-1)w(m)$, then $w^R = w(m)w(m-1)\dots w(1)$, where $w(i) \in \{0,1\}$ for $i \in \{1,\dots,m\}$.

Theorem 1. *The language PAL*$= \{ww^R : w \in \{0,1\}^*\}$ *does not belong to CRL.*

In our proof we use the fact that shrinking TPDA are „equivalent" to simpler, length-reducing TPDA (This was shown by Niemann [12] for both nondeterministic and deterministic versions. We are not aware of the „official" publication of this result, thus let us mention that our proof is applicable with slight modification to the characterization of CRL by shrinking deterministic TPDA [14]). We show that such an automaton working on words from a certain set of palindromes has to fulfill two contradictory conditions. On the one hand it has to move the contents of one pushdown to the other very often, on the second hand it must not lose too much information about any part of the input, if we ensure that these palindromes are build out of blocks of high Kolmogorov complexity (see [10,11] for other results that apply Kolmogorov complexity). This is impossible as the automaton is length reducing. As a technical tool we use a kind of pumping lemma, which is of independent interest. Note that even CRL languages over one-letter alphabet need not to be semilinear (e.g. $\{a^{2^n} | n \in \mathbb{N}\} \in$ CRL, [9]), so a pumping technique has to be used in a rather sophisticated way. In fact, this is the first application of pumping technique to non context-free languages known to the authors.

Observe that co-PAL, the complement of PAL, is also context-free. Moreover, PAL (as well as co-PAL) is an unambiguous context-free language (UCFL), i.e. there exists a context-free grammar for PAL such that every word from PAL has only one derivation tree (see [8]). Therefore, as a conclusion of the previous result, we obtain the following theorem, answering the open question from [1].

Corollary 1. *The classes CFL∩ co-CFL and UCFL∩ co-UCFL are not included in CRL.*

Note that this is quite a tight separation. It is known ([8]) that inside CFL there exists a hierarchy with respect to the degree of ambiguity and the unambiguous languages define its last but one level (see Theorem 7.3.1 in [8]). At the lowest level there is the class of deterministic context-free languages, which is strictly included in CRL.

The rest of the paper is devoted to the proof of Theorem 1. In Section 3 we introduce basic notions and present a formal definition of length-reducing two-pushdown automata. In Section 4 we introduce the notion of a computation graph and show some of its properties. Section 5 is devoted to the notion of a mutant which is crucial for our proof method. In Section 6 we present the proof by contradiction. Due to space limitations we omit many details, in particular several technical lemmata are given without proofs.

3 Basic Definitions and Properties

Let $|x|$ be the length of x if x is a word or a path in a graph (i.e. number of edges in the path), and the number of elements if x is a set. If x is a sequence (or a word) then $x(i)$ denotes the ith element of x (ith symbol). Let shift(u, l) denote a left cyclic shift of the word u by l positions, i.e. shift$(u, l) = u(l+1) \ldots u(|u|)u(1) \ldots u(l)$. Let $w = xyz$ be a word over a finite alphabet. Then, the position of y in w is equal to $|x|$.

The set of Church-Rosser languages (CRL) was originally defined by finite, length-reducing and confluent string-rewriting systems [9]. We do not make use of this characterization in this paper, so we omit the formal definition. Instead we use the characterization by two-pushdown automata defined below.

Definition 1. *A **two-pushdown automaton** (TPDA) $M = (Q, \Sigma, \Gamma, q_0, \perp, F, \delta)$ with window of length $k = 2j$ is a nondeterministic automaton with two pushdown stores. Formally, it is defined by a set of states Q, an input alphabet Σ, a tape alphabet Γ ($\Sigma \subseteq \Gamma$), an initial state $q_0 \in Q$, a bottom marker of the pushdown stores $\perp \in \Gamma \backslash \Sigma$, a set of final states $F \subseteq Q$ and a transition relation $\delta : Q \times \Gamma_{\perp,j} \times \Gamma_{j,\perp} \to \mathcal{P}(Q \times \Gamma^* \times \Gamma^*)$ where $\Gamma_{\perp,j} = \Gamma^j \cup \{\perp v : |v| \leq j-1, v \in \Gamma^*\}$, $\Gamma_{j,\perp} = \Gamma^j \cup \{v \perp : |v| \leq j-1, v \in \Gamma^*\}$, and $\mathcal{P}(Q \times \Gamma^* \times \Gamma^*)$ denotes the set of finite subsets of $Q \times \Gamma^* \times \Gamma^*$. M is a **deterministic two-pushdown automaton** (DTPDA) if δ is a (partial) function from $Q \times \Gamma_{\perp,j} \times \Gamma_{j,\perp}$ into $Q \times \Gamma^* \times \Gamma^*$. A (D)TPDA is called **length-reducing** if $(p, u', v') \in \delta(q, u, v)$ implies $|u'v'| < |uv|$, for all $q \in Q$, $u \in \Gamma_{\perp,j}$, and $v \in \Gamma_{j,\perp}$.*

A **configuration** of a TPDA M, in which the automaton is in a state q_i and has u and v written down on its pushdowns will be described by the word uq_iv^R. Note that v is written from the right to the left, so the symbols $\perp$ are on the both ends of uq_iv^R. Let the **cursor position** of a configuration uq_iv be the length of the left pushdown, i.e. $|u|$. We say that the configuration $C_1 = uz'q'x'v$ follows the configuration $C_0 = uzqxv$, shortly $C_0 \vdash_M C_1$, if M can move in one step from C_0 to C_1, i.e. if $(q', z', x') \in \delta(q, z, x)$. For an input word $x \in \Sigma^*$, the corresponding **initial configuration** is $\perp q_0 x \perp$. M **accepts** by empty pushdown stores which means that $L(M) = \{x \in \Sigma^* : \exists_{q \in F} \perp q_0 x \perp \vdash_M^* q\}$.

We also require the transition function to be such that during the whole computation the symbols $\perp$ occur only on bottoms of the pushdowns. In particular they can be removed from the pushdowns only in the last step.

Both GCSL and CRL may be characterized by classes of languages recognized by two-pushdown length-reducing automata:

Theorem 2 ([12,14]). *A language is accepted by a length-reducing TPDA if and only if it is a growing context-sensitive language. A language is accepted by a length-reducing DTPDA if and only if it is a Church-Rosser language.*

4   Computation Graphs and Their Properties

We introduce the notion of a computation graph, similar to the derivation graphs from [6,5]. It describes a computation of a two-pushdown automaton on a particular input.

Definition 2. *Let $M = (Q, \Sigma, \Gamma, q_0, \perp, F, \delta)$ be a two-pushdown deterministic automaton, let $x = x(1)x(2)\ldots x(n)$ be an input word and let $C_0 = \perp q_0 x \perp \vdash C_1 \vdash \ldots \vdash C_j$ be a computation of M on x. A **computation graph** $\mathcal{G}_M(C_0, C_j)$ corresponding to the configuration C_j is described the triple $\langle G_j, \prec, \mathcal{L} \rangle$, where $G_j = (V_j, E_j)$ is a graph, $\prec$ is a partial order relation in the set V_j and $\mathcal{L} : V_j \to \Gamma \cup Q \cup \delta$ is a labelling function that divides the vertices into three categories: **symbol**, **state** and **instruction** vertices. We define $\mathcal{G}_M(C_0, C_j)$ inductively:*

$j = 0$: $V_0 = \{s_0\} \cup \{\rho_{-1}, \rho_0, \ldots, \rho_{n+2}\}$, $\mathcal{L}(\rho_i) = x(i)$ for every $1 \leq i \leq n$, $\mathcal{L}(\rho_i) = \perp$ for $i < 1$ or $i > n$, s_0 is labeled by the initial state of M, $\rho_{-1} \prec \rho_0 \prec s_0 \prec \rho_1 \prec \ldots \prec \rho_{n+2}$ and $E_0 = \emptyset$.

$j > 0$: Assume that $C_{j-1} = \beta_1 z_1 \ldots z_{p_0} q_{j-1} z_{p_0+1} \ldots z_p \beta_2$, $C_j = \beta_1 z'_1 \ldots z'_{r_0} q_j z'_{r_0+1} \ldots z'_r \beta_2$, and an instruction $I = \langle \delta(q_{j-1}, z_1 \ldots z_{p_0}, z_{p_0+1} \ldots z_p) = (q_j, z'_1 \ldots z'_{r_0}, z'_{r_0+1} \ldots z'_r) \rangle$ is executed in the jth step. The set V_j is obtained by adding the set $\{D_j, \pi'_1, \ldots, \pi'_r, s_j\}$ to V_{j-1}, where D_j is an instruction vertex labeled by I, each π'_i is a symbol vertex labeled by z'_i $(i = 1, \ldots, r)$ and s_j is a state vertex labeled by q_j. Let $\pi_1, \ldots, \pi_p$ be consecutive (with respect to $\prec$) symbol vertices corresponding to the symbols $z_1, \ldots, z_p$ (i.e. symbols "replaced" in the jth step by $z'_1 \ldots z'_r$) and let s_{j-1} be the state vertex corresponding to q_{j-1}. The set E_j is obtained by adding the set of directed edges $\{(D_j, \pi'_1), \ldots, (D_j, \pi'_r), (D_j, s_j)\} \cup \{(\pi_1, D_j), \ldots, (\pi_p, D_j), (s_{j-1}, D_j)\}$ to E_{j-1}. Finally, we extend $\prec$ by setting $\pi \prec \pi'_1 \prec \ldots \prec \pi'_{r_0} \prec s_j \prec \pi'_{r_0+1} \prec \ldots \prec \pi'_r \prec \bar{\pi}$, for each two vertices $\pi, \bar{\pi} \in V_{j-1}$ with $\pi \prec \pi_1$ and $\pi_p \prec \bar{\pi}$.

If it does not lead to confusion we shall sometimes use the term computation graph to the graphs G_j. We say that π is an **alive vertex** in G_j if it has no outcoming edges in G_j. This means that the symbol corresponding to π **does** occur on one of the pushdown stores in the configuration C_j. The only exceptions are ρ_{-1} and ρ_{n+2} that are artificial vertices introduced for technical reasons. The vertices from V_0 we call **ground vertices**.

They correspond to the symbols of the initial configuration (except ρ_{-1} and ρ_{n+2}) and have no incoming edges. Observe that computation graphs are planar and acyclic, the relation $\prec$ describes a „left-to-right" ordering of alive as well as ground vertices in planar representations of the graph. If $\pi_1 \preceq \pi_2$ then we say that π_1 is to the left of π_2.

We extend the notion $\vdash$ for computation graphs in the following way. It holds that $G \vdash H$ for computation graphs G, H iff there exist an initial configuration C_0 and configurations C, C' such that $G = \mathcal{G}(C_0, C)$, $H = \mathcal{G}(C_0, C')$ and $C \vdash C'$.

Let $init(u)$ for $u \in \Sigma^*$ denote the computation graph corresponding to the initial configuration on the input word u.

Let G be a computation graph. Then $Gr(G)$ is the sequence of labels of all ground vertices in G except ρ_{-1} and ρ_{n+2}, and $Cur(G)$ is the sequence of labels of all alive vertices except ρ_{-1} and ρ_{n+2}. In other words, $Gr(G)$ describes the initial configuration and $Cur(G)$ describes the last configuration of the computation described by G, i.e. $G = \mathcal{G}(Gr(G), Cur(G))$.

In this paper we apply the term *path* exclusively to paths that start in a ground vertex and finish in an alive vertex. The ground vertex in a path is called its **source** and the alive vertex is the **sink** of the path. Let σ be a path in G with a sink π. We say that σ is **short** if there is no path with sink π that is shorter than σ. The relation $\prec$ induces a left-to-right partial ordering of paths. A path σ_1 is to the left (right, resp.) of σ_2 iff none of vertices of σ_1 lays to the right (left, resp.) of a vertex of σ_2. Note that σ_1 and σ_2 may have common vertices, however, they must not cross. A path σ is the leftmost (rightmost) in a set S of paths if it is to the left (right) of every path $\sigma' \in S$.

The **height** of a vertex π is the number of instruction vertices in any short path with the sink in π. We say that an alive vertex π is i-**successor** if one of its short paths starts in ρ_i, the ith ground vertex.

Let us fix a length-reducing DTPDA $M = (Q, \Sigma, \Gamma, q_0, \bot, F, \delta)$ with the window of length $k > 2$ (one can simulate an automaton with the window of size 2 by an automaton with longer window). The rest of this section concerns computation graphs of M.

Proposition 1. *Let G be a computation graph.*

(a) Let π be an alive vertex in G. Then, there exists $-1 \leq i \leq n+2$ such that π is i-successor.

(b) Let D be an instruction vertex in G and let π be a child of D. If π is i-successor then there exists π', a parent of D such that π' is an i-successor in some graph G' such that $G' \vdash^* G$.

(c) Let σ_1, σ_2 be short paths in G and let $\pi_1, \ldots, \pi_p$ be all common vertices of σ_1 and σ_2 written in the order in which they occur in the paths. Let σ_i^j be a subpath of σ_i that starts in π_{j-1} and finishes in π_j for $i = 1, 2$ and $j = 1, \ldots, p+1$ (with two exceptions: σ_i^1 starts in a source of σ_i and σ_i^{p+1} finishes in a sink of σ_i). Then, for every sequence $a_1, \ldots, a_p \in \{1, 2\}$, the path $\sigma_{a_1}^1 \sigma_{a_2}^2 \ldots \sigma_{a_{p+1}}^{p+1}$ is also a short path in G.

(d) Let π, π' be vertices of G, let $\pi \prec \pi'$. If π is j_1-successor then π' is j_2-successor for some $j_2 \geq j_1$. Similarly, if π' is j_3-successor then π is j_4-successor for some $j_4 \leq j_3$.

Note here that an alive vertex may be i-successor for many i's (and at least one). On the other hand, it is possible that there is no alive vertex that is j-successor for some

j. The following lemma shows the dependency between the maximal number of alive vertices of some height and the number of ground vertices (i.e. the length of the input word) in a computation graph.

Lemma 1 (Heights Lemma). *Let G be a computation graph of M for some configuration during the computation on an input word of length $n > 0$. Let p_h be the number of alive vertices in G with height greater or equal to h. Then $p_h \leq 5n \left(\frac{k}{k+1}\right)^h$, where k is the window size of M.*

Corollary 2 (Maximal Paths Property). *Let G be a computation graph corresponding to a configuration of M during the computation on an input word of length n. Then, there are no vertices of height (and no short paths with length) greater than or equal to $(\frac{1}{\log((k+1)/k)} + 1)\log(5n)$.*

Now, we introduce some notions that help us to determine the impact of the context-sensitivity of automata on the results of computations „inside" subwords of the input word.

Definition 3 (Description of a path). *Let $\sigma = \pi_1, \pi_2 \ldots, \pi_{2l+1}$ be a path in a computation graph $G = (V, E)$. A **description of the path** σ, $desc(\sigma)$, consists of two sequences: the sequence $(lab_1, \ldots, lab_{2l+1})$ of labels of consecutive vertices on the path and the sequence $p_1, p_2, \ldots, p_{2l}$ of numbers, such that for each even j, π_j is an instruction vertex, π_{j-1} is the (p_{j-1})st parent of π_j and π_{j+1} is the (p_j)th child of π_j, where the numeration is according to $\prec$. We say that a path $\sigma \in G$ is γ-**path** if $desc(\sigma) = \gamma$.*

Definition 4. *Descriptions $\gamma_1, \ldots, \gamma_{l-1}$ decompose a computation graph G into subgraphs $G_1, \ldots G_l$ if the following conditions are satisfied:*
1. There are paths $\sigma_1, \ldots, \sigma_{l-1}$ in G such that $desc(\sigma_i) = \gamma_i$ for $i = 1, \ldots, l$ and σ_i is to the left of σ_{i+1} for $i = 1, \ldots, l-2$.
2. G_1 is the subgraph of G induced by all vertices to the left of σ_1 or inside σ_1, G_l is the subgraph of G induced by all vertices to the right of σ_l or inside σ_l and G_i (for $1 < i < l$) is the subgraph of G induced by all vertices to the right of σ_{i-1} and to the left of σ_i or inside σ_{i-1} or inside σ_i.

If there are descriptions $\gamma_1, \ldots, \gamma_{l-1}$ that decompose G into subgraphs $G_1, \ldots, G_l$ then we write $G = G_1 \ldots G_l$.

We extend the notions Gr and Cur into subgraphs of computations graphs. For a subgraph H, $Gr(H)$ is the sequence of labels of all (but the rightmost) ground vertices in H. In a similar way we define $Cur(H)$. Obviously, if $G = G_1, \ldots, G_l$ then $Gr(G) = Gr(G_1) \ldots Gr(G_l)$ and $Cur(G) = Cur(G_1) \ldots Cur(G_l)$.

Now, using „cut and paste" technique, we provide some conditions under which „pumping" is possible. In the following we shall write $\langle \tau \rangle_j$ as an abbreviation for $\tau, \ldots, \tau$, where a τ is taken j times.

Lemma 2 (General Pumping Lemma). *Assume that descriptions $\langle \gamma \rangle_2$ decompose a computation graph G into G_1, G_2, G_3 and $x_j = Gr(G_j)$, $x'_j = Cur(G_j)$ for $j = 1, 2, 3$.*

Then $x_1 x_2^i x_3$ for $i > 0$ is the initial configuration of M for some input word. Moreover, if we start a computation in $x_1 x_2^i x_3$ then we get a configuration $x_1' x_2'^i x_3'$ and $\langle \gamma \rangle_{i+1}$ decompose $\mathcal{G}(x_1 x_2^i x_3, x_1' x_2'^i x_3')$ into $G_1, \langle G_2 \rangle_i, G_3$.

Let $G_1 G_2 G_3$ be a computation graph of a DTPDA M with window of length k. We say that the window **touches** a subgraph G_2 of $G_1 G_2 G_3$ if and only if $G_1 G_2 G_3 \vdash G'$ for some G' and at least one alive vertex of G_2 becomes non-alive in G' (i.e. is „used" in a step). Let $G_1 G_2 G_3$ and G' be computation graphs, $G_1 G_2 G_3 \vdash_M^* G'$, and the window does not touch G_2 in $G_1 G_2 G_3$. The window *touches G_2 for the first time* after $G_1 G_2 G_3$ in G' if the window does not touch G_2 in $G_1 G_2 G_3$ and in all graphs following $G_1 G_2 G_3$ and preceding G' and it touches G_2 in G'.

5   Mutants

A crucial property of context-free languages, that underlies pumping lemma is that given derivation uniquely determines a derivative of each part of the starting phrase. We would like to have an analog of this property for Church-Rosser languages that would allow us to treat parts of pushdown contents as representatives of subwords of a given input word, i.e. as result of its reduction. Unfortunately, it is not so simple because the work of DTPDA is context-sensitive. As a poor substitute for such a representative we introduce a notion of *mutant*. It might seem that a good candidate for mutant of a given subword v could be a word stored between two alive vertices which are sinks for short paths starting from the left and right ends of v, respectively. However, we should care about two additional circumstances. First, we would like to avoid a situation that „representations" of two disjoint words overlap by many symbols. Second, we should care about the fact that although there exist short paths to every alive vertex, there might exist ground vertices in which no short path starts. This is a reason why the notion of *mutant*, defined below, is a little bit complicated.

Definition 5. *Let G be a computation graph corresponding to a computation on an input word of length n. Then, a **rightmost-left** pseudo-successor of i in G, shortly $RL_G(i)$, is the rightmost symbol vertex in the set $Left_G(i) = \{\pi \mid \exists_{j \leq i} \ \pi \text{ is alive } j\text{-successor}\}$. A **rightmost-left** i-path is the rightmost short path with a sink $RL_G(i)$ and a source $\pi \preceq \rho_i$.*

*Similarly, a **leftmost-right** pseudo-successor of i in G, shortly $LR_G(i)$, is the leftmost symbol vertex in the set $Right_G(i) = \{\pi \mid \exists_{j \geq i} \ \pi \text{ is alive } j\text{-successor}\}$. A **leftmost-right** i-path is the leftmost short path with a sink $LR_G(i)$ and a source $\pi \succeq \rho_i$.*

Definition 6 (Mutant). *Let G be a computation graph corresponding to a computation on an input word of length n, $0 \leq l < r \leq n+1$. Let $\pi_l = RL_G(l)$, $\pi_r = LR_G(r)$. If there is no alive symbol vertex π such that $\pi_l \prec \pi \prec \pi_r$ then (l,r)-mutant is undefined in G, $\mathcal{M}(l,r,G) = \emptyset$. Otherwise (l,r)-mutant is defined in G and it is equal to $\mathcal{M}(l,r,G) = (\sigma_l, \sigma_r, \tau)$, where σ_l is a rightmost-left l-path, σ_r is a leftmost-right r-path and $\tau = \tau_1 \tau_2 \ldots \tau_p$ is a sequence of all alive vertices between π_l and π_r, i.e. $\pi_l = \tau_1 \prec \tau_2 \ldots \prec \tau_p = \pi_r$.*

*A **type** of a mutant $\mathcal{M}(l,r,G) = (\sigma_l, \sigma_r, \tau_1 \ldots \tau_p)$ is equal to $(desc(\sigma_l), desc(\sigma_r), \mathcal{L}(\tau_1) \ldots \mathcal{L}(\tau_p))$. Two mutants are equivalent if they have equal types. A **length** of a mutant $(\sigma_l, \sigma_r, \tau)$ is equal to $|\tau|$.*

At first sight it may seem to be more natural to define a (l,r)-mutant whenever $RL_G(l) \preceq LR_G(r)$. However, because of technical reasons we leave it undefined when there are no vertices between $RL_G(l)$ and $LR_G(r)$. In particular, now we have got a notion which satisfies properties stated below.

Lemma 3. *Let G, G' be computation graphs corresponding to a computation on an input word of length n, $G \vdash G'$, and $0 \le l < r \le n+1$. Assume that (l,r)-mutant is defined in G and $\mathcal{M}(l,r,G) = (\sigma_l, \sigma_r, \tau)$. Then*

(a) *For every $1 < i < |\tau|$, if $\tau(i)$ is j-successor then $l < j < r$.*

(b) *If the window does not touch τ in G then the (l,r)-mutant remains unchanged in G', i.e. $\mathcal{M}(l,r,G) = \mathcal{M}(l,r,G')$.*

(c) *If $|\tau| > 2k$ then the (l,r)-mutant is defined in G'.*

(d) *If the (l,r)-mutant is defined in G' then its length is not smaller than $|\tau| - k$ and not bigger than $|\tau| + k$.*

(e) *Assume that the (l',r')-mutant is defined in G for $r < l' < r' \le n+1$ and $\mathcal{M}(l',r',G) = (\sigma_{l'}', \sigma_{r'}', \tau')$. Then, $|\tau \cap \tau'| \le 2$.*

(f) *The paths σ_l and σ_r are disjoint.*

6   Proof of the Main Theorem

Our proof exploits the notion of Kolmogorov complexity (cf. [10]). Recall that Kolmogorov complexity of a binary word x (denoted by $K(x)$), is the length of the shortest program (written binary) that prints x. We use the fact that for every natural n there are binary words of length n with Kolmogorov complexity greater than $n - 1$. We call them *hard* words.

For the sake of contradiction assume that the length-reducing DTPDA $M = (Q, \Sigma, \Gamma, q_0, \perp, F, \delta)$ with the window of size k recognizes the language PAL. We will analyse computations of M on inputs which consists of many repetitions of a palindrome. Let $\mathcal{W}_d = \{(ww^R)^{2i+1} \mid |w| = m, 2i + 1 \le m^d, K(w) > m - 1\}$. Note that elements of $(ww^R)^*$ are palindromes for every $w \in \{0,1\}^*$. The copies of ww^R are called **blocks**, by the ith block we mean the ith occurence (from left side) of the word ww^R in the input word that belongs to $(ww^R)^*$ (note that two copies of ww^R cannot overlap in the word from $(ww^R)^*$ if w is hard). By a **middle block** of the word $(ww^R)^{2i+1}$ we mean the $(i+1)$-st block. A mutant of the j-th block of $(ww^R)^i$ in a graph G is the $(2|w|(j-1)+1, 2|w|j)$-mutant in G. Let $e = |\Gamma| + |Q|$.

First let us present a high level sketch of the proof. Many arguments are based on the fact that short paths have lengths $O(\log n)$ and on the intuition that only $O(s)$ bits can be transmitted across paths of length s. The computation of M on some inputs from $(ww^R)^*$ (for hard w) may be divided into phases such that the contents of pushdowns at the end of each phase preserve periodic structure. Intuitively, this periodicity makes M unable to determine the position of the middle block and to compare all symmetric positions of the input. The formal proof is divided in few steps:

Step 1. First, we restrict our attention to inputs that contain polynomial number of blocks (with respect to the length of the block), i.e. to inputs from $\mathcal{W}_d$ for fixed d. For such inputs we show that as long as the mutant of the middle block is defined, the mutants of all other blocks are also defined and „long" [Middle Block Lemma]. It follows from the fact that if the mutant of the middle block is defined then the amount of information exchanged between the left and the right part of the input word is small (because there exists a short path „between" these parts). And if the mutant of the block is short then we are not able to store all information about its contents.

Step 2. Using a technical notion of a *periodicity property* we show that starting in a „periodic" configuration (i.e. graph) $G = G_1 G_2^i G_3$ in which one of pushdowns is short (the window is inside G_1 or G_3), the automaton achieves another „periodic" configuration $G' = G_1'(G_2')^*(G_3')$ such that the window touches G_3' for the first time in G' (if it touches G_1 in G) [Pumping Lemma]. This property is satisfied under the condition that the mutants of all blocks are in G defined and long. It guarantees also that the mutants of all blocks are defined and long also in G'. We show this lemma in two stages. First, we restrict our attention to inputs in which the number of blocks is polynomial with respect to the length of the block and we make use of Middle Block Lemma. The second stage is based on the „periodicity" of the graphs G and G' and it gives a result that is independent of the number of blocks.

Step 3. Finally observe that, for each input word $u \in (ww^R)^*$, w satisfies periodicity property for appropriate parameters concerning the initial configuration for the input word u. Thus if w is hard and long enough then starting from the initial configuration (of the input word = $u \in (ww^R)^*$) it is possible to apply Pumping Lemma l times (for every $l \in \mathbb{N}$). On the other hand each „application" of Pumping Lemma denotes a part of the computation in which the automaton has to make many steps (in order to „move" the window through all the subgraph G_2^*). And each step shortens the length of the configuration. So after a constant number of „applications" of Pumping Lemma we get a configuration that is too short to contain „long" mutants of all blocks (recall that mutants of disjoint parts of the input may overlap by at most two symbols).

Now, we provide a useful lemma saying that (for inputs from $\mathcal{W}_d$) M should store "unique" description of every block, as long as the mutant of the middle block is defined.

Lemma 4 (Middle Block Lemma). *For every $d > 0$ there exists a number $mbl(d)$ such that the following condition is satisfied for every computation graph G describing a computation on an input from $\mathcal{W}_d$ with $|w| = m > mbl(d)$: if a mutant of a middle block is defined in G then mutants of all other blocks are also defined in G and the length of the mutant of each block (but the middle block, possibly) is greater than $c'm$ (where $c' = \frac{1}{4\lceil \log e \rceil}$).*

Proof. (Sketch) The proof of Middle Block Lemma uses an incompressibility method and the „cut and paste" technique (applied to computation graphs). Assuming that the lemma is false we conclude that during a computation the mutant of the ith block becomes short (for some i) while the mutant of the middle block is still defined. W.l.o.g we may assume that the ith block lies in the first half of the input word. Then any "cut and paste" operation replacing the subgraph of the computation graph that corresponds to the mutant of the ith block by another subgraph with the same alive vertices and the same paths on its borders does not change the second half of the input. This implies that the mutant of the ith block should describe this block uniquely, otherwise we would accept an input word that is not a palindrome. But this contradicts the assumption that blocks have large Kolmogorov complexity (because the mutant is short). □

Now, we introduce a notion of **periodicity property**, and next we present Pumping Lemma which provides an "inductive step" for "generating" consecutive phases of computations. Finally, by applying this lemma, we get a contradiction.

Definition 7. *A word $w \in \Sigma^m$ satisfies **periodicity property** with a parameter $(G_1, G_2, G_3, \alpha, r, j)$, where $G_1 G_2 G_3$ is a computation graph and $\alpha, r, j \in \mathbb{N}$ iff*
$\forall_{i>0}$ $init((ww^R)^{\alpha+ir}) \vdash^*_M G_1 G_2^i = G_3$, *and*

- $2m\alpha \leq r \leq m^j$, α *is odd*, r *is even.*
- $Gr(G_2) = (shift(ww^R, a))^r$ *for some $a \in \mathbb{N}$.*
- *There exists a description γ such that $= \langle \gamma \rangle_{i+1}$ decompose $G_1 G_2^i G_3$ into $G_1, \langle G_2 \rangle_i G_3$ for every $i > 0$.*
- *A path on the right border of $G_1 G_2^l$ is short in $G_1 G_2^i G_3 = $ for every $0 \leq l \leq i$ and $i > 0$.*

*The word $w \in \Sigma^m$ satisfies **strong periodicity property** with the parameter $(G_1, G_2, G_3, \alpha, r, j)$ if it satisfies periodicity property with this parameter and for every $i > 0$, the mutant of each block is defined in $G_1 G_2^i G_3$ and its length is greater than $c'm$ in $G_1 G_2^i G_3$ ($c' = \frac{1}{4\lceil \log e \rceil}$ as in Middle Block Lemma).*

Proposition 2. *Assume that w satisfies **periodicity property** with a parameter $(G_1, G_2, G_3, \alpha, r, j)$, $|w| = m > 4$. Then:*

(a) $|Cur(G_a)| \leq m^{j+2}$ *for $a = 1, 3$.*
(b) *The lth block of the input $(ww^R)^{\alpha+ir}$ is included in the infix subgraph G_2^i of the computation graph $G_1 G_2^i G_3$ for every $i > 0$ and $\lceil (\alpha+r)/2 \rceil < l < \alpha + ir - \lfloor (\alpha + r)/2 \rfloor$.*
(c) *Assume that the bth block and the $(b+2r)$th block of $(ww^R)^{\alpha+ir}$ are contained in the infix subgraph G_2^i of the graph $G_1 G_2^i G_3$. Then mutants of blocks b and $b+2r$ in $G_1 G_2^i G_3$ are equivalent.*
(d) *If there exists $a > 1$ such that a mutant of every block of $(ww^R)^{\alpha+ar}$ is (defined and) longer than $c'm$ in $G_1 G_2^a G_3$ then w satisfies strong periodicity property with the parameter $(G_1, G_2, G_3, \alpha, r, j)$.*

Lemma 5 (Pumping Lemma). *For every $j \geq 1$ exists $m_0 \in \mathbb{N}$ such that for every $w \in \Sigma^m$, if w satisfies strong periodicity property with a parameter $(G_1, G_2, G_3, \alpha, r, j)$, $K(w) > m-1$ and $m > m_0$ then w satisfies strong periodicity property with a parameter $(H_1, H_2, H_3, \alpha', r', 20j)$ such that:*

- $\alpha' = \alpha + gr$, $r' = fr$ *for some $g, f \in \mathbb{N}$,*
- $init((ww^R)^{\alpha'+ir}) \vdash^*_M G_1 G_2^{g+if} G_3 \vdash^*_M H_1(H_2)^i H_3$ *for every $i > 0$ where $H_1 H_2^i H_3$ is **the first graph after** $G_1 G_2^{g+if} G_3$ in which the window touches G_3 (if the window touches G_1 in $G_1 G_2^{g+if} G_3$) or G_1 (otherwise).*

Proof. (Sketch) We show that the lemma is satisfied for $m_0 = mbl(80j)$. Let $w \in \Sigma^m$ satisfy strong periodicity property with the parameter $(G_1, G_2, G_3, \alpha, r, j)$, $m > m_0$, $K(w) > m-1$. Observe that the window in a graph of the type $G_1 G_2^* G_3$ touches G_1 or G_3 (because there is only one symbol from Q in each configuration, so appropriate state vertex cannot be in G_2). W.l.o.g. we may assume that it touches G_1 (the second case may be = proved in the same way). Let $NSC(i)$, for $i > 0$, denote a first graph after $G_1 G_2^i G_3$ in which the window touches G_3.

Let G be a graph describing a computation on an input word $v \in (ww^R)^*$. We say that paths σ_1, σ_2 (contained in G) are *parallel* in G iff $\mathrm{desc}(\sigma_1) = \mathrm{desc}(\sigma_2)$ and $p_2 - p_1$ is divisible by $2m$, where $p_1 < p_2$ are positions of the sources of σ_1 and σ_2 (i.e. positions of the sources of σ_1, σ_2 in appropriate blocks of v are equal).

Let $u = ww^R$. The outline of our proof is following:

1. We show that the mutant of the middle block of $u^{\alpha + ir}$ is defined in $NSC(i)$ for every $i > 0$ such that $\alpha + ir \leq m^{40j}$. The proof is based on Middle Block Lemma and the „periodicity" of the configuration $G_1 G_2^i G_3$.

2. By counting arguments we show that there exist two parallel short paths in $NSC(i)$ for some $i > 1$ that satisfies $\alpha + ir \leq m^{5j}$.

3. Using General Pumping Lemma and the existense of parallel paths in $NSC(i)$ we define subgraphs H_1, H_2, H_3 and numbers α', r', g, f such that w satisfies periodicity property with a parameter $(H_1, H_2, H_3, \alpha', r', 20j)$ where $H_1 H_2^i H_3$ is a first graph after $G_1 G_2^{g+if} G_3$ (for every $i > 0$), in which the window touches the suffix G_3, where $\alpha' = \alpha + gr, r' = fr$.

4. Finally, using item 1. above and Proposition 2 (in particular the facts (c)-(d) saying about „parallelity" of mutants of blocks), we show that w satisfies strong periodicity property with the parameter $(H_1, H_2, H_3, \alpha', r', 20j)$. □

If we ensure that strong periodicity property (with appropriate parameters) is satisfied in the initial configuration then we may apply Pumping Lemma many times, obtaining consecutive "semi-border" configurations (i.e. configurations in which one pushdown is relatively short; interchangeably the left and the right one) in which periodic structure of pushdowns (and graphs) are preserved, and the mutant of each block is longer than $c'm$ in these configurations. On the other hand, M shortens pushdowns in each step, what in consequence makes unable to preserve "long" mutants of blocks (required by Pumping Lemma) and gives contradiction.

Let us take w such that $|w| = m > mbl(80^{l+1})$ for a constant l, $K(w) > m - 1$. Let $\alpha_0 = 3, r_0 = 6m, j_0 = 20$. Let $v = (ww^R)^{r_0 + 3}$, $G = init(v)$, i.e. G is the computation graph corresponding to the initial configuration on the input word v. Let $G_{1,0}, G_{2,0}, G_{3,0}$ be subgraphs of G corresponding respectively to the first block ww^R of v, the next r_0 blocks and the last two blocks. One can easily verify that w satisfies strong periodicity property with $(G_{1,0}, G_{2,0}, G_{3,0}, \alpha_0, r_0, j_0)$. Now, we can apply Pumping Lemma l times obtaining $G_{1,i}, G_{2,i}, G_{3,i}, \alpha_i, r_i$ for $i \in \{1, \ldots, l\}$ such that w satisfies strong periodicity property with the parameter $(G_{1,i}, G_{2,i}, G_{3,i}, \alpha_i, r_i, j_i)$, i.e.:

A. for any natural number p, $init((ww^R)^{\alpha_l + pr_l}) = G_{1,0} G_{2,0}^{p_0} G_{3,0} \vdash_M^* G_{1,1} G_{2,1}^{p_1} G_{3,1} \vdash_M^* \cdots \vdash_M^* G_{1,l} G_{2,l}^{p_l} G_{3,l}$, where $p_l = p$ and $p_1 \geq \ldots p_{l-1} \geq p$ are some natural numbers.

B. the mutant of each block is longer than $c'm$ in configurations $\{G_{1,i} G_{2,i}^{p_i} G_{3,i}\}_{i=0,\ldots,l}$

C. the window touches $G_{1,i}$ for every even i and $G_{3,i}$ for every odd i.

Let $Cur(G_{1,i}) = x_i$, $Cur(G_{2,i}) = w_i$, $Cur(G_{3,i}) = y_i$ for $0 \leq i \leq l$. Let $p = \max_{i=0,\ldots,l}(|x_i| + |y_i|)$. This guarantees that "margins" $G_{1,i}$, $G_{3,i}$ contain relatively short parts of pushdowns what forces M to make "many" steps in every „transition" $G_{1,i}G_{2,i}^{p_i}G_{3,i} \vdash_M^* G_{1,i+1}G_{2,i+1}^{p_{i+1}}G_{3,i+1}$. Condition B implies that $|x_i w_i^{p_i} y_i| \geq (\alpha_l + pr_l)c'm/2$. (It is a consequence of the fact that the number of blocks is equal to $\alpha_l + pr_l$, mutant of every block is longer than $c'm$ and the mutants of two consecutive blocks may overlap by at most two symbols, by Lemma 3(e).) So, $|w_i^{p_i}| \geq (\alpha_l + pr_l)c'm/2 - (|x_i| + |y_i|) \geq (\alpha_l + pr_l)c'm/2 - p \geq (\alpha_l + pr_l)c'm/4$ for every i. Observe that the position of the cursor has to move through all the part $G_{2,i}^{p_i}$ during the computation $G_{1,i}G_{2,i}^{p_i}G_{3,i} \vdash_M^* G_{1,i+1}G_{2,i+1}^{p_{i+1}}G_{3,i}$; so M makes at least $(\alpha_l + pr_l)c'm/4k$ steps, since the cursor position may move by at most k in one step. Consequently, M makes at least $l(\alpha_l + pr_l)c'm/4k$ steps during the computation: $init((ww^R)^{\alpha_l + pr_l}) = G_{1,0}G_{2,0}^{p_0}G_{3,0} \vdash_M^* \cdots \vdash_M^* G_{1,l}G_{2,l}^{p_l}G_{3,l}$. This implies that $|x_l w_l^{p_l} y_l| \leq 2(\alpha_l + pr_l)m - l(\alpha_l + pr_l)c'm/4k$, because each step shortens the length of pushdowns by at least one. Comparing it with condition B (and Lemma 3(e)), we have $2(\alpha_l + pr_l)m - l(\alpha_l + pr_l)c'm/4k > (\alpha_l + pr_l)c'm/4$, i.e. $2 - l\frac{c'}{4k} > c'/4$. However, for $l \geq \lceil 8k/c' \rceil$, this inequality is false. Contradiction – M does not recognize PAL. This finishes the proof of Theorem 1.

Acknowledgements. We thank F. Otto and G. Niemann for helpful comments.

References

1. M. Beaudry, M. Holzer, G. Niemann, F. Otto, *McNaughton Languages*, Mathematische Schriften Kassel 26/00 (conference version: *On the Relationship between the McNaughton Families of Languages and the Chomsky Hierarchy*, DLT 2001, LNCS 2295, 340–348).
2. G. Buntrock, *Wachsende Kontextsensitive Sprachen*, Habilitationsschrift, Würzburg, 1995.
3. G. Buntrock, K. Loryś, *On growing context-sensitive languages*, Proc. of International Colloquium on Automata, Languages and Programming, 1992, LNCS 623, 77–88.
4. G. Buntrock, K. Loryś, *The variable membership problem: Succintness versus complexity*, Proc. of STACS, 1994, LNCS 775, Springer-Verlag, 595–606.
5. G. Buntrock, F. Otto, *Growing Context-Sensitive Languages and Church-Rosser Languages*, Information and Computation 141 (1998), 1–36.
6. E. Dahlhaus, M.K. Warmuth, *Membership for growing context-sensitive grammars is polynomial*, Journal of Computer Systems Sciences , 33 (1986), 456–472.
7. A. Gladkij, *On the complexity of derivations for context-sensitive grammars*, Algebri i Logika 3, 1964, 29–44. [In Russian]
8. M. A. Harrison, *Introduction to Formal Language Theory*, Addison-Wesley, 1978.
9. R. McNaughton, P. Narendran, F. Otto, *Church-Rosser Thue systems and formal languages*, Journal of the Association Computing Machinery, 35 (1988), 324–344.
10. M. Li, P. Vitanyi, *An Introduction to Kolmogorov Complexity and its Applications*, Springer-Verlag 1993.
11. M. Li, P. Vitanyi, *A New Approach to Formal Language Theory by Kolmogorov Complexity*, SIAM J. on Comp., 24:2(1995), 398–410.
12. G. Niemann, *CRL, CDRL und verkuerzende Zweikellerautomaten.*, Unpublished note, 1997.
13. G. Niemann, F. Otto, *Restarting automata, Church-Rosser languages, and confluent internal contextual languages*, Proc. of DLT 99, Aachener Informatik-Berichte 99-5, 49–62.
14. G. Niemann, F. Otto, *The Church-Rosser Languages Are the Deterministic Variants of the Growing Context-Sensitive Languages*, FoSSaCC 1998, LNCS 1378, 243–257.

Intersection of Regular Languages and Star Hierarchy

Sebastian Bala

Institute of Computer Science
University of Wrocław
Przesmyckiego 20, 51151 Wrocław, Poland

Abstract. In this paper we consider the intersection-nonemptiness problem for languages given by regular expression without binary +. We prove that for regular expressions of star height 2 this problem is $PSPACE$-complete and it is NP-complete for regular expressions of star height at most 1.

1 Introduction

In [K77] D.Kozen proved that the nonemptiness problem for intersection of regular languages given by finite automata is $PSPACE$-complete. This problem is a convenient tool applied to prove $PSPACE$-hardness. It has been applied to prove $PSPACE$-hardness of many problems, for example: (1) simultaneous rigid E-unification with one variable [GV97], (2) problem of deciding purity for finitely generated subgroups of free groups [BM00], (3) satisfiability problem of quadric word equations with regular constraints [DR99]. Kozen's result is also of independent interest and it has been investigated with various restriction. It was proved in [BM00] that problem remains $PSPACE$-complete for languages specified by injective or inverse deterministic finite automata. However, putting constraints on the transition function of the automata is not only way to restrict the problem. In [LR92] was considered similar problem, where number of given automata is bounded by some function $g(n)$ in the input length n, which is the length of the encodings of these automata. The nonemptiness problem for intersection languages for such restriction is $NSPACE(g(n)log(n))$-complete with respect to logspace reduction.

In this paper we consider the problem given by regular expressions without the binary + operator. The removal of + slightly restricts expressive power of the regular expressions. This remark concerns the + operators which do not occur under a star operator. However, if all plus occurrences appear under star operator, we can convert it to equivalent expression without +. It just goes to show that, among the three operators of regular expression, the star operator is the most essential one.

One natural measure of the complexity of a regular expression is the number of nested stars in the expression, which is called the star height of the expression. Star height gives us, even for expressions without +, a strict hierarchy for regular

P. Widmayer et al. (Eds.): ICALP 2002, LNAI 2380, pp. 159–169, 2002.

languages – it means that for each integer $k \geq 0$ there exists a regular language that is denoted by regular expression of star height k, but not $k-1$ [Yu], [DS66], [E63]. Kozen's proof [K77] operates on regular expressions, not on automaton given by transition function. After removal of $+$, from Kozen's proof [K77], we can easily obtain star height 3. We prove that if the star height of a regular expressions is restricted to 2, then the problem remains $PSPACE$-complete. But if star height equals 1, the problem is NP-complete.

2   Notations and Preliminaries

The *star height* of a regular expression e over the alphabet Σ , denoted by $h(e)$, is nonnegative integer defined recursively as follows:

1. $h(e) = 0$, if $e = \emptyset$ or a for $a \in \Sigma$,
2. $h(e) = \max\{h(e_1), h(e_2)\}$, if $e = (e_1 + e_2)$ or $e = (e_1 \cdot e_2)$, where e_1 and e_2 are regular expressions over Σ,
3. $h(e) = h(e_1) + 1$ if $e = (e_1)^*$ and e_1 is a regular expression over Σ.

We consider *one-tape Turing machine*. A Turing machine M is a structure $(\Sigma, Q, \delta, \{\textbf{Left}, \textbf{Right}\}, q_{start}, q_{stop})$; where Σ is a tape alphabet; Q is a set of states; **Left, Right** denote direction in which the head moves; $\delta : Q \times \Sigma \longrightarrow Q \times \Sigma \times \{\textbf{Left}, \textbf{Right}\}$ is a transition function; q_{start} is a start state; and q_{stop} is the only accepting state.

We will use a string $\#CONF_0\#CONF_1\# \ldots \#CONF_{stop}\#\#$ for denoting a *accepting computation* of M with some input x. $CONF_i$ is a description of ith configuration of this computation and it consists of the tape content after the ith step of M, together with blank symbols $\flat$) at the end and the only state symbol stuck just before a tape symbol observed after the ith step.

A pair (u, v) is a *valid transition* if u and v are three lettered words over the $\Sigma \cup Q \cup \{\flat, \#\}$ alphabet, and if the transition function of M permits, u and v as the ith, $(i+1)^{th}$ and $(i+2)^{th}$ letters of the consecutive configurations. Further are considered the Turing machines with space bounded by some polynomial p. Therefore, the positions $i^{th}, (i+1)^{th}$ and $(i+2)^{th}$ are taken $mod\, p(|x|)$ and we allow to single $\#$ occurrence in v or u. The set of valid transitions we denote by $ValTrans$. In the remaining part of this paper we consider, without any loss of generality, only Turing machines which halt on an empty tape with the head at the first position of its tape.

In this paper we will use the *Chinese Remainder Theorem*. It is provided below

Lemma 1 *Let $m_1, \ldots, m_n$ and $x_1, \ldots, x_n$ be integers. Assume that for every pair (i, j) we have*

$$x_i \equiv x_j \pmod{gcd(m_i, m_j)}. \tag{1}$$

Then there exists a nonnegative integer x such that

$$x \equiv x_i \pmod{m_i} \quad for\ 1 \leq i \leq n. \tag{2}$$

Furthermore, x is unique modulo the least common multiple of $m_1, \ldots, m_k$ and condition (1) is necessary for the existence of x.

$LCM(x, y)$ is an abbreviation for the least common multiple of x and y.

3   The Intersection-Nonemptiness Problem for Star Height Two

Let $s_1, \ldots, s_k \in \Sigma^+$. SUM_i denotes regular expressions of the form $(\sum_{i=1}^{k} s_i)^*$ and $L(SUM_i)$ be a language represented by SUM_i. $s_1, \ldots, s_k \in \Sigma^+$ are called *components* of SUM_i. Given SUM_i, we define its *size* as $Size(SUM_i) = \sum_{i=1}^{k} |s_i|$.

In this section we consider the following decision problem:

Problem 2 *(INTERSECTION SUM)*
INSTANCE: An alphabet Σ, regular expressions $SUM_1, \ldots, SUM_p$ over Σ.
QUESTION: Is there $w \in \Sigma^+$, such that $w \in \bigcap_{i=1}^{p} L(SUM_i)$?

The intersection-nonemptiness problem for regular expressions in general is in $PSPACE$ [K77]. Therefore, the $INTERSECTION\ SUM$, as subproblem of the general case, is in $PSPACE$. Below, we show $PSPACE - hardness$ of the $INTERSECTION\ SUM$ problem by encoding a accepting computation of a one tape deterministic Turing machine.

Let M be any deterministic machine, defined as in chapter 2, with space bound p, where p is a polynomial and $p(n) \geq n$. Let $\Delta = |\{\, u \mid (u, y)\ or\ (y, u)\ is\ a\ valid\ transition\ of\ M,\ for\ some\ word\ y\}|$.

Let $\Phi : \Delta \mapsto \{1, \ldots, |\Delta|\}$ be a bijection. For every word from Δ, Φ assign a unique nonnegative integer number. Symbols $P_1, \ldots P_{p(|x|)}$ mark number of a cell on the workspace of M. Without loss of generality, we assume that $p(|x|)$ is even. The $\#$ symbol separates consecutive configurations, 1 symbol is used for unary encoding of nonnegative integer numbers, $\flat$ is a blank symbol and it appears on those positions of a configuration where no symbols from $\Sigma \cup Q$ were written . $\overline{CONF_k} = 1^{|\Delta|} a_1 P_1 1^{|\Delta|} a_2 \ldots 1^{|\Delta|} a_{p(|x|)} P_{p(|x|)} 1^{|\Delta|}$ is a *thin and marked* version of $CONF_k = a_1 a_2 \ldots a_{p(|x|)}$.

We will construct a set of regular expressions $SUM_1, .., SUM_{p(|x|)+2}$. The intersection of languages denoted by SUM_i-expressions will be a set of words representing a valid, thin, and marked computation of M. In other words, $w \in \bigcap_{i=1}^{p(|x|)+2} L(SUM_i)$ if and only if w is a valid, thin, marked sequence of consecutive configurations of M on the input x. Besides, the sequence of configuration begins with the initial configuration and ends with the accepting configuration. There exists at most one such word because M is deterministic.

Let $\overline{CONF_0} = q_{start} x\, \flat^{p(|x|)-|x|-1}$ and $\overline{CONF_{stop}} = q_{stop}\, \flat^{p(|x|)-1}$. We define following sets of words:

$$A_{p(|x|)+1} = \{1^{|\Delta|} a P_i 1^{|\Delta|} b P_{i+1} \mid 1 \leq i < p(|x|)\ \text{is a odd number}; a, b \in \Sigma \cup Q \cup \{\flat\}\},$$

$$B_{p(|x|)+1} = \{1^{|\Delta|} \#\},$$

$$C_{p(|x|)+1} = \{\#\overline{CONF_0}\#, 1^{|\Delta|} \#\overline{CONF_{stop}}\#\#\},$$

$$Odd = A_{p(|x|)+1} \cup B_{p(|x|)+1} \cup C_{p(|x|)+1},$$

$$A_{p(|x|)+2} = \{1^{|\Delta|}aP_i1^{|\Delta|}bP_{i+1} \mid 1 < i < p(|x|) \text{ is a even number}; a, b \in \Sigma \cup Q \cup \{\flat\}\},$$

$$B_{p(|x|)+2} = \{1^{|\Delta|}aP_{p(|x|)}1^{|\Delta|}\#1^{|\Delta|}bP_1 \mid a, b \in \Sigma \cup Q \cup \{\flat\}\},$$

$$C_{p(|x|)+2} = \{\#\overline{CONF_0}\#1^{|\Delta|}aP_1 \mid a \in \Sigma \cup Q \cup \{\flat\}\},$$

$$D_{p(|x|)+2} = \{1^{|\Delta|}\flat P_{p(|x|)}1^{|\Delta|}\#\overline{CONF_{p(|x|)}}\#\#\},$$

$$Even = A_{p(|x|)+2} \cup B_{p(|x|)+2} \cup C_{p(|x|)+2} \cup D_{p(|x|)+2}.$$

Below we define regular expressions $ODD = SUM_{p(|x|)+1}$ and $EVEN = SUM_{p(|x|)+2}$:

$$ODD = (\sum_{s \in Odd} s)^*,$$

$$EVEN = (\sum_{s \in Even} s)^*.$$

Fact 3   *1.* $|\Delta| = O((|\Sigma| + |Q|)^3)$,
   2. $Size(ODD)$ *and* $Size(EVEN)$ *are* $O((|\Sigma| + |Q|)^2|\Delta|p(|x|))$ *therefore both are* $O(p(|x|))$.

Proposition 4 *If a word* $w \in L(ODD) \cap L(EVEN)$ *then* w *has the form*

$$\#\overline{CONF_0}\#WORD_1\# \dots \#WORD_l\#\overline{CONF_{stop}}\#\# \tag{3}$$

and for every $1 \leq i \leq l$

$$WORD_i = 1^{|\Delta|}a_1P_11^{|\Delta|}a_2P_2 \dots 1^{|\Delta|}a_{p(|x|)}P_{p(|x|)}1^{|\Delta|}$$

where $a_1, \dots, a_{p(|x|)} \in \Sigma \cup Q \cup \{\flat\}$.

Expressions $SUM_1, \dots, SUM_{p(|x|)}$ will be constructed in a such way that their intersection forces valid transitions between consecutive configurations. Let

$$A_i = \{1^k aP_j1^{|\Delta|-k} \mid a \in \Sigma \cup Q \cup \{\flat\}; 1 \leq j \leq p(|x|); j \neq i; 1 \leq k \leq |\Delta|\},$$

$$B_i = \{1^k\#1^{|\Delta|-k} \mid 1 \leq k \leq |\Delta|\},$$

$$C_i = \{\#\overline{CONF_0^{1-}}1^\tau \mid \tau = |\Delta| - \Phi(x_ix_{i+1}x_{i+2})\},$$

where $\overline{CONF_0^{1-}}$ is the $\overline{CONF_0}$ devoid of the last block of 1, x_i, x_{i+1}, x_{i+2} are i^{th}, $(i+1)^{th}$ and $(i+2)^{th}$ letter of string $\#CONF_0\#$,

$$D_i = \{1^\alpha \overline{CONF_{stop}^{-1}} \mid \alpha = \Phi(x_ix_{i+1}x_{i+2})\},$$

where $\overline{CONF_{p(|x|)}^{-1}}$ is the string $CONF_{stop}$ without of the first block of ones, x_i, x_{i+1}, x_{i+2} are the ith, $(i+1)^{th}$ and $(i+2)^{th}$ letters of string $\#CONF_{stop}\#$.

It remains to define sets which will be responsible for transitions of M. For $1 < i < p(|x|)$

$$E_i = \{1^k a P_{i-1} 1^{|\Delta|} b P_i 1^{|\Delta|} c P_{i+1} 1^{|\Delta|-l} \mid a, b, c \in \Sigma \cup Q \cup \{\flat\}; \text{ there exists}$$

$$a \text{ word } def \text{ such that } (abc, def) \in ValTrans, \; k = \Phi(abc), \; l = \Phi(def)\}.$$

$$E_1 = \{1^k \#1^{|\Delta|} a P_1 1^{|\Delta|} b P_2 1^{|\Delta|-l} \mid a, b \in \Sigma \cup Q \cup \{\flat\}; \text{ there exists } \#cd$$

$$\text{such that } (\#ab, \#cd) \in ValTrans, \; k = \Phi(\#ab), \; l = \Phi(\#cd)\}.$$

$$E_{p(|x|)} = \{1^k a P_{p(|x|)-1} 1^{|\Delta|} b P_{p(|x|)} 1^{|\Delta|} \#1^{|\Delta|-l} \mid a, b \in \Sigma \cup Q \cup \{\flat\}; \text{ there exists}$$

$$cd\# \text{ such that } (ab\#, cd\#) \in ValTrans, \; k = \Phi(ab\#), \; l = \Phi(cd\#)\}.$$

At the end of this construction we put

$$Sum_i = A_i \cup B_i \cup C_i \cup D_i \cup E_i$$

and

$$SUM_i = \left(\sum_{s \in Sum_i} s \right)^*.$$

Above construction guarantees that, if the starting configuration contains only one state symbol then remaining contain only one state symbol, too. Moreover the form of E_i ensures that the coded Turing Machine M works according to the definition of the transition function. Note that if we want to create a word w of the form *(3)* from Sum_i, then every occurrence of P_i in w comes from E_i. The suffix of ones of a word from E_i determines exactly which letter will be put on the ith position of next configuration.

Proposition 5 *A word $w \in \bigcap_{i=1}^{p(|x|)+2} L(SUM_i)$ if and only if w is a valid, thin and marked version of accepting computation of M on the input x.*

Fact 6 1. $|ValTrans| = O(|\Sigma|^2 |Q||\Delta|)$,
  2. *For every $1 \le i \le p(|x|)$, $Size(SUM_i)$ is $O((|\Sigma|+|Q|)|\Delta|^2 p(|x|))$ and therefore $O(p(|x|))$.*
  3. *The total size of all SUM_i expressions is $O(p^2(|x|))$.*

It is left to the reader to verify that the above construction can be done in the logarithmic space.

By Kozen's result [K77], Facts 3 and 6, and by Propositions 4 and 5 we obtain the following lemma

Lemma 7 *The INTERSECTION SUM problem is PSPACE-complete.*

Remark 8 *The INTERSECTION SUM problem is PSPACE-hard even if it is considered over a binary alphabet.*
Proof. It is easy to remark that symbols from $\{P_1, \ldots, P_{p(|x|)}\} \cup Q \cup \Sigma \cup \{b, \#\}$ can to be unary encoded in two letter alphabet $\{0, 1\}$. If we encode each of the symbols as a word of the form $1^i 0$ then the PSPACE-hardness proof of the *INTERSECTION SUM* problem remains correct. Besides, total size of the regular expressions increases at most $p(|x|) + |\Sigma| + |Q| + 2$ times. ∎

Problem 9 *(STAR HEIGHT TWO)*
INSTANCE: An alphabet Σ, regular expressions $SUM_1, \ldots, SUM_p$ over Σ. The $SUM_1, \ldots, SUM_p$ have star height at most two and they are based on the operators · (concatenation) and ∗ (Kleene-closure).
QUESTION: Is there $w \in \Sigma^+$, such that $w \in \bigcap_{i=1}^{p} L(SUM_i)$?

Note that an expression $(s_1 + s_2 + \ldots + s_k)^*$ is equivalent to $((s_1)^*(s_2)^* \ldots (s_k)^*)^*$. Therefore, by Lemma 7 we obtain

Theorem 10 *The STAR HEIGHT TWO problem is PSPACE-complete.*

4   The Intersection-Nonemptiness Problem for Star Height One

Let k be any positive integer number, $w_1, \ldots, w_{k+1}, c_1, \ldots, c_k \in \Sigma^*$ for some alphabet Σ. Notation ONE_i is used for denoting regular expressions of the form

$$w_1(c_1)^* w_2 \ldots w_k (c_k)^* w_{k+1} \tag{4}$$

In this section we consider the following decision problem:

Problem 11 *(INTERSECTION STAR ONE)*
INSTANCE: An alphabet Σ, Regular expressions $ONE_1, \ldots, ONE_p$ over Σ.
QUESTION: Is there $w \in \Sigma^+$, such that $w \in \bigcap_{i=1}^{p} L(ONE_i)$?

We prove that *INTERSECTION STAR ONE* is $NP-complete$. First, we show how to reduce the *3SAT-problem* to *INTERSECTION STAR ONE*. Let ϕ be any boolean expression in conjunctive normal form with m variables $(x_1, \ldots, x_m)$ and n clauses. Without any loss of generality, we assume that each clause has exactly three literals (ϕ is in $3CNF$ form). Let Σ be an alphabet containing a symbol X_i and $\neg X_i$ if and only if some clause of the ϕ formula has an occurrence of x_i or $\neg x_i$. Besides, Σ contains special symbols $\#$ and $\uparrow$. Every $\#$ occurrence separates consecutive clause representations. Between every i^{th} and $(i+1)^{th}$ occurrence of the $\#$ symbol there is a fragment of the expression that correspond to the ith clause.
Let

$$ONE_{m+1} = \uparrow \#(\uparrow X_1^1)^*(\uparrow X_2^1)^*(\uparrow X_3^1)^* \uparrow \# \ldots$$

$$\ldots \uparrow \#(\uparrow X_1^n)^*(\uparrow X_2^n)^*(\uparrow X_3^n)^* \uparrow \# \uparrow,$$

$$ONE_{m+2} = \uparrow \# \uparrow (X_1^1)^*(X_2^1)^*(X_3^1)^* \uparrow \# \ldots$$

$$\ldots \uparrow \# \uparrow (X_1^n)^*(X_2^n)^*(X_3^n)^* \uparrow \# \uparrow .$$

We use the X_j^i symbol for representing the jth literal of the ith clause.

Remark 12 $L(ONE_{m+1}) \cap L(ONE_{m+2})$ *contain all of words in form*

$$\uparrow \# \uparrow X_{i_1}^1 \uparrow \# \ldots \# \uparrow X_{i_k}^k \uparrow \# \ldots \# \uparrow X_{i_n}^n \uparrow \# \uparrow,$$

where $i_1, \ldots, i_n \in \{1,2,3\}$, *and does not contain any another.*

Let us create expressions ONE_i for $0 \le i \le m$. Each ONE_i will be responsible for that only one symbol, either X_i or $\neg X_i$, which can appear in the word w, for any $w \in \bigcap_{i=1}^{m+2} ONE_i$.

In the following, we describe how ONE_i is constructed:

- We put $(\uparrow)^*(\uparrow \#)^*(\# \uparrow)^*$ and $(\uparrow \#)^*(\# \uparrow)^*(\uparrow)^*$ respectively at the beginning and at the end of the expression.
- Every pair of adjoining clause representations is separated by $(\uparrow \#)^*(\# \uparrow)^*$.
- For every occurrence of X_k^l in ONE_i, provided $X_k^l \neq X_i$ and $X_k^l \neq \neg X_i$, we put $(\uparrow X_k^l)^*(X_k^l \uparrow)^*$. Else, if $X_k^l = X_i$, we put $\uparrow X_i$ in the place of the appearance of X_k^l, else if $X_k^l = \neg X_i$ then we put $\neg X_i \uparrow$.

Example Given

$$(x_1 \vee \neg x_2 \vee x_3) \wedge (\neg x_3 \vee \neg x_4 \vee x_1) \wedge (x_4 \vee \neg x_1 \vee x_2)$$

$$ONE_2 = (\uparrow)^*(\uparrow \#)^*(\# \uparrow)^*(\uparrow X_1)^*(X_1 \uparrow)^*(\neg X_2 \uparrow)^*(\uparrow X_3)^*(X_3 \uparrow)^*(\uparrow \#)^*$$

$$(\# \uparrow)^*(\uparrow \neg X_3)^*(\neg X_3 \uparrow)^*(\uparrow \neg X_4)^*(\neg X_4 \uparrow)^*(\uparrow X_1)^*(X_1 \uparrow)^*(\uparrow \#)^*$$

$$(\# \uparrow)^*(\uparrow X_4)^*(X_4 \uparrow)^*(\uparrow \neg X_1)^*(\neg X_1 \uparrow)^*(\uparrow X_2)^*(\uparrow \#)^*(\# \uparrow)^*(\uparrow)^*.$$

Remark 13 *If* $w \in L(ONE_{m+1}) \cap L(ONE_{m+2}) \cap L(ONE_i)$ *then* w *has not any occurrences* X_i *or* $\neg X_i$. *Which one of the* X_i *and the* $\neg X_i$ *don't appear in* w *it depend on that the first occurrence of star in* ONE_i *(* $(\uparrow)^*$ *) was 'executed' or not.*

Lemma 14 *The INTERSECTION STAR ONE problem is* $NP-hard$.
Proof. Let $\phi = C_1 \wedge \ldots \wedge C_n$ be a $3CNF$ formula, where $C_1, \ldots, C_n$ are clauses. An ordered set of literals $\{y_1, \ldots y_n\}$ is called a *witness* for ϕ, if y_i has an occurrence in C_i, for $i = 1, \ldots, n$. By Remark 12 we conclude that each of $w \in \bigcap_{i=1}^{m+2} L(ONE_i)$ is a counterpart of the witness for ϕ. If $\{y_1, \ldots y_n\}$ is consistent then it is called *witness of satisfiability* for ϕ. If ϕ has not any witness of satisfiability then every witnesses for ϕ is inconsistent. Hence, by Remark 13, if ϕ is not satisfiable then $\bigcap_{i=1}^{m+2} L(ONE_i) = \emptyset$. Finally, the construction of ONE_i expressions enables putting a X_j symbol between the consecutive $\#$ symbols, accordingly to true literals under any v assignment. ∎

Remark 15 *The INTERSECTION STAR ONE problem remains $NP-hard$ if it is considered over binary alphabet.*
Proof. We use the same argument as in the proof of Remark 8. ∎

It remains to show that the *INTERSECTION STAR ONE* problem is in *NP*. But first, we consider intersection-nonemptiness problem for simpler expressions – with at most single star occurrence.

Problem 16 *(SINGLE STAR)*
INSTANCE: An alphabet Σ, set of regular expressions

$$\{a_1(b_1)^*c_1, a_2(b_2)^*c_2, \ldots, a_n(b_n)^*c_n\}$$

where $a_i, b_i, c_i \in \Sigma^*$.
QUESTION: Is there $w \in \Sigma^+$, such that $w \in \bigcap_{i=1}^{n} L(a_i(b_i)^*c_i)$?

Note that for every number d there exists at most one word w with the length d, contained in $L(a_i(b_i)^*c_i)$.

Let $E_{i,j} = \{w|\ w \in L(a_i(b_i)^*c_i) \ \wedge \exists_u(u \in L(a_j(b_j)^*c_j \wedge |u| = |w|)\}$ and $INT_{i,j} = L(a_i(b_i)^*c_i) \cap L(a_j(b_j)^*c_j)$.

Lemma 17 *(i)   For the language $INT_{i,j}$ to have infinitely many words, it is necessary and suffices that the words respectively the first shortest and the second shortest word from $E_{i,j}$ are equal to the first shortest and the second shortest word from $E_{j,i}$.*
(ii)   The length of these words does not exceed $MAX_{i,j} + 2 \cdot LCM(|b_i|, |b_j|)$, where $MAX_{i,j} = \max\{|a_i| + |c_i|, |a_j| + |c_j|\}$.
(iii)   There are only three possibilities:
- *$INT_{i,j}$ is empty,*
- *$INT_{i,j}$ contain only one word,*
- *$INT_{i,j}$ is infinite.*

Proof. The length x of word $w \in INT_{i,j}$ satisfies

$$x \equiv |a_i| + |c_i| \pmod{b_i}$$
$$x \equiv |a_j| + |c_j| \pmod{b_j} \tag{5}$$

Let x' be the smallest solution of (5) provided that any solution exists. By Theorem 1, solutions of (5) have the following form:

$$\{x', x' + LCM(|b_i|, |b_j|), x' + 2 \cdot LCM(|b_i|, |b_j|), x' + 3 \cdot LCM(|b_i|, |b_j|), \ldots\}$$

Let $PE_{i,j} = \{k \mid k = |w| \wedge w \in E_{i,j}\}$. It easy to see that:
- $PE_{i,j}$, like $E_{i,j}$, is either empty or infinite,
- $PE_{i,j} = PE_{j,i}$,
- $PE_{i,j}$ is the set of some solutions of (5),
- if we take $y = \min(PE_{i,j})$ then

$$PE_{i,j} = \{y, y + LCM(|b_i|, |b_j|), y + 2 \cdot LCM(|b_i|, |b_j|),$$
$$y + 3 \cdot LCM(|b_i|, |b_j|), \ldots\}, \tag{6}$$

- $MAX_{i,j} \leq y \leq MAX_{i,j} + LCM(|b_i|, |b_j|)$ and hence *(ii)* is satisfied.

Assume that the condition in Lemma 17 is satisfied. Let u be the second short-est word of $INT_{i,j}$. We can choose such a word $v_{i,j} = z_1(b_i)^{(\frac{LCM(|b_i|,|b_j|)}{|b_i|}-1)}z_2$ ($z_2 z_1 = b_i$) that $u = u_1 v_{i,j} u_2$, and by pumping $v_{i,j}$ we can obtain all words from $INT_{i,j}$. Hence, $INT_{i,j}$ is infinite. Note that, for every $z \in PEL_{i,j}$ there exists $w \in INT_{i,j}$ of length z. Besides, there exists $v_{j,i} = z_1'(b_j)^{(\frac{LCM(|b_i|,|b_j|)}{|b_j|}-1)}z_2'$, such that $z_2' z_1' = b_j$ and $v_{j,i} = v_{i,j}$.

On the other hand, if condition in Lemma 17 is not satisfied, we have two cases:

- first – $E_{i,j}$ is empty, therefore $INT_{i,j}$ is empty,
- second – the first or the second shortest word from $E_{i,j}$ and $E_{j,i}$ differ on some symbol. In such a situation pumping maintains this difference. Hence, either $INT_{i,j}$ is empty or contain only one nonempty word. ∎

Problem 18 *(PAIR INTERSECTION)*
INSTANCE: The alphabet Σ, two regular expressions $\{a_i(b_i)^* c_i, a_j(b_j)^* c_j\}$, where $a_i, b_i, c_i \in \Sigma^*$.
QUESTION: Is there $w \in \Sigma^+$, such that $w \in L(a_i(b_i)^* c_i) \cap L(a_j(b_j)^* c_j)$?

Let $Pair(i,j)$ be a function which solves the *PAIR INTERSECTION* problem. Let us assume that $Pair(i,j)$ returns

- message 'empty', if $INT_{i,j}$ is empty,
- message 'infinite', if $INT_{i,j}$ is infinite,
- a word w, if w is only word in $INT_{i,j}$.

By Lemma 17 we can the write such a function that it will work in $O(LCM(|b_i|, |b_j|)^2 + MAX_{i,j})$ steps.

Below, we show the algorithm how to decide the *SINGLE STAR* problem. Let us call it *PAIR VERIFY*:

status :='infinite'; i:=1; j:=0;
while (*status* ='infinite' **and** $i \leq n$) **do**
    $i := i + 1$;
    while (*status* ='infinite' **and** $j < i$) **do**
        $j := j + 1$;
        status := $Pair(i,j)$;
if *status*='infinite' **then return** 'true';
if *status*='empty' **then return** 'false';
if *status* $\in \bigcap\limits_{i=1}^{n} L(a_i(b_i)^* c_i)$ **then return** 'true';
else return 'false';

The soundness of *PAIR VERIFY* follows easily from the Chinese Remain-der Theorem.

Given $B_\cup = \{|b_1|, \ldots, |b_n|\} \cup \{|a_1| + |c_1|, \ldots, |a_n| + |c_n|\}$, $C_{max} = \max B_\cup$ we obtain the following fact concerning complexity of *PAIR VERIFY*.

Fact 19 *PAIR VERIFY works in $O(C_{max}^4)$ steps. Therefore the SINGLE STAR problem is in polynomial time decidable.*

Now we show NP-reduction of the *INTERSECTION STAR ONE* to the *SINGLE STAR* problem. It will be called *DECOMP*. The NP-reduction and *PAIR VERIFY* (as polynomial time verifier) give us nondeterministic polynomial algorithm for the *INTERSECTION STAR ONE* problem.

Let *Stack* be a family of sets of regular expressions of the same form like on input the *SINGLE STAR* problem. Inp_r denotes the set of expressions in the rth step of reduction. Writing *loop* we mean a part of the expression of the form $(u)^*$. Functions $pref(v(u)^*)$ and $suff(v(u)^*)$ denotes respectively some prefix and some suffix of the expression $v(u)^*$. For example, given expression $S = v_1 v_2 (u_1 u_2)^*$ for any $v_1, v_2, u_1, u_2 \in \Sigma^*$, it may happen that $pref(S) = v_1$, or $pref(S) = v_1 v_2$, or $pref(S) = v_1 v_2 (u_1 u_2)^* u_1$, or $pref(S) = v_1 v_2 (u_1 u_2)^*$. Similarly for the $suff$ map.

DECOMP-r^{th} step: $Inp_r = \{a_{ir1}(b_{ir1})^* a_{ir1} \cdots a_{ire}(b_{ire})^* a_{ire+1} | 1 \le i \le n\}$

1. guess number k where $1 \le k \le n$, assuming that the leftmost loop of the kth expression 'ends' earliest;
2. for $1 \le l \le n$ and $l \ne k$ guess the partition of $a_{lr1}(b_{lr1})^*$:

$$a_{lr1}(b_{lr1})^* = pref(a_{lr1}(b_{lr1})^*) \cdot suff(a_{lr1}(b_{lr1})^*);$$

3.

$$Stack := Stack \cup \{\{pref(a_{1r1}(b_{1r1})^*), \ldots, pref(a_{(k-1)r1}(b_{(k-1)r1})^*),$$

$$a_{kr1}(b_{kr1})^*, pref(a_{(k+1)r1}(b_{(k+1)r1})^*), \ldots, pref(a_{nr1}(b_{nr1})^*))\}\};$$

4. set next input

$$Inp_{r+1} := \{suff(a_{ir1}(b_{ir1})^*)a_{ir1} \cdots a_{ire}(b_{ire})^* a_{ire+1} | 1 \le i \le n \text{ and } i \ne k\} \cup$$

$$\cup \{a_{kr2}(b_{kr2})^* \cdots a_{kre}(b_{kre})^* a_{kre+1}\}$$

Fact 20 *DECOMP works in $\sum_{i=1}^{n} ire +1$ steps and after the last step of decomposition $|Stack| = \sum_{i=1}^{n} ire +1$.*

In order to decide *INTERSECTION STAR ONE*, after nondeterministic decomposition, *PAIR VERIFY* is applied to every member of *Stack*.

Lemma 21 *The INTERSECTION STAR ONE problem is in NP.*

Problem 22 *(STAR HEIGHT ONE)*
INSTANCE: An alphabet Σ, regular expressions $SUM_1, \ldots, SUM_p$ over Σ. The $SUM_1, \ldots, SUM_p$ have star height at most one and they are based on the

*operators · (concatenation) and * (Kleene-closure).*
QUESTION: Is there $w \in \Sigma^+$, such that $w \in \bigcap_{i=1}^{p} L(SUM_i)$?

Note that every regular expression of star height at most one which does not contain any occurrence of binary + operator has a form such as *(4)*. Thus, the *INTERSECTION STAR ONE* problem is equivalent to the *STAR HEIGHT ONE* problem . Therefore,

Theorem 23 *it The STAR HEIGHT ONE problem is $NP-$ complete.*

Acknowledgements. *Leszek Pacholski, Kasia Paluch, especially Krzysiek Loryś.*

References

[BM94]   J.-C. Birget, S. Margolis, J. Meakin, P. Weil, *PSPACE-completness of Certain Algorithm Problems On The Subgroups of Free Groups*, Proc. LNCS. Springer-Verlag, ICALP'94.

[BM00]   J.-C. Birget, S. Margolis, J. Meakin, P. Weil, *PSPACE-complete Problems for subgroups of free groups and inverse automata*, Theoretical Computer Science 242(1-2), 247-281, 2000.

[CH91]   S. Cho, D. Huynh, *Finite-automaton aperiodicity is PSPACE-complete*, Theoretical Computer Science 88 (1991) 411-424.

[DR99]   V. Diekert, J.M. Robson, *On Quadric Word Equations*, STACS'99, Proc. LNCS. Springer-Verlag, Berlin, 1999.

[DS66]   F. Dejean and M.P. Schützenberger, *On Question of Eggan*, Information and Control 9, 1966.

[E63]   L.C. Eggan, *Transition Graphs and Star Height of Regular Events*, Michigan Math. J. 10, 1963.

[GV97]   Y. Gurevich, A. Voronkov, *Monadic Simultaneous Rigid E-Unification and Related Problems* Proc. LNCS. Springer-Verlag, ICALP'97.

[HU]   J. Hopcroft, J. Ullman, *Introduction to Automata Theory, Formal Languages and Computation* Addison-Wesley. 1979.

[JR91]   T. Jiang, B. Ravikumar A Note On The Space Complexity of Some Decision Problems for Finite Automata, Information Processing Letters, 40(1991) 25-31.

[K77]   D. Kozen, *Lower Bounds For Natural Proof Systems*, Proc. 18-th Symp. Fundations of Comp. Sci. 1977.

[LR92]   K.-J. Lange, P. Rossmanith, *The Emptiness Problem for Intersection of Regular Languages*, Conference on Mathematical Foundations of Computer Science, Proc. LNCS. Springer-Verlag, 1992.

[P]   C.H. Papadimitriou, *Computational Complexity*, Adison-Wesley Publ. 1994.

[Sch]   K.U. Schultz, *Makanin's Algorithm for Word Equations: two improvements and a generalizations*, IWWERT'90, LNCS 572, 85-150, 1992.

[Yu]   S. Yu, Regular Language in G.Rozenber A.Salomaa Eds. *Handbook of Formal Language*, vol.1, Springer-Verlag Berlin Heidelberg, 1997.

On the Construction of Reversible Automata for Reversible Languages

Sylvain Lombardy

Laboratoire Traitement et Communication de l'Information
Ecole Nationale Supérieure des Télécommunications
46, rue Barrault   75634 Paris Cedex 13, France
lombardy@enst.fr

Abstract. Reversible languages occur in many different domains. Although the decision for the membership of reversible languages was solved in 1992 by Pin, an effective construction of a reversible automaton for a reversible language was still unknown. We give in this paper a method to compute a reversible automaton from the minimal automaton of a reversible language. With this intention, we use the universal automaton of the language that can be obtained from the minimal automaton and that contains an equivalent automaton which is quasi-reversible. This quasi-reversible automaton has nearly the same properties as a reversible one and can easily be turned into a reversible automaton.

Keywords: Finite automata, reversible languages, reversible automata, universal automata

Introduction

Reversible languages are a class of rational languages that stands at the junction of several domains. P. Silva [13] has considered them in the study of inversive semigroups. J.-E. Pin [10] and P.-C. Héam [6] have given some topological properties of reversible languages. They constitute also a positive variety of rational languages and are in correpondence with a variety of ordered monoids: $\mathbf{E}_{\mathbf{com}}^{-}$ [11]. Among rational languages, they are a natural generalization to the notion of group languages. It is the reason why, in order to extend results proved on group languages, studying this class can prove to be wise. C. Nicaud [9] has studied average complexity of some operations on reversible automata. S. Lombardy and J. Sakarovitch [7] have given a new effective proof to compute their star height.

J.-E. Pin has given some characterizations of reversible languages in [10]. He gives also an algorithm to decide whether a rational language is reversible. Unfortunately, his paper does not provide any effective method to build a reversible automaton.

We present here such a construction from the minimal automaton of a reversible language. We use a canonical automaton, called *universal automaton*, attached to every rational language. On the one hand, we give an effective new

P. Widmayer et al. (Eds.): ICALP 2002, LNCS 2380, pp. 170–182, 2002.

method to compute this automaton from the minimal automaton of the language; on the other hand, we prove that this automaton contains a *quasi-reversible* automaton that accepts the language. As it is very easy to turn quasi-reversible automata into reversible ones, this provides a construction for reversible automata.

In a first part, we give some basic definitions about automata, that will be used in the course of this paper. In the second part, we define reversible languages. We recall the characterization given by J.-E. Pin[10]. Then, we introduce quasi-reversible automata, that accept reversible languages, are smaller, and give easily reversible automata.

In a third part, we present the universal automaton of a rational language, that is based on an idea of J. H. Conway [4]. This automaton has been studied by A. Arnold, A. Dicky, and M. Nivat [1], O. Matz and A. Potthoff [8] and J. Sakarovitch [12]. The main property of this finite automaton is that there exists a morphism from any automaton that accepts the same language, into the universal automaton. We give an effective construction of it.

The fourth part is devoted to the study of universal automata of reversible languages. We show that their strongly connected components are reversible. This result is the key of the theorem of the fifth part which claims that the universal automaton of a reversible language contains a quasi-reversible automaton that accepts this language.

1   Definitions

We denote by A^* the free monoid generated by a set A. Elements of A^* are words, the identity of this monoid is the empty word 1_{A^*}.

Definition 1. *We denote a (finite) automaton by a 5-tuple $\langle Q, A, E, I, T \rangle$, where Q is a finite set of states, A is a finite set of letters, E, the set of transitions, is a subset of $Q \times A \times Q$, and I (resp. T), the set of initial states (resp. terminal states), is a subset of Q.*

Definition 2. *Let $\mathcal{A}$ be an automaton and p a state of $\mathcal{A}$. The **past** of p in $\mathcal{A}$ is the set of words that label a path from an initial state of $\mathcal{A}$ to p. The **future** of p in $\mathcal{A}$ is the set of words that label a path from p to a terminal state of $\mathcal{A}$. We respectively denote the past and the future of p in $\mathcal{A}$ by $\mathsf{Past}\mathcal{A}p$ and $\mathsf{Fut}\mathcal{A}p$.*

Definition 3. *Let $\mathcal{A} = \langle Q, A, E, I, T \rangle$ and $\mathcal{B} = \langle R, A, F, J, U \rangle$ be two automata. A mapping μ from Q into R is a **morphism** of automata if and only if:*

$$p \in I \Rightarrow p\mu \in J, \qquad p \in T \Rightarrow p\mu \in U,$$
$$\text{and} \quad (p, a, q) \in E \Rightarrow (p\mu, a, q\mu) \in F. \tag{1}$$

Proposition 1. *Let μ be a morphism from an automaton $\mathcal{A}$ into an automaton $\mathcal{B}$. Then, for every state p of $\mathcal{A}$,*

$$\text{Past}\mathcal{A}p \subseteq \text{Past}\mathcal{B}p\mu, \quad \text{Fut}\mathcal{A}p \subseteq \text{Fut}\mathcal{B}p\mu. \tag{2}$$

The proof is by induction on the length of words.   □

Definition 4. *Let $\mathcal{A} = \langle Q, A, E, I, T \rangle$ be an automaton. For every state p in Q, for every letter a, we denote $p \cdot a$ the set $\{q \in Q \mid (p, a, q) \in E\}$. With the convention $p \cdot 1_{A^*} = \{p\}$, we can extend this definition to subsets of Q and to words:*

$$\forall u = u'a \in A^+, p \cdot u = (p \cdot u') \cdot a, \quad \forall P \subseteq Q, P \cdot u = \bigcup_{p \in P} p \cdot u. \tag{3}$$

Symetrically, for every subset P of Q, for every word u,

$$u \cdot P = \{p \in Q \mid p \cdot u \cap P \neq \emptyset\}. \tag{4}$$

There is a path labelled by a word u between two states p and q if and only if q belongs to $p \cdot u$.

Definition 5. *A **strongly connected component** (SCC for short) of an automaton is a maximal subautomaton such that there is a path between every pair of states.*

2   Reversible Languages: Definition and Realization

Definition 6. *An automaton $\mathcal{A} = \langle Q, A, E, I, T \rangle$ is **reversible** if, for every state p in Q, for every letter a in A there exists at most one transition in E that comes from p (resp. goes to p) with label a.*
A rational language is reversible if there exists a reversible automaton that accepts it.

Remark 1. A reversible automaton may have several initial or terminal states. As a consequence, the minimal automaton of a reversible language may be not reversible.

Theorem 1. *[10] There is a polynomial time algorithm for testing whether the language accepted by a minimal automaton can be accepted by a reversible automaton.*

In this paper, J.-E. Pin gives a construction for reversible automata. This construction is based on a decomposition of words with respect to a lemma due to Ash [2].

We give now a new representation for reversible languages.

Definition 7. *An automaton* $\mathcal{A} = \langle Q, A, E, I, T \rangle$ *is **quasi-reversible** if, in case two transitions of E have the same label and arrive in or come from the same state, none of them belongs to any SCC.*

Example 1. Among the three automata of Figure 1, the first automaton is reversible. The second one is quasi-reversible: both transitions that come from p with label a do not belong to any SCC. The last one is not quasi-reversible: one of both transitions that come from q with label b does belong to an SCC (it is a loop).

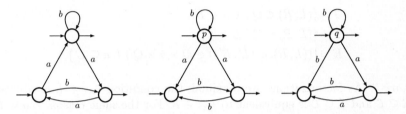

Fig. 1. A reversible, a quasi-reversible and a non-reversible automata.

The following proposition proves that languages that are accepted by quasi-reversible automata are reversible.

Proposition 2. *If $\mathcal{A}$ is a quasi-reversible automaton, one can build a reversible automaton that is equivalent to $\mathcal{A}$.*

Proof. We call sensitive pair a pair of transitions that is in contradiction with reversibility. The proof is by induction on the number of sensitive pair in $\mathcal{A}$. If there is no such a pair, then $\mathcal{A}$ is reversible. If there are n sensitive pairs, let $\{e_1, e_2\}$ be one of them. As neither e_1 nor e_2 belongs to any SCC, there is no path that contains these two transitions. One creates two copies $\mathcal{A}_1$ and $\mathcal{A}_2$ of $\mathcal{A}$ and removes e_1 from $\mathcal{A}_1$ and e_2 from $\mathcal{A}_2$. The number of sensitive pair is smaller in each of this two quasi-reversible automata. Thus, by induction, they can be turned into reversible automata.   □

Remark 2. One also could transform the quasi-reversible automaton into a set of reversible "string-like"-automata, as we shall see later.

Quasi-reversible automaton can be much smaller than reversible ones. P.-C. Héam [5] has given the following example: the language $(aa + ab + bb)^{2n}$ is finite and is therefore reversible. The minimal automaton of this language is quasi-reversible and has $6n+1$ states, whereas the smallest reversible automaton that accepts this language has at least $(3\sqrt{2}/4)^n$ states.

3   Universal Automaton

The universal automaton of a language has been defined by Conway [4] as a "factor matrix". A complete study of properties of this automaton can be found in [12]. We recall here the definition and some basic properties of this automaton. Lastly, we introduce a new construction for the universal automaton.

Definition 8. *Let $\mathcal{L}$ be a rational language of A^*. A **factorization** of $\mathcal{L}$ in A^* is a maximal couple (with respect to the inclusion) of languages (L,R) such that $L.R \subseteq \mathcal{L}$.*
*The **universal automaton** of $\mathcal{L}$ is $\mathcal{U}_\mathcal{L} = \langle Q, A, E, I, T \rangle$, where Q is the set of factorizations of $\mathcal{L}$, and*

$$
\begin{aligned}
I &= \{(L,R) \in Q \mid 1_{A^*} \in L\}, \\
T &= \{(L,R) \in Q \mid L \subseteq \mathcal{L}\}, \\
E &= \{((L,R), a, (L',R')) \in Q \times a \times Q \mid L.a \subseteq L'\}.
\end{aligned}
\tag{5}
$$

Remark 3. As factorizations are maximal, the condition $1_{A^*} \in L$ is equivalent to $R \subseteq \mathcal{L}$ and $L \subseteq \mathcal{L}$ is equivalent to $1_{A^*} \in R$. For the same reason, $L.a \subseteq L'$ is equivalent to $a.R' \subseteq R$ and to $L.a.R' \subseteq \mathcal{L}$.

Proposition 3. *The universal automaton of a rational language is a finite automaton.*

Factorizations of a rational language are recognized by the syntactic monoid of this language. Thus there is a finite number of factorizations.   □

Example 2. Let $\mathcal{L}_1 = b^*ab((a+b)b^* + 1) + b^*(a+1)$. Figure 2 shows the minimal automaton of this language.

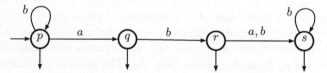

Fig. 2. Minimal automaton of $\mathcal{L}_1$.

By computing the syntactic monoid, we can get the factorizations of $\mathcal{L}_1$:

$$
\begin{aligned}
&(b^*, \mathcal{L}_1), \qquad (b^*(a+1), (ba+1)b^*) \\
&(b^*(ab+1), (a+1)b^*), \qquad (\mathcal{L}_1, b^*).
\end{aligned}
\tag{6}
$$

Hence, the universal automaton of $\mathcal{L}_1$ has four states. They are all initial, because 1_{A^*} belongs to the left factor of every factorization and they are all terminal because the left factor of every factorization is a subset of $\mathcal{L}_1$. We show at the end of this section that we can build the universal automaton without computing these factorizations.

Proposition 4. *Let* (L, R) *be a state of a universal automaton* $\mathcal{U}_{\mathcal{L}}$. *Then:*

$$\mathsf{Past}\mathcal{U}_{\mathcal{L}}(L, R) = L \quad \mathsf{Fut}\mathcal{U}_{\mathcal{L}}(L, R) = R. \tag{7}$$

The proof is by induction on the length of words. □

Corollary 1. *The universal automaton of a rational language* $\mathcal{L}$ *accepts* $\mathcal{L}$.

Proof. For every terminal state (L, R) of the universal automaton, $\mathsf{Past}\mathcal{U}_{\mathcal{L}}(L, R)$ is a subset of $\mathcal{L}$. Thus, the language accepted by $\mathcal{U}_{\mathcal{L}}$ is a subset of $\mathcal{L}$. Let $L = \{u \in A^* \mid u.\mathcal{L} \subseteq \mathcal{L}\}$. Then, $(L, \mathcal{L})$ is a factorization which is an initial state of $\mathcal{U}_{\mathcal{L}}$ and is future is $\mathcal{L}$. Hence, $\mathcal{U}_{\mathcal{L}}$ accepts $\mathcal{L}$. □

We are here especially interested in the following property of the universal automaton.

Proposition 5. *Let* $\mathcal{L}$ *be a rational language, and* $\mathcal{A}$ *be a trim automaton that accepts* $\mathcal{L}$. *Then there exists a morphism from* $\mathcal{A}$ *into the universal automaton of* $\mathcal{L}$.

We build a mapping μ from states of $\mathcal{A}$ into factorizations of $\mathcal{L}$: $p\mu = (L_p, R_p)$, with:

$$R_p = \{v \in A^* \mid \mathsf{Past}\mathcal{A}p.v \subseteq \mathcal{L}\}, \qquad L_p = \{u \in A^* \mid u.R_p \subseteq \mathcal{L}\}. \tag{8}$$

We check then that μ is a morphism between automata. □

The following proposition gives an effective method to build the universal automaton of a rational language from its minimal automaton.

Proposition 6. *Let* $\mathcal{A} = \langle Q, A, E, \{i\}, T \rangle$ *be the minimal automaton of* $\mathcal{L}$ *and* P *be the set of states of the codeterminized automaton of* $\mathcal{A}$:

$$P = \{X \subseteq Q \mid \exists u \in A^*, X = u \cdot T\}. \tag{9}$$

Let $P_\cap$ *be the closure of* P *under intersection (without the empty set):*

$$X, Y \in P_\cap, X \cap Y \neq \emptyset \Longrightarrow X \cap Y \in P_\cap. \tag{10}$$

Then, the universal automaton $\mathcal{U}_{\mathcal{L}}$ *is isomorphic to* $\langle P_\cap, A, F, J, U \rangle$, *with:*

$$
\begin{aligned}
J &= \{X \mid i \in X\} \\
U &= \{X \mid X \subseteq T\} \\
F &= \{(X, a, Y) \mid X \cdot a \subseteq Y \text{ and} \forall p \in X, p \cdot a \neq \emptyset\}.
\end{aligned} \tag{11}
$$

Proof. (*sketch*) Let φ be the mapping from $P_\cap$ into $A^* \times A^*$ defined by $X\varphi = (L_X, R_X)$, with:

$$L_X = \bigcup_{p \in X} \mathsf{Past}\mathcal{A}p, \qquad R_X = \bigcap_{p \in X} \mathsf{Fut}\mathcal{A}p. \tag{12}$$

For every X *in* $P_\cap$, $X = \bigcap_{u \in R_X} u \cdot T$. *Hence,* (L_X, R_X) *is a factorization, and, for every factorization* (L, R), *the set* $X = \bigcap_{u \in R} u \cdot T$ *belongs to* $P_\cap$ *and* $X\varphi = (L, R)$. *Thus* φ *is a one-to-one mapping between* $P_\cap$ *and factorizations of* $\mathcal{L}$. *Then, we check that* φ *is an isomorphism of automata.* □

Remark 4. The minimality of $\mathcal{A}$ is not required in this construction. We only assume that it is deterministic. We can show that the cardinality of $P_\cap$ is not affected by the minimality of $\mathcal{A}$.

Example 3. The minimal automaton of $\mathcal{L}_1$ is shown on Figure 2. Figure 3 shows its codeterminized automaton. The set $\{\{p\}, \{p,q\}, \{p,r\}, \{p,q,r,s\}\}$ is closed

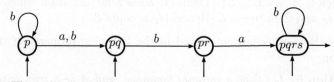

Fig. 3. The codeterminized automaton of the minimal automaton of $\mathcal{L}_1$.

under intersection. Hence, the automaton defined Proposition 6 (and isomorphic to the universal automaton) has the following transition table:

	$\{p\}$	$\{p,q\}$	$\{p,r\}$	$\{p,q,r,s\}$
$\{p\}$	b	a,b	b	a,b
$\{p,q\}$			b	b
$\{p,r\}$				a,b
$\{p,q,r,s\}$				b

(13)

All the states are initial and terminal. The universal automaton of $\mathcal{L}_1$ is drawn on Figure 4.

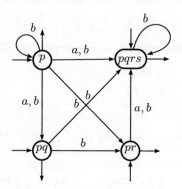

Fig. 4. Universal automaton of $\mathcal{L}_1$.

4   Strongly Connected Components of the Universal Automaton of a Reversible Language

We show in this part that every SCC of the universal automaton of a reversible language is reversible. This result is an important step to prove that there is a quasi-reversible automaton that accepts the language in the universal automaton. We define first the subset expansion of an automaton. We show that the universal automaton is one of its subautomata and that the SCCs of the subset expansion of a reversible automaton are reversible.

Definition 9. *Let $\mathcal{A} = \langle Q, A, E, I, T \rangle$ be an automaton that accepts $\mathcal{L}$. Let S be the set of antichains of $\mathcal{P}(Q)$ (the set of subsets of Q, partially ordered by inclusion). The **subset expansion** of $\mathcal{A}$ is the automaton $V_{\mathcal{A}} = \langle S, A, F, J, U \rangle$, with:*

$$\begin{aligned} J =& \{X \in S \mid \exists Y \in X, Y \subseteq I\}, \\ U =& \{X \in S \mid \forall Y \in X, Y \cap T \neq \emptyset\}, \\ F =& \{(X, a, X') \mid \forall Y \in X, \exists Y' \in X', Y' \subseteq Y \cdot a\}. \end{aligned} \tag{14}$$

Remark 5. If $\mathcal{A}$ is a minimal automaton, the subset expansion of $\mathcal{A}$ is exactly the *subset automaton of order* 0 introduced by R. S. Cohen and J.A. Brzozoski [3]. In this case, the construction is the same as Proposition 6, except that the set of states is isomorphic to $\mathcal{P}(Q) \setminus \{\emptyset\}$ instead of being isomorphic to $P_{\cap}$ (which is smaller) defined in the proposition.

Lemma 1. *The subset expansion of an automaton $\mathcal{A}$ accepts the same language as $\mathcal{A}$.*

Proof. Let I (*resp.* T) be the set of initial (*resp.* terminal) states of $\mathcal{A}$. We prove by induction on the length of words, that there is a path labelled by u between X and X' if and only if:

$$\forall Y \in X, \exists Y' \in X', Y' \subseteq Y \cdot u \tag{15}$$

If a word u is accepted by $V_{\mathcal{A}}$, there exists a path labelled by u between an initial state X and a terminal state X'. There exists Y in X such that $Y \subseteq I$. Hence, there exists Y' in X' such that $Y' \subseteq Y \cdot u \subseteq I \cdot u$. As X' is terminal, $Y' \cap T \neq \emptyset$ and $I \cdot u \cap T \neq \emptyset$. Therefore, the word u is accepted by $\mathcal{A}$.
Conversely, if u is accepted by $\mathcal{A}$, $I \cdot u \cap T \neq \emptyset$, hence, there is a path labelled by u between the initial state $\{I\}$ and the terminal state $\{\{p\} \mid p \in T\}$.   □

Proposition 7. *Let $\mathcal{A}$ be an automaton that accepts $\mathcal{L}$. The universal automaton of $\mathcal{L}$ is a subautomaton of the subset expansion of $\mathcal{A}$.*

We build a mapping from states of $\mathcal{U}_\mathcal{L}$ into states of $\mathcal{V}_\mathcal{A}$:

$$(L, R) \longmapsto \min\{I \cdot u \mid u \in L\}, \tag{16}$$

where I is the set of initial states of $\mathcal{A}$. We show then that this is an injective morphism of automata.   □

Proposition 8. *Let $\mathcal{A}$ be a reversible automaton. The SCCs of the subset expansion of $\mathcal{A}$ are reversible.*

Proof. Let $\mathcal{V}_\mathcal{A}$ be the subset expansion of $\mathcal{A}$. We show that in every SCC of $\mathcal{V}_\mathcal{A}$, every word induces a one-to-one mapping between sets that characterize states. Let p and q be two states of the same SCC in $\mathcal{V}_\mathcal{A}$. p and q are both antichains of subsets of $\mathcal{P}(Q)$, where Q is the set of states of $\mathcal{A}$. There exist two words u and v that respectively label paths from p to q and from q to p.

We show by induction on the cardinal k of the elements of p (*resp. q*) that u (*resp. v*) induces a bijection from p onto q (*resp.* from q onto p) between elements of the same cardinal.

There is no element of cardinal zero, thus the base of the induction is trivial. Let Y be an element of p of cardinal k.

i) As $\mathcal{A}$ has a deterministic transition system, the cardinal of $Y \cdot u$ is smaller than or equal to k. If it is smaller, by induction hypothesis, there exists Y' in p with the same cardinal as $Y \cdot u$ such that $Y' \cdot u = Y \cdot u$. As $\mathcal{A}$ is codeterministic, $Y' \subset Y$, which is in contradiction with the fact that p is an antichain. Therefore, $Y \cdot u$ has the same cardinal as Y and u is a bijection from elements of Y onto elements of $Y \cdot u$.

ii) If $Y \cdot u$ does not belong to q, there is a smaller element Z in q in bijection with an element Y' in p such that $Y' \cdot u = Z \subset Y \cdot u$. Once more, it implies $Y' \subset Y$, which is impossible, because p is an antichain.

iii) If $Y \cdot u = Y' \cdot u$, as Y is in bijection with $Y \cdot u$, it means that $Y \subseteq Y'$. If Y' belongs to p, $Y = Y'$. Thus u induces an injective function from elements of p of cardinal k into elements of q of same cardinal. As v induces a similar injection from q into p, these functions are one-to-one.

Therefore, from a state p, given a word u, there is at most one state in the same SCC, that can be reached by a path labelled by u. Thus, SCCs are deterministic and, symetrically, codeterministic.   □

Corollary 2. *The SCCs of the universal automaton of a reversible language are reversible.*

5   Universal Automaton of a Reversible Language

We prove in this part that the universal automaton of a reversible language contains a quasi-reversible subautomaton that accepts the language. For this purpose, we show that, for every reversible language, there exists a reversible automaton such that its image, in the universal automaton of the language, is quasi-reversible. This reversible automaton has a particular form. It is a union of "string-like"-automata:

Definition 10. *An automaton is a **string-like automaton** (SL-automaton for short) if*
    i) it has one initial and one terminal state,
    ii) there is at most one transition that comes from (resp. goes to) every SCC,
    iii) it is connected,
    iv) it is trim.

Fig. 5. Shape of an SL-automaton.

The general shape of an SL-automaton is presented in Figure 5, where every ellipse is an SCC (that may be a single state).

Proposition 9. *Every reversible automaton is equivalent to a finite union of reversible SL-automata.*

The construction is illustrated on Figure 6. If the automaton is reversible (or merely quasi-reversible) every SL-automaton is reversible. □

Definition 11. *A union of SL-automata is minimal if one can not erase any SL-automaton without changing the language accepted by the union.*

In other words, every SL-automaton of a minimal union accepts a word which is not accepted by the other components.

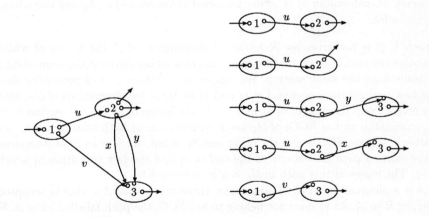

Fig. 6. From an automaton to a set of SL-automata.

Proposition 10. *Let $\mathcal{L}$ be a reversible language and $\mathcal{A}$ an automaton that accepts $\mathcal{L}$ and that is a minimal finite union of reversible SL-automata. Let φ be a morphism from $\mathcal{A}$ into $\mathcal{U}_{\mathcal{L}}$. Then, the image of every SCC of $\mathcal{A}$ is an SCC of $\mathcal{U}_{\mathcal{L}}$.*

Proof. We show that, if there is an SCC $\mathcal{S}$ for which it is not true, every word accepted by the SL-automaton $\mathcal{R}$, that contains $\mathcal{S}$, is accepted by another SL-automaton of $\mathcal{A}$, which is in contradiction with the minimality of the set of SL-automata.

Let φ be a morphism from $\mathcal{A}$ into $\mathcal{U}_{\mathcal{L}}$. Let w be a word accepted only by $\mathcal{R}$ in $\mathcal{A}$. As $\mathcal{R}$ is string-like, in $\mathcal{R}$, every path crosses every SCC. Thus, there exists a state r in $\mathcal{S}$ that belongs to the path labelled by w. One can write $w = u.v$, where u (*resp.* v) is in Past$\mathcal{R}r$ (*resp.* in Fut$\mathcal{R}r$).

As $\mathcal{S}\varphi$ is not an SCC, there exists a transition e in the SCC that contains $\mathcal{S}\varphi$ but that does not belong to $\mathcal{S}\varphi$. Let x be a word labelling a loop around $r\varphi$ that passes through e. As u (*resp.* v) is in the past (*resp.* the future) of r and as $\mathcal{U}_{\mathcal{L}}$ recognizes $\mathcal{L}$, for every k in $\mathbb{N}$, the word $u.x^k.v$ is in $\mathcal{L}$. As $\mathcal{A}$ is reversible, there exists an integer n such that, for every state of $\mathcal{A}$, either x^n labels a loop around the state or there is no path starting in this state and labelled by x^n. The word $u.x^n.v$ is in $\mathcal{L}$. If it was accepted by $\mathcal{R}$, there would be a loop labelled by x^n around r, and, as $\mathcal{S}\varphi$ is deterministic, the transition e would belong to the image of this loop. Thus, this word is not accepted by $\mathcal{R}$ and there exists another SL-automaton of $\mathcal{A}$ that accepts it, and, as x^n labels a loop, that accepts $u.v$ too. Which is in contradiction with the assumption.

Therefore, the image of every SCC of every SL-automaton of $\mathcal{A}$ is an SCC. □

Proposition 11. *Let $\mathcal{L}$ be a reversible language and $\mathcal{A}$ an automaton that accepts $\mathcal{L}$ and that is a minimal union of reversible SL-automata. Then the image of every SL-automaton of $\mathcal{A}$ in the universal automaton of $\mathcal{L}$, by any morphism, is reversible.*

Proof. If it is not true, let $\mathcal{R}$ be an SL-automaton of $\mathcal{A}$ the image of which contains two transitions with the same label that either arrive at the same state p or come from the same state p. We can assume without loss of generality that the first configuration occurs. Let t_1 and t_2 be these two transitions of $\mathcal{U}_{\mathcal{L}}$, that are respectively the images of s_1 and s_2. As the image of an SL-automaton is an SL-automaton and as SCCs of $\mathcal{U}_{\mathcal{L}}$ are reversible, one of both transitions, t_1, for instance, is in an SCC $\mathcal{S}$ and the other one, t_2, is not. As $\mathcal{R}$ is an SL-automaton, there exists a path that starts at the end of s_2 and the last transition of which is s_1. The image of this path in $\mathcal{U}_{\mathcal{L}}$ is a loop labeled by x.

If $\mathcal{A}$ is a minimal union of SL-automata, there exists a word w that is accepted only by $\mathcal{R}$ in $\mathcal{A}$. As s_2 does not belong to any SCC, the path labelled by w in $\mathcal{R}$ passes through s_2. Let u be the label of the part of this path which ends with s_2 and v the label of the second part. Words u and v respectively belong to the past and the future of p in $\mathcal{U}_{\mathcal{L}}$. Thus $u.x^k.v$ belongs to $\mathcal{L}$ for every k, and, as in the proof of Proposition 10, with the same notations, we prove that there exists an SL-automaton of $\mathcal{A}$ that accepts $u.x^n.v$ (and that can therefore not be $\mathcal{R}$) and $u.v = w$, which is in contradiction with the assumption. □

Definition 12. *Let $\mathcal{U}_\mathcal{L}$ be the universal automaton of a reversible language $\mathcal{L}$. The maximum quasi-reversible subautomaton of $\mathcal{U}_\mathcal{L}$ is the largest subautomaton of $\mathcal{U}_\mathcal{L}$*
  i) which is quasi-reversible,
  ii) which has the same SCCs as $\mathcal{U}_\mathcal{L}$.

This automaton exists because the SCCs of $\mathcal{U}_\mathcal{L}$ are reversible. It is obtained from $\mathcal{U}_\mathcal{L}$ by deleting transitions that do not belong to any SCC and that are in contradiction with the assumption of quasi-reversibility.

Theorem 2. *The maximum quasi-reversible subautomaton of the universal automaton of a reversible language accepts this language.*

Proof. Let $\mathcal{A}$ be a minimal finite union of reversible SL-automata that accepts $\mathcal{L}$. The image of every SL-automaton in $\mathcal{U}_\mathcal{L}$ by any morphism is a reversible automaton which covers SCCs of $\mathcal{U}_\mathcal{L}$ that are intersected. Thus, the image of $\mathcal{A}$ is a subautomaton of the maximum quasi-reversible subautomaton of $\mathcal{U}_\mathcal{L}$. Therefore, this one accepts the language $\mathcal{L}$. □

Example 4. The language $\mathcal{L}_1$ is reversible. (This can be check using its syntactic monoid [10]). Figure 7 shows the maximum quasi-reversible subautomaton of its universal automaton.

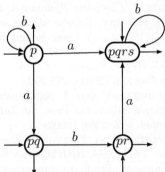

Fig. 7. A quasi-reversible automaton for $\mathcal{L}_1$.

6   Complexity of the Algorithm

The size of the universal automaton may be exponential. Thus, the complexity of the algorithm that is described in this paper will be at least exponential.

The first step is the construction of the universal automaton. The computation of the codeterminized automaton and the closure of its states under intersection is polynomial in the size of the output (that may be exponential in the size of the input).

The second step is the computation of the maximal quasi-reversible subautomaton of the universal automaton. This requires the computation of the SCCs, that is linear in the number of transitions. The number of transitions is quadratic in the number of states, but the square of an exponential number remains exponential. Then, the criterium of quasi-reversibility is local and the construction is polynomial in the size of the universal automaton.

Therefore, this algorithm is exponential in the size of the minimal automaton, but it is polynomial in the size of the output.

The transformation of a quasi-reversible automaton into a reversible one may be exponential, that makes the algorithm doubly exponential.

Acknowledgement. The author thanks Jacques Sakarovitch for his pertinent advices. He is grateful to Pierre-Cyrille Héam for invaluable conversations on this topic and others.

References

1. A. ARNOLD, A. DICKY, AND M. NIVAT, A note about minimal non-deterministic automata. *Bull. of E.A.T.C.S.* **47** (1992), 166–169.
2. C. J. ASH, Finite semigroups with commuting idempotents, J. Austral. Math. Soc. (Series A) **43** (1987), 81–90.
3. R. S. COHEN AND J. A. BRZOZOWSKI, General properties of star height of regular events, *J. Computer System Sci.* **4** (1970), 260–280.
4. J. H. CONWAY, *Regular algebra and finite machines*, Chapman and Hall, 1971.
5. P.-C. HÉAM, A lower bound for reversible automata, *Theoret. Informatics Appl.* **34** (2000), 331–341.
6. P.-C. HÉAM, Some topological properties of rational sets, *J. of Automata, Lang. and Comb.* **6** (2001), 275-290.
7. S. LOMBARDY AND J. SAKAROVITCH, Star height of reversible languages and universal automata, *Proc. 5th Latin Conf., (S. Rajsbaum, Ed.), Lecture Notes in Comput. Sci.* **2286** (2002).
8. O. MATZ AND A. POTTHOFF, Computing small nondeterministic finite automata. *Proc. TACAS'95, BRICS Notes Series* (1995), 74–88.
9. C. NICAUD, Étude du comportement en moyenne des automates finis et des langages rationnels, *Thèse de doctorat*, Université Paris 7, 2000.
10. J.-E. PIN, On reversible automata, *Proc. 1st LATIN Conf., (I. Simon, Ed.), Lecture Notes in Comput. Sci.* **583** (1992), 401–416.
11. J.-E. PIN, A variety theorem without complementation, *Russian mathematics* **39** (1995), 80–90.
12. J. SAKAROVITCH, *Éléments de théorie des automates*, Vuibert, to appear.
13. P.V. SILVA, On free inverse monoid languages, *Theoret. Informatics and Appl.* **30** (1996), 349–378.

Priority Queues, Pairing, and Adaptive Sorting

Amr Elmasry

Computer Science Department, Rutgers University, New Brunswick, NJ 08903
elmasry@cs.rutgers.edu

Abstract. We introduce Binomialsort, an adaptive sorting algorithm that is optimal with respect to the number of inversions. The number of comparisons performed by Binomialsort, on an input sequence of length n that has I inversions, is at most $2n \log \frac{I}{n} + O(n)$ [1]. The bound on the number of comparisons is further reduced to $1.89n \log \frac{I}{n} + O(n)$ by using a new structure, which we call trinomial queues. The fact that the algorithm is simple and relies on fairly simple structures makes it a good candidate in practice.

1 Introduction

In many applications the lists to be sorted do not consist of randomly distributed elements, but are already partially sorted. An adaptive sorting algorithm benefits from the existing presortedness in the input sequence. Mannila [10] formalized the concept of presortedness. He studied several measures of presortedness and introduced the concept of optimality with respect to these measures. One of the main measures of presortedness is the number of inversions in the input sequence [6]. The number of inversions is the number of pairs in the wrong order. More precisely, for an input sequence X of length n, the number of inversions is defined

$$Inv(X) = |\{(i,j) \mid 1 \leq i < j \leq n \text{ and } x_i > x_j\}|.$$

We assume that x_i is to the left of x_j, whenever i is less than j. For an adaptive sorting algorithm to be optimal with respect to the number of inversions, its running time should be in $O(n \log \frac{Inv(X)}{n} + n)$ [5].

Finger trees [5,2,12] was the first data structure that utilizes the existing presortedness in the input sequence. Using a finger tree, the time required to sort a sequence X of length n is $O(n \log \frac{Inv(X)}{n} + n)$. The fact that the underlying data structure is pretty complicated, and requires a lot of pointer manipulations, makes these algorithms impractical. Mehlhorn [11] introduced a sorting algorithm that achieves the above bound, but his algorithm is not fully practical as well. Other algorithms which are optimal with respect to the number of inversions, include Blocksort [9] which runs in-place, and the tree-based Mergesort [14] which is optimal with respect to several other measures of presortedness. Among the sorting algorithms which are optimal with respect to the number of

---

[1] We define $\log x$ to be $\max(0, \log_2 x)$

P. Widmayer et al. (Eds.): ICALP 2002, LNCS 2380, pp. 183–194, 2002.
© Springer-Verlag Berlin Heidelberg 2002

inversions, Splitsort [7] and adaptive Heapsort [8] are the most promising from the practical point of view. Both algorithms require at most $2.5n \log n$ comparisons. Splaysort, sorting by repeated insertion in a splay tree [15], was proved to be optimal with respect to the number of inversions [3]. Moffat et al. [13] performed experimental results showing that Splaysort is practically efficient. See [4] for a nice survey of adaptive sorting algorithms.

We introduce Binomialsort, an optimal sorting algorithm with respect to the number of inversions. The number of comparisons performed by Binomialsort is at most $2n \log \frac{Inv(X)}{n} + O(n)$. Next, we introduce trinomial queues, an alternative to binomial queues, and use it to reduce the bound on the number of comparisons performed by the algorithm to $1.89n \log \frac{Inv(X)}{n} + O(n)$. The fact that our algorithm is simple and uses simple structures makes it practically efficient, while having a smaller constant, for the leading term of the bound on the number of comparisons, than other known algorithms. The question about the existence of a selection sort algorithm, that uses a number of comparisons that matches the information theoretical lower bound of $n \log \frac{Inv(X)}{n} + O(n)$, is open.

2 Adaptive Transformations on Binomial Queues

A binomial tree [1,16] of rank r is constructed recursively by making the root of a binomial tree of rank $r - 1$ the rightmost child of the root of another binomial tree of rank $r - 1$. A binomial tree of rank 0 consists of a single node. The following properties follow from the definition:

- The rank of an n-node (assume n is a power of 2) binomial tree is $\log n$.
- The root of a binomial tree, with rank r, has r sub-trees each of which is a binomial tree, having respective ranks $0, 1, \ldots, r - 1$ from left to right.
- Other than the root, there are 2^i nodes with rank $\log n - i - 1$, for all i from 0 to $\log n - 1$.

A binomial queue is a heap-ordered binomial tree. It satisfies the additional constraint that every node contains a data value smaller than or equal to those stored in its children. We use a binary implementation of binomial queues. In such an implementation, every node has two pointers, One pointing to its right sibling and the other to its rightmost child. The right pointer of a rightmost child points to the leftmost child to form a circular list. Given a pointer to a node, both its rightmost and leftmost children can be accessed in constant time. The list of its children can be sequentially accessed from left to right.

Given a forest F, each node of which contains a single value, the sequence of values obtained by a pre-order traversal of F is called the corresponding sequence of F and denoted by $Pre(F)$. Our traversal gives precedence to the trees of the forest in left-to-right order. Also, the precedence ordering of the sub-trees of a given node proceeds from left to right. A transformation on F that results in a forest F' shall be considered Inv-adaptive provided that $Inv(Pre(F')) \leq Inv(Pre(F))$. Our sorting algorithm will be composed of a series of *Inv-adaptive* transformations, applied to an initial forest consisting of a sequence of singleton-node trees (the input sequence). The idea is to maintain the presortedness in the input sequence, by keeping the number of inversions in the corresponding sequence of the forest non-increasing with time.

The following transformations are needed throughout the algorithm:

Heapifying Binomial Queues

Given an n-node binomial queue, T, such that the value at its root is not the smallest value, we want to restore the heap property. Applying the standard *heapify* operation will do. Moreover, this transformation is *Inv-adaptive*. Recall that the *heapify* operation proceeds by finding the node, say x, with the smallest value among the children of the root and swapping its value with that of the root. This step is repeated with the node x as the current root, until either a leaf or a node that has a value smaller than or equal to all the values of its children is reached. To show that the *heapify* operation is *Inv-adaptive*, consider any two elements $x_i, x_j \in Pre(T)$, where $i < j$. If these two elements were swapped during the *heapify* operation, then $x_i > x_j$. Since x_i appears before (to the left of) x_j in $Pre(T)$, we conclude that this swap decreases the number of inversions.

It remains to investigate how the *heapify* operation is implemented. Finding the minimum value within a linked list requires linear time. This may lead to an $O(\log^2 n)$ time for the *heapify* operation. We can do better, however, by maintaining with every node an extra pointer that points to the node with the smallest value among all its left siblings, including itself. We call this pointer, the pointer for the prefix minimum (pm). The pm pointer of the rightmost child of a node will, therefore, point to the node with the smallest value among all the children of the parent node. To maintain the correct values in the pm pointers, whenever the value of a node is updated all the pm pointers of its right siblings, including itself, have to be updated. This is accomplished by proceeding from left-to-right; the pm pointer of a given node, x, is updated to point to the smaller of the value of x and the value of the node pointed to by the pm pointer of the left sibling of x. A *heapify* at a node with rank r_1 reduces to a *heapify* at its child with the smallest value whose rank is $r2$, after one comparison plus at most $r_1 - r_2$ comparisons to maintain the pm pointers. Let $H(r)$ be the number of comparisons used by the above algorithm to *heapify* a binomial queue whose root has rank r, then

$$H(0) = 0, \quad H(r_1) \le H(r_2) + r_1 - r_2 + 1.$$

The number of comparisons is, therefore, at most $2 \log n$.

The number of comparisons can still be reduced as follows. First, the path from the root to a leaf, where every node has the smallest value among its siblings, is determined by utilizing the pm pointers. No comparisons are required for this step. Next, the value at the root is compared with the values of the nodes of this path bottom up, until the correct position of the root is determined. The value at the root is then inserted at this position, and all the values at the nodes above this position are shifted up. The *pm* pointers of the nodes whose values moved up and those of all their right siblings are updated. The savings are due to the fact that at each level of the queue (except possibly for the level of the final destination of the old value of the root), either a comparison with the old value of the root takes place or the *pm* pointers are updated, but not both. Then, the number of comparisons is at most $\log n + 1$. In the sequel it is to be understood that our binomial queues will be augmented with *pm* pointers as described above.

Lemma 1. *The number of comparisons needed to heapify an n-node binomial queue is at most* $\log n + 1$.

Inv-Adaptive Construction of Binomial Queues

Given an input sequence of length n, we want to build a corresponding binomial queue *Inv-adaptively*. This is done by recursively building a binomial queue with the rightmost $\frac{n}{2}$ elements, and another binomial queue with the leftmost $\frac{n}{2}$ elements. The root of the right queue is linked to the root of the left queue as its rightmost child. If the value at the root of the right queue was smaller than the value at the root of the left queue, the two values are swapped. A *heapify* operation is then called to maintain the heap property for the right queue, and the *pm* pointer of the root of the right queue is updated. Note that swapping the values at the roots of the two queues is an *Inv-adaptive* transformation. Let $B(n)$ be the number of comparisons needed to build a queue of n nodes as above, then

$$B(2) = 1, \quad B(n) \le 2B(\frac{n}{2}) + \log n + 2.$$

Lemma 2. *Given an input sequence of length n, a binomial queue can be Inv-adaptively built with less than 3n comparisons.*

Proof. Consider the recursion tree corresponding to the above recursive relations. There are $\frac{n}{2^i}$ nodes each representing a sub-problem of size 2^i, for all i from 2 to $\log n$. The number of comparisons associated with each of these sub-problems is two plus the logarithm of the size of the sub-problem, for a total of $\sum_{i=2}^{\log n}(i+2)\frac{n}{2^i} < 2.5n$. Also, there are $\frac{n}{2}$ sub-problems each of size 2, representing the initial conditions, for a total of another $\frac{n}{2}$ comparisons. Note that, a bound of $2.75n$ can be achieved by replacing the initial condition with $B(4) = 5$. □

3   Adaptive Pairing

For any two queues q_1 and q_2, where q_1 is to the left of q_2, a pairing operation is defined as follows. If the value of the root of q_2 is the smaller, perform a left-link by making q_1 the leftmost child of the root of q_2. Otherwise, perform a right-link by making q_2 the rightmost child of the root of q_1. See Fig. 1. It is straightforward to verify that this way of pairing is *Inv-adaptive*. Consider the corresponding sequence of the heap before and after the pairing. This sequence will not change by a right-link. A left-link changes the position of the root of the right queue, by moving it before (to the left of) all the elements of the left queue. Since the value of the root of the right queue is smaller than all the elements of the left queue, the number of inversions of the corresponding sequence decreases. The *Inv-adaptive* property follows.

Given a forest of priority queues, we combine these queues into a single queue as follows. Starting from the rightmost queue, each queue is paired with the result of the linkings of the queues to its right. We refer to this right-to-left sequence of node pairings as a *right-to-left incremental pairing* pass. The intuition behind doing the pairing from right to left is to keep the number of right-links throughout the algorithm linear, as will be clear later.

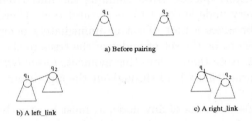

a) Before pairing

b) A left_link

c) A right_link

Fig. 1. Adaptive pairing.

4   The Binomialsort Algorithm

Given an input sequence X of length n. If n is not a power of 2 we split the input list into sub-lists of consecutive elements whose lengths are powers of 2. The lengths of the sub-lists have to be strictly increasing from left to right. Note that this splitting is unique. The sorting algorithm is then applied on each sub-list separately. We then do a merging phase to combine the sorted sub-lists. We start with the leftmost (shortest) two sub-lists and merge them. Then, the result of the merging is repeatedly merged with the next list to the right (next list in size), until all the elements are merged.

Lemma 3. *The number of comparisons spent in merging is less than $2n$.*

Proof. Let the lengths of the k sub-lists from left to right be $2^{j_1}, 2^{j_2}, \ldots 2^{j_k}$, where $j_1 < j_2 < \ldots < j_k$. Let C_i be the number of comparisons needed to merge the leftmost i sub-lists, where $i \le k$. We prove by induction that $C_i < \sum_{l=1}^{i} 2^{j_l+1}$. The base case follows, since $C_2 < 2^{j_1} + 2^{j_2}$. We also have $C_i < C_{i-1} + \sum_{l=1}^{i} 2^{j_l}$. Assuming the hypothesis is true for $i - 1$, then $C_i < \sum_{l=1}^{i-1} 2^{j_l+1} + \sum_{l=1}^{i} 2^{j_l}$. Using the fact that $\sum_{l=1}^{i-1} 2^{j_l} < 2^{j_i}$, then $C_i < \sum_{l=1}^{i-1} 2^{j_l+1} + 2^{j_i+1}$, and the hypothesis is true for all i. Since $n = \sum_{l=1}^{k} 2^{j_l}$, then $C_k < 2n$.   □

Throughout the algorithm, we assume without loss of generality that n is a power of 2. The algorithm in a nutshell is:

> Build an initial binomial queue *Inv-adaptively*.
> **While** *the heap is not empty* **do**
>   Delete the root of the heap and output its value.
>   Do a *fix-up* step.
>   Do a *right-to-left incremental pairing* pass.
> **enddo**

To implement the *fix-up* step efficiently, we associate a pseudo-rank with every node. This rank field may be different from the natural rank defined in Sect. 2 for binomial queues. After building the initial binomial queue, the pseudo-rank of every node is equal to its natural rank. Consider any node, x, the first time it becomes a root of a tree immediately before a pairing pass. Define *to-the-left(x)* to be the set of nodes of the trees to the left of the tree of x, at this moment. Note that, before this moment, *to-the-left(x)* is not defined. The following invariant must hold throughout the algorithm:

Invariant 1. The pseudo-rank of any node, x, must be strictly greater than the pseudo-rank of all the nodes in *to-the-left(x)*.

For a given node, call the children that were left-linked to that node, the L-children. Call the, at most one, child that may be right-linked to that node, the R-child (We will show later that, as a result of our pairing strategy, any node may gain at most one right child). And call the other children the I-children. The following two properties, concerning the children of a given node, are a direct consequence of Invariant 1:

Property 1. The R-child is the rightmost child and its pseudo-rank is the largest pseudo-rank among its siblings and parent.

Property 2. The pseudo-ranks of the L-children are smaller than that of their parent. Moreover, they form an increasing sequence from left to right.

If the pseudo-rank of the root of a binomial queue is equal to its natural rank, this binomial queue is a heavy binomial queue. If the pseudo-rank of this root is one more than its natural rank, the binomial queue is a light binomial queue. The number of nodes of a light binomial queue, whose pseudo-rank is r, is 2^{r-1}. We call the root of a light binomial queue, a light node. All other nodes are called normal nodes. The special case of a light binomial queue with pseudo-rank 0 is not defined. In other words, a binomial queue with pseudo-rank 0 is a single node, and is always heavy.

Invariant 2.1. The I-children of a normal node, whose pseudo-rank is r, are the roots of heavy or light binomial queues. The pseudo-ranks of these I-children, from left to right, form a consecutive sequence from 0 up to either $r - 1$ or r.

Invariant 2.2. Immediately before the pairing pass, all the roots of the queues are normal nodes.

For the initial binomial queue, all the children are I-children, and both Invariant 1 and Invariant 2 hold. When time comes for a node to leave the heap, the pseudo-ranks of its children form at most two runs of strictly increasing values from left to right. One is formed from the L-children, and the other is formed from the I-children and the R-child. The purpose of the *fix-up* step is to reconstruct such a forest of queues to form one single run of increasing pseudo-ranks from left to right, while maintaining Invariant 1 and Invariant 2. Subsequent to the initial construction of the binomial queue, the pseudo-ranks are updated during the *fix-up* step as illustrated below.

The Fix-Up Step

Let w be the node that has just been deleted before the *fix-up* step. We start with the leftmost queue and traverse the roots of the queues from left to right, until a node whose pseudo-rank is greater than or equal to the pseudo-rank of its right neighbor is encountered (if at all such a node exists). Call this node x and its pseudo-rank r_x. This node must have been the first L-child linked to w, and its right neighbor must be the leftmost I-child (By Invariant 2, this queue is a single node whose pseudo-rank is 0). Property 2 implies that, before w leaves the heap, its pseudo-rank must have been strictly greater than r_x. Property 1 implies that if w has an R-child, the pseudo-rank of this child must be greater than $r_x + 1$. Starting from the right neighbor of x, we continue traversing the roots of the queues from left to right, while the pseudo-ranks of the traversed nodes are less than or equal to $r_x + 1$. Call the forest of queues defined by these nodes the working-forest. The above analysis implies that the roots of the queues of the working-forest are I-children of w. Invariant 2 implies that the pseudo-rank of the root of the rightmost queue of the working-forest is either r_x or $r_x + 1$.

A trivial case arises if there is only one queue in the working forest. By Invariant 2, The pseudo-rank of this node, y, must be 0. This implies that r_x is equal to 0, and x is a single node as well. The fix-up step, in this case, ends by making the node that has the smaller value of x and y, the parent of the other node, and assigning a pseudo-rank 1 to it. If the number of queues of the working-forest is more than one, we perform a *combine* step.

a) Combine. The *combine* starts by promoting the single node queue representing the leftmost queue of the working forest, giving it a pseudo-rank of $r_x + 1$, and making it the root of the combined queue. In other words, we make the other roots of the queues of the working-forest I-children of this promoted node. The binomial queues of the working forest are traversed from left to right, until a heavy binomial queue, say h, is encountered. The pseudo-rank of the root of each of the light binomial queues, to the left of h, is decreased by 1, making these binomial queues heavy. Let r_h be the pseudo-rank of the root of h. We proceed by splitting h into two binomial queues; the right queue is the queue whose root was the rightmost child of the root of h before the split, while the rest of h form the left queue. The left binomial queue is assigned a pseudo-rank $r_h - 1$, indicating that it is a heavy binomial queue, while the right binomial queue is assigned a pseudo-rank r_h, indicating that it is a light binomial queue. After this step, the pseudo-ranks of the roots of the queues of the working-forest still form a consecutive sequence from left to right, and Invariant 2.1 holds. Note that this process is similar to subtracting 1 from the binary representation of an integer. Finally, perform a *heapify* operation on the combined queue. See Fig. 2. It is straightforward to verify that the *combine* step is *Inv-adaptive*.

Consider the case where all the queues of the working-forest are light queues. The pseudo-rank of the root of each of these queues is decreased by 1, making it a heavy binomial queue. If the pseudo-rank of the root of the rightmost queue was r_x and now becomes $r_x - 1$, for Invariant 2 to hold, the promoted root is assigned a pseudo-rank r_x, instead of r_x+1, and a *fusion* step is to be performed.

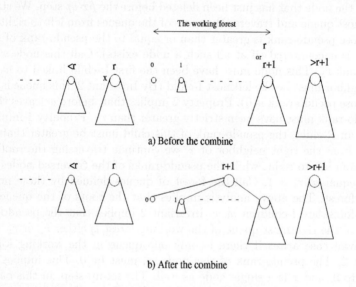

Fig. 2. The combine step. Symbols beside nodes represent pseudo-ranks.

b) Fusion. If the pseudo-rank of the root of the combined queue is r_x, a *fusion* step is performed. Call the root of the combined queue y. The way the *combine* step is done insures that the combined queue, in this case, is a binomial queue. The queue of y is fused with the queue of x as follows. If the value of y is smaller than the value of x, the two values are swapped and a *heapify* step is performed on the queue of y. The pseudo-rank of x is incremented by 1. Let z be the rightmost child of x and its pseudo-rank r_z. If $r_z < r_x$, the fusion ends by making the queue of y the rightmost I-child of x. In this case, Invariant 2 implies $r_z = r_x - 1$, and Invariant 2.1 still holds after the *fusion*. Otherwise ($r_z = r_z$, a case that happens only if x was promoted in a previous *combine* step), the value of y is compared with the value of z. If the value of y is smaller, the two values are swapped and a *heapify* step is performed on the queue of y. The queue of y becomes the rightmost I-child of z, and the pseudo-rank of z is incremented by 1. Since the queue of y is a binomial queue, Invariant 2.1 again holds in this case. See Fig. 3 for an example. Since the *heapify* operation is *Inv-adaptive*, it follows that the *fusion* step is *Inv-adaptive* as well.

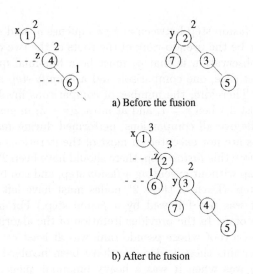

a) Before the fusion

b) After the fusion

Fig. 3. The fusion step. Numbers beside nodes represent pseudo-ranks.

c) Gluing. The last step of the fix-up is to maintain Invariant 2.2. The roots of the queues are traversed from left to right, searching for light nodes. Light nodes may have been I-children of w, that were not included in the combined queue of the working forest. When a light node is encountered its rank is decreased by 1, making it a normal node. We may now face another problem; Invariant 1 may no longer be true, since there may be two sibling nodes having the same rank. The first observation is that, for a given rank, there may be at most two roots that have this rank. The second observation is that the queues of any two roots having the same rank must have been one queue, which was split by a *combine* step at some point of the algorithm. This follows from the fact that the right queue, among any of these pairs, was a light queue, indicating that it was split in a previous step of the algorithm. Each pair of such queues is glued together by making the right queue the rightmost child of the root of the left queue, and increasing the rank of the root of the glued queue by 1. Note that, during the *gluing* step, no comparisons are performed between the values of the nodes.

Analysis

Lemma 4. *The number of right-links in all the pairing passes is less than n.*

Proof. The *right-to-left incremental pairing* pass ensures that a node can gain at most one R-child. When a node gains an R-child it becomes the root of the rightmost queue. From this point on this node will never have a right sibling and will always be a node on the right spine of the rightmost queue. Hence, it will never gain another R-child in any pairing pass.   □

Lemma 5. *The number of comparisons involved in all the fusion steps is at most $2n$.*

Proof. Consider a *fusion* step between any two queues q_1 and q_2, where q_1 is to the left of q_2. Let r be the pseudo-rank of the roots of the two queues before the *fusion*. The first observation is that q_2 must be a binomial queue. During the *fusion*, in the worst case, one comparison and a *heapify* step are performed on q_2, at most twice. Therefore, the number of comparisons involved is at most 2 when $r = 0$, at most 4 when $r = 1$, and at most $2(r + 2)$ in general.

What makes the overall comparisons, performed during *fusion* steps, linear is that *fusion* steps are not executed in most of the iterations of the algorithm. More precisely, Before this *fusion* step, there should have been 2^r nodes that have already left the heap without executing a *fusion* step, and can be associated only with this *fusion* step. (Each of these 2^r nodes must have left the heap with a *combine* step that was not followed by a *fusion* step.) For more illustration, consider the nodes of q_2. In the previous iteration of the algorithm, these nodes were descendents of a root whose pseudo-rank was at least $r + 1$. The last time the queue that contains these nodes may have been involved in a *fusion* step (as a right queue) was when it was a heavy binomial queue, i.e. it had 2^{r+1} descendents. From this point, a total of 2^r nodes, of the descendents of this queue, have left the heap, and each time the *fix-up* did not include a *fusion* step.

We charge each of these 2^r nodes twice. Since, for all values of r, 2^{r+1} credits cover the cost of the current *fusion* step, the lemma follows.   □

Lemma 6. *The pseudo-rank of a normal node, that has s descendants, is less than $\log s + 1$.*

Proof. Let r be the pseudo-rank of a normal node that has s descendants. To establish a lower bound on s, we use Invariant 2. The node itself contributes 1 to s. Consider the I-children of this node. The leftmost I-child is a single node. The other children are roots of binomial queues whose pseudo-ranks form a consecutive sequence from 1 to either $r - 1$ or r. Even if these binomial queues are light, they contribute at least $2^0, 2^1, \ldots, 2^{r-2}$ to s, then $s \geq 2 + \sum_{i=0}^{r-2} 2^i > 2^{r-1}$.□

Lemma 7. *The number of comparisons involved in the heapify operations in all the combine steps is less than $n \log \frac{Inv(X)}{n} + 3n$.*

Proof. Consider the moment when a node w is leaving the heap. Let f_w be the number of descendents of the L-children of w. Each of these nodes represents an element that is greater than w and was appearing before (to the left of) w in the input sequence. Since all the steps of the algorithm are *Inv-adaptive*, taking the summation over all the nodes, then $\sum_x f_x \leq Inv(X)$.

Assume that the pseudo-rank of the first L-child gained by w is r. Using Lemma 6, then $r < \log f_w + 1$. During the *combine* step that follows the deletion of w, the accompanying *heapify* operation is performed on a queue whose pseudo-rank is at most $r + 1$, for a total of at most $r + 2$ comparisons.

Let CH be the number of comparisons involved in all the *heapify* operations of the *combine* steps. Using Jensen's inequality, and taking the summation over all the nodes of the heap, then $CH < \sum_x (\log f_x + 3) \leq n \log \frac{Inv(X)}{n} + 3n$.   □

Lemma 8. *The number of left-links in all the pairing passes is less than* $n \log \frac{Inv(X)}{n} + n$.

Proof. Consider the moment when a node w is leaving the heap. Let f_w be as defined in Lemma 7. Let l_w be the number of L-children of w. Using Property 2, and applying Lemma 6 on each of the sub-trees defined by these nodes, then $f_w > \sum_{i=1}^{l_w - 1} 2^{i-1} + 1$, and hence $l_w < \log f_w + 1$. Let L be the number of left-links. Using Jensen's inequality, then $L < \sum_x (\log f_x + 1) \leq n \log \frac{Inv(X)}{n} + n$.   □

The following theorem follows, as a consequence of the above lemmas.

Theorem 1. *Binomialsort sorts a sequence X of length n, in at most* $2n \log \frac{Inv(X)}{n} + O(n)$ *comparisons and* $O(n \log \frac{Inv(X)}{n} + n)$ *time.*

5   Trinomial Queues

A trinomial tree of rank r is constructed recursively using 3 trinomial trees of rank $r - 1$. The first is linked as the rightmost child of the second. Then, the second is linked as the rightmost child of the third. A trinomial tree of rank 0 consists of a single node. The following properties hold for trinomial trees: (Assuming that the number of nodes n is a power of 3)

- The rank of an n-node trinomial tree is $\log_3 n$.
- The root of a trinomial tree, with rank r, has r children, having respective ranks $0, 1, \ldots, r - 1$ from left to right.
- Other than the root, there are 2.3^i nodes with rank $\log_3 n - i - 1$, for all i from 0 to $\log_3 n - 1$.

A trinomial queue is a heap ordered trinomial tree. Applying the same techniques we used for *heapifying* binomial queues on trinomial queues, the number of comparisons needed to do the *heapify* on a trinomial queue of rank r is at most $2r + 1$. In a similar fashion, building an initial trinomial queue *Inv-adaptively* can be done in linear time.

A trinomial queue is considered heavy if the pseudo-rank of its root is equal to its natural rank. A light trinomial queue, with pseudo-rank r, consists of two heavy trinomial queues, with pseudo-ranks $r - 1$, one of them is the rightmost child of the other. This means that a light trinomial queue, with rank r, has 2.3^{r-1} nodes. During the *combine* step, when a heavy trinomial queue, with pseudo-rank r, is split, its rightmost child becomes the right queue, which is a light queue with pseudo-rank r, and the rest of the queue becomes the left queue, which is a heavy queue with pseudo-rank $r - 1$.

Using trinomial queues instead of binomial queues throughout the algorithm, it can be easily verified that the number of comparisons involved in all the *heapify* operations during the *fix-up* steps is at most $\frac{2}{\log 3} n \log \frac{Inv(X)}{n} + O(n)$, and the number of left-links is at most $\frac{1}{\log 3} n \log \frac{Inv(X)}{n} + O(n)$. All the other operations require a linear number of comparisons, for a total of at most $\frac{3}{\log 3} n \log \frac{Inv(X)}{n} + O(n)$ comparisons.

Theorem 2. *Trinomialsort sorts a sequence X of length n, in at most* $1.89n \log \frac{Inv(X)}{n} + O(n)$ *comparisons and* $O(n \log \frac{Inv(X)}{n} + n)$ *time.*

References

1. M. Brown. *Implementation and analysis of binomial queue algorithms.* SIAM J. Comput. 7 (1978), 298-319.
2. M. Brown and R. Tarjan. *Design and analysis of data structures for representing sorted lists.* SIAM J. Comput. 9 (1980), 594-614.
3. R. Cole. *On the dynamic finger conjecture for splay trees. Part II: The proof.* SIAM J. Comput. 30 (2000), 44-85.
4. V. Estivill-Castro. *A survey of adaptive sorting algorithms.* ACM Comput. Surv. vol 24(4) (1992), 441-476.
5. L. Guibas, E. McCreight, M. Plass and J. Roberts. *A new representation of linear lists.* ACM Symp. on Theory of Computing 9 (1977), 49-60.
6. D. Knuth. *The Art of Computer Programming. Vol III: Sorting and Searching.* Addison-wesley, second edition (1998).
7. C. Levcopoulos and O. Petersson. *Splitsort - An adaptive sorting algorithm.* Information Processing Letters 39 (1991), 205-211.
8. C. Levcopoulos and O. Petersson. *Adaptive Heapsort.* J. of Alg. 14 (1993), 395-413.
9. C. Levcopoulos and O. Petersson. *Exploiting few inversions when sorting: Sequential and parallel algorithms.* Theoretical Computer Science 163 (1996), 211-238.
10. H. Mannila. *Measures of presortedness and optimal sorting algorithms.* IEEE Trans. Comput. C-34 (1985), 318-325.
11. K. Mehlhorn. *Sorting presorted files.* Proc. Of the 4th GI Conference on Theory of Computer Science. LNCS 67 (1979), 199-212.
12. K. Mehlhorn *Data Structures and Algorithms. Vol. 1. Sorting and Searching.* Springer-Verlag, Berlin/Heidelberg. (1984).
13. A. Moffat, G. Eddy and O. Petersson. *Splaysort: fast, verstile, practical.* Softw. Pract. and Exper. Vol 126(7) (1996), 781-797.
14. A. Moffat, O. Petersson and N. Wormald *A tree-based Mergesort.* Acta Informatica, Springer-Verlag (1998), 775-793.
15. D. Sleator and R. Tarjan. *Self-adjusting binary search trees.* J. ACM 32(3) (1985), 652-686.
16. J. Vuillemin. *A data structure for manipulating priority queues.* Comm. ACM 21(4) (1978), 309-314.

Exponential Structures for Efficient Cache-Oblivious Algorithms[*]

Michael A. Bender[1], Richard Cole[2], and Rajeev Raman[3]

[1] Computer Science Department, SUNY Stony Brook, Stony Brook, NY 11794, USA.
bender@cs.sunysb.edu.
[2] Computer Science Department, Courant Institute, New York University, New York, NY
10012, USA. cole@cs.nyu.edu.
[3] Department of Mathematics and Computer Science, University of Leicester,
Leicester LE1 7RH, UK. R.Raman@mcs.le.ac.uk

Abstract. We present *cache-oblivious* data structures based upon *exponential structures*. These data structures perform well on a hierarchical memory but do not depend on any parameters of the hierarchy, including the block sizes and number of blocks at each level. The problems we consider are searching, partial persistence and planar point location. On a hierarchical memory where data is transferred in blocks of size B, some of the results we achieve are:

- We give a linear-space data structure for dynamic searching that supports searches and updates in optimal $O(\log_B N)$ *worst-case* I/Os, eliminating amortization from the result of Bender, Demaine, and Farach-Colton (FOCS '00). We also consider finger searches and updates and batched searches.
- We support partially-persistent operations on an ordered set, namely, we allow searches in any previous version of the set and updates to the latest version of the set (an update creates a new version of the set). All operations take an optimal $O(\log_B(m+N))$ amortized I/Os, where N is the size of the version being searched/updated, and m is the number of versions.
- We solve the planar point location problem in linear space, taking optimal $O(\log_B N)$ I/Os for point location queries, where N is the number of line segments specifying the partition of the plane. The pre-processing requires $O((N/B)\log_{M/B} N)$ I/Os, where M is the size of the 'inner' memory.

1 Introduction

A modern computer has a hierarchical memory consisting of a sequence of levels – machine registers, several levels of on-chip cache, main memory, and disk – where the farther the level is from the CPU, the larger the capacity and the longer the access time. On present-day machines, the cache is 2 orders of magnitude faster than main memory, and 6 - 7 orders of magnitude faster than disk. In order to amortize the cost of a memory access, data is transferred between levels in *blocks* of contiguous locations; we refer to each such transfer as an I/O. Due to the cost of an I/O, it is important to minimise I/Os.

[*] This work was supported in part by HRL Laboratories, NSF Grant EIA-0112849 and Sandia National Laboratories (Bender); in part by NSF grants CCR-9800085 and CCR-0105678 (Cole); and in part by EPSRC grant GR L/92150 and UISTRF project 2001.04/IT (Raman).

P. Widmayer et al. (Eds.): ICALP 2002, LNCS 2380, pp. 195–207, 2002.
© Springer-Verlag Berlin Heidelberg 2002

Unfortunately, a multilevel memory hierarchy is complex, and it can be unwieldy or impossible to write programs tuned to the parameters of each level of the memory hierarchy. Also, this parameterization leads to inflexible algorithms that are only tuned for one memory platform. The traditional alternative is to assume a two-level memory rather than a multilevel memory. The best-known two-level memory model was defined by Aggarwal and Vitter [1] (see [24] for others). In this model the memory hierarchy is composed of an arbitrarily large disk divided into blocks, and an internal memory. There are three parameters: the problem size N, the block size B, and the main memory size M. An I/O transfers one block between main memory and disk. The goal is to minimize the number of I/Os, expressed as a function of N, M, and B. In recent years many algorithms have been designed for the Aggarwal-Vitter and related models, many of which are reviewed in some recent surveys [24,5].

One limitation of the Aggarwal-Vitter model is that it only applies to two-level memory hierarchies. In contrast, it is becoming increasingly important to achieve data locality on *all* levels of the hierarchy. For example, Ailamaki et al. [2] report that in many standard database applications, the bottleneck is main-memory computation, caused by poor cache performance. Even in main memory, simultaneous locality at cache and disk block levels improves performance by reducing address translation misses [19,20]. Furthermore, the memory hierarchy is rapidly growing steeper because in recent years, processing speeds have been increasing at a faster rate than memory speeds (60% per year versus 10% per year). This trend is entrenched and appears unlikely to change in the next few years if not longer.

Frigo et al. [16,18] proposed the elegant *cache-oblivious model*, which allows one to reason about a two-level memory hierarchy, but prove results about an unknown multilevel memory hierarchy. Cache-oblivious algorithms assume a standard (flat) memory as in the RAM model, and are unaware of the structure of the memory hierarchy, including the capacities, block sizes, or access times of the levels. I/Os occur transparently to the algorithm, and I/O performance is analyzed in the Aggarwal-Vitter model [1] with arbitrary block and main memory sizes. The model assumes an *ideal cache*, where I/Os are performed by an omniscient off-line algorithm (Frigo et al. note that any reasonable block-replacement strategy approximates the omniscient strategy to within a constant factor). The key observation is that if the algorithms perform few I/Os in this scenario, then they perform well for all block and main memory sizes on all levels of the memory hierarchy simultaneously, for any memory hierarchy. By way of contrast, henceforth we refer to the Aggarwal-Vitter model as the *cache-aware* model.

Frigo et al. [16,18] gave optimal algorithms for sorting, FFT, and matrix transpose. Optimal cache-oblivious algorithms have also been found for LU decomposition [11, 23]. These algorithms make the (generally reasonable) *tall cache* assumption that $M = \Omega(B^2)$. Prokop [18] observed how to layout a complete binary tree in linear space so as to optimize root to leaf traversals in the I/O cost model. We call his layout a *height partitioning* layout, and review it later. Bender, Demaine and Farach-Colton [8] considered the dynamic predecessor searching problem, and obtained a linear space data structure with amortized $O(\log_B N)$ update cost and worst case $O(\log_B N)$ search cost. They also consider the problem of traversing the leaves of the tree. Their result, like most of our results, does not use the tall cache assumption.

Our approach. In this paper we present a new general approach for solving irregular and dynamic problems cache-obliviously: *the exponential structure*, which is based on the exponential tree previously developed and used by Andersson and Thorup [3,22,4] for designing fast data structures in the transdichotomous model [15]. An exponential tree is a tree of $O(\log \log N)$ levels where the degrees of nodes descending from the root level decrease doubly exponentially, e.g. as in the series $N^{1/2}, N^{1/4}, N^{1/8}, \cdots, 2$.

We show that exponential trees, and more generally exponential structures, are similarly powerful in a hierarchical memory. We use exponential trees to devise alternative solutions to the cache-oblivious B-tree, and we provide the first worst-case solution to this problem. Then we generalize these techniques to devise new cache-oblivious data structures, such as persistent search trees (which are not trees), search trees that permit finger searches, search trees for batched searches, and a planar point location structure. By using exponential structures to solve a variety of problems, we introduce a powerful tool for designing algorithms that are optimized for hierarchical memories.

Our results are as follows:

Dynamic predecessor searching: The problem here is to store a set of N totally ordered items, while supporting predecessor queries, insertions and deletions. We give a series of data structures, culminating in one that takes $O(N)$ space and supports all operations in $O(\log_B N)$ worst-case I/Os, which is optimal. This eliminates amortization from previous cache-oblivious solutions to this problem [8,12,10].

The main difficulty faced in supporting dynamic search structures cache-obliviously is the need to maintain data locality (at all levels of granularity) in the face of updates. Arbitrary updates appear to require data sparsity. However, sparsity reduces locality. To balance these constraints, Bender et al. [8] developed strongly weight-balanced trees and the *packed-memory* structure. Two other recent solutions with the same amortized complexity also use the packed memory structure or similar techniques [12,10]. In essence, the packed-memory problem is to maintain an ordered list of length n in an array of size $O(n)$, preserving the order, under insertions and deletions. Although simple amortized algorithms to the packed-memory problem in the RAM model yield good cache-oblivious algorithms, it is far from clear if the same holds for Willard's complex worst-case RAM solution [25].

As we will see, exponential structures are particularly effective at addressing the tension between locality and sparsity. Our solution uses a new worst-case approach to dynamizing exponential trees. In contrast to an earlier worst-case dynamization by Andersson and Thorup [4] in the context of dynamic searching in the RAM model, the main difficulty we face is to balance data sparsity and locality for dynamically changing "fat" nodes in the exponential tree. As a result, our solutions, unlike Andersson and Thorup's, will use superlinear space at the deeper levels of the search structure. One of our dynamizations seems to be simpler than that of Andersson and Thorup, and we believe would somewhat simplify their constructions.

Finger search: This is the variant of dynamic predecessor searching in which successive accesses are nearby. More precisely, suppose two successive accesses to an ordered set are separated by d elements. We give a linear size data structure such that the second access performs $O(\log^* d + \log_B d)$ amortized and worst-case I/Os respectively, depending on whether it is an update or a search.

By way of comparison, in the standard RAM model there are solutions using $O(\log d)$ operations [13,17], and this extends in a straightforward way to the cache-aware model with bounds of $O(\log_B d)$. The cache-oblivious data structures of [8,10] do not support finger searches, and that of [12] has a higher update time.

Partial persistence: Here one seeks to support queries in the past on a changing collection of totally ordered items. Specifically, suppose we start with an ordered set of N items at time t_0, and then perform a series of m updates at times $t_1 < t_2 < \cdots < t_m$ ($t_0 < t_1$, of course). At any time t'_j, $t_j < t'_j < t_{j+1}$, one can perform a query of the form: "what was the largest item $x < y$ at time t_i?", where $i \leq j$. We support updates and queries in $O(\log_B(N + m))$ I/Os and use $O(N + m)$ space. Note that the structure here is a dag. A key issue is to give an appropriate definition of the weight of a fat node.

By way of comparison we note that in the standard RAM model there are solutions using $O(\log(N + m))$ operations [21], and our I/O bound matches the cache-aware result of [14].

Planar point location: The input comprises N arbitrarily-oriented line segments in the plane which are non-intersecting, except possibly at the endpoints. The task is to preprocess the line segments so as to rapidly answer queries of the form: which line segment is directly above query point p? (The traditional planar point location problem boils down to solving this problem.)

This can be solved using a structure for partial persistence that supports predecessor queries on items drawn from a partially ordered universe, but with the requirement that the items present at any given time t_i be totally ordered; however, there is no requirement that items have a relative ordering outside their time span of existence. Sarnak and Tarjan [21] gave a linear space solution for this problem supporting queries in $O(\log N)$ time. As our partial persistence data structure does not support partial orders, we are obliged to take a different approach. We build a search dag similar to that used for our partial persistence data structure, but build it offline; it has linear size, is built using $O((N/B) \log_{M/B} N)$ I/Os, and supports queries using $O(\log_B N)$ I/Os. This is a worst case result, and depends on the tall cache assumption. This data structure also supports batched queries efficiently.

By way of comparison, several linear-space data structures are known in the RAM model that support queries in $O(\log N)$ time and can be constructed in $O(N \log N)$ time. In the two-level model, Goodrich et al. [14] give a linear space data structure that supports queries in $O(\log_B N)$ I/Os but whose construction takes $O(N \log_B N)$ I/Os. Arge et al. [6] consider a batched version of this problem, but their data structure requires superlinear space.

Batched search: We give a linear-sized structure for performing simultaneous predecessor queries on r query items among N base items. This is challenging when there is only a partial order on the r items (so that they cannot necessarily be sorted); however, the N base items are totally ordered and each query item is fully ordered with respect to the base items. (An example scenario has as base items non-intersecting lines crossing a vertical strip in the plane, and as query items points in the vertical strip.) The I/O cost for the search is $O(r \log_B(N/r) + (r/B) \cdot \log_{M/B} r)$. This result uses the tall cache assumption (and hence $\log_{M/B} r = O(\log_M r)$). We give an off-line algorithm for constructing the search structure; it uses $O((N/B) \log_{M/B} N)$ I/Os. We believe we can also maintain the structure dynamically with the same I/O cost as for searches, but have yet to check all the details.

In the rest of this abstract we outline our solutions to some of the above problems. Proofs of the results on batched searching and finger searching, persistent search, as well as details of the strongest version of the planar point location result, are omitted from this abstract and may be found in [9].

2   Dynamic Searching

We do not discuss deletions, which can essentially be handled by standard lazy approaches. Our solution uses a known optimal layout [18] for complete binary search trees as a building block. This layout places an N node tree in a size N array as follows. The tree is partitioned into height $\frac{1}{2} \log N$ subtrees, each of which is laid out recursively in an array segment of length $\sqrt{N}$. We call this the *height-partitioning* layout. It is straightforward to extend this construction to trees in which all the leaves are at the same level, where this is the first level in which this number of leaves would fit. The search time is unchanged and the space is still linear.

Lemma 1. *(Prokop). A search on a tree stored in a height-partitioning layout uses* $O(\log_B N)$ *I/Os.*

Our dynamic solution uses a similar structure, but in order to handle updates efficiently we need something more flexible. In our search trees, as with B-trees, all items are at the leaves and only keys are stored at internal nodes. As suggested in the introduction, internal nodes may have many children; to emphasize their large size we will call them *fat nodes* henceforth. Likewise, we will refer to a *layer* of fat nodes rather than a level.

We define the *volume* of a fat node in an exponential tree to be the number of items stored in its descendant leaves. We extend the definition to subtrees: the volume of a subtree is simply the volume of its root. Let T be a layer i fat node. An update that changes T's volume will be called an update in T. Note that an update will be in one fat node in each layer.

2.1   The Amortized $O(\log_B N + \log \log N)$ Solution

We parameterize the fat nodes by layer, the leaves forming layer 0. A layer i fat node, $i \geq 1$, will have a volume in the range $[2^{2^i} - 2^{2^{i-1}}, 2 \cdot 2^{2^i})$, except for the topmost fat node, where the range is given by $[2 \cdot 2^{2^{k-1}}, 2 \cdot 2^{2^k})$. The reason for the term "$-2^{2^{i-1}}$" will become clear later. Each layer 0 fat node contains a single item. Loosely speaking, the volumes of the fat nodes square at each successive layer.

Each fat node is implemented as a complete binary tree with all its leaves on the same level; appropriate keys are stored at the leaves and internal vertices. This binary tree is then stored in a height partitioning layout. Each leaf of this binary tree, except perhaps the rightmost one, will have two pointers, one to each of its two children fat nodes. We note that a layer i fat node, $i \geq 1$, has $O(2^{2^{i-1}})$ leaves and internal vertices.

Lemma 2. *A search takes* $O(\log_B N + \log \log N)$ *I/Os.*

Proof. Going from one layer to another takes one I/O, and there are $\log \log N$ such transfers in all. The searches within layer i consume $O(\log_B 2^{2^{i-1}})$ I/Os. Summed over all the layers, this comes to $O(\log_B N + \log \log N)$.   □

Next we explain when a fat node needs updating and how it is done. If a layer i fat node T acquires volume $2 \cdot 2^{2^i}$ it is split as evenly as possible into two subtrees T_1 and T_2, each of size roughly 2^{2^i}. However, we wish to perform this split without changing T's children fat nodes. As the children have volumes up to $2 \cdot 2^{2^{i-1}}$, we can only achieve a split with the subtrees T_1 and T_2 having sizes in the range $[2^{2^i} - 2^{2^{i-1}}, 2^{2^i} + 2^{2^{i-1}}]$. Such a split can readily be implemented in $O(|T|) = O(2^{2^{i-1}})$ time and I/Os.

When T splits this adds one to the number of its parent's children. This is accommodated by completely rebuilding the parent, which can be done in time and I/Os linear in the size of the parent, i.e. $O(2^{2^i})$ time and I/Os. We call this operation *refreshing* the parent. A standard argument shows that the amortized times and I/Os spent by an insertion on splitting and refreshing the layer i fat node T it traverses are $O(1/2^{2^{i-1}})$ and $O(1)$ respectively. Thus, we obtain:

Theorem 1. *There is a cache oblivious linear size data structure for an ordered set of* N *items supporting searches in worst case* $O(\log_B N + \log \log N)$ *I/Os and updates in amortized* $O(\log_B N + \log \log N)$ *I/Os.*

The simplicity and consequent practicality of the above solution has been confirmed in preliminary experimental work [20].

2.2   The Amortized $O(\log_B N)$ I/O Solution

The main change here is to keep a level i layer fat node and all its descendant layer fat nodes in a contiguous portion of the array. This will increase the splitting time and as we will see increases the space needed.

Now when splitting layer i fat node T into subtrees T_1 and T_2, besides creating T_1 and T_2 we need to copy all of T's descendant fat nodes into either the portion of the array being used for T_1 and its descendants or that being used for T_2 and its descendants. This will take time proportional to the size of T and its descendants. Thus the cost of splitting T becomes $O(2^{2^i})$. However, amortized time and I/Os spent by an insertion on splitting the layer i fat node T it traverses is still $O(1)$. Let us turn to the space needed.

Lemma 3. *Layer* i *fat node* T *has at most* $2 \cdot 2^{2^i}/(2^{2^{i-1}} - 2^{2^{i-2}})$ *subtrees.*

We need to leave sufficient space for all of T's possible children and their descendants, i.e. $2 \cdot (2^{2^{i-1}} + 2^{2^{i-2}} + 2)$ layer $i - 1$ fat nodes and their descendants. Thus T and its descendant fat nodes will need space $O(2^i \cdot 2^{2^i})$, and so the whole exponential tree uses space $O(N \log N)$.

Lemma 4. *The search time is $O(\log_B N)$.*

Proof. Consider layer i fat nodes, where $\sqrt{B} < 2^i \cdot 2^{2^i} \leq B$. Searching such a layer i fat node and all its descendant fat nodes takes $O(1)$ I/Os, since together they occupy $O(1)$ pages. Searching the higher portion of the tree, above layer i, takes $O(\log_B N + \log \log N - \log \log B)$ I/Os, for there are $O(\log \log N - \log \log B)$ layers in this portion of the exponential tree. But this is $O(\log_B N)$ I/Os in total.   □

To reduce the space, each leaf is replaced by a record that stores between $\log N$ and $2 \log N$ items in sorted order. This reduces the number of items in the exponential tree to $O(N/\log N)$ and hence the overall space used is reduced to $O(N)$.

Theorem 2. *There is a cache oblivious linear size data structure for an ordered set of N items supporting searches in worst case $O(\log_B N)$ I/Os and updates in amortized $O(\log_B N)$ I/Os.*

2.3   The Worst Case $O(\log_B N)$ Solution

The solution is quite similar to the solution with the same amortized complexity bound. The main changes to a layer i fat node tree T are the following:

- The implementation of the interface between fat nodes is slightly changed.
- The refreshing is performed incrementally over $\Theta(2^{2^{i-1}})$ insertions, and is initiated every $\Theta(2^{2^{i-1}})$ insertions in T.
- The splitting is performed incrementally over $\Theta(2^{2^i})$ insertions, and is initiated every $\Theta(2^{2^i})$ insertions in T.

The fat node interface is changed so that when a child fat node of T splits, T need not be immediately refreshed. In order to link to the new children generated by such a split, T is provided with space for an extra level of vertices: each link to a child fat node may be replaced by a vertex and, emanating from the vertex, links to the two new children fat nodes. However, only one additional layer of vertices will be allowed: we will ensure that T has been refreshed before any of the new child fat nodes splits.

A refreshing proceeds by copying all the leaf vertices in fat node T and then building internal vertices above them. Until a refreshing is complete searches and updates continue to be performed on the old version of T. We cross-link the copies of each leaf in the old and new versions of T so that any update resulting from a split of a child of T can be performed in both versions.

A split task proceeds by first splitting the top layer fat node, and then copying descendant fat nodes one by one in a depth first ordering. As a fat node is copied it is refreshed and split if need be. When the copying of a fat node is complete, searches will proceed through the new version(s) of the fat node(s). Consider the task that is splitting a layer j tree T, and suppose that it has just completed copying a descendant layer i fat node S. Now suppose that a search goes through S; it then follows a pointer to a layer $i - 1$ fat node in the old version of T. Following this pointer could result in an I/O, even if S and its descendants would fit in a single block once their copying is complete. Call such a pointer a j-long pointer, and note that the span of a j-long pointer is contained within a layer $j + 1$ tree. To guarantee a good search performance, we schedule tasks in such a way that if a search follows a k-long pointer, it will subsequently only follow j-long pointers for $j < k$. From this it follows that searches require only $O(\log_B N)$ I/Os.

We face the difficulty that, unfortunately, tasks could interfere with each other. We ensure that only one task at any time will be copying a fat node. Consider a split task $\mathcal{T}$ associated with fat node T. Suppose $\mathcal{T}$ is about to copy a descendant fat node S, but it finds S is already in the process of being copied by another task associated with a proper descendant of T; then $\mathcal{T}$ takes over the task, completes it, and finally does its own copying. This is readily implemented with a copy ID at the root of each fat node tree S identifying the name and layer of the task currently copying S, if any.

Interference also occurs if a layer h tree S is to be split but it is in the process of being copied by task $\mathcal{T}$ which is splitting a higher layer tree T. Then the task $\mathcal{S}$ for S will wait until $\mathcal{T}$ completes its copying of S and will then proceed with its own copying of S, unless taken over by some higher level task.

The handling of space is less obvious now. Each fat node is provided with double the space it needs; at any time it is stored in one half of the space and refreshing occurs in the other half. Also, lists of free space are needed within each layer tree to provide space for its children when they split. When the task $\mathcal{T}$ splitting a layer i tree finishes copying a layer $h < i$ subtree S and S's descendant subtrees, $\mathcal{T}$ releases the space for S, and if S had been in the process of splitting, the space for the old version of S.

Lemma 5. *A split takes $O(2^{2^i})$ time and I/Os.*

Next, we specify the choices of parameters. Now a layer i tree T has volume in the range $[2^{2^i}, 4 \cdot 2^{2^i})$. A split is initiated when its volume reaches $3 \cdot 2^{2^i}$ and will complete following at most 2^{2^i} insertions to T. The resulting trees have volumes in the range $[1.5 \cdot 2^{2^i} - 2 \cdot 2^{2^{i-1}}, 2.5 \cdot 2^{2^i} + 2 \cdot 2^{2^{i-1}}]$. It is readily checked that for $i \geq 2$, the trees created by the split have volumes in the range $[2^{2^i}, 3 \cdot 2^{2^i}]$.

Lemma 6. *The space used is $O(N \log^2 N)$.*

Proof. A layer i tree T has at most $4 \cdot 2^{2^i}/2^{2^{i-1}}$ child subtrees. $\qquad\square$

To achieve space $O(N)$ we use buckets of size $\Theta(\log^2 N)$, implemented as two layers of records of size in the range $[\log N, 2 \log N]$. Thus, we have shown:

Theorem 3. *There is a cache oblivious linear size data structure that supports searches and updates on an ordered set of items using $O(\log_B N)$ I/Os in the worst case.*

3 Planar Point Location

In this part of the extended abstract we only describe two results: both use preprocessing of $O(N \log_B N)$ I/Os and support queries using $O(\log_B N)$ I/Os; but the first result uses $O(N \log \log N)$ space, while the second uses only $O(N)$ space. We also briefly outline our stronger result which achieves linear space, $O((N/B) \log_{M/B} N)$ I/Os for preprocessing and $O(\log_B N)$ I/Os for queries, but requires the tall cache assumption.

We use the standard *trapezoidal decomposition*: vertical lines are extended up and down from each segment endpoint, terminating when they reach another segment. This partitions the plane into a collection of trapezoids, with vertical sides, and at most two non-vertical edges per trapezoid. A key tool in our preprocessing will be the *partition procedure*: given a parameter $m < N$, it selects $O(N/m)$ segments and forms the trapezoidal decomposition they define; each trapezoid will contain $\Theta(m)$ segments or part segments of the N input segments, and for each trapezoid these segments are identified. Such a trapezoidal decomposition is said to have *size* $\Theta(m)$.

The basic idea is to do a plane sweep with a vertical line from left to right, maintaining the collection of segments currently crossing its sweep line in sets of contiguous segments with sizes at least $m/7$ and at most m. It is helpful to set $m' = m/7$. Each pair of adjacent sets is separated by a segment already added to the set of selected segments. Each set is associated with a trapezoid T of the partition; T's left vertical boundary is known; its top and bottom boundaries are the selected segments bounding the set; its right boundary has yet to be specified.

We define the *current* size of a set to be the number of segments it contains (i.e. that cross the sweep line), and the *full* size to be the number of segments contained in the associated trapezoid. We ensure that the current and full sizes of a set are always in the range $[m', 7m']$. We further ensure that on creation, a set's sizes are in the range $[2m', 6m']$.

As left and right segment endpoints are swept over, the corresponding segments are respectively inserted into, or deleted from, the appropriate sets. When a set S's full or current size reaches an extreme value (due to the processing of an endpoint p, say) the vertical line through p provides the right boundary of the associated trapezoid. If S's current size lies outside $[2m', 6m']$, S is either split or merged with a neighbor set, as appropriate. A merge with set S' also results in the trapezoid associated with S' receiving the same right boundary. If the current size of the merged set is greater than $6m'$, the new set is immediately split. Following the splits and merges due to the processing of point p, for each remaining new set, a new trapezoid is instantiated, with left boundary the vertical line through p.

The splitting proceeds a little unusually and is explained next. The set of segments to be split all intersect the sweep line. Consequently they can be sorted along the sweep line. The middle m' segments are considered, and among these the one with the rightmost endpoint is chosen as the separator. The sets resulting from the split will thus have sizes in the range $[2.5m', 5m']$.

It is also possible that a point p swept over is the right endpoint of a selected segment σ. In this case, the two trapezoids bounded by σ are completed with right boundaries through p, and the corresponding sets are merged. If this new set has size more than $6m'$ it is split, and if has size more than $11m'$ it is partitioned by a three-way split, so as to obtain sets in the range $[3m', 6m']$ (two groups of m' segments each are considered, located in the one third and two thirds positions, rather than the middle).

The sweepline procedure works as follows. First, the segment endpoints are sorted according to their x values. Each endpoint is then considered in increasing x order. If it is a left endpoint, we locate the set containing this endpoint and add the incident segment to the set. Similarly, a right endpoint results in the deletion of the incident segment.

The locating is done by means of a search over the selected segments that intersect the sweep line. These segments are maintained in a dynamic search tree. Thus the search uses $O(\log_B m)$ I/Os and maintaining the search tree over the selected segments takes $O(m \log_B m)$ I/Os.

Lemma 7. *The partition procedure forms at most* $11N/m'$ *trapezoids.*

Proof. We count how many right vertical boundaries are created. There are a few ways for this to occur. First, a set can reach full size $7m'$; this requires at least m' insertions into the set since its creation. Second, a set can reach current size m'. This requires at least m' deletions from the set since its creation. Third, a set can be a neighbor of a set whose size drops to m'. Fourth, a set can be bounded by a selected segment whose right endpoint is reached.

There are at most N/m' sets in each of the first three categories. We show that there are at most $8N/m'$ sets in the fourth category by a credit argument. Each segment endpoint is given two credits. When processed, it gives its credits to the two selected segments bounding its set. Consider a selected segment σ whose right endpoint is reached. We show it has received at least m' credits. Consider the m' segments from which σ was selected. Let $\tau \neq \sigma$ be one of these segments. Suppose that when the sweepline reaches τ's right endpoint, τ is no longer in a set bounded by σ; then there must have been at least m' updates to the set containing τ that was created when σ was selected (so that τ could go to a different set). Each of these m' updates contributed one credit to σ. Otherwise, on reaching τ's right endpoint, the endpoint gives a credit to σ. Including the credit from σ's right endpoint, we see that σ always receives at least m' credits.   □

Our first result applies the partition procedure with $m = \sqrt{N} \log N$. Within each trapezoid two sets of segments are identified: those fully contained in the trapezoid and those that touch or cross a vertical boundary (the only boundaries that can be crossed by segments). Each trapezoid and its fully contained segments are handled recursively. For each vertical boundary, we create a "side" search structure for the segments that cross it. Recall that we seek the segment immediately above a query point p.

The selected segments will be used to define the top layer search structure. Deeper layers are defined by the recursion on the trapezoids. The goal of the top search structure is two-fold: first to identify the segment among the selected segments which is directly above the query point; second, to identify the trapezoid defined by the selected segments containing the query point, and thereby enable a recursive continuation of the search. This search structure comprises two levels: the first level is a search tree over the x values given by the endpoints of the selected segments. For each pair of adjacent x values, the second level has a search structure on the selected segments spanning that pair of x values. Both of these can be implemented using binary search trees, laid out in a height-partitioned manner. Thus the space used in the top layer is $O(N/\log^2 N)$ and the search time within the top layer is $O(\log_B N)$.

The side search structure is simply a priority search tree. This can be implemented using a complete binary tree, which is then stored using the height-partitioning layout. In each layer, a query will search two side structures, one for each vertical boundary of the trapezoid being searched; each search will seek the first edge, if any, in the side structure on or above the query point. The search in layer i takes $O(\log_B 2^{2^i})$ I/Os, which summed over all layers is $O(\log_B N)$ I/Os. In order to avoid additive terms of $O(\log \log N)$ in the search time we allocate space for each subproblem recursively. Note that the above structure uses space $O(N \log \log N)$.

Theorem 4. *There is a cache oblivious data structure of size $O(N \log \log N)$ for planar point location that can be constructed using $O(N \log_B N)$ I/Os and supports queries using $O(\log_B N)$ I/Os.*

Now, we obtain a linear space solution. The basic idea is to create a bottom level comprising trapezoids holding $O(\log N)$ segments each. These segments can be stored contiguously, but in no particular order within each trapezoid; such a trapezoid can be searched with $O((\log N)/B)$ I/Os by means of a sequential scan. The bottom level is created by applying the partition procedure with $m = \log N$. Next, a planar point location search structure is build on the selected $O(N/\log N)$ segments, using our first solution. This structure is used to find the segment, among the selected segments, immediately above the query point. Finally, we have to connect the selected segment identified by the query to a particular trapezoid. The difficulty is that a single segment may bound multiple trapezoids. Thus, we need a third part to the data structure. Each segment is partitioned at the boundaries of the trapezoids it bounds, and a search tree is constructed over these boundary points. Having found a segment, the search continues by finding the segment portion immediately above the query point, and then proceeds to search the one trapezoid immediately beneath this segment portion.

Theorem 5. *There is a cache oblivious linear size data structure for planar point location that can be build using $O(N \log_B N)$ I/Os and supports queries using $O(\log_B N)$ I/Os.*

In our strongest result, we seek to build a trapezoidal decomposition of size $a \log N$ for some constant $a \geq 1$. In building the size $a \log N$ decomposition we also build a search structure that in $\Theta((N/B) \log_{M/B} N)$ I/Os can locate N query points, by returning the trapezoid containing the query point. More generally it can also locate $r < N$ query points using $\Theta((r/B) \log_{M/B} r + r \frac{\log(N/r)}{\log B})$ I/Os.

The basic approach is to build in turn trapezoidal decompositions of sizes $aN^{3/4}$, $aN^{(3/4)^2}, \ldots, a \log N$, using each decomposition as a stepping stone to the next one, a being a suitable constant. However, the most natural approach, which is to build each decomposition in turn appears to require re-reading of all the line segments to compute each decomposition, without being able to maintain locality at a sufficiently large granularity; this has an unacceptable cost of $\Omega((N/B) \log \log N)$ I/Os. A second approach, to proceed recursively, i.e. as a trapezoid in one decomposition is found, to immediately compute its partitioning in the smaller decomposition, appears to lead to segment duplication (of those segments that cross multiple trapezoids in a given decomposition, of which there can be many); this appears to entail unacceptable space (and work) $\Theta(N \cdot \text{polylog}(N))$.

Instead, our approach is to seek to identify individual trapezoids containing $\Theta(\log N)$ segments as early as possible, and then to remove them from the series of trapezoidal decompositions, putting them in the final size $a \log N$ decomposition right away. Thus our algorithm simultaneously builds the series of decreasing-size trapezoidal decompositions and the size $a \log N$ decomposition. Not surprisingly, some care is needed to choreograph this activity. Details can be found in [9].

Theorem 6. *There is a batched search structure for planar point location that uses space $O(N)$ and can be built with $O((N/B) \log_{M/B} N)$ I/Os and supports batched search of $r \leq N$ items using $O((r/B) \log_{M/B} r + r \log_B(N/r))$ I/Os.*

References

1. A. Aggarwal and J. S. Vitter. The I/O complexity of sorting and related problems. *Communications of the ACM* **31**, 1116–1127, 1988.
2. A. Ailamaki, D. J. DeWitt, M. D. Hill, and D. A. Wood. DBMSs on a Modern Processor: Where Does Time Go? In *Proc. 25th VLDB Conference* (1999), pp 266–277.
3. A. Andersson. Faster Deterministic Sorting and Searching in Linear Space. In *Proc. 37th IEEE FOCS* (1996), pp. 135–141.
4. A. Andersson and M. Thorup. Tight(er) worst-case bounds on dynamic searching and priority queues. In *Proc. 31st ACM STOC* (2000), pp. 335–342.
5. L. Arge. External memory data structures. In J. Abello, P. M. Pardalos, and M. G. C. Resende, eds, *Handbook of Massive Data Sets*. Kluwer Academic, 2002.
6. L. Arge, D. E. Vengroff and J. S. Vitter. External-Memory Algorithms for Processing Line Segments in Geographic Information Systems (Extended Abstract). In *Proc. 6th European Symposium on Algorithms* (1995), LNCS 979, 295-310.
7. R. Bayer and E. M. McCreight. Organization and maintenance of large ordered indexes. *Acta Informatica*, 1(3):173–189, February 1972.
8. M. A. Bender, E. Demaine and M. Farach-Colton Cache-oblivious B-trees. In *Proc. 41st IEEE FOCS* (2000), pp. 399–409.
9. M. A. Bender, R. Cole and R. Raman. Exponential structures for efficient cache-oblivious algorithms. University of Leicester TR 2002/19, 2002.
10. M. A. Bender, Z. Duan, J. Iacono and J. Wu. A locality-preserving cache-oblivious dynamic dictionary. In *Proc. 13th ACM-SIAM SODA* (2002), pp. 29–38.
11. R. D. Blumofe, M. Frigo, C. F. Joerg, C. E. Leiserson, and K. H. Randall. An analysis of dag-consistent distributed shared-memory algorithms. In *Proc. 8th ACM SPAA*, pp. 297–308, 1996.
12. G. S. Brodal, R. Fagerberg and R. Jacob. Cache-oblivious search trees via binary trees of small height. In *Proc. 13th ACM-SIAM SODA* (2002), pp. 39–48.
13. M. R. Brown, R. E. Tarjan. A data structure for representing sorted lists. *SIAM J. Comput.* **9** (1980), pp. 594–614.
14. M. T. Goodrich, J-J. Tsay, D. E. Vengroff and J. S. Vitter. External-Memory Computational Geometry (Preliminary Version). In *Proc. 34th IEEE FOCS* (1993), pp. 714–723.
15. M. L. Fredman and D. E. Willard. Surpassing the information theoretic bound with fusion trees. *J. Comput. System Sci.*, 47(3):424–436, 1993.
16. M. Frigo, C. E. Leiserson, H. Prokop and S. Ramachandran Cache-oblivious algorithms. In *Proc. 40th IEEE FOCS* (1999), pp. 285–298.
17. K. Mehlhorn. *Data structures and algorithms, 1. Sorting and searching.* Springer, 1984.
18. H. Prokop. Cache-oblivious algorithms. MS Thesis, MIT, 1999.
19. N. Rahman and R. Raman. Adapting radix sort to the memory hierarchy. TR 00-02, King's College London, 2000. Prel. vers. in *Proc. ALENEX 2000*.

20. N. Rahman, R. Cole and R. Raman. Optimised predecessor data structures for internal memory. In *Proc. 5th Workshop on Algorithm Engg.*, LNCS 2141, pp. 67–78, 2001.
21. N. Sarnak and R. E. Tarjan. Planar point location using persistent search trees. *Communications of the ACM* **29**, 1986, 669–679.
22. M. Thorup. Faster Deterministic Sorting and Priority Queues in Linear Space. *Proc. 9th ACM-SIAM SODA* (1998), pp. 550-555.
23. S. Toledo. Locality of reference in *LU* decomposition with partial pivoting. *SIAM Journal on Matrix Analysis and Applications*, 18(4):1065–1081, Oct. 1997.
24. J. S. Vitter. External memory algorithms and data structures: Dealing with MASSIVE data. *ACM Computing Surveys* **33** (2001) pp. 209–271.
25. D. E. Willard. A density control algorithm for doing insertions and deletions in a sequentially ordered file in good worst-case time. *Information and Computation*, 97(2):150–204, 1992.

Bounded-Depth Frege Systems with Counting Axioms Polynomially Simulate Nullstellensatz Refutations

(Extended Abstract)

Russell Impagliazzo[*] and Nathan Segerlind[**]

Department of Computer Science
University of California, San Diego
La Jolla, CA 92093
{russell,nsegerli}@cs.ucsd.edu

Abstract. We show that bounded-depth Frege systems with counting axioms modulo m polynomially simulate Nullstellensatz refutations modulo m. When combined with a previous result of the authors, this establishes the first size (as opposed to degree) separation between Nullstellensatz and polynomial calculus refutations.

1 Introduction

In this paper we study proof sizes in propositional systems that utilize modular counting in limited ways. The complexity of propositional proofs has strong connections to computational and circuit complexity [10,12,15,5]. In particular, NP equals $coNP$ if and only if there exists a propositional proof system that has a polynomial-size proof of every tautology [10]. But before we can prove lower bounds for all proof systems, it seems we should be able to prove lower bounds for specific proof systems.

Lower bounds for proof systems often correspond to lower bounds for associated classes of circuits. Since some of the strongest circuit lower bounds are for bounded-depth circuits with modular gates [16,19], proof systems that allow modular reasoning are of considerable interest. Three such systems are: bounded-depth Frege systems augmented with counting axioms (**BDF+CA**) [1,2,4,18], which add to bounded-depth systems a family of axioms stating that a set of size N cannot be partitioned into set of size m when N is indivisible by m, the Nullstellensatz system (**NS**) [4,7], which captures static polynomial reasoning, and the polynomial calculus [9,17] which captures iterative polynomial reasoning. These systems are inter-related; in particular, the **NS** system was introduced as a technical tool for proving lower bounds for **BDF+CA**.

[*] Research Supported by NSF Award CCR-9734911, grant #93025 of the joint US-Czechoslovak Science and Technology Program, NSF Award CCR-0098197, and USA-Israel BSF Grant 97-00188
[**] Partially supported by NSF grant DMS-9803515

P. Widmayer et al. (Eds.): ICALP 2002, LNCS 2380, pp. 208–219, 2002.
© Springer-Verlag Berlin Heidelberg 2002

We show that bounded-depth Frege systems with counting axioms modulo m polynomially simulate Nullstellensatz refutations modulo m. This shows that earlier lower bounds [1,2] for **BDF+CA** implicitly contained lower bounds for **NS**. (Krajíček [13] shows that the algebraic techniques similar to the those of these papers can be used to obtain **NS** lower bounds, and later papers [4,7,6,11] explicitly use **NS** lower bounds as a tool to prove **BDF+CA** lower bounds. We show that **BDF+CA** lower bounds imply **NS** lower bounds.) From this simulation, we derive new upper bounds for **BDF+CA**, and new size lower bounds for **NS**. In particular, we establish the first superpolynomial size separation between Nullstellensatz and polynomial calculus refutations.

Bounded-depth Frege systems work with propositional formulas and Boolean connectives, whereas the Nullstellensatz and polynomial calculus systems work with polynomials. So in order to simulate one system by the other, we first need to give a translation of Boolean formulas to systems of polynomials. Such translations for specific formulas were implicitly made in the papers using Nullstellensatz as a tool for lower bounds for **BDF+CA** [4,6]. The most direct translation, converting the formula to an equivalent $3 - CNF$ and writing each clause as a degree 3 equation, would give a translation that is (provably) weaker than these implicit translations. We give a translation so that our translations of studied formulas are at least as strong as the historical ones. Thus, our simulation shows that **BDF+CA** lower bounds for these formulas imply matching **NS** lower bounds for their historical translations. As an intermediate step, we give a general notion of reduction between formulas and systems of polynomials.

Outline: In section 2, we give basic definitions and notation. In section 3, we introduce modified counting axioms and show that it can be proved from the normal counting axioms. It will be easier simulate **NS** proofs with these modified counting axioms than to do so directly. In section 4, we give a notion of reduction from propositional formulas to systems of polynomials, and prove that such reductions translate **NS** proofs to **BDF+CA** proofs. In section 5 we give the translation from a formula f to a system of polynomials, and show that f is always reducible to its translation. We explore some applications of this simulation in section 6, such as **BDF+CA** upper bounds for Tseitin tautologies [20], induction on sums [11] and "tau tautologies" [3,14]; as well as a superpolynomial size separation between **NS** and **PC**.

2   Definitions, Notation, and Conventions

In this paper, we make much use of partitions of sets into pieces of a fixed size. We use the following definitions:

Definition 1. *Let S be a set. The set $[S]^m$ is the collection of m element subsets of S. For $e, f \in [S]^m$, we say that e conflicts with f, $e \perp f$, if $e \neq f$ and $e \cap f \neq \emptyset$.*

When N is a positive integer, we let $[N]$ denote the set of integers $\{i \mid 1 \leq i \leq N\}$.

Definition 2. *A* monomial *is a product of variables. A* term *is scalar multiple of a monomial. For a term* $t = c\prod_{i \in I} x_i^{\alpha_i}$, *its* multilinearization *is* $\bar{t} = c\prod_{i \in I} x_i$. *Let* $f = \sum_{i=1}^{k} t_i$ *be a polynomial. The* multilinearization *of* f *is* $\bar{f} = \sum_{i=1}^{k} \bar{t}_i$. *A polynomial* f *is* multilinear *if* $f = \bar{f}$.

Definition 3. *Let* $n > 0$ *be given, and let* $x_1, \ldots x_n$ *be variables. Let* $I \subseteq [n]$ *be given. The monomial* x_I *is defined to be* $\prod_{i \in I} x_i$.

We think of a multilinear polynomial f in $x_1, \ldots x_n$ in the normal form:
$f = \sum_{I \subseteq [n]} a_I x_I$
Technically, we will consider refutation systems rather than proof systems, in that the systems show a set of hypotheses to be unsatisfiable by deriving FALSE from the hypotheses and axioms. The Nullstellensatz and polynomial calculus systems demonstrate that sets of polynomials have no common solution, and are inherently refutation systems. Frege systems can be viewed either as deriving tautologies, or refuting inconsistent sets of hypotheses; for ease of comparison, we will only consider their use as refutation systems.

Frege systems are sound, implicationally complete propositional proof systems over a finite set of connectives with a finite number of axiom schema and inference rules. Cook and Reckhow prove [10] that any two Frege systems polynomially simulate one another, in a way that preserves depth up to constant factors. For this reason, our results hold for any bounded depth Frege system. For concreteness, the reader can keep in mind the following Frege system whose connectives are NOT gates, and unbounded fan-in OR gates, and whose inference rules are: (1) Axioms $\overline{A \vee \neg A}$, (2)Weakening $\frac{A}{A \vee B}$ (3) Cut $\frac{A \vee B \quad (\neg A) \vee C}{B \vee C}$ (4)Merging $\frac{\bigvee X \vee \bigvee Y}{\bigvee (X \cup Y)}$ and (5)Unmerging $\frac{\bigvee (X \cup Y)}{\bigvee X \vee \bigvee Y}$.

For $m \geq 2$, bounded-depth Frege with counting axioms modulo m (written as $BDF + CA_m$) is an extension of bounded-depth Frege systems that has axioms that state for any integer N with $N \not\equiv_m 0$, it is impossible to partition a set of N elements into pieces of size m.

Definition 4. *Let* $m \geq 2$ *and* $N \not\equiv_m 0$ *be given. Let* V *be a set of* N *elements. For each* $e \in [V]^m$, *introduce a variable* x_e, *representing that the elements of* e *are grouped together in a partition.* $Count_m^V$ *is the negation of the conjunction of the following set of formulas:*

1. *For each* $v \in V : \bigvee_{e \ni v} x_e$ *(all elements are covered by some set in the partition)*
2. *For each* e, f *with* $e \perp f : \neg(x_e \wedge x_f)$ *(all sets in the partition are disjoint)*

$BDF + CA_m$ derivations are Frege derivations that allow the use of substitution instances (where each variable is replaced by a formula) of $Count_m^{[N]}$, with $N \not\equiv_m 0$, as axioms.

To prove that a system of polynomials $f_1, \ldots f_k$ has no common roots, one could give a list of polynomials $p_1, \ldots p_k$ so that $\sum_{i=1}^{k} p_i f_i = 1$. Because we are interested in Boolean variables, we add the polynomials $x^2 - x$ as hypotheses to guarantee all roots are zero-one roots.

Definition 5. *For a system of polynomials $f_1, \ldots f_k$ in variables $x_1, \ldots x_n$ over a field F, a* Nullstellensatz refutation *is a list of polynomials $p_1, \ldots p_k, r_1, \ldots r_n$ satisfying the following equation:*

$$\sum_{i=1}^{k} p_i f_i + \sum_{j=1}^{n} r_j \left(x_j^2 - x_j\right) = 1$$

The degree of the refutation is the maximum degree of the polynomials $p_i f_i$, $r_j \left(x_j^2 - x_j\right)$. We define the *size* of a Nullstellensatz refutation to be the total number of (non-zero) monomials appearing in the polynomials $p_1, \ldots p_k$, $f_1, \ldots f_k$ and $r_1, \ldots r_n$. **NS** is defined over any ring, but is complete only when working in a field. In this paper, we will work over $\mathbb{Z}_m$ for any fixed $m \geq 2$, and do not assume m is prime.

One of the corollaries of the simulation is a superpolynomial size separation between polynomial calculus refutations and Nullstellensatz refutations. However, our contribution here is the lower bound for **NS**; the polynomial calculus upper bound was known [11]. Since we never use the polynomial calculus directly in this paper, we omit a definition; see [9,17,7].

3   Contradictory Partitions of Satisfied Variables

In this section, we introduce a variant counting principle that will serve as an intermediate point in our simulation. We show that this variant principle is provable from the standard counting principle. Our simulation will then convert the Nullstellensatz refutation to a bounded depth Frege proof of the negation of this variant principle, yielding a bounded depth Frege refutation from the standard counting principle. The variant says that for any assignment to a set of variables, it is impossible to to have two m-partitions, one which perfectly covers the satisfied variables, and another which covers the satisfied variables and k new points with $1 \leq k \leq m - 1$.

Definition 6. *Let $m \geq 2, 1 \leq k \leq m - 1$ and $n \geq 1$ be integers. Let $u_1, \ldots u_n$ be a set of boolean variables. For each $e \in [n]^m$, let y_e be a variable, and for each $e \in [n+k]^m$, let z_e be a variable.*

$CP_m^{n,k}(\boldsymbol{u}, \boldsymbol{y}, \boldsymbol{z})$ is the negation of the conjunction of the following formulas:

1. *For each $e \in [n]^m$, and $i \in e$: $y_e \to u_i$. Every variable covered by the first partition is satisfied.*
2. *For each $i \in [n]$: $u_i \to \bigvee_{i \in e} y_e$. Every satisfied variable is covered by the first partition.*
3. *For each $e, f \in [n]^m$ with $e \perp f$: $\neg y_e \vee \neg y_f$. Sets in the first partition are disjoint.*
4. *For each $e \in [n+k]^m, i \in e, i \leq n$: $z_e \to u_i$ Every variable covered by the second partition is satisfied.*
5. *For each $i \in [n]$: $u_i \to \bigvee_{i \in e} z_e$. Every satisfied variable is covered by the second partition.*

6. For each i, $n + 1 \leq i \leq n + k$: $\bigvee_{i \in e} z_e$ Every extra point is covered by the second partition.

7. For each $e, f \in [n + k]^m$ with $e \perp f$: $\neg z_e \vee \neg z_f$. Sets in the second partition are disjoint.

Lemma 1. *Fix m and k. For all n, the tautology $CP_m^{n,k}$ has a constant depth, size $O(n^m)$ proof in bounded-depth Frege from a counting modulo m axiom.*

Proof. Fix m, n and k. The proof of $CP_m^{n,k}$ is by contradiction. We define a set U of size $mn + k$ and formulas ϕ_e for each $e \in [U]^m$ so that we can derive $\left(\neg \text{Count}_m^U\right)[x_e \leftarrow \phi_e]$ in size $O(n^m)$ from the hypothesis $\neg CP_m^{n,k}$.

U contains m copies of each variable u_i and k extra points; more precisely, U contains $p_{r,i}$, $r \in [m]$, $i \in [n]$ (the r'th copy of u_i), and $p_{m,i}$, $n + 1 \leq i \leq k$ (the extra points.) We define the partition formulas, ϕ_e, for $e \in [U]^m$, as follows:

1. When a variable is false, we group together its m copies: For each $i \in [n]$,
   $\phi_{\{p_{1,i},\ldots p_{m,i}\}} = \neg u_i$.
2. Use the first partition to group the first $m - 1$ copies of all the true variables: For each $r \in [m - 1]$, each $i_1, \ldots i_m \in [n]$, $\phi_{\{p_{r,i_1},\ldots p_{r,i_m}\}} = y_{\{i_1,\ldots i_m\}}$.
3. Use the the second partition to group the last copy of each true variable, and the extra points: For each $i_1, \ldots i_m \in [n + k]$, $\phi_{\{p_{m,i_1},\ldots p_{m,i_m}\}} = z_{\{i_1,\ldots i_m\}}$.
4. No other groups are used: For all other $e \in [U]^m$, $\phi_e = 0$.

Note that the above formulas are literals or constants, so the above substitution does not increase formula depth.

Now we sketch the derivation of $\left(\neg \text{Count}_m^U\right)[x_e \leftarrow \phi_e]$ from $\neg CP_m^{n,k}$. It is easily verified that the derivation has constant depth and size $O((mn + k)^m) = O(n^m)$.

We need to prove that every point of U is covered by the partition defined by the ϕ's. First, for any $p_{r,i} \in U$ with $i \in [n]$, $r \in [m - 1]$, observe that $\neg u_i$ implies $\phi_{\{p_{1,i},\ldots p_{m,i}\}}$. On the other hand, u_i implies $\bigvee_{i \in e} y_e = \bigvee_{i_1,\ldots i_{m-1}} \phi_{\{p_{r,i},p_{r,i_1}\ldots p_{r,i_{m-1}}\}}$. In either case $p_{r,i}$ is covered. (Frege proofs of this would involve using the cut rule on u_i to do the case analysis, and then weakening.) If $r = m$ and $1 \leq i \leq n$, we simply replace the y_e with z_e in the second case above. If $r = m, i > n$ then there is only the second case, using the z_e.

Next we need to prove that no overlapping edges are used. Let $e_1, e_2 \in [U]^m$ be given so that $e_1 \perp e_2$, and neither ϕ_{e_1} nor ϕ_{e_2} is identically 0. If $\phi_{e_1} = \neg u_i$ and $\phi_{e_2} = y_f$, then e_1 is $\{p_{r,i} \mid r \in [m]\}$ and e_2 is $\{p_{r,j} \mid j \in f\}$ for some $r \in [m]$ and $f \in [n]^m$ so that $i \in f$. $\neg CP_m^{n,k}$ includes the clause $y_f \to u_i$, which is equivalent to $\neg\neg u_i \vee \neg y_f = \neg\phi_{e_1} \vee \neg\phi_{e_2}$.

If $\phi_{e_1} = y_{f_1}$ and $\phi_{e_2} = y_{f_2}$, then e_1 is $\{p_{r_1,i} \mid i \in f_1\}$ and e_2 is $\{p_{r_2,i} \mid i \in f_2\}$ with $r_1 = r_2$ and $f_1 \perp f_2$. From $\neg CP_m^{n,k}$ derive $\neg y_{f_1} \vee \neg y_{f_2} = \neg\phi_{e_1} \vee \neg\phi_{e_2}$.

The only other cases are when $\phi_{e_1} = \neg u_i$ and $\phi_{e_2} = z_f$ or $\phi_{e_1} = z_{f_1}$ and $\phi_{e_2} = z_{f_2}$. These are handled similarly to the preceding cases.

4   The Simulation

In this section, we define a notion of reduction from a set of Boolean formulas to a system of polynomials. If Γ reduces to P, then any Nullstellensatz refutation of P can be converted to a **BDF+CA** refutation of Γ of roughly the same size.

For a polynomial f, treat a monomial with coefficient a in f as having a distinct copies. On a given assignment, f is zero modulo m if and only if there is an m-partition on its satisfied monomials. To reduce a formula to a system of polynomials, we will define such a partition for each polynomial, and prove that defined partition does indeed partition the satisfied monomials. We can restrict our attention to multilinear polynomials.

Definition 7. *Let $f = \sum_{I \subseteq [n]} a_I x_I$ be a multilinear polynomial over $\mathbb{Z}_m$. The set of monomials of f, denoted M_f, is $\{m_{c,I} \mid I \subseteq [n], \ c \in [a_I]\}$*

Think of $m_{c,I}$ as the c'th copy of the monomial x_I. (Do not confuse the monomial $m_{c,I}$ with the modulus m.)

Definition 8. *Let $x_1, \ldots x_n$ be variables. Let f be a multilinear polynomial in the variables $x_1, \ldots x_n$.*

Let $y_1, \ldots y_N$ be boolean variables. For each $i \in [n]$, let δ_i be a formula in $\boldsymbol{y}$. For each $E \in [M_f]^m$, let θ_E be a formula in $\boldsymbol{y}$.

We say that the θ's form an m-partition of the satisfied monomials of f with definitions δ if the following hold:

1. *For each $E \in [M_f]^m, I \in E$: $\theta_E \to \bigwedge_{k \in I} \delta_k$. Only satisfied monomials are covered.*
2. *For each $I \in M_f, c \in [a_I]$: $\left(\bigwedge_{k \in I} \delta_k \right) \to \bigvee_{\substack{E \in [M_f]^m \\ E \ni m_{c,I}}} \theta_E$. All satisfied monomials are covered.*
3. *For each $E, F \in [M_f]^m, E \perp F$, $\neg\theta_E \vee \neg\theta_F$ All sets in the partition are disjoint.*

Definition 9. *Let $x_1, \ldots x_n$ and $y_1, \ldots y_N$ be boolean variables. Let $\Gamma(\boldsymbol{y})$ be a propositional formula. Let $F = \{f_1, \ldots f_k\}$ be a system of polynomials from $\mathbb{Z}_m[\boldsymbol{x}]$.*

A (d, T)-reduction from Γ to F is a set of depth d formulas $\delta_i(\boldsymbol{y})$, for $i \in [n]$, and $\beta_E^i(\boldsymbol{y})$, for $i \in [k]$, $E \in \left[M_{\bar{f}_i}\right]^m$, so that there is a size T, depth d Frege derivation from $\Gamma(\boldsymbol{y})$ that, for each $i \in [k]$, the β^i's form an m-partition on the satisfied monomials of $\bar{f}_i$ with definitions δ.

The following is our main technical result, showing that the above notion of reduction translates $\mathbf{NS_m}$ refutations into $\mathbf{BDF + CA_m}$ refutations.

Theorem 1. *Let $m > 1$ be an integer. Let $x_1, \ldots x_n$ and $y_1, \ldots y_N$ be boolean variables.*

Let $\Gamma(\boldsymbol{y})$ be a propositional formula, and let F be a system of polynomials over $\mathbb{Z}_m$ in $\boldsymbol{x}$ so that Γ reduces to F in depth d and size T.

If there is a Nullstellensatz refutation of F with size S, then there is a depth $O(d)$ Frege with counting axioms modulo m refutation of $\Gamma(\boldsymbol{x})$ with size $O(S^{2m}T)$.

Proof. Let $p_1, \ldots p_k, r_1, \ldots r_n$ be a size S Nullstellensatz refutation of F. Let $\delta_k(\boldsymbol{y})$, for $k \in [n]$, and $\beta_E^i(\boldsymbol{y})$, for $i \in [k]$, $E \in [M_{\bar{f}_i}]^m$, be formulas so that from Γ there is a size T, depth d proof that for each $i \in [k]$ the $\beta_E^i(\boldsymbol{x})$'s form an m-partition on the satisfied monomials of $\bar{f}_i$ with definitions δ.

We will derive the negation of the contradictory partitions of satisfied variables principles from Γ. The variables will correspond to all the monomials that appear in the expansion of $\sum_{i=1}^k \bar{p}_i \bar{f}_i$ when we cross-multiply terms and multilinearize, but do not cancel like terms. One partition of these variables will be based on extending the m-partition on the satisfied monomials of each f_i to a partition on the satisfied monomials in $p_i f_i$. On the other hand, if we do combine like terms, all terms but the constant term have coefficients that are 0 modulo m. The second partition will thus group like terms, along with $m-1$ extra point to group the remaining constant term with.

More precisely, the set of variables will be the collection, over $i \in [k]$, of all pairs of monomials from $\bar{p}_i$ and $\bar{f}_i$.

$$V = \bigcup_{i=1}^k \{(m_{c,I}, m_{d,J}, i) \mid m_{c,I} \in M_{\bar{p}_i}, \; m_{d,J} \in M_{\bar{f}_i}\}$$

Notice that $|V| = O(S^2)$.

For each $v \in V$, with $v = (m_{c,I}, m_{d,J}, i)$, let $\gamma_v = \bigwedge_{k \in I \cup J} \delta_k$. Think of this as an occurrence of the monomial $x_{I \cup J}$. We will give formulas θ_E, that define a partition on the satisfied monomials with $m-1$ many extra points, and η_E, that define a partition on the satisfied monomials with no extra points. Using these definitions, we give a size $O(|V|^m + T) = O(S^{2m} + T)$, depth $O(d)$ derivation from Γ of the following substitution instance of $\neg CP_m^{|V|, m-1}$:

$$\neg CP_m^{|V|, m-1} [u_v \leftarrow \gamma_v, \; y_E \leftarrow \eta_E, \; z_E \leftarrow \theta_E]$$

On the other hand, by lemma 1, $CP_m^{|V|, m-1}$ has constant depth Frege proofs of size $O(|V|^m)$, so $CP_m^{|V|, m-1} [u_v \leftarrow \gamma_v, \; y_E \leftarrow \eta_E, \; z_E \leftarrow \theta_E]$ has a depth $O(d)$ Frege with counting axioms modulo m proof of size $O(|V|^m T)$.

Therefore, Γ has an $O(d)$ depth Frege with counting axioms modulo m refutation of size $O(S^{2m}T)$.

The Partition with $m-1$ Extra Points

Notice that we have the following equation:

$$\sum_{i=1}^k \overline{\bar{p}_i \bar{f}_i} = \overline{\sum_{i=1}^k p_i f_i + \sum_{j=1}^n r_j(x_j^2 - x_j) = 1}$$

Therefore, when we collect terms after cross-multiplying the terms of $\sum_{i=1}^k \bar{p}_i \bar{f}_i$ and multilinearizing, the coefficient of every nonconstant term is 0 modulo m, and the constant term is 1 modulo m.

For each $S \subseteq [n]$, let $V_S = \{(m_{c,I}, m_{d,J}, i)) \in V \mid I \cup J = S\}$. Think of these as the occurrences of x_S in the multilinearized expansion. For each $S \subseteq [n]$, $S \neq \emptyset$, there is an m-partition on V_S, call it $\mathcal{P}_S$. Likewise, there is an m-partition on $V_\emptyset \cup [m-1]$, call it $\mathcal{P}_\emptyset$. Define the formulas θ_E as follows: for each $E \in ([V] \cup [m-1])^m$, if $E \in \mathcal{P}_S$ for some $S \subseteq [n]$, then $\theta_E = \bigwedge_{k \in S} \delta_k$, otherwise $\theta_E = 0$.

We now sketch a size $O(|V|^m)$ and depth $O(1)$ Frege with counting axioms modulo m proof that these formulas define an m-partition of the satisfied monomials of $\sum_{i=1}^k \bar{p}_i \bar{f}_i$ with $m - 1$ extra points. It is trivial from the definition of θ_E that the edges cover only satisfied monomials. That every satisfied monomial $\bigwedge_{k \in S} \delta_k$ is covered is also trivial: the edge from $\mathcal{P}_S$ is used if and only if the monomial $x_S (\bigwedge_{k \in S} \delta_k)$ is satisfied. Finally, it easily shown that the formulas for two overlapping edges are never both satisfied: only edges from $\mathcal{P}_S$ are used (regardless of the values of the x's), so for any pair of overlapping edges, $E \perp F$, one of the two formulas θ_E or θ_F is identically 0.

The Partition with No Extra Points

The idea is that an m-partition on the satisfied monomials of $\bar{f}_i$ can be used to build an m-partition on the satisfied monomials of $\bar{p}_i \bar{f}_i$.

For each $E \in [V]^m$, define η_E as follows: if $E = \{(m_{c,I}, m_{d_l,J_l}, i) \mid l \in [m]\}$ for some $i \in [k]$, $m_{c,I} \in M_{f_i}$, then $\eta_E = \bigwedge_{k \in I} \delta_k \wedge \beta_{\{m_{d_l,J_l} \mid l \in [m]\}}$, otherwise, $\eta_E = 0$.

There is a size $O(S + |V|^m)$, depth $O(d)$ Frege derivation from Γ that the η_E's form an m-partition on the satisfied monomials of $\sum_{i=1}^k \bar{p}_i \bar{f}_i$ with definitions δ. We briefly sketch how to construct the proof. Begin by deriving from Γ, for each i, that the β_E^i's form an m-partition on the satisfied monomials of $\bar{f}_i$.

"Every satisfied monomial is covered." Let $(m_{c,I}, m_{d,J}, i) \in V$ be given. If $\bigwedge_{k \in I \cup J} \delta_k$ holds, then so do $\bigwedge_{k \in I} \delta_k$ and $\bigwedge_{k \in J} \delta_k$. Because the β^i's form an m-partition on the satisfied monomials of $\bar{f}_i$ with definitions δ, we may derive $\bigvee_{F \in [M_{f_i}]^m} \beta_F^i$. From this and $\bigwedge_{k \in I} \delta_k$, derive $\bigvee_{F \in [M_{f_i}]^m} (\bigwedge_{k \in I} \delta_k \wedge \beta_F^i)$. A weakening inference applied to this yields $\bigvee_{E \in [V]^m} \eta_E$.

"Every monomial covered is satisfied." Let $v = (m_{c,I}, m_{d,J}, i) \in V$ be given so that $v \in E$ and η_E holds. For this to happen, $E = \{(m_{c,I}, m_{d_l,J_l}, i) \mid l \in [m]\}$, for some $m_{c,I} \in M_{\bar{p}_i}$. By definition,, $\eta_E = \bigwedge_{k \in I} \delta_k \wedge \beta_{\{m_{d_l,J_l} \mid l \in [m]\}}^i$, and therefore $\bigwedge_{k \in I} \delta_k$ holds. Because the β^i's form an m-partition on the satisfied monomials of $\bar{f}_i$, we have that $\bigwedge_{k \in J} \delta_k$ holds. Therefore $\bigwedge_{k \in I \cup J} \delta_k$ holds.

"No two conflicting edges E and F can have η_E and η_F simultaneously satisfied." If $E \perp F$, and neither θ_E nor θ_F is identically 0, then they share the same $\bar{p}_i$ component. That is, there exists i, $m_{c,I} \in M_{\bar{p}_i}$ so that $E = \{(m_{c,I}, m_{d_l,J_l}, i) \mid l \in [m]\}$, and $F = \{(m_{c,I}, m_{d'_l,J'_l}, i) \mid l \in [m]\}$. Because $E \perp F$, we have $\{m_{d_l,J_l} \mid l \in [m]\} \perp \{m_{d'_l,J'_l} \mid l \in [m]\}$. Because the β^i's form an m-partition on the satisfied monomials of $\bar{f}_i$, we can derive $\neg\beta_{\{m_{d_l,J_l} \mid l \in [m]\}}^i \vee \neg\beta_{\{m_{d'_l,J'_l} \mid l \in [m]\}}^i$. We weaken this formula to obtain $\neg\beta_{\{m_{d_l,J_l} \mid l \in [m]\}}^i \vee \neg\beta_{\{m_{d'_l,J'_l} \mid l \in [m]\}}^i \vee \bigvee_{k \in I} \neg\delta_k$, and from this derive $\neg\left(\bigwedge_{k\in I}\delta_k\wedge\beta^i_{\{m_{d_l,J_l}|l\in[m]\}}\right)\vee\neg\left(\bigwedge_{k\in I}\delta_k\wedge\beta^i_{\{m_{d'_l,J'_l}|l\in[m]\}}\right)=\neg\eta_E\vee\neg\eta_F$.

5   Translations of Formulas into Polynomials

To apply the results in the last section, we need to provide a way of translating formulas into sets of polynomials, and give a reduction between the formulas and their translations.

t-**CNFs:** The simplest type of translation is when Γ is a t-CNF for t constant. Then for each clause γ_i, we can express γ_i as a multilinear polynomial f_i. For every assignment satisfying γ_i, the number of satisfied monomials in f_i is divisible by m, so we can partition them into sets of size m. Since γ_i and f_i depend on at most t variables, the number of such monomials is $O(2^t)$ and we can define and prove valid this partition with proof of size at most $O(2^t)$. Thus, if $t\in O(\log n)$, this gives us a depth 1, polynomial-size reduction from Γ to the f_i.

Translation to 3-CNF: More involved translations of formulas into sets of polynomials use extension variables that represent sub-formulas. The simplest way of doing this would be to reduce an unbounded fan-in formula Γ to a bounded fan-in formula, and then introduce one new variable y_g per gate g, with the polynomial that says y_g is computed correctly from its inputs. It is easy to give a reduction from Γ to this translation, of depth $depth(\Gamma)$ and size $poly(|\Gamma|)$. (We would let δ_{y_g} be the subformula rooted at g.) However, there would usually be no small degree Nullstellensatz refutation of the resulting system of equations, even for trivial Γ. For example, say that we translated the formula $x_1,\neg(((((x_1\vee x_2)\vee\ldots x_n)$ this way. It can be shown that the resulting system of polynomials is weaker than the induction principles, which require $\Omega(\log n)$ degree **NS** refutations [8].

A robust translation: Instead, we give a translation that is at least as strong as the traditional ways of translating formulas into polynomials. For large fan-in gates, we introduce variables that say whether an input is the first satisfied input. By making these input variables mutually exclusive, we can write the value of a gate as the sum of the variables for each input.

Definition 10. *Let Γ be a formula in the variables $x_1,\ldots x_n$ and the connectives $\{\bigvee,\neg\}$. For each pair of subformulas g_1 and g_2 of Γ, we write $g_1\to g_2$ if g_1 is an input to g_2. Order the subformulas of f, and write $g_1<g_2$ if g_1 precedes g_2 in this ordering. In addition to $x_1,\ldots x_n$, for each (non-input) subformula g of f, introduce a variable y_g - "the value of g". For each $\bigvee$ gate g_2 of f, and each input to g_2, $g_1\to g_2$, introduce a variable z_{g_1,g_2}- "g_1 is the first satisfied input of g_2".*

The polynomial translation of f, $POLY(f)$, is the following set of polynomials:

1. *We restrict to boolean values. For all g, $g_1\to g$: $y_g^2-y_g$ and $z_{g_1,g}^2-z_{g_1,g}$.*
2. *If g_1 is a satisfied input to OR gate g, and g_1 precedes g_2, then g_2 is not the first satisfied input to g: $y_{g_1}z_{g_2,g}$.*

3. If g_1 is the first satisfied input of g, then g_1 is satisfied: $z_{g_1,g}y_{g_1} - z_{g_1,g}$.
4. If an OR gate g is false, then for each input g_1 of g, g_1 is false: $(y_g - 1)y_{g_1}$
5. An OR gate g is satisfied if and only if some input to g is the first satisfied input of g: $y_g - \sum_{g_1 \to g} z_{g_1,g}$.
6. A not gate g is the negation of its input g_1: $y_{g_1} + y_g - 1$.
7. The formula Γ is satisfied: $y_\Gamma - 1$

One can easily show by induction that Γ is satisfiable if and only if POLY(Γ) has a common root.

Theorem 2. *If Γ is a formula in the variables $x_1, \ldots x_n$ and the connectives $\{\bigvee, \neg\}$, then Γ is reducible to POLY(Γ) in depth $O(depth(\Gamma))$ and size polynomial in $|\Gamma|$.*

The definition of a variable y_g is simply $\delta_{y_g} = g(\boldsymbol{x})$. For a variables $z_{g_1,g}$, the definition is $\delta_{z_{g_1,g}} = g_1(\boldsymbol{x}) \wedge \bigwedge_{\substack{g_2 \to g \\ g_2 < g_1}} \neg g_2(\boldsymbol{x})$.

For the partitions, the only interesting case is that of the fifth class of polynomials. Consider a polynomial $y_g - \sum_{g_1 \to g} z_{g_1,g}$. (Note that, in normal form, each -1 is replaced by $m - 1$.) We define the partition on the satisfied monomials as follows: For each $g_1 \to g$, we group together the monomial y_g with the $m - 1$ copies of the monomial $z_{g_1,g}$ if and only if $\delta_{z_{g_1,g}}$. No other sets are part of the partition. This automatically ensures that each monomial $z_{g_1,g}$ is covered if satisfied, covered at most once, and not covered if not satisfied. To prove that this is a valid partition on satisfied monomials, we need this also to be satisfied for the monomial y_g. From $\delta_{z_{g_1,g}}$ we can easily derive by weakening $\neg\delta_{y_{g_2}}$ for each $g_2 < g_1, g_2 \to g$ and $\delta_{y_{g_1}}$. It then follows that $\delta z_{g_1,g} \to \neg\delta_{z_{g_2,g}}$ for any $g_2 \to g, g_2 \neq g_1$. Thus, y_g is covered at most once. Then one can derive $\delta_{y_g} \equiv \bigvee \delta_{z_{g_1,g}}$ by repeatedly applying cut rules. This proves that y_g is covered if and only if it is satisfied.

Example: We illustrate the above translation with the counting principles themselves. The variables introduced in the clauses $\neg y_e \vee \neg y_f$ are easily solved for, yielding the equations $y_e y_f = 0$. Each clause $\bigvee_{e \ni v} y_e$ introduces n new variables $z_{i,e}$, with $\sum z_{i,e} = 1$, and $z_{i,e}y_e = z_{i,e}$. One can verify that the equations $z_{i,e} = y_e$ are expressible as low degree linear combinations of the above. Substituting, the equations become the standard translation: $\sum_{e \ni v} y_e = 1$ and $y_e y_f = 0, e \perp f$.

6   Applications

We summarize some applications of the simulation. To save space, the proofs are omitted. Interested parties can find the proofs in the full version of the paper.

Many tautologies studied in propositional proof complexity, such as Tseitin's tautologies and the τ formulas of Nisan-Wigderson generators built from parity, can be expressed as inconsistent systems of linear equations over a field $\mathbb{Z}_q$ in which each equation involves only a small number of variables. An equation involving w variables can be explicitly encoded by a DNF of size $O(w2^w)$, and the system can be explicitly encoded by taking the conjunction of the explicit encodings of the equations.

Theorem 3. *Fix a prime number* q. *Let* A *be an* $m \times n$ *matrix, let* $x_1, \ldots x_n$ *be variables and let* $b \in \mathbb{Z}_q^m$ *be so that* $Ax = b$ *has no solutions. Let* w *be the maximum number of non-zero entries in any row of* A.

There is a constant depth Frege with counting axioms modulo q *refutation of the explicit encoding of* $Ax = b$ *of size polynomial in* m,n *and* 2^w.

In another paper [11], the authors introduced a family of unsatisfiable sets of clauses based on the following principle: Let N be a positive integer. Suppose that we have N rows of N boolean variables. There is no assignments to these variables with the following properties: (1) The parity of the first row is 0 (2) The parity of the final row is 1 (3)For each r from 2 to N, the parity of row r is equal to the parity of row r times the parity of row $r - 1$.

The details of this formula, call it $IS(N)$, and its algebraic translation, $AIS(N)$, can be found in [11]. Their properties that concern us here are: (1) $IS(N)$ has no refutation of polynomial size in bounded-depth Frege with counting axioms modulo two (2) $AIS(N)$ has a polynomial size refutation in the polynomial calculus modulo two (3) $AIS(N)$ has a quasipolynomial size refutation in Nullstellensatz modulo two (4) the principles $IS(N)$ reduces to $AIS(N)$ is polynomial size and constant depth.

Combining these facts with theorem 1, shows that there is no polynomial size Nullstellensatz refutation of the principles $AIS(N)$ and that there is a quasipolynomial size refutation of $IS(N)$ in bounded-depth Frege with counting axioms modulo two.

Acknowledgements. The authors would like to thank Sam Buss for many useful conversations and insightful suggestions, and the anonymous referees for comments that led to many improvements.

References

1. M. Ajtai. Parity and the pigeonhole principle. In *FEASMATH: Feasible Mathematics: A Mathematical Sciences Institute Workshop*. Birkhauser, 1990.
2. M. Ajtai. The independence of the modulo p counting principles. In *Proceedings of the Twenty-Sixth Annual ACM Symposium on the Theory of Computing*, pages 402–411, 23–25 May 1994.
3. M. Alekhnovich, E. Ben-Sasson, A. Razborov, and A. Wigderson. Pseudorandom generators in propositional proof complexity. In *41st Annual Symposium on Foundations of Computer Science: proceedings: 12–14 November, 2000, Redondo Beach, California*, pages 43–53, 2000.
4. P. Beame, R. Impagliazzo, J. Krajíček, T. Pitassi, and P. Pudlák. Lower bound on Hilbert's Nullstellensatz and propositional proofs. In *35th Annual Symposium on Foundations of Computer Science*, pages 794–806. IEEE, 20–22 November 1994.
5. P. Beame and T. Pitassi. Propositional proof complexity: Past, present, and future. *Bulletin of the EATCS*, 65:66–89, 1998.
6. P. Beame and S. Riis. More one the relative strength of counting principles. In P. Beame and S. Buss, editors, *Proof Complexity and Feasible Arithmetics*, pages 13–35. American Mathematical Society, 1998.

7. S. Buss, R. Impagliazzo, J. Krajíček, P. Pudlák, R. Razborov, and J. Sgall. Proof complexity in algebraic systems and bounded depth frege systems with modular counting. *Computational Complexity*, 6, 1997.

8. S. Buss and T. Pitassi. Good degree bounds on nullstellensatz refutations of the induction principle. In *Proceedings of the Eleventh Annual Conference on Computational Complexity*, 1996.

9. M. Clegg, J. Edmonds, and R. Impagliazzo. Using the Groebner basis algorithm to find proofs of unsatisfiability. In *Proceedings of the Twenty-Eighth Annual ACM Symposium on the Theory of Computing*, pages 174–183, 22–24 May 1996.

10. S. Cook and R. Reckhow. The relative efficiency of propositional proof systems. *The Journal of Symbolic Logic*, 44(1):36 – 50, March 1979.

11. R. Impagliazzo and N. Segerlind. Counting axioms do not polynomially simulate counting gates (extended abstract). In *Proceedings of the Forty-Second Annual Symposium on Foundations of Computer Science*, pages 200–209. IEEE Computer Society, 2001.

12. J. Krajíček. *Bounded Arithmetic, Propositional Logic and Complexity Theory*. Cambridge University Press, 1995.

13. J. Krajíček. Uniform families of polynomial equations over a finite field and structures admitting and euler characteristic of definable sets. *Proc. London Mathematical Society*, 3(81):257–284, 2000.

14. J. Krajíček. Tautologies from pseudo-random generators. *The Bulletin of Symbolic Logic*, 7(2):197–212, 2001.

15. P. Pudlák. The lengths of proofs. In S. R. Buss, editor, *Handbook of Proof Theory*, pages 547–637. Elsevier North-Holland, 1998.

16. A. Razborov. On the method of approximations. In *Proceedings of the Twenty First Annual ACM Symposium on Theory of Computing*, pages 167–176, 15–17 May 1989.

17. A. Razborov. Lower bounds for the polynomial calculus. *CMPCMPL: Computational Complexity*, 7, 1998.

18. S. Riis. Count(q) does not imply Count(p). *Annals of Pure and Applied Logic*, 90(1–3):1–56, 15 December 1997.

19. R. Smolensky. Algebraic methods in the theory of lower bounds for Boolean circuit complexity. In *Proceedings of the Nineteenth Annual ACM Symposium on Theory of Computing*, pages 77–82, New York City, 25–27 May 1987.

20. G. Tseitin. On the complexity of proofs in propositional logics. *Seminars in Mathematics*, 8, 1970.

On the Complexity of Resolution with Bounded Conjunctions*

(Extended Abstract)

Juan Luis Esteban[1], Nicola Galesi[1,2], and Jochen Messner[3]

[1] Universitat Politècnica de Catalunya, esteban@lsi.upc.es.
[2] University of Toronto, galesi@cs.toronto.edu.
[3] Universität Ulm, messner@informatik.uni-ulm.de.

Abstract. We analyze size and space complexity of $Res(k)$, a family of proof systems introduced by Krajíček in [16] which extend Resolution by allowing disjunctions of conjunctions of up to $k \geq 1$ literals. We prove the following results: (1) The treelike $Res(k)$ proof systems form a strict hierarchy with respect to proof size and also with respect to space. (2) Resolution polynomially simulates treelike $Res(k)$, and is almost exponentially separated from treelike $Res(k)$. (3) The space lower bounds known for Resolution also carry over to $Res(k)$. We obtain almost optimal space lower bounds for PHP_n, GT_n, Random Formulas, CT_n, and Tseitin Tautologies.

1 Introduction

A central theme in Computational Complexity is whether there is an *efficient* propositional proof system, i.e. a proof system that for any tautology provides a proof polynomial in the size of the tautology. As observed in [13] this corresponds to the question whether NP = coNP. Hence the investigation of the complexity of proof systems, can be seen as a way to tackle NP $\neq$ coNP: prove that for every propositional proof system there are tautologies that require long proofs. The approach that one follows is to consider proof systems of increasing power and to investigate their limitations.

Among the most studied proof systems are those related to Resolution. While there are several lower bounds for the complexity of proofs in propositional Resolution [15,25,5,9], Resolution-based proof systems are still a subject of research [16,4,3]. On the one hand one is interested in finding more and more powerful combinatorial lower bounds techniques that hopefully can be applied to stronger systems [22,23]. On the other hand Resolution-based proof systems are of practical interest in the field of Automated Theorem Proving.

* The authors were partly supported by the DAAD project *Acciones integradas Hispano-Alemanas* no. 70116. J.L. Esteban was also partly supported by MEC through grant PB 98-0937-C04 (FRESCO Project). N. Galesi was also partly supported by Spanish grant TIC2001-1577-C03-02 and by CSERC Canadian funds.

P. Widmayer et al. (Eds.): ICALP 2002, LNCS 2380, pp. 220–231, 2002.

Given that no polynomial time algorithm can exist to find proofs for a non-efficient proof system like Resolution, Bonet *et al.* in [12] proposed the following approach: for a proof system P, find algorithms running in time polynomial in the size of the shortest P-proof of the formula we are seeking proofs for. If such an algorithm exists we say that P is *automatizable*. Despite that many proof systems, including Resolution, are not automatizable conditioned to plausible complexity assumptions ([2,18,12]), there are examples of Resolution-based proof systems known to be subexponentially automatizable [5,9,3].

This work focuses on the family of refutation systems $Res(k)$, $k \geq 1$, introduced by Krajíček in [16] as a generalization of Resolution. Instead of clauses, $Res(k)$ allows to infer k-clauses, i.e. disjunctions of k-bounded conjunctions. In [4] Atserias *et al.* gave exponential lower bounds for $Res(2)$ refutations of random formulas. Moreover, generalizing the algorithm of Beame *et al.* in [11], Atserias and Bonet in [3] provided a quasi-polynomial automatization for treelike $Res(k)$, the restricted system in which the proof is a tree.

In the following we say that a proof system P *dominates* a system Q if (1) P polynomially simulates Q, i.e. for any Q-proof there is P-proof of the same formula that is polynomially related in size. And (2) P is almost exponentially separated from Q, i.e. there exists a class of formulas having polynomial size P-proofs, but requiring Q-proofs of almost exponential size. Let us note here that throughout the paper we will consider a lower bound of $2^{\Omega(n/\log n)}$ as *almost exponential*. It is known that Resolution dominates treelike Resolution [10,8]. We improve this result by proving that Resolution dominates treelike $Res(k)$, for k constant. In fact we prove the stronger result that treelike $Res(k+1)$ dominates treelike $Res(k)$ and we show that Resolution simulates treelike $Res(k)$ by only a linear increase in size (see also [17,19]). To prove the separations we extend to $Res(k)$ a technique based on games introduced by Pudlák and Impagliazzo in [21] and used by Ben-Sasson et al. in [8] to give lower bounds for treelike Resolution. Then we define a generalization of the Pebbling Contradictions (see [8]) and we generalize the technique of [8] based on pebbling of graphs, to work for $Res(k)$. In particular we show a combinatorial Lemma (Lemma 7) about pebbling of graphs which might be interesting in its own right and could also be applied to other areas.

Concerning automatizability, our result and the algorithm from [3] show the quasipolynomial automatizability for a hierarchy of almost exponentially separated proof systems. Moreover, since the Generalized Pebbling Contradictions have Resolution refutations using constant size clauses, it says that there are examples for which the algorithm of [9] can be exponentially faster in finding Resolution proofs than the algorithm of [3] in finding treelike $Res(k)$ proofs.

Proof size is not the only complexity measure for proof systems. The definition of space proof complexity was given for Resolution by Esteban and Torán in [14] and generalized to other proof systems by Aleckhnovich *et al.* in [1]. Lower bounds for space complexity of Resolution are known for several classes of contradictions ([14,1,7]). Moreover [14,1,6] provided relationships between size and space for Resolution refutations. Here we prove a size-space relationship for treelike $Res(k)$ which allows us to translate the previous size lower bounds to almost optimal space lower bounds. Moreover we obtain that treelike $Res(k)$ also forms a strict hierarchy with respect to space. Finally we investigate space complexity for daglike $Res(k)$. As a consequence of a simple $Res(k)$ *Locality Lemma*, similar to that of [1] for Resolution, we prove linear space lower bounds for PHP_n, GT_n, Tseitin Tautologies and Random Formulas in $Res(k)$.

The rest of the paper is structured as follows: In Section 2 we give preliminary definitions. In Section 3 we introduce the Generalized Pebbling Tautologies and prove the size upper bounds for the separation of the treelike $Res(k)$ hierarchy. Section 4 is devoted to the size lower bounds for treelike $Res(k)$ and to the statements of the separations. Section 5 is concerned with space complexity for $Res(k)$. Due to the limited space, most proofs had to be omitted in this extended abstract.

2   Preliminary Definitions

A literal l is a variable or its negation. As usual we denote by $\neg l$ the opposite literal. A *k-term* is a conjunction of up to $k \geq 1$ literals. A *k-clause* is an unbounded disjunction of k-terms. A set of k-clauses or *configuration* means the conjunction of the k-clauses contained in it. We use calligraphic letter to denote configurations and $|\mathcal{F}|$ denote the number of k-clauses in $\mathcal{F}$. Assignments (possibly total) on the variables of a formula or of a set of k-clauses are usually denoted by ρ. The size $|\rho|$ of an assignment ρ, is the number of different variables to which ρ is giving a truth value. We call an assignment ρ' a *sub-assignment* of ρ if any variable that is assigned by ρ' is assigned by ρ to the same value.

$Res(k)$ is a refutation system, introduced in [16]. It is defined by the following rules:(i) Weakening, (ii) $\wedge$-Introduction, and (iii) k-Cut.

$$(i)\ \frac{A}{A \vee \bigwedge_{l \in L} l} \quad (ii)\ \frac{A \vee \bigwedge_{l \in L} l \quad B \vee \bigwedge_{l \in K} l}{A \vee B \vee \bigwedge_{l \in L \cup K} l} \quad (iii)\ \frac{A \vee \bigwedge_{l \in L} l \quad B \vee \bigvee_{l \in L} \neg l}{A \vee B}$$

where A and B are k-clauses, and L, K are sets of literals such that $|L \cup K| \leq k$. Notice that $Res(1)$ is Resolution with a Weakening rule.

Refutations can be seen as directed acyclic graphs (*dags*). We say that a proof is *treelike* if the underlying graph is a tree. Treelike $Res(k)$ is restricted to treelike refutations.

The *size* of a $Res(k)$ refutation is the number of k-clauses forming the proof. Given a refutation P we use $|P|$ to denote the size of P. Sometimes we use $Size_k(\mathcal{F})$ (resp. $Size_k^*(\mathcal{F})$) to denote the minimal size of a refutation of $\mathcal{F}$ in $Res(k)$ (resp. in treelike $Res(k)$).

We consider a well-known *Pebbling Game* on dags. In a dag G let us call a node v a *source* if it has no predecessor, and let us call v a *target* if it has no successor. The aim of the Pebbling Game is to put a pebble on a target node of the dag using the following rules: (1) a pebble can be put on any source node; (2) a pebble can be taken away from any node; (3) a pebble can be put on any node, provided all its predecessors are pebbled. The *Pebbling* Number of a dag G (denoted by $pn(G)$), is the minimal number of pebbles needed to pebble a target node in G, following the rules of the game.

Using the Pebbling Game, in [14] the space of a refutation is defined as the minimal number of pebbles needed to pebble its underlying graph. Here we extend this definition to $Res(k)$. Using the equivalent formulation from [1], we can view a $Res(k)$ refutation of a formula $\mathcal{F}$ as a set of configurations $\mathcal{C}_0, \dots, \mathcal{C}_s$ such that $\mathcal{C}_0 = \emptyset$, $\mathcal{C}_s$ is the empty k-clause and each $\mathcal{C}_t$ for $t = 1, \dots, s$ is obtained by one of the following rules: (1) *Axiom Download*: $\mathcal{C}_t := \mathcal{C}_{t-1} \cup \{C\}$ for some clause $C \in \mathcal{F}$; (2) *Memory erasing*: $\mathcal{C}_t := \mathcal{C}_{t-1} - \{C\}$ for some $C \in \mathcal{C}_{t-1}$; (3) *Inference Adding*: $\mathcal{C}_t := \mathcal{C}_{t-1} \cup \{C\}$, for some C obtained by one of the rules of $Res(k)$ applied to two clauses in $\mathcal{C}_{t-1}$. Given a refutation P as a set of configurations, the *space* of P is the maximal size of a configuration in P. The *space* of refuting an unsatisfiable formula $\mathcal{F}$ in $Res(k)$ (resp in treelike $Res(k)$), denoted by $Space_k(\mathcal{F})$ (resp. $Space_k^*(\mathcal{F})$), is the minimal space of a $Res(k)$ refutation (resp. treelike refutation) of $\mathcal{F}$.

3   Treelike $Res(k)$ Refutations for Pebbling Contradictions

Ben-Sasson *et al.* considered in [8] Pebbling Contradictions associated to the Pebbling Game on dags G. They proved that treelike Resolution refutations of these formulas require exponential size in $pn(G)$, which gives a $2^{\Omega(n/\log n)}$ lower bound using a family of graphs from [20] with pebbling number $\Omega(n/\log n)$.

We prove that these formulas have $O(n)$ size refutations in treelike $Res(2)$ (Theorem 2). Therefore they give an almost exponential separation between treelike Resolution and treelike $Res(2)$. In this section we define a generalization of the Pebbling Contradictions to extend the separation to successive levels of treelike $Res(k)$.

3.1   The Basic Pebbling Contradictions

For any node w of a given dag $G = (V, E)$, let $x(w)$ mean that node w can be pebbled. The Pebbling Game is described using a Horn formula that for any node w contains the clause $\neg x(v_1) \vee \cdots \vee \neg x(v_k) \vee x(w)$ where $v_1, \dots, v_k$ ($k \geq 0$) are all the predecessors of w. If w is a source, the clause is just $x(w)$ and we call it a *source clause*, otherwise it is called a *pebbling clause*. In order to obtain a contradiction we add for each target node $t \in V$ the *target clause* $\neg x(t)$. We denote this contradiction by Peb_G. For our purpose it actually suffices to consider dags G where every non-source node has in-degree 2. For such a graph G any pebbling clause in G is of the form $\neg x(u) \vee \neg x(v) \vee x(w)$ where u and v are the parents of w. Since Peb_G is a Horn formula it has a very simple treelike Resolution refutation.

Proposition 1 *Let $\mathcal{F}$ be an unsatisfiable Horn formula. Then there is a treelike Resolution refutation of $\mathcal{F}$ that uses any input clause from $\mathcal{F}$ at most once.*

3.2   Generalized Pebbling Contradictions

The contradiction $\text{Peb}_{G,k,k'}$ $(k, k' \geq 1)$ is obtained from Peb_G by introducing $k \cdot k'$ variables $x(v)_{i,j}$, $i \in [k']$, $j \in [k]$ for each propositional variable $x(v)$ in Peb_G. Each variable $x(v)$ is replaced by

$$\bigwedge_{i \in [k']} \bigvee_{j \in [k]} x(v)_{i,j}.$$

The resulting formula is then transformed into CNF using de Morgan's laws, and distributivity. Hence, each source clause $x(s)$ in Peb_G will correspond to the $\text{Peb}_{G,k,k'}$-*source clauses*

$$x(s)_{i,1} \vee \cdots \vee x(s)_{i,k}$$

for $i \in [k']$. Each target clause $\neg x(t)$ in Peb_G will correspond to the $\text{Peb}_{G,k,k'}$-*target clauses*

$$\neg x(t)_{1,j_1} \vee \cdots \vee \neg x(t)_{k',j_{k'}}$$

for $j_1, \ldots, j_{k'} \in [k]$. And each pebbling clause $\neg x(u) \vee \neg x(v) \vee x(w)$ in Peb_G will correspond to the $\text{Peb}_{G,k,k'}$-*pebbling clauses*

$$\neg x(u)_{1,j_1} \vee \cdots \vee \neg x(u)_{k',j_{k'}} \vee \neg x(v)_{1,l_1} \vee \cdots \vee \neg x(v)_{k',l_{k'}} \vee x(w)_{i,1} \vee \cdots \vee x(w)_{i,k}$$

for $j_1, \ldots, j_{k'}, l_1, \ldots, l_{k'} \in [k]$, $i \in [k']$. Clearly, $\text{Peb}_{G,k,k'}$ is a contradiction since Peb_G is. Moreover $\text{Peb}_{G,k,k'}$ has a small Resolution refutation in treelike $Res(k)$.

Theorem 2 *There is a treelike $Res(k)$ refutation of $\text{Peb}_{G,k,k'}$ that involves less than twice the number of clauses in $\text{Peb}_{G,k,k'}$.*

Note that $\text{Peb}_{G,2,1}$ are the Pebbling Contradictions in [8]. By Theorem 2 and the lower bound of [8] we get an almost exponential separation between treelike Resolution and treelike $Res(2)$.

4   Lower Bounds for Generalized Pebbling Contradictions

In this section we show that any treelike $Res(k)$ refutation of $\text{Peb}_{G,k+1,k'}$ with $k' \geq k$ is of size at least $2^{(pn(G)-3)/k}$. To obtain the lower bound we generalize a game introduced in [21] to prove lower bounds for treelike Resolution. It is a 2-Player game where the two players build a partial assignment, one variable per round. Here we extend the rules of this game such that at each round the two players can play with up to k variables at once.

4.1   A Game on Contradictions

The game $G_k(\mathcal{F})$ is a 2-Player game played on the unsatisfiable CNF formula $\mathcal{F}$. The aim of the first player, the Prover, is to build an assignment that falsifies an initial clause of $\mathcal{F}$. The aim of the second player, the Delayer, is to get the maximal number of points. At each round the Prover asks for a set L of up to k

yet unassigned literals in $\mathcal{F}$. The Delayer answers with a partial (possibly total) assignment ρ to the variables in L. If ρ falsifies either the conjunction or the disjunction of the literals in L, then the round is over. Otherwise the Prover extends ρ to a total assignment over the variables in L and the Delayer scores one point.

We show that each treelike $Res(k)$ refutation yields a strategy for the Prover in which the Delayer scores a number of points at most logarithmic in the size of the refutation. Actually already a special type of decision tree (called k-decision tree, here) for $\mathcal{F}$ can be used by the Prover to obtain a good strategy.

It is well known (see [8]) that a treelike Resolution refutation of a CNF formula $\mathcal{F}$ can be transformed into a binary decision tree T of the same size such that for any assignment to $\mathcal{F}$, T yields a falsified clause of $\mathcal{F}$. In T each inner node is labeled by a variable and the decision how to continue the path at an inner node is determined by the assignment to its variable. So any total assignment will lead to a leaf node of T associated to a clause that is falsified by that assignment. Here we consider binary decision trees where each inner node is labeled by a k-term. The decision how to continue a path at an inner node is determined by the value of its k-term under the assignment. We call such a tree a k-decision tree for $\mathcal{F}$. Similar to the well known result for $k = 1$ one obtains the following result for any $k \geq 1$.

Proposition 3 *If $\mathcal{F}$ has a treelike $Res(k)$ refutation of size S, then $\mathcal{F}$ has k-decision tree of size $\leq S$.*

For $k = 1$ also the reverse inequality holds (see [8]). Since for any contradiction $\mathcal{F}$ in k-CNF there is a trivial k-decision tree of linear size (just check for each clause whether it is falsified), we obtain the following separation between the size of k-decision trees and the size of treelike $Res(k)$ refutations.

Proposition 4 *There is a family of contradictions $\mathcal{F}_m$ with 3-decision trees of size $O(m)$ that require refutations of size $2^{\Omega(m)}$ in treelike $Res(k)$.*

Proof. Since by Theorem 13 treelike $Res(k)$ is simulated by Resolution, the lower bound is given by the lower bounds for 3-CNF contradictions (see, e.g., [9]).

The following proposition provides a useful relation between the size of a k-decision tree for a contradiction $\mathcal{F}$ and the number of points the Delayer can score in $G_k(\mathcal{F})$.

Proposition 5 *If $\mathcal{F}$ has a k-decision tree of size S, then the Prover has a strategy for $G_k(\mathcal{F})$ such that the Delayer scores at most $\lfloor \log S \rfloor$ points.*

As a consequence we obtain the following corollary. that we will use to prove lower bounds for treelike $Res(k)$ refutations.

Corollary 6 *If the Delayer in $G_k(\mathcal{F})$ has a strategy that yields at least p points, then any k-decision tree for F, as well as any treelike $Res(k)$ refutation for $\mathcal{F}$, is of size at least 2^p.*

Notice however that this method will not allow us to prove directly lower bounds for treelike $Res(k)$ refutations of formulas in k-CNF.

4.2   The Delayer's Strategy

Let us in the following fix a dag $G = (V, E)$ where each non-source node has in-degree 2, fix further constants k, k' with $k' \geq k \geq 1$. We will describe a strategy for the delayer that yields at least $(pn(G) - 3)/k$ points in the game $G_k(\text{Peb}_{G,k+1,k'})$.

For sets $S, T \subseteq V$ let us denote by $pn(S, T)$ the pebbling number of the graph $G' = (V, E')$ where $E' = E \setminus ((V \times S) \cup (T \times V))$. In other words we obtain G' from G by additionally making each node in S to a source node, and each node in T to a target node.

To describe the strategy of the Delayer we will need Lemma 7.

Lemma 7 *For any disjoint sets $W, S, T \subseteq V$, there exists a partition X, Y of W ($X \cup Y = W$ and $X \cap Y = \emptyset$) such that: $pn(S, T) \leq |X| + pn(S \cup X, T \cup Y)$.*

Proof (sketch). By induction on $|W|$. The base case, $|W| = 1$, is from [8]. ∎

Now we are ready to describe the strategy of the Delayer for the game $G_k(\text{Peb}_{G,k+1,k'})$. She keeps two sets of source and target nodes that she (eventually) modifies at each round. At the beginning $S_0 = T_0 = \emptyset$. Let S_r and T_r be the sets built after round r. Assume that at round $r + 1$ the Prover asks for a set L of at most k literals. Let us denote by W the set of nodes associated with the variables in L. W is divided into the four sets $W \cap S_r$, $W \cap T_r$, $W_= = \{w \in W \setminus (S_r \cup T_r) \mid pn(S_r, T_r \cup \{w\}) = pn(S_r, T_r)\}$, and $W_> = W \setminus (S \cup T \cup W_=)$. Now the Delayer assigns 1 to every variable in L that is associated with a node in $W \cap S_r$; she assigns 0 to every variable in L associated with a node in $(W \cap T_r) \cup W_=$. If either the conjunction or the disjunction of L is falsified by the so constructed assignment, the round is over, and the Delayer sets $T_{r+1} = T_r \cup W_=$, and $S_{r+1} = S_r$, in this case the pebbling number remains the same $pn(S_r, T_r) = pn(S_{r+1}, T_{r+1})$. Otherwise the Prover assigns a value to the remaining variables in L, the Delayer scores one point and defines S_{r+1} and T_{r+1} as follows: by Lemma 7, she chooses a partition X, Y of $W_>$ s.t. $pn(S_r, T_r \cup W_=) \leq pn(S_r \cup X, T_r \cup W_= \cup Y) + |X|$. Now $S_{r+1} = S_r \cup X$, and $T_{r+1} = T_r \cup W_= \cup Y$ (in this case the pebbling number decreases by at most $|X| \leq k$).

Assuming that the Delayer follows this strategy, she maintains the following invariants: (I1) If a variable $x(v)_{j,i}$ is assigned a value in round r or before then the associated node v is in $S_r \cup T_r$. (I2) If $v \in S_r$ then there are at most k associated variables $x(v)_{j,i}$ that are assigned to 0. (I3) If $v \in T_r$ then there are at most $k - 1$ associated variables $x(v)_{j,i}$ that are assigned to 1. (I4) $pn(G) \leq pn(S_r, T_r) + |S_r|$.

First we claim that at the end of the game $G_k(\text{Peb}_{G,k+1,k'})$, say at round e, the pebbling number is considerably reduced. Namely we have:

Lemma 8 $pn(S_e, T_e) \leq 3$.

Proof. Let $G' = (V, E')$ where $E' = E \setminus ((V \times S_e) \cup (T_e \times V))$. Remember that $pn(S_e, T_e)$ was defined to be the pebbling number of G'. The game ends when the constructed partial assignment falsifies a clause of $\mathrm{Peb}_{G,k+1,k'}$. If a source clause $x(s)_{i,1} \vee \cdots \vee x(s)_{i,k+1}$ associated to a source s in G is falsified then $s \in T_e$ due to $(I1)$ and $(I2)$. Hence s is both a source and a target node in G', which shows that one pebble suffices for a pebbling of G'. Similarly, when a target clause $\neg x(t)_{1,j_1} \vee \cdots \vee \neg x(t)_{k',j_{k'}}$ is falsified then $t \in S_e$ by $(I1)$ and $(I3)$ (since $k' \geq k$) and the pebbling number of G' is one. Finally assume that a pebbling clause associated to a node w with predecessors u and v is falsified. Similar to the previous considerations we obtain that $u, v \in S_e$, and $w \in T_e$. Hence, for a pebbling of G' it suffices to use three pebbles.

Due to Invariant $(I4)$ this implies that $|S_e| \geq pn(G) - 3$. Moreover we have

Lemma 9 *The Delayer scores at least $|S_e|/k$ points.*

Combining this with Corollary 6 we obtain

Theorem 10 *If G is a dag where any non-source node has in-degree 2, and $k' \geq k \geq 1$, then the Delayer can score at least $(pn(G) - 3)/k$ points in the game $G_k(\mathrm{Peb}_{G,k+1,k'})$.*

4.3   Almost Exponential Separations for Treelike *Res(k)*

It is shown in [20] that there is an infinite family of graphs G (where each non-source node in G has in-degree 2) such that $pn(G) = \Omega(n/\log n)$, where n is the number of nodes in G. Combining Theorem 10 with Corollary 6 this shows that for such a graph G, any treelike $Res(k)$ refutation for $\mathrm{Peb}_{G,k+1,k}$ has size $2^{\Omega(n/k \log n)}$. On the other hand $\mathrm{Peb}_{G,k+1,k}$ consists of at most $O(n)$ clauses. Hence, by Theorem 2 there is a treelike $Res(k+1)$ refutation of $\mathrm{Peb}_{G,k+1,k}$ of size $O(n)$. This yields an almost exponential separation between treelike $Res(k)$ and $Res(k+1)$.

Corollary 11 *Let $k > 0$. There is a family of formulas $\mathcal{F}$ with a treelike $Res(k+1)$ refutation of size s such that any treelike $Res(k)$ refutation has size $2^{s/\log s}$.*

Corollary 12 *Treelike $Res(k+1)$ dominates treelike $Res(k)$.*

Moreover we have that Resolution simulates treelike $Res(k)$.

Theorem 13 *Resolution simulates treelike $Res(k)$ with an at most linear increase in size.*

An immediate corollary of previous Theorem and lower bounds for each level of the treelike $Res(k)$ hierarchy is the following

Corollary 14 *Resolution dominates treelike $Res(k)$ for $k \geq 1$.*

5   Space Complexity in $Res(k)$

We turn now on space complexity for proofs in $Res(k)$. In this section we show that, as for the size, the space hierarchy for treelike $Res(k)$ also forms a strict hierarchy for constant k. Moreover we obtain space lower bounds in $Res(k)$ for all those contradictions for which Resolution space lower bounds are known.

5.1   Space Separations for the Treelike $Res(k)$ Hierarchy

Consider the following definition from [3]. Given a formula $\mathcal{F}$ over variables in X, and a $k \in N$, define a new formula $\mathcal{F}_k$ this way: for any set of literals L over X, with $|L| \leq k$, introduce a new literal z_L meaning $\bigwedge_{l \in L} l$. Let $\mathrm{Ex}(X, k)$ be the set of clauses $\neg z_L \vee l$, for $l \in L$ and $z_L \vee \bigvee_{l \in L} \bar{l}$. Then $\mathcal{F}_k$ is the union of $\mathcal{F}$ and $Ex(X, k)$.

We obtain the following two lemmas by adapting to space complexity a result from [3] that was proved for size complexity. In [3] it was shown that the proof size in (treelike) Resolution for $\mathcal{F}_k$ and the proof size in (treelike) $Res(k)$ for $\mathcal{F}$ are linearly related.

Lemma 15 *For any $\mathcal{F}$ and $k \in N$, if $\mathcal{F}$ has a (treelike) $Res(k)$ refutation in space S, then $\mathcal{F}_k$ has a (treelike) Resolution refutation in space $S + 3$.*

Lemma 16 *For any $\mathcal{F}$ and $k \in N$, if $\mathcal{F}_k$ has a (treelike) Resolution refutation in space S, then $\mathcal{F}$ has a (treelike) $Res(k)$ in space $S + 3$.*

For treelike Resolution [14] proved the following relationship.

Lemma 17 ([14]) *If a formula on n variables has treelike Resolution refutations in space S, then it has treelike Resolution refutations of size at most $\binom{n+S}{S}$.*

Extending the this lemma to treelike $Res(k)$ we obtain

Theorem 18 *For any formula $\mathcal{F}$ over n variables and $k \in N$, if $Size_k^*(\mathcal{F}) \geq S$, then $Space_k^*(\mathcal{F}) \geq \Omega(\frac{\log S}{\log n})$.*

Proof. Let $\mathcal{F}$ be a contradiction over n variables such that $Size_k^*(\mathcal{F}) \geq S$. According to a result in [3] this implies that $Size_1^*(\mathcal{F}_k) \geq \frac{S}{k}$. Since the space in Resolution is upper bounded by the number of variables, it is easy to see that Lemma 17 in turn implies that $Space_1^*(\mathcal{F}_k) \geq \Omega(\log S/\log n)$, which implies the claim by Lemma 15.

As a corollary of the previous theorem and the size lower bound of Corollary 12, we obtain a space lower bound for $Peb_{G,k+1,k}$.

Corollary 19 $Space_k^*(Peb_{G,k+1,k}) \geq \Omega(n/\log^2 n)$.

On the other hand, by adapting the proof of the size upper bound in Theorem 2 for $Peb_{G,k+1,k}$ in treelike $Res(k+1)$, we obtain the following space upper bound.

Lemma 20 $\text{Peb}_{G,k+1,k}$ *has a treelike* $\text{Res}(k+1)$ *refutation in constant space.*

Therefore the treelike $\text{Res}(k)$ space hierarchy, is strict.

Corollary 21 *For* $k \geq 1$ *there is a family of formulas* $\mathcal{F}$ *over* n *variables that have constant space refutations in treelike* $\text{Res}(k+1)$ *but require space* $\Omega(n/\log^2 n)$ *in treelike* $\text{Res}(k)$.

5.2   Space Lower Bounds for $\text{Res}(k)$

The space lower bounds in [14] are based on minimal assignments satisfying sets of pebbled clauses. If a set of clauses has a minimal satisfying assignment of size n, then the set has at least n clauses. In [1] this is rephrased in terms of the *Resolution Locality Lemma*. We study lower bounds for $\text{Res}(k)$ introducing and proving a $\text{Res}(k)$ *Locality Lemma*. We prove that all known lower bounds for Resolution still hold for $\text{Res}(k)$.

We start by proving the extension of the Locality Lemma for $\text{Res}(k)$. As in [1] for Resolution, the Locality Lemma will give us lower bounds for the class of n-semiwide tautologies, including PHP_n, GT_n, CT_n, and also for Random CNF and Tseitin Tautologies.

Lemma 22 *Let* $\mathcal{C}$ *be a set of* k-*clauses and* ρ *a partial assignment that satisfies* $\mathcal{C}$. *Then there exists a sub-assignment* ρ' *of* ρ *that satisfies* $\mathcal{C}$ *with* $|\rho'| \leq |\mathcal{C}| \cdot k$.

Proof. Let $\mathcal{C}$ be $C_1 \wedge \ldots \wedge C_n$. Since ρ satisfies $\mathcal{C}$ then for each k-clause $C_i \in \mathcal{C}$ there exists at least a k-term t_i such that ρ satisfies t_i. Take ρ' the sub-assignment of ρ satisfying exactly one k-term in each k-clause. Since each k-term is the conjunction of at most k literals, then $|\rho'| \leq k \cdot |\mathcal{C}|$.

We extend the concept of semiwideness in [1] to sets of k-clauses.

Definition 1 *Given a partial assignment* ρ *and a set of* k-*clauses* $\mathcal{C}$ *we say that* ρ *is* $\mathcal{C}$-*consistent if* ρ *can be extended to an assignment* ρ' *satisfying* $\mathcal{C}$.

Definition 2 *We say that a contradictive set* $\mathcal{C}$ *of* k-*clauses is* n-*semiwide if there is a partition* $\mathcal{C}'$, $\mathcal{C}''$ *of* $\mathcal{C}$ *with* (1) $\mathcal{C}'$ *is satisfiable; and* (2) *every* $\mathcal{C}'$-*consistent assignment of size less than* n *is* $\mathcal{C}' \cup \{C\}$-*consistent for every* $C \in \mathcal{C}''$.

Theorem 23 *If a set* $\mathcal{C}$ *of* k-*clauses is* n-*semiwide then* $\text{Space}_k(\mathcal{C}) > n/k$.

Proof. Let $\mathcal{C}'$, $\mathcal{C}''$ be the partition guaranteed by the n-sewideness of $\mathcal{C}$. Assume by contradiction that $\text{Space}_k(\mathcal{C}) \leq n/k$. Then $\mathcal{C}$ has a $\text{Res}(k)$ refutation $\mathcal{C}_1, \ldots, \mathcal{C}_l$ consisting of configurations $\mathcal{C}_i$ with at most n/k k-clauses. We show now by induction on $i = 1, \ldots, l$ that there exist an assignment ρ_i such that (1) ρ_i satisfies $\mathcal{C}_i$, (2) ρ_i is $\mathcal{C}'$-consistent, and (3) $|\rho_i| \leq k \cdot |\mathcal{C}_i|$. This is a contradiction since the last configuration contains the empty clause.

The base case $i = 1$ is trivial since $\mathcal{C}_1$ is empty. Let now $i \geq 1$. By the inductive assumption there is an assignment ρ_i fulfilling (1)-(3). $\mathcal{C}_{i+1}$ is obtained from $\mathcal{C}_i$ by one of the following three cases.

Axiom Download: $\mathcal{C}_{i+1} = \mathcal{C}_i \cup \{C\}$ for an axiom $C \in \mathcal{C}$. Since $|\rho_i| \leq k \cdot (|\mathcal{C}_{i+1}| - 1) < n$ we can use n-semiwideness to obtain that in both cases $C \in \mathcal{C}'$ and $C \in \mathcal{C}''$, ρ_i can be extended to an assignment ρ' that satisfies $\mathcal{C}' \cup \{C\}$ (and still $\mathcal{C}_i$). Using the Locality Lemma we take ρ_{i+1} to be a sub-assignment of ρ' that satisfies $\mathcal{C}_{i+1}$ with $|\rho_{i+1}| \leq k \cdot |\mathcal{C}_{i+1}|$.

Inference Step: $\mathcal{C}_{i+1}$ is obtained from $\mathcal{C}_i$ by adding a k-clause obtained by a rule of $Res(k)$. By soundness of $Res(k)$, ρ_i also satisfies $\mathcal{C}_{i+1}$. Let $\rho_{i+1} = \rho_i$.

Memory Erasing: $\mathcal{C}_{i+1} = \mathcal{C}_i - \{C\}$. By the Locality Lemma we obtain a sub-assignment ρ_{i+1} of ρ_i that satisfies $\mathcal{C}_{i+1}$ with $|\rho_{i+1}| \leq k \cdot |\mathcal{C}_{i+1}|$

From the $\Omega(n)$-semiwideness of PHP_n, CT_n and GT_n in [1] we get:

Theorem 24 $Space_k(GT_n) > \frac{n}{2k}$, $Space_k(PHP_n) > \frac{n}{k}$, $Space_k(CT_n) > \frac{n}{k}$.

We can prove $Res(k)$ space lower bounds for Tseitin Tautologies $T(G)$ associated to a connected graphs G with high expansion (see [25]), using the $Res(k)$ Locality Lemma and modifying the proofs of [1].

Theorem 25 *Let G be connected graph over n vertices and with expansion factor $\Omega(n)$. Then for any $k \in N$, $Space_k(T(G)) > \Omega(n)$.*

We reduce $Res(k)$ space lower bounds for Random 3-CNF to lower bounds for the *Matching Space* of a game, introduced by [7], played on bipartite graphs. Given a 3-CNF $\mathcal{F}$, let $G_{\mathcal{F}}$ be the standard bipartite graph associated to $\mathcal{F}$. Using the $Res(k)$ Locality Lemma we prove that:

Theorem 26 *For any contradictive 3-CNF $\mathcal{F}$, $Space_k(\mathcal{F}) \geq MSpace(G_{\mathcal{F}})/k$.*

As random 3-CNF define bipartite expander graphs, by Theorem 26 we have:

Theorem 27 *Let $\mathcal{F} \sim \mathbf{F}^n_{\Delta \cdot n}$, with $\Delta > 4.6$. Then for any $0 < \epsilon < \frac{1}{2}$, for each $k \in N$, $Space_k(\mathcal{F}) > \Omega(\frac{n \cdot \Delta^{\frac{1+\epsilon}{1-\epsilon}}}{k})$.*

6 Open Problems

Some problems directly related to this work are the following: Do the daglike $Res(k)$ systems also form a strict hierarchy with respect to size or to space? Is there a relation between size and space in $Res(k)$ similar to that of Theorem 18? The latter would also solve the problem for the space complexity of the Pebbling Contradictions in Resolution (see also [1,8,6]).

Acknowledgments. We would like to thank Jan Krajíček for some remarks about consequences of our separations in Bounded Arithmetic, Maria Luisa Bonet for helpful discussions, and Albert Atserias for the hint that allowed us to improve the simulation in Theorem 13.

References

1. M. Alekhnovich, E. Ben-Sasson, A. Razborov, A. Wigderson. Space complexity in propositional calculus. *STOC 2000* pp. 358–367. To appear in *SIAM J. Comput.*
2. M. Alekhnovich, A. Razborov. Resolution is not automatizable unless $W[P]$ is not tractable. *FOCS 2001*. pp.190-199.
3. A. Atserias, M. L. Bonet. On the automatizability of Resolution and related propositional proof systems. TR02-010, *ECCC*, 2002.
4. A. Atserias, M.L. Bonet, J.L. Esteban. Lower Bounds for the Weak Pigeonhole Principle Beyond Resolution. *ICALP 2001*, pp. 1005–1016.
5. P. Beame, T. Pitassi. Simplified and Improved Resolution Lower Bounds. *FOCS 1996*, pp. 274–282.
6. E. Ben-Sasson. Size-Space tradeoffs for Resolution. To appear in,*STOC 2002*.
7. E. Ben-Sasson, N. Galesi. Space Complexity of Random Formulae in Resolution. *CCC 2001*, pp. 42–51.
8. E. Ben-Sasson, R. Impagliazzo, A. Wigderson. Near optimal separation of treelike and general Resolution. TR00-005, *ECCC*, 2000.
9. E. Ben-Sasson, A. Wigderson. Short Proofs Are Narrow—Resolution Made Simple. *J. ACM* 48(2) pp. 149–168, 2001.
10. M.L. Bonet, J.L. Esteban, N. Galesi, J. Johannsen. On the Relative Complexity of Resolution Refinements and Cutting Planes Proof Systems. *SIAM J. Comput.* 30(5) pp. 1462–1484, 2000.
11. P. Beame, R.M. Karp, T. Pitassi, M.E. Saks. On the Complexity of Unsatisfiability Proofs for Random k-CNF Formulas. *STOC 1998*, pp. 561–571.
12. M.L. Bonet, T. Pitassi, R. Raz On Interpolation and Automatization for Frege Systems. *SIAM J. Comput.* 29(6) pp. 1939–1967, 2000.
13. S. Cook, R. Reckhow. The relative efficiency of propositional proof systems. *J. Symbolic Logic* 44 pp. 36–50, 1979.
14. J.L. Esteban, J. Torán. Space bounds for resolution. *Inform. and Comput.* 171 (1) pp. 84–97, 2001.
15. A. Haken. The Intractability of Resolution. *Theoret. Comp. Sci.* 39, 297–308, 1985.
16. J. Krajíček. On the weak pigeonhole principle. *Fund. Math.* 170(1-3) pp. 123–140, 2001.
17. J. Krajíček. Lower Bounds to the Size of Constant-Depth Propositional Proofs. *J. Symbolic Logic* 59(1) pp. 73–86, 1994.
18. J. Krajíček, P. Pudlák. Some Consequences of Cryptographical Conjectures for S_2^1 and EF. *Inform. and Comput.* 140(1) pp. 82–94, 1998.
19. A. Maciel, T. Pitassi, A.R. Woods. A new proof of the weak pigeonhole principle. *STOC 2000* pp. 368–377.
20. W.J. Paul, R.E. Tarjan, J.R. Celoni. Space bounds for a game on graphs. *Math. Systems Theory*, 10 pp. 239–251, 1977.
21. P. Pudlák, R. Impagliazzo. Lower bounds for DLL algorithms. *SODA 2000* pp. 128–136.
22. R. Raz. Resolution Lower Bounds for the Weak Pigeonhole Principle. TR01-21. *ECCC*, 2001.
23. A. Razborov. Resolution Lower Bounds for Perfect Matching Principles. MS, 2001.
24. G.S. Tseitin. On the complexity of derivation in propositional calculus. *Studies in Constructive Mathematics and Mathematical Logic*, Part 2. pp. 115-125. Consultants Bureau, New York-London, 1968.
25. A. Urquhart. Hard examples for resolution. *J. ACM* 34(1) pp. 209-219, 1987.

Cryptographic Hardness Based on the Decoding of Reed-Solomon Codes

Aggelos Kiayias[1] and Moti Yung[2]

[1] Graduate Center, CUNY, NY USA, akiayias@gc.cuny.edu
[2] CertCo, NY USA moti@cs.columbia.edu

Abstract. We investigate the decoding problem of Reed-Solomon Codes (aka: the Polynomial Reconstruction Problem – PR) from a cryptographic hardness perspective. Following the standard methodology for constructing cryptographically strong primitives, we formulate a decisional intractability assumption related to the PR problem. Then, based on this assumption we show: (i) *hardness of partial information extraction* and (ii) *pseudorandomness*. This lays the theoretical framework for the exploitation of PR as a basic cryptographic tool which, as it turns out, possesses unique properties. One such property is the fact that in PR, the size of the corrupted codeword (which corresponds to the size of a ciphertext and the plaintext) and the size of the index of error locations (which corresponds to the size of the key) are independent and can even be super-polynomially related. We then demonstrate the power of PR-based cryptographic design by constructing a stateful cipher.

1   Introduction

Finding new problems based on which we can design cryptographic primitives is an important research area. Given a presumably hard problem it is usually non-trivial to exploit it directly in cryptography. Many times, in order to serve as the base for secure cryptographic primitives, we need to find related hard decision problems (predicates). This is the fundamental methodology initiated by Goldwasser and Micali in [GM84] where they started the quest for formal notions and proofs of security in cryptography. The decision problem's hardness, typically seems related to (or at times proved in some sense related or, even better, reducible from) the hardness of the original problem. Hard predicate assumptions allow formal security proofs (in the form of reductions) for advanced cryptographic primitives such as pseudorandomness and semantically secure encryption. Examples of decisions problems are Quadratic-Residuosity (related to factoring) and Decisional Diffie-Hellman (related to the Diffie-Hellman problem).

In this work, our goal is to investigate the possibility of cryptographic primitives whose security is based on the problem of *Polynomial Reconstruction* (PR). Recall that the problem of Polynomial Reconstruction is defined as follows: Given n points over a (large) finite field $\mathbb{F}$, such that at least t of them belong to the graph of a polynomial p of degree less than k, recover such a polynomial (where $n > t > k$). For small t the problem is believed to be hard [BW86,GS98,Sud97].

P. Widmayer et al. (Eds.): ICALP 2002, LNCS 2380, pp. 232–243, 2002.
© Springer-Verlag Berlin Heidelberg 2002

Note that Polynomial Reconstruction as is, does not seem to be ready for direct cryptographic exploitation: even if presumed hard, it is not at all clear how to build advanced cryptographic primitives whose security can be reduced to it. Indeed, when Naor and Pinkas [NP99] first employed the problem cryptographically in a context of protocol design, they actually introduced a related pseudorandomness assumption (which itself, motivates further investigation).

In this work, we first identify a decisional problem naturally related to PR. This problem is based on the following basic question: given a PR-instance that contains n points and an index $i \in \{1, \ldots, n\}$, does the i-th point of the instance belong in the graph of the polynomial solution or not? (note that in the range of our parameters, a PR-instance has a unique solution with very high probability). We formalize the hardness of this predicate for all indices i as the "Decisional-PR-Assumption" (DPR).

Based on the DPR-Assumption we show: (i) *hardness of partial information extraction*: an adversary with access to a PR-instance who wishes to predict the value of some computable function on a new point of the polynomial-solution, gains only negligible advantage compared to an adversary who wishes to predict the same value without seeing the instance — this holds true even if the point follows an adversarially chosen probability distribution; also: (ii) *pseudorandomness*: PR-instances are pseudorandom in the sense that they are indistinguishable from random sets of points, for any poly-time observer. These results suggest that PR is quite robust in the cryptographic sense and is suitable for employment in cryptographic constructions.

There are several possible advantages of the PR problem which can be exploited by cryptographic primitives built on it, for example: (i) The natural dichotomy and independence exhibited between the key-size (index of error locations) and the size of Reed-Solomon encoded messages (or concealed information in PR-based systems) allows key-sizes to be selected independently of (and possibly super-polynomially smaller than) the message size; we know of no other problem that allows such a property in the cryptographic literature. (ii) The PR problem enjoys a unique algebraic structure. (iii) The operation of polynomial interpolation which is basic in PR cryptographic primitives can be implemented quite efficiently (especially in special purpose hardware).

With the above advantages in mind, we apply our results to the design of PR-based cryptographic primitives. As an application of our methodology, we introduce a new semantically-secure stateful cipher based on Polynomial Reconstruction. Our cipher demonstrates the exploitation of structural properties possessed by the PR problem and exhibits unique properties: (1) *Forward Secrecy*: if the system is broken into at a certain time, this only affects the security of future messages; (2) *Computational perfect secrecy*: deciphering a single ciphertext is equivalent to getting the system private key; (3) *Superpolynomial message size*: the plaintext can be *superpolynomial* in the key-size; (4) *Built-in error correction*; and (5) *Key-equivalence*: all keys have equal strength.

A more complete technical account of the present work is included in [KY02]. **Notation.** All computations are performed in a (large) finite field $\mathbb{F}$. Tuples in $\mathbb{F}^n$ are denoted by $\mathbf{x}$ and $(\mathbf{x})_i$ denotes the i-th coordinate of $\mathbf{x}$. Denote by

$(n)_k := n(n-1)\ldots(n-k+1)$, and if A is a set denote by $(A)_k$ the set of all k-tuples over A without repetitions. PPT stands for "probabilistic polynomial-time." All algorithms mentioned in the paper are PPT Turing Machines, and denoted by $\mathcal{A}, \mathcal{B}$ etc. For any PPT $\mathcal{A}$ defined over a space D, and input $x \in D$, if y is in the range of $\mathcal{A}$ we will denote by $\mathbf{Prob}_{x \in_U D}[\mathcal{A}(x) = y]$ the probability that $\mathcal{A}$ returns y when x is uniformly distributed over D; the probability is also taken over the internal coin tosses of $\mathcal{A}$ (note that y may be a function of x). A function $\alpha(n) : \mathbb{N} \to \mathbb{R}$ is negligible if for all c it holds that $\alpha(n) < n^{-c}$ for sufficiently large n. A function $\beta(n) : \mathbb{N} \to \mathbb{R}$ is called non-negligible if it is not negligible for all large enough inputs, namely there is a c s.t. $\beta(n) \geq n^{-c}$ for all n sufficiently large. When the probability of an event is greater equal to $1 - \epsilon(n)$ where $\epsilon(n)$ is negligible, then we write that the event happens "with overwhelming probability." A probability distribution $\mathcal{D}$ over some space R of objects of size polynomial in n is called (polynomial time) samplable if there is a PPT $S_\mathcal{D}$ so that for all $y \in R$, $\mathbf{Prob}_\mathcal{D}[y] = \mathbf{Prob}[S_\mathcal{D}(1^n) = y]$.

2 The Problem

Definition 1. Polynomial Reconstruction (PR). *Given $n, k, t \in \mathbb{N}$ and $\{\langle z_1, y_1 \rangle, \ldots, \langle z_n, y_n \rangle\}$ with $z_i \neq z_j$ for $i \neq j$, output all $\langle p(x), I \rangle$ such that $p \in \mathbb{F}[x]$, $\mathrm{degree}(p) < k$, $I \subseteq \{1, \ldots, n\}$, $|I| \geq t$ and $\forall i \in I(p(z_i) = y_i)$.*

PR, from a coding theoretic perspective, asks for all messages that agree with at least t positions of the received, partially corrupted, Reed-Solomon codeword. For a general treatment on the subject the interested reader is referred to [Ber68] or [MS77]. Note that $k < n$ since k/n is the message rate of the code, and that we further require that at least one solution $\langle p(x), I \rangle$ exists.

STRUCTURE OF THE INSTANCE SPACE. An instance of PR will be denoted by $X := \{\langle z_i, y_i \rangle\}_{i=1}^n$; the set of all instances with parameters n, k, t will be denoted by $\mathcal{S}_{n,k,t}$. In order to refer to PR with parameters n, k, t we will write $\mathrm{PR}[n, k, t]$. Note that unless stated otherwise we assume that n is polynomially related to $\log |\mathbb{F}|$. Let $I \subseteq \{1, \ldots, n\}$ with $|I| = t$. We denote by $\mathcal{S}_{n,k,t}(I)$ the subset of $\mathcal{S}_{n,k,t}$ so that for any $X \in \mathcal{S}_{n,k,t}(I)$ it holds that X has a solution of the form $\langle p, I \rangle$. It is clear that $\mathcal{S}_{n,k,t} = \cup_{|I|=t} \mathcal{S}_{n,k,t}(I)$.

Some basic facts about the structure of the instance space $\mathcal{S}_{n,k,t}$ are the following: (i) The number of elements of $\mathcal{S}_{n,k,t}$ can be approximated within negligible error by $\binom{n}{t}(|\mathbb{F}|)_n |\mathbb{F}|^{2n-t+k}$; (ii) There is a straightforward way to sample the uniform distribution over $\mathcal{S}_{n,k,t}$: sample an element of $\mathcal{S}_{n,k,t}(\{1, \ldots, t\})$ and then permute the points according to a random permutation; (iii) Provided that the finite field is large ($\log |\mathbb{F}| \geq 2n$) the ratio of the number of instances in $\mathcal{S}_{n,k,t}$ with more than solution, over $\#\mathcal{S}_{n,k,t}$ is less than 2^{-n}. As a result a randomly chosen instance has, with overwhelming probability, a unique solution.

Consequently any instance $X \in \mathcal{S}_{n,k,t}$ uniquely defines a polynomial p (with overwhelming probability) such that $\mathrm{degree}(p) < k$. We denote this polynomial by s_X (for solution of X). The set of indices that corresponds to the graph of p which we call "the index-solution set" is denoted by $I(X)$. Obviously, the recovery of s_X implies the recovery of $I(X)$ and vice-versa.

SOLVABILITY OF PR. When $t \geq \frac{n+k}{2}$ then $PR[n, k, t]$ has only one solution and it can be found with the algorithm of Berlekamp and Welch [BW86] ($\frac{n+k}{2}$ is the error-correction bound of the Reed-Solomon codes). When t is beyond the error-correction bound then having more than one solution is possible. Sudan proposed an algorithm that solves the PR beyond the error-correction bound when $t \geq \sqrt{2kn}$ in [Sud97] and later in [GS98], Guruswami and Sudan presented an algorithm that solves the PR for $t > \sqrt{kn}$. In [GSR95] it was proven that when $t > \sqrt{kn}$ the number of solutions is bounded by a polynomial. In [GS98] it is pointed out that the possibility of an algorithm that solves instances for smaller values of t might be limited. As a matter of fact, a certain decisional version of PR was shown to be NP-Complete (for very small values of t), see [GSR95]. We note here that the solvability of PR (and related problems) was also studied in the context of lattices, see [BN00]. As a result, the current state of knowledge implies that $PR[n, k, t]$ is hard for the choice of parameters $t < \sqrt{kn}$.

SECURITY PARAMETERS. In our exposition we will use n as be the security parameter. The parameters k, t are functions in n, so that $k < t < n$ and $t < \sqrt{nk}$. The straightforward brute-force algorithm for solving $PR[n, k, t]$ requires checking all possibilities and as a result has complexity proportional to $\min(\binom{n}{k}, \binom{n}{t})$. The parameters $[n, k(n), t(n)]$ are called *sound* for $PR[n, k, t]$ if $k(n)$ and $t(n)$ are chosen so that $t < \sqrt{kn}$ and $\min(\binom{n}{k}, \binom{n}{t})$ is exponential in n. Note that we will suppress (n) in $k(n), t(n)$. Observe that if $[n, k, t]$ are sound parameters then it also holds that $[n, k+1, t]$ are sound parameters (provided that $k+1 < t$). Intuitively this means that allowing the degree of the solution-polynomial to be greater without changing the other parameters it cannot make the problem easier. We will assume sound parameters throughout.

ALTERING THE DISTRIBUTION OF PR-INSTANCE SOLUTIONS. Apart from the uniform distribution over $\mathcal{S}_{n,k,t}$ we also consider distributions where h points of the polynomial solution of a PR-instance follow a non-uniform probability distribution $\mathcal{D}_h$ over $\mathbb{F}^h$ (where $0 < h < k$). Such "modified solution distributions" over $\mathcal{S}_{n,k,t}$ will be denoted by $\mathcal{D}_h^{w_1,\ldots,w_h}$ where $w_1, \ldots, w_h \in \mathbb{F}$. A certain PR-instance X distributed according to $\mathcal{D}_h^{w_1,\ldots,w_h}$ follows the uniform distribution with the exception of the values $s_X(w_1), \ldots, s_X(w_h)$ which are distributed according to $\mathcal{D}_h$. It is easy to see that $\mathcal{D}_h^{w_1,\ldots,w_h}$ is samplable provided that $\mathcal{D}_h$ is samplable.

Remark 1. A basic fact about PR-instances is that in any probability statement that involves a PPT $\mathcal{A}$ and the uniform distribution over the instance space $\mathcal{S}_{n,k,t}$ (or even a modified solution distribution as defined above), it is possible to *fix* the index-solution-set without significantly affecting the probability: $\forall \mathcal{A} \, \exists \mathcal{A}' \, \forall I \subseteq \{1, \ldots, n\}, |I| = t$, $\mathbf{Prob}_{X \in_U \mathcal{S}_{n,k,t}}[\mathcal{A}(X) = y] \simeq \mathbf{Prob}_{X \in_U \mathcal{S}_{n,k,t}(I)}[\mathcal{A}'(X) = y]$ (this is because one can randomize the index-solution-set of a given PR-instance by randomly permuting the points — note that "$\simeq$" means that the distance of the two probabilities is negligible in n).

2.1   The Intractability Assumption

A decision problem that relates naturally to the hardness of solving an instance X of $\mathrm{PR}[n, k, t]$ is the following: given X and an index $i \in \{1, \ldots, n\}$ decide whether $i \in I(X)$. We postulate that such decision is computationally hard to make whenever PR is hard. Since this has to hold true for all indices we will use a counter-positive argument to formalize the related decisional intractability assumption. In the definition below we describe a pair of predicates that refutes the assumption by "revealing" one of the points that belongs in the graph of the solution-polynomial (note that we formulate probabilities independently of the index-solution-set since given any PR-instance the index-solution-set can be randomized — see remark 1 above:

Definition 2. *A pair of* PPT *predicates* $\mathcal{A}_1, \mathcal{A}_2$ *is called a gap-predicate-pair for the parameters* n, k, t *if for all* $I \subseteq \{1, \ldots, n\}$ *with* $|I| = t$ *it holds that:*

$$| \mathbf{Prob}[\mathcal{A}_1(i, X) = 1] - \mathbf{Prob}[\mathcal{A}_2(i, X) = 1] | = \begin{cases} \text{negligible} & \forall i \notin I \\ \text{non–negligible for some } i_0 \in I \end{cases}$$

where the probabilities are taken over all choices of $X \in \mathcal{S}_{n,k,t}(I)$ *and internal coin-tosses of the predicates* $\mathcal{A}_1, \mathcal{A}_2$. *We further require that* $i_0 \leq n - k$.

A gap-predicate-pair when given a PR instance X and $i \in \{1, \ldots, n\}$ exhibits a measurable difference for at least one $i \in I(X)$, where at the same time it exhibits no measurable difference for indices outside $I(X)$. Using this, we formulate the Decisional-PR-Assumption as follows:

Decisional-PR-Assumption. $(\mathrm{DPR}[n, k, t])$
For any sound parameters $[n, k, t]$ there does not exist a gap-predicate-pair.

It is easy to see that solving PR for some parameters violates the DPR assumption. On the other hand, one can show that if DPR is violated by a gap-predicate-pair that has a certain *samplability* property for each $\mathcal{S}_{n,k,t}(I)$ then PR is in probabilistic polynomial-time (this uses the fact that $i_0 \leq n - k$). Although this is not a full reduction of the functional to the decisional version of the problem (as it could be the case that there exist gap-predicate-pairs that do not possess the samplability property) it is an indication that the decisional-PR-Assumption is connected to the assumed hardness of PR.

3   Hardness of Recovering Partial Information of Any Specific Polynomial Value

In this section we show that $\mathrm{PR}[n, k, t]$ "leaks no partial information" about any specific polynomial value under the DPR-Assumption. In particular, we show that for some fixed value $w \in \mathbb{F}$, given an instance $X := \{\langle z_i, y_i \rangle\}_{i=1}^n \in \mathcal{S}_{n,k,t}$ with $w \notin \{z_1, \ldots, z_n\}$, we get no polynomial advantage in predicting the value of *any* function g over the polynomial value $s_X(w)$ for $s_X(w)$ drawn from any polynomially samplable probability distribution $\mathcal{D}_1$, unless the DPR fails for parameters $[n, k-1, t]$. In the remaining of the section we will fix $w \in \mathbb{F}$ and we will assume that $\mathcal{S}_{n,k,t}$ does not contain instances with w among the z-values (which is a negligible probability event). The generality of the proof stems from the fact that we can map a $\mathrm{PR}[n, k-1, t]$-instance X into a $\mathrm{PR}[n, k, t]$-instance X' of which we can select the value $s_{X'}(w)$. Then, we can use any algorithm that makes a non-negligible prediction regarding some property of $s_{X'}(w)$ to extract a parameterized predicate that is sensitive to a parameter choice inside the index-solution-set. This predicate yields a gap-predicate-pair that violates $\mathrm{DPR}[n, k-1, t]$.

For the rest of the section fix some value $w \in \mathbb{F}$. Next, we formalize the concept of "leaking no partial information." Informally, we can describe the definition as follows: for any PPT that predicts the value of $g(s_X(w))$ given a PR instance, there is another algorithm with essentially the same functionality that operates *without* the PR instance.

Definition 3. $\mathrm{PR}[n, k, t]$ *leaks no partial information means that for all poly-time computable* $g : \mathbb{F} \to R$ *and all polynomial-time samplable probability distributions* $\mathcal{D}_1$ *over* $\mathbb{F}$ *it holds: for all* PPT $\mathcal{A}$ *there exists a* PPT $\mathcal{A}'$ *such that the following is negligible in* n:

$$| \mathbf{Prob}_{X \in \mathcal{D}_1^w \mathcal{S}_{n,k,t}}[\mathcal{A}(X) = g(s_X(w))] - \mathbf{Prob}_{u \in \mathcal{D}_1 \mathbb{F}}[\mathcal{A}'(1^n) = g(u)] |$$

A consequence of remark 1 is that the definition above can be made more specific so that: for all PPT $\mathcal{A}$ there exists a PPT $\mathcal{A}'$ so that for all $I \subseteq \{1, \ldots, n\}$ with $|I| = t$ it holds that the following is negligible in n:

$$| \mathbf{Prob}_{X \in \mathcal{D}_1^w \mathcal{S}_{n,k,t}(I)}[\mathcal{A}(X) = g(s_X(w))] - \mathbf{Prob}_{u \in \mathcal{D}_1 \mathbb{F}}[\mathcal{A}'(1^n) = g(u)] |$$

So, the probability of success of any PPT $\mathcal{A}$ is taken over $\mathcal{S}_{n,k,t}(I)$ following the distribution $\mathcal{D}_1^w$, independently of the index-solution-set I. The core of the proof that PR leaks no partial information is the following lemma:

Lemma 1. *Suppose that there is a poly-time computable* $g : \mathbb{F} \to R$ *and a probability distribution* $\mathcal{D}_1$ *for which* $\mathrm{PR}[n, k, t]$ *leaks partial information. Then there exists a* PPT $\mathcal{B}$ *such that for all* $I \subseteq \{1, \ldots, n\}$ *with* $|I| = t$, *if* $\beta_i(n) :=$ $\mathbf{Prob}_{X \in_U \mathcal{S}_{n,k-1,t}(I)} [\mathcal{B}(i, X) = 1]$ *with* $i \in \{0, \ldots, n\}$ *it holds that*
1. *For all* $i \notin I$ $|\beta_{i-1}(n) - \beta_i(n)|$ *is negligible.*
2. *There exists an* $i_0 \in I$ *such that* $|\beta_{i_0-1}(n) - \beta_{i_0}(n)|$ *is non-negligible and* $i_0 \leq n - k + 1$.

The proof of this Lemma is a crucial contribution. It exhibits the two main proof-techniques used throughout; one technique involves controlling portions of the instance's solution, whereas the other technique involves a "walking argument" over the points of the instance. Now observe that if $\mathcal{A}_1(i, X) := \mathcal{B}(i, X)$ and $\mathcal{A}_2(i, X) := \mathcal{B}(i-1, X)$, it follows easily that $\mathcal{A}_1, \mathcal{A}_2$ is a gap-predicate-pair. As a result,

Theorem 1. *Suppose that there is a poly-time computable* $g : \mathbb{F} \to R$ *and a probability distribution* $\mathcal{D}_1$ *for which* $\mathrm{PR}[n, k, t]$ *leaks partial information. Then the DPR-Assumption fails for parameters* $[n, k-1, t]$.

In the rest of the section we present special cases of the above Theorem which appear frequently in cryptographic settings. Let us assume that the distribution $\mathcal{D}_1$ is uniform. Let $g : \mathbb{F} \to R$ be a poly-time computable function. Define $\mathbb{F}_a = \{u \mid g(u) = a; u \in \mathbb{F}\}$ for any $a \in R$. We say that g is *balanced* if for all $a \in R$ and all polynomials q it holds that $\mid \frac{|\mathbb{F}_a|}{|\mathbb{F}|} - \frac{1}{|R|} \mid < \frac{1}{q(\log |\mathbb{F}|)}$ (for sufficiently large $|\mathbb{F}|$). The balanced property means that any image under g corresponds to roughly the same number of pre-images. This is a very general condition that applies to individual bits of elements of $\mathbb{F}$ as well as to various length bit-sequences of elements of $\mathbb{F}$.

Naturally, guessing an unknown value of a balanced function with a uniformly distributed pre-image cannot be done with probability significantly greater than $1/|R|$. This and theorem 1 imply:

Corollary 1. *For any balanced $g : \mathbb{F} \to R$, the success of any PPT $\mathcal{A}$ that given $X \in \mathcal{S}_{n,k,t}$, computes the value $g(s_X(w))$ is only by a negligible fraction different than $1/|R|$ unless the* DPR*-Assumption fails for parameters $[n, k-1, t]$.*

More specifically we can give the following examples of balanced predicates/functions that are hard to compute given a $\mathrm{PR}[n, k, t]$-instance:

Proposition 1. *The following problems are hard under the* DPR$[n, k-1, t]$*:*

1. *Let $\mathrm{BIT}_l(a)$ denote the l-th least significant bit of $a \in \mathbb{F}$. Given $X \in \mathcal{S}_{n,k,t}$ predict $\mathrm{BIT}_l(s_X(w))$ with non-negligible advantage where l represents any bit, except the $\log\log |\mathbb{F}|$ most significant — in particular l as a function of $\log |\mathbb{F}|$ should satisfy that for any $c \in \mathbb{N}$, $l < \log |\mathbb{F}| - c \log\log |\mathbb{F}|$ for sufficiently large $\log |\mathbb{F}|$.*
2. *Let $\mathrm{BITS}_l(a)$ denote the sequence of the l least significant bits of $a \in \mathbb{F}$. Given $X \in \mathcal{S}_{n,k,t}$ predict $\mathrm{BITS}_l(s_X(w))$ with probability $\frac{1}{2^l} + \alpha(n)$ where $\alpha(n)$ is non-negligible.*
3. *Let $\mathrm{QR}(a)$ be 1 iff $a \in \mathbb{F}$ is a quadratic residue, where $\mathbb{F}$ is of prime order. Given $X \in \mathcal{S}_{n,k,t}$ predict $\mathrm{QR}(s_X(w))$ with non-negligible advantage.*

We note that the exclusion of the $\log\log |\mathbb{F}|$ most significant bits from the item (1) above is independent of our treatment as depending on the order of the field they may be easy to guess, and as a result BIT_l might not be balanced. Note that if the finite field is chosen appropriately all bits of $s_X(w)$ will be hard: e.g. if we restrict to finite fields $\mathbb{F}$ such that there is a $c \in \mathbb{N}$: $|\mathbb{F}| - 2^{\lfloor \log |\mathbb{F}| \rfloor} \leq (\log |\mathbb{F}|)^c$ then all bits will be hard (e.g. a field of numbers modulo a Mersenne prime).

We complete this section by pointing out that theorem 1 can be extended to any h points of the solution-polynomial. In particular, if h points of the solution polynomial of a PR-instance follow a probability distribution $\mathcal{D}_h$ over $\mathbb{F}^h$ (where $1 \leq h < k$) then provided that $\mathrm{DPR}[n, k-h, t]$ holds, no probabilistic polynomial-time bounded adversary can extract any non-trivial information about these points.

4   Pseudorandomness

In this section we will show that distinguishing instances of $\mathrm{PR}[n, k, t]$ from random elements of $\mathcal{S}_n := (\mathbb{F})_n \times \mathbb{F}^n$ is hard under the DPR-Assumption (which essentially amounts to saying that instances of $\mathrm{PR}[n, k, t]$ are pseudorandom under the DPR). We start with a standard definition:

Definition 4. *Let $\{\mathcal{F}_n\}_{n \in \mathbb{N}}$ be a family of sets, such that $\mathcal{F}_n$ contains all possible choices of elements of size n. Note that for simplicity we will use $\mathcal{F}_n$ to refer to the family $\{\mathcal{F}_n\}_{n \in \mathbb{N}}$. Two families of sets $A_n, B_n \subseteq \mathcal{F}_n$ are* (polynomial-time, computationally) *indistinguishable if for any PPT predicate $\mathcal{A}$,*

$$| \mathbf{Prob}_{X \in_U A_n}[\mathcal{A}(X) = 1] - \mathbf{Prob}_{X \in_U B_n}[\mathcal{A}(X) = 1] |$$

is negligible in n. If on the other hand there is an $\mathcal{A}$ for which the probability above is non-negligible in n, we will say that $\mathcal{A}$ is a distinguisher for A_n, B_n. A family of sets A_n is called pseudorandom *if it is indistinguishable from $\mathcal{F}_n$.*

Note that for this section we consider $B_n = \mathcal{F}_n := \mathcal{S}_n = (\mathbb{F})_n \times \mathbb{F}^n$ and $A_n := \mathcal{S}_{n,k,t}$ (the set of $\mathrm{PR}[n, k, t]$ instances). Let $\mathcal{A}$ be a distinguisher for $\mathcal{S}_{n,k,t}$ and $\mathcal{S}_n$. Because of remark 1 it holds that the particular choice of the index-solution set I is essentially independent of the distinguishing probability, i.e. for some $\mathcal{A}'$ and for all $I \subseteq \{1, \ldots, n\}$, $|I| = t$, it holds that the following is non-negligible in n:

$$| \mathbf{Prob}_{X \in_U \mathcal{S}_{n,k,t}(I)}[\mathcal{A}'(X) = 1] - \mathbf{Prob}_{X \in_U \mathcal{S}_n}[\mathcal{A}(X) = 1] |$$

In other words, remark 1 suggests that any distinguisher between $\mathcal{S}_{n,k,t}$ and $\mathcal{S}_n$ can essentially serve as a distinguisher between $\mathcal{S}_{n,k,t}(I)$ and $\mathcal{S}_n$ for all subsets I (by permuting the points of the given PR-instance first).

The core of the pseudorandomness proof is the next lemma which shows how such a distinguisher can be used to extract a parameterized over $\{0, \ldots, n\}$ predicate $\mathcal{B}$ that its behavior is sensitive to some choice of the parameter that belongs in the index-solution-set of a given PR-instance.

Lemma 2. *Let $\mathcal{A}$ be a PPT predicate s.t. for all $I \subseteq \{1, \ldots, n\}$ with $|I| = t$, $\mathcal{A}$ is a distinguisher for $\mathcal{S}_{n,k,t}(I)$ and $\mathcal{S}_n$. Then there exists a PPT $\mathcal{B}$, for which it holds that for all $I \subseteq \{1, \ldots, n\}$ with $|I| = t$, there exists an $i_0 \in I$ with $i_0 \leq n - k$, such that if*

$$\beta_i(n) := \mathbf{Prob}_{X \in_U \mathcal{S}_{n,k,t}(I)}[\mathcal{B}(i, X) = 1] \quad \text{for} \quad i \in \{0, \ldots, n\}$$

it holds that $|\beta_{i-1}(n) - \beta_i(n)|$ is negligible for any $i \notin I$ and non-negligible for i_0.

Now observe that if $\mathcal{A}_1(i, X) := \mathcal{B}(i, X)$ and $\mathcal{A}_2(i, X) := \mathcal{B}(i - 1, X)$, it follows easily that $\mathcal{A}_1, \mathcal{A}_2$ is a gap-predicate-pair. As a result,

Theorem 2. *Under the DPR-Assumption for $[n, k, t]$, the family of PR instances $\mathcal{S}_{n,k,t}$ is pseudorandom.*

5   Application: A Stateful Cipher

A cipher design involves two parties, who share some common random input (the key). The goal of a cipher design is the secure transmission of a sequence of messages. Suppose that I denotes the shared randomness between the sender and the receiver. A cipher is defined by two probabilistic functions $f_I : \mathcal{K} \times \mathbb{P} \to \mathcal{K} \times \mathbb{C}$ and $g_I : \mathcal{K} \times \mathbb{C} \to \mathcal{K} \times \mathbb{P}$. The spaces $\mathcal{K}, \mathbb{P}, \mathbb{C}$ denote the state-space, plaintext-space and ciphertext-space respectively. The functions f, g have the property that if $f_I(s, m) = (s', c)$ (encryption) it holds that $g_I(s, c) = (s', m)$ (decryption); note that s' (given by both f, g) is the state that succeeds the state s.

Stream-ciphers use public state sequences of the form $\langle 0, 1, 2, 3, \ldots \rangle$. The reader is referred to [Lub96] for more details on stream ciphers and how they can be built based on pseudorandom number generators. Block-ciphers encrypt messages of size equal to some fixed security parameter which are called blocks. Such ciphers are typically at the same state throughout and this state is considered to be secret (it coincides with the secret shared random key). The reader is referred to [Gol98] for further details on block-ciphers and generic constructions.

If a cipher, which operates on blocks, employs a "secret state-sequence update" and uses the shared randomness (the key) only as the initial state of the state-sequence, it is called a *stateful* cipher, see the figure below; (note that in a stateful cipher we suppress the subscript I from the functions f, g).

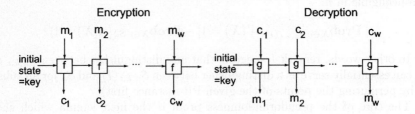

Fig. 1.

In the remaining of this section we introduce a stateful cipher that is based on PR and possesses unique properties.

DESCRIPTION OF THE PR-CIPHER. Let $[n, \frac{k-1}{2}, t]$ with $k \leq t$ be sound parameters for the PR problem. We work in a finite field $\mathbb{F}$ with $\log |\mathbb{F}| \geq 3n$. The state-space $\mathcal{K}$ is defined to be the set of n-bitstrings with Hamming weight t. For some $s \in \mathcal{K}$ we define I_s to be the corresponding subset of $\{1, \ldots, n\}$, and v_s be the corresponding integer that has s as its binary representation. We denote by $V_{\mathcal{K}}$ the set of all numbers that their binary representation belongs in $\mathcal{K}$. Let $\mathbb{P} := \mathbb{F}^{\frac{k-1}{2}}$ and $\mathbb{C} := (\mathbb{F})_n \times \mathbb{F}^n$. The shared randomness between the two parties is a random $s_0 \in \mathcal{K}$, that is the initial state of the cipher.

The encryption function f of the cipher is defined as follows: $\langle s, \mathbf{m} \rangle$ is mapped to a random instance Y of $\mathcal{S}_{n,k,t}(I_s)$ so that $s_Y(0)$ is a random element of $V_{\mathcal{K}}$ (the next key to be used), and $s_Y(1) = (\mathbf{m})_1, \ldots, s_Y(\frac{k-1}{2}) = (\mathbf{m})_{\frac{k-1}{2}}$. The decryption function g is defined as follows: given $\langle s, Y \rangle \in \mathcal{K} \times \mathbb{C}$, the polynomial p that corresponds to the pairs of Y whose index is in I_s is interpolated. The decrypted message is set to be $\langle p(1), \ldots, p(\frac{k-1}{2}) \rangle$ and the next state is set to the binary representation of $p(0)$.

SEMANTIC-SECURITY. A semantic-security adversary $\mathcal{A}$ for a stateful cipher is a PPT that takes the following steps: (i) queries a polynomial number of times the encryption-mechanism, (ii) generates two messages M_1, M_2 and obtains the ciphertext that corresponds to the encryption of M_b where b is selected at random from $\{1, 2\}$, (iii) queries the encryption-mechanism a polynomial number of times. Finally the adversary predicts the value of b. This is illustrated in the figure below. A cipher is said to be semantically secure if any semantic-security adversary predicts b with negligible advantage in the security parameter n. For more details regarding semantically secure symmetric encryption, see [Lub96, KY00].

Operation of a Semantic Security Adversary

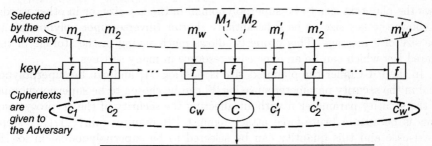

The Adversary decides whether C is an encryption of M_1 or M_2

Fig. 2.

FORWARD SECRECY. A cipher is said to satisfy forward secrecy if in the case of a total security breach at some point of its operation (i.e. the internal state is revealed) the adversary is unable to extract any information about the previously communicated messages.

This is formalized by two chosen plaintext security adversaries who are submitting adaptively messages to the encryption oracle. The encryption oracle flips a coin and answers by encrypting the plaintexts submitted by one of the two adversaries (the same adversary throughout). At some point the internal state of the system is revealed to the adversaries. Forward secrecy is violated if the adversaries can tell with probability significantly better than one half whose messages the encryption oracle was returning.

COMPUTATIONAL PERFECT SECRECY. A generic chosen plaintext adversary for a stateful cipher is is an adversary who is allowed to query an encryption oracle a number of times and then required to meet a certain (unspecified) challenge. For some stateful-cipher we consider the following two attacks that can be launched by a generic chosen plaintext adversary: (i) "existential" where the generic chosen plaintext adversary is allowed to query the encryption oracle a number of times and then is asked to decrypt the next ciphertext (which encrypts a random secret message) (ii) "universal" where a generic chosen plaintext adversary is allowed to query the encryption oracle a number of times and then is asked to recover the state of the cipher (something that allows the recovery of all future messages from that point on).

It is clear that for any cipher an existential attack reduces to a universal attack. Nevertheless it is not at all apparent if the opposite direction in the reduction holds. A stateful-cipher for which it holds that a generic chosen plaintext adversary launching an existential attack implies the existence of a generic chosen plaintext adversary launching a universal attack is said to satisfy *computational perfect secrecy*. The equivalence of attacks that recover the message to attacks that recover the key has been postulated by Shannon as "perfect secrecy." Blum and Goldwasser [BG84] designed a factoring based public-key system where they reduced semantic security of a message to breaking the key (i.e. factoring the composite). They coined the notion of "computational perfect secrecy," a variant of which we define above.

SUPERPOLYNOMIAL MESSAGE-SIZE. A cryptosystem that has this property allows the plaintext size to be superpolynomial in the key-size, or in other words, it allows the key-size to be substantially shorter (inverse-super-polynomial) in the size of a message. This property allows much saving in the storage of the shared key which can be an expensive resource in many settings.

In the PR-Cipher the plaintext size is $\frac{k-1}{2}\lfloor \log |\mathbb{F}| \rfloor$ and can be superpolynomial in the security parameter since $\log |\mathbb{F}|$ can be chosen to be superpolynomial in the security parameter n without affecting the security of the cryptosystem. This is because a brute-force attack against PR requires $\min\{\binom{n}{t}, \binom{n}{k}\}$ steps worst-case and this quantity can be selected to be superpolynomial in $\log |\mathbb{F}|$ even if $\log |\mathbb{F}|$ is superpolynomial in n.

ERROR-CORRECTING DECRYPTION. A cryptosystem is said to allow error correcting decryption if the decryption procedure is able to correct errors that are introduced during the transmission (possibly by an adversary). This combines the decryption operation with the error-correction operation (that is important to apply independently in any setting where two parties communicate). A cryptosystem that transmits plaintexts consisting of d blocks is called d'-error-correcting if up to d' corrupted blocks can be corrected for each transmitted plaintext. The PR-cipher (which transmits plaintexts that consist of $\frac{k-1}{2}$ "blocks": elements of the underlying finite field $\mathbb{F}$) is $\frac{t-k}{2}$-error-correcting since the interpolation step during decryption can be substituted by the [BW86] polynomial-reconstruction algorithm that can withstand up to $\frac{t-k}{2}$ errors (in the worst-case).

KEY-EQUIVALENCE. A symmetric cryptosystem is said to satisfy the key-equivalence property if there are no families of keys of measurable size that are susceptible to attacks that do not apply to the key-space in general. By "measurable-size" we mean that the ratio of the size of the family of keys over the key-space size is a non-negligible function. The key-equivalence property is an important security aspect for a symmetric cryptosystem as it suggests that there are no "weak" keys.

The following theorem summarizes the properties of the PR-Cipher:

Theorem 3. *The PR-Cipher with parameters* n, k, t,
(i) is semantically secure under $\mathrm{DPR}[n, \frac{k-1}{2}, t]$.

(ii) satisfies forward secrecy under DPR$[n, \frac{k-1}{2}, t]$.
(iii) satisfies computational perfect secrecy.
(iv) allows super-polynomial in n message-size.
(v) is $\frac{t-k}{2}$-error-correcting.
(vi) satisfies the key-equivalence property.

References

[Ber68] Elwyn R. Berlekamp, *Algebraic Coding Theory*. McGraw-Hill, 1968.

[BW86] Elwyn R. Berlekamp and L. Welch, *Error Correction of Algebraic Block Codes*. U.S. Patent, Number 4,633,470, 1986.

[BN00] Daniel Bleichenbacher and Phong Nguyen, *Noisy Polynomial Interpolation and Noisy Chinese Remaindering,* In Advances in Cryptology — Eurocrypt 2000, Lecture Notes in Computer Science, Springer-Verlag, vol. 1807, pp. 53–69, May 2000.

[BG84] Manuel Blum and Shafi Goldwasser, *An Efficient Probabilistic Public-Key Encryption Scheme Which Hides All Partial Information*, In Advances in Cryptology — Crypto 1984, Lecture Notes in Computer Science, Springer-Verlag, vol. 196, pp. 289–302, 1985.

[Gol98] Oded Goldreich, *Foundations of Cryptography: Fragments of a Book*, manuscript 1998.

[GSR95] Oded Goldreich, Madhu Sudan and Ronitt Rubinfeld, *Learning Polynomials with Queries: The Highly Noisy Case,* in the Proceedings of the 36th Annual Symposium on Foundations of Computer Science, IEEE Computer Society, pp. 294–303, 1995. (ECCC Technical Report: TR98-060).

[GM84] Shafi Goldwasser and Silvio Micali, *Probabilistic encryption*, Journal of Computer and System Sciences, vol. 28(2), pp. 270-299, April 1984.

[GS98] Venkatesan Guruswami and Madhu Sudan, *Improved Decoding of Reed-Solomon and Algebraic-Geometric Codes*. In the Proceedings of the 39th Annual Symposium on Foundations of Computer Science, IEEE Computer Society, pp. 28–39, 1998.

[KY02] Aggelos Kiayias and Moti Yung, *Cryptographic Hardness based on the Decoding of Reed-Solomon Codes with Applications*, ECCC Technical report TR02-017, 2002.

[KY00] Jonathan Katz and Moti Yung, *Complete Characterization of Security Notions for Probabilistic Private-key Encryption*, in the Proceedings of the 32nd Annual ACM Symposium on Theory of Computing, ACM, pp. 245–254, 2000.

[Lub96] Michael Luby, *Pseudorandomness and Cryptographic Applications*, Princeton University Press, 1996.

[MS77] F. J. MacWilliams and N. Sloane, *The Theory of Error Correcting Codes*. North Holland, Amsterdam, 1977.

[NP99] Moni Naor and Benny Pinkas, *Oblivious Transfer and Polynomial Evaluation*. in the Proceedings of the 31st Annual ACM Symposium on Theory of Computing, ACM, pp. 245–254, 1999. (Full Version *Oblivious Polynomial Evaluation*, available at
http://www.wisdom.weizmann.ac.il/~naor/onpub.html.)

[Sud97] Madhu Sudan, *Decoding of Reed Solomon Codes beyond the Error-Correction Bound*. Journal of Complexity 13(1), pp. 180–193, 1997.

Perfect Constant-Round Secure Computation via Perfect Randomizing Polynomials

Yuval Ishai[1]* and Eyal Kushilevitz[2]**

[1] Princeton University, USA. yishai@cs.princeton.edu.
[2] Technion, Israel. eyalk@cs.technion.ac.il.

Abstract. Various information-theoretic constant-round secure multiparty protocols are known for classes such as NC^1 and polynomial-size branching programs [1,13,18,3,19,10]. All these protocols have a small probability of failure, or alternatively use an *expected* constant number of rounds, suggesting that this might be an inherent phenomenon. In this paper we prove that this is not the case by presenting several constructions of *perfect* constant-round protocols.

Our protocols are obtained using *randomizing polynomials* – a recently introduced representation [19], which naturally relaxes the standard polynomial representation of boolean functions. Randomizing polynomials represent a function f by a low-degree mapping from its inputs and independent random inputs to a vector of outputs, whose distribution depends only on the value of f. We obtain several constructions of degree-optimal *perfect* randomizing polynomials, whose distinct output distributions are perfectly separated. These results on randomizing polynomials are of independent complexity-theoretic interest.

1 Introduction

Representation of functions by low-degree multivariate polynomials has proved to be a surprisingly powerful tool in complexity theory. Such a representation is also useful in the context of *secure multiparty computation*; in particular, most general-purpose protocols for secure multiparty computation can be used to evaluate constant-degree polynomials in a constant number of rounds.[1] A major difficulty, however, is that not many functions can be evaluated using low-degree polynomials, and even some very simple functions, like the logical OR of n bits, require polynomials of degree n.

A natural relaxation of the standard representation notion which gets around this obstacle was recently suggested in [19]. *Randomizing polynomials* extend the standard representation by incorporating randomness and by allowing *several* polynomials to simultaneously act on the same inputs and random inputs. Instead of directly outputting the value of the represented function, a randomizing polynomials vector is required to produce an *output distribution* which directly corresponds to this value. This is best illustrated by the next example, which shows a degree-2 representation of the OR function by randomizing polynomials.

---

* Work done while at AT&T Labs – Research and DIMACS.
** Work done in part while at IBM T.J. Watson Research Center. Supported in part by the Mitchell-Schoref program at the Technion and MANLAM Fund 120-044.

[1] More generally, the round complexity of these protocols is proportional to the *multiplicative depth* of an arithmetic circuit computing the function f of interest, where multiplicative depth is defined similarly to ordinary circuit depth except that addition gates are ignored.

Let $\mathbb{F}$ be some finite field, and $p = (p_1, \ldots, p_s)$ a vector of degree-2 polynomials over $\mathbb{F}$ in the n inputs $x = (x_1, \ldots, x_n)$ and the sn random inputs $r = (r_{ij})$, $1 \leq i \leq s$, $1 \leq j \leq n$, defined by $p(x, r) = (\sum_{j=1}^{n} x_j r_{1j}, \ldots, \sum_{j=1}^{n} x_j r_{sj})$. For any input $x \in \{0, 1\}^n$, let $P(x)$ denote the output distribution of p on x, i.e., the distribution over $\mathbb{F}^s$ induced by a uniform choice of r from $\mathbb{F}^{sn}$. It is not hard to verify that the above polynomial vector p satisfies the following two properties: (1) If $\mathrm{OR}(x) = \mathrm{OR}(y)$ then the output distributions $P(x)$ and $P(y)$ are *identical*; in other words, there exist probability distributions D_0, D_1 such that for any $x \in \{0, 1\}^n$, $P(x)$ is equal to $D_{\mathrm{OR}(x)}$ (specifically, D_0 is concentrated on the zero vector, and D_1 is uniform over $\mathbb{F}^s$). (2) The statistical distance between D_0, D_1 is close to 1 (more precisely, it is $1 - |\mathbb{F}|^{-s}$).

Property (1) guarantees that from a sample of $P(x)$ it is impossible to learn *anything* about x except, perhaps, $\mathrm{OR}(x)$; Property (2) guarantees that from such a sample it is indeed possible to correctly compute $\mathrm{OR}(x)$ with high probability. Thus, learning a sample from $P(x)$ is, in a sense, information-theoretically equivalent to learning $\mathrm{OR}(x)$. Consequently, the task of securely computing the OR function may be reduced to the task of securely sampling from $P(x)$, which in turn can be reduced to that of securely evaluating a related vector of *deterministic* degree-2 polynomials over $\mathbb{F}$.

The application to secure computation will be more thoroughly discussed in Section 1.1. For the time being, however, we point out the fact that the above two output distributions D_0 and D_1 are not perfectly separated, and note that this does come at a cost: even if a perfect protocol is used for the evaluation of degree-2 polynomials, the resultant secure protocol for OR will have a nonzero error probability. This issue, which is further motivated below, stands in the center of the current work.

A general definition of randomizing polynomials $p(x, r)$ representing a boolean function $f(x)$ can be easily derived from the above example.[2] In [19] it was shown that *any* boolean function f can be represented by degree-3 randomizing polynomials, where the complexity of this representation (defined as the total number of inputs and outputs) is at most quadratic in the branching program size of f. It was also shown that almost all functions, with the exception of functions which are "similar" to the OR function, do not admit a degree-2 representation. However, the general degree-3 construction of [19] suffers from the same deficiency as the above example: the two output distributions D_0, D_1 are not completely disjoint. Thus, the general reduction it provides in the context of secure computation introduces a small probability of error. This raises the question whether *perfect* low-degree randomizing polynomials, for which the output distributions are perfectly separated, can be constructed, and if so at what cost.

Before describing our results we give some background on secure multiparty computation and motivate the above question in this context.

1.1   Secure Multiparty Computation and Its Round Complexity

A secure multiparty computation protocol allows k parties to evaluate a function of their inputs in a distributed way, so that both the *privacy* of their inputs and the *correctness* of the outputs are maintained. These properties should hold in the presence of an *adversary* which may corrupt at most t parties. The main focus of this work is on the *information-theoretic* setting for secure computation, in which security should hold

---

[2] For concreteness, we set a constant threshold of $1/2$ on the statistical distance between D_0, D_1; this distance can be amplified, without increasing the degree, by concatenating several copies of p having disjoint sets of random inputs.

against a computationally unbounded adversary. Nonetheless, our results are also useful in the alternative *computational* setting, as discussed in Section 1.2.

The *round complexity* of interactive protocols is one of their most important complexity measures. Indeed, substantial research efforts have been invested into characterizing the round complexity of various distributed tasks, such as zero-knowledge proofs and Byzantine Agreement. This is the case also for the general task of secure computation. Following the initial plausibility results [26,17,6,9], much of the research in this area has shifted to various *complexity* aspects of secure computation. In particular, the problem of obtaining constant-round secure protocols has attracted a considerable amount of attention [1,5,4,13,18,24,3,7,19,10,21]. Our work continues this line of research, and focuses on the following question: can *perfectly* secure computation be realized with a constant number of rounds in the *worst case*?[3]

In the computational setting for secure computation, any function that can be (efficiently) computed can also be securely computed in a constant number of rounds [26, 5,21]. The situation is not as well understood in the information-theoretic setting. Several (efficient) constant-round protocols are known in this setting for function classes such as NC^1, polynomial-size branching programs, and related linear algebra classes [1, 13,18,3,19,10]. All these protocols have a small probability of failure, or alternatively use an *expected* constant number of rounds, suggesting that this might be an inherent phenomenon. In the current work we show that this is not the case: we obtain *perfect* constant-round protocols which typically match or beat their previous non-perfect (information-theoretic) counterparts in every efficiency aspect.

1.2   Our Results

We present two main constructions of perfect randomizing polynomials, which in turn can be transformed via general-purpose protocols for perfectly secure computation (e.g., [6,11,14]) to perfect constant-round protocols. (This transformation is outlined in Section 2.3.) The communication complexity of the resultant protocols is proportional to the complexity of the underlying randomizing polynomials. Their exact number of rounds depends on the specific notion of security. For instance, t-security against a passive adversary can be achieved in 2 rounds if $t < k/3$ or in 3 rounds if $t < k/2$ (see [19]), and t-security against an active adversary can be achieved in 3 rounds with $t = \Omega(k)$ using a 2-round VSS protocol from [14] (assuming a broadcast channel is available). From now on, we describe the results in terms of randomizing polynomials and do not spell out the specific consequences for constant-round secure computation.

A combinatorial construction. Our first construction of perfect randomizing polynomials is combinatorial in nature, and is based on a boolean formula representation. We first derive a natural information-theoretic analogue of Yao's *garbled circuit* construction [26], which is originally cast in the computational setting, and observe that it gives rise to a representation by perfect degree-3 randomizing polynomials over $GF(2)$. The complexity of this construction is at least quadratic (and always polynomial) in the formula size. We then present an optimization which allows a significant complexity improvement. We demonstrate this improvement for the function OR, for which the complexity of the optimized representation is $2^{O(\sqrt{\log n})} \cdot n$.

An algebraic construction. Our second construction is linear-algebraic in nature, and is based on a branching program representation. (The relevant branching program models

---

[3] The term "perfect security" binds together perfect privacy and correctness requirements; see, e.g., [8] for its formal definition.

are defined in Section 2.2.) We obtain, for any *counting* branching program of size ℓ, a perfect degree-3 representation of complexity ℓ^2. This construction applies to deterministic branching programs and logspace computation as special cases, but does not apply to the nondeterministic model.[4] In the full version of this paper we show that by settling for *statistical privacy* (a relaxation of property (1) of randomizing polynomials) it is possible to get an efficient representation also in the nondeterministic case. This yields perfectly-correct (yet statistically-private) constant-round protocols for NL.

We note that since branching programs can simulate formulas, the latter constructions can be efficiently applied to a presumably larger class of functions than the former. However, as the OR example demonstrates, the complexity of the (optimized) formula-based construction can be significantly better.

Efficiency advantages. In addition to providing perfect security in a strictly-constant number of rounds, our results also offer some independent efficiency advantages. In the information-theoretic setting, all previous constant-round protocols required either a significant parallel repetition of certain subprotocols or computation over large fields to make their failure probability small. (In the context of randomizing polynomials, this is needed to amplify the separation between the two output distributions.) Our constructions avoid this overhead. For instance, our perfect solution for branching programs over $\mathbb{F} = GF(2)$ is slightly more efficient than a similar solution from [19] that can only achieve a small constant separation (which then needs to be amplified).

In the computational setting for secure computation, Yao's garbled circuit technique gives rise to constant-round protocols whose efficiency is *linear* in the *circuit* size of the function being computed and a security parameter [26,5,21]. Since none of the alternative information-theoretic techniques (and ours in particular) efficiently applies to circuits, they are generally considered less appealing. However, they do offer some efficiency advantages in the computational setting as well. Most notably, an advantage of most of these techniques over the garbled circuit approach is that they efficiently scale to *arithmetic* computation over large moduli. Our solution for *counting* branching programs, which applies to arithmetic formulas as a special case, is especially appealing in this context. This application and its generalization to arbitrary *rings* are further addressed in [12]. Finally, our constructions can be beneficial also in the boolean case; since their complexity does not involve a cryptographic security parameter (nor does its analysis hide large constants), they may be preferable for small branching programs or formulas which arise in specific applications (such as the "millionaire's problem" [25]).

2   Preliminaries

Notation. We let $\mathbb{F}$ denote a finite field; the underlying field $\mathbb{F}$ is often omitted when it can be understood from the context. The *statistical distance* between two probability distributions Y_0 and Y_1 is defined as $\mathrm{SD}(Y_0, Y_1) = \max_E |\Pr[Y_0 \in E] - \Pr[Y_1 \in E]|$.

2.1   Randomizing Polynomials

Syntax. A *randomizing polynomials vector* $p = (p_1, \ldots, p_s)$ is a vector of polynomials in the $n + m$ variables $x = (x_1, \ldots, x_n)$ and $r = (r_1, \ldots, r_m)$ over some finite field $\mathbb{F}$. The variables $x_1, \ldots, x_n$ will be referred to as the *inputs* of p and $r_1, \ldots, r_m$ as its *random inputs*. The *output complexity* of p is the number of polynomials s, its *randomness*

---

[4] Counting is usually more powerful than nondeterminism. However, in the context of perfect randomizing polynomials it is not clear how to eliminate the extra information provided by the exact count.

complexity is the number of random inputs m, and its *complexity* is the total number of all inputs and outputs $s + n + m$. Finally, the *degree* of p is defined as the (total) degree of its maximum-degree entry, where both ordinary inputs and random inputs count towards the degree.[5] For instance, if $p = (p_1, p_2)$ where $p_1(x, r) = x_1 r_1^2$ and $p_2(x, r) = x_1 x_2 + r_1 + r_2 + r_3$, then the degree of p is 3.

Semantics. The following semantics of randomizing polynomials generalize the original ones from [19], and incorporate computational efficiency requirements which were not considered in the original definitions. However, the default notion of randomizing polynomials remains the same.

For any $x \in \mathbb{F}^n$ and $r \in \mathbb{F}^m$, the *output* $p(x, r) = (p_1(x, r), \ldots, p_s(x, r))$ is an s-tuple over $\mathbb{F}$. For any $x \in \mathbb{F}^n$, let $P(x)$ denote the *output distribution* of p on input x, induced by a uniform choice of $r \in \mathbb{F}^m$. Thus, randomizing polynomials may be thought of as computing a function from inputs to output distributions. We say that p represents a function f if the output distribution $P(x)$ "corresponds" to the function value $f(x)$. Motivated by the application to secure computation, we break this condition into two requirements, termed *privacy* and *correctness*.

Definition 1. *A polynomial vector $p(x, r)$ over $\mathbb{F}$ is an ϵ-private, δ-correct randomizing polynomials representation for a function $f : A^n \to B$, where $A \subseteq \mathbb{F}$ and B is an arbitrary set, if it has the following two properties:*

- ϵ-CORRECTNESS. *There exists a randomized simulator algorithm S such that for any input $x \in A^n$, $\mathrm{SD}(S(f(x)), P(x)) \le \epsilon$.*
- δ-PRIVACY. *There exists a reconstruction algorithm C such that for any $x \in A^n$, $\Pr[C(P(x)) \ne f(x)] \le \delta$.*

When referring to a uniform family of randomizing polynomials, parameterized by n, we require that the simulator and the reconstruction algorithms be efficient in n.

By default, we define randomizing polynomials to be 0-private, 1/4-correct (similarly to [19]). The main new variant considered in this paper is that of *perfect* randomizing polynomials, defined as 0-private, 0-correct randomizing polynomials. Finally, we will also consider a third variant of *ϵ-private, perfectly-correct* randomizing polynomials.

2.2 Formulas and Branching Programs

Formulas. A formula is a single-output boolean circuit in n input variables in which each gate has a fan-out of 1. Specifically, a formula F is a directed binary tree. Each of its leaves is labeled by a *literal* which is either a variable from $x_1, \ldots, x_n$ or a negated variable from $\bar{x}_1, \ldots, \bar{x}_n$. Each internal node of the tree, referred to as a *gate*, has two incoming edges, referred to as *wires*, and is labeled by either AND or OR. This includes the node which is the root of the tree; we think of this node as also having an outgoing wire which is called the *output* wire of F. Any input $x \in \{0, 1\}^n$ naturally assigns a unique *value* to each wire. The value of the formula F, denoted $F(x)$, is the value of its output wire.

Branching programs. Syntactically, a branching program (BP for short) is defined by a directed acyclic graph $G(V, E)$, two special vertices $s, t \in V$, and a labeling function ϕ assigning to each edge in E a literal (i.e., x_i or $\bar{x}_i$) or the constant 1. Its *size* is defined as $|V| - 1$. Each input assignment $x = (x_1, \ldots, x_n)$ naturally induces an unlabeled subgraph G_x, whose edges include every $e \in E$ such that $\phi(e)$ is satisfied by x. An

---

[5] This convention is crucial for the application to secure multiparty computation.

accepting path on input x is a directed $s - t$ path in the graph G_x. We attach two main semantics to branching programs. A mod-q *counting* BP (CBP for short), where $q \geq 2$ is prime, computes the function $f : \{0,1\}^n \to \mathrm{GF}(q)$ such that $f(x)$ is the *number* of accepting paths on x modulo q. A mod-q *nondeterministic* BP (NBP for short) computes the boolean function $f : \{0,1\}^n \to \{0,1\}$, such that $f(x) = 1$ iff the number of accepting paths on x is *nonzero* modulo q. By setting q to be larger than the possible number of paths, we get the usual notion of nondeterminism over the integers. Finally, perhaps the most useful notion of BP is the special case of *deterministic BP*, where each input induces at most *one* accepting paths. A deterministic BP may be viewed as a mod-q CBP or NBP with an arbitrary $q \geq 2$.

2.3   Secure Multiparty Computation

The reader is referred to [8,16,22,15,2] for formal definitions of secure computation. We emphasize though that the issues addressed in this paper are quite insensitive to the exact notion of security. Our results provide *information-theoretic reductions* from the task of securely computing a general function f, represented by a formula or a branching program, to that of securely computing a vector of degree-3 polynomials. Such a reduction, originally described in [19] for the non-perfect case, proceeds as follows. Given a representation of $f(x)$ by $p(x, r)$, the secure computation of f (whose n inputs are arbitrarily partitioned among the k players) can be reduced to the secure computation of the randomized function $P(x)$. The latter, in turn, reduces to the secure computation of the *deterministic* function $p'(x, r^1, \ldots, r^{t+1}) \stackrel{\text{def}}{=} p(x, r^1 + \ldots + r^{t+1})$, where t is the security threshold, by assigning each input vector r^j to a distinct party and instructing it to pick it at random. Note that the degree of p' is the same as that of p. The above reduction preserves perfectness: if $p(x, r)$ is a perfect representation for f and if a perfectly secure protocol is used for evaluating p', then the resultant protocol for f is also perfectly secure.[6] In the remainder of this paper we will phrase our results in terms of randomizing polynomials and will not state the corollaries to secure computation.

3   Perfect Randomizing Polynomials from Formulas

In this section we construct perfect degree-3 randomizing polynomials from a boolean formula representation. The construction works over $\mathbb{F} = \mathrm{GF}(2)$, which we take to be the underlying field throughout this section. We start with a basic construction, which may be viewed as the natural information-theoretic analogue of Yao's garbled circuit construction. (Our notation for this section closely follows the presentation of Yao's construction from [23].) We later present an optimization of this basic construction.

Let F be a boolean formula of size s computing the function $f : \{0,1\}^n \to \{0,1\}$. It may be assumed, without loss of generality, that F has depth $d = O(\log s)$. We will efficiently transform F to a perfect degree-3 randomizing polynomials representation for f, whose complexity is polynomial in s. As usual, denote by x the input for F and by $x_1, \ldots, x_n$ its individual n variables. Let m be the number of wires in F, where the m-th wire is the output wire. For $i \in [m]$, denote by $b_i(x)$ (or simply b_i) the value of the i-th wire induced by the input x. In constructing the randomizing polynomials vector $p_F(x, r)$ we use random inputs of two types: m bits denoted $r_1, \ldots, r_m$ corresponding

---

[6] In the case of security against an *active* adversary, it is important that the input domain of p' be defined so that x is taken from A^n, the input domain of f, rather than from $\mathbb{F}^n$ (if they are different). In the default boolean case ($A = \{0,1\}$) standard protocols from the literature can be modified to handle such a restriction on the inputs of p' with little or no efficiency overhead.

to the m wires of F, and m pairs of strings W_i^0, W_i^1 again in correspondence with the m wires. The length of the strings W_i^b is defined inductively (from top to bottom) as follows: $|W_m^b| = 0$ (for $b \in \{0, 1\}$) and if k is the output wire of some gate g and i, j are the input wires of this gate then $|W_j^b| = |W_i^b| = 2(|W_k^b| + 1)$ (for $b \in \{0, 1\}$); therefore the length of each of these strings is at most $O(2^d)$ =poly(s). We view each string W_i^b as if it is broken into two equal-size halves denoted $W_i^{b,0}, W_i^{b,1}$. We use c_i to denote the value of wire i masked by r_i; namely, $c_i = b_i \oplus r_i$.

To define the polynomial vector p_F, we specify several polynomials for each wire. In what follows $\oplus$ denotes bitwise-xor among strings; when we want to emphasize that the operation is applied to single bits we will usually denote it by either $+$ or $-$. We call *meta-polynomial* a polynomial that involves strings. Each meta-polynomial has a simple transformation into a vector of polynomials that operate bit-by-bit (e.g., for $a \in \mathrm{GF}(2)$ and strings A, B of length t the expression $a \cdot A \oplus B$ is a meta-polynomial that represents the length-t polynomial vector $(a \cdot A_1 + B_1, \ldots, a \cdot A_t + B_t)$). When we want to emphasize that a term T is actually a string we write $\langle T \rangle$. We use $\circ$ to denote concatenation. We now describe the polynomial vector associated with each wire.

Input wires: For an input wire i, labeled by a literal ℓ, we use the following meta-polynomial $\langle W_i^\ell \circ (\ell + r_i) \rangle$. Note that ℓ is either some variable x_u or its negation, which can be written as the degree-1 polynomial $1 - x_u$. Also, note that each term W_i^ℓ can be represented by $\ell \cdot W_i^1 \oplus (1 - \ell) \cdot W_i^0$. All together, this is a degree-2 meta-polynomial which, as described above, is just a short writing for a vector of degree-2 polynomials over the boolean inputs x and boolean random inputs r, W.

Output wires of gates: Let g be a gate with input wires i, j and output wire k. We associate with this wire 4 meta-polynomials. Specifically, for each of the 4 choices of $c_i, c_j \in \{0, 1\}$, we define a corresponding meta-polynomial on strings of length $|W_k^b| + 1$. This degree-3 meta-polynomial can be thought of as the garbled table entry indexed by (c_i, c_j) and is defined as follows:

$$Q_k^{c_i,c_j}(x, r) \overset{\text{def}}{=} W_i^{c_i-r_i,c_j} \oplus W_j^{c_j-r_j,c_i} \oplus \langle W_k^{g(c_i-r_i,c_j-r_j)} \circ (g(c_i - r_i, c_j - r_j) + r_k) \rangle \quad (1)$$

Note that $Q_k^{c_i,c_j}$ actually depends only on the random inputs. Also note that g is either an AND gate, in which case $g(a, b) = a \cdot b$, or an OR gate, in which case $g(a, b) = 1 - (1 - a) \cdot (1 - b)$; hence all occurrences of g in the above expression can be replaced by degree-2 polynomials. Moreover, as above, each expression of the form W_i^h can be represented by $h \cdot W_i^1 \oplus (1 - h) \cdot W_i^0$. The degree of the meta-polynomial $Q_k^{c_i,c_j}$ is 3.

Output wire of the formula: With this wire we associate a single degree-1 polynomial (in addition to the 4 meta-polynomials $Q_m^{c_i,c_j}$, as described above) which is simply the random input r_m.

We now show that the above construction is perfectly correct and private.

Correctness. Given $\alpha = p_F(x, r)$, it is possible to go over the formula from bottom to top and compute for each wire i the value $\langle W_i^{b_i} \circ c_i \rangle$. Applying this to the output wire m, and since for this wire we also have r_m in the output of p_F (by the construction), one can compute $f(x) = b_m = c_m + r_m$ as required.

Privacy. Consider any output vector, α, of p_F. This output consists of an output bit for each of the polynomials described above. Let α' be the vector α excluding its last bit (i.e., the value r_m); we claim that given any possible input x the output α' is obtained with the same probability. As a first step, by the correctness proof, for each wire i of F, the vector α' completely determines the string $W_i^{b_i}$ and the bit c_i (also note that the b_i's are determined by x). Therefore, if indeed α' is equally possible given any x, and since $r_m = c_m + f(x)$, then α reveals nothing but $f(x)$.

It remains to prove that for every input x the vector α' has the same probability to appear in the output. For this, consider the values $W_i^{b_i}, c_i$ for any i (which, as argued, are totally determined given α'). We show, by induction from top to bottom, that the number of choices for the strings $W_i^{1-b_i}$ that are consistent with α' is independent of x. This is clearly true for the output wire since $|W_m^{1-b_m}| = 0$. For the induction step consider an output wire k of a gate g whose input wires are i, j and assume, without loss of generality, that $c_i = c_j = 0$ (otherwise we just need to permute the 4 polynomials below accordingly). In this case α contains the output of the following 4 meta-polynomials:

$$
\begin{aligned}
Q_k^{0,0}(x,r) &= W_i^{b_i,0} \;\oplus\; W_j^{b_j,0} \;\oplus\; \langle W_k^{g(b_i,b_j)} \circ (g(b_i,b_j) + r_k)\rangle \\
Q_k^{0,1}(x,r) &= W_i^{b_i,1} \;\oplus\; W_j^{1-b_j,0} \;\oplus\; \langle W_k^{g(b_i,1-b_j)} \circ (g(b_i,1-b_j) + r_k)\rangle \\
Q_k^{1,0}(x,r) &= W_i^{1-b_i,0} \;\oplus\; W_j^{b_j,1} \;\oplus\; \langle W_k^{g(1-b_i,b_j)} \circ (g(1-b_i,b_j) + r_k)\rangle \\
Q_k^{1,1}(x,r) &= W_i^{1-b_i,1} \;\oplus\; W_j^{1-b_j,1} \;\oplus\; \langle W_k^{g(1-b_i,1-b_j)} \circ (g(1-b_i,1-b_j) + r_k)\rangle
\end{aligned}
$$

Note that at this stage we already assigned values to both strings corresponding to the output wire k of this gate, i.e. to $W_k^{b_k}, W_k^{1-b_k}$ (and clearly $g(b_i,b_j) = b_k$; together with c_k this determines r_k). Hence, the third summand in each of the above meta-polynomials is already fixed. On the other hand, for the input wires i, j we are only committed to the values $W_i^{b_i}, W_j^{b_j}$ and still have the freedom to choose $W_i^{1-b_i}, W_j^{1-b_j}$. Examining the 4 equations above we make the following observations: (a) in the first equation there are no unknowns (in fact, when we choose $W_k^{b_k}$ in the first part of the proof, we choose it so that this equation holds). (b) in the second and third equations there is a unique choice for $W_j^{1-b_j,0}, W_i^{1-b_i,0}$ that satisfies the equation. (c) in the fourth equation we have a constraint on what the string $W_i^{1-b_i,1} \oplus W_j^{1-b_j,1}$ should be. The number of choices that satisfy this constraint is clearly independent of x, as needed.

Efficiency. The complexity of the above construction is dominated by the total size of the strings W_i^b for all input wires i. The length of each string W_i^b depends only on the *depth* of wire i. For a wire of depth d, this length is $O(2^d)$. For example, if F is a balanced formula of size s and depth $\log_2 s$, then the complexity of p_F is $O(s^2)$.

3.1 An Efficiency Improvement

The above proof of privacy leaves some freedom in choosing the random strings (case (c) above). We can get rid of this freedom and thereby obtain some efficiency improvements over the basic construction. To do this, we break the symmetry between the 2 input wires i, j of each gate. We associate with one of these two wires, say j, a shorter string by letting $W_j^{b,1} = W_j^{b,0}$ for $b \in \{0,1\}$. The proof of correctness remains as before. In the proof of privacy, we no longer have freedom in (c), as $W_j^{1-b_j,1}$ was already fixed (because $W_j^{1-b_j,1} = W_j^{1-b_j,0}$); hence, there is a unique way to choose $W_i^{1-b_i,1}$ so as to satisfy the fourth equation (and this is, again, independent of x). The efficiency improvement is obtained since we do not have $|W_i^b| = |W_j^b| = 2(|W_k^b| + 1)$ as before, but rather we can do with $|W_j^b| = |W_k^b| + 1$ and only $|W_i^b| = 2(|W_k^b| + 1)$.

This simple observation already leads to some significant efficiency improvements. If F is a completely balanced formula of size s and depth $d = \log_2 s$, then the total length of the strings W_i^b (which dominates the complexity of p_F) is now only $O(3^d) = O(s^{\log_2 3})$.

For instance, since the OR_n function (OR of n bits) admits such a balanced formula, this basic optimization applies to OR_n yields complexity of $O(n^{\log_2 3})$ (compared with $O(n^2)$ given by the basic construction).

A further efficiency gain may be obtained by skewing the formula tree so that each of its leaves has roughly the same contribution to the overall complexity. We leave open the question of converting a general formula into an optimal equivalent form, and proceed with the interesting special case of the function OR_n (or, equivalently, AND_n). A useful feature of this function is that there is a complete freedom in choosing the shape of its formula: *any* binary tree T with n leaves naturally induces a formula F_T for OR_n, where the leaves are labeled by the n distinct inputs and the internal nodes by binary OR gates. Hence, our problem is equivalent to finding a binary tree T of size n which minimizes the total weight of its n vertices subject to the following constraints: (1) the root has weight 0; (2) if v is an internal node of weight w, then the weights of its two sons are $w + 1$ and $2(w + 1)$. It can be shown that such a tree T of total weight $n \cdot 2^{O(\sqrt{\log n})}$ can be efficiently constructed (details omitted for lack of space). Hence, there is a perfect degree-3 representation for OR_n of complexity $n \cdot 2^{O(\sqrt{\log n})}$.

4   Perfect Randomizing Polynomials from Branching Programs

In this section we construct perfect randomizing polynomials from a branching program representation. First, in Section 4.1, we give an overview of our solutions and provide some background and intuition. Then, in Section 4.2, we present a construction for CBP (which includes *deterministic* branching programs as a special case) and state our result for NBP whose proof is omitted from this version.

4.1   Overview of Constructions

Before describing our new solutions, it is instructive to review some previous ideas and techniques from [18,19] on which we rely. The previous construction of (nonperfect) randomizing polynomials from branching programs [19] uses an NBP representation, and is based on the following two facts.

Fact 1. [18] Given a mod-q NBP of size ℓ computing f, there exists a function $L(x)$, mapping an input x to an $\ell \times \ell$ matrix over $\mathbb{F} = GF(q)$, such that:

- Each entry of $L(x)$ is a degree-1 polynomial in a single variable x_i.
- $f(x)$ is in one-to-one correspondence with $\text{rank}(L(x))$. Specifically, if $f(x) = 1$ then $L(x)$ is of full rank, and if $f(x) = 0$ then its rank is one less than full.

Fact 2. [19] Let M be an arbitrary square matrix over $\mathbb{F}$, and R_1, R_2 be independent uniformly random matrices of the same dimension. Then, the distribution of the random variable $R_1 M R_2$ depends only on $\text{rank}(M)$. Moreover, if $\text{rank}(M) \neq \text{rank}(M')$ then $SD(R_1 M R_2, R_1 M' R_2) > \epsilon_q$, where ϵ_q is a constant depending only on q.

The final degree-3 representation, based on the above two facts, has the form $p(x, R_1, R_2) = R_1 L(x) R_2$, where both the random inputs and the output vector are parsed as matrices. Randomizing polynomials of this form cannot possibly achieve perfect correctness: setting all of the random inputs to 0 will always yield an all-0 output vector. A natural solution that comes to mind is to replace R_1 and R_2 by uniform *nonsingular* matrices. It is easy to see that perfect correctness will be achieved, and it can also be argued that the perfect privacy will not be violated. However, it is not clear how to incorporate random nonsingular matrices into the randomizing polynomials framework; in fact, it follows by a divisibility argument that it is *impossible* to generate a uniformly random nonsingular matrix over a finite field $\mathbb{F}$ from (finitely many) uniform random elements of $\mathbb{F}$, regardless of the complexity or the degree of such a generation.

To get around this problem, we take a closer look at the matrices $L(x)$ generated by the mapping L. It turns out that, for matrices *of this special form*, it is possible to pick R_1 and R_2 from carefully chosen *subgroups* of nonsingular matrices, so that: (1) random matrices from these subgroups can be generated by degree-1 polynomials; (2) $R_1 L(x) R_2$ achieves perfect privacy and correctness with respect to the *number* of accepting paths (mod q) on input x. This approach gives our solution for CBP, presented next. However, it falls short of providing an efficient solution for their nondeterministic counterparts, except when the modulus q is small.

4.2   Counting Branching Programs

Throughout this section let $\mathbb{F} = \mathrm{GF}(q)$, where q is an arbitrary prime. Our goal is to convert a mod-q counting branching program computing $f : \{0, 1\}^n \to \mathbb{F}$ into an efficient representation of f by perfect degree-3 randomizing polynomials. As outlined above, we start with a refined version of Fact 1. Its proof is similar to that of Fact 1 (from [18]) and is omitted.

Lemma 1. *Suppose there is a mod-q CBP of size ℓ computing f. Then, there exists a function $L(x)$, mapping an input x to an $\ell \times \ell$ matrix over $\mathbb{F} = \mathrm{GF}(q)$, such that:*

- *Each entry of $L(x)$ is a degree-1 polynomial in a single input variable x_i.*
- *$L(x)$ contains the constant -1 in each entry of its second diagonal (the one below the main diagonal) and the constant 0 below this diagonal.*
- *$f(x) = \det(L(x))$.*

Our variant of Fact 2 relies on the following simple randomization lemma.

Lemma 2. *Let $\mathcal{H}$ be a set of square matrices over $\mathbb{F}$, and $\mathcal{G}_1, \mathcal{G}_2$ be multiplicative groups of matrices of the same dimension as $\mathcal{H}$. Denote by '$\sim$' the equivalence relation on $\mathcal{H}$ defined by: $H \sim H'$ iff there exist $G_1 \in \mathcal{G}_1, G_2 \in \mathcal{G}_2$ such that $H = G_1 H' G_2$. Let R_1, R_2 be uniformly and independently distributed matrices from $\mathcal{G}_1, \mathcal{G}_2$, respectively. Then, for any H, H' such that $H \sim H'$, the random variables $R_1 H R_2$ and $R_1 H' R_2$ are identically distributed.*

Lemma 2 will be instantiated with the following matrix sets.

Definition 2. *Let $\mathcal{H}$ be the set of $\ell \times \ell$ matrices over $\mathbb{F} = \mathrm{GF}(q)$ containing only -1's in their second diagonal (the diagonal below the main diagonal), and 0's below the second diagonal. Define two matrix groups $\mathcal{G}_1$ and $\mathcal{G}_2$ as follows:*

- *$\mathcal{G}_1$ consists of all matrices with 1's on the main diagonal and 0's below it.*
- *$\mathcal{G}_2$ consists of all matrices with 1's on the main diagonal and 0's in all of the remaining entries except, perhaps, those of the rightmost column.*

From now on, '$\sim$' denotes the equivalence relation on $\mathcal{H}$ induced by $\mathcal{G}_1, \mathcal{G}_2$, as defined in Lemma 2.

The following lemma shows that a matrix from $\mathcal{H}$ can be brought into a canonical form, uniquely defined by its determinant, by multiplying it from the left by some $G_1 \in \mathcal{G}_1$ and from the right by some $G_2 \in \mathcal{G}_2$.

Lemma 3. *For any $H \in \mathcal{H}$ there exist $G_1 \in \mathcal{G}_1$ and $G_2 \in \mathcal{G}_2$ such that $G_1 H G_2$ contains -1's in its second diagonal, $\det(H)$ in its top-right entry, and 0's elsewhere.*

Proof. Consider two types of matrix operations: (a) Add to row i some multiple of row $i' > i$; (b) Add to the last column a multiple of some other column. As illustrated in Figure 1, a matrix $H \in \mathcal{H}$ can be transformed, using a sequence of (a) and (b) operations, to a matrix H_0 containing -1's in its second diagonal, an arbitrary value in its top-right entry, and 0's elsewhere. Note that none of these operations changes the determinant, and hence $\det(H_0) = \det(H)$. It follows that the top-right entry of H_0 must be equal to its determinant. We conclude the proof by observing that each operation of type (a) is a left-multiplication by a matrix from $\mathcal{G}_1$, and each operation of type (b) is a right-multiplication by a matrix from $\mathcal{G}_2$.   □

$$
\begin{pmatrix}
* & * & * & * & * & * \\
-1 & * & * & * & * & * \\
0 & -1 & * & * & * & * \\
0 & 0 & -1 & * & * & * \\
0 & 0 & 0 & -1 & * & * \\
0 & 0 & 0 & 0 & -1 & *
\end{pmatrix}
\overset{(a)}{\Longrightarrow}
\begin{pmatrix}
0 & 0 & 0 & 0 & 0 & * \\
-1 & 0 & 0 & 0 & 0 & * \\
0 & -1 & 0 & 0 & 0 & * \\
0 & 0 & -1 & 0 & 0 & * \\
0 & 0 & 0 & -1 & 0 & * \\
0 & 0 & 0 & 0 & -1 & *
\end{pmatrix}
\overset{(b)}{\Longrightarrow}
\begin{pmatrix}
0 & 0 & 0 & 0 & 0 & * \\
-1 & 0 & 0 & 0 & 0 & 0 \\
0 & -1 & 0 & 0 & 0 & 0 \\
0 & 0 & -1 & 0 & 0 & 0 \\
0 & 0 & 0 & -1 & 0 & 0 \\
0 & 0 & 0 & 0 & -1 & 0
\end{pmatrix}
$$

Fig. 1. Bringing a matrix $H \in \mathcal{H}$ to a canonical form H_0

The following is an easy corollary.

Lemma 4. *Let $\mathcal{H}, \mathcal{G}_1, \mathcal{G}_2$ be as in Definition 2. Then, for any $H, H' \in \mathcal{H}$, $\det(H) = \det(H')$ implies $H \sim H'$.*

Our final randomizing polynomials construction is given in the next theorem.

Theorem 1. *Suppose that f can be computed by a mod-q counting branching program of size ℓ. Then, there exists (constructively) a representation of f by perfect degree-3 randomizing polynomials over $\mathbb{F} = \mathrm{GF}(q)$, with output complexity $\binom{\ell+1}{2}$ and randomness complexity $\binom{\ell}{2} + \ell - 1$.*

Proof. Consider the polynomial vector $p(x, r^1, r^2) = R_1(r^1)L(x)R_2(r^2)$, where L is as promised by Lemma 1, and R_1 (resp., R_2) is a degree-1 mapping of the random inputs r^1 (resp., r^2) to a uniformly random matrix from $\mathcal{G}_1$ (resp., $\mathcal{G}_2$). Note that the number of random field elements that r^1 and r^2 should contain is $\binom{\ell}{2}$ and $\ell - 1$, respectively. Hence the specified randomness complexity. For the output complexity, note that it is enough to include the $\binom{\ell+1}{2}$ entries of the output matrix on or above the main diagonal. The perfect privacy of p follows from Lemmas 1,2, and 4. Finally, its perfect correctness follows from the fact that the determinant of the output matrix is always equal to $\det(L(x))$, which by Lemma 1 is equal to $f(x)$.   □

Due to lack of space, we omit the full treatment of the nondeterministic case. If the modulus q is small, the problem reduces to the previous counting case. In general, however, we are only able to obtain perfect correctness by relaxing the privacy requirement. The main relevant theorem is the following:

Theorem 2. *Given a mod-q NBP of size ℓ computing f and a security parameter k, it is possible to compute in time $poly(\ell, \log q, k)$ a perfectly-correct, 2^{-k}-private, degree-3 randomizing polynomials representation of f over $\mathrm{GF}(2)$.*

Remark. An alternative approach for obtaining perfect constant-round protocols for CBP is to reduce the problem to an *inversion* of a *triangular* matrix. Specifically, the value of the CBP may be obtained as the top-right entry of $(I - A_x)^{-1}$, where A_x is the adjacency matrix of the graph G_x (cf. [18]). Since $I - A_x \in \mathcal{G}_1$, a straightforward modification of the secure inversion protocol from [1] (essentially replacing random nonsingular matrices by random matrices from $\mathcal{G}_1$) can be used to compute the desired entry with perfect security and a strictly constant number of rounds. A disadvantage of this solution is that it requires more rounds than a protocol based on degree-3 randomizing polynomials.

Acknowledgments. We thank Amos Beimel, Ronald Cramer, Ivan Damgrard, and Serge Fehr for helpful discussions and comments.

References

1. J. Bar-Ilan and D. Beaver. Non-cryptographic fault-tolerant computing in a constant number of rounds. In *Proc. of 8th PODC*, pages 201–209, 1989.
2. D. Beaver. Secure Multi-party Protocols and Zero-Knowledge Proof Systems Tolerating a Faulty Minority. *J. Cryptology*, Springer-Verlag, (1991) 4: 75-122.
3. D. Beaver. Minimal-latency secure function evaluation. EUROCRYPT 2000.
4. D. Beaver, J. Feigenbaum, J. Kilian, and P. Rogaway. Security with low communication overhead. In *Proc. of CRYPTO '90*, pages 62–76.
5. D. Beaver, S. Micali, and P. Rogaway. The round complexity of secure protocols (extended abstract). In *Proc. of 22nd STOC*, pages 503–513, 1990.
6. M. Ben-Or, S. Goldwasser, and A. Wigderson. Completeness theorems for non-cryptographic fault-tolerant distributed computation. STOC, 1988.
7. C. Cachin, J. Camenisch, J. Kilian, and J. Muller. One-round secure computation and secure autonomous mobile agents. In *ICALP 2000*.
8. R. Canetti. Security and composition of multiparty cryptographic protocols. *J. of Cryptology*, 13(1), 2000.
9. D. Chaum, C. Crépeau, and I. Damgrard. Multiparty unconditionally secure protocols (extended abstract). In *Proc. of 20th STOC*, pages 11–19, 1988.
10. R. Cramer and I. Damgrard. Secure distributed linear algebra in a constant number of rounds. In *Proc. Crypto 2001*.
11. R. Cramer, I. Damgrard, and U. Maurer. General secure multi-party computation from any linear secret-sharing scheme. In *Proc. of EUROCRYPT 2000*.
12. R. Cramer, S. Fehr, Y. Ishai, and E. Kushilevitz. Efficient Multi-Party Computation over Rings. Manuscript, 2002.
13. U. Feige, J. Kilian, and M. Naor. A minimal model for secure computation (extended abstract). In *Proc. of 26th STOC*, pages 554–563, 1994.
14. R. Gennaro, Y. Ishai, E. Kushilevitz and T. Rabin. The Round Complexity of Verifiable Secret Sharing and Secure Multicast. In *Proc. 33rd STOC*, 2001.
15. S. Goldwasser and L. Levin. Fair Computation of General Functions in Presence of Immoral Majority. In *CRYPTO '90, LNCS 537*, Springer-Verlag, 1990.
16. O. Goldreich. Secure multi-party computation.
    www.wisdom.weizmann.ac.il/~oded/pp.html, 2000.
17. O. Goldreich, S. Micali, and A. Wigderson. How to play any mental game (extended abstract). In *Proc. of 19th STOC*, pages 218–229, 1987.
18. Y. Ishai and E. Kushilevitz. Private simultaneous messages protocols with applications. In *Proc. of ISTCS '97*, pp. 174-183, 1997.
19. Y. Ishai and E. Kushilevitz. Randomizing Polynomials: A New Representation with Applications to Round-Efficient Secure Computation. In *Proc. of FOCS '00*.

20. J. Kilian. Basing cryptography on oblivious transfer. STOC '98, pp. 20-31, 1988.
21. Y. Lindell. Parallel Coin-Tossing and Constant-Round Secure Two-Party Computation. In Prof. of Crypto '01.
22. S. Micali and P. Rogaway. Secure computation. In *Proc. of CRYPTO '91*.
23. M. Naor, B. Pinkas, and R. Sumner. Privacy Preserving Auctions and Mechanism Design. In *Proc. ACM Conference on Electronic Commerce 1999*, pages 129-139.
24. T. Sandler, A. Young, and M. Yung. Non-interactive cryptocomputing for NC^1. In *Proc. of 40th FOCS*, pages 554–566, 1999.
25. A. C. Yao. Protocols for secure computations (extended abstract). In Proc. of FOCS 1982.
26. A. C. Yao. How to generate and exchange secrets. In Proc. of FOCS 1986.

Exponential Lower Bound for Static Semi-algebraic Proofs

Dima Grigoriev[1], Edward A. Hirsch[2*], and Dmitrii V. Pasechnik[3]

[1] IRMAR, Université de Rennes, Campus de Beaulieu, 35042 Rennes, cedex France.
dima@maths.univ-rennes1.fr .
http://www.maths.univ-rennes1.fr/~dima/

[2] Steklov Institute of Mathematics at St.Petersburg, 27 Fontanka, 191011
St.Petersburg, Russia.
hirsch@pdmi.ras.ru . http://logic.pdmi.ras.ru/~hirsch/

[3] Department of Technical Mathematics and Informatics, Faculty ITS, Delft
University of Technology, Mekelweg 4, 2628 CD Delft, The Netherlands.
d.pasechnik@its.tudelft.nl.
http://ssor.twi.tudelft.nl/~dima/

Abstract. Semi-algebraic proof systems were introduced in [1] as extensions of Lovász-Schrijver proof systems [2,3]. These systems are very strong; in particular, they have short proofs of Tseitin's tautologies, the pigeonhole principle, the symmetric knapsack problem and the clique-coloring tautologies [1].

In this paper we study *static* versions of these systems. We prove an exponential lower bound on the length of proofs in one such system. The same bound for two tree-like (dynamic) systems follows. The proof is based on a lower bound on the *"Boolean degree"* of Positivstellensatz Calculus refutations of the symmetric knapsack problem.

1 Introduction

Algebraic proof systems. An observation that a propositional formula can be written as a system of polynomial equations has lead to considering *algebraic* proof systems, in particular, the Nullstellensatz (NS) and the Polynomial Calculus (PC) proof systems, see Subsection 2.2 below (we do not dwell much here on the history of this rich area, several nice historical overviews one could find in e.g., [4,5,6,7,8,9]).

For these proof systems several interesting complexity lower bounds on the degrees of the derived polynomials were obtained [6,7,9]. When the degree is close enough to linear (in fact, greater than the square root), these bounds imply exponential lower bounds on the proof complexity (more precisely, on the number of monomials in the derived polynomials) [7]. If polynomials are given by formulas rather than by sums of monomials as in NS or in PC, then the complexity could decrease significantly. Several gaps between these two kinds of proof systems are demonstrated in [10].

---

* Partially supported by grant of RAS contest-expertise of young scientists projects and grants from CRDF (RM1-2409-ST-02), RFBR (02-01-00089), and NATO.

P. Widmayer et al. (Eds.): ICALP 2002, LNCS 2380, pp. 257–268, 2002.

Semi-algebraic proof systems. In [1], we have introduced several *semi-algebraic* proof systems. In these system, one deals with polynomial inequalities, and new inequalities can be derived by algebraic operations like the sum, the multiplication and the division. The simplest semi-algebraic systems are the so-called Lovász-Schrijver calculi (see [2,3], cf. also [11] and Subsection 2.3 below), where the polynomials are restricted to quadratic ones. No exponential lower bounds are known so far even for these restricted systems (a number of lower bounds on the number of steps of Lovász-Schrijver *procedure* is known [12,13,14,15,1], but they do not imply exponential lower bounds on the size of proofs [1]). Moreover, general semi-algebraic proof systems, (where one allows polynomials of arbitary degree, see [1] and Subsection 2.3 below), appear to be very strong. In [1], it is proved that such systems have short proofs of Tseitin's tautologies, the pigeonhole principle, clique-coloring tautologies and the symmetric knapsack problem. They also polynomially simulate the Cutting Planes proof system [16, 17,18,19] with polynomially bounded coefficients. Another (and much stronger) kind of semi-algebraic proof system was introduced in [20] with no focus on the complexity.

Static systems and our results. Another proof system manipulating polynomial inequalities called the Positivstellensatz Calculus was introduced in [21]. Lower bounds on the degree in this system were established for the parity principle, for Tseitin's tautologies [22] and for the knapsack problem [23]. Lower bounds on the Positivstellensatz Calculus degree are possible because its "dynamic" part is restricted to an ideal and an element of a cone is obtained from an element of ideal by adding the sum of squares to it. On the contrary, the semi-algebraic proof systems introduced in [2,3,1] are completely "dynamic" proof systems. (The discussion on static and dynamic proof systems can be found in [21]. Briefly, the difference is that in the dynamic semi-algebraic proof systems a derivation constructs gradually an element of the cone generated by the input system of inequalities, while in the Positivstellensatz Calculus the sum of squares is given explicitly.) We consider a static version of Lovász-Schrijver calculi and prove an exponential lower bound on the size of refutation of the symmetric knapsack problem (Section 4); this bound also translates into the bound for the tree-like version of (dynamic) LS. The key ingredient of the proof is a linear lower bound on the "Boolean degree" of Positivstellensatz Calculus refutations (Section 3). Note that exponential lower bounds on the size of (static!) Positivstellensatz refutations are still unknown.

Organization of the paper. We start with the definitions of proof systems in general and the particular proof systems we use in our paper (Section 2). We then prove a lower bound on the "Boolean degree" of Positivstellensatz Calculus refutations of the symmetric knapsack problem (Section 3), and derive from it an exponential lower bound on the size of proofs in a static semi-algebraic proof system and in the tree-like versions of two dynamic semi-algebraic proof systems (Section 4). Finally, we formulate open questions (Section 5).

2   Definitions

2.1   Proof Systems

A *proof system* [24] for a language L is a polynomial-time computable function mapping words (proof candidates) onto L (whose elements are considered as theorems).

A *propositional proof system* is a proof system for any fixed co-NP-complete language of Boolean tautologies (e.g., tautologies in DNF).

When we have two proof systems Π_1 and Π_2 for the same language L, we can compare them. We say that Π_1 *polynomially simulates* Π_2, if there is a function g mapping proof candidates of Π_2 to proof candidates of Π_1 so that for every proof candidate π for Π_2, one has $\Pi_1(g(\pi)) = \Pi_2(\pi)$ and $g(\pi)$ is at most polynomially longer than π.

Proof system Π_1 is *exponentially separated* from Π_2, if there is an infinite sequence of words $t_1, t_2, \ldots \in L$ such that the length of the shortest Π_1-proof of t_i is polynomial in the length of t_i, and the length of the shortest Π_2-proof of t_i is exponential.

Proof system Π_1 is *exponentially stronger* than Π_2, if Π_1 polynomially simulates Π_2 and is exponentially separated from it.

When we have two proof systems for different languages L_1 and L_2, we can also compare them if we fix a reduction between these languages. However, it can be the case that the result of the comparison is more due to the reduction than to the systems themselves. Therefore, if we have propositional proof systems for languages L_1 and L_2, and the intersection $L = L_1 \cap L_2$ of these languages is co-NP-complete, we will compare these systems as systems[1] for L.

2.2   Algebraic Proof Systems

There is a series of proof systems for languages consisting of unsolvable systems of polynomial equations. To transform such a proof system into a propositional proof system, one needs to translate Boolean tautologies into systems of polynomial equations.

To translate a formula F in k-DNF, we take its negation $\neg F$ in k-CNF and translate each clause of $\neg F$ into a polynomial equation. A clause containing variables $v_{j_1}, \ldots, v_{j_t}$ $(t \leq k)$ is translated into an equation

$$(1 - l_1) \cdot \ldots \cdot (1 - l_t) = 0, \tag{1}$$

where $l_i = v_{j_i}$ if variable v_{j_i} occurs positively in the clause, and $l_i = (1 - v_{j_i})$ if it occurs negatively. For each variable v_i, we also add the equation $v_i^2 - v_i = 0$ to this system.

---

[1] If one can decide in polynomial time for $x \in L_1$, whether $x \in L$, then any proof system for L_1 can be restricted to $L \subseteq L_1$ by mapping proofs of elements of $L_1 \setminus L$ into any fixed element of L. For example, this is the case for L_1 consisting of all tautologies in DNF and L consisting of all tautologies in k-DNF.

Remark 1. Observe that it does not make sense to consider this translation for formulas in general DNF (rather than k-DNF for constant k), because an exponential lower bound for any system using such encoding would be trivial (note that $(1 - v_1)(1 - v_2) \ldots (1 - v_n)$ denotes a polynomial with exponentially many monomials).

Note that F is a tautology if and only if the obtained system S of polynomial equations $f_1 = 0$, $f_2 = 0$, ..., $f_m = 0$ has no solutions. Therefore, to prove F it suffices to derive a contradiction from S.

Nullstellensatz (NS) [4]. A proof in this system is a collection of polynomials $g_1, \ldots, g_m$ such that

$$\sum_i f_i g_i = 1.$$

Polynomial Calculus (PC) [8]. This system has two derivation rules:

$$\frac{p_1 = 0;\ p_2 = 0}{p_1 + p_2 = 0} \quad \text{and} \quad \frac{p = 0}{p \cdot q = 0}. \tag{2}$$

I.e., one can take a sum[2] of two already derived equations $p_1 = 0$ and $p_2 = 0$, or multiply an already derived equation $p = 0$ by an arbitrary polynomial q. The proof in this system is a derivation of $1 = 0$ from S using these rules.

Positivstellensatz [21]. A proof in this system consists of polynomials $g_1, \ldots, g_m$ and $h_1, \ldots, h_l$ such that

$$\sum_i f_i g_i = 1 + \sum_j h_j^2 \tag{3}$$

Positivstellensatz Calculus [21]. A proof in this system consists of polynomials $h_1, \ldots, h_l$ and a derivation of $1 + \sum_j h_j^2 = 0$ from S using the rules (2).

2.3 Dynamic Semi-algebraic Proof Systems

To define a propositional proof system manipulating with inequalities, we again translate each formula $\neg F$ in CNF into a system S of linear inequalities, such that F is a tautology if and only if S has no 0-1 solutions. Given a Boolean formula in CNF, we translate each its clause containing variables $v_{j_1}, \ldots, v_{j_t}$ into the inequality

$$l_1 + \ldots + l_t \geq 1, \tag{4}$$

where $l_i = v_{j_i}$ if the variable v_{j_i} occurs positively in the clause, and $l_i = 1 - v_{j_i}$ if v_{j_i} occurs negatively. We also add to S the inequalities

$$x \geq 0, \tag{5}$$
$$x \leq 1 \tag{6}$$

for every variable x.

---

[2] Usually, an arbitrary linear combination is allowed, but clearly it can be replaced by two multiplications and one addition.

Lovász-Schrijver calculus (LS) [2,3] (cf. also [11]). In the weakest of Lovász-Schrijver proof systems, the contradiction must be obtained using the rule

$$\frac{f_1 \geq 0; \ \ldots; \ f_t \geq 0}{\sum_{i=1}^{t} \lambda_i f_i \geq 0} \quad \text{(where } \lambda_i \geq 0\text{),} \tag{7}$$

applied to linear or quadratic f_i's and the rules

$$\frac{f \geq 0}{fx \geq 0}; \quad \frac{f \geq 0}{f(1-x) \geq 0} \quad \text{(where } f \text{ is linear, } x \text{ is a variable).} \tag{8}$$

Also, the system S is extended by the axioms

$$x^2 - x \geq 0, \quad x - x^2 \geq 0 \tag{9}$$

for every variable x.

LS$_+$ [2,3,11]. This system has the same axioms and derivation rules as LS, and also has the axiom

$$l^2 \geq 0 \tag{10}$$

for every linear l.

Note that the Lovász-Schrijver systems described above deal either with linear or quadratic inequalities. In [1], several extensions of Lovász and Schrijver proof systems are introduced. The main idea is to allow a proof to contain monomials of degree up to d.

LS^d. This system is an extension of LS. The difference is that rule (8) is now restricted to f of degree at most $d-1$ rather than to linear inequalities. Rule (7) can be applied to any collection of inequalities of degree at most d.

Remark 2. Note that LS=LS^2.

2.4 Static Semi-algebraic Proof Systems

Nullstellensatz is a "static" version of Polynomial Calculus; Positivstellensatz is a "static" version of Positivstellensatz Calculus. Similarly, we define "static" versions of the semi-algebraic proof systems defined in the previous subsection.

Static LS^n. A proof in this system is a a refutation of a system of inequalities $S = \{s_i \geq 0\}_{i=1}^{t}$, where each $s_i \geq 0$ is either an inequality given by the translation (4), an inequality of the form $x_j \geq 0$ or $1 - x_j \geq 0$, or an inequality of the form $x_j^2 - x_j \geq 0$. The refutation consists of positive real coefficients $\omega_{i,l}$ and multisets $U_{i,l}^+$ and $U_{i,l}^-$ defining the polynomials

$$u_{i,l} = \omega_{i,l} \cdot \prod_{k \in U_{i,l}^+} x_k \cdot \prod_{k \in U_{i,l}^-} (1 - x_k)$$

such that

$$\sum_{i=1}^{t} s_i \sum_l u_{i,l} = -1. \tag{11}$$

Static LS$_+^n$. The difference from the previous system is that S is extended by inequalities $s_{t+1} \geq 0, \ldots, s_{t'} \geq 0$, where each polynomial s_j ($j \in [t+1..t']$) is a square of another polynomial s_j'. The requirement (11) transforms into

$$\sum_{i=1}^{t'} s_i \sum_l u_{i,l} = -1. \tag{12}$$

Static LS$_+$. The same as static LS$_+^n$, but the polynomials s_i' can be only linear.

Remark 3. Note that static LS$_+$ includes static LS^n.

Remark 4. Note that these static systems are not propositional proof systems in the sense of Cook and Reckhow [24], but are something more general, since there is no clear way to verify (11) in deterministic polynomial time (cf. [25]). However, they can be easily augmented to match the definition of Cook and Reckhow, e.g., by including a proof of the equality (11) or (12) using axioms of a ring (cf. F-NS of [10]). Clearly, if we prove a lower bound for the original system, the lower bound will be valid for any augmented system as well.

Remark 5. The size of a refutation in these systems is the length of a reasonable bit representation of all polynomials $u_{i,l}$, s_i (for $i \in [1..t]$) and s_j' (for $j \in [t+1..t']$) and is thus at least the number of $u_{i,l}$'s.

Example 1. We now present a very simple static LS$_+$ proof of the propositional pigeonhole principle. The negation of this tautology is given by the following system of inequalities:

$$\sum_{\ell=1}^{m-1} x_{k\ell} \geq 1; \qquad 1 \leq k \leq m; \tag{13}$$

$$x_{k\ell} + x_{k'\ell} \leq 1; \qquad 1 \leq k < k' \leq m; \ 1 \leq \ell \leq m-1. \tag{14}$$

(That says that the k-th pigeon must get into a hole, while two pigeons k and k' cannot share the same hole ℓ.)

Here is the static LS$_+$ proof:

$$\sum_{k=1}^{m} \left(\sum_{\ell=1}^{m-1} x_{k\ell} - 1 \right) +$$

$$\sum_{\ell=1}^{m-1} \left(\sum_{k=1}^{m} x_{k\ell} - 1 \right)^2 +$$

$$\sum_{\ell=1}^{m-1} \sum_{k=1}^{m} \sum_{k \neq k'=1}^{m} (1 - x_{k\ell} - x_{k'\ell}) x_{k\ell} +$$

$$\sum_{\ell=1}^{m-1} \sum_{k=1}^{m} (x_{k\ell}^2 - x_{k\ell})(m-1)$$

$$= -1. \qquad \qquad \square$$

3 Linear Lower Bound on the "Boolean Degree" of Positivstellensatz Calculus Refutations of the Knapsack

We use the following notation from [7,23]. For a polynomial f, its *multilineariza-tion* $\overline{f}$ is a polynomial obtained by the reduction of f modulo $(x - x^2)$ for every variable x, i.e., f is the unique multilinear polynomial equivalent to f modulo these ("Boolean") polynomials. When $f = \overline{f}$ we say that f is reduced.

For a monomial t one can define its *Boolean degree* $\mathrm{Bdeg}(t)$ as $\deg(\overline{t})$, in other words, the number of occurring variables; then one extends the concept of Bdeg to polynomials: $\mathrm{Bdeg}(f) = \max \mathrm{Bdeg}(t_i)$, where the maximum is taken over all non-zero monomials t_i occurring in f. Thereby, one can define Bdeg of a deriva-tion in PC and subsequently in Positivstellensatz and Positivstellensatz Calculus as maximum Bdeg of *all* polynomials in the derivation (in Positivstellensatz and Positivstellensatz Calculus, this includes polynomials h_j^2, cf. definition in Sub-section 2.2).

The following lemma extends the argument in the proof of [7, Theorem 5.1] from deg to Bdeg.

Lemma 1. *Let* $f(x_1, \ldots, x_n) = c_1 x_1 + \ldots + c_n x_n - m$, *where* $c_1, \ldots, c_n \in \mathbb{R} \setminus \{0\}$. *Let* q *be deducible in PC from the knapsack problem* $f = 0$ *with* $\mathrm{Bdeg} \leq \lceil (n-1)/2 \rceil$. *Then one can represent*

$$q = \sum_{i=1}^{n} (x_i - x_i^2) g_i + fg, \tag{15}$$

where $\deg(fg) \leq \mathrm{Bdeg}(q)$.

Proof. Similarly to the proof of [7, Theorem 5.1], we conduct the induction along a (fixed) deduction in PC. Assume (15) and consider a polynomial qx_1 obtained from q by multiplying it by a variable x_1. W.l.o.g. one can suppose that g is reduced. Then $\overline{qx_1} = \overline{fgx_1}$; denote $h = \overline{gx_1}$. Let $d = \deg(h) - 1$. We need to verify that $d + 2 = \deg(fh) \leq \mathrm{Bdeg}(qx_1)$. Taking into account that

$$d + 1 = \deg(h) \leq \deg(g) + 1 = \deg(fg) \leq \mathrm{Bdeg}(q) \leq \mathrm{Bdeg}(qx_1),$$

the mere case to be brought to a contradiction is when $\mathrm{Bdeg}(qx_1) = \mathrm{Bdeg}(q) = \deg(g) + 1 = d + 1$.

We write $g = p + x_1 p_1$ where all the terms of g not containing x_1 are gathered in p. Clearly, $\deg(p) \leq \deg(g) = d$. Moreover, $\deg(p) = d$ because if $\deg(p) < d$, we would have $d + 1 = \deg(h) \leq \mathrm{Bdeg}(gx_1) \leq \max(\mathrm{Bdeg}(x_1 p), \mathrm{Bdeg}(x_1^2 p_1)) \leq d$.

On the other hand, $d = \mathrm{Bdeg}(q) - 1 \leq \lceil (n-1)/2 \rceil - 1$. Therefore, [7, Lemma 5.2] applied to the instance $c_2 x_2 + \ldots + c_n x_n - 0$ of symmetric knapsack states that

$$\deg(\overline{(c_2 x_2 + \ldots + c_n x_n) p}) = \deg(p) + 1 = d + 1$$

(one should add to the formulation of [7, Lemma 5.2] the condition that p is reduced).

Hence there exists a monomial $x^J = \prod_{j \in J} x_j$ occurring in p for a certain $J \subseteq \{2, \ldots, n\}$, $|J| = d$, and besides, there exists $i \in [2..n]$ such that the monomial $x_i x^J$, being of the degree $d + 1$, occurs in the polynomial $(c_2 x_2 + \ldots + c_n x_n)p$, in particular $i \notin J$.

Because of that the monomial $T = x_i x^J x_1$ with $\deg(T) = d + 2$ occurs in

$$p' = \overline{(c_2 x_2 + \ldots + c_n x_n)p x_1}.$$

Furthermore, T occurs in

$$\overline{fg x_1} = \overline{((c_2 x_2 + \ldots + c_n x_n) + (c_1 x_1 - m))(p + x_1 p_1) x_1}$$

since after opening the parenthesis in the right-hand side of the latter expression we obtain only p' and two subexpressions

$$\overline{(c_1 x_1 - m)(p + x_1 p_1) x_1} = \overline{(c_1 - m) g x_1} \quad \text{and} \quad \overline{(c_2 x_2 + \ldots + c_n x_n) x_1 p_1 x_1}$$

of Boolean degree at most $d + 1$ (thereby, any monomial from these subexpressions cannot be equal to the *reduced* monomial T). Finally, due to the equality $\overline{qx_1} = \overline{fg x_1}$, we conclude that $\mathrm{Bdeg}(qx_1) \geq \deg(\overline{qx_1}) = \deg(\overline{fg x_1}) \geq d + 2$; the achieved contradiction proves the induction hypothesis for the case of the rule of the multiplication by a variable (note that the second rule in (2) can be replaced by the multiplication by a variable with a multiplicative constant).

Now we proceed to the consideration of the rule of taking the sum of two polynomials q and r. By the induction hypothesis we have

$$r = \sum_{i=1}^{n} (x_i - x_i^2) u_i + f u,$$

where u is reduced and $\deg(fu) \leq \mathrm{Bdeg}(r)$. Then making use of (15) we get $\overline{r + q} = \overline{fv}$ where $v = \overline{g + u}$. The inequality

$$\deg(v) \leq \max\{\deg(g), \deg(u)\} \leq \max\{\mathrm{Bdeg}(q), \mathrm{Bdeg}(r)\} - 1$$
$$\leq \lceil (n-1)/2 \rceil - 1 \leq \lceil n/2 \rceil - 1$$

enables us to apply [7, Lemma 5.2] to v, this implies that $\deg(\overline{fv}) = \deg(v) + 1 = \deg(fv)$. Therefore, $\mathrm{Bdeg}(r + q) \geq \deg(\overline{r + q}) = \deg(\overline{fv}) = \deg(fv)$. $\qquad \square$

The next corollary extends [7, Theorem 5.1].

Corollary 1. *Any PC refutation of the knapsack f has Bdeg greater than $\lceil (n - 1)/2 \rceil$.*

Now we can formulate the following theorem extending the theorem of [23] from deg to Bdeg. Denote by δ a stairs-form function which equals to 2 out of the interval $(0, n)$ and which equals to $2k + 4$ on the intervals $(k, k + 1)$ and $(n - k - 1, n - k)$ for all integers $0 \leq k < n/2$.

Theorem 1. *Any Positivstellensatz Calculus refutation of the symmetric knapsack problem $f = x_1 + \ldots + x_n - m$ has Bdeg greater or equal to $\min\{\delta(m), \lceil (n - 1)/2 \rceil + 1\}$.*

Proof. The proof of the theorem follows the proof of the theorem [23]. First, we apply Lemma 1 to the deduction in PC being an ingredient of the deduction in Positivstellensatz Calculus (see definitions in 2.2). This provides a refutation in Positivstellensatz Calculus of the form

$$-1 = \sum_{i=1}^{n}(x_i - x_i^2)g_i + fg + \sum_j h_j^2. \tag{16}$$

The rest of the proof follows literally the proof from [23] which consists in applying to (16) the homomorphism B introduced in [23]. It is worthwhile to mention that B is defined on the quotient algebra $\mathbb{R}[x_1, \ldots, x_n]/(x_1 - x_1^2, \ldots, x_n - x_n^2)$, thereby, the proof in [23] actually, estimates Bdeg rather than just deg. $\square$

Remark 6. A shorter self-contained proof of Theorem 1 is given in [26].

4   Exponential Lower Bound on the Size of Static LS$_+$ Refutations of the Symmetric Knapsack

In this section we apply the results of Section 3 to obtain an exponential lower bound on the size of static LS$_+$ refutations of the symmetric knapsack. We follow the notation introduced in Subsection 2.4 and Section 3. The *Boolean degree of a static LS (LS$_+$) refutation* is the maximum Boolean degree of the polynomials $u_{i,l}$ in Subsection 2.4.

Let us fix for the time being a certain (threshold) d.

Lemma 2. *Denote by M the number of $u_{i,l}$'s occurring in (12) that have Boolean degrees at least d. Then there is a variable x and a value $a \in \{0,1\}$ such that the result of substituting $x = a$ in (12) contains at most $M(1 - d/(2n))$ non-zero polynomials $u_{i,l}|_{x=a}$ of Boolean degrees at least d. (Note that by substituting in (12) a value a for x we obtain a valid static LS$_+$ refutation of the system $S|_{x=a}$).*

Proof. Since there are at least Md polynomials $u_{i,l}$ of Boolean degrees at least d containing either x or $1 - x$, there is a variable x such that either x or $1 - x$ occurs in at least $Md/(2n)$ of these polynomials. Therefore, after substituting the appropriate value for x, at least $Md/(2n)$ polynomials $u_{i,l}$ vanish from (12). $\square$

For the symmetric knapsack problem

$$x_1 + x_2 + \ldots + x_n - m = 0 \tag{17}$$

we can rewrite its static LS$_+$ refutation in the following way. Denote

$$\begin{aligned} f_0 &= x_1 + \ldots + x_n - m, \\ f_i &= x_i - x_i^2 \quad (1 \le i \le n), \\ f_i &= (s_i')^2 \quad (n+1 \le i \le n') \end{aligned}$$

(m is not an integer). The refutation can be represented in the form

$$\sum_{i=0}^{t} f_i \sum_{l} g_{i,l} + \sum_{j=n+1}^{n'} f_j t_j + \sum_{j=n'+1}^{n''} t_j = -1, \qquad (18)$$

where

$$g_{i,l} = \gamma_{i,l} \cdot \prod_{k \in G_{i,l}^+} x_k \cdot \prod_{k \in G_{i,l}^-} (1 - x_k),$$

$$t_j = \tau_j \cdot \prod_{k \in T_j^+} x_k \cdot \prod_{k \in T_j^-} (1 - x_k)$$

for appropriate multisets $G_{i,l}^-$, $G_{i,l}^+$, T_j^- and T_j^+, positive real τ_j and *arbitrary* real $\gamma_{i,l}$.

Lemma 3. *If $n/4 < m < 3n/4$, then the Boolean degree D of any static LS_+ refutation of the symmetric knapsack problem is at least $n/4$.*

Proof. Replacing in t_j each occurrence of x_i by $f_i + x_i^2$ and each occurrence of $1 - x_i$ by $f_i + (1 - x_i)^2$ and subsequently opening the parentheses in t_j, one can gather all the terms containing at least one of f_i and separately the products of squares of the form x_i^2, $(1 - x_i)^2$. As a result one gets a representation of the form

$$\sum_{i=0}^{n} f_i g_i + \sum_{j=1}^{n'''} h_j^2 = -1$$

for appropriate polynomials g_i, h_j of Boolean degrees $\mathrm{Bdeg}(g_i), \mathrm{Bdeg}(h_j^2) \leq D$, thereby a Positivstellensatz (and Positivstellensatz Calculus) refutation of the symmetric knapsack of Boolean degree at most $D + 2$. Then Theorem 1 implies that $D \geq \lceil (n-1)/2 \rceil - 1 \geq n/4$. $\square$

Theorem 2. *For $m = (2n + 1)/4$ the number of $g_{i,l}$'s and t_j's in (18) is $\exp(\Omega(n))$.*

Proof. Now we set $d = \lceil n/8 \rceil$ and apply Lemma 2 consecutively $\kappa = \lfloor n/4 \rfloor$ times. The result of all these substitutions in (18) we denote by (18'), it contains $n - \kappa$ variables; denote by $u_{i,l}'$ the polynomial we thus get from $u_{i,l}$. We denote by f_0' the result of substitutions applied to f_0. Note that after all substitutions we obtain again an instance of the knapsack problem. Taking into account that the free term m' of f_0' ranges in the interval $[m - \kappa, m]$ and since $(n - \kappa)/4 < m - \kappa < m < 3(n - \kappa)/4$, we are able to apply Lemma 3 to (18'). Thus, the degree of (18') is at least $(n - \kappa)/4 > d$.

Denote by M_0 the number of $u_{i,l}$'s of the degrees at least d in (18). By Lemma 2 the refutation (18') contains at most $M_0(1 - d/(2n))^\kappa \leq M_0(1 - 1/16)^{n/4}$ non-zero polynomials $u_{i,l}'$ of degrees at least d. Since there is at least one polynomial $u_{i,l}'$ of such degree, we have $M_0(1 - 1/16)^{n/4} \geq 1$, i.e. $M_0 \geq (16/15)^{n/4}$, which proves the theorem. $\square$

Corollary 2. *Any static LS_+ refutation of (17) for $m = (2n+1)/4$ must have size $\exp(\Omega(n))$.*

Corollary 3. *Any treelike LS_+ (or LS^n) refutation of (17) for $m = (2n+1)/4$ must have size $\exp(\Omega(n))$.*

Proof. The size of such treelike refutation (even the numer of instances of axioms f_i used in the refutation) is at least the number of polynomials $u_{i,l}$.   □

Remark 7. The value $m = (2n+1)/4$ in Theorem 2 and its corollaries above can be changed to any non-integer value between $\lceil n/4 \rceil$ and $\lfloor 3n/4 \rfloor$ by tuning the constants in the proofs (and in the $\Omega(n)$ in the exponent).

5   Open Questions

1. Prove an exponential lower bound for a static semi-algebraic *propositional* proof system. Note that we have only proved an exponential lower bound for static LS_+ as a proof system for the co-NP-complete language of *systems of 0-1 linear inequalities*, because the symmetric knapsack problem is not obtained as a translation of a Boolean formula in DNF.
2. Prove an exponential lower bound for a dynamic semi-algebraic proof system, e.g., for LS.
3. Can static LS be polynomially simulated by a certain version of the Cutting Planes proof system?

References

1. Grigoriev, D., Hirsch, E.A., Pasechnik, D.V.: Complexity of semi-algebraic proofs. In: Proceedings of the 19th International Symposium on Theoretical Aspects of Computer Science, STACS 2002. Volume 2285 of Lecture Notes in Computer Science., Springer (2002) 419–430
2. Lovász, L., Schrijver, A.: Cones of matrices and set-functions and 0–1 optimization. SIAM Journal on Optimization **1** (1991) 166–190
3. Lovász, L.: Stable sets and polynomials. Discrete Mathematics **124** (1994) 137–153
4. Beame, P., Impagliazzo, R., Krajíček, J., Pitassi, T., Pudlák, P.: Lower bounds on Hilbert's Nullstellensatz and propositional proofs. Proc. London Math. Soc. **73** (1996) 1–26
5. Beame, P., Impagliazzo, R., Krajíček, J., Pudlák, P., Razborov, A.A., Sgall, J.: Proof complexity in algebraic systems and bounded depth Frege systems with modular counting. Computational Complexity **6** (1996/97) 256–298
6. Razborov, A.A.: Lower bounds for the polynomial calculus. Computational Complexity **7** (1998) 291–324
7. Impagliazzo, R., Pudlák, P., Sgall, J.: Lower bounds for the polynomial calculus. Computational Complexity **8** (1999) 127–144

8. Clegg, M., Edmonds, J., Impagliazzo, R.: Using the Groebner basis algorithm to find proofs of unsatisfiability. In: Proceedings of the 28th Annual ACM Symposium on Theory of Computing, STOC'96, ACM (1996) 174–183

9. Buss, S., Grigoriev, D., Impagliazzo, R., Pitassi, T.: Linear gaps between degrees for the polynomial calculus modulo distinct primes. Journal of Computer and System Sciences **62** (2001) 267–289

10. Grigoriev, D., Hirsch, E.A.: Algebraic proof systems over formulas. Technical Report 01-011, Electronic Colloquim on Computational Complexity (2001) ftp://ftp.eccc.uni-trier.de/pub/eccc/reports/2001/TR01-011/index.html.

11. Pudlák, P.: On the complexity of propositional calculus. In: Sets and Proofs: Invited papers from Logic Colloquium'97. Cambridge University Press (1999) 197–218

12. Stephen, T., Tunçel, L.: On a representation of the matching polytope via semidefinite liftings. Math. Oper. Res. **24** (1999) 1–7

13. Cook, W., Dash, S.: On the matrix-cut rank of polyhedra. Math. Oper. Res. **26** (2001) 19–30

14. Dash, S.: On the Matrix Cuts of Lovász and Schrijver and their use in Integer Programming. Technical report tr01-08, Rice University (2001) http://www.caam.rice.edu/caam/trs/2001/TR01-08.ps.

15. Goemans, M.X., Tunçel, L.: When does the positive semidefiniteness constraint help in lifting procedures. Mathematics of Operations Research (2001) to appear.

16. Gomory, R.E.: An algorithm for integer solutions of linear programs. In: Recent Advances in Mathematical Programming. McGraw-Hill (1963) 269–302

17. Chvátal, V.: Edmonds polytopes and a hierarchy of combinatorial problems. Discrete Math. **4** (1973) 305–337

18. Cook, W., Coullard, C.R., Turán, G.: On the complexity of cutting-plane proofs. Discrete Appl. Math. **18** (1987) 25–38

19. Chvátal, V., Cook, W., Hartmann, M.: On cutting-plane proofs in combinatorial optimization. Linear Algebra Appl. **114/115** (1989) 455–499

20. Lombardi, H., Mnev, N., Roy, M.F.: The Positivstellensatz and small deduction rules for systems of inequalities. Mathematische Nachrichten **181** (1996) 245–259

21. Grigoriev, D., Vorobjov, N.: Complexity of Null- and Positivstellensatz proofs. Annals of Pure and Applied Logic **113** (2001) 153–160

22. Grigoriev, D.: Linear lower bound on degrees of Positivstellensatz calculus proofs for the parity. Theoretical Computer Science **259** (2001) 613–622

23. Grigoriev, D.: Complexity of Positivstellensatz proofs for the knapsack. Computational Complexity **10** (2001) 139–154

24. Cook, S.A., Reckhow, A.R.: The relative efficiency of propositional proof systems. Journal of Symbolic Logic **44** (1979) 36–50

25. Pitassi, T.: Algebraic propositional proof systems. In Immerman, N., Kolaitis, P.G., eds.: Descriptive Complexity and Finite Models. Volume 31 of DIMACS Series in Discrete Mathematics and Theoretical Computer Science. American Mathematical Society (1997)

26. Grigoriev, D., Hirsch, E.A., Pasechnik, D.V.: Complexity of semi-algebraic proofs. Technical Report 01-103, Revision 01, Electronic Colloquim on Computational Complexity (2002) ftp://ftp.eccc.uni-trier.de/pub/eccc/reports/2001/TR01-103/index.html#R01

Paths Problems in Symmetric Logarithmic Space

Andreas Jakoby and Maciej Liśkiewicz*

Universität zu Lübeck
Inst. für Theoretische Informatik, Wallstr. 40, D-23560 Lübeck, Germany
{jakoby/liskiewi}@tcs.mu-luebeck.de

Abstract. This paper studies space complexity of some basic problems for undirected graphs. In particular, the two disjoint paths problem (TPP) is investigated that asks whether for a given undirected graph G and nodes s_1, s_2, t_1, t_2, G admits two node disjoint paths connecting s_1 with t_1 and s_2 with t_2. The solving of this problem belongs to the fundamental tasks of routing algorithms and VLSI design, where two pairs of nodes have to be connected via disjoint paths in a given network. One of the most important results of this paper says that TPP can be solved by a symmetric nondeterministic Turing machine that works in logarithmic space. It is well known that switching from undirected to directed graphs, TPP becomes intractable.

Furthermore, the space complexity of minor detections is discussed. We show that testing for K_4-minor can be reduced (via log-space many-one reduction) to planarity testing and that detecting K_5-minor is hard for $\mathcal{SL}$. As a corollary we obtain that series-parallel graphs can be recognised in $\mathcal{SL}$.

Finally, the problem to determine the number of self avoiding walks in undirected series-parallel graphs is considered. It is proved that this problem can be solved in $\mathcal{FL}^{\langle \mathcal{SL} \rangle}$.

1 Introduction

The st-connectivity problem for undirected graphs is a generic problem for the complexity class $\mathcal{SL}$. This class, introduced by Lewis and Papadimitriou [18], covers all problems that can be solved by logarithmic space bounded symmetric NTMs, i.e. nondeterministic Turing machines M for which the relation that M starting in configuration C yields in one step a configuration C', is symmetric. A natural generalisation of the st-connectivity problem is the k-disjoint paths problem: for a given undirected graph G and sequence of pairs $(s_1, t_1), \ldots, (s_k, t_k)$ of nodes of G decide whether there exist paths $\Pi_1, \ldots, \Pi_k$ in G, mutually node-disjoint, such that Π_i joins s_i and t_i, with $1 \le i \le k$. This problem has been studied intensively in the literature and it is well known that k-disjoint paths is $\mathcal{NP}$-complete, if the number of pairs k is a variable of the input [15]. On the other hand, in [21] Robertson and Seymour have given an algorithm solving the k-disjoint paths problem in polynomial time for any constant k. In [24] Shiloach has shown that for $k = 2$ the problem can be solved even by an $O(n \cdot m)$ algorithm, where n, resp. m, is the number of nodes, resp. edges, of the input graph.

For directed graphs st-connectivity is the basic complete problem for the class $\mathcal{NL}$. However, if one switches to the directed k-disjoint paths problem the complexity changes

---

* On leave from Instytut Informatyki, Uniwersytet Wroclawski, Poland.

P. Widmayer et al. (Eds.): ICALP 2002, LNCS 2380, pp. 269–280, 2002.

drastically: in [11] Fortune et al. prove that this problem becomes $\mathcal{NP}$-complete even for $k = 2$. One of the main results of our paper shows that the 2-disjoint paths problem (TPP) for undirected graphs remains in $\mathcal{SL}$ and hence it is complete for this class. We leave as an open question if the k-disjoint paths problem remains in $\mathcal{SL}$ for any constant $k \geq 3$. Note that in the literature one also considers a different variant of the paths problem, namely the so called k-vertex disjoint paths problem. In this case we ask whether for a given undirected G and two nodes s and t there are k node disjoint paths from s to t. It is well known that this problem is $\mathcal{SL}$-complete for any constant k (see [20]). Moreover one can easily reduce this problem to the k-disjoint paths problem by replacing in a given graph G the nodes s and t by k copies $s_1, \ldots, s_k$ and $t_1, \ldots, t_k$ and then connecting the copies with the remaining nodes of G in the same way as s, resp. t (w.l.o.g. we can assume that there is no edge (s, t) in G). On the other hand, for $k \geq 3$ we do not know at all if a log-space reduction from the k-disjoint paths problem to k-vertex disjoint paths problem exists.

This paper investigates also the complexity of H-minor testing problem that is related in a natural way to the k-disjoint paths problems. For some particular graphs H the H-minor testing problem is to decide if a given undirected graph G has H as minor (i.e. a subgraph which is homeomorphic to H). From [21] we obtain that for every fixed graph H, the problem of H-minor testing can be solved in time $O(n^3)$ for an input graph G of size n. For some specific H even faster algorithms are known. In [16] Kézdy and McGuinness show that K_5-minor can be find in G in time $O(n^2)$. In this paper we investigate space complexity of the H-minor testing. It is easy to see that the problem is $\mathcal{L}$-hard for any $H = K_\ell$, with $\ell \geq 3$. Hence e.g. recognition of undirected series-parallel graphs, which is equivalent with testing whether a given graph G contains K_4 as a minor, is $\mathcal{L}$-hard, too, but it is open whether the problem is $\mathcal{SL}$-hard. We prove that the K_4-minor testing problem is log-space many-one reducible to planarity testing and that the K_5-minor testing is $\mathcal{SL}$-hard. On the other hand Allender and Mahajan [2] have shown that checking if a graph is planar can be done in $\mathcal{SL}$ leaving however as an open question whether the problem is hard for $\mathcal{SL}$. Note that planarity testing is equivalent with deciding if a given graph has $K_{3,3}$ or K_5 as a minor. Hence one can conclude: K_4-minor testing $\leq_m^{\log}$ planarity testing $\leq_m^{\log}$ K_5-minor testing. As a corollary we obtain that recognition of undirected series-parallel graphs is in $\mathcal{SL}$ but we leave as an open question whether K_5-minor testing can be solved in $\mathcal{SL}$, too.

For series-parallel graphs or equivalently for K_4-free graphs we will investigate the problem to generate the decomposition tree and to count the number of self avoiding walks (SAWs for short). A walk is called self avoiding if it follows a path which does not intersect itself (see e.g. [17]). We prove that decomposition of series-parallel graphs and to count SAWs in such graphs can be done in $\mathcal{FL}^{\mathcal{SL}}$. Changing from series-parallel graphs to planar graphs the problem of counting SAWs increases its complexity significantly: it is known that counting SAWs even in planar graph is $\#\mathcal{P}$-complete ([17]) whereas counting SAWs in directed series-parallel graphs with two terminals is in $\mathcal{FL}$ ([14]). We leave it as an open problem whether this problem for series-parallel graphs can be solved in $\#\mathcal{L}$.

This paper is organised as follows. In the next section we will give some basic definitions. In the third section we will show how to solve the 2-disjoint path problem in $\mathcal{SL}$. In section 4 we will investigate the graph minor problem. Section 5 and 6 are concerned with series-parallel graphs. In section 5 the basic ideas for computing a decomposition tree are presented and in section 6 we investigate the counting problem of SAWs. The paper ends with some conclusions and open problems.

2   Preliminaries

Definition 1. 2-disjoint paths problem (TPP): *for a given undirected graph $G = (V, E)$ and four nodes $s_1, s_2, t_1, t_2 \in V$ decide whether there are two node disjoint paths from s_1 to t_1 and from s_2 to t_2 in G.*

$\quad$ *H-**minor testing:** for a given undirected graph G decide whether G has H as minor i.e. a subgraph which is homeomorphic to H.*

The class $\mathcal{SL}$ contains all languages which can be reduced via a log-space many-one reduction to the undirected st-connectivity problem or equivalently all languages which can be recognised by symmetric NTMs that run within log-space. Denote by $\mathcal{L}^{\langle \mathcal{SL} \rangle}$ the class of all languages accepted by deterministic log-space oracle Turing machines (log-space oracle DTMs for short) with an oracle from $\mathcal{SL}$. An oracle DTM has a work tape and a write-only query tape (with unlimited length) which is initialised after every query. Furthermore, a symmetric log-space bounded oracle NTM M with oracle B is a symmetric NTM working within log-space that generates its queries to B by its current instantaneous description where B has also access to the input word of M. That means that B is a subset of $\{(d, X) : |d| = \log |X|\}$, where d is an instantaneous description of M and X is an input word. Now let $\mathcal{SL}^{\langle B \rangle}$ be the class of all languages accepted by symmetric log-space bounded oracle NTMs with oracle B and let $\mathcal{SL}^{\langle \mathcal{SL} \rangle} := \bigcup_{B \in \mathcal{SL}} \mathcal{SL}^{\langle B \rangle}$. This definition has been given by [3]. In [19] Nisan and Ta-Shma proved that the three classes above are equal: $\mathcal{SL} = \mathcal{L}^{\langle \mathcal{SL} \rangle} = \mathcal{SL}^{\langle \mathcal{SL} \rangle}$. Finally let us define $\mathcal{FSL} = \mathcal{FL}^{\langle \mathcal{SL} \rangle}$ as a set of all functions computed by log-space oracle DTMs with an oracle from $\mathcal{SL}$ that generate outputs on write-only output-tape.

Now let us give a formal definition of undirected series-parallel graphs.

Definition 2. $G = (V, E)$ *with two distinguished nodes $s, t \in V$, called the terminals of G (s can be equal to t) is an undirected series-parallel (s, t)-graph, (s, t)-USP graph for short, if either G is a single node v (i.e. $|V| = 1$ and $|E| = 0$) with $s = t = v$, or a single loop (v, v) (i.e. $|V| = 1$ and $|E| = 1$) with $s = t = v$, or a single edge (v, u) (i.e. $|V| = 2$ and $|E| = 1$) with $s = v$ and $t = u$, or there are two undirected series-parallel graphs $G_1 = (V_1, E_1)$ with terminals s_1, t_1 and $G_2 = (V_2, E_2)$ with terminals s_2, t_2 such that $V = V_1 \cup V_2$, $E = E_1 \cup E_2$, and either G is a*
*(A) **parallel composition** of G_1 and G_2: $s = s_1 = s_2$ and $t = t_1 = t_2$, or*
*(B) **series composition** of G_1 and G_2: $s = s_1$ and $t_1 = s_2$ and $t = t_2$, or*
*(C) **tree composition** of G_1 and G_2: $s = s_1$, $t = t_1$, and $|V_1 \cap V_2| = 1$. We call the node in $V_1 \cap V_2$ a t-node.*

We say that G is undirected series-parallel (USP for short) if there exist two nodes s, t of G such that G is an (s, t)-USP graph. Note that there are several alternative definitions for undirected series-parallel graphs known from the literature. We will use the definition above, because it leads to a natural definition of the decomposition tree of a USP graph that will be useful for counting the number of SAWs. To solve the recognition problem for USP graphs we will use the following result from Duffin [9]: An undirected graph G is USP iff it is K_4-free, i.e. it has no K_4 as a minor.

Note that because of the parallel composition step USP graphs are multigraphs, i.e. there might be more than one edge connecting a pair of nodes. On the other hand, changing form common graphs to multigraphs does not influence the computational complexity of TPP. Hence in the following, graphs will always refer to multigraphs.

3   The Undirected Two Paths Problem

To solve most problems discussed in this paper we give deterministic log-space algorithms which use an $\mathcal{SL}$ oracle for deciding the following version of the reachability problem: *Given an undirected graph $G = (V, E)$, a subset of nodes $U \subset V$ of a constant size, and two nodes $u, v \in V \setminus U$ decide, whether u is reachable from v without passing a node in U.* Obviously, the problem can be solved in $\mathcal{SL}$. To denote that an algorithm uses the oracle above for a particular input we write $\mathrm{USTCON}(G \setminus U, u, v)$.

Now we are ready to focus on the TPP problem. Because the common st-connectivity problem is $\mathcal{SL}$-hard hence TPP remains hard for $\mathcal{SL}$ as well. The main result of this section says that TPP can be solved in $\mathcal{SL}$. To prove this let us first consider the case that G is 3-connected and planar. Analogously to [27] one can show:

Theorem 1. *Let $G = (V, E)$ be a 3-connected planar graph and $s_1, s_2, t_1, t_2 \in V$. Then there exists no solution for the TPP problem iff there exists a planar drawing of G and a face F such that s_1, s_2, t_1, t_2 belong to F and traveling along the border of F we pass the terminal nodes in the order s_1, s_2, t_1, t_2.*

Hence for a 3-connected planar graph G one can solve TPP in $\mathcal{SL}$ just by adding a new node z and eight edges $(s_1, z), (s_2, z), (t_1, z), (t_2, z)$ and $(s_1, s_2), (s_1, t_2), (t_1, s_2), (t_1, t_2)$ to G and testing if the modified graph is not planar. One can show that it is planar iff the original graph is planar, all terminal nodes belong to the same face, and the ordering of the terminal nodes is s_1, s_2, t_1, t_2. There are some (small) counterexamples proving that this construction cannot be used for 2-connected planar graphs.

Before solving the TPP problem for the general case we will modify G by adding four auxiliary edges $(s_1, s_2), (s_1, t_2), (t_1, s_2), (t_1, t_2)$. Note that these edges do not change the solution of the TPP problem. Furthermore, we will assume that t_1 is reachable from s_1 without passing through the nodes s_2 and t_2 and t_2 is reachable from s_2 without passing through s_1 and t_1. Note that if one of these assumptions is not fulfilled the TPP problem has a negative solution.

Recall that in [24] Shiloach has shown that TPP has a positive solution for any choice of terminal nodes, if the graph is 4-connected and not planar. This result can be extended to reduced graphs [23,26,25].

Definition 3. *For a graph $G = (V, E)$ and four nodes s_1, s_2, t_1, t_2 let A, B, and C be three subsets of V such that $V = A \cup B$, $C = A \cap B$, $|C| \leq 3$, $s_1, s_2, t_1, t_2 \in A$, and $E \cap ((A \setminus C) \times (B \setminus C)) = \emptyset$. A graph $G' = (V', E')$ is called a reduction of G if* *for such subsets A, B, C it holds $V' = A$, $E' = (E \setminus (B \times B)) \cup (C \times C)$, and $|C|$ is minimum over all choices of $A', B',$ and C' fulfilling the conditions above. G is called* **reduced** *iff there exists no reduction G' of G.*

Note that for each solution of the TPP for a reduced graph each path that enters the node set $B \setminus C$ has also to leave the set B. Hence, there exists at most one path in each solution that enters $B \setminus C$. Since C has a minimum cardinality the nodes in B are connected. Therefore it follows that an instance of the TPP problem has a positive solution iff the instance has also a positive solution after performing one reduction step.

Theorem A [23,26,25] *For a 3-connected reduced graph $G = (V, E)$ and four nodes $s_1, s_2, t_1, t_2 \in V$ there exists no positive solution of the TPP problem iff G is a planar graph with a face containing s_1, s_2, t_1, t_2 in order.*

This result is closely related to Theorem 1. Unfortunately it is not known, whether a graph can be transformed into a reduced graph via an $\mathcal{SL}$ algorithm. Below we show an operation which can be done in $\mathcal{SL}$ and which has the following properties: it transforms an arbitrary graph into a so called *quasi 4-connected graph* such that the TPP problem on the resulting instance has a positive solution iff the original one has a positive solution. Furthermore, we will show that our quasi 4-connected graph is 3-connected and planar if its reduced graph has a negative solution for the TPP problem. Hence a solution for the original TPP problem can be obtained from Theorems 1 and A. Till the end of this section we will focus on quasi 4-connected graphs.

Definition 4. *Graph $G = (V, E)$ is called quasi $(k+1)$-connected iff G is k-connected and deleting k nodes of G results in at most two connected components $G_1 = (V_1, E_1)$ and $G_2 = (V_1, E_1)$ such that G_1 is not connected to G_2 and either $|V_1| = 1$ or $|V_2| = 1$. $G = (V, E)$ is called quasi $(k+1)$-connected with respect to a terminal set T iff $G' = (V, E \cup (T \times T))$ is quasi $(k+1)$-connected.*

For a node v and a constant k define $\text{sep}_k(v) \subseteq V$ with $|\text{sep}_k(v)| = k$ such that:

1) deleting $\text{sep}_k(v)$ from G separates v from all terminal nodes
2) $\forall u \in \text{sep}_k(v) :$ deleting $\text{sep}_k(v) \setminus \{u\}$ does not separate v from any terminal node
3) $\forall V' \subset V$ with $|V'| = k$ that fulfills the two conditions above for $\text{sep}_k(v) :$
   deleting V' does not separate a node of $\text{sep}_k(v)$ from any terminal node.

For a node v of G that can not be separated from the terminal nodes let $\text{sep}_k(v) = \emptyset$.

Lemma 1. *For a constant k let $G = (V, E)$ be a k-connected graph and $T \subseteq V$ be the set of terminal nodes of G and $v \in V$ a non terminal node. If v can be separated from all terminal nodes by deleting k nodes, then $\text{sep}_k(v) \neq \emptyset$ is well defined, unique, and can be computed in log-space by using USTCON as an oracle.*

Instead of investigating G we will now consider the graph $G' = (V', E')$ that can be generated from G and the terminal nodes in 3 steps: Assume that G is a connected graph, then define $G_1 = (V_1, E_1)$ and $G_2 = (V_2, E_2)$ as follows:

$V_1 := \{ v \in V \mid v$ can not be separated from the terminals by deleting one node $\}$,
$E_1 := \{ (u, v) \mid u, v \in V_1$ and $(u, v) \in E \}$,
$V_2 := \{ v \in V_1 \mid v$ can not be separated from the terminals by deleting two nodes $\}$,
$E_2 := \{ (u, v) \mid u, v \in V_2$ and $(u, v) \in E_1$ or $\exists w \in V_1 \setminus V_2 : (u, v) = \text{sep}_2(w) \}$.

For $u \in V$ define: $R[u] \subset V$ as the set of nodes v such that v is not reachable from a terminal without passing through a node in $sep_3(u)$. We assume that $sep_3(u) \cap R[u] = \emptyset$ for all $u \in V$. Let $r[u]$ be the lexicographically first node in $R[u]$ if such a node exists and let $r[u] = u$ otherwise. Note that $R[u] = \emptyset$ if $sep_3(v) = \emptyset$. Now we define G':

$V_3 := \{\, v \in V_2 \mid v$ can not be separated from the terminals by deleting three nodes $\,\}$,
$V' := V_3 \cup \{\, v \in V_2 \mid \exists u \in V_2 \; : \; sep_3(u) \neq \emptyset$ and $v = r[u] \,\}$,
$E' := (E_2 \cap (V_3 \times V_3)) \cup \{\, (r[u], r[v]) \mid u, v \in V_2$ and $u \in sep_3(v) \,\}$.

Note that $R[v]$ and $r[v]$ can be computed in log-space by using USTCON as an oracle. Furthermore, the question whether v belongs to V' or e belongs to E' can be decided in $\mathcal{SL}$ on input G, v, and the terminal nodes. From the construction it follows:

Lemma 2. *The graph G' as constructed above is quasi 4-connected with respect to the terminal nodes. Moreover TPP has a positive solution for G, s_1, s_2, t_1, t_2 iff it has a positive solution for G', s_1, s_2, t_1, t_2.*

It remains to show how G' can be used to solve the TPP problem. A reduced graph of G' is generated by a sequence of reduction steps (Definition 3). By induction it follows:

Lemma 3. *Assume that a reduced graph of G' has a negative solution for the corresponding TPP problem. Then G' is planar.*

Now we are ready to give an Algorithm solving the TPP problem:

```
input: G = (V, E), s1, t1, s2, t2;
   if ¬USTCON(G \ {s1, t1}, s2, t2) ∨ ¬USTCON(G \ {s2, t2}, s1, t1)
      then return (G, s1, t1, s2, t2) ∉ TPP;
E := E ∪ {(s1, s2), (s1, t2), (t1, s2), (t1, t2)};
let  G' = (V', E') be the quasi 4-connected graph obtained from (V, E);
V' := V' ∪ {z};  E' := E' ∪ {(s1, z), (s2, z), (t1, z), (t2, z)};
   if (V', E') is planar
      then return (G, s1, t1, s2, t2) ∉ TPP
      else return (G, s1, t1, s2, t2) ∈ TPP;
```

The correctness of the algorithm follows from Theorem 1 as well as Lemma 3 and Theorem A. Moreover one can show that using the oracle USTCON the algorithm works in log-space. Hence we obtain:

Theorem 2. *The TPP problem is in $\mathcal{SL}$.*

4   The Complexity of Detecting Some Graph Minors

In this section we investigate the complexity of H-minor testing problem for some specific small graphs H. For the only nontrivial minor H of 3 nodes, namely for K_3, the H-minor testing problem is equivalent to decide whether a given undirected graph contains a cycle. Since the last problem is complete for $\mathcal{L}$ ([8,13]) the K_3-minor problem is $\mathcal{L}$-complete, too. Testing whether a given graph contains K_4 as a minor, what is equivalent to recognition of undirected series-parallel graphs (see [9]), seems to be more complicated. One of the main results of this section says that K_4-minor problem is at most as difficult as planarity testing.

Theorem 3. *K_4-minor testing is log-space many-one reducible to planarity testing.*

Proof: Let $G = (V, E)$ be a given undirected graph, with n nodes and m edges. We construct a sequence $G_1, G_2, \ldots, G_t$ of graphs with $t = \binom{m}{2}$, such that G is series-parallel iff all G_i are planar. The construction goes as follows: let $e = (u, w)$ and $e' = (u', w')$ be the i-th pair of edges of G (with respect to a fixed enumeration of such pairs). Then define $G_i = (V \cup \{v_e, v_{e'}\}, (E \setminus \{(u, w), (u', w')\}) \cup \{(u, v_e), (v_e, w), (u', v_{e'}), (v_{e'}, w'), (v_e, v_{e'})\})$, i.e. we split e and e' by inserting the new nodes v_e and $v_{e'}$ and connect v_e with $v_{e'}$.

To show that our reduction gives the sequence of graphs with the property above let us assume first that G contains a subgraph H homeomorphic to K_4. Let $a, b, c, d \in V$ be the nodes of H which induce K_4 and $\Pi_{a,b}, \Pi_{a,c}, \Pi_{a,d}, \Pi_{b,c}, \Pi_{b,d}$, and $\Pi_{c,d}$ the corresponding paths in G. Let a' be the immediate successor of a on $\Pi_{a,c}$ and b' be the immediate successor of b on $\Pi_{b,d}$. Note that it can happen that $a' = c$ or $b' = d$. Now let G_i be the graph constructed with respect to the edges $e = (a, a')$ and $e' = (b, b')$. Then the nodes $a, b, c, d, v_e, v_{e'}$ determine a subgraph of G_i which is homeomorphic to $K_{3,3}$. Hence we conclude that G_i is not planar.

Now assume that G does not have a K_4 as a minor and let us assume to the contrary that there exists some i such that G_i is not planar. This means that in G_i there exists a subgraph H homeomorphic to K_5 or to $K_{3,3}$. If at most one of the new nodes v_e and $v_{e'}$ is in H then one can obtain from H a subgraph in G which is homeomorphic to K_4 by reversing the construction of G_i – a contradiction.

Otherwise, if H is homeomorphic to $K_{3,3}$ we can show that G has to contain a subgraph homeomorphic to K_4, too. Let a, b, c, a', b', c' be the nodes of G which induce a $K_{3,3}$ such that $\{a, b, c\}$ and $\{a', b', c'\}$ are the bipartite components. Because each new node of G_i has degree 3 it is impossible that both belong to the same bipartite component. But it can happen that one of them belongs to the first component and the second belongs to the other bipartite component. Let us consider this case. W.l.o.g. we assume that

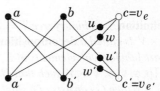

Fig. 1. To obtain from $K_{3,3}$ in G_i a K_4 in G one needs to remove v_e, $v_{e'}$, and the incident edges and to add the edges (u, w) and (u', w').

c and c' are the new nodes, i.e. let $c = v_e$ and $c' = v_{e'}$, with $e = (u, w)$ and $e' = (u', w')$. It is easy to prove (see e.g. Fig. 1) that the nodes a, b, a', b' induce a subgraph in G which is homeomorphic to K_4 – a contradiction.

Let assume that H is homeomorphic to K_5 and both new nodes v_e and $v_{e'}$, with $e = (u, w)$ and $e' = (u', w')$, belong to H. Since the nodes of K_5 have degree 4, all the nodes which induce K_5 in G_i (let us call them z_1, z_2, z_3, z_4, z_5) belong to V. This means that v_e and $v_{e'}$ lay on paths Π_{z_j, z_k}, resp. $\Pi_{z_{j'}, z_{k'}}$, connecting z_j and z_k, resp. $z_{j'}$ and $z_{k'}$ in H. If $j = j'$ and $k = k'$ the new nodes lay on the same path and the nodes of the remaining paths belong to V. Hence $\{z_1, z_2, z_3, z_4, z_5\} \setminus \{z_j\}$ induces a K_4 minor in G – a contradiction. Let us now focus on the case that Π_{z_j, z_k} and $\Pi_{z_{j'}, z_{k'}}$ are different.

Because the degree of v_e and $v_{e'}$ is 3 the nodes u and w have to lay on Π_{z_j,z_k} and u' and w' have to lay on $\Pi_{z_{j'},z_{k'}}$. Then replacing the edges $(u, v_e), (v_e, w)$ on Π_{z_j,z_k} by (u, w) and on $\Pi_{z_{j'},z_{k'}}$ the edges $(u', v_{e'}), (v_{e'}, w')$ by (u', w') we get a subgraph of G homeomorphic to K_5, and hence to K_4 – a contradiction. ∎

Using the theorem above and the result of Allender and Mahajan [2] we obtain

Corollary 1 *The problem of recognition of undirected series-parallel graphs is in $\mathcal{SL}$.*

Testing whether a graph G has a K_5 minor might be more difficult than testing whether G contains K_4 or $K_{3,3}$ minor. The result below seems to confirm this conjecture.

Theorem 4. *If H has a cycle with a node of degree at least 4 then H-minor testing is $\mathcal{SL}$-hard.*

Corollary 2 *The problem to test whether a graph has a K_5 as a minor is $\mathcal{SL}$-hard.*

Note that from this claim it does not follow that planarity testing is $\mathcal{SL}$-hard.

5 Decomposing Undirected Series-Parallel Graphs

Recognition and decomposition of directed series-parallel graphs are well studied topics in the literature [5,10,14,12,28]. In this paper we will investigate the space complexity of these problems for the undirected case. One of the basic tools used to examine series-parallel graphs is the decomposition tree. We will apply such a tree later to count the number of paths in an (s, t)-USP graph.

Definition 5. *A binary tree $T = (V_T, E_T)$ with a labeling $\sigma : V_T \to (\{p, s, t\} \cup E) \times V \times V$ is called a **decomposition tree** of an (s, t)-USP graph $G = (V, E)$ iff leaves of T are labeled with elements of $E \times V \times V$ and internal nodes are labeled with elements of $\{p, s, t\} \times V \times V$ such that G can be generated recursively using T as follows: 1. if $V_T = \emptyset$ then G consists of a single node; 2. if T is a single node v then G consists of the single edge (s, t) and $\sigma(v) = ((s, t), s, t)$; 3. otherwise, let T_1 (resp. T_2) be the right (resp. left) subtree of T, with the root v, and for $i \in \{1, 2\}$ let T_i be the decomposition tree of G_i and*

(a) if $\sigma(v) = (p, s, t)$ then G is the parallel composition of G_1 and G_2,
(b) if $\sigma(v) = (s, s, t)$ then G is the serial composition of G_1 and G_2,
(c) if $\sigma(v) = (t, s, t)$ then G is the tree composition of G_1 and G_2.

Our algorithm which generates the decomposition tree for a given (s, t)-USP graph G is based on a deterministic log-space bounded method using the $\mathcal{SL}$ oracle USTCON which determines whether a given graph G is an (s, t)-USP graph. This algorithm will efficiently traverse an USP graph in a decomposition tree like order and it will generate the tree nodes during this traverse. For the rest of this section we will present the basic ideas for this method.

We start with the following observation. From the definition of a USP graph G it follows, that each subgraph G' of G of at least 4 nodes is at most 2-connected, i.e there are 2 nodes that separates G' into at least two disconnected subgraphs. This property leads us to the following strategy for the traversing of G:

Verify that for a given pair s, t and each node v there exists a sequence $G_0, G_1, ..,$ G_k with $G_0 = G$ and $G_k = (\{v\}, \emptyset)$, such that for each $i \in \{1, .., k\}$ G_{i-1} can be constructed by parallel, series or tree composition of G_i with an other graph.

To specify G_i we will use a triple (s_i, t_i, h_i) where s_i and t_i are two corresponding terminal nodes and h_i is an inner node of G_i. If G_{i-1} is a series or tree composition of G_i and an other graph then we assign $h_i := s_i$. Otherwise, if G_{i-1} is parallel composed, we choose h_i as the nearest bridge node to s_i in G_i decomposing G_i into two disconnected subgraphs G'_i and G''_i where s_i belongs to G'_i and t_i belongs to G''_i.

For $s_i, t_i,$ and v let $V' \subset V \setminus \{s_i, t_i\}$ be the set of nodes reachable from v without passing through s_i or t_i. Define $G_{s_i, t_i}[v] = (V_{s_i, t_i}[v], E_{s_i, t_i}[v])$ as the empty graph if $V' = \emptyset$ and as the subgraph of G induced by the node set $V' \cup \{s_i, t_i\}$ otherwise. Our strategy to solve the recognition problems for (s, t)-USP graphs is based on the following fact:

Lemma 4. *For a given USP graph G and three nodes $s, t,$ and v the following holds: If $G_{s,t}[v]$ is an (s, t)-USP graph then either $s = v$, or $t = v$, or $G_{s,t}[v]$ is not 2-connected, i.e. there exists a bridge node u in $G_{s,t}[v]$. Furthermore, either $u = v$, or*

- $v \in V_{s,u}[v]$ *and thus $V_{u,t}[v]$ is empty and $G_{s,u}[v]$ is an (s, u)-USP graph, or*
- $v \in V_{u,t}[v]$ *and thus $V_{s,u}[v]$ is empty and $G_{u,t}[v]$ is an (u, t)-USP graph.*

On the other hand, if u is a bridge node in a graph G, then G is a series (resp. tree) composition of two subgraphs, composed at u. Note that inside a 2-connected USP graph we do not have to consider a tree composition step. Moreover, if G can be decomposed into two or more disconnected subgraphs by deleting two nodes s and t then G is a parallel composition of these subgraphs composed at s and t.

Definition 6. *For a given graph $G = (V, E)$ and three nodes s, t, v, G is called (s, t)-series-consistent according to v iff there exists a sequence of pairs $(s_0, t_0), .., (s_k, t_k)$ such that $s_0 = s$, $t_0 = t$, and for all $i \in \{1, \ldots, k\}$ it holds that $v = s_k$ or $v = t_k$ and either $s_i = s_{i-1}$ and t_i is a bridge node of $G_{s_{i-1}, t_{i-1}}[v]$ and v is a node in $G_{s_i, t_i}[v]$ or $t_i = t_{i-1}$ and s_i is a bridge node of $G_{s_{i-1}, t_{i-1}}[v]$ and v is a node in $G_{s_i, t_i}[v]$.*

To decide whether a graph is an USP graph we use the following lemma:

Lemma 5. *A 2-connected graph $G = (V, E)$ is an (s, t)-USP graph iff G is (s, t)-series-consistent according to each node $v \in V$.*

We construct a decomposition tree by following the decomposition described above for some specific nodes in G. Then we can conclude:

Theorem 5. *A decomposition tree for a (s, t)-USP graph can be computed in $\mathcal{FSL}$.*

6   Counting Paths

We investigate the problem of counting the number of SAWs (1) from the source to a specific point, (2) from the source to any point, and (3) between any two points. From [17] it is known that all these problems are complete for $\#\mathcal{P}$ even for grid graphs.

Theorem 6. *Each of the three SAW counting problems in USP graphs is in $\mathcal{FSL}$.*

Till the end of this section we will sketch a proof for this theorem. We start with the following result that can be proved by induction on the structure of a given USP graph.

Theorem 7. *For each USP graph G and each node s of G there exists a node t such that G is an (s, t)-USP graph.*

Hence, to show Theorem 6 it is enough to give an $\mathcal{FSL}$ algorithm only for counting the number of SAWs from the source to a specific point. Below we describe such an algorithm for a given (s, t)-USP graph $G = (V, E)$, with $n = |V|$ and $m = |E|$ and a point $v \in V$. The decomposition tree generated by the algorithm discussed in the previous section has the following properties. (i) Any series step which composes two subgraphs such that terminals of the resulting graph are not used for a parallel composition step later on is replaced by a tree composition step. So our algorithm prefers tree composition when ever possible. Furthermore (ii) t-node u associated with a node w of the decomposition tree, with $\sigma(w) = (\mathsf{t}, s, t)$, belongs always to the same 2-connected component as s and t.

Let $T = (V_T, E_T)$ of size $z \leq n + m$ be a decomposition tree of G fulfilling (i) and (ii). We denote the number of SAWs from the source s to the point v by #SAW. By induction one can show that #SAW $\leq 2^{n+m}$. The prime number theorem implies $\prod_{p_i \leq n+m} p_i = e^{(n+m)(1+o(1))} >$ #SAW, where $p_1 < p_2 < \ldots$ denotes the standard enumeration of primes. By this inequality, taking all $p_i \leq n + m$ the values #SAW mod p_i give a Chinese Remainder Representation of #SAW. In the rest of this section we will show how to compute such a representation in $\mathcal{FSL}$. To convert it to the ordinary binary representation in log-space one can use the result of [7].

Let $p \leq n + m$ be a prime. For (s, t)-USP graph G let $[st]$ denote the number of SAWs mod p from source s to sink t. To compute this value we proceed inductively. If G is a parallel composition of (s_1, t_1)-USP graph G_1 and (s_2, t_2)-USP graph G_2, where s_1, t_1, s_2, and t_2 are terminals with $s_1 = s_2 = s$ and $t_1 = t_2 = t$, we have: $[st] = [s_1 t_1]_1 + [s_2 t_2]_2$. The index i in the $[.]_i$ notation stresses that we count the corresponding paths in G_i only. If G is a series composition of G_1 and G_2 then $[st] = [s_1 t_1]_1 \cdot [s_2 t_2]_2$. Finally for a tree composition of (s, t)-USP G_1 and (s_2, t_2)-USP G_2 we have $[st] = [s_1 t_1]_1$.

To evaluate $[st]$ we consider the decomposition tree T of G and we interpret it as an arithmetic expression as follows. Every leaf represents the integer 1. An internal node v of T labeled by s (resp. p) corresponds to a multiplication (resp. addition) of the expressions given by the sons of w. If an internal node v is labeled by t then we just ignore this node and the right son of w that corresponds to G_2 and consider the father of w as a direct predecessor of G_1. The value of the root of T equals $[st]$. Now using the log-space algorithm of [6] one can transform T into a binary tree T' of depth $O(\log z)$ representing an arithmetic expression with the same value as T. We evaluate T' using the algorithm in [4]. This algorithm works in space $O(\log z + \log(n + m)) \leq O(\log n)$.

To compute the number of SAWs from a terminal s to a non-terminal v we will first consider the case that s and v belong to the same 2-connected component of G. Note that such a component is a USP graph generated by series and parallel composition steps only. Let G' be such a 2-connected component of G. To compute #SAW we associate with each node w of the decomposition tree T of G' a tuple $\langle \alpha, \beta, \gamma, \delta \rangle$ with the following meaning. Let G'' be an (s', t')-USP graph represented by subtree rooted in w. Then α

is the number of SAWs mod p in G'' joining s' and v (note that if v does not belong G'' then $\alpha = 0$) and β is the number of SAWs mod p in G'' joining t' and v. Next γ denotes the number of SAWs mod p in G'' joining s' and v without going through t' and analogously δ denotes the number of SAWs mod p in G'' joining t' and v without going through s'. The values α, β, γ, and δ of a USP graph can be computed recursively by using the following equalities.

(1) If G'' is a series composition of (s_1, t_1)-USP G_1 and (s_2, t_2)-USP G_2, then $\alpha = \alpha_1 + [s_1 t_1]_1 \cdot \alpha_2, \beta = \beta_2 + [s_2 t_2]_2 \cdot \beta_1, \gamma = \alpha_1 + [s_1 t_1]_1 \cdot \gamma_2$, and $\delta = \beta_2 + [s_2 t_2]_2 \cdot \delta_1$.

(2) If G is a parallel composition of G_1 and G_2 then $\alpha = \alpha_1 + [s_2 t_2]_2 \cdot \delta_1 + \alpha_2 + [s_1 t_1]_1 \cdot \delta_2$, $\beta = \beta_1 + [t_2 s_2]_2 \cdot \gamma_1 + \beta_2 + [t_1 s_1]_1 \cdot \gamma_2, \gamma = \gamma_1 + \gamma_2$, and $\delta = \delta_1 + \delta_2$.

Now we search in T for a node w_0 which corresponds to the subgraph $G^{(0)}$ such that $G^{(0)}$ is a series composition of $(s_{0,1}, t_{0,1})$-USP $G_1^{(0)}$ and $(s_{0,2}, t_{0,2})$-USP $G_2^{(0)}$ with $t_{0,1} = s_{0,2} = v$. Let $w_{0,1}$, resp. $w_{0,2}$ be the nodes corresponding to $G_1^{(0)}$, resp. $G_2^{(0)}$ in the decomposition tree. If such node w_0 does not exist the input graph G is just (s, t)-USP with $t = v$ and hence it holds $\#SAW = [st]$ (recall $s \neq v$). Then we prune away from T all subtrees rooted in $w_{0,1}$ and in $w_{0,2}$. The node w_0 is a new leaf of a tree that we will denote by T'. We associate $\langle \alpha = [s_{0,1} t_{0,1}]_1, \beta = [s_{0,2} t_{0,2}]_2, \gamma = [s_{0,1} t_{0,1}]_1, \delta = [s_{0,2} t_{0,2}]_2 \rangle$ to w_0. For the remaining leafs we choose $\langle \alpha = 0, \beta = 0, \gamma = 0, \delta = 0 \rangle$. Finally, we define for each inner node w of T' an operation on the tuples as follows: if w corresponds to a series composition then we use operation (1) and for a parallel composition we use operation (2). We claim that if one computes the value $\langle \alpha, \beta, \gamma, \delta \rangle$ for the root of the tree T' then $\alpha = \#SAW$. Moreover this tuple can be computed in $\mathcal{FSL}$. From the properties (i) and (ii) of the tree T it follows that a subtree of T' induced by non-zero tuples is a path from w_0 to the root of T'.

If v does not belong to the same 2-connected component as s we proceed analogously.

7   Conclusions and Open Problems

In this paper we have presented an $\mathcal{SL}$-algorithm for the k-disjoint paths problem, with $k = 2$. Finding disjoint paths can be a helpful tool for minor detection because an edge of a minor represents a path in the original graph. But nothing is known about the space complexity for $k > 2$. Note that if the k-disjoint paths problem can be solved in $\mathcal{SL}$, with $k = 10$, then the K_5-minor testing problem becomes $\mathcal{SL}$-complete. We conjecture that the k-disjoint paths problem can be solved in $\mathcal{SC}$ for any constant k.

In a second part of this paper we have investigated paths problems for series-parallel graphs. We have shown that testing, decomposition, or counting SAWs for USP graphs can be done in log-space using only an oracle for st-connectivity. Since $\mathcal{SL}/\text{poly} = \mathcal{L}/\text{poly}$ [1] all these problems can be solved in deterministic log-space using an advice of polynomial length. It remains open, whether the counting problems remains solvable in $\mathcal{FSL}$ if we consider SAWs of specific length.

Acknowledgement. Thanks are due to Eric Allender, Markus Bläser, and some unknown referees for helpful comments and pointers to the literature.

References

1. R. Aleliunas, R. Karp, R. Lipton, L. Lovasz, C. Rackoff, *Random Walks, Universal Sequences and the Complexity of Maze Problems,* FOCS, 1979, 218-223.
2. E. Allender, M. Mahajan, *The Complexity of Planarity Testing,* STACS, 2000, 87-98.
3. Y. Ben-Asher, K.-J. Lange, D. Peleg, and A. Schuster, *The Complexity of Reconfiguring Network Models,* Inf. & Comp. 121, 1995, 41-58.
4. M. Ben-Or, R. Cleve *Computing Algebraic Formulas Using a Constant Number of Registers,* SIAM J. Comput. 21, 1992, 54-58.
5. H. Bodlaender, B. de Fluiter, *Parallel Algorithms for Series Parallel Graphs,* ESA, 1996, 277-289.
6. S. Buss, S. Cook, A. Gupta, V. Ramachandran, *An Optimal Parallel Algorithm for Formula Evaluation,* SIAM J. Comput. 21, 1992, 755-780.
7. A. Chiu, G. Davida, B. Litow, *Division in logspace-uniform NC^1,* Theoretical Informatics and Applications, 35, 2001, 259-275.
8. S. Cook, *Towards a complexity theory of synchronous parallel computation,* L'Enseignement Mathematique, XXVIII, 1981, 99-124.
9. R. Duffin, *Topology of Series-Parallel Networks,* J. Math. Analysis Appl. 10, 1965, 303-318.
10. D. Eppstein, *Parallel Recognition of Series-Parallel Graphs,* Inf. & Comp. 98, 1992, 41-55.
11. S. Fortune, J. Hopcroft and J. Wyllie, *The directed subgraph homomorphism problem,* Theoretical Computer Science 10, 1980, 111-121.
12. X. He, Y. Yesha, *Parallel Recognition and Decomposition of Two Terminal Series Parallel Graphs,* Inf. & Comp. 75, 1987, 15-38.
13. B. Jenner, K.-J. Lange, P. McKenzie, *Tree Isomorphism and Some Other Complete Problems for Deterministic Logspace,* publication #1059, DIRO, Université de Montréal, 1997.
14. A. Jakoby, M. Liśkiewicz, R. Reischuk, *Space Efficient Algorithms for Series-Parallel Graphs* STACS, 2001, 339-352. See also ECCC Report TR02-021, 2002.
15. R. Karp, *On the complexity of the combinatorial problems,* Networks 5, 1975, 45-48.
16. A. Kézdy and P. McGuinness, *Sequential and Parallel Algorithms to Find a K_5 Minor,* SODA, 1992, 345-356.
17. M. Liśkiewicz, M. Ogihara, and S. Toda, *The Complexity of Counting Self-Avoiding Walks in Two-Dimensional Grid Graphs and in Hypercube Graphs,* ECCC Report TR01-061, 2001.
18. H. R. Lewis and C. H. Papadimitriou, *Symmetric space-bounded computations,* Theoretical Computer Science 19, 1982, 161-187.
19. N. Nisan and A. Ta-Shma, *Symmetric logspace is closed under complement,* FoCS, 1995, 140-146.
20. J. Reif, *Symmetric Complementation,* Journal of the ACM, vol. 31(2), 1984, 401-421.
21. N. Robertson, P. D. Seymour, *Graph Minors XIII. The Disjoint Paths Problems,* J. of Combinatorial Theory, Series B 63, 1995, 65-110.
22. N. Robertson, P. D. Seymour, R. Thomas, *Non-Planar Extensions Of Planar Graphs,* http://www.math.gatech.edu/~thomas/ext.ps
23. P. D. Seymour, *Disjoint paths in Graphs,* Discrete Math. 29, 1980, 293-309.
24. Y. Shiloach, *A Polynomial Solution to the Undirected Two Paths Problem,* J. of the ACM, Vol. 27, 1980, 445-456.
25. R. Thomas, *Graph Planarity and Related Topics,* Graph Drawing, 1999, 137-144.
26. C. Thomassen, *2-linked Graphs,* Europ. J. Combinatorics 1, 1980, 371-378.
27. G. Woeginger, *A simple solution to the two paths problem in planar graphs,* Inform. Process. Lett. 36, 1990, No. 4, 191-192.
28. J. Valdes, R. Tarjan, E. Lawlers *The Recognition of Series Parallel Digraphs,* SIAM J. Comput. 11, 1982, 298-313.

Scheduling Search Procedures

Peter Damaschke

Chalmers University, Computing Sciences, 41296 Göteborg, Sweden
ptr@cs.chalmers.se

Abstract. We analyze preemptive on-line scheduling against random-ized adversaries, where the goal is only to finish an unknown distin-guished target job. The problem is motivated by a clinical gene search project, but it leads to theoretical questions of broader interest, includ-ing some natural but unusual probabilistic models. For some versions of our problem we get optimal competitive ratios, expressed by given parameters of instances.

1 Introduction

1.1 A Motivation from Clinical Genetics

This work was originally motivated by a tumor suppressor gene search project [4]. In many cells from a certain type of tumor, end segments of chromosome 1 of varying lengths are deleted. This suggests the conjecture of a tumor sup-pressor gene located in the shortest missing segment. The putative suppressor gene is knocked out only if it is absent or damaged on the partner chromosome, too. Therefore one wishes to identify a gene in that region which exhibits fatal mutations on the partner chromosome in the available tumor cells but works properly in healthy control cells.

Let us represent the chromosome as unit interval $[0, 1]$, where some right end segment is deleted in every such clinical case. A deletion is given by its break-point, i.e. the new endpoint of the damaged chromosome. Their locations are known in all cases, subject to some discretization. However, detecting mutations is a time-consuming *job*, as it requires sequencing of the DNA material, and analysis and interpretation of the results (alignment, comparison, checking the effect on the encoded protein and its function in the cell, etc.).

It is natural to assume that breakpoints follow some fixed distribution and are independent in different examples. Under the conjecture of a single suppressor gene, this *target* gene must be to the right of all observed breakpoints.

One aspect of this gene search problem is: Given a set of candidate genes, how should we distribute simultaneous work among these candidates so as to optimize the chances of early success? Let us model the problem in a more abstract language that also allows generalization.

In a set of n objects (here: the candidate genes), one object is the target (here: the suppressor gene). Our goal is to identify the unknown target as early as possible. In order to check whether or not an object is the target, one has to perform some job on that object. In view of our application we allow preemptive scheduling without preemption penalty. Thus one can even run a mix of jobs, where each job permanently gets some fraction of time. (In practice one has to approximate such continuous schedules by short time slots according to the prescribed fractions.) The search stops if the target job has been finished, but since the target is unknown, it is unavoidable to partly run the other jobs as well. Thus we get an on-line problem, with the target completion time as the cost to minimize.

A crucial point here is that the on-line player (searcher) has some prior knowledge about probable target positions that she should take advantage of. In our case, the target is most likely close to the rightmost breakpoint, since otherwise we should have observed more breakpoints to the right, i.e. smaller deletions not affecting the suppressor gene. However the on-line player cannot simply assign probabilities to the target candidates given the rightmost breakpoint. Rather the target is given by nature, and the observed rightmost breakpoint, in the following called the *signal*, obeys some distribution depending on the target. We call this scenario the *adversarial instance random signal (AIRS)* framework. Note the similarity to Bayesian decision making (cf. [3, Chapter 15]). The difference is that our goal is not prediction but scheduling jobs to identify the true target. We believe that the AIRS framework is natural and might be of interest also beyond scheduling search procedures.

In our particular case, the target candidates are n points in interval $[0,1]$, and the signal is sampled by the following random experiment: For a number m (the number of clinical cases in a population monitored during several years), m breakpoints in $[0,1]$ have been independently sampled according to the even distribution. We define the rightmost breakpoint to the left of the target as our signal. (The breakpoints could be sampled from an arbitrary continuous distribution, but then we could map it to $[0,1]$ such that the new coordinate of a point is the probability that the breakpoint falls to the left of this point. So we can w.l.o.g. consider the even distribution on the unit interval.)

If the target is point t, the signal density near point t is approximately $me^{(s-t)m}$ where $s \leq t$ is the signal. If m is sufficiently large and the target candidates are in a relatively small region of the unit interval, we can therefore adopt a more elegant formulation that says essentially the same: The possible signals s are points on the real axis, and the possible targets t are n fixed points. The signal density is e^{s-t} for $s \leq t$, and 0 for $s > t$.

Before we derive scheduling results and discuss their conclusions for the above real-world problem, we introduce our notion and the general models, to provide a language in which our scheduling problems can be defined.

1.2   Online Problems with Randomized Input

Choice tree, instances, and signals. For the basics of competitive analysis we refer to [3,8]. It is well-known that the definition of competitive ratio and hence the "best" strategy depends on the on-line player's prior knowledge and the off-line player's power. We use off-line player and adversary as synonymes.

Traditionally, the adversary is supposed to have the full control of input, in the sense that she can freely choose all parts of input being unknown to the on-line player. This corresponds to the classical competitive ratio of a strategy for an on-line minimization problem which is the cost of the on-line solution, divided by the cost of the best off-line solution, maximized over all instances. Some criticism against this malicious assumption is discussed in [5], and generalized models, called diffuse adversary and comparative analysis, have been proposed. In particular, in the diffuse adversary setting an instance may have probabilistic components the adversary has no influence on. The motivation in [5] was that classical worst-case competitive analysis is often unable to discriminate among several strategies of apparently different quality. We were led to such generalizations by the application reported above.

On-line problems can be considered as games. One part of this game is a mechanism that determines a *state* of the problem. We split every state into an *instance*, which alone determines the costs that any strategy will incur, and a *signal* that captures all a priori information a player gets about the instance. By this abstract notion of signal one can model any kind of a priori knowledge (cf. the idea of comparative analysis in [5]). In the worst case for the on-line player, the signal is just constant or independent of instances such that it is not informative.

The adversary generates a state by decisions we can think of as a rooted tree. Every inner node including the root is either a *chance node* or an *adversary node*. Edges going from a chance node to its children are labeled by probabilities that sum up to 1. The adversary starts her decision process at the root. If the current node is an adversary node, she can proceed to an arbitrary child, if it is a chance node, the adversary must proceed to a child according to the given probabilities. Every leaf is labeled by an instance (hidden from the on-line player) and a signal (forwarded to the on-line player). Both players know the tree and all its labels completely. Intuitively, the tree encodes the on-line player's background knowledge.

This tree model resembles [3, Section 6.1.1], but here we use a tree only to describe the generation of states, whereas the players' strategies are considered separately. Note that the fan-out of nodes can be uncountably infinite, such that we must define a probability density on the children of a chance node, rather than probabilities of children.

If an adversary (chance) node is child of an adversary (chance) node then it can obviously be merged with the parent node. Thus we may w.l.o.g. consider trees consisting of alternating layers of chance and adversary nodes. Moreover, the adversary may make all her deterministic decisions first, conditional on the outcomes of chance nodes. Thus we can transform any such tree into an equivalent two-layer tree, merely consisting of an adversarial root, chance nodes as its children, and leaves. This is the diffuse adversary from [5] who selects a distribution of states from a class of distributions. (However these normalizations might not always be natural and blow up the tree.)

Costs, regret, and competitive ratio. An *action* assigns a cost to every instance. Here we assume that both players have the same actions available. A strategy is to deterministically choose and apply an action. (In the present paper we do not consider randomized strategies.) The on-line player may take the signal as a parameter for her choice.

Next we define the competitive ratio of a strategy, due to the above normalization restricted to the two-layer case. Once we have defined the competitive ratio of a strategy for every chance node, we just take the maximum as the "overall" competitive ratio of the strategy, since the adversary has the free choice among the chance nodes. We adopt the following definition of competitive ratio for any chance node:

Definition 1. *ER is the expected ratio of the on-line cost and the cost of a best action for the same instance.*

This is justified particularly if the on-line player sees in retrospect what the instance was, and imagines that the adversary would have applied the cheapest action on this instance, and if the game is invariant under scaling, i.e. if multiplying all costs by any positive factor is an isomorphism of the labeled tree. Then ER measures the on-line player's expected regret, maximized over all adversarial choices.

In [5], the competitive ratio for the diffuse adversary has been defined differently. To distinguish this measure from ER we denote it by RE (ratio of expectations):

Definition 2. *RE is the ratio of the on-line player's expected costs and the expected cost of best actions for all instances.*

Rather obviously but amazingly, RE is the limit of ER when the same game is played many times and the total costs are compared. (Due to lack of space we state this only informally here.) Our interpretation of this fact is that an on-line player who aims at minimizing ER has a long-run view, maybe because she must permanently solve problems of this type, whereas an on-line player faced with a unique instance of a problem should prefer ER. In this sense, ER seems to be more fundamental. Both $ER \leq RE$ and $RE \leq ER$ is possible. We always have to state which version of the competitive ratio a result refers to.

Finally, we may extend our model towards comparative analysis and relax the assumption of an omniscient adversary, assuming that she also gets a signal, but a more informative one as the on-line player. (A formal definition is straightforward.) This allows to evaluate the power of certain parts of information.

1.3   Contributions and Related Work

We will discuss several natural versions of on-line scheduling of search procedures, depending on the power of on-line player and adversary. According to our general assumption that both players have the same actions available at the same costs, the adversary is always supposed at least to verify the target.

For the sake of a presentation being consistent in itself we use some notation that may deviate from standard terminology (like ER, RE defined above, denotation of job parameters etc.). Let n be the number of objects, one of them being the target. The amount of time to recognize whether the ith object is the target is called the *decision time* of this object/job and is denoted c_i.

In order to get in easier, we will first consider simple models where the jobs have target probabilities p_i, where $\sum_{i=1}^{n} p_i = 1$. It will then turn out that the main results carry over to the AIRS setting where the on-line player gets a random signal correlated to the adversarial target. We also briefly consider the use of partially known decision times of objects.

The study of on-line scheduling was initiated in article [7] which also contains some results for bounded execution times of jobs. For more recent results and further pointers we refer e.g. to [2]. The versions considered in the present paper are static in that no further jobs appear over time. Only parameters of the initially given jobs are unknown. Scheduling against randomized adveraries is also addressed in [6], but there the execution times are random variables, while the objective functions are the classical ones. Thus there is no overlap with our work. Geometric search problems such as in [1] are of similar nature, but differ in the costs of "preemptions", i.e. changing directions of search.

2   Models with Target Probabilities

2.1   Minimizing the Expected Completion Time

An obvious goal for any player who knows the p_i and c_i (but not the actual target) is to minimize the expected completion time of the target job. Define $d_i = c_i/p_i$. W.l.o.g. let the jobs be sorted by non-decreasing d_i. Then the optimal solution is simply to execute the jobs one-by-one in this ordering (known as Smith's Rule). Define $e_i = d_i - d_{i-1}$ (where $d_0 := 0$). Then the expected target completion time in the optimal solution is

$$OPT = \sum_j p_j \sum_{i \leq j} c_i = \sum_{i \leq j} p_i p_j d_i = \sum_{k \leq i \leq j} p_i p_j e_k.$$

Now we start comparing different players. First consider an adversary who selects the p_i and c_i but not the target. This adversary has to use Smith's Rule to minimize her expected costs, such that OPT becomes the denominator in RE.

Suppose that the on-line player also knows the p_i but not the c_i. For the special case $p_i = 1/n$, [7] have shown that the simple Round Robin algorithm that devotes equal time fractions to all jobs not finished yet gives the optimal competitive ratio $2 - \frac{2}{n+1}$. (The result in [7] refers to minimizing the sum of completion times which is obviously an equivalent formulation.) We generalize the asymptotic competitive ratio 2 to arbitrary "weights" p_i, by a strategy that we call Weighted Round Robin (WRR): Assign fractions of time proportional to the p_i to the jobs not completed yet.

Theorem 1. *WRR has competitive ratio $RE < 2$ against an adversary who knows the target probabilities and decision times but not the target.*

Proof. From the definition of the d_i, and $d_1 \leq \ldots \leq d_n$, it follows immediately that WRR completes the jobs in the ordering of indices. By this observation we can calculate the expected completion time WRR of the target job:

$$WRR = \sum_j p_j \sum_{k \leq j} e_k \sum_{i \geq k} p_i = \sum_{k \geq j} p_j e_k (1 - \sum_{i < k} p_i) = \sum_{k \geq j} p_j e_k - \sum_{i < k \leq j} p_i p_j e_k,$$

which finally gives $WRR = \sum_{k \leq i,j} p_i p_j e_k$.
For any fixed k we have $\sum_{i,j \geq k} p_i p_j e_k = 2 \sum_{j \geq i \geq k} p_i p_j e_k - \sum_{i \geq k} p_i^2 e_k$.
It follows $WRR = 2 \cdot OPT - \sum_i p_i^2 d_i = 2 \cdot OPT - \sum_i p_i c_i$.   □

In contrast to RE, one can only get poor ER:

Theorem 2. *WRR achieves $ER \leq n$, and any deterministic strategy has competitive ratio $ER \geq n/2$.*

Proof. The upper bound is trivial. For the lower bound, consider instances with $p_i = 1/n$ and c_i rapidly increasing with i. If job i is the target then the adversary needs slightly more than c_i time. Since the on-line player does not know the c_i, we may assume instead that she is told the fact $c_1 < \ldots < c_n$, but not the numbering of jobs. Given any deterministic on-line strategy, we construct an instance that fools it: At an arbitrary moment, the strategy has devoted at most $1/n$ of the time to some job. Decide that this is job 1, and let c_1 be the time the strategy has spent on it. Then wait much longer than c_1 time units. The on-line player has devoted at most $1/(n-1)$ of the time to some of the remaining jobs. Define c_2 similarly as above, and so on. Hence the on-line player finds target i not earlier than at time $(n - i + 1)c_i$. The ratio averaged over all targets is roughly $n/2$.   □

2.2   Online Completion Time vs. Target Decision Time

From now on we consider an omniscient adversary who even knows the target, corresponding to the "retrospective regret" discussed in the Introduction. That means, the adversary only needs time c_i if the ith object is the target. An on-line strategy that completes the ith job at time T_i achieves $ER = \sum_i p_i T_i / c_i$ and $RE = \sum_i p_i T_i / \sum_i p_i c_i$. Now we are going to study on-line players with different levels of prior knowledge.

First suppose that the on-line player knows the p_i and c_i. Then a one-by-one schedule where the jobs are ordered by non-decreasing c_i^2 / p_i is optimal with respect to ER. (This follows immediately from Smith's Rule if we look at the weighted sum of completion times with weights p_i / c_i.) On the other hand, the objective to minimize the expected target completion time, and thus RE, requires to sort the jobs by non-decreasing c_i / p_i.

However, we assume in the remainder of this section that the on-line player is ignorant of the c_i (but still knows the p_i). Then she has to start with a WRR type strategy, otherwise the competitive ratio can be arbitrarily large. W.l.o.g., the weights in a deterministic strategy remain constant in time, as long as no candidate has been discarded.

We restrict attention to fixed-weight strategies where the on-line player ignores the fact that some non-targets may have already expired. Any such strategy is completely characterized by the constant time fraction x_i devoted to the ith candidate $(i = 1, \ldots, n)$. Since $T_i = c_i/x_i$, we have $ER = \sum_{i=1}^n p_i/x_i$ and $RE = (\sum_{i=1}^n p_i c_i/x_i)/(\sum_{i=1}^n p_i c_i)$.

Moreover, fixed-weight strategies are not a proper restriction in the following problem version: The target j is chosen according to the probabilities p_j, and the adversary fixes the c_i conditional on the target. Trivially, it is always good for the adversary to make the non-target decision times much larger than c_j; we may simply consider them as infinite. This version is an appropriate model in cases where the on-line player has a verification criterion for the target, but no falsification criteria for non-targets. We aim at a minimum competitive ratio ER. (Recall the single-instance motivation.)

Lemma 1. *Given positive numbers $p_1, \ldots, p_n$, the $x_1, \ldots, x_n$ that minimize expression $\sum_{i=1}^n p_i/x_i$ under the constraint $\sum_{i=1}^n x_i = X$, for some fixed X, are $x_i = X\sqrt{p_i}/\sum_{i=1}^n \sqrt{p_i}$, which results in $\sum_{i=1}^n p_i/x_i = (\sum_{i=1}^n \sqrt{p_i})^2/X$.*

Proof. For any positive constants p, q, a, the function $p/x + q/(a - x)$ attains its minimum if $x^2/(a - x)^2 = p/q$. It follows easily that our objective function is minimized if the x_i are proportional to the $\sqrt{p_i}$. □

Since we have $X = 1$ in our problem, the on-line player has to choose $x_i = \sqrt{p_i}/\sum_{i=1}^n \sqrt{p_i}$ to get the best ER which equals $U := (\sum_{i=1}^n \sqrt{p_i})^2$. We call this strategy ROOT, and we call U the *uniformity* of distribution $p_1, \ldots, p_n$. It satisfies $1 \le U \le n$, with $U = 1$ iff some $p_i = 1$, and $U = n$ iff all p_i are equal. U plays a similar role in our search problem as entropy in searching by comparisons. (Due to space limitations we only mention this analogy as a hint.)

Theorem 3. *ROOT satisfies $ER = U$ against an off-line player who knows the target, and this is the optimal competitive ratio for fixed-weight strategies.* □

The result suggests to extend ROOT straightforwardly to the model where the adversary has to fix the c_i before the target is selected: Always keep the x_i proportional to the $\sqrt{p_i}$, such that $\sum x_i = 1$, where the sum is taken over all jobs not finished yet. Since the fixed-weight strategy cannot be better, we get $ER \le U$.

2.3   Monotone But Unknown Target Probabilities

In all the previous considerations, the on-line player was aware of the p_i. If, opposed to this, she is completely ignorant of the p_i and c_i then RR with equal fractions is the optimal strategy. If she knows the c_i but not the p_i then her best strategy against any adversary is to process the jobs one-by-one in order of non-decreasing c_i. Due to lack of space we omit the straighforward proofs. In either case, the competitive ratio can be up to n.

Interestingly, the on-line player can do better if the ordering of the p_i is known, though not their values. W.l.o.g. assume that she knows $p_1 \geq \ldots \geq p_n$. A motivation for this case is given later, prior to Corollary 1.

The on-line player is in the worst position if she does not know the c_i, and the adversary even knows the target. Then the following strategy, called HAR-MONIC, performs well. As usual, $H_n = \sum_{i=1}^{n} 1/i$ denotes the n-th harmonic number. We fix the fractions of a round-robin schedule by $x_i = 1/(iH_n)$. The idea is simply to prefer more likely targets. (We remark that HARMONIC can be approximately realized in a nice incremental way: In rounds $r = 1, 2, 3, \ldots$, a time slot is devoted to all divisors of r.) Define $I := \sum_{i=1}^{n} ip_i$, which is the expected index of the target if objects are sorted by non-increasing target probability. HARMONIC achieves $ER \leq \sum_{i=1}^{n} p_i/x_i = \sum_{i=1}^{n} p_i iH_n = H_n I$. The expected index I is also related to the uniformity U we defined earlier:

Lemma 2. *For any $p_1 \geq \ldots \geq p_n > 0$ we have:*
$2 \sum_{i=1}^{n} ip_i - \sum_{i=1}^{n} p_i \leq (\sum_{i=1}^{n} \sqrt{p_i})^2.$

Proof. By monotonicity, it is $\sqrt{p_i p_j} \geq p_i$ for $j \leq i$, and $\sqrt{p_i p_j}$ appears twice in the right hand side if $i \neq j$ and once if $i = j$. From these observations we obtain $(\sum_{i=1}^{n} \sqrt{p_i})^2 \geq \sum_{i=1}^{n} (2i - 1)p_i.$ □

It follows $2I - 1 \leq U$. Together with the preceding calculation this shows the following result which is within an $O(\log n)$ factor of the lower bound U for a stronger on-line player.

Theorem 4. *For monotone p_i, HARMONIC achieves competitive ratio $ER \leq H_n I < H_n(U + 1)/2 = O(U \log n)$ against an adversary who knows the target.*
□

3   Targets That Give Random Signals

3.1   Adversarial Instances and Random Signals (AIRS)

Now we consider the scenario where the target is freely chosen in the adversarial root of the tree. If object t is the target then, in the chance node for t, signal s is forwarded to the on-line player with probability $p(s|t)$ (where $\sum_s p(s|t) = 1$ for every t). We restrict our presentation to the case of finite sets of targets and signals, although extensions are possible.

Here we consider only the scheduling problem version with the powerful adversary who also fixes decision time c_t along with the target. For the same reasons as discussed earlier, the adversary would make all non-target decision times infinite, and the on-line player can restrict herself to fixed-weight Round Robin strategies.

Let T_{st} be the time that the on-line player needs if the target is t and signal s is received. (Note that costs of every action depend on t only, but the on-line strategy may depend on s, hence the completion time depends on both.) We defined the competitive ratio ER as $\max_t \sum_s p(s|t)T_{st}/c_t$. Note that T_{st}/c_t is the on-line player's multiplicative regret for the particular state (t, s). Since the chance nodes below the root correspond to the targets (i.e. instances), we have $RE = ER$.

Let $x(t|s)$ be the time fraction devoted by the on-line strategy to object t if the signal was s. Since $T_{st} = c_t/x(t|s)$ we get $ER = \max_t \sum_s p(s|t)/x(t|s)$. We wish to fix the $x(t|s)$ so as to minimize ER, under the constraints $\forall s : \sum_t x(t|s) = 1$.

3.2   The Optimal Solution

Lemma 3. *In an optimal solution to the above scheduling problem, all terms $\sum_s p(s|t)/x(t|s)$ are equal.*

Proof. Consider any solution, some s, and different targets t, t'. If we keep $x(t|s) + x(t'|s)$ constant but change the summands, we can lower the sum for one target and raise it arbitrarily for the other target. After a sequence of such changes we can raise the sums for all targets except one to the current maximum. Finally raise the unique smallest sum and let the others equally decrease, until all sums are equal. Details are straightforward.   □

Lemma 3 and the following conclusion can be generalized to any AIRS problems where, in the set of available strategies, costs can be continuously exchanged between targets in this way.

Lemma 4. *Let t_1 be one of the target objects. Minimizing $\sum_s p(s|t_1)/x(t_1|s)$, under the original constraints $\forall s : \sum_t x(t|s) = 1$ and the further constraints $\forall t : \sum_s p(s|t)/x(t|s) = \sum_s p(s|t_1)/x(t_1|s)$, is equivalent to the problem of minimizing ER.*

Proof. Since any feasible solution to the new problem has to satisfy the original constraints, it is also a feasible solution to the original problem. Conversely, by Lemma 3, an optimal solution to the original problem is among the feasible solutions to the new problem.   □

Since the new objective function from Lemma 4 is differentiable, we can apply Lagrange's method. The Lagrange function is

$$F = \sum_s \frac{p(s|t_1)}{x(t_1|s)} + \sum_s \lambda_s \left(\sum_t x(t|s) - 1 \right) + \sum_{t \neq t_1} \mu_t \left(\sum_s \frac{p(s|t)}{x(t|s)} - \sum_s \frac{p(s|t_1)}{x(t_1|s)} \right).$$

Defining an extra factor $\mu_{t_1} := 1 - \sum_{t \neq t_1} \mu_t$, we can write this more compactly:

$$F = \sum_s \lambda_s \left(\sum_t x(t|s) - 1 \right) + \sum_t \mu_t \sum_s \frac{p(s|t)}{x(t|s)},$$

which also takes away the artificial distinction of one target, at cost of an additional constraint $\sum_t \mu_t = 1$. The derivatives are

$$\frac{\partial F}{\partial x(t|s)} = \lambda_s - \mu_t p(s|t)/x(t|s)^2.$$

For a system of $x(t|s)$ to be an extremum, all these derivatives must be 0. This yields $x(t|s)^2 = p(s|t)\mu_t/\lambda_s$.

The signal-wise constraints $\sum_t x(t|s) = 1$ imply $\sqrt{\lambda_s} = \sum_t \sqrt{\mu_t}\sqrt{p(s|t)}$. Using abbreviation $w_t := \sqrt{\mu_t}$, we may now write the time fractions of the on-line schedule as $x(t|s) = w_t\sqrt{p(s|t)}/\sum_r w_r\sqrt{p(s|r)}$. Thus, in an optimal solution, the $x(t|s)$ for any fixed s are proportional to weighted $\sqrt{p(s|t)}$, where the weights w_t depend on the target only, in other words, they are equal for all signals. The weights must satisfy $\sum_t w_t^2 = 1$. Using the introduced notion, the competitive ratio is

$$ER = \frac{1}{n}\sum_{s,t} p(s|t)/x(t|s) = \frac{1}{n}\sum_s \left(\sum_t \sqrt{p(s|t)}/w_t \cdot \sum_r w_r\sqrt{p(s|r)}\right).$$

Let us minimize the inner product for any fixed s, for the moment ignoring the demand that the weights have to be independent of s. By the Cauchy-Schwarz inequality, this product is at least $(\sum_t \sqrt{p(s|t)})^2$. On the other hand, this lower bound is achieved if all w_t are equal. Therefore the weights minimizing each summand of the outer sum are already independent of s, such that the whole expression is minimized if weights are equal. This shows:

Theorem 5. *Taking $x(t|s) = \sqrt{p(s|t)}/\sum_r \sqrt{p(s|r)}$ gives the best ER which is* $\frac{1}{n}\sum_s(\sum_t \sqrt{p(s|t)})^2$. $\qquad\square$

This resembles the ROOT strategy from Theorem 3. The terms defined by $U(s) := (\sum_t \sqrt{p(s|t)})^2$ can be interpreted as uniformities of target likelihoods for a given signal, although the $p(s|t)$ for given s do not sum up to 1. Then the competitive ratio is the average uniformity.

3.3   Totally Ordered Targets and One-Sided Signals

We apply the result to the special problem described in the introduction.

If we do not distinguish signals between the same consecutive targets, that is, if we only use n signals corresponding to the $n-1$ intervals between targets and the infinite leftmost interval, we may enumerate the signals and targets $1, \ldots, n$ from the left to the right. For convenience we use denotations $p(s|t)$ and $x(t|s)$ with numbers s, t. We define $L_i := \sqrt{e^{-(t_i - t_{i-1})}}$, where $t_0 = -\infty$, and the other t_i are the target points. Thus, for $s \le t$ we can write $p(s|t) = (1-L_s^2)(L_{s+1}\cdot\ldots\cdot L_t)^2$. (Let the product be 0 if $s > t$, and 1 if $s = t$.)

Applying the optimal strategy we get $x(t|s) = L_{s+1}\cdot\ldots\cdot L_t/\sum_r L_{s+1}\cdot\ldots\cdot L_r$. Straightforward calculations that we omit here show for a regular pattern of targets (with $L = L_i$ for all i):

Proposition 1. *If equidistant targets are given then the competitive ratio is not larger than* $(1 + L)/(1 - L)$. $\square$

The above strategy remains nearly optimal if our information on the breakpoint distribution is somewhat inaccurate. However if the on-line player completely lacks such information, targets can no longer be mapped to specific points on the axis. Nevertheless, due to the linear ordering there still remains some a-priori knowledge, namely the fact that, for any s, the $p(s|t)$ decrease with t, whereas their values may be unknown! Thus we may still apply HARMONIC. Similarly as in Theorem 4 we get:

Corollary 1. *HARMONIC guarantees, for unknown $p(s|t)$ decreasing with t, competitive ratio $O(\max_t(\sum_s \sqrt{p(s|t)})^2 \log n)$.* $\square$

4   The Case of Known Bounds on Decision Times

We add one result where the on-line player knows intervals $[a_i, b_i] \ni c_i$, but in a simple non-probabilistic setting only: the adversary can freely choose a target i and the c_i in the given intervals, and the signal forwarded to the on-line player is constant.

We propose the following strategy called Interval Round Robin (IRR): Let t be a positive real parameter growing in time, and devote time to the jobs in such a way that the following is true at every moment and for every job i:

- If $t < a_i$ then the job has not been started yet.
- If $a_i \leq t \leq c_i$ then the time already spent on the job is t.

Note that this invariant implies that IRR proceeds exclusively with the ith job for a time a_i when t reaches a_i (and t remains constant during this period), and that IRR always assigns equal time fractions to all running jobs. If we visualize work on the t-axis, the real time runs by an integer factor faster than t grows, namely by the number of currently running jobs. These are the jobs with $[a_i, c_i] \ni t$. Moreover t may jump instantaneously from some b_i to the next larger a_j if there is no given interval in between.

Theorem 6. *IRR gives the optimal competitive ratio for every interval system $[a_i, b_i]$, against an adversary who can choose the decision times within the given intervals, and the target.*

Proof. For any choice of the c_j, the competitive ratio of IRR is $\max_j \sum_{b_i < c_j} b_i/c_j + \sum_{a_i \leq c_j \leq b_i} 1$. The adversary may adjust the c_j in such a way that this expression is maximized. (Note that the c_j are easy to compute: W.l.o.g. every c_j is one of the $a_i \in [a_j, b_j]$.)

We show that the resulting competitive ratio is optimal for deterministic on-line players, by an adversary strategy: Consider j and $c_j = a_i$ where our competitive ratio attains its maximum for the given interval system. Pre-select the jobs i with $b_i < c_j$ or $a_i \leq c_j \leq b_i$ as possible targets. Call them *short jobs* and *long jobs*, respectively. Given an on-line strategy, consider the first moment when the time spent on every long job has reached c_j. Call the long job that reached c_j last the *slow job*. If all short jobs have already reached their respective b_i then take the slow job as target. In the other case, if some short job did not reach its b_i yet then wait until the last short job did so, and then choose this short job as target. In both cases, the competitive ratio exceeds that of IRR.   □

It remains to study scheduling with bounded decision times in the other models. Searching for more than one target would be another natural generalization.

Acknowledgements. Thanks to: Tommy Martinsson (Clinical Genetics, Sahlgrenska University Hospital, Gothenburg) for vivid discussions about the biological side, to Olle Nerman (Mathematical Statistics, Chalmers) who brought us together, and to the referees who suggested some reorganization of the original manuscript making the motivation of the various models more obvious.

Conclusions of our results for application to the concrete data are omitted, as they are certainly not of theoretical interest.

References

1. R. Baeza-Yates, J. Culberson, G. Rawlins: Searching in the plane, *Info. and Comp.* 106 (1993), 234-252
2. L. Becchetti, S. Leonardi: Non-clairvoyant scheduling to minimize average flow time on single and parallel machines, *33rd ACM STOC'2001*, 94-103
3. A. Borodin, R. El-Yaniv: *Online Computation and Competitive Analysis*, Cambridge Univ. Press 1998
4. K. Ejeskär: Genetic alterations in Scandinavian neuroblastoma tumors, PhD thesis, Gothenburg University 2000
5. E. Koutsoupias, C. Papadimitriou: Beyond competitive analysis, *SIAM J. Computing* 30 (2000), 300-317
6. R.H. Möhring: Scheduling under uncertainty: Optimizing against a randomizing adversary, Technical Report 681-2000, Dept. of Mathematics, TU Berlin
7. R. Motwani, S. Phillips, E. Torng: Nonclairvoyant scheduling, *Theoretical Computer Science* 130 (1994), 17-47
8. J. Sgall: On-line scheduling – a survey, in *LNCS* 1442 (1997), 196-231

Removable Online Knapsack Problems

Kazuo Iwama* and Shiro Taketomi

School of Informatics, Kyoto University, Kyoto, Japan
{iwama, taketomi}@kuis.kyoto-u.ac.jp

Abstract. We introduce an on-line model for a class of hand-making games such as Rummy and Mah-Jang. An input is a sequence of items, $u_1, \cdots, u_i, \cdots$ such that $0 < |u_i| \leq 1.0$. When u_i is given, the on-line player puts it into the bin and can discard any selected items currently in the bin (including u_i) under the condition that the total size of the remaining items is at most one. The goal is to make this total size as close to 1.0 as possible when the game ends. We also discuss the multi-bin model, where the player can select a bin out of the k ones which u_i is put into. We prove tight bounds for the competitive ratio of this problem, both for $k = 1$ and $k \geq 2$.

1   Introduction

There is a class of popular games, such as Rummy, Mah-Jong, Scrabble, etc., where the goal of each player is to make a "good hand" by repeating the draw and discard of pieces of cards or tiles. In Rummy, for example, the player is given seven cards at the beginning and subsequently, in each round, he/she draws one card from the top of the stock and discards a most useless card from the hand. The goal is, for example, to develop a (consecutive) sequence of three or more cards of the same suit. After the play ends, each player is given a score (or a penalty) calculated by a scoring-function of the final hand.

In this paper we introduce a simple on-line model for this kind of games and prove its tight competitive ratio. Our model is as follows: An input is a sequence of items $u_1, u_2, \cdots, u_i, \cdots$ where each u_i has its size $|u_i|$ such that $0 < |u_i| \leq 1$. When u_i is given, the on-line player puts it into the bin and discards zero or more items currently in the bin (including u_i). The sum of the sizes of the items remaining in the bin must be at most 1.0 and our goal is to make it as close to 1.0 as possible when the game ends. As a simple example, suppose that items are given as 0.5, 0.6, 0.5 in this order. Then the player, for example, can keep the first 0.5, replace it by the second 0.6 and discard the third 0.5, which results in 0.6 as the final cost. However, the better action is to keep 0.5, discards 0.6 and keep 0.5, which achieves 1.0, as the final result. Thus the previous action can only achieve the competitive ratio (the optimal cost over the cost achieved by the algorithm) of $\frac{1.0}{0.6} \approx 1.66$, at best.

We also consider the multi-bin model, where the player can use at most k bins. Namely, in each round, he/she can put u_i into one of the k bins and can

---

* supported in part by Scientific Research Grant, Ministry of Japan, No. 13480081.

P. Widmayer et al. (Eds.): ICALP 2002, LNCS 2380, pp. 293–305, 2002.

discard arbitrary items currently existing in the k bins. The goal is to make the cost of the best bin as close to 1.0 as possible. (One can think of a Rummy hand including two unrelated sequences of length two. The player wants to extend either of the sequences by keeping both.) We call our problem k-*bin removable on-line knapsack problem* (k-*bin ROK*).

Related Results. The most important feature of our current model is to be able to discard items once put into the bin. If we prohibit this action, then the problem becomes the on-line knapsack problem [6,7]. Now the player can only choose whether or not the current item is put into the bin and easily gets in trouble if he/she receives an input such as $0.01, 0.01, \cdots$. If the player keeps not taking these items at all, then the competitive ratio gets worse and worse. If the player takes one item at any moment, then the adversary gives an item of size 1.0. Thus there is no competitive algorithm for this model [7]. In [6,7], this difficulty was overcome using stochastic models, where they obtained almost optimal algorithms. For our new model, the previous input is no harm, either, since the player can discard items in the bin when he/she receives the item of size 1.0.

Another related model is called on-line bin-packing, where the player has to keep every item in some bin of size one. The goal is to minimize the number of nonempty bins. Apparently this problem is closely related to the approximate bin-packing which has a long history of research. (Typical approximation algorithms, such as Best-Fit [5], are also on-line algorithms and therefore competitive ratio is equal to approximation ratio.) Unfortunately there still remains a gap: the best upper and lower bounds of the competitive ratio are 1.589 [8] and 1.540 [10], respectively. For the monotone on-line bin-packing, where items are given in the decreasing order, those are 1.222 [5] and 1.143 [1], respectively. For other related models, see [2,9].

Recall that our model can use two or more bins, which can be viewed as a relaxation of the restriction against the on-line player by allowing him/her to hold multiple solutions. This notion was first introduced in [3] for the on-line independent-set problem. [4] shows that if the player can hold up to a polynomial number of solutions, then the competitive ratio for the same problem is $\theta(\frac{n}{\log n})$, improved from $n - 1$ for the single-solution model.

Our Results. We prove tight bounds for the competitive ratios of 1-bin ROK and k-bin ROK for $k \geq 2$: (i) There exist on-line, polynomial time algorithms for 1-bin ROK and 2-bin ROK which achieve ratios of less than $\frac{\sqrt{5}+1}{2}(\approx 1.618)$ and $\alpha(\approx 1.3815)$, respectively, where α is one of the roots of equation $2x^3 + 4x^2 - 5x - 6 = 0$. (ii) Any algorithms for 1-bin ROK and k-bin ROK, $k \geq 2$, have competitive ratios of at least $\frac{\sqrt{5}+1}{2} - \varepsilon$ and $\alpha - \varepsilon$, for any positive constant ε, respectively. Note that the algorithm for 2-bin ROK obviously works for k-bin ROK, $k \geq 3$, with the same upper bound of competitive ratio and the off-line optimal cost for k-bin ROK is the same for all $k \geq 1$. Therefore our results show that using two bins is better than one bin but three or more bins does not help.

In what follows, we give formal definitions of the problems in Section 2, upper and lower bounds for 1-bin ROK in Section 3, and upper and lower bounds for

k-bin ROK, $k \geq 2$, in Section 4. Finally in Section 5, we briefly state possible extensions of the model.

2   Definitions

An *instance* σ of k-bin ROK is a sequence $u_1, u_2, \cdots, u_i, \cdots, u_n$ of *items* such that $0 < |u_i| \leq 1$, where $|u_i|$ is the *size of item* u_i. In each round i, u_i is given and the on-line player has to decide (i) which bin out of the k bins, $B_1, B_2, \cdots, B_k$, u_i is put into and (ii) which (zero or more) items among the items currently in the bins (including u_i) are discarded so that the total size, denoted by $|B_j|$, of items in B_j will be at most 1.0 for all $1 \leq j \leq k$. We often call $|B_j|$ the *size of bin* B_j.

Let $\mathcal{A}$ be an algorithm for k-bin ROK. For an instance σ, $|\mathcal{A}(\sigma)|$ denotes the *cost* achieved by $\mathcal{A}$, which is defined as the largest cost of the bins (i.e., $\max\{|B_1|, \cdots, |B_k|\}$) after the final round for input σ is completed. $|\text{OPT}(\sigma)|$ is the cost achieved by the off-line optimal algorithm. $\frac{|\text{OPT}(\sigma)|}{|\mathcal{A}(\sigma)|}$ is called the *competitive ratio* (*CR*) of algorithm $\mathcal{A}$ for input σ, and its worst-case value, i.e., $CR(\mathcal{A}) = \max\limits_{\sigma} \frac{|\text{OPT}(\sigma)|}{|\mathcal{A}(\sigma)|}$, is called CR of $\mathcal{A}$.

3   1-bin ROK

Let $t = \frac{\sqrt{5}-1}{2} (\approx 0.618)$ be one of the roots of equation $\frac{x}{1-x} = \frac{1}{x}$ and let $\alpha = \frac{1}{t} (\approx 1.618)$. In this section, we use only one bin, so B_1 is simply denoted by B. We first prove the upper bound, i.e., we give an optimal on-line algorithm for 1-bin ROK. Our basic idea is quite simple: (i) If the current item can fit the bin without discarding any existing item, then we take it. (ii) Otherwise, we maintain the cost of the bin so that it will be the *lowest* possible as long as the current CR (= the CR supposing that the input ends with the current item) is below our goal CR, i.e., α. Recall the simple example given in the previous section. When the on-line player receives 0.5 and 0.6, the current optimal cost is 0.6. Since $0.6/0.5 = 1.2 < \alpha$, the player's action should have been not to take 0.6 but to keep 0.5. This basic strategy does not change in the k-bin case, although its implementation is much more complicated.

Theorem 1. There exists an algorithm $\mathcal{A}$ for 1-bin ROK such that $CR(\mathcal{A}) < \alpha$.

Proof. To describe the algorithm $\mathcal{A}$, we need the following classification on the size of an item. An item u is said to be in class S, M, L, and X if $0 < |u| < 2t-1$, $2t - 1 \leq |u| \leq 1 - t$, $1 - t < |u| \leq t$, and $t < |u| \leq 1$, respectively (see Fig. 1). An item in S (M, L, X, respectively) is often written as s (m, ℓ, x, respectively). Suppose that an item u is given in the ith round. B denotes the set of items held by the bin at the beginning of this round and $|B|$ its cost. It should be noted that if $|B| > t$ then $\mathcal{A}$ can achieve the CR less than α by discarding all the following items since the optimal cost is at most 1.0, namely, the game is over. $\mathcal{A}$ takes the following action in each round:

if $|B| > t$
then $\mathcal{A}$ just discards u. – (1)
else if $|u| + |B| \le 1$
  then $\mathcal{A}$ puts u into the bin and discards nothing. – (2)
  else ($|B| \le t$ and $|u| + |B| > 1$ and hence $|u| > 1 - t$, i.e., $u \in L$ or $u \in X$)
    if $u \in X$
    then holds u and discards all the others. – (3)
    else (i.e., $u \in L$)
      if B includes only s's and m's
      then holds u and discards s's and m's, one by one, in an arbitrary
        order until the size of the bin first becomes 1.0 or less. – (4)
      else (As shown below, B contains a single ℓ and zero or more s's)
        if $|u| + |\ell| \le 1$
        then makes the bin of only u and ℓ. – (5)
        else if $|u| < |\ell|$
          then holds u and discards ℓ, else discards u. – (6)

Now we prove that $\frac{|\text{OPT}(\sigma)|}{|\mathcal{A}(\sigma)|} < \alpha$ for any input σ by induction. When $|\sigma| = 1$, then the single item u must be in B by action (2) above. Therefore the cost ratio is obviously 1.0. Let $|B|$ and $|\text{OPT}|$ be the costs before round i achieved by $\mathcal{A}$ and by the optimal algorithm, respectively, and $|B'|$ and $|\text{OPT}'|$ be the similar costs after round i. What we have to do is to prove that if $\frac{|\text{OPT}|}{|B|} < \alpha$, then $\frac{|\text{OPT}'|}{|B'|} < \alpha$.

Suppose that action (1) or (3) is taken by $\mathcal{A}$ in round i. Then we have $|B'| > t$. Since $|\text{OPT}'| \le 1$, $\frac{|\text{OPT}'|}{|B'|} < \frac{1}{t} = \alpha$. It is also easy when action (2) is taken, since $\frac{|\text{OPT}'|}{|B'|} \le \frac{|\text{OPT}|+|u|}{|B|+|u|} < \frac{|\text{OPT}|}{|B|} < \alpha$. Suppose that action (4) is taken. Let $B = \{b_1, b_2, \cdots, b_m\}$ and suppose that $b_1, b_2, \cdots, b_m$ are discarded in this order. Then there must be $1 \le i \le m$ such that $|b_i| + |b_{i+1}| + \cdots + |b_m| + |u| > 1$ and $|b_{i+1}| + \cdots + |b_m| + |u| = |B'| \le 1$. Since all b_i is in S or M, $|B'| > 1 - |b_i| \ge t$. If action (5) occurs, then we have $|u| > 1 - t$ and $|\ell| > 1 - t$. Consequently, $|B'| = |u| + |\ell| > 2(1 - t) = 3 - \sqrt{5} > \frac{\sqrt{5}-1}{2} = t$.

Our analysis is a bit more complicated if action (6) occurs. If action (6) occurs, the following (i) through (vi) hold: (i) The actions taken by $\mathcal{A}$ so far are only (2) or (6). (Reason: If (1),(3),(4) or (5) is taken, the cost exceeds t at that moment and action (1) is always taken in later rounds.) (ii) All s's given so far are in B. (If the size of the bin is at most t, then s can always enter the bin by

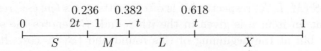

Fig. 1. Classification of item u.

(2).) (iii) No m has come so far. (Otherwise, these (one or more) m's must be in B for the same reason as above. Now consider the earliest round (maybe this round) in which ℓ is first given. Since $t < |m| + |\ell| \leq 1.0$, if (2) happens then (1) always happens later and otherwise (4) must happen.)

Therefore the player has so far received some s's $(s_1, s_2, \cdots, s_h)$ and some ℓ's $(\ell_1, \ell_2, \cdots, \ell_k = u$, in this order). (iv) The first ℓ_1 must enter B by action (2) due to (i) above and because action (4) did not happen in that round. Hence the number of ℓ's $(= k)$ is at least two. (v) After ℓ_i is given, B contains all s's given so far and a single ℓ which is the *smallest* one in $\{\ell_1, \cdots, \ell_i\}$ due to (i), (iv) above and action (6), which is a key point of $\mathcal{A}$ as mentioned at the beginning of this section.. (vi) Let ℓ_{min} and ℓ'_{min} be the smallest and the second smallest items in $\{\ell_1, \cdots, \ell_k\}$, respectively. Then $|\ell_{min}| + |\ell'_{min}| > 1$. (If ℓ_{min} and ℓ'_{min} have arrived in this order, then B contains ℓ_{min} when ℓ'_{min} arrived by (v). If $|\ell_{min}| + |\ell'_{min}| \leq 1$, then action (5) must have happened. Similarly for the opposite order.)

Now let ℓ_{max} be the largest one in $\{\ell_1, \cdots, \ell_k\}$. Because of (vi), no two items in $\{\ell_1, \cdots, \ell_k\}$ can enter B, which means $|\text{OPT}'| \leq |s_1| + \cdots + |s_h| + |\ell_{max}|$. On the other hand, $|B'| = |s_1| + \cdots + |s_h| + |\ell_{min}|$ due to (v). Hence, $\frac{|\text{OPT}'|}{|B'|} = \frac{|s_1| + \cdots + |s_h| + |\ell_{max}|}{|s_1| + \cdots + |s_h| + |\ell_{min}|} \leq \frac{|\ell_{max}|}{|\ell_{min}|} < \frac{t}{1-t} = \alpha$.

Thus no matter which action is taken in this round, $\frac{|\text{OPT}'|}{|B'|} < \alpha$.

We next show the lower bound.

Theorem 2. Let $\mathcal{A}$ be any on-line algorithm for 1-bin ROK. Then $CR(\mathcal{A}) > \alpha - \varepsilon$ for any $\varepsilon > 0$.

Proof. Let $t = \frac{1}{\alpha} = \frac{\sqrt{5}-1}{2}$ as before and the adversary gives two items u_1, u_2 such that $|u_1| = t + \varepsilon'$, for a small $\varepsilon' > 0$, and $|u_2| = 1 - t$. Since $|u_1| + |u_2| > 1$, either $|u_1|$ or $|u_2|$ can stay in B. If B contains u_2 then the adversary stops the input at this moment. Then the CR is $\frac{t+\varepsilon'}{1-t} > \alpha$. Otherwise, i.e., if B contains u_1, then the adversary gives u_3 such that $|u_3| = t$. The optimal cost is $1.0(= |u_2| + |u_3|)$ but what the player can do the best is to leave u_1 in the bin. Hence the CR is at least $\frac{1}{t+\varepsilon'}$. For any small $\varepsilon > 0$, we can select sufficiently small $\varepsilon' > 0$ such that $\frac{1}{t+\varepsilon'} > \frac{1}{t} - \varepsilon = \alpha - \varepsilon$.

4 k-bin ROK

Suppose that an item u is given in the ith round. Then the on-line player puts u into one of the k bins and discards an arbitrary (zero or more) number of items currently in the bins. Let α (≈ 1.3815) be one of the roots of equation $2x^3 + 4x^2 - 5x - 6 = 0$. (Note that we use the same α although the value is different from the previous section.) Let $t = 1/\alpha$ (≈ 0.724) as before.

4.1   Lower Bounds

We first prove the lower bound.

Theorem 3. Let $\mathcal{A}$ be any on-line algorithm for k-bin ROK, $k \geq 2$. Then $CR(\mathcal{A}) > \alpha - \varepsilon$ for any $\varepsilon > 0$.

Proof. The adversary uses items s, m, m' and ℓ of four different sizes, where $|s|, |m|$ and $|\ell|$ satisfy the following equations:

$$|s| + |m| + |\ell| = 1,$$

$$\frac{1}{|m| + |\ell|} = \frac{3|m|}{|s| + |\ell|} = \frac{2|\ell|}{|s| + |m|} = \alpha.$$

By a simple calculation, one can see that $|m|$ (≈ 0.3153) is one of the roots of equation $3x^3 + 16x^2 + x - 2 = 0$, $|\ell|$ (≈ 0.4085) is one of the roots of equation $2x^3 - 9x^2 - 4x + 3 = 0$, and $|s| = 1 - |m| - |\ell|$ (≈ 0.2762). Also, let $|m'| = |m| + \varepsilon'$ for a small constant $\varepsilon' > 0$. Although the on-line player (denoted by $\mathcal{A}$) can use an arbitrary number of bins, it is useless to use three or more bins as shown below.

Now the adversary first gives m' and ℓ. $\mathcal{A}$ can hold these items in different bins or in a single bin. ($\mathcal{A}$ can also discard one or both items but such actions are obviously useless since $\mathcal{A}$ can do so later.) In the former case, the adversary stops the input. Then the CR is $\frac{|m'| + |\ell|}{|\ell|} > \frac{0.31 + 0.40}{0.41} > 1.7 > \alpha$. (Note that $0.40 < |\ell| < 0.41$ and $|m'| > 0.31$.) Thus we can assume that $\mathcal{A}$ now holds m' and ℓ in a single bin.

The adversary then gives s. There are three possibilities for $\mathcal{A}$'s action, i.e., holding s in a different bin (Case 1), replacing m' by s (Case 2) and replacing ℓ by s (Case 3). Again it should be noted that $\mathcal{A}$ can take other actions such as (i) holding only one of s, m' and ℓ, (ii) just discarding s and (iii) holding s in the second bin and discarding m' or ℓ. However, the configurations of the bins after these actions are equal to the configurations which can be obtained by removing some items from the bins of Case 1. Since $\mathcal{A}$ can remove items later, it is obviously useless to take these actions.

The following actions of the adversary and of $\mathcal{A}$ are summarized in Fig. 2. In Case 1, for example, the adversary gives m as the fourth item. Since $|m| + |m'| + |\ell| > 1$ and $|m'| + |\ell| > |s| + |m|$, $\mathcal{A}$ can get at most $|m'| + |\ell|$ as its cost, while the optimal cost is $|s| + |m| + |\ell| = 1$. Thus the adversary stops the input at this moment and the CR is at least $\frac{1}{|m'| + |\ell|}$. Similarly for Cases 2 and 3: In Case 2, the next item by the adversary is m'. If $\mathcal{A}$ replaces ℓ by this m', the game ends and the $CR > \frac{3}{2}$. If $\mathcal{A}$ puts m' into the new bin, then the adversary gives another m'. Here one can see that it is useless to use the third bin to hold this m' since we already have the bin holding a single m'. Thus the whole information of the actions by the adversary and by $\mathcal{A}$ is included in Fig. 2; details may be omitted. One can see the CR is at least $\min\left\{ \frac{3}{2}, \frac{1}{|m'| + |\ell|}, \frac{3|m'|}{|s| + |\ell|}, \frac{2|\ell|}{|s| + |m'|} \right\}$, which becomes larger than $\alpha - \varepsilon$ by selecting sufficiently small ε'.

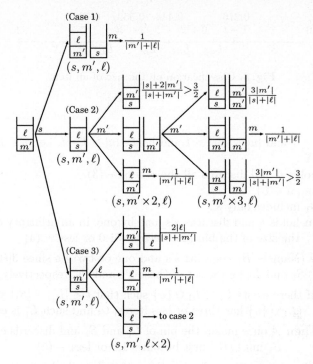

Fig. 2. The adversary for the lower bound.

4.2   Upper Bounds

Now we prove the upper bound, which is the most important result in this paper. Note that we can achieve the best possible bound (matching to the lower bound of Theorem 3) by using only two bins.

Theorem 4. *There is an algorithm $\mathcal{A}$ for 2-bin ROK such that $CR(\mathcal{A}) < \alpha$.*

Proof. We divide each item into six different classes $S, L_{SS}, L_{SL}, L_M, L_L$, and X. An item in each class is denoted by $s, \ell_{SS}, \ell_{SL}, \ell_M, \ell_L$, and x, respectively. Each item satisfies: $0 < |s| \leq 1 - t$, $1 - t < |\ell_{SS}| < 0.4$, $0.4 \leq |\ell_{SL}| < 2t - 1$, $2t - 1 \leq |\ell_M| \leq 2 - 2t$, $2 - 2t < |\ell_L| \leq t$, and $t < |x| \leq 1$ (see Fig. 3). L_S denotes $L_{SS} \cup L_{SL}$ and L denotes $L_S \cup L_M \cup L_L$. Our algorithm $\mathcal{A}$ uses two bins B_1 and B_2 and consists of two stages, Stage I and Stage II.

Stage I of the Algorithm. In Stage I, $\mathcal{A}$ uses only B_1, whose single round is described as follows:

if $|B_1| > t$
then $\mathcal{A}$ just discards u. – (1)
else if $|u| + |B_1| \leq 1$
   then puts u into B_1 and discards nothing. – (2)

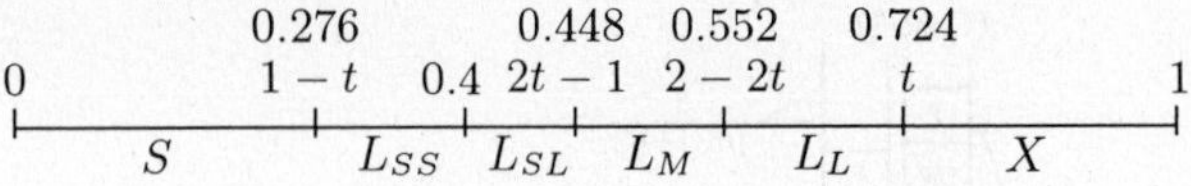

Fig. 3. Classification on the size of an item

else ($|B_1| \leq t$ and $|u| + |B_1| > 1$ and hence $|u| > 1 - t$, i.e., $u \in L$ or $u \in X$)
    if $u \in X$
    then holds u and discards all the others. – (3)
    else (i.e., $u \in L$)
        if B_1 includes only s's
        then holds u and discards s's, one by one, in an arbitrary order until
            the size of the bin first becomes 1.0 or less. – (4)

        else (Namely B_1 contains s's and one or two ℓ's since $3|\ell| > t$. Let
            S_1 and L_1 be the set of those s's and ℓ's, respectively)

        if there exists $L_1' \subseteq L_1 \cup \{u\}$ such that $t < |L_1'| + |S_1| \leq 1 + |S_1|$
            ($L_1 \cup \{u\}$ has three or less items, to find such L_1' is easy)

        then $\mathcal{A}$ once makes the bin of L_1' and S_1 and discards elements of
            S_1 until $|B_1|$ first becomes 1.0 or less. – (5)

        else goes to Stage II after the action described below. – (6)

Thus, actions (1) through (5) are similar to the previous ones for the 1-bin model; if B_1 already has enough cost, then just discards u (action (1)), just puts u into B_1 if possible (action (2)) and makes the bin of cost larger than t if possible (actions (3) through (5)). One can easily prove that the CR is less than α if one of these actions (other than (6)) is taken, using a similar argument as before. Now we describe action (6):

Action (6): We first summarize the situation before $\mathcal{A}$ executes this action: (i) $u \in L$, (ii) $|u| + |B_1| > 1$ and B_2 is empty, (iii) B_1 includes zero or more s's and one or two ℓ's (if B_1 includes two ℓ's, the both ℓ's are in L_S since $|B_1| \leq t$), and (iv) The items given so far are all in B_1 (i.e., only action (2) has been taken so far). Considering this situation, we are introducing six different cases: The first two cases apply when B_1 includes only one ℓ, i.e., Case 1 is for $|\ell| \geq |u|$ and Case 2 for $|\ell| < |u|$. The third case applies when B_1 includes two ℓ's ($\in L_S$) and the current u satisfies that $u \in L_L$. The remaining three cases are for B_1's including two ℓ's and $u \in L_S$. (u cannot be in L_M since if so, action (5) should have happened because $|u| + |\ell_S| > t$.) Let those three ℓ_S's (two in B_1 and u) be $\ell_{S1}, \ell_{S2}, \ell_{S3}$ such that $|\ell_{S1}| \leq |\ell_{S2}| \leq |\ell_{S3}|$. Then one can claim that for any two of those three ℓ_S's, the sum of their sizes is at most t (if more than t, the action (5) should have happened since the sum of the sizes is at most 1.0). Therefore at most one of these three ℓ_S's can be in L_{SL} and we introduce two cases, the case that no ℓ_S's are in L_{SL} (i.e., $\ell_{S3} \in L_{SS}$) and the case that $\ell_{S3} \in L_{SL}$. The latter case is further divided into two cases, $\frac{3(t-|\ell_{S3}|)}{1-|\ell_{S2}|} > \alpha$ and $\frac{3(t-|\ell_{S3}|)}{1-|\ell_{S2}|} \leq \alpha$, for technical reason.

$\mathcal{A}$'s action for each case, described later, consists of two actions; one for handling u and the other for moving to the initial state for Stage II. As given later, $\mathcal{A}$ uses 18 different *States*, State (1) through State (17) and a special state denoted by State F. If $\mathcal{A}$ is in State F, then it has already achieved the CR α, i.e., it has a bin of size more than t. Other states are denoted by (W_1, W_2, I). Here, $W_1(W_2)$ shows the items in $B_1(B_2)$ and I the items so far given. For example, State (13) is denoted by $(\{S, \ell_{S1}, \ell_{S2}\}, \phi; \{s, \ell_{SS}, \ell_{SL}^*\})$, namely, B_1 includes all the items in S so far given, ℓ_{S1} and ℓ_{S2} (described above). B_2 is empty. Furthermore, we can see from $I = \{s, \ell_{SS}, \ell_{SL}^*\}$ that $\mathcal{A}$ has so far received zero or more s's and ℓ_{SS}'s, and exactly one ℓ_{SL} ($*$ and $**$ show explicitly the number of items, one and two, respectively). Occasionally, we use such a description as ℓ_{Lmax}, which means the largest item among ℓ_L's so far given. Similarly for ℓ_{Lmin}. Also, each state includes "Condition" about I. For example, State (5) has $I = \{s, \ell_S^*, \ell_L\}$, which shows that we have so far received zero or more s's and ℓ_L's, and a single ℓ_S. Its Condition says that the sum of the size of the single ℓ_S and the size of any one of the ℓ_L's is more than 1.0. Note that this cannot be guaranteed simply by the definitions of L_S and L_L. However, we will show that Condition is always met when $\mathcal{A}$ enters that state. Also, as one can see later, (W_1, W_2, I) and Condition have enough information to calculate the CR of $\mathcal{A}$ at that moment.

Now we are ready to describe Cases 1-6:

Case 1. (B_1 includes a single ℓ and $|\ell| \geq |u|$) $\mathcal{A}$ puts u into B_2. As for the initial state for Stage II, there are four possibilities: Since $|\ell| \geq |u|$, we must have (i) $\ell, u \in L_L$, (ii) $\ell \in L_L$ and $u \in L_M$, (iii) $\ell \in L_L$ and $u \in L_S$, or (iv) $\ell, u \in L_M$. (No other possibilities for the following reason: If $\ell \in L_M$ and $u \in L_S$, for example, then their total size exceeds t and action (5) should have happened. Also, suppose that both ℓ and u are in L_S. Then $|\ell| + |u| < 1.0$ and since action (2) did not happen, we have $|u| + |B_1| = |u| + |\ell| + |S_1| > 1$. It then follows that $t < |\ell| + |u| + |S_1| < 1 + |S_1|$, which means that action (5) should have happened, too.) In the case (i), we have so far received some s's and exactly two items in L_L (i.e., the ℓ in B_1 and the current u). Those two ℓ_L's can be written as ℓ_{Lmax} and ℓ_{Lmin}, and $\mathcal{A}$'s initial state for Stage II is $(\{S, \ell_{Lmax}\}, \{\ell_{Lmin}\}; \{s, \ell_L\})$ (= State (1)). Similarly for (ii) to (iv). Case (ii): $\mathcal{A}$ moves to the initial state $(\{S, \ell_L\}, \{\ell_{Mmin}\}; \{s, \ell_M, \ell_L\})$ = State (8). Note that $\mathcal{A}$ has received only one item in L_M (= the current u) and therefore Condition of State (8) is met. Case (iii): $\mathcal{A}$ goes to $(\{S, \ell_{Lmin}\}, \{\ell_S\}; \{s, \ell_S^*, \ell_L\})$ = State (5). So far we have received one ℓ_L and one $\ell_S(= u)$. If the sum of their sizes is at most 1, then $\mathcal{A}$ should not have come to action (6), i.e., Condition of State (5) is met. Case (iv): $\mathcal{A}$ goes to $(\{S, \ell_{Mmax}\}, \{\ell_{Mmin}\}; \{s, \ell_M\})$ = State (7). We have received two ℓ_M's. For the same reason as Case (iii), Condition of State (7) is met.

Case 2. (B_1 includes a single ℓ and $|\ell| < |u|$) $\mathcal{A}$ puts u into B_2. We consider four different cases similar to Case 1 above: (i) $\ell, u \in L_L$, (ii) $\ell \in L_M$ and $u \in L_L$, (iii) $\ell \in L_S$ and $u \in L_L$, and (iv) $\ell, u \in L_M$. No other cases can happen for the same reason as before. $\mathcal{A}$'s initial states for Stage II corresponding to these (i) to

(iv) are States (4), (10), (2), and (9), respectively. Similar arguments as before can be done to show that Conditions for those initial states are met.

Case 3. (B_1 includes two ℓ_S's and $u \in L_L$) u is put into B_2. $\mathcal{A}$'s initial state for Stage II is State (3). We have received only two ℓ_S's so far. If u and one of those ℓ_S's can enter a single bin then $\mathcal{A}$ should not have come to action (6). Also, we can write $|B_1| = |\ell_{Smin}| + |\ell_{Smax}| + |S|$, which is at most t. Thus Condition of State (3) is met.

Case 4. ($\ell_{S3} \in L_{SS}$) $\mathcal{A}$ puts u into B_1 and discards ℓ_{S3} (maybe u) so that B_1 will contain all the s's, ℓ_{S1} and ℓ_{S2}. The sum of these items is at most t, since otherwise, $t < |\ell_{S1}| + |\ell_{S2}| + |S_1| \leq 1 + |S_1|$ holds and $\mathcal{A}$ should not have come to action (6). $\mathcal{A}$ goes to State (11). It turns out that $|\ell_{S1}| + |\ell_{S2}| + |\ell_{S3}| > 1.0$. (If the sum ≤ 1.0, action (5) should have happened since $|\ell_{S1}| + |\ell_{S2}| + |\ell_{S3}| > 3(1 - t) > 3 \cdot (1 - 0.774) > t$.) Also, as shown before, the size of no two items among those ℓ_{S1}, ℓ_{S2}, and ℓ_{S3} exceeds t. Finally, what $\mathcal{A}$ has discarded so far is only ℓ_{S3}. Thus Condition of State (11) is met.

Case 5. ($\ell_{S3} \in L_{SL}$ and $\frac{3(t-|\ell_{S3}|)}{1-|\ell_{S2}|} > \alpha$.) $\mathcal{A}$ puts u into B_1 and discards ℓ_{S3}. $\mathcal{A}$ goes to State (13), where Condition is met for the same reason as Case 4 and by the condition $\frac{3(t-|\ell_{S3}|)}{1-|\ell_{S2}|} > \alpha$.

Case 6. ($\ell_{S3} \in L_{SL}$ and $\frac{3(t-|\ell_{S3}|)}{1-|\ell_{S2}|} \leq \alpha$.) $\mathcal{A}$ puts u into B_1 and discards ℓ_{S2}. The initial state is State (16). Note that we have only two ℓ_{SS}'s (ℓ_{S1} and ℓ_{S2}) and only one ℓ_{SL} ($= \ell_{S3}$) so far. Thus Condition of State (16) is met.

Stage II of the Algorithm. Now $\mathcal{A}$ has finished action (6) and is moving to Stage II. As for the competitive ratio at this moment, one can refer to the description in each state given later. We shall then describe each round (an item u is given) of Stage II. If $\mathcal{A}$ is in State F, then $\mathcal{A}$ just discards u. Otherwise, i.e., if $\mathcal{A}$ is in State (i), $1 \leq i \leq 17$, then:

if $u \in X$
then $\mathcal{A}$ makes a bin containing only u and goes to State F. – (a)
else if $t < |u| + |B_1| \leq 1$ or $t < |u| + |B_2| \leq 1$
    then $\mathcal{A}$ puts u into B_1 or B_2 to make a bin of size more than t and goes
        to State F. – (b)
    else (Since B_1 includes at most two ℓ's (if three then their total size exceeds
        t), $B_1 \cup \{u\}$ includes at most three ℓ's. Also $B_1 \cup \{u\}$ may includes
        some s's. Similarly for B_2. Let S_1 (S_2, resp.) be the set of those s's in
        $B_1 \cup \{u\}$ ($B_2 \cup \{u\}$, resp.) and L_1 (L_2, resp.) be the set of those ℓ's in
        $B_1 \cup \{u\}$ ($B_2 \cup \{u\}$, resp.))
        if there is $L_1' \subseteq L_1$ ($L_2' \subseteq L_2$, resp.) such that $t < |L_1'| + |S_1| \leq 1 + |S_1|$
        ($t < |L_2'| + |S_2| \leq 1 + |S_2|$, resp.)
        then $\mathcal{A}$ once makes the bin of $L_1' \cup S_1$ ($L_2' \cup S_2$, resp.) and discards s's
            in an arbitrary order until its size first becomes 1.0 or less. Then
            $\mathcal{A}$ goes to State F. – (c)
        else $\mathcal{A}$ follows Action of State (i) – (d).

In what follows, we give the description of each state, which consists of (W_1, W_2, I), Condition, CR where we prove that the current competitive ratio when $\mathcal{A}$ comes to that state is less than α, and Action. Note that Action includes which state $\mathcal{A}$ goes to next, but if that is omitted then $\mathcal{A}$ stays in the same state. State transitions are summarized in Fig. 4. As one can see, for any state once $\mathcal{A}$ leaves that state then $\mathcal{A}$ never comes back there. We will occasionally evaluate the values $\frac{t}{2-2t}$, $4 - \frac{2}{t}$, $\frac{2-2t}{2t-1}$, and $\frac{8}{2-t} - 5$. Since $0.723 < t < 0.724$, these values are bounded as $\frac{t}{2-2t} < \frac{0.724}{0.552} < 1.32$, $4 - \frac{2}{t} < 4 - \frac{2}{0.724} < 1.24$, $\frac{2-2t}{2t-1} < \frac{0.554}{0.446} < 1.25$, and $\frac{8}{2-t} - 5 < \frac{8}{1.276} - 5 < 1.27$. Thus all the values are less than $\alpha(> 1.38)$. In the following, it is said that a bin is *good* if its cost is more than t.

State (1) $(\{S, \ell_{Lmax}\}, \{\ell_{Lmin}\}; \{s, \ell_L\})$

Condition: none

CR: No two ℓ_L's can enter one bin and all the items in S so far given are in B_1. Hence $CR = \frac{|\ell_{Lmax}| + |S|}{|\ell_{Lmax}| + |S|} = 1$.

Action: (i) (Recall that $\mathcal{A}$ did not take action (b) or (c).) If $u \in S$, then one can see that both $|B_1| + |u|$ and $|B_2| + |u|$ are at most t. In this case $\mathcal{A}$ puts u into B_1 (and stays in this state). (ii) If $u \in L_S$, $\mathcal{A}$ exchanges u and ℓ_{Lmax}, and moves to State (2). Since u is the first item in L_S, and u and ℓ_{Lmin} cannot enter one bin (otherwise, $\mathcal{A}$ could have made a good bin with u and ℓ_{Lmin}, which means that action (b) or (c) must have happened), Condition of State (2) is met. (iii) If $u \in L_M$, $\mathcal{A}$ exchanges u and ℓ_{Lmin}, and moves to State (8). Since u is the first item in L_M, Condition of State (8) is met. (iv) Suppose that $u \in L_L$. Then, if $|u| > |\ell_{Lmax}|$, then $\mathcal{A}$ exchanges u and ℓ_{Lmax}, else if $|u| < |\ell_{Lmin}|$, $\mathcal{A}$ exchanges u and ℓ_{Lmin}. Otherwise discards u. **Note:** The action in (iv) will often appear below and we will simply say that "$\mathcal{A}$ updates ℓ_{Lmax} or ℓ_{Lmin}."

State (2) $(\{S_1, \ell_S\}, \{S_2, \ell_{Lmin}\}; \{s, \ell_S^*, \ell_L\})$
S_1 denotes the set of s's which are given before $\mathcal{A}$ enters this state for the first time. Also, S_2 denotes the set of s's given after that.

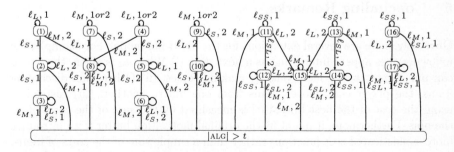

Fig. 4. The action in Stage II

Condition: The single ℓ_S and any one of ℓ_L's so far given cannot enter one bin. $|S_1| + |\ell_{Lmax}| \leq t$. (The second condition is obvious since a good bin should have been made otherwise.)

CR: Because of Condition, the optimal bin can include only one ℓ. Therefore $CR = \frac{|\ell_{Lmax}| + |S_1| + |S_2|}{|\ell_{Lmin}| + |S_2|} \leq \frac{|\ell_{Lmax}| + |S_1|}{|\ell_{Lmin}|} < \frac{t}{2-2t} < \alpha$

Action: (i) If $u \in S$, puts u into B_2. (ii) Again recall that $\mathcal{A}$ did not take action (b) or (c). So, if $u \in L_S$, we can put u into B_1. $\mathcal{A}$ does so and moves to State (3). Since u is the second item in L_S, we can write that $|B_1| + |u| = |\ell_{Smin}| + |\ell_{Smax}| + |S_1| \leq t$. Also, u and ℓ_{Lmin} cannot enter one bin. (Otherwise, $\mathcal{A}$ could have made a good bin with u and ℓ_{Lmin}, and action (b) or (c) should have happened.) For these reasons and by the condition of this state, Condition of State (3) is met. (iii) If $u \in L_M$, $\mathcal{A}$ could have made a good bin with u and ℓ_S (i.e., this cannot happen.) (iv) If $u \in L_L$, $\mathcal{A}$ updates ℓ_{Lmin}. **Note:** $\mathcal{A}$ stays in this state for (i) and (iv). Although we have to show that the current Condition is met after the actions in (i) and (iv), it is usually obvious and omitted besides a few exceptions like State (16).

State (3) ($\{S_1, \ell_{Smin1}, \ell_{Smin2}\}, \{S_2, \ell_{Lmin}\}; \{s, \ell_S, \ell_L\}$)
S_1 and S_2 are the same as in State (2). Let ℓ_{Smin1} and ℓ_{Smin2} be the smallest and the second smallest items in L_S so far given.
Condition: No ℓ_S and ℓ_L can enter one bin. No three ℓ_S's can enter one bin. $|\ell_{Smin}| + |\ell_{Smax}| + |S_1| \leq t$.
CR: Let $|\ell_{Smin1}| = x$ ($1 - t < x \leq \frac{t}{2}$). Due to the condition, $|\ell_{Lmin}| > 1 - x$ and $|\ell_{Smax}| \leq t - x - |S_1|$. $CR \leq \max\left\{ \frac{2|\ell_{Smax}| + |S_1| + |S_2|}{|\ell_{Lmin}| + |S_2|}, \frac{|\ell_{Lmax}| + |S_1| + |S_2|}{|\ell_{Lmin}| + |S_2|} \right\} \leq \max\left\{ \frac{2t - 2x - |S_1|}{1-x}, \frac{t + (t-2x)}{1-x} \right\} = \frac{2t - 2x}{1-x} = 2 - \frac{2-2t}{1-x} < 4 - \frac{2}{t} < \alpha$

Action: (i) If $u \in S$, puts u into B_2. (ii) If $u \in L_S$, $\mathcal{A}$ updates ℓ_{Smin2}. (iii) If $u \in L_M$, $\mathcal{A}$ should have made a good bin with u and ℓ_{Smin1}. (iv) If $u \in L_L$, $\mathcal{A}$ updates ℓ_{Lmin}.

Due to space limitation, State (4) - State (17) are omitted. See http://www.lab2.kuis.kyoto-u.ac.jp/ iwama/ICALP02.pdf. □

5   Concluding Remarks

Obviously there are several extensions for the multi-bin model. For example, the following ones appear interesting as a next step: (i) When u_i is given, the player can put it into two or more bins by making copies. (ii) Other than discarding items, the player can "move" items from one bin to another. (iii) Instead of using the cost of the best bin, $\mathcal{A}(\sigma)$ is measured by the *sum* of the costs of all the bins. Furthermore, it might fit the original motivation better if we introduce more sophisticated cost functions rather than a simple sum of the costs of items. An apparent first step along this line is to allow each item to have its own value (as suggested by the name of the problem). Average-case analysis will also be interesting.

References

1. J. Csirik, G. Galambos, and G. Turán, "A lower bound on on-line algorithms for decreasing lists," *Proc. EURO VI*, 1984.
2. J. Csirik, and G. Woeginger, "Resource argumentation for online bounded space bin packing," *Proc. ICALP 2000*, pp. 296-304, 2000.
3. M. M. Halldórsson, "Online coloring known graphs," *Electronic J. Combinatorics*, Vol.7, R7, 2000. www.combinatorics.org.
4. M. M. Halldórsson, K. Iwama, S. Miyazaki, and S. Taketomi, "Online independent sets," *Proc. COCOON 2000*, pp. 202-209, 2000.
5. D. S. Johnson, A. Demers, J. D. Ullman, M. R. Garey, and R. L. Graham, "Worst-case performance bounds for simple one-dimensional packing algorithms," *SIAM Journal on Computing*, Vol. 3(4), pp. 299-325, 1974.
6. G. S. Lueker, "Average-case analysis of off-line and on-line knapsack problems," *Proc. Sixth Annual ACM-SIAM SODA*, pp. 179-188, 1995.
7. A. Marchetti-Spaccamela and C. Vercellis, "Stochastic on-line knapsack problems," *Math. Programming*, Vol. 68(1, Ser. A), pp. 73-104, 1995.
8. S. S. Seiden, "On the online bin packing problem," *Proc. ICALP 2001*, pp. 237-248, 2001.
9. S. S. Seiden, "An optimal online algorithm for bounded space variable-sized bin packing," *Proc. ICALP 2000*, pp. 283-295, 2000.
10. A. van Vliet, "On the asymptotic worst case behavior of harmonic fit," *J. Algorithms*, Vol. 20, pp. 113-136, 1996.

New Bounds for Variable-Sized and Resource Augmented Online Bin Packing

Leah Epstein[1]*, Steve Seiden[2]**, and Rob van Stee[3]***

[1] School of Computer Science, The Interdisciplinary Center, Herzliya, Israel.
lea@idc.ac.il.
[2] Department of Computer Science, 298 Coates Hall, Louisiana State University,
Baton Rouge, LA 70803, U.S.A. sseiden@acm.org.
[3] Institut für Informatik, Albert-Ludwigs-Universität, Georges-Köhler-Allee, 79110
Freiburg, Germany. vanstee@informatik.uni-freiburg.de.

Abstract. In the variable-sized online bin packing problem, one has to assign items to bins one by one. The bins are drawn from some fixed set of sizes, and the goal is to minimize the sum of the sizes of the bins used. We present new algorithms for this problem and show upper bounds for them which improve on the best previous upper bounds. We also show the first general lower bounds for this problem. The case where bins of two sizes, 1 and $\alpha \in (0, 1)$, are used is studied in detail. This investigation leads us to the discovery of several interesting fractal-like curves. Our techniques are also applicable to the closely related resource augmented online bin packing problem, where we have also obtained the first general lower bounds.

1 Introduction

In this paper we investigate the bin packing problem, one of the oldest and most thoroughly studied problems in computer science [3,5]. The influence and importance of this problem are witnessed by the fact that it has spawned off whole areas of research, including the fields of online algorithms and approximation algorithms. In particular, we investigate a natural generalization of the classical online bin packing problem known as online variable-sized bin packing. We show improved upper bounds and the first lower bounds for this problem, and in the process encounter several strange fractal-like curves.

Problem Definition: In the *classical bin packing* problem, we receive a sequence σ of *pieces* $p_1, p_2, \ldots, p_N$. Each piece has a fixed *size* in $(0, 1]$. In a slight abuse of notation, we use p_i to indicate both the ith piece and its size. We have an infinite number of *bins* each with *capacity* 1. Each piece must be assigned to

* Research supported by Israel Science Foundation (grant no. 250/01).
** Research supported by the Louisiana Board of Regents Research Competitiveness Subprogram.
*** This work done while the author was at the CWI, The Netherlands. Research supported by the Netherlands Organization for Scientific Research (NWO), project number SION 612-30-002.

P. Widmayer et al. (Eds.): ICALP 2002, LNCS 2380, pp. 306–317, 2002.
© Springer-Verlag Berlin Heidelberg 2002

a bin. Further, the sum of the sizes of the pieces assigned to any bin may not exceed its capacity. A bin is *empty* if no piece is assigned to it, otherwise it is *used*. The goal is to minimize the number of bins used.

The *variable-sized bin packing* problem differs from the classical one in that bins do not all have the same capacity. There are an infinite number of bins of each capacity $\alpha_1 < \alpha_2 < \cdots < \alpha_m = 1$. The goal now is to minimize the sum of the capacities of the bins used.

In the *resource augmented bin packing* problem, one compares the performance of a particular algorithm $\mathcal{A}$ to that of the optimal offline algorithm in an unfair way. The optimal offline algorithm uses bins of capacity one, where $\mathcal{A}$ is allowed to use bins of capacity $\theta > 1$.

In the *online* versions of these problems, each piece must be assigned in turn, without knowledge of the next pieces. Since it is impossible in general to produce the best possible solution when computation occurs online, we consider approximation algorithms. Basically, we want to find an algorithm which incurs cost which is within a constant factor of the minimum possible cost, no matter what the input is. This factor is known as the asymptotic performance ratio.

A bin-packing algorithm uses *bounded space* if it has only a constant number of bins available to accept items at any point during processing. These bins are called *open* bins. Bins which have already accepted some items, but which the algorithm no longer considers for packing are *closed* bins. While bounded space algorithms are sometimes desirable, it is often the case that unbounded space algorithms can achieve lower performance ratios.

We define the asymptotic performance ratio more precisely. For a given input sequence σ, let $\mathrm{cost}_{\mathcal{A}}(\sigma)$ be the sum of the capacities of the bins used by algorithm $\mathcal{A}$ on σ. Let $\mathrm{cost}(\sigma)$ be the minimum possible cost to pack pieces in σ. The *asymptotic performance ratio* for an algorithm $\mathcal{A}$ is defined to be $R_{\mathcal{A}}^{\infty} = \limsup_{n \to \infty} \max_{\sigma} \{ \frac{\mathrm{cost}_{\mathcal{A}}(\sigma)}{\mathrm{cost}(\sigma)} | \mathrm{cost}(\sigma) = n \}$. The *optimal asymptotic performance ratio* is defined to be $R_{\mathrm{OPT}}^{\infty} = \inf_{\mathcal{A}} R_{\mathcal{A}}^{\infty}$. Our goal is to find an algorithm with asymptotic performance ratio close to $R_{\mathrm{OPT}}^{\infty}$.

Previous Results: The online bin packing problem was first investigated by Johnson [10,11]. He showed that the NEXT FIT algorithm has performance ratio 2. Subsequently, it was shown by Johnson, Demers, Ullman, Garey and Graham that the FIRST FIT algorithm has performance ratio $\frac{17}{10}$ [12]. Yao showed that REVISED FIRST FIT has performance ratio $\frac{5}{3}$, and further showed that no online algorithm has performance ratio less than $\frac{3}{2}$ [21]. Brown and Liang independently improved this lower bound to 1.53635 [1,15]. This was subsequently improved by van Vliet to 1.54014 [20]. Chandra [2] shows that the preceding lower bounds also apply to randomized algorithms.

Define $u_{i+1} = u_i(u_i - 1) + 1$, $u_1 = 2$, and $h_{\infty} = \sum_{i=1}^{\infty} 1/(u_i - 1) \approx 1.69103$. Lee and Lee showed that the HARMONIC algorithm, which uses bounded space, achieves a performance ratio arbitrarily close to h_{∞} [14]. They further showed that no bounded space online algorithm achieves a performance ratio less than h_{∞} [14]. A sequence of further results has brought the upper bound down to 1.58889 [14,16,17,19].

The variable-sized bin packing problem was first investigated by Frieson and Langston [8,9]. Kinnerly and Langston gave an online algorithm with performance ratio $\frac{7}{4}$ [13]. Csirik proposed the VARIABLE HARMONIC algorithm, and showed that it has performance ratio at most h_∞ [4]. This algorithm is based on the HARMONIC algorithm of Lee and Lee [14]. Like HARMONIC, it uses bounded space. Csirik also showed that if the algorithm has two bin sizes 1 and $\alpha < 1$, and that if it is allowed to pick α, then a performance ratio of $\frac{7}{5}$ is possible [4]. Seiden has recently shown that VARIABLE HARMONIC is an optimal bounded-space algorithm [18]. The resource augmented problem was first investigated by Csirik and Woeginger [6].

Our Results: In this paper, we present new algorithms for the variable-sized online bin packing problem. By combining the upper bounds for these algorithms, we give improve the upper bound for this problem from 1.69103 to 1.63597. Our technique extends the general packing algorithm analysis technique developed by Seiden [19]. We also show the first lower bounds for variable-sized online bin packing and resource augmented online bin packing. Due to space considerations, we focus on the case were there are two bins sizes. However, our techniques are applicable to the general case. We think that our results are particularly interesting because of the unusual fractal-like curves that arise in the investigation of our algorithms and lower bounds.

2   Upper Bounds

To begin, we present two different online algorithms for variable-sized bin packing. We focus in on the case where there are two bin sizes, $\alpha_1 < 1$ and $\alpha_2 = 1$, and examine how the performance ratios of our algorithms change as a function of α_1. Since it is understood that $m = 2$, we abbreviate α_1 using α. Both of our algorithms are combinations of the HARMONIC and REFINED HARMONIC algorithms. Both have a real parameter $\mu \in (\frac{1}{3}, \frac{1}{2})$. We call these algorithms VRH1(μ) and VRH2(μ). VRH1(μ) is defined for all $\alpha \in (0,1)$, but VRH2(μ) is only defined for

$$\alpha > \max\left\{\frac{1}{2(1-\mu)}, \frac{1}{3\mu}\right\}. \tag{1}$$

First we describe VRH1(μ). Define $n_1 = 50$, $n_2 = \lfloor n_1\alpha \rfloor$, $\epsilon = 1/n_1$ and

$$T = \left\{\frac{1}{i} \,\middle|\, 1 \le i \le n_1\right\} \cup \left\{\frac{\alpha}{i} \,\middle|\, 1 \le i \le n_2\right\} \cup \{\mu, 1-\mu\}.$$

Define $n = |T|$. Note that it may be that $n < n_1 + n_2 + 2$, since T is not a multi-set. Rename the members of T as $t_1 = 1 > t_2 > t_3 > \cdots > t_n = \epsilon$. For convenience, define $t_{n+1} = 0$. The interval I_j is defined to be $(t_{j+1}, t_j]$ for $j = 1, \ldots, n$. Note that these intervals are disjoint and that they cover $(0, 1]$. A piece of size s has *type* j if $s \in I_j$. Define the *class* of an interval I_j to be α if $t_j = \alpha/k$ for some positive integer k, otherwise the class is 1.

The basic idea of VRH1 is as follows: When each piece arrives, we determine the interval I_j to which it belongs. If this is a class 1 interval, we pack the item in a size 1 bin using a variant of REFINED HARMONIC. If it is a class α interval, we pack the item in a size α bin using a variant of HARMONIC.

VRH1 packs bins in *groups*. All the bins in a group are packed in a similar fashion. The groups are determined by the set T. We define $g = 3$ if $\alpha > 1 - \mu$, otherwise $g = 2$. We also define $h = 6$ if $\alpha/2 > \mu$, $h = 5$ if $\alpha/2 \leq \mu < \alpha$ and $h = 4$ otherwise. Note that these functions are defined so that $t_g = 1 - \mu$ and $t_h = \mu$. The groups are named $(g, h), 1, \ldots, g - 1, g + 1, g + 2, \ldots, n$.

Bins in group $j \in \{1, 2, , \ldots, n\} \setminus \{g\}$ contain only type j pieces. Bins in group (g, h) all have capacity 1. Closed bins in this group contain one type g piece and one type h piece. Bins in group n all have capacity 1 and are packed using the NEXT FIT algorithm. I.e. there is one open bin in group n. When a type n piece arrives, if the piece fits in the open bin, it is placed there. If not, the open bin is closed, the piece is placed in a newly allocated open group n bin.

For group $j \in \{1, 2, , \ldots, n - 1\} \setminus \{g\}$, the capacity of bins in the group depends on the class of I_j. If I_j has class 1, then each bin has capacity one, and each closed bin contains $\lfloor 1/t_j \rfloor$ items of type j. Note that t_j is the reciprocal of an integer for $j \neq h$ and therefore $\lfloor 1/t_j \rfloor = 1/t_j$. If I_j has class α, then each bin has capacity α, and each closed bin contains $\lfloor \alpha/t_j \rfloor$ items of type j. Similar to before, t_j/α is the reciprocal of an integer and therefore $\lfloor \alpha/t_j \rfloor = \alpha/t_j$. For each of these groups, there is at most one open bin.

The algorithm has a real parameter $\tau \in [0, 1]$, which for now we fix to be $\frac{1}{7}$. Essentially, a proportion τ of the type h items are reserved for placement with type g items. A precise definition of VRH1 appears below. The algorithm uses the sub-routine $\text{PUT}(p, G)$, where p is an item and G is a group.

VRH1

Initialize $x \longleftarrow 0$ and $y \longleftarrow 0$.
For each item p:
  $j \longleftarrow$ type of p.
  If $j = n$ then pack p using
  NEXT FIT in a group n bin.
  Else, if $j = g$ then $\text{PUT}(p, (g, h))$.
  Else, if $j = h$:
    $x \longleftarrow x + 1$.
    If $y < \lfloor \tau x \rfloor$:
      $y \longleftarrow y + 1$.
      $\text{PUT}(p, (g, h))$.
    Else $\text{PUT}(p, h)$.
  Else $\text{PUT}(p, j)$.

$\text{PUT}(p, G)$

If there is no open bin in G
  allocate a new bin b.
Else,
  let b be an arbitrary
  open bin in G.
Pack p in b.

We analyze VRH1 using the technique of *weighting systems* introduced in [19]. A weighting system is a tuple $(\mathbb{R}^\ell, \mathbf{w}, \xi)$, where $\mathbb{R}^\ell$ is a real vector space, $\mathbf{w}$ is a *weighting function*, and ξ is a *consolidation function*. We shall simply describe the weighting system for VRH1, and assure the reader that our definitions meet the requirements put forth in [19]. We use $\ell = 3$, and define $\mathbf{a}$, $\mathbf{b}$ and $\mathbf{c}$ to be orthogonal unit basis vectors. The weighting function is:

$$\mathbf{w}(x) = \begin{cases} \mathbf{b} & \text{if } x \in I_g; \\ (1-\tau)\dfrac{\mathbf{a}}{2} + \tau\,\mathbf{c} & \text{if } x \in I_h; \\ \dfrac{\mathbf{a}\,x}{1-\epsilon} & \text{if } x \in I_n; \\ \mathbf{a}\,t_i & \text{otherwise.} \end{cases}$$

The consolidation function is $\xi(x\,\mathbf{a} + y\,\mathbf{b} + z\,\mathbf{c}) = x + \max\{y, z\}$.

The following lemmas are proved in [7] and [19], respectively.

Lemma 1. *For all input sequences σ, $\mathrm{cost}_{vrh1}(\sigma) \leq \xi\left(\sum_{i=1}^{n} \mathbf{w}(p_i)\right) + O(1)$.*

Lemma 2. *For any input σ on which VRH1 achieves a performance ratio of c, there exists an input σ' where 1) VRH1 achieves a performance ratio of at least c, 2) every bin in an optimal solution is full, and 3) every bin in some optimal solution is packed identically.*

Given these two lemmas, the problem of upper bounding the performance ratio of VRH1 is reduced to that of finding the single packing of an optimal bin with maximal weight/size ratio. Details of this reduction can be found in [7], however, the basic ideas are in [19]. We consider the following integer program: Maximize $\xi(\mathbf{x})/\beta$ subject to

$$\mathbf{x} = \mathbf{w}(y) + \sum_{j=1}^{n-1} q_j \mathbf{w}(t_j); \quad y = \beta - \sum_{j=1}^{n-1} q_j\, t_{j+1}; \quad y > 0, \tag{2}$$

$$q_j \in \mathbb{N} \text{ for } 1 \leq j \leq n-1, \quad \beta \in \{1, \alpha\}; \tag{3}$$

over variables $\mathbf{x}, y, \beta, q_1, \ldots, q_{n-1}$. Intuitively, q_j is the number of type j pieces in an optimal bin. y is an upper bound on space available for type n pieces. We require $y > 0$ because a type j piece is strictly larger than t_{j+1}. Call this integer linear program $\mathcal{P}$. The value of $\mathcal{P}$ upper bounds the asymptotic performance ratio of VRH1. It is easily determined using a branch and bound procedure very similar to those in [19,18].

Now we describe VRH2(μ). Redefine

$$T = \left\{ \frac{1}{i} \,\middle|\, 1 \leq i \leq n_1 \right\} \cup \left\{ \frac{\alpha}{i} \,\middle|\, 1 \leq i \leq n_2 \right\} \cup \{\alpha\mu,\ \alpha(1-\mu)\}.$$

Define n_1, n_2, ϵ and n as for VRH1. Again, rename the members of T as $t_1 = 1 > t_2 > t_3 > \cdots > t_n = \epsilon$. (1) guarantees that $1/2 < \alpha(1-\mu) < \alpha < 1$ and $1/3 < \alpha\mu < \alpha/2 < 1/2$, so we have $g = 3$ and $h = 6$. The only difference from VRH1 is that (g, h) bins have capacity α. Otherwise, the two algorithms are identical. We therefore omit a detailed description and analysis of VRH2. We display the upper bound on the performance ratio achieved using the best of VRH1(μ), VRH2(μ) and VARIABLE HARMONIC in Figure 2. This upper bound is achieved by optimizing μ for each choice of α. Our upper bound is at most $\frac{373}{228} < 1.63597$ for all α, which is the performance ratio of REFINED HARMONIC in the classic bin packing context.

3   Lower Bounds

We now consider the question of lower bounds. Prior to this work, no general lower bounds for either variable-sized or resource augmented online bin packing were known. We present our results in terms of the variable-sized problem. Due to space limitations, we can only mention the resource augmented results.

Our method follows along the lines laid down by Liang, Brown and van Vliet [1,15,20]. We give some unknown online bin packing algorithm $\mathcal{A}$ one of k possible different inputs. These inputs are defined as follows: Let $\varrho = s_1, s_2, \ldots, s_k$ be a sequence of *item sizes* such that $0 < s_1 < s_2 < \cdots < s_k \leq 1$. Let ϵ be a small positive constant. We define σ_0 to be the empty input. Input σ_i consists of σ_{i-1} followed by n items of size $s_i + \epsilon$. Algorithm $\mathcal{A}$ is given σ_i for some $i \in \{1, \ldots, k\}$.

A *pattern* with respect to ϱ is a tuple $p = \langle \text{size}(p), p_1, \ldots, p_k \rangle$ where $\text{size}(p)$ is a positive real number and $p_i, 1 \leq i \leq k$ are non-negative integers such that $\sum_{i=1}^{k} p_i\, s_i < \text{size}(p)$. Intuitively, a pattern describes the contents of some bin of capacity $\text{size}(p)$. Define $\mathcal{P}(\varrho, \beta)$ to be the set of all patterns p with respect to ϱ with $\text{size}(p) = \beta$. Further define $\mathcal{P}(\varrho) = \bigcup_{i=1}^{m} \mathcal{P}(\varrho, \alpha_i)$. Note that $\mathcal{P}(\varrho)$ is necessarily finite. Given an input sequence of items, an algorithm is defined by the numbers and types of items it places in each of the bins it uses. Specifically, any algorithm is defined by a function $\Phi : \mathcal{P}(\varrho) \mapsto \mathbb{R}_{\geq 0}$. The algorithm uses $\Phi(p)$ bins containing items as described by the pattern p. We define $\phi(p) = \Phi(p)/n$.

Consider the function Φ that determines the packing used by online algorithm $\mathcal{A}$ uses for σ_k. Since $\mathcal{A}$ is online, the packings it uses for $\sigma_1, \ldots, \sigma_{k-1}$ are completely determined by Φ. We assign to each pattern a *class*, which is defined $\text{class}(p) = \min\{i \mid p_i \neq 0\}$. Intuitively, the class tells us the first sequence σ_i which results in some item being placed into a bin packed according to this pattern. I.e. if the algorithm packs some bins according to a pattern which has class i, then these bins will contain one or more items after σ_i. Define $\mathcal{P}_i(\varrho) = \{p \in \mathcal{P}(\varrho) \mid \text{class}(p) \leq i\}$. Then if $\mathcal{A}$ is determined by Φ, its cost for σ_i is simply $n \sum_{p \in \mathcal{P}_i(\varrho)} \text{size}(p)\phi(p)$. Since the algorithm must pack every item, we have the following constraints: $n \sum_{p \in \mathcal{P}(\varrho)} \phi(p)\, p_i \geq n$, for $1 \leq i \leq k$. For a fixed n, define $\chi_i(n)$ to be the optimal offline cost for packing the items in σ_i. The following lemma gives us a method of computing the optimal offline cost for each sequence:

Lemma 3. *For $1 \leq i \leq k$, $\chi^* = \lim_{n \to \infty} \chi_i(n)/n$ exists and is the value of the linear program:*

$$\min \left\{ \sum_{p \in \mathcal{P}_i(\varrho)} \text{size}(p)\phi(p) \,\middle|\, 1 \leq \sum_{p \in \mathcal{P}(\varrho)} \phi(p)\, p_j, \quad \text{for } 1 \leq j \leq i; \right\} \tag{4}$$

over variables χ_i and $\phi(p), p \in \mathcal{P}(\varrho)$.

Proof Clearly, the LP always has a finite value between $\sum_{j=1}^{i} s_j$ and i. For any fixed n, the optimal offline solution is determined by some ϕ. It must satisfy the constraints of the LP, and the objective value is exactly the cost incurred. Therefore the LP lower bounds the optimal offline cost. The LP is a relaxation in that it allows a fractional number of bins of any pattern, whereas a legitimate solution must have an integral number. Rounding the relaxed solution up to get a legitimate one, the change in the objective value is at most $|\mathcal{P}(\varrho)|/n$. ∎

Suppose we are given a sequence of item sizes ρ, then we are interested in the value $c = \min_{\mathcal{A}} \max_{i=1,\dots,k} \limsup_{n\to\infty} \text{cost}_{\mathcal{A}}(\sigma_i)/\chi_i(n)$. As $n \to \infty$, we can replace $\chi_i(n)/n$ by χ_i^*. Once we have the values $\chi_1^*, \dots, \chi_k^*$, we can readily compute a lower bound for our online algorithm:

Lemma 4. *The optimal value of the linear program: Minimize c subject to*

$$c \geq \frac{1}{\chi_i^*} \sum_{p \in \mathcal{P}_i(\varrho)} \text{size}(p)\phi(p), \qquad \text{for } 1 \leq i \leq k;$$

$$1 \leq \sum_{p \in \mathcal{P}(\varrho)} \phi(p)\, p_i, \qquad \text{for } 1 \leq i \leq k;$$

(5)

over variables c and $\phi(p), p \in \mathcal{P}(\varrho)$, is a lower bound on the asymptotic performance ratio of any online bin packing algorithm.

Proof For any fixed n, any algorithm $\mathcal{A}$ has some $\varPhi$ which must satisfy the second constraint. Further, $\varPhi$ should assign an integral number of bins to each pattern. However, this integrality constraint is relaxed, and $\sum_{p \in \mathcal{P}_i(\varrho)} \text{size}(p)\phi(p)$ is $1/n$ times the cost to $\mathcal{A}$ for σ_i as $n \to \infty$. The value of c is just the maximum of the performance ratios achieved on $\sigma_1, \dots, \sigma_k$. ∎

Although this is essentially the result we seek, a number of issues are left to be resolved. The first is that these linear programs have a variable for each possible pattern. The number of such patterns is potentially quite large, and we would like to reduce the LP size if possible. We show that this goal is indeed achievable. We say that a pattern p of class i is *dominant* if $s_i + \sum_{j=1}^k p_j\, s_j > \text{size}(p)$. Let p be a non-dominant pattern with class i. There exists a unique dominant pattern q of class i such that $p_j = q_j$ for all $i \neq j$. We call q the *dominator* of p with respect to class i.

Lemma 5. *In computing the values of the linear programs in Lemmas 3 and 4, it suffices to consider only dominant patterns.*

Proof We transform an LP solution by applying the following operation to each non-dominant pattern p of class i: Let $x = \phi(p)$ in the original solution. We set $\phi(p) = 0$ and increment $\phi(q)$ by x, where q is the dominator of p with respect to i. The new solution remains feasible, and its objective value has not changed. Further, the value of $\phi(p)$ is zero for every non-dominant p, therefore these variables can be safely deleted. ∎

Given a sequence of item sizes ϱ, we can compute a lower bound $L_m(\varrho, \alpha_1, \dots, \alpha_{m-1})$ using the following algorithm: 1) Enumerate the dominant patterns. 2)

For $1 \leq i \leq k$, compute χ_i via the LP given in Lemma 3. 3) Compute and return the value of the LP given in Lemma 4. Step one is most easily accomplished via a simple recursive function. Our concern in the remainder of the paper shall be to study the behavior of $L_m(\varrho, \alpha_1, \ldots, \alpha_{m-1})$ as a function of ϱ and $\alpha_1, \ldots, \alpha_{m-1}$.

4   Lower Bound Sequences

Up to this point, we have assumed that we were given some fixed item sequence ϱ. We consider now the question of choosing ϱ. We again focus in on the case where there are two bin sizes, and examine properties of $L_2(\varrho, \alpha_1)$. We again abbreviate α_1 using α and L_2 using L.

To begin, we define the idea of a *greedy* sequence. Let ϵ denote the empty sequence, and $\wedge$ the sequence concatenation operator. The greedy sequence $\Gamma_\tau(\beta)$ for capacity β with cutoff τ is defined by

$$\gamma(\beta) = \frac{1}{\left\lfloor \frac{1}{\beta} \right\rfloor + 1}; \qquad \Gamma_\tau(\beta) = \begin{cases} \epsilon & \text{if } \beta < \tau, \\ \gamma(\beta) \wedge \Gamma_\tau(\beta - \gamma(\beta)) & \text{otherwise.} \end{cases}$$

The sequence defines the item sizes which would be used if we packed a bin of capacity β using the following procedure: At each step, we determine the remaining capacity in our bin. We choose as the next item the largest reciprocal of an integer which fits without using the remaining capacity completely. We stop when the remaining capacity is smaller than τ. Note that for $\tau = 0$, we get the infinite sequence. We shall use Γ as a shorthand for Γ_0.

The recurrence u_i described in Section 1, which is found in connection with bounded-space bin packing [14], gives rise to the sequence $\frac{1}{u_i} = \frac{1}{2}, \frac{1}{3}, \frac{1}{7}, \frac{1}{43}, \frac{1}{1807}$, ... This turns out to be the infinite greedy sequence $\Gamma(1)$. Somewhat surprisingly, it is also the sequence used by Brown, Liang and van Vliet in the construction of their lower bounds [1,15,20]. In essence, they analytically determine the value of $L_1(\Gamma_\tau(1))$. Liang and Brown lower bound the value, while van Vliet determines it exactly. This well-known sequence is our first candidate. Actually, we use the first k items sizes in it, and we re-sort them so that the algorithm is confronted with items from smallest to largest. In general, this re-sorting seems to be good heuristic, since the algorithm has the most decisions to make about how the smallest items are packed, but on the other hand has the least information about which further items will be received. The results are shown in Figure 1.

Examining Figure 1, one immediately notices that $L(\Gamma_\tau(1), \alpha)$ exhibits some very strange behavior. The curve is highly discontinuous. Suppose we have a finite sequence ϱ, where each item size is a continuous function of $\alpha \in (0, 1)$. Tuple p is a *potential pattern* if there exists an $\alpha \in (0, 1)$ such that p is a pattern. The set of breakpoints of p with respect to ϱ is defined to be $B(p, \varrho) = \{\alpha \in (0, 1) | \sum_{i=1}^{k} p_i \, s_i = \text{size}(p)\}$. Let $\mathcal{P}^*$ be the set of all potential patterns. The set of all breakpoints is $B(\varrho) = \bigcup_{p \in \mathcal{P}^*} B(p, \varrho)$. Intuitively, at each breakpoint some combinatorial change occurs, and the curve may jump. In the intervals between breakpoints, the curve behaves nicely as summarized by the following lemma:

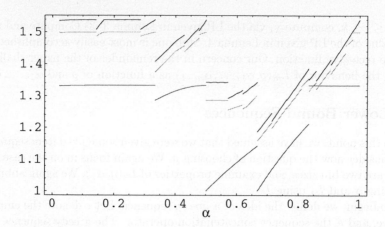

Fig. 1. The evolution of the curves given by the greedy item sequence. The lowest curve is $\frac{1}{2}, \frac{1}{3}$; the middle curve is $\frac{1}{2}, \frac{1}{3}, \frac{1}{7}$; the highest curve is $\frac{1}{2}, \frac{1}{3}, \frac{1}{7}, \frac{1}{43}$.

Lemma 6. *Let ϱ be a finite item sequence, with each item size a continuous function of $\alpha \in (0,1)$. In any interval $I = (\ell, h)$ which does not contain a breakpoint, $L(\varrho, \alpha)$ is continuous. Furthermore, for all $\alpha \in I$,*

$$L(\varrho, \alpha) \geq \min \left\{ \frac{\ell + h}{2h}, \frac{2\ell}{\ell + h} \right\} L\left(\varrho, \tfrac{1}{2}(\ell + h)\right).$$

This lemma is proved in [7].

Considering Figure 1 again, there are sharp drops in the lower bound near the points $\frac{1}{3}$ $\frac{1}{2}$ and $\frac{2}{3}$. It is not hard to see why the bound drops so sharply at those points. For instance, if α is just larger than $\frac{1}{2} + \epsilon$, then the largest items in $\Gamma(1)$ can each be put in their own bin of size α. If $\alpha \geq \frac{2}{3} + 2\epsilon$, two items of size $\frac{1}{3} + \epsilon$ can be put pairwise in bins of size α. In short, in such cases the online algorithm can pack some of the largest elements in the list with very little wasted space, hence the low resulting bound.

This observation leads us to try other sequences, in which the last items cannot be packed well. A first candidate is the sequence $\alpha, \Gamma(1-\alpha)$. As expected, this sequence performs much better than $\Gamma(1)$ in the areas described above.

It is possible to find further improvements for certain values of α. As a general guideline for finding sequences, items should not fit too well in either bin size. If an item has size x, then $\min\left\{1 - \lfloor\frac{1}{x}\rfloor x, \alpha - \lfloor\frac{\alpha}{x}\rfloor x\right\}$ should be as large as possible. In areas where a certain item in a sequence fits very well, that item should be adjusted (e.g. use an item $1/(j+1)$ instead of using the item $1/j$) or a completely different sequence should be used. (This helps explain why the algorithms have a low competitive ratio for α close to 0.7 where this minimum is never very large.)

Furthermore, as in the classical bin packing problem, sequences that are bad for the online algorithm should have very different optimal solutions for each prefix sequence. Finally, the item sizes should not increase too fast or slow: If items are very small, the smallest items do not affect the online performance much, while if items are close in size, the sequence is easy because the optimal solutions for the prefixes are alike.

In addition to the greedy sequence and $\alpha, \Gamma(1 - \alpha)$, we have found that the following sequences yield good results in restricted areas: $\alpha/2, \Gamma(1 - \alpha/2)$; $\alpha, \frac{1}{3}, \frac{1}{7}, \frac{1}{43}; \frac{1}{2}, \frac{1}{4}, \frac{1}{5}, \frac{1}{21}$; and $\frac{1}{2}, \frac{\alpha}{2}, \frac{1}{9}, \Gamma_\tau(\frac{7}{18} - \frac{\alpha}{2})$.

Using Lemma 6 we obtain the main theorem of this section:

Theorem 1. *Any online algorithm for the variable sized bin packing problem with $m = 2$ has asymptotic performance ratio at least 1.33561.*

Proof First note that for $\alpha \in (0, 1/43]$, the sequence $\frac{1}{2}, \frac{1}{3}, \frac{1}{7}, \frac{1}{43}$ yields a lower bound of $217/141 > 1.53900$ as in the classic problem: Bins of size α are of no use. For $\alpha > 1/43$, we use the sequences described in the preceding paragraphs. For each sequence ϱ, we compute a lower bound on $(1/43, 1)$ using the following procedure. Define $\varepsilon = 1/10000$. We break the interval $(0,1)$ into subintervals using the lattice points $\varepsilon, 2\varepsilon, \ldots, 1 - \varepsilon$. To simplify the determination of break-points, we use a constant sequence for each sub-interval. This constant sequence is fixed at the upper limit of the interval. I.e. throughout the interval $[\ell\varepsilon, \ell\varepsilon + \varepsilon)$ we use the sequence $\varrho|_{\alpha = \ell\varepsilon + \varepsilon}$. Since the sequence is constant, a lower bound on the performance ratio of any online bin packing algorithm with $\alpha \in [\ell\varepsilon, \ell\varepsilon + \varepsilon)$ can be determined by the following algorithm:

1. $\varrho' \leftarrow \varrho|_{\alpha = \ell\varepsilon + \varepsilon}$.
2. Initialize $B \leftarrow \{\ell\varepsilon, \ell\varepsilon + \varepsilon\}$.
3. Enumerate all the patterns for ϱ' at $\alpha = \ell\varepsilon + \varepsilon$.
4. For each pattern: (a) $z \leftarrow \sum_{i=1}^{k} p_i s_i$. (b) If $z \in (\ell\varepsilon, \ell\varepsilon + \varepsilon)$ then $B \leftarrow B \cup \{z\}$.
5. Sort B to get $b_1, b_2, , \ldots, b_j$.
6. Calculate and return the value:

$$\min_{1 \le i < j} \min \left\{ \frac{b_i + b_{i+1}}{2 b_{i+1}}, \frac{2 b_i}{b_i + b_{i+1}} \right\} L \left(\varrho', \tfrac{1}{2}(b_i + b_{i+1}) \right).$$

We implemented this algorithm in `Mathematica`, and used it to find lower bounds for each of the aforementioned sequences. The lowest lower bound is $495176908800/370749511199$, in the interval $[0.7196, 0.7197)$.   ∎

An overview of the lower bounds can be seen in Figure 2.

Using the same techniques, we also obtain the first general lower bound for resource augmented online bin packing. Define θ to be the size of the online algorithm's bins. Our bound is displayed in Figure 3. Notably, For $\theta < \frac{3}{2}$, online algorithms are strictly worse than optimal.

5   Conclusions

We have shown new algorithms and lower bounds for variable-sized online bin packing with two bin sizes. By combining these algorithms with VARIABLE HAR-MONIC, choosing for each size of the second bin α the best algorithm for that size,

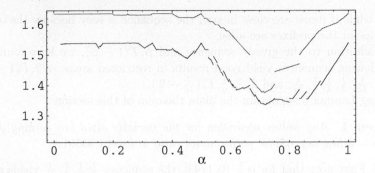

Fig. 2. The best upper and lower bounds for variable sized online bin packing. The upper bound is best of the VRH1, VRH2 and VARIABLE HARMONIC algorithms.

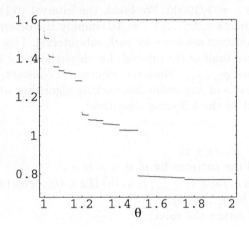

Fig. 3. The lower bound for resource augmented online bin packing.

we find an algorithm with asymptotic performance ratio of at most $\frac{373}{228} < 1.63597$ for all α. The best previous upper bound was $h_\infty \approx 1.69103$.

The largest gap between the performance of the algorithm and the lower bound is 0.18193, achieved for $\alpha = 0.9071$. The smallest gap is 0.03371 achieved for $\alpha = 0.6667$. Note that for $\alpha \le \frac{1}{2}$, there is not much difference with the classical problem: having the extra bin size does not help the online algorithm much. To be more precise, it helps about as much as it helps the offline algorithm.

Our work raises the following questions: is there a value of α where it is possible to design a better algorithm and show a matching lower bound? Or, can a lower bound be shown anywhere that matches an existing algorithm? Note that at the moment there is also a small gap between the competitive ratio of the best algorithm and the lower bound in the classical bin packing problem.

Another interesting open problem is analyzing variable-sized bin packing with an arbitrary number of bin sizes.

References

1. BROWN, D. J. A lower bound for on-line one-dimensional bin packing algorithms. Tech. Rep. R-864, Coordinated Sci. Lab., University of Illinois at Urbana-Champaign, 1979.

2. CHANDRA, B. Does randomization help in on-line bin packing? *Information Processing Letters 43*, 1 (Aug 1992), 15–19.

3. COFFMAN, E. G., GAREY, M. R., AND JOHNSON, D. S. Approximation algorithms for bin packing: A survey. In *Approximation Algorithms for NP-hard Problems*, D. Hochbaum, Ed. PWS Publishing Company, 1997, ch. 2.

4. CSIRIK, J. An on-line algorithm for variable-sized bin packing. *Acta Informatica 26*, 8 (1989), 697–709.

5. CSIRIK, J., AND WOEGINGER, G. On-line packing and covering problems. In *On-Line Algorithms—The State of the Art*, A. Fiat and G. Woeginger, Eds., Lecture Notes in Computer Science. Springer-Verlag, 1998, ch. 7.

6. CSIRIK, J., AND WOEGINGER, G. Resource augmentation for online bounded space bin packing. In *Proceedings of the 27th International Colloquium on Automata, Languages and Programming* (Jul 2000), pp. 296–304.

7. EPSTEIN, L., SEIDEN, S. S., AND VAN STEE, R. On the fractal beauty of bin-packing. Tech. Rep. SEN-R0104, CWI, Amsterdam, 2001.

8. FRIESEN, D. K., AND LANGSTON, M. A. A storage-size selection problem. *Information Processing Letters 18* (1984), 295–296.

9. FRIESEN, D. K., AND LANGSTON, M. A. Variable sized bin packing. *SIAM Journal on Computing 15* (1986), 222–230.

10. JOHNSON, D. S. *Near-optimal bin packing algorithms*. PhD thesis, Massachusetts Institute of Technology, Cambridge, Massachusetts, 1973.

11. JOHNSON, D. S. Fast algorithms for bin packing. *Journal Computer Systems Science 8* (1974), 272–314.

12. JOHNSON, D. S., DEMERS, A., ULLMAN, J. D., GAREY, M. R., AND GRAHAM, R. L. Worst-case performance bounds for simple one-dimensional packing algorithms. *SIAM Journal on Computing 3* (1974), 256–278.

13. KINNERSLEY, N., AND LANGSTON, M. Online variable-sized bin packing. *Discrete Applied Mathematics 22*, 2 (Feb 1989), 143–148.

14. LEE, C., AND LEE, D. A simple on-line bin-packing algorithm. *Journal of the ACM 32*, 3 (Jul 1985), 562–572.

15. LIANG, F. M. A lower bound for online bin packing. *Information Processing Letters 10* (1980), 76–79.

16. RAMANAN, P., BROWN, D., LEE, C., AND LEE, D. On-line bin packing in linear time. *Journal of Algorithms 10*, 3 (Sep 1989), 305–326.

17. RICHEY, M. B. Improved bounds for harmonic-based bin packing algorithms. *Discrete Applied Mathematics 34* (1991), 203–227.

18. SEIDEN, S. S. An optimal online algorithm for bounded space variable-sized bin packing. In *Proceedings of the 27th International Colloquium on Automata, Languages and Programming* (Jul 2000), pp. 283–295.

19. SEIDEN, S. S. On the online bin packing problem. In *Proceedings of the 28th International Colloquium on Automata, Languages and Programming* (Jul 2001), pp. 237–249.

20. VAN VLIET, A. An improved lower bound for online bin packing algorithms. *Information Processing Letters 43*, 5 (Oct 1992), 277–284.

21. YAO, A. C. C. New algorithms for bin packing. *Journal of the ACM 27* (1980), 207–227.

The Quest for Small Universal Cellular Automata

Nicolas Ollinger

LIP, École Normale Supérieure de Lyon, 46, allée d'Italie
69 364 Lyon Cedex 07, France, `Nicolas.Ollinger@ens-lyon.fr`

Abstract. We formalize the idea of intrinsically universal cellular automata, which is strictly stronger than classical computational universality. Thanks to this uniform notion, we construct a new one-dimensional universal automaton with von Neumann neighborhood and only 6 states, thus improving the best known lower bound both for computational and intrinsic universality.

1   Why Study Small Universal Machines?

Designing very small universal machines is an old and fascinating challenge, introduced by Shannon in 1956 [10], which usually involves tricky encodings. This problem has been also explored for other computational machines like Post machines [4]. As an abstract computing model, cellular automata provide the same concerns. Because of its uniformity – there is no separation between control and data – and parallelism, the cellular automata model also provides a kind of intrinsic universality.

An intrinsically universal cellular automaton can simulate, using macro-cells to encode single cells and a linear time slowdown, any given cellular automaton. Understanding how to construct very small automata of this kind involves understanding the way to structure data in both space and time, the way complex computation can be transfered from the local rule to the global one.

In his pioneer study of self-reproduction, Von Neumann [12] introduced a computationally universal (unformally, able to simulate any Turing machine) automaton – in fact, it can be proved intrinsically universal. As far as we know, the question to find very small universal automata was first studied by Smith III [11] and Banks [2] was the first to consider intrinsic universality. Banks closed the problem in dimension 2 and higher. In the case of one-dimensional cellular automata, the problem is more difficult. Until now, there was a gap between the smallest computationally universal automaton and the smallest intrinsically universal one. The different results are reviewed on Table 1.

In the present paper, we fill the gap by exhibiting an intrinsically universal cellular automaton with first-neighbors neighborhood and only 6 states. More material, like simulation programs and big colorfull space-time diagrams, is available on the Web at the url `http://www.ens-lyon.fr/~nollinge/6st/`.

P. Widmayer et al. (Eds.): ICALP 2002, LNCS 2380, pp. 318–329, 2002.

Table 1. Some previously known universal cellular automata

year	author	d	ν	states	universality
1966	von Neumann [12]	2	5	29	intrinsic
1968	Codd [3]	2	5	8	intrinsic
1970	Banks [2]	**2**	**5**	**2**	**intrinsic**
		1	3	18	intrinsic
1971	Smith III [11]	2	7	7	computation
		1	3	18	computation
1987	Albert and Čulik II [1]	1	3	14	intrinsic
1990	Lindgren and Nordhal [5]	**1**	**3**	**7**	**computation**

here d is the dimension of the cellular automaton and ν its neighborhood size: 5 corresponds to the von Neumann neighborhood and 3 to the first-neighbors neighborhood.

2   Definitions

A *cellular automaton* $\mathcal{A}$ is a quadruple $\left(\mathbb{Z}^d, S, N, \delta\right)$ such that $\mathbb{Z}^d$ is the d-dimensional regular grid, S is a finite set of states, N is a finite set of ν vectors of $\mathbb{Z}^d$ called the neighborhood of $\mathcal{A}$ and δ is the local transition function of $\mathcal{A}$ which maps S^ν to S.

A *configuration* $\mathcal{C}$ of a cellular automaton $\mathcal{A}$ maps $\mathbb{Z}^d$ to the set of states of $\mathcal{A}$. The state of the i-th cell of $\mathcal{C}$ is denoted as $\mathcal{C}_i$. The local transition function δ of $\mathcal{A}$ is naturally extended to a *global transition function* $G_\mathcal{A}$ which maps a configuration $\mathcal{C}$ of $\mathcal{A}$ to a configuration $\mathcal{C}'$ of $\mathcal{A}$ satisfying, for each cell i, the equation $\mathcal{C}'_i = \delta\left(\mathcal{C}_{i+v_1}, \ldots, \mathcal{C}_{i+v_\nu}\right)$, where $\{v_1, \ldots, v_\nu\}$ is the neighborhood of $\mathcal{A}$. A *space-time diagram* of a cellular automaton $\mathcal{A}$ is an infinite sequence of configurations $(\mathcal{C}_t)_{t \in \mathbb{N}}$ such that, for every time t, $\mathcal{C}_{t+1} = G_\mathcal{A}(\mathcal{C}_t)$. The usual way to represent space-time diagrams is to draw the sequence of configurations successively, from bottom to top.

A *sub-automaton*[1] of a cellular automaton corresponds to a stable restriction on the states set. A cellular automaton is a sub-automaton of another cellular automaton if (up to a renaming of states) the space-time diagrams of the first one are space-time diagrams of the second one. To compare cellular automata, we introduce a notion of rescaling space-time diagrams. To formalize this idea, we introduce the following notations:

σ^k. Let S be a finite state set and k be a vector of $\mathbb{Z}^d$. The shift σ^k is the bijective map from $S^{\mathbb{Z}^d}$ onto $S^{\mathbb{Z}^d}$ which maps a configuration $\mathcal{C}$ to the configuration $\mathcal{C}'$ such that, for each cell i, the equation $\mathcal{C}'_{i+k} = \mathcal{C}_i$ is satisfied.

---

[1] The prefix *sub* emphasizes the fact that $(S'^{\mathbb{Z}}, G)$ is an (algebraic) sub-structure of $(S^{\mathbb{Z}}, G)$. One could have also used the terminology *divisor* as the set of space-time diagrams of one automaton is included into the one of the other.

o^m. Let S be a finite state set and $m = (m_1, \dots, m_d)$ be a finite sequence of strictly positive integers. The *packing map* o^m is the bijective map from $S^{\mathbb{Z}^d}$ onto $(S^{m_1 \cdots m_d})^{\mathbb{Z}^d}$ which maps a configuration $\mathcal{C}$ to the configuration $\mathcal{C}'$ such that, for each cell i, the equation $\mathcal{C}'_i = (\mathcal{C}_{mi}, \dots, \mathcal{C}_{m(i+1)-1})$ is satisfied. The principle of $o^{(3,2)}$ is depicted on Fig. 1.

Fig. 1. The way $o^{(3,2)}$ cuts a two-dimensional configuration

Definition 1. *Let $\mathcal{A}$ be a d-dimensional cellular automaton with states set S. A $\langle m, n, k \rangle$-rescaling of $\mathcal{A}$ is a cellular automaton $\mathcal{A}^{\langle m,n,k \rangle}$ with states set $S^{m_1 \cdots m_d}$ and global transition function $G_{\mathcal{A}}^{\langle m,n,k \rangle} = \sigma^k \circ o^m \circ G_{\mathcal{A}}^n \circ o^{-m}$.*

Definition 2. *Let $\mathcal{A}$ and $\mathcal{B}$ be two cellular automata. Then $\mathcal{B}$ simulates $\mathcal{A}$ if there exists a rescaling of $\mathcal{A}$ which is a sub-automaton of a rescaling of $\mathcal{B}$.*

The relation of simulation is a quasi-order on cellular automata. It is a generalization of the order introduced by Mazoyer and Rapaport [6]. In [8], we motivate the introduction of this relation and discuss its main properties. In particular, it induces a maximal equivalence class which exactly corresponds to the set of intrinsically universal cellular automata as described by Banks [2] and Albert and Čulik II [1].

Definition 3. *A cellular automaton $\mathcal{A}$ is intrinsically universal if, for each cellular automaton $\mathcal{B}$, there exists a rescaling of $\mathcal{A}$ of which $\mathcal{B}$ is a sub-automaton.*

In the remaining part of this paper, we will especially consider cellular automata of dimension 1. As any cellular automaton can be simulated by a one-way cellular automaton, that is a cellular automaton with neighborhood $\{0, -1\}$, there exist intrinsically universal one-way cellular automata. Therefore, to prove that a particular cellular automaton is intrinsically universal, it is sufficient to prove that it can simulate any one-way cellular automaton.

3 A Simple 8 State Universal Cellular Automaton

Describing in details the behavior of a particular cellular automaton is not an easy task. The following text attempts to give a feeling of the way things work. To verify the correctness of the universality, one only needs the local transition function and the rules to encode a particular cellular automaton. These details will also be given.

We present the automaton in two steps. First, we describe the macroscopic idea behind the simulation. Secondly, we consider the microscopic encoding of those ideas into a cellular automaton.

3.1   Macroscopic Considerations

To simulate a cellular automaton, one has to iterate and compute in parallel a local transition function $\delta : S^N \to S$. Let Ξ be the binary alphabet $\{0, 1\}$ and $n = \lceil \log_2 |S| \rceil$. We can encode S on Ξ^n and decompose δ into n boolean functions $\delta_i : \Xi^{Nn} \to \Xi$. A sample cellular automaton, which will be used to illustrate our simulation, is presented on Fig. 2. As the single boolean operator NAND, denoted $|$, is a complete basis[2] for boolean functions computation, we can consider the functions (δ_i) as boolean circuits involving only the NAND operator. So, to simulate a cellular automaton, it is sufficient to compute n boolean circuits in parallel in constant time.

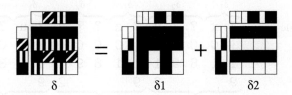

$$\delta \qquad = \qquad \delta_1 \qquad + \qquad \delta_2$$

Fig. 2. A sample cellular automaton and its decomposition

Rather than usual circuits, we consider leveled circuits with two kind of gates: copy and NAND. The gates of a leveled boolean circuit are partitioned into a finite number of levels $L_0, L_1, \dots , L_k$, such that every variable is on level L_0, the output is on level L_k and any gate on level L_{i+1} takes its inputs on level L_i. Sample representations of boolean functions by leveled circuits are given on Fig. 3. To simulate cellular automata, we want to simulate such circuits. In order to do so, we must flatten the circuit to encode it on some $S^{\mathbb{Z}}$. We introduce the idea of a boolean circuit cellular automata simulator.

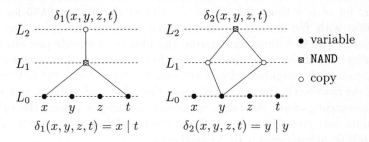

Fig. 3. Boolean functions and their leveled circuit representations

Definition 4. *A boolean circuit cellular automata simulator is a discrete dynamical model. Each cell of the model contains a stored boolean value which is updated at each time step according to some read-only information: an operator (copy or NAND) and the relative position of the operands.*

---

[2] A complete basis B for boolean functions is a set of boolean functions such that any boolean function can be obtained by circuits which doors compute functions from B. For an introduction on boolean functions, see [13].

As the operands can be arbitrarily far away, a boolean circuit cellular automata simulator is not a cellular automaton but it can simulate any cellular automaton, by local simulation of cellular automata cells and local transition function. We simply build a leveled circuit for each δ_i. Each circuit, padded with copy-only levels if necessary, must have the same height T. A cell s_j of the cellular automaton is encoded on the simulator by a macro-cell: a block of cells consisting of concatenation of the circuits of the (δ_i). Variables are extracted from the circuits outputs of the other macro-cells. In T time steps, each macro-cell s_j achieves exactly one transition of the simulated automaton. A sample encoding is depicted in Fig. 4.

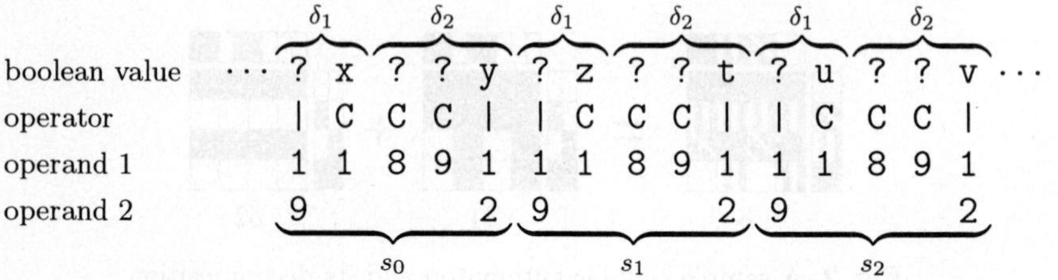

		δ_1		δ_2		δ_1		δ_2		δ_1		δ_2	
boolean value	$\cdots$?	x	?	?	y	?	z	?	?	t	?	u	?	? v $\cdots$
operator			C	C	C				C	C	C				C	C	C		
operand 1	1	1	8	9	1	1	1	8	9	1	1	1	8	9	1
operand 2	9				2	9				2	9				2

$$s_0 \qquad\qquad s_1 \qquad\qquad s_2$$

In this example, the macro-cells s_j encode the cells of the sample cellular automaton. Only the encoding boolean values are given, ? correspond to noise. For the operators encoding, | corresponds to NAND and C to copy. One step of the simulation is achieved by 2 time steps of the simulator (one step by level).

Fig. 4. Encoding cells using the boolean circuit cellular automata simulator

When simulating a cellular automaton, the simulator uses only finitely many different kind of cells. So, it accesses operands at a bounded distance m (an upper bound for m is n times the size of the biggest circuit over NAND for a boolean function with Nn inputs). To simulate the boolean circuit cellular automata simulator with a cellular automaton, the idea is to encode one circuit cell by a macro-cell consisting of circa m cells. Every circuit cell will send its boolean value to each m neighbors circuit cells on its right. When the i-th information cross the macro-cell, the macro-cell reads the i-th cell of its block and, depending on its contents, stores the value, applies an operand or ignores the value. When m values have crossed a macro-cell, the macro-cell has got its new boolean value and will be able to send it to its neighbors.

3.2   Microscopic Encoding

Background layer. We choose to represent a circuit cell by $m + 2$ cells of our universal automaton, where m is the maximal distance to look at. We use three kinds of states. The first kind, Op state, concerns m cells and memorizes the way to handle incoming information, as described above. One state cell is of the second kind Val and is used to store the boolean value of the macro-cell. The last cell state is of the third kind Sig and transmits a boolean value from macro-cell

to macro-cell. A computation is done each time the three kinds of state meet.
The details of the interactions are depicted on Fig. 5.

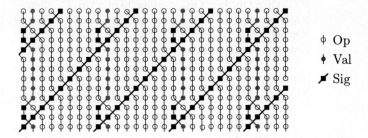

ϕ Op

$\blacklozenge$ Val

$\nearrow$ Sig

Fig. 5. Three kinds of states to move the information

On the figure, time goes from bottom to top. The Val states are the ones
which stay on a same column. The Sig states are the ones which go at real time
to the right. The Op states go slowly to the left when crossing Sig cells. We just
defined a cellular automaton with von Neumann neighborhood and 3 states. The
local transition function is given on Table 2.

Table 2. Local transition function of the 3 state automaton

s_l	s_m	s_r	$\delta(s_l, s_m, s_r)$	s_l	s_m	s_r	$\delta(s_l, s_m, s_r)$
Op	Op	Op	Op	Sig	Op	Op	Sig
Op	Op	Sig	Op	Sig	Op	Sig	Val
Op	Op	Val	Op	Sig	Op	Val	Sig
Op	Sig	Op	Op	Sig	Sig	Op	Op
Op	Sig	Sig	Sig	Sig	Val	Op	Sig
Op	Sig	Val	Sig	Val	Op	Op	Op
Op	Val	Op	Val	Val	Op	Sig	Op

Computational layer. To achieve the construction, we split the three kinds
of states into sub-kinds. The Sig kind is split in Sig_0 and Sig_1 boolean signals.
The Val kind is also split in Val_0 and Val_1 boolean values. The Op kind is split
in $Op_\sharp$, Op_C, Op_F, and $Op_|$ to encode respectively, the end of a macro cell,
the copy of the incoming signal, the absence of operation, and a NAND between
incoming and stored values. We obtain an 8 state intrinsically universal cellular
automaton, whose transition function is completely described on Table 3.

Encoding of a circuit cell. The exact encoding of a circuit cell, starting from
a macro-cell of the boolean circuit cellular automata simulator, is the following:

Let m be the maximum relative position of an operand used in the macro-cell
operators (for example $m = 9$ on Fig. 4). The basic shape for cells-encoding is

$$\omega(x, d_1, \ldots, d_{m-1}) = Val_x\, Op_{d_1} \cdots Op_{d_{m-4}}\, Sig_x\, Op_{d_{m-3}}\, Op_{d_{m-2}}\, Op_{d_{m-1}}\, Op_\sharp .$$

Table 3. Local transition function of the 8 state universal automaton

s_l	s_m	s_r	$\delta(s_l, s_m, s_r)$	s_l	s_m	s_r	$\delta(s_l, s_m, s_r)$
Op_i	Op_j	Op_k	Op_j	Sig_v	Op_i	Op_j	Sig_v
Op_i	Op_j	Sig_v	Op_j	Sig_v	Op_i	$\mathrm{Sig}_{v'}$	$\mathrm{Val}_{\varphi(v,i,v')}$
Op_i	Op_j	Val_v	Op_j	Sig_v	Op_i	$\mathrm{Val}_{v'}$	Sig_v
Op_i	Sig_v	Op_j	Op_j	Sig_v	$\mathrm{Sig}_{v'}$	Op_i	Op_i
Op_i	Sig_v	$\mathrm{Sig}_{v'}$	Sig_v	Sig_v	$\mathrm{Val}_{v'}$	Op_i	$\mathrm{Sig}_{\psi(v,v',i)}$
Op_j	Sig_v	$\mathrm{Val}_{v'}$	$\mathrm{Sig}_{v'}$	Val_v	Op_i	Op_j	Op_i
Op_i	Val_v	Op_j	Val_v	Val_v	Op_i	$\mathrm{Sig}_{v'}$	Op_i

$$\text{where} \quad \varphi(v, i, v') = \begin{cases} v' & \text{if } i = C, \\ v \mid v' & \text{if } i = \mid, \\ v & \text{else,} \end{cases} \quad \text{and} \quad \psi(v, v', i) = \begin{cases} v' & \text{if } i = \sharp, \\ v & \text{else.} \end{cases}$$

Cells are encoded as instanciations of the operators of these basic shape :

copy cell $\overset{C}{\underset{1}{}}$,
$$\omega_1^C(x) = \omega(x, F, \dots, F)$$

copy cell $\overset{C}{\underset{k}{}}$ with $k > 1$,
$$\omega_k^C(x) = \omega(x, \underbrace{F, \dots, F}_{k-2}, C, F, \dots, F)$$

NAND cell $\overset{\mid}{\underset{k}{}}$,
$$\omega_1^{\mid}(x) = \omega(x, \underbrace{F, \dots, F}_{k-2}, \mid, F, \dots, F)$$

NAND cell $\overset{\mid}{\underset{k}{}}$ with $k > 1$,
$$\omega_k^{\mid}(x) = \omega(x, \underbrace{F, \dots, F}_{k-2}, C, \underbrace{F, \dots, F}_{l-k-1}, \mid, F, \dots, F)$$

Let $\omega_1, \dots, \omega_n$ be the sequence of encoding cells corresponding to the macro-cell we want to simulate. Let $v_1, \dots, v_n$ be the initial boolean values to put into the circuit to encode a particular state. The macro-cell to use on the 8 state cellular automaton is then, shifting to ensure good synchronization:

$$\omega_2(v_1)\omega_3(v_2) \cdots \omega_n(v_{n-1})\omega_1(v_n).$$

One step of the simulation is then achieved in $m(m+2)T$ steps of the simulator where T is the number of levels of the circuit. This transformation was applied to the configuration on Fig. 4 to obtain the configuration on Fig. 6.

4 Tuning the Number of States

In the previous section, an 8 state universal cellular automaton was built. Now, by a careful analysis of the previous automaton, we first show that only 7 of the 8 states are really necessary to achieve universality. Secondly, we briefly explain how the number of Op states can be reduced to 2 sub-kinds, leading to a 6 state universal cellular automaton.

```
(0,0) 1-----%---#0-----/-+-#0-----/--+#0|----/---#0-----/--|#
(0,1) 1-----%---#0-----/-+-#0-----/--+#0|----/---#1-----%--|#
(1,0) 1-----%---#1-----%-+-#0-----/--+#0|----/---#0-----/--|#
(1,1) 1-----%---#1-----%-+-#0-----/--+#0|----/---#1-----%--|#
```

The symbols -, +, #, and | respectively encode the operators Op_F, Op_C, $Op_\sharp$, and $Op_|$. The symbols 0 and 1 encode the values Val_0 and Val_1. The symbols / and % encode the signals Sig_0 and Sig_1. One step of the simulation is achieved in 198 time steps of the simulator.

Fig. 6. Encoding of the sample automaton on the 8 state universal automaton

4.1   Emulate Copy: From 8 to 7 States

When used with constants, the NAND operator acts as follows: $x \mid 0 = 1$ and $x \mid 1 = \neg x$. In the previous simulation, we replace each signal Sig_x by a triple of signals Sig_1, Sig_x, Sig_0. We can then use these constants to emulate Op_C because of the equality

$$1 \mid (x \mid (0 \mid y)) = x \ .$$

The Op_C cells are replaced by the sequence $Op_|$, $Op_|$, $Op_|$ and the three other sub-kinds of Op cells are kept but guarded by an Op_F on the left and on the right to ignore the constant signals. This transformation was applied to the configurations on Fig. 6 to obtain the configurations on Fig. 7. When applying these transformation, the time to simulate one time step on the simulator grows from $m(m+2)T$ to $m(3m+4)T$ time steps of the simulator.

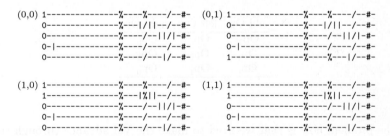

The symbols -, #, and | respectively encode the operators Op_F, $Op_\sharp$, and $Op_|$. The symbols 0 and 1 encode the values Val_0 and Val_1. The symbols / and % encode the signals Sig_0 and Sig_1. One step of the simulation is achieved in 558 time steps of the simulator.

Fig. 7. Encoding of the sample automaton on the 7 state universal automaton

So, we obtain a 7 state intrinsically universal cellular automaton by removing the Op_C state from the previous 8 state intrinsically universal automaton.

4.2   Split the Operators: From 7 to 6 States

To obtain a 6 state intrinsically universal cellular automaton, we will now show how to modify the preceding automaton to use only 2 Op sub-kinds.

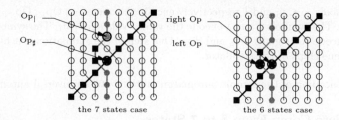

the 7 states case   the 6 states case

Fig. 8. Location of the computations in the universal cellular automata

The trick is displayed on Fig. 8. In the case of the 7 state automaton, the only active computations concern $Op_\sharp$ and $Op_|$ and are applied by looking at the Op on the right of the collision between Val and Sig. We now split the Op in a left Op and a right Op. The two sub-kinds of Op are the boolean values Op_0 and Op_1. The left Op controls the modification of the Val value: if it is equal to Op_1, a NAND between the Sig and the Val is applied and stored in the Val; if it is equal to Op_0, the Val keeps its value. The right Op controls the modification of the Sig value: if it is equal to Op_1, the Val value is stored in the Sig; if it is equal to Op_0, the Sig keeps it value. The 3 old Op are emulated as explained on Table 4.

Table 4. Emulation of the 7 state Op by the 6 state Op

old Op	left Op	right Op
Op_F	Op_0	Op_0
$Op_\sharp$	Op_0	Op_1
$Op_|$	Op_1	Op_0

The problem now is that we need to make the Val advance through the Op two by two. To achieve this, we add a new Sig between every two consecutive Sig. The purpose of this garbage signal is not to carry useful information, it only moves the Op.

By doing this, we introduce two new problems. First, we must take care that the garbage signals do not perturb the computation. Secondly, in order to preserve the injectivity of the simulation, we need to clean the information of the garbage signals at the end of every macro-cell transition.

Perturbation of the computation. We first consider the problem of the perturbation of the computation. On Fig. 9 is represented the typical sequence

$$F_F F_F F_F \cdots F_F F_F F_{\#} \mid_{\#} \mid_{\#} \mid_F F_F F_F F_F \cdots F_F F_F F_F \left(F_{\#} \mid_F F_F\right) F_F F_F F_F \cdots F_{\#} \#_{\mid} F_F$$

Fig. 9. Encoding of a cell with in subscript the operations of the garbage signals

of Op cells for an encoding using the 7 state universal cellular automaton. In subscript are displayed the operations performed by the garbage signals between the useful signals. The only case where the Val value is modified is by applying a NAND at the end. This will erase the information. But, the time before, the value of the Val has been saved on a Sig signal. In fact, just after the $Op_{\#}$ is executed, the next starting configuration is entirely contained in the sequence of useful signals. We will put everything back in the Val when cleaning the garbage.

Cleaning of the garbage. The cleaning of the garbage is rather technical. The main trick is the partial configuration given on Fig. 10. After going through these configuration, the garbage signal of the Sig_0 and Sig_1 guards contain the value Sig_1, the garbage signal of the Sig_x signal contains $Sig_{\bar{x}}$ and the value cell on its left contains Val_x. Thus, to avoid synchronization problems, the encoding algorithm goes as follows, starting from the 7 state macro-cell configuration.

signal		0	x	1	0	y	1
main op		|	F	F	|	|	|
garbage op	#	F	F	#	#	#	F
cell value	?	1	1	1	1	$\bar{y}$	y

Fig. 10. Sequence of operations to clean the garbage

First, for each subword encoding a cell of the boolean circuit cellular automata simulator, which consists of $3m$ cells of the kind Op, concatenate $n-2$ times the configuration $Op_F\, Op_F\, Op_F$ where n is the size of the boolean circuit cellular automata simulator macro-cell. Then, concatenate the cleaning pattern of Fig. 10, that is $Op_{\mid}\, Op_F\, Op_F\, Op_{\mid}\, Op_{\mid}\, Op_{\mid}$. Finally, apply the 7 states to 6 states conversion as described on Table 4. This will let enough space between signals to insert the garbage signals at a distance at least 4 after each coding signal. Finally, the time to simulate one time step on the simulator grows from $m(3m+4)T$ to $(m+n)(6(m+n)+7)T$ time steps of the simulator.

Once again, we apologize for the difficulty to read such statements. We hope it will at least help the interested reader to understand the ideas and path which led to this universal 6 state cellular automaton. For a formal proof, it is sufficient to extract the encoding of the meta-cells from the text and use the local transition function given in Table 5 to simulate another intrinsically universal cellular automaton.

5   The Quest Just Begins

Usually the construction of very small computationally universal cellular automata involves very tricky encodings. The main problem is to encode, and be

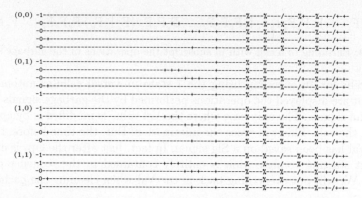

The symbols - and + respectively encode the operators Op_0 and Op_1. The symbols 0 and 1 encode the values Val_0 and Val_1. The symbols / and % encode the signals Sig_0 and Sig_1. One step of the simulation is achieved in 2548 time steps of the simulator.

Fig. 11. Encoding of the sample automaton on the 6 state universal automaton

Table 5. Local transition function of the 6 state universal automaton

s_l	s_m	s_r	$\delta\left(s_l, s_m, s_r\right)$	s_l	s_m	s_r	$\delta\left(s_l, s_m, s_r\right)$
Op_i	Op_j	Op_k	Op_j	Sig_v	Op_i	Op_j	Sig_v
Op_i	Op_j	Sig_v	Op_j	Sig_v	Op_i	$Sig_{v'}$	Val_v
Op_i	Op_j	Val_v	Op_j	Sig_v	Op_i	$Val_{v'}$	Sig_v
Op_i	Sig_v	Op_j	Op_j	Sig_v	$Sig_{v'}$	Op_i	Op_i
Op_i	Sig_v	$Sig_{v'}$	Sig_v	Sig_v	$Val_{v'}$	Op_i	$Sig_{\psi(v,v',i)}$
Op_j	Sig_v	$Val_{v'}$	$Sig_{\varphi(j,v,v')}$	Val_v	Op_i	Op_j	Op_i
Op_i	Val_v	Op_j	Val_v	Val_v	Op_i	$Sig_{v'}$	Op_i

where $\quad \varphi(j, v, v') = \begin{cases} v' & \text{if } j = 0, \\ v \mid v' & \text{if } j = 1, \end{cases} \quad$ and $\quad \psi(v, v', i) = \begin{cases} v & \text{if } i = 0, \\ v' & \text{if } i = 1. \end{cases}$

able to differentiate between both the control part of the sequential device and the passive data collection with as few states as possible. Therefore, the universal sequential device which is encoded into a cellular automaton must already be in some sense minimal, as in Lindgren and Nordhal [5] automaton. Those two levels of optimization lead to difficulties to understand how to effectively compute with the obtained cellular automata and the choice of the minimal sequential device turns back to problems of non-uniformity like transfer between control and data, between letters and states.

If, as we believe, cellular automata must be considered as an acceptable programming system, that is as a clean computing model, universality investigations must be done inside the model itself and not through doubtful notions of extrinsic simulation. The notion of intrinsic universality try to respond to this necessity. As it is defined on top of a notion of intrinsic simulation which respects both the locality and the uniformity of cellular automata, the notion of intrinsically universal cellular automaton is formal (thus allowing to prove that some cellular automata are not universal which is impossible without a formal definition), uniform and involves full parallelism. This helped us to design an 8 states universal cellular automaton without any tricky encoding. The trick to go from 8 states to 7 states was also very simple. Thus, we were able to do as good as [5] without beginning to really encode. Our tricky encoding lies into the transformation from 7 to 6 states.

The problem to find the smallest intrinsically universal cellular automaton is still open. We think that, using very tricky encodings, it should be possible to transform our automaton into an intrinsically universal automaton with 4 states.

References

1. J. Albert and K. Čulik II, A simple universal cellular automaton and its one-way and totalistic version, *Complex Systems*, **1**(1987), no. 1, 1–16.
2. E. R. Banks, Universality in cellular automata, in *Conference Record of 1970 Eleventh Annual Symposium on Switching and Automata Theory*, pages 194–215, IEEE, 1970.
3. E. Codd, *Cellular Automata*, Academic Press, 1968.
4. M. Kudlek and Y. Rogozhin, New small universal circular Post machines, in R. Freivalds, editor, *Proceedings of FCT'2001*, volume 2138 of *LNCS*, pages 217–226, 2001.
5. K. Lindgren and M. G. Nordahl, Universal computation in simple one-dimensional cellular automata, *Complex Systems*, **4**(1990), 299–318.
6. J. Mazoyer and I. Rapaport, Inducing an order on cellular automata by a grouping operation, *Discrete Appl. Math.*, **218**(1999), 177–196.
7. M. Minsky, *Finite and Infinite Machines*, Prentice Hall, 1967.
8. N. Ollinger, Toward an algorithmic classification of cellular automata dynamics, 2001, LIP RR2001-10, http://www.ens-lyon.fr/LIP.
9. N. Ollinger, Two-states bilinear intrinsically universal cellular automata, in R. Freivalds, editor, *Proceedings of FCT'2001*, volume 2138 of *LNCS*, pages 396–399, 2001.
10. C. E. Shannon, A universal Turing machine with two internal states, *Ann. Math. Stud.*, **34**(1956), 157–165.
11. A. R. Smith III, Simple computation-universal cellular spaces, *J. ACM*, **18**(1971), no. 3, 339–353.
12. J. von Neumann, *Theory of Self-reproducing Automata*, University of Illinois Press, Chicago, 1966.
13. I. Wegener, *The complexity of boolean functions*, Wiley-Teubner, 1987.

Hyperbolic Recognition by Graph Automata

Christophe Papazian and Eric Rémila

Laboratoire de l'Informatique du Parallélisme CNRS UMR 5668
Ecole Normale Supérieure de Lyon
46 Allée d'Italie, 69364 Cedex 07 Lyon, France
{Christophe.Papazian,Eric.Remila}@ens-lyon.fr

Introduction

Graph automata were first introduced by P. Rosenstiehl [10], under the name of *intelligent graphs*, surely because a network of finite automata is able to know some properties about its own structure. Hence P. Rosenstiehl created some algorithms that find Eulerian paths or Hamiltonian cycles in those graphs, with the condition that every vertex has a fixed degree [11]. Those algorithms are called "myopic" since each automaton has only the knowledge of the state of its immediate neighborhood. This hypothesis seems to be absolutely essential for the modelisation of the physical reality. A. Wu and A. Rosenfeld ([12] [13]) developed ideas of P. Rosenstiehl, using a simpler and more general formalism: the *d*-graphs. For each of these authors, a graph automata is formed by synchronous finite automata exchanging information according to an adjacent graph.

A. Wu and A. Rosenfeld have been especially interested in the problem of graph recognition. What is the power of knowledge of a parallel machine about its own architecture ? The motivation for this problem in parallelism is clear (can we control the structure of a parallel computer just using this computer ?) but this problem can also be interpreted in Biology or Physics as study of dynamic systems with local rules, or as a theoretical model of computer or mobile networks. In a general point of view, this is also a method to study the local algorithms on regular structures.

There is a general difficulty of this task, due to the finite nature of the automaton. As it has a finite memory, an arbitrary large number can not be stored in such a memory or in the signals exchanged by automata, . Hence, for example, coordinates can not be stored. Thus appropriate techniques have to be used: we use time differences to compute coordinates, and we will define signal flows to do comparisons between coordinates.

The first non-trivial result on this framework is due to A. Wu and A. Rosenfeld, who gave a linear algorithm allowing a graph automata to know if its graph is a rectangle or not. Then, E. Rémila [8] extended this result to other geometrical structures, with a system of cutting along certain lines, by using some methods of signal transmission in a fixed direction. In a recent paper [5], we gave a very general method that allows to recognize a large class of finite subgraphs of $\mathbb{Z}^2$ (and a lot of subclasses like classes of convex figures, of cell compatible

P. Widmayer et al. (Eds.): ICALP 2002, LNCS 2380, pp. 330–342, 2002.
© Springer-Verlag Berlin Heidelberg 2002

figures ...) by putting orientations on edges and computing coordinates for each vertex.

In the present paper, we are interested in the problem of recognition of finite subgraphs of infinite regular hyperbolic networks. A new difficulty arises, due to the hyperbolic structure which is not intuitive, in our opinion: working in a hyperbolic plane is not as natural a working in $\mathbb{Z}^2$. Thus, we limit ourselves to networks constructed from Dehn groups (i. e. fundamental groups of compact orientable 2-manifolds, see [1]), which are natural hyperbolic generalizations of the planar grid. Thus, the group structure can be used. Moreover, these groups are said *automatic* (in the sense of [4]), and the automatic structure gives us a powerful tool for our study.

The main results are a linear time (linear with the size of the studied graph) algorithm of recognition for a very large and natural class of finite subgraphs of hyperbolic regular networks and a quadratic time algorithm for the most general case. We even show that the result could be extended to larger classes.

1   Definitions

Notation: We will use a, b, c for constant on letters, u, v, w for words (and ϵ for the empty word), and x, y, z for variables on letters. We will use d, i, j, k for integers, e, f for edges, and μ, ν for vertices.

1.1   d-Graph

A *graph* is a set $G = (V, E)$ with V a finite set of vertices, and E a subset of V^2, whose elements are called "edges". We will only consider graphs without loops (no (ν, ν) edges) and symmetric (if (μ, ν) is in E, so is (ν, μ)). A vertex μ is *a neighbor* of another vertex ν if (ν, μ) is in E. The *degree* of a vertex is the number of its neighbors; the degree of the graph is the maximum degree among its vertices. A vertex of degree d is called a d-vertex. A *path* of G from ν to μ is a finite sequence $(\nu_0, \nu_1, \ldots, \nu_p)$ (with $\nu = \nu_0$ and $\nu_p = \mu$) such that two consecutive vertices of the sequence are neighbors. A *cycle* is a path such that $\mu = \nu$. We say that G is connected if for each pair (ν, μ) of V^2, there exists a path of G from ν to μ. We will only consider connected graphs.

A *cyclic conjugate* of a word $w = x_1...x_n$ is a word $w_i = x_i...x_n x_1...x_{i-1}$ for all $1 \leq i \leq n$ (see [1]).

A *labelling* λ is a mapping from E to a finite alphabet Σ. Given any path of G, the labelling of this path is the word built from the sequence of labels of successive edges of the path. Given any cycle of G, a word w is a *contour word* of this cycle iff w or w^{-1} is a cyclic conjugate of the labelling of this cycle.

Let d be a fixed integer such that $d \geq 2$. A d-graph is a 3-tuple $(G, \nu_0, \mathfrak{h})$, where $G = (V, E)$ is a symmetric connected graph with only two kinds of vertices: vertices of degree 1 (which are called #-vertices by Rosenfeld) and vertices of degree d, ν_0 is a d-vertex of G (which is called the leader, or the general), $\mathfrak{h}$ is a mapping from E to $\{1, 2, ..., d\}$ such that, for each d-vertex ν of V, the partial mapping $\mathfrak{h}(\nu,.)$ is bijective (injective if we do not consider the #-vertices). The subgraph G' of G induced by the set of d-vertices is called the underlying graph of G.

From any graph G' of degree at most d, we can construct a d-graph $(G, \nu_0, \mathfrak{h})$ whose underlying graph is G': we add some #-vertices until the degree of each (non #) vertex is d, arbitrarily choose a leader and, for each vertex, build an order on its neighbors.

For each vertex, we call an up-edge, an edge that links this vertex to another vertex that is closer to the leader than itself. We call down-edge in the opposite case.

1.2   Graph Automata

Definition. A *finite d-automaton* is a pair (Q, δ) such that Q is a finite set of states with # in Q, and δ is a function from $Q \times (\mathbb{Z}_d)^d \times Q^d$ to Q such that, for each element Υ of $Q \times (\mathbb{Z}_d)^d \times Q^d$, $\delta(\Upsilon) = \#$ if and only if the first component of Υ is #.

A *graph automaton* is a set $M = (G_d, A)$ with G_d a d-graph, and A a finite d-automaton (Q, δ).

A *configuration* C of M is a mapping from the set of vertices of G_d to the set of states Q, such that $C(\nu) = \#$ iff ν is a #-vertex.

We define, for each vertex ν, the *neighborhood vector* $H(\nu) \in \mathbb{Z}_d{}^d$ in the following way: for each integer i of $\{1, 2, \dots, d\}$, let μ_i denote the vertex such that $\mathfrak{h}(\nu, \mu_i) = i$. The ith component of $H(\nu)$ is $\mathfrak{h}(\mu_i, \nu)$.

We compute a new configuration from a previous one by applying the transition function δ simultaneously to each vertex of G_d, computing a new state for each (non-#) vertex by reading its state and the states of its neighbors, using the vector of neighborhood:

$$\mathcal{C}_{new}(\nu) = \delta(\mathcal{C}(\nu), H(\nu), \mathcal{C}(\mu_1), \dots \mathcal{C}(\mu_d))$$

(#-vertices always stay in the # state). Hence, we have a synchronous model, with local finite memory. The use of the neighborhood vector is necessary for this computation (see [9] for details).

There is a special state q_0 in Q that we call the *quiescent state*. If a vertex ν and all its (non-#) neighbors are in this state, ν will stay in this same state in the next configuration.

The *initial configuration* (the one from which we will begin the computation) is a configuration where all vertices are in the state q_0, except the leader, which is in a state q_{init}.

We say that a configuration C of G_d can be *reached* in M if C is obtained from the initial configuration after a finite number of steps of computations described above.

Similarity with a network. To simplify our explanations, we will consider that each automaton sends and receives signals to/from its neighbors at each step of computation, instead of looking at their entire states. It is easily possible, considering their states as sets of $d+1$ memory registers, one for keeping tracks about its own characteristics, and d others that contain messages for the neighbors, one register for each neighbor. Hence, at each step of computation, a vertex only looks at the messages computed for itself in the state of its neighbors.

To allow simultaneous exchanges, we consider a message as *a set of signals*.

1.3   The Problem of Recognition

Now, we will only use automata with two particular states: the accepting state and the reject state.

Definition 1 The class of accepted graphs for one d-automaton A *is the class of d-graphs G_d such that there is a configuration, where the leader is in the accepting state, that can be reached in (G_d, A).*

In the same way, we can define the class of rejected graphs.

Definition 2 A recognizer *is an automaton that rejects all d-graphs that it does not accept.*

In fact, we have a decision problem: As input, we have graphs, and for each automaton, we can compute if the graph is accepted or rejected. This natural problem has been formalized by A. Wu and A. Rosenfeld.

So the aim of the recognition of d-graphs is to build a recognizer for a given class of d-graphs.

1.4   Hyperbolic Regular Network

We call *a cell*, a cycle of a graph such that it is not possible to extract a shorter cycle of the graph from the cycle ("no vertex nor edges inside").Two different cells are said *adjacent* if they share one edge. A finite subgraph $G = (V, E)$ is said *cell compatible* if E is a union of cells. If, moreover, for each pair (C, C') of cells included in E with a common vertex ν, either C and C' are adjacent, or there exists cells $(C = C_0, C_1, ..., C_i = C')(i < d)$ included in E, such that each C_i contains the vertex ν and C_i is adjacent to C_{i+1}, then we say that G is *locally cell connected*.

Definition 3 *The* cell-neighborhood *of a vertex ν of a d-graph $G = (V, E)$ of $\Gamma(k, d)$ is the sub-graph of G composed by the cells containing the vertex ν.*

A free group $\mathcal{G} =< S >=< S; \emptyset >$ (with S a finite set) is the set of finite words of Σ^* with the natural reduction $xx^{-1} = \epsilon$, where $\Sigma = S \cup S^{-1}$. A group $\mathcal{G} =< S; R >$, is the quotient group $< S > /d(< R >)$, with $d(< R >)$, the group generated by the set of words xrx^{-1} ($x \in S$, $r \in R$). Informally, G is the quotient group of $< S >$ where all words of R (and the symmetric words of words of R and the cyclic permutations of words of R) are equal to ϵ, the empty word.

So we can have two different words of Σ^* that will represent a same element of $< S; R >$ (for example any element of R^* is equal to ϵ).

We will now consider, for $g \geq 2$, the alphabet $\Sigma_g = \{a_1, ..., a_{2g}, a_1^{-1}, ..., a_{2g}^{-1}\}$, the fundamental group (of the orientable compact 2-manifold of genus g) $\mathcal{G}_g = < \Sigma_g ; w_\perp = a_1 a_2 ... a_{2g} a_1^{-1} a_2^{-1} ... a_{2g}^{-1} >$ and the Cayley graph associated to such a group. Remember that a Cayley graph $G = (V, E)$ of a given group $\mathcal{G} = < S, R >$ is a graph such that V is the set of elements of $\mathcal{G}$ and $(v, w) \in E$ if and only if there is a generator $a \in S \cup S^{-1}$ such that va and w are the same element of $\mathcal{G}$. The symmetric unlabeled graph induced by the Cayley graph of $\mathcal{G}_g$ is $\Gamma(4g, 4g)$ defined as follows:

Definition 4 *A network $\Gamma(k, d)$ is the unique undirected planar graph, such that every cell has k edges, every vertex has d neighbors, and $\Gamma(k, d)$ can be drawn in such a way that each bounded region contains a finite number of vertices. It corresponds to the regular tessellations of Euclidean or hyperbolic planes.*

So, $\Gamma(4, 4)$ is the infinite grid isomorphic to $\mathbb{Z}^2$, and $\Gamma(4g, 4g)$ is a natural extension in hyperbolic geometries.

Hence, the cycles of length $4g$ of $\Gamma(4g, 4g)$ are cells. A canonical labelling of $\Gamma(4g, 4g)$ is a labelling λ from the set of edges to Σ_g such that the cells admit $w_\perp$ as a contour word, and for all edges (ν, μ) and (ν, μ') we have $(\lambda(\nu, \mu) = \lambda(\nu, \mu')) \Rightarrow (\mu = \mu')$, and we have $\lambda(\nu, \mu) = \lambda(\mu, \nu)^{-1}$.

A finite graph G is called *a manifold on $\mathcal{G}_g$* if we have a valid labelling λ on G to Σ_g such that every cell of G admits $w_\perp$ as a contour word, every vertex does not have two outgoing edges with the same label, and we have $\lambda(\nu, \mu) = \lambda(\mu, \nu)^{-1}$.

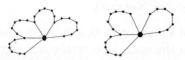

Fig. 1. One possible structure of local cell-neighborhood for $\Gamma(8, 8)$ on the left. On the right, a cell-neighborhood not allowed in locally cell connected manifold.

As we can see from figure 1, the definition of locally cell connected graph forces the local structure of the cell-neighborhood of each vertex, and, conversely, each connected subgraph of $\Gamma(4g, 4g)$, such that the cell-neighborhood of each vertex is like the figure 1 (on the left), is locally cell connected.

Note that two isomorphic locally cell connected subgraphs of $\Gamma(4g, 4g)$ also are isometric (from an Hyperbolic point of view). This is not true for all cell compatible subgraphs of $\Gamma(4g, 4g)$. There are some kneecap effects around vertices that induce several possible local labellings (see figure 2).

Each subgraph of $\Gamma(4g, 4g)$ obviously admits a valid labelling, but some other graphs (for example: tori, cylinders, spheres ...) also do. Informally, we can say that each graph accepting a valid labeling can be locally mapped to the

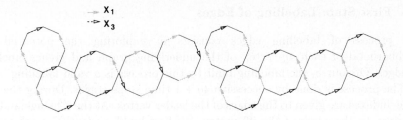

Fig. 2. Two possible labellings of the same graph, due to the kneecap effect.

graph $\Gamma(4g, 4g)$ and, moreover, all the local mappings have a kind of weak local coherence.

Moreover, a labelling of a locally cell connected subgraph of $\Gamma(4g, 4g)$ is completely determined by the labels of the edges outgoing from a fixed vertex. This is not true for (non-locally) connected cell compatible subgraphs of $\Gamma(4g, 4g)$. This fact is important for the determination of a possible isomorphism from a given graph to a subgraph of $\Gamma(4g, 4g)$. It is the reason why we limit ourselves in studying the recognition of the class of $4g$-graphs whose underlying graph is isomorphic to a locally cell connected subgraph.

Finally, we say that a manifold $G = (V, E)$ (with a labelling λ) is mappable if a morphism φ exists from V onto the vertices of $\Gamma(4g, 4g)$ such that the labelling is unchanged: $\lambda(\varphi(\nu), (\varphi(\mu)) = \lambda(\nu, \mu)$. Note that if φ is injective, then G is isomorphic to a subgraph of $\Gamma(4g, 4g)$.

2   The Automaton

Now, we will construct a $4g$-automaton A_{4g} that recognizes the class of $4g$-graphs whose underlying graph is a cell locally connected figure of $\Gamma(4g, 4g)$. The process of recognition is divided into three steps:

First, each automaton tries to put coherent labellings on its edges. By doing this process, we recognize the locally cell connected manifolds of $\mathcal{G}_g$.

Afterwards, it (informally) puts words of the finite group $\mathcal{G}_g$ on its vertices, that must be coherent with the previously chosen labelling. Hence, we have an injection from the vertices of the graph to the elements of the group. By doing this second process, we recognize the mappable locally cell connected manifolds of $\mathcal{G}_g$.

Finally, automata do some processing on the border to detect the possible contradictions. By doing this last process, we recognize the subgraphs of $\Gamma(4g, 4g)$.

We will often say that *a vertex* sends a signal or computes a result to mean that *the copy of the automaton on this vertex* does such a task.

2.1   First Step: Labelling of Edges

The process of labelling edges consists in exploring the potential $2g$-neighborhood of each $4g$-vertex of the underlying graph and giving labels to its edges. Of course, the labelling built by this process is a valid labelling.

The process is done by successive $4g + 1$ time steps stages. During the first stage, indexes are given to the edges of the leader vertex. At the ith stage, indexes are given to the edges of the ith vertex. We first build a (depth) search tree in the graph, to have a natural order on the different vertices. Such a tree can easily be constructed by a graph automaton, see [12] and [13] for details.

Each stage is computed in $4g + 1$ time steps using signals exploring the $2g$-neighborhoods of vertices.

$t = 0$, **Beginning**

The vertex ν sends a signal A_i^1 with $1 \leq i \leq 4g$ through each of its edges e_i (we use $\mathfrak{h}(\nu, .)$ to enumerate the edges of ν).

$1 \leq t < 2g$ **Transmission**

Each time a vertex receives a signal A_i^k $(k < 2g)$ through one of its edge, it sends a signal A_i^{k+1} through all of its other edges.

$t = 2g$ **Identification of cells**

When a vertex receives two signals A_i^{2g} by the edge e and A_j^{2j} by the edge f, it sends back a signal $B_{i,j}$ through e and $B_{j,i}$ through f.

$2g < t < 4g$ **Back transmission**

When a vertex receives a signal $B_{i,j}$, it sends it back through the edge from it previously received a signal A_i^k.

$t = 4g$, **Final computation**

The vertex ν must receive one or two signal $B_{i,j}$ (and $B_{i,j'}$) through its edge e_i. Informally, it allows to know that a cell of the $2g$-neighborhood exists between the edges e_i and e_j (and e_i and e'_j). As we know the neighbor edges of any given edge, we can put labels on all the edges of ν, knowing at least one label previously given in a precedent stage.

$t = 4g + 1$, **Next step**

We send a signal to the next vertex ν', to indicate that ν' must begin the labelling process and to give the label of the edge (ν', ν).

Remark 1 *During this process, each vertex can moreover store, in its own state, the number of its border edges, i. e. edges which belong to only one (potential) cell. Each vertex has 0 or 2 border edges issued from itself due to local cell connectivity. Hence, the set of (undirected) border edges is a set of cycles.*

This first step recognizes the locally cell connected manifold of $\mathcal{G}_g$ in $\mathcal{O}(g|V| + r)$ time steps and build a valid labelling for the manifolds recognized.

2.2   Second Step: Coordinates of Vertices

It is the main part of the work that must be done by the recognizer.

Now, as all edges are labeled, we would like to place coordinates from $\mathcal{G}$ on each vertex. Each finite automaton has a fixed memory, thus words of Σ_g^* of arbitrarily large length cannot be stored in such an automaton. So we will encode those coordinates into the time-space diagram of the automata, using a particular process.

In the Euclidean networks, we used coordinates of the form $a^n b^m$, in fact as words of $\{ab\}^*$, to represent the elements of the group $< a, b \mid aba^{-1}b^{-1} > = \mathbb{Z}^2$. In the hyperbolic problem, we will use coordinates of Σ_g^* to represent the elements of $\Gamma(4g, 4g)$

The coordinates (the words) have to be consistent with the labelling of edges. If we assign the word w_1 to the vertex ν_1 and the word w_2 to the neighbor vertex ν_2, and the label of (ν_1, ν_2) is a, then $w_1 a = w_2$ in $\mathcal{G}$. As we need to know if the words we put on vertices are coherent with the labelling, we have to solve the word problem: given any two words w and w', are they equal?

The Word Problem on Hyperbolic Groups: Building of a Tree

We explain here a new method to solve the word problem on hyperbolic Cayley graph, using finite automata. Remember that we could easily solve the word problem on Euclidean groups, by using Euclidean coordinates (that is the consequence of the Abelian properties of these groups). This is not the case for hyperbolic groups.

To solve the word problem on $\mathcal{G}_g =< \Sigma_g, w_\perp >\sim \Gamma(4g, 4g)$, we build a sublanguage L_T (a set of words) of S^* in bijection with $\mathcal{G}$, such that any words w of S^* could be easily transformed into the word of L_T that represents the same element of $\mathcal{G}$. To compare two given words consists now to transform them into the corresponding words of L_T and verify they are the same word.

Now, we explain how to build T, the graph of L_T. Given a $\Gamma(4g, 4g)$ and a given vertex $\odot$ (the origin) in this network, we can remark that every cell has one vertex that is closest to $\odot$ (distance n) and one vertex that is farthest from $\odot$ (distance $n + 2g$). To each cell C, we can associate a unique word w_C that is a cyclic conjugate of $w_\perp = a_1 a_2 ... a_{2g} a_1^{-1} ... a_{2g}^{-1}$ and w_C describes the successive labels of the edges in a contour word of the cell beginning at the closest vertex to $\odot$.

As we only take cyclic conjugates of w_0, and not words obtained by order reversing, words w_C induce an orientation of cells which gives a global orientation of Γ, as we can see on the figure 3.

Hence, for each cell C, the word $w_C = x_1 x_2 ... x_{4g}$ implies a natural numbering of the edges of the cell, from the first edge that is labelled with x_1 to the $4g^{th}$ edge that is labelled with x_{4g}. The first edge and the last edge are connected to the vertex of the cell C that is the closest to $\odot$. The edges labelled with x_{2g} and x_{2g+1} are connected to the farthest (from $\odot$) vertex of C.

Theorem 1 *If we consider the graph $\Gamma(4g, 4g)$ and the given vertex $\odot$, and, for each cell, we remove the $2g + 1^{th}$ edge, we transform the graph into a tree T, without changing the distance (from $\odot$) on the network.*

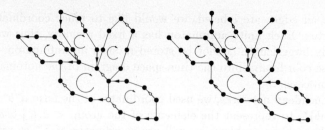

Fig. 3. A part of $\Gamma(8,8)$, with labelling on the left, and the tree that we obtained on the right.

Proof. The proof is easy by induction. Consider the cells that contain $\odot$. By removing one of the farthest edge of each of this cell, we do not change the distance between a vertex of these cells and $\odot$. So, the theorem holds for vertex at distance 1 from $\odot$.

Suppose, by induction hypothesis, that this is true for all vertices at distance n from $\odot$. All vertices at distance $n+1$ was connected to vertex at distance n. If we do not removed an edge from a vertex, its distance is not changed. But if we remove one edge, it means that this vertex was connected to two distinct vertices at distance n (from $\odot$). Now, it is connected to only one of this two vertex. So we do not change the distance (from $\odot$) for all these vertices: the induction hypothetis is verified for $n+1$.

And the subnetwork obtained is a tree, as we can see that, for each vertex ν, there is only one neighbor of ν that is closer to $\odot$ than ν. It is a consequence of the fact that, in any infinite $\Gamma(k,d)$, given two separate geodesics (they do not share any edge) from ν to $\odot$, the first edge of the first geodesic is on the same cell that the first edge of the second one: the cell where ν is the farthest vertex (from $\odot$). But in T, we remove one of these two edges, so we can not have such a configuration: there is only one path from any ν to $\odot$ (see [6] for structural details on $\Gamma(k,d)$).   □

Using the Tree to Compute Coordinates Words

So, the tree T that we build implies a language L_T of geodesic words. We will use this language to compute the geodesic words that we will use as coordinates.

If we have a word u of L_T and consider the word ub ($b \in \Sigma_g$), the word ub is in L_T except if this is the obvious case $u = vb^{-1}$ (that implies $ub = v \in L_T$) or if b corresponds to a forbidden (removed in T) edge. In this case, the subword composed by the $2g$ last letters of ub correspond to a half cell of the form $x_1x_2...x_{2g} = w_{end}x_{2g}$ such that $w_{end}x_{2g}$ is a subword of a cyclic conjugate of $\overline{w_{\perp}} = a_{2g}...a_2a_1a_{2g}^{-1}...a_1^{-1}$. We have to change it into into the other half of the same cell corresponding to the word $x_{2g}x_{2g-1}...x_1 = x_{2g}\overline{w_{end}}$.

By this substitution, we transform $u_0 = ub = u'w_{end}x_{2g}$ into the word $u_1 = u'x_{2g}\overline{w_{end}} = u'b\overline{w_{end}}$. A new subword w_{half}, of length $2g$, corresponding to another forbidden half-cell, can be created in $u'x_{2g}\overline{w_{end}}$, but the only possibility is that the last letter of w_{half} is $x_{2g} = b$, since $x_{i+1}x_i$ cannot be a subword of a cyclic conjugate of $\underline{w_\perp}$. This enforces: $w_{half} = w_{end}x_{2g}$.

So if we can state $u' = u''w_{end}$, then we transform $u_1 = u'x_{2g}\overline{w_{end}}$ into $u_2 = u''x_{2g}(\overline{w_{end}})^2$.

Repeating this process until there is no possible substitution, a word $u_{last} = u'''x_{2g}(\overline{w_{end}})^j$ of L_T is obtained, in $\mathcal{O}(\text{length of } w)$, such that ub and u_{last} represent the same element of G_g.

Hence, we have a linear time algorithm θ such that given a word u of L_T and a letter $b \in \Sigma_g$ computes the word $v = ub$ (in G_g) such that $v \in L_T$.

If we consider a word $v = y_1...y_n$ of Σ^*, to find the corresponding word u into L_T, just apply this algorithm on each prefix of v: $u_0 = \epsilon$ and $u_i = \theta(u_{i-1}y_i)$. Hence $u = u_n$.

Hence, we have an algorithm to compute the geodesics in quadratic time on one automaton. It will be linear on graph automata, due to the fact we have as many automata as the size of the words we need to transform.

The Algorithm of Coordinates on Graph Automata

Now, we will use the properties on L_T to compute coordinates for each vertex. These coordinates will be encoded in the space time of each vertex, using a sequence of signals L_{a_i} that represents the sequence of letters a_i that will be a word of L_T corresponding to a particular element of G_g. The sequence of signals is finished by a signal E (for "End").

Initialization. So the leader vertex sends signals L_x to each of its neighbors, where $x \in S$, is the label of the edge trough which the signal L is sent. The next time step, it sends a signal E to each of its neighbors.

Main process. Intuitively, each vertex will receive one or two words as a sequence of signals L_{x_i}. If the vertex receives only one word, there is no problem. If it receives several words, the vertex has to verify that the words are the same after a possible transformation. If the two words are not equal, there is a rejection.

When a vertex receives a word, letters of the word are stacked on a buffer of size $2g$. But at any time, the vertex verifies if the buffer can be simplified, simulating the algorithm of computing geodesics of T. Hence, the words received by next vertices are always in T except the last letter. $2g$ time steps after receiving the first letter (signal L), and one time step before sending the first letter of the word to a particular neighbor, the considered vertex sends a signal L_x where x is the label of the edge that leads to this neighbor, to complete the word.

To finalize this process, each time a vertex does not have down-edges and is assigned one single element $\mathcal{G}$ by this process, it will send a "Final" signal through its up-edges. When a vertex received the "Final" signal through all its down-edges, it will send this signal through its up-edges. It is obvious that when the leader received this signal from all its edges, the process of coordinates succeeded.

This second step recognizes the localy cell connected mappable (for the labelling built by the first step) manifold of $\mathcal{G}_g$ in $\mathcal{O}(g.r)$ time steps, by building a morphism φ from the manifold to $\mathcal{G}_g$.

2.3   Final Step: Non-overlap Verification

A mappable manifold is a subset of $\Gamma(4g, 4g)$ if and only if there are no two vertices with the same coordinates. So we have to verify the injectivity of φ. This is the last problem of *the overlaps*.

The General Case of Locally Cell Connected Subgraphs of $\Gamma(4g, 4g)$

We have to compare the coordinates of each vertex to the coordinates of other vertices. The method is easy. We just build a search tree and for each vertex ν on the tree, we make the process of puting coordinates on other vertices, as if ν were the leader. If another vertex receives the empty word (in L_T), we know there is an overlap.

This process takes $\mathcal{O}(g.|V|.r)$ time steps. So, it is quadric in the worst case. As we have to compare $\mathcal{O}(|V|^2)$ different coordinates of size $\mathcal{O}(r)$ with a space of size $\mathcal{O}(|V|)$, the time needed is at least $\mathcal{O}(|V|.r)$, using this method of resolution. We obtained $\mathcal{O}(r.\ln(r))$ for Euclidean networks, using the same arguments. Note that we can limit the verification to border vertices, but it does not actually change the complexity.

Theorem 2 *The class of locally cell connected subgraphs of $\Gamma(4g, 4g)$ can be recognized in quadratic time $\mathcal{O}(g.|V|.r)$.*

Linear Time for the Large Subclass of Convex Figures

Now, we can carry out different processes on the border to find geometrical properties using linear time , especially the convexity property.

Definition 5 *A cell compatible subgraph, a mappable graph or a manifold G (of $\Gamma(4g, 4g)$, $g > 1$) is convex iff there is only one single border cycle (the set of vertices of degree less than $4g$ defined a single connected cycle) and for any path p of $2g + 1$ consecutive vertices on the border such that p is a subsequence of a cell c, c is in G.*

This definition of convexity could be seen as *a local convexity property*. In fact, this definition of convexity that we give here is directly linked to the theorem of small cancellations ([3], for more details about this important theorem in hyperbolic groups). Due to this theorem (and more accurately, due to inherent property of hyperbolic networks), this local convexity implies strong consequences on graphs.

Theorem 3 (Theorem of small cancellations) *(Dehn 1911)*
For any $\mathcal{G}_g = <S, R>$, if w is a word of S^* that is equal to ϵ (null word) in $\mathcal{G}_g$, then there exists a word y, subword of w, with $|y| \leq 2g+1$, such that y can be reduced to $y' = y$ with $|y'| < |y|$.

Theorem 4 *The class of convex cell connected subgraphs of $\Gamma(4g, 4g)$ can be recognized in linear time.*

In fact, it is very easy to verify such a property of convexity. To begin any process on the border, the leader begins a depth first search in the graph ([12]), until it finds a border vertex. Then, this vertex becomes a local leader on its border cycle. When the process on this cycle is finished, the depth first search resumes until it finds another border vertex on a cycle not yet reached or it explored all the graph. Hence, the complexity of border algorithm is based on the number of vertices of the underlying graph. Then, just send a word trough the border cycle, like in the precedent part, and verify that the last $2g+1$ letters do not define a single cell. And if it is the case, verify that the cell is in the graph by looking at the other (non border) edges of the border vertices. Hence, we have a linear time algorithm to decide if a graph is convex. And from the theorem of small cancellations, we can directly deduce the following lemma:

Lemma 1 *When a mappable graph is convex, there is no possible overlap: a mappable convex graph is a subgraph of $\Gamma(4g, 4g)$*

3   Extensions

The algorithm proposed here can be extended in many different ways. But we can see that we do not need any property of planarity in our algorithm, nor compass.

We give a solution for $\Gamma(4g, 4g)$, but this result is true for all $\Gamma(k, d)$ with k and d enough large to have the property of hyperbolicity.

The locally cell connected property is not necessary for linear time complexity: we can use a global cell connected property, with more state in the automaton. We must finally note that there is a trivial algorithm for non-cell connected graphs in exponential time.

References

1. R. C. Lyndon, P. E. Schupp, *Combinatorial Group Theory*, Springer, (1977)
2. M. Gromov, *Hyperbolic groups*, in S.M. Gersten (ed.), *Essays in Group Theory*, Mathematical Sciences Research Institute Publications 8, Berlin (1987), 75–263
3. E. Ghys, P. de la Harpe *Sur les Groupes Hyperboliques d'après Mikhael Gromov* Progress in Mathematics, Birkhäuser, (1990)
4. David B. A. Epstein *Word Processing in Groups* J&B Publishers, (1992)
5. C. Papazian, E. Rémila *Linear time recognizer for subsets of Z^2* Proceedings of Fondamentals of Computation Theory (FCT), (2001), LNCS 2138, 400-403

6. C.Papazian, E. Rémila *Some Properties of Hyperbolic Networks* Proceeding of Discrete Geometry for Computer Imagery (DGCI), (2000), LNCS 1953, 149-158

7. J. Mazoyer, C. Nichitiu, E. Rémila, *Compass permits leader election*, Proceedings of Symposium on Discrete Algorithms (SODA), SIAM Editor (1999), 948-949

8. E. Rémila, *Recognition of graphs by automata*, Theoretical Computer Science 136, (1994), 291-332

9. E. Rémila, *An introduction to automata on graphs*, Cellular Automata, M. Delorme and J. Mazoyer (eds.), Kluwer Academic Publishers, Mathematics and Its Applications 460, (1999), 345-352

10. P. Rosenstiehl, *Existence d'automates finis capables de s'accorder bien qu'arbitrairement connectés et nombreux*, Internat. Comp. Centre 5 (1966), 245-261

11. P. Rosensthiel, J.R Fiksel and A. Holliger, *Intelligent graphs: Networks of finite automata capable of solving graph problems* , R. C. Reed, Graph Theory and computing, Academic Press, New-York, (1973), 210-265

12. A. Wu, A. Rosenfeld, *Cellular graph automata I*, Information and Control 42 (1979) 305-329

13. A. Wu, A. Rosenfeld, *Cellular graph automata II*, Information and Control 42 (1979) 330-353

Quantum and Stochastic Branching Programs of Bounded Width
(Track A)

Farid Ablayev[1], Cristopher Moore[2], and Christopher Pollett[3]

[1] Dept. of Theoretical Cybernetics
Kazan State University
420008 Kazan, Russia
ablayev@ksu.ru
[2] Computer Science Department
University of New Mexico
Albuquerque, New Mexico 87131
moore@cs.unm.edu
[3] Dept. of Math and Computer Science
San Jose State University
One Washington Square
San Jose, California 95192
pollett@mathcs.sjsu.edu

Abstract. We prove upper and lower bounds on the power of quantum and stochastic branching programs of bounded width. We show any NC^1 language can be accepted exactly by a width-2 quantum branching program of polynomial length, in contrast to the classical case where width 5 is necessary unless $NC^1 = ACC$. This separates width-2 quantum programs from width-2 doubly stochastic programs as we show the latter cannot compute the middle bit of multiplication. Finally, we show that bounded-width quantum and stochastic programs can be simulated by classical programs of larger but bounded width, and thus are in NC^1.

1 Introduction

Interest in quantum computation has been steadily increasing since Shor's discovery of a polynomial time quantum algorithm for factoring [14]. A number of models of quantum computation have been considered, including quantum versions of Turing machines, simple automata, circuits, and decision trees. The goal of much of this research has been to understand in what ways quantum algorithms do and do not offer a speed-up over the classical case, and to understand what classical techniques for proving upper and lower complexity bounds transfer to the quantum setting.

Branching programs have proven useful in a variety of domains, such as hardware verification, model checking, and other CAD applications [15]. Recently, several models of quantum branching programs have been proposed [1, 5,11]. Ablayev, Gainutdinova, and Karpinski [1] gave a matrix-based definition

P. Widmayer et al. (Eds.): ICALP 2002, LNCS 2380, pp. 343–354, 2002.
© Springer-Verlag Berlin Heidelberg 2002

of quantum branching programs as a natural generalization of quantum finite automata [8,9]. In contrast to what had been shown about one-way quantum automata [4], they showed that arbitrary boolean functions can be computed by one-way quantum branching programs. They exhibited a symmetric boolean function which could be computed with log-width, leveled, oblivious read-once quantum programs which require linear width for classical programs of this type. Finally, they gave a lower bound on the width of a read-once quantum program computing a boolean function in terms of the minimal width of a classical ordered binary decision diagram (OBDD) for the same function. Nakanishi, Hamaguchi, and Kashiwabara [11] took a graph-based approach to defining quantum branching programs. They give a language L_{Half} which can be recognized by ordered bounded-width quantum branching programs of polynomial length but which cannot be recognized by probabilistic programs of this type. Lastly, Ambainis, Schulman, and Vazirani [5] point out that 3-qubit quantum computers can simulate width-5 permutation branching programs and hence contain NC^1.

In this paper we prove several new results for quantum branching programs of bounded width. After reviewing the definition of [1], we show that width-2 quantum programs are more powerful than width-2 doubly stochastic programs, and are as strong as deterministic branching programs of width 5. Specifically, we show that polynomial-length, width-2 quantum branching programs can recognize any NC^1 language exactly. This is surprising, since such programs act on a single qubit. On the other hand, we show that polynomial-length, width-2 doubly stochastic programs cannot compute the middle bit of the multiplication function. In the classical case, Yao [16] showed that width-2 deterministic programs require superpolynomial length to compute the majority function, and Barrington [6] showed that width 5 is sufficient for deterministic programs to capture NC^1.

Finally, we improve the result of Ablayev, Gainutdinova, and Karpinski [1] by showing that bounded-probability quantum and stochastic programs can be simulated by deterministic programs of the same length and larger, but still bounded, width. Therefore these classes are contained in NC^1, and in fact for bounded-width quantum programs exact acceptance is just as strong as acceptance with bounded probability. We use the techniques of this result to show that polynomial-length width-2 stochastic programs accepting with error margin more than $\epsilon = 1/4$ cannot compute majority. In addition, we show that polynomial-length width-2 stochastic programs accepting with error margin more than $\epsilon = 1/8$ and polynomial-length width-3 stochastic programs accepting with error margin more than $\epsilon = 1/\sqrt{5}$ must compute functions in ACC.

2   Preliminaries

We begin by discussing the classical model of branching programs and then show how to quantize it. A good source of information on branching programs is Wegener's book [15], and for an introduction to quantum computation see Nielsen and Chuang [12].

Definition 1 *A* branching program *is a finite directed acyclic graph which recognizes some subset of* $\{0,1\}^n$. *Each node (except for the sink nodes) is labelled with an integer* $1 \leq i \leq n$ *and has two outgoing arrows labelled* 0 *and* 1. *This corresponds to querying the ith bit* x_i *of the input, and making a transition along one outgoing edge or the other depending on the value of* x_i. *There is a single source node corresponding to the start state, and there is a subset* A *of the sink nodes corresponding to accepting states. An input* x *is* accepted *if and only if it induces a chain of transitions leading to a sink node in* A.

A branching program is oblivious *if the nodes can be partitioned into levels* $V_1, \ldots, V_l$ *such that the nodes in* V_l *are the sink nodes, nodes in each level* V_j *with* $j < l$ *have outgoing edges only to nodes in the next level* V_{j+1}, *and all nodes in a given level* V_j *query the same bit* x_{i_j} *of the input. Such a program is said to have* length l, *and* width k *if each level has at most* k *nodes.*

Oblivious branching programs have an elegant algebraic definition. Recall that a *monoid* is a set with an associative binary operation · and an identity 1 such that $1 \cdot a = a \cdot 1 = a$ for all a.

Definition 2 *Let* M *be a monoid and* $S \subset M$ *an accepting set. Let* x_i, $1 \leq i \leq n$ *be a set of Boolean variables. A* branching program over M *of length* l *is a string of* l *instructions; the jth instruction is a triple* $(i_j, a_j, b_j) \in \{1, \ldots, n\} \times M \times M$, *which we interpret as* a_j *if* $x_{i_j} = 0$ *and* b_j *if* $x_{i_j} = 1$. *Given an input* x, *the* yield $Y(x)$ *of the program is the product in* M *of all its instructions. We say that the input* x *is* accepted *if* $Y(x) \in S$, *and the set of such inputs is the language* L *recognized by the program.*

Such programs are often called *non-uniform deterministic finite automata* (NUDFAs); a computation over a deterministic finite automaton consists of taking a product in its syntactic monoid, while in a NUDFA we allow the same variable to be queried many times, and for "true" and "false" to be mapped into a different pair of monoid elements in each query.

A common monoid is T_k, the set of functions from a set of k objects into itself. Then the program makes transitions among k states, and we can equivalently define oblivious, width-k branching programs by choosing an initial state and a set of accepting states, where the k states correspond, according to an arbitrary ordering, to the k vertices in each level V_j.

Definition 3 *An oblivious width-k branching program is a branching program over* T_k, *where the accepting set* $S \subset T_k$ *consists of those elements of* T_k *that map an initial state* $s \in \{1, \ldots, k\}$ *to a final state* $t \in A$ *for some subset* $A \subset \{1, \ldots, k\}$.

We define language classes recognized by (non-uniform) families of bounded-width branching programs whose length increases polynomially with n:

Definition 4 k-BWBP *is the class of languages recognized by polynomial-length branching programs of width* k, *and* BWBP $= \cup_k k$-BWBP.

Recall that a *group* is a monoid where every element has an inverse, and a group is *Abelian* if $ab = ba$ for all a, b. A subgroup $H \subseteq G$ is *normal* if the left and right cosets coincide, $aH = Ha$ for all $a \in G$. A group is *simple* if it has no normal subgroups other than itself and $\{1\}$.

Barrington [6] studied branching programs over the permutation group on k objects $S_k \subset T_k$; such programs are called *permutation programs*. He showed that polynomial-length programs over S_5, and therefore width-5 branching programs, can recognize any language in NC^1, the class of languages recognizable by boolean circuits of polynomial width and logarithmic depth. The version of Barrington's result that we will use is:

Theorem 1 ([6,10]). *Let G be a non-Abelian simple group, and let $a \neq 1$ be any non-identity element. Then any language L in NC^1 can be recognized by a family of polynomial-length branching programs over G such that their yield is $Y(x) = a$ if $x \in L$ and 1 otherwise.*

Since the smallest non-Abelian simple group is $A_5 \subset S_5$, the group of even permutations of 5 objects, and since we can choose a to map some initial state s to some other final state t, width 5 suffices. Conversely, note that we can model a width-k branching program as a boolean product of l transition matrices of dimension k, and a simple divide-and-conquer algorithm allows us to calculate this product in $O(\log l)$ depth. Thus $BWBP \subset NC^1$, so we have

$$5\text{-}BWBP = BWBP = NC^1 .$$

To define stochastic and quantum branching programs, we write the probability of acceptance as an inner product. Let e_s and e_t be k-dimensional vectors whose entries are 1 for the initial state and accepting final states respectively and 0 otherwise, and M_j the matrix corresponding to the jth instruction. Then write

$$P(x) = \left\langle e_s \left| \prod_{j=1}^{l} M_j \right| e_t \right\rangle$$

In the deterministic case $P = 1$ or $P = 0$ for all x, since the transition matrix corresponding to an element of T_k has exactly one 1 in each column. For a group this is true of the rows as well, in which case the M_j are permutation matrices. We can generalize this by letting the M_j be *stochastic* matrices, i.e. matrices with non-negative entries where each column sums to 1, and letting e_s be an initial probability distribution over the set of states. Then P is the probability that the program accepts. If the transpose of a stochastic matrix is also stochastic the matrix is called *doubly stochastic*. If all the matrices in a program are doubly stochastic then we say the program is a *doubly stochastic* program.

In the quantum case, we let the M_j be complex-valued and *unitary*, i.e. $M_j^{-1} = M_j^\dagger$ where $\dagger$ denotes the Hermitian conjugate, and e_s and e_t be initial and final state vectors with $|e_s|^2 = |e_t|^2 = 1$. Then the program accepts with probability

$$P(x) = \left| \left\langle e_s \left| \prod_{j=1}^{l} M_j \right| e_t \right\rangle \right|^2$$

For the purposes of this paper we will only need one possible final state vector, however, we could easily generalize the above to allow for many possible orthogonal final state vectors by summing over terms of the above type one for each possible final state vector. Note also that this is a "measure-once" model analogous to the quantum finite automata of [9], in which the system evolves unitarily except for a single measurement at the end. We could also allow multiple measurements during the computation, by representing the state as a density matrix and making the M_i superoperators instead of purely unitary operators; we do not do this here.

We can define recognition in several ways for the quantum case. We say that a language L is accepted *with bounded probability* if there is some $\epsilon > 0$ such that $P(x) > 1/2 + \epsilon$ if $x \in L$ and $P(x) < 1/2 - \epsilon$ if $x \notin L$, and accepted *exactly* if $P(x) = 1$ if $x \in L$ and $P(x) = 0$ if $x \notin L$ as in the deterministic case.

We denote by $B\cdot$ (where we will drop the '$\cdot$' when clear) the language classes recognized with bounded probability, and $E\cdot$ those recognized exactly. Writing SBP and QBP for stochastic and quantum branching programs respectively, we define the classes of languages recognized by width-k stochastic and quantum programs of polynomial length k-BSBP, k-BQBP, and k-EQBP. Note that we remove "BW" to avoid acronym overload. We write BSBP for $\cup_k k$-BSBP and define BQBP and EQBP similarly. We have

$$\text{BWBP} \subseteq \text{EQBP} \subseteq \text{BQBP}$$

and

$$\text{BWBP} \subseteq \text{BSBP}$$

but in principle k-BSBP could be incomparable with k-EQBP or k-BQBP.

3   Width-2 Doubly Stochastic and Quantum Programs

In this section we show that width-2 quantum programs with exact acceptance contain NC^1 and also show that these programs are stronger than width-2 doubly stochastic programs.

First we note that stochastic programs are stronger than permutation programs for width 2. It is easy to see that any program over $\mathbb{Z}_2$ simply yields the parity of some subset of the x_i. The AND_n function, which accepts only the input with $x_i = 1$ for all i, is not of this form, and so this language cannot be recognized by a width-2 permutation program. However, it can easily be recognized by a stochastic program P with bounded error which queries each variable once as follows: for $i < n$ it maps $x_i = 1$ and 0 to the identity $\begin{pmatrix} 1 & 0 \\ 0 & 1 \end{pmatrix}$ and the matrix $\begin{pmatrix} 1/2 \ 1/2 \\ 1/2 \ 1/2 \end{pmatrix}$ respectively, and for x_n it maps 1 and 0 to $\begin{pmatrix} 3/4 \ 1/4 \\ 0 \ 1 \end{pmatrix}$ and $\begin{pmatrix} 3/8 \ 5/8 \\ 3/8 \ 5/8 \end{pmatrix}$ respectively. Taking the first state to be both the initial and final state, P accepts with probability 3/4 if $x_i = 1$ for all i and 3/8 otherwise. Note that except for one matrix this is in fact a doubly stochastic program. If we had treated the variable x_n in the same fashion as the other variables we would have gotten a doubly stochastic program accepting AND_n with one-sided error.

Despite being stronger than their permutation counterparts, the next result shows width-2 doubly stochastic branching programs are not that strong. Let $MULT_k^n$ be the boolean function which computes the kth bit of the product of two n-bit integers. Define $MULT^n$ to be $MULT_{n-1}^n$. i.e., the middle bit of the product. We will argue that this function requires at least exponential lengthed width 2 stochastic programs.

Lemma 1. *Any width-2 doubly stochastic program on n variables is equivalent to one which queries each variable once and in the order 1,2,3, ... ,n.*

Proof. Any 2×2 stochastic matrix can be written as $\begin{pmatrix} p & 1-p \\ 1-p & p \end{pmatrix}$ for some $p \in [0,1]$. It is easy to verify that matrices of this kind commute. Hence, if we have a product of such matrices $\prod_j^n M_{x_{j_i}}$ we can rewrite it so that we first take the product of all the matrices that depend on x_1, then those that depend on x_2, and so on. To finish the proof we note that products of doubly stochastic matrices are again doubly stochastic, so we can use a single doubly stochastic matrix for the product of all the matrices that depend on a given x_i.

The above lemma shows we can convert any width-2 doubly stochastic program into one which is read-once and with a fixed variable ordering. i.e. a randomized ordered binary decision diagram (OBDD). The next result is proved in Ablayev and Karpinski [2].

Theorem 2. *A BP-OBDD that correctly computes $MULT^n$ has length at least $2^{\Omega(n/\log n)}$.*

So by Lemma 1 we have immediately:

Corollary 1 *Any width 2 doubly stochastic program correctly computing $MULT^n$ with bounded error has length at least $2^{\Omega(n/\log n)}$.*

So width-2 stochastic programs are not that strong. However, width-2 *quantum* programs are surprisingly strong, as the next result shows. Note that a width-2 quantum program has a state space equivalent to a single qubit such as a single spin-1/2 particle [12].

Theorem 3. NC^1 *is contained in 2-EQBP.*

Proof. First, recall that A_5, the smallest non-Abelian simple group, is the set of rotations of the icosahedron. Therefore, the group $SO(3)$ of rotations of $\mathbb{R}^3$, i.e. the $3{\times}3$ orthogonal matrices with determinant 1, contains a subgroup isomorphic to A_5.

There is a well-known 2-to-1 mapping from $SU(2)$, the group of 2×2 unitary matrices with determinant 1, to $SO(3)$. Consider a qubit $a|0\rangle + b|1\rangle$ with $|a|^2 + |b|^2 = 1$; we can make a real by multiplying by an overall phase. The *Bloch sphere* representation (see e.g. [12]) views this state as the point on the unit sphere with latitude θ and longitude ϕ , i.e. $(\cos\phi\cos\theta, \sin\phi\cos\theta, \sin\theta)$, where $a = \cos\theta/2$ and $b = e^{i\phi}\sin\theta/2$.

Given this representation an element of $SU(2)$ is equivalent to some rotation of the unit sphere. Recall the *Pauli matrices*

$$\sigma_x = \begin{pmatrix} 0 & 1 \\ 1 & 0 \end{pmatrix}, \quad \sigma_y = \begin{pmatrix} 0 & i \\ -i & 0 \end{pmatrix}, \quad \sigma_z = \begin{pmatrix} 1 & 0 \\ 0 & -1 \end{pmatrix}$$

Then we can rotate an angle α around the x, y or z axes with the following operators:

$$R_x(\alpha) = e^{i(\alpha/2)\sigma_x} = \begin{pmatrix} \cos\alpha/2 & -i\sin\alpha/2 \\ -i\sin\alpha/2 & \cos\alpha/2 \end{pmatrix}$$

$$R_y(\alpha) = e^{i(\alpha/2)\sigma_y} = \begin{pmatrix} \cos\alpha/2 & -\sin\alpha/2 \\ \sin\alpha/2 & \cos\alpha/2 \end{pmatrix}, \quad \text{and}$$

$$R_z(\alpha) = e^{i(\alpha/2)\sigma_z} = \begin{pmatrix} e^{-i\alpha/2} & 0 \\ 0 & e^{i\alpha/2} \end{pmatrix} .$$

This makes $SU(2)$ a *double cover* of $SO(3)$, where each element of $SO(3)$ corresponds to two elements $\pm U$ in $SU(2)$. (Note that angles get halved by this mapping.) Therefore, $SU(2)$ has a subgroup which is a double cover of A_5. One way to generate this subgroup is with $2\pi/5$ rotations around two adjacent vertices of an icosahedron. Since two such vertices are an angle $\tan^{-1} 2$ apart, if one is pierced by the z axis and the other lies in the x-z plane we have

$$a = R_z(2\pi/5) = \begin{pmatrix} e^{i\pi/5} & 0 \\ 0 & e^{-i\pi/5} \end{pmatrix}$$

$$b = R_y(\tan^{-1} 2) \cdot a \cdot R_y(-\tan^{-1} 2)$$

$$= \frac{1}{\sqrt{5}} \begin{pmatrix} e^{i\pi/5}\tau + e^{-i\pi/5}\tau^{-1} & -2i\sin\pi/5 \\ -2i\sin\pi/5 & e^{-i\pi/5}\tau + e^{i\pi/5}\tau^{-1} \end{pmatrix}$$

where $\tau = (1+\sqrt{5})/2$ is the golden ratio. Now consider the group element $c = a \cdot b \cdot a$; this rotates the icosahedron by π around the midpoint of the edge connecting these two vertices. In $SU(2)$, this maps each of the eigenvectors of σ_y to the other times an overall phase. Taking these as the initial and final state,

$$e_s = \frac{|0\rangle + i|1\rangle}{\sqrt{2}}, \quad e_t = \frac{|0\rangle - i|1\rangle}{\sqrt{2}}$$

we have

$$|\langle e_s|c|e_t\rangle|^2 = 1$$

while, since the two eigenvectors are orthogonal,

$$|\langle e_s|1|e_t\rangle|^2 = 0 \ .$$

Now, Theorem 1 tells us that for any language in NC^1 we can construct a polynomial-length program over A_5 that yields the element equivalent to c if the input is in the language and 1 otherwise. Mapping this language to $SU(2)$ gives a program which yields $\pm c$ or 1, and accepts with probability 1 or 0.

4   Classical Simulations of Stochastic and Quantum Branching Programs

In this section we give general results on simulating stochastic and quantum programs by classical ones. We use this to show that width-2 quantum programs can be simulated by bounded-width deterministic programs, and to show that if $\epsilon > 1/4$ then width-2 stochastic programs for majority require super-polynomial length. We also get that polynomial-length width-2 stochastic programs accepting with error margin more than $\epsilon = 1/8$ and polynomial-length width-3 stochastic programs accepting with error margin more than $\epsilon = 1/\sqrt{5}$ must compute functions in ACC.

Theorem 4. *If a language is recognized with bounded probability $1/2 + \epsilon$ by a width-w stochastic branching program, then it is also recognized by a deterministic branching program of the same length, and width*

$$w_C \le \lfloor \epsilon^{-(w-1)} \rfloor$$

Similarly, if a language is recognized with bounded probability $1/2 + \epsilon$ by a width-w_Q quantum branching program, it is recognized by a deterministic program of the same length, and width

$$w_C \le \lfloor (\epsilon/2)^{-(2w-1)} \rfloor$$

The proof uses essentially the same techniques as were used for Theorem 3 of [13] and Proposition 6 of [8], which show that stochastic and quantum finite state automata that accept with bounded probability can be simulated by deterministic finite state automata and therefore accept regular languages. Here we are extending this method to branching programs where the automaton is non-uniform; this was done for read-once branching programs in [1].

The idea is that if two state vectors are within a distance θ of each other, where θ depends on ϵ, then if the same operators are applied to both they must either both accept or both reject, and so are equivalent. Since only a finite number of balls of radius $\theta/2$ can fit into the state space we end up with a finite number of states. We start with the following lemmas. Note that we use the L_1

norm $|v|_1 = \sum_i |v_i|$ in the stochastic case, and the L_2 norm $|v|_2 = \sqrt{\sum_i |v_i|^2}$ in the quantum case.

As shorthand, say that a final probability distribution or state vector is *accepting* if it causes the program to accept with probability at least $1/2 + \epsilon$ and *rejecting* if it causes the program to accept with probability at most $1/2 - \epsilon$.

Lemma 2. *Consider a stochastic or quantum branching program. Let v, v' be two probability distributions or state vectors such that v is accepting and v' is rejecting. Then $|v - v'|_1 \geq 4\epsilon$ in the stochastic case, or $|v - v'|_2 \geq 2\epsilon$ in the quantum case.*

Proof. We prove the stochastic case first. As in the definition above, let $A \subset \{1, \dots, k\}$ be the set of accepting states. By hypothesis $\sum_{i \in A} v_i \geq 1/2 + \epsilon$ and $\sum_{i \notin A} v_i \leq 1/2 - \epsilon$, and vice versa for v'. Then

$$|v - v'|_1 = \sum_i |v_i - v'_i| \geq \left| \sum_{i \in A} v_i - \sum_{i \in A} v'_i \right| + \left| \sum_{i \notin A} v_i - \sum_{i \notin A} v'_i \right| \geq 4\epsilon \ .$$

In the quantum case, let e_t be the accepting final state. Write $v_{\mathrm{acc}} = \langle v | e_t \rangle e_t$ and $v_{\mathrm{rej}} = v - v_{\mathrm{acc}}$ for the components of v parallel to e_t and orthogonal to it, and similarly for v'. By hypothesis $|v_{\mathrm{acc}}|_2^2 \geq 1/2 + \epsilon$ and $|v_{\mathrm{rej}}|_2^2 \leq 1/2 - \epsilon$, and vice versa for v'. Then

$$|v_{\mathrm{acc}} - v'_{\mathrm{acc}}|, \ |v_{\mathrm{rej}} - v'_{\mathrm{rej}}| \geq \sqrt{1/2 + \epsilon} - \sqrt{1/2 - \epsilon}$$

so

$$|v - v'|_2 = \sqrt{|v_{\mathrm{acc}} - v'_{\mathrm{acc}}|_2^2 + |v_{\mathrm{rej}} - v'_{\mathrm{rej}}|_2^2} = \sqrt{1 + 2\epsilon} - \sqrt{1 - 2\epsilon} \geq 2\epsilon$$

where the final inequality comes from some simple algebra.

Given a set D in a metric space (see e.g. [3]) with metric ρ, we say two points x, y in D are *θ-equivalent* if they are connected by a chain of points where each step has distance less than θ; that is, $x \sim y$ if there are $z_0, \dots, z_m \in D$ such that $x = z_0$, $y = z_m$, and $\rho(z_i, z_{i+1}) < \theta$ for all $0 \leq i < m$. Call the equivalence classes of this relation the *θ-components* of D. Then:

Lemma 3. *Suppose a stochastic or quantum branching program P accepts a language with bounded probability $1/2 + \epsilon$. Then it is equivalent to a deterministic branching program of the same length where the states are the θ-components of the set of possible states at each step, where θ is given by Lemma 2.*

Proof. Since stochastic matrices never increase the L_1 distance, and unitary matrices preserve the L_2 distance, the θ-component of the P's state on a given step is a function only of the θ-component it was in on the previous step, and the value of variable queried on that step. This defines a deterministic branching program D on the set of θ-components of the same length as P. Lemma 2 shows that if two final states are in the same θ-component where $\theta = 4\epsilon$ or 2ϵ as applicable, they either both accept or both reject. Therefore, whether the final state is accepting or rejecting depends only on its θ-component, and D accepts if and only if P does.

Now all that remains is to bound the number of θ-components, or equivalently the maximum number of balls of radius $\theta/2$, that can fit in the state space of a stochastic or quantum program of width w. The w-dimensional probability distributions form a flat $(w-1)$-dimensional manifold of volume V and diameter 2, and a ball of radius $\theta/2 = 2\epsilon$ has volume at least $\epsilon^{w-1}V$. Similarly, the set of w-dimensional state vectors v with $|v|^2 = 1$ forms the complex sphere C^w, a $(2w-1)$-dimensional manifold of volume V and diameter 2. Due to the positive curvature of C^w, a ball of radius $\theta/2 = \epsilon$ has volume at least $(\epsilon/2)^{2w-1}V$. This completes the proof of Theorem 4 (note that both these bounds can be significantly improved with a little more work).

While we may need an exponentially larger width to simulate a stochastic or quantum branching program, the width is still constant. Thus bounded-width programs of all three types are equivalent, and since bounded-width classical programs are contained in NC^1 they all are. Conversely, we showed in Theorem 3 that NC^1 is contained in width-2 quantum programs, so we have

Corollary 2.

$$2\text{-EQBP} = 2\text{-BQBP} = \mathrm{BQBP} = \mathrm{BSBP} = \mathrm{BWBP} = \mathrm{NC}^1 \ .$$

In other words, width-2 quantum programs with exact acceptance are exactly as strong as quantum or stochastic programs of arbitrary bounded width, with exact or bounded-probability acceptance.

A more careful analysis of the number of θ-components can also be used to derive lower bounds on the capabilities of width-2 stochastic programs. For instance:

Corollary 3. *Any width-2 stochastic program recognizing majority with bounded probability $1/2 + \epsilon$ where $\epsilon > 1/4$ must have superpolynomial length.*

Proof. In the width-2 case the state space of a stochastic program consists of the pairs of points on the line from $(1,0)$ to $(0,1)$ in the plane. This line has L_1-length 2. If we allow θ-chains which might be centered on the end points, the maximum number of θ-components that can fit into this space is bounded by $\lfloor (2+4\epsilon)/4\epsilon \rfloor = \lfloor 1+1/2\epsilon \rfloor$. When $\epsilon > 1/4$ this gives 2, so using Lemma 3 a width-2 stochastic program with $\epsilon > 1/4$ can be simulated by a width-2 deterministic program of the same length. Finally, Yao [16] showed that deterministic width-2 programs for majority require super-polynomial length.

Recall that $\mathrm{ACC}[k]$ is the class of languages computed by families of polynomial-sized, unbounded fan-in, AND, OR, NOT, MOD_k circuits of constant depth. The class ACC is $\cup_k \mathrm{ACC}[k]$. Our techniques above can also be used to get the following result.

Corollary 4. *Polynomial-length width-2 stochastic programs accepting with error margin more than $\epsilon = 1/8$ and polynomial-length width-3 stochastic programs accepting with error margin more than $\epsilon = 1/\sqrt{5}$ must compute functions in* ACC.

Proof. By Barrington and Therien [7], we know that polynomial length width-4 deterministic programs compute functions in ACC. By our analysis in Corollary 3, if $\lfloor 1 + 1/2\epsilon \rfloor \leq 4$ then a polynomial-length width-2 stochastic program can be simulated by a width-4 deterministic program of the same length. This inequality holds provided $\epsilon > 1/8$. For the width-3 case we apply Theorem 4 to get the bound on simulating width-3 stochastic programs by width-4 deterministic ones.

Acknowledgments. We are grateful to Alexander Russell, Eric Allender, David Mix Barrington, and Azaria Paz for helpful conversations and e-mails. C.M. is supported by NSF grant PHY-0071139 and the Sandia University Research Program.

References

1. F. Ablayev, A. Gainutdinova, and M. Karpinski. On computational Power of quantum branching programs. *Proc. FCT 2001*, Lecture Notes in Computer Science 2138: 59–70, 2001.
2. F. Ablayev and M. Karpinski. A lower bound for integer multiplication on randomized read-once branching programs. *Electronic Colloquium on Computational Complexity* TR 98-011, 1998. http://www.eccc.uni-trier.de/eccc
3. P. Alexandrov. *Introduction to set theory and general topology.* Berlin, 1984.
4. A. Ambainis and R. Freivalds. 1-way quantum finite automata: strengths, weakness, and generalizations. *Proc. 39th IEEE Symp. on Foundations of Computer Science (FOCS)*, 332–342, 1998.
5. A. Ambainis, L. Schulman, and U. Vazirani. Computing with Highly Mixed States. *Proc. 32nd Annual ACM Symp. on Theory of Computing (STOC)*, 697–704, 2000.
6. D.A. Barrington. Bounded-width polynomial branching programs recognize exactly those languages in NC^1. *Journal of Computer and System Sciences* 38(1): 150–164, 1989.
7. D.A. Barrington and D. Therien. Finite Monoids and the Fine Structure of NC^1 *Journal of the ACM* 35(4): 941–952, 1988.
8. A. Kondacs and J. Watrous On the power of quantum finite automata. *Proc. of the 38th IEEE Symp. on Foundations of Computer Science (FOCS)*, 66–75, 1997.
9. C. Moore and J.P. Crutchfield. Quantum automata and quantum grammars. *Theoretical Computer Science* 237: 275–306, 2000.
10. C. Moore, D. Thérien, F. Lemieux, J. Berman, and A. Drisko. Circuits and Expressions with Non-Associative Gates. *Journal of Computer and System Sciences* 60: 368–394, 2000.
11. M. Nakanishi, K. Hamaguchi, and T. Kashiwabara. Ordered quantum branching programs are more powerful than ordered probabilistic branching programs under a bounded-width restriction. *Proc. 6th Annual International Conference on Computing and Combinatorics (COCOON)* Lecture Notes in Computer Science 1858: 467–476, 2000.

354   F. Ablayev, C. Moore, and C. Pollett

12. M.A. Nielson and I.L. Chuang. *Quantum Computation and Quantum Information.* Cambridge University Press. 2000.
13. M. Rabin. Probabilistic automata. *Information and Control* 6: 230–245, 1963.
14. P. Shor. Polynomial-time algorithms for prime factorization and discrete logarithms on a quantum computer. *SIAM Journal on Computing* 26(5): 1484–1509, 1997.
15. Ingo Wegener. *Branching Programs and Binary Decision Diagrams.* SIAM Monographs on Discrete Mathematics and Applications. 2000.
16. A.C. Yao. Lower Bounds by Probabilistic Arguments *Proc. of the 24th IEEE Symp. on Foundations of Computer Science (FOCS)*, 420–428, 1983.

Spanning Trees with Bounded Number of Branch Vertices[*]

Luisa Gargano[1], Pavol Hell[2], Ladislav Stacho[2], and Ugo Vaccaro[1]

[1] Dipartimento di Informatica ed Applicazioni,
Università di Salerno,
84081 Baronissi, Italy,
{lg,uv}@dia.unisa.it
[2] School of Computing Science,
Simon Fraser University,
Burnaby BC, V5A 1S6 Canada,
{pavol,lstacho}@cs.sfu.ca

Abstract. We introduce the following combinatorial optimization problem: Given a connected graph G, find a spanning tree T of G with the smallest number of branch vertices (vertices of degree 3 or more in T). The problem is motivated by new technologies in the realm of optical networks. We investigate algorithmic and combinatorial aspects of the problem.

1 Introduction

The existence of a Hamilton path in a given graph G is a much studied problem, both from the algorithmic and the graph–theoretic point of view. It is known that deciding if such a path exists is an NP-complete problem, even in cubic graphs G [10]. On the other hand, if the graph G satisfies any of a number of density conditions, a Hamilton path is guaranteed to exist. The best known of these density conditions, due to Dirac [6], requires each vertex of G to have a degree of at least $n/2$. (We shall reserve n to always mean the number of vertices of G, and to avoid trivialities we shall always assume that $n \geq 3$.) Other conditions relax the degree constraint somewhat, while requiring at the same time that $K_{1,3}$ (or sometimes $K_{1,4}$) is not an induced subgraph of G. Excluding these subgraphs has the effect of forcing each neighbourhood of a vertex to have many edges, allowing us to guarantee the existence of a Hamilton path with somewhat decreased degree condition.

---

[*] Research of the first and fourth authors was partially supported by the European Community under the RTN project: "APPROXIMATION AND RANDOMIZED ALGORITHMS IN COMMUNICATION NETWORKS (ARACNE)", and by the Italian Ministry of Education, University, and Research under the PRIN project: "RESOURCE ALLOCATION IN WIRELESS NETWORKS (REALWINE)"; the third author is on a Pacific Institute of the Mathematical Sciences postdoctoral fellowship at the School of Computing Science, Simon Fraser University.

P. Widmayer et al. (Eds.): ICALP 2002, LNCS 2380, pp. 355–365, 2002.

There are several natural optimization versions of the Hamilton path problem. For instance, one may want to minimize the number of leaves [15], or minimize the maximum degree, in a spanning tree of G [11,12,14,20,22]; either of these numbers is equal to two if and only if G has a Hamilton path. The best known optimization problem of this sort is the longest path problem [7,13,2]. (G has a Hamilton path if and only if the longest path has n vertices.) It is known that, unless $P = NP$, there is no polynomial time constant ratio approximation algorithm for the longest path problem, even when restricted to cubic graphs G which have a Hamilton path, cf. [2] and [7] where a number of other nonapproximability results are also discussed. In this paper, we introduce another possible optimization problem - minimizing the number of branch vertices in a spanning tree of G.

A *branch vertex* of G is a vertex of degree greater than two. If G is a connected graph, we let $s(G)$ denote the smallest number of branch vertices in any spanning tree of G. Since a spanning tree without branch vertices is a Hamilton path of G, we have $s(G) = 0$ if and only if G admits a Hamilton path. A tree with at most one branch vertex will be called a *spider*. Note that a spider may in fact be a path, i.e., have no branch vertices. Thus a graph G with $s(G) \leq 1$ admits a spanning subgraph that is a spider; we will say that G admits a *spanning spider*. There is an interesting intermediate possibility: We will call a graph G *arachnoid*, if it admits a spanning spider centred at each vertex of G. (A spider with a branch vertex is said to be *centred* at the branch vertex; a spider without branch vertices, i.e., a path, is viewed as centred at any vertex.) It follows from these definitions that a graph G with $s(G) = 0$ (i.e., a graph with a Hamilton path) is arachnoid, and that every arachnoid graph has $s(G) \leq 1$.

Our interest in the problem of minimizing the number of branch vertices arose from a problem in optical networks [19,25]. The wavelength division multiplexing (WDM) technology of optical communication supports the propagation of multiple laser beams through a single optical fiber, as long as each beam has a different wavelength. A *lightpath* connects two nodes of the network by a sequence of fiber links, with a fixed wavelength. Thus two lightpaths using the same link must use different wavelengths. This situation gives rise to many interesting combinatorial problems, cf. [3,8].

We consider a different situation, resulting from a new technology allowing a switch to replicate the optical signal by splitting light. *Light–trees* extend the lightpath concept by incorporating optical multicasting capability. Multicast is the ability to transmit information from a single source node to multiple destination nodes. Many bandwidth–intensive applications, such as worldwide web browsing, video conferencing, video on demand services, etc., require multicasting for efficiency purposes. Multicast has been extensively studied in the electronic networking community and has recently received much attention in the optical networking community [23,24,28,30].

The multicast can be supported at the WDM layer by letting WDM switches make copies of data packets directly in the optical domain via light splitting. Thus a light–tree enables all-optical communication from a source node to a set of destination nodes (including the possibility of the set of destinations consisting of all other nodes) [30]. The switches which correspond to the nodes of degree greater than two have to be able to split light (except for the source of the multicast, which can transmit to any number of neighbours). Each node with splitting capability can forward a number of copies equal to the number of its neighbours, while each other node can support only "drop and continue", that enables the node to receive the data and forward one copy of it; for the rationale behind this assumption see [30,1]. It should be noticed that optical multicasting (which can be implemented via light–trees) has improved characteristics over electronic multicasting since splitting light is "easier" than copying a packet in an electronic buffer [23].

However, typical optical networks will have a limited number of these more sophisticated switches, and one has to position them in such a way that all possible multicasts can be performed. Thus we are lead to the problem of finding spanning trees with as few branch vertices as possible.

Specifically, let G be the graph whose vertices are the switches of the network, and whose edges are the fiber links. With $s(G)$ light-splitting switches, placed at the branch vertices of an optimal spanning tree, we can perform all possible multicasts. In particular, if $s(G) = 1$, i.e., if G has a spanning spider, we can perform all possible multicasts with just one special switch. If G is a arachnoid graph, no light–splitting switches are needed. (Recall that the source of the multicast can transmit to any number of neighbours.) If $s(G) > 0$, the minimum number of light-splitting switches needed for all possible multicasts in G, is in fact equal to $s(G)$. Indeed, if k vertices of G are allowed to be branch vertices, then multicasting from one of these vertices results in a spanning tree of G with at most k branch vertices, thus $k \geq s(G)$.

In this paper, we investigate the parameter $s(G)$, with emphasis on graphs which admit a spanning spider, or which are arachnoid, in analogy with the study of graphs which admit a Hamilton path. We show that an efficient algorithm to recognize these graphs or a nice characterization for them is unlikely to exist, as the recognition problems are all NP-complete. In fact, we show that $s(G)$ is even hard to approximate. We explore several density conditions, similar to those for Hamilton paths, which are sufficient to give interesting upper bounds on $s(G)$. Finally, we also relate the parameter $s(G)$ to a number of other well studied graph parameters, such as connectivity, independence number, and the length of a longest path.

In dealing with branch vertices, it is helpful to observe that a cut vertex v of a graph G such that $G - v$ has at least three components must be a branch vertex of any spanning tree of G. We will use this observation throughout our arguments.

1.1   Summary of the Paper

The rest of the paper is organized as follows. In Section 2 we study the computational complexity of the problems we introduced and show they are all NP-complete. We also give nonapproximability results for $s(G)$. In Section 3 we explore density conditions which imply strong upper bounds on $s(G)$. In Section

4 we relate the parameter $s(G)$ to other important graph parameters; we also show that there is a polynomial time algorithm to find a spanning tree of a connected graph G with at most $\alpha(G) - 2$ branch vertices, where $\alpha(G)$ is the independence number of the graph. Finally, in Section 5 we conclude and give some open problems.

2   Complexity Results

We shall show that all of the problems we have introduced are NP-complete. Since all of them, when appropriately stated, belong to NP, we shall only exhibit the corresponding reductions. Recall that it is NP-complete to decide whether, given a graph G and a vertex v, there exists a Hamilton path in G starting at v [10].

We begin with the case of graphs G that admit a spanning spider, i.e., with $s(G) \leq 1$.

Proposition 1 *It is NP-complete to decide whether a graph G admits a spanning spider.*

Proof. Suppose G is a given graph, and v a given vertex of G. Construct a new graph G' which consists of three copies of G and one additional vertex adjacent to the vertex v of all three copies of G. It is then easy to see that G' has a spanning spider (necessarily centred at the additional vertex), if and only if G admits a Hamilton path starting at v.

More generally, we prove that it is NP-complete to decide whether a given graph admits a spanning tree with at most k branch vertices:

Proposition 2 *Let k be a fixed non-negative integer. It is NP-complete to decide whether a given graph G satisfies $s(G) \leq k$.*

Proof. In view of the preceding result, we may assume that $k \geq 2$. (Recall that $s(G) = 0$ is equivalent to G admitting a Hamilton path, a well known NP-complete problem.) Let again G be a given graph, and v a given vertex of G. This time we construct a graph G' from $2k$ disjoint copies of G and a complete graph on k additional vertices, by making each additional vertex adjacent to the vertex v of its own two copies of G. It is again easy to check that G' admits a spanning tree with at most k branch vertices (necessarily among the vertices of the additional complete graph), if and only if G admits a Hamilton path starting at v.

The problem of recognizing arachnoid graphs is also intractable:

Proposition 3 *It is NP-complete to decide whether a given graph G is arachnoid.*

Proof. We show that the problem of deciding whether a graph has a spanning spider can be reduced to the problem of deciding whether a graph is arachnoid. Given a graph G on n vertices, we construct new graph G' as follows. Take n copies of G, denoted by $G^1, G^2, \ldots, G^n$. Let $\{v_1^i, v_2^i, \ldots, v_n^i\}$ be the vertex set of G^i $(i = 1, 2, \ldots, n)$. The graph G' is constructed from $G^1, G^2, \ldots, G^n$ by identifying all vertices v_j^i $(j = 1, 2, \ldots, n)$ into one vertex called w. The vertex w is a cut-vertex in G', and hence if G' has a spanning spider, then w must be its center. Moreover, it is clear that such a spider exists if and only if G is arachnoid.

We show now that even approximating $s(G)$ seems to be an intractable problem. (More results on nonapproximability can be obtained by the same technique from other results of [2].)

Proposition 4 *Let k be any fixed positive integer. Unless $P = NP$, there is no polynomial time algorithm to check if $s(G) \leq k$, even among cubic graphs with $s(G) = 0$.*

Proof. This will follow from [2], and the following observation. Let $\ell(G)$ denote the maximum length of a path in G. Thus $\ell(G) = n$ if and only if G admits a Hamilton path, that is, $s(G) = 0$. We claim that in any cubic graph G with $s(G) > 0$

$$\ell(G) \geq \frac{2}{2s(G) + 1}(n - s(G)). \tag{1}$$

Consider an optimal spanning tree T of G, and denote by S its set of branch vertices. Thus $|S| = s(G)$. The graph induced on $T - S$ is a union of at most $2s(G) + 1$ paths. Indeed, there are at most $s(G) - 1$ paths joining two branch vertices (these are called *trunks* of T), and at most $s(G) + 2$ paths joining a branch vertex to a leaf (these are called *branches* of T). Thus the two longest paths of $T - S$ have at least the number of vertices counted by the right hand side of the above inequality. Since some path of G must contain both these paths of $T - S$, the inequality follows.

It is shown in [2] that there is no polynomial time algorithm guaranteed to find $\ell(G) \geq n/k$, even among cubic graphs with $\ell(G) = n$. It follows from our inequality that (for large enough n) $\ell(G) \geq \frac{1}{s(G)+1}n$, and the proposition follows.

We close this section by noticing that the inequality (1) can be generalized as follows (thus allowing corresponding generalizations of Proposition 4):

$$\ell(G) \geq \frac{2}{s(G)(\Delta - 1) + 1}(n - s(G)),$$

where Δ is the maximum degree of G.

3   Density

Of the several possible density conditions that assure the existence of a Hamilton path [9], we focus on the following result from [18] and [16]:

Theorem 1 [18][16] *Let G be a connected graph that does not contain $K_{1,3}$ as an induced subgraph.*

If each vertex of G has degree at least $(n-2)/3$ (or, more generally, if the sum of the degrees of any three independent vertices is at least $n-2$), then G has a Hamilton path.

We shall prove a similar result for spanning spiders:

Theorem 2 *Let G be a connected graph that does not contain $K_{1,3}$ as an induced subgraph.*

If each vertex of G has degree at least $(n-3)/4$ (or, more generally, if the sum of the degrees of any four independent vertices is at least $n-3$), then G has a spanning spider.

In fact, we have the following general result which treats the existence of a spanning tree with at most k vertices, and includes both the above theorems (corresponding to $k = 0$ and $k = 1$):

Theorem 3 *Let G be a connected graph that does not contain $K_{1,3}$ as an induced subgraph, and let k be a nonnegative integer.*

If each vertex of G has degree at least $\frac{n-k-2}{k+3}$ (or, more generally, if the sum of the degrees of any $k+3$ independent vertices is at least $n-k-2$), then $s(G) \le k$.

The proof of Theorem 3 is omitted from this extended abstract, but we offer the following observations.

Remark 1 We first notice that there is no new result of this type for arachnoid graphs, because the smallest lower bound on the degrees of a connected graph G without $K_{1,3}$ which would guarantee that G is arachnoid is $\frac{n-2}{3}$: The graph R_p obtained from a triangle abc by attaching a separate copy of $K_p, p \ge 3$, to each vertex a, b, c, is a $K_{1,3}$-free graph with minimum degree $\frac{n-2}{3}$, which is not arachnoid (all spanning spiders must have center in the triangle abc). However, if all degrees of a connected graph G without $K_{1,3}$ are at least $\frac{n-2}{3}$, then already Theorem 1 implies that G has a Hamilton path (and hence is arachnoid).

We also remark that Theorem 3 does not hold if $K_{1,3}$ is not excluded as an induced subgraph: For $k = 0$ this is well-known, and easily seen by considering, say the complete bipartite graph $K_{p,p+2}$ $(p \ge 1)$. For $k \ge 1$ we can take a path on $k+1$ vertices and attach a K_p $(p \ge 2)$ to every vertex of the path. Moreover, we attach an extra K_p to the first and the last vertex of the path. The resulting graph is not $K_{1,3}$-free, and has no spanning tree with k branch vertices. However, the degree sum of any $k+3$ independent vertices in the graph is at least $n-k-1$, where $n = (k+3)p - 2$.

Finally, we note that the bound $n - k - 2$ in the theorem is nearly best possible. For $k = 0$, this is again well-known, and can be seen by considering, say, the above graph R_p. For $k \ge 1$, we consider following example. Take two copies of R_p, where $p = k+1$. Shrink one K_p back to a vertex in one of the two copies of R_p, and attach the vertex to a vertex of degree $p-1$ in the other copy of R_p. The resulting graph has four copies of K_p, each with $p - 1$ vertices of degree $p-1$, and one copy K_p, denoted by K, with $p-2$ vertices of degree $p-1$. Take $k - 1$ vertices of degree $p - 1$ in K, and attach a copy of K_p to each. The resulting graph is $K_{1,3}$-free, and has no spanning tree with at most k branch vertices. The degree sum of any $k + 3$ independent vertices is at least $n - k - 5$, where $n = (k + 3)p + 2$.

We can prove a similar result for graphs that do not contain $K_{1,4}$ as an induced subgraph. (For the existence of Hamilton paths and cycles, such results can be found in [17,4].)

Naturally, there is a tradeoff between this weaker assumption and the minimum degrees in G one has to assume. We only state it here for $k \geq 1$, the main emphasis of our paper. The proof is omitted from this extended abstract.

Theorem 4 *Let G be a connected graph that does not contain $K_{1,4}$ as an induced subgraph, and let k be a nonnegative integer.*

If each vertex of G has degree at least $\frac{n+3k+1}{k+3}$ (or, more generally, if the sum of the degrees of any $k + 3$ independent vertices is at least $n + 3k + 1$), then $s(G) \leq k$.

We have also attempted to generalize the best known density theorem on hamiltonicity - the theorem of Dirac [6], which implies, in particular, that a graph in which all degrees are at least $\frac{n-1}{2}$ has a Hamilton path. We have the following conjecture:

Conjecture *Let G be a connected graph and k a nonnegative integer.*

If each vertex of G has degree at least $\frac{n-1}{k+2}$ (or, more generally, if the sum of the degrees of any $k + 2$ independent vertices is at least $n - 1$), then $s(G) \leq k$. We believe the conjecture, at least for $k = 1$. (Note that it is true for $k = 0$ by Dirac's theorem.) We have proved, for instance, that any connected graph G with all degrees at least $\frac{n-1}{3}$ has either a spanning spider with at most three branches, or has a spanning caterpillar. We are hoping to prove that the spanning caterpillar can be chosen so that all leaves except one are incident with the same vertex, i.e., that the caterpillar is also a spider. This would imply the conjecture for $k = 1$, and would also give additional information on the kind of spiders that are guaranteed to exist - that is, either a spider with at most three branches, or a spider with many branches but all except one having just one edge. We know that having all degrees at least $\frac{n-1}{3}$ is not sufficient to imply the existence of a spider with at most three branches - the graph $K_{p,2p}$ is a counterexample.

4   Relation to Other Problems

We relate our parameter $s(G)$ to other classical graph theoretic parameters.

First we make some additional remarks about arachnoid graphs.

Proposition 5 *If G is an arachnoid graph, then, for any set S of vertices, the graph $G - S$ has at most $|S| + 1$ components.*

Proof. If the deletion of S leaves at least $|S| + 2$ components, then no spider centred in one of the components can be spanning.

The condition in the proposition is a well-known necessary condition for a graph to have a Hamilton path. Recall that we have also observed in Section 3, that we don't have a density condition which implies that a graph is arachnoid without also implying that it has a Hamilton path. Thus we are lead to ask whether or not every arachnoid graph must have a Hamilton path. This is, in fact, not the case, but examples are not easy to find. One can, for instance, take a *hypotraceable graph* G, that is a graph which does not have a Hamilton path, but such that for each vertex v, the graph $G - v$ has a Hamilton path. Hypotraceable graphs are constructed in [26,27].

Proposition 6 *Every hypotraceable graph is arachnoid.*

Proof. For any vertex x, consider the Hamilton path in $G - v$, where v is any neighbour of x. Adding the edge xv yields a spanning spider of G centred at x.

The following observation shows the relationship between path coverings and $s(G)$.

Proposition 7 *If G is a connected graph whose vertices can be covered by k disjoint paths, then G has a spanning tree with at most $2k - 2$ branch vertices, that is,*

$$s(G) \leq 2k - 2.$$

Let $\alpha(G)$ denote the *independence number* of G, i.e. maximum size of an independent set of vertices in G, and let $\kappa(G)$ denote the *connectivity* of G, i.e. minimum number of vertices removal of which disconnects G or results in an empty graph. Chvátal and Erdős [5] proved that vertices of any graph G can be covered by at most $\lceil \alpha(G)/\kappa(G) \rceil$ vertex disjoint paths. Using the previous proposition we have

Theorem 5 *Let G be a connected graph. Then*

$$s(G) \leq 2 \lceil \alpha(G)/\kappa(G) \rceil - 2.$$

Thus, for 1-connected graphs, previous theorem gives $s(G) \leq 2\alpha(G) - 2$. One may in fact do a little bit better as shown by the following Theorem 6. A caterpillar in which all branch vertices are of degree 3 shows that Theorem 6 is best possible.

Theorem 6 *Let G be a connected graph which is not complete.*
 Then

$$s(G) \leq \alpha(G) - 2,$$

and a spanning tree with at most $\alpha(G) - 2$ branch vertices can be found in polynomial time.

Proof. We shall prove a more general statement. Let $T \subseteq G$ be a tree and let L be the set of all leaves and S be the set of all branch vertices of T. For every leaf $u \in L$, there is unique branch vertex $s(u) \in S$ (closest to u). These two vertices are joined by unique $s(u) - u$ path in T called the *branch of u*. Recall that the path joining two branch vertices of T is called a trunk of T. Note that a tree with l branch vertices has at least $l+2$ branches (and hence at least $l+2$ leaves).

Suppose that G is a counterexample to our theorem, i.e., it has no spanning tree with at most $\alpha(G) - 2$ branch vertices. Let $T \subseteq G$ be a tree of G with at most $\alpha(G) - 2$ branch vertices. We assume T is chosen so that it contains the greatest number of vertices, and, subject to this condition, so that

(i) the sum of the degrees of the branch vertices of T is as small as possible.

We may assume that T has exactly $\alpha(G) - 2$ branch vertices, by the connectivity of G and the maximality of T. It follows that T has at least $\alpha(G)$ leaves. If any two of these, say $u, v \in L$ are adjacent, then the tree $T + uv - us(u)$ violates (i). Thus, the set of leaves of T is an independent set of cardinality $\alpha(G)$. Since G is a counterexample, there is a vertex r in $V(G) - V(T)$. As the maximum independent set in G has cardinality $\alpha(G)$, the vertex r must be adjacent to at least one leaf of T, contradicting the maximality of T.

To see that the above process is constructive, it is enough to start with any tree T and then to modify T so that:

1. It has at most $\alpha(G) - 2$ branch vertices. This can be done, since if T has more than $\alpha(G) - 2$ branch vertices, then it has more that $\alpha(G)$ leaves, and at least two of them, say u, v are adjacent. Now, we modify T to $T + uv - us(u)$. This either results in $\alpha(G) - 2$ branch vertices in T, or we can repeat the process.

2. It has at most $\alpha(G)$ leaves. This can be accomplished by a procedure similar to the above.

3. It contains all the vertices of G. Indeed, if T has fewer than $\alpha(G) - 2$ branch vertices, then we can simply add any new vertex into T. Otherwise T has at least $\alpha(G) - 2$ branch vertices thus, at least $\alpha(G)$ leaves, and any vertex not in T must be adjacent to a leaf. This obviously allows for an extension.

To check which case applies takes polynomial time, and each of the cases is applied at most $|V(G)|$ times.

5 Concluding Remarks

In Section 2 we have proved that, unless $P = NP$, there is no polynomial time algorithm to approximate the minimum number of branch vertices $s(G)$ of any spanning tree of G, to within a constant factor. Under stronger complexity assumptions, stronger nonapproximability results can be derived. The first natural question to ask is which kind of positive results on the approximability of $s(G)$ one can obtain.

In Section 4 we proved that there exists a polynomial time algorithm to find a spanning tree of a graph G with at most $\alpha(G) - 2$ branch vertices, for any connected graph G. A (better) nonconstructive bound is given in Theorem 5 depending on the connectivity $\kappa(G)$ of G. An interesting problem is to improve Theorem 6 constructively taking into account the connectivity $\kappa(G)$ of G.

We have observed that every graph with a Hamilton path (such graphs are sometimes called *traceable*) is arachnoid, and proved in Proposition 6 that every hypotraceable graph is arachnoid, but these are the only ways we have of constructing arachnoid graphs. It would be interesting to find constructions of arachnoid graphs that are neither traceable (do not have a Hamilton path), nor hypotraceable.

In the same vein, all our arachnoid graphs have the property that each vertex is the center of a spider with at most three branches. We propose the problem of constructing arachnoid graphs in which this is not so, i.e., in which some vertex v is the center of only spiders with more than three branches.

Acknowledgements. We would like to thank Bruno Beauquier for his interest and inspiration.

References

1. M. Ali, *Transmission Efficient Design and Management of Wavelength–Routed Optical Networks*, Kluwer Academic Publishing, 2001.
2. C. Bazgan, M. Shanta, Z. Tuza, "On the Approximability of Finding A(nother) Hamiltonian Cycle in Cubic Hamiltonian Graphs", *Proc. 15th Annual Symposium on Theoretical Aspects of Computer Science*, LNCS, Vol. 1373, 276-286, Springer, (1998).
3. B. Beauquier, J-C. Bermond, L. Gargano, P. Hell, S. Perennes, U. Vaccaro, "Graph Problems arising from Wavelength-Routing in All-Optical Networks", *Proc. of WOCS*, Geneve, Switzerland, 1997.
4. G. Chen, R. H. Schelp, "Hamiltonicity for $K_{1,r}$-free graphs", *J. Graph Th.*, 20, 423–439, 1995.
5. V. Chvátal, P. Erdős. "A note on Hamiltonian circuits", *Discr. Math.*, 2:111–113, 1972.
6. G. A. Dirac, "Some theorems on abstract graphs", *Proc. London Math. Soc.*, 2:69–81, 1952.
7. T. Feder, R. Motwani, C. Subi, "Finding Long Paths and Cycles in Sparse Hamiltonian Graphs", *Proc. Thirty second annual ACM Symposium on Theory of Computing (STOC'00)* Portland, Oregon, May 21–23, 524-529, ACM Press, 2000.
8. L. Gargano, U. Vaccaro, "Routing in All–Optical Networks: Algorithmic and Graph–Theoretic Problems", in: *Numbers, Information and Complexity*, I. Althofer et al. (Eds.), Kluwer Academic Publisher, pp. 555-578, Feb. 2000.
9. R. J. Gould. Updating the Hamiltonian problem—a survey. *J. Graph Theory*, 15(2):121–157, 1991.
10. R. M. Karp, "Reducibility among combinatorial problems", *Complexity of Computer Computations*, R.E. Miller and J.W.Thatcher (eds.), Plenum Press, (1972), 85-103.

11. S. Khuller, B. Raghavachari, N. Young, "Low degree spanning trees of small weight", *SIAM J. Comp.*, 25 (1996), 335–368.
12. J. Könemann, R. Ravi, "A Matter of Degree: Improved Approximation Algorithms for Degree-Bounded Minimum Spanning Trees", *Proc. Thirty second annual ACM Symp. on Theory of Computing (STOC'00)*, Portland, Oregon, 537–546, (2000).
13. D. Krager, R. Motwani, D.S. Ramkumar, "On Approximating the Longest Path in a Graph", *Algorithmica*, 18 (1997), 82–98.
14. A. Kyaw, "A Sufficient Condition for a Graph to have a k Tree", *Graphs and Combinatorics*, 17: 113–121, 2001.
15. M. Las Vergnas, "Sur une proprieté des arbres maximaux dans un graphe", *Compte Rendus Acad. Sc. Paris*, 271, serie A, 1297–1300, 1971.
16. Y. Liu, F. Tian, and Z. Wu, "Some results on longest paths and cycles in $K_{1,3}$-free graphs", *J. Changsha Railway Inst.*, 4:105–106, 1986.
17. L. R. Markus. Hamiltonian results in $K_{1,r}$-free graphs. In *Proc. of the Twenty-fourth Southeastern International Conference on Combinatorics, Graph Theory, and Computing (Boca Raton,FL, 1993)*, volume 98, pages 143–149, 1993.
18. M. M. Matthews, D. P. Sumner. Longest paths and cycles in $K_{1,3}$-free graphs. *J. Graph Theory*, 9(2):269–277, 1985.
19. B. Mukherjee, *Optical Communication Networks*, McGraw–Hill, New York, 1997.
20. V. Neumann Lara, E. Rivera-Campo, "Spanning trees with bounded degrees", *Combinatorica*, 11(1):55–61, 1991.
21. O. Ore. Note on Hamilton circuits. *Amer. Math. Monthly*, 67:55, 1960.
22. B. Raghavari, "Algorithms for finding low–degree structures", in: *Approximation Algorithms for NP–Hard Problems*, D.S. Hochbaum (Ed.) PWS Publishing Company, Boston, 266–295, 1997.
23. L. Sahasrabuddhe, N. Singhal, B. Mukherjee, "Light-Trees for WDM Optical Networks: Optimization Problem Formulations for Unicast and Broadcast Traffic," *Proc. Int. Conf. on Communications, Computers, and Devices (ICCD-2000)*, IIT Kharagpur, India, (2000).
24. L. H. Sahasrabuddhe, B. Mukherjee, "Light Trees: Optical Multicasting for Improved Performance in Wavelength–Routed Networks", *IEEE Comm. Mag.*, 37: 67–73, 1999.
25. T.E. Sterne, K. Bala, *MultiWavelength Optical Networks*, Addison–Wesley, 1999.
26. C. Thomassen, "Hypohamiltonian and hypotraceable graphs", *Disc. Math.*, 9(1974), 91–96.
27. C. Thomassen, "Planar cubic hypo-Hamiltonian and hypotraceable graphs", *J. Combin. Theory Ser. B*, 30(1), 36–44, 1981.
28. Y. Wang, Y. Yang, "Multicasting in a Class of Multicast–Capable WDM Networks", *Journal of Lightwave Technology*, 20 (2002), 350–359.
29. S. Win, "Existenz von Gerüsten mit vorgeschriebenem Maximalgrad in Graphen", *Abh. Mat. Sem. Univ. Hamburg*, 43: 263–267, 1975.
30. X. Zhang, J. Wei, C. Qiao, "Costrained Multicast Routing in WDM Networks with Sparse Light Splitting", *Proc. of IEEE INFOCOM 2000*, vol. 3: 1781–1790, Mar. 2000.

Energy Optimal Routing in Radio Networks Using Geometric Data Structures*

René Beier, Peter Sanders, and Naveen Sivadasan

Max Planck Insitut für Informatik
Saarbrücken, Germany
{rbeier,sanders,ns}@mpi-sb.mpg.de

Abstract. Given the current position of n sites in a radio network, we discuss the problem of finding routes between pairs of sites such that the energy consumption for this communication is minimized. Though this can be done using Dijkstra's algorithm on the complete graph in quadratic time, it is less clear how to do it in near linear time. We present such algorithms for the important case where the transmission cost between two sites is the square of their Euclidean distance plus a constant offset. We give an $\mathcal{O}(kn \log n)$ time algorithm that finds an optimal path with at most k hops, and an $\mathcal{O}(n^{1+\epsilon})$ time algorithm for the case of an unrestricted number of hops. The algorithms are based on geometric data structures ranging from simple 2-dimensional Delaunay triangulations to more sophisticated proximity data structures that exploit the special structure of the problem.

1 Introduction

Networks between a number of sites can be set up without additional infrastructure using radio communication between the sites. Such an approach has recently gained considerable interest. We discuss the fundamental problem of finding good routes between pairs of sites.

Since the sites have often only limited energy reserves from batteries or power from solar panels, a prime optimization criterion is the energy consumption. We model this as follows. When two sites, say u and v, communicate directly, the sender node u needs a transmission energy $C_u + |uv|^2$. The *cost offset* C_u accounts for distance independent energy consumption like the energy consumption of the signal processing during sending and receiving. $|uv|^2$ denotes the square of the Euclidean distance of u and v. Although many cost functions for energy consumptions have been proposed this quadratic behavior is the most fundamental one since it accurately describes the behavior of electromagnetic waves in free space.

---

* This work was partially supported by DFG grant SA 933/1-1 and the Future and Emerging Technologies programme of the EU under contract number IST-1999-14186 (ALCOM-FT).

P. Widmayer et al. (Eds.): ICALP 2002, LNCS 2380, pp. 366–376, 2002.

In addition to energy consumption, another expensive resource is bandwidth[1] — transmitting a data packet using range r inhibits[2] communication in an area of size $\Theta(r^2)$. Hence, our quadratic cost measure can also be viewed as an estimate for the degree to which a route hinders other connections from using the same frequencies. A third cost measure is the reliability of a connection. The more nodes are involved, the higher is the probability of a failure. This cost measure is indirectly captured by the offset C_u.

We model the network as a complete geometric graph where the nodes of the graphs corresponds to the sites of the network and an edge from node u to v in the graph has weight $|uv|^2 + C_u$. In principle, an optimal path can be easily found in time $\mathcal{O}(n^2)$ using Dijkstra's shortest path algorithm in this graph.

The subject of this paper is to exploit the geometric structure of the problem to find energy optimal routes in time closer to $\mathcal{O}(n)$ than $\mathcal{O}(n^2)$. After developing a number of algorithms for this problem, we were made aware of previous work for geometric shortest path problems by Chan, Efrat, and Har-Peled [6,4] motivated by routing airplanes while taking into account that fuel consumption grows superlinearly with distance. It turned out that we rediscovered a very simple algorithm for cost functions without offset and at least quadratic growth. Here it suffices to search for a path in the Delaunay triangulation and hence there is a simple $\mathcal{O}(n \log n)$ algorithm. This is rather easy to see since a non-Delaunay edge e on a shortest path contains points inside the smallest circle enclosing the edge. Replacing e by two edges going via such a point leads to a smaller sum of squared lengths. The problem of this cost function for radio networks is that in large networks it leads to paths with an unrealistically large number of hops (edges).

For general cost functions, Chan, Efrat, and Har-Peled obtain an algorithm using time and space $\mathcal{O}(n^{4/3+\epsilon})$ for any positive constant ϵ.

Although this is a remarkable theoretical achievement, we believe that faster algorithms are so important for many applications that they are worth investigating even for more specialized classes of cost functions. Quadratic cost functions with offset seem a good compromise for radio networks. The quadratic component implies nice geometric properties and models energy consumption in free space and bandwidth consumption. The offsets model energy not invested in radio waves and allow us to penalize paths with too many hops.

1.1   New Results

In Section 2 we start with a very simple algorithm based on Voronoi diagrams that finds optimal 2-hop paths for uniform offset costs in time $\mathcal{O}(\log n)$ after a preprocessing time of $\mathcal{O}(n \log n)$ and space $\mathcal{O}(n)$. Using the dynamic programming principle this can easily be extended to an $\mathcal{O}(n \log n)$ algorithm for optimal

---

[1] 120 MHz of UMTS frequency range in Germany cost about $50 \cdot 10^9$ Euro

[2] In channel based systems like GSM, inhibition basically means blocking the channel for other concurrent communication. In CDMA systems like UMTS the inhibition is more subtle but nevertheless present.

paths with at most four hops. We then give a more general algorithm for finding optimal routes with up to k hops and non-uniform offset costs with running time $\mathcal{O}(kn \log n)$ and $\mathcal{O}(n)$ space by reducing it to a special case of 3D nearest neighbor problem and solving it efficiently. All the above algorithms are quite practical. Finally Section 3 gives a more theoretical solution for the general case of non-uniform costs and arbitrary number of hops with running time and space $\mathcal{O}(n^{1+\epsilon})$ for any constant $\epsilon > 0$. The solution is based on 3D closest bichromatic pair queries [7] similar to the general algorithm in [4]. We mainly improve the nearest neighbor queries involved to take advantage of the special structure of the problem.

1.2   Further Related Work

Energy efficient communication in radio network is a widely studied problem. Besides simple constant cost models [9] the most common simple model is our model with $f(|uv|) = |uv|^\rho$ for some constant $\rho \geq 2$ [12,10].[3] Rabaey et al. [8] mention sensor and monitoring networks as a natural application for radio networks where positions of stations are known because they are also needed on the application level. Energy-efficient communication by solving network optimization problems in a graph with transmission power as edge weights is an intensively studied approach. However, so far more complex problems like broadcasting and multi-cast routing have been studied (e.g. [15,14]) whereas the quadratic complexity of solving the shortest path problem in a complete graph has not been challenged.

An intensive area of research is distributed routing in dynamic situations without complete information.[4] Unfortunately, the worst case performance of such systems is not encouraging. Therefore we believe that research should be driven from both sides, namely : making pessimistic distributed models more powerful and making static models with complete information more dynamic. We view our work as a contribution in the latter direction because faster routing means that a centralized scheduler can handle more routing requests.

Although we are mainly applying computational geometry to radio communication, one could also take the more abstract point of view that we are interested in solving fundamental network problems in implicitly defined (geometric) graphs. There are results in this direction on spanning trees [11] and matchings [13].

2   Bounded Number of Hops

There are several reasons why one might prefer routes that traverse only a small number of edges (hops) in the communication graph. If there are more hops,

---

[3] Powers $\rho < 2$ are only relevant for nowadays rather special cases like short wave transmission with reflections on the ionosphere.

[4] A list of papers can be found at
http://www1.ics.uci.edu/~atm/adhoc/paper-collection/papers.html.

latency might get too large, reliability becomes an issue and distance independent energy consumption in the intermediate nodes may become too large. Therefore, we might be interested in the energy minimal path with the additional constraint that at most k hops should be used. Section 2.1 presents a simple algorithm for the case of two-hop paths with uniform offsets based on 2-dimensional closest point queries and extends this approach to paths with three and four-hops. Then Section 2.2 gives a generalization for k-hop paths with non-uniform offsets that needs more sophisticated and less standard data structures.

2.1   Two, Three, or Four Hops

Theorem 1. *In a complete geometric graph defined by n points in the plane with edge weights $C + |uv|^2$, queries for optimal 2-hop paths can be answered in time $\mathcal{O}(\log n)$ after preprocessing time $\mathcal{O}(n \log n)$ using $\mathcal{O}(n)$ space.*

Proof. Given source s and target t we can express the cost μ_w of the 2-hop path (s, w, t) as a function of $d = |st|$ and the distance r between w and the mid-point M of segment st. Fig. 1 illustrates this situation. We have

$$h^2 = r^2 - (d/2 - c)^2 = r^2 - d^2/4 - c^2 + dc,$$
$$\mu_w = a^2 + b^2 = c^2 + h^2 + (d - c)^2 + h^2 = d^2 + 2c^2 - 2dc + 2h^2$$
$$= d^2/2 + 2r^2.$$

Hence, we can minimize the path length by picking the intermediate node with the smallest Euclidean distance from M. Finding the nearest neighbor is a standard problem from computational geometry and can be solved in time $\mathcal{O}(\log n)$ after $\mathcal{O}(n \log n)$ preprocessing. For example we can build a Voronoi diagram of the nodes and a point location data structure for locating w in the Voronoi diagram [5, Section 7].   ∎

The 2-hop algorithm from Theorem 1 can be generalized for a dynamically changing set of transmitters. Using the dynamic closest point data structure [1] we can answer two-hop route request in time $\mathcal{O}(\text{Polylog } n)$. For any given $\epsilon > 0$, the data structure can be maintained dynamically in amortized time $\mathcal{O}(n^{1+\epsilon})$

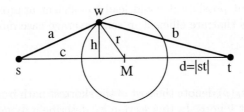

Fig. 1. Points with equal cost for two-hop routing lie on a circle around the mid point M between source s and target t.

per insert/delete operation. If we are content with approximate results within a factor $(1 + \delta)$, an answer can be obtained in time $\mathcal{O}(\log(n)/\delta)$ [3]. Although these worst case bounds are only obtained for rather complicated theoretical algorithms we expect that simple data structures will work in practical situations. For example, if transmitter positions are uniformly distributed, all operations run in constant expected time using a simple grid based data structure [2].

We now explain how the 2-hop algorithm leads to an efficient algorithm for three and four hops.

Theorem 2. *In a complete geometric graph defined by n points in the plane with edge weights $C + |uv|^2$, an optimal path with three or four hops between nodes s and t can be found in time $\mathcal{O}(n \log n)$ using space $\mathcal{O}(n)$.*

Proof. We first explain the method for a three-hop path (s, u, v, t). We build a data structure for answering nearest neighbor queries as in the case of two-hop in time $\mathcal{O}(n \log n)$. Then we consider all $n - 2$ possibilities for choosing v. We can find the shortest three-hop path (s, u, v, t) by adding $3C + |vt|^2$ to the shortest two-hop path from s to v. This is possible in time $\mathcal{O}(\log n)$ by Theorem 1. We get an overall execution time of $\mathcal{O}(n \log n)$.

To find a four-hop path (s, u, v, w, t) we proceed in a similar way only that we now have to find two such 2-hop paths ($s \to v$ and $t \to v$) for each possible choice of v. ∎

For a practical implementation of the above four-hop algorithm, several optimizations suggest themselves that might improve the constant factors involved. One could exploit that all the $2(n - 2)$ nearest neighbor queries are independent of each other. Hence, it suffices to implement a *batched* nearest neighbor algorithm that runs in time $\mathcal{O}(n \log n)$. Another more heuristic approach abandons batched access and considers candidate points close to the mid point M first. Let l be the length of the shortest path found so far. It can be shown that points outside the disc centered at M with radius $\sqrt{l - |st|^2/4}$ cannot possibly yield improved routes and the search can be stopped.

The basic approach from Section 2.1 for 2–4 hop routing can also be generalized for non-quadratic cost functions. We first locate "ideal" relays position in the Voronoi diagram or a grid file and then search in the neighborhood for good points using a more accurate cost measure. We have not analyzed the complexity of this approach but possibly it could lead to efficient approximate algorithms or exact algorithms that are efficient in some average case model.

2.2 k Hops

For any $p \in P$, let $\mu_k(p)$ denote the cost of the shortest path from the source node s to p using at most k hops. In this section we describe a dynamic programming algorithm that determines $\mu_k(t)$ in time $\mathcal{O}(kn \log n)$ for any $k \leq n - 1$. Clearly, $\mu_1(p) = C_s + |sp|^2$. For $i > 1$ the function μ_i can be computed recursively as follows:

$$\mu_{i+1}(p) = \min\{\mu_i(p), \min_{q \in P}\{\mu_i(q) + C_q + |qp|^2\}\}. \tag{1}$$

A trivial implementation of this formula yields an algorithm with running time $\mathcal{O}(kn^2)$: We start with μ_1 and iteratively determine μ_i for i increasing from 2 to k. In each of the $k - 1$ phases, equation 1 is applied for all $p \in P$. We can exploit the geometric structure of the problem to speed up the computation.

Theorem 3. *In a complete geometric graph defined by n points in the plane with edge weights $C_u + |uv|^2$, the single source shortest path problem restricted to paths with at most k-hops can be solved in time $\mathcal{O}(kn \log n)$ using space $\mathcal{O}(n)$.*

Proof. Applying Lemmas 1 and 3 below, $\mu_{i+1}(p)$ can be computed from μ_i for all $p \in P$ in time $\mathcal{O}(n \log n)$. This corresponds to one phase of the algorithm. For all $k-1$ phases, a total of $\mathcal{O}(kn \log n)$ time is sufficient. ∎

The problem of finding the point $q \in P$ that minimizes the expression $\mu_i(q) + C_q + |pq|^2$ on the right hand side of equation 1 can be reduced to a 3D nearest neighbor query. Consider the points P embedded into the plane $z = 0$ in 3-dimensional space. Thus a point $p \in P$ has the form $p = (x_p, y_p, 0) \in \mathbb{R}^3$. We code the cost of a i-hop path from s to p as well as the offset cost C_p using the third dimension: Define $f_i : P \to \mathbb{R}^3$; $f_i(x_p, y_p, 0) := (x_p, y_p, \sqrt{\mu_i(p) + C_p})$. Let $f_i(P) := \{f_i(p) : p \in P\}$. Since the vector from p to $f_i(p)$ is orthogonal to the plane $z = 0$ we have for each pair $p, q \in P$:

$$|f_i(q)p|^2 = |qp|^2 + \left(\sqrt{\mu_i(q) + C_q}\right)^2 = \mu_i(q) + C_q + |qp|^2.$$

Let $p \in P$ be fixed and choose $q \in P$ such that $f_i(q)$ is closest to p among the points in $f_i(P)$. Then q minimizes the expression $\mu_i(q) + C_q + |pq|^2$. Using equation 1 we can compute $\mu_{i+1}(p)$ by answering one nearest neighbor query for p. This justifies the following lemma:

Lemma 1. *The problem of computing $\mu_{i+1}(p)$ from $\mu_i(p)$ reduces to an instance of the nearest neighbor problem with sites $f_i(P)$ and query point $p = (p_x, p_y, 0)$.*

A standard approach for solving the nearest neighbor problem uses Voronoi diagrams together with a point location algorithm. The worst case combinatorial complexity of the Voronoi diagram $\mathcal{V}(S)$ for a 3D point set with n elements is $\Theta(n^2)$. Therefore, a straightforward application does not improve the asymptotic running time of our algorithm. Note however that for our purposes all query points lie in the plane $e := (z = 0)$. Hence it suffices to compute the subdivision of plane e that is induced by the Voronoi diagram $\mathcal{V}(f_i(P))$. Let $\mathcal{V}_e$ denote this subdivision of e, which can be regarded as the intersection of $\mathcal{V}(f_i(P))$ and plane e. It is very similar to a 2D Voronoi diagram except that there might be less than n cells (faces). The worst case combinatorial complexity of $\mathcal{V}_e$ is only linear in the number of points in P. This can easily be verified using Euler's formula for planar graphs.

In the following we give an algorithm for computing $\mathcal{V}_e(S)$ in time $\mathcal{O}(n \log n)$. We use the well known relationship between Voronoi diagrams and the intersection of upper half spaces [5]. Given a set S of n points in $\mathbb{R}^3$ we map each point $p = (p_x, p_y, p_z) \in S$ to the hyperplane h_p in $\mathbb{R}^4$ which is the graph of the 3-variate linear function

$$h_p(x, y, z) = 2p_x x + 2p_y y + 2p_z z - (p_x^2 + p_y^2 + p_z^2).$$

A given query point $q = (q_x, q_y, q_z)$ is closest to $p \in S$ (among all points in S) if and only if

$$p = \arg\max_{s \in S} h_s(q_x, q_y, q_z).$$

This can be interpreted geometrically in 4-dimensional space with dimensions x, y, z and v as the height[5]: Consider the half spaces bounded from below (with respect to v) by the hyperplanes $H(S) := \{h_s : s \in S\}$. The upper envelope $\mathcal{UE}(H(S))$ of $H(S)$ is defined as the boundary of the intersection of these half spaces which is the graph of the 3-variate function $f(x, y, z) = \max_{s \in S} h_s(x, y, z)$. Hence $p \in S$ is a nearest neighbor to query point q if and only if $f(q) = h_p(q)$. In this case h_p is a supporting hyperplane that contributes a facet to $\mathcal{UE}(H(S))$. Consequently, the orthogonal projection of $\mathcal{UE}(H(S))$ to the hyperplane $v = 0$ is the Voronoi diagram $\mathcal{V}(S)$. For our purposes we need to intersect the Voronoi diagram of the 3D point set $f_i(P)$ with the hyperplane $(z = 0)$. The worst case complexity of the resulting subdivision $\mathcal{V}_e$ and $\mathcal{V}(f_i(P))$ differ by a factor of n. To avoid computing unnecessary information, we schedule the transition to the subspace $z = 0$ before computing the upper envelope and obtain the following algorithm:

Algorithm 1

1. Compute the set of hyperplanes $H = H(f_i(P))$.
2. Intersect each hyperplane $h \in H$ with the hyperplane $z = 0$ to get a set of 2D hyperplanes H_z.
3. Compute the upper envelope $\mathcal{UE}(H_z)$.
4. Project $\mathcal{UE}(H_z)$ onto the xy-plane ($v = z = 0$).

Lemma 2. *Algorithm 1 computes* $\mathcal{V}_e(f_i(P))$ *correctly and has running time* $\mathcal{O}(n \log n)$.

Proof. Let $q = (q_x, q_y, 0)$ be any query point on the xy-plane lying in the Voronoi cell of some site $p \in f_i(P) = S$ (i.e, p is a nearest neighbor for q). So we have $h_p(q) = \max_{s \in S} h_s(q)$. Note that for any $s \in S$ the 2D hyperplane $h_s^z \in H_z$ is the graph of a 2-variate $h_s^z(x, y) = h_s(x, y, 0)$. Hence, for q, $h_p(q) = h_p^z(q_x, q_y) = \max_{s \in S} \{h_s^z(q_x, q_y)\}$. This means that h_p^z contributes a facet to $\mathcal{UE}(H_z)$. Moreover, in the projection of the $\mathcal{UE}(H_z)$ to the xy-plane, q is part of the Voronoi region that corresponds to site p.

---

[5] So we can use the notion of above and below.

Since the worst case combinatorial complexity of $\mathcal{UE}(H_z)$ is $\mathcal{O}(n)$, it can be computed in time $\mathcal{O}(n \log n)$ [5]. It is easy to verify that all other steps of algorithm 1 can be done in linear time.   ∎

Having computed $\mathcal{V}_e(f_i(P))$, we can make use of a point location algorithm to answer 2D nearest neighbor queries efficiently.

Lemma 3. *Let P be a set of n points in $\mathbb{R}^3$. Allowing $\mathcal{O}(n \log n)$ preprocessing time, the nearest neighbor problem for sites P and query points lying in the xy-plane can be answered in time $\mathcal{O}(\log n)$.*

3   The General Case

In this section we consider the case of computing the shortest path without any restrictions in the number of hops used. Since all edge weights are non-negative, this corresponds to computing $\mu_{n-1}(t)$ for target node t. Using the results of the last section one could solve the problem in time $\mathcal{O}(n^2 \log n)$ which is even worse than applying Dijkstra's algorithm to the complete graph. Therefore, we apply a different shortest path algorithm which was already used in [4]. It is similar to Dijkstra's algorithm in that it settles nodes in order of the distance from the source. But edges are never explicitly relaxed. Instead a powerful geometric data structure replaces both edge relaxation and the priority queue data structure of Dijkstra's algorithm.

Theorem 4. *In a complete geometric graph defined by n points in the plane with edge weights $C_u + |uv|^2$, the single source shortest path problem can be solved in $\mathcal{O}(n^{1+\epsilon})$ time and space for any constant $\epsilon > 0$.*

Let $\mu(v) = \mu_{n-1}(v)$ denote the length of a shortest path from source s to $v \in V$. During the execution of the algorithm each node is either reached (red) or unreached (blue). Initially all nodes except the source node are blue. Nodes change their color from blue to red in order of increasing $\mu()$ values. At the time a node becomes red, its $\mu()$ value is known. Let R and B denote the set of red and blue nodes respectively. In each iteration, we determine a blue node with smallest distance to the source:

$$v = \arg\min_{b \in B} \min_{r \in R} \{\mu(r) + |rb|^2 + C_r\} \tag{2}$$

This can be modeled as a closest bichromatic point problem with Euclidean distance function using one additional dimension. We define

$$f : R \longrightarrow \mathbb{R}^3; \qquad f(r_x, r_y, 0) = (r_x, r_y, \sqrt{\mu(r) + C_r}).$$

For any red-blue pair $(r, b) \in R \times B$, we have $|f(r)b|^2 = |rb|^2 + \sqrt{\mu(r) + C_r}^2$. Let $f(R) = \{f(r) : r \in R\}$. Then the closest bichromatic pair between the sets $f(R)$ and B gives the solution to Equation 2. Algorithm 2 states the pseudo code for the general case.

Algorithm 2

$B \leftarrow V\backslash\{s\}$, $R \leftarrow \{s\}$, $\mu[s] \leftarrow 0$;
while $(B \neq \emptyset)$

 Find closest bichromatic pair (r, b) between sets $f(R)$ and B.
 $R \leftarrow R \cup \{b\}$, $B \leftarrow B\backslash\{b\}$, $\mu[b] \leftarrow |f(r)b|^2$;

Correctness can be shown by induction with the following invariants: For any red-blue pair (r, b) we have $\mu(r) \leq \mu(b)$ and the $\mu()$-values of the red nodes are determined correctly. When a point p is discovered then Equation 2 ensures that all the remaining blue points have a $\mu()$-value at least as large as $\mu(p)$. Since edge weights are strictly positive, there cannot be a blue point on a shortest $s \to p$ path and hence $\mu(p)$ is computed correctly. The dynamic bichromatic closest points problems we have to solve are of special type: all blue points are lying in the xy-plane and the distance function is simply the Euclidean distance. This enables us to speed up the computation compared to the general 3D dynamic closest pair problem.

Corollary 1. *In a complete geometric graph defined by n points in the plane with edge weights $C_u + |uv|^2$, the single source shortest path problem can be reduced to answer a sequence of $n - 1$ closest bichromatic neighbor queries for dynamically changing point sets $R, B \in \mathbb{R}^3$ of cumulative size n where all points in B are lying in the xy-plane.*

A general technique for solving the dynamic bichromatic closest pair problem is given by Eppstein [7], who reduces the problem to the *dynamic nearest neighbors problem* for dynamically changing sets of sites R and B. For blue sites B the query points are red ($q \in R$) and vice versa. Since in our application the blue set is basically a 2D point set, we distinguish two types of nearest neighbor queries:

1. Sites S are restricted to the xy-plane ($S \subset \mathbb{R}^2$, query $q \in \mathbb{R}^3$).
2. Queries q are restricted to the xy-plane ($S \subset \mathbb{R}^3$, $q \in \mathbb{R}^2$).

We show that these special types of the 3D problem can be regarded as 2D nearest neighbor problems. Consider the first type. Given a query point $q = (q_x, q_y, q_z)$ we simply find the nearest neighbor for $q' = (q_x, q_y, 0)$. It is easy to verify that this is also a solution for q.

 The second type of nearest neighbor queries we considered already in the last subsection for a static set of sites. Here we again exploit the relation between Voronoi diagrams and the upper envelope of hyperplanes. Instead of pre-computing the upper envelope, we determine for each individual query point $q = (q_x, q_y, 0, 0)$ the first hyperplane that is intersected by the ray emanating from the point $(q_x, q_y, 0, +\infty)$ in $-v$ direction (see [1]). Since all query points lie in xy-plane, the corresponding query rays are part of the subspace $z = 0$. Hence it suffices to restrict the calculations to 3D space (with dimensions x, y, v). We maintain a set of 2D planes in $\mathbb{R}^3$ which are obtained by intersecting the hyper-planes $H(f(R)) = \{h_p : p \in f(R)\}$ (using the notation from section 2.2) with hyperplane $z = 0$. Agarwal and Matoušek [1] describe algorithms for dynamic ray shooting problems and state the following lemma.

Lemma 4. *Given a convex polytope in $\mathbb{R}^3$ described as the intersection of n half-spaces, one can process it in time $\mathcal{O}(n^{1+\epsilon})$ into a data structure of size $\mathcal{O}(n^{1+\epsilon})$, so that the first point of the polytope boundary hit by a query ray can be determined in $\mathcal{O}(\log^5 n)$ time. The data structure can be maintained dynamically in amortized time $\mathcal{O}(n^\epsilon)$ per insert/delete operation.*

Combined with the result of Eppstein [7] for maintaining the bichromatic closest pair, this proves Theorem 4. In contrast to the work of [4], we considered a more restrictive cost function which leads to a better running time. They used the lower envelope of a set of bivariate functions to solve the nearest neighbor problem whereas we maintain a set of linear object, namely 2D planes.

We showed how to solve the problem by answering a sequence of closest bichromatic point queries for dynamic sets R and B. These queries have more special properties that we have not exploited:

1. We never insert points into set B.
2. We never delete points from set R.

Our application is even more specific when we view a deletion followed by an insertion as one operation, which is then the only operation we need: A node turns its color from blue to red and is lifted up in the third dimension keeping its x- and y-coordinates. This might be a starting points for finding more practical algorithms.

4   Discussion

We have proposed efficient algorithms for energy and bandwidth efficient communication in a model designed for simplicity and resemblance to the actual physical conditions. The problems of communication in real radio networks are vastly more complicated and probably do not allow similarly clean solutions. However, we believe that the algorithmic principles developed here may nevertheless remain useful. In particular, the 2–4 hop routing algorithms from Section 2.1 are so simple and flexible that they might be useful in practice.

Acknowledgments. We would like to thank Edgar Ramos for contributing valuable geometric know-how.

References

1. P. D. Agarwal and J. Matousek. Ray shooting and parametric search. In N. Alon, editor, *Proceedings of the 24th Annual ACM Symposium on the Theory of Computing*, pages 517–526, Victoria, B.C., Canada, May 1992. ACM Press.
2. J. Bentley, B. W. Weide, and A. C. Yao. Optimal expected-time algorithms for closest point problems. *ACM Transactions on Mathematical Software*, 6(4):563–580, December 1980.

3. T. Chan. Approximate nearest neighbor queries revisited. In *Proceedings of the 13th International Annual Symposium on Computational Geometry (SCG-97)*, pages 352–358, New York, June 4–6 1997. ACM Press.

4. T. Chan and A. Efrat. Fly cheaply: On the minimum fuel consumption problem. *Journal of Algorithms*, 41(2):330–337, November 2001.

5. M. de Berg, M. van Kreveld, M. Overmars, and O. Schwarzkopf. *Computational Geometry Algorithms and Applications*. Springer-Verlag, Berlin Heidelberg, 2., rev. ed. edition, 2000.

6. A. Efrat and S. Har-Peled. Fly cheaply: On the minimum fuel-consumption problem. In *Proceedings of the Fourteenth Annual Symposium on Computational Geometry (SCG'98)*, pages 143–145, New York, June 1998. Association for Computing Machinery.

7. D. Eppstein. Dynamic euclidean minimum spanning trees and extrema of binary functions. *Disc. Comp. Geom.*, 13:111–122, 1995.

8. J. M. Rabaey et al. Picoradio supports ad hoc ultra-low power wireless networking. *IEEE Computer Magazine*, pages 42–48, July 2000.

9. K. Nakano, S. Olariu, and A. Y. Zomaya. Energy-efficient permutation routing in radio networks. *IEEE Transactions on Parallel and Distributed Systems*, 12(6):544–557, 2001.

10. D. Patel. Energy in ad-hoc networking for the picoradio. Master's thesis, UC Berkeley, 2000.

11. F. P. Preparata and M. I. Shamos. *Computational Geometry*. Springer, 1985.

12. T. S. Rappaport. *Wireless Communication*. Prentice Hall, 1996.

13. K. R. Varadarajan and P. K. Agarwal. Approximation algorithms for bipartite and non-bipartite matching in the plane. In *SODA*, pages 805–814, 1999.

14. Peng-Jun Wan, G. Calinescu, Xiang-Yang Li, and Ophir Frieder. Minimum-energy broadcast routing in static ad hoc wireless networks. In *IEEE Infocom*, 2001.

15. J. E. Wieselthier, G. D. Nguyen, and A. Ephremides. On the construction of energy-efficient broadcast and multicast trees in wireless networks. In *IEEE Infocom*, volume 2, pages 585–594. IEEE, 2000.

Gossiping with Bounded Size Messages in *ad hoc* Radio Networks

(Extended Abstract)

Malin Christersson[1], Leszek Gąsieniec[2*], and Andrzej Lingas[1]

[1] Department of Computer Science, Lund University, Box 118, S-221 00 Lund, Sweden, {Malin.Christersson,Andrzej.Lingas}@cs.lth.se
[2] Department of Computer Science, University of Liverpool, Chadwick Building, Peach Street, Liverpool L69 7ZF, UK, leszek@csc.liv.ac.uk.

Abstract. We study deterministic algorithms for the gossiping problem in *ad hoc* radio networks under the assumption that each combined message contains at most $b(n)$ single messages or bits of auxiliary information, where b is an integer function and n is the number of nodes in the network. We term such a restricted gossiping problem $b(n)$-*gossiping*. We show that $\sqrt{n}$-gossiping in an *ad hoc* radio network on n nodes can be done deterministically in time $\tilde{O}(n^{3/2})$ which asymptotically matches the best known upper bound on the time complexity of unrestricted deterministic gossiping[†]. Our upper bound on $\sqrt{n}$-gossiping is tight up to a poly-logarithmic factor and it implies similarly tight upper bounds on $b(n)$-gossiping where function b is computable and $1 \leq b(n) \leq \sqrt{n}$ holds. For symmetric *ad hoc* radio networks, we show that even 1-gossiping can be done deterministically in time $\tilde{O}(n^{3/2})$. We also demonstrate that $O(n^t)$-gossiping in a symmetric *ad hoc* radio network on n nodes can be done in time $O(n^{2-t})$. Note that the latter upper bound is $o(n^{3/2})$ when the size of a combined message is $\omega(n^{1/2})$. Furthermore, by adopting known results on repeated randomized broadcasting in symmetric *ad hoc* radio networks, we derive a randomized protocol for 1-gossiping in these networks running in time $\tilde{O}(n)$ on the average. Finally, we observe that when a collision detection mechanism is available, even deterministic 1-gossiping in symmetric *ad hoc* radio networks can be performed in time $\tilde{O}(n)$.

1 Introduction

An *ad hoc* radio network is usually modeled as a directed graph. The nodes of the graph are interpreted as distributed processors that communicate and work in synchronous time-steps. There is a directed edge between two nodes if the successor node is in the radio range of the predecessor one. If the relation of

---

* Work of this author is supported in part by EPSRC grants GR/N09855 and GR/R85921.
† Notation $\tilde{O}(f(n))$ stands for $O(f(n) \cdot \log^c n)$, for any constant $c > 0$.

P. Widmayer et al. (Eds.): ICALP 2002, LNCS 2380, pp. 377–389, 2002.

being in the range is symmetric then each edge is accompanied by the reverse edge and the network is called *symmetric*. In every time-step, each processor can either transmit or receive a message as well as perform local computations. A transmitted message is always sent to all the neighboring processors in one time-step. A neighboring processor however, can only receive the message if it is the only message sent to it at this time-step. All our results (with exception of section 3.4) are presented in a model with a single radio frequency where collision detection is not available, e.g., see [3,5].

There are two basic tasks in network communication, *broadcasting* and *gossiping*. The broadcasting operation consists in distributing a single message from one source node to all other nodes. The gossiping operation consists in distributing a unique message from each node to all other nodes in the network. In order to make the gossiping problem feasible, the network must be strongly connected. We shall study the gossiping problem in *ad hoc* radio networks, i.e., networks where the topology of the network is unknown to the processors, each processor knows only the total number of nodes n and its own identification number drawn from set $\{1, 2, .., cn\}$, for some constant c.

In the literature on the gossiping problem in *ad hoc* radio networks, one always assumes, with few exceptions [3,10] that each processor can transmit all its current knowledge in a single step of the algorithm. Such assumption may not be realistic in case of very large networks where nodes can acquire very large knowledge. And indeed it is very natural to impose an upper bound on the size of a combined message that a single processor can transmit in a single step. In [3] Bar-Yehuda *et al.* and in [10] Clementi *et al.* consider a radio network model with the size of combine messages limited to a logarithmic number of bits. In this paper we rather express this limit in terms of an integer function $b(n)$ where n is the number of processors in the network. The upper bound on the size of a combined message is interpreted in terms of the number of unit messages (i.e., the ones originally held at nodes of the network, if they are of the same size) included in it. More precisely, such a combined message to transmit is allowed to contain up to $b(n)$ single messages or alternatively single bits of auxiliary information (if size of single messages varies from node to node). We term so restricted problem as the $b(n)$-*gossiping* problem.

Similarly as the authors of previous works on the gossiping problem in *ad hoc* radio networks, we are primarily interested in communication complexity of our gossiping algorithms (e.g., we do not consider the cost of local computations at each node).

1.1   Previous Work

The standard collision-free communication procedure for *ad hoc* radio networks is called Round Robin. The procedure runs in rounds, in the i-th step of each round the node with identity i transmits its whole knowledge to all its neighbors. Round Robin is used as a subroutine in many broadcasting and gossiping algorithms. It completes gossiping in D rounds where D is the network eccentricity, i.e., the maximum of the lengths of shortest directed paths from a node u to a node v, taken over all ordered pairs (u, v) of nodes in the network. Hence, it requires $O(nD)$ time-steps to complete gossiping.

In general directed *ad hoc* radio networks Chrobak, Gąsieniec and Rytter have shown in [9] that there exists a deterministic algorithm for broadcasting that works in time $O(n \log^2 n)$. Using their broadcasting procedure, they also developed a gossiping algorithm that runs in time $O(n^{3/2} \log^2 n)$. Both algorithms are non-constructive. Later Indyk [16] developed a constructive solutions with similar complexities $\tilde{O}(n)$ and $\tilde{O}(n^{3/2})$. In [8], Chrobak, Gąsieniec and Rytter presented an $O(n \log^4 n)$-time randomized algorithm for gossiping in *ad hoc* radio networks. In [11], Clementi, Monti and Silvestri presented algorithms whose time performance is expressed in terms of the maximum in-degree Δ of the network as well as its eccentricity D. Their algorithms for gossiping and broadcasting in *ad hoc* radio network run in time $O(D\Delta \log^3 n)$ and $O(D\Delta \log^2 n)$, respectively. The same authors have also studied deterministic broadcasting in known radio networks in the presence of both static as well as dynamic faults [12]. Also recently an alternative deterministic $\tilde{O}(n\sqrt{D})$-time gossiping algorithm was proposed by Gąsieniec and Lingas in [14].

For symmetric *ad hoc* radio networks, Chlebus *et al.* have presented an optimal linear-time broadcasting protocol in [5]. The protocol yields also linear-time gossiping in these networks.

Recently, Chlebus *et al.* established several upper and lower bounds on the time complexity of the so called *oblivious* gossiping algorithms in *ad hoc* radio networks [6]. In an oblivious multiple-communication algorithm, the fact that a processor transmits or not at a given time-step depends solely on the identification number of the processor, n and the number of the time-step.

In symmetric radio networks with combined messages of size limited to $O(\log n)$ bits and partially known (e.g., a local neighborhood is known) networks [3], Bar-Yehuda *et al.* introduced a randomized distributed communication protocol that performs broadcast of k messages in time $O((D + k) \log \Delta \log n)$ admitting a randomized gossiping algorithm in time $O(n \log \Delta \log n)$. Another interesting study, by Clementi *et al.*, of both randomized and deterministic multi-broadcast protocols in directed *ad-hoc* radio networks with messages of size $O(\log n)$ can be found in [10]. Their approach imposes a deterministic gossiping algorithm that runs in time $O(n\Delta^2 \log^3 n)$. Recently, the issue of gossiping in entirely known radio network with messages of a unit size has been studied by Gąsieniec and Potapov in [15]. In this paper we investigate (mainly) deterministic communication algorithms in both symmetric as well as general **ad hoc** radio networks with bounded size messages.

For the survey of numerous earlier results on broadcasting and gossiping in *ad hoc* radio networks (e.g., [2,17]), the reader is referred to [4,18].

1.2   This Work

Initially we present (time) upper bounds on $b(n)$-gossiping in *ad hoc* networks for non-trivial values of $b(n)$. Our main result is that $\sqrt{n}$-gossiping in an *ad hoc* radio network on n nodes can be done deterministically in time $\tilde{O}(n^{3/2})$ which asymptotically matches the best known upper bound on the time complexity of unrestricted deterministic gossiping. We also observe that our upper bound on $\sqrt{n}$-gossiping is tight up to a poly-logarithmic factor and that it implies similarly tight upper bounds on $b(n)$-gossiping where function b is computable and $1 \leq b(n) \leq \sqrt{n}$ holds.

For symmetric *ad hoc* radio networks, we prove that even 1-gossiping can be done deterministically in time $\tilde{O}(n^{3/2})$. We also show that $O(n^t)$-gossiping in a symmetric *ad hoc* radio network on n nodes can be done in time $O(n^{2-t})$. Note that the latter upper bound is $o(n^{3/2})$ when the size of a combined message is $\omega(n^{1/2})$. Furthermore, by adopting known results on repeated randomized broadcasting in symmetric *ad hoc* radio networks, we derive a randomized protocol for 1-gossiping in these networks running in time $\tilde{O}(n)$ on the average. Finally, we observe that when a collision detection mechanism is available, even deterministic 1-gossiping in symmetric *ad hoc* radio networks can be performed in time $\tilde{O}(n)$.

2   Bounded Gossiping in Directed *ad hoc* Radio Networks

2.1   Bounded Broadcasting and Round-Robin

In order to specify our algorithm for $\sqrt{n}$-gossiping in an *ad hoc* radio network on n nodes, we adopt the following conventions.

- For an integer function b, a $b(n)$-*broadcast* from a node v means deterministically broadcasting from v in such a way that only combined messages containing up to $b(n)$ selected single messages (known to v) or bits of auxiliary information are transmitted.
- For an integer function $b(n)$, $b(n)$-ROUNDROBIN means the Round-Robin procedure operating with combined messages containing no more than $b(n)$ single (original) messages known to the transmitting node **at the beginning** of the current round of Round Robin.

The broadcasting algorithm due to Chrobak *et al.* [9] and its alternative constructive variant due to Indyk [16] imply the following fact.

Fact 1. *For any computable integer function $b(n)$, a constructive $b(n)$-broadcast from a node of an ad hoc radio network on n nodes can be performed deterministically in time $\tilde{O}(n)$.*

The next fact is straightforward.

Fact 2. *For any computable integer function $b(n)$, an execution of a single round of $b(n)$-ROUNDROBIN in a network with n nodes takes time $O(n)$.*

2.2   The $\sqrt{n}$-Gossiping Algorithm

Following notation from [9], at any stage of our algorithm each node v keeps a set $K(v)$ of messages. Initially $K(v)$ is a singleton containing a message originating from v. Whenever v receives a combined message, all singleton messages included in it that are not known to v are inserted into $K(v)$. During the execution of the algorithm, some singleton messages will be deleted from $K(v)$.

Similarly as in the gossiping algorithm of Chrobak *et al.* in [9], our algorithm for $\sqrt{n}$-gossiping relies on efficient selection of a node that maximizes value $|K(v)|$ (procedure FINDMAX) and efficient broadcasting of $K(v)$ from node v (procedure DOBROADCAST). In order to stay within the limit on the size of combined messages, the broadcasting will be implemented as a sequence of $\sqrt{n}$-broadcasts (see Fact 1) in our algorithm. In order to specify our algorithm shortly, we consider the following $\sqrt{n}$-BLK procedure consisting in the identification of a maximum and the subsequent broadcasting phase.

Procedure $\sqrt{n}$-BLK

1. Determine a node v_{max} having the maximum number among the nodes maximizing $|K(v)|$.
2. Distribute $K(v_{max})$ from v to all other nodes in the network by repeating $\sqrt{n}$-broadcast from v_{max} $\lceil |K(v_{max})|/\sqrt{n} \rceil$ times.
3. For each node w set $K(w)$ to $K(w) \setminus K(v)$.

By [9] and Fact 2, we have the following lemma.

Lemma 1.
Step 1 in $\sqrt{n}$-BLK can be performed in time $\tilde{O}(n)$ using $O(\log n)$- bit messages. Step 2 in $\sqrt{n}$-BLK takes time $\tilde{O}(tn)$ where $t = \lceil |K(v_{max})|/\sqrt{n} \rceil$.

Let c be a constant at $\sqrt{n} \log^2 n$ specifying the number of rounds of Round-Robin in the first phase of the gossiping algorithm of Chrobak *et al.*, see [9]. In what follows we present our algorithm for $\sqrt{n}$-gossiping.

Algorithm $\sqrt{n}$-GOSSIP

Phase 1: for $i = 1, ..., c\lceil \sqrt{n} \log^2 n \rceil$ **do**
         Perform a single round of $\sqrt{n}$-ROUNDROBIN.
         repeat $\sqrt{n}$-BLK **until** $\max_v |K(v)| \leq \sqrt{n}$.
     od
Phase 2: repeat $\sqrt{n}$-BLK **until** $\max_v |K(v)| = 0$.

Theorem 1. *Algorithm $\sqrt{n}$-GOSSIP performs $\sqrt{n}$-gossiping in time $\tilde{O}(n^{3/2})$.*

Proof. At the entry to Phase 2, for each node v the inequality $|K(v)| \leq \sqrt{n}$ holds. Hence, all the $\sqrt{n}$-broadcastings performed in Phase 2 are equivalent to unrestricted broadcasting. Moreover each original message has been transmitted to at least $c\lceil \sqrt{n} \log^2 n \rceil$ other nodes. This is done during $c\lceil \sqrt{n} \log^2 n \rceil$ rounds of $\sqrt{n}$-ROUNDROBIN. During each round every node transmits (to all its neighbors) its whole share of information (a number $< \sqrt{n}$ of messages) possessed at the beginning of the round. This means that after i rounds each message has been transmitted to all (at least i) nodes within distance i.

For the above reasons, Phase 2 is analogous to the second phase in the gossiping algorithm (DoGossip) in [9] and consequently an analogous time complexity analysis can be applied. We conclude that Phase 2 completes gossiping in time $\tilde{O}(n^{3/2})$. It remains to estimate the time complexity of Phase 1.

The $c\lceil\sqrt{n}\log^2 n\rceil$ rounds of $\sqrt{n}$-ROUNDROBIN are executed in total time $O(n^{3/2}\log^2 n)$ by Fact 2. Consider the first calls of $\sqrt{n}$-BLK procedure after completion of each $\sqrt{n}$-ROUNDROBIN round in Phase 1. We may assume, w.l.o.g., that $\sqrt{n} \geq d\log n$ where d is the constant at $\log n$ in in Lemma 1. By this lemma, Step 1, the first $\sqrt{n}$-broadcastings in Step 2 and Step 3 take in total $\tilde{O}(n \times \sqrt{n}\log^2 n)$ time. The total number of the remaining $\sqrt{n}$-broadcasting performed in Phase 1 and thus in particular the number of the remaining calls of $\sqrt{n}$-BLK in Phase 1 is clearly $O(\sqrt{n})$. Hence, the overall time taken by Phase 1 is $\tilde{O}(n^{3/2})$ by Lemma 1.

2.3   Extensions to $b(n)$-Gossiping for $b(n) \leq \sqrt{n}$

By the following lemma, our efficient algorithm for $\sqrt{n}$-gossiping can be easily transformed into an efficient $b(n)$-gossiping where b is a computable integer function satisfying $b(n) \leq \sqrt{n}$ for $n \in N$.

Lemma 2. *Let b and f be computable integer functions satisfying $1 \leq f(n) \leq b(n)$ for $n \in N$. Any deterministic algorithm for $b(n)$-gossiping in ad hoc radio networks running in time $T(n)$ can be transformed into another deterministic algorithm for $f(n)$-gossiping in such networks running in time $O(T(n)b(n)/f(n))$.*

Proof. Replace each transmission of a combined message of size (i.e., the number of single messages or the number of auxiliary bits included) not exceeding $b(n)$ by at most $\lceil b(n)/f(n)\rceil$ transmissions of combined messages of size not exceeding $f(n)$.

Combining Theorem 1 with Lemma 2, i.e., adapting algorithm $\sqrt{n}$-GOSSIP according to Lemma 2, we obtain the following generalization of Theorem 1.

Theorem 2. *Let b be a computable integer function satisfying $1 \leq b(n) \leq \sqrt{n}$ for $n \in N$. Algorithm $\sqrt{n}$-GOSSIP can be constructively transformed into into a deterministic algorithm for $b(n)$-gossiping running in time $\tilde{O}(n^2/b(n))$.*

The tightness of the upper time-bound on $\sqrt{n}$-gossiping given in Theorem 1 as well as its generalization given in Theorem 2 follows from the following fact.

Fact 3[15] *Let b be an integer function. Any algorithm for $b(n)$-gossiping in known radio network requires $\Omega(n^2/b(n))$ time in the worst case.*

Proof. Consider a directed network composed of a directed line on n vertices and all directed backward edges. Note that each unit message from the first half of the line has to pass through all the nodes in the second half of the line in order to reach the last node on the line. Thus, the sum of sizes of combined messages successfully transmitted by the nodes in the second half has to be $\Omega(n/2 \times n/2)$.

On the other hand, because of the presence of backward edges, only **one** node on the line can successfully transmit at the same time-step, and the size of a combined transmitted message is clearly bounded by $b(n)$ from above. The lower bound follows.

Corollary 1. *The upper time-bounds on $\sqrt{n}$-gossiping and $b(n)$-gossiping in general in* ad hoc *radio networks given in Theorems 1, 2 are tight up to a polylogarithmic factor.*

3   Bounded Gossiping in Symmetric *ad hoc* Radio Networks

Here, we consider symmetric *ad hoc* radio networks modeled as graphs with bidirectional edges. By Fact 3, $O(1)$-gossiping in directed *ad hoc* networks requires $\Omega(n^2)$ time. On the other hand we know that 1-gossiping in these networks can be performed in time $O(n^2 \log^2 n)$ by multiple application of the broadcasting procedure due to Chrobak *et al.* [9]. In this section, we show that there exists an adaptive deterministic algorithm that performs 1-gossiping in symmetric *ad hoc* radio networks in time $\tilde{O}(n^{3/2})$. Our algorithm breaks the $\tilde{O}(n^2)$ bound achieved by multiple application of an $\tilde{O}(n)$-time broadcasting procedure for first time. Recall here that even in the case of known symmetric networks, $O(1)$-gossiping requires $\Omega(n \log n)$ time [15].

3.1   The 1-Gossiping Algorithm for Symmetric Networks

The high level specification of our algorithm for 1-gossiping in symmetric networks is as follows.

Algorithm DO-GOSSIP

Step 1 Distinguish one node as a leader λ.
Step 2 Build a BFS tree T rooted in λ.
Step 3 Every node at level i in T sends to λ its source message and labels of all (or at most $\sqrt{n}$ if the number is larger) neighbors at level $i - 1$.
Step 4 Leader λ processes whole information.
Step 5 Leader λ distributes along consecutive levels of BFS tree the source messages and additional control packets.

Theorem 3. *Algorithm DO-GOSSIP performs gossiping in time $\tilde{O}(n^{3/2})$.*

Proof. The algorithm consists of five steps. We show here that each of the steps can be performed in time $O(n^{3/2})$. Note that we are primarily focused on communication complexity, i.e., we ignore the cost of local computations performed by nodes.

Step 1 This can be done via binary search (in the set $\{1, ..., cn\}$) in $O(\log n)$ iterations, where each iteration employs a single use of a broadcasting procedure, for details see [9]. The time-cost of this step is $\tilde{O}(n)$.

Step 2 We build a BFS tree rooted in λ. The construction is performed as follows. We adopt a broadcasting procedure due to Chlebus et al. (see [5]) that is based on use of strongly selective families. More precisely, in our algorithm each node follows exactly the same pattern of transmission as in [5], however nodes perform simple local calculations and send more complex messages. In our setting, each node v transmits and receives pairs of integers (note that since we are allowed to send only bits there is a logarithmic slowdown comparing to speed achieved by Chlebus et al. [5]). Each pair is formed by a label of a sender followed by a currently computed shortest distance of the sender from the leader. Each node upon arrival of a new pair updates its current record (distance from the leader and label of its predecessor) if the new distance is smaller than the old one.

Consider a shortest path $P(v) = < \lambda = v_0, v_1, .., v_l = v >$ between leader λ and some other node v. A message sent along $P(v)$ can be delayed at every intermediate node v_i, for $i \in \{0, .., l-1\}$, for some time due to the collision effect caused by incident edges. Let d_i be an in-degree of nod v_i. It is known that in an undirected graph $\sum_{i=0}^{l-1} d_i = O(n)$. We also know that a delay in transmission at node v_i is not larger than $\min(d_i^2 \log n, n)$, due to a property of strongly selective families [7]. This implies that message initially sent by a leader will reach v before time $\tilde{O}(n^{3/2})$, for any node v. Thus in time $\tilde{O}(n^{3/2})$ every node will learn its distance from the leader (root of the BFS tree) and its predecessor (a parent in BFS tree). The whole process is completed by one round of Round Robin to to let every node to learn about its children in BFS tree as well as to learn about its neighbors placed at its predecessor level. This completes the proof that BFS tree can be computed in time $\tilde{O}(n^{3/2})$.

Step 3 During this round every node will use a structure of a BFS tree to deliver to the leader its: label v, BFS level $l(v)$, and information about its neighbors at level $l(v) - 1$. Note that information on neighbors is complete if there are at most $\sqrt{n}$ neighbors at level $l(v) - 1$, otherwise information on arbitrary $\sqrt{n}$ neighbors is delivered. The delivery process is organized in the form of a pipeline. The information to be sent from each node is transported by a pipeline of nodes that always form a path $P(w)$ from the root to some node w. We assume every node but w in path $P(w)$ is currently holding only one message to be transmitted to the root. Whenever the number of messages to be transmitted from w to the root drops to one, node w extends pipeline by one of its "not yet pipelined" children. If all children have been explored or w is a leaf the end of a pipeline gets closer to the root in order to find another branch of BFS tree to be explored. Note that each node v taking part in the pipeline perform transmissions in time t iff $t = l(v) \bmod 3$ to avoid collisions. Since every 3rd round root λ gets a new message, and a total number of messages is bounded by $n^{3/2}$ the time required by this round is bounded by $O(n^{3/2})$.

Step 4 The leader computes for each level in the BFS tree and each node at this level the sequence of transmission steps to be performed. Each two consecutive levels and all existing edges between those levels in the BFS tree form a bipartite graph. The multiple broadcasting to be performed in step 5 is performed also in a form of a pipeline, where a uniform information available at all nodes of level l is to be transmitted to all nodes at level $l+1$, for all levels l in the BFS tree in time $\tilde{O}(n^{3/2})$. Three are four cases to be handled:

A The size s_l of level l is $\leq \sqrt{n}$. In this case each node at level l transmits separately, i.e., the number of control messages transmitted to level l is equal to s_l. Note also that in the remaining cases B,C, and D we assume that the size s_l of level l is $> \sqrt{n}$.

B All nodes at level $l+1$ with connected to at most $\sqrt{n}$ nodes at level l can be informed in $O(\log^2 n)$ rounds since the leader knows everything about their neighborhood.

C If the number of nodes at level $l+1$ with neighborhood $> \sqrt{n}$ is at most $\sqrt{n}$ then each of them can be informed in a separate call from level l.

D We can assume now that the size of level $l+1$ is $s_{l+1} > \sqrt{n}$ and that each node at level $l+1$ has at least $\sqrt{n}$ neighbors at level l. In this case the leader computes a sequence of transmissions, performed at level l, of length $O(\sqrt{n}\log n)$ as follows. Assume that the size of level l is s_l. Note that the total (out)degree at level l is $\geq s_{l+1} \cdot \sqrt{n}$. This means that there exist node v at level l that can inform $\geq \frac{s_{l+1} \cdot \sqrt{n}}{s_l} \geq \frac{s_{l+1}}{\sqrt{n}}$ nodes at level $l+1$. Thus the number of $U(t)$ uninformed nodes at level $l+1$ at time t follows the recurrence:

$$U(0) = s_{l+1} \quad \text{and} \quad U(t+1) \leq U(t)(1 - \frac{1}{\sqrt{n}}).$$

In particular $U(t+\sqrt{n}) = U(t)(1-\frac{1}{\sqrt{n}})^{\sqrt{n}} \leq \frac{1}{2} \cdot U(t)$. It follows that after $\sqrt{n}$ calls from level l the number of uniformed nodes at level $l+1$ drops down at least by half. Thus to inform all nodes at level $l+1$ the number of calls from (nodes delivered to the leader) level l can be bounded by $O(\sqrt{n}\log n)$.

Note that since there are at most $\sqrt{n}$ levels of size $\geq \sqrt{n}$ which receive $\tilde{O}(\sqrt{n})$ control messages and every other level l receives s_l control messages, the total number of control messages is $\tilde{O}(n)$. It also true that the total number of messages to be delivered to levels 1 through l is $\tilde{O}(s_1 + ..s_l)$.

Step 5 A distribution of the original messages supported by control messages is also performed in the form of a pipeline. Initially only control messages are sent to enable efficient transmission at every consecutive level starting with control messages destined at level 1, then $2, 3, 4$ *etc.* And just after all control messages at level i are installed transmission delay at this layer is set to some general delay (established by the leader) of size $\tilde{O}(\sqrt{n})$. We will prove now that all control messages will delivered in time $\tilde{O}(n^{3/2})$. Assume that levels 1 to l received all of their control messages in time $\tilde{O}((s_1 + .. + s_l)\sqrt{n})$, and that the control messages for the next levels are already available (in the right order) at levels 1 through l, one control message at each level. Note that level $l+1$ awaits $\min(s_{l+1}, \tilde{O}(\sqrt{n}))$ control messages. Each control message can be delivered to level $l + 1$ every $\tilde{O}(\sqrt{n})$th round. Thus during next $\min(s_{l+1}, \tilde{O}(\sqrt{n})) \cdot \tilde{O}(\sqrt{n}) = \tilde{O}(s_{l+1}\sqrt{n})$ rounds all control messages are delivered at level $l+1$. Also in the mean time the beginning of a pipeline is fed with appropriate control messages. Following this reasoning we prove that control messages are delivered in time $\tilde{O}((s_1 + .. + s_{last})\sqrt{n}) = \tilde{O}(n^{3/2})$. Now all original messages can be sent also in time $\tilde{O}(n^{3/2})$.

3.2 n^t-Gossiping in Time $O(n^{2-t})$

In this section we present a n^t-gossiping algorithm that works in time $O(n^{2-t})$. Note that this algorithm improves the bound stated in the previous section when the size of a combined message is $\omega(n^{1/2})$.

The algorithm works as follows:

1. Build DFS tree T_D in linear time applying technique use in [5].
2. Enumerate nodes of T_D in postfix order.
3. For each node in T_D find its neighbor with the largest postorder label and include this edge to the new spanning tree T_P.

In rooted spanning tree T, if node w is a descendant of node v and w is not a child of v an edge that connects two nodes v and w is called a shortcut in T.

Lemma 3. *Tree T_P is free of shortcuts.*

Proof. Lemma follows from the definition of tree T_P.

Since tree T_P has no shortcuts all messages stored originally in all nodes of the network can be collated in the root of the tree in linear time and constant size messages, similarly as it was done in previous section.

A distribution of all original messages from the root of T_P is performed in two stages. Initially we distribute all messages to selected nodes, called a backbone of the network, such that every node is at most at distance n^{1-t} from some node belonging to the backbone. Later messages available in the backbone nodes are distributed to all other nodes in the network.

Stage 1 Informing nodes in the backbone (n^{1-t}-covering). This stage works in greedy fashion. Initially a backbone is formed by a root of T_P. At any stage the algorithm checks (using e.g. prefix traversal of T_P) whether there exists a paths of length n^{1-t} sticking out of a backbone part. If it does all messages are pipelined along this path in time $O(n^{1-t})$ (the length of the path is equal to the number of messages of size n^t.) Since the algorithm will visit at most $\frac{n}{n^{1-t}}$ paths to be visited the total cost of this stage is $O(n)$.

Stage 2 During this stage messages available in backbone nodes will be distributed to all other nodes in the network. This is done by $n^{1-t} + n^t$ rounds of Round Robin. Messages stored in backbone nodes are numbered from 1 to n^{1-t}.

At the beginning of every round of Round Robin each node finds a messages with the largest index i received from its neighbors during the last round. And then it sends message this messages to all its neighbors. The following invariant holds. Any node at distance i from the closest backbone node receives message with index k in round $i + k$.

3.3   Fast Randomized 1-Gossiping

Bar-Yehuda *et al.* devised fast randomized multiple communication protocols for symmetric *ad hoc* networks in [3]. A basic setup phase in their protocols is a construction of a BFS tree in expected time $O(n \log^2 n)$. It uses messages of size at most $O(\log n)$. By slowing down their BFS method by a logarithmic factor, we can modify it to use only one bit messages. Once the BFS tree is available, in particular they show how k broadcasts from the root of the tree can be performed in expected time $O(k + n \log^2 n)$. Given the BFS tree, we can collect all the original messages in its root, analogously as in step 3 of DO-GOSSIP, in time proportional to the total number of original messages and auxiliary bits to be delivered to the root, i.e., $\tilde{O}(n)$. Now, it is sufficient to perform $k = n$ broadcasts of the original messages from the root in order to complete 1-gossiping. In conclusion, we obtain the following theorem.

Theorem 4. *A randomized* 1*-gossiping in a symmetric ad hoc radio network can be performed in expected time* $\tilde{O}(n)$.

3.4   Fast Deterministic 1-Gossiping in the Presence of Collision Detection

A collision detection mechanism simplifies the construction of a BFS tree even more than randomisation. We utilize this in the following three-step algorithm (similar to the randomized one) for 1-gossiping in the symmetric case.

1. Build BFS tree T in linear time using collision detection.
2. Collect all the original messages in the root of T.
3. Repetitively broadcast bits of a $0 - 1$-string encoding the original messages from the root of T through T using collision detection.

Theorem 5. *The three-step algorithm can perform gossiping in a symmetric ad hoc radio network in time* $\tilde{O}(n^{3/2})$.

Proof. To implement step 1, i.e., to construct a BFS tree, a node, say of identity 1, sends a one bit message. In the next time-step, all nodes that got a message or indication of collision for first time in the previous time-step send a 1 bit message. The latter step is iterated $n - 2$ times. The time-step in which a node gets a message or indication of collision for first time specifies its distance from the root of the BFS tree, i.e., the node of identity 1. The BFS construction is completed by $O(\log n)$ rounds of Round Robin in which each node learns about distances of its neighbors to the root and after chooses a neighbor with the distance by one smaller as its parent in the tree.

Given the BFS tree, step 2 can be implemented analogously as step 3 in DO-GOSSIP in time proportional to the total number of original messages and auxiliary bits to be delivered to the root, i.e., $\tilde{O}(n)$.

Finally, to implement step 3, the root encodes the messages collected in step 2 into a $0-1$ string of length $\tilde{O}(n)$, and broadcasts its consecutive bits through the BFS tree using pipelining. In the broadcast, receiving 1 or indication of a collision is interpreted as 1, a silence as 0.

4   Conclusion

Following [3,10] we studied here a more realistic model of gossiping in *ad hoc* radio networks assuming an upper bound on the size of combined message that can be transmitted in a single step. For upper bounds up to $\sqrt{n}$, we have provided tight (up to a poly-logarithmic factor) upper time-bounds on gossiping in this model. However the main open problem for directed networks of whether or not the known $\tilde{O}(n^{3/2})$ upper bound for (unrestricted) gossiping in such networks is tight up to a poly-logarithmic factor remains open. The tightness of our results imply that any algorithm for unrestricted gossiping substantially improving the $\tilde{O}(n^{3/2})$ bound has to transmit combined messages of size substantially exceeding $\sqrt{n}$.

As for the symmetric case, although our time-bounds on bounded gossiping are generally better than the corresponding ones in the directed case, it is an interesting open problem whether they are tight up to a logarithmic factor (in the randomized model and the model with collision detection this is true).

Acknowledgments. We are grateful to Anders Dessmark and Aris Pagourtzis for valuable discussion and comments.

References

1. N. Alon, A. Bar-Noy, N. Linial, and D. Peleg, Single round simulation of radio networks, *J. Algorithms*, 13 (1992), pp. 188–210.
2. R. Bar-Yehuda, O. Goldreich, and A. Itai, On the time complexity of broadcast in radio networks: An exponential gap between determinism and randomisation, *J. Computer and System Sciences*, 45 (1992), pp. 104–126.
3. R. Bar-Yehuda, A. Israeli, and A. Itai, Multiple communication in multi-hop radio networks, *SIAM J. on Computing*, 22 (1993), pp. 875–887.
4. B.S. Chlebus, Randomized communication in radio networks, a chapter in "Handbook on Randomized Computing," P.M. Pardalos, S. Rajasekaran, J.H. Reif, and J.D.P. Rolim, (Eds.), Kluwer Academic Publishers, to appear.
5. B.S. Chlebus, L. Gąsieniec, A.M. Gibbons, A. Pelc, and W. Rytter, Deterministic broadcasting in unknown radio networks, in *Proc. 11th ACM-SIAM Symp. on Discrete Algorithms*, San Francisco, California, 2000, pp. 861–870.

6. B. Chlebus, L. Gąsieniec, A. Lingas and A.T. Pagourtzis. Oblivious gossiping in ad hoc radio networks in *Proc. 5th International Workshop on Discrete Algorithms and Methods for Mobile Computing and Communications (DIALM 2001)*, pp. 44-51.

7. B.S. Chlebus, L. Gąsieniec, A. Östlin, and J.M. Robson, Deterministic radio broadcasting, in *Proc. 27th Int. Colloquium on Automata, Languages and Programming*, Geneva, Switzerland, 2000, Springer LNCS 1853, pp. 717–728.

8. M. Chrobak, L. Gąsieniec, and W. Rytter, A randomized algorithm for gossiping in radio networks, In *Proc. COCOON'2001*, pp. 483–492.

9. M. Chrobak, L. Gąsieniec, and W. Rytter, Fast broadcasting and gossiping in radio networks, in *Proc. 41st IEEE Symp. on Foundations of Computer Science*, Redondo Beach, California, 2000, pp. 575–581.

10. A.E.F. Clementi, A. Monti, and R. Silvestri, Distributed multi-broadcast in unknown radio networks, in *Proc., 20th ACM Symp. on Principles of Distributed Computing*, 2001, Newport, Rhode Island, pp. 255–264.

11. A.E.F. Clementi, A. Monti, and R. Silvestri, Selective families, superimposed codes, and broadcasting in unknown radio networks, in *Proc., 12th ACM-SIAM Symp. on Discrete Algorithms*, Washington, DC, 2001, pp. 709–718.

12. A.E.F. Clementi, A. Monti, and R. Silvestri, Round robin is optimal for fault-tolerant broadcasting on wireless networks, In *Proc., 9th Ann. European Symposium on Algorithms*, BRICS, University of Aarhus, Denmark, (ESA'2001), pp. 452–463.

13. G. De Marco, and A. Pelc, Faster broadcasting in unknown radio networks, *Information Processing Letters*, 79(2), pp. 53–56.

14. L. Gąsieniec and A. Lingas, On adaptive deterministic gossiping in *ad hoc* radio networks, In *Proc., 13th ACM-SIAM Symp. on Discrete Algorithms*, San Francisco, 2002, pp. 689–690.

15. L. Gąsieniec and I. Potapov, Gossiping with unit messages in known radio networks, to appear in Proc. *2nd IFIP International Conference on Theoretical Computer Science*, Montreal, August 2002.

16. P. Indyk, Explicit construction of selectors and related combinatorial structures with applications, In *Proc., 13th ACM-SIAM Symp. on Discrete Algorithms*, San Francisco, 2002, pp. 697–704.

17. E. Kushilevitz, and Y. Mansour, An $\Omega(D \log(N/D))$ lower bound for broadcast in radio networks, *SIAM J. on Computing*, 27 (1998), pp. 702–712.

18. A. Pelc, Broadcasting in radio networks, a chapter in [20].

19. D. Peleg, Deterministic radio broadcast with no topological knowledge, 2000, a manuscript.

20. I. Stojmenovic, (Ed.), "Handbook of Wireless Networks and Mobile Computing," John Wiley, February 2002.

The Kolmogorov-Loveland Stochastic Sequences Are Not Closed under Selecting Subsequences

Wolfgang Merkle

Ruprecht-Karls-Universität Heidelberg
Mathematisches Institut
Im Neuenheimer Feld 294
D-69120 Heidelberg, Germany
merkle@math.uni-heidelberg.de

Abstract. It is shown that the class of Kolmogorov-Loveland stochastic sequences is not closed under selecting subsequences by monotonic computable selection rules. This result gives a strong negative answer to the question whether the Kolmogorov-Loveland stochastic sequences are closed under selecting subsequences by Kolmogorov-Loveland selection rules, i.e., by not necessarily monotonic, partially computable selection rules. The following previously known results are obtained as corollaries. The Mises-Wald-Church stochastic sequences are not closed under computable permutations, hence in particular they form a strict superclass of the class of Kolmogorov-Loveland stochastic sequences. The Kolmogorov-Loveland selection rules are not closed under composition.

1 Introduction

We start with a very brief review of the developments that led to the concept of a Kolmogorov-Loveland stochastic sequence. Our account follows the survey by Ambos-Spies and Kučera [2], see there for further details and references. Around 1920, von Mises proposed a concept of randomness that is based on considering limiting frequencies in an infinite sequence. In the setting of uniformly distributed binary sequences, von Mises proposed to call a sequence random if the sequence itself and all subsequences selected by admissible selection rules behave like a typical sequence behaves according to the strong law of large numbers, i.e., the frequencies of 1's in the prefixes of the sequence converges to 1/2. Von Mises did not make precise which selection rules should be considered as being admissible, however, he postulated that whether a certain place is to be selected should depend only on the already observed bits of the sequence and, in particular, must not depend on the value of the sequence at this place.

For the moment and until the concept is formally introduced in Definition 2, we use the following informal notion of a selection rule. A selection rule specifies a process that scans bits of a given infinite binary sequence. Based on the already scanned bits of the sequence, the selection rules determines a yet unscanned bit that is to be scanned next and, before actually inspecting this bit, specifies whether this bit is to be selected. A sequence S is called stochastic with respect

P. Widmayer et al. (Eds.): ICALP 2002, LNCS 2380, pp. 390–400, 2002.

to a given set of admissible selection rules if for any admissible selection rule
the sequence of bits selected from S is either finite or the frequency of 1's in the
prefixes of the sequence converges to $1/2$.

Around 1940, at a time when formalizations of the concept of computability
had just been developed, Church [4] proposed to admit just the computable
selection rules. According to custom at that time, Church actually considered
only selection rules that are monotonic, i.e., where always the place to be scanned
next is strictly larger than all the places scanned before. Church argued that
according to a result of Wald there are sequences that are stochastic with respect
to this class of selection rules.

Ville [22] demonstrated that for any countable class of monotonic selection
rules there is a stochastic sequence such that every prefix of the sequence contains more 1's than 0's, i.e., stochastic sequences may lack certain statistical
properties that are associated with the intuitive understanding of randomness.
In order to overcome such shortcomings, later on more restrictive concepts of
randomness were proposed. Most of these concepts can be defined in terms of
betting strategies as follows. A sequence is random if by betting successively on
the places of the sequence according to an admissible betting strategy, the gain
achieved in the course of the game is bounded. In this context, most attention
has been received by the concepts of a Martin-Löf random set and of a computably random set. These concepts have been introduced by Martin-Löf and
Schnorr, respectively, and can be defined in terms of betting strategies where the
payoff function can be effectively approximated from below and by computable
betting strategies.

Kolmogorov [7,8] and Loveland [11,12] independently introduced a more liberal version of selection rule, leading to a more restricted concept of stochastic sequence. They proposed to admit partially computable selection rules that
are not required to be monotonic. These selection rules are called Kolmogorov-
Loveland selection rules and a sequence that is stochastic with respect to such selection rules is called Kolmogorov-Loveland stochastic. Furthermore, a sequence
is called Mises-Wald-Church stochastic if it is stochastic with respect to monotonic Kolmogorov-Loveland selection rules; such selection rules were considered
by Daley [5]. Already Loveland [11] demonstrated that the class of Mises-Wald-
Church stochastic sequences properly contains the class of Kolmogorov-Loveland
stochastic sequences, while it took over 20 years until Shen [19] showed that in
turn the latter class properly contains the class of Martin-Löf random sequences.

The question of whether Kolmogorov-Loveland random sequences are closed
under selecting subsequences by Kolmogorov-Loveland selection rules suggests
itself and has been been reported as an open problem by Shen' [18], by Li and
Vitányi Ê[10, p. 150], by Ambos-Spies and Kučera [2, Open Problem 2.8] and
by others [8,21]. In what follows, we answer the question negatively and in a
form that is stronger than might have been expected. There is a Kolmogorov-
Loveland stochastic sequence S such that even a monotonic computable selection
rule suffices to select from S a subsequence T that is not Kolmogorov-Loveland
stochastic. Moreover, the latter property of T holds in the strong form that there

is a Kolmogorov-Loveland selection rule that selects from T a sequence that is not just unbalanced, but consists only of 1's.

2   Notation and Basic Facts

Our notation is mostly standard, for unexplained terms and further details we refer to the textbooks and surveys cited in the bibliography [1,2,3,10,16].

Unless explicitly stated otherwise, the terms set and sequence refer to a subset of the natural numbers $\mathbb{N}$ and to an infinite binary sequence, respectively. We identify sets and sequences by identifying a set S with its characteristic sequence $S(0)S(1)S(2)\dots$. We refer to the binary value $S(i)$ as bit i of the sequence S; the bit $S(i)$ is equal to 1 if and only if i is in the set S.

A word is a finite binary sequence; the empty word is denoted by λ. An assignment is a function from a subset I of $\hat{\mathbb{E}}\mathbb{N}$ to $\{0,1\}$. For an assignment σ with finite domain $\{z_0 < \dots < z_{n-1}\}$, the WORD ASSOCIATED WITH σ is the (unique) word w of length n that satisfies $w(i) = \sigma(z_i)$ for $i = 0, \dots, n-1$. A sequence S can be viewed as a function $i \mapsto S(i)$, hence restricting a sequence S to a set I yields an assignment with domain I.

In what follows, we consider random experiments where the bits of a sequence are determined by independent tosses of biased coins or, formally, are determined by an infinite series of independent Bernoulli trials. In this connection, we employ the following fact from probability theory.

Remark 1. For any given natural number k and rational number $\varepsilon > 0$, we can compute a natural number $m = m(k, \varepsilon)$ such that in m independent tosses of a coin with bias of at least ε, the majority of the outcomes will reflect the bias with probability of at least $1 - 1/2^k$.

A coin is called biased if the two sides of the coin do not come up with equal probabilities. It is plausible that if a biased coin is tossed sufficiently often, then with high probability the majority of the outcomes will reflect the bias, i.e., the side that is more likely will indeed come up more often than the other side. A formal proof of this statement can be given by means of Chernoff bounds [17, Lemma 11.9]. By using such bounds it is straightforward to show that given k and $\varepsilon > 0$, one can compute a natural number $m = m(k, \varepsilon)$ such that in m independent tosses of a coin where the more likely side comes up with probability of at least $1/2 + \varepsilon$, the probability that the majority of the outcomes does not reflect the bias is at most $1/2^k$.

3   Stochastic Sequences

In this section, we review the definitions of Mises-Wald-Church stochastic and Kolmogorov-Loveland stochastic sequences and we discuss some basic facts related to these concepts. Both concepts are defined in terms of partially computable selection rules, where the difference is that in the case of a Kolmogorov-Loveland stochastic sequences the selection rules are not necessarily monotonic.

Intuitively speaking, a selection rule defines a process that scans bits of a given sequence A. More precisely, the selection rule determines a sequence of mutually distinct places $x_0, x_1, \ldots$ at which A is scanned and specifies which of these places are not just scanned but are in fact selected. The place x_{i+1} and the decision whether this place shall be selected depends solely on the previously scanned bits $A(x_0)$ through $A(x_i)$. Formally, a selection rule is a partial function that receives as input the information $w = A(x_0) \ldots A(x_i)$ that has been obtained so far by scanning the unknown sequence A, and outputs a place $x(w)$ to be scanned next and a bit $b(w)$ that indicates whether $x(w)$ is to be selected.

Definition 2. *A* SELECTION RULE *is a not necessarily total function*

$$s \colon \{0,1\}^* \to \mathbb{N} \times \{0,1\}$$
$$w \mapsto (x(w), b(w))$$

such that whenever $s(w)$ is defined then for all proper prefixes v of w, the value $s(v)$ is defined and $x(w)$ differs from $x(v)$.

   Consider the situation where a selection rule s as above is applied to a sequence A. The SEQUENCE OF SCANNED PLACES $x_0, x_1, \ldots$ *is defined inductively by*

$$x_0 = x(\lambda), \qquad x_{i+1} = x(A(x_0) A(x_1) \ldots A(x_i)) \, .$$

The SEQUENCE OF SELECTED PLACES $z_0, z_1, \ldots$ *is the subsequence of the sequence of scanned places that is obtained by cancelling all elements where the corresponding value of b is equal to 0, i.e., x_0 is cancelled if $b(\lambda) = 0$ and*

$$x_{i+1} \text{ is cancelled if } b(A(x_0) A(x_1) \ldots A(x_i)) = 0 \, .$$

Finally, the sequence that is SELECTED *by s from A is $A(z_0) A(z_1) \ldots$.*

   A selection rule s is called MONOTONIC *if for all sequences the sequence of scanned places is strictly ascending. (Equivalently, one may require $x(v) < x(w)$ whenever both values are defined and v is a proper prefix of w.)*

In connection with Definition 2, consider the situation where a selection rule s scans successively the places $x_0, x_1, \ldots$ when applied to the sequence A. In case the value of $s(A(x_0) \ldots A(x_n))$ is undefined for some n, the inductive definition of the sequence of scanned places gets stuck, hence the latter sequence and accordingly the sequence selected from A are both finite.

Definition 3. *A sequence S is* STOCHASTIC *with respect to a given set of admissible selection rules if for any admissible selection rule the sequence of selected places $z_0, z_1, \ldots$ is either finite or in the limit the frequency of 1's (hence also of 0's) in the selected sequence converges to $1/2$, i.e.,*

$$\lim_{n \to \infty} \frac{\{i < n | S(z_i) = 1\}}{n} = \frac{1}{2} \, . \tag{1}$$

A KOLMOGOROV-LOVELAND SELECTION RULE *is a selection rule that is partially computable. A sequence is* KOLMOGOROV-LOVELAND STOCHASTIC *if it is stochastic with respect to Kolmogorov-Loveland selection rules. A sequence is* MISES-WALD-CHURCH STOCHASTIC *if it is stochastic with respect to monotonic Kolmogorov-Loveland selection rules.*

By definition, every Kolmogorov-Loveland stochastic sequence is Mises-Wald-Church stochastic. The reverse implication does not hold, see Loveland [11] and Corollary 10 below. If we determine the bits of a sequence by independent tosses of a fair coin, then with probability $\hat{E}1$ we obtain a sequence that is Kolmogorov-Loveland stochastic, hence in particular there are such sequences; for a proof see Remark 7 below.

The concept of Kolmogorov-Loveland stochasticity from Definition 3 is the usual one considered in the literature [2,8,10,12,13,21]. The concept is robust in so far as we can replace in its definition partially computable selection rules by computable ones without changing the concept. The latter and a related observation on nonmonotonic betting strategies are stated in Remark 4. A similar phenomenon occurs in connection with Martin-Löf random sequences. They can be defined equivalently in terms of sequential tests that are determined by enumerable or by computable functions [10, Definition 2.5.1].

Remark 4. The concept of a Kolmogorov-Loveland stochastic sequence remains the same if we replace in its definition partially computable selection rules by computable ones.

For a proof, fix any sequence S. If S is Kolmogorov-Loveland stochastic, then it is in particular stochastic with respect to computable selection rules. Next suppose that S is not Kolmogorov-Loveland stochastic. Then there is a Kolmogorov-Loveland selection rule s that selects from S a sequence of places $z_0, z_1, \ldots$ such that the sequence $T = S(z_0), S(z_1), \ldots$ is biased, i.e., where (1) is false. Decompose T into the subsequences T_{even} and T_{odd} that contain exactly the bits $S(z_i)$ where z_i is even and where z_i is odd, respectively. Then at least one of these subsequences must be biased, too; we assume that this is the case for T_{even} and we omit the virtually identical considerations for the case where T_{odd} is biased. We argue that we can obtain a computable selection rule s' that selects the biased sequence T_{even} from S, which then finished the proof. The new selection rule works basically by simulating s. However, s' only scans but never selects odd numbers. Furthermore, if the simulation of s takes to many computation steps (or if s wants to scan an odd place that has already been scanned), the new selection rule simply scans the next previously unscanned odd number. We omit the technical details of the definition of s' due to lack of space.

Muchnik, Semenov and Uspensky [15] consider sequences that are random with respect to not necessarily monotonic, partially computable betting strategies. By an argument similar to the one just given it can be shown that this concept of a random sequence remains the same if we replace in its definition partially computable betting strategies by computable ones.

Remarks 5 and 6 state two well-known facts about stochastic sequences. Remark 5 is false for the class of Kolmogorov-Loveland stochastic sequences; this is just the assertion of Theorem 8. Furthermore, from the proof of Theorem 8 we obtain as a corollary the known fact that Remark 6 does not extend to the class of Mises-Wald-Church stochastic sequences.

Remark 5. The class of Mises-Wald-Church stochastic sequences is closed under selecting sequences by monotonic partially computable selection rules.

For a proof by contradiction, assume that there is a Mises-Wald-Church stochastic sequence S_0 from which we can select by a monotonic partially computable selection rule s_0 a sequence S_1 that is not Mises-Wald-Church stochastic. The latter means that from S_1 we can select by a monotonic partially computable selection rule s_1 a sequence S_2 where the frequency of 1's does not converge to $1/2$. But then contrary to our assumption S_0 can not be Mises-Wald-Church stochastic because we could select S_2 from S_0 directly by a monotonic partially computable selection rule obtained by combining s_1 and s_2 in the obvious way.

Remark 6. The class of Kolmogorov-Loveland stochastic sequences is closed under computable permutations.

For a proof by contradiction, assume that there is a sequence $S_0 = b_0 b_1 \ldots$ and a computable permutation $\pi : \mathbb{N} \to \mathbb{N}$ such that S_0 but not the permuted sequence $S_1 = b_{\pi(0)} b_{\pi(1)} \ldots$ is Kolmogorov-Loveland stochastic. Then there is a partially computable selection rule $s \mapsto (x(w), b(w))$ which selects from S_1 a sequence S_2 for which the frequency of 1's does not converge to $1/2$. But S_2 is selected from S_0 by the partially computable selection rule

$$w \mapsto (\pi^{-1}(x(w)), b(w)) \,.$$

Hence S_0 is not Kolmogorov-Loveland stochastic, contrary to our assumption.

4   Main Results

Remark 7 describes a probabilistic construction of stochastic sets due to van Lambalgen [9]. In a nutshell, the construction works by determining the bits of a sequence by independent coin tosses where the probabilities for a 1 converge to $1/2$.

Remark 7. Let $\beta_0, \beta_1, \ldots \ldots$ be a sequence of reals that converges to $1/2$ and consider the random experiment where the bits of a sequence $\widehat{S}$ are chosen by independent tosses of biased coins such that bit i is 1 with probability β_i. Then with probability 1 the sequence $\widehat{S}$ will be Kolmogorov-Loveland stochastic. In fact, with probability 1 the sequence $\widehat{S}$ will be stochastic with respect to any fixed countable set of admissible selection rules.

For a proof, consider triples of a Kolmogorov-Loveland selection rule s, a binary value r, and a rational number ε. The sequence $\widehat{S}$ is *not* Kolmogorov-Loveland stochastic if and only if there is such a triple where s selects from $\widehat{S}$ a sequence $z_0, z_1, \ldots$ with

$$\limsup_{i \in \mathbb{N}} \frac{\{i < n | \widehat{S}(z_i) = r\}}{n} \geq \frac{1}{2} + \varepsilon \,; \tag{2}$$

i.e., we have to show that with probability 1 condition (2) is false for all such triples. By σ-additivity of the underlying probability measure it suffices to show that (2) is false with probability 1 for any such triple.

Fix any triple s, r, and ε as above. Let $\widehat{T}$ be the sequence selected by s from $\widehat{S}$. The key observation is that the probability distribution with respect to $\widehat{T}$ remains the same if we change the random experiment such that we do not determine the whole sequence $\widehat{S}$ in advance but instead we simulate the scanning procedure given by s and in case the next place to be scanned is x, the corresponding bit is obtained by tossing a biased coin which yields a 1 with probability β_x. But β_x is less than $1/2 + \varepsilon/2$ for almost all x, hence by the strong law of large numbers the probability that (2) is false is 1.

The technique described in Remark 7 has been used by van Lambalgen in connection with sequences that are stochastic with respect to countable sets of monotonic selection rules. Shen' [19] uses the technique for proving that the class of Martin-Löf-random sequences is properly contained in the class of Kolmogorov-Loveland stochastic sequences. Lutz and Schweizer [13] show that every sequence can be reduced to a Kolmogorov-Loveland stochastic sequence via a truth-table reduction computable in polynomial time. Their proof is based on a special form of the construction from Remark 7. If we keep a given nonzero bias on an interval that is sufficiently long, then with high probability the majority of the bits assigned on this interval will reflect the bias; hence we can code a single bit of information into such an interval by choosing the probability for a 1 either above or below $1/2$ according to the given bias. This idea is also employed in the proof of our main result, which is stated as the following theorem.

Theorem 8. *The class of Kolmogorov-Loveland stochastic sequences is not closed under selection of subsequences by monotonic computable selection rules.*

Proof. In the proof we employ the probabilistic construction from Remark 7. That is, we define a sequence $\beta_0, \beta_1, \ldots$ of rational numbers that converges to $1/2$; then we consider the chance experiment where the bits of a sequence are determined by independent tosses of biased coins such that bit i is equal to 1 with probability β_i. According to Remark 7, with probability 1 we obtain a sequence that is Kolmogorov-Loveland stochastic. Furthermore, we argue that the β_i can be chosen such that again with probability 1 we obtain a sequence from which a subsequence that is not Kolmogorov-Loveland stochastic can be selected by a monotonic computable selection rule. By additivity of the underlying probability measure, with probability 1 we obtain a sequence that witnesses the assertion of the theorem; thus in particular there is such a sequence and the theorem holds.

Recall the definition of $m(k, \varepsilon)$ from Remark 1 and for all $k > 0$, let

$$\varepsilon_k = \frac{1}{k+1} \quad \text{and} \quad m_k = m(k, \varepsilon_k) \ .$$

Partition the natural numbers into successive intervals $J_1, J_2, \ldots$ where J_k has length $2m_k + 1$. For each interval J_k, let z_k be the least number in the interval, let I_k^0 contain the next m_k numbers in the interval (i.e., the least m_k numbers different from z_k) and let I^1_k contain the remaining m_k numbers in J_k. Then every number x is either equal to some z_k or is in a uniquely determined interval of the form I^i_k, hence we can define probabilities β_x by

$$\beta_x = \begin{cases} \frac{1}{2} & \text{if } x = z_k \text{ for some } k, \\ \frac{1}{2} - \varepsilon_k & \text{if } x \in I^0_k \text{ for some } k, \\ \frac{1}{2} + \varepsilon_k & \text{if } x \in I^1_k \text{ for some } k. \end{cases}$$

Now consider the random experiment where the bits of a sequence are chosen according to the probabilities β_x. For given k and i, we may ask whether the bias ε_k is reflected by the majority of the random values assigned on interval I^i_k. By choice of the β_x and the m_k, the probability that the majority is not reflected is at most $1/2^k$. The sum over these "error probabilities" over all intervals I^i_k converges, hence with probability 1 at most a finite number of "errors" will occur according to the Borel-Cantelli lemma [6]. That is, with probability 1 we obtain a sequence such that for almost all intervals the majority of the assigned bits reflects the bias. By the introductory remarks then it suffices to show that from any such sequence we can select as required a sequence that is not Kolmogorov-Loveland stochastic. So fix any such sequence S.

Consider the following monotonic computable selection rule, which successively selects numbers from the intervals $J_1, J_2, \ldots$. While working on interval J_k, for a start the least number z_k in this interval is selected. If the value assigned to z_k is 0, then all numbers in I^0_k are selected in their natural order and otherwise, if the value is 1, likewise all numbers in I^1_k are selected.

Apply this selection rule to the sequence S in order to select a sequence T. Partition the natural numbers into consecutive intervals of length $m_k + 1$ and let z'_k be the first number in the kth interval and let I'_k contain the remaining m_k numbers. Then for all k,

$$T(z'_k) = S(z_k) \quad \text{and} \quad w'_k = w_k , \tag{3}$$

where w'_k and w_k are the words associated with the restriction of T to I'_k and of S to $I^{S(z_k)}_k$, respectively. That is, the application of the selection rule to S can be pictured as transferring the assignment on z_k to z'_k and, depending on $S(z_k)$, the assignments on either I^0_k or I^1_k to I'_k.

By choice of S, for almost all k the majority of the values $T(j)$ where j is in I'_k agrees with the value of $T(z'_k)$, hence for all such intervals we can easily determine $T(z'_k)$ by scanning the restriction of T to the interval I'_k. As a consequence there is a nonmonotonic computable selection rule s that selects from T exactly the z'_k such that $T(z'_k)$ equals 1. By (3) and because S is Kolmogorov-Loveland stochastic, there are infinitely many such z'_k, hence T is not Kolmogorov-Loveland stochastic. □

Remark 9. For any given computable, unbounded, and non-decreasing function $b: \mathbb{N} \mapsto \mathbb{N}$, we can adjust the proof of Theorem 8 such that for all k the length of the intervals I_k^0 and I_k^1 is $b(k)$. Expressed in terms of density of the places selected from the sequence T, this means that for any computable rational-valued function r that converges non-ascendingly to 0, we can arrange that the places that are selected from T form a subset of $\mathbb{N}$ that contains a fraction of at least $r(n)$ of the first n natural numbers.

Inspection of the proof of Theorem 8 reveals that it is not necessary to obtain a sequence S such that for almost all k the intervals I_k^0 and I_k^1 both reflect the corresponding bias, but it is sufficient if this happens in the limit for a fraction of $\delta > 1/2$ of all indices k. So in order to obtain intervals I_k^i of length $b(k)$, we simply choose the β_x for the numbers x in these intervals so close to 1 that the probability that both intervals reflect the bias is at least δ_0 for some fixed $\delta_0 > \delta$. Since b is unbounded and non-decreasing, this is compatible with the requirement that the β_x converge to $1/2$, hence the construction goes through. With the new construction, we can still select from T a sequence that is biased; however, we cannot assume as before that this sequence consist only of 1's.

Shen' [20] observed that the proof of Theorem 8 yields a simplified proof of the following result due to Loveland [11].

Corollary 10. *(i) The class of Mises-Wald-Church stochastic sequences is not closed under computable permutations.*

(ii) The Kolmogorov-Loveland stochastic sequences form a proper subclass of the class of Mises-Wald-Church stochastic sequences.

Proof. By Remark 6 the class of Kolmogorov-Loveland stochastic sequences is closed under computable permutations, hence (i) implies (ii).

A counterexample that proves (i) is given by the sequence $T = b_0 b_1 \ldots$ from the proof of Theorem 8. The sequence T is Mises-Wald-Church stochastic. This follows by Remark 5 because T is selected by a monotonic computable selection rule from the sequence S, where the latter sequence is Kolmogorov-Loveland stochastic and hence is a fortiori Mises-Wald-Church stochastic. On the other hand, the sequence $T_\pi = b_{\pi(0)} b_{\pi(1)} \ldots$ is not Mises-Wald-Church stochastic if we let π be the computable permutation that for each k exchanges z_k' and the largest number in I_k' (while mapping all other numbers to themselves). For a proof, observe that we obtain a monotonic computable selection rule that selects from T_π an infinite sequence that contains only 1's if we combine the permutation π and the selection rule s constructed in the proof of Theorem 8 in the obvious way.   □

Lutz [14] pointed out that Theorem 8 implies the result of Shen' [18] that Kolmogorov-Loveland selection rules are not closed under composition.

Remark 11. Kolmogorov-Loveland selection rules are not closed under composition if viewed as mappings from sequences to sequences.

The latter assertion was demonstrated by Shen'[18,20]; it also follows easily from Theorem 8. If the Kolmogorov-Loveland selection rules were closed under composition, then the argument given in Remark 5 could be extended to show that the Kolmogorov-Loveland stochastic sequences are closed under selecting sequences by Kolmogorov-Loveland selection rules.

Acknowledgements. The implications of Theorem 8 stated in Corollary 10 and Remark 11 have been pointed out by Alexander Shen' and Jack Lutz, respectively; we would like to thank them for their contributions and for the permission to include the material in this article. Furthermore, we are grateful to Klaus Ambos-Spies, Jack Lutz, Jan Reimann, Alexander Shen', Nicolai Vereshchagin, and Paul Vitányi for helpful discussions and to the anonymous referees for comments and corrections.

References

1. K. Ambos-Spies and E. Mayordomo. Resource-bounded measure and randomness. In *Complexity, Logic, and Recursion Theory*, Marcel Dekker, Inc, 1997.
2. K. Ambos-Spies and A. Kučera. Randomness in Computability Theory. In P. Cholak et al. (eds.), *Computability Theory: Current Trends and Open Problems*, Contemporary Mathematics 257:1–14, American Mathematical Society, 2000.
3. J.L. Balcázar, J. Díaz and J. Gabarró. *Structural Complexity*, Vol. I and II. Springer, 1995 and 1990.
4. A. Church. On the concept of a random number. *Bulletin of the AMS*, 46:130–135,1940.
5. R. P. Daley. Minimal-program complexity of pseudo-recursive and pseudo-random sequences. *Mathematical Systems Theory*, 9:83–94,1975.
6. W. Feller. *An Introduction to Probability Theory and Its Applications. Vol. I*, third edition, revised printing. John Wiley, 1968.
7. A. N. Kolmogorov. On tables of random numbers. *Sankhyā*, Ser. A, 25:369–376, 1963.
8. A. N. Kolmogorov and V. A. Uspensky. Algorithms and randomness. *Theory of Probability and its Applications* 32:389–412, 1987.
9. M. van Lambalgen. *Random sequences*, Doctoral dissertation, University of Amsterdam, Amsterdam, 1987.
10. M. Li and P. Vitányi *An Introduction to Kolmogorov Complexity and Its Applications*, second edition, Springer, 1997.
11. D. W. Loveland. A new interpretation of the von Mises' concept of random sequence. *Zeitschrift für mathematische Logik und Grundlagen der Mathematik*, 12:279–294, 1966.
12. D. W. Loveland. The Kleene hierarchy classification of recursively random sequences. *Transactions of the AMS*, 125:497–510, 1966.
13. J. H. Lutz and D. L. Schweizer. Feasible reductions to Kolmogorov-Loveland stochastic sequences. *Theoretical Computer Science*, 225:185–194, 1999.
14. J. H. Lutz. *Private Communication*, March 2002.
15. An. A. Muchnik, A. L. Semenov, and V. A. Uspensky. Mathematical metaphysics of randomness. *Theoretical Computer Science*, 207:263–317, 1998.
16. P. Odifreddi. *Classical Recursion Theory. Vol. I.* North-Holland, Amsterdam, 1989.
17. C. H. Papadimitriou. *Computational Complexity.* Addison-Wesley Publishing Company, 1994.
18. A. Kh. Shen'. The frequency approach to the definition of the notion of a random sequence (in Russian). *Semiotika i Informatika*, 18:14–41, 1982.
19. A. Kh. Shen'. On relations between different algorithmic definitions of randomness. *Soviet Mathematics Doklady*, 38:316–319, 1988.

20. A. Kh. Shen'. *Private Communication*, November 2001.
21. V. A. Uspensky, A. L. Semenov, and A. Kh. Shen'. Can an individual sequence of zeros and ones be random? *Russian Math. Surveys*, 45:121–189, 1990.
22. J. Ville, *Étude Critique de la Notion de Collectif*. Gauthiers-Villars, Paris, 1939.

The Nondeterministic Constraint Logic Model of Computation: Reductions and Applications

Robert A. Hearn[1] and Erik D. Demaine[2]

[1] Artificial Intelligence Laboratory, Massachusetts Institute of Technology,
200 Technology Square, Cambridge, MA 02139, USA, rah@ai.mit.edu
[2] Laboratory for Computer Science, Massachusetts Institute of Technology,
200 Technology Square, Cambridge, MA 02139, USA, edemaine@mit.edu

Abstract. We present a nondeterministic model of computation based on reversing edge directions in weighted directed graphs with minimum in-flow constraints on vertices. Deciding whether this simple graph model can be manipulated in order to reverse the direction of a particular edge is shown to be PSPACE-complete by a reduction from Quantified Boolean Formulas. We prove this result in a variety of special cases including planar graphs and highly restricted vertex configurations, some of which correspond to a kind of passive constraint logic. Our framework is inspired by (and indeed a generalization of) the "Generalized Rush Hour Logic" developed by Flake and Baum [2].

We illustrate the importance of our model of computation by giving simple reductions to show that multiple motion-planning problems are PSPACE-hard. Our main result along these lines is that classic unrestricted sliding-block puzzles are PSPACE-hard, even if the pieces are restricted to be all dominoes (1×2 blocks) and the goal is simply to move a particular piece. No prior complexity results were known about these puzzles. This result can be seen as a strengthening of the existing result that the restricted Rush Hour™ puzzles are PSPACE-complete [2], of which we also give a simpler proof. Finally, we strengthen the existing result that the pushing-blocks puzzle Sokoban is PSPACE-complete [1], by showing that it is PSPACE-complete even if no barriers are allowed.

1 Introduction

Motivating Application: Sliding Blocks. Motion planning of rigid objects is concerned with whether a collection of objects can be moved (translated and rotated), without intersection among the objects, to reach a goal configuration with certain properties. Typically, one object is distinguished, the remaining objects serving as obstacles, and the goal is for that object to reach a particular position. This general problem arises in a variety of applied contexts such as robotics and graphics. In addition, this

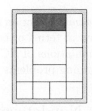

Fig. 1. Move the large square to the bottom center.

problem arises in the recreational context of *sliding-block puzzles* [5], where the pieces are typically integral rectangles, L shapes, etc., and the goal is simply to move a particular piece to a specified target position. See Fig. 1 for an example.

P. Widmayer et al. (Eds.): ICALP 2002, LNCS 2380, pp. 401–413, 2002.

The *Warehouseman's Problem* [4] is a particular formulation of this problem in which the objects are rectangles of arbitrary side lengths, packed inside a rectangular box. In 1984, Hopcroft, Schwartz, and Sharir [4] proved that deciding whether the rectangular objects can be moved so that each object is at its specified final position is PSPACE-hard. Their construction critically requires that some rectangular objects have dimensions that are proportional to the box dimensions. Although not mentioned in [4], the Warehouseman's Problem captures a particular form of sliding-block puzzles in which all pieces are rectangles. However, two differences between the two problems are that sliding-block puzzles typically require only a particular piece to reach a position, instead of the entire configuration, and that sliding-block puzzles involve blocks of only constant size.

In this paper, we prove that the Warehouseman's Problem and sliding-block puzzles are PSPACE-hard even for 1×2 rectangles (dominoes) packed in a rectangle. In contrast, there is a simple polynomial-time algorithm for 1×1 rectangles packed in a rectangle. Thus our results are tight.

Hardness Framework. To prove that sliding blocks and other problems are PSPACE-hard, this paper builds a general framework for proving PSPACE-hardness which simply requires the construction of a couple of gadgets that can be connected together in a planar graph. Our framework is inspired by the one developed by Flake and Baum [2], but is simpler and more powerful. We prove that several different models of increasing simplicity are equivalent, permitting simple constructions of PSPACE-hardness. In particular, we derive simple constructions for sliding blocks, Rush Hour [2], and a restricted form of Sokoban [1].

Nondeterministic Constraint Logic Model of Computation. Our framework can also be viewed as a model of computation in its own right, and that is the focus of this paper. We show that a Nondeterministic Constraint Logic (NCL) machine has the same computational power as a space-bounded Turing machine. Yet, it has a more concise formal description, and has a natural interpretation as a kind of logic network. Thus, it is reasonable to view NCL as a simple computational model that corresponds to the class PSPACE, just as, for example, deterministic finite automata correspond to the regular expressions.

Roadmap. Section 2 describes our model of computation in more detail, formulated in terms of both graphs and circuits. Section 3 proves increasingly simple formulations of NCL to be PSPACE-complete. Section 4 proves various motion-planning problems to be PSPACE-hard using the restricted forms of NCL.

2   Nondeterministic Constraint Logic

2.1   Graph Formulation

The simplest description of NCL arises as reversal of edges in a directed graph. A "machine" is specified by a *constraint graph*: an undirected graph together with an assignment of nonnegative integers (*weights*) to edges and integers (*minimum in-flow constraints*) to vertices. A configuration of this machine is an orientation (direction) of the edges such that the sum of incoming edge weights at each vertex is at least the minimum in-flow constraint of that vertex. A move from one configuration to another configuration is simply the reversal of a single edge such that the minimum in-flow constraints remain satisfied. The standard decision question from a particular NCL machine and configuration is whether a specified edge can be eventually reversed by a sequence of moves. We can view such a sequence as a nondeterministic computation.

Equivalent Forms. A constraint graph G_2 is an *equivalent form* of constraint graph G_1 if every configuration of G_1 can be reached if and only if a corresponding configuration of G_2 can be reached, and the configuration map preserves identity of non-loop edges: that is, every non-loop edge in G_1 may be assigned an edge in G_2 such that reversing one always corresponds to reversing the other. Note that any loop edge can trivially be reversed; see, e.g., Fig. 2.

Normal Form. We say that a constraint graph is in *normal form* if all edge weights are 1 or 2, all minimum in-flow constraints are 2, and all vertices have degree 3. This is the form of NCL that we shall be primarily concerned with. In all graph diagrams, we adopt the convention that red (light gray) edges have weight 1, blue (dark gray) edges have weight 2, and unlabeled vertices have a minimum in-flow constraint of 2.

Fig. 2 shows one translation to normal form that we will use frequently. In fact, every constraint graph has an equivalent normal form which can be computed in polynomial time. We omit the proof in this abstract; it is not needed for our main results.

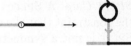

Fig. 2.   Normalizing red-to-blue conversion.

2.2   Circuit Formulation

In this Section and the following we show that a normal-form constraint graph may be viewed as a kind of circuit made up of various kinds of logic gates wired together. The circuit model is useful for visualizing how some graphs work, and is also useful for reductions to various other problems.

Gates. A *gate* is an object with a set of *ports* (each of which is either an *input* or an *output*), possibly an internal state, and a set of constraints relating the port states and the internal state. A port may be either *active* or *inactive*.

Circuits. A *circuit* is a collection of gates togther with a one-to-one pairing of all of their ports. The pairs are called *wires*. We do not require that wires connect inputs to outputs; in fact, much of the special character of NCL circuits results from constraints induced by wiring inputs to inputs or outputs to outputs. Actually, the input/output labeling is not necessary, and merely serves to place the gates in a familiar digital logic context.

We require consistency of ports connected by wires, as follows: (1) an inactive output may not be connected to an active input, (2) two active inputs may not be connected, and (3) two inactive outputs may not be connected. That is, the input/output distinction has the effect of reversing the notions of active and inactive; indeed, this is the only effect of the input/output labeling.

We give circuits a "kinematics" by allowing any sequence of individual changes to port or internal gate states consistent with the constraints. We do not, however, give circuits a "dynamics"; a circuit's state evolution is nondeterministic.

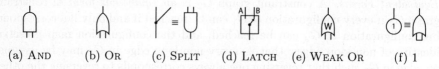

(a) AND   (b) OR   (c) SPLIT   (d) LATCH   (e) WEAK OR   (f) 1

Fig. 3.   Gates. Inputs are at bottom; outputs are at top.

AND and OR Gates. An AND gate (Fig. 3(a)) is a gate with two inputs and one output, and the constraint that the output may be active only if the inputs are both active. Similarly, an OR gate (Fig. 3(b)) has the constraint that its output may be active only if at least one input is active.

SPLIT Gate. A SPLIT (Fig. 3(c)) is a gate with one input and two outputs, and the constraint that the outputs may be active only if the input is active. Because SPLIT is not a symmetric gate, we are careful to distinguish the input side by drawing the outputs at a 45° angle. In fact, SPLIT is equivalent to AND with the inputs and outputs reversed. That is, if we replace an AND in a circuit by a SPLIT, using the SPLIT outputs in place of the AND inputs and vice-versa, then the SPLIT imposes the same constraints on the surrounding circuit behavior as the AND gate did. However, we will often use SPLITs in circuit diagrams, to indicate the normal direction of information flow.

LATCH Gate. A LATCH (Fig. 3(d)) is a gate with one input, two outputs, and one Boolean internal state variable. The internal state can change only while the input is active. One output (A) may be active only if the input is active or the internal state is false; and the other output (B) may be active only if the input is active or the internal state is true. As it turns out, a LATCH is often easier to construct than an OR gate as a gadget used in a reduction from NCL, and we will show that it is just as useful.

WEAK OR Gate. A WEAK OR gate (Fig. 3(e)) is identical to an OR gate, except that we require that any circuit containing a WEAK OR must make it impossible for both inputs to be active at once. Like LATCH, we will show that WEAK OR is just as useful as OR; it is often easier to construct for reductions, because a WEAK OR gadget built out of something else (such as sliding blocks) need not function correctly in all the cases an OR must.

1 Gate. A 1 gate (Fig. 3(f)) has a single output, which is unconstrained (and thus may serve to supply an active input to another gate). This is merely a shorthand for an OR with the inputs wired together.

INVERTER Gate. Although it is not needed for our construction, we point out for comparison that it is impossible to make an inverter, that is, a gate whose output is active exactly when its input is inactive. The idea of an inverter does not map onto the passive nature of NCL: ports are *permitted*, but not required, to change state when their constraints are satisfied.

The approach in [2] requires inverters in a similar computational context, and Flake and Baum show how to construct inverters by using a kind of dual-rail logic. However, our reductions have no need of inverters, so we may omit this step, and view individual wires as representing our logic values.

2.3   Universal Gate Sets

Here we show that circuits made with AND and OR gates are equivalent to normal-form constraint graphs: graphs and AND/OR circuits are merely two different languages for describing the same computational processes. We define equivalence as for constraint graphs, substituting "port" for "edge" where appropriate.

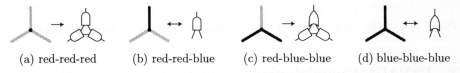

(a) red-red-red   (b) red-red-blue   (c) red-blue-blue   (d) blue-blue-blue

Fig. 4. Converting normal-form vertices into AND/OR subcircuits.

Lemma 1. *Normal-form constraint graphs and* AND *and* OR *circuits are polynomial-time equivalent.*

Proof sketch. Apply the conversions in Fig. 4.   □

Lemma 1 shows that AND and OR are universal gates. As we will show in Section 3.1, this means that we may show a problem to be PSPACE-hard by showing how to construct an AND and OR circuit as an instance of the problem. In comparison, our AND and OR gates have essentially the same properties as the "both" and "either"

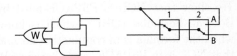

(a) LATCH built from AND and WEAK OR

(b) OR built from SPLIT and LATCH

Fig. 5. Gate emulations. Inputs are on the left; outputs are on the right.

gates in [2], but their Generalized Rush Hour Logic requires additional machinery to build Boolean "and" and "or" operations because of their use of dual-rail logic. Furthermore, here we show that two other sets of gates, which are often easier to construct, work just as well. In Section 4, we show three different problems PSPACE-hard; in each one, a different set of gates proves most convenient.

Lemma 2. LATCH *may be emulated with* AND *and* WEAK OR; WEAK OR *may be emulated with* OR; OR *may be emulated with* SPLIT *and* LATCH.

Proof sketch. Fig. 5 shows the LATCH and the OR constructions; OR may substitute directly for WEAK OR. □

We summarize these results with the following theorem, recalling that AND is equivalent to SPLIT:

Theorem 1. *The following are polynomial-time equivalent: normal-form constraint graphs,* AND/OR *circuits,* AND/LATCH *circuits, and* AND/WEAK OR *circuits.*

Proof. Lemmas 1 and 2. □

Corollary 1. *The following are polynomial-time equivalent: planar normal-form constraint graphs, planar* AND/OR *circuits, planar* AND/LATCH *circuits, and planar* AND/WEAK OR *circuits.*

Proof. All of the relevant reductions use planar subcircuits and subgraphs. □

3　PSPACE-Completeness

In this section, we show that NCL is PSPACE-complete, and provide reductions showing that some simplified forms of NCL are also PSPACE-complete.

3.1　Nondeterministic Constraint Logic

We show that NCL is PSPACE-hard by giving a reduction from Quantified Boolean Formulas (QBF), which is known to be PSPACE-complete [3], even when the formula is in conjunctive normal form. A simple argument then shows that NCL is in PSPACE, and therefore PSPACE-complete.

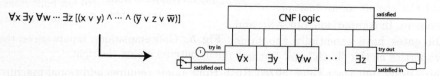

Fig. 6. Schematic of the reduction from Quantified Boolean Formulas to NCL.

Reduction. First we will give an overview of the reduction and the gadgets we will need; then we will analyze the gadgets' properties. We use the circuit form of NCL. The reduction is illustrated schematically in Fig. 6. We translate a given quantified Boolean formula ϕ into an instance of NCL, so that a particular gate in the resulting circuit may be activated if and only if ϕ is true.

One way to determine the truth of a quantified Boolean formula is as follows: Consider the initial quantifier in the formula. Assign its variable first to false and then to true, and for each assignment, recursively ask whether the remaining formula is true under that assignment. For an existential quantifier, return true if either assignment succeeds; for a universal quantifer, return true only if both assignments succeed. For the base case, all variables are assigned, and we only need to test whether the CNF formula is true under the current assignment.

This is essentially the strategy our reduction shall employ. We define *variable gadgets* and *quantifier gadgets* (Fig. 7). The quantifier gadgets are connected together into a string, one per quantifier in the formula. Each quantifier gadget is connected to its own variable gadget. The variable gadgets feed into the CNF network, which corresponds to the unquantified formula. The output from the CNF network connects to the rightmost quantifier gadget; the output of our overall circuit is the satisified out port from the leftmost quantifier gadget. (We use the attached LATCH to show a related result.)

When a quantifier gadget is activated, all quantifier gadgets to its left have fixed particular variable assignments, and only this quantifier gadget and those to the right are free to change their variable assignments. The activated quantifier gadget can declare itself "satisfied" if and only if the Boolean formula read from here to the right is true given the variable assignments on the left.

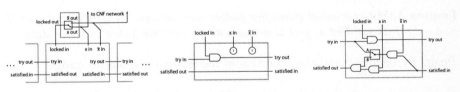

(a) Variable; connections   (b) Existential quantifier   (c) Universal quantifier

Fig. 7. QBF reduction gadgets.

Variable Gadget. A variable gadget (shown in Fig. 7(a)) is simply a LATCH, with the input port used to lock or release the variable state, and the output ports used to indicate that the variable is either true or false. The LATCH input port serves as the variable locked out port. This input/output switch reverses the sense of activation: for locked out to activate, the LATCH input must be inactive, locking it.

Quantifier Gadgets. A quantifier gadget is activated by activating its try in port. Its try out port is enabled to activate only if try in is active, and its variable gadget is locked. Thus, a quantifier gadget may nondeterministically "choose"

a variable assignment, and recursively "try" the rest of the formula under that assignment and those that are locked by quantifiers to its left. For satisfied out to activate, indicating that the formula from this quantifier on is currently satisfied, we require (at least) that satisfied in is active.

We need both existential and universal quantifier gadgets, described below.

CNF Formula. In order to evaulate the formula for a particular variable assignment, we construct an AND and OR network corresponding to the unquantified part of the formula, fed inputs from the variable gadgets, and feeding into the satisfied in port of the rightmost quantifier gadget, as in Fig. 6. The satisfied in port of the rightmost quantifier gadget is further protected by an AND gate, so it may activate only if try out is active and the formula is currently satisfied.

Lemma 3. *A quantifier gadget's satisfied in port may not activate unless its try out port is active.*

Proof sketch. Induct from right to left, using properties of quantifier gadgets. □

Existential Quantifier. For an existential quantifier gadget (Fig. 7(b)) we use the basic circuitry required to meet the definition of a quantifier gadget; we leave the variable ports unconstrained by connecting them to 1 gates. If the formula is true under some assignment of an existentially quantified variable, then its quantifier gadget may lock the variable gadget in the corresponding state, and recursively receive the satisfied in signal, releasing its satisfied out port. Here we exploit the nondeterminism in the model to choose between true and false.

Lemma 4. *An existential quantifier gadget may activate its satisfied out port if and only if its satisfied in port is active with its variable locked in some state.*

Proof. By Lemma 3 and the definition of the existential quantifer gadget.   □

Universal Quantifier. A universal quantifer gadget (Fig. 7(c)) may only enable satisfied out if the formula is true under both variable assignments. We use a LATCH as a memory bit to record that one assignmnent has been successfully tried, and then enable satisfied out only if the bit so indicates, and the other assignment is currently satisfied. To ensure that the bit resets appropriately, the other LATCH state is constrained to be active when try in is inactive.

Lemma 5. *A universal quantifier gadget may activate its satisfied out port if and only if its satisfied in port is at one time active with its variable locked in the false ($\overline{x}$) state, and at a later time is again active with its variable locked in the true (x) state, with try in remaining active throughout.*

Proof sketch. The constraints on the gates essentially force this route.   □

We summarize the behavior of both types of quantifiers with the following:

Lemma 6. *A quantifier gadget may activate its satisfied out port if and only if its try in port is active, and the formula read from the corresponding quantifier to the right is true given the variable assignments that are fixed by the quantifier gadgets to the left.*

Proof sketch. By induction from right to left using Lemmas 4 and 5.   □

Theorem 2. *NCL is PSPACE-complete.*

Proof. Lemma 6 establishes PSPACE-hardness. A simple nondeterministic algorithm traverses the state space, maintaining only the current state, so NCL is in NPSPACE, and Savitch's Theorem [6] says that NPSPACE = PSPACE.   □

Corollary 2. *Deciding whether a specified configuration of an NCL graph is reachable is PSPACE-complete.*

Proof sketch. The configuration which is identical to the initial configuration, but with the attached LATCH state switched, is reachable just when ϕ is true.   □

3.2   Planar Nondeterministic Constraint Logic

The result obtained in the previous section used particular constraint graphs (represented as circuits), which turn out to be nonplanar. Thus, reductions from NCL to other problems must provide a way to encode arbitrary graph connections into their particular structure. For 2D motion-planning kinds of problems, such a reduction would typically require some kind of crossover gadget. Crossover gadgets are

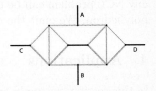

Fig. 8. Planar crossover.

a common requirement in complexity results for these kinds of problems, and can be among the most difficult gadgets to design. For example, the crossover gadget used in the proof that Sokoban is PSPACE-complete [1] is quite intricate. A crossover gadget is also among those used in the Rush Hour proof [2].

In this section we show that any normal-form NCL graph can be translated into an equivalent normal-form planar NCL (PNCL) graph, obviating the need for crossover gadgets in reductions from NCL. Fig. 8 illustrates the reduction. All vertices have minimum in-flow constraints of 2, so the blue-red-red vertices need either the blue edge or both red edges to be directed inward. The degree-4 vertices need two edges to be directed inward.

Lemma 7. *In a crossover gadget, each of the edges A and B may face outward if and only if the other faces inward, and each of the edges C and D may face outward if and only if the other faces inward.*

Proof sketch. The constraints on the vertices force this behavior.   □

The crossover subgraph is not in normal form, and Corollary 1 only applies to graphs in normal form. To solve this problem, we replace each degree-4 vertex in Fig. 8 with the equivalent subgraph in Fig. 9.

Lemma 8. *In a half-crossover gadget, at least two of the edges A, B, C, and D must face inward; any two may face outward.*

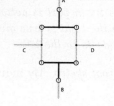

Fig. 9. Half-crossover.

Proof sketch. A similar but simpler constraint analysis as in Lemma 7.   □

Theorem 3. *Every normal-form constraint graph has an equivalent planar normal-form constraint graph which can be computed in polynomial time.*

Proof. Lemmas 7 and 8.   □

3.3 Nondeterministic Constraint Logic on a Polyhedron

Nondeterministic Constraint Logic has a particularly simple geometric form. Any NCL graph can be translated into an equivalent simple planar 3-connected graph, which is isomorphic to the edges of a convex polyhedron in 3D. Therefore, any NCL problem can be thought of as an edge redirection problem on a convex polyhedron. We omit the construction and the proof in this abstract.

4 Applications

In this section, we apply our results from the previous section to various puzzles and motion-planning problems. One result (sliding blocks) is completely new, and provides a tight bound; one (Rush Hour) reproduces an existing result, with a simpler construction; the last (Sokoban) strengthens an existing result.

4.1 Sliding Blocks

We define the *Sliding Blocks* problem as follows: given a configuration of rectangles (*blocks*) of constant sizes in a rectangular 2-dimensional box, can the blocks be translated and rotated, without intersection among the objects, so as to move a particular block? We give a reduction from PNCL showing that Sliding Blocks is PSPACE-hard even when all the blocks are 1×2 rectangles (dominoes). In contrast, there is a simple polynomial-time algorithm for 1×1 blocks; thus, our results are tight. The *Warehouseman's Problem* [4] is a related problem in which there are no restrictions on block size, and the goal is to achieve a particular total configuration. Its PSPACE-hardness also follows from our result.

Fig. 10 shows a schematic of our reduction and the two required gate gadgets. A and B are (inactive) inputs; C is the (inactive) output. Activation proceeds by moving "holes" forward: A activates by moving out, C by moving in.

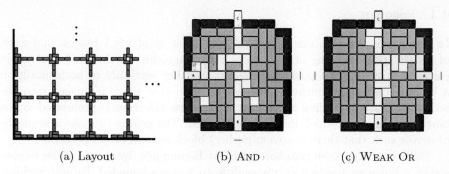

(a) Layout   (b) AND   (c) WEAK OR

Fig. 10. Sliding Blocks layout and gates.

To build arbitrary planar circuits, we also need "straight" and "turn" blocks. These may be formed from (5×5)-gate combinations of AND gates. The construction is omitted in this abstract.

4.2 Rush Hour

In the puzzle *Rush Hour*, one is given a sliding-block configuration with the additional constraint that each block is constrained to move only horizontally or vertically on a grid. The goal is to move a particular block to a particular location at the edge of the grid. In the commercial version of the puzzle, the grid is 6×6, the blocks are all 1×2 or 1×3 ("cars" and "trucks"), and each block constraint direction is the same as its lengthwise orientation.

Flake and Baum [2] showed that the generalized problem is PSPACE-complete, by showing how to build a kind of reversible computer from Rush Hour gadgets that work like our AND and OR gates, as well as a crossover gadget. Tromp [7] strengthened their result by showing that Rush Hour is PSPACE-complete even if the blocks are all 1×2.

Here we give a simpler construction showing that Rush Hour is PSPACE-complete, again using the traditional 1×2 and 1×3 blocks which must slide lengthwise. We only need an AND and a LATCH, as shown in Fig. 11.

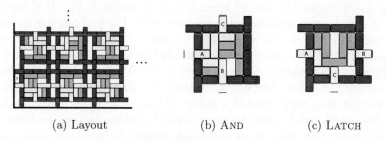

(a) Layout   (b) AND   (c) LATCH

Fig. 11. Rush Hour layout and gadgets.

4.3 Sokoban

In the puzzle *Sokoban*, one is given a configuration of 1×1 blocks, and a set of target positions. One of the blocks is distinguished as the *pusher*. A move consists of moving the pusher a single unit either vertically or horizontally; if a block occupies the pusher's destination, then that block is pushed into the adjoining space, providing it is empty. Otherwise, the move is prohibited. Some blocks are *barriers*, which may not be pushed. The goal is to make a sequence of moves such that there is a (non-pusher) block in each target position.

Culberson [1] proved Sokoban is PSPACE-complete, by showing how to construct a Sokoban position corresponding to a space-bounded Turing machine. Using PNCL, we give an alternate construction. Our result applies even if there are no barriers allowed, thus strengthening Culberson's result.

Figs. 12(a) and 12(b) illustrate the AND and OR gates. In each, block C may be reversibly pushed left only when A and/or B have been pushed left/up, as appropriate. Fig. 12(c) shows gadgets for basic wiring (A can move right only if D has), parity switching (D to E), wire flipping and pusher pass-through (H to J), and turning corners (F to G). These gadgets permit arbitrary planar circuits.

The target position corresponds to a desired PNCL configuration; this ensures that moves that violate the above conditions do not permit solution.

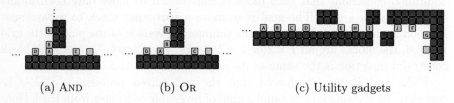

(a) AND   (b) OR   (c) Utility gadgets

Fig. 12. Sokoban gadgets.

5   Conclusion

We proved that one of the simplest possible forms of motion planning, involving sliding 1×2 blocks (dominoes) around in a rectangle, is PSPACE-hard. This result is a major strengthening of previous results. The problem has no artificial constraints, such as the movement restrictions of Rush Hour; it has object size constraints which are tightly bounded, unlike the unbounded object sizes in the Warehouseman's Problem. Also compared to the Warehouseman's Problem, the task is simply to move a block at all, rather than to reach a total configuration.

Along the way, we presented a model of computation of interest in its own right, and which can be used to prove several motion-planning problems to be PSPACE-hard. Our hope is to apply this approach to several other motion-planning problems whose complexity remains open, for example:

1. 1×1 **Rush Hour.** While 1×1 sliding blocks can be solved in polynomial time, if we enforce horizontal or vertical motion constraints as in Rush Hour,

does the problem become PSPACE-complete? Deciding whether a block may move at all is in P, but how hard is moving a given block to a given position?

2. **Lunar Lockout.** In this puzzle, robots are placed on a grid, and each move slides a robot in one direction until it hits another robot; robots are not allowed to fly off to infinity. The goal is to bring a particular robot to a specified position. Is this problem PSPACE-complete?

Acknowledgment. We thank John Tromp for several useful suggestions.

References

1. Joseph Culberson. Sokoban is PSPACE-complete. In *Proceedings of the International Conference on Fun with Algorithms*, pages 65–76, Elba, Italy, June 1998.
2. Gary William Flake and Eric B. Baum. *Rush Hour* is PSPACE-complete, or "Why you should generously tip parking lot attendants". *Theoretical Computer Science*, 270(1–2):895–911, January 2002.
3. Michael R. Garey and David S. Johnson. *Computers and Intractability: A Guide to the Theory of NP-Completeness*. W. H. Freeman & Co., 1979.
4. J. E. Hopcroft, J. T. Schwartz, and M. Sharir. On the complexity of motion planning for multiple independent objects: PSPACE-hardness of the 'Warehouseman's Problem'. *International Journal of Robotics Research*, 3(4):76–88, 1984.
5. Edward Hordern. *Sliding Piece Puzzles*. Oxford University Press, 1986.
6. Walter J. Savitch. Relationships between nondeterministic and deterministic tape complexities. *Journal of Computer and System Sciences*, 4(2):177–192, 1970.
7. John Tromp. On size 2 Rush Hour logic. Manuscript, December 2000.

Constraint Satisfaction Problems in Non-deterministic Logarithmic Space

Víctor Dalmau*

Departament de Tecnologia, Universitat Pompeu Fabra ,
Estació de França, Passeig de la Circumval.lacio 8. Barcelona 08003, Spain
victor.dalmau@tecn.upf.es
http://lsi.upc.es/~dalmau

Abstract. We study which constraint satisfaction problems (CSPs) are solvable in NL. In particular, we identify a general condition called bounded path duality, that explains all the families of CSPs previously known to be in NL. Bounded path duality captures the class of constraint satisfaction problems that can be solved by linear Datalog programs, i.e., Datalog programs with at most one IDB in the body of each rule. We obtain several alternative characterizations of bounded path duality. We also address the problem of deciding which constraint satisfaction problems have bounded path duality. In this direction we identify a subclass of bounded path duality problems, called (1,k)-path duality problems for which membership is decidable. Finally, we study which closure operations guarantee bounded path duality. We show that closure under any operation in the pseudovariety generated by the class of dual discriminator operations is a sufficient condition for bounded path duality.

1 Introduction

The constraint satisfaction problem provides a framework in which it is possible to express, in a natural way, many combinatorial problems encountered in artificial intelligence and elsewhere. A constraint satisfaction problem is represented by a set of variables, a domain of values for each variable, and a set of constraints between variables. The aim of a constraint satisfaction problem is then to find an assignment of values to the variables that satisfies the constraints.

Solving a general constraint satisfaction problem is known [4,23] to be NP-complete. One of the main approaches pursued by researchers in artificial intelligence to tackle this problem has been the identification of tractable cases obtained by imposing restrictions in the constraints (see [5,6,8,9,11,12,18,19,20, 25,26,32,33]).

Recently [11] (see also [17]) it has been observed that the constraint satisfaction problem can be recast as the following fundamental algebraic problem: given two finite relational structures $\mathcal{A}$ and $\mathcal{B}$, is there a homomorphism $\mathcal{A} \longrightarrow \mathcal{B}$? In

---

* Research conducted whilst the author was visiting the University of California, Santa Cruz, supported by NSF grant CCR–9610257.

P. Widmayer et al. (Eds.): ICALP 2002, LNCS 2380, pp. 414–425, 2002.

this framework, the problem of identifying which restrictions in the constraints guarantee tractability is equivalent to deciding for which structures $\mathcal{B}$, the homomorphism problem $\mathcal{A} \longrightarrow \mathcal{B}$, when only $\mathcal{A}$ is part of the input, denoted by $\mathrm{CSP}(\mathcal{B})$, is solvable in polynomial time. In this paper we adopt completely this framework.

The class NL of problems solvable in non-deterministic logarithmic space has received a lot of interest in complexity theory. It is known that $\mathrm{NL} \subseteq \mathrm{NC}$ and, therefore, problems in NL are highly parallelizable. Despite the large amount of tractable cases of constraint satisfaction problems identified so far, very few subclasses of constraint satisfaction problems are known to be in NL. To our knowledge, the only families of CSP problems known to be in NL are the class of bijunctive satisfiability problems [29], including 2-SAT, which later on was generalized to the class of implicational constraints [20] (see also [17]) and the class of implicative Hitting-Set Bounded (see [27]) originally defined (although with a different name) in [29].

In this paper we address the problem of classifying which constraint satisfaction problems are solvable in NL, that is, our target is to identify for which structures $\mathcal{B}$, $\mathrm{CSP}(\mathcal{B})$ is solvable in NL. Whereas the paper does not achieve this goal, it isolates a condition, called *bounded path duality* which guarantees membership to NL. We notice here that at present, at the best of our knowledge, all familes of constraint satisfaction problems known to be solvable in NL belong to the class of bounded path duality problems.

The class of bounded path duality problems, besides being solvable in NL, possesses a remarkable feature: it allows several equivalent formalizations, some of them coming from different domains. In particular, it can be defined (1) in terms of expressibility in certain classes of logics (existential positive fragment of finite-variable infinitary logic with restricted conjunction, restricted Krom monotone SNP, linear Datalog), (2) in terms of a particular type of games, called pebble-relation games, and (3) as structures generating an ideal with an obstruction set of bounded pathwidth.

Deciding which structures have bounded path duality is not known to be decidable. We identify a subclass of bounded path duality problems, called $(1, k)$-path duality problems, for which membership is decidable. Finally, we attack the problem of identifying bounded path duality problems via algebraic conditions. The algebraic approach to CSP, started in [19], pursues the identification of tractable families of constraints satisfaction problems, by means of some algebraic condition, that of preservation, on the set of constraints (see [1,2,3,5,6,8,18, 19]). Our findings in this direction is a collection of algebras: the pseudovariety generated by dual discriminator operations, that guarantees bounded path duality. This class of problems contains, as particular case, the class of implicational relations.

For space restrictions most of the proofs are not included in this extended abstract. The reader is referred to the full version of the paper which can be downloaded from the author's web page.

2   Basic Definitions

A *vocabulary* is a finite set of relation symbols or predicates. In the following τ always denotes a vocabulary. Every relation symbol R in τ has an *arity* $r \geq 0$ associated to it. We also say that R is a r-ary relation symbol.

A τ-structure $\mathcal{A}$ consists of a set A, called the *universe* of $\mathcal{A}$, and a relation $R^{\mathcal{A}} \subseteq A^r$ for every relation symbol $R \in \tau$ where r is the arity of R. Unless otherwise stated we will assume that we are dealing with *finite* structures.

If $\mathcal{A}$ is a τ-structure and $B \subseteq A$, then $\mathcal{A}_{|B}$ denotes the substructure induced by $\mathcal{A}$ on B, i.e., the τ-structure $\mathcal{B}$ with universe B and $R^{\mathcal{B}} = R^{\mathcal{A}} \cap B^r$ for every r-ary $R \in \tau$.

An *homomorphism* from a τ-structure $\mathcal{A}$ to a τ-structure $\mathcal{B}$ is a mapping $h : A \to B$ such that for every r-ary $R \in \tau$ and every $\langle a_1, \ldots, a_r \rangle \in R^{\mathcal{A}}$, we have $\langle h(a_1), \ldots, h(a_r) \rangle \in R^{\mathcal{B}}$. We denote it by $\mathcal{A} \xrightarrow{h} \mathcal{B}$. We say that $\mathcal{A}$ maps homomorphically to $\mathcal{B}$, and denote it by $\mathcal{A} \longrightarrow \mathcal{B}$ iff there exists some homomorphism from $\mathcal{A}$ to $\mathcal{B}$. We denote by $\hom(\mathcal{A}, \mathcal{B})$ the set of all homomorphisms from $\mathcal{A}$ to $\mathcal{B}$. We will assume by convention that for every set B there exists one mapping $\lambda : \emptyset \to B$. Consequently, if $\mathcal{A}$ is a structure with an empty universe then $\{\lambda\} = \hom(\mathcal{A}, \mathcal{B})$.

Let $\mathcal{A}$ be a τ structure and let $\tau' \subseteq \tau$. We denote by $\mathcal{A}[\tau']$ the τ'-structure such that for every $R \in \tau'$, $R^{\mathcal{A}[\tau']} = R^{\mathcal{A}}$. Similarly, if $\mathcal{C}$ is a collection of τ-structures we denote by $\mathcal{C}[\tau']$ the set $\{\mathcal{A}[\tau'] : \mathcal{A} \in \mathcal{C}\}$. STR denotes the class of all structures and consequently, STR$[\tau]$ denotes the class of all τ-structures.

Following [11], CSP$(\mathcal{B})$ is defined to be the set of all structures $\mathcal{A}$ such that $\mathcal{A} \longrightarrow \mathcal{B}$.

3   Infinitary Logic

The following definition is borrowed from [21]. Let τ be a vocabulary consisting of relational symbols and let $\{v_1, \ldots, v_n, \ldots\}$ be a countable set of variables. The class $\mathcal{L}_{\infty,\omega}$ of *infinitary formulas* over τ is the smallest collection of formulas such that

- it contains all first-order formulas over τ.
- if φ is a formula of $\mathcal{L}_{\infty,\omega}$ then so is $\neg\varphi$.
- if φ is a formula of $\mathcal{L}_{\infty,\omega}$ and v_i is a variable, then $(\forall v_i)\varphi$ and $(\exists v_i)\varphi$ are also formulas of $\mathcal{L}_{\infty,\omega}$.
- if Ψ is a set of $\mathcal{L}_{\infty,\omega}$ formulas, then $\bigvee \Psi$ and $\bigwedge \Psi$ are also formulas of $\mathcal{L}_{\infty,\omega}$.

Remark: Unless otherwise explicitly stated, we do not allow the use of equalities ($=$) in the formulas. This restriction is not essential and simplifies matters slightly. The main reason for this is that the notion of homomorphism, when applied over structures that contain the relation equality, or some other built-in predicate, must be modified in order to apply only to non-built-in predicates. Furthermore, we will be mainly interested in classes of structures closed under homomorphism, for which equality does not add, in general, expressiveness.

The concept of a *free variable* in a formula of $\mathcal{L}_{\infty,\omega}$ is defined in the same way as for first-order logic. A *sentence* of $\mathcal{L}_{\infty,\omega}$ is a formula φ of $\mathcal{L}_{\infty,\omega}$ with no free variables. The semantics of $\mathcal{L}_{\infty,\omega}$ is a direct extension of the semantics of first-order logic, with $\bigvee \Psi$ interpreted as a disjunction over all formulas in Ψ and $\bigwedge \Psi$ interpreted as a conjunction.

Let k be a non-negative integer. The *infinitary logic with k variables*, denoted by $\mathcal{L}^k_{\infty,\omega}$ consists of all formulas of $\mathcal{L}_{\infty,\omega}$ with at most k distinct variables. Let $\exists \mathcal{L}^k_{\infty,\omega}$ be the collection of all formulas of $\mathcal{L}^k_{\infty,\omega}, k \geq 0$, that are obtained from atomic formulas using infinitary disjunctions, infinitary conjunctions, and existential quantification only.

Let j be a non-negative integer. We will denote by *j-restricted* infinitary conjunction the infinitary conjunction $\bigwedge \Psi$ when Ψ is a collection of $\mathcal{L}_{\infty,\omega}$ formulas such that (a) every formula with more than j free variables is quantifier-free and (b) at most one formula in Ψ has quantifiers and is not a sentence. If furthermore the set Ψ is finite then we will call it *j-restricted conjunction*

Let $0 \leq j \leq k$ be non-negative integers. Let $M^{j,k}$ ($N^{j,k}$) be the collection of all formulas of $\mathcal{L}^k_{\infty,\omega}$, that are obtained from atomic formulas using infinitary disjunction, *j-restricted* infinitary conjunction (*j-restricted* conjunction), and existential quantification only.

4   Quasi-Orderings and Pathwidth of Relational Structures

A *quasi-ordering* on a set S is a reflexive and transitive relation $\leq$ on S. Let $\langle S, \leq \rangle$ be a quasi-ordered set. Let S' and S'' be subsets of S. We say that S' is a *filter* if it is closed under $\leq$ upward; that is, if $x \in S'$ and $x \leq y$, then $y \in S'$. The *filter generated* by S'' is the set $F(S'') = \{y \in S : \exists x \in S''\ x \leq y\}$. We say that S' is an *ideal* if it is closed under $\leq$ downward; that is, if $x \in S'$ and $y \leq x$, then $y \in S'$. The *ideal generated* by S'' is the set $I(S') = \{y \in S : \exists x \in S''\ y \leq x\}$.

Let I be an ideal of $\langle S, \leq \rangle$. We say that a set $O \subseteq S$ forms an *obstruction set* for I if

$$x \in I \text{ iff } \forall y \in O(y \not\leq x)$$

That is, O is an obstruction set for I if I is the complement of $F(O)$

Let τ be a vocabulary. The set of τ-structures, $\mathrm{STR}[\tau]$, is quasi-ordered by the homomorphism relation. In consequence, a set C of τ-structures is an ideal of $\langle \mathrm{STR}[\tau], \longrightarrow \rangle$ if

$$\mathcal{B} \in C, \mathcal{A} \longrightarrow \mathcal{B} \implies \mathcal{A} \in C$$

Let us define a notion of pathwidth relative to relational structures, which is the natural generalization of the notion of pathwidth over graphs. We follow the lines of previous generalizations of similar notions as treewidth. For reasons that will be made clear later it is desirable to parameterize the ordinary notion of pathwidth to capture a finer structure. For this purpose we will consider not only the maximum size of any set of the path-decomposition but also the maximum size of its pairwise intersection.

Definition 1. *Let $\mathcal{A}$ be a τ-structure. A path-decomposition of $\mathcal{A}$ is a collection $S_1, \ldots, S_n$ $S_i \subseteq A$ such that:*

1. *for every r-ary relation R in τ and every $\langle a_1, \ldots, a_r \rangle \in R^{\mathcal{A}}$, there exists $1 \leq i \leq n$ such that $\{a_1, \ldots, a_r\} \subseteq S_i$.*
2. *if $a \in S_i \cap S_j$, then $a \in S_l$ for all $i \leq l \leq j$.*

The width of the path-decomposition is defined to be the pair $\langle \max\{|S_i \cap S_{i+1}| : 1 \leq i \leq n-1\}, \max\{|S_i| : 1 \leq i \leq n\}\rangle$. We say that a structure $\mathcal{A}$ has pathwidth (j, k) if it has a path decomposition of width (j, k). We say that a set of structures C has pathwidth (j, k) if every structure $\mathcal{A}$ in C has pathwidth (j, k).

5   Pebble-Relation Games

In this section we will introduce a game, called (j, k)-pebble-relation game, that captures expressibility in $M^{j,k}$.

Let S_1 and S_2 be two (not necessarily finite) subsets. A *relation* with domain S_1 and range S_2 is a collection of functions with domain S_1 and range S_2. Remark: some confusion can arise from the fact that generally (and in this paper) the name relation is used with another meaning, for example, a r-ary relation over B is a subset of B^r. Both concepts are perfectly consistent, since a r-ary relation over B is, indeed, a relation with domain $\{1, \ldots, r\}$ and range B, in our sense.

Let f be a function of domain S_1 and range S_2, and let S_1' be a subset of its domain S_1. We will denote by $f_{|S_1'}$ the restriction of f to S_1. Similarly, let T be a relation of domain S_1 and range S_2, and let S_1' be a subset of its domain S_1. We will denote by $T_{|S_1'}$ the relation with domain S_1' and range S_2 that contains $f_{|S_1'}$ for every $f \in T$. For every relation T we denote by $\mathrm{dom}(T)$ the domain of T. We have two relations of domain $\emptyset$: the relation $\{\lambda\}$ denoted by $\overline{T}$ and the relation $\emptyset$ denoted by $\underline{T}$.

Let $0 \leq j \leq k$ be non-negative integers and let $\mathcal{A}$ and $\mathcal{B}$ be (not necessarily finite) τ-structures. The (j, k)-pebble-relation "(j, k)-PR" game on $\mathcal{A}$ and $\mathcal{B}$ is played between two players, the *Spoiler* and the *Duplicator*. A configuration of the game consists of a relation T with domain $I = \{a_1, \ldots, a_{k'}\} \subseteq A$, $k' \leq k$ and range B such that every function f in T is an homomorphism from $\mathcal{A}_{|I}$ to $\mathcal{B}$.

Initially $I = \emptyset$ and T contains the (unique) homomorphism from $\mathcal{A}_{|\emptyset}$ to $\mathcal{B}$, that is, λ. Each round of the game consists of a move from the Spoiler and a move from the Duplicator. Intuitively, the Spoiler has control on the domain I of T, which can be regarded as placing some pebbles on the elements of A that constitute I, whereas the Duplicator decides the content of T after the domain I has been set by the Spoiler. There are two types of rounds: *shrinking* rounds and *blowing* rounds.

Let T^n be the configuration after the n-th round. The Spoiler decides whether the following round is a shrinking or a blowing round. If the $(n + 1)$-th round is a shrinking round, the Spoiler sets I^{n+1} (the domain of T^{n+1}) to be a subset of the domain I^n of T^n. The Duplicator responds by projecting every function in T^n into the subdomain defined by I^{n+1}, that is, $T^{n+1} = T^n_{|I^{n+1}}$. A blowing turn only can be performed if $|I^n| \leq j$. In this case the Spoiler sets I^{n+1} to be a superset of I^n with $|I^{n+1}| \leq k$. The duplicator responds by providing a T^{n+1} with domain I^{n+1} such that $T^{n+1}_{|I^n} \subseteq T^n$. That is, T^{n+1} should contain some extensions of functions in T^n over the domain I^{n+1} (recall that any such extension must be an homomorphism from $\mathcal{A}_{|I^{n+1}}$ to $\mathcal{B}$). The Spoiler wins the game if the response of the Duplicator sets T^{n+1} to $\emptyset$, i.e., the Duplicator could not extend successfully any of the functions. Otherwise, the game resumes. The Duplicator wins the game if he has an strategy that allows him to continue playing "forever", i.e., if the Spoiler can never win a round of the game.

Now, we will present an algebraic characterization of the (j,k)-PR game.

Definition 2. *Let $0 \leq j \leq k$ be non-negative integers and let $\mathcal{A}$ and $\mathcal{B}$ be (not necessarily finite) τ-structures. We say that the Duplicator has a winning strategy for the (j,k)-pebble-relation game on $\mathcal{A}$ and $\mathcal{B}$ if there is a nonempty family $\mathcal{H}$ of relations such that:*

(a) *every relation T has range B and domain I for some $I \subseteq A$ with $|I| \leq k$.*
(b) *for every relation T in $\mathcal{H}$ of domain I, $\emptyset \neq T \subseteq \mathrm{hom}(\mathcal{A}_{|I}, \mathcal{B})$*
(c) *$\mathcal{H}$ is closed under restrictions: for every T in $\mathcal{H}$ with domain I and every $I' \subseteq I$, we have that $T_{|I'} \in \mathcal{H}$.*
(d) *$\mathcal{H}$ has the (j,k)-forth property: for every relation T in $\mathcal{H}$ with domain I with $|I| \leq j$ and every superset I' of I with $|I'| \leq k$, there exists some relation T' in $\mathcal{H}$ with domain I' such that $T'_{|I} \subseteq T$.*

The following results show that pebble-relation games, expressiveness in the existential positive fragment of finite-variable infinitary logic with restricted conjunction, and obstruction sets of bounded pathwidth are equivalent mathematical embodiments of the same concept.

Theorem 1. *Let $0 \leq j \leq k$ be non-negative integers and let $\mathcal{A}$ and $\mathcal{B}$ be (not necessarily finite) τ-structures. The following statements are equivalent:*

1. *The Duplicator has a winning strategy $\mathcal{H}$ for the (j,k)-PR game on $\mathcal{A}$ and $\mathcal{B}$.*
2. *For every sentence φ in $M^{j,k}$ such that $\mathcal{A} \models \varphi$ we have that $\mathcal{B} \models \varphi$.*
3. *For every sentence φ in $N^{j,k}$ such that $\mathcal{A} \models \varphi$ we have that $\mathcal{B} \models \varphi$.*
4. *Every τ-structure $\mathcal{P}$ with pathwidth (j,k) that maps homomorphically to $\mathcal{A}$ also maps homomorphically to $\mathcal{B}$.*

Theorem 2. *Let $0 \leq j \leq k$ be non-negative integers and let $\mathcal{C}$ be a class of τ-structures. The following statements are equivalent:*

1. *The class $\mathcal{C}$ is $M^{j,k}$-definable, i.e., there is a sentence φ of $M^{j,k}$ such that for every τ-structure we have that $\mathcal{A} \in \mathcal{C}$ iff $\mathcal{A} \models \varphi$.*
2. *If $\mathcal{A}$ and $\mathcal{B}$ are τ-structures such that $\mathcal{A} \in \mathcal{C}$ and the Duplicator has a winning strategy for the (j,k)-PR game on $\mathcal{A}$ and $\mathcal{B}$, then $\mathcal{B} \in \mathcal{C}$.*

Furthermore, if $\neg C$ is an ideal finitely generated we also have the following equivalences:

(a) *The class C is $M^{j,k}$-definable.*
(b) *The class C is $N^{j,k}$-definable.*
(c) *If A and B are τ-structures such that $A \in C$ and the Duplicator has a winning strategy for the (j,k)-PR game on A and B, then $B \in C$.*
(d) *The class $\neg C$ has an obstruction set of pathwidth (j,k)*

6   Datalog Programs

Let τ be a vocabulary consisting of relational symbols. The class SNP [22,28] is the set of all existential second-order sentences with a universal first-order part, i.e., sentences of the form $\exists S_1, \ldots, S_l \forall v_1, \ldots, v_m \varphi(v_1, \ldots, v_m)$ where φ is a quantifier-free first-order formula over the vocabulary $\tau \cup \{S_1, \ldots, S_l\}$ with variables among $v_1, \ldots, v_m$ (we will assume that φ is a CNF). We consider some restrictions that can be enforced on the class SNP. For monotone SNP [11], we require that every occurrence of a relation symbol from τ to have negative polarity, i.e., a negation applied to it. For j-adic SNP, we require that every second-order relation $S_i, 1 \le i \le l$ to have arity at most j. For k-ary, we require that the number of variables universally quantified m to be at most k. For Krom SNP we require that every clause of the quantifier-free first-order part φ to have at most two occurrences of a second-order relation symbol.

Every Krom SNP formula φ over the vocabulary τ defines a class of τ-structures, namely, the set containing every τ-structure A such that $A \models \varphi$. Furthermore, the problem of deciding, given a τ-structure A, whether $A \models \varphi$ is solvable in NL [13]. Thus expressibility in Krom SNP is a sufficient condition for membership in NL.

For restricted Krom SNP we additionally require that every clause of the quantifier-free first-order part φ to have at most one positive occurrence of a second-order relation symbol and at most one negative occurrence of a second-order relation symbol.

Theorem 3. *Let $0 \le j \le k$ be non-negative integers and let B be a τ-structure. There exists a sentence φ in j-adic k-ary restricted Krom monotone SNP with equalities such that for every τ-structure A, $A \models \varphi$ iff the Duplicator has a winning strategy for the (j,k)-PR game on A and B.*

Restricted Krom SNP formulas can be regarded alternatively as a particular type of Datalog programs called, linear Datalog Programs.

Let τ be a vocabulary. A *Datalog Program* over τ is a finite set of rules of the form

$$t_0 : -t_1, \ldots, t_m$$

where each t_i is an atomic formula $R(v_1, \ldots, v_m)$. The relational predicates that occur in the heads of the rules are the *intensional database* predicates (IDBs), while all others are the *extensional database* predicates (EDBs) and must belong to τ. One of the IDBs is designated as the *goal* of the program. Note that IDBs may occur in the bodies of rules, and, thus, a Datalog program is a recursive specification of the IDBs with semantics obtained via least fixed-points of monotone operators (see [31]).

A Datalog Program is called linear if every rule contains at most one occurrence of a IDB in its body. Let $0 \le j \le k$ be non-negative integers, (j, k)-Datalog is said to be the collection of all Datalog programs in which every rule has at most k variables and at most j variables in the head.

Lemma 1. *Let $0 \le j \le k$ be non-negative integers and let $\mathcal{C}$ be a collection of τ-structures. The two following sentences are equivalent:*

1. *$\mathcal{C}$ is definable in linear (j, k)-Datalog.*
2. *$\neg\mathcal{C}$ is definable in j-adic k-ary restricted Krom monotone SNP*

Furthermore, if $\neg\mathcal{C}$ is an ideal we also have the following equivalences:

1. *$\mathcal{C}$ is definable in linear (j, k)-Datalog.*
2. *$\neg\mathcal{C}$ is definable in j-adic k-ary restricted Krom monotone SNP*
3. *$\neg\mathcal{C}$ is definable in j-adic k-ary restricted Krom SNP with equalities.*

Theorem 4. *Let $0 \le j \le k$ be non-negative integers. Every linear (j, k)-Datalog Program is expressible in $M^{j,k}$*

If furthermore $\neg\mathcal{C} = I(\mathcal{B})$ for some τ-structure $\mathcal{B}$, then we can combine Theorem 2, Lemma 1, Theorem 3, and Theorem 4, obtaining the main result of this paper.

Theorem 5. *Let $0 \le j \le k$ be non-negative integer, let $\mathcal{E}$ be a τ-structure, and let $\neg\mathcal{C} = \mathrm{CSP}(\mathcal{E}) = I(\mathcal{E})$. The following sentences are equivalent:*

1. *The class $\mathcal{C}$ is $M^{j,k}$-definable.*
2. *The class $\mathcal{C}$ is $N^{j,k}$-definable.*
3. *The class $\mathcal{C}$ is definable in linear (j, k)-Datalog.*
4. *If $\mathcal{A}$ and $\mathcal{B}$ are finite structures such that $\mathcal{A} \in \mathcal{C}$ and Duplicator has a winning strategy for the (j, k)-PR game on $\mathcal{A}$ and $\mathcal{B}$, then $\mathcal{B} \in \mathcal{C}$.*
5. *The class $\neg\mathcal{C}$ has an obstruction set of pathwidth (j, k).*
6. *The class $\neg\mathcal{C}$ is definable in j-adic k-ary restricted Krom SNP with equalities.*
7. *The class $\neg\mathcal{C}$ is definable in j-adic k-ary restricted Krom monotone SNP.*

Let $\mathcal{E}$ be a τ-structure. If $\neg\mathcal{C} = I(\mathcal{E})$ satisfies any of the conditions of Theorem 5 we say that $\mathcal{E}$ has (j, k)-path duality. Furthermore, we say that $\mathcal{E}$ has j-path duality if $\mathcal{E}$ has (j, k)-path duality for some $j \le k$. Finally, we say that $\mathcal{E}$ has bounded path duality if $\mathcal{E}$ has j-path duality for some $0 \le j$. Recall that if a structure $\mathcal{E}$ has bounded path duality then $\mathrm{CSP}(\mathcal{E})$ is in NL.

7   Deciding Bounded Path Duality

In this section we consider the problem of deciding whether a structure has bounded path duality. Although we do not give a complete answer to the problem we provide some decidable conditions that guarantee that a structure has bounded path duality.

7.1   1-Path Duality

In the particular case $j = 1$, it is possible to obtain some decidability results, by parallelizing some results originally proven in [11].

Theorem 6. *For every $1 \leq k$, the problem of deciding whether a τ-structure $\mathcal{B}$ has $(1, k)$-path duality is decidable.*

Proof. By Theorem 3, there exists a sentence φ in 1-adic (also called monadic) monotone SNP such that for every τ-structure $\mathcal{A}$, $\mathcal{A} \models \varphi$ iff the Duplicator has a winning strategy for the $(1, k)$-PR game on $\mathcal{A}$ and $\mathcal{B}$. It is widely known (see [11]) that there exists a sentence ψ is monadic monotone SNP such that for every τ-structure $\mathcal{A}$, $\mathcal{A} \models \psi$ iff $\mathcal{A} \in \mathrm{CSP}(\mathcal{B})$. Thus deciding $(1, k)$-path duality is equivalent to decide whether $(\varphi \implies \psi)$ is a tautology for finite structures. The later is decidable since so is containment for monadic monotone SNP [11]. ∎

7.2   2-Path Duality. Algebraic Approach

One of the most successful approaches [19] in studying the complexity of constraint satisfaction problems has been the use of universal algebra, or more precisely clone theory. The algebraic approach (see [7] for an exposition) relies in the concept of polymorphism, or preservation by an operation, defined as follows: Let $\mathcal{B}$ be a τ-structure with universe B and let $\gamma : B^l \to B$ be an l-ary operation on B. We say that γ preserves $\mathcal{B}$ if for every r-ary relation symbol R in τ and every $t_1, \ldots, t_l$ (not necessarily different) tuples in $R^{\mathcal{B}}$, we have that the tuple

$$\langle \gamma(t_1[1], \ldots, t_l[1]), \ldots, \gamma(t_1[r], \ldots, t_l[r]) \rangle$$

belongs to $R^{\mathcal{B}}$.

We assume familiarity with the basic notions of algebra (see [24] for example) such as algebra, subalgebra, homomorphic image or direct product.

For a class $\mathcal{J}$ of similar algebras, $I\mathcal{J}, S\mathcal{J}, H\mathcal{J}, P\mathcal{J}, P_{\mathrm{fin}}\mathcal{J}$ denote the class of all algebras isomorphic to algebras from $\mathcal{J}$, the class of subalgebras, homomorphic images, direct products, finite direct products of algebras from $\mathcal{J}$ respectively.

A class of algebras is called a *variety* if it is closed under I, H, S, P. It is called a *pseudovariety* if it is closed under $I, H, S, P_{\mathrm{fin}}$. The variety generated by the class $\mathcal{J}$ i.e., the smallest variety containing $\mathcal{J}$ is denoted, by var $\mathcal{J}$. The pseudovariety generated by $\mathcal{J}$ is denoted by pvar $\mathcal{J}$. Also $\mathcal{J}_{\mathrm{fin}}$ denotes the class of all finite algebras from $\mathcal{J}$.

Definition 3. *Let $0 \leq j \leq k$ be non-negative integers and let (B, F) be a finite algebra. We say that (B, F) has (j, k)-path duality (j-path duality, bounded path-duality) if every relational structure $\mathcal{B}$ with universe B preserved by every operation in F has (j, k)-path duality (j-path duality, bounded path duality).*

Proposition 1. *Let $\mathcal{J}$ be a class of finite algebras with (j, k)-path duality. Then every algebra in $I\mathcal{J}$, $S\mathcal{J}$, $H\mathcal{J}$ has (j, k)-path duality.*

It is currently open whether Proposition 1 can be extended in order to include $P_{\text{fin}}\mathcal{J}$. Nevertheless, the next result shows that for a certain class of algebras $\mathcal{H}$, $P_{\text{fin}}\mathcal{H}$ has 2-path duality. In order to introduce $\mathcal{H}$ we need some definitions.

For every set B, the dual discriminator μ_B is the 3-ary operation defined as [30]:

$$\mu_B(x, y, z) = \begin{cases} y \text{ if } y = z \\ x \text{ otherwise} \end{cases}$$

Let $\mathcal{H}$ be the set containing all algebras (B, μ_B) such that μ_B is the dual discriminator operation on B.

Lemma 2. *Every algebra (B, γ) in $P_{\text{fin}}(\mathcal{H})$ has 2-path duality.*

Proposition 2. *For every set $\mathcal{J}$ of similar finite algebras,*

$$\text{pvar}(\mathcal{J}) = (\text{var}(\mathcal{J}))_{\text{fin}} = HSP_{\text{fin}}(\mathcal{J})$$

Finally, putting all the pieces together we have

Theorem 7. *Every algebra in $\text{pvar}(\mathcal{H})$ has 2-path duality.*

It has been shown that the class of implicational constraints [20], known to be solvable in NL, corresponds to the class of structures preserved by a dual discriminator operation [18]. Therefore, the class of implicational constraints, containing as a particular example 2-SAT, have 2-path duality. It is possible to show that the class of implicative Hitting-Set Bounded [27] has also bounded path duality (although is not contained in the class of structures preserved by an operation in $\text{pvar}(\mathcal{H})$). More examples of constraint satisfaction problems $\text{CSP}(\mathcal{B})$, such that $\mathcal{B}$ has bounded path duality can be found in the literature about H-coloring, which can be reformulated, as the constraint satisfaction problem $\text{CSP}(\mathcal{B})$, when $\mathcal{B}$ is a graph. For undirected graphs, the only case solvable in polynomial time corresponds to bipartite graphs [14], which have 2-path duality. For undirected graphs, it has been shown that if $\mathcal{B}$ is an oriented path, a directed cycle, or and unbalanced oriented cycle [15,16] then it has bounded path duality.

8   Open Problems

It is unknown whether bounded path duality captures the class of CSPs solvable in NL, i.e, whether there exists a structure that does not have bounded path duality such that $\text{CSP}(\mathcal{B})$ is solvable in NL. In this direction, it is not difficult to show, using some results in [13], that for every structure $\mathcal{B}$, if $\text{CSP}(\mathcal{B})$ is in NL, then it can be expressed in restricted Krom monadic SNP with equality, $s(x, y)$ (true when y is the immediate successor of x in some total ordering), 0 and max (first and last elements of the total ordering respectively). Here we understand that the homomorphism condition only applies to non-built-in predicates.

Another interesting problem is to extend the class of structures $\mathcal{B}$ known to have bounded path duality via closure conditions. With that respect, it would be interesting to decide whether majority operations, or even more generally, near-unanimity operations [30] (see also [18]) guarantee bounded path duality. This result is true, in particular, for structures with domain of size at most 2.

Acknowledgments. I am grateful to Phokion Kolaitis for stimulating discussions and for sharing with me his expertise on logic and games. I also want to thank Andrei Krokhin for many comments on this paper and its presentation.

References

1. A. Bulatov, P. Jeavons, and M. Volkov. Finite Semigroups Imposing Tractable Constraints. In *School on Algorithmic Aspects of the Theory of Semigroups*, 2001.
2. A. Bulatov, A. Krokhin, and P. Jeavons. Constraint Satisfaction Problems and Finite Algebras. In *27th International Colloquium on Automata Languages, and Programming, ICALP'00*, volume 1853 of *Lecture Notes in Computer Science*, pages 160–171, 2000.
3. A. Bulatov, A. Krokhin, and P. Jeavons. The Complexity of Maximal Constraint Languages. In *33rd Annual ACM Symposium on Theory of Computing, STOC'01*, pages 667–674, 2001.
4. S.A. Cook. The Complexity of Theorem-Proving Procedures. In *3rd Annual ACM Symposium on Theory of Computing STOC'71*, pages 151–158, 1971.
5. M.C. Cooper, D.A. Cohen, and P.G. Jeavons. Characterizing Tractable Constraints. *Artificial Intelligence*, 65:347–361, 1994.
6. V. Dalmau. A New Tractable Class of Constraint Satisfaction Problems. In *6th International Symposium on Artificial Intelligence and Mathematics*, 2000.
7. V. Dalmau. *Computational Complexity of Problems over Generalized Formulas.* PhD thesis, Universitat Politècnica de Catalunya, 2001.
8. V. Dalmau and J. Pearson. Set Functions and Width 1. In *5th International Conference on Principles and Practice of Constraint Programming, CP'99*, volume 1713 of *Lecture Notes in Computer Science,* pages 159–173, Berlin/New York, 1999. Springer-Verlag.
9. R. Dechter and J. Pearl. Network-based Heuristics for Constraint Satisfaction Problems. *Artificial Intelligence*, 34(1):1–38, 1988.
10. T. Feder. Removing Inequalities and Negation for Homomorphism-Closed Problems. manuscript.
11. T. Feder and M.Y. Vardi. The Computational Structure of Monotone Monadic SNP and Contraint Satisfaction: A Study through Datalog and Group Theory. *SIAM J. Computing*, 28(1):57–104, 1998.
12. E.C. Freuder. A Sufficient Condition for Backtrack-bounded Search. *Journal of the ACM*, 32:755–761, 1985.

13. E. Grädel. Capturing Complexity Classes by Fragments of Second-Order Logic. *Theoretical Computer Science*, 101(1):35–57, 1992.
14. P. Hell and J. Nešetřil. On the Complexity of H-coloring. *J. Comb. Theory, Series B*, 48:92–110, 1990.
15. P. Hell and X. Zhu. Homomorphisms to oriented paths. *Discrete Mathematics*, 132:107–114, 1994.
16. P. Hell and X. Zhu. The Existence of Homomorphisms to Oriented Cycles. *SIAM J. Discrete Math.*, 8, 1995.
17. P. Jeavons. On the Algebraic Structure of Combinatorial Problems. *Theoretical Computer Science*, 200:185–204, 1998.
18. P. Jeavons, D. Cohen, and M.C. Cooper. Constraints, Consistency and Closure. *Artificial Intelligence*, 101:251–265, 1998.
19. P. Jeavons, D. Cohen, and M. Gyssens. Closure Properties of Constraints. *Journal of the ACM*, 44(4):527–548, July 1997.
20. L. Kirousis. Fast Parallel Constraint Satisfaction. *Artificial Intelligence*, 64:147–160, 1993.
21. P. G. Kolaitis and M. Vardi. On the Expressive Power of Datalog: Tools and a Case Study. *Journal of Computer and System Sciences*, 51(1):110–134, 1995.
22. P.G. Kolaitis and M. Vardi. The Decision Problem for the Probabilities of Higher-Order Properties. In *19th Annual ACM Symposium on Theory of Computing*, pages 425–435, 1987.
23. A. K. Mackworth. Consistency in networks of relations. *Artificial Intelligence*, 8:99–118, 1977.
24. R.N. McKenzie, G.F. McNulty, and W.F. Taylor. *Algebras, Lattices and Varieties*, volume 1. Wadsworth and Brooks, 1987.
25. U. Montanari. Networks of Constraints: Fundamental Properties and Applications to Picture Processing. *Information Sciences*, 7:95–132, 1974.
26. U. Montanari and F. Rossi. Constraint Relaxation may be Perfect. *Artificial Intelligence*, 48:143–170, 1991.
27. S. Khanna N. Creignou and M. Sudan. *Complexity Classification of Boolean Constraint Satisfaction Problems*, volume 7 of *Monographs on Discrete Mathematics and Applications*. SIAM, 2001.
28. C. H. Papadimitriou and M. Yannakakis. Optimization, Approximation, and Complexity Classes. *Journal of Computer and System Sciences*, 43:425–440, 1991.
29. T.J. Schaefer. The Complexity of Satisfiability Problems. In *10th Annual ACM Symposium on Theory of Computing*, pages 216–226, 1978.
30. A. Szendrei. Idempotent algebras with restrictions in subalgebras. *Acta Sci. Math.*, 51:251–268, 1987.
31. J.D. Ullman. *Principles of Database and Knowledge-Base Systems*, volume II. Computer Science Press, 1989.
32. P. van Beek and R. Dechter. On the Minimality and Decomposability of Row-convex Constraint Networks. *Journal of the ACM*, 42:543–561, 1995.
33. P. van Hentenryck, Y. Deville, and C-M. Teng. A Generic Arc-consistency Algorithm and its Specializations. *Artificial Intelligence*, 1992.

Cache Oblivious Distribution Sweeping

Gerth Stølting Brodal*,** and Rolf Fagerberg*

BRICS***, Department of Computer Science, University of Aarhus, Ny Munkegade,
DK-8000 Århus C, Denmark. {gerth,rolf}@brics.dk

Abstract. We adapt the distribution sweeping method to the cache oblivious model. Distribution sweeping is the name used for a general approach for divide-and-conquer algorithms where the combination of solved subproblems can be viewed as a merging process of streams. We demonstrate by a series of algorithms for specific problems the feasibility of the method in a cache oblivious setting. The problems all come from computational geometry, and are: orthogonal line segment intersection reporting, the all nearest neighbors problem, the 3D maxima problem, computing the measure of a set of axis-parallel rectangles, computing the visibility of a set of line segments from a point, batched orthogonal range queries, and reporting pairwise intersections of axis-parallel rectangles. Our basic building block is a simplified version of the cache oblivious sorting algorithm Funnelsort of Frigo et al., which is of independent interest.

1 Introduction

Modern computers contain a hierarchy of memory levels, with each level acting as a cache for the next. Typical components of the memory hierarchy are: registers, level 1 cache, level 2 cache, main memory, and disk. The time for accessing a level in the memory hierarchy increases from one cycle for registers and level 1 cache to figures around 10, 100, and 100,000 cycles for level 2 cache, main memory, and disk, respectively [15, p. 471], making the cost of a memory access depend highly on what is the current lowest memory level containing the element accessed. The evolution in CPU speed and memory access time indicates that these differences are likely to increase in the future [15, pp. 7 and 429].

As a consequence, the memory access pattern of an algorithm has become a key component in determining its running time in practice. Since classic asymptotic analysis of algorithms in the RAM model is unable to capture this, a number of more elaborate models for analysis have been proposed. The most widely used of these is the I/O model of Aggarwal and Vitter [1], which assumes a memory hierarchy containing two levels, the lower level having size M and the transfer

* Partially supported by the Future and Emerging Technologies programme of the EU under contract number IST-1999-14186 (ALCOM-FT).
** Supported by the Carlsberg Foundation (contract number ANS-0257/20).
*** Basic Research in Computer Science, www.brics.dk, funded by the Danish National Research Foundation.

P. Widmayer et al. (Eds.): ICALP 2002, LNCS 2380, pp. 426–438, 2002.
© Springer-Verlag Berlin Heidelberg 2002

between the two levels taking place in blocks of B elements. The cost of the computation in the I/O model is the number of blocks transferred. The model is adequate when the memory transfer between two levels of the memory hierarchy dominates the running time, which is often the case when the size of the data significantly exceeds the size of main memory, as the access time is very large for disks compared to the remaining levels of the memory hierarchy. By now, a large number of results for the I/O model exists—see e.g. the survey by Vitter [18]. A significant part of these results are for problems within computational geometry.

Recently, the concept of *cache oblivious* algorithms has been introduced by Frigo et al. [12]. In essence, this designates algorithms optimized in the I/O model, except that one optimizes to a block size B and a memory size M which are *unknown*. I/Os are assumed to be performed automatically by an off-line optimal cache replacement strategy. This seemingly simple change has significant consequences: since the analysis holds for any block and memory size, it holds for *all* levels of the memory hierarchy. In other words, by optimizing an algorithm to one unknown level of the memory hierarchy, it is optimized to each level automatically. Furthermore, the characteristics of the memory hierarchy do not need to be known, and do not need to be hardwired into the algorithm for the analysis to hold. This increases the portability of implementations of the algorithm, which is important in many situations, including production of software libraries and code delivered over the web. For further details on the concept of cache obliviousness, see [12].

Frigo et al. introduced the concept of cache oblivious algorithms and presented optimal cache oblivious algorithms for matrix transposition, FFT, and sorting [12]. Bender et al. [6], gave a proposal for cache oblivious search trees with search cost matching that of standard (cache aware) B-trees [4]. Simpler cache oblivious search trees with complexities matching that of [6] were presented in [7, 10]. Cache-oblivious data structures based on on exponential structures are presented in [5]. Recently, a cache-oblivious priority queue has been developed [2], which in turn gives rise to several cache-oblivious graph algorithms.

We consider cache oblivious algorithms within the field of computational geometry. Existing algorithms may have straightforward cache oblivious implementation—this is for example the case for the algorithm know as Graham's scan [14] for computing the convex hull of a point set [11]. This algorithm first sorts the points, and then scans them while maintaining a stack containing the points on the convex hull of the points visited so far. Since the sorting step can be done by the Funnelsort algorithm of Frigo et al. [12] and a simple array is an efficient cache oblivious implementation of a stack, we immediately get a cache oblivious convex hull algorithm performing optimal $O(\text{Sort}(N))$ I/Os, where $\text{Sort}(N)$ is the optimal number of I/Os required for sorting. In this paper, we devise nontrivial cache oblivious algorithms for a number of problems within computational geometry.

In Section 2 we first present a version of the cache oblivious sorting algorithm Funnelsort of Frigo et al., which will be the basic component of our cache oblivious algorithms and which seems of independent interest due to its simplicity. In

Section 3 we develop cache oblivious algorithms based on Lazy Funnelsort for a sequence of problems in computational geometry. Common to these problems is that there exist external memory algorithms for these problems based on the *distribution sweeping* approach of Goodrich et al. [13].

Goodrich et al. introduced distribution sweeping as a general approach for developing external memory algorithms for problems which in internal memory can be solved by a divide-and-conquer algorithm based on a plane sweep. Through a sequence of examples they demonstrated the validity of their approach. The examples mentioned in [13, Section 2] are: orthogonal line segment intersection reporting, the all nearest neighbors problem [19], the 3D maxima problem [16], computing the measure of a set of axis-parallel rectangles [8], computing the visibility of a set of line segments from a point [3], batched orthogonal range queries, and reporting pairwise intersections of axis-parallel rectangles.

We investigate if the distribution sweeping approach can be adapted to the cache oblivious model, and answer this in the affirmative by developing optimal cache oblivious algorithms for each of the above mentioned problems. Theorem 1 summarizes our results. These bounds are known to be optimal in the I/O model [13] and therefore are also optimal in the cache oblivious model.

Due to lack of space, we in this paper only give the details of two of the algorithms, and refer to [9] for the rest.

Theorem 1. *In the cache oblivious model the 3D maxima problem on a set of points, computing the measure of a set of axis-parallel rectangles, the all nearest neighbors problem, and computing the visibility of a set of non-intersecting line segments from a point can be solved using optimal $O(\text{Sort}(N))$ I/Os, and the orthogonal line segment intersection reporting problem, batched orthogonal range queries, and reporting pairwise intersections of axis-parallel rectangles can be solved using optimal $O(\text{Sort}(N) + \frac{T}{B})$ I/Os, where N is the input size, T the output size, and $\text{Sort}(N)$ the number of I/Os required to sort N elements.*

Goodrich et al. described distribution sweeping as a top-down approach. We instead describe it bottom-up, which facilitates our use of Funnelsort as a basic building block. The basic idea of the distribution sweeping approach is to sort the geometric objects, e.g. points and endpoints of line segments, w.r.t. one dimension and then apply a divide-and-conquer approach on this dimension where solutions to adjacent *strips* are merged to a solution for the union of the strips. This merging may be viewed as a sweep of the strips along another dimension. The details of the merging step is unique for each specific problem to be solved, but the overall structure of the method resembles Mergesort.

We note that the general method is not confined to problems within computational geometry—rather, any divide-and-conquer algorithm that combines solutions to subproblems in a merge-like fashion seems like a candidate for using the method, provided that the divide phase of the algorithm can be done as a separate preprocessing step by e.g. sorting. For such an algorithm, the applicability of the method in a cache oblivious setting is linked to the degree of locality of the information needed in each merge step, a point we elaborate on in the beginning of Sect. 3.

Preliminaries. By a *binary tree* we denote a rooted tree where nodes are either *internal* and have two children, or are *leaves* and have no children. The *size* $|T|$ of a tree T is its number of leaves. The *depth* $d(v)$ of a node v is the number of nodes (including v) on the path from v to the root. By *level i* in the tree we denote all nodes of depth i. We use $\log_x y$ as a shorthand for $\max\{1, \log_x y\}$.

2   Lazy Funnelsort

Frigo et al. in [12] gave an optimal cache oblivious sorting algorithm called *Funnelsort*, which may be seen as a cache oblivious version of Mergesort. In this section, we present a new version of the algorithm, termed *Lazy Funnelsort*. The benefit of the new version is twofold. First, its description, analysis, and implementation are, we feel, simpler than the original—features which are important for a problem as basic as sorting. Second, this simplicity facilitates the changes to the algorithm needed for our cache oblivious algorithms for problems in computational geometry. We also generalize Funnelsort slightly by introducing a parameter d which allows a trade-off between the constants in the time bound for Funnelsort and the strength of the "tall cache assumption" [12]. The choice $d = 3$ corresponds to the description in [12].

Central to Funnelsort is the concept of a *k-merger*, which for each invocation merges the next k^d elements from k sorted streams of elements. As a k-merger takes up space super-linear in k, it is not feasible to merge all N elements by an N-merger. Instead, Funnelsort recursively produces $N^{1/d}$ sorted streams of size $N^{1-1/d}$ and then merges these using an $N^{1/d}$-merger. In [12], a k-merger is defined recursively in terms of $k^{1/2}$-mergers and buffers, and the invocation of a k-merger involves a scheduling of its sub-mergers, driven by a check for fullness of all of its buffers at appropriate intervals.

Our modification lies in relaxing the requirement that all buffers of a merger should be checked (and, if necessary, filled) at the same time. Rather, a buffer is simply filled when it runs empty. This change allows us to "fold out" the recursive definition of a k-merger to a tree of binary mergers with buffers on the edges, and, more importantly, to define the merging algorithm in a k-merger directly in terms of nodes of this tree.

We define a k-merger as a perfectly balanced binary tree with k leaves. Each leaf contains a sorted input stream, and each internal node contains a standard binary merger. The output of the root is the output stream of the entire k-merger. Each edge between two internal nodes contains a buffer, which is the output stream of the merger in the lower node and is one of the two input streams of the merger in the upper node. The sizes of the buffers are defined recursively: Let $D_0 = \lceil \log(k)/2 \rceil$ denote the number of the middle level in the tree, let the *top tree* be the subtree consisting of all nodes of depth at most D_0, and let the subtrees rooted by nodes at depth $D_0 + 1$ be the *bottom trees*. The edges between nodes at depth D_0 and depth $D_0 + 1$ have associated buffers of size $\lceil k^{d/2} \rceil$, and the sizes of the remaining buffers is defined by recursion on the top tree and the bottom trees. For consistency, we think of the output stream of the root of the k-merger as a buffer of size k^d.

A k-merger, including the buffers associated with its middle edges, is laid out in memory in contiguous locations. This statement holds recursively for the top tree and for the bottom trees of the k-merger. In effect, a k-merger is the same as the van Emde Boas layout of a binary tree [17], except that edges now are buffers and take up more than constant space.

Procedure FILL(v)
    while v's output buffer is not full
        if left input buffer empty
            FILL(left child of v)
        if right input buffer empty
            FILL(right child of v)
        perform one merge step

Fig. 1. The merging algorithm

In Figure 1 our algorithm is shown for the binary merge process in each internal node of a k-merger. The last line means moving the smallest of the two elements in the fronts of the input buffers to the rear of the output buffer. The entire k-merger is simply invoked by a call FILL(r) on the root r of the merger. This will output k^d merged elements to the output buffer of the merger.

Concerning implementation details, we note that the input buffers of the merger may run empty during the merging. Exhausting of input elements should be propagated upward in the merger, marking a buffer as exhausted when both of its corresponding input buffers are exhausted. We also note that buffers are emptied completely before they are filled, so they need not be implemented as circular arrays, in contrast to [12].

Lemma 1. *Let $d \geq 2$. The size of a k-merger (excluding its output buffer) is bounded by $c \cdot k^{(d+1)/2}$ for a constant $c \geq 1$. Assuming $B^{(d+1)/(d-1)} \leq M/2c$, a k-merger performs $O(\frac{k^d}{B} \log_M(k^d) + k)$ I/O's during an invocation.*

Proof. The space is given by the recursion formula $S(k) = k^{1/2} \cdot k^{d/2} + (k^{1/2} + 1) \cdot S(k^{1/2})$, which has a solution as stated.

For the I/O bound, we consider the recursive definition of buffer sizes in a k-merger, and follow the recursion until the space bound for the subtree (top tree or bottom tree) to recurse on is less than $M/2$, i.e. until $\bar{k}^{(d+1)/2} \leq M/2c$, where $\bar{k}$ is the number of leaves of the subtree. As $\bar{k}$ is the first such value, we know that $(\bar{k}^2)^{(d+1)/2} = \bar{k}^{d+1} > M/2c$. The buffers whose sizes will be determined during this partial recursion we denote *large* buffers. Removing the edges containing large buffers will partition the tree of the merger into a set of connected subtrees, which we denote *base trees*. By the tall cache assumption, a base tree and one block for each of the $\bar{k}$ buffers in its edges to leaves can be contained in memory, as $\bar{k} \cdot B \le (M/2c)^{2/(d+1)} \cdot (M/2c)^{(d-1)/(d+1)} \le M/2c$.

If the k-merger itself is a base tree, the merger and one block for each input stream will fit in memory, and the number of I/Os for outputting the k^d elements during an invocation is $O(k^d/B + k)$, as claimed. Otherwise, consider a call FILL(v) to the root v of a base tree. This call will output $\Omega(\bar{k}^d)$ elements to the output buffer of v. Loading the base tree and one block for each of the $\bar{k}$ buffers just below the base tree into memory will incur $O(\bar{k}^{(d+1)/2}/B + \bar{k})$ I/Os. This is $O(1/B)$ I/Os per element output, since $\bar{k}^{d+1} > M/2c$ implies $\bar{k}^{d-1} > (M/2c)^{(d-1)/(d+1)} \ge B$ and hence $\bar{k} \le \bar{k}^d/B$. During the call FILL(v), the buffers just below the base tree may run empty, which will trigger calls to the nodes below these buffers. Such a call may evict the base tree from memory, leading to its reloading when the call finishes. However, a buffer of size $\Omega(\bar{k}^d)$ has been filled during this call, so the same calculation as above shows that the reloading of the base tree incurs $O(1/B)$ I/Os per element inserted into the buffer. The last time a buffer is filled, it may not be filled completely due to exhaustion. This happens only once for each buffer, so we can instead charge $O(1/B)$ I/Os to each position in the buffer in the argument above. As the large buffers are part of the space used by the entire k-merger, and as this space is sublinear in the output of the k-merger, this is $O(1/B)$ I/O per element merged.

In summary, charging an element $O(1/B)$ I/Os each time it is inserted into a large buffer will account for the I/Os performed. As $F = (M/2c)^{1/(d+1)}$ is the minimal number of leaves for a base tree, each element can be inserted in at most $\log_F k = O(d \log_M k) = O(\log_M k^d)$ large buffers, including the output buffer of the k-merger. From this the stated I/O bound follows. $\square$

Theorem 2. *Under the assumptions in Lemma 1, Lazy Funnelsort uses $O(d\frac{N}{B} \log_M N)$ I/Os to sort N elements.*

Proof. The algorithm recursively sorts $N^{1/d}$ segments of size $N^{1-1/d}$ of the input and then merges these using an $N^{1/d}$-merger. When the size of a segment in a recursive call gets below $M/2$, the blocks in this segment only needs to be loaded once into memory during the sorting of the segment, as the space consumption of a merger is linearly bounded in its output. For the k-mergers used at the remaining higher levels in the recursion tree, we have $k^d \ge M/2c \ge B^{(d+1)/(d-1)}$, which implies $k^{d-1} \ge B^{(d+1)/d} > B$ and hence $k^d/B > k$. By Lemma 1, the number of I/Os during a merge involving n' elements is $O(\log_M(n')/B)$ per element. Hence, the total number of I/Os per element is

$$O\left(\frac{1}{B} \left(1 + \sum_{i=0}^{\infty} \log_M N^{(1-1/d)^i} \right) \right) = O\left(d \log_M(N)/B \right) .$$

$\square$

3   Distribution Sweeping

Before going into the details for the various geometric problems, we below summarize the main technical differences between applying the distribution sweeping approach in the I/O model and in the cache oblivious model.

- In the I/O model, distribution sweeping uses $\Theta(M/B)$-ary merging. For cache oblivious algorithms, we do not know the parameters M and B, and instead use on binary merging. This is a simplification of the approach.
- In the I/O model, an entire merging process is completed before another merging process is started. In the cache oblivious model, we are building on (Lazy) Funnelsort, so this does not hold. Rather, a scheduling of the various merging processes takes place, and the intermediate outputs of merging processes are stored in buffers of limited size and used as input for other merging processes. This is a complication of the approach.

To illustrate the latter point, we note that in the distribution sweeping algorithms for batched orthogonal range queries, for orthogonal line segment intersection reporting, and for finding pairwise rectangle intersections, the merging process at a node needs to access the already merged part like a stack when generating the required output. In the I/O model this is not a problem, since there is always only one output stream present. In the cache oblivious model, the access to already merged parts is a fundamental obstacle, since this information may already have been removed by the merger at the parent node. Similar complications arise in the algorithm for all nearest neighbors. The solutions to these problems form a major part of the contribution of this paper.

On the other hand, for the 3D maxima problem and for computing the measure of a set of axis-parallel rectangles, this problem does not show up. The only difference from the merging performed in Lazy Funnelsort is that each input and output element is labeled with constant additional information, and that computing the labeling of an output element requires information of constant size to be maintained at the nodes of the merging process. For computing the visibility of a set of line segments from a point the situation is basically the same, except that some input points to a node in the merging process are removed during the merging.

Due to lack of space we only give the algorithms for batched orthogonal range queries and the all nearest neighbors problem, and refer to [9] for the rest.

3.1   Batched Orthogonal Range Queries

Problem 1. Given N points in the plane and K axis-parallel rectangles, report for each rectangle R all points which are contained in R.

The basic distribution sweeping algorithm for range queries proceeds as follows. First all N points and the $2K$ upper left and upper right rectangle corners are sorted on the first coordinate. Each corner point contains a full description of the rectangle. After having sorted the points we use a divide-and-conquer approach on the first coordinate, where we merge the sequences of points from two adjacent strips A and B to the sequence of points in the strip $A \cup B$. All sequences are sorted on the second coordinate, and the merging is performed as a bottom-up sweep of the strip $A \cup B$. The property maintained is that if a rectangle corner is output for a strip, then we have reported all points in the strip that are contained in the rectangle.

While merging strips A and B, two lists L_A and L_B of points are generated: L_A (L_B) contains the input points from A (B), which are by now below the sweep line. If the next point p is an input point from A (B), we insert p into L_A (L_B) and output p. If p is a rectangle corner from A, and p is the upper left corner of a rectangle R that spans B completely in the first dimension, then the points in $L_B \cap R$ are reported by scanning L_B until the first point below the rectangle is found (if R only spans B partly, then the upper right corner of R is contained in B, i.e. $L_B \cap R$ has already been reported). On the RAM this immediately gives an $O(N \log N)$ time algorithm. The space usage is $O(N)$, since it is sufficient to store the L lists for the single merging process in progress. In the I/O model, a merging degree of $\Theta(\frac{M}{B})$ gives an $O(\mathrm{Sort}(N))$ time algorithm with a space usage of $O(\frac{N}{B})$ blocks.

Unfortunately, this approach does not immediately give an optimal cache oblivious algorithm. One problem is that the interleaved scheduling of the merge processes at nodes in a k-merger seems to force us to use $\Theta(n \log n)$ space for storing each input point in an L list at each level in the worst case. This space consumption is sub-optimal, and is also a problem in the proof of Theorem 2, where we for the case $N \leq M/2$ use that the space is linearly bounded.

We solve this problem in three phases: First we calculate for each node of a k-merger how many points will actually be reported against some query rectangle— without maintaining the L lists. By a simple change in the algorithm, we can then reduce the space needed at a node to be bounded by the reporting done at the node. Finally, we reduce the space consumption to $O(\frac{N}{B})$ blocks by changing the scheduling of the merging processes such that we force the entire merging process at certain nodes to complete before returning to the parent node.

In the following we consider a k-merger where the k input streams are available in k arrays holding a total of N points, and where $k = N^{1/d}$. In the first phase we do no reporting, but only compute how much reporting will happen at each of the $k - 1$ nodes. We do so by considering a slightly different distribution sweeping algorithm. We now consider all N input points and all $4K$ corners of the rectangles. When merging the points from two strips A and B, we maintain the number a (b) of rectangles intersecting the current sweep line that span strip A (B) completely and have two corners in B (A). We also maintain the number of points r_A (r_B) in A (B) below the sweep line which cause at least one reporting at the node when applying the above algorithm. Whenever the next point is the lower left (right) corner of a rectangle spanning B (A) completely, b (a) is increased. Similarly we decrease the counter when a corresponding top-most corner is the next point. If the next point is an input point from A (B), we increase r_A (r_B) by one if and only if a (b) is nonzero. Since the information needed at each node is constant, we can apply the Lazy Funnelsort scheduling and the analysis from Lemma 1 for this first phase.

By including the lower rectangle corner points in the basic reporting algorithm, we can simultaneously with inserting points into L_A and L_B keep track of a and b, and avoid inserting a point from A (B) into L_A (B) if the point will not be reported, i.e. if a (b) is zero. This implies that all points inserted into L_A and L_B will be reported at least once, so the space $O(r_A + r_B)$ required for L_a and L_b is bounded by the amount of reporting generated at the node.

Finally, to achieve space linear in the total input N of the k-merger, we will avoid allocating the L lists for all nodes simultaneously if this will require more than linear space. The reporting generated by a k-merger will be partitioned into iterations, each of which (except the last) will generate $\Omega(N)$ reporting using space $O(N)$. The details are as follows. First we apply the above algorithm for computing the r_A and r_B values of each node of the k-merger. In each iteration we identify (using a post-order traversal principle) a node v in the k-merger where the sum of the r_A and r_B values at the descendants is at least N, and at most $3N$ (note: for each node we have $r_A + r_B \leq N$). If no such node exists, we let v be the root. We first allocate an array of size $3N$ to hold all the L_A and L_B lists for the descendants of v. We now complete the entire merging process at node v, by repeatedly applying FILL(v) until the input buffers of v are exhausted. We move the content of the output buffer of v to a temporary array of size N, and when the merging at v finished we move the output to a global array of size N which holds the final merged lists of several nodes simultaneously. If the k input streams have size $N_1, \ldots, N_k$, and node v spans streams $i..j$, the merged output of v is stored at positions $1 + \sum_{\ell=1}^{i-1} N_i$ and onward. When the merging of v is finished, we set r_A and r_B of all descendants of v to zero.

For the analysis, we follow the proof of Lemma 1. We first note that by construction, we use space $\Theta(N)$ and in each iteration (except the last) generate $\Omega(N)$ reporting. If $N \leq M/2c$, all computation will be done in internal memory, when the input streams first have been loaded into memory, i.e. the number of I/Os used is $O(\frac{N}{B} + \frac{T}{B})$. For the case $N > M/2c$, i.e. $k < \frac{N}{B}$, we observe that each base tree invoked only needs to store $O(1)$ blocks from the head of each L list in the nodes of the base tree. Writing a point to an L list can then be charged to the later reporting of the point. Reading the first blocks of the L lists in a base tree has the same cost as reading the first blocks of each of the input streams to the base tree. We conclude that the I/Os needed to handle the L lists can either be charged to the reporting or to the reading of the input streams of a base tree. The total number of I/Os used in an iteration is $O(k + \frac{N}{B} + \frac{T'}{B})$, where T' is the amount of reporting, plus the number of I/Os used to move points from a base tree to the base next. Over all iterations, the latter number of I/Os is at most $O(\frac{N}{B} \log_M N)$. We conclude that the k-merger in total uses $O(\frac{N}{B} \log_M N + \frac{T}{B})$ I/Os and uses $O(\frac{N}{B})$ blocks of space. Analogous to the proof of Theorem 2 it follows that the entire algorithm uses $O(d\frac{N}{B} \log_M N + \frac{T}{B})$ I/Os.

3.2   All Nearest Neighbors

Problem 2. Given N points in the plane, compute for each point which other point is the closest.

We solve the problem in two phases. After the first phase, each point p will be annotated by another point p_1 which is at least as close to p as the closest among all points lying *below* p. The point p_1 itself does not need to lie below p. If no points exist below p, the annotation may be empty. The second phase is symmetric, with *above* substituted for *below*, and will not be described further. The final result for a point p is the closest of p_1 and the corresponding annotation from the second phase.

In the first phase, we sort the points on the first dimension and apply a divide-and-conquer approach from [19] on this dimension. For each vertical strip S, we will produce a stream containing the points in S in decreasing order w.r.t. the second dimension, with each point annotated by *some* other point p_1 from S (or having empty annotation). The divide-and-conquer approach will be patterned after Lazy Funnelsort, and for streams analogous to output streams of k-mergers, the annotation will fulfill an invariant as above, namely that p_1 is at least as close as the closest among the points from S lying below p (for streams internal to k-mergers, this invariant will not hold).

The base case is a strip containing a single point with empty annotation. For a strip being the union of two strips A and B, we merge the streams for A and B by a downward plane sweep, during which we maintain two *active sets* S_A and S_B of copies of points from A and B, respectively. For clarity, we in the discussion below refer to such a copy as an *element* x, and reserve the term *point* p for the original points in the streams being merged.

These active sets are updated each time the sweepline passes a point p. The maintenance of the sets are based on the following definition: Let c denote the intersection of the horizontal sweepline and the vertical line separating A and B, let p be a point from S, let p_1 be the point with which p is annotated, and let d denote Euclidean distance. By $U(p)$ we denote the condition $d(p, p_1) \leq d(p, c)$, where $d(p, p_1)$ is taken as infinity if p has empty annotation. If $U(p)$ holds and p is in A (B), then no point in B (A) lying below the sweepline can be closer to p than p_1.

We now describe how the active sets are updated when the sweepline passes a point $p \in A$. The case $p \in B$ is symmetric. We first calculate the distance $d(p, x)$ for all elements x in $S_A \cup S_B$. If this is smaller than the distance of the current annotation of x (or p, or both), we update the annotation of x (or p, or both). A copy of the point p is now added to S_A if condition $U(p)$ does not hold. In all cases, p is inserted in the output stream of the merge process. Finally, if for any x in $S_A \cup S_B$ condition $U(x)$ is now true, we remove x from its active set. When the sweepline passes the last point of S, we remove any remaining elements in S_A and S_B.

By induction on the number of points passed by the sweepline, all elements of S_A are annotated by a point at least as close as any other element currently in S_A. Also, $U(x)$ is false for all $x \in S_A$. As observed in [19], this implies that for any two elements x_1 and x_2 from S_A, the longest side of the triangle $\triangle x_1 c x_2$ is the side $x_1 x_2$, so by the law of cosines, the angle $\angle x_1 c x_2$ is at least $\pi/3$. Therefore S_A can contain at most two elements, since the existence of three elements x_1, x_2, and x_3 would imply an angle $\angle x_i c x_j$ of at least $2\pi/3$ between two of these. By the same argument we also have $|S_B| \leq 2$ at all times.

Let $I(p, X)$ denote the condition that the annotation of p is a point at least as close as the closest point among the points lying below p in the strip X. Clearly, if a point $p \in A$ is passed by the sweepline without having a copy inserted into S_A, we know that $I(p, B)$ holds. If a copy x of p is inserted into S_A, it follows by induction on the number of points passed by the sweepline that $I(x, B)$ holds when x is removed from S_A. Similar statements with A and B interchanged also hold.

As said, our divide-and-conquer algorithm for phase one is analogous to Lazy Funnelsort, except that the merge process in a binary node of a k-merger will be the sweep line process described above. We allocate $O(1)$ extra space at each node to store the at most four copies contained in the two active sets of the merge process. We will maintain the following invariant: when a k-merger spanning a strip S finishes its merge process, condition $I(p, S)$ holds for all points p in output stream of the k-merger. Correctness of the algorithm follows immediately from this invariant. From the statements in the previous paragraph we see that the invariant is maintained if we ensure that when a k-merger finishes, the annotations of points p in its output stream have been updated to be at least as close as the annotations of *all* copies of p removed from active sets during the invocation of the k-merger.

To ensure this, we keep the copies of a point p in a doubly linked list along the path toward the root of the k-merger. The list contains all copies currently contained in active sets, and has p itself as the last element. Note that in a k-merger, the merge processes at nodes are interleaved—part of the output of one process is used as input for another before the first process has finished—so the length of this list can in the worst case be the height of the k-merger.

Consider a merge step in a node v in a k-merger which moves a point p from an input buffer of v to its output buffer. During the step, several types of updates of the linked list may be needed:

1. If p is currently contained in a list, the forward pointer of the next-to-last element needs to be updated to p's new position.
2. A copy x of p may be inserted in an active set of v. If p is currently not in a list, a new list containing x and p is made. Otherwise, x is inserted before p in p's list.
3. Elements of active sets of v may be removed from the sets. Each such element x should be deleted from its linked list.

This updating is part of the algorithm. Additionally, in the third case the annotation of x is propagated to the next element y in the list, i.e. the annotation of y is set to the closest of the annotations of x and y. This ensures the invariant discussed above.

What remains is to analyze the I/O complexity of the algorithm. As the space usage of a node in a k-merger is still $O(1)$, the analysis of Lazy Funnelsort is still valid, except that we need to account for the I/Os incurred during updates of the linked lists. In the full version of the paper [9], we prove that the I/Os for these updates is also bounded by the number of I/Os performed by the Lazy Funnelsort.

References

1. A. Aggarwal and J. S. Vitter. The input/output complexity of sorting and related problems. *Communications of the ACM*, 31(9):1116–1127, Sept. 1988.
2. L. Arge, M. A. Bender, E. D. Demaine, B. Holland-Minkley, and J. I. Munro. Cache-oblivious priority queue and graph algorithm applications. In *Proc. 34th Ann. ACM Symp. on Theory of Computing*. ACM Press, 2002. To appear.
3. M. J. Atallah and J.-J. Tsay. On the parallel-decomposability of geometric problems. *Algorithmica*, 8:209–231, 1992.
4. R. Bayer and E. McCreight. Organization and maintenance of large ordered indexes. *Acta Informatica*, 1:173–189, 1972.
5. M. A. Bender, R. Cole, and R. Raman. Exponential structures for efficient cache-oblivious algorithms. In *Proc. 29th International Colloquium on Automata, Languages, and Programming (ICALP)*, 2002. These proceedings.
6. M. A. Bender, E. Demaine, and M. Farach-Colton. Cache-oblivious B-trees. In *Proc. 41st Ann. Symp. on Foundations of Computer Science*, pages 399–409, 2000.
7. M. A. Bender, Z. Duan, J. Iacono, and J. Wu. A locality-preserving cache-oblivious dynamic dictionary. In *Proc. 13th Ann. ACM-SIAM Symp. on Discrete Algorithms*, pages 29–39, 2002.
8. J. L. Bentley. Algorithms for Klee's rectangle problems. Carnegie-Mellon University, Pittsburgh, Penn., Department of Computer Science, unpublished notes, 1977.
9. G. S. Brodal and R. Fagerberg. Cache oblivious distribution sweeping. Technical Report RS-02-18, BRICS, Dept. of Computer Science, University of Aarhus, 2002.
10. G. S. Brodal, R. Fagerberg, and R. Jacob. Cache oblivious search trees via binary trees of small height. In *Proc. 13th Ann. ACM-SIAM Symp. on Discrete Algorithms*, pages 39–48, 2002.
11. M. de Berg, M. van Kreveld, M. Overmars, and O. Schwarzkopf. *Computational Geometry: Algorithms and Applications*. Springer Verlag, Berlin, 1997.
12. M. Frigo, C. E. Leiserson, H. Prokop, and S. Ramachandran. Cache-oblivious algorithms. In *40th Annual Symposium on Foundations of Computer Science*, pages 285–297, 1999.
13. M. T. Goodrich, J.-J. Tsay, D. E. Vengroff, and J. S. Vitter. External-memory computational geometry. In *Proc. 34th Ann. Symp. on Foundations of Computer Science*, pages 714–723, 1993.
14. R. L. Graham. An efficient algorithm for determining the convex hull of a finite planar set. *Inf. Process. Lett.*, 1:132–133, 1972.
15. J. L. Hennessy and D. A. Patterson. *Computer Architecture: A Quantitative Approach*. Morgan Kaufmann, second edition, 1996.
16. H. T. Kung, F. Luccio, and F. P. Preparata. On finding the maxima of a set of vectors. *Journal of the ACM*, 22(4):469–476, Oct. 1975.

17. H. Prokop. Cache-oblivious algorithms. Master's thesis, Massachusetts Institute of Technology, June 1999.
18. J. S. Vitter. External memory algorithms and data structures: Dealing with massive data. *ACM Computing Surveys*, 33(2):209–271, June 2001.
19. D. E. Willard and Y. C. Wee. Quasi-valid range querying and its implications for nearest neighbor problems. In *Proceedings of the Fourth Annual Symposium on Computational Geometry*, pages 34–43. ACM Press, 1988.

One-Probe Search*

Anna Östlin and Rasmus Pagh

BRICS** Department of Computer Science
University of Aarhus, Denmark
{annao,pagh}@brics.dk

Abstract. We consider dictionaries that perform lookups by probing a *single word* of memory, knowing only the size of the data structure. We describe a randomized dictionary where a lookup returns the correct answer with probability $1 - \epsilon$, and otherwise returns "don't know". The lookup procedure uses an expander graph to select the memory location to probe. Recent explicit expander constructions are shown to yield space usage far smaller than what would be required using a deterministic lookup procedure. Our data structure supports efficient *deterministic* updates, exhibiting new probabilistic guarantees on dictionary running time.

1 Introduction

The *dictionary* is one of the most well-studied data structures. A dictionary represents a set S of elements (called *keys*) from some universe U, along with information associated with each key in the set. Any $x \in U$ can be looked up, i.e., it can be reported whether $x \in S$, and if so, what information is associated with x. We consider this problem on a unit cost word RAM in the case where keys and associated information have fixed size and are not too big (see below). The most straightforward implementation, an array indexed by the keys, has the disadvantage that the space usage is proportional to the size of U rather than to the size of S. On the other hand, arrays are extremely time efficient: A single memory probe suffices to retrieve or update an entry.

It is easy to see that there exists no better deterministic one-probe dictionary than an array. In this paper we investigate *randomized* one-probe search strategies, and show that it is possible, using much less space than an array implementation, to look up a given key with probability arbitrarily close to 1. The probability is over coin tosses performed by the lookup procedure. In case the memory probe did not supply enough information to answer the query, this is realized by the lookup procedure, and it produces the answer "don't know". In particular, by iterating until an answer is found, we get a Las Vegas lookup procedure that can have an expected number of probes arbitrarily close to 1.

---

* Partially supported by the IST Programme of the EU under contract number IST-1999-14186 (ALCOM-FT).

** Basic Research in Computer Science (www.brics.dk), funded by the Danish National Research Foundation.

It should be noted that one-probe search is impossible if one has no idea how much data is stored. We assume that the query algorithm knows the size of the data structure – a number that only changes when the size of the key set changes by a constant factor. The fact that the size, which may rarely or never change, is the only kind of global information needed to query the data structure means that it is well suited to support concurrent lookups (in parallel or distributed settings). In contrast, all known hash function based lookup schemes have some kind of global hash function that must be changed regularly. Even concurrent lookup of the *same* key, without accessing the same memory location, is supported to some extent by our dictionary. This is due to the fact that two lookups of the same key are not very likely to probe the same memory location.

A curious feature of our lookup procedure is that it makes its decision based on a constant number of *equality* tests – in this sense it is comparison-based. However, the data structure is not *implicit* in the sense of Munro and Suwanda [11], as it stores keys not in S.

Our studies were inspired by recent work of Buhrman et al. [4] on randomized analogs of bit vectors. They presented a Monte Carlo data structure where one bit probe suffices to retrieve a given bit with probability arbitrarily close to 1. When storing a sparse bit vector (few 1s) the space usage is much smaller than that of a bit vector. When storing no associated information, a dictionary solves the *membership* problem, which can also be seen as the problem of storing a bit vector. Our Las Vegas lookup procedure is stronger than the Monte Carlo lookup procedure in [4], as a wrong answer is never returned. The price paid for this is an *expected* bound on the number of probes, a slightly higher space usage, and, of course, that we look up one word rather than one bit. The connection to [4] is also found in the underlying technique: We employ the same kind of unbalanced bipartite expander graph as is used there. Recently, explicit constructions[1] of such graphs with near-optimal parameters have been found [14,15].

Let $u = |U|$ and $n = |S|$. We assume that one word is large enough to hold one of $2u + 1$ different symbols plus the information associated with a key. (Note that if this is not the case, it can be simulated by accessing a number of consecutive words rather than one word – an efficient operation in many memory models.) Our main theorem is the following:

Theorem 1. *For any constant $\epsilon > 0$ there exists a nonexplicit one-probe dictionary with success probability $1 - \epsilon$, using $O(n \log \frac{2u}{n})$ words of memory. Also, there is an explicit construction using $n \cdot 2^{O((\log \log u)^3)}$ words of memory.*

Note that the space overhead for the nonexplicit scheme, a factor of $\log \frac{2u}{n}$, is exponentially smaller than that of an array implementation.

In the second part of the paper we consider dynamic updates to the dictionary (insertions and deletions of keys). The fastest known dynamic dictionaries use hashing, i.e., they select at random a number of functions from suitable families, which are stored and subsequently used deterministically to direct searches.

---

[1] Where a given neighbor of a vertex can be computed in time polylogarithmic in the number of vertices.

A main point in this paper is that a fixed structure with random properties (the expander graph) can be used to move random choices from the data structure itself to the lookup procedure. The absence of hash functions in our data structure has the consequence that updates can be performed in a very local manner. We show how to deterministically perform updates by probing and changing a number of words that is nearly linear in the degree of the expander graph (which, for optimal expanders, is at most logarithmic in the size of the universe). Current explicit expanders are not fast enough for our dynamic data structure to improve known results in a standard RAM model. However, if we augment the RAM with an instruction for computing neighbors in an optimal expander graph with given numbers of vertices, an efficient dynamic dictionary can be implemented.

Theorem 2. *In the expander-augmented RAM model, there is a dictionary where a sequence of a insertions/deletions and b lookups in a key set of size at most n takes time $O(a(\log \frac{2u}{n})^{1+o(1)} + b + t)$ with probability $1 - 2^{-\Omega(a+t/(\log \frac{2u}{n})^{1+o(1)})}$. The space usage is $O(n \log \frac{2u}{n})$ words.*

When the ratio between the number of updates and lookups is small, the expected average time per dictionary operation is constant. Indeed, if the fraction of updates is between $(\log \frac{2u}{n})^{-1-\Omega(1)}$ and $n^{-\omega(1)}$, and if $u = 2^{n^{1-\Omega(1)}}$, the above yields the best known probability, using space polynomial in n, that a sequence of dictionary operations take average constant time. The intuitive reason why the probability bound is so good, is that time consuming behavior requires bad random choices in many invocations of the lookup procedure, and that the random bits used in different invocations are *independent*.

1.1 Related Work

As described above, this paper is related to [4], in scope as well as in tools. The use of expander graphs in connection with the membership problem was earlier suggested by Fiat and Naor [6], as a tool for constructing an efficient implicit dictionary.

Yao [16] showed an $\Omega(\log n)$ worst case lower bound on the time for dictionary lookups on a restricted RAM model allowing words to contain only keys of S or special symbols from a fixed set whose size is a function of n (e.g., pointers). The lower bound holds when space is bounded by a function of n, and u is sufficiently large. It extends to give an $\Omega(\log n)$ lower bound for the expected time of randomized Las Vegas lookups.

Our data structure violates Yao's lower bound model in two ways: 1. We allow words to contain certain keys not in S (accessed only through equality tests); 2. We allow space depending on u. The second violation is the important one, as Yao's lower bound can be extended to allow 1. Yao also considered deterministic one-probe schemes in his model, showing that, for $n \leq u/2$, a space usage of $u/2 + O(1)$ words is necessary and sufficient for them to exist.

The worst case optimal number of word probes for membership was studied by Pagh in [13] in the case where U equals the set of machine words. It was shown that *three* word probes are necessary when using m words of space, unless $u = 2^{\Omega(n^2/m)}$ or $u \leq n^{2+o(1)}$. Sufficiency of three probes was shown for all parameters (in most cases it followed by the classic dictionary of Fredman et al. [7]). In the expected sense, most hashing based dictionaries can be made to use arbitrarily close to 2 probes per lookup by expanding the size of the hash table by a constant factor.

Dictionaries with sublogarithmic lookup time that also allow efficient deterministic updates have been developed in a number of papers [1,2,8,9,12]. Let n denote an upper bound on the number of keys in a dynamic dictionary. For lookup time $t = o(\log \log n)$, the best known update time is $n^{O(1/t)}$, achieved by Hagerup et al. [9]. The currently best *probabilistic* guarantee on dynamic dictionary performance, first achieved by Dietzfelbinger and Meyer auf der Heide in [5], is that each operation takes constant time with probability $1 - O(m^{-c})$, where c is any constant and m is the space usage in words (which must be some constant factor larger than n). This implies that a sequence of a insertions/deletions and b lookups takes time $O(a + b + t)$ with probability $1 - O(m^{-t/n})$.

2   Preliminaries

In this section we define (n, d, ϵ)-expander graphs and state some results concerning these graphs. For the rest of this paper we will assume ϵ to be a multiple of $1/d$, as this makes statements and proofs simpler. This will be without loss of generality, as the statements we show do not change when rounding ϵ down to the nearest multiple of $1/d$.

Let $G = (U, V, E)$ be a bipartite graph with left vertex set U, right vertex set V, and edge set E. We denote the set of neighbors of a set $S \subseteq U$ by $\Gamma(S) = \bigcup_{s \in S} \{v \mid (s, v) \in E\}$. We use $\Gamma(x)$ as a shorthand for $\Gamma(\{x\})$, $x \in U$.

Definition 3. *A bipartite graph $G = (U, V, E)$ is d-regular if the degree of all nodes in U is d. A bipartite d-regular graph $G = (U, V, E)$ is an (n, d, ϵ)-expander if for each $S \subseteq U$ with $|S| \leq n$ it holds that $|\Gamma(S)| \geq (1 - \epsilon)d|S|$.*

Lemma 4. *For $0 < \epsilon < 1$ and $d \geq 1$, if $|V| \geq (1-\epsilon)dn(2u/n)^{1/\epsilon d}e^{1/\epsilon}$ then there exists an (n, d, ϵ)-expander graph $G = (U, V, E)$, where $|U| = u$.*

Proof. Our proof is a standard application of the probabilistic method. Let $G = (U, V, E)$ be a randomly generated graph created by the following procedure. For each $u \in U$ choose d neighbors with replacement, i.e., an edge can be chosen more than once, but then the double edges are removed. We will argue that the probability that this graph fails to be a (n, d, ϵ)-expander graph is less than 1 for the choices of $|V|$ and d as stated in the lemma. The degrees of the nodes in U in this graph may be less than d, but if there exists a graph that is expanding with degree at most d for all nodes, then there clearly exists a graph that is expanding with exactly degree d as well.

We must bound the probability that some subset of $i \leq n$ vertices from U has fewer than $(1 - \epsilon)di$ neighbors. A subset $S \subseteq U$ of size i can be chosen in $\binom{u}{i}$ ways and a set $V' \subseteq V$ of size $(1 - \epsilon)di$ can be chosen in $\binom{|V|}{(1-\epsilon)di}$ ways. (Note that $|V| \geq (1 - \epsilon)di$.) The probability that such a set V' contains all of the neighbors for S is $(\frac{(1-\epsilon)di}{|V|})^{di}$. Thus, the probability that some subset of U of size $i \leq n$ has fewer than $(1 - \epsilon)di$ neighbors is at most

$$\sum_{i=1}^{n} \binom{u}{i} \binom{|V|}{(1-\epsilon)di} \left(\frac{(1-\epsilon)di}{|V|} \right)^{di}$$

$$< \sum_{i=1}^{n} \left(\frac{ue}{i} \right)^{i} \left(\frac{|V|e}{(1-\epsilon)di} \right)^{(1-\epsilon)di} \left(\frac{(1-\epsilon)di}{|V|} \right)^{di}$$

$$\leq \sum_{i=1}^{n} \left(\left(\frac{(1-\epsilon)di}{|V|} \right)^{\epsilon d} e^{d} u/i \right)^{i}.$$

If the term in the outermost parentheses is bounded by $1/2$, the sum is less than 1. This is the case when $|V|$ fulfills the requirement stated in the lemma. □

Corollary 5. *For any constants $\alpha, \epsilon > 0$ there exist an (n, d, ϵ)-expander $G = (U, V, E)$ for the following parameters:*

- *$|U| = u$, $d = O(\log(2u/n))$ and $|V| = O(n \log(2u/n))$.*
- *$|U| = u$, $d = O(1)$ and $|V| = O(n (2u/n)^{\alpha})$.*

Theorem 6. *(Ta-Shma [14]) For any constant $\epsilon > 0$ and for $d = 2^{O((\log\log u)^3)}$, there exists an explicit (n, d, ϵ)-expander $G = (U, V, E)$ with $|U| = u$ and $|V| = n \cdot 2^{O((\log\log u)^3)}$.*

3   Static Data Structure

Let $S \subseteq U$ denote the key set we wish to store. Our data structure is an array denoted by T. Its entries may contain the symbol x for keys $x \in S$, the symbol $\neg x$ for keys $x \in U \backslash S$, or the special symbol $\perp \notin U$. (Recall our assumption that one of these symbols plus associated information fits into one word.) For simplicity we will consider the case where there is no information associated with keys, i.e., we solve just the membership problem. Extending this to allow associated information is straightforward. We make use of a $(2n + 1, d, \epsilon/2)$-expander with neighbor function Γ. Given that a random element in the set $\Gamma(x)$ can be computed quickly for $x \in U$, the one-probe lookup procedure is very efficient.

```
procedure lookup_ε(x)
    choose v ∈ Γ(x) at random;
    if T[v] = x then return 'yes'
    else if T[v] ∈ {¬x, ⊥} then return 'no'
    else return 'don't know';
end;
```

The corresponding Las Vegas lookup algorithm is the following:

> **procedure** lookup(x)
>   **repeat**
>   choose $v \in \Gamma(x)$ at random;
>   **until** $T[v] \in \{x, \neg x, \bot\}$;
>   **if** $T[v] = x$ **then return** *'yes'* **else return** *'no'*;
> **end**;

3.1   Requirements to the Data Structure

The success probability of $\text{lookup}_\epsilon(x)$ and the expected time of lookup(x) depends on the content of the entries indexed by $\Gamma(x)$ in T. To guarantee correctness and success probability $1 - \epsilon$ in each probe for x, the following conditions should hold:

1. If $x \in S$, at least a fraction $1 - \epsilon$ of the entries $T[v]$, $v \in \Gamma(x)$, contain x, and none contain $\neg x$ or $\bot$.
2. If $x \notin S$, at least a fraction $1 - \epsilon$ of the entries $T[v]$, $v \in \Gamma(x)$, contain either $\neg x$ or $\bot$, and none contain x.

By inserting $\bot$ in all entries of T except the entries in $\Gamma(S)$, condition 2 will be satisfied for all $x \notin S$ with $|\Gamma(x) \cap \Gamma(S)| \leq \epsilon d$. A key notion in this paper is the set of ϵ-*ghosts* for a set S, which are the keys of U that have many neighbors in common with S. For each ϵ-ghost x we will need some entries in T with content $\neg x$.

Definition 7. *Given a bipartite graph $G = (U, V, E)$, a key $x \in U$ is an ϵ-ghost for the set $S \subseteq U$ if $|\Gamma(x) \cap \Gamma(S)| > \epsilon|\Gamma(x)|$ and $x \notin S$.*

Lemma 8. (Buhrman et al. [4]) *There are at most n ϵ-ghosts for a set S of size n in a $(2n + 1, d, \epsilon/2)$-expander graph.*

In order to fulfill conditions 1 and 2, we need to assign entries in T to the keys in S and to the ϵ-ghosts for S.

Definition 9. *Let $G = (U, V, E)$ be a bipartite d-regular graph and let $0 < \epsilon < 1$. An assignment for a set $S \subseteq U$, is a subset $A \subseteq E \cap (S \times \Gamma(S))$ such that for $v \in \Gamma(S)$, $|A \cap (S \times \{v\})| = 1$. A $(1 - \epsilon)$-balanced assignment for S is an assignment A, where for each $s \in S$ it holds that $|A \cap (\{s\} \times \Gamma(s))| \geq (1 - \epsilon)d$.*

Lemma 10. *If a graph $G = (U, V, E)$ is an (n, d, ϵ)-expander then there exists a $(1 - \epsilon)$-balanced assignment for every set $S \subseteq U$ of size at most n.*

To show the lemma we will use Hall's theorem [10]. A *perfect matching* in a bipartite graph (U, V, E) is a set of $|U|$ edges such that for each $x \in U$ there is an edge $(x, v) \in E$, and for each $v \in V$ there is at most one edge $(x, v) \in E$.

Theorem 11. (Hall's theorem) *In any bipartite graph $G = (U, V, E)$, where for each subset $U' \subseteq U$ it holds that $|U'| \leq |\Gamma(U')|$, there exists a perfect matching.*

Proof of Lemma 10. Let S be an arbitrary subset of U of size n. Let $G' = (S, \Gamma(S), E')$ be the subgraph of G induced by the nodes S and $\Gamma(S)$, i.e., $E' = \{(s, v) \in E \mid s \in S\}$. To prove the lemma we want to show that there exists an assignment A such that for each $s \in S$, $|A \cap (\{s\} \times \Gamma(s))| \geq (1 - \epsilon)d$. The idea is to use Hall's theorem $(1 - \epsilon)d$ times by repeatedly finding a perfect matching and removing the nodes from $\Gamma(S)$ in the matching.

Since G is an (n, d, ϵ)-expander we know that for each subset $S' \subseteq S$ it holds that $|\Gamma(S')| \geq (1 - \epsilon)d|S'|$. Assume that we have i perfect matchings from S to non-overlapping subsets of $\Gamma(S)$ and denote by M the nodes from $\Gamma(S)$ in the matchings. For each subset $S' \subseteq S$ it holds that $|\Gamma(S')\backslash M| \geq ((1 - \epsilon)d - i)|S'|$. If $(1 - \epsilon)d - i \geq 1$ then the condition in Hall's theorem holds for the graph $G_i = (S, (\Gamma(S)\backslash M), E'\backslash E_i)$, where E_i is the set of edges incident to nodes in M, and there exists a perfect matching in G_i. From this it follows that at least $(1 - \epsilon)d$ non-overlapping (in $\Gamma(S)$) perfect matchings can be found in G'. The edges in the matchings define a $(1 - \epsilon)$-balanced assignment. □

3.2   Construction

We store the set S as follows:

1. Write $\perp$ in all entries not in $\Gamma(S)$.
2. Find the set $\bar{S}$ of ϵ-ghosts for S.
3. Find a $(1 - \epsilon)$-balanced assignment for the set $S \cup \bar{S}$.
4. For $x \in S$ write x in entries assigned to x.
   For $x \in \bar{S}$ write $\neg x$ in entries assigned to x in $\Gamma(S)$.

By Lemma 8 the set $\bar{S}$ found in step 2 contains at most n keys, and by Lemma 10 it is possible to carry out step 3. Together with the results on expanders in Section 2, this concludes the proof of Theorem 1.

We note that step 2 takes time $\Omega(|U\backslash S|)$ if we have only oracle access to Γ. When the graph has some structure it is sometimes possible to do much better. Ta-Shma shows in [14] that this step can be performed for his class of graphs in time polynomial in the size of the right vertex set, i.e., polynomial in the space usage. All other steps are clearly also polynomial time in the size of the array.

In the dynamic setting, covered in Section 4, we will take an entirely different approach to ghosts, namely, we care about them only if we see them. We then argue that the time spent looking for a particular ghost before it is detected is not too large, and that there will not be too many different ghosts.

4   Dynamic Updates

In this section we show how to implement efficient dynamic insertions and deletions of keys in our dictionary. We will use a slightly stronger expander graph than in the static case, namely a $(4n', d, \epsilon/3)$-expander where n' is an upper bound on the size of the set that can be handled. The parameter n' is assumed to be known to the query algorithm. Note that n' can be kept in the range, say, n to $2n$ at no asymptotic cost, using standard global rebuilding techniques. Our dynamic dictionary essentially maintains the static data structure described in the previous section. Additionally, we maintain the following auxiliary data structures:

- A priority queue with all keys in S plus some set $\bar{S}$ of keys that appear negated in T. Each key has as priority the size of its assignment, which is always at least $(1 - \epsilon)d$.
- Each entry $T[v]$ in T is augmented with
  - A pointer $T_p[v]$ which, if entry v is assigned to a key, points to that key in the priority queue.
  - A counter $T_c[v]$ that at any time stores the number of keys in S that have v as a neighbor.

Since all keys in the priority queue are assigned $(1 - \epsilon)d$ entries in T, the performance of the lookup procedure is the desired one, except when searching for ϵ-ghosts not in $\bar{S}$. We will discuss this in more detail later.

4.1   Performing Updates

We first note that it is easy to maintain the data structure during deletions. All that is needed when deleting $x \in S$ is decreasing the counters $T_c[v]$, $v \in \Gamma(x)$, and replacing x with $\neg x$ or $\perp$ (the latter if the counter reaches 0). Finally, x should be removed from the priority queue. We use a simple priority queue that requires space $O(d + n)$, supports insert in $O(d)$ time, and increasekey, decreasekey, findmin and delete in $O(1)$ time. The total time for a deletion in our dictionary is $O(d)$.

When doing insertions we have to worry about maintaining a $(1 - \epsilon)$-balanced assignment. The idea of our insertion algorithm is to assign all neighbors to the key being inserted. In case this makes the assignment of other keys too small (easily seen using the priority queue), we repeat assigning all neighbors to them, and so forth. Every time an entry in T is reassigned to a new key, the priority of the old and new key are adjusted in the priority queue. The time for an insertion is $O(d)$, if one does not count the associated cost of maintaining assignments of other keys. The analysis in Section 4.2 will bound this cost. Note that a priori it is not even clear whether the insertion procedure terminates.

A final aspect that we have to deal with is ghosts. Ideally we would like $\bar{S}$ to be at all times the current set of ϵ-ghosts for S, such that a $(1 - \epsilon)$-balanced assignment was maintained for all ghosts. However, this leaves us with the hard problem of finding new ghosts as they appear. We circumvent this problem by only including keys in $\bar{S}$ if they are selected for examination and found to be ϵ-ghosts. A key is selected for examination if a lookup of that key takes more than $\log_{1/\epsilon} d$ iterations. The time spent on examinations and on lookups of a ghost before it is found, is bounded in the next section.

The sequence of operations is divided up into stages, where each stage (except possibly the last) contains n' insert operations. After the last insertion in a stage, all keys in $\bar{S}$ that are no longer ϵ-ghosts are deleted. This is done by going through all keys in the priority queue. Keys of $\bar{S}$ with at least $(1-\epsilon)d$ neighbors containing $\perp$ are removed from the priority queue. Hence, when a new stage starts, $\bar{S}$ will only contain ϵ-ghosts.

4.2   Analysis

We now sketch the analysis of our dynamic dictionary. First, the total work spent doing assignments and reassignments is analyzed. Recall that the algorithm maintains a $(1 - \epsilon)$-balanced assignment for the set $S \cup \bar{S}$ of keys in the priority queue. Keys enter the priority queue when they are inserted in S, and they may enter it when they are ϵ-ghosts for the current set. It clearly suffices to bound the total work in connection with insertions in the priority queue, as the total work for deletions cannot be larger than this. We will first show a bound on the number of keys in $\bar{S}$.

Lemma 12. *The number of keys in the set $\bar{S}$ never exceeds $2n'$.*

Proof. Let S be the set stored at the beginning of a stage. $\bar{S}$ only contains ϵ-ghosts for S at this point. Let S' denote the keys inserted during the stage. New keys inserted into $\bar{S}$ have to be ϵ-ghosts for $S \cup S'$. According to Lemma 8, the fact that $|S \cup S'| \leq 2n'$ implies that there are at most $2n'$ ϵ-ghosts for $S \cup S'$ (including the ϵ-ghosts for S). Thus, the number of keys in $\bar{S}$ during any stage is at most $2n'$. □

It follows from the lemma that the number of insertions in the priority queue is bounded by 3 times the number of insertions performed in the dictionary. The remainder of our analysis of the number of reassignments has two parts: We first show that our algorithm performs a number of reassignments (in connection with insertions) that is within a constant factor of *any* scheme maintaining a $(1-\epsilon/3)$-balanced assignment. The scheme we compare ourselves to may be *off-line*, i.e., know the sequence of operations in advance. Secondly, we give an off-line strategy for maintaining a $(1 - \epsilon/3)$-balanced assignment using $O(d)$ reassignments per update. This proof strategy was previously used for an assignment problem by Brodal and Fagerberg [3].

In the following lemmas, the set M is the set for which a balanced assignment is maintained, and the insert and delete operations are insertions and deletions in this set. In our data structure M corresponds to $S \cup \bar{S}$.

Lemma 13. *Let $G = (U, V, E)$ be a d-regular bipartite graph. Suppose O is a sequence of insert and delete operations on a dynamic set $M \subseteq U$. Let B be an algorithm that maintains a $(1 - \frac{\epsilon}{3})$-balanced assignment for M, and let C be our "assign all" scheme for maintaining a $(1 - \epsilon)$-balanced assignment for M. If B makes at most k reassignments during O, then C assigns all neighbors to a key at most $\frac{3}{\epsilon}(k/d + |M|_{\mathrm{start}})$ times, where $|M|_{\mathrm{start}}$ is the initial size of M.*

Proof. To show the lemma we will argue that the assignment of C, denoted A_C, will become significantly "less different" from the assignment of B, denoted A_B, each time C assigns all neighbors of a key to that key. At the beginning $|A_B \backslash A_C| \leq d|M|_{\text{start}}$, since $|A_B| \leq d|M|_{\text{start}}$. Each of the k reassignments B performs causes $|A_B \backslash A_C|$ to increase by at most one. This means that the reassignments made by C during O can decrease $|A_B \backslash A_C|$ by at most $k + d|M|_{\text{start}}$ in total.

Each time C assigns all entries in $\Gamma(x)$ to a key x, at least ϵd reassignments are done, since the assignment for x had size less than $(1 - \epsilon)d$ before the reassignment. At this point at least $(1 - \frac{\epsilon}{3})d$ pairs (x, e) are included in A_B, i.e., at most $\frac{\epsilon}{3}d$ of the neighbors of x are not assigned to x in A_B. This means that at least $\frac{2\epsilon}{3}d$ of the reassignments made by C decrease $|A_B \backslash A_C|$, while at most $\frac{\epsilon}{3}d$ reassignments may increase $|A_B \backslash A_C|$. In total, $|A_B \backslash A_C|$ is decreased by at least $\frac{\epsilon}{3}d$ when C assigns all neighbors to a key. The lemma now follows, as $|A_B \backslash A_C|$ can decrease by $\frac{\epsilon}{3}d$ at most $(k + d|M|_{\text{start}})/(\frac{\epsilon}{3}d)$ times. □

Lemma 14. *Let $G = (U, V, E)$ be a $(4n', d, \epsilon/3)$-expander. There exists an off-line algorithm maintaining a $(1 - \frac{\epsilon}{3})$-balanced assignment for a dynamic set $M \subseteq U$, during a stage of $3n'$ insertions, by performing at most $4dn'$ reassignments, where $|M| \leq n'$ at the beginning of the stage.*

Proof. Let M' be the set of $3n'$ keys to insert. Define $\tilde{M} = M \cup M'$; we have $|\tilde{M}| \leq 4n'$. Let $A_{\tilde{M}}$ be a $(1 - \frac{\epsilon}{3})$-balanced assignment for $\tilde{M}$ (shown to exist in Lemma 10).

The off-line algorithm knows the set M' of keys to insert from the start, and does the following. First, it assigns neighbors to the keys in M according to the assignment $A_{\tilde{M}}$, which requires at most dn' reassignments. Secondly, for each insertion of a key $x \in M'$, it assigns neighbors to x according to $A_{\tilde{M}}$, which requires at most d reassignments. This will not cause any key already in the set to lose an assigned neighbor, hence no further reassignments are needed to keep the assignment $(1 - \frac{\epsilon}{3})$-balanced. It follows that he total number of reassignments during the $3n'$ insertions is at most $4dn'$, proving the lemma. □

The above two lemmas show that in a sequence of a updates to the dictionary there are $O(a)$ insertions in the priority queue, each of which gives rise to $O(d)$ reassignments in a certain off-line algorithm, meaning that our algorithm uses $O(ad)$ time for maintaining a $(1 - \epsilon)$-balanced assignment for the set $S \cup \bar{S}$ in the priority queue.

We now turn to analyzing the work done in the lookup procedure. First we will bound the number of iterations in all searches for keys that are *not* undetected ϵ-ghosts. Each iteration has probability at least $1 - \epsilon$ of succeeding, independently of all other events, so we can bound the probability of many iterations using Chernoff bounds. In particular, the probability that the total number of iterations used in the b searches exceeds $\frac{2}{1-\epsilon}b + t$ is less than $e^{-\frac{1-\epsilon}{4}t}$.

When searching for a key that is not an undetected ϵ-ghost, the probability of selecting it for examination is bounded from above by $1/d$. In particular, by Chernoff bounds we get that, for $k > 0$, the total number of examinations during all b lookups is at most $b/d + k$ with probability $1 - (\frac{e}{1+kd/b})^{b/d+k}$. For $k = (2e - 1)b/d + t/d$ we get that the probability of more than $2eb/d + t/d$ examinations is bounded by $2^{-t/d}$. Each examination costs time $O(d)$, so the probability of spending $O(b+t)$ time on such examinations is at least $1 - 2^{-t/d}$.

We now bound the work spent on finding ϵ-ghosts. Recall that an ϵ-ghost is detected if it is looked up, and the number of iterations used by the lookup procedure exceeds $\log_{1/\epsilon} d$. Since we have an ϵ-ghost, the probability that a single lookup selects the ghost for examination is at least $\epsilon^{\log_{1/\epsilon} d - 1} = \Omega(1/d)$. We define $d' = O(d)$ by $1/d' = \epsilon^{\log_{1/\epsilon} d - 1}$. Recall that there are at most $2n'$ ϵ-ghosts in a stage, and hence at most $2a$ in total. We bound the probability that more than $4ad' + k$ lookups are made on undetected ϵ-ghosts, for $k > 0$. By Chernoff bounds the probability is at most $e^{-k/4d'}$. Each lookup costs $O(\log d)$ time, so the probability of using time $O(ad\log d + t)$ is at least $1 - e^{-t/4d' \log d}$.

In summary, we have bounded the time spent on four different tasks in our dictionary:

- The time spent looking up keys that are not undetected ϵ-ghosts is $O(b+t)$ with probability $1 - 2^{-\Omega(t)}$.
- The time spent examining keys that are not undetected ϵ-ghosts is $O(b+t)$ with probability $1 - 2^{-\Omega(t/d)}$.
- The time spent looking up ϵ-ghosts before they are detected is $O(ad\log d + t)$ with probability $1 - 2^{-\Omega(t/d\log d)}$.
- The time spent assigning, reassigning and doing bookkeeping is $O(ad)$.

Using the above with the first expander of Corollary 5, having degree $d = O(\log \frac{2u}{n'})$, we get the performance bound stated in Theorem 2. Using the constant degree expander of Corollary 5 we get a data structure with constant time updates. This can also be achieved in this space with a trie, but a trie would use around $1/\alpha$ word probes for lookups of keys in the set, rather than close to 1 word probe, expected.

5   Conclusion and Open Problems

In this paper we studied dictionaries for which a single word probe with good probability suffices to retrieve any given key with associated information. The main open problem we leave is whether the space usage of our dictionary is the best possible for one-probe search.

It is known that three word probes are necessary and sufficient in the worst case for lookups in dictionaries, even when using superlinear space. An obvious open question is how well one can do using two word probes and a randomized lookup procedure. Can the space utilization be substantially improved? Another point is that we bypass Yao's lower bound by using space dependent on u. An interesting question is: How large a dependence on u is necessary to get around Yao's lower bound. Will space $n \log^* u$ do, for example?

Acknowledgment. We thank Thore Husfeldt for helpful comments.

References

[1] Arne Andersson and Mikkel Thorup. Tight(er) worst-case bounds on dynamic searching and priority queues. In *Proceedings of the 32nd Annual ACM Symposium on Theory of Computing (STOC '00)*, pages 335–342. ACM Press, 2000.

[2] Paul Beame and Faith Fich. Optimal bounds for the predecessor problem. In *Proceedings of the 31st Annual ACM Symposium on Theory of Computing (STOC '99)*, pages 295–304. ACM Press, 1999.

[3] Gerth Stølting Brodal and Rolf Fagerberg. Dynamic representations of sparse graphs. In *Proceedings of the 6th International Workshop on Algorithms and Data Structures (WADS '99)*, volume 1663 of *Lecture Notes in Computer Science*, pages 342–351. Springer-Verlag, 1999.

[4] Harry Buhrman, Peter Bro Miltersen, Jaikumar Radhakrishnan, and S. Venkatesh. Are bitvectors optimal? In *Proceedings of the 32nd Annual ACM Symposium on Theory of Computing (STOC '00)*, pages 449–458. ACM Press, 2000.

[5] Martin Dietzfelbinger and Friedhelm Meyer auf der Heide. A new universal class of hash functions and dynamic hashing in real time. In *Proceedings of the 17th International Colloquium on Automata, Languages and Programming (ICALP '90)*, volume 443 of *Lecture Notes in Computer Science*, pages 6–19. Springer-Verlag, 1990.

[6] Amos Fiat and Moni Naor. Implicit $O(1)$ probe search. *SIAM J. Comput.*, 22(1):1–10, 1993.

[7] Michael L. Fredman, János Komlós, and Endre Szemerédi. Storing a sparse table with $O(1)$ worst case access time. *J. Assoc. Comput. Mach.*, 31(3):538–544, 1984.

[8] Michael L. Fredman and Dan E. Willard. Surpassing the information theoretic bound with fusion trees. *J. Comput. System Sci.*, 47:424–436, 1993.

[9] Torben Hagerup, Peter Bro Miltersen, and Rasmus Pagh. Deterministic dictionaries. *J. Algorithms*, 41(1):69–85, 2001.

[10] Philip Hall. On representatives of subsets. *J. London Math. Soc.*, 10:26–30, 1935.

[11] J. Ian Munro and Hendra Suwanda. Implicit data structures for fast search and update. *J. Comput. System Sci.*, 21(2):236–250, 1980.

[12] Rasmus Pagh. A trade-off for worst-case efficient dictionaries. *Nordic J. Comput.*, 7(3):151–163, 2000.

[13] Rasmus Pagh. On the Cell Probe Complexity of Membership and Perfect Hashing. In *Proceedings of the 33rd Annual ACM Symposium on Theory of Computing (STOC '01)*, pages 425–432. ACM Press, 2001.
http://www.brics.dk/ pagh/papers/probe.pdf

[14] Amnon Ta-Shma. Storing information with extractors. To appear in Information Processing Letters.

[15] Amnon Ta-Shma, Christopher Umans, and David Zuckerman. Loss-less condensers, unbalanced expanders, and extractors. In *Proceedings of the 33rd Annual ACM Symposium on Theory of Computing (STOC '01)*, pages 143–152. ACM Press, 2001.

[16] Andrew C.-C. Yao. Should tables be sorted? *J. Assoc. Comput. Mach.*, 28(3):615–628, 1981.

New Algorithms for Subset Query, Partial Match, Orthogonal Range Searching, and Related Problems

Moses Charikar[1], Piotr Indyk[2], and Rina Panigrahy[3]

[1] Princeton University
moses@cs.princeton.edu
[2] MIT
indyk@theory.lcs.mit.edu
[3] Cisco Systems
rinap@cisco.com

Abstract. We consider the *subset query* problem, defined as follows: given a set $\mathcal{P}$ of N subsets of a universe U, $|U| = m$, build a data structure, which for any *query* set $Q \subset U$ detects if there is any $P \in \mathcal{P}$ such that $Q \subset P$. This is essentially equivalent to the partial match problem and is a fundamental problem in many areas. In this paper we present the first (to our knowledge) algorithms, which achieve non-trivial space and query time bounds for $m = \omega(\log N)$. In particular, we present two algorithms with the following tradeoffs:

- $N \cdot 2^{O(m \log^2 m \sqrt{c/\log N})}$ space, and $O(N/2^c)$ time, for any c
- Nm^c space and $O(mN/c)$ query time, for any $c \leq N$

We extend these results to the more general problem of orthogonal range searching (both exact and approximate versions), approximate orthogonal range intersection and the exact and approximate versions of the nearest neighbor problem in ℓ_∞.

1 Introduction

The *subset query* problem is defined as follows: given a set $\mathcal{P}$ of N subsets of a universe U, $|U| = m$, build a data structure, which for any *query* set $Q \subset U$ detects if there is any $P \in \mathcal{P}$ such that $Q \subset P$. This problem is of fundamental importance in many areas. In information retrieval, it occurs in applications which allow the user to search for documents containing a given set of words (e.g., as in Google). In the field of databases, it corresponds to the *partial match problem*, which involves searching for records which satisfy specified equality constraints[1] (e.g., "Sex=Male and City=Boston"). The partial match problem is also of large interest for IP packet classification, where the goal is to classify a packet depending on its parameters (source and destination address, length etc) [1,6,7,8,16].

---

[1] For an easy equivalence between the partial match problem and the subset query problem, see Preliminaries.

P. Widmayer et al. (Eds.): ICALP 2002, LNCS 2380, pp. 451–462, 2002.
© Springer-Verlag Berlin Heidelberg 2002

Due to its high importance, the subset query and partial match problems have been investigated for quite a while. It is easy to see that this problem has two fairly trivial solutions: (a) store all answers to all queries (requiring 2^m space and $O(m)$ query time), or (b) scan the whole database for the answer (requiring linear storage and $O(Nm)$ query time). Unfortunately, both solutions are quite unsatisfactory in most applications. The first non-trivial result for this problem has been obtained by Rivest [14,15]. He showed that the 2^m space of the trivial "exhaustive storage" solution can be somewhat improved for $m \leq 2 \log N$. He also presented a trie-based algorithm using linear storage, which achieved sublinear query time when the database content is generated at *random.*

Unfortunately, since those early results, there has been essentially no progress on this problem to this date[2]. Consequently, it is believed (e.g., see [5] or [10] and the references therein), that the problem inherently suffers from the "curse of dimensionality" , i.e., that there is no algorithm for this problem which achieves *both* "fast" query time and "small" space. Various variants of the conjecture exist, depending on the definition of "fast" and "small". Recently it has been shown [5] that the problem requires space superpolynomial in n if the query time (in the cell-probe model) is $o(\log m)$. However, there is an exponential gap between the lower bound and the upper bounds.

In sharp contrast to these problems, recent research has yielded fairly good bounds for the approximate nearest neighbor problem in Euclidean spaces, a fundamental problem that is very challenging for points in high dimensions. In particular, [12] and [11] provided the first algorithm for $(1 + \epsilon)$-approximate nearest neighbor in l_2^d or l_1^d, with polynomial storage and $(1/\epsilon + d + \log N)^{O(1)}$ query time. For l_∞^d, [9,10] gave a 3-approximate nearest neighbor algorithm using $N^{O(\log d)}$ storage and having $(d + \log N)^{O(1)}$ query time.

In this paper we present the first (to our knowledge) algorithms, which achieve non-trivial space and query time bounds for $m = \omega(\log N)$. In particular, we present two algorithms with the following tradeoffs:

- $N \cdot 2^{O(m \log^2 m \sqrt{c/\log N})}$ space, and $O(N/2^c)$ time, for any c
- Nm^c space and $O(mN/c)$ query time, for any $c \leq N$

For reporting queries, our algorithms output a collection of pointers to pre-computed lists. Thus they do not need to output each individual point. The first algorithm is interesting when the goal is to achieve a $o(N)$ query time. For example, if we set $c = \log N / poly \log m$, then we can achieve an algorithm with space $N2^{m/poly \log m}$ (i.e., subexponential in m), with query time $N^{1-1/poly \log m}$. For $m = O(\log^{1.4} N)$ (and a different value of c), the first algorithm yields a data structure with space $N^{1+o(1)}$ and query time sublinear in N. On the other hand, the second algorithm is interesting when the goal is to achieve a $o(mN)$ query time. This becomes interesting if m is fairly large compared to N, e.g. when $m > N^{(1+\alpha)/2}$ for $\alpha > 0$. In this case, by setting $c = \sqrt{N}$ we get significantly subexponential space requirements (i.e., $2^{O(m^{1/(1+\alpha)} \log m)}$), while achieving significantly sublinear query time of $O(m\sqrt{N})$.

---

[2] From the theoretical perspective - many data structures has been designed which improve the space or the query time in practice.

We also mention that the first result shows that the strongest version of the curse-of-dimensionality conjecture (which interprets $2^{o(m)}$ space as "small" and $o(N)$ query time as "fast", with $m = \omega(\log N)$) is *false*.

We extend the above results to a more general problem of *orthogonal range searching* [2], which involves searching for points contained in a d-dimensional query rectangle. Range searching is one of the most investigated problems in computational geometry. It is known [13] that one can construct a (range-tree) data structure of size $O(N \log^d N)$ which answers range queries in time $O(\log^{d-1} N)$; sub-logarithmic improvements to these bounds are also known.

We show that the second of the aforementioned subset query algorithms can be extended to solve orthogonal range search within the same bounds (with m replaced by d). This yields an algorithm with Nd^c space and $O(dN/c)$ query time. The algorithm can be immediately used to solve the nearest neighbor problem in l_∞^d with space bound unchanged and query time multiplied by $\log n$. Moreover, a special case of orthogonal range search with *aligned* query rectangles (i.e. whose projection on each dimension is of the form $[x2^i, (x+1)2^i]$) reduces directly to the partial match problem with $m = s$, where s is the number of bits required to specify each point in the database. This result is interesting for range queries in databases where queries require exact matches on most fields and specify ranges in a few fields. In addition, we show that $(1+\epsilon)$-approximate orthogonal range searching can be reduced to the partial match problem with $m = O(s/\epsilon)$. This reduction gives interesting trade-offs when combined with the first of the aforementioned subset query algorithms. The results also apply to the $(1+\epsilon)$-approximate nearest neighbor problem in l_∞. We can also show that a $(1+\epsilon)$-approximate range intersection can be reduced to a partial match problem with $m = \tilde{O}(s/\epsilon^4)$.[3]

In addition, we show that orthogonal range searching in dimension d can be reduced to the subset query problem with $m = O(d^2 \log^2 N)$. This demonstrates the close relationship between the two problems. Moreover, combined with the result of [9] (who showed that the subset query problem can be reduced to c-approximate nearest neighbor in l_∞^d for $c < 3$) it implies an interesting fact: if we efficiently solve the c-approximate nearest neighbor in l_∞^d for $c < 3$, then we can solve this problem *exactly* with similar resources.

2   Preliminaries

In this section we show that several superficially different problems are essentially equivalent to the subset query problem. These easy equivalences are widely known. In this paper we switch between these problems whenever convenient.

Partial match problem. Given a set $\mathcal{D}$ of N vectors in $\{0, 1\}$, build a data structure, which for any *query* $q \in \{0, 1, *\}$, detects if there is any $p \in \mathcal{D}$ such that q matches p. The symbol "$*$" acts as a "don't care" symbol, i.e., it matches

---

[3] We use $\tilde{O}(n)$ to denote $O(n\,poly\log n)$.

both 0 and 1. Note that one can extend any algorithm solving this problem to handle non-binary symbols in vectors and queries, by replacing them with their binary representations.

It is easy to reduce partial match problem to the subset query problem. To this end, we replace each $p \in \mathcal{D}$ by a set of all pairs (i, p_i), for all $i = 1 \ldots m$. In addition, we replace each query q by a set of all pairs (i, q_i) such that $q_i \neq *$. The correctness is immediate.

It is also see how to reduce subset query to the partial match problem. This is done by replacing each database set P by its characteristic vector, and replacing query set Q by its characteristic vector in which 0's are replaced with $*$'s.

Containment query problem. This problem is almost the same as the subset query problem, except that we seek a set P such that $P \subset Q$. Since P is a subset of Q iff its complement is a superset of the complement of Q, the subset and the containment query problems are equivalent as well.

3   Algorithm for Set Containment

In this section, we present an algorithm for set containment. A naive algorithm to answer set containment queries on a universe of size m is to write the query set as a bit vector and index it into an array of size 2^m. Here each array location points to a list L of sets from the database contained in the set corresponding to the array location.

To reduce the space complexity one can use random sampling to hash the query sets into a smaller hash table by focusing on a random subset of the universe. Say we random sample with probability p and obtain pm elements of the universe. One can now construct an array of size 2^{pm} based only on these sampled elements. Again, the query set maps to an array location which contains a list L of database sets. However, we can only guarantee that a set in list L is contained in Q with respect to the sampled elements. In fact, one can show that the sets in L are *almost* contained in Q. In other words the difference from Q of a set in L is likely to be small. (A set with a large difference from Q is likely to be eliminated since the random sample would contain an element from the difference.) The fact that sets in L have small differences from Q implies a bound on the number of distinct differences from Q. We choose our sampling probability such that the number of distinct differences from Q of sets in L is small (sublinear in N). However this property by itself does not help in yielding a query time sublinear in N. This is because even though the number of possible differences from Q is small, the number of sets in L may be quite large as there could be several sets that have the same difference from Q.

However, the fact that the sets in L have small differences from Q suggests that the sets in L could be grouped according to their difference from Q. It is not clear how to do this since the differences depend on the query set Q and a number of query sets Q map onto the same location in the array. However, this idea can be adapted to yield an efficient data structure. For the list L we construct a baseline set and work with differences from that set. This baseline set R which we call a *representative set* is such that sets in L have small differences from R. We group the sets in L by their differences from R. Now assuming a query set Q contains R we only need to look at which of the differences from R are contained in Q.

However, we cannot ensure that every query set Q that hashes to a list L contains its representative set R. It turns out that we can relax this condition to say that Q almost contains R, i.e. the difference $R - Q$ is small.

We now build a second level hash table to organize the collection L. Entries in this second level table are indexed by small subsets of R. The query set Q maps onto the location corresponding to $R - Q$ (note that this is a small subset by the properties of the representative set). This location is associated with the subcollection $L' \subseteq L$ of sets that do not intersect $R - Q$. Further, the sets in L' are grouped according to their difference from R. Since the sets in the original collection L had small differences from R, the sets in the subcollections L' also have small differences. For each group of sets with the same difference from R, we check to see if the difference from R is completely contained in Q. If so, all the sets in the group are contained in Q; if not, all sets are not contained in Q. This is depicted in Figure 1. The query time is determined by the number of groups (i.e. number of differences from R) that must be examined by the algorithm. The number of differences can be bounded by the fact the sets have small differences from R; we choose our parameters so that this is sublinear in N.

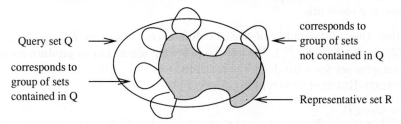

Fig. 1. *The second level hash table in the containment query data structure. The "ears" around R represent collections of sets that have the same difference from R*

3.1   Technical Details

We start by defining relaxed notions of containment and disjointness:

Definition 1 (Almost contains). *A set Q is said to x-almost contain set P if $|P - Q| \leq x$.*

Definition 2 (Almost disjoint). *A set Q is said to be x-almost disjoint from set P if $|P \cap Q| \leq x$.*

We will use parameters p and x in presenting the algorithm details; values for these parameters will be picked later. For brevity, we will use *almost contains* and *almost disjoint* to mean x-almost contains and x-almost disjoint respectively.

We first select a subset $S \subseteq U$ by picking $p \cdot m$ elements uniformly and at random from U. The first level hash table is indexed by subsets of S. Query set Q is mapped to the location indexed by $Q \cap S$. The size of this table is 2^{pm}. The table location corresponding to subset $Q' \subseteq S$ has associated with it a list L of sets such that $\forall P \in L, P \cap S \subseteq Q'$. (These are sets that seem to be contained in Q with respect to the sampled elements.) Note that $P \in L \Leftrightarrow S \cap (P - Q) = \emptyset$. If a set P is not almost contained in Q (i.e. $|P - Q| > x$) then $\mathbf{Pr}[P \in L] \le 2^{-px}$. The expected number of sets in L that are not x-almost contained in Q is at most $N \cdot 2^{-px}$. So with probability at least $1/2$, there are at most $2N \cdot 2^{-px}$ sets in the list that are not almost contained in Q (i.e. very few sets have large differences from Q). We will say that a query set Q *is good for* a list of sets L from the database if at most $f = 2N \cdot 2^{-px}$ sets in L are not almost contained in Q. By the preceding discussion, if the query set Q maps onto an array location associated with a list L, with probability at least $1/2$, Q *is good for* L. For the remainder of the discussion, we will assume that this is indeed the case and show how this can be ensured in the end.

Recall that we associate with each list L, a representative set R so that sets in L can be examined efficiently. We now describe the precise properties that R satisfies and an algorithm to construct such a representative set. A set R is said to be *a representative set* for a list L if

1. For any query set Q that is good for L, $|R - Q| \le g = p \cdot m$.
2. Except for at most $(2/p) \cdot N \cdot 2^{-px}$ sets in L, the others differ from R in at most x/p elements.

Note that the total number of distinct differences of sets in L from R is at most $(2/p) \cdot N \cdot 2^{-px} + \binom{m}{x/p}$. We use the following algorithm to produce such a representative set for a list L.

CONSTRUCT-REPRESENTATIVE(L)

1. $R = \cup_{P \in L} P$.
2. Check if there exists a subset $G \subseteq R, |G| = g$ that is almost disjoint from all sets in L, except for at most f sets.
3. If no such set G exists, stop and output R, else $R \leftarrow R - G$.
4. Go back to step 2.

Claim. CONSTRUCT-REPRESENTATIVE-SET(L) produces a valid representative set for L.

Proof. Consider the final value of R produced by the algorithm. We claim that for all query sets Q that are good for L, $|R - Q| \le g$. Suppose in fact, there is a query set Q that is good for L such that $|R - Q| > g$. Since Q is good for L, except for at most f sets, the sets in L are almost contained in Q (i.e. have at most x elements outside Q). Let G be any g element subset of $R - Q$. Except for at most f sets in L, the sets in L must be almost disjoint from G. (i.e. have intersection of size at most x). But the algorithm would have detected this subset in Step 2, and would not have terminated in Step 3. By contradiction, R must satisfy the first condition required of a representative set.

Now we establish the second condition. Initially, all sets in L are completely contained in R. For a given set in L, the elements outside the final set R are precisely those elements which are removed from R as a result of discarding the set G in each iteration. The number of iterations is at most m/g since the initial set R has at most m elements and g elements are removed in each iteration. In each iteration at most f sets in L are not almost disjoint from G. This gives a total of $(m/g) \cdot f \leq (2/p) \cdot N \cdot 2^{-px}$ such sets. For the remaining sets, the intersection with the set G in each iteration has size at most x. Thus, in removing G, at most x elements are removed from such a set. So the total number of elements from such a set outside the final R is at most $(m/g)\cdot x \leq x/p$. This proves that R satisfies the second condition required of a representative set.

Recall that to check whether Q contains a set P, we check that **(1)** Q contains P with respect to the elements in R, and **(2)** Q contains P with respect to the elements in $\bar{R}$. To perform the first check, we need to eliminate sets that contains any elements in $R - Q$. Note that $|R - Q| \leq g$ by the properties of the representative set. To facilitate the first check, we keep a second level array with each list L indexed by all possible subsets of R of size at most g. The size of this array is at most $\binom{m}{g} \leq 2^{pm \log m}$. Query set Q indexes into the array location indexed by $R - Q$. This second level array indexes into a sublist of L that contains sets in L disjoint from $R - Q$. To perform the second check, this sublist is grouped by the differences from R, i.e. sets with the same difference from R are grouped together.

The query time for this algorithm is determined by the number of distinct differences from R that the algorithm must examine. This number is at most $(2/p) \cdot N/2^{px} + \binom{m}{x/p}$. Further, the space requirement is $2^{pm} \cdot 2^{pm \log m} \cdot N$. Setting $px = t$ where $t < (\log N)/2$, and solving for $\binom{m}{x/p} = N/2^t$, we get $p = \sqrt{t \log m / \log N}$ The number of distinct differences is now at most $(2/p) \cdot N/2^{px} + \binom{m}{x/p} = 2m \cdot N/2^t + N/2^t = (2m + 1)N/2^t$. We also know that with probability $1/2$, Q is *good for* L. So if we have $O(m)$ such arrays every set Q will be good for some list L in one of the arrays with very high probability. Setting $t = 2c \log m$, this gives us a query time of $O(N/2^c)$ and space of $N \cdot 2^{O(m \log^2 m \sqrt{c/\log N})}$. It can be verified that the time to construct the data structure is $N \cdot 2^{O(m \log^2 m \sqrt{c/\log N})}$.

We can obtain better space bounds when the size of the query set is bounded by q. In this case, we can get a query time of $O(N/2^c)$ using space $N \cdot 2^{O(q \log^2 m \sqrt{c/\log N})}$. The basic idea is to replace the random sampling step by a certain construct called a *Bloom filter* [4], used for succinct representation of sets to support membership queries. The details will appear in the full paper.

4   Partial Match and Orthogonal Range Search in Very High Dimensions

In this section we describe the algorithm which achieves $Nm^{O(c)}$ space and $O(mN/c)$ query time, for any $c \leq N$. We first describe in the the context of the subset query problem, and then modify it to work for general orthogonal range search.

The main part of the data structure is a "tree of subsets" of $\mathcal{P}$ (call it T). The root of T is formed by the set $\mathcal{P}$, and each internal node corresponds to a subset of $\mathcal{P}$. We show first how to construct the children of the root. For any $i \in U$, we say that the *attribute i is selective* for $\mathcal{P}' \subset \mathcal{P}$, if the number of sets in the family which do not contain i is at least N/c. Let I be the set of attributes which are selective for $\mathcal{P}$. For each $i \in I$ we form a child $\mathcal{P}_i$ of $\mathcal{P}$ by removing from $\mathcal{P}$ all sets which do not contain i. We then apply the same procedure to the new nodes. The procedure stops when there are no further nodes to expand.

Given a query Q, the search proceeds as follows. We start from the root and then proceed further down the tree. At any node (corresponding to a set $\mathcal{P}'$), we check if there is any attribute $i \in Q$ which is selective for $\mathcal{P}'$. If so, we move to $\mathcal{P}'$. Otherwise, we enumerate all pairs (i, j), such that $i \in Q$ and $P_j \in \mathcal{P}'$ does not contain i. Note that we can enumerate all such pair in time linear in their number. For every such pair, we mark the set P_j as *not* an answer to the query Q. At the end, we verify if the number of marked nodes in less than N; if so, we output YES, otherwise we output NO.

The complexity bounds for the above data structure can obtained as follows. Firstly, observe that any parent-child pair $\mathcal{P}', \mathcal{P}''$ we have $|\mathcal{P}''| \leq \mathcal{P}'| - N/c$. Thus, the depth of the tree T is at most c, and thus its size is bounded by m^c (in fact, slightly better bound of $\binom{m}{c}$ can be shown). Moreover, the tree can be traversed from the root to any (final) node in $O(m + N/c)$ time. This is due to the fact that the selectivity property is monotone, and thus once an attribute $i \in Q$ ceases to be selective, we never have to check it again. Finally, the cost of enumerating the pairs (i, j) is at most mN/c, since the numbers of j's per each $i \in Q$ is at most N/c. The bounds follow.

The data structure can be generalized to orthogonal range queries in the following way. Firstly, observe that by increasing the query time by an additive factor of $O(d \log n)$ we can assume that all coordinates of all database points are distinct numbers from $\{1 \ldots nd\}$ (this is based on folklore technique of replacing each number by its rank in a sorted order). We create a tree T in a similar way as before. Moreover, it is sufficient to consider query rectangles which are products of left-infinite intervals $[-\infty, a_i]$ for $i = 1 \ldots d$. Starting from the root node, for each dimension i, we find a "threshold" t_i such that the number of points with the i-th coordinate greater than t_i is exactly N/c. Then we construct the children sets $\mathcal{P}_i$ by removing from $\mathcal{P}$ all points with the i-th coordinate greater than t_i, and apply the same procedure recursively to the newly created nodes.

In order to answer a query $\prod_i [-\infty, a_i]$, we proceed top-down as before. At any node $\mathcal{P}'$ we check if there is any i such that $a_i < t_i$. If so, can move to the node $\mathcal{P}'_i$. Otherwise, we have $a_i > t_i$ for all i's (we can assume there are not ties). In this case we enumerate all pairs (i, j) such that the i-th coordinate of the j-th point is greater than t_i. Among such pairs, we choose those for which the i-th coordinate is greater than a_i, and mark the j-th point as not-an-answer.

The analysis of space and time requirements is as for the first algorithm.

5   Approximate Orthogonal Range Search

The special case of orthogonal range search where the coordinates in each dimension can only take on values $\{0, 1\}$ is exactly the partial match problem. In this case, points in a d dimensional space can be represented by bit vectors of length d and a query rectangle can be represented as a ternary vector with don't cares or '*' in certain dimensions. As explained in Section 2, the partial match problem can be reduced to set containment. The general case of d dimensional orthogonal range search also reduces to the partial match problem when the query rectangles are well aligned (i.e. the projection on each dimension is of the form $[i2^j, (i+1)2^j]$.) In this case too, points can be represented as bit vectors and a query rectangle can be represented as a ternary vector (with coordinates in $\{0, 1, *\}$), creating an instance of the partial match problem. This gives us an algorithm for high dimensional orthogonal range search where the query rectangle is well aligned. Specifically if s bits are required to represent a point in the space then an aligned orthogonal range search can be performed in sublinear time by using $2^{(s \log^2 s / \sqrt{\log N})}$ space. Note that most algorithms for range search use a tree based recursive space partitioning scheme whereas our algorithm is based on hashing.

We now extend this result to a $(1 + \epsilon)$-approximation algorithm for a general d dimensional orthogonal range search problem where the query rectangle may not be aligned. By a $(1 + \epsilon)$-approximation we mean that we will approximate the query rectangle by larger rectangle containing the query rectangle where each side is blown up by a factor of at most $(1 + \epsilon)$.

Theorem 1. *A $(1+\epsilon)$-approximation of an instance of orthogonal range search can be reduced to an instance of set containment with $m = O(s/\epsilon)$ elements in the universe where s is the number of bits in a binary representation of a single point in the high dimensional space.*

To achieve this we first generalize the partial match problem from binary vectors to k-ary vectors. The query now consists of a range (subinterval) in $[0..k-1]$ in each dimension. The database now consists of d dimensional k-ary vectors. A vector is set to match the query if it satisfies the range specified in the query for every dimension. We can reduce this problem to set containment by using k elements per dimension, one for each possible value. We construct the query set by looking at the range specified in each dimension and including only the elements that fall in the range. As for a point in the database for each dimension we include the element corresponding to the value for that dimension. This reduction gives a total of $m = k \cdot d$ elements.

Now look at the range specified in a query rectangle on any one dimension. This range consists of a lower bound, l and an upper bound, u. Express these bounds in k-ary representation (base k). Look at the most significant digit where they differ. Let b_l and b_u be the values of that digit in the two bounds, l and u. This gives us a subinterval $[b_l, b_u]$ contained in $[0..k-1]$. If the length of this subinterval is large say at least $k/2$ then by ignoring the later digits we get a

$4/k$ approximation to the range specified in the query (we get an error of one on each side). If this condition holds for all dimensions we get a $2/k$ approximation by reducing it to the k-ary partial match problem.

However this condition may not be true in every dimension. We force this condition by having multiple representations of every coordinate. Say p bits are required to represent a coordinate in binary. By grouping these p bits into blocks of size $\log k$ we get a k-ary representation (assume k is a power of 2 and is > 4). Now look at a range $[l, u]$ of length $r = u - l$ specified on this coordinate. Let the ith digit be the most significant digit of the k-ary representation of r. We want this digit to be large. Further we want the ith digit of l to be small so that by adding r to l we do not affect the previous digit. So we will insist that the ith digit of l is at most $k/2$ and the ith digit of r is between $k/8$ and $k/4 - 1$. The first condition can be achieved by having two representations of l, that is l and $l + D$ where $D = k/2.k/2 \ldots k/2$ repeated $p/\log k$ times (in k-ary representation). Clearly either l or $l + D$ has its ith digit at most $k/2$. We can ensure that the ith digit of r is large by having $\log k$ different representations; that is shift the groupings of bits into k-ary digits by $1, 2, \ldots \log k$ bits. In at least one of these representations the most significant digit of r is at least $k/8$ and $< k/4$. So we have a total of $2 \log k$ representation obtained by the different shifts and the displacement. In at least one of these by ignoring all digits after the most significant digit of r we get a $16/k$ approximation to the length of the range. We concatenate these $2 \log k$ different representations of the range $[l, u]$. For representations that do not satisfy the above conditions we place the don't care range, $[0..k - 1]$, for all its digits. We thus get a k-ary vector consisting of $(p/\log k) \cdot 2 \log k = 2p$ digits for each coordinate. Similarly each coordinate value of any point in database requires $2p$ k-ary digits. So if a d dimensional point requires s bits to specify it, we will get $2s$ k-ary digits after concatenating the $2 \log k$ different representations for each coordinate. After reducing this to set containment we get $m = 2s \cdot k$ elements. The amount of space required to answer this in sub linear time is $2^{2s \cdot k \cdot poly \log(sk)/\sqrt{\log N}}$. The approximation factor $\epsilon = 16/k$. The space required in terms of ϵ is $2^{\tilde{O}(s/\epsilon)/\sqrt{\log N}}$.

Alternatively one could have $(2 \log k)^d$ sets of data structures and after trying out the $2 \log k$ different representations for each coordinate use the appropriate data structure. Each data structure would now correspond to a universe of size $m = 2sk/\log k$. In the full paper, we will discuss applications to the approximate nearest neighbor problem in ℓ_∞ and explain how the ideas for orthogonal range search can be extended to orthogonal range intersection and containment.

6   Reduction from Orthogonal Range Search to the Subset Query Problem

Let $G = (X, E)$ and $G' = (X', E')$ be a directed graph, $|X| \leq |X'|$. We say that $f, g : X \to X'$ embed G into G', if for all $p, q \in X$ we have $(p, q) \in E$ iff $(f(p), g(q)) \in E'$; this definition can be extended to randomized embeddings in a natural way. Let $G = (X, \subset)$ be a poset defined over the set X of all intervals $I \subset \{1 \ldots u\}$, and let $G' = (X', \mathcal{E})$ be a directed graph over the set $X' = P_s(\{1 \ldots v\})$ ($P_s(Z)$ here denotes all s-subsets of Z) such that for any $A, B \in X'$ we have $(A, B) \in \mathcal{E}$ iff $A \cap B = \emptyset$.

Lemma 1. *There is an embedding of G into G' such that $s = O(\log u)$ and $v = O(u)$.*

Proof. For simplicity we assume u is a power of 2. We start from solving the following problem: design mappings $h_1 : \{1 \ldots u\} \to P_s(\{1 \ldots O(u)\})$ and $h_2 : X \to P_s(\{1 \ldots O(u)\})$ such that $i \in I$ iff $h_1(i) \cap h_2(I) \neq \emptyset$ (intuitively, I will play the role of a "forbidden interval"). To this end, consider a binary tree T with the set of leafs corresponding to the elements of $\{1 \ldots u\}$ and the set of internal nodes corresponding to the dyadic intervals over $\{1 \ldots u\}$ (i.e., we have nodes $\{1, 2\}$, $\{3, 4\}$ etc. on the second level, $\{1 \ldots 4\}$, $\{5 \ldots 8\}$ etc on the third level and so on). We define $h_1(i)$ to be the set of all nodes on the path from the root to the leaf i. We also define $h_2(I)$ to be the set of nodes in T corresponding to intervals from the dyadic decomposition of I (i.e., the unique minimum cardinality partitioning of I into dyadic intervals). Note that $|T| = O(u)$.

The correctness of h_1 and h_2 follows from the fact that $i \in I$ iff there is a node on the path from the root to i corresponding to a dyadic interval contained in I.

Now, we observe that $\{a \ldots b\} \not\subset \{c \ldots d\}$ iff $a \in \{1 \ldots c - 1\}$ or $a \in \{d + 1 \ldots u\}$ or $b \in \{1 \ldots c - 1\}$ or $b \in \{d + 1 \ldots u\}$. Therefore, we can define $f(\{a \ldots b\}) = h_1(a) \cup h_1(b)$ and $g(\{c \ldots d\}) = h_2(\{1 \ldots c-1\}) \cup h_2(\{d+1 \ldots u\})$.

Lemma 2. *There is a randomized embedding of G' into $(P_s(\{1 \ldots s^2/\delta\}, \mathcal{E})$ which preserves non-intersection between any two elements of G' with probability $1 - \delta$, for $0 < \delta < 1$. The intersection is always preserved.*

Proof. It is sufficient to use a random function $h : \{1 \ldots v\} \to \{1 \ldots s^2/\delta\}$ and extend it to sets.

Consider now the orthogonal range query problem in d-dimensional space. By using the above lemmas we can reduce this problem to the subset query problem for N subsets of $\{1 \ldots d^2 \log^2 |P|/\delta\}$ in the following way. For $i = 1 \ldots d$ use Lemma 1 to obtain functions f_i and g_i. For simplicity we assume that the values of f_i and f_j do not intersect for $i \neq j$; we make the same assumption for g_i's. For any point $p = (x_1 \ldots x_d)$ define $L(p) = \cup_i f_i(\{x_i, x_i\})$. Also, for any box $B = \{a_1 \ldots b_1\} \times \{a_2 \ldots b_2\} \times \ldots \times \{a_d \ldots b_d\}$ define $R(B) = \cup_i g_i(\{a_i \ldots b_i\})$. We have that $p \in B$ iff $L(p) \cap R(B) = \emptyset$. Since $L(p)$ and $R(B)$ are of size $O(d \log u)$, we can apply Lemma 2 to reduce the universe size to $O(d^2 \log^2 u)$ (say via function h). The condition $A \cap B = \emptyset$ is equivalent to $A \subset \overline{B}$, where $\overline{B}$ is the complement of B. Therefore, with a constant probability, there exists $p \in P$ which belongs to B iff $h(R(B)) \subset \overline{h(L(p))}$. Thus, we proved the following theorem.

Theorem 2. *If there is a data structure which solves the subset query problem for n subsets of* $\{1 \ldots k\}$ *having query time* $Q(n, k)$ *and using space* $S(n, k)$, *then there exists a randomized algorithm for solving the orthogonal range query problem for n points in* $\{1 \ldots u\}^d$ *with query time* $Q(n, O(d^2 \log^2 u))$ *and using space* $S(n, O(d^2 \log^2 u))$.

Note that one can always reduce u to $O(n)$ by sorting all coordinates first and replacing them by their ranks.

References

1. H. Adiseshu, S. Suri, and G. Parulkar. Packet Filter Management for Layer 4 Switching. *Proceedings of IEEE INFOCOM*, 1999.
2. P. Agarwal and J. Erickson. Geometric range searching and it's relatives. *Advances in Discrete and Computational Geometry*, B. Chazelle, J. Goodman, and R. Pollack, eds., Contemporary Mathematics 223, AMS Press, pp. 1-56, 1999.
3. S. Arya and D. Mount. Approximate Range Searching *Proceedings of 11th Annual ACM Symposium on Computational Geometry*, pp. 172-181, 1995.
4. B. Bloom. Space/time tradeoffs in hash coding with allowable errors. *Communications of the ACM*, 13(7):422–426, July 1970.
5. A. Borodin, R. Ostrovsky, and Y. Rabani. Lower bounds for high dimensional nearest neighbor search and related problems. *Proceedings of the Symposium on Theory of Computing*, 1999.
6. D. Eppstein and S. Muthukrishnan. Internet Packet Filter Management and Rectangle Geometry. *Proceedings of 12th ACM-SIAM Symp. Discrete Algorithms* 2001, pp. 827-835.
7. A. Feldmann and S. Muthukrishnan, Tradeoffs for Packet classification. *Proceedings of IEEE INFOCOM*, 3:1193–1202. IEEE, March 2000.
8. P. Gupta and N. McKeown. Algorithms for Packet Classification. *IEEE Network Special Issue*, March/April 2001, 15(2):24–32.
9. P. Indyk. On approximate nearest neighbors in non-euclidean spaces. *Proceedings of the 39th IEEE Symposium on Foundations of Computer Science*, pp. 148-155, 1998.
10. P. Indyk. High-dimensional computational geometry. *Ph.D. thesis, Stanford University*, 2001.
11. P. Indyk and R. Motwani. Approximate nearest neighbor: towards removing the curse of dimensionality. *Proceedings of the 30th ACM Symposium on Theory of Computing*, pp. 604–613, 1998.
12. E. Kushilevitz, R. Ostrovsky, and Y. Rabani. Efficient search for approximate nearest neighbor in high dimensional spaces. *Proceedings of the Thirtieth ACM Symposium on Theory of Computing*, pages 614–623, 1998.
13. G. S. Lueker. A data structure for orthogonal range queries. *Proceedings of the Symposium on Foundations of Computer Science*, 1978.
14. R. L. Rivest. *Analysis of Associative Retrieval Algorithms*. Ph.D. thesis, Stanford University, 1974.
15. R. L. Rivest. *Partial match retrieval algorithms*. SIAM Journal on Computing 5 (1976), pp. 19-50.
16. V. Srinivasan, S. Suri, and G. Varghese. Packet classification using tuple space search. *ACM Computer Communication Review* 1999. ACM SIGCOMM'99, Sept. 1999.

Measuring the Probabilistic Powerdomain

Keye Martin[1], Michael Mislove[2] and James Worrell[2]

[1] Programming Research Group, Wolfson Building, University of Oxford, UK
[2] Department of Mathematics, Tulane University,
6823 St Charles Avenue, New Orleans LA 70118, USA

Abstract. In this paper we initiate the study of measurements on the probabilistic powerdomain. We show how measurements on the underlying domain naturally extend to the probabilistic powerdomain, so that the kernel of the extension consists of exactly those normalized valuations on the kernel of the measurement on the underlying domain. This result is combined with now-standard results from the theory of measurements to obtain a new proof that the fixed point associated with a weakly hyperbolic IFS with probabilities is the unique invariant measure whose support is the attractor of the underlying IFS.

1   Introduction

A relatively recent discovery [12] in domain theory is that most domains come equipped with a natural measurement: A Scott continuous map into the non-negative reals which encodes the Scott topology. The existence of measurements was exploited by Martin [12,13,14] to study the space of maximal elements of a domain, and to formulate various fixed point theorems for domains, including fixed point theorems for non-monotonic maps.

The theory of measurements meshes particularly fruitfully with the idea of domains as models of classical spaces. Here we say that a domain D is a model of a topological space X if the set of maximal elements of D equipped with the relative Scott topology is homeomorphic to X. Under quite mild conditions on D the set of normalized Borel measures on X, equipped with the weak topology, can be embedded into the set of maximal elements of the probabilistic powerdomain $\mathbf{P}D$. This construction was utilized by Edalat [2,3] to provide new results on the existence of attractors for iterated function systems, and to define a generalization of the Riemann integral to functions on metric spaces.

In this paper we show that each measurement $m \colon D \to [0,1]$ (satisfying a suitable condition) has a natural extension to a measurement $M \colon \mathbf{P}D \to [0,1]$. Moreover we show that the kernel of M, equipped with the relative Scott topology, is homeomorphic to the space of continuous valuations on the kernel of m

---

[1,2]The support of the US Office of Naval Research is gratefully acknowledged.
[2] The support of the National Science Foundation is gratefully acknowledged.

P. Widmayer et al. (Eds.): ICALP 2002, LNCS 2380, pp. 463–475, 2002.
© Springer-Verlag Berlin Heidelberg 2002

equipped with the weak topology. As a consequence we find that if D is an ω-continuous model of a regular space X, then the set of normalized valuations on X, equipped with the weak topology, can be embedded into the set of maximal elements of $\mathbf{P}D$. We also use the measurement on $\mathbf{P}D$ to provide a new proof that the fixed point associated with a weakly hyperbolic iterated function system with probabilities is the unique measure whose support is the attractor of the underlying iterated function system.

2   Background

2.1   Domain Theory

A *poset* $(P, \sqsubseteq)$ is a set P endowed with a partial order $\sqsubseteq$. The least element of P (if it exists) is denoted $\bot$, and the set of maximal elements of P is written $\max P$. Given $A \subseteq P$, we write $\uparrow A$ for the set $\{x \in P \mid (\exists a \in A) a \sqsubseteq x\}$. A function $f \colon P \to Q$ between posets P and Q is *monotone* if $x \sqsubseteq y$ implies $f(x) \sqsubseteq f(y)$ for all $x, y \in P$. A subset $A \subseteq P$ is *directed* if each finite subset $F \subseteq A$ has an upper bound in A.

Note that since $F = \emptyset$ is a possibility, a directed subset must be non-empty. A *(directed) complete partial order (dcpo)* is a poset P in which each directed set $A \subseteq P$ has a least upper bound, denoted $\bigsqcup A$.

If D is a dcpo, and $x, y \in D$, then we write $x \ll y$ if for each directed subset $A \subseteq D$, if $y \sqsubseteq \bigsqcup A$, then $\uparrow x \cap A \neq \emptyset$. We then say x *is way-below* y. Let $\downarrow y = \{x \in D \mid x \ll y\}$; we say that D is *continuous* if it has a *basis*, i.e., a subset $B \subseteq D$ such that for each $y \in D$, $\downarrow y \cap B$ is directed with supremum y. If D has a countable basis then we say it is ω-*continuous*. The way-below relation on a continuous dcpo has the *interpolation property*: if $x \ll y$ then there exists a basis element z such that $x \ll z \ll y$.

A subset U of a dcpo D is *Scott-open* if it is an upper set (i.e., $U = \uparrow U$) and is inaccessible by directed suprema (i.e., for each directed set $A \subseteq D$, if $\bigsqcup A \in U$ then $A \cap U \neq \emptyset$). The collection ΣD of all Scott-open subsets of D is called the *Scott topology* on D. If D is continuous, then the Scott topology on D is locally compact, and the sets $\uparrow x$ where $x \in D$ form a basis for the topology. Given dcpos D and E, a function $f \colon D \to E$ is continuous with respect the Scott topologies on D and E iff it is monotone and preserves directed suprema (i.e., for each directed $A \subseteq D$, $f(\bigsqcup A) = \bigsqcup f(A)$).

2.2   Valuations and the Probabilistic Powerdomain

We briefly recall some basic definitions and results about valuations and the probabilistic powerdomain.

Definition 1. *Let X be a topological space. A* continuous valuation *on X is a mapping $\mu \colon (\Omega X, \subseteq) \to ([0, 1], \leq)$ satisfying:*

1. Strictness: $\mu(\emptyset) = 0$,

2. *Monotonicity:* $U \subseteq V \Rightarrow \mu(U) \leq \mu(V)$.
3. *Modularity: for all* $U, V \in \Omega X$, $\mu(U \cup V) + \mu(U \cap V) = \mu(U) + \mu(V)$.
4. *Continuity: for every directed family* $\{U_i\}_{i \in I}$, $\mu(\bigcup_{i \in I} U_i) = \sup_{i \in I} \mu(U_i)$.

Each element $x \in X$ gives rise to a valuation defined by

$$\delta_x(U) = \begin{cases} 1 & \text{if } x \in U, \\ 0 & \text{otherwise.} \end{cases}$$

A *simple valuation* has the form $\sum_{a \in A} r_a \delta_a$ where A is a finite subset of X, $r_a \geq 0$, and $\sum_{a \in A} r_a \leq 1$. A valuation μ is *normalized* if $\mu(X) = 1$.

For the most part we will consider continuous valuations defined on the Scott topology ΣD of a dcpo D. The set of all such valuations, ordered by $\mu \sqsubseteq \nu$ if and only if $\mu(U) \leq \nu(U)$ for all $U \in \Sigma D$, forms a dcpo $\mathbf{P}D$: the probabilistic powerdomain of D. Our main reference for the probabilistic powerdomain is the PhD thesis of Jones [9] from which the following result is taken.

Theorem 1 (Jones [9]). *If D is a continuous domain then $\mathbf{P}D$ is continuous with a basis* $\mathcal{B} = \{\Sigma_{i=1}^n r_i \delta_{p_i} \mid p_i \in B\}$, *where* $B \subseteq D$ *is a basis for* D.

Proof. (Sketch) Define a dissection of D to be a disjoint family of crescents $\mathcal{D} = \{C_i\}_{i \in I}$, where $C_i = \uparrow x_i \setminus U_i$ for some $x_i \in \mathcal{B}$ and $U_i \in \Sigma D$. Given $\mu \in \mathbf{P}D$ and $0 < r < 1$ define

$$\mu_{\mathcal{D}, r} = \sum_{i \in I} r \mu(C_i) \delta_{x_i}.$$

The substantial part of the proof, which is elided here, is to show that the set of $\mu_{\mathcal{D}, r}$ for all $\mathcal{D}$ and r is directed with join μ. □

Obviously valuations bear a close resemblance to measures. Lawson [11] showed that any valuation on an ω-continuous dcpo D extends uniquely to a measure on the Borel σ-algebra generated by the Scott topology (equivalently by the Lawson topology) on D. This result was generalized to continuous dcpos by Alvarez-Manilla, Edalat and Saheb-Djahromi [1]. Both these results depend heavily on the axiom of choice. In this paper, we avoid using either theorem. We do use the elementary result that each valuation on a dcpo D extends uniquely to a finitely additive set function on the field $\mathcal{F}D$ generated by ΣD. Each member R of this field can be written as a disjoint union of crescents, i.e., $R = \bigcup_{i=1}^n U_i \setminus V_i$ for $U_i, V_i \in \Sigma D$. The extension of a valuation ν to $\mathcal{F}D$ assigns R the value

$$\sum_{i=1}^n (\nu(U_i) - \nu(U_i \cap V_i)).$$

Also we recall from [7, Section 3.2] that if $E \in \mathcal{F}D$ and $\mu \in \mathbf{P}D$ then we may define $\mu|_E \in \mathbf{P}D$ by $\mu|_E (O) = \mu(O \cap E)$ for all $O \in \Sigma D$.

In Section 6 we quote the result of Pettis [16] that any continuous valuation on a locally compact Hausdorff space has a unique extension to a measure. But this is only used to mediate between the formulation of the main result of that section and the results of Hutchinson [8] which are stated for measures.

3 Measurements

Definition 2. *Let D and E be continuous dcpos, and $m: D \to E$ a Scott continuous map. Define $\ker m = \{x \in D \mid m(x) \in \max E\}$. We say that m is a measurement if for all $x \in \ker m$ and Scott-open U containing x, there exists $\varepsilon \in E$ such that $\{y \in D \mid y \sqsubseteq x \text{ and } \varepsilon \ll m(y)\} \subseteq U$.*

In the words of [12] the last condition in the definition says that m *induces the Scott topology near* $\ker m$. This implies that $\ker m \subseteq \max D$. The condition becomes more transparent in the (typical) instance that E is the dcpo $[0, \infty)^*$ of non-negative reals in their opposite order. Then it is equivalent to requiring that there exists $\varepsilon > 0$ such that $\{y \in D : y \sqsubseteq x \text{ and } m(y) < \varepsilon\} \subseteq U$.

Example 1. The following computational models for classical spaces yield natural measurements into $[0, \infty)^*$.

1. If $\langle X, d \rangle$ is a locally compact metric space, then its *upper space*

$$\mathbf{U}X = \{\emptyset \neq K \subseteq X : K \text{ is compact}\}$$

ordered by reverse inclusion is a continuous dcpo. The supremum of a directed set $S \subseteq \mathbf{U}X$ is $\bigcap S$, and the way-below relation is given by $A \ll B$ iff $B \subseteq \text{int } A$. Given $K \in \mathbf{U}X$, defining the diameter of K by

$$|K| = \sup\{d(x, y) : x, y \in K\},$$

it is readily verified that $K \mapsto |K|$ is a measurement on $\mathbf{U}X$ whose kernel is $\max \mathbf{U}X = \{\{x\} : x \in X\}$.

2. Given a metric space $\langle X, d \rangle$, the *formal ball model* [5] $\mathbf{B}X = X \times [0, \infty)$ is a poset ordered by

$$(x, r) \sqsubseteq (y, s) \text{ iff } d(x, y) \leq r - s.$$

The way-below relation is characterized by

$$(x, r) \ll (y, s) \text{ iff } d(x, y) < r - s.$$

The poset $\mathbf{B}X$ is a continuous dcpo iff the metric d is complete. Moreover $\mathbf{B}X$ has a countable basis iff X is separable. A natural measurement on π on $\mathbf{B}X$ is given by $\pi(x, r) = r$. Then $\ker \pi = \max \mathbf{B}X = \{(x, 0) : x \in X\}$.

One of the motivations behind the introduction of measurements in [12] was to facilitate the formulation of sharper fixed point theorems. The following is a basic example of one such result.

Theorem 2. *Let $f: D \to D$ be a monotone map on a pointed continuous dcpo D equipped with a measurement $m: D \to E$. Furthermore, suppose that for each $d \in D$ the set $\{m(f^n(d)) : n \in \mathbb{N}\}$ is directed with supremum in $\max E$. Then*

$$x^\star = \bigsqcup_{n \geq 0} f^n(\bot) \in \max D$$

is the unique fixed point of f. Moreover $x^\star$ is an attractor in the sense that for all $y \in \ker m$, $f^n(y) \to x^\star$ in the relative Scott topology on $\ker m$.

Martin [14] gives a necessary and sufficient condition for a measurement $m\colon D \to [0,\infty)^*$ to extend to a measurement $M\colon \mathbf{C}D \to [0,\infty)^*$, where $\mathbf{C}$ is the convex powerdomain. This result is used to show that for any ω-continuous dcpo D, with $\max D$ regular, the Vietoris hyperspace of $\max D$ embeds into $\max \mathbf{C}D$ (as the kernel of a measurement). Furthermore, applying standard results from the theory of measurements, Edalat's domain-theoretic construction of attractors of iterated function systems is recovered. Here we seek analogous results with the probabilistic powerdomain in place of the convex powerdomain. In particular, we provide a sufficient condition for a measurement m on D to extend to a measurement on the probabilistic powerdomain $\mathbf{P}D$.

As we mentioned earlier, it is conventional to consider measurements into $[0,\infty)^*$. However in the rest of this paper we find it convenient to consider measurements into $[0,1]$. Any measurement of the former type can, by composition with $f(x) = 2^{-x}$, be turned into a measurement of the latter type. To ensure that such measurements extend to the probabilistic powerdomain we strengthen the condition that m induces the Scott topology near $\ker m$ by requiring that for any $U \in \Sigma D$

$$(\forall x \in U \cap \ker m)(\exists V \in \Sigma D)(\exists \varepsilon > 0)(x \in V \text{ and } m_\varepsilon(V \cap \ker m) \subseteq U) \qquad (\ddagger)$$

where $m_\varepsilon(S) = \{y : y \in \downarrow S \text{ and } 1 - m(y) < \varepsilon\}$. Note that both measurements in Example 1 (when turned into measurements into $[0,1]$) satisfy $(\ddagger)$. In Example 2 we exhibit a measurement not satisfying $(\ddagger)$ which fails to extend to the probabilistic powerdomain.

We have the following characterization of those ω-continuous dcpos admitting measurements satisfying $(\ddagger)$.

Theorem 3. *If D is an ω-continuous dcpo, then $\max D$ is regular in the relative Scott topology iff there is a measurement $m\colon D \to [0,1]$ satisfying $(\ddagger)$ such that $\ker m = \max D$.*

Proof. Since the condition $(\ddagger)$ implies that m is a *Lebesgue measurement* [14], if D admits such a measurement then $\max D$ is regular by [14, Theorem 11.6] (in fact $\max D$ is Polish in this case). Conversely if $\max D$ is regular, then the construction given in the proof of [14, Theorem 11.6] yields the required measurement. $\qquad\square$

Our main result, Theorem 5, implies that if D is a continuous dcpo admitting a measurement m satisfying $(\ddagger)$, then the set of normalized valuations on $\ker m$ embeds into $\max \mathbf{P}D$ as the kernel of a measurement. Combining this with Theorem 3 we obtain Corollary 1. This result was first proved, in a different way, in Martin [14, Theorem 11.8].

Corollary 1. *If D is an ω-continuous dcpo with $\max D$ regular, then the set of normalized valuations on $\max D$ embeds into $\max \mathbf{P}D$.*

4   Comparing Valuations

One of the most elegant results about the probabilistic powerdomain is the Splitting Lemma. This bears a close relationship to a classic problem in probability theory: find a joint distribution with given marginals.

Lemma 1 (Jones [9]). *Let* $\mu = \sum_{a \in A} r_a \delta_a$ *and* $\nu = \sum_{b \in B} s_b \delta_b$ *be simple valuations. Then* $\mu \ll \nu$ *if and only if there are* $\{t_{a,b} \mid a \in A, b \in B\} \subseteq [0,1]$ *satisfying*

1. *For each* $a \in A$, $\sum_{b \in B} t_{a,b} = r_a$,
2. *For each* $b \in B$, $\sum_{a \in A} t_{a,b} < s_b$, *and*
3. $t_{a,b} \neq 0$ *implies* $a \ll b$.

In the remainder of this section we give a characterization of when a simple valuation lies way-below an arbitrary continuous valuation.

Proposition 1 (Kirch [10]). *Let* ν *be a continuous valuation on* D, *then* $\sigma = \sum_{a \in A} r_a \delta_a \ll \nu$ *if and only if* $\forall J \subseteq A$, $\sum_{a \in J} r_a < \nu(\uparrow\! J)$.

Definition 3. *Fix a finite subset* $A \subseteq D$, *and for each* $J \subseteq A$ *define*

$$D_J^A = \bigcap_{a \in J} \uparrow\! a \setminus \bigcup_{a' \in A \setminus J} \uparrow\! a'.$$

Observe that $\{D_J^A\}_{J \subseteq A}$ is a family of crescents partitioning D.

Proposition 2. *Let* ν *be a continuous valuation on* D, $\sum_{a \in A} r_a \delta_a$ *a simple valuation on* D, *and* $\{E_i\}_{i \in I} \subseteq \mathcal{F}D$ *a partition of* D *refining* $\{D_J^A\}_{J \subseteq A}$. *Then* $\sum_{a \in A} r_a \delta_a \ll \nu$ *iff there exists a relation* $R \subseteq A \times I$ *such that*

(i) $(a, i) \in R$ *implies* $E_i \subseteq \uparrow\! a$.
(ii) *For all* $J \subseteq A$, $\sum_{a \in J} r_a < \sum_{i \in R(J)} \nu(E_i)$.

Proof. ($\Rightarrow$) Suppose $\sum_{a \in A} r_a \delta_a \ll \nu$. Define R by $R(a, i)$ just in case $E_i \subseteq \uparrow\! a$. Then, given $J \subseteq A$, by Proposition 1,

$$\sum_{a \in J} r_a < \nu(\uparrow\! J) = \sum_{i \in R(J)} \nu(E_i).$$

($\Leftarrow$) Given a relation R satisfying conditions (i) and (ii) above, then for all $J \subseteq A$ we have

$$\sum_{a \in J} r_a < \sum_{i \in R(J)} \nu(E_i) \leq \nu(\uparrow\! J).$$

Thus $\sum_{a \in A} r_a \delta_a \ll \nu$ by Proposition 1.   □

Next we give an alternative characterization of the way-below relation on PD. This is a slight generalization of the Splitting Lemma, and should be seen as dual to Proposition 2.

Proposition 3. *Suppose $\sum_{a \in A} r_a \delta_a$ and ν are continuous valuations on D and $\{E_i\}_{i \in I} \subseteq \mathcal{F}D$ is a partition of D refining $\{D_j^A\}_{J \subseteq A}$. Then $\sum_{a \in A} r_a \delta_a \ll \nu$ iff there exists a family of 'transport numbers' $\{t_{a,i}\}_{a \in A, i \in I}$ where*

1. *For each $a \in A$, $\sum_{i \in I} t_{a,i} = r_a$*
2. *For each $i \in I$, $\sum_{a \in A} t_{a,i} < \nu(E_i)$*
3. *$t_{a,i} > 0$ implies $E_i \subseteq \uparrow a$.*

Proof. ($\Leftarrow$) Given the existence of a family of transport numbers $\{t_{a,i}\}$, define $R \subseteq A \times I$ by $R(a, i)$ iff $t_{a,i} > 0$. Then R satisfies (i) and (ii) in Proposition 2.

($\Rightarrow$) By Proposition 2 there exists a relation $R \subseteq A \times I$ satisfying conditions (i) and (ii) thereof. The proof that such a relation yields transport numbers as required uses the max-flow min-cut theorem from graph theory. The basic idea is due to Jones [9], but we refer the reader to the formulation of Heckmann [7, Lemma 2.7] which is general enough to apply to the present setting. □

For our main results, we can equally-well use Proposition 2 or (the dual form) Proposition 3 to characterize the way-below relation. Next we define a operation $\star$ for composing splittings with a common index set by 'projecting out that index'. Suppose $s = \{s_{i,j}\}_{i \in I, j \in J}$ and $t = \{t_{j,k}\}_{j \in J, k \in K}$ are families of non-negative real numbers where I, J and K are finite. Assuming that $\sum_{k \in K} t_{j,k} > 0$ for each $j \in J$ we define $t \star s$ to be an $I \times K$-indexed family where

$$(t \star s)_{i,k} = \sum_{j \in J} s_{i,j} \left(\frac{t_{j,k}}{\sum_{k' \in K} t_{j,k'}} \right).$$

Proposition 4. *Let s and t be as above, then for each $i \in I$,*

$$\sum_{k \in K} (t \star s)_{i,k} = \sum_{j \in J} s_{i,j}. \tag{1}$$

Furthermore, if $\sum_{i \in I} s_{i,j} < \sum_{k \in K} t_{j,k}$ for each $j \in J$, it follows that

$$\sum_{i \in I} (t \star s)_{i,k} < \sum_{j \in J} t_{j,k} \tag{2}$$

for each $k \in K$.

Proof. Simple algebra. □

5 Measuring the Probabilistic Powerdomain

Definition 4. *If* $m: D \to [0,1]$ *is a measurement on a continuous dcpo* D, *then we define* $M: \mathbf{P}D \to [0,1]$ *by* $M(\mu) = \int m\,d\mu$, *where the integral is that defined by Jones [9]. In particular, by the continuity properties of this integral we get that* $M(\mu) = \sup\{\sum_{a \in A} r_a m(a) \mid \sum_{a \in A} r_a \delta_a \ll \mu\}$.

To motivate condition (‡), consider the following example where M, as defined above, fails to be a measurement.

Example 2. Let P be the dcpo obtained by adding a top element ∞ to the naturals in their usual order. Let $P' = \{n' \mid n \in \mathbb{N}\} \cup \{\infty'\}$ be a disjoint copy of P. Finally write D for the dcpo consisting of the sum of P and P' together with a copy of the naturals in the discrete order $\{n'' \mid n \in \mathbb{N}\}$, with additionally $n, n' \sqsubseteq n''$ for all $n \in \mathbb{N}$.

Define a measurement $m: D \to [0,1]$ by $m(\infty) = m(\infty') = m(n'') = 0$ for all $n \in \mathbb{N}$, and $m(n) = m(n') = 2^{-n}$ for all $n \in \mathbb{N}$. Now the valuation $\mu = \sum_{n \in \mathbb{N}} 2^{-(n+1)} \delta_{n''}$ is in $\ker M$, and $\delta_0 \ll \mu$. Furthermore, defining $\nu_N = \sum_{n \le N} 2^{-(n+1)} \delta_{n''} + \sum_{n > N} 2^{-(n+1)} \delta_{n'}$ we have that $\nu_N \sqsubseteq \mu$ but not $\delta_0 \ll \nu_N$. However by choosing N large enough we can make $M(\nu_N)$ arbitrarily close to 1. Thus M is not a measurement on $\mathbf{P}D$, cf. Definition 2.

Proposition 5. *Let* $\mu \in \ker M$, *i.e.,* $\int m\,d\mu = 1$. *Then for a crescent* $E = U \setminus V$, *where* $U, V \in \Sigma D$, *we have that* $\mu(E) > 0$ *implies* $E \cap \ker m \ne \emptyset$.

Proof. We define a decreasing sequence of crescents $\langle E_n \mid n \in \mathbb{N}\rangle$ with $\mu(E_n) > 0$ for all $n \in \mathbb{N}$. First, $E_0 = E$. Next, assuming E_n is defined, let $\rho = \frac{1}{\mu(E_n)} \mu|_{E_n}$. Since

$$\mu = \mu|_{E_n} + \mu|_{E_n^c},$$

the inequality $M(\mu|_{E_n^c}) \le \mu(E_n^c)$ forces $M(\mu|_{E_n}) = \mu(E_n)$, whence $M(\rho) = 1$. By the proof of Theorem 1 there is a dissection $\mathcal{D}$ of D and $0 < r < 1$ such that $M(\rho_{\mathcal{D},r}) > 1 - 1/n$. In particular, there exists $x_n \in E_n$, namely one of the mass points of $\rho_{\mathcal{D},r}$, such that $m(x_n) \ge 1 - 1/n$ and $\mu(E_n \cap \uparrow x_n) > 0$. Now set $E_{n+1} = E_n \cap \uparrow x_n$.

The proposition now follows since $\langle x_n \mid n \in \mathbb{N}\rangle$ is an increasing sequence in E, and so $\bigsqcup x_n \in E \cap \ker m$.   □

Proposition 6. *Let* $\mu \in \ker M$. *If* $U_1, U_2 \in \Sigma D$ *with* $U_1 \cap \ker m = U_2 \cap \ker m$, *then* $\mu(U_1) = \mu(U_2)$.

Proof. Since neither of the crescents $U_1 \setminus U_2$ and $U_2 \setminus U_1$ meets $\ker m$ it follows that

$$
\begin{aligned}
\mu(U_1) &= \mu(U_1 \cap U_2) + \mu(U_1 \setminus U_2) \\
&= \mu(U_1 \cap U_2) \text{ (by Proposition 5)} \\
&= \mu(U_1 \cap U_2) + \mu(U_2 \setminus U_1) \text{ (by Proposition 5)} \\
&= \mu(U_2).
\end{aligned}
$$

Theorem 4. *The set of normalized valuations on* ker m *equipped with the weak topology is homeomorphic to* ker M *with the relative Scott topology.*

Proof. Suppose μ is a valuation on ker m with total mass 1. Then we easily see that $\mu^*\colon \Sigma D \to [0,1]$ defined by $\mu^*(O) = \mu(O \cap \ker m)$ is a valuation on ΣD. Since

$$\mu^*\{x : m(x) \geq 1 - 1/n\} = \mu(\ker m) = 1,$$

then $M(\mu^*) \geq 1 - 1/n$ for all $n > 0$. Thus $\mu^* \in \ker M$.

Conversely, suppose $\mu \in \ker M$. We define a valuation μ_* on the open sets of ker m as follows. For an open set $O \subseteq \ker m$ we define $\mu_*(O) = \mu(O^\dagger)$ where $O^\dagger$ is the greatest Scott open subset of D such that $O^\dagger \cap \ker m = O$. Now for all open subsets O_1, O_2 of ker m,

$$\begin{aligned}
\mu_*(O_1 \cup O_2) + \mu_*(O_1 \cap O_2) &= \mu((O_1 \cup O_2)^\dagger) + \mu((O_1 \cap O_2)^\dagger) \\
&= \mu(O_1^\dagger \cup O_2^\dagger) + \mu(O_1^\dagger \cap O_2^\dagger) \quad \text{(by Proposition 6)} \\
&= \mu(O_1^\dagger) + \mu(O_2^\dagger) \\
&= \mu_*(O_1) + \mu_*(O_2).
\end{aligned}$$

Thus μ_* is modular. By similar reasoning it also follows that μ_* is Scott continuous.

One easily sees that the maps $\mu \mapsto \mu^*$ and $\mu \mapsto \mu_*$ are inverse. The fact that they are both continuous follows from the characterization of convergence in the Scott topology in [4, Theorem 4.1]. This requires the metrizability of ker m, which holds whenever m satisfies (‡). ◻

The following theorem is the main result of the paper. It says that if the measurement $m\colon D \to [0,1]$ satisfies (‡), then the induced map $M\colon \mathbf{P}D \to [0,1]$ is a measurement. The proof is a little technical, though not difficult. Here we just outline the structure of the argument, omitting the technical details.

Theorem 5. *Let* $\nu \in \ker M$ *and* $\sigma \ll \nu$. *Then there exists* $\varepsilon > 0$ *such that whenever* $\rho \sqsubseteq \nu$ *and* $M(\rho) > 1 - \varepsilon$, *then* $\sigma \ll \rho$.

Proof (sketch). Since the simple valuations form a basis of $\mathbf{P}D$, using the interpolation property of $\ll$ on $\mathbf{P}D$ and the continuity of M it may be assumed, without loss of generality, that both σ and ρ are simple — say $\sigma = \sum_{a \in A} r_a \delta_a$ and $\rho = \sum_{b \in B} s_b \delta_b$.

Applying Proposition 3 with the partition $\{D_J^A\}_{J \subseteq A}$ we obtain a splitting $u = \{u_{a,J}\}$ between σ and ν. Applying again Proposition 3, this time with the partition $\{D_K^{A \cup B}\}_{K \subseteq A \cup B}$, we obtain a splitting $v = \{v_{b,K}\}$ between ρ and ν. Below we illustrate the valuations and splittings under consideration.

(3)

Our strategy is to obtain a splitting between σ and ρ, in the sense of Lemma 1, by combining u and (a modification of) v using Proposition 4. In fact, to apply Proposition 4 we have to re-index v to make it composable with u. Specifically we define a $P(A) \times B$-indexed set of transport numbers $w = \{w_{J,b}\}$ by

$$w_{J,b} = \sum \{v_{b,K} : K \cap A = J\}.$$

The delicate part of the proof is to show that the composition $w \star u$ really defines a splitting between σ and ρ, i.e., that it satisfies (1)-(3) in Lemma 1. As one would expect, this depends on a careful choice of the value of ε. It is here that we crucially use the fact that m satisfies condition ($\ddagger$). $\qquad\square$

6   Iterated Function Systems

Definition 5. *An* iterated function system *(IFS) on a complete metric space X is given by a collection of continuous maps $f_i : X \to X$ indexed over a finite set I. Such an IFS is denoted $\langle X, \{f_i\}_{i \in I}\rangle$. If each map f_i is contracting then the IFS is said to be* hyperbolic.

A hyperbolic IFS induces a contraction F on the complete metric space of non-empty compact subsets of X equipped with the Hausdorff metric. F is defined by

$$F(K) = \bigcup_{i \in I} f_i(K).$$

By Banach's contraction mapping theorem F has a unique fixed point: the *attractor* of the IFS. An alternate domain-theoretic proof this result, due to Hayashi [6], involves considering F as a continuous selfmap of $\mathbf{U}X$ and deducing that the least fixed point of F is maximal in $\mathbf{U}X$, and therefore is a unique fixed point. Many different fractal sets arise as, or can be approximated by, attractors of IFS's.

Definition 6. *A* weighted IFS *$\langle X, \{(f_i, p_i)\}_{i \in I}\rangle$ consists of an IFS $\langle X, \{f_i\}_{i \in I}\rangle$ and a family of weights $0 < p_i < 1$, where $\sum_{i \in I} p_i = 1$. This data induces a so-called Markov operator $G : \mathcal{M}X \to \mathcal{M}X$ on the set $\mathcal{M}X$ of Borel probability measures on X, given by*

$$G(\mu)(B) = \sum_{i \in I} p_i \mu(f_i^{-1}(B)) \qquad (4)$$

for each Borel subset $B \subseteq X$.

The space $\mathcal{M}X$ equipped with the weak topology can be metrized by the *Hutchinson metric* [8]. Furthermore, if a weighted IFS is hyperbolic then the map G is contracting with respect to the Hutchinson metric. In this case the unique fixed point of G, obtained by the contraction mapping theorem, defines a probability measure called an *invariant measure* for the IFS. The support of the invariant measure is the attractor of the underlying IFS. This construction is an important method of defining fractal measures. Next we outline a domain-theoretic construction, due to Edalat [2], of invariant measures for so-called weakly hyperbolic IFS's on compact metric spaces.

If X is a locally compact metric space, then a weighted IFS $\langle X, \{(f_i, p_i)\}_{i \in I} \rangle$ induces a continuous map $T \colon \mathbf{P}\mathbf{U}X \to \mathbf{P}\mathbf{U}X$ – the domain theoretic analog of the Markov operator – defined by

$$T(\mu)(O) = \sum_{i \in I} p_i \mu((\mathbf{U}f_i)^{-1}(O)) \tag{5}$$

where $\mathbf{U}f_i \colon \mathbf{U}X \to \mathbf{U}X$ is the map $K \mapsto f_i(K)$.

Specializing now to the case where X is compact we have $T(\delta_X) = \sum_{i \in I} p_i \delta_{f_i(X)}$. Iterating, it follows that

$$T^n(\delta_X) = \sum_{i_1, \dots, i_n \in I} p_{i_1} \dots p_{i_n} \delta_{f_{i_1} \dots f_{i_n}(X)}. \tag{6}$$

Thus $M(T^n(\delta_X))$, the measurement of the n-th iterate, is equal to 2^{-r_n}, where

$$r_n = \sum_{i_1, \dots, i_n \in I} p_{i_1} \dots p_{i_n} |f_{i_1} \dots f_{i_n}(X)|.$$

A sufficient condition ensuring that $M(T^n(\delta_X)) \to 1$ as $n \to \infty$ is to require that for all $\varepsilon > 0$, there exists $n \geq 0$ such that $|f_{i_1} \dots f_{i_n}(X)| < \varepsilon$ for all sequences $i_1 i_2 \dots i_n \in I^n$. In fact, by König's lemma, it is sufficient that for each infinite sequence $i_1 i_2 \dots \in I^\omega$, $|f_{i_1} \dots f_{i_n}(X)| \to 0$ an $n \to \infty$. Edalat calls an IFS satisfying the latter condition *weakly hyperbolic*. It is clearly the case that every hyperbolic IFS is weakly hyperbolic.

Theorem 6 (Edalat [2]). *A weakly hyperbolic weighted IFS* $\langle X, \{(f_i, p_i)\}_{i \in I} \rangle$ *on a compact metric space X has a unique invariant measure which is moreover an attractor for the Markov operator (4).*

Proof. Every continuous valuation on a compact metric space extends to a Borel measure, and conversely every Borel measure restricts to a continuous valuation, see [16]. Thus, to prove the existence of a unique invariant measure, it will suffice to prove that there is a unique continuous valuation ν on X such that $\nu(O) = \sum_{i \in I} p_i \nu(f_i^{-1}(O))$ for all open $O \subseteq X$.

Let D be the sub-dcpo of $\mathbf{P}\mathbf{U}X$ consisting of valuations with mass 1. Then D is pointed and continuous, and T restricts to a monotone map $D \to D$. Thus we may apply Theorem 2 to deduce that T has a unique fixed point on D. Next we show that $\mu \in \max \mathbf{P}\mathbf{U}X$ is a fixed point of T iff the corresponding valuation μ_* on $\max \mathbf{U}X = X$, as defined in Theorem 4, is invariant for the IFS.

Suppose $\mu = T(\mu)$, then for any open set $O \subseteq X$ we have (using the notation of Theorem 4):

$$
\begin{aligned}
\mu_*(O) &= \mu(O^\dagger) \\
&= T(\mu)(O^\dagger) \\
&= \sum_{i \in I} p_i \mu((\mathbf{U}f_i)^{-1}(O^\dagger)) \quad \text{by (5)} \\
&= \sum_{i \in I} p_i \mu((f_i^{-1}(O))^\dagger) \quad \text{by Proposition 6} \\
&= \sum_{i \in I} p_i \mu_*(f_i^{-1}(O)).
\end{aligned}
$$

Thus μ_* is invariant. The converse follows similarly.

Finally, from Theorem 2 we may conclude that the unique invariant measure is an attractor for the Markov operator, since the latter agrees with T on max($\mathbf{PU}D$). □

The construction of the unique invariant measure here is essentially the same as in Edalat [2]. However it is justified in a different way. Edalat deduces that the least fixed point of T is a unique fixed point by proving that it is maximal. This observation depends on a characterization of the maximal elements of $\mathbf{PU}X$ in terms of their supports. This last requires some more measure-theoretic machinery than we have used here: in particular he uses the result of Lawson [11] on extending continuous valuations on ω-continuous dcpos to Borel measures over the Lawson topology.

References

1. M. Alvarez-Manilla, A. Edalat, and N. Saheb-Djahromi. An extension result for continuous valuations. *Journal of the London Mathematical Society*, 61(2):629–640, 2000.
2. A. Edalat. Power Domains and Iterated Function Systems. *Information and Computation*, 124:182–197, 1996.
3. A. Edalat. Domain theory and integration. *Theoretical Computer Science*, 151:163–193, 1995.
4. A. Edalat. When Scott is weak at the top. *Mathematical Structures in Computer Science*, 7:401–417, 1997.
5. A. Edalat and R. Heckmann. A computational model for metric spaces. *Theoretical Computer Science*, 193:53–78, 1998.
6. S. Hayashi. Self-similar sets as Tarski's fixed points. *Publications of the Research Institute for Mathematical Sciences*, 21(5):1059–1066, 1985.
7. R. Heckmann. Spaces of valuations. *Papers on General Topology and Applications: 11th Summer Conference at the University of Southern Maine, Vol. 806, Annals of the New York Academy of Sciences*, pp. 174–200. New York, 1996.
8. J.E. Hutchinson. Fractals and self-similarity. *Indiana University Mathematics Journal*, 30:713–747, 1981.
9. C. Jones. *Probabilistic nondeterminism*, PhD Thesis, Univ. of Edinburgh, 1990.
10. O. Kirch. Bereiche und Bewertungen. Master's thesis. Technische Hochschule Darmstadt, 1993.

11. J. Lawson. Valuations on Continuous Lattices. In Rudolf-Eberhard Hoffmann, editor, *Continuous Lattices and Related Topics*, volume 27 of *Mathematik Arbeitspapiere*. Universität Bremen 1982.
12. K. Martin. The measurement process in domain theory. *Proc. 27th International Colloquium on Automata, Languages and Programming (ICALP)*, Lecture Notes in Computer Science, Vol. 1853, Springer-Verlag, 2000.
13. K. Martin. Unique fixed points in domain theory. *Proceedings of 17th Annual Conference on Mathematical Foundations of Programming Semantics*, BRICS Notes Series, NS-01-2. May 2001.
14. K. Martin. Fractals and domain theory. Submitted. http://web.comlab.ox.ac.uk/oucl/work/keye.martin/
15. M.W. Mislove. Topology, domain theory and theoretical computer science. *Topology and its Applications*, 89:3–59, 1998.
16. B.J. Pettis. On the extension of measures. *Annals of Mathematics*, 54(1):186–197, 1951.

Games Characterizing Levy-Longo Trees

C.-H.L. Ong and P. Di Gianantonio

[1] Oxford University Computing Laboratory, Wolfson Building, Parks Road, Oxford
OX1 3QD, United Kingdom. Fax: +44 1865 273839. lo@comlab.ox.ac.uk
[2] Dipartimento di Matematica e Informatica, Universitá de Udine, via delle Scienze,
206, 33100 Udine, Italy. Fax +39 0432558499. digianantonio@dimi.uniud.it

Abstract. We present a simple *strongly universal* innocent game model
for Levy-Longo trees i.e. every point in the model is the denotation
of a unique Levy-Longo tree. The observational quotient of the model
then gives a universal, and hence fully abstract, model of the pure Lazy
Lambda Calculus.

1 Introduction

This paper presents a *strongly universal* innocent game model for Levy-Longo
trees [11,12] (i.e. every point in the model is the denotation of a unique Levy-
Longo tree). We consider arenas in the sense of [8,14] in which questions may
justify either questions or answers, but answers may only justify questions; and
we say that an answer (respectively question) is *pending* in a justified sequence
if no question (respectively answer) is explicitly justified by it. Plays are justified
sequences that satisfy the standard conditions of Visibility and Well-Bracketing,
and a new condition, which is a *dual* of Well-Bracketing, called

> *Persistence*: If an odd-length (respectively even-length) play *s* has a
> pending O-answer (respectively P-answer) – let *a* be the last such in *s*,
> and if *s* is followed by a question *q*, then *q* must be explicitly justified
> by *a*.

We then consider *conditionally copycat* strategies, which are *innocent* strategies
(in the sense of [8]) that behave in a *copycat* fashion as soon as an O-answer
is followed by a P-answer. Together with a *relevance* condition, we prove that
the recursive such strategies give a *strongly universal* model of Levy-Longo trees
i.e. every strategy is the denotation of a unique Levy-Longo tree. To our knowledge, this is the first universal model of Levy-Longo trees. The observational
quotient of the model then gives a universal and fully abstract model of the pure
Lazy Lambda Calculus [15,3].

Related work. Universal models for the Lazy Lambda Calculus with convergence test were first presented in [2] and [13]. The model studied in the former
is in the AJM style [1], while that in the latter, by McCusker, is based on an
innocent-strategy [8] universal model for call-by-name FPC, and is obtained via
a universal and fully abstract translation from the Lazy Lambda Calculus into

P. Widmayer et al. (Eds.): ICALP 2002, LNCS 2380, pp. 476–487, 2002.
© Springer-Verlag Berlin Heidelberg 2002

call-by-name FPC. The present paper considers the *pure* (i.e. without any constant) Lazy Lambda Calculus. Our model is the same as McCusker's, except that it has three additional constraints: Persistence, which is a constraint on plays, and Conditional Copycat and Relevance, which are constraints on strategies. Since Persistence constrains Opponent as well as Player, the model presented here is not simply a submodel of McCusker's.

An AJM-style game model of the the pure Lazy Lambda Calculus was presented by the second author in [6]. The strategies therein are history-free and satisfy a *monotonicity* condition. Though fully abstract for the language, the model is not universal (there are finite monotone strategies that are not denotable). However we believe it is possible to achieve universality by introducing a condition similar to relevance. In [10,9] game models based on *effectively almost-everywhere copycat* (or EAC) strategies are constructed which are strongly universal for Nakajima trees and Böhm trees respectively. Several local structure results for AJM-style game models can be found in [7].

2   Arenas and Nested Levels

An **arena** is a triple $A = \langle M_A, \lambda_A, \vdash_A \rangle$ where M_A is a set of moves; $\lambda_A : M_A \to \{PQ, PA, OQ, OA\}$ is a labelling function which, for given a move, indicates which of P or O may make the move and whether it is a question (Q) or an answer (A); and $\vdash_A \subseteq (A + \{*\}) \times A$, where $*$ is a dummy move, is called the *justification relation* (we read $m_1 \vdash_A m_2$ as "m_1 justifies m_2") satisfying the following axioms: we let m, m', m_i range over M_A

1. Either $* \vdash_A m$ (in which case we call m an *initial move*) or else $m^- \vdash_A m$ for some m^-.
2. Every initial move is an O-question.
3. If $m \vdash_A m'$ then m and m' are moves by different players.
4. If $m_1 \vdash_A m$ and $m_2 \vdash_A m$ then $m_1 = m_2$.
5. If $m \vdash_A m'$ and m is an answer then m' is a question ("Answers may only justify questions.").

In the following we shall refer to standard arena constructions such as product $A \times B$, function space $A \Rightarrow B$ and lifting $A_\perp$; the reader may wish to consult e.g. [8,14] for a definition. We use square and round parentheses in bold type as meta-variables for moves as follows:

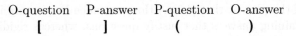

A **justified sequence** over an arena A is a finite sequence of alternating moves such that, except the first move which is initial, every move m has a *justification pointer* (or simply *pointer*) to some earlier move m^- whereby $m^- \vdash_A m$; we say that m is *explicitly justified* by m^-. A question (respectively answer) in a justified sequence s is said to be **pending** just in case no answer (respectively question) in s is explicitly justified by it. This extends the standard meaning of "pending questions" to "pending answers". Recall the definition of the **P-view** $\ulcorner s \urcorner$ of a justified sequence s:

$$\ulcorner \epsilon \urcorner = \epsilon$$
$$\ulcorner s\, m \urcorner = \ulcorner s \urcorner m \qquad \text{if } m \text{ is a P-move}$$
$$\ulcorner s\, m \urcorner = m \qquad \text{if } m \text{ is initial}$$
$$\ulcorner s\, m_0\, u\, m \urcorner = \ulcorner s \urcorner m_0\, m \qquad \text{if the O-move } m \text{ is explicitly justified by } m_0$$

In $\ulcorner s\, m_0\, u\, m \urcorner$ the pointer from m to m_0 is retained, similarly for the pointer from m in $\ulcorner s\, m \urcorner$ in case m is a P-move.

Definition 1. A justified sequence s over A is said to be a *legal position* (or *play*) just in case it satisfies:

1. *Visibility*: Every P-move (respective non-initial O-move) is explicitly justified by some move that appears in the P-view (respectively O-view) at that point.
2. *Well-Bracketing*: Every P-answer (respectively O-answer) is an answer to (i.e. explicitly justified by) the last pending O-question (respectively P-question).
3. *Persistence*: If an odd-length (respectively even-length) s has a pending O-answer (respectively P-answer) – let a be the last such in s, and if s is followed by a question q, then q must be explicitly justified by a.

Remark 1. Except for Persistence, all that we have introduced so far are standard notions of the *innocent* approach to Game Semantics in the sense of [8]. Note that there can be at most one pending O-answer (respectively P-answer) in a P-view (respectively O-view). It is an immediate consequence of Well-Bracketing that no question may be answered more than once in a legal position.

Persistence may be regarded as a *dual* of Well-Bracketing: it is to questions what Well-Bracketing is to answers. The effect of Persistence is that in certain situations, namely when there is a pending O-answer, a strategy has no choice over which question it can ask, or equivalently over which argument it can interrogate, *at that point* (of course it may decide instead to answer an O-question). An apparently similar restriction on the behaviour of strategies is imposed by the *rigidity* condition introduced by Danos and Harmer [5]. (For any legal position of a rigid strategy, the pointer from a question is to some move that appears in the *R-view* of the play at that point.) However since Persistence is a constraint on *plays* containing answers that justify questions, whereas rigidity is a condition on *strategies* over arenas whose answers do *not* justify any move, it is not immediately obvious how the two notions are related.

Nested levels. Take any set M that is equipped with a function $\lambda : M \rightarrow \{Q, A\}$ which labels elements as either questions or answers. Let s be a finite sequence of elements from M – call s a *dialogue*. Set $\#_{\mathrm{qn}}(s)$ and $\#_{\mathrm{ans}}(s)$ respectively to be the number of questions and the number of answers in s. Following

[6], we define the **nested level** at sm (or simply the **level** of m whenever s is understood) to be

$$\mathsf{NL}(sm) = \begin{cases} \delta - 1 & \text{if } m \text{ is a question} \\ \delta & \text{if } m \text{ is an answer} \end{cases}$$

where $\delta = \#_{\mathrm{qn}}(sm) - \#_{\mathrm{ans}}(sm)$; we define $\mathsf{NL}(\epsilon) = 0$. For example, the nested levels of the moves in the dialogue $[\,()\,([\,]\,[\,()\,]\,)\,()\,]\,[\,($ are:

```
Nested Level
     3                      ()
     2                 []  [   ]
     1              () (          ) ()     (
     0           [                      ] [
```

For $l \geq 0$, we write $s \upharpoonright l$ to mean the subsequence of s consisting of moves at level l. We say that an answer a in a dialogue t is *closed* if it is the last move in t at level l, where l is the level of a in t.

We state some basic properties of nested levels of dialogues.

Lemma 1. *In the following, we let s range over dialogues.*

1. *For any $s = umm'$, if m and m' are at different levels l and l' respectively, then m and m' are either both questions (in which case $l' = l + 1$) or both answers (in which case $l' = l - 1$). As a corollary we have:*
2. *If a and b in a dialogue are at levels l_1 and l_2 respectively, then for any $l_1 \leq l \leq l_2$, there is some move between a and b (inclusive) at level l.*
3. *For any $l \geq 0$, if $l < \mathsf{NL}(s)$ (respectively $l > \mathsf{NL}(s)$) then the last move in s at level l, if it exists, is a question (respectively answer).*
4. *Suppose s begins with a question. For each l, if $s \upharpoonright l$ is non-empty, the first element is a question, thereafter the elements alternate strictly between answers and questions.*
5. *Take any dialogue sq, where q is a question. Suppose $\mathsf{NL}(sq) = l$ then an answer a is the last occurring closed answer in s if and only if a is the last move at level l in s.*

The notion of nested level is useful for proving that the composition of strategies is well-defined. Note that Lemma 1 holds for dialogues in general – there is no assumption of justification relation or pointers, nor of the distinction between P and O.

3   Conditionally Copycat Strategies and Relevance

Recall that a *P-strategy* (or simply *strategy*) σ for a game A is defined to be a non-empty, prefix-closed set of legal positions of A satisfying:

1. For any even-length $s \in \sigma$, if sm is a legal position then $sm \in \sigma$.
2. (*Determinacy*). For any odd-length s, if sm and sm' are in σ then $m = m'$.

A strategy is said to be ***innocent*** [8] if whenever even-length $sm \in \sigma$ then for any odd-length $s' \in \sigma$ such that $\ulcorner s \urcorner = \ulcorner s' \urcorner$, we have $s'm \in \sigma$. That is to say, σ is completely determined by a partial function f (say) that maps P-views p to *justified P-moves* (i.e. $f(p)$ is a P-move together with a pointer to some move in p). We write f_σ for the minimal such function that defines σ. We say that an innocent strategy σ is ***compact*** just in case f_σ is a finite function (or equivalently σ contains only finitely many P-views).

Definition 2. We say that an innocent strategy σ is ***conditionally copycat*** (or simply ***CC***) if for any odd-length P-view $p \in \sigma$ in which there is an O-answer which is immediately followed by a P-answer (i.e. p has the shape "$\cdots$)] $\cdots$"), then $pm \in \sigma$ for some P-move m which is explicitly justified by the penultimate O-move in p.

CC strategies can be characterized as follows.

Lemma 2 (CC). *An innocent strategy σ is CC if and only if for every even-length P-view p in σ that has the shape u $)_0$ $]_0$ v*

1. *for any O-move m, if $pm \in \sigma$ then $pmm' \in \sigma$ for some P-move m', and*
2. *if v is a non-empty segment, then v is a **copycat block** of moves. I.e. v has the shape*

$$a_1 \, b_1 \, a_2 \, b_2 \, \cdots \, a_n \, b_n$$

   where $n \geq 1$ such that
   a) for each i, the P-move b_i is a question iff the preceding O-move a_i is a question
   b) $]_0$ explicitly justifies a_1 uniquely, $)_0$ explicitly justifies b_1 uniquely, and for each $i \geq 1$, b_i explicitly justifies a_{i+1} uniquely, and each a_i explicitly justifies b_{i+1} uniquely.
   In other words v is an interleaving of two sequences v_1 and v_2, such that in each v_i, each element (except the first) is explicitly justified by the preceding element in the other sequence.

Composition of strategies. Suppose σ and τ are strategies over arenas $A \Rightarrow B$ and $B \Rightarrow C$ respectively. The set of ***interaction sequences*** arising from σ and τ is defined as follows:

$$\mathbf{ISeq}(\sigma, \tau) \;=\; \{u \in \mathcal{L}(A, B, C) : u \upharpoonright (A, B, b) \in \sigma, u \upharpoonright (B, C) \in \tau\}$$

where $\mathcal{L}(A, B, C)$ is the set of local sequences (see [8,14]) over (A, B, C), and where b ranges over occurrences of initial B-moves in u, $u \upharpoonright (A, B, b)$ is the subsequence of u consisting of moves from the arenas A and B that are hereditarily justified by the occurrence b (note that the subsequence inherits the pointers associated with the moves), and similarly for $u \upharpoonright (B, C)$. We can now define the composite strategy $\sigma; \tau$ over $A \Rightarrow C$ as $\sigma; \tau = \{u \upharpoonright (A, C) : u \in \mathbf{ISeq}(\sigma, \tau\}$. In $u \upharpoonright (A, C)$ the pointer of every initial A-move is to the unique initial C-move.

The nested level of an interaction sequence is well-defined, since an interaction sequence is a dialogue. It is useful to establish a basic property about nested levels of interaction sequences.

Lemma 3. *For any m_1 and m_2 in $u \in \mathbf{ISeq}(\sigma, \tau)$ and for any $l \geq 0$, if the segment $m_1 m_2$ appears in $u \upharpoonright l$, then m_1 explicitly justifies m_2 in u.*

Remark 2. The proof of the Lemma appeals to the assumption that $u \upharpoonright (B, C)$ and $u \upharpoonright (A, B, b)$ satisfy Persistence, and to the structure of interaction sequences (in particular, Locality and the Switching Convention). If legal positions are not required to satisfy Persistence, then the Lemma does not hold.

A notion of relevance. We consider a notion of relevance whereby P is not allowed to respond to an O-question by engaging O indefinitely in a dialogue at one level higher, nor is P allowed to "give up"; instead he must eventually answer the O-question.

Definition 3. We say that a CC strategy σ is ***relevant*** if whenever $f_\sigma : p\,[\ \mapsto (_0$, then there is some $b \geq 0$, and there are moves $)_0, (_1,)_1, \cdots, (_b,)_b$, and $]$ such that
$$f_\sigma : p\,[\,(_0\,)_0 \cdots (_b\,)_b \mapsto\,].$$
We call b the *branching factor* of σ at the P-view $p\,[$. (The reason behind the name is explained in the proof of Lemma 6.)

Theorem 1. *If σ and τ are relevant CC strategies over arenas $A \Rightarrow B$ and $B \Rightarrow C$ respectively then the composite $\sigma;\tau$ is also a relevant CC strategy.*

The category $\mathbb{L}$. We define a category called $\mathbb{L}$ whose objects are arenas and whose maps $A \to B$ are relevant CC strategies of the arena $A \Rightarrow B$. It is completely straightforward to verify that $\mathbb{L}$ is cartesian closed (see e.g. [8] for a very similar proof): the terminal object is the empty arena; for any arenas A and B, their cartesian product is given by the standard product construction $A \times B$, and the function space arena is $A \Rightarrow B$. However lifting $(-)_\perp$ is *not* functorial. We write $\mathbb{L}_{\mathrm{rec}}$ for the subcategory whose objects are arenas but whose maps are *recursive* (in the sense of [8, §5.6]), relevant, CC strategies.

Remark 3. (i) There is no way lifting can be functorial in a category of conditionally copycat strategies. Take a CC strategy $\sigma : A \to B$. Since $\mathrm{id}_\perp = \mathrm{id} : A_\perp \to B_\perp$, $\sigma_\perp$ is forced to respond to the initial move q_B in $B_\perp$ with the initial move q_A in $A_\perp$, and to respond to the P-view $q_B q_A a_A$ with the move a_B. Now almost all P-views in $\sigma_\perp$ contain an O-answer a_A immediately followed by a P-answer a_B, and so, by Lemma 2, $\sigma_\perp$ is almost always constrained to play copycat, whereas σ may not be restricted in the same way. (It is easy to construct concrete instances of σ and $\sigma_\perp$.)

(ii) Functoriality of lifting is not necessary for the construction of our game models of the Lazy Lambda Calculus. The domain equation $D = [D \Rightarrow D]_\perp$ is solved in an auxiliary category of games whose maps are the subgame relations (see e.g. [2]), and lifting *is* functorial in this category. All we need are two (relevant, CC) strategies, $\mathsf{up}_D : D \to D_\perp$ and $\mathsf{dn}_D : D_\perp \to D$, such that $\mathsf{dn}_D \circ \mathsf{up}_D = \mathsf{id}_D$, which are easily constructible for any arena D.

(iii) Indeed functoriality of lifting is *inconsistent* with our model being fully abstract. A feature of our model is that there are "few" denotable strategies that are compact-innocent; indeed the innocent strategy denoted by a closed term is compact if and only if the term is unsolvable of a finite order. Now we know from [3, Lemma 9.2.8] that projections on the finite approximations $\mathcal{D}_n$ of the fully abstract model $\mathcal{D}$ of the Lazy Lambda Calculus are not λ-definable. If *all* the domain constructions involved in the domain equation $D = [D \Rightarrow D]_\perp$ were functorial, these projections would be maps that are definable *categorically*, which would imply that our model is not fully abstract.

4   Universality and Full Abstraction

The model. We denote the initial solution of the recursive domain equation $D = [D \Rightarrow D]_\perp$ in the category $\mathbb{L}$ as the arena $\mathcal{D}$. The arena $\mathcal{D}$ satisfies the properties:

1. Every question justifies a unique answer, and at most one question.
2. Every answer justifies a unique question.

With respect to the justification relation, $\mathcal{D}$ has the structure of a finitely-branching tree in which every node has either one or two descendants; see Figure 1 for a picture of $\mathcal{D}$.

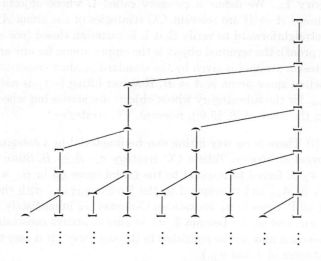

Fig. 1. A picture of $\mathcal{D}$

Note that $[\mathcal{D} \Rightarrow \mathcal{D}]_\perp$ and $\mathcal{D}$ are *identical* (not just isomorphic) arenas. For any closed λ-term s, we shall write $[\![s]\!]$ for its denotation in the model given by

$\mathcal{D}$ in $\mathbb{L}$ (so that $[\![s]\!]$ is a relevant, CC strategy over $\mathcal{D}$). By adapting a standard method in [4] based on an approximation theorem, we have the following result:

Lemma 4 (Adequacy). *For any closed term s, we have $[\![s]\!] = \bot$, the strategy that has no response to the opening move, if and only if s is strongly unsolvable (i.e. s is not β-convertible to a λ-abstraction).*

Structure of P-views. We aim to describe P-views of $\mathcal{D}$ in terms of blocks (of moves) of two kinds, called α and β respectively.

For $n \geq 0$, an α_n-*block* is an alternating sequence of O-questions and P-answers of length $2n + 1$, beginning with an O-question, such that each element except the first is explicitly justified by the preceding element, as follows:

$$\llbracket_0 \;\rrbracket \llbracket_1 \;\rrbracket \;\cdots\; \llbracket_{n-1} \;\rrbracket \llbracket_n$$

We call $\llbracket_i$ the i-th question of the block.

For $m \geq 0, i \geq 0$ and $j \geq 1$, a $\beta_m^{(i,j)}$-*block* is an alternating sequence of P-questions and O-answers of length $2m + 1$, beginning with a P-question, such that each element except the first is explicitly justified by the preceding element, as follows:

$$(_0 \;) (_1 \;) \;\cdots\; (_{m-1} \;) (_m$$

We call $(_i$ the i-th question of the block. The superscript (i,j) in $\beta_m^{(i,j)}$ encodes the target of the justification pointer of $(_0$ relative to the P-view of which the $\beta_m^{(i,j)}$-block is a part (about which more anon). A $\overline{\beta}_m^{(i,j)}$-block is just a $\beta_m^{(i,j)}$-block followed by a $)$, which is explicitly justified by the last question $(_m$. An α-block is just an α_n-block, for some n; similarly for a β-block.

Suppose we have a P-view of the form

$$p \;\; = \;\; A_1 \, B_1 \, A_2 \, B_2 \;\cdots\; A_k \, B_k \;\cdots$$

where each A_k is an α_{n_k}-block and each B_k is a $\beta_{l_k}^{(i_k, j_k)}$-block. The superscript (i_k, j_k) encodes the fact that the 0-th question of the block B_k is explicitly justified by the j_k-th question of the block A_{k-i_k}. Thus we have the following constraints: for each $k \geq 1$

$$0 \leq i_k < k \qquad \wedge \qquad 1 \leq j_k \leq n_{k-i_k} \tag{1}$$

The lower bound of j_k is 1 rather than 0 because, by definition of $\mathcal{D}$ (see Figure 1), the only move that the 0-th question of any α-block can justify is an answer. Note that since p is a P-view by assumption, for each $k \geq 2$, the 0-th question of the α-block A_k is explicitly justified by the last question of the preceding β-block.

Remark 4. It is straightforward to see that given any finite alternating sequence γ of α- and β-blocks

$$\gamma \;\; = \;\; \alpha_{n_1} \, \beta_{l_1}^{(i_1, j_1)} \;\cdots\; \alpha_{n_k} \, \beta_{l_k}^{(i_k, j_k)} \;\cdots$$

subject to the constraints (1), there is exactly one P-view p of $\mathcal{D}$ that has the shape γ. Therefore there is no harm in referring to the P-view p simply as γ, and we shall often do so in the following.

Lemma 5 (P-view Characterization). *Suppose the even-length P-view*

$$W = \alpha_{n_1} \beta_{l_1} \cdots \alpha_{n_m} \beta_{l_m}$$

is in a relevant CC strategy σ over $\mathcal{D}$ for some $m \geq 0$. Then exactly one of the following holds:

(1) For each $j \geq 0$, $W \alpha_j \in \mathrm{dom}(f_\sigma)$.

(2) There is some $n \geq 0$ such that $W \alpha_n \in \sigma \setminus \mathrm{dom}(f_\sigma)$.

(3) There are some $n_{m+1} \geq 0$, some $0 \leq i < m+1$ and some $1 \leq j \leq n_{m+1-i}$ such that $f_\sigma : W \alpha_{n_{m+1}} \mapsto (^{(i,j)}$; further by relevance, for some $l \geq 0$, we have

$$f_\sigma : W \alpha_{n_{m+1}} \overline{\beta}_l^{(i,j)} \mapsto].$$

Moreover by CC we have $W \alpha_{n_{m+1}} \overline{\beta}_l^{(i,j)}] C \in \mathrm{dom}(f_\sigma)$, for each (odd-length) copycat block C, as defined in Lemma 2.

For any λ-term s, if the set $\{i \geq 0 : \exists t. \lambda\beta \vdash s = \lambda x_1 \cdots x_i t\}$ has no supremum in $\mathbb{N}$, we say that s has *order infinity*; otherwise if the supremum is n, we say that s has *order n*. A term that has order infinity is unsolvable (e.g. **yk**, for any fixpoint combinator **y**). We give an informal definition of $\mathrm{LT}(s)$, the **Levy-Longo tree** [11,12] of a λ-term s, as follows:

- Suppose s is unsolvable: If s has order infinity then $\mathrm{LT}(s)$ is the singleton tree $\top$; if s has order $n \geq 0$ then $\mathrm{LT}(s)$ is the singleton tree $\bot_n$.
- Suppose $s =_\beta \lambda x_1 \cdots x_m y s_1 \cdots s_n$ where $m, n \geq 0$. Then $\mathrm{LT}(s)$ is the tree:

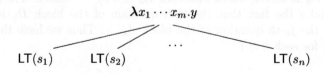

It is useful to fix a *variable-free representation* of Levy-Longo trees. We write $\mathbb{N} = \{0, 1, 2, \cdots\}$ and $\mathbb{N}_+ = \{1, 2, 3, \cdots\}$. A **Levy-Longo pre-tree** is a partial function T from the set $(\mathbb{N}_+)^*$ of *occurrences* to the following set of *labels*

$$\mathbb{N} \times (\mathbb{N} \times \mathbb{N}_+) \times \mathbb{N} \quad \cup \quad \{\bot_i : i \geq 0\} \quad \cup \quad \{\top\}$$

such that

1. $\mathrm{dom}(T)$ is prefix-closed.

2. Every occurrence that is labelled by any of $\perp_i$ and $\top$ is maximal in $\mathrm{dom}(T)$.
3. If $T(l_1 \cdots l_m) = \langle n, (i,j), b \rangle$ then:
   a) $l_1 \cdots l_m l \in \mathrm{dom}(T) \iff 1 \le l \le b$, and
   b) $0 \le i \le m+1$, and
   c) If $i \le m$ then $T(l_1 \cdots l_{m-i})$ is a triple, the first component of which is at least j.

(The case of $i = m+1$ corresponds to the head variable at $l_1 \cdots l_m$ being a free variable.) We say that the pre-tree is *closed* if $T(l_1 \cdots l_m) = \langle n, (i,j), b \rangle \Rightarrow i \le m$. A **Levy-Longo tree** is the Levy-Longo pre-tree given by $\mathrm{LT}(s)$ for some λ-term s. In the following, we shall only consider closed pre-trees and trees.

To illustrate the variable-free representation, consider the following (running) example.

Example 1. Set $s = \lambda x_1 x_2 x_1 \perp_1 (\lambda y_1 y_2 y_3 y_2 (\lambda z x_1)) \top$. The Levy-Longo tree $\mathrm{LT}(s)$ as shown in the figure below

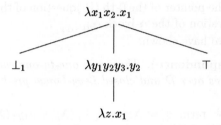

In variable-free form, $\mathrm{LT}(s)$ is the following partial function:

$$\begin{cases} \epsilon \mapsto \langle 2, (0,1), 3 \rangle & 2 \mapsto \langle 3, (0,2), 1 \rangle & 21 \mapsto \langle 1, (2,1), 0 \rangle \\ 1 \mapsto \perp_1 & 3 \mapsto \top \end{cases}$$

We Take $\mathrm{LT}(s) : 21 \mapsto \langle 1, (2,1), 0 \rangle$ which encodes the label $\lambda z x_1$ of the tree at occurrence 21: the first component is the *nested depth* of the λ-abstraction: in this case it is a 1-deep λ-abstraction (i.e. of order one); the second component (i,j) says that the head variable (x_1 in this case) is a copy of the j-th (in this case, first) variable bound at the occurrence i (in this case, two) levels up; and the third component is the *branching factor* at the occurrence, which is 0 in this case i.e. the occurrence 21 has 0 children.

Thanks to Lemma 5, we can now explain the correspondence between relevant CC strategies over $\mathcal{D}$ and closed Levy-Longo pre-trees; we shall write the pre-tree corresponding to the strategy σ as T_σ. Using the notation of Lemma 5, the action of the strategy σ on a P-view $p \in \sigma$ of the shape $\alpha_{n_1} \beta_{l_1}^{(i_1, j_1)} \cdots \alpha_{n_m} \beta_{l_m}^{(i_m, j_m)} [$ determines precisely the label of T_σ at the occurrence $l_1 \cdots l_m$. Corresponding to each of the three cases in Lemma 5, the label defined at the occurrence is as follows:

1. $\top$
2. $\perp_n$ where $n \geq 0$, and
3. $\langle n, (i,j), b \rangle$.

It is easy to see the occurrence in question is maximal in $\text{dom}(T_\sigma)$ in cases 1 and 2. Suppose case 3 i.e. $T_\sigma(l_1 \cdots l_m) = \langle n, (i,j), b \rangle$. From the P-view p, we can work out the label of T_σ at each prefix $l_1 \cdots l_k$ (where $k \leq m$) of the corresponding occurrence, which is $\langle n_{k+1}, (i_{k+1}, j_{k+1}), b_{k+1} \rangle$, as determined by

$$f_\sigma : \alpha_{n_1} \, \beta_{l_1}^{(i_1, j_1)} \; \cdots \; \alpha_{n_k} \, \beta_{l_k}^{(i_k, j_k)} \; \alpha_{n_{k+1}} \, \overline{\beta}_{b_{k+1}}^{(i_{k+1}, j_{k+1})} \mapsto \mathbf{]}$$

we set $\langle n_{m+1}, (i_{m+1}, j_{m+1}), b_{m+1} \rangle = \langle n, (i,j), b \rangle$. Note that b_{k+1} is well-defined because of relevance. Thus the domain of T_σ is prefix-closed. Take any $k \leq m$. For each $1 \leq l \leq b_{k+1}$, we have the odd-length P-view

$$\alpha_{n_1} \, \beta_{l_1}^{(i_1, j_1)} \; \cdots \; \alpha_{n_k} \, \beta_{l_k}^{(i_k, j_k)} \; \alpha_{n_{k+1}} \, \beta_l^{(i_{k+1}, j_{k+1})} \, \mathbf{[} \; \in \; \sigma$$

and so, we have $l_1 \cdots l_k l \in \text{dom}(T_\sigma) \iff 1 \leq l \leq b_{k+1}$. Finally, we must have $j_{k+1} \leq n_{k-i_{k+1}}$, as the pointer of the 0-th (P-)question of the β-block $\beta_l^{(i_{k+1}, j_{k+1})}$ is to the j_{k+1}-th question of the α-block $\alpha_{n_{k-i_{k+1}}}$.

To summarize, we have shown:

Lemma 6 (Correspondence). *There is a one-to-one correspondence between relevant CC strategies over $\mathcal{D}$ and closed Levy-Longo pre-trees.*

Example 2. Take the term $s = \lambda x_1 x_2 x_1 \perp_1 (\lambda y_1 y_2 y_3 y_2 (\lambda z x_1)) \top$ in the preceding example. In the following table, we illustrate the exact correspondence between the relevant CC strategy $[\![s]\!]$ denoted by s on the one hand, and the Levy-Longo tree $\mathsf{LT}(s)$ of the term on the other.

P-views in $[\![s]\!]$		occurrences	labels of $\mathsf{LT}(s)$
$\boxed{\alpha_2 \, \overline{\beta}_3^{(0,1)}}$	$\mapsto \mathbf{]}$	ϵ	$\langle 2, (0,1), 3 \rangle$
$\alpha_2 \, \beta_1^{(0,1)} \, \alpha_1$	$\in \sigma \setminus \text{dom}(f_\sigma)$	1	$\perp_1$
$\alpha_2 \, \beta_1^{(0,1)} \, \boxed{\alpha_3 \, \overline{\beta}_1^{(0,2)}}$	$\mapsto \mathbf{]}$	2	$\langle 3, (0,2), 1 \rangle$
$\alpha_2 \, \beta_3^{(0,1)} \, \alpha_n$	$\mapsto \mathbf{]}$ for $n \geq 0$	3	$\top$
$\alpha_2 \, \beta_2^{(0,1)} \, \alpha_3 \, \beta_1^{(0,2)} \, \boxed{\alpha_1 \, \overline{\beta}_0^{(2,1)}}$	$\mapsto \mathbf{]}$	21	$\langle 1, (2,1), 0 \rangle$

For each P-view shown above, note that the subscripts in bold give the corresponding occurrence in the Levy-Longo tree, and the label at that occurrence is specified by the (subscripts and the superscript in the) block that is framed. The first, third and fifth P-views define the "boundary" beyond which the copycat response sets in.

Using an argument similar to the proof of [4, Thm 10.1.23], we can show that every *recursive* closed Levy-Longo pre-tree T is the Levy-Longo tree of some closed λ-term. Thus we have:

Theorem 2 (Universality).

1. *The denotation of a closed λ-term s is a recursive, relevant, CC strategy which corresponds to $LT(s)$ in the sense of Lemma 6.*
2. *Every recursive, relevant, CC strategy over $\mathcal{D}$ is the denotation of a closed λ-term. I.e. for every $\sigma \in \mathbb{L}_{rec}(1, \mathcal{D})$ there is some $s \in \Lambda^o$ such that $[\![s]\!] = \sigma$.*

It follows that two closed λ-terms have the same denotation in $\mathcal{D}$ iff they have the same Levy-Longo tree. As a straightforward corollary, the observational quotient of the model then gives a universal, and hence fully abstract, model of the pure Lazy Lambda Calculus.

Acknowledgements.   The authors are grateful to EU TMR LINEAR, Merton College, and Oxford University Computing Laboratory for their support.

References

1. S. Abramsky, R. Jagadeesan, and P. Malacaria. Full abstraction for PCF. *Information and Computation*, 163, 2000.
2. S. Abramsky and G. McCusker. Games and full abstraction for the Lazy Lambda Calculus. In *Proc. LICS*, pages 234–243. The Computer Society, 1995.
3. S. Abramsky and C.-H. L. Ong. Full abstraction in the Lazy Lambda Calculus. *Information and Computation*, 105:159–267, 1993.
4. H. Barendregt. *The Lambda Calculus*. North-Holland, revised edition, 1984.
5. V. Danos and R. Harmer. The anatomy of innocence. In *Proc. CSL 2001*, pages 188–202. Springer Verlag, 2001. LNCS Vol. 2142.
6. P. Di Gianantonio. Game semantics for the Pure Lazy Lambda Calculus. In *Proc. TLCA 2001*, pages 106–120. Springer-Verlag, 2001.
7. P. Di Gianantonio, G. Franco, and F. Honsell. Game semantics for the untyped $\lambda\beta\eta$-calculus. In *Proc. TLCA 1999*, pages 114–128. Springer-Verlag, 1999. LNCS Vol. 1591.
8. J. M. E. Hyland and C.-H. L. Ong. On Full Abstraction for PCF: I. Models, observables and the full abstraction problem, II. Dialogue games and innocent strategies, III. A fully abstract and universal game model. *Information and Computation*, 163:285–408, 2000.
9. A. D. Ker, H. Nickau, and C.-H. L. Ong. A universal innocent game model for the Böhm tree lambda theory. In *Proc. CSL 1999*, pages 405 – 419. Springer-Verlag, 1999. LNCS Volume 1683.
10. A. D. Ker, H. Nickau, and C.-H. L. Ong. Innocent game models of untyped λ-calculus. *Theoretical Computer Science*, 272:247–292, 2002.
11. J.-J. Levy. An algebraic interpretation of equality in some models of the lambda calculus. In C. Böhm, editor, *Lambda Calculus and Computer Science Theory*, pages 147–165. Springer-Verlag, 1975. *LNCS* No. 37.
12. G. Longo. Set-theoretical models of lambda calculus: Theories, expansions and isomorphisms. *Annals of Pure and Applied Logic*, 24:153–188, 1983.
13. G. McCusker. Full abstraction by translation. In *Advances in Theory and Formal Methods of Computing*. IC Press, 1996.
14. G. McCusker. *Games for recursive types*. BCS Distinguished Dissertation. Cambridge University Press, 1998.
15. G. D. Plotkin. Call-by-name, call-by-value and the lambda calculus. *Theoretical Computer Science*, 1:125–159, 1975.

Comparing Functional Paradigms for Exact Real-Number Computation

Andrej Bauer[1]*, Martín Hötzel Escardó[2], and Alex Simpson[3]**

[1] IMFM, University of Ljubljana, Slovenia
[2] School of Computer Science, University of Birmingham, England
[3] LFCS, Division of Informatics, University of Edinburgh, Scotland

Abstract. We compare the definability of total functionals over the reals in two functional-programming approaches to exact real-number computation: the *extensional* approach, in which one has an abstract datatype of real numbers; and the *intensional* approach, in which one encodes real numbers using ordinary datatypes. We show that the type hierarchies coincide up to second-order types, and we relate this fact to an analogous comparison of type hierarchies over the *external* and *internal* real numbers in Dana Scott's category of equilogical spaces. We do not know whether similar coincidences hold at third-order types. However, we relate this question to a purely topological conjecture about the Kleene-Kreisel continuous functionals over the natural numbers. Finally, although it is known that, in the extensional approach, parallel primitives are necessary for programming total first-order functions, we demonstrate that, in the intensional approach, such primitives are not needed for second-order types and below.

1 Introduction

In functional programming, there are two main approaches to exact real-number computation. One is to use a specialist functional programming language that contains the real numbers as an abstract datatype. This approach is *extensional* in the sense that the data structures representating real numbers are hidden from view and one may only manipulate reals via representation-independent operations upon them. A second approach is to use an ordinary functional language, and to encode real numbers using standard infinite data structures, for example, streams. This approach is *intensional* in the sense that one has direct access to the encodings of reals, allowing the possibility of distinguishing between different representations of the same real number. In recent years, the extensional approach has been the subject of much theoretical investigation via the study of specialist languages, such as Di Gianantonio's **RL** [Di 93] and Escardó's **RealPCF** [Esc96]. On the other hand, the intensional approach is the one that is actually used when exact real-number computation is implemented in practice—see, for example, [GL01].

* Research supported by the Slovene Ministry of Science grant Z1-3138-0101-01
** Research supported by an EPSRC Advanced Research Fellowship.

P. Widmayer et al. (Eds.): ICALP 2002, LNCS 2380, pp. 488–500, 2002.
© Springer-Verlag Berlin Heidelberg 2002

This paper presents preliminary results in a general investigation relating the two approaches. Specifically, we address the question of how the programmability of higher-type total functionals over the real numbers compares between the two approaches. To this end, we consider two type hierarchies built using function space and product over a single base type, real. The first hierarchy is constructed by interpreting each type σ as the set $[\sigma]_E$ of extensionally programmable total functionals of that type, and the second by interpreting σ as the set $[\sigma]_I$ of intensionally programmable total functionals. As our first main result, Theorem 1, we prove that for all second-order (and below) types σ, the sets $[\sigma]_E$ and $[\sigma]_I$ coincide, thus a second-order functional is extensionally programmable if and only if it is intensionally programmable. This result thus applies at the type level at which many interesting functionals, including definite integration

$$(f, a, b) \mapsto \int_a^b f(x)\, \mathrm{d}x \quad : \quad (\text{real} \to \text{real}) \times \text{real} \times \text{real} \ \to \ \text{real} \,,$$

reside. See [EE00] and [Sim98] for accounts of integration within the extensional and intensional approaches respectively.

We prove Theorem 1 by relating it to an analogous question of the coincidence of type hierarchies in the setting of Dana Scott's category of equilogical spaces [Sco96,BBS02]. In that setting there is an *external* type hierarchy $(\sigma)_E$, built over Euclidean space, and there is an *internal* hierarchy $(\sigma)_I$, built over the object of real numbers as defined in the internal logic of the category. Again, we show that $(\sigma)_E$ and $(\sigma)_I$ coincide up to second-order types, Theorem 2.

It is of course natural to ask whether the above type hierarchies also coincide for third-order σ and above. We do not know the answer to this question, but a further contribution of this paper is to relate the agreement of the hierarchies at higher types to a purely topological conjecture about the Kleene-Kreisel continuous functionals [Kle59,Kre59] of second-order type, see Sect. 5. However, regarding the extension to third-order types, we remark that we lack examples of genuinely interesting *total* functionals of type three to which such generalisations of our results would apply.

Our methodology for studying the two approaches to exact real-number computation is to consider a paradigmatic programming language for each. For the extensional approach, we use Escardó's **RealPCF+**, which is **RealPCF** [Esc96] extended by a parallel existential operator—a language that enjoys the merit of being universal with respect to its domain-theoretic semantics [ES99]. For the intensional approach, we encode real numbers within Plotkin's **PCF++**, which is **PCF** extended by parallel-conditional and existential operators [Plo77]. Again, **PCF++** enjoys a universality property with respect to its denotational semantics [Plo77].

Admittedly, both **RealPCF+** and **PCF++** are idealized languages, distant from real-world functional languages such as Haskell [Has]. As such, they provide the perfect vehicles for a theoretical investigation into programmability questions such as ours. Nevertheless, it is our desire that our results should relate to the practice of exact real-number computation. There is one main obstacle to such a transference of the results: the parallel features of **PCF++** do not appear in Haskell and related languages. We address this issue in Sect. 7, where we show that, again for second-order σ and below, the parallel features of **PCF++** are nowhere required to program functionals in $[\sigma]_I$, Theorem 3. Thus a second-order total functional over the reals is programmable in an ordinary sequential functional language if and only if it is programmable in the idealized, specialist and highly parallel language **RealPCF+**. Again, we do not know whether this result extends to third-order types and above.

Although our investigation is one into questions of programmability (i.e. of definability) within **RealPCF+** and **PCF++**, we carry out the investigation purely at the denotational level, relying on known universality results to infer definability consequences from the semantic correspondences we establish. In doing so, there is one major way in which the results presented in this paper depart from the outline presented above. A full investigation would show that the *computable* (and hence definable) total functionals coincide between the denotational interpretations of the extensional and intensional approaches. Instead, we establish the coincidence for *arbitrary* continuous functionals, whether computable or not. We remark that the results we prove, although computability free, do nonetheless have definability consequences relative to functional languages with programs given by infinite syntax trees, or, equivalently, relative to languages extended with oracles for all set-theoretic functions from $\mathbb{N}$ to $\mathbb{N}$.

Our reason for ignoring computability questions is that the results we establish already require significant technical machinery from domain theory and, especially, topology. Although we believe that it should be possible to prove effective versions of the results by effectivizing the topological lemmas that we use, it is certainly not a triviality to do so. We leave this as a task for future research. Only once this task is completed will the original programming questions that motivated the research in this paper be fully resolved. Nevertheless, the results in this paper provide a strong indication of the outcome of these questions, and, moreover, introduce techniques that are likely to be useful in addressing them.

For lack of space, proofs are only outlined in this conference version of the paper. In this version, our main goal is to convey the flavour of how mathematical tools from domain theory, topology and category theory may be combined to attack seemingly innocuous questions that originate in functional programming. In doing so, we assume some familiarity with these three subjects, for which our basic references are [AJ94,Dug89,Mac71] respectively.

2   Domains for Real-Number Computation

We first fix terminology—see [AJ94] for definitions. We write *dcppo* to mean directed-complete pointed partial order, i.e. one with least element, and we typically use $\sqsubseteq$ for the partial order. We call a dcppo ω-*continuous* if it has a countable basis. For us, a *domain* is an ω-continuous bounded-complete dcppo. We write $\omega\mathbf{BC}$ for the category of domains and (directed-)continuous functions, and we write $\omega\mathbf{L}$ for its full subcategory of ω-continuous lattices. Both categories are cartesian closed with exponentials given by the dcppo of all continuous functions.

Our main interest will be in two particular domains, one for each of the two approaches to exact real-number computation mentioned in the introduction. The *interval domain* $\mathcal{I}$ has underlying set $\{\mathbb{R}\} \cup \{[a,b] \mid a \le b \in \mathbb{R}\}$, with its order defined by $\delta \sqsubseteq \delta'$ if and only if $\delta \supseteq \delta'$. This is indeed a domain.

The interval domain is intimately connected with the extensional approach to exact real-number computation. Indeed, the abstract datatype of real numbers in **RealPCF** [Esc96] is specifically designed to have $\mathcal{I}$ as its denotational interpretation. Furthermore, Escardó and Streicher [ES99] have established a universality result with respect to the domain-theoretic semantics: every computable element in the domain interpreting a **RealPCF** type is definable, by a term of that type, in the language **RealPCF+**, which is **RealPCF** extended with a parallel existential operator. In this paper, although we are motivated by definability questions, we do not wish to entangle ourselves in computability issues. Thus we remark on the following modified version of Escardó and Streicher's result. Every element (computable or not) in the domain interpreting a **RealPCF** type is definable in the language Ω**RealPCF+**, which is **RealPCF+** extended with an oracle for every set-theoretic function from $\mathbb{N}$ to $\mathbb{N}$.

Under the intensional approach to exact real-number computation, one needs to select a computationally *admissible* representation of real numbers [WK87]. There are many equivalent choices. For simplicity, we use a mantissa-exponent representation, where the mantissa, a real number in the interval $[-1, 1]$, is represented using *signed-binary* expansions. Specifically, a real number is represented by a pair (n, α) where the mantissa $\alpha \in \{-1, 0, 1\}^\omega$ represents the number $0.\alpha_0\alpha_1\alpha_2\ldots$, i.e. $\sum_{i=0}^{\infty} 2^{-(i+1)}\alpha_i$, and the exponent $n \in \mathbb{N}$ gives a multiplier of 2^n, thus the pair (n, α) represents the real number $\sum_{i=0}^{\infty} 2^{n-(i+1)}\alpha_i$.

To implement the above representation in a functional programming language, one would most conveniently encode a real number as a pair consisting of a natural number followed by a stream. However, in order to fix on as simple a language as possible, we use instead a direct implementation in Plotkin's **PCF** [Plo77] extended with product types. In **PCF**, the base type, nat, is interpreted as the flat domain $\mathbb{N}_\perp = \{\perp\} \cup \mathbb{N}$ with least element $\perp$. Function space and product are interpreted using the cartesian-closed structure of ω**BC**. As we are interested in definability, we mention Plotkin's universality result: every computable element in the domain interpreting a **PCF** type is definable in the language **PCF++**, which is **PCF** extended with parallel-conditional and existential operators. Again, there is a computability-free version of this result. Every element (computable or not) in the domain interpreting a **PCF** type is definable in the language Ω**PCF++**, which is **PCF++** extended with an oracle for every set-theoretic function from $\mathbb{N}$ to $\mathbb{N}$.

We represent real numbers, in **PCF**, using the type nat $\to$ nat whose denotational interpretation is the *function domain* $\mathcal{J} = \mathbb{N}_\perp^{\mathbb{N}_\perp}$. We say that a function $f \in \mathcal{J}$ is *real representing* if $f(0) \neq \perp$ and if $f(x) \in \{0, 1, 2\}$ when $x > 0$. Any such real-representing f encodes the real number $\sum_{i=1}^{\infty} 2^{f(0)-i}(f(i) - 1)$.

3   Two Type Hierarchies of Assemblies

Our goal is to investigate the type hierarchies of total functionals on reals programmable in the two approaches to exact real-number computation. We consider simple types over a base type of real numbers, with types given by:

$$\sigma ::= \mathsf{real} \mid \sigma \times \sigma' \mid \sigma \to \sigma' .$$

The *order* of a type is: $\mathrm{order}(\mathsf{real}) = 0$; $\mathrm{order}(\sigma \times \sigma') = \max(\mathrm{order}(\sigma), \mathrm{order}(\sigma'))$; and $\mathrm{order}(\sigma \to \sigma') = \max(1 + \mathrm{order}(\sigma), \mathrm{order}(\sigma'))$.

For the extensional approach, we study the total functionals on reals programmable in the language Ω**RealPCF+**. Every such functionals is represented by an element in the type hierarchy over $\mathcal{I}$ in ω**BC**. However, the type hierarchy over $\mathcal{I}$ contains both superfluous elements and redundancies. For example, $\mathcal{I}$ itself contains "partial" real numbers (proper intervals) in addition to "total" reals (singleton intervals). At first-order types, such as $\mathcal{I}^{\mathcal{I}}$, there are elements that do not represent total functions on reals because they fail to preserve total reals. Furthermore, at the same type, it is possible to have two different functions $f, g \colon \mathcal{I} \to \mathcal{I}$ that represent the same total function on reals, because, although they behave identically on total reals, they differ in their behaviour on partial reals.

For the intensional approach, we study the functionals programmable in Ω**PCF++**, using the representation described in Sect. 2. This time, every such functional is represented by an element in the type hierarchy over $\mathcal{J}$ in ω**BC**. Again, there is superfluity and redundancy. Within $\mathcal{J}$, we singled out the real-representing elements in Sect. 2, and in fact each real number has infinitely many different representations. Because of this, there are two ways that a function from $\mathcal{J}$ to $\mathcal{J}$ may fail to represent a function on real numbers: either it may map some real-representing element to a non-real-representing element; or it may map two different representations of the same real number to representations of different real numbers.

Assemblies offer a convenient way of identifying the elements of the hierarchies over $\mathcal{I}$ and $\mathcal{J}$ in ω**BC** that represent total functionals on reals. An *assembly* is a triple $A = (|A|, \|A\|, \Vdash_A)$ where $|A|$ is a set, $\|A\|$ is a domain, and $\Vdash_A$ is a binary relation between $\|A\|$ and $|A|$ such that, for all $a \in |A|$, there exists $x \in \|A\|$ such that $x \Vdash_A a$. A morphism from one assembly A to another B is simply a function $f \colon |A| \to |B|$ for which there exists a continuous $g \colon \|A\| \to \|B\|$ such that $x \Vdash_A a$ implies $g(x) \Vdash_B f(a)$, in which case we say that g *tracks* f. We write **Asm**$(\omega$**BC**$)$ for the category of assemblies over domains, and **Asm**$(\omega$**L**$)$ for the full subcategory of assemblies over ω-continuous lattices. Again, both categories are cartesian closed, with the exponential B^A given by

$$|B^A| = \{f \colon |A| \to |B| \mid f \text{ is a morphism from } A \text{ to } B\}$$

$$\|B^A\| = \|B\|^{\|A\|} \text{ in } \omega\mathbf{BC}$$

$$g \Vdash_{B^A} f \iff g \text{ tracks } f .$$

We use $\mathbf{Asm}(\omega\mathbf{BC})$ to define the two type hierarchies of total functionals we are interested in. For the extensional approach, we define an assembly $[\![\sigma]\!]_E$ for each type σ. For the base type, real, this is given by:

$$|[\![\mathsf{real}]\!]_E| = \mathbb{R} \qquad \|[\![\mathsf{real}]\!]_E\| = \mathcal{I} \qquad \delta \Vdash_{[\![\mathsf{real}]\!]_E} x \iff \delta = \{x\}$$

from which $[\![\sigma]\!]_E$ is defined using the cartesian-closed structure of $\mathbf{Asm}(\omega\mathbf{BC})$.

The hierarchy of extensional functionals over $\mathbb{R}$ is given by the sets $|[\![\sigma]\!]_E|$, for which we henceforth use the less cluttered $[\sigma]_E$. We have that $[\mathsf{real}]_E = \mathbb{R}$; $[\sigma \times \sigma']_E = [\sigma]_E \times [\sigma']_E$; and $[\sigma \to \sigma']_E$ is a set of (total) functions from $[\sigma]_E$ to $[\sigma']_E$. In fact, by [Nor00b], $[\sigma \to \sigma']_E$ is exactly the set of continuous functions with respect to the interpretation of the σ type hierarchy over $\mathbb{R}$ in the cartesian-closed category of sequential topological spaces (see Sect. 5). For each type σ, the set $|[\![\sigma]\!]_E|$, for which we henceforth use the less cluttered notation $[\sigma]_E$, is a set of total functionals over the reals. By the universality of $\Omega\mathbf{RealPCF+}$, the functionals $f \in [\sigma]_E$ are exactly those for which there exists an $\Omega\mathbf{RealPCF+}$ program P of type σ such that P computes f, as witnessed by the relation $[\![P]\!] \Vdash_{[\sigma]_E} f$, where $[\![P]\!]$ is the denotational interpretation of P.

Similarly, for the intensional approach, the assembly $[\![\mathsf{real}]\!]_I$ is defined by:

$$|[\![\mathsf{real}]\!]_I| = \mathbb{R} \quad \|[\![\mathsf{real}]\!]_I\| = \mathcal{J} \quad f \Vdash_{[\![\mathsf{real}]\!]_I} x \iff \begin{array}{l} f \text{ is real representing and} \\ x = \sum_{i=1}^{\infty} 2^{f(0)-i} \cdot (f(i) - 1) \end{array}$$

and again $[\![\sigma]\!]_I$ is induced for arbitrary σ using the cartesian-closed structure of $\mathbf{Asm}(\omega\mathbf{BC})$. This time the set $|[\![\sigma]\!]_I|$, for which we henceforth write $[\sigma]_I$, is the set of those total functionals f for which there exists an $\Omega\mathbf{PCF++}$ program P of type σ^* (where $\mathsf{real}^* = \mathsf{nat} \to \mathsf{nat}$ and $(\cdot)^*$ commutes with function space and product) such that P computes f, as witnessed by the relation $[\![P]\!] \Vdash_{[\sigma]_I} f$.

Theorem 1. *For any type σ with* $\mathrm{order}(\sigma) \leq 2$, *it holds that* $[\sigma]_E = [\sigma]_I$.

By the coincidence of the $[\sigma]_E$ hierarchy with the hierarchy in the cartesian-closed category of sequential topological spaces [Nor00b], one can charcterise the sets $[\sigma]_E$, for σ with $\mathrm{order}(\sigma) \leq 2$, as the continuous functionals with respect to the compact-open topology on function spaces. Thus, as a consequence of Theorem 1, we obtain a purely topological description of the $[\sigma]_I$ hierarchy for σ with $\mathrm{order}(\sigma) \leq 2$.

4   Two Type Hierarchies of Equilogical Spaces

We prove Theorem 1 by relating it to another situation in which there are competing hierarchies of total functionals over the reals, but in which the problem of comparing the two hierarchies is more tractable. This second pair of hierarchies arises in Dana Scott's category of *equilogical spaces* [Sco96,BBS02], a cartesian-closed extension of the category of topological spaces.

In the present paper we only consider countably-based equilogical spaces, and we do not impose Scott's T_0 condition. For our purposes then, an *equilogical space* is a triple $X = (|X|, \|X\|, q_X)$ where $|X|$ is a set, $\|X\|$ is a countably-based topological space and $q_X : \|X\| \to |X|$ is a surjective function. A morphism from one equilogical space X to another Y is simply a function $f : |X| \to |Y|$ for which there exists a continuous $g : \|X\| \to \|Y\|$ such that $q_Y \circ g = f \circ q_X$. Again we say that g *tracks* f. We write $\omega\mathbf{Equ}$ for the category of equilogical spaces.

We write $\omega\mathbf{Top}$ for the category of countably-based topological spaces. There is a full and faithful functor from $\omega\mathbf{Top}$ to $\omega\mathbf{Equ}$, mapping a countably-based space S to (S, S, id_S). A remarkable fact is that is $\omega\mathbf{Equ}$ is equivalent to the category $\mathbf{Asm}(\omega\mathbf{L})$ [BBS02]. Thus $\omega\mathbf{Equ}$ is cartesian closed.

There are two non-isomorphic equilogical spaces, each with good claims to be *the* equilogical space of real numbers. The *external* reals, $\mathsf{R_E}$, is the inclusion of the topological Euclidean reals as the object $(\mathbb{R}, \mathbb{R}, \mathrm{id}_\mathbb{R})$. The *internal* reals, $\mathsf{R_I}$, is the object $(\mathbb{R}, \mathbb{N} \times 3^\omega, \mathsf{r})$, where $3 = \{-1, 0, 1\}$ with the discrete topology, both 3^ω and $\mathbb{N} \times 3^\omega$ are given the product topologies, and r is:

$$\mathsf{r}(n, \alpha) = \sum_{i=0}^{\infty} 2^{n-(i+1)} \alpha_i . \tag{1}$$

Thus, the internal reals are again based on the intensional signed-digit notation. The reason for the terminology is that the internal reals are given as the object of Cauchy reals as defined in the internal logic of $\mathbf{Asm}(\omega\mathbf{L})$.

We use the cartesian-closed structure to determine two type hierarchies, the *external* $([\sigma])_E$, and the *internal* $([\sigma])_I$ in $\omega\mathbf{Equ}$, defined at base type by:

$$([\mathsf{real}])_E = \mathsf{R_E} \qquad\qquad ([\mathsf{real}])_I = \mathsf{R_I} .$$

We write $(\sigma)_E$ as an abbreviation for $|([\sigma])_E|$, and $(\sigma)_I$ for $|([\sigma])_I|$.

Theorem 2. *For any type σ with $\mathrm{order}(\sigma) \leq 2$, it holds that $(\sigma)_E = (\sigma)_I$.*

5   Proofs of Theorems 1 and 2

In this section, we outline the proof of Theorem 2 and the derivation of Theorem 1 from Theorem 2.

We shall need to consider various types of topological spaces. A space is said to be *zero-dimensional* if every neighbourhood of a point has a clopen subneighbourhood, where a *clopen* set is one that is both open and closed.

In a topological space T, an infinite sequence $(x_i)_{i \geq 0}$ *converges* to a point x, notation $(x_i) \to x$, if, for all neighbourhoods $U \ni x$, the sequence (x_i) is eventually in U (i.e., there exists $l \geq 0$ such that $x_j \in U$ for all $j \geq l$). A subset $X \subseteq T$ is *sequentially open* if, whenever $(x_i) \to x \in X$, it holds that (x_i) is eventually in X. Every open set is sequentially open. A space T is said to be *sequential* if every sequentially open subset is open. We write $\mathbf{Seq}$ for the category of sequential spaces. This category is known to be cartesian closed. If S and T are sequential then the exponential T^S is given by the set of all continuous functions endowed with the unique sequential topology that induces the convergence relation $(f_i) \to f$ if and only if, whenever $(x_i) \to x$ in S, it holds that $(f_i(x_i)) \to f(x)$ in T.

We write $\omega\mathbf{qTop}$ for the category of all quotient spaces of countably-based spaces, i.e. a topological space T is an object of $\omega\mathbf{qTop}$ if and only if there exists a countably-based space S with a topological quotient $q\colon S \twoheadrightarrow T$. There are subcategory inclusions $\omega\mathbf{Top} \hookrightarrow \omega\mathbf{qTop} \hookrightarrow \mathbf{Seq}$. Importantly, the category $\omega\mathbf{qTop}$ is cartesian closed with its cartesian-closed structure inherited from $\mathbf{Seq}$ [MS02].

A topological space is said to be *hereditarily Lindelöf* if, for every family $\{U_i\}_{i \in I}$ of open sets, there is a countable subfamily $\{U_j\}_{j \in J}$ (i.e. where $J \subseteq I$ is countable) such that $\bigcup_{j \in J} U_j = \bigcup_{i \in I} U_i$. It is easily shown that every space in $\omega\mathbf{qTop}$ is hereditarily Lindelöf.

The next proposition relates the above notions to an important property of the function $r\colon \mathbb{N} \times 3^\omega \to \mathbb{R}$, defined in (1), which is a topological quotient. We first introduce terminology that makes sense in an arbitrary category. Given an object Z and a morphism $g\colon X \to Y$ we say that Z is *g-projective*, or equivalently that g *projects* Z, if, for every $f\colon Z \to Y$, there exists $\overline{f}\colon Z \to X$ such that the left-hand diagram below commutes. Dually, we say that Z is *g-injective*, or equivalently that g *injects* Z, if, for every $f\colon X \to Z$, there exists $\overline{f}\colon Y \to Z$ such that the right-hand diagram commutes.

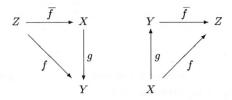

Proposition 1. *Zero-dimensional hereditarily-Lindelöf spaces are r-projective.*

Consider the full subcategory $\omega\mathbf{0Equ}$ of $\omega\mathbf{Equ}$ consisting of those equilogical spaces that are isomorphic to one X for which $\|X\|$ is zero-dimensional. Easily, $\omega\mathbf{0Equ}$ is closed under finite products, and it contains every countably-based zero-dimensional space under the inclusion of $\omega\mathbf{Top}$ in $\omega\mathbf{Equ}$. Moreover, using Proposition 1, $\omega\mathbf{0Equ}$ contains the objects $([\sigma])_I$ for σ with $\mathrm{order}(\sigma) \leq 1$.

We say that a morphism $e\colon X \to Y$ in $\omega\mathbf{Equ}$ is *tight* if it is mono and it projects every space in $\omega\mathbf{0Equ}$. Every tight morphism is also epi. We say that an equilogical space is *tight-injective* if it is injective with respect to every tight map. It can be proved that the full subcategory $\omega\mathbf{Equ_{ti}}$ of tight-injective objects in $\omega\mathbf{Equ}$ is cartesian closed and contains every countably-based space. Thus every object $([\sigma])_E$ is tight-injective.

We prove Theorem 2 by constructing tight morphisms $([\sigma])_I \to ([\sigma])_E$ when $\mathrm{order}(\sigma) \leq 2$. The first lemma gives the crucial construction for function types.

Lemma 1. *Given tight maps $e\colon X \to Z$ and $f\colon Y \to W$, where Y is in $\omega\mathbf{0Equ}$ and Z, W are in $\omega\mathbf{Equ_{ti}}$, then the function $f^{e^{-1}}\colon |Y|^{|X|} \to |W|^{|Z|}$ restricts to a function $g\colon |Y^X| \to |W^Z|$ giving a tight morphism $g\colon Y^X \to W^Z$.*

Lemma 2. *If* $\mathrm{order}(\sigma) \leq 2$ *then* $(\sigma)_I = (\sigma)_E$ *and the identity function gives a tight morphism* $([\sigma])_I \to ([\sigma])_E$.

Theorem 2 is an immediate consequence.

We next consider how Theorem 2 might be extended to higher types. Certainly, the proof above does not extend directly, because one can show that $([(\mathrm{real} \to \mathrm{real}) \to \mathrm{real}])_I$ is not in $\omega\mathbf{0Equ}$. However, this leaves open the possibility of replacing the use of $\omega\mathbf{0Equ}$ with that of another category.

Proposition 2. *Suppose there exists a full subcategory of* $\omega\mathbf{Equ}$ *satisfying four conditions: (i) it is closed under finite products; (ii) it contains the 1-point compactification of* $\mathbb{N}$*; (iii) it contains every object* $([\sigma])_I$*; (iv) every object in the subcategory is projective with respect to the "identity"* $\mathsf{R}_I \to \mathsf{R}_E$*. Then* $(\sigma)_E = (\sigma)_I$ *for all types* σ.

We do not know whether such a subcategory exists. The difficult conditions to reconcile are (iii) and (iv). Let us pinpoint our ignorance more exactly by considering the "pure" second- and third-order types:

$$\mathsf{real}_2 \equiv (\mathsf{real} \to \mathsf{real}) \to \mathsf{real} \qquad\qquad \mathsf{real}_3 \equiv \mathsf{real}_2 \to \mathsf{real}$$

Proposition 3.

1. $(\mathsf{real}_3)_E \supseteq (\mathsf{real}_3)_I$.
2. $(\mathsf{real}_3)_E = (\mathsf{real}_3)_I$ *if and only if the object* $([\mathsf{real}_2])_I$ *is projective with respect to the identity* $\mathsf{R}_I \to \mathsf{R}_E$ *(tracked by* r*)*.

We have not succeeded in establishing whether $([\mathsf{real}_2])_I$ is projective with respect to the identity $\mathsf{R}_I \to \mathsf{R}_E$. However, we have managed to reduce this condition to a conjecture concerning the topology of the Kleene-Kreisel continuous functionals over $\mathbb{N}$ [Kle59,Kre59]. Many presentations of the continuous functionals are known, but, for our conjecture, the simplest description is as the hierarchy of simple types over $\mathbb{N}$ in the cartesian-closed category $\omega\mathbf{qTop}$, or equivalently in $\mathbf{Seq}$, or equivalently in the cartesian-closed category of compactly-generated Hausdorff spaces, see [Nor80].

Conjecture 1. The sequential space $\mathbb{N}^{\mathbb{B}}$, where $\mathbb{B} = \mathbb{N}^{\mathbb{N}}$, is zero dimensional.

Proposition 4. *If Conjecture 1 holds then* $([\mathsf{real}_2])_I$ *is projective with respect to the identity* $\mathsf{R}_I \to \mathsf{R}_E$ *and hence* $(\mathsf{real}_3)_E = (\mathsf{real}_3)_I$.

We prove Theorem 1 by reducing it to Theorem 2. This requires some work. Although we have stated that $\omega\mathbf{Equ}$ is equivalent to the full subcategory $\mathbf{Asm}(\omega\mathbf{L})$ of $\mathbf{Asm}(\omega\mathbf{BC})$, this is of no immediate help because neither $[\mathsf{real}]_E$ nor $[\mathsf{real}]_I$ resides in this subcategory. However, following [Bau00], there is a second way of viewing $\mathbf{Asm}(\omega\mathbf{L})$, and hence $\omega\mathbf{Equ}$, as (equivalent to) a full subcategory of $\mathbf{Asm}(\omega\mathbf{BC})$, under which $[\mathsf{real}]_E$ and $[\mathsf{real}]_I$ are included.

We say that an assembly A in $\mathbf{Asm}(\omega\mathbf{BC})$ is *dense* if $\mathrm{supp}(A)$ is dense in $\|A\|$ under the Scott topology, where:

$$\mathrm{supp}(A) = \{x \in \|A\| \mid \text{there exists } a \in |A| \text{ such that } x \Vdash a\}\ .$$

An assembly is *essentially dense* if it is isomorphic to a dense assembly. It holds that $\mathbf{Asm}_{\mathbf{ed}}(\omega\mathbf{BC}) \simeq \omega\mathbf{Equ}$, where $\mathbf{Asm}_{\mathbf{ed}}(\omega\mathbf{BC})$ is the full subcategory of essentially dense assemblies in $\mathbf{Asm}(\omega\mathbf{BC})$ [BBS02,Bau00].

Proposition 5. *For any type σ,*

1. $[\![\sigma]\!]_E$ *is a dense assembly, and*
2. *if* $\mathrm{order}(\sigma) \leq 1$ *then* $[\![\sigma]\!]_I$ *is an essentially dense assembly.*

Statement 1 follows from Normann's density theorem for an ω-algebraic variant of the interval domain [Nor00b]. Statement 2 will be addressed in Sect. 6.

Lemma 3. *For any type σ,*

1. $[\![\sigma]\!]_E = (\sigma)_E$, *and*
2. *if* $\mathrm{order}(\sigma) \leq 2$ *then* $[\![\sigma]\!]_I = (\sigma)_I$.

Theorem 1 follows immediately from Lemma 3 and Theorem 2.

6   Extensionalization

In this section, we prove Proposition 5.2. Our proof establishes a property of the domains underlying $[\![\sigma]\!]_I$, for first-order σ, that we call *extensionalization*. This property is of interest independent of its application to Proposition 5.2.

Following [Ber93], we define the set of *total elements* $\mathcal{T}_\tau \subseteq \|\tau\|$, where $\|\tau\|$ is the domain interpreting a **PCF** type τ. This is by:

$$\mathcal{T}_{\mathsf{nat}} = \mathbb{N} \subseteq \mathbb{N}_\perp$$
$$\mathcal{T}_{\tau_1 \times \tau_2} = \mathcal{T}_{\tau_1} \times \mathcal{T}_{\tau_2}$$
$$\mathcal{T}_{\tau_1 \to \tau_2} = \{f \in \|\tau_1 \to \tau_2\| \mid \text{for all } x \in \mathcal{T}_{\tau_1},\, f(x) \in \mathcal{T}_{\tau_2}\}$$

Recall also that, for a type σ over real, $\|[\![\sigma]\!]_I\| = \|\sigma^*\|$, where $(\cdot)^*$ is the translation to **PCF** types from Sect. 3. The proof of the next proposition uses Berger's generalization of the "KLS Theorem" [Ber93] together with a purely topological lemma.

Lemma 4. *If S is a nonempty closed subspace of a countably-based zero-dimensional space T then S is a retract of T.*

Proposition 6 (Extensionalization). *For any σ with $\mathrm{order}(\sigma) \leq 1$, the identity function on $[\![\sigma]\!]_I$ is tracked by a function $i : \|[\![\sigma]\!]_I\| \to \|[\![\sigma]\!]_I\|$ with the property that, for all $x \in \mathcal{T}_{\sigma^*}$, $i(x) \in \mathcal{T}_{\sigma^*} \cap \mathrm{supp}([\![\sigma]\!]_I)$.*

We call this result extensionalization for the following reason. As in Sect. 3, in order for an element $f \in \|[\![\text{real} \to \text{real}]\!]_I\|$ to track a morphism from $[\![\text{real}]\!]_I$ to $[\![\text{real}]\!]_I$ it must both preserve real-representing elements (i.e. it must preserve $\text{supp}([\![\text{real}]\!]_I)$) and it must also preserve the equivalence between such representations; thus one might say that it must behave "extensionally". The proposition relates such "extensional" elements of $\|[\![\text{real} \to \text{real}]\!]_I\|$ to total ones. Firstly, because i tracks the identity, it maps every extensional element f to an equivalent total extensional one. Secondly, every non-extensional but total f is mapped to an arbitrary extensional and still total element. Thus the total elements of $\|[\![\text{real} \to \text{real}]\!]_I\|$ are all "extensionalized" by i. Again we do not know whether such a process of extensionalization is also available for second-order σ and above.

Corollary 1. *If* $\text{order}(\sigma) \leq 1$ *then the identity on* $[\![\sigma]\!]_I$ *is tracked by a function* i *such that, for all* $x \in \|[\![\sigma]\!]_I\|$, $i(x)$ *is in the Scott-closure of* $\text{supp}([\![\sigma]\!]_I)$.

Proposition 5.2 follows, as the property stated in the corollary is easily seen to be sufficient to establish that $[\![\sigma]\!]_I$ is essentially dense.

7   Eliminating Parallelism

To conclude the paper, we return to our original motivation for studying the $[\![\sigma]\!]_E$ and $[\![\sigma]\!]_I$ hierarchies, namely that they correspond to the total functionals on reals definable in the two approaches to exact real-number computation. As discussed in Sect. 3, the $[\![\sigma]\!]_E$ functionals are exactly those programmable in $\Omega\textbf{RealPCF+}$, and the $[\![\sigma]\!]_I$ functionals are those programmable in $\Omega\textbf{PCF++}$. Both these languages contain parallel primitives.

   In the context of **PCF**, Normann has proved that the type hierarchies of total functionals over $\mathbb{N}$ programmable in **PCF** and **PCF++** are identical for arbitrary types [Nor00a]. By the same proof, the hierarchies of $\mathbb{N}$-functionals programmable in $\Omega\textbf{PCF}$ and $\Omega\textbf{PCF++}$ are identical. In other words, parallel primitives are unnecessary as far as programming total functionals over $\mathbb{N}$ is concerned. It is natural to ask whether a similar phenomenon of elimination of parallelism occurs also for total functionals over $\mathbb{R}$.

   For the extensional approach, the situation is unsatisfactory. In [EHS99], it is proved that there is no sequential way of implementing even the first-order function of binary addition. For this reason, core **RealPCF** contains a primitive parallel-conditional operation. However, one may still question whether the parallel existential of **RealPCF+** is required for programming total functionals. The only known result is that all second-order functionals can be defined in languages strictly weaker than **RealPCF+** [Nor02].

   Our final result is that, in the intensional approach, parallelism is eliminable up to type two. Recall, from Sect. 3, our notation for **PCF** and its semantics.

Theorem 3. *If* $\text{order}(\sigma) \leq 2$ *then, for any* $f \in [\![\sigma]\!]_I$, *there exists an* $\Omega\textbf{PCF}$ *program* P *of type* σ^* *such that* $[\![P]\!] \Vdash_{[\![\sigma]\!]_I} f$.

The proof uses extensionalization, Proposition 6, to reduce the result to Normann's result for third-order **PCF** types. The type restriction on Proposition 6 is the only obstacle to extending Theorem 3 to higher-order types.

To ease comparison with the results for the extensional case discussed above, we remark that we have also proved a version of Theorem 3 for the standard (oracle free) versions of **PCF** and **PCF++**. Specifically, a total functional on $\mathbb{R}$ is definable in **PCF** if and only if it is definable in **PCF++**. The proof involves writing **PCF** programs for the extensionalization functions i of Proposition 6. The coding details of these functions are interesting, and may appear elsewhere.

References

[AJ94]   S. Abramsky and A. Jung. Domain theory. In *Handbook of Logic in Computer Science*, volume 3, pages 1–168. OUP, 1994.

[Bau00] A. Bauer. *The Realizability Approach to Computable Analysis and Topology.* PhD thesis, Carnegie Mellon University, 2000.

[BBS02] A. Bauer, L. Birkedal, and D.S. Scott. Equilogical spaces. *Theoretical Computer Science*, to appear, 2002.

[Ber93] U. Berger. Total sets and objects in domain theory. *Annals of Pure and Applied Logic*, 60:91–117, 1993.

[Di 93] P. Di Gianantonio. *A Functional Approach to Computability on Real Numbers.* PhD thesis, Università di Pisa, 1993.

[Dug89] J. Dugundji. *Topology.* Wm. C. Brown, 1989.

[EE00]   M.H. Escardó and A. Edalat. Integration in real PCF. *Information and Computation*, 160:128–166, 2000.

[EHS99] M.H. Escardó, M. Hofmann, and Th. Streicher. *Mediation is inherently parallel.* Talk at Workshop on Domains V, Darmstadt, 1999.

[ES99]   M.H Escardó and Th. Streicher. Induction and recursion on the partial real line with applications to real PCF. *Theoretical Computer Science*, 210:121–157, 1999.

[Esc96] M.H. Escardó. PCF extended with real numbers. *Theoretical Computer Science*, 162:79–115, 1996.

[GL01]   P. Gowland and D. Lester. A survey of exact computer arithmetic. In *Computability and Complexity in Analysis*. Springer LNCS 2064, 2001.

[Has]   `http://www.haskell.org/`.

[Kle59] S.C. Kleene. Countable functionals. In *Constructivity in Mathematics*, pages 81–100. North-Holland, 1959.

[Kre59] G. Kreisel. Interpretation of analysis by means of functionals of finite type. In *Constructivity in Mathematics*, pages 101–128. North-Holland, 1959.

[Mac71] S. Mac Lane. *Categories for the Working Mathematician.* Springer-Verlag, 1971.

[MS02]   M. Menni and A.K. Simpson. Topological and limit-space subcategories of countably-based equilogical spaces. *Mathematical Structures in Computer Science*, to appear, 2002.

[Nor80] D. Normann. *Recursion on the countable functionals.* Springer LNM 811, 1980.

[Nor00a] D. Normann. Computabilty over the partial continuous functionals. *Journal of Symbolic Logic*, 65:1133–1142, 2000.

[Nor00b]   D. Normann. The continuous functionals of finite types over the reals. *Electronic Notes in Theoretical Computer Science*, 35, 2000.

[Nor02]   D. Normann. Exact real number computations relative to hereditarily total functionals. *Theoretical Computer Science*, to appear, 2002.

[Plo77]   G.D. Plotkin. LCF considered as a programming language. *Theoretical Computer Science*, 5(3):223–255, 1977.

[Sco96]   D.S. Scott. A new category? Unpublished Manuscript. Available at http://www.cs.cmu.edu/Groups/LTC/, 1996.

[Sim98]   A.K. Simpson. Lazy functional algorithms for exact real functionals. In *Mathematical Foundations of Computer Science*, pages 456–464. Springer LNCS 1450, 1998.

[WK87]   K. Weihrauch and C. Kreitz. Representations of the real numbers and of the open subsets of the set of real numbers. *Annals of Pure and Applied Logic*, 35(3):247–260, 1987.

Random Sampling from Boltzmann Principles

Philippe Duchon[1], Philippe Flajolet[2], Guy Louchard[3], and Gilles Schaeffer[4]

[1] Université Bordeaux I, 351 Cours de la Libération, F-33405 Talence, France
[2] Algorithms Project, INRIA-Rocquencourt, F-78153 Le Chesnay, France
[3] Université Libre de Bruxelles, Département d'informatique,
Boulevard du Triomphe, B-1050 Bruxelles, Belgium
[4] ADAGE Group, LORIA, F-54000 Villers-les-Nancy, France

Abstract. This extended abstract proposes a surprisingly simple framework for the random generation of combinatorial configurations based on *Boltzmann models*. Random generation of possibly complex structured objects is performed by placing an appropriate measure spread over the whole of a combinatorial class. The resulting algorithms can be implemented easily within a computer algebra system, be analysed mathematically with great precision, and, when suitably tuned, tend to be efficient in practice, as they often operate in linear time.

1 Introduction

In this text, *Boltzmann models* are proposed as a framework for the random generation of structured combinatorial configurations, like words, trees, permutations, constrained graphs, and so on. A Boltzmann model relative to a combinatorial class $\mathcal{C}$ depends on a control parameter $x > 0$ and places an appropriate measure that is spread over the whole of $\mathcal{C}$. Random objets under a Boltzmann model then have a fluctuating size, but objects with the same size invariably occur with the same probability. In particular, a *Boltzmann sampler* (i.e., a random generator that obeys a Boltzmann model), with the size of its output conditioned to be a fixed value n, draws *uniformly* at random an object of size n.

As we demonstrate in this article, Boltzmann samplers can be derived systematically (and simply) for classes that are specified in terms of a basic collection of general-purpose combinatorial constructions. These constructions are precisely the ones that surface recurrently in modern theories of combinatorial analysis; see, e.g., [2,7,9] and references therein. As a consequence, one obtains with surprising ease Boltzmann samplers covering an extremely wide range of combinatorial types.

Fixed-size generation is the standard paradigm in the random generation of combinatorial structures, and a vast literature exists on the subject. There, either specific bijections are exploited or general combinatorial decompositions are put to use in order to generate objects at random based on possibility counts—this has come to be known as the "recursive method" originating with Nijenhuis and Wilf [12] and formalized by Flajolet, Zimmermann, and Van Cutsem in [8]. In contrast, the basic principle of Boltzmann sampling is to *relax* the constraint

P. Widmayer et al. (Eds.): ICALP 2002, LNCS 2380, pp. 501–513, 2002.

of generating objects of a strictly fixed size, and prefer to draw objects with a randomly varying size. As we shall see, normally, one can *tune* the value of the control parameter x in order to favour objects of a size in the vicinity of a target value n. If needed, one can pile up a filter that rejects objects whose size is out of range. In this way, Boltzmann samplers may also serve for approximate-size as well as exact-size random generation.

We propose Boltzmann samplers as an attractive alternative to standard combinatorial generators based on the recursive method and implemented in packages like Combstruct (under the computer algebra system MAPLE) and CS (under MuPAD). Boltzmann algorithms are expected to be competitive when compared to many existing combinatorial methods: they only necessitate a small *fixed* number of multiprecision constants that are fairly easy to compute; when suitably optimized, they operate in low polynomial time—often even in linear time. Accordingly, uniform generation of objects with sizes in the range of millions is becoming a possibility, whenever the approach is applicable.

2   Boltzmann Models and Generators

We consider a class $\mathcal{C}$ of combinatorial objects of sorts, with $|\cdot|$ the size function from $\mathcal{C}$ to $\mathbb{Z}_{\geq 0}$. By $\mathcal{C}_n$ is meant the subclass of $\mathcal{C}$ comprising all the objects in $\mathcal{C}$ having size n. Each $\mathcal{C}_n$ is assumed to be finite. One may think of binary words (with size defined as length), permutations, graphs and trees of various types (with size defined as number of vertices), and so on. Any set $\mathcal{C}$ endowed with a size function and satisfying the finiteness axiom will henceforth be called a *combinatorial class*.

Definition 1. *The Boltzmann models of parameter x exist in two varieties, the ordinary version and the exponential version. They assign to any object $\gamma \in \mathcal{C}$ the following probability:*

$$\text{Ordinary/Unlabelled case:} \quad \mathbb{P}_x(\gamma) = \frac{1}{C(x)} \cdot x^{|\gamma|} \quad \text{with} \quad C(x) = \sum_{\gamma \in \mathcal{C}} x^{|\gamma|},$$

$$\text{Exponential/Labelled case:} \quad \mathbb{P}_x(\gamma) = \frac{1}{\widehat{C}(x)} \cdot \frac{x^{|\gamma|}}{|\gamma|!} \quad \text{with} \quad \widehat{C}(x) = \sum_{\gamma \in \mathcal{C}} \frac{x^{|\gamma|}}{|\gamma|!}.$$

A Boltzmann generator (or sampler) $\Gamma C(x)$ for a class $\mathcal{C}$ is a process that produces objects from $\mathcal{C}$ according to a Boltzmann model.

The normalization coefficients are nothing but the counting generating functions of ordinary type (OGF) for $C(x) := \sum_n C_n x^n$ and exponential type (EGF) for $\widehat{C}(x) := \sum_n C_n x^n/n!$. Only *coherent* values of x defined to be such that $0 < x < \rho_C$ (or $\rho_{\widehat{C}}$), with ρ_f the radius of convergence of f are to be considered.

The name "Boltzmann model" comes from the great statistical physicist Boltzmann whose works (together with those of Gibbs) led to enounce the following principle: *Statistical mechanical configurations of energy equal to E in a system have a probability of* *occurrence proportional to* $e^{-\beta E}$. (There, β is an inverse temperature.) If one identifies size of a combinatorial configuration with energy of a thermodynamical system and sets $x = e^{-\beta}$, then what we term the ordinary Boltzmann models become the true model of statistical mechanics. The counting generating function in the combinatorial world then coincides with the normalization constant in the statistical mechanics world where it is known as the *partition function* and is often denoted by Z. Under this perhaps artificial dictionary, Boltzmann models and random combinatorics become united.

For reasons which will become apparent, we have also defined the exponential Boltzmann model. These are appropriate for handling *labelled* combinatorial structures while the ordinary models are to be used for *unlabelled* combinatorial models. In the unlabelled universe, all elementary components of objects ("atoms") are indistinguishable, while in the labelled universe, they are all distinguished from one another by bearing a distinctive mark, say one of the integers between 1 and n if the object considered has size n. (This terminology is standard in combinatorial enumeration and graph theory [2,7,9].)

The size of the resulting object under a Boltzmann model is a random variable, denoted throughout by N, whose law is quantified by the following lemma.

Proposition 1. *The random size of the object produced under the ordinary Boltzmann model of parameter x satisfies*

$$\mathbb{E}_x(N) = x\frac{C'(x)}{C(x)}, \qquad \mathbb{E}_x(N^2) = \frac{x^2 C''(x) + x C'(x)}{C(x)}. \tag{1}$$

Proof. By construction the probability of drawing an object of size n is $\mathbb{P}_x(N = n) = C_n x^n / C(x)$. Consequently, the probability generating function of N is $C(xz)/C(x)$ and the result follows.

In the next two sections (Sections 3 and 4), we develop a collection of rules by which one can assemble Boltzmann generators from simpler ones. The combinatorial classes considered are built on a small set of constructions that have a wide expressive power. The language in which classes are specified is in essence the same as the one underlying the recursive method [6,8]: it consists of the constructions of union, product, sequence, set, and cycle. For each allowable class, a Boltzmann sampler can be built in an entirely systematic manner.

3   Ordinary Boltzmann Generators

A *combinatorial construction* builds a new class $\mathcal{C}$ from structurally simpler classes $\mathcal{A}, \mathcal{B}$, in such a way that $\mathcal{C}_n$ is determined from objects in $\{\mathcal{A}_j\}_{j=0}^n, \{\mathcal{B}_j\}_{j=0}^n$. Constructions considered here are disjoint *union* (+), cartesian *product* ($\times$), and *sequence* formation ($\mathfrak{S}$). We define these in turn and concurrently build the corresponding Boltzmann sampler ΓC for the composite class $\mathcal{C}$, given random generators $\Gamma A, \Gamma B$ for the ingredients and assuming the values of intervening generating functions $A(x), B(x)$ at x to be known exactly.

Disjoint union. Write $C = A + B$ if C is the union of disjoint copies of A and B, while size on C is inherited from A, B. One has $C(x) = A(x) + B(x)$. The Boltzmann model corresponding to $C(x)$ is then a mixture of the models associated to $A(x)$ and $B(x)$, with the probability of selecting a particular γ in C being $\mathbb{P}(\gamma \in A) = A(x)/C(x)$, $\mathbb{P}(\gamma \in B) = A(x)/C(x)$. Let us be given a generator for a Bernoulli variable Bern(p) defined as follows: Bern$(p) = 1$ with probability p; Bern$(p) = 0$ with probability $1 - p$; a sampler ΓC given ΓA and ΓB is simply obtained by

function $\Gamma C(x : \text{real})$; let $p_A := A(x)/(A(x) + B(x))$;
if Bern(p_A) then return$(\Gamma A(x))$ else return$(\Gamma B(x))$ fi.

Cartesian Product. Write $C = A \times B$ if C is the set of ordered pairs from A and B, and size on C is inherited additively from A, B. For generating functions, one finds $C(x) = A(x) \cdot B(x)$. A random element of $C(x)$ is then obtained by forming a pair $\langle \alpha, \beta \rangle$ with α, β drawn *independently* from the Boltzmann models $A(x), B(x)$, respectively:

function $\Gamma C(x : \text{real})$; return$(\langle \Gamma A(x), \Gamma B(x) \rangle)$ {independent calls}.

Sequences. Write $C = \mathfrak{S}(A)$ if C is composed of all the finite sequences of elements of A. The sequence class C is also the solution to the symbolic equation $C = 1 + AC$ (with **1** the empty sequence), which only involves unions and products. Consequently, one has $C(x) = (1 - A(x))^{-1}$. This gives rise to a recursive generator for sequences. Once recursion is unwound, the resulting generator assumes a particularly simple form:

function $\Gamma C(x : \text{real})$; let $A(x)$ be the value of the OGF of A;
draw K according to Geometric$(A(x))$;
return the K-tuple $\langle \Gamma A(x), \Gamma A(x), ..., \Gamma A(x) \rangle$ {independent calls}.

Finite sets. There finally remains to discuss initialization (when and how do we stop?). Clearly if C is finite (and in practice small), one can generate a random element of C by selecting it according to the finite probability distribution given explicitly by the definition of the Boltzmann model.

Proposition 2. *Define as* specifiable *an unlabelled class that can be specified (in a possibly recursive way) from finite sets by means of disjoint unions, cartesian products, and the sequence construction. Let C be an unlabelled specifiable class. Let x be a "coherent" parameter in $(0, \rho_C)$, and let ΓC be the generator compiled from the definition of C by means of the three rules above. Then ΓC correctly draws elements from C according to the ordinary Boltzmann model. It halts with probability 1 and in finite expected time.*

Example 1. Words without long runs. Consider the collection $\mathcal{R}$ of binary words over the alphabet $A = \{a, b\}$ such that they never have more than m consecutive occurrences of any letter. The set W of all words is expressible by a regular expression written in our notation $W = \mathfrak{S}(b) \times \mathfrak{S}(a\mathfrak{S}(a)b\mathfrak{S}(b)) \times \mathfrak{S}(a)$. This expresses the fact that any word has a "core" formed with blocks of a's and blocks of b's in alternation that is bordered by a header of b's and a trailer of a's. The decomposition serves for $\mathcal{R}$: e.g., replace any internal $a\mathfrak{S}(a)$ by $\mathfrak{S}_{1..m}(a)$ and any $b\mathfrak{S}(b)$ by $\mathfrak{S}_{1..m}(b)$, where $\mathfrak{S}_{1..m}$ means a sequence of between 1 and m elements. The composition rules given above give rise to a generator for $\mathcal{R}$ of the following form: two generators produce sequences of a's or b's according to a truncated geometric law; a generator for the product $\mathcal{C} := (\mathfrak{S}_{1..m}(a)\mathfrak{S}_{1..m}(b))$ is built according to the product rule; a generator for the "core" sequence $\mathcal{D} :=$ $\mathfrak{S}(\mathcal{C})$ is constructed according to the sequence rule. The generator assembled *automatically* from the general rules is then

$$\mathrm{Geom}_{\leq m}(x; b)\left\{ \mathrm{Geom}\left[\tfrac{x^2(1-x^m)^2}{(1-x)^2}\right] \circ \left\langle \mathrm{Geom}_{1..m}(x; a), \mathrm{Geom}_{1..m}(x; b)\right\rangle \right\} \mathrm{Geom}_{\leq m}(x; a).$$

Example 2. Trees (rooted, plane). Take first the class $\mathcal{B}$ of binary trees defined by the recursive specification $\mathcal{B} = \mathcal{Z} + (\mathcal{Z} \times \mathcal{B} \times \mathcal{B})$, where $\mathcal{Z}$ is the class comprising the generic node. The generator $\Gamma\mathcal{Z}$ is deterministic and consists simply of the instruction "output a node" (since $\mathcal{Z}$ is finite and in fact has only one element). The Boltzmann generator $\Gamma\mathcal{B}$ calls $\Gamma\mathcal{Z}$ (and halts) with probability $x/B(x)$ where $B(x)$ is the OGF of binary trees, $B(x) = (1 - \sqrt{1 - 4x^2})/(2x)$. With the complementary probability corresponding to the strict binary case, it will make a call to $\Gamma\mathcal{Z}$ and two recursive calls to itself. In other words: *the Boltzmann generator for binary trees as constructed automatically from the composition rules produces a random sample of the (subcritical) branching process with probabilities $x/B(x)$, $xB(x)^2/B(x)$.* Unbalanced 2-3 trees are similarly produced from $\mathcal{U} = \mathcal{Z} + \mathcal{U}^2 + \mathcal{U}^3$, unary-binary trees from $\mathcal{V} = \mathcal{Z}(1 + \mathcal{V} + \mathcal{V}^2)$, etc.

Example 3. Secondary structures. This example is inspired by the works of Waterman *et al.*, themselves motivated by the problem of enumerating secondary RNA structures. To fix ideas, consider rooted binary trees where edges contain 2 or 3 atoms and leaves ("loops") contain 4 or 5 atoms. A specification is $\mathcal{S} = (\mathcal{Z}^4 + \mathcal{Z}^5) + (\mathcal{Z}^2 + \mathcal{Z}^3)^2 \times (\mathcal{S} \times \mathcal{S})$. A Bernoulli switch will decide whether to halt or not, two independent recursive calls being made in case it is decided to continue, with the algorithm being sugared with suitable Bernoulli draws. The method is clearly universal for this entire class of problems.

4   Exponential Boltzmann Generators

We consider here *labelled structures* in the precise technical sense of combinatorial theory; see, e.g., [7]. A labelled object of size n is then composed of n distinguishable atom, each bearing a distinctive label that is an integer in the interval $[1, n]$. Labelled combinatorial classes can be subjected to the *labelled product* defined as follows: if $\mathcal{A}$ and $\mathcal{B}$ are labelled classes, the product $\mathcal{C} = \mathcal{A} \star \mathcal{B}$ is obtained by forming all ordered pairs $\langle \alpha, \beta \rangle$ with $\alpha \in \mathcal{A}$ and $\beta \in \mathcal{B}$ and relabelling them in all possible order-consistent ways. From the definition, a binomial convolution $C_n = \sum_{k=0}^{n} \binom{n}{k} A_k B_{n-k}$, takes care of relabellings. In terms of exponential generating functions, this becomes $\widehat{C}(z) = \widehat{A}(z) \cdot \widehat{B}(z)$.

Like in the ordinary case, we proceed by assembling Boltzmann generators for structured objects from simpler ones.

Disjoint union. The unlabelled construction carries over verbatim.

Labelled product. The cartesian product construction adapts to this case: in order to produce an element from $C = \mathcal{A} \star \mathcal{B}$, simply produce an independent pair by the cartesian product rule, but using the EGF values $\widehat{A}(x), \widehat{B}(x)$.

Sequences. In the labelled universe, $\mathcal{C}$ is the sequence class of $\mathcal{A}$, written $\mathcal{C} = \mathfrak{S}(\mathcal{A})$ iff it is composed of all the sequences of elements from A up to order-consistent relabellings. Then, the EGF relation $\widehat{C}(x) = (1 - \widehat{A}(x))^{-1}$ holds, and the sequence construction of the generator ΓC from ΓA given in Section 3 and based on the geometric law is applicable.

Sets. This is a new construction that we did not consider in the unlabelled case. The class $\mathcal{C}$ is the set-class of $\mathcal{A}$, written $\mathcal{C} = \mathfrak{P}(\mathcal{A})$ ($\mathfrak{P}$ is reminiscent of "powerset") if $\mathcal{C}$ is the quotient of $\mathfrak{S}\{\mathcal{A}\}$ by the relation that declares two sequences as equivalent if one derives from the other by an arbitrary permutation of the components. It is then easily seen that the EGFs are related by $\widehat{C}(x) = \sum_{k \geq 0} \widehat{A}(x)^k / k! = e^{\widehat{A}(x)}$, where the factor $1/k!$ "kills" the order present in sequences. A moment of reflection shows that, under the exponential Boltzmann model, the probability for a set in $\mathcal{C}$ to have k components is $e^{-\widehat{A}(x)} \widehat{A}(x)^k / k!$, that is, a Poisson law of rate $\widehat{A}(x)$. This gives rise to a simple algorithm for generating sets (analogous to the geometric algorithm for sequences):

> function $\Gamma C(x : \text{real})$; let $\widehat{A}(x)$ be the value of the EGF of $\mathcal{A}$;
> draw K according to Poisson$(\widehat{A}(x))$;
> return the K-tuple $\langle \Gamma A(x), \Gamma A(x), ..., \Gamma A(x) \rangle$ {independent calls}.

Cycles. This construction, written $\mathcal{C} = \mathfrak{C}(\mathcal{A})$, is defined like sets but with two sequences being identified if one is a cyclic shift of the other. The EGFs satisfy $\widehat{C}(x) = \sum_{k \geq 0} \widehat{A}(x)^k / k = \log(1 - \widehat{A}(x))^{-1}$. The log-law (also known as "logarithmic series distribution") of rate $\lambda < 1$, is defined by $\mathbb{P}(X = k) = (-\log(1 - \lambda))^{-1} \lambda^k / k$. Then cycles under the exponential Boltzmann model can be drawn like in the case of sets upon replacing the Poisson law by the log-law.

Proposition 3. *Define as* specifiable *a labelled class that can be specified (in a possibly recursive way) from finite sets by means of disjoint unions, cartesian products, as well as sequence, set and cycle constructions. Let $\mathcal{C}$ be a labelled specifiable class. Let x be a "coherent" parameter in $(0, \rho_{\widehat{C}})$, and let ΓC be the generator compiled from the definition of $\mathcal{C}$ by means of the five rules above. Then ΓC correctly draws elements from $\mathcal{C}$ according to the exponential Boltzmann model. It halts with probability 1 and in finite expected time.*

Example 4. Set partitions. A set partition of size n is a partition of the integer interval $[1, n]$ into a certain number of nonempty classes, also called blocks, the blocks being by definition unordered between themselves. Let $\mathfrak{P}_{\geq 1}$ represent the powerset construction where the number of components is constrained to be ≥ 1. The labelled class of all set partitions is then definable as $\mathcal{S} = \mathfrak{P}(\mathfrak{P}_{\geq 1}(\mathcal{Z}))$, where $\mathcal{Z}$ consists of a single labelled atom, $\mathcal{Z} = \{1\}$. The EGF of $\mathcal{S}$ is the well-known generating function of the Bell numbers, $\widehat{S}(x) = e^{e^x - 1}$. By the composition rules, a random generator is as follows: *Choose the number K of blocks as $Poisson(e^x - 1)$. Draw K independent copies $X_1, X_2, \ldots, X_K$ from the Poisson law of rate x, each conditioned to be at least 1.*

Example 5. Random surjections (or ordered set partitions). These may be defined as functions from $[1, n]$ to $[1, n]$ such that the image of f is an initial segment of $[1, n]$ (i.e., there are no "gaps"). One has for the class $\mathcal{Q}$ of surjections $\mathcal{Q} = \mathfrak{S}(\mathfrak{P}_{\geq 1}(\mathcal{Z}))$. Thus a random generator for $\mathcal{Q}$ first chooses a number of components $K \in \mathrm{Geom}\,(e^x - 1)$ and then launches K Poisson generators.

Example 6. Cycles in permutations. This corresponds to $\mathcal{P} = \mathfrak{P}(\mathfrak{C}_{\geq 1}(\mathcal{Z}))$ and is obtained by a (Poisson∘Log) process. (This example is loosely related to the Shepp–Lloyd model that generates permutations by ordered cycle lengths.)

Example 7. Assemblies of filaments in a liquid. We may model these as sets of sequences, $\mathcal{F} = \mathfrak{P}(\mathfrak{S}_{\geq 1}(\mathcal{Z}))$. The EGF is $\exp(z/(1 - z))$. The random generation algorithm is a compound of the form (Poisson∘Geometric), with appropriate parameters. (See A000262 in Sloane's encyclopedia [14].)

5   The Realization of Boltzmann Samplers

In this section, we examine the way Boltzmann sampling can be implemented and sketch a discussion of complexity issues involved. In this abstract, only the *real-arithmetic model* ($\mathbb{R}$) is considered. There, what is assumed to be given is a random-access machine with unit cost for (exact) real arithmetic operations and elementary transcendental functions over the real numbers.

By definition, a Boltzmann sampler requires as input the value of the control parameter x that defines the Boltzmann model of use. As seen in previous sections, it also needs the finite collection of values at x of the generating functions that intervene in a specification. We assume these values to be provided by what we call the (generating function) *"oracle"*. Such constants, which need only be precomputed *once*, are likely to be provided by a multiprecision package or a computer algebra system used as coroutine.

First one has to specify fully generators for the probabilistic laws Geom (λ), Pois (λ), Loga (λ), as well as the Bernoulli generator Bern (p), where the latter outputs 1 with probability p and 0 otherwise. A random generator 'uniform ()'

produces at unit cost a random variable uniformly distributed over the real interval $(0, 1)$.

Bernoulli generator. The Bernoulli generator is simply
$$\mathrm{Bern}\,(p) := \text{if uniform}\,() \le p \text{ then return}(1) \text{ else return}(0) \text{ fi.}$$
This generator serves in particular to draw from unions of classes.

Geometric, Poisson, and Logarithmic generators. For the remaining laws, we let p_k be the probability that a random variable with the desired distribution has value k, namely,

$$\mathrm{Geom}\,(\lambda): \; (1 - \lambda)\lambda^k; \quad \mathrm{Pois}\,(\lambda): \; e^{-\lambda}\frac{\lambda^k}{k!}; \quad \mathrm{Loga}\,(\lambda): \; \frac{1}{\log(1 - \lambda)^{-1}}\frac{\lambda^k}{k}.$$

The general scheme that goes well with real-arithmetic models is the *sequential algorithm*:

$U := \text{uniform}\,()$; $S := 0$; $k := 0$;
while $U < S$ do $S := S + p_k$; $k := k + 1$; od; return(k).

This scheme is nothing but a straightforward implementation based on inversion of distribution functions (see [4, Sec. 2.1]). For the three distributions under consideration, the probabilities p_k can themselves be computed recurrently on the fly. In particular, under the model that has unit cost for real arithmetic operations and functions, the sequential generators have a useful property: *a variable with outcome k is drawn with a number of operations that is $O(k + 1)$*. This has immediate consequences for all classes that are specifiable in the sense of Propositions 2 and 3.

Theorem 1. *Consider a specifiable class $\mathcal{C}$, either labelled or unlabelled. Assume as given an oracle that provides the finite collection of exact values of the intervening generating functions at a coherent value x. Then, the Boltzmann generator $\Gamma C(x)$ has a complexity in the number of real-arithmetic operations that is linear in the size of its output object.*

The linear complexity in the abstract model $\mathbb{R}$, as expressed in Theorem 1, provides an indication of the broad type of complexity behaviour one may aim for in practice, namely linear-time complexity. For instance, one may realize a Boltzmann sampler by truncating real numbers to some fixed precision, say using floating point numbers represented on 64 bits or 128 bits. The resulting samplers operate in time linear in the size of the output, though they may fail (by lack of digits in values of generating functions) in a small number of cases, and accordingly must deviate (slightly) from uniformity. Pragmatically, such samplers are likely to suffice for most medium-size simulations.

A sensitivity analysis of truncated Boltzmann samplers would be feasible, though rather heavy to carry out. One could even correct perfectly the lack of uniformity by appealing to an adaptive precision strategy based on guaranteed multiprecision floating point arithmetic. (The reader may get a feeling of the type of analysis involved by referring to the papers by Denise, Zimmermann, and Dutour, e.g., [3], where a thorough examination of the recursive method under this angle has been conducted.) In the full paper [5], we shall discuss bit-level implementations of Boltzmann samplers (see Knuth and Yao's insightful work [10] for context), as well as implementation issues raised by the oracle.

6   Exact-Size and Approximate-Size Sampling

Our primary objective in this article is the fast random generation of objects of some large size. Two types of constraints on size are considered. In *exact-size* random sampling, objects of $\mathcal{C}$ should be drawn uniformly at random from the subclass $\mathcal{C}_n$ of objects of size *exactly* n. In *approximate-size* random sampling, objects should be drawn with a size in an interval of the form $[n(1-\varepsilon), n(1+\varepsilon)]$, for some quantity $\varepsilon \geq 0$ called the (relative) *tolerance*, with two objects of the same size still being equally likely to occur. The conditions of exact and approximate-size sampling are immediately satisfied if one filters the output a Boltzmann generator by *rejecting* the elements that do not obey the desired size constraint. The main question is when and how can this rejection process be made reasonably efficient. The major conclusion from this and the next section is as follows: in many cases, including all the examples seen so far, *approximate-size sampling is achievable in linear time under the real-arithmetic model* of Theorem 1. The constants appear to be not too large if a "reasonable" tolerance on size is allowed.

The outcome of a basic Boltzmann sampler has a random size N whose distribution is exactly described by Proposition 1. First, for the rejection sampler tuned at the "natural" value $x = x_n$ such that $\mathbb{E}_{x_n}(N) = n$, a direct application of Chebyshev's inequalities gives:

Theorem 2. *Let $\mathcal{C}$ be a specifiable class and ε a fixed nonzero tolerance on size. Assume the following Mean Value and Variance Conditions,*

$$\lim_{x \to \rho^-} \mathbb{E}_x(N) = +\infty, \qquad \lim_{x \to \rho^-} \frac{\sqrt{\mathbb{E}_x(N^2) - \mathbb{E}_x(N)^2}}{\mathbb{E}_x(N)} = 0. \qquad (2)$$

Then, the rejection sampler equipped with the value $x = x_n$ defined by the inversion relation $x_n C'(x_n)/C(x_n) = n$ succeeds in one trial with probability tending to 1 as $n \to \infty$. Its total cost is $O(n)$ on average.

The mean and variance conditions *are* satisfied by the class $\mathcal{S}$ of set partitions (Example 4) and the class $\mathcal{F}$ of assemblies of filaments (Example 7).

It is possible to discuss at a fair level of generality cases where rejection sampling is efficient, even though the strong moment conditions of Theorem 2 may not hold. The discussion is fundamentally based on the types of singularities that the generating functions exhibit. This is an otherwise well-researched topic as it is central to asymptotic enumeration [7,13].

Theorem 3. *Let $\mathcal{C}$ be a combinatorial class that is specifiable. Assume that the generating function $C(z)$ (for $z \in \mathbb{C}$) has an isolated singularity at ρ, which is the*

*unique dominant singularity. Assume also that the singular expansion of $C(z)$
at ρ is of the form (with P a polynomial)*

$$C(z) \underset{z \to \rho}{\sim} P(z) + c_0(1 - z/\rho)^{-\alpha} + o((1 - z/\rho)^{-\alpha}). \qquad (3)$$

When the exponent $-\alpha$ is negative, for any fixed nonzero tolerance ε, the rejection sampler corresponding to $x = x_n$ succeeds in an expected number of trials asymptotic to the constant

$$\frac{1}{\xi_\alpha(\varepsilon)}, \quad where \quad \xi_\alpha(\varepsilon) = \frac{\alpha^\alpha}{\Gamma(\alpha)} \int_{-\varepsilon}^{\varepsilon} (1 + s)^{\alpha-1} e^{\alpha(1+s)} \, ds.$$

Moreover the total cost of this rejection sampler is $O(n)$ on average.

Words without long runs, surjections, and permutations (Examples 1, 5, and 6) have generating functions with a polar singularity, corresponding to the singular exponent -1, and hence satisfy the conditions above.

We note here that a condition $-\alpha < 0$ can often be ensured by successive differentiations of generating functions. Combinatorially, this corresponds to a *"pointing"* construction. Boltzmann sampling combined with pointing and rejection is developed in the full article [5] as a viable optimization technique.

7   Singular Boltzmann Samplers

We now discuss two infinite categories of models, where it is of advantage to place oneself right at the singularity $x = \rho_C$ in order to develop a rejection sampler from a Boltzmann model for C. One category covers several of the sequence constructions, the other one corresponds to a wide set of recursive specifications.

Singular samplers for sequences. Define a sequence construction $C = \mathfrak{S}(A)$ to be supercritical if $\rho_A > \rho_C$. The generating function of C and A satisfy $C(x) = (1 - A(x))^{-1}$, so that the supercriticality condition corresponds to $A(\rho_C) = 1$, with the (dominant) singularity ρ_C of $C(x)$ being necessarily a pole.

Theorem 4. *Consider a sequence construction $C = \mathfrak{S}(A)$ that is supercritical. Generate objects from A sequentially according to $\Gamma A(\rho_C)$ until the total size becomes at least n. With probability tending to 1 as $n \to \infty$, this produces a random C object of size $n + O(1)$ in one trial. Exact-size random generation is achievable from this generator by rejection in expected time $O(n)$.*

This theorem applies to "cores" of words without long runs (from Example 1) and surjections (Example 5), for which exact-size generation become possible in linear time. It also provides a global setting for a variety of *ad hoc* algorithms developed by Louchard in the context of efficient generation of certain types (directed, convex) of random planar diagrams known as "animals" and "polyominos".

Example 8. Coin fountains (𝒪). These were enumerated by Odlyzko and Wilf. They correspond to Dyck paths taken according to area (disregarding length). The OGF is the continued fraction $O(z) = 1 \big/ (1 - z \big/ (1 - z^2 \big/ (1 - z^3 \big/ (\cdots))))$. At top level, the singular Boltzmann sampler of Theorem 4 applies (write $\mathcal{O} = \mathfrak{S}(\mathcal{Q})$ and $O(z) = (1 - Q(z))^{-1}$). The root ρ of $Q(z) = 1$ is easily found to high precision as $\rho = 0.5761487691 \cdots$. The objects of $\mathcal{Q}$ needed are with high probability of size at most $O(\log n)$, so that they can be generated by whichever subexponential method is convenient. The overall (theoretical and practical) complexity is $O(n)$ with *very* low implementation constants. Random generation well in the range of millions is now easy thanks to the singular Boltzmann generator.

Singular samplers for recursive structures. What we call a recursive class $\mathcal{C}$ is the component $\mathcal{C} = \mathcal{F}_1$ of a system of mutually dependent equations:

$$\{\mathcal{F}_1 = \Psi_1(\mathcal{Z}; \mathcal{F}_1, \ldots, \mathcal{F}_m), \ldots, \mathcal{F}_m = \Psi_m(\mathcal{Z}; \mathcal{F}_1, \ldots, \mathcal{F}_m)\}$$

where the Ψ's are *any* functional term involving *any* constructor defined previously ('+', '×' or '⋆', and $\mathfrak{S}, \mathfrak{P}, \mathfrak{C}$) The system is said to be irreducible if the dependency graph between the $\mathcal{F}_j$ is strongly connected (everybody depends on everybody else). In such a case, the singular type of the generating functions is a square-root, as follows from a mild generalization of a famous theorem by Drmota, Lalley, and Woods; see [7, Ch. 8] and references therein. A consequence is that coefficients of generating functions are of the universal form $\rho^{-n}n^{-3/2}$. In particular objects of a small size are likely to be produced by the singular generator $\Gamma C(\rho_C)$ whereas the expectation of size $\mathbb{E}_{\rho_C}(N)$ is infinite. (In other words, a very high dispersion of sizes is observed.) The singular sampler considered here simply uses the singular value $\rho = \rho_C$ together with an "early-abort" strategy: it aborts its execution as soon as the size of the partially generated object exceeds the tolerance upper bound. The process is repeated till an object within the tolerance bounds is obtained.

Theorem 5. *Let $\mathcal{C}$ be a combinatorial class given by a recursive specification that is irreducible and aperiodic. For any fixed nonzero tolerance ε, the "early-abort" rejection sampler succeeds in a number of trials that is $O(n^{1/2})$ on average. Furthermore, the total cost K_n of this sampler satisfies*

$$\mathbb{E}(K_n) \sim \frac{n}{\varepsilon}\left((1 - \varepsilon)^{1/2} + (1 + \varepsilon)^{1/2}\right). \tag{4}$$

For exact-size generation, the "early-abort" rejection sampler has complexity $O(n^2)$.

The early-abort sampler thus gives linear-time approximate-size random generation for all the simple varieties of trees of Example 2 (including binary trees, unary-binary trees, 2–3 trees, and so on) and for secondary structures (Example 3). For all these cases, exact-size is also achievable in quadratic time. The method is roughly comparable to drawing from a suitably dimensioned critical branching process in combination with abortion and rejection.

The rejection algorithm above is akin to the "Florentine algorithm" invented by Barcucci–Pinzani–Sprugnoli [1] to generate prefixes of Motzkin words and certain directed plane animals. The cost analysis is related to Louchard's work [11].

8   Conclusions

As shown here, combinatorial decompositions allow for random generation in low polynomial time. In particular, approximate-size random generation is often of a linear time complexity. Given the large number of combinatorial decompositions that have been gathered over the past two decades (see, e.g., [2,7,9], we estimate to perhaps a hundred the number of classical combinatorial structures that are amenable to efficient Boltzmann sampling. In contrast with the recursive method [3,8,12], memory requirements are kept to a minimum since only a table of constants of size $O(1)$ is required.

In forthcoming works starting with [5], we propose to demonstrate the versatility of Boltzmann sampling including: the generation of unlabelled multisets and powersets, the encapsulation of constructions like substitution and pointing, and the realization of Boltzmann samplers at bit-level. (Linear boolean complexity seems to be achievable in many cases of practical interest.)

Acknowledgements. The authors are grateful to Brigitte Vallée for several architectural comments on an early version of this manuscript. Thanks also to Bernard Ycart, Jim Fill, Marni Mishna, and Paul Zimmermann for encouragements and constructive observations. This work was supported in part by the ALCOM-FT Project IST-1999-14186 of the European Union.

References

1. BARCUCCI, E., PINZANI, R., AND SPRUGNOLI, R. The random generation of directed animals. *Theoretical Computer Science 127*, 2 (1994), 333–350.
2. BERGERON, F., LABELLE, G., AND LEROUX, P. *Combinatorial species and tree-like structures.* Cambridge University Press, Cambridge, 1998. Translated from the 1994 French original by Margaret Readdy, With a foreword by Gian-Carlo Rota.
3. DENISE, A., AND ZIMMERMANN, P. Uniform random generation of decomposable structures using floating-point arithmetic. *Theoretical Computer Science 218*, 2 (1999), 233–248.
4. DEVROYE, L. *Non-Uniform Random Variate Generation.* Springer Verlag, 1986.
5. DUCHON, P., FLAJOLET, P., LOUCHARD, G., AND SCHAEFFER, G. Boltzmann samplers for random combinatorial generation. In preparation, 2002.
6. FLAJOLET, P., SALVY, B., AND ZIMMERMANN, P. Automatic average–case analysis of algorithms. *Theoretical Computer Science 79*, 1 (Feb. 1991), 37–109.
7. FLAJOLET, P., AND SEDGEWICK, R. *Analytic Combinatorics.* 2001. Book in preparation: Individual chapters are available as INRIA Research Reports 1888, 2026, 2376, 2956, 3162, 4103 and electronically under
http://algo.inria.fr/flajolet/Publications/books.html.

8. FLAJOLET, P., ZIMMERMANN, P., AND VAN CUTSEM, B. A calculus for the random generation of labelled combinatorial structures. *Theoretical Computer Science 132*, 1-2 (1994), 1–35.

9. GOULDEN, I. P., AND JACKSON, D. M. *Combinatorial Enumeration*. John Wiley, New York, 1983.

10. KNUTH, D. E., AND YAO, A. C. The complexity of nonuniform random number generation. In *Algorithms and complexity (Proc. Sympos., Carnegie-Mellon Univ., Pittsburgh, Pa., 1976)*. Academic Press, New York, 1976, pp. 357–428.

11. LOUCHARD, G. Asymptotic properties of some underdiagonal walks generation algorithms. *Theoretical Computer Science 218*, 2 (1999), 249–262.

12. NIJENHUIS, A., AND WILF, H. S. *Combinatorial Algorithms*, second ed. Academic Press, 1978.

13. ODLYZKO, A. M. Asymptotic enumeration methods. In *Handbook of Combinatorics*, R. Graham, M. Grötschel, and L. Lovász, Eds., vol. II. Elsevier, Amsterdam, 1995, pp. 1063–1229.

14. SLOANE, N. J. A. *The On-Line Encyclopedia of Integer Sequences*. 2000. Published electronically at http://www.research.att.com/~njas/sequences/.

On the Average Performance of Orthogonal Range Search in Multidimensional Data Structures*

Amalia Duch and Conrado Martínez

[1] Laboratorio Nacional de Informática Avanzada (LANIA), Xalapa, Mexico.
amalia@lania.mx
[2] Departament de Llenguatges i Sistemes Informàtics, Universitat Politècnica de Catalunya, E-08034 Barcelona, Spain. conrado@lsi.upc.es.

Abstract. In this work we present the average-case analysis of orthogonal range search for several multidimensional data structures. We first consider random relaxed K-d trees as a prototypical example. Later we extend these results to many different multidimensional data structures. We show that the performance of range searches is related to the performance of a variant of partial matches using a mixture of geometric and combinatorial arguments. This reduction simplifies the analysis and allows us to give exact lower and upper bounds for the performance of range searches. Furthermore, under suitable conditions ("small range queries"), we can also get a very precise asymptotic estimate for the expected cost of range searches.

1 Introduction

Orthogonal range search appears frequently in applications of large databases, geographical information systems, multimedia databases and computer graphics, among others [1]. Given a collection of multidimensional data points and a query rectangle, the goal of an orthogonal range search (range search, for short) is to retrieve all the data points in the collection that fall inside the given rectangle. Apart from the applications of range search as such, it is implicitly involved in more complex region queries and other associative queries.

Many data structures have been proposed for the management of multidimensional data and specifically for range search (see for instance [2]). Among these, K-d trees [3], quadtrees [4], K-d tries [5] and multiple variants of these.

However, the mathematical analysis of the performance of range searches has proven a difficult task. The original analysis by Bentley *et al.* and most subsequent work (e.g. [6,7]) rely on the unrealistic assumption that the considered tree data structure is perfectly balanced, which often yields unduly optimistic

---

* This research was supported by project DGES PB98-0926 (AEDRI) of the Spanish Ministery for Education and Science. The second author was also supported the Future and Emergent Technologies programme of the EU under contract IST-1999-14186 (ALCOM-FT).

P. Widmayer et al. (Eds.): ICALP 2002, LNCS 2380, pp. 514–524, 2002.

results. Only recently, there has been remarkable progresses on this direction with two recent papers [8,9] that provide upper (Ω) and lower bounds (big-Oh) for the average performance of range search in standard K-d trees, squarish K-d trees (a variant of the former introduced in [9]) and other multidimensional data structures.

In this work we analyze the average cost of range queries using the same random model as in [8,9] and obtain sharper results: in particular, exact upper and lower bounds (Theorems 1 and 2) on the performance of range searches, and tight asymptotic estimates of the average performance when query hyperrectangles are "small" enough (Theorem 4). We obtain these results using a combination of geometric and combinatorial arguments. Our proof techniques are rather different from those in [8,9], but they are also easily applicable to many multidimensional data structures. We analyze first the average cost of range search in randomized K-d trees [10] and later discuss how our results generalize to other multidimensional data structures. We begin reviewing random(ized) relaxed K-d trees and orthogonal range queries in Section 2, as well as the random models used in the sequel. In Sect. 3 we introduce *sliced partial matches* and relate the performance of range search with the performance of sliced partial matches; we use this relationship to provide a tight asymptotic analysis of the average cost of range search. In Sect. 4 we show that the results for randomized K-d trees can be easily extended to most tree-like multidimensional data structures, namely, standard K-d trees, squarish K-d trees, K-d-t trees, standard and relaxed K-d tries, quadtrees and quadtries.

Furthermore, we have conducted a preliminary experimental study whose results are in total accordance with the theoretical analysis described in this paper. For details about this experimental study please see [11]; due to space limitations, we do not include them in this extended abstract. We conclude in Sect. 5 with some final remarks and future work.

2   Basic Definitions

Let $F = \{x^{(1)}, x^{(2)}, \ldots, x^{(n)}\}$, $n \geq 0$, be the file of K-dimensional data points. We assume that each $x \in F$ is a K-tuple $x = (x_0, \ldots, x_{K-1})$ in $[0,1]^K$.

A *relaxed K-d tree* [10] for a set F of K-dimensional data points is a binary tree in which: (a) each node contains a K-dimensional data point and has an associated discriminant $j \in \{0, 1, \ldots, K-1\}$; (b) for each node x with discriminant j, the following invariant is true: any data point y in the left subtree satisfies $y_j < x_j$ and any data point z in the right subtree satisfies $z_j \geq x_j$.

Notice that the sequence of discriminants in a path from the root to any leaf is arbitrary. On the contrary, the definition of standard K-d trees [3] requires the sequence of discriminants along any path to be cyclic, starting with $j = 0$. Thus the root of the tree at level 0 discriminates with respect to the first coordinate ($j = 0$), its sons at level 1 discriminate w.r.t. the second coordinate ($j = 1$), and in general, all nodes at level m discriminate w.r.t. coordinate $j = m \mod K$.

Our average-case analysis of range searches over relaxed K-d trees in Sect. 3 will assume that trees are *random*. We say that a relaxed K-d tree of size n is *random* if it is built by n insertions where the points are independently drawn from a continuous distribution in $[0,1]^K$ and the discriminants are uniformly and independently drawn from $\{0, \ldots, K-1\}$.

In random relaxed K-d trees, these assumptions about the distribution of the input imply that the $n!^K K^n$ possible configurations of input file and discriminant sequences are equiprobable [10]. For standard K-d trees, since there is a fixed rule for discriminants, the assumption of random insertions (producing random standard K-d trees) implies that all $n!^K$ input sequences are equiprobable [3, 12]. In particular, in a random relaxed K-d tree each of the $n \cdot K$ possibilities of (key, discriminant) pairs are equally likely to appear in the root and once the root is fixed, the left and right subtrees are independent random relaxed K-d trees.

A *range query* is a K-dimensional hyperrectangle Q. We shall write $Q = [\ell_0, u_0] \times [\ell_1, u_1] \times \cdots \times [\ell_{K-1}, u_{K-1}]$, with $\ell_i \leq u_i$, for $0 \leq i < K$.

We will use the probabilistic model of random range queries introduced in [8, 9]. In this model, the center of a *random range query* Q is an independently drawn point z in $[0,1]^K$, sampled from the same continuous distribution as the data points, and the edges of the query have given lengths $\Delta_0, \Delta_1, \ldots \Delta_{K-1}$, with $0 \leq \Delta_i \leq 1$, for $0 \leq i < K$. Therefore, $\ell_i = z_i - \Delta_i/2$ and $u_i = z_i + \Delta_i/2$, for $0 \leq i < K$.

Range searching in any variant of K-d trees is straightforward. When visiting a node x that discriminates w.r.t. the j-th coordinate, we must compare x_j with the j-th range $[\ell_j, u_j]$ of the query. If the query range is totally above (or below) that value, we must search only the right subtree (respectively, left) of that node. If, on the contrary, $\ell_j \leq x_j \leq u_j$ then both subtrees must be searched; additionally, we must check whether x falls or not inside the query hyperrectangle. This procedure continues recursively until empty subtrees are reached.

We will measure the cost of range queries by the number of nodes of the K-d tree visited during the search. If the number of points to be reported by the range search is P then the cost R_n of the range search will be of the form $\Theta(P + W_n)$, where W_n is the *overhead*.

3   Analysis of the Cost of Range Searches

3.1   Bounding Rectangles, Slices, and Sliced Partial Match

The *bounding rectangle* $B(x) = [l_0(x), u_0(x)] \times \ldots \times [l_{K-1}(x), u_{K-1}(x)]$ of a point $x = (x_0, \ldots, x_{K-1})$ in a K-d tree t is the region of $[0,1]^K$ corresponding to the leaf replaced by x when x was inserted into t. Formally, it is defined as follows: (a) if x is the root of t then $B(x) = [0,1]^K$; (b) if $y = (y_0, \ldots, y_{K-1})$ is the father of x in t and y discriminates w.r.t. the j-th coordinate then: (b.1) if $x_j < y_j$

then $B(x) = [l_0(y), u_0(y)] \times \ldots \times [l_j(y), y_j] \times \ldots \times [l_{K-1}(y), u_{K-1}(y)]$, and (b.2)
if $x_j \geq y_j$ then $B(x) = [l_0(y), u_0(y)] \times \ldots \times [y_j, u_j(y)] \times \ldots \times [l_{K-1}(y), u_{K-1}(y)]$.

Lemma 1. *A point x with bounding rectangle $B(x)$ is visited by a range search with query hyperrectangle Q if and only if $B(x)$ intersects Q.*

Proof. See [8,9].

In order to relate the performance of range searches with the performance of partial matches, we need to introduce several notions, beginning with that of *slice*. Given a bitstring $w = (w_0, \ldots, w_{K-1})$ of length K, the slice Q_w is the K-dimensional hyperrectangle defined by

$$Q_w = [\ell'_0, u'_0] \times [\ell'_1, u'_1] \times \cdots \times [\ell'_{K-1}, u'_{K-1}],$$

where $[\ell'_i, u'_i] = [\ell_i, u_i]$ if $w_i = 0$ and $[\ell'_i, u'_i] = [0,1]$ if $w_i = 1$. Notice that $Q_{00\ldots0} = Q$ and $Q_{11\ldots1} = [0,1]^K$.

Another useful notion is that of *proper slice*. The proper slice $\hat{Q}_w$ is the hyperregion defined by

$$\hat{Q}_w = Q_w - \bigcup_{v < w} Q_v,$$

that is, the region that results when all the slices properly contained within Q_w are subtracted from it (see Fig. 3.1). The only proper slice consisting of a simple connected region is $\hat{Q}_{00\ldots0} = Q_{00\ldots0} = Q$; in general, $\hat{Q}_w$ consists of $2^{\text{order}(w)}$ connected subregions, where order(w) is the number of 1s in the bitstring w.

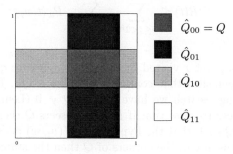

Fig. 1. Example of the proper slices induced by a query Q

The most important concept in this subsection is that of *sliced partial match*. Given a query hyperrectangle Q, a bitstring w and a point $y \in [0,1]^K$, a sliced partial match acts as a standard partial match with query $q = (q_0, q_1, \ldots, q_{K-1})$ where $q_i = y_i$ if $w_i = 1$ and $q_i = *$ if $w_i = 0$ (hence the specification pattern of the partial match is w), but contrary to a standard partial match it only counts ("reports") the visited points x in the data structure such that $x \in \hat{Q}_w$.

To every sliced partial match with point y and specification pattern w we associate the hyperplane $H(y, w)$ defined by

$$H(y, w) = \{x \in [0, 1]^K \mid \forall i : w_i = 1 \implies x_i = y_i\}.$$

For instance, if $K = 2$ then $H(y, 00) = [0, 1]^K$, $H(y, 11) = \{y\}$, $H(y, 01)$ is the line passing through y parallel to the horizontal axis and $H(x, 10)$ is the line passing through x parallel to the vertical axis.

Lemma 2. *A point x with bounding rectangle $B(x)$ is visited and reported by a sliced partial match with query hyperrectangle Q, specification w, and point y if and only if $x \in \hat{Q}_w$ and the bounding rectangle $B(x)$ intersects the hyperplane $H(y, w)$.*

Proof. The proof is immediate from Lemma 1. Notice that a sliced partial match behaves as a range query in which the hyperrectangle query "degenerates" to the hyperplane $H(y, w)$. And, by definition, the point must also belong to $\hat{Q}_w$ to be reported.

3.2   The Combinatorial Characterizations

In this subsection we state several relations between the cost $R(t)$ of an orthogonal range search in a K-d tree t and the performance $P_w(t, y)$ of a sliced partial match with specification pattern w and query point y in a K-d tree t. The implicit query hyperrectangle Q is the same for both the range search and the sliced partial match.

Theorem 1. *Given a query Q with corners $v_0, v_1, \ldots, v_{2^K - 1}$ and a K-d tree t,*

$$R(t) \leq \sum_{0 \leq j < 2^K} \sum_{w \in (0+1)^K} P_w(t, v_j).$$

Proof. Consider a point x visited by a range search with query Q. Let w be the index of the proper slice that contains x, i.e., $x \in \hat{Q}_w$. Recall that since x is visited by the range search we have $B(x) \cap Q \neq \emptyset$ (Lemma 1). Therefore, by Lemma 2, it suffices to show that if $B(x)$ intersects Q then there exists at least one corner v_j of Q such that the hyperplane $H(v_j, w)$ does intersect $B(x)$.

If $B(x)$ contains any of the corners of Q then the statement above is clearly true: since the hyperplane $H(v_j, w)$ contains v_j, it must intersect $B(x)$. If $B(x)$ does not contain any corner of Q, there are two possibilities to consider: either $B(x)$ is entirely within Q, or $B(x)$ intersects one or more faces of Q. If $B(x)$ is totally inside Q then $w = 0 \ldots 0$ and indeed $H(v_j, 0 \ldots 0) = [0, 1]^K$ intersects $B(x)$ for any corner v_j. On the other hand, if $B(x)$ intersects one or more faces of Q but does not contain a corner nor it is contained inside Q then $w \neq 11 \ldots 1$ and $\hat{Q}_w$ must "contact" one of the intersected faces, in the sense that the face is a boundary of $\hat{Q}_w$. Let f be such face. Now, the hyperplane $H(v_j, w)$ contains this face (and hence it intersects $B(x)$), provided that v_j is any corner of the face f.

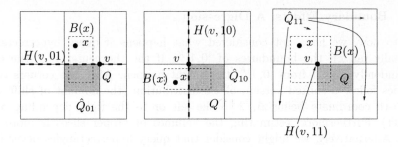

Fig. 2. Graphical illustration of the proof of Theorem 1

Theorem 2. *Given a query Q with center at z and a K-d tree t,*

$$R(t) \geq \sum_{w \in (0+1)^K} P_w(t, z).$$

Proof. The statement of the theorem is immediate, once we show that whenever a point x is reported by a sliced partial match with parameters w and z then it is also visited by the range search with query Q. Formally, we have to show that if $x \in \hat{Q}_w$ and $H(z, w)$ intersects the bounding rectangle $B(x)$ of x then $B(x)$ also intersects Q. Indeed, if $w = 00 \ldots 0$ then the statement trivially holds since $x \in Q$. On the other hand, if $w \neq 00 \ldots 0$ then $H(z, w)$ and $\hat{Q}_w$ are disjoint. Since $B(x)$ intersects $\hat{Q}_w$ (x is part of both by hypothesis) the only way for $B(x)$ to intersect $H(z, w)$ is to intersect Q too (then, because of Lemma 1, x must be visited by the range search).

Theorem 3. *Given a query Q with center at z divide $[0, 1]^K$ into 2^K quadrants $C_0, C_1, \ldots, C_{2^K - 1}$. Let $R^{(i)}(t)$ denote the number of points of the i−th quadrant visited by a range search with query Q in the K-d tree t. Similarly, let $P_w^{(i)}(t, y)$ denote the number of points of the i−th quadrant reported by sliced partial match with pattern w and point y in the K-d tree t. Let v_i be the unique corner of Q belonging to the i−th quadrant. Then*

$$R^{(i)}(t) = \sum_{w \in (0+1)^K} P_w^{(i)}(t, v_i).$$

Proof (Sketch of proof). For a point x belonging to the i-th quadrant and the proper slice $\hat{Q}_w$, the intersection of $H(v_i, w)$ with $B(x)$ implies the intersection of Q with $B(x)$. On the other hand, if $B(x)$ intersects Q there is at least one corner v such that $H(v, w)$ intersects $B(x)$; it is not difficult to see that one of these corners must be v_i, the unique corner of Q in the i-th quadrant.

3.3   Boundary Effects: A Digression

Up to now, we have not considered what happens if query hyperrectangles partially fall off the boundaries of $[0,1]^K$. If the the center z of the query is randomly drawn from $[0,1]^K$ this might happen. Also, the corners of the queries follow a shifted version of z's distribution (the amount of shift w.r.t. the i-th coordinate being $\Delta_i/2$ to the left or to the right depending of the corner). Furthermore, estimating the volumes of proper slices is more difficult. Alternatively, we might consider that query hyperrectangles never intersect the $[0,1]^K$ boundary by imposing that z is randomly drawn from the distribution that we obtain by mapping the continuous distribution on $[0,1]^K$ to $[\Delta_0/2, 1 - \Delta_0/2] \times \cdots \times [\Delta_{K-1}/2, 1 - \Delta_{K-1}/2]$. But again, the corners are distributed according to a shifted version of z's distribution. In any case, the corners of the queries are not necessarily distributed as the data points and the center of the query, and they cannot fall in some regions of $[0,1]^K$ (e.g., for $K = 2$ if queries are allowed to fall partially off the boundaries the lower left corner falls in $[-\Delta_0/2, 1 - \Delta_0/2] \times [-\Delta_1/2, 1 - \Delta_1/2]$; if queries are not allowed to fall off the boundaries then the lower left corner can only fall in $[0, 1 - \Delta_0] \times [0, 1 - \Delta_1]$).

Thus the developments in the next subsection are only valid when the Δ's are small enough, since the perturbation due to the boundary and shifting effects is then negligible.

3.4   The Expected Cost of Range Search in Relaxed K-d Trees

The theorems of Subsection 3.2 show that the analysis of range search reduces to the analysis of sliced partial matches.

Theorem 4. *Let $\mathbb{E}[R_n]$ be the expected cost of a range search with a random query in a random K-d tree of size n. Let $\mathbb{E}[P_{n,w}]$ be the expected cost of a sliced partial match with pattern w in a random K-d tree of size n, with respect to a random query and a random point. If the data points are uniformly distributed, the centers are uniformly distributed in $[\Delta_0/2, 1 - \Delta_0/2] \times \cdots \times [\Delta_{K-1}/2, 1 - \Delta_{K-1}/2]$, i.e., no query falls partially off the boundaries and the Δ's are small enough then*

$$\mathbb{E}[R_n] = \sum_{w \in (0+1)^K} \mathbb{E}[P_{n,w}].$$

Proof. Clearly $R(t) = \sum_{0 \le i < 2^K} R^{(i)}(t)$. Hence,

$$R(t) = \sum_{0 \le i < 2^K} \sum_{w \in (0+1)^K} P_w^{(i)}(t, v_i). \tag{1}$$

The next step is to take expectations on both sides of the equation above. Two points are worth noting: (1) since the center of the query z is (almost) uniformly distributed, each v_i is (almost) uniformly distributed (see Subsection 3.3); (2) the expected value $\mathbb{E}\left[P_{n,w}^{(i)}\right] = \mathrm{Vol}(C_i) \cdot \mathbb{E}[P_{n,w}]$, where $\mathrm{Vol}(C_i)$ is the probability that, given a randomly drawn point z, a second random point falls in the i-th quadrant defined by z. Since $\sum_{0 \leq i < 2^K} \text{Vol}(C_i) = 1$ the result follows. It is worth noting that the equality given in (2) does not hold if the length of the edges of the query is not small enoug, for then neither the center of the query nor its corners are uniformly distributed in $[0,1]^K$.

Although (1) gives a precise relationship between the cost of range search and the cost of sliced partial matches, we cannot use it to get results about the variance of R_n or its probability distribution since the costs of the sliced partial matches (the random variables $P_{n,w}^{(i)}$'s) are not independent.

Theorems 1 and 2 can be used as we have done in Theorem 3 to get corresponding lower and upper bounds for $\mathbb{E}[R_n]$ (but they cannot be used to bound the variance, for the same reason as before).

Now we need to analyze the expected cost of sliced partial matches in random(ized) relaxed K-d trees. It easily follows from the analysis of the expected cost of standard partial matches in random relaxed K-d trees [13]. Our next theorem gives the expected cost of sliced partial matches in random relaxed K-d trees.

Theorem 5. *If $w \neq 00 \ldots 0$ and $w \neq 11 \ldots 1$, the expected cost $\mathbb{E}[P_{n,w}]$ of a sliced partial match in a random relaxed K-d tree of size n w.r.t. a random point and the pattern w is*

$$\mathbb{E}[P_{n,w}] = \text{Vol}(\hat{Q}_w) \cdot \beta(\rho) \cdot n^{\alpha(\rho)} + \mathcal{O}(1),$$

where $\rho = \text{order}(w)/K$, $\alpha \equiv \alpha(x) = (\sqrt{9 - 8x} - 1)/2$, $\beta(x) = \Gamma(2\alpha + 1)/((1 - x)(\alpha + 1)\alpha^3 \Gamma^3(\alpha))$, and $\text{Vol}(\hat{Q}_w)$ is the probability that a point falls inside the proper slice $\hat{Q}_w$ of a randomly centered query. Furthermore, $\mathbb{E}[P_{n,00\ldots0}] = \text{Vol}(\hat{Q}_{00\ldots0}) \cdot n$ and $\mathbb{E}[P_{n,11\ldots1}] = 2 \cdot \text{Vol}(\hat{Q}_{11\ldots1}) \cdot (H_{n+1} - 1)$, where $H_n = \sum_{1 \leq j \leq n} 1/j = \log n + \gamma + \mathcal{O}(1/n)$ denotes the n-th harmonic number.

Proof (Sketch of proof). It is relatively easy to establish that $\mathbb{E}[P_{n,w}] = \text{Vol}(\hat{Q}_w) \cdot \mathbb{E}\left[P_{n,w}^{[\text{std}]}\right]$, where $\mathbb{E}\left[P_{n,w}^{[\text{std}]}\right]$ is the expected number of visited nodes in a standard partial match with query pattern w on a random K-d tree of size n. The statement of the theorem follows from the known asymptotic estimates of $\mathbb{E}\left[P_{n,w}^{[\text{std}]}\right]$ (see [13]).

Corollary 1. *The expected cost of a random centered range search of sides Δ_0, $\ldots$, Δ_{K-1} in a random relaxed K-d tree of size n satisfies, for uniformly and independently drawn data points and centers of the queries (with queries not falling off the boundaries), and if the values of the Δ's are small enough,*

$$\mathbb{E}[R_n] = \Delta_0 \cdots \Delta_{K-1} \cdot n + \sum_{1 \leq j < K} c_j \cdot n^{\alpha(j/K)}$$
$$+ 2 \cdot (1 - \Delta_0) \cdots (1 - \Delta_{K-1}) \cdot \log n + \mathcal{O}(1),$$

where

$$c_j = \beta(j/K) \cdot \sum_{w:order(w)=j} \left(\prod_{i:w_i=0} \Delta_i \right) \cdot \left(\prod_{i:w_i=1} (1 - \Delta_i) \right).$$

Proof. The corollary easily follows from Theorem 4 and the asymptotic estimates given in Theorem 5. Finally, observe that for the uniform distribution we have

$$\text{Vol}(\hat{Q}_w) = \left(\prod_{i:w_i=0} \Delta_i \right) \cdot \left(\prod_{i:w_i=1} (1 - \Delta_i) \right).$$

since we do not allow query hyperrectangles to intersect the $[0,1]^K$ boundaries. If the queries were allowed to fall off the boundaries of $[0,1]^K$ then the values given above for the volumes of proper slices are not valid, but they are still a good approximation if Δ's are small enough, and thus the asymptotic estimate for $\mathbb{E}[R_n]$ is reasonably good even for the model where queries may fall partially off the boundaries.

Notice that the term $\Delta_0 \cdots \Delta_{K-1} \cdot n$ in $\mathbb{E}[R_n]$ is the expected number of reported points and hence the overhead is $\mathcal{O}(n^{\alpha(1/K)})$.

4   Other Multidimensional Data Structures

It is important to stress that no assumptions were made with respect to the way that discriminants are assigned during the construction of the tree, so all theorems of subsection 3.2 apply to standard K-d trees [3], squarish K-d trees [9], K-d-t trees [14] and other variants. It turns out that the theorems also apply to quadtrees [4] without change.

Furthermore, similar arguments to that of Theorem 5 are also valid, relating the expected cost of sliced partial matches to the expected cost of standard partial matches. In general, when $w \neq 00\ldots0$ and $w \neq 11\ldots1$ for the data structures mentioned above we have

$$\mathbb{E}[P_{n,w}] = \beta_w \cdot \text{Vol}(\hat{Q}_w) \cdot n^{\alpha(\rho)} + \mathcal{O}(1),$$

where $\rho = order(w)/K$, $\alpha(x) = 1 - x + \phi(x)$ and β_w is a constant depending on w. The expected cost of range search takes hence the form

$$\mathbb{E}[R_n] = \Delta_0 \cdots \Delta_{K-1} \cdot n + \sum_{1 \le j < K} c_j \cdot n^{\alpha(j/K)} \tag{2}$$
$$+ 2 \cdot (1 - \Delta_0) \cdots (1 - \Delta_{K-1}) \cdot \log n + \mathcal{O}(1),$$

where $c_j = \sum_{w:order(w)=j} \beta_w \cdot \text{Vol}(\hat{Q}_w)$.

Different data structures are characterized by different α's and β's. For standard K-d trees the necessary analysis is from [15] where it was shown that $\phi(x) < 0.07$ and that it is the unique real solution of

$$(\phi(x) + 3 - x)^x (\phi(x) + 2 - x)^{1-x} - 2 = 0.$$

No closed expression for the β's is given, but their values can be explicitly computed with some effort; for $K \leq 4$ all the numerical values are given in [15]. Squarish K-d trees [9] have optimal performance since $\phi(x) = 0$; however, the values of the β's are not yet known and the only way for the moment to compute them would be through experimental measurement. For quadtrees the analysis of partial match can be found in [16]; $\alpha(x)$ is the same as for K-d trees. But the β_w's depend only on K and the order of w. For $K = 2$, we have $\beta_{01} = \beta_{10} = \Gamma(2\alpha + 2)/(2\alpha^3 \Gamma^3(\alpha)) \approx 1.5950991$, but no explicit form is given for higher dimensions.

Last, but not least, the theorems in Sect. 3 also apply to K-d tries, relaxed K-d tries and quadtries. These multidimensional data structures are space-oriented rather than data-oriented, since they induce partitions of the space that are independent of the data points (except that the recursive subdivision of the space stops when the corresponding region contains only one or no points). In order to apply the theorems of Sect. 3, we only need to assume that each internal node "contains" the middle point of the hyperplane(s) associated to the internal node, to meaningfully define slice partial matches in these data structures. The average cost of range searches in these digital data structures satisfies (2), but the β_w's involve a fluctuating periodic term (depending on n) of small amplitude and bounded by a constant. For instance, for relaxed K-d tries [13] we have $\alpha = \alpha(x) = \log_2(2 - x)$ and $\beta_w = \beta(\mathrm{order}(w)/K)$ with

$$\beta(x) = \frac{1}{\log 2}\Big((\alpha - 1)\Gamma(-\alpha) + \delta(\log_2 n)\Big),$$

and $\delta(\cdot)$ a periodic function of period 1, mean 0 and small amplitude that also depends on α.

5   Final Remarks

In this work we have studied the average case analysis of orthogonal range search in relaxed K-d trees. We have also shown how this analysis can be extended to several hierarchical multidimensional data structures such as K-d trees, quadtrees, K-d tries and other variants of these (basically, all data structures which induce axis-parallel partitions of the search space, either for data-oriented partitions as in K-d trees or quadtrees, or for space-oriented partitions as in K-d tries and quadtries). As a result, we have proved that several multidimensional data structures exhibit low overhead for range searches (of sublinear time complexity with small constant factors).

In our analysis we have used a mixture of combinatorial and geometric arguments that allows us to relate the cost of range search to the cost of partial match searches. The analysis of partial match searches has already been conducted for many multidimensional data structures using purely analytical/combinatorial techniques; our approach exploits these known results, overcoming the apparent inapplicability of these techniques to the analysis of range search.

Furthermore, since range search is on the basis of more complex associative queries, it seems likely that our approach could be successfully applied to the analysis of the performance of more complex region queries and to nearest neighbor search. We are already working on these open problems and conjecture that the results will be rather similar to those that we have obtained for range search.

References

1. Samet, H.: The Design and Analysis of Spatial Data Structures. Addison-Wesley (1990)
2. Bentley, J., Friedman, J.: Data structures for range searching. ACM Computing Surveys **11** (1979) 397–409
3. Bentley, J.: Multidimensional binary search trees used for associative retrieval. Communications of the ACM **18** (1975) 509–517
4. Bentley, J., Finkel, R.: Quad trees: A data structure for retrieval on composites keys. Acta Informatica **4** (1974) 1–9
5. Rivest, R.L.: Partial-match retrieval algorithms. SIAM Journal on Computing **5** (1976) 19–50
6. Bentley, J., Stanat, D.: Analysis of range searches in quad trees. Information Processing Letters **3** (1975) 170–173
7. Silva-Filho, Y.: Average case analysis of region search in balanced k-d trees. Information Processing Letters **8** (1979) 219–223
8. Chanzy, P., Devroye, L., Zamora-Cura, C.: Analysis of range search for random k-d trees. Acta Informatica **37** (2001) 355–383
9. Devroye, L., Jabbour, J., Zamora-Cura, C.: Squarish k-d trees. SIAM Journal on Computing **30** (2000) 1678–1700
10. Duch, A., Estivill-Castro, V., Martínez, C.: Randomized k-dimensional binary search trees. In Chwa, K.Y., Ibarra, O.H., eds.: Int. Symposium on Algorithms and Computation (ISAAC'98). Volume 1533 of Lecture Notes in Comput. Sci., Springer (1998) 199–208
11. Duch, A., Martínez, C.: On the average performance of orthogonal range search in multidimensional data structures. Technical Report ALCOMFT-TR-01-185, ALCOM-FT (2001) Also published as technical report LSI-01-54-R, LSI-UPC. Available from http://www.brics.dk/~alcomft/TR/ALCOMFT-TR-01-185.ps and http://www.lsi.upc.es/~techreps/ps/R01-54.ps.gz.
12. Mahmoud, H.: Evolution of Random Search Trees. J. Wiley & Sons (1992)
13. Martínez, C., Panholzer, A., Prodinger, H.: Partial match queries in relaxed multidimensional search trees. Algorithmica **29** (2001) 181–204
14. Cunto, W., Lau, G., Flajolet, P.: Analysis of kdt-trees: kd-trees improved by local reorganisations. In Dehne, F., Sack, J.R., Santoro, N., eds.: Workshop on Algorithms and Data Structures (WADS'89). Volume 382 of Lecture Notes in Comput. Sci., Springer-Verlag (1989) 24–38
15. Flajolet, P., Puech., C.: Partial match retrieval of multidimensional data. Journal of the ACM **33** (1986) 371–407
16. Flajolet, P., Gonnet, G., Puech, C., Robson, J.M.: Analytic variations on quadtrees. Algorithmica **10** (1993) 473–500

Bialgebraic Modelling of Timed Processes

Marco Kick*

LFCS, University of Edinburgh, Edinburgh EH9 3JZ, Scotland
mk@dcs.ed.ac.uk

Abstract. We give an abstract axiomatic account of timed processes using monoids and their (partial and total) actions. Subsequently, we present categorical formulations thereof, including a novel characterisation of partial monoid actions as coalgebras for an *evolution comonad*. Adapting the approach of Turi and Plotkin [24], we then exhibit an abstract theory of well-behaved operational rules suitable for timed processes and, for discrete time, also derive a concrete syntactic format encompassing all rules we found in the literature.

Introduction

Over the past decade, much research effort has been directed towards establishing a theory of real-time systems. Amongst other approaches, extensions of standard process algebras [6,10,16] with timing features have been considered. The meaning of such *timed processes* is usually given by *structural operational semantics (SOS)* [20]: one inductively defines a *labelled transition system (LTS)* [20] on the set of processes, describing the behaviour of programs as (sequences of) transitions inferred from syntax-directed rules.

In this paper, we attempt to unify the plethora of studied languages, e.g., [4, 17,19,22,25], by axiomatising shared fundamental mathematical properties; we model timed processes by *timed transition systems (TTSs)* – specific LTSs incorporating intuitive properties of time passing. In particular, we adopt the widely accepted design principle of separating computation from time passing, using *action* and *time* transitions to denote, respectively, instantaneous computations and pure time passing. We only consider the case of action and time transitions defined independently of each other, with Temporal CCS (TeCCS) [17] as the paradigmatic example.

In the study of processes, strong emphasis has been put on *behavioural equivalences* (see [8] for an overview) and on proving *congruence results* for them. This has led to *rule formats*: syntactic constraints on SOS rules whose satisfaction automatically establishes a congruence result. Arguably, the most well-known format is GSOS [7] which ensures that strong bisimulation [16] is a congruence. To our knowledge, there is no corresponding format for timed processes. Paving the way for such a format, we present a mathematical theory of well-behaved SOS rules for timed processes; furthermore, for the case of *discrete* time, we derive a syntactic format which, we believe, captures all examples in the literature.

---

* This work is supported by EPSRC grant GR/M56333.

P. Widmayer et al. (Eds.): ICALP 2002, LNCS 2380, pp. 525–536, 2002.

To achieve these last two goals, our approach builds on the framework of [24]. There, *abstract operational rules* are a suitable kind of natural transformation, parametric in functorial notions Σ and B of *signature* and (one-step) *behaviour*, specifying how to derive the meaning of compound expressions from the meaning of their arguments (as traditional SOS rules do). Mathematically, the core of the theory is that such abstract rules inductively define a *distributive law* of a monad over a comonad [21] (respectively (co)freely generated from Σ and B). Using the *bialgebras* [24] of such distributive law, and under a mild condition on B (preservation of weak pullbacks), a congruence result for coalgebraic B-bisimulation [1] follows.

GSOS arises by considering abstract rules for the behaviour functor

$$(1) \qquad\qquad B_A X = \mathcal{P}_{\mathrm{fi}}(X)^A$$

on **Set** (see [24] for definitions): B_A-coalgebras are (image finite) LTSs, the corresponding coalgebraic bisimulation is strong bisimulation, and B_A preserves weak pullbacks. Hence, [24] yields, apart from a new proof, a conceptual reason *why* bisimulation is a congruence for GSOS languages. We want to apply these techniques to our model of timed processes, i.e., TTSs, by fitting it into the general framework; as we will see, to successfully do so we need to slightly generalise [24].

The structure of the paper is as follows. In §1, stressing the crucial concepts, we consider *time domains* (special monoids) and TTSs to model time and timed processes. We then show that TTSs are partial monoid actions of a time domain; total monoid actions play a prominent role as *delay operators*; both kinds of actions are then suitably combined in the new notion of *biaction*. In preparation for applying [24], we give categorical characterisations of these notions in §2, in particular of partial monoid actions (i.e., TTSs) as coalgebras of a novel *evolution comonad*, and of biactions as bialgebras for a distributive law (of the monad for total actions over the evolution comonad).

Finally, in §3, we adapt the framework of [24] to accommodate our categorical description of timed processes: rather than using behaviour functors (as in [24]), we need to begin with a behaviour *comonad* to account for TTSs (as coalgebras for the evolution comonad). Thus, a correspondingly different type of natural transformation modelling abstract rules must be used: still inducing a distributive law, yet removing the restriction to cofreely generated comonads. In some sense, behaviour comonads model 'big-step' semantics as they usually describe 'complete' computations, rather than just one step. For the case of discrete time, where the evolution comonad is actually cofreely generated from a functor, thus enabling us to directly apply [24], we then derive the aforementioned syntactic format.

Benefits from the approach of the present paper are: a more unified view of timed processes by isolating fundamental mathematical properties underlying many calculi; highlighting important mathematical structures in languages for timed processes; and providing another case study documenting the universality and flexibility of the categorical approach to operational semantics of [24].

1   A Mathematical Model of Timed Processes

In this section we present our abstract model of timed processes based on an axiomatic treatment of time transitions: time is modelled by special monoids, timed processes by LTSs with their transition relations restricted to incorporate special properties of time passing[1]. The following definitions of time domain and TTS are synthesised from the (slightly differing) ones in [12,13,18,25].

Definition 1. *A* time domain *is a commutative monoid* $(\mathcal{T}, +, 0)$ *(abbreviated as $\mathcal{T}$) whose induced preorder $\leq$, defined by $s \leq t \Leftrightarrow (\exists u). \ s + u = t$, is a linear order, and which satisfies the* cancellation rule

$$(2) \qquad\qquad t + u = t + v \Rightarrow u = v$$

Note that 0 is the least element with respect to $\leq$, $+$ is monotone with respect to $\leq$ (implied by (2)), and whenever $s \leq t$, there is a unique u such that $s + u = t$, written as $t - s$. This *partial subtraction* (only defined for $s \leq t$) can be extended to the total *truncated subtraction* $\dot{-}$, setting $t \dot{-} s = 0$ if $s > t$, making time domains *closed* [15]. Examples of time domains include the naturals $\mathbb{N}$ for *discrete time*, the non-negative reals $\mathbb{R}_{\geq 0}$ for *continuous time*, and the trivial domain $\{0\}$; note that $\mathcal{T}$ is either trivial or infinite (cf. [13]).

Definition 2. *A* TTS *is an LTS* $(P, \mathcal{T}, \rightsquigarrow)$ *where P is a set of* processes, $\mathcal{T}$ *is a* time domain, *and the transition relation $\rightsquigarrow \subseteq P \times \mathcal{T} \times P$ satisfies the following axioms, writing $p \xrightarrow{t} p'$ for $(p, t, p') \in \rightsquigarrow$:*

(Determinacy) $p \overset{t}{\rightsquigarrow} p' \wedge p \overset{t}{\rightsquigarrow} p'' \Rightarrow p' = p''$

(Zero-Delay) $p \overset{0}{\rightsquigarrow} p$

(Continuity) $p \overset{s+t}{\rightsquigarrow} p' \Leftrightarrow (\exists p''). \ p \overset{s}{\rightsquigarrow} p'' \overset{t}{\rightsquigarrow} p'$

Axiom (Determinacy) states that no choices are resolved by the passage of time. Axioms (Determinacy) and (Continuity) (*additivity* in [18]) are widely accepted, e.g., in [17,18,25] (sometimes only one implication of (Continuity) is used, e.g. in [12,13]). Axiom (Zero-Delay) is usually not assumed since most languages allow only non-zero time transitions; yet it is intuitively clear and can be added without inconsistencies.

Definition 3. *Given TTSs $(P_i, \mathcal{T}, \rightsquigarrow_i)$, $i \in \{1, 2\}$, a relation $R \subseteq P_1 \times P_2$ is a* (strong) time bisimulation *(over $\mathcal{T}$) if $(p_1, p_2) \in R$ implies for all $t \in \mathcal{T}$ that*

$$p_1 \overset{t}{\rightsquigarrow}_1 p_1' \Rightarrow (\exists p_2'). \ p_2 \overset{t}{\rightsquigarrow}_2 p_2' \wedge (p_1', p_2') \in R$$
$$p_2 \overset{t}{\rightsquigarrow}_2 p_2' \Rightarrow (\exists p_1'). \ p_1 \overset{t}{\rightsquigarrow}_1 p_1' \wedge (p_1', p_2') \in R$$

We write $p_1 \sim p_2$ if there exists a strong time bisimulation containing (p_1, p_2).

---

[1] Since action transitions are defined independently by GSOS rules, we omit their treatment, as it is just an instance of [24], using the behaviour functor B_A.

Note that calculi like TeCCS [17] combine time bisimulation with standard action bisimulation. Axiom (Determinacy) forces the transition relation $\rightsquigarrow$ of a TTS $(P, \mathcal{T}, \rightsquigarrow)$ to be a binary partial function $* : P \times \mathcal{T} \rightharpoonup P$ (written in infix notation), (Zero-Delay) and (Continuity) ensure the following equations up to Kleene equality ($\simeq$):

$$(3) \qquad p * 0 \simeq p \qquad\qquad p * (s + t) \simeq (p * s) * t$$

In other words, $*$ has to be a *partial monoid action* of $\mathcal{T}$ on P. Note that the TTS is completely determined by $*$:

$$(4) \qquad p \overset{t}{\rightsquigarrow} p' \Leftrightarrow p' \simeq p * t$$

Definition 4. *Let $\mathcal{T}$ be a time domain and $*_i$ partial $\mathcal{T}$-actions on P_i, $i \in \{1, 2\}$. A homomorphism from $*_1$ to $*_2$ is a (total) function $f : P_1 \rightarrow P_2$ such that for $p \in P_1$ and $t \in \mathcal{T}$, $f(p *_1 t) \simeq (fp) *_2 t$.*

Using (4), $R \subseteq P_1 \times P_2$ is a time bisimulation if $(p_1, p_2) \in R$ implies for all $t \in \mathcal{T}$, with $(\ldots)\!\downarrow$ denoting that the expression $(\ldots)$ is defined (dually, undefinedness will be expressed as $(\ldots)\!\uparrow$):

$$(p_1 *_1 t)\!\downarrow \Leftrightarrow (p_2 *_2 t)\!\downarrow \qquad\qquad (p_1 *_1 t)\!\downarrow \Rightarrow (p_1 *_1 t, p_2 *_2 t) \in R$$

Thus, R is *closed* under partial actions. Note how the rewritten definition incorporates (Determinacy): $p_1 \overset{t}{\rightsquigarrow}_1 p_1'$ can only hold for $p_1' = p_1 *_1 t$, hence existential quantification is no longer required. Furthermore, a time bisimulation R can be endowed with a canonical partial action, viz., $(p_1, p_2)*t \simeq (p_1 *_1 t, p_2 *_2 t)$, making the *projections* $\pi_i : R \rightarrow P_i$ homomorphisms from $*$ to $*_i$.

Consider TeCCS now; after adding transitions according to (Zero-Delay)[2], its operational rules define a TTS on the set $\text{TeCCS}_{/\sim}$ of processes modulo time bisimulation. An important operator is *time prefixing*: intuitively, $(t).p$ represents the process p *delayed* by $t > 0$ units of time; formally, its behaviour is given by the following SOS rules[3]:

$$(5) \qquad \frac{}{(s+t).p \overset{s}{\rightsquigarrow} (t).p} \ (t > 0) \qquad \frac{}{(t).p \overset{t}{\rightsquigarrow} p} \qquad \frac{p \overset{s}{\rightsquigarrow} p'}{(t).p \overset{s+t}{\rightsquigarrow} p'}$$

Extending for $t = 0$, (Zero-Delay) forces $(0).p = p$ since $(0).p \overset{0}{\rightsquigarrow} p$ is derivable from the rules, hence $(0).p \sim p$; one also easily shows $(s).(t).p \sim (s+t).p$. These equations, plus the delay intuition, inspire the following definition:

Definition 5. *Let $(P, \mathcal{T}, \rightsquigarrow)$ be a TTS; a delay operator on it is a binary total function $\bullet : \mathcal{T} \times P \rightarrow P$ (written in infix notation) satisfying the equations*

$$(6) \qquad 0 \bullet p = p \qquad\qquad s \bullet (t \bullet p) = (s + t) \bullet p$$

---

[2] TeCCS does not define $\overset{0}{\rightsquigarrow}$-transitions.

[3] The side condition $t > 0$ was implicit in [17] since $(0).p$ is not a TeCCS process.

Hence, a delay operator is a *total monoid action* of $\mathcal{T}$ on P; for example, the extended time prefixing is a delay operator on $\mathrm{TeCCS}_{/\sim}$. Analysing the interaction of time transitions with time prefixing or, more abstractly, of a partial action with a total one, bearing in mind that $\leq$ is a total order so that at least one of $s \dot{-} t$ and $t \dot{-} s$ is equal to 0, we introduce the following notion.

Definition 6. *A $\mathcal{T}$-biaction is a set P with a partial action $*$ and a delay operator $\bullet$ satisfying*

$$(7) \qquad (t \bullet p) * s \simeq (t \dot{-} s) \bullet (p * (s \dot{-} t))$$

A $\mathcal{T}$-biaction gives a minimal account of timed processes which can be delayed and perform time transitions, (7) linking these actions in the 'right' way. From our previous remarks, it follows that $\mathrm{TeCCS}_{/\sim}$ is an example of a biaction.

2   Timed Processes Categorically

In order to fit the previous considerations into the framework of [24], we now give categorical formulations of (total and partial) actions and biactions; in the remainder of this section, let $\mathcal{T}$ be a time domain. For partial actions, driven by their equivalence with TTSs, and the necessity to account for time bisimulation, we will present a coalgebraic description; for dual reasons, we use the following folklore description of total $\mathcal{T}$-actions as algebras for a monad:

Proposition 1. *The map $X \mapsto \mathcal{T} \times X$ induces a monad on* **Set** *whose algebras are the $\mathcal{T}$-actions.*

Instead of viewing partial actions as partial functions $X \times \mathcal{T} \rightharpoonup X$, we will regard them as *total* functions $X \to E_{\mathcal{T}}X$ where $E_{\mathcal{T}}X$ contains $\mathcal{T}$-*evolutions* – certain partial functions $\mathcal{T} \rightharpoonup X$ with properties mimicking (3):

Definition 7. *A $\mathcal{T}$-evolution (on X) is a partial function $e : \mathcal{T} \rightharpoonup X$ such that*

$$(8) \qquad e(0){\downarrow} \qquad\qquad e(s+t){\downarrow} \Rightarrow e(s){\downarrow}$$

We denote the set of all $\mathcal{T}$-evolutions on X by $E_{\mathcal{T}}X$ (or simply EX).

Intuitively, an evolution e in EX describes a timed process whose time transitions are defined by $e \overset{t}{\rightsquigarrow} x \overset{\mathrm{df}}{\Leftrightarrow} e(t) \simeq x$; using this notation, and abbreviating $(\exists x).\, e \overset{t}{\rightsquigarrow} x$ by $e \overset{t}{\rightsquigarrow}$, the two axioms (8) become $e \overset{0}{\rightsquigarrow}$ and $e \overset{s+t}{\rightsquigarrow} \Rightarrow e \overset{s}{\rightsquigarrow}$, i.e., very basic versions of (Zero-Delay) and (one direction of) (Continuity).

Given a function $f : X \to Y$, defining $Ef : EX \to EY$ by $e \mapsto f \circ e$ makes E an endofunctor on **Set**, as is routinely verified. Note that, since f is a total map, the domains of e and $f \circ e$ are equal; E is not only a functor:

Proposition 2. *E is a comonad, with counit ε and comultiplication δ given by*

$$\varepsilon_X : EX \to X, \ e \mapsto e(0)$$

$$\delta_X : EX \to E^2X, \ e \mapsto \begin{cases} \left(\lambda t. \begin{cases} e(s+t) & \textit{if } e(s+t){\downarrow} \\ \mathrm{undef} & \textit{if } e(s+t){\uparrow} \end{cases} \right) & \textit{if } e(s){\downarrow} \\ \mathrm{undef} & \textit{if } e(s){\uparrow} \end{cases}$$

Definition 8. *We call* (E, ε, δ) *the* evolution comonad *on* **Set**.

Note that δ transforms an evolution e on X into an evolution $\delta(e)$ on EX by acting as a *parameterised shift* or *lookahead*: $\delta(e)(t)$ is equal to $e + t$, i.e., the evolution e after $t \in \mathcal{T}$ units of time have passed, the case distinction merely taking care of potential undefinedness. Furthermore, the comonad law $\varepsilon_E \circ \delta = id_E$ states $\delta(e)(0) = e + 0 = e$, i.e., shifting by 0 is the same as not shifting at all.

Standard (image finite) LTSs are coalgebras for the behaviour functor B_A; in contrast, TTSs are partial $\mathcal{T}$-actions and need the notion of coalgebras for a comonad to account for the axioms in (3):

Proposition 3. *E-coalgebras are partial $\mathcal{T}$-actions, which in turn are TTSs; the passage from an E-coalgebra $k : P \to EP$ to a TTS $(P, \mathcal{T}, \rightsquigarrow)$, and vice versa, is given by*

$$(9) \qquad\qquad p \overset{t}{\rightsquigarrow} p' \;\Leftrightarrow\; k(p)(t) \simeq p'$$

A comonad D is *cofreely generated* by an endofunctor B if each DX is the carrier of the final $(X \times B)$-coalgebra (see e.g. [23]). One important consequence is then that D–Coalg $\cong B$–Coalg; this can be interpreted as big-step transitions (from D-coalgebras) being completely determined by one-step transitions (from B-coalgebras). For the evolution comonad and discrete time, we note the following well-known fact:

Proposition 4. $E_\mathbb{N}$ *is cofreely generated by* $B_\mathbb{N} \overset{\text{df}}{=} 1 + Id : \textbf{Set} \to \textbf{Set}$.

Hence $E_\mathbb{N}$–Coalg $\cong B_\mathbb{N}$–Coalg, i.e., any partial $\mathbb{N}$-action on some processes P corresponds to a unique function $P \to 1 + P$: for $t \in \mathbb{N}$ we have $t = 1 + \ldots + 1$ (t times); hence, knowing the 'next step' (via a $B_\mathbb{N}$-coalgebra) is by (Continuity) sufficient to define all time transitions (via an $E_\mathbb{N}$-coalgebra). We can use $B_\mathbb{N}$ to model the *qualitative* notions of time in [9,19]: a special deterministic action (χ or σ) denoting the passage of an (unspecified) amount of time (which might be thought of as the duration of a clock cycle).

For a behaviour functor B, there is a notion of *coalgebraic B-bisimulation* [1]: a *B-bisimulation* between B-coalgebras $k_1 : X_1 \to BX_1$ and $k_2 : X_2 \to BX_2$ is a B-coalgebra $k : X \to BX$ and B-coalgebra homomorphisms $f_i : X \to X_i$. For example, for B_A, one obtains strong bisimulation in this way. This definition readily generalises to comonads, yielding for the evolution comonad, via the equivalence of TTSs and partial monoid actions:

Proposition 5. *E-bisimulation is strong time bisimulation.*

For $\mathcal{T} = \mathbb{N}$, we obtain strong time bisimulation over $\mathbb{N}$; yet, such a bisimulation is completely determined by matching the next step, which is exactly given by $B_\mathbb{N}$-bisimulation, and which is also the appropriate notion of (time) bisimulation for qualitative time. The following proposition will be needed later.

Proposition 6. *E preserves pullbacks.*

Finally, we can now give a categorical description of the $\mathcal{T}$-biactions introduced above. A *distributive law of a monad* (T, η, μ) *over a comonad* (D, ε, δ) [21] is a natural transformation $\ell : TD \Rightarrow DT$ subject to the equations

$$\ell \circ \eta_D = D\eta \qquad\qquad \ell \circ \mu_D = D\mu \circ \ell_T \circ T\ell$$

and their duals

$$\varepsilon_T \circ \ell = T\varepsilon \qquad\qquad \delta_T \circ \ell = D\ell \circ \ell_D \circ T\delta$$

The ℓ-bialgebras [24] are then structures $TX \xrightarrow{h} X \xrightarrow{k} DX$, consisting of a T-algebra h and a D-coalgebra k such that $k \circ h = Dh \circ \ell \circ Tk$.

Proposition 7. *The map*

$$(10) \qquad \ell_X : \mathcal{T} \times EX \to E(\mathcal{T} \times X), \quad (t, e) \mapsto \lambda s.(t \doteq s, e(s \doteq t))$$

induces a distributive law whose bialgebras are biactions.

Hence, biactions are obtained as bialgebras, distributing total over partial actions. As total $\mathcal{T}$-actions can alternatively be described as coalgebras for the exponential comonad $(\_)^{\mathcal{T}}$ (dually to Prop. 1), an interesting open question is whether there is an entirely coalgebraic description of biactions, using a distributive law of comonads in order to combine $(\_)^{\mathcal{T}}$ with E.

3   Categorical Operational Semantics for Timed Processes

We now turn our attention to abstract operational rules for timed processes or, more abstractly, behaviour comonads. In [24], *abstract GSOS rules* were given by a natural transformation of type

$$(11) \qquad\qquad \Sigma(Id \times B) \Rightarrow BT$$

for functorial notions Σ and B of signature and (local) behaviour, with Σ freely generating the syntax monad T. Such rules then inductively induce a distributive law $TD \Rightarrow DT$ of T over the comonad D cofreely generated by B, hence distributing free syntax over cofree behaviour.

For $B_{\mathcal{A}}$, we can translate (11) into concrete rules as follows: for each set X of variables and each n-ary operator σ in the signature Σ, there is a map

$$[\![\sigma]\!] : (X \times \mathcal{P}_{\mathrm{fi}}(X)^{\mathcal{A}})^n \to \mathcal{P}_{\mathrm{fi}}(TX)^{\mathcal{A}}$$

Its arguments are n pairs of variables $x_i \in X$ and 'behaviours' $\beta_i \in \mathcal{P}_{\mathrm{fi}}(X)^{\mathcal{A}}$, interpreted as process names and transitions (described by mapping labels to targets of transitions with that label); its result is a behaviour encoding the transitions of the compound process $\sigma(x_1, \dots, x_n)$. Careful analysis, in particular of naturality, shows that this is actually equivalent to defining GSOS rules.

The Id-component in the premises of (11) is needed to associate names to behaviours; for E, with its counit $\varepsilon : E \Rightarrow Id$ available for that purpose, it suffices to use ΣE instead of $\Sigma(Id \times E)$. Inspection of rules in the literature allows even further simplification, as shown by the following proposition.

Proposition 8. *The rules of TeCCS induce a natural transformation of type*

(12) $$\Sigma E \Rightarrow E(Id + \Sigma)$$

Proof. We will only illustrate how to define some rules, leaving the proof of naturality to the reader. Consider weak choice $\oplus$ *with its (slightly adapted) rules*

$$\frac{p_1 \overset{t}{\rightsquigarrow} p_1', \; p_2 \overset{t}{\rightsquigarrow} p_2'}{p_1 \oplus p_2 \overset{t}{\rightsquigarrow} p_1' \oplus p_2'} \qquad \frac{p_1 \overset{t}{\rightsquigarrow} p_1', \; p_2 \overset{t}{\not\rightsquigarrow}}{p_1 \oplus p_2 \overset{t}{\rightsquigarrow} p_1'} \qquad \frac{p_1 \overset{t}{\not\rightsquigarrow}, \; p_2 \overset{t}{\rightsquigarrow} p_2'}{p_1 \oplus p_2 \overset{t}{\rightsquigarrow} p_2'}$$

Recalling (9), in particular $p \overset{t}{\not\rightsquigarrow}$ being equivalent to $k(p)(t)\uparrow$, we obtain the map $[\![\oplus]\!] : (EX)^2 \to E(X + \Sigma X)$,

$$[\![e_1 \oplus e_2]\!] = \lambda t. \begin{cases} e_1(t) \oplus e_2(t) & \textit{if } e_1(t)\downarrow \wedge \, e_2(t)\downarrow \\ e_1(t) & \textit{if } e_1(t)\downarrow \wedge \, e_2(t)\uparrow \\ e_2(t) & \textit{if } e_1(t)\uparrow \wedge \, e_2(t)\downarrow \\ \text{undef} & \textit{if } e_1(t)\uparrow \wedge \, e_2(t)\uparrow \end{cases}$$

Note how the cases match the rules. Writing $\mathcal{T}^+ \overset{\text{df}}{=} \mathcal{T} \setminus \{0\}$, time prefixing, with rules given in (5), yields the map $[\![(\_).\_]\!] : \mathcal{T}^+ \times EX \to E(X + \Sigma X)$, recalling $\varepsilon(e) = e(0)$:

$$[\![(t).e]\!] = \lambda s. \begin{cases} (t - s).\varepsilon(e) & \textit{if } s < t \\ \varepsilon(e) & \textit{if } s = t \\ e(s - t) & \textit{if } s > t \; \wedge \; e(s - t)\downarrow \\ \text{undef} & \textit{if } s > t \; \wedge \; e(s - t)\uparrow \end{cases}$$

More precisely, $[\![(\_).\_]\!] : \mathcal{T}^+ \times EX \to E(X + (\mathcal{T}^+ \times X)) \cong E(\mathcal{T} \times X)$, since $X + (\mathcal{T}^+ \times X) \cong (X \times \{0\}) + (X \times \mathcal{T}^+) \cong X \times (\{0\} + \mathcal{T}^+) \cong X \times \mathcal{T}$, and also $[\![(\_).\_]\!] = \ell|_{(\mathcal{T}^+ \times EX)}$, i.e., (10) restricted to arguments of type $\mathcal{T}^+ \times EX$. Note that this definition also makes sense for $t = 0$, yielding $[\![(0).e]\!] = e$ since $s < 0$ never holds, hence resulting in $[\![(\_).\_]\!] = \ell$.

We will now show how to obtain a distributive law $TE \Rightarrow ET$ from (12). Since E is no longer cofreely generated from a behaviour functor, we need to place some conditions on the rules relating them to the operations ε and δ of E (which were void in the cofree case). Despite stating the following theorem for E, it holds for arbitrary comonads (D, ε, δ).

Theorem 1. *Let Σ be a functor freely generating a monad T; furthermore suppose ρ is a natural transformation of type (12) satisfying*

(13)

$$\begin{array}{ccc} \Sigma E & \overset{\rho}{\Longrightarrow} & E(Id + \Sigma) \\ \Sigma\epsilon \downarrow\downarrow & & \downarrow\downarrow \epsilon_{Id+\Sigma} \\ \Sigma & \underset{inr}{\Longrightarrow} & Id + \Sigma \end{array}$$

$$\begin{array}{ccc} \Sigma E & \overset{\rho}{=\!=\!=\!=\!=\!\Longrightarrow} & E(Id + \Sigma) \\ \Sigma\delta \downarrow\downarrow & & \downarrow\downarrow \delta_{Id+\Sigma} \\ \Sigma E^2 & \underset{\rho E}{\Longrightarrow} E(E + \Sigma E) \underset{E[Einl, \rho]}{\Longrightarrow} & E^2(Id + \Sigma) \end{array}$$

Then this induces a distributive law $TE \Rightarrow ET$.

Note that (12) is the simplest case in a hierarchy of natural transformations describing operational rules for behaviour comonads, distinguished by the complexity of terms allowed in the right-hand side of rule conclusions; the other extreme is $\Sigma E \Rightarrow ET$, permitting arbitrary terms, as opposed to (12) allowing at most one function symbol. Increasing expressivity is traded off against the complexity of constraints to be satisfied in order to respect the operations of the comonad, with (13) also being the simplest case. For TeCCS, we obtain:

Proposition 9. *The natural transformation of Prop. 8 satisfies* (13).

Corollary 1. *Time bisimulation is a congruence for TeCCS.*

Proof. This follows from Thm. 1, Props. 5, 6, and 9, and [24, Cor. 7.5].

Note that this is a weaker congruence result than what was shown in [17]: there, the bisimulation obtained by *combining* action and time bisimulation was a congruence; here, we only obtain that each of the two on its own is a congruence (by [24], since the action rules are GSOS, and by the preceding corollary).

Discrete Time. Let us now consider the case $\mathcal{T} = \mathbb{N}$. As shown in Prop. (4), $E_\mathbb{N}$ is cofreely generated by the functor $B_\mathbb{N}$, enabling us to use (11) to define the rules and to introduce the following syntactic rule format. Let Σ be a signature and fix an enumeration (without repetitions) $\{x_k \mid k \geq 1\}$ of a countable set X of variables. Our rule format will use GSOS [7] rules of the form

$$(14) \qquad \frac{\{x_i \rightsquigarrow x_{n+i}\}_{i \in I},\ \{x_j \not\rightsquigarrow\}_{j \in J}}{\sigma(x_1, \ldots, x_n) \rightsquigarrow \theta}$$

where $\sigma \in \Sigma$ is an n-ary function symbol, $I, J \subseteq \{1, \ldots, n\}$, and θ a term over Σ and X. The fact that we consider GSOS rules means that θ contains no fresh variables, i.e., all variables in θ must occur somewhere else in the rule[4]. Moreover, for each x_k, $1 \leq k \leq n$, there is at most one positive and one negative premise, and the targets of the positive premises (if there are any) are fixed to be the next $|I|$ variables after x_n in the order of the enumeration.

A rule (14) is *consistent* if $I \cap J = \emptyset$, *complete* if $I \cup J = \{1, \ldots, n\}$, and it has *type* (I, J); two such rules, with respective types (I_k, J_k), are *mutually exclusive* if $(I_1 \cap J_2) \cup (I_2 \cap J_1) \neq \emptyset$, i.e., there is at least one variable occurring both positively (i.e., in a positive premise) in one rule and negatively in the other; for consistent rules, note that this is equivalent to $I_1 \neq I_2$ (since $J_k = \{1, \ldots, n\} \setminus I_k$). Say then that a set of consistent and complete rules (14) is in *deterministic single-label GSOS (dslGSOS)* format over X if any two rules for the same operator $\sigma \in \Sigma$ are mutually exclusive.

Consistent rules have no conflicting premises, i.e., no variable occurs both positively and negatively in the same rule; yet, since we only use complete rules

---

[4] Also all variables in $\{x_k \mid 1 \leq k \leq n + |I|\}$ must be pairwise distinct, but that is already guaranteed by our use of the enumeration.

for the dslGSOS format, each x_k must occur inside the premises, therefore, it occurs in *exactly one* premise. Mutual exclusion then ensures for each operator that there is at most one rule applicable at a time (assuming each x_k has either no or exactly one next step, as defined by a $B_\mathbb{N}$-coalgebra). Note that there can only be finitely many rules for each operator $\sigma \in \Sigma$ without violating mutual exclusion; hence, if Σ is finite, each set of rules in dslGSOS format is automatically finite. Furthermore, mutual exclusion is a *global* condition on *sets* of rules, unlike the *local* variable conditions of the GSOS format, or consistency and completeness, which refer only to single rules; also the restriction to image-finite sets of GSOS rules in [24], ensuring the one-to-one correspondence between such sets of rules and natural transformations (11) for the functor B_A, is a weak but nevertheless global condition.

Theorem 2. *There is a one-to-one correspondence between rules in dslGSOS format and natural transformations of type*

$$\Sigma(Id \times B_\mathbb{N}) \Rightarrow B_\mathbb{N} T$$

Proof. The proof is done along the lines of the proof of [24, Thm 1.1]. We obtain an exact correspondence (not just up to equivalence of sets of rules as in [24]) since the dslGSOS format, deploying the variable enumeration, unambiguously prescribes which variables occur at which places in the rules.

An example of rules in the above format is given by the time rules of the language ATP [19]. We have also defined a new one-step version of the time rules of TeCCS fitting in dslGSOS. Using the above theorem and the equivalence between $E_\mathbb{N}$- and $B_\mathbb{N}$-coalgebras, we were then able to prove categorically that the one-step rules induce the same TTS as the original (big-step) rules.

Note that for both the above languages, the action rules are GSOS, hence we can describe them using the behaviour functor B_A and (11). For the full languages, combining the two sets of rules given by two natural transformations $\rho_A : \Sigma(Id \times B_A) \Rightarrow B_A T$ and $\rho_\mathbb{N} : \Sigma(Id \times B_\mathbb{N}) \Rightarrow B_\mathbb{N} T$, precomposing ρ_A and $\rho_\mathbb{N}$ with the respective projections yields natural transformations

(15) $\Sigma(Id \times (B_A \times B_\mathbb{N})) \Rightarrow B_A T$ $\Sigma(Id \times (B_A \times B_\mathbb{N})) \Rightarrow B_\mathbb{N} T$

Since $(B_A \times B_\mathbb{N})T \cong B_A T \times B_\mathbb{N} T$, (15) is equivalent to (11) instantiated with $B = B_A \times B_\mathbb{N}$. Thus, by [24], the combined bisimulation is a congruence for both languages since it corresponds exactly to coalgebraic bisimulation for $B_A \times B_\mathbb{N}$.

Future Work

Most pressingly, we plan to give a syntactic characterisation of the abstract format (12) for timed processes for the general case: this is as yet beyond the scope of this paper but we envisage a format corresponding to (12) on which to impose (global) side conditions corresponding to (13).

For discrete time, (15) should also allow to treat such calculi where action and time transitions are no longer defined independently, in particular by adopting the *maximal progress assumption* [11], e.g., (discrete time) TiCCS [25] and TPL [9]. We want to extend this to the case of an arbitrary time domain, also to obtain a congruence result for the combined bisimulation for (big-step) TeCCS. Hopefully, we can achieve this by instantiating (12) with products of comonads (alike to using products of functors in the discrete case).

In a different direction, we will investigate the idea of *normal forms*: for instance, the normal forms of $(0).p$ and $(s).(t).p$ would be p and $(s + t).p$. In order to achieve this, we could use a retraction $T_{NF} \lhd T$ for two (syntax) monads, corresponding to rewriting terms into normal forms and the inclusion of normal forms into terms, coupled with a distributive law $T_{NF}E \Rightarrow ET_{NF}$ (induced by rules like (12)) to model the operational semantics of normal forms only; the retraction should then induce a distributive law for the full language.

The last point seems closely connected to defining rules corresponding to (12) in the category of $\mathcal{T}$-actions, using a more complex syntax with a 'built-in' total $\mathcal{T}$-action corresponding to time prefixing; as behaviour, we would use the *lifting* (see [24]) of E obtained by the distributive law (10). This might also clear up the somewhat mysterious rôle of said law, since currently it is only used in a restricted form (by the rules for time prefixing, see Prop. 8), although clearly an important mathematical structure.

Finally, we hope to be able to deal with operational semantics for timed automata [2], a very prominent approach to formalising real-time systems. Treating also this quite different (from the process-algebraic point of view) approach within the same or a similar framework as in the present paper would even further emphasise the flexibility of the categorical framework.

Acknowledgements. The author would like to thank Daniele Turi and Gordon Plotkin for numerous helpful discussions.

References

1. P. Aczel and P. F. Mendler. A final coalgebra theorem. In D. H. Pitt, D. E. Rydeheard, P. Dybjer, A. M. Pitts, and A. Poigné, editors, *Category Theory and Computer Science, LNCS* 389, pp. 357–365, 1989. Springer.
2. R. Alur and D. L. Dill A theory of timed automata. *Theoretical Computer Science* **25(2)**, pp. 183–235, 1994.
3. J.C.M. Baeten and J.W. Klop, editors. *Concurrency Theory (CONCUR '90), LNCS* 458, 1990. Springer.
4. J.C.M. Baeten and C.A. Middelburg. Process algebra with timing: real time and discrete time. In Bergstra et al. [5], chapter 10.
5. J. A. Bergstra, A. Ponse, and S. A. Smolka, editors. *Handbook of Process Algebra.* North-Holland, 2001.
6. J.A. Bergstra and J.W. Klop. Process algebra for synchronous communication. *Information and Computation* **60**, pp. 109–137, 1984.
7. B. Bloom, S. Istrail, and A. R. Meyer. Bisimulation can't be traced. *Journal of the ACM* **42(1)**, pp. 232–268, 1995.

8. R.J. van Glabbeek. The linear time – branching time spectrum I; the semantics of concrete, sequential processes. In Bergstra et al. [5], chapter 1, pages 3–99.

9. M. Hennessy and T. Regan. A process algebra for timed systems. *Information and Computation* **117**, pp. 221–239, 1995.

10. C.A.R. Hoare. *Communicating Sequential Processes.* Prentice Hall, 1985.

11. J.J.M. Hooman and W.P. de Roever. Design and verification in real-time distributed computing: an introduction to compositional methods. In E. Brinksma, G. Scollo, and Chris A. Vissers, editors. *International Conference on Protocol Specification, Testing and Verification,* 1989. North-Holland.

12. A. Jeffrey. A linear time process algebra. In Larsen and Skou [14], pp. 432–442.

13. A. S. A. Jeffrey, S. A. Schneider, and F. W. Vaandrager. A comparison of additivity axioms in timed transition systems. Technical Report CS-R9366, CWI, 1993.

14. K.G. Larsen and A. Skou, editors. *Computer Aided Verification (CAV '91), LNCS* 575, 1991. Springer.

15. W. Lawvere. Metric spaces, generalized logic, and closed categories. In *Rendiconti del Seminario Matematico e Fisico di Milano, XLIII.* Tipografia Fusi, 1973.

16. R. Milner. *Communication and Concurrency.* Prentice Hall, 1989.

17. F. Moller and C. Tofts. A temporal calculus of communicating systems. In Baeten and Klop [3], pp. 401–415.

18. X. Nicollin and J. Sifakis. An overview and synthesis on timed process algebras. In Larsen and Skou [14], pp. 376–398.

19. X. Nicollin and J. Sifakis. The algebra of timed processes, ATP: Theory and application. *Information and Computation* **114**, pp. 131–178, 1994.

20. G. D. Plotkin. A structural approach to operational semantics. Technical Report DAIMI FN-19, Computer Science Department, Aarhus University, 1981.

21. J. Power and H. Watanabe. Distributivity for a monad and a comonad. In B. Jacobs and J. Rutten, editors, *Second Workshop on Coalgebraic Methods in Computer Science (CMCS'1999),* volume 19 of *ENTCS,* 1999.

22. S. A. Schneider. An operational semantics for timed CSP. Technical Report PRG-TR-1-91, Oxford University, 1991.

23. D. Turi. *Functorial Operational Semantics and its Denotational Dual.* PhD thesis, Free University, Amsterdam, 1996.

24. D. Turi and G. Plotkin. Towards a mathematical operational semantics. In *Twelfth Annual Symposium on Logic in Computer Science (LICS '97),* pp. 280–291, 1997. IEEE Computer Society Press.

25. Y. Wang. Real-time behaviour of asynchronous agents. In Baeten and Klop [3], pp. 502–520.

Testing Labelled Markov Processes

Franck van Breugel[1], Steven Shalit[2], and James Worrell[2]

[1] York University, Department of Computer Science
4700 Keele Street, Toronto, M3J 1P3, Canada
[2] Tulane University, Department of Mathematics
6823 St Charles Avenue, New Orleans LA 70118, USA

Abstract. Larsen and Skou introduced a notion of bisimulation for probabilistic transition systems. They characterized probabilistic bisimilarity in terms of a probabilistic modal logic and also in terms of 'button pressing' tests. Desharnais et al. extended the notion of probabilistic bisimulation and the logical characterization of probabilistic bisimilarity to labelled Markov processes. These processes generalize probabilistic transition systems in that they also allow continuous state spaces. We extend the characterization of probabilistic bisimilarity in terms of testing to labelled Markov processes. One of our main technical contributions is the construction of a final object in a category of labelled Markov processes and the identification of a natural metric on the state space of the final labelled Markov process. This metric provides us with another characterization of probabilistic bisimilarity: states are probabilistic bisimilar if and only if they have distance 0.

Introduction

In their influential paper [19], Larsen and Skou present an elegant theory of probabilistic transition systems. We will not discuss the importance of such a theory here, but we refer the reader to op. cit. for such a discussion. A probabilistic transition system reacts to actions of the environment. These reactions are specified by means of probability distributions on the set of states of the system. Larsen and Skou adapt the notion of a bisimulation, the key behavioural equivalence for ordinary transition systems due to Milner and Park, to the probabilistic setting. Their paper contains two fundamental characterizations of probabilistic bisimilarity. First of all, they introduce a probabilistic variant of Hennessy Milner logic [18]. They show that

> states are probabilistic bisimilar if and only if they satisfy the same probabilistic modal logic formulae.

This result is very similar to the logical characterization of ordinary bisimilarity in terms of Hennessy Milner logic presented in op. cit. To prove their logical characterization of probabilistic bisimilarity, Larsen and Skou restrict themselves

---

[1] Supported by the Natural Sciences and Engineering Research Council of Canada.
[2] Supported by the US Office of Naval Research.

P. Widmayer et al. (Eds.): ICALP 2002, LNCS 2380, pp. 537–548, 2002.

to the class of systems that satisfy the minimal deviation assumption. This assumption roughly states that the probability distributions, which specify the reactions of the system, have finite support. Secondly, Larsen and Skou present a collection of 'button pressing' tests for probabilistic transition systems, similar to the ones introduced by Abramsky [1] for ordinary transition systems. They prove that

> states are probabilistic bisimilar if and only if they yield the same probability distribution for each test.

Abramsky proves a similar result for ordinary transition systems in op. cit. Larsen and Skou's proof of the above result also relies on the minimal deviation assumption.

In [8,12,13], Desharnais, Edalat and Panangaden present a foundational study of labelled Markov processes. These processes generalize probabilistic transition systems in that they also allow continuous state spaces. To deal with such state spaces, probability distributions are replaced with probability measures. We will not discuss the importance of this generalization here, but we refer the reader to the work of Desharnais et al. where an extensive discussion can be found. In [8] they introduce the notion of probabilistic bisimilarity for labelled Markov processes. For probabilistic transition systems, this notion coincides with the original notion introduced by Larsen and Skou. In [12] they extend the logical characterization of probabilistic bisimularity to labelled Markov processes. Surprisingly, they show that the probabilistic modal logic of Larsen and Skou can be simplified considerably. Furthermore, their proof demonstrates that the minimal deviation assumption is not needed. The nature of their proof is quite different from the proof of Larsen and Skou in that it uses some of the tools of measure theory.

In this paper we generalize the characterization of probabilistic bisimilarity in terms of testing from probabilistic transition systems to labelled Markov processes. In contrast to Larsen and Skou's proof, ours does not rely on the minimal deviation assumption and uses some measure theory. This result is interesting because the presentation of probabilistic bisimilarity for labelled Markov processes is hedged around measurability conditions, and is correspondingly less intuitive as a notion of observational equivalence.

In [14], Desharnais, Gupta, Jagadeesan and Panangaden present yet another characterization of probabilistic bisimilarity. They define a pseudometric[3] on the states of a probabilistic transition system using functional expressions. Roughly speaking, these functional expressions arise from a non-standard real-valued semantics for Larsen and Skou's probabilistic modal logic. The idea is that the pseudometric measures the behavioural proximity of states of a probabilistic transition system. In particular, they demonstrate that

> states are probabilistic bisimilar if and only if they have distance 0.

---

[3] A pseudometric satisfies the same axioms as a metric, except that different elements can have distance 0.

They extend their results to labelled Markov processes in [15]. Besides giving rise to the above characterization, such a pseudometric is attractive as it provides us with a robust notion of approximate behavioural equivalence. Probabilistic bisimilarity is not robust in the sense that a very small perturbation in the probabilities associated with a process makes bisimilar states non-bisimilar and vice versa. For a more detailed discussion we refer the reader to [11].

One of the main technical contributions of this paper is the construction of a final object in the category of labelled Markov processes and zig-zag maps. A zig-zag map between labelled Markov processes may be thought of as a functional bisimulation. By definition, there exists a unique zig-zag map from a given labelled Markov process to the final labelled Markov process. This map provides another characterization of probabilistic bisimilarity, since

> states are probabilistic bisimilar if and only if the unique zig-zag map from the labelled Markov process to the final labelled Markov process identifies them.

We view labelled Markov processes as coalgebras of the Giry functor on the category of measurable spaces [17]. In our previous papers [9,10,11] we worked with coalgebras of a functor on the category of metric spaces whose definition was based on the Hutchinson metric. The construction of the final labelled Markov process involves an interplay between these two viewpoints and allows us to transfer results proved in our previous papers to the present more general setting. In particular, we show that the state space of the final labelled Markov process carries a natural metric, and prove that this metric is closely related to the pseudometric of Desharnais et al. As a consequence we obtain a new proof of the fact, first shown in [14], that states are probabilistic bisimilar if and only if they take the same value on all functional expressions. The link between functional expressions and the final labelled Markov process plays a key role in our proof that probabilistic bisimilarity coincides with testing equivalence.

1   Labelled Markov Processes and Bisimulation

A *labelled Markov process* consists of a set X of states, a σ-field Σ on X, a finite[4] set Act of actions and a transition probability function $\mu : X \times Act \times \Sigma \to [0, 1]$ such that

1. for all $x \in X$ and $a \in Act$, the function $\mu(x, a, \cdot) : \Sigma \to [0, 1]$ is a subprobability measure, and
2. for all $a \in Act$ and $A \in \Sigma$, the function $\mu(\cdot, a, A) : X \to [0, 1]$ is measurable.

Instead of $\mu(x, a, A)$ we will often write $\mu_{x,a}(A)$ or $\mu_a(x, A)$. The function μ_a describes the reaction of the process to the action a selected by the environment. This represents a reactive model of probabilistic processes. Given the process is

---

[4] We restrict to finite action sets for ease of exposition, but our main results still hold for countable action sets.

in state x and reacts to action a chosen by the environment, $\mu_{x,a}(A)$ is the probability that the process makes a transition to a state in the set A. Note that this is a conditional probability. Also notice that we consider *sub*probability measures, i.e., measures whose total mass may be smaller than 1. This allows for the possibility that the process may refuse an action. The probability of refusal of the action a given the process is in state x is $1 - \mu_{x,a}(X)$.

An important special case is when the σ-field Σ is taken to be the powerset of X and, for all states x and actions a, the subprobability measure $\mu_{x,a}$ is completely determined by a discrete probability distribution. This case corresponds to the original model of Larsen and Skou.

Given labelled Markov processes $\langle X, \Sigma, \mu \rangle$ and $\langle X', \Sigma', \mu' \rangle$, a measurable function $f : X \to X'$ is a *zig-zag map* if for all $x \in X$, $a \in Act$ and $A \in \Sigma'$,

$$\mu_a(x, f^{-1}(A)) = \mu'_a(f(x), A).$$

Let $\langle X, \Sigma, \mu \rangle$ be a labelled Markov process. An equivalence relation R on X is a *probabilistic bisimulation* if $x \, R \, y$ implies $\mu_{x,a}(A) = \mu_{y,a}(A)$ for all $a \in Act$ and R-closed and measurable sets A of X. A set A is R-closed if $x \in A$ and $x \, R \, y$ implies $y \in A$. States are *probabilistic bisimilar* if they are related by a probabilistic bisimulation.

Probabilistic bisimulations are the relational counterparts of zig-zag maps. As shown in [19], they can also be seen in a very precise way as the probabilistic analogs of the strong bisimulations of Milner and Park. A probabilistic bisimulation divides the state space into blocks, such that for all states in the same block, the probability of making a transition to any measurable amalgamation of blocks is the same. The presence of the measurability conditions make this notion less intuitively compelling than the discrete version. However, in Section 4, we characterize probabilistic bisimilarity as indistinguishability in the simple 'button pressing' testing framework of Larsen and Skou.

2   A Final Labelled Markov Process

Labelled Markov processes and zig-zag maps form a category. In this section we construct the final object of this category. Furthermore, we characterize probabilistic bisimilarity on the states of a given labelled Markov process as the kernel of the unique zig-zag map to the final labelled Markov process.

First, we observe that the category of labelled Markov processes is equal to the category of $\mathcal{G}$-coalgebras, where $\mathcal{G}$ is an endofunctor on the category $\mathbb{M}es$ of measurable spaces and measurable functions. Hence, it suffices to construct a final $\mathcal{G}$-coalgebra. Second, we consider the category of $\mathcal{H}$-coalgebras, where $\mathcal{H}$ is an endofunctor on the category $\mathbb{K}\mathbb{M}et$ of compact metric spaces and nonexpansive functions. A function is nonexpansive if it does not increase any distances. The category of $\mathcal{H}$-coalgebras consists of labelled Markov processes whose state space is endowed with a metric, and whose zig-zag maps are nonexpansive. In [9] we showed that this category has a final object exploiting the metric final coalgebra theorem of Rutten and Turi [21]. Here we present an alternative construction of the final $\mathcal{H}$-coalgebra. Third, we consider the functor $\mathcal{B}$ from $\mathbb{KMet}$ to $\mathbb{Mes}$, which maps a metric space $\langle X, d \rangle$ to the Borel measurable space on X generated by the metric topology induced by the metric d. We show that this functor $\mathcal{B}$ maps the constructed final $\mathcal{H}$-coalgebra to a final $\mathcal{G}$-coalgebra. Although a direct construction of the final labelled Markov process may be feasible, by using the Kolmogorov consistency theorem [20, Theorem V.5.1], the present construction allows us to transfer the metric on the final $\mathcal{H}$-coalgebra to the final $\mathcal{G}$-coalgebra. This metric plays a crucial role in the rest of this paper.

Next, we introduce the category of coalgebras and homomorphisms. Let $\mathbb{C}$ be a category and $\mathcal{F} : \mathbb{C} \to \mathbb{C}$ a functor. An $\mathcal{F}$-*coalgebra* $\langle C, f \rangle$ consists of an object C in $\mathbb{C}$ together with an arrow $f : C \to \mathcal{F}(C)$ in $\mathbb{C}$. An $\mathcal{F}$-*homomorphism* from $\langle C, f \rangle$ to an $\mathcal{F}$-coalgebra $\langle D, g \rangle$ is an arrow $h : C \to D$ in $\mathbb{C}$ such that $\mathcal{F}(h) \circ f = g \circ h$.

We write $\mathcal{G}\langle X, \Sigma \rangle$ for the set of subprobability measures on a measurable space $\langle X, \Sigma \rangle$. For each $A \in \Sigma$ we have a projection function $p_A : \mathcal{G}\langle X, \Sigma \rangle \to [0, 1]$ sending measure μ to $\mu(A)$. We take $\mathcal{G}\langle X, \Sigma \rangle$ to be a measurable space by giving it the smallest σ-field such that all the projections p_A are measurable. Next, $\mathcal{G}$ is turned into an endofunctor on the category $\mathbb{Mes}$ by defining $\mathcal{G}(f)(\mu) = \mu \cdot f^{-1}$. This functor is studied in detail by $\mathcal{G}$iry [17].

Let $c \in (0, 1)$ be fixed once and for all. Given a metric space $\langle X, d \rangle$, we write $\langle X, d \rangle \xrightarrow{c} [0, 1]$ for the set of c-contractive functions from X to $[0, 1]$. A function is c-contractive if it decreases all distances by at least a factor c. We write $\mathcal{H}\langle X, d \rangle$ for the set of Borel subprobability measures on $\langle X, d \rangle$ equipped with the metric d_H defined by

$$d_H(\mu, \nu) = \sup \left\{ \int f d\mu - \int f d\nu \mid f \in \langle X, d \rangle \xrightarrow{c} [0, 1] \right\}.$$

This metric is variously known as the $\mathcal{H}$utchinson metric, the Kantorovich metric and also as the Wasserstein metric. The metric metrizes the weak topology on $\mathcal{H}\langle X, d \rangle$, viz. the weakest topology such that for all continuous $f : \langle X, d \rangle \to [0, 1]$ the map $\mu \mapsto \int f d\mu$ is itself continuous [16, Theorem 2.5.17]. Also, recall that $\mathcal{H}\langle X, d \rangle$ is compact whenever $\langle X, d \rangle$ is a compact metric space [5, Theorem 9.5.1]. We extend $\mathcal{H}\langle X, d \rangle$ to an endofunctor $\mathcal{H}$ on the category $\mathbb{KMet}$ by defining $\mathcal{H}(f)(\mu) = \mu \cdot f^{-1}$. This functor has the important property of being locally contractive, as we show in [9, Proposition 1].

Given a labelled Markov process $\langle X, \Sigma, \mu \rangle$, μ may be regarded as a measurable map from $\langle X, \Sigma \rangle$ to $\mathcal{G}\langle X, \Sigma \rangle^{Act}$, where $(\cdot)^{Act}$ denotes the Act-fold product. That is, labelled Markov processes are nothing but coalgebras of the functor $\mathcal{G}(\cdot)^{Act} : \mathbb{Mes} \to \mathbb{Mes}$. Furthermore, the coalgebra homomorphisms in this case are just the zig-zag maps. In our previous papers [9,10,11] we consider the class of coalgebras of the functor $\mathcal{H}(\cdot)^{Act} : \mathbb{KMet} \to \mathbb{KMet}$. For such a coalgebra $\langle \langle X, d \rangle, \mu \rangle$, the map $\mu : \langle X, d \rangle \to \mathcal{H}\langle X, d \rangle^{Act}$ is nonexpansive. This requirement is trivially satisfied if d is a discrete metric; in the continuous case it is a significant restriction. However, in the present paper we show how to transfer results from the more restrictive setting of $\mathcal{H}(\cdot)^{Act}$-coalgebras to the general setting of $\mathcal{G}(\cdot)^{Act}$-coalgebras. For simplicity, in the remaining part of this section we consider coalgebras of the functors $\mathcal{H}$ and $\mathcal{G}$. That is, we restrict ourselves to the unlabelled case. All of our results can easily be extended to deal with labels.

Theorem 1. *There is a final $\mathcal{H}$-coalgebra $\langle \mathcal{H}^\omega(1), \iota \rangle$.*

Proof sketch. Starting with the final object 1 of $\mathbb{KMet}$, we construct the chain

$$1 \xleftarrow{\;!\;} \mathcal{H}(1) \xleftarrow{\mathcal{H}(!)} \mathcal{H}^2(1) \xleftarrow{\mathcal{H}^2(!)} \cdots \tag{1}$$

where ! is the unique map from $\mathcal{H}(1)$ to 1. Since $\mathcal{H}$ is locally contractive, the chain (1) has a limit in $\mathbb{KMet}$, denoted $\mathcal{H}^\omega(1)$, and $\mathcal{H}$ preserves this limit according to [3, Lemma 3.13]. As a consequence the canonical map from the limit object $\mathcal{H}(\mathcal{H}^\omega(1))$ to the limit object $\mathcal{H}^\omega(1)$ is an isomorphism, whose inverse we denote ι. According to (the dual of) [2, Proposition 5], $\langle \mathcal{H}^\omega(1), \iota \rangle$ is a final coalgebra.   □

Next, we show that the above mentioned functor $\mathcal{B}$ maps the final $\mathcal{H}$-coalgebra to a final $\mathcal{G}$-coalgebra.

Suppose $\langle X, d \rangle$ is a separable metric space[5]. Let $\mathcal{O}$ be a countable basis of open subsets of $\langle X, d \rangle$ which is a π-system, i.e., closed under finite intersections. Every open subset of $\langle X, d \rangle$ is a countable union of sets from $\mathcal{O}$, so $\sigma(\mathcal{O})$, the σ-field generated by $\mathcal{O}$, equals $\mathcal{B}\langle X, d \rangle$. The set of subprobability measures on $\mathcal{B}\langle X, d \rangle$ can be given a σ-field structure in the following ways.

- Σ_B is the smallest σ-field such that p_A is measurable for each $A \in \mathcal{B}\langle X, d \rangle$.
- Σ_O is the smallest σ-field such that p_A is measurable for each $A \in \mathcal{O}$.
- Σ_H is the Borel σ-field on $\mathcal{H}\langle X, d \rangle$.

Lemma 1. $\Sigma_B = \Sigma_O = \Sigma_H$.

Proof sketch. (i) $\Sigma_B = \Sigma_O$. Clearly $\Sigma_O \subseteq \Sigma_B$. For the converse, consider

$$\mathcal{L} = \{\, A \in \mathcal{B}\langle X, d \rangle \mid p_A \text{ is } \Sigma_O\text{-measurable} \,\}.$$

$\mathcal{L}$ is a λ-system, i.e., it is closed under countable disjoint unions, complements and contains the empty set. Also, by definition of Σ_O, we have that the π-system $\mathcal{O}$ is contained in $\mathcal{L}$. By the $\lambda - \pi$ theorem [6, Theorem 3.2] we have $\sigma(\mathcal{O}) \subseteq \mathcal{L}$. Since $\mathcal{B}\langle X, d \rangle = \sigma(\mathcal{O})$ and $\mathcal{L} \subseteq \mathcal{B}\langle X, d \rangle$, we have that $\mathcal{L} = \mathcal{B}\langle X, d \rangle$. Thus $\Sigma_B \subseteq \Sigma_O$ by minimality of Σ_B.

(ii) $\Sigma_O = \Sigma_H$. Σ_O is countably generated by sets of the form $p_A^{-1}(q, 1]$ for $q \in [0, 1] \cap \mathbb{Q}$ and $A \in \mathcal{O}$. However $p_A^{-1}(q, 1] \in \Sigma_H$ since this set is actually open in the metric topology on $\mathcal{H}\langle X, d \rangle$, i.e., the weak topology[6]. Thus $\Sigma_O \subseteq \Sigma_H$. Now the unique structure theorem [4, Theorem 3.3.5] says that any sub-σ-field of a Borel σ-field on a separable space which is countably generated and separates points is in fact equal to the whole Borel σ-field. This concludes our argument.   □

---

[5] A compact metric space is separable.

[6] According to [20, Theorem II.6.1], a net $\{\mu_\alpha\}_\alpha$ converges to μ in the weak topology if and only if for every open set A of X, $\liminf_\alpha \mu_\alpha(A) \geq \mu(A)$.

Corollary 1. $\mathcal{G} \circ \mathcal{B} = \mathcal{B} \circ \mathcal{H}$.

Lemma 2. *The functor* $\mathcal{B}$

1. maps the chain (1) to the chain

$$1 \xleftarrow{\ !\ } \mathcal{G}(1) \xleftarrow{\mathcal{G}(!)} \mathcal{G}^2(1) \xleftarrow{\mathcal{G}^2(!)} \cdots \qquad (2)$$

where 1 is the final object in Mes *and* ! *is the unique map from* $\mathcal{G}(1)$ *to* 1.
2. maps the limit object $\mathcal{H}^\omega(1)$ *to the limit object* $\mathcal{G}^\omega(1)$ *of (2).*

Theorem 2. *The functor* $\mathcal{B}$ *maps the final* $\mathcal{H}$*-coalgebra* $\langle \mathcal{H}^\omega(1), \iota \rangle$ *to a final* $\mathcal{G}$*-coalgebra.*

Proof sketch. Clearly, the $\mathcal{B}$-image of ι is an isomorphism as well. Since

$$\mathcal{B}\left(\mathcal{H}\left(\mathcal{H}^\omega(1)\right)\right) = \mathcal{G}\left(\mathcal{B}\left(\mathcal{H}^\omega(1)\right)\right) \quad \text{[Corollary 1]}$$
$$= \mathcal{G}\left(\mathcal{G}^\omega(1)\right) \quad \text{[Lemma 2.2]}$$

the result follows from Lemma 2.1 and (the dual of) [2, Proposition 5].   □

Let us denote the compact metric space $\mathcal{H}^\omega(1)$ by $\langle X_\omega, d_\omega \rangle$. Then the measurable space $\mathcal{G}^\omega(1)$ consists of the set X_ω and the Borel σ-field Σ_ω on X_ω induced by the metric d_ω. As a consequence, we can transfer the metric d_ω from the final $\mathcal{H}$-coalgebra to the final $\mathcal{G}$-coalgebra.

Let $\langle X, \Sigma, \mu \rangle$ be a labelled Markov process, whose state space is analytic and whose σ-field is the Borel σ-field on X. The importance of this restriction is discussed in [12]. We denote the unique map from this labelled Markov process to the final labelled Markov process $\langle X_\omega, \Sigma_\omega, \iota \rangle$ by $[\![\cdot]\!]$. Probabilistic bisimilarity can be characterized as the kernel of this map.

Theorem 3. *For all* $x, y \in X$, *the states* x *and* y *are probabilistic bisimilar if and only if* $[\![x]\!] = [\![y]\!]$.

3   A Pseudometric

As we already mentioned in the introduction, another characterization of probabilistic bisimilarity is that states are probabilistic bisimilar if and only if they satisfy the same formulae of the probabilistic modal logic of Larsen and Skou. The formulae of this logic are given by

$$\phi ::= \top \mid \phi \wedge \phi \mid \neg \phi \mid \langle a \rangle_q \phi$$

where $a \in Act$ and $q \in [0,1] \cap \mathbb{Q}$. The semantics of this logic is given by a satisfaction relation between the set of formulae and the set X of states of a labelled Markov process. In particular, state x satisfies formula $\langle a \rangle_q \phi$ if the mass of the set of states satisfying ϕ with respect to the measure $\mu_{x,a}$ exceeds q.

Motivated by this idea, Desharnais et al. [14,15] introduce a 'logic' for labelled Markov processes based on a syntactic class of functional expressions. These functional expressions get interpreted as $[0, 1]$-valued measurable functions on the state space of a labelled Markov process. The set $\mathcal{F}$ of functional expressions is generated by

$$f ::= 1 \mid \min(f, f) \mid 1 - f \mid \langle a \rangle f \mid f \overset{.}{-} q$$

where $a \in Act$ and $q \in [0, 1] \cap \mathbb{Q}$. Informally, there is the following correspondence between functional expressions and probabilistic modal logic formulae. $\top$ is represented by 1, conjunction by min, negation by $1-$, and $\langle a \rangle_q$ splits into $\langle a \rangle$ and $\overset{.}{-} q$. There is a little poetic license in this analogy, however we will see that functional expressions play an analogous role to probabilistic modal logic formulae in characterizing probabilistic bisimilarity in terms of testing.

The semantics of functional expressions is parameterized by $c \in (0, 1)$. This corresponds to the contraction in the definition of the functor $\mathcal{H}$; its role here is to give greater weight to observations made at smaller modal depth. Given a labelled Markov process $\langle X, \Sigma, \mu \rangle$, with X an analytic space and Σ the Borel σ-field on X, to each functional expression f there corresponds a map $f_{\langle X, \Sigma, \mu \rangle}$ from X to $[0, 1]$ defined by structural induction on f. We define $1_{\langle X, \Sigma, \mu \rangle}(x) = 1$ and

$$(\langle a \rangle f)_{\langle X, \Sigma, \mu \rangle}(x) = c \cdot \int f_{\langle X, \Sigma, \mu \rangle} \, d\mu_{x, a}.$$

The other clauses are interpreted in the obvious manner[7], and it is straightforward to verify by structural induction on $f \in \mathcal{F}$ that $f_{\langle X, \Sigma, \mu \rangle}$ is measurable.

Now, following Desharnais et al., we define a pseudometric $d_{\langle X, \Sigma, \mu \rangle}$ on the state space X of the labelled Markov process by

$$d_{\langle X, \Sigma, \mu \rangle}(x, y) = \sup \{ f_{\langle X, \Sigma, \mu \rangle}(x) - f_{\langle X, \Sigma, \mu \rangle}(y) \mid f \in \mathcal{F} \}.$$

The main result in this section says that the distance between states in the pseudometric $d_{\langle X, \Sigma, \mu \rangle}$ is given by the distance of their images in the final labelled Markov process. In particular, by Theorem 3, states are probabilistic bisimilar just in case they take the same values under all functional expressions. The latter result was first proven in [14, Theorem 1].

The pseudometric $d_{\langle X, \Sigma, \mu \rangle}$ and the metric d_ω on the final labelled Markov process are related via the unique zig-zag map $[\![\cdot]\!]$ as follows.

Theorem 4. *For all* $x, y \in X$, $d_{\langle X, \Sigma, \mu \rangle}(x, y) = c \cdot d_\omega([\![x]\!], [\![y]\!])$.

This theorem extends [9, Theorem 3], where the labelled Markov process $\langle X, \Sigma, \mu \rangle$ is assumed to be an $\mathcal{H}(\cdot)^{Act}$-coalgebra. The extension of this result to the present setting relies on the fact that the final labelled Markov process can be seen as an $\mathcal{H}(\cdot)^{Act}$-coalgebra. Once this reduction is made, the key idea is that since the class of functional expressions is closed under min and max, we can use the Stone-Weierstrass theorem to show that any nonexpansive map $\langle X_\omega, d_\omega \rangle \to [0, 1]$ can be uniformly approximated by functional expressions.

---

[7] $r \overset{.}{-} q$ is equal to $r - q$ if $r \geq q$ and 0 otherwise.

Using Theorem 4, in conjunction with some ideas presented in [10, Section 5], one can show that $d_{\langle X, \Sigma, \mu \rangle}$ can be uniformly approximated to within $2 \cdot c^n$ by considering functional expressions of modal depth smaller than n. This provides an alternative proof of [14, Lemma 5]. This lemma, reproduced below, is used in the proof of the coincidence of probabilistic bisimilarity and testing equivalence in the next section.

Lemma 3. *Given $\varepsilon > 0$, there exists a finite set $\mathcal{F}_F$ of functional expressions such that for any labelled Markov process $\langle X, \Sigma, \mu \rangle$ and for all x, $y \in X$,*

$$0 \le d_{\langle X, \Sigma, \mu \rangle}(x, y) - \sup \left\{ f_{\langle X, \Sigma, \mu \rangle}(x) - f_{\langle X, \Sigma, \mu \rangle}(y) \mid f \in \mathcal{F}_F \right\} < \varepsilon.$$

4 Testing

As we mentioned earlier, 0 distance in our pseudometric coincides with probabilistic bisimilarity. Now we consider the correspondence with probabilistic testing as formalized by Larsen and Skou. We extend their result that probabilistic bisimilarity coincides with testing equivalence to the more general setting of labelled Markov processes. Furthermore, we give substance to the intuition that the pseudometric measures the behavioural proximity of states by showing that the pseudometric metrizes the topology induced by the probability distributions of the tests.

There is an extensive literature on testing processes, including Bloom and Meyer's [7]. A typical intuition is that a process is a black box whose only interface to the outside world consists of buttons (corresponding to actions). The most basic kind of test is to try and press one of the buttons: either the button will go down and the process will make an invisible state change (corresponding to a labelled transition), or the button doesn't go down (corresponding to the absence of a labelled transition). An important question arises as to which mechanisms are allowed to combine the basic button pushing experiments. Here, following Larsen and Skou, we suppose that the tester is allowed to make multiple copies of a process in order to experiment independently on one copy at a time. This feature is crucial in capturing branching-time equivalences. The set $\mathcal{T}$ of tests is given by the following syntax:

$$t ::= \omega \mid a.t \mid (t, \ldots, t)$$

where $a \in Act$. The test ω does nothing but successfully terminate, while $a.t$ specifies the test: press the a-button and in case of success proceed with t. Finally, $(t_1, \ldots, t_n)$ specifies the test: make n copies of (the current state of) the process and perform the test t_i on the i-th copy for $i = 1, \ldots, n$. With each test t we associate a set O_t of possible observations as follows:

$$O_\omega = \{\omega^\checkmark\} \qquad O_{a.t} = \{a^\times\} \cup \{a^\checkmark.e \mid e \in O_t\} \qquad O_{(t_1, \ldots, t_n)} = O_{t_1} \times \cdots \times O_{t_n}$$

The only observation of the test ω is successful termination: $\omega^\checkmark$. Upon performing $a.t$ one possibility, denoted by $a^\times$, is that the a-button fails to go down (and so the test terminates unsuccessfully). Otherwise, the a-button goes down and we proceed to observe e by running t in the next state; this is denoted by $a^\vee.e$.

For a given test t, each state x of a labelled Markov process $\langle X, \Sigma, \mu \rangle$ induces a probability distribution $P_{t,x}$ on O_t. The definition of $P_{t,x}$ is by structural induction on t as follows.

$$P_{\omega,x}(\omega^\vee) = 1 \qquad P_{a.t,x}(a^\times) = 1 - \mu_{x,a}(X) \qquad P_{a.t,x}(a^\vee.e) = \int (P_{t,\cdot}(e))\, d\mu_{x,a}$$
$$P_{(t_1,\dots,t_n),x}(e_1,\dots,e_n) = P_{t_1,x}(e_1) \times \cdots \times P_{t_n,x}(e_n)$$

We are now in a position to state the main result of this section.

Theorem 5. *Let $\langle X, \Sigma, \mu \rangle$ be a labelled Markov process. Then a sequence $\langle x_n \rangle_{n \in \mathbb{N}}$ of states converges to the state x in the pseudometric $d_{\langle X, \Sigma, \mu \rangle}$ if and only if for each test t the sequence of distributions $\langle P_{t,x_n} \rangle_{n \in \mathbb{N}}$ converges to $P_{t,x}$ pointwise.*

As a consequence, the pseudometric $d_{\langle X, \Sigma, \mu \rangle}$ metrizes the topology on the states of a labelled Markov process induced by the probability distributions of the tests. Furthermore, we can also conclude from the above result that states are probabilistic bisimilar if and only if they yield the exact same probability distribution on the observation set of any test. The latter result generalizes [19, Theorem 6.5] and does not rely on the minimal deviation assumption[8].

We sketch a proof of Theorem 5, starting with the only if direction. Observe that n-fold multiplication, considered as a map $\prod_{i=1}^{n}[0,1] \to [0,1]$, is n-Lipschitz. We augment the set of functional expressions with an n-fold product construct for each $n \in \mathbb{N}$, where $(\prod_{i=1}^{n} f_i)_{\langle X, \Sigma, \mu \rangle} = \frac{1}{n} \cdot \prod_{i=1}^{n} (f_i)_{\langle X, \Sigma, \mu \rangle}$.

In Section 3 we showed that any nonexpansive function on the final labelled Markov process can be uniformly approximated by functional expressions. We deduce that, since this extra construct is nonexpansive, it does not change the definition of the pseudometric $d_{\langle X, \Sigma, \mu \rangle}$ based on functional expressions.

With each test-observation pair (t, e), where $e \in O_t$, we associate a functional expression $f_{t,e}$. The idea is that $(f_{t,e})_{\langle X, \Sigma, \mu \rangle}(x)$ is proportionate to the probability of observing e when the test t is run from state x.

$$f_{\omega,\omega^\vee} = 1 \qquad f_{a.t,a^\times} = (1 - \langle a \rangle 1) \mathbin{\dot{-}} (1 - c) \qquad f_{a.t,a^\vee.e} = \langle a \rangle f_{t,e}$$
$$f_{(t_1,\dots,t_n),(e_1,\dots,e_n)} = \prod_{i=1}^{n} f_{t_i,e_i}$$

In fact, it is easy to see that for each test t and observation $e \in O_t$ there is a constant $k_{t,e}$ such that $(f_{t,e})_{\langle X, \Sigma, \mu \rangle}(x) = k_{t,e} \cdot P_{t,x}(e)$. Intuitively, $k_{t,e}$ is proportionate to the 'width' and 'depth' of the test t.

From this we immediately conclude that if $\langle x_n \rangle_{n \in \mathbb{N}}$ converges to x in the pseudometric $d_{\langle X, \Sigma, \mu \rangle}$, then for each test t the sequence of probability distributions $\langle P_{t,x_n} \rangle_{n \in \mathbb{N}}$ converges pointwise to $P_{t,x}$. Now we turn our attention to proving the converse.

The idea behind this part of the proof is to show how we can determine the approximate value of a functional expression through testing. The reader is

---

[8] A discrete labelled Markov process $\langle X, \Sigma, \mu \rangle$ satisfies the minimal deviation assumption if the set $\{\mu_{x,a}(y) \mid x, y \in X\}$ is finite for each $a \in Act$.

invited to compare this with the proof of [19, Theorem 8.4] where Larsen and Skou show how to test the truth or falsity of a probabilistic modal logic formula. In particular, we can avoid the minimum deviation assumption precisely because we only test for approximate 'truth values'.

Lemma 4. *Let $f \in \mathcal{F}$, $0 \leq \alpha < \beta \leq 1$ and $\delta > 0$. Then there exist $t \in \mathcal{T}$ and $E \subseteq O_t$ such that for all $x \in X$,*

1. *whenever $f_{\langle X, \Sigma, \mu \rangle}(x) \geq \beta$ then $P_{t,x}(E) \geq 1 - \delta$, and*
2. *whenever $f_{\langle X, \Sigma, \mu \rangle}(x) \leq \alpha$ then $P_{t,x}(E) \leq \delta$,*

where $P_{t,x}(E) = \sum_{e \in E} P_{t,x}(e)$.

Thus, if we run test t in state x and observe $e \in E$ then with high confidence we can assert that $f_{\langle X, \Sigma, \mu \rangle}(x) > \alpha$. On the other hand, if we observe $e \notin E$ then with high confidence we can assert that $f_{\langle X, \Sigma, \mu \rangle}(x) < \beta$.

Proof sketch. The proof proceeds by structural induction on $f \in \mathcal{F}$. We sketch a few cases.

$1 - f$ Assume for $\langle f, 1 - \beta, 1 - \alpha, \delta \rangle$ we have test t with evidence set E. Then we take test t with evidence set $O_t \setminus E$ for $\langle 1 - f, \alpha, \beta, \delta \rangle$.

$\langle a \rangle f$ Pick an $N \in \mathbb{N}$ and $\delta' > 0$. By the induction hypothesis, for $i = 1, \ldots, N$, we have a test t_i with evidence set E_i for $\langle f, \frac{i-1}{N}, \frac{i}{N}, \delta' \rangle$. We take $t = (t_1, \ldots, t_N)$ as a candidate test for $\langle f, \alpha, \beta, \delta \rangle$. Fix $x \in X$ and define a random variable θ on the probability space $\langle O_t, P_{t,x} \rangle$ by

$$\theta(e_1, \ldots, e_N) = \tfrac{1}{N} \cdot \#\{\, e_i \mid e_i = a^\vee.e_i' \wedge e_i' \in E_i \,\}.$$

By taking suitably large N and small δ' one can make $E[\theta]$, the expected value of θ, arbitrarily close to $\int f \, d\mu_{x,a}$. Moreover for large N, θ is close to $E[\theta]$ with high probability, and thus close to $\int f \, d\mu_{x,a}$ with high probability. Finally, this estimation can be made independently of the original choice of $x \in X$. Thus for large N and small δ', t is a test for $\langle \langle a \rangle f, \alpha, \beta, \delta \rangle$ with set of evidence $E = \{\, e \in O_t \mid \theta(e) \geq \frac{\beta + \alpha}{2c} \,\}$. $\square$

We can now conclude the proof of Theorem 5. Suppose $\langle x_n \rangle_{n \in \mathbb{N}}$ is a sequence of states and x is a state such that for each test t the sequence of probability distributions $\langle P_{t,x_n} \rangle$ converges to $P_{t,x}$. Then from the lemma above we may conclude that for each individual functional expression f, the sequence $\langle f_{\langle X, \Sigma, \mu \rangle}(x_n) - f_{\langle X, \Sigma, \mu \rangle}(x) \rangle_{n \in \mathbb{N}}$ converges to 0. By Lemma 3 this implies that the sequence $\langle x_n \rangle_{n \in \mathbb{N}}$ converges to x in the pseudometric $d_{\langle X, \Sigma, \mu \rangle}$.

References

1. S. Abramsky. Observation equivalence as a testing equivalence. *Theoretical Computer Science*, 53(2/3):225–242, 1987.
2. J. Adámek and V. Koubek. Least fixed point of a functor. *Journal of Computer System Sciences*, 19(2):163–178, October 1979.

3. P. America and J.J.M.M. Rutten. Solving reflexive domain equations in a category of complete metric spaces. *Journal of Computer and System Sciences*, 39(3):343–375, December 1989.
4. W. Arveson. *An Invitation to C*-Algebras*. Springer-Verlag, New York, 1976.
5. M.F. Barnsley. *Fractals Everywhere*. Academic Press, Boston, second edition, 1993.
6. P. Billingsley. *Probability and Measure*. Wiley, New York, third edition, 1995.
7. B. Bloom and A. Meyer. Experimenting with process equivalence. *Theoretical Computer Science*, 101(2):223–237, July 1992.
8. R. Blute, J. Desharnais, A. Edalat, and P. Panangaden. Bisimulation for labelled Markov processes. In *Proceedings of the 12th Annual IEEE Symposium on Logic in Computer Science*, pages 149–158, Warsaw, June/July 1997. IEEE.
9. F. van Breugel and J. Worrell. Towards quantitative verification of probabilistic transition systems. In F. Orejas, P.G. Spirakis, and J. van Leeuwen, editors, *Proceedings of the 28th International Colloquium on Automata, Languages, and Programming*, volume 2076 of *Lecture Notes in Computer Science*, pages 421–432, Crete, July 2001. Springer-Verlag.
10. F. van Breugel and J. Worrell. An algorithm for quantitative verification of probabilistic transition systems. In K.G. Larsen and M. Nielsen, editors, *Proceedings of the 12th International Conference on Concurrency Theory*, volume 2154 of *Lecture Notes in Computer Science*, pages 336–350, Aalborg, August 2001. Springer-Verlag.
11. F. van Breugel and J. Worrell. A behavioural pseudometric for probabilistic transition systems. Available at www.cs.yorku.ca/~franck, October 2001.
12. J. Desharnais, A. Edalat, and P. Panangaden. A logical characterization of bisimulation for labeled Markov processes. In *Proceedings of the 13th Annual IEEE Symposium on Logic in Computer Science*, pages 478–487, Indianapolis, June 1998. IEEE.
13. J. Desharnais, A. Edalat, and P. Panangaden. Bisimulation for labelled Markov processes. To appear in *Information and Computation*. Available at www-acaps.cs.mcgill.ca/~prakash, 1999.
14. J. Desharnais, V. Gupta, R. Jagadeesan, and P. Panangaden. Metrics for labeled Markov systems. In J.C.M. Baeten and S. Mauw, editors, *Proceedings of the 10th International Conference on Concurrency Theory*, volume 1664 of *Lecture Notes in Computer Science*, pages 258–273, Eindhoven, August 1999. Springer-Verlag.
15. J. Desharnais, V. Gupta, R. Jagadeesan, and P. Panangaden. Metrics for labeled Markov systems. Available at www-acaps.cs.mcgill.ca/~prakash, November 2001.
16. G.A. Edgar. *Integral, Probability, and Fractal Measures*. Springer-Verlag, New York, 1998.
17. M. Giry. A categorical approach to probability theory. In B. Banaschewski, editor, *Proceedings of the International Conference on Categorical Aspects of Topology and Analysis*, volume 915 of *Lecture Notes in Mathematics*, pages 68–85, Ottawa, August 1981. Springer-Verlag.
18. M. Hennessy and R. Milner. Algebraic laws for nondeterminism and concurrency. *Journal of the ACM*, 32(1):137–161, January 1985.
19. K.G. Larsen and A. Skou. Bisimulation through probabilistic testing. *Information and Computation*, 94(1):1–28, September 1991.
20. K.R. Parthasarathy. *Probability Measures on Metric Spaces*. Academic Press, New York, 1967.
21. D. Turi and J.J.M.M. Rutten. On the foundations of final semantics: non-standard sets, metric spaces, partial orders. *Mathematical Structures in Computer Science*, 8(5):481–540, October 1998.

Why Computational Complexity Requires Stricter Martingales*

John M. Hitchcock and Jack H. Lutz

Department of Computer Science
Iowa State University
{jhitchco,lutz}@cs.iastate.edu

Abstract. The word "martingale" has related, but different, meanings in probability theory and theoretical computer science. In computational complexity and algorithmic information theory, a martingale is typically a function d on strings such that $E(d(wb)|w) = d(w)$ for all strings w, where the conditional expectation is computed over all possible values of the next symbol b. In modern probability theory a martingale is typically a sequence $\xi_0, \xi_1, \xi_2, \ldots$ of random variables such that $E(\xi_{n+1}|\xi_0, \ldots, \xi_n) = \xi_n$ for all n.

This paper elucidates the relationship between these notions and proves that the latter notion is too weak for many purposes in computational complexity, because under this definition every computable martingale can be simulated by a polynomial-time computable martingale.

1 Introduction

Since martingales were introduced by Ville [21] in 1939 (having been implicit in earlier works of Lévy [9,10]), they have followed two largely disjoint paths of scientific development and application. Along the larger and, to date, more significant path, Doob developed them into a powerful tool of probability theory that, especially following his influential 1953 book [6], has become central to many areas of research, including probability, stochastic processes, functional analysis, fractal geometry, statistical mechanics, and mathematical finance. Along the smaller and more recent path, effective martingales (martingales satisfying various computability conditions) have been used in theoretical computer science, first in the 1970's by Schnorr [17,18,19,20] in his investigations of Martin-Löf's definition of randomness [14] and variants thereof, and then in the 1990's by Lutz [11,13] in the development of resource-bounded measure. Many researchers have extended these developments, and effective martingales are now an active research topic that makes frequent contributions to our understanding of computational complexity, randomness, and algorithmic information.

A curious thing about these two paths of research is that they interpret the word "martingale" differently. In computational complexity and algorithmic

---

* This research was supported in part by National Science Foundation Grant 9988483.

P. Widmayer et al. (Eds.): ICALP 2002, LNCS 2380, pp. 549–560, 2002.

information theory, a martingale is typically a real-valued function d on $\{0,1\}^*$ such that

$$E[d(wb)|w] = d(w) \tag{1.1}$$

for all strings w, where the expectation is conditioned on the *bit history* w (the string seen thus far) and computed over the two possible values of the next bit b. When the underlying probability measure is uniform (0 and 1 equally likely, independent of prior history), equation (1.1) becomes the familiar identity

$$d(w) = \frac{d(w0) + d(w1)}{2}. \tag{1.2}$$

Intuitively, a martingale d is a strategy for betting on the successive bits of an infinite binary sequence, and $d(w)$ is the amount of capital that a gambler using d will have after w if the sequence starts with w. Thus $d(\lambda)$ is the initial capital, and equation (1.1) says that the payoffs are fair.

On the other hand, in probability theory, a martingale is typically a sequence $\xi_0, \xi_1, \xi_2, \ldots$ of random variables such that

$$E[\xi_{n+1}|\xi_0, \ldots, \xi_n] = \xi_n \tag{1.3}$$

for all $n \in \mathbb{N}$. Such a sequence is also called a martingale sequence or a *martingale process*, and we exclusively use the latter term here in order to distinguish the two notions under discussion.

To understand the essential difference between martingales and martingale processes, we first need to dispose of three inessential differences. First a martingale is a function from $\{0,1\}^*$ to $\mathbb{R}$, while a martingale process is a sequence of random variables. To see that this is only a difference in notation, let $\mathbf{C}$ be the Cantor space, consisting of all infinite binary sequences. Then we can identify each martingale d with the sequence $\xi_0, \xi_1, \xi_2, \ldots$ of functions $\xi_n : \mathbf{C} \to \mathbb{R}$ defined by

$$\xi_n(S) = d(S[0..n-1]),$$

where $S[0..n-1]$ is the n-bit prefix of S. Then $\xi_0, \xi_1, \xi_2, \ldots$ is a sequence of random variables and equation (1.1) says that

$$E[\xi_{n+1}|w] = \xi_n \tag{1.4}$$

for all $n \in \mathbb{N}$ and $w \in \{0,1\}^n$. (See sections 2 and 3 for a precise treatment of this and other ideas developed intuitively in this introduction.)

The other two inessential differences are that martingales, unlike martingale processes, are typically required to be nonnegative and to have $\mathbf{C}$ as their underlying sample space (i.e., as the domain of each of the random variables ξ_n). To date it has been convenient to include nonnegativity in the martingale definition because most applications have required martingales that are nonnegative (or, equivalently, bounded below). Similarly, it has been convenient to have $\mathbf{C}$ –

or some similar sequence space – as the underlying sample space because martingales have been used to investigate the structures of such spaces. However, neither of these requirements is essential or likely to persist into the future. (E.g., as the use of martingales in computational complexity expands, it is likely that "nonnegative" will be moved from the definition to the theorems where it is still needed.) In this paper, in order to facilitate our comparison, we ignore the nonnegativity requirement on martingales, and for both martingales and martingale processes, we focus on the case where the underlying sample space is **C**.

The essential difference between the martingale processes of probability theory and the martingales of theoretical computer science is thus the difference between equations (1.3) and (1.4). Translating our remarks following (1.1) into the notation of (1.4), ξ_n denotes the gambler's capital after n bets, and equation (1.4) says that for each bit history $w \in \{0,1\}^n$, the expected value of the gambler's capital ξ_{n+1} after the next bet, *conditioned on the bit history w*, is the gambler's capital ξ_n before the next bet. In contrast, equation (1.3) says that for each *capital history* $c_0, \ldots, c_n$, the expected value of the gambler's capital ξ_{n+1} after the next bet, *conditioned on the capital history $\xi_0 = c_0, \ldots, \xi_n = c_n$*, is the gambler's capital ξ_n before the next bet. As we shall see, it is clear that (1.3) holds if (1.4) holds, but if two or more bit histories correspond to the same capital history, then it is possible to satisfy (1.3) without satisfying (1.4). Thus the martingale requirement of theoretical computer science is *stricter* than the martingale process requirement of probability theory.

In this paper we prove that this strictness is essential for computational complexity in the sense that martingale processes cannot be used in place of martingales as a basis for resource-bounded measure or resource-bounded randomness.

Resource-bounded measure uses resource-bounded martingales to define measure in complexity classes [11,12,13]. For example, a set X of decision problems has *measure 0 in* the complexity class E = DTIME(2^{linear}), and we write $\mu(X|\text{E}) = 0$, if there is a polynomial time computable nonnegative martingale that *succeeds*, i.e., wins an unbounded amount of money on, every element of $X \cap \text{E}$. An essential condition for this definition to be nontrivial is that E does not have measure 0 in itself, i.e., that there is no polynomial-time nonnegative martingale that succeeds on every element of E. This is indeed true by the Measure Conservation Theorem [11].

In contrast, we show here that there *is* a polynomial-time nonnegative martingale *process* that succeeds on every element of E. In fact, our main theorem says that for *any* computable nonnegative martingale process d, there is a polynomial-time nonnegative martingale process d' that succeeds on every sequence that d succeeds on. That is, computable nonnegative martingale processes cannot use time beyond polynomial to succeed on additional sequences. It follows that for every computably presentable class $\mathcal{C}$ of decision problems – and hence for every reasonable complexity class $\mathcal{C}$ – there is a polynomial-time nonnegative martingale process that succeeds on every element of $\mathcal{C}$. Thus martingale processes cannot be used as a basis for resource-bounded measure.

Martingale processes are similarly inadequate for resource-bounded random-ness. For example, a sequence $S \in \mathbf{C}$ is p-*random* if there is no polynomial-time nonnegative martingale that succeeds on it [19,11]. An essential feature of resource-bounded randomness is the existence [19], in fact abundance [11, 2], of decidable sequences that are random with respect to a given resource bound. For example, although no element of E can be p-random, almost every element of the complexity class EXP = DTIME($2^{\text{polynomial}}$) is p-random [11, 2]. However, the preceding paragraph implies that for every decidable sequence S there is a polynomial-time nonnegative martingale process that succeeds on S, so *no* decidable sequence could be p-random if we used martingale processes in place of martingales in defining p-randomness. Moreover, we also show that there exist computably random sequences (sequences on which no computable nonnegative martingale succeeds) on which polynomial-time nonnegative martingale processes can succeed.

Historically, the 1939 martingale definition of Ville [21] was the strict definition (1.4) now used in theoretical computer science. It was Doob [5] who in 1940 relaxed Ville's definition to the form (1.3) that is now so common in probability theory [7,16,1]. Of course the difference in usage between these two fields is not at all a dichotomy. The relaxed definition (1.3) is used in randomized algorithms [15] and other areas of theoretical computer science where the complexities of the martingales are not an issue, and probability theory also uses the more abstract notion of an $\mathcal{F}$-martingale process (also formulated by Doob [5] and described in section 3 below), of which martingales and martingale processes are the two extreme cases.

Our results show that resource-bounded measure and randomness do in fact require martingales that are stricter than the martingale processes used so commonly in probability theory. However, these results do not disparage the latter notion. Quite to the contrary, it is to be anticipated that theoretical computer science will avail itself of and effectivize increasingly sophisticated aspects of martingales and measure-theoretic probability in the coming years. Our results and the arguments by which we prove them are to be regarded as steps toward expanding the interface between these two fields.

2   Preliminaries

A *decision problem* (a.k.a. *language*) is a set $A \subseteq \{0,1\}^*$. We identify each language with its characteristic sequence $[\![s_0 \in A]\!][\![s_1 \in A]\!][\![s_2 \in A]\!] \cdots$, where $s_0, s_1, s_2, \ldots$ is the standard enumeration of $\{0,1\}^*$ and $[\![\phi]\!] =$ if ϕ then 1 else 0. We write $A[i..j]$ for the string consisting of the i-th through j-th bits of (the characteristic sequence of) A.

A class $\mathcal{C}$ of languages is *computably presentable* (a.k.a. *recursively presentable* [3]) if there is an effective enumeration $M_0, M_1, \ldots$ of deterministic Turing machines, each of which halts on all inputs, such that $\mathcal{C} = \{L(M_i) | i \in \mathbb{N}\}$, where $L(M_i)$ is the language decided by M_i.

A *prefix set* is a language A such that no element of A is a prefix of any other element of A. If A is a language and $n \in \mathbb{N}$, then we write $A_{=n} = A \cap \{0,1\}^n$ and $A_{\leq n} = A \cap \{0,1\}^{\leq n}$.

The Cantor space $\mathbf{C}$ is the set of all infinite binary sequences. If $w \in \{0,1\}^*$ and $x \in \{0,1\}^* \cup \mathbf{C}$, then $w \sqsubseteq x$ means that w is a prefix of x. The *cylinder* generated by a string $w \in \{0,1\}^*$ is $\mathbf{C}_w = \{A \in \mathbf{C} \mid w \sqsubseteq A\}$.

A *σ-algebra* on $\mathbf{C}$ is a nonempty collection $\mathcal{F}$ of subsets of $\mathbf{C}$ that is closed under complements and under countable unions. For any collection $\mathcal{A}$ of subsets of $\mathbf{C}$ there is a unique smallest σ-algebra $\sigma(\mathcal{A})$ on $\mathbf{C}$ that contains $\mathcal{A}$. The Borel σ-algebra on $\mathbf{C}$ is $\mathcal{B} = \sigma(\{\mathbf{C}_w \mid w \in \{0,1\}^*\})$. We use the uniform probability measure on $\mathbf{C}$, which is the function $\mu : \mathcal{B} \to [0,1]$ determined by the values $\mu(w) = \mu(\mathbf{C}_w) = 2^{-|w|}$ for all $w \in \{0,1\}^*$.

Let $\mathcal{F}$ be a σ-algebra on $\mathbf{C}$. We say that a function $f : \mathbf{C} \to \mathbb{R}$ is *$\mathcal{F}$-measurable* if for all $t \in \mathbb{R}$,

$$\{S \in \mathbf{C} \mid \xi(S) \leq t\} \in \mathcal{F}.$$

A *random variable* on $\mathbf{C}$ is a function $\xi : \mathbf{C} \to \mathbb{R}$ that is $\mathcal{B}$-measurable. We write $\mathrm{E}[\xi]$ for the *expectation* of a random variable ξ. The *indicator function* of a set $A \subseteq \mathbf{C}$ is the function

$$\mathbf{1}_A : \mathbf{C} \to \{0,1\}$$

$$\mathbf{1}_A(S) = \begin{cases} 1 & \text{if } S \in A \\ 0 & \text{if } S \notin A. \end{cases}$$

If ξ is a random variable and $A \subseteq \mathbf{C}$ satisfies $\mu(A) > 0$, then the *conditional expectation of ξ given A* is

$$\mathrm{E}[\xi|A] = \frac{\mathrm{E}[\xi \cdot \mathbf{1}_A]}{\mu(A)}.$$

If $\xi_0, \ldots, \xi_{n+1}$ are random variables and $t_0, \ldots, t_n \in \mathbb{R}$, we write $E[\xi_{n+1}|\xi_0 = t_0, \ldots, \xi_n = t_n]$ for $E[\xi_{n+1}|\{S \in \mathbf{C}|\xi_0(S) = t_0, \ldots, \xi_n(S) = t_n\}]$. If ξ is a random variable, $\mathcal{A}$ is a countable partition of $\mathbf{C}$, and $\mathcal{F} = \sigma(\mathcal{A})$, then the *conditional expectation of ξ given $\mathcal{F}$* is the random variable

$$E[\xi|\mathcal{F}] : \mathbf{C} \to \mathbb{R}$$

$$E[\xi|\mathcal{F}](S) = \mathrm{E}[\xi|A] \text{ where } S \in A \in \mathcal{A}$$

which is defined for μ-almost all S.

We say that a real-valued function $f : \{0,1\}^* \to \mathbb{R}$ is *computable* if there is a computable function $\hat{f} : \mathbb{N} \times \{0,1\}^* \to \mathbb{Q}$ such that for all $n \in \mathbb{N}$ and $w \in \{0,1\}^*$, $|f(w) - \hat{f}(n,w)| \leq 2^{-n}$. We say that $\hat{f}$ is a *computation* of f. We often write $\hat{f}_n(w)$ for $\hat{f}(n,w)$. If $\hat{f}$ is computable in polynomial-time (where n is input in unary), then f is *polynomial-time computable*. If $f : \{0,1\}^* \to \mathbb{Q}$ is itself a (polynomial-time) computable function, then we say that f is *(polynomial-time) exactly computable*.

We say that $f : \{0,1\}^* \to \mathbb{R}$ is *constructive* (a.k.a. *lower semicomputable*) if there is a computable function $h : \mathbb{N} \times \{0,1\}^* \to \mathbb{Q}$ such that for any $w \in \{0,1\}^*$, $h(n,w) \le h(n+1,w) < f(w)$ for all $n \in \mathbb{N}$ and $\lim_{n \to \infty} h(n,w) = f(w)$.

3   Varieties of Martingales

In this section we introduce the different notions of martingales used in theoretical computer science and probability theory. As noted in the introduction, we use the terms "martingale" for the former and "martingale process" for the latter. We begin with the martingale definition commonly used in the theory of computing.

Definition. A function $d : \{0,1\}^* \to \mathbb{R}$ is a *martingale* if

$$d(w) = \frac{d(w0) + d(w1)}{2} \tag{3.1}$$

for all $w \in \{0,1\}^*$.

Intuitively, a martingale d represents a strategy in a betting game. The gambler begins with $d(\lambda)$ of capital and is betting on an unknown sequence $S \in \mathbf{C}$. The gambler places a wager on the first bit of S being 0 or 1. If the first bit of S is 0, the gambler then holds $d(0)$ capital; otherwise, the first bit is 1 and the gambler holds $d(1)$ capital. The gambler then bets on the second bit of S possibly using his knowledge of the first bit of S. In general, after n rounds of this game, the gambler knows that the first n bits of S are $w = S[0..n-1]$. Using this knowledge he wagers on the $(n+1)$-st bit of S. Equation (3.1) says that this is a fair gambling game. That is, the payoffs are fair: if S is chosen uniformly at random, the gambler can expect to have the same amount of capital after each stage of the game.

We will use random variables and conditional expectations to make this idea of fair gambling more precise. Let $d : \{0,1\}^* \to \mathbb{R}$ be an arbitrary function. For each $n \in \mathbb{N}$, we define the function

$$\xi_{d,n} : \mathbf{C} \to \mathbb{R}$$

$$\xi_{d,n}(S) = d(S[0..n-1]).$$

Observe that each $\xi_{d,n}$ is a discrete random variable on $(\mathbf{C}, \mathcal{B}, \mu)$. We associate the sequence of random variables $\boldsymbol{\xi_d} = (\xi_{d,0}, \xi_{d,1}, \dots)$ with d. We can now interpret the martingale condition (3.1) as a conditional expectation.

Observation 3.1. *A function $d : \{0,1\}^* \to \mathbb{R}$ is a martingale if and only if*

$$\mathrm{E}[\xi_{d,|w|+1}|\mathbf{C}_w] = \xi_{d,|w|} \tag{3.2}$$

for all $w \in \{0,1\}^*$.

In probability theory, martingales are typically defined in the following more general form.

Definition. Let $\xi = (\xi_0, \xi_1, \dots)$ be a sequence of random variables. We say that ξ is a *martingale process* if for all $n \in \mathbb{N}$, $\mathrm{E}[\xi_n] < \infty$ and

$$\mathrm{E}[\xi_{n+1} | \xi_0 = c_0, \dots, \xi_n = c_n] = c_n \qquad (3.3)$$

for all values of $c_0, \dots, c_n \in \mathbb{R}$. (As we shall see below, condition (3.3) can also be stated more concisely using a conditional expectation given a σ-algebra.)

We can also view a martingale process ξ as a gambling game. Again the gambler is wagering on an unknown sequence $S \in \mathbf{C}$. The initial capital is $\xi_0(S)$. After the nth stage of the game, the gambler has capital $\xi_n(S)$. The condition (3.3) says that the payoffs are fair in this game. This notion of fairness is more relaxed than the martingale condition (3.2). In order to make a precise comparison we extend the definition of martingale processes.

Definition. A function $d : \{0,1\}^* \to \mathbb{R}$ is a *martingale process* if ξ_d is a martingale process.

The martingale process condition for a function d is

$$\mathrm{E}[\xi_{d,n+1} | \xi_{d,0} = c_0, \dots, \xi_{d,n} = c_n] = c_n. \qquad (3.4)$$

This fairness condition involves the *capital history* of the gambling game rather than revealed *bit history* of the sequence S. In (3.2), the conditioning is done on the bit history w. The conditioning in (3.4) is done on the capital history. Intuitively, the martingale condition is more "local" than the martingale process condition.

We now give a more concrete characterization of which functions $d : \{0,1\}^* \to \mathbb{R}$ are martingale processes. Define an equivalence relation $\approx_d$ on $\{0,1\}^*$ by

$$x \approx_d y \iff |x| = |y| \text{ and } (\forall 1 \le i \le n) \; d(x[0..i-1]) = d(y[0..i-1]).$$

For each $w \in \{0,1\}^*$ we define the equivalence class $[w]_d = \{v \in \{0,1\}^* | w \approx_d v\}$.

Observation 3.2. *A function $d : \{0,1\}^* \to \mathbb{R}$ is a martingale process if and only if*

$$2 \Big| [w]_d \Big| d(w) = \sum_{v \in [w]_d} d(v0) + d(v1) \qquad (3.5)$$

for all $w \in \{0,1\}^*$.

Any martingale d is also a martingale process; the following example shows that the converse is not true.

Example 3.3. Define for all $u \in \{0,1\}^*$

$$d(\lambda) = d(0) = d(1) = 1,$$
$$d(0u) = 0,$$
$$d(1u) = 2.$$

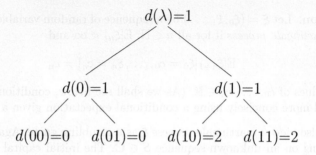

Fig. 1. The martingale process d of Example 3.3.

Then d is not a martingale, but d is a martingale process. Because the strings 0 and 1 have the same capital histories, $[0]_d = \{0,1\}$, so the averaging condition (3.5) allows the capital to "shift" in a manner not allowed by a martingale.

We now discuss a more general formulation of martingales that is used in probability theory. See also [5,6,15,4] for discussions of this notion. The following definition will yield the martingales and martingale processes defined above as special cases.

Definition. 1. A *filtration on* $\mathbf{C}$ is a sequence of σ-algebras $\mathcal{F} = (\mathcal{F}_0, \mathcal{F}_1, \dots)$ on $\mathbf{C}$ such that $\mathcal{F}_n \subseteq \mathcal{F}_{n+1}$ for all $n \in \mathbb{N}$.
2. Let $\boldsymbol{\xi}$ be a sequence of random variables and let $\mathcal{F}$ be a filtration on C. Then $\boldsymbol{\xi}$ is an $\mathcal{F}$-*martingale process* if the following conditions hold.
    (i) For all $n \in \mathbb{N}$, ξ_n is $\mathcal{F}_n$ measurable and $\mathrm{E}[\xi_n] < \infty$.
    (ii) For all $n \in \mathbb{N}$,

$$\mathrm{E}[\xi_{n+1}|\mathcal{F}_n] = \xi_n. \tag{3.6}$$

We also say that $\boldsymbol{\xi}$ is a *martingale relative to* $\mathcal{F}$.

The conditional expectation (3.6) can be viewed as a more generalized notion of fairness in the gambling game. For example, if $\boldsymbol{\xi}$ is an $\mathcal{F}$-martingale process for some filtration $\mathcal{F}$, then $\boldsymbol{\xi}$ is also a martingale process. Before we make any further comparisons we extend the filtration definition to functions.

Definition. Let $\mathcal{F}$ be a filtration. A function $d : \{0,1\}^* \to \mathbb{R}$ is an $\mathcal{F}$-*martingale process* if $\boldsymbol{\xi}_d$ is an $\mathcal{F}$-martingale process.

For each $n \in \mathbb{N}$, let

$$\mathcal{M}_n = \sigma(\{\mathbf{C}_w | w \in \{0,1\}^n\}).$$

We let $\boldsymbol{\mathcal{M}} = (\mathcal{M}_0, \mathcal{M}_1, \dots)$.

Observation 3.4. *A function* $d : \{0,1\}^* \to \mathbb{R}$ *is a martingale if and only if* d *is an* $\mathcal{M}$-*martingale process.*

Let $d : \{0,1\}^* \to \mathbb{R}$ be arbitrary. For each $n \in \mathbb{N}$, define

$$B_{d,n} = \{[w]_d \mid w \in \{0,1\}^n\},$$

$$C_{d,n} = \left\{ \bigcup_{w \in A} C_w \,\middle|\, A \in B_{d,n} \right\}, \text{ and}$$

$$\mathcal{F}_{d,n} = \sigma(C_{d,n}).$$

We let $\mathcal{F}_d = (\mathcal{F}_{d,0}, \mathcal{F}_{d,1}, \dots)$.

Observation 3.5. *A function* $d : \{0,1\}^* \to \mathbb{R}$ *is a martingale process if and only if* d *is an* $\mathcal{F}_d$-*martingale process.*

If d is a martingale relative to some filtration $\mathcal{F}$, then d is also an $\mathcal{F}_d$-martingale process. That is, the martingale process requirement uses the coarsest filtration possible. On the other hand, the martingale requirement uses the essentially finest filtration $\mathcal{M}$. (If $\mathcal{F}$ is a finer filtration than $\mathcal{M}$, then d is an $\mathcal{F}$-martingale process if and only if d is an $\mathcal{M}$-martingale process.)

A very useful property of martingales in theoretical computer science is that the sum of two martingales is a martingale. For any filtration $\mathcal{F}$, the analogous fact also holds for $\mathcal{F}$-martingale processes. In contrast, it is well known [4] that the sum of two martingale processes need not be a martingale process. We include an example for completeness.

Example 3.6. Define for all $u \in \{0,1\}^*$ and $v \in \{0,1\}^+$

$$d_1(\lambda) = d_1(0) = d_1(1) = 1,$$
$$d_1(00u) = 3,$$
$$d_1(01u) = d_1(10u) = 0,$$
$$d_1(11u) = 1,$$

and

$$d_2(\lambda) = d_2(0) = d_2(1) = 1,$$
$$d_2(0v) = 0,$$
$$d_2(10) = d_2(11) = 2,$$
$$d_2(10v) = 4,$$
$$d_2(11v) = 0.$$

Then d_1 and d_2 are martingale processes. Let $d = d_1 + d_2$. Then $[00]_d = \{00, 11\}$, and

$$2d(00) = 6 \neq 8 = d(000) + d(001) + d(110) + d(111),$$

so d is not a martingale process.

4   Martingale Processes and Complexity

In this section we present our results, all of which concern the complexities and success sets of martingale processes.

Definition. Let $d : \{0,1\}^* \to \mathbb{R}$.

1. We say that d *succeeds on* a sequence $S \in \mathbf{C}$ if $\limsup_{n \to \infty} d(S[0..n-1]) = \infty$.
2. The *success set* of d is $S^{\infty}[d] = \{S \in \mathbf{C} | d \text{ succeeds on } S\}$.

The following technical lemma is crucial for our main theorem.

Lemma 4.1. (Exact Computation Lemma) *For every computable martingale process d and every $m \in \mathbb{N}$, there is an exactly computable martingale process d' such that for all $w \in \{0,1\}^*$, $|d'(w) - d(w)| < 2^{-m}$.*

Our next lemma can be regarded as a "speedup" theorem for exactly computable martingale processes, but its proof uses a very slow simulation technique analogous to slow diagonalization.

Lemma 4.2. *For every exactly computable nonnegative martingale process d there is a polynomial-time exactly computable nonnegative martingale process d' such that $S^{\infty}[d] = S^{\infty}[d']$.*

We now have the main theorem of this paper, which says that polynomial-time computable martingale processes are equivalent to arbitrary computable martingale processes.

Theorem 4.3. *For every computable nonnegative martingale process d there is a polynomial-time exactly computable nonnegative martingale process d' such that $S^{\infty}[d] = S^{\infty}[d']$.*

Proof. This follows immediately from Lemmas 4.1 and 4.2. $\qquad\square$

Theorem 4.3 has the following consequence for resource-bounded measure.

Corollary 4.4. *For every computably presentable class $\mathcal{C}$, there is a polynomial-time exactly computable nonnegative martingale process d such that $\mathcal{C} \subseteq S^{\infty}[d]$.*

Proof. Lutz [11] has shown that for every computably presentable class $\mathcal{C}$ (called "rec-countable" in the terminology of [11]) there is a computable nonnegative martingale d such that $\mathcal{C} \subseteq S^{\infty}[d]$. Since d is a computable martingale process, the conclusion of the corollary follows by Theorem 4.3. $\qquad\square$

Since complexity classes such as $\mathbf{E}, \mathbf{EXP}, \mathbf{ESPACE}$, etc. are all computably presentable, Corollary 4.4 implies that martingale processes cannot be used in place of martingales as a basis for resource-bounded measure.

We now prove a generalized Kraft inequality that enables us to establish an upper bound on the power of computable martingale processes. For any function $d : \{0,1\}^* \to \mathbb{R}$ and $A \subseteq \{0,1\}^*$, we say that A is *closed under* $\approx_d$ if for all $w \in \{0,1\}^*$,

$$w \in A \Rightarrow [w]_d \subseteq A.$$

Lemma 4.5. *If d is a nonnegative martingale process and $A \subseteq \{0,1\}^*$ is a prefix set that is closed under $\approx_d$, then*

$$\sum_{w \in A} d(w) 2^{-|w|} \leq d(\lambda).$$

Theorem 4.6. *For every computable nonnegative martingale process d there is a constructive nonnegative martingale d' such that $S^\infty[d] \subseteq S^\infty[d']$.*

Theorem 4.6 implies that no computable nonnegative martingale process can succeed on a sequence that is random in the sense of Martin-Löf [14]. In contrast, we now show that there exist computably random sequences (sequences on which no computable nonnegative martingale succeeds [19]) on which polynomial time nonnegative martingale processes can succeed.

Theorem 4.7. *For every $S \in \mathbf{C}$ satisfying $K(S[0..n-1]) < n - \log n$ almost everywhere, there is a polynomial-time nonnegative martingale process that succeeds on S.*

Corollary 4.8. *There exist a computably random sequence S and a polynomial-time nonnegative martingale d such that d succeeds on S.*

Proof. Lathrop and Lutz [8] proved that there is a computably random sequence $S \in \mathbf{C}$ that is *ultracompressible* in the sense that for every computable, non-decreasing, unbounded function $g : \mathbb{N} \to \mathbb{N}$, for all but finitely many $n \in \mathbb{N}$, $K(S[0..n-1]) < K(n) + g(n)$. Such a sequence S clearly satisfies the hypothesis of Theorem 4.7. □

References

1. N. Alon and J. H. Spencer. *The Probabilistic Method*. Wiley, 1992.
2. K. Ambos-Spies and E. Mayordomo. Resource-bounded measure and randomness. In A. Sorbi, editor, *Complexity, Logic and Recursion Theory*, Lecture Notes in Pure and Applied Mathematics, pages 1–47. Marcel Dekker, New York, N.Y., 1997.
3. J. L. Balcázar, J. Díaz, and J. Gabarró. *Structural Complexity I (second edition)*. Springer-Verlag, Berlin, 1995.
4. K. L. Chung. *A Course in Probability Theory*. Academic Press, third edition, 2001.
5. J. L. Doob. Regularity properties of certain families of chance variables. *Transactions of the American Mathematical Society*, 47:455–486, 1940.
6. J. L. Doob. *Stochastic Processes*. Wiley, New York, N.Y., 1953.
7. R. Durrett. *Essentials of Stochastic Processes*. Springer, 1999.
8. J. I. Lathrop and J. H. Lutz. Recursive computational depth. *Information and Computation*, 153:139–172, 1999.
9. P. Lévy. Propriétés asymptotiques des sommes de variables indépendantes ou enchaînées. *Journal des mathématiques pures et appliquées. Series 9.*, 14(4):347–402, 1935.

10. P. Lévy. *Théorie de l'Addition des Variables Aleatoires*. Gauthier-Villars, 1937 (second edition 1954).
11. J. H. Lutz. Almost everywhere high nonuniform complexity. *Journal of Computer and System Sciences*, 44:220–258, 1992.
12. J. H. Lutz. The quantitative structure of exponential time. In L.A. Hemaspaandra and A.L. Selman, editors, *Complexity Theory Retrospective II*, pages 225–254. Springer-Verlag, 1997.
13. J. H. Lutz. Resource-bounded measure. In *Proceedings of the 13th IEEE Conference on Computational Complexity*, pages 236–248, New York, 1998. IEEE Computer Society Press.
14. P. Martin-Löf. The definition of random sequences. *Information and Control*, 9:602–619, 1966.
15. R. Motwani and P. Raghavan. *Randomized Algorithms*. Cambridge University Press, 1995.
16. S. M. Ross. *Stochastic Processes*. Wiley, 1983.
17. C. P. Schnorr. Klassifikation der Zufallsgesetze nach Komplexität und Ordnung. *Z. Wahrscheinlichkeitstheorie verw. Geb.*, 16:1–21, 1970.
18. C. P. Schnorr. A unified approach to the definition of random sequences. *Mathematical Systems Theory*, 5:246–258, 1971.
19. C. P. Schnorr. Zufälligkeit und Wahrscheinlichkeit. *Lecture Notes in Mathematics*, 218, 1971.
20. C. P. Schnorr. Process complexity and effective random tests. *Journal of Computer and System Sciences*, 7:376–388, 1973.
21. J. Ville. *Étude Critique de la Notion de Collectif*. Gauthier–Villars, Paris, 1939.

Correspondence Principles for Effective Dimensions[*]

John M. Hitchcock

Department of Computer Science
Iowa State University
jhitchco@cs.iastate.edu

Abstract. We show that the classical Hausdorff and constructive dimensions of any union of Π_1^0-definable sets of binary sequences are equal. If the union is effective, that is, the set of sequences is Σ_2^0-definable, then the computable dimension also equals the Hausdorff dimension. This second result is implicit in the work of Staiger (1998).
Staiger also proved related results using entropy rates of decidable languages. We show that Staiger's computable entropy rate provides an equivalent definition of computable dimension. We also prove that a constructive version of Staiger's entropy rate coincides with constructive dimension.

1 Introduction

Lutz has recently effectivized classical Hausdorff dimension to define the constructive and computable dimensions of sets of infinite binary sequences [1,3]. In early lectures on these effective dimensions [2], Lutz conjectured that there should be a *correspondence principle* stating that the constructive dimension of every sufficiently simple set X coincides with its classical Hausdorff dimension. In this paper we provide such a principle, along with an analogous correspondence principle for computable dimension. Specifically, given a set X of infinite binary sequences, let $\dim_H(X)$ be the Hausdorff dimension of X, $\mathrm{cdim}(X)$ be the constructive dimension of X, and $\dim_{\mathrm{comp}}(X)$ be the computable dimension of X. Our correspondence principle for constructive dimension says that for every set X that is an *arbitrary* union of Π_1^0-definable sets of sequences, $\mathrm{cdim}(X) = \dim_H(X)$. The correspondence principle for computable dimension says that for every Σ_2^0-definable set X of sequences, $\dim_{\mathrm{comp}}(X) = \dim_H(X)$. We show that these results are optimal in the arithmetical hierarchy. Staiger [6] has proven closely related results in his investigations of Kolmogorov complexity and Hausdorff dimension. The correspondence principle for computable dimension is implicit in his results on martingale exponents of increase. In addition, for each set X of sequences he defined a kind of *entropy rate* that coincides with classical Hausdorff dimension. Staiger proved that a computable version of this entropy rate is equal to the classical Hausdorff dimension for Σ_2^0-definable sets. We show

[*] This research was supported in part by National Science Foundation Grant 9988483.

P. Widmayer et al. (Eds.): ICALP 2002, LNCS 2380, pp. 561–572, 2002.
© Springer-Verlag Berlin Heidelberg 2002

here that for every set X, Staiger's computable entropy rate of X coincides with the computable dimension of X. This provides a second proof of the correspondence principle for computable dimension. We also show that a constructive version of Staiger's entropy rate coincides with constructive dimension.

This paper is organized as follows. Section 2 contains the necessary preliminaries. In section 3 we review Hausdorff, computable, and constructive dimensions. The correspondence principles are presented in section 4. The comparison of effective dimensions with effective entropy rates is given in section 5.

2   Preliminaries

We write $\{0,1\}^*$ for the set of all finite binary *strings* and $\mathbf{C}$ for the Cantor space of all infinite binary *sequences*. For $\omega \in \{0,1\}^* \cup \mathbf{C}$ and $i,j \in \mathbb{N}$, $\omega[i..j]$ is the string consisting of bits i through j of ω. For $w,v \in \{0,1\}^*$, we write $w \sqsubseteq v$ if w is a prefix of v and $w \sqsubset v$ if w is a proper prefix of v. A *prefix set* is a language $A \subseteq \{0,1\}^*$ such that no element of A is a prefix of any other element of A. The sets of strings of length n and of length less than n are $\{0,1\}^n$ and $\{0,1\}^{<n}$. For any $n \in \mathbb{N}$ and $A \subseteq \{0,1\}^*$, $A_{=n} = A \cap \{0,1\}^n$ and $A_{<n} = A \cap \{0,1\}^{<n}$.

We write DEC for the class of *decidable* languages and CE for the class of *computably enumerable* languages.

We will define the first two levels of the arithmetical hierarchy of subsets of $\mathbf{C}$. For each $w \in \{0,1\}^*$, the basic open set $\mathbf{C}_w$ is the set of all sequences in $\mathbf{C}$ that begin with prefix w. We let $\mathbf{C}_\top = \emptyset$.

Definition. Let $X \subseteq \mathbf{C}$.

- $X \in \Sigma_1^0$ if there is a computable function $h : \mathbb{N} \to \{0,1\}^* \cup \{\top\}$ such that

$$X = \bigcup_{i=0}^{\infty} \mathbf{C}_{h(i)}.$$

- $X \in \Pi_1^0$ if $X^c \in \Sigma_1^0$.
- $X \in \Sigma_2^0$ if there is a computable function $h : \mathbb{N} \times \mathbb{N} \to \{0,1\}^* \cup \{\top\}$ such that

$$X = \bigcup_{i=0}^{\infty} \bigcap_{j=0}^{\infty} \mathbf{C}_{h(i,j)}^c.$$

- $X \in \Pi_2^0$ if $X^c \in \Sigma_2^0$.

Note that every $X \in \Sigma_1^0$ is open and every $X \in \Pi_1^0$ is closed in the standard (product) topology on $\mathbf{C}$. In fact, Σ_1^0 and Π_1^0 are the *computably open* and *computably closed* subsets of $\mathbf{C}$, respectively. Recall that $\mathbf{C}$ is *compact*. This implies that for any closed set $X \subseteq \mathbf{C}$ and any collection of strings $A \subseteq \{0,1\}^*$ such that $X \subseteq \bigcup_{w \in A} \mathbf{C}_w$, there is a finite subcollection $A' \subseteq A$ such that $X \subseteq \bigcup_{w \in A'} \mathbf{C}_w$.

We say that a real-valued function $f : \{0,1\}^* \to [0,\infty)$ is *computable* if there is a computable function $\hat{f} : \mathbb{N} \times \{0,1\}^* \to [0,\infty) \cap \mathbb{Q}$ such that for all

$n \in \mathbb{N}$ and $w \in \{0,1\}^*$, $|f(w) - \hat{f}(n,w)| \leq 2^{-n}$. A rational-valued function $f : \{0,1\}^* \to [0,\infty) \cap \mathbb{Q}$ that is itself computable is called *exactly computable*. We say that $f : \{0,1\}^* \to [0,\infty)$ is *lower semicomputable* if there is a computable function $g : \mathbb{N} \times \{0,1\}^* \to [0,\infty) \cap \mathbb{Q}$ such that for any $w \in \{0,1\}^*$, $g(n,w) \leq g(n+1,w) < f(w)$ for all $n \in \mathbb{N}$ and and $f(w) = \lim_{n\to\infty} g(n,w)$.

3   Hausdorff, Constructive, and Computable Dimensions

In this section we briefly review classical Hausdorff dimension, constructive dimension, and computable dimension. Further details are available in Lutz's introductory papers [1,3].

For each $k \in \mathbb{N}$, let $\mathcal{A}_k$ be the set of all prefix sets $A \subseteq \{0,1\}^*$ such that $A_{<k} = \emptyset$. For each $X \subseteq \mathbf{C}$, $s \in [0,\infty)$, and $k \in \mathbb{N}$, we define

$$H_k^s(X) = \inf\left\{ \sum_{w \in A} 2^{-s|w|} \,\middle|\, A \in \mathcal{A}_k \text{ and } X \subseteq \bigcup_{w \in A} \mathbf{C}_w \right\}$$

and

$$H^s(X) = \lim_{k\to\infty} H_k^s(X).$$

Definition. The *Hausdorff dimension* of a set $X \subseteq \mathbf{C}$ is

$$\dim_{\mathrm{H}}(X) = \inf\left\{ s \in [0,\infty) \,\middle|\, H^s(X) = 0 \right\}.$$

Lutz [1] proved an alternative characterization of Hausdorff dimension using functions called gales and supergales. Gales and supergales are generalizations of martingales and supermartingales.

Definition. Let $s \in [0,\infty)$. A function $d : \{0,1\}^* \to [0,\infty)$ is an *s-supergale* if for all $w \in \{0,1\}^*$,

$$d(w) \geq \frac{d(w0) + d(w1)}{2^s}. \tag{3.1}$$

If equality holds in (3.1) for all $w \in \{0,1\}^*$, then d is an *s-gale*. A martingale is a 1-gale and a supermartingale is a 1-supergale.

Intuitively, a supergale is viewed as a function betting on an unknown binary sequence. If w is a prefix of the sequence, then the capital of the supergale after placing its first $|w|$ bets is given by $d(w)$. Assuming that w is a prefix of the sequence, the supergale places bets on $w0$ and $w1$ also being prefixes. The parameter s determines the fairness of the betting; as s decreases the betting is less fair. The goal of a supergale is to bet successfully on sequences.

Definition. Let $s \in [0,\infty)$ and let d be an s-supergale.

1. We say d *succeeds on* a sequence $S \in \mathbf{C}$ if

$$\limsup_{n\to\infty} d(S[0..n-1]) = \infty.$$

2. The *success set* of d is

$$S^\infty[d] = \{S \in \mathbf{C} \mid d \text{ succeeds on } S\}.$$

Theorem 3.1. (Lutz [1]) *For any $X \subseteq \mathbf{C}$,*

$$\dim_\mathrm{H}(X) = \inf\left\{s \,\middle|\, \begin{array}{l} \text{there exists a } s\text{-gale } d \\ \text{for which } X \subseteq S^\infty[d] \end{array}\right\}.$$

This characterization of Hausdorff dimension motivates the following definitions of computable and constructive dimensions. We say that an s-supergale d is *constructive* if it is lower-semicomputable.

Definition. Let $X \subseteq \mathbf{C}$.

1. The *computable dimension* of X is

$$\dim_\mathrm{comp}(X) = \inf\left\{s \,\middle|\, \begin{array}{l} \text{there exists a computable} \\ s\text{-gale } d \text{ for which } X \subseteq S^\infty[d] \end{array}\right\}.$$

2. The *constructive dimension* of X is

$$\mathrm{cdim}(X) = \inf\left\{s \,\middle|\, \begin{array}{l} \text{there exists a constructive} \\ s\text{-supergale } d \text{ for which } X \subseteq S^\infty[d] \end{array}\right\}.$$

3. The *constructive dimension* of a sequence $S \in \mathbf{C}$ is $\mathrm{cdim}(S) = \mathrm{cdim}(\{S\})$.

Observe that for any set $X \subseteq \mathbf{C}$,

$$0 \leq \dim_\mathrm{H}(X) \leq \mathrm{cdim}(X) \leq \dim_\mathrm{comp}(X) \leq 1.$$

An important property of constructive dimension is the following *pointwise stability* property.

Lemma 3.2. (Lutz [3]) *For all $X \subseteq \mathbf{C}$, $\mathrm{cdim}(X) = \sup_{S \in X} \mathrm{cdim}(S)$.*

The following exact computation lemma shows that computable dimension can be equivalently defined using exactly computable gales in place of computable gales.

Lemma 3.3. (Lutz [1]) *For any computable s-gale d where 2^s is rational, there is an exactly computable s-gale d' such that $S^\infty[d] \subseteq S^\infty[d']$.*

4   Correspondence Principles

In this section we will prove that that $\mathrm{cdim}(X) = \dim_\mathrm{H}(X)$ for any X that is an *arbitrary* union of Π_1^0-definable sets. We will also show that $\dim_\mathrm{comp}(X) = \dim_\mathrm{H}(X)$ if X is Σ_2^0-definable.

Lemma 4.1. *If $X \in \Pi_1^0$, then $\dim_\mathrm{H}(X) = \dim_\mathrm{comp}(X)$.*

Proof. Let $X \in \Pi_1^0$. Since $\dim_{\mathrm{comp}}(X) \geq \dim_{\mathrm{H}}(X)$, it is enough to prove that $\dim_{\mathrm{comp}}(X) \leq \dim_{\mathrm{H}}(X)$. For this, let $s > \dim_{\mathrm{H}}(X)$ be such that 2^s is rational.

Since $s > \dim_{\mathrm{H}}(X)$, for each $r \in \mathbb{N}$, there is a prefix set $A_r \subseteq \{0,1\}^*$ such that

$$\sum_{w \in A_r} 2^{-s|w|} \leq 2^{-r} \text{ and } X \subseteq \bigcup_{w \in A_r} \mathbf{C}_w.$$

Because $\mathbf{C}$ is compact and X is closed, X is compact. Thus each A_r may be taken finite.

Because $X \in \Pi_1^0$, there is a computable function $h : \mathbb{N} \to \{0,1\}^* \cup \{\top\}$ such that

$$X = \bigcap_{i=0}^{\infty} \mathbf{C}_{h(i)}^c.$$

For each $k \in \mathbb{N}$, let

$$X_k = \bigcap_{i=0}^{k} \mathbf{C}_{h(i)}^c.$$

Then for each $k \in \mathbb{N}$, it is easy to compute a finite prefix set B_k such that

$$\sum_{w \in B_k} 2^{-s|w|} \text{ is minimal and } X_k = \bigcup_{w \in B_k} \mathbf{C}_w.$$

For each $r \in \mathbb{N}$, let

$$k_r = \min \left\{ k \;\middle|\; \sum_{w \in B_k} 2^{-s|w|} \leq 2^{-r} \right\}.$$

We know that each k_r exists because of the existence of the finite prefix sets A_r that satisfy the condition. Also, each k_r can be computed by computing the finite sets B_k until the condition is satisfied.

The rest of the proof is based on a construction used in characterizing Hausdorff dimension in terms of gales [1]. There the prefix sets A_r mentioned above are used to give an s-gale that succeeds on X. Here we use the finite, computable prefix sets B_{k_r} in the same manner to give a computable s-gale that succeeds on X.

Define for each $r \in \mathbb{N}$ a function $d_r : \{0,1\}^* \to [0, \infty)$ by

$$d_r(w) = \begin{cases} 2^{(s-1)(|w|-|v|)} & \text{if } (\exists v \sqsubseteq w) v \in B_{k_r} \\ \displaystyle\sum_{\substack{u \in \{0,1\}^* \\ wu \in B_{k_r}}} 2^{-s|u|} & \text{otherwise.} \end{cases}$$

Then each d_r is an s-gale. Also, $d_r(\lambda) \leq 2^{-r}$ and $d_r(w) = 1$ for all $w \in B_{k_r}$. Next define a function d on $\{0,1\}^*$ by $d = \sum_{r=0}^{\infty} 2^r d_{2r}$. Then

$$d(\lambda) = \sum_{r=0}^{\infty} 2^r d_{2r}(\lambda) \leq \sum_{r=0}^{\infty} 2^r 2^{-2r} = 2,$$

so by induction it follows that $d(w) < \infty$ for all strings w. Therefore, by linearity, d is an s-gale.

Let $S \in X$. Then $S \in X_{k_{2r}}$ for all $r \in \mathbb{N}$. This means that S has some prefix $S[0..n-1] \in B_{k_{2r}}$, and then

$$d(S[0..n-1]) \geq 2^r d_{2r}(S[0..n-1]) = 2^r.$$

Therefore d succeeds on S, so $X \subseteq S^\infty[d]$.

To see that d is computable, define $\hat{d} : \mathbb{N} \times \{0,1\}^* \to [0,\infty)$ by

$$\hat{d}(i,w) = \sum_{r=0}^{\lceil s|w| \rceil + i} 2^r d_{2r}(w).$$

We can exactly compute $\hat{d}$ by using the function h to uniformly compute the sets B_{k_r}. Then

$$
\begin{aligned}
\left| d(w) - \hat{d}(i,w) \right| &= \sum_{r=\lceil s|w| \rceil + i + 1}^{\infty} 2^r d_{2r}(w) \\
&\leq \sum_{r=\lceil s|w| \rceil + i + 1}^{\infty} 2^r 2^{s|w|} d_{2r}(\lambda) \\
&\leq \sum_{r=\lceil s|w| \rceil + i + 1}^{\infty} 2^{r+s|w|} 2^{-2r} \\
&= 2^{s|w|} \sum_{r=\lceil s|w| \rceil + i + 1}^{\infty} 2^{-r} \\
&= 2^{s|w|} 2^{-\lceil s|w| \rceil - i} \\
&\leq 2^{-i},
\end{aligned}
$$

so $\hat{d}$ is a computable approximation of d. Therefore d is computable, so it witnesses that $\dim_{\text{comp}}(X) \leq s$. Since $s > \dim_{\mathrm{H}}(X)$ is arbitrary with 2^s rational and the rationals are dense, it follows that $\dim_{\text{comp}}(X) \leq \dim_{\mathrm{H}}(X)$. $\square$

We now use the preceding lemma to give our correspondence principle for constructive dimension.

Theorem 4.2. *If $X \subseteq \mathbf{C}$ is a union of Π_1^0 sets, then $\dim_{\mathrm{H}}(X) = \text{cdim}(X)$.*

Proof. Let $\mathcal{I}$ be an arbitrary index set, $X_\alpha \in \Pi_1^0$ for each $\alpha \in \mathcal{I}$, and $X = \bigcup_{\alpha \in \mathcal{I}} X_\alpha$. By definition, $\dim_{\mathrm{H}}(X) \leq \text{cdim}(X)$. Using Lemma 3.2 (the pointwise stability of constructive dimension), Lemma 4.1, and the monotonicity of Hausdorff dimension, we have

$$
\begin{aligned}
\text{cdim}(X) &= \sup_{\alpha \in \mathcal{I}} \text{cdim}(X_\alpha) \\
&= \sup_{\alpha \in \mathcal{I}} \dim_{\mathrm{H}}(X_\alpha) \\
&\leq \dim_{\mathrm{H}}(X).
\end{aligned}
$$

$\square$

Theorem 4.2 yields a pointwise characterization of the classical Hausdorff dimension of unions of Π_1^0 sets.

Corollary 4.3. *If $X \subseteq \mathbf{C}$ is a union of Π_1^0 sets, then*

$$\dim_H(X) = \sup_{S \in X} \operatorname{cdim}(S).$$

Proof. This follows immediately from Theorem 4.2 and Lemma 3.2. $\square$

If we require that the union in Theorem 4.2 be effective, we arrive at the following correspondence principle for computable dimension. This result also follows implicitly from Staiger's work on martingale exponents of increase [6].

Theorem 4.4. *If $X \in \Sigma_2^0$, then $\dim_H(X) = \dim_{\operatorname{comp}}(X)$.*

Proof. Let $X \in \Sigma_2^0$. Since $\dim_{\operatorname{comp}}(X) \geq \dim_H(X)$, it is enough to prove that $\dim_{\operatorname{comp}}(X) \leq \dim_H(X)$. For this, let $s > \dim_H(X)$ be such that 2^s is rational. As in the proof of the Lemma 4.1, it suffices to give a computable s-gale d that succeeds on X.

Since $X \in \Sigma_2^0$, there is a computable function $h : \mathbb{N} \times \mathbb{N} \to \{0,1\}^* \cup \{\top\}$ such that

$$X = \bigcup_{j=0}^{\infty} \bigcap_{i=0}^{\infty} \mathbf{C}_{h(i,j)}^c.$$

For each $j \in \mathbb{N}$, let

$$X_j = \bigcap_{i=0}^{\infty} \mathbf{C}_{h(i,j)}^c.$$

Since each $X_j \subseteq X$, $\dim_H(X_j) \leq \dim_H(X) < s$. Each $X_j \in \Pi_1^0$, so by Lemma 4.1, for each $j \in \mathbb{N}$, there is a computable s-gale d_j with $d_j(\lambda) \leq 1$ that succeeds on X_j. Let $d = \sum_{j=0}^{\infty} 2^{-j} d_j$. Then d is an s-gale, d is computable by using h to uniformly compute the d_j, and $X \subseteq S^\infty[d]$. $\square$

We note that Theorems 4.2 and 4.4 cannot be extended to higher levels of the arithmetical hierarchy.

Observation 4.5. *There is a set $X \in \Pi_2^0$ such that $\dim_H(X) \neq \operatorname{cdim}(X)$.*

Proof. It is well known that there exists a Martin-Löf random sequence $S \in \Delta_2^0$. (A sequence S is in Δ_2^0 if S is decidable relative to an oracle for the halting problem.) Let $X = \{S\}$. Since $S \in \Delta_2^0$, we have $X \in \Pi_2^0$. Lutz [3] observed that all random sequences have constructive dimension 1, so $\operatorname{cdim}(X) = 1$. But any singleton has Hausdorff dimension 0, so $\dim_H(X) = 0$. $\square$

5 Dimension and Entropy Rates

In this section we compare our correspondence principles to related work of Staiger [6] on entropy rates. This comparison yields a new characterization of constructive dimension.

Definition. Let $A \subseteq \{0,1\}^*$. The *entropy rate of A* is

$$H_A = \limsup_{n \to \infty} \frac{\log |A_{=n}|}{n}.$$

(Here the logarithm is base 2 and we use the convention that $\log 0 = 0$.) Staiger observed that this entropy rate has a useful alternate characterization.

Lemma 5.1. (Staiger [5]) *For any $A \subseteq \{0,1\}^*$,*

$$H_A = \inf \left\{ s \left| \sum_{w \in A} 2^{-s|w|} < \infty \right. \right\}.$$

Definition. Let $A \subseteq \{0,1\}^*$. The *δ-limit of A* is

$$A^\delta = \{S \in \mathbf{C} | (\exists^\infty n) S[0..n-1] \in A\}.$$

For any $X \subseteq \mathbf{C}$, define

$$\mathcal{H}(X) = \{H_A | A \subseteq \{0,1\}^* \text{ and } X \subseteq A^\delta\},$$

$$\mathcal{H}_{\mathrm{DEC}}(X) = \{H_A | A \in \mathrm{DEC} \text{ and } X \subseteq A^\delta\},$$

and

$$\mathcal{H}_{\mathrm{CE}}(X) = \{H_A | A \in \mathrm{CE} \text{ and } X \subseteq A^\delta\}.$$

We call the infima of $\mathcal{H}(X), \mathcal{H}_{\mathrm{DEC}}(X)$, and $\mathcal{H}_{\mathrm{CE}}(X)$ the *entropy rate of X*, the *computable entropy rate of X*, and the *constructive entropy rate of X*, respectively.

Classical Hausdorff dimension may be characterized in terms of entropy rates.

Theorem 5.2. *For any $X \subseteq \mathbf{C}$, $\dim_{\mathrm{H}}(X) = \inf \mathcal{H}(X)$.*

A proof of Theorem 5.2 can be found in [5]; it also follows from Theorem 32 of [4].

Staiger proved the following relationship between computable entropy rates and Hausdorff dimension.

Theorem 5.3. (Staiger [6]) *For any $X \in \Sigma_2^0$, $\dim_{\mathrm{H}}(X) = \inf \mathcal{H}_{\mathrm{DEC}}(X)$.*

Putting Theorems 4.4, 5.2, and 5.3 together, for any $X \in \Sigma_2^0$ we have

$$
\begin{array}{ccc}
\dim_{\mathrm{H}}(X) = & \mathrm{cdim}(X) & = \dim_{\mathrm{comp}}(X) \\
\| & \| & \| \\
\inf \mathcal{H}(X) = & \inf \mathcal{H}_{\mathrm{CE}}(X) = & \inf \mathcal{H}_{\mathrm{DEC}}(X).
\end{array}
$$

We will extend this to show that $\mathrm{cdim}(X) = \inf \mathcal{H}_{\mathrm{CE}}(X)$ and $\mathrm{dim}_{\mathrm{comp}}(X) = \inf \mathcal{H}_{\mathrm{DEC}}(X)$ hold for *arbitrary* $X \subseteq \mathbf{C}$. Note that the latter together with Theorem 5.3 provides a second proof of Theorem 4.4.

First we show that the dimensions are lower bounds of the entropy rates.

Lemma 5.4. *For any $X \subseteq \mathbf{C}$,*

$$\mathrm{cdim}(X) \leq \inf \mathcal{H}_{\mathrm{CE}}(X)$$

and

$$\mathrm{dim}_{\mathrm{comp}}(X) \leq \inf \mathcal{H}_{\mathrm{DEC}}(X).$$

Proof. We begin with a general construction that will be used to prove both inequalities. Let $A \subseteq \{0,1\}^*$ and let $t > s > H_A$ such that 2^t and 2^s are rational. For each $n \in \mathbb{N}$, define a function $d_n : \{0,1\}^* \to [0, \infty)$ by

$$d_n(w) = \begin{cases} 2^{-t(n-|w|)} \cdot \left| \{v \in A_{=n} | w \sqsubseteq v\} \right| & \text{if } |w| \leq n \\ 2^{(t-1)(|w|-n)} d(w[0..n-1]) & \text{if } |w| > n. \end{cases}$$

Then each d_n is a t-gale. Define a function d on $\{0,1\}^*$ by $d = \sum_{n=0}^{\infty} 2^{(t-s)} d_n$. Then

$$d(\lambda) = \sum_{n=0}^{\infty} 2^{(t-s)n} 2^{-tn} |A_{=n}| = \sum_{w \in A} 2^{-s|w|} < \infty$$

because $s > H_A$. By induction, $d(w) < \infty$ for all strings w, so $d : \{0,1\}^* \to [0, \infty)$. By linearity, d is also a t-gale. For any $w \in A$, we have

$$d(w) \geq 2^{(t-s)|w|} d_{|w|}(w) = 2^{(t-s)|w|},$$

so it follows that $A^\delta \subseteq S^\infty[d]$.

Let $r > \inf \mathcal{H}_{\mathrm{CE}}(X)$ be arbitrary. Then there is a computably enumerable A with $X \subseteq A^\delta$ and $H_A < r$. We can also choose 2^t and 2^s rational so that $H_A < s < t < r$. Because A is computably enumerable, the t-gale d defined above is constructive. Since $X \subseteq A^\delta \subseteq S^\infty[d]$, we have $\mathrm{cdim}(X) \leq t < r$. As this holds for all $r > \inf \mathcal{H}_{\mathrm{CE}}(X)$, we have $\mathrm{cdim}(X) \leq \inf \mathcal{H}_{\mathrm{CE}}(X)$.

Similarly, let $r > \inf \mathcal{H}_{\mathrm{DEC}}(X)$ be arbitrary. Take a decidable A and 2^s, 2^t rational such that $X \subseteq A^\delta$ and $H_A < s < t < r$. We will show that the t-gale d defined above is computable. For this, choose a natural number $k > \frac{1}{t-s}$. Define a function $\hat{d} : \{0,1\}^* \times \mathbb{N} \to [0, \infty) \cap \mathbb{Q}$ by

$$\hat{d}(w, r) = \sum_{n=0}^{kr} 2^{(s-t)n} d_n(w).$$

Then $\hat{d}$ is exactly computable. For any string w, $d_n(w) \leq 1$ for all n, so for any precision $r \in \mathbb{N}$,

$$
\begin{aligned}
|d(w) - \hat{d}(w, r)| &= \sum_{n=kr+1}^{\infty} 2^{(s-t)n} d_n(w) \\
&\leq \sum_{n=kr+1}^{\infty} 2^{(s-t)n} \\
&= 2^{(s-t)(kr)} \\
&< 2^{-r}.
\end{aligned}
$$

Therefore $\hat{d}$ demonstrates that d is computable. Then $\dim_{\mathrm{comp}}(X) \leq t < r$ because $X \subseteq A^\delta \subseteq S^\infty[d]$. It follows that $\dim_{\mathrm{comp}}(X) \leq \inf \mathcal{H}_{\mathrm{DEC}}(X)$ because $r > \inf \mathcal{H}_{\mathrm{DEC}}(X)$ is arbitrary. $\qquad\square$

Next we give lower bounds for constructive and computable dimension by entropy rates.

Lemma 5.5. *For all $X \subseteq \mathbf{C}$,*

$$
\inf \mathcal{H}_{\mathrm{CE}}(X) \leq \mathrm{cdim}(X)
$$

and

$$
\inf \mathcal{H}_{\mathrm{DEC}}(X) \leq \dim_{\mathrm{comp}}(X).
$$

Proof. Suppose that d is an s-supergale with $X \subseteq S^\infty[d]$. Assume without loss of generality that $d(\lambda) < 1$ and let $A = \{w \mid d(w) > 1\}$. Then for all $n \in \mathbb{N}$, $\sum_{w \in \{0,1\}^n} d(w) \leq 2^{sn}$ and $|A_{=n}| \leq 2^{sn}$. Also, $X \subseteq S^\infty[d] \subseteq A^\delta$. For any $t > s$,

$$
\sum_{w \in A} 2^{-t|w|} = \sum_{n=0}^{\infty} 2^{-tn} |A_{=n}| \leq \sum_{n=0}^{\infty} 2^{(s-t)n} < \infty,
$$

so $H_A \leq t$. Therefore $H_A \leq s$.

If we let $s > \mathrm{cdim}(X)$ such that 2^s is rational, then there is a constructive s-supergale d succeeding on A. Then the set A defined above is computably enumerable, so $H_A \in \mathcal{H}_{\mathrm{CE}}(X)$. We showed that $H_A \leq s$, so $\inf \mathcal{H}_{\mathrm{CE}}(X) \leq s$. Therefore $\inf \mathcal{H}_{\mathrm{CE}}(X) \leq \mathrm{cdim}(X)$.

If $s > \dim_{\mathrm{comp}}(X)$ is rational, then by Lemma 3.3 there is an exactly computable s-gale d succeeding on A. Then the set A above is decidable, and analogously we obtain $\inf \mathcal{H}_{\mathrm{DEC}}(X) \leq \mathrm{cdim}(X)$. $\qquad\square$

Combining Lemmas 5.4 and 5.5 yields new characterizations of constructive and computable dimension.

Theorem 5.6. *For all* $X \subseteq \mathbf{C}$,

$$\mathrm{cdim}(X) = \inf \mathcal{H}_{\mathrm{CE}}(X)$$

and

$$\mathrm{dim}_{\mathrm{comp}}(X) = \inf \mathcal{H}_{\mathrm{DEC}}(X).$$

We observe that some resource-bounded analogues of these results hold. For example, if we define the similar concepts for polynomial-time and polynomial-space computability [1], the proofs of Lemmas 5.4 and 5.5 can be extended to show that

$$\mathrm{dim}_{\mathrm{p}}(X) \geq \inf \mathcal{H}_{\mathrm{P}}(X)$$

and

$$\mathrm{dim}_{\mathrm{pspace}}(X) = \inf \mathcal{H}_{\mathrm{PSPACE}}(X)$$

hold for all $X \subseteq \mathbf{C}$.

Acknowledgments. The author thanks Jack Lutz, Elvira Mayordomo, and an anonymous referee for helpful suggestions and remarks.

References

1. J. H. Lutz. Dimension in complexity classes. In *Proceedings of the Fifteenth Annual IEEE Conference on Computational Complexity*, pages 158–169. IEEE Computer Society Press, 2000.
2. J. H. Lutz. Information and computation seminar. Iowa State University, 2000. Unpublished lectures.
3. J. H. Lutz. The dimensions of individual strings and sequences. Technical Report cs.CC/0203016, ACM Computing Research Repository, 2002.
4. C. A. Rogers. *Hausdorff Measures*. Cambridge University Press, 1998. Originally published in 1970.
5. L. Staiger. Kolmogorov complexity and Hausdorff dimension. *Information and Control*, 103:159–94, 1993.
6. L. Staiger. A tight upper bound on Kolmogorov complexity and uniformly optimal prediction. *Theory of Computing Systems*, 31:215–29, 1998.

A Total Approach to Partial Algebraic Specification

José Meseguer[1] and Grigore Roşu[2]

[1] University of Illinois at Urbana-Champaign
[2] NASA Ames Research Center - RIACS

Abstract. Partiality is a fact of life, but at present explicitly partial algebraic specifications lack tools and have limited proof methods. We propose a sound and complete way to support execution and formal reasoning of explicitly partial algebraic specifications within the total framework of membership equational logic (MEL) which has a high-performance interpreter (Maude) and proving tools. This is accomplished by a sound and complete mapping PMEL → MEL of *partial* membership equational (PMEL) theories into total ones. Furthermore, we characterize and give proof methods for a practical class of theories for which this mapping has "almost-zero representational distance," in that the partial theory and its total translation are *identical* up to minor syntactic sugar conventions. This then supports very direct execution of, and formal reasoning about, partial theories at the total level. In conjunction with tools like Maude and its proving tools, our methods can be used to execute and reason about partial specifications such as those in CASL.

1 Introduction

Any algebraic specification formalism worth its salt has somehow to deal with partiality, because in practice many functions are partial. This can be done within *explicitly partial* formalisms, or, alternatively, within *total* formalisms supporting notions of subtype (subsort) so that partial functions then appear as appropriate restrictions to given subtypes of corresponding total functions.

Various explicitly partial specification formalisms were explored from the early days of algebraic specification by different authors, e.g., [13,3] (see [1,9] for surveys and bibliographies on equational specification, including partial approaches). Reichel's book [20] gives a systematic study of algebraic specification in an explicitly partial way using *existence equations*. The comparison between explicitly partial formalisms has been facilitated by Mossakowski's work using maps of institutions [17]. The interest in explicitly partial specifications has been recently revived by the CASL specification language [7], which supports subsorts and partial operations, with both *existence* and *strong* equations.

At present, the main practical limitations of explicitly partial equational specifications appear in the areas of *execution, formal reasoning*, and *tools*, where the

P. Widmayer et al. (Eds.): ICALP 2002, LNCS 2380, pp. 572–584, 2002.

state of the art is considerably less developed than for total approaches. Execution requires adequate notions of confluence and termination. As far as we know, the only theoretical study of such matters at the explicitly partial level is the work of Hintermeier, C. Kirchner, and H. Kirchner in the context of *galactic algebras* [12]; but we are not aware of any systems supporting execution of explicitly partial specifications, galactic or otherwise. In fact, we are not aware of any tools that *directly support* either execution or formal reasoning for explicitly partial equational specifications. The only tool that we know of supporting theorem proving for CASL specifications is the HOL-CASL tool [18], where partial specifications are translated into total ones in higher-order logic [18,19]; in particular, inductive reasoning is supported using second order logic in HOL. This can be viewed as an *indirect* method of supporting theorem proving for explicitly partial specifications: the reasoning happens in a different logic, and the original specifications are modified considerably by the translation.

Developing well-engineered interpreters and theorem provers that directly support execution and formal reasoning for explicitly partial specifications seems a worthwhile goal; however, this will certainly require a strong commitment along several years by several research groups to carry out the new theoretical developments required and to build appropriate tools. In the meantime, are there any viable alternatives to support both execution and reasoning for partial specifications? The main goal of this paper is to propose one such alternative, based on *total* equational specifications.

An important advantage of our proposal is that there are several interpreters and proving tools for total equational formalisms with varying degrees of support for partiality via subsorts: two based on order-sorted algebra (OBJ3 [11] and CafeOBJ [8]) and one based on membership equational logic (Maude [5]). Specifically, our proposal uses a *logical framework* approach based on the ideas:

1. *unification* of different explicitly partial equational specification formalisms (including all partial formalisms in the so-called Mossakowski's web [17]) within the *partial membership equational logic framework* (PMEL) [15];

2. a *conservative map of logics* embedding PMEL within the *(total) membership equational logic* framework (MEL) [15]; this embedding preserves initial algebras, free algebras, and left adjoints to forgetful functors induced by theory maps in both directions; furthermore, one can *borrow* the MEL inference system to get a complete PMEL inference system [4,15];

3. characterization of a class of theories of practical interest for which the above embedding PMEL $\rightarrow$ MEL becomes a mapping with *almost-zero representational distance*; that is, except for minor syntactic sugar differences, the partial specification remains *unchanged* when translated to a total one;

4. execution in a tool supporting MEL specifications such as Maude of the corresponding total versions of the explicitly partial equational specifications, which can be reinterpreted as execution of the original PMEL specifications; similarly, inductive theorem proving support for PMEL using tools such as the Maude inductive theorem prover (ITP) and other Maude tools [6].

The theoretical foundations underlying 1–2 were already developed in [15]. Based on such foundations it is possible to execute and reason about PMEL specifications via their corresponding MEL translation. However, for *arbitrary* PMEL specifications the MEL-based execution and reasoning are somewhat indirect, in the sense that the original PMEL specification *has to be modified*. This is because *definedness* in PMEL is expressed as *having a sort* in MEL, but definedness is always *inherited by subterms*. For example, a PMEL existence equation $f(g(x, y)) = k(h(x))$ (both terms are defined and are equal) with x, y of sort s, must be translated into the MEL axioms (its *envelope*) $f(g(x, y)) = k(h(x))$, $f(g(x, y)) : \top_1$, $g(x, y) : \top_2$, $k(h(x)) : \top_1$, and $h(x) : \top_3$, where $\top_1, \top_2, \top_3$ are the *top sorts* in the appropriate connected components of the sort poset.

In this work, we give conditions and proof techniques allowing us to embed, execute, and reason about explicitly partial specifications in the MEL total framework *without modifying them*, without having to add envelopes (ideas 3–4). Thus, because of the almost-zero representational distance thus obtained and the "borrowing" of the MEL logic to get a sound and complete PMEL proof system, we can support execution and reasoning for PMEL specifications in a *very direct way* within a total framework. In particular, since the PMEL specifications that pass the test do not have to be modified, for *execution purposes* we can rely on the well-developed theory of MEL execution by rewriting presented in [2].

2   Total Membership Equational Logic

We next recall some (total) membership equational logic (MEL) definitions needed in the paper. The interested reader is referred to [15] for more on MEL.

A *signature* Ω in MEL is a triple (K, Σ, π) where K is a set of *kinds*, Σ is a K-sorted (or, more precisely, K-kinded) algebraic signature, and $\pi : S \to K$ is a function that assigns to each element in its domain, called a *sort*, a kind. Given a signature Ω in MEL, an Ω-*(membership) algebra* A is a Σ-algebra together with a set $A_s \subseteq A_{\pi(s)}$ for each sort $s \in S$, and an Ω-*homomorphism* $h : A \to B$ is a Σ-morphism such that for each $s \in S$ we have $h_{\pi(s)}(A_s) \subseteq B_s$. We let $\mathbf{Alg}_\Omega$ denote the category of Ω-algebras and Ω-homomorphisms.

Given a signature Ω and a K-indexed set of *variables*, an *atomic* (Ω, X)-*equation* has the form $t = t'$ where $t, t' \in T_{\Sigma,k}(X)$, and an *atomic* (Ω, X)-*membership* has the form $t : s$ where s is a sort and $t \in T_{\Sigma, \pi(s)}(X)$. An Ω-*sentence* in MEL has the form $(\forall X)\ a\ \text{if}\ a_1 \wedge \ldots \wedge a_n$, where $a, a_1, \ldots, a_n$ are atomic (Ω, X)-equations or (Ω, X)-memberships, and $\{a_1, \ldots, a_n\}$ is a set (no duplications). If $n = 0$ then the Ω-sentence is called *unconditional* and written $(\forall X)\ a$. Given an Ω-algebra A and a map $\theta : X \to A$, then $A, \theta \models_\Omega t = t'$ iff $A_{\theta(t)} = A_{\theta(t')}$, and $A, \theta \models_\Omega t : s$ iff $A_{\theta(t)} \in A_s$. A satisfies $(\forall X)\ a\ \text{if}\ a_1 \wedge \ldots \wedge a_n$, written $A \models_\Omega (\forall X)\ a\ \text{if}\ a_1 \wedge \ldots \wedge a_n$, iff for each $\theta : X \to A$, if $A, \theta \models_\Omega a_1$ and ... and $A, \theta \models_\Omega a_n$ then $A, \theta \models_\Omega a$. A *specification* (or *theory*) $T = (\Omega, \Gamma)$ in MEL consists of a signature Ω and a set of Ω-sentences Γ. An Ω-algebra A satisfies (or is a model of) $T = (\Omega, \Gamma)$, written $A \models T$, iff it satisfies each sentence in Γ. We let $\mathbf{Alg}(T)$ denote the full subcategory of $\mathbf{Alg}_\Omega$ of Ω-algebras satisfying T.

We let $\mathbf{Th}_{\mathrm{MEL}}$ denote the category of theories in MEL (see [15] for the definition of morphisms). Theories T and T' are semantically equivalent in MEL, written $T \equiv_{\mathrm{MEL}} T'$, iff they have the same models, i.e., iff $\mathbf{Alg}(T) = \mathbf{Alg}(T')$. It is known [15] that MEL admits complete deduction, so $T \equiv_{\mathrm{MEL}} T'$ is equivalent to $T \Leftrightarrow T'$, where $(\Omega, \Gamma) \Rightarrow (\Omega', \Gamma')$ means that $\Gamma \vdash_\Omega \varphi'$ for each $\varphi' \in \Gamma'$.

To make specifications easier to read and to emphasize that order-sorted specifications are a special case of membership equational ones, the following syntactic sugar conventions are widely accepted and supported by Maude [5]:

Subsorts. Given sorts s, s' with $\pi(s) = \pi(s') = k$, the declaration $s < s'$ is syntactic sugar for the conditional membership $(\forall x : k)\ x : s'$ **if** $x : s$.

Operations. If $\sigma \in \Omega_{k_1 \ldots k_n, k}$ and $s_1, \ldots, s_n, s \in S$ with $\pi(s_1) = k_1, \ldots, \pi(s_n) = k_n$, $\pi(s) = k$, then the declaration $\sigma : s_1 \times \cdots \times s_n \to s$ is syntactic sugar for $(\forall x_1 : k_1, \ldots, x_n : k_n)\ \sigma(x_1, \ldots, x_n) : s$ **if** $x_1 : s_1 \wedge \ldots \wedge x_n : s_n$.

Variables. $(\forall x : s, X)\ a$ **if** $a_1 \wedge \ldots \wedge a_n$ is syntactic sugar for the Ω-sentence $(\forall x : \pi(s), X)\ a$ **if** $a_1 \wedge \ldots \wedge a_n \wedge x : s$. With this, the operation declaration $\sigma : s_1 \times \cdots \times s_n \to s$ is equivalent to $(\forall x_1 : s_1, \ldots, x_n : s_n)\ \sigma(x_1, \ldots, x_n) : s$.

3   Partial Membership Equational Logic

Partial membership equational logic (abbreviated PMEL), as shown in Section 15 of [15], has good properties as logical framework for partial algebraic specification. It generalizes in a straightforward way several well-known partial specification formalisms and furthermore can specify up to isomorphism all the categories of models for the different partial specification formalisms in [17].

If $(S, \leq)$ is a poset then $\hat{S} = S/\equiv_\leq$, with $\equiv_\leq$ the smallest equivalence relation containing $\leq$, is the set of its connected components and we will call them its *kinds* for convenience and let k or $[s]$ denote one of them. A *signature* in partial membership equational logic is a triple $\Omega = (S, \leq, \Sigma)$ with $(S, \leq)$ a poset of *sorts* s.t. each of its connected components k has a top element $\top_k$, and Σ is an $\hat{S}$-sorted signature. A *partial Ω-algebra* is an S-indexed set with $A_s \subseteq A_{s'}$ for each $s \leq s'$, together with a *partial* function $A_f : A_{\top_{k_1}} \times \cdots \times A_{\top_{k_n}} \dashrightarrow A_{\top_k}$ for each $f : k_1 \ldots k_n \to k$ in Σ. $\mathbf{PAlg}_\Omega$ is the category of partial Ω-algebras [15].

Given an S-indexed set of variables X, we let $\hat{X}$ denote the $\hat{S}$-indexed set associated to X in the obvious way ($\hat{X}_k = \bigcup_{s \in k} X_s$). Notice that the same variable can occur in more than one of the sets X_s, but that $\hat{X}_{[s]}$ will have only one occurrence of it. We often write $x : s$ to denote the fact that $x \in X_s$, and X is often given by a list of pairs (variable:sort) separated by comma. Then an *atomic (Ω, X)-equation* has the form $t = t'$ where $t, t' \in T_{\Sigma, k}(\hat{X})$, and an *atomic (Ω, X)-membership* has the form $t : s$, where $t \in T_{\Sigma, [s]}(\hat{X})$. An *$\Omega$-sentence* in PMEL has the form $(\forall X)\ a$ **if** $a_1 \wedge \ldots \wedge a_n$, where $a, a_1, \ldots, a_n$ are atomic (Ω, X)-equations or atomic (Ω, X)-memberships, and $a_1, \ldots, a_n$ form a set (no duplications). If $n = 0$ then the Ω-sentence is called *unconditional* and written $(\forall X)\ a$. As in membership equational logic, (total) operation declarations $\sigma : s_1 \ldots s_m \to s$ are syntactic sugar for formulae[1] $(\forall x_1 : s_1, ..., x_m : s_m)\,\sigma(x_1, ..., x_m) : s$. Given a partial Ω-algebra A and a map $\theta\colon X \to A$, then $A, \theta \models^{\underline{o}}_\Omega t = t'$ iff $A_{\theta(t)}$ and $A_{\theta(t')}$ are both defined and $A_{\theta(t)} = A_{\theta(t')}$, and $A, \theta \models^{\underline{o}}_\Omega t : s$ iff $A_{\theta(t)}$ defined and $A_{\theta(t)} \in A_s$. *A satisfies* $(\forall X)$ a **if** $a_1 \wedge ... \wedge a_n$, written $A \models^{\underline{o}}_\Omega (\forall X)$ a **if** $a_1 \wedge ... \wedge a_n$, if and only if for each $\theta\colon X \to A$, if $A, \theta \models^{\underline{o}}_\Omega a_1$ and ... and $A, \theta \models^{\underline{o}}_\Omega a_n$ then $A, \theta \models^{\underline{o}}_\Omega a$. In this paper, we consider therefore what is called *existential* satisfaction in the literature on partiality. Other types of satisfaction (weak, strong) can be defined using the existential one in most practical situations; a more comprehensive study of these will be given elsewhere in a work dedicated entirely to PMEL.

A *specification* (or *theory*) $P = (\Omega, \Gamma)$ in PMEL consists of a signature Ω and a set of Ω-sentences Γ. A partial Ω-algebra A satisfies (or is a model of) $P = (\Omega, \Gamma)$, written $A \models^{\underline{o}} P$, iff it satisfies each sentence in Γ. Given a partial Ω-theory P, we let $\mathbf{PAlg}(P)$ denote the full subcategory of $\mathbf{PAlg}_\Omega$ of partial algebras satisfying P. Ω-theories P and P' are semantically equivalent in PMEL, written $P \equiv_{\mathrm{PMEL}} P'$, iff they have the same models, i.e., $\mathbf{PAlg}(P) = \mathbf{PAlg}(P')$.

Example 1. We give a partial membership equational specification for (small) categories. Let the kind k have two sorts, *Object* and *Arrow*, such that *Object* < *Arrow* (the identity arrow is identified with its object), and two total operation declarations $s, t\colon Arrow \to Object$, which are just syntactic sugar for the formulae $(\forall x : k)\, s(x) : Object$ **if** $x : Arrow$ and $(\forall x : k)\, t(x) : Object$ **if** $x : Arrow$. Consider also an operation $\_;\_ : k \times k \to k$ and the following formulae:

$$
\begin{aligned}
&(\forall O : Object)\ s(O) = O &&(1)\\
&(\forall O : Object)\ t(O) = O &&(2)\\
&(\forall O : Object, A : Arrow)\ O; A = A \text{ if } s(A) = O &&(3)\\
&(\forall O : Object, A : Arrow)\ A; O = A \text{ if } t(A) = O &&(4)\\
&(\forall A, A' : Arrow)\ A; A' : Arrow \text{ if } t(A) = s(A') &&(5)\\
&(\forall A, A' : Arrow)\ t(A) = s(A') \text{ if } A; A' : Arrow &&(6)\\
&(\forall A, A' : Arrow)\ s(A; A') = s(A) \text{ if } t(A) = s(A') &&(7)\\
&(\forall A, A' : Arrow)\ t(A; A') = t(A') \text{ if } t(A) = s(A') &&(8)\\
&(\forall A, A', A'' : Arrow)\ (A; A'); A'' = A; (A'; A'') \text{ if } t(A) = s(A') \wedge t(A') = s(A'') &&(9)
\end{aligned}
$$

Equations (1), (2) give the source and the target of an identity, (3) and (4) are the usual identity axioms, (5) and (6) define composition of arrows as a partial operation, (7) and (8) give the source and the target of the composition of two arrows when defined, and (9) states the associativity of composition. The partial algebras satisfying the theory above are exactly the (small) categories.

We recall the notion of Ω-*envelope* of a PMEL sentence [15]:

$$
\begin{aligned}
env_\Omega(t = t') &= t = t' \wedge \bigwedge_{u \in NVST(t,t')} u : \top_{k(u)}\\
env_\Omega(t : s) &= t : s \wedge \bigwedge_{u \in NVST(t)-\{t\}} u : \top_{k(u)}\\
env_\Omega((\forall X)\ a \text{ if } a_1 \wedge ... \wedge a_n) &= \{(\forall X)\ a' \text{ if } env_\Omega(a_1, ..., a_n) \mid a' \in env_\Omega(a)\},
\end{aligned}
$$

where $NVST(t)$ and $NVST(t,t')$ are the sets of all non-variable subterms of t and of t and t', respectively, $env_\Omega(a_1, ..., a_n)$ is a shorthand for $env_\Omega(a_1) \wedge ... \wedge env_\Omega(a_n)$, and $a' \in env_\Omega(a)$ means that a' is among the conjuncts in $env_\Omega(a)$.

---

[1] Together with an appropriate operation declaration on kinds.

Being concerned with defining the institution of PMEL, [15] actually extended the definition of sentences to allow conjunctions as conclusions. The envelope map extends trivially to sets of sentences and moreover, given a specification $P = (\Omega, \Gamma)$ in PMEL, the *envelope of P* is the specification $env(P) = (\Omega, env_\Omega(\Gamma))$. Notice that $env(env(P)) = env(P)$ and that $P \equiv_{PMEL} env(P)$.

Example 2. The envelope of the specification of categories in Example 1, written in sugared notation and omitting the declarations for the total operations, is:

$$
\begin{array}{ll}
(\forall O : Object) \; s(O) = O & (\; 1 \;) \\
(\forall O : Object) \; s(O) : Arrow & (\; 1.1 \;) \\
(\forall O : Object) \; t(O) = O & (\; 2 \;) \\
(\forall O : Object) \; t(O) : Arrow & (\; 2.1 \;) \\
(\forall O : Object, A : Arrow) \; O; A = A \text{ if } s(A) = O \wedge s(A) : Arrow & (\; 3 \;) \\
(\forall O : Object, A : Arrow) \; O; A : Arrow \text{ if } s(A) = O \wedge s(A) : Arrow & (\; 3.1 \;) \\
(\forall O : Object, A : Arrow) \; A; O = A \text{ if } t(A) = O \wedge t(A) : Arrow & (\; 4 \;) \\
(\forall O : Object, A : Arrow) \; A; O : Arrow \text{ if } t(A) = O \wedge t(A) : Arrow & (\; 4.1 \;) \\
(\forall A, A' : Arrow) \; A; A' : Arrow \text{ if } C(A, A') & (\; 5 \;) \\
(\forall A, A' : Arrow) \; t(A) = s(A') \text{ if } A; A' : Arrow & (\; 6 \;) \\
(\forall A, A' : Arrow) \; t(A) : Arrow \text{ if } A; A' : Arrow & (\; 6.1 \;) \\
(\forall A, A' : Arrow) \; s(A') : Arrow \text{ if } A; A' : Arrow & (\; 6.2 \;) \\
(\forall A, A' : Arrow) \; s(A; A') = s(A) \text{ if } C(A, A') & (\; 7 \;) \\
(\forall A, A' : Arrow) \; s(A; A') : Arrow \text{ if } C(A, A') & (\; 7.1 \;) \\
(\forall A, A' : Arrow) \; s(A) : Arrow \text{ if } C(A, A') & (\; 7.2 \;)
\end{array}
$$

$$
\begin{array}{ll}
(\forall A, A' : Arrow) \; t(A; A') = t(A') \text{ if } C(A, A') & (\; 8 \;) \\
(\forall A, A' : Arrow) \; t(A; A') : Arrow \text{ if } C(A, A') & (\; 8.1 \;) \\
(\forall A, A' : Arrow) \; t(A') : Arrow \text{ if } C(A, A') & (\; 8.2 \;) \\
(\forall A, A', A'' : Arrow) \; (A; A'); A'' = A; (A'; A'') \text{ if } C(A, A') \wedge C(A', A'') & (\; 9 \;) \\
(\forall A, A', A'' : Arrow) \; (A; A'); A'' : Arrow \text{ if } C(A, A') \wedge C(A', A'') & (\; 9.1 \;) \\
(\forall A, A', A'' : Arrow) \; A; (A'; A'') : Arrow \text{ if } C(A, A') \wedge C(A', A'') & (\; 9.2 \;) \\
(\forall A, A', A'' : Arrow) \; A; A' : Arrow \text{ if } C(A, A') \wedge C(A', A'') & (\; 9.3 \;) \\
(\forall A, A', A'' : Arrow) \; A'; A'' : Arrow \text{ if } C(A, A') \wedge C(A', A'') & (\; 9.4 \;)
\end{array}
$$

where $C(X, Y)$ is a shorthand for "$t(X) = s(Y) \wedge t(X) : Arrow \wedge s(Y) : Arrow$".

4   From Partial to Total Membership Equational Logic

Since MEL admits complete deduction and since it is already efficiently implemented, one would want to reduce PMEL proof tasks to MEL proof tasks. More precisely, one would like to have an efficient and automatic translator, say tr, taking Ω-specifications P and Ω-sentences φ in PMEL to specifications and sentences $tr(P)$ and $tr(\varphi)$ such that $P \models^\circ \varphi$ if and only if $tr(P) \models tr(\varphi)$. Notice that such a translator would automatically produce complete deduction for PMEL. It is worth mentioning that such translators that "borrow" other logics are usually obtained from (generalized) maps of institutions [4,15].

There is a straightforward translation of specifications in PMEL to specifications in MEL, in which (almost) nothing changes. Unfortunately, this translation is not always correct as seen shortly. There is another translation, based on the construction of envelopes, which is always correct but usually unnecessarily long. We show conditions under which the two translations are equivalent, thus providing a simple way to use an executable specification system like Maude [5] and its proving tools [6] to execute and verify properties of specifications in PMEL.

4.1 A Straightforward Translation

The simplest possible translation of PMEL sentences and specifications is to let them (almost) unchanged. More precisely, given a specification $P = (\Omega, \Gamma)$ in PMEL, let $\overline{P}$ be $(\overline{\Omega}, Ax_\Omega \cup \overline{\Gamma})$ in MEL, where if $\Omega = (S, <, \Sigma)$ and φ is an Ω sentence in PMEL, then (in order to keep the translation clear, we do not use sugared notation below)

- $\overline{\Omega}$ is the signature (K, Σ, π) in MEL with $K = \hat{S}$ and $\pi(s) = [s]$;
- Ax_Ω is the set of $\overline{\Omega}$-sentences associated to Ω by the expected syntactic sugar conventions, i.e., containing a sentence $(\forall x : k)\ x : s'$ if $x : s$ for each pair $s < s'$ in connected component k;
- $\overline{\varphi}$ is exactly φ under the appropriate syntactic sugar conventions for MEL and the appropriate reinterpretation of operations in Ω into operations in $\overline{\Omega}$; this operation can be trivially extended to sets of Ω-sentences.

We call $\overline{(\_)} \colon \mathbf{Th_{PMEL}} \to \mathbf{Th_{MEL}}$ the *desugaring* map, let $\mathbf{Th_{MEL}^{\top}}$ denote its image, and call the theories in $\mathbf{Th_{MEL}^{\top}}$ *topped*. Because of the consistent syntactic sugar conventions in both PMEL and MEL, there should be no syntactic difference between P and $\overline{P}$ in practice.

Intuitively, the straightforward translation from PMEL to MEL presented in this subsection is *the* desired translation, because it has what it can be called "almost-zero representation distance" between P and $\overline{P}$. A natural question is, however, whether it is correct or not. As seen later in the paper, it is correct for the example above but, as shown by concrete examples in the extended version of this paper [16], it is neither complete nor sound in general, that is, there are Ω-theories P, P' and Ω-sentences φ, φ' in PMEL such that $P \models \varphi$ but $\overline{P} \not\models \overline{\varphi}$, and $P' \not\models \varphi'$ but $\overline{P'} \models \overline{\varphi'}$. It is a major goal of this paper to give conditions under which the straightforward translation is both sound and complete.

4.2 The Correct Translation

Given a signature $\Omega = (S, <, \Sigma)$ and an Ω-sentence $\varphi = (\forall X)\ a$ if $a_1 \wedge \ldots \wedge a_n$ in PMEL, let $\overline{\Omega}$ and Ax_Ω be defined as in Subsection 4.1, and let $\overline{env}_\Omega(\varphi)$ be the set of $\overline{\Omega}$-sentences $\{(\forall X)\ a'$ if $env_\Omega(a_1, \ldots, a_n) \mid a' \in env_\Omega(a)\}$ (we use the same notational conventions as in Section 3), with the appropriate interpretation of kinds and operations in Ω into kinds and operations in $\overline{\Omega}$. $\overline{env}_\Omega$ can be extended to sets of Ω-formulae in the obvious way. Given a specification $P = (\Omega, \Gamma)$ in PMEL, let $\overline{env}(P)$ be the specification in MEL $(\overline{\Omega}, Ax_\Omega \cup \overline{env}_\Omega(\Gamma))$. Note that $\overline{env}(P) = \overline{env(P)}$. For a topped theory $\overline{P} \in \mathbf{Th_{MEL}^{\top}}$, let $env^{\top}(\overline{P})$ be $\overline{env}(P)$.

It is important to notice that despite the fact that P is semantically equivalent to $env(P)$ in PMEL, it is *not* the case that $\overline{P}$ and $\overline{env}(P)$ are equivalent in MEL, because the straightforward translation does not preserve the semantical equivalence of specifications in PMEL (see Subsection 4.1).

As shown in [15], for any signature Ω in PMEL, any $\overline{\Omega}$-algebra A and any partial Ω-algebra B, one can build a partial Ω-algebra A° by forgetting all the elements that don't have a sort and letting the operations undefined accordingly, and one can build an $\overline{\Omega}$-algebra $B^\bullet$ by freely adding to B all the undefined, or "error", values. Moreover, given an Ω-theory P in PMEL then $(\_)^\circ$ and $(\_)^\bullet$ can be organized as functors $(\_)^\circ \colon \mathbf{Alg}(env^\top(\overline{P})) \to \mathbf{PAlg}(P)$ and $(\_)^\bullet \colon \mathbf{PAlg}(P) \to \mathbf{Alg}(env^\top(\overline{P}))$. The following results are proved in Section 14 in [15]:

Proposition 1. *Given an Ω-theory P and an Ω-sentence φ in PMEL, and a (total) $\overline{\Omega}$-algebra A, then*

1. *The functor $(\_)^\bullet \colon \mathbf{PAlg}(P) \to \mathbf{Alg}(\overline{env}(P))$ is a left adjoint left inverse to $(\_)^\circ \colon \mathbf{Alg}(\overline{env}(P)) \to \mathbf{PAlg}(P)$,*
2. *$(A \models_{\overline{\Omega}} \overline{env}_\Omega(\varphi)$ iff $A^\circ \models_\Omega \varphi)$ and $(A \models \overline{env}(P)$ iff $A^\circ \models P)$,*
3. *$P \models \varphi$ iff $\overline{env}(P) \models \overline{env}_\Omega(\varphi)$.*

Thus, "$P \vdash \varphi$ iff $\overline{env}(P) \vdash \overline{env}_\Omega(\varphi)$" gives sound complete deduction for PMEL.

$\overline{env}$ can be actually organized as a generalized map of institutions [15]. Note that, due to 1. in Proposition 1, $(\mathbf{Alg}(\overline{env}(P)))^\circ = \mathbf{PAlg}(P)$ and there is in fact a perfect correspondence between free algebras F_P in $\mathbf{PAlg}(P)$ and free algebras $F_{\overline{env}(P)}$ in $\mathbf{Alg}(\overline{env}(P))$, since we have $F_P^\bullet = F_{\overline{env}(P)}$ and $F_{\overline{env}(P)}^\circ = F_P$. In particular, there is a perfect correspondence between initial algebras I_P in $\mathbf{PAlg}(P)$ and initial algebras $I_{\overline{env}(P)}$ in $\mathbf{Alg}(\overline{env}(P))$.

4.3 Putting All the Maps Together

There is also a map S from $\mathbf{Th}_{\mathrm{MEL}}^\top$ to $\mathbf{Th}_{\mathrm{PMEL}}$, called *sugaring*, which, due to the syntactic sugar conventions in both PMEL and MEL, is also just syntactic identity in practice. Given $T = (\Omega, \Gamma) \in \mathbf{Th}_{\mathrm{MEL}}^\top$, let $S(T)$ be the theory $(S(\Omega), S(\Gamma))$ in $\mathbf{Th}_{\mathrm{PMEL}}$, where:

- $S(\Omega)$ is the PMEL signature obtained from Ω by forgetting its kinds (but keeping its sorts) and adding $s < s'$ for each sorts s, s' such that the sentence $(\forall x : k)\ x : s'$ if $x : s$ is in Γ;
- $S((\forall X)\ a$ **if** $C)$ be the PMEL $S(\Omega)$-sentence $(\forall \tilde{X})\ a$ **if** $\tilde{C}$, where $\tilde{X}$ is the set of (sorted) variables $\bigcup_{x:k \in X}\{x : s \mid x : s \in C\}$ and $\tilde{C}$ is the conjunction of all the atomic sentences in $C - \{x : s \mid x : s \in C\}$.

The following picture shows all the maps defined so far in the paper:

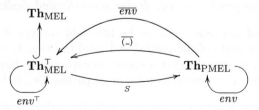

Notice that S and $\overline{(\_)}$ are inverse of each other, that $env^\top; env^\top = env^\top$ and $env; env = env$, and that $env; \overline{(\_)} = \overline{(\_)}; env^\top = \overline{env}$. Due to these, Proposition 1 is often used in the sequel in a slightly modified version, where $\overline{env}(P)$ is replaced by $env^\top(T)$ and P by $S(T)$ for some topped theory T in $\mathbf{Th}_{\mathrm{MEL}}^\top$.

5   Strict and Duplex Theories

In this section we define and study four categories of total membership equational logic theories that are strongly related to our goal of representing partial theories as total theories with almost-zero representational distance. Various subsequent results and notions, such as strictness, work for any MEL theories, but, since most results in this paper only make sense in the context of topped MEL theories, in order to keep the presentation simple we assume that all the (total) MEL theories from now on are topped.

A crucial aspect of partiality is that if a term is defined then *any* subterm of it is also defined. This motivates the following category of total MEL theories:

Definition 1. *Given any theory* T *in* $\mathbf{Th}_{\mathrm{MEL}}^{\top}$, *a term* $t \in T_{\Sigma}(X)$ *is* **T-strict** *iff for any subterm* u *of* t *there is some sort* s_u *such that* $T \models (\forall X)\, u{:}s_u$. T *is* **strict** *iff for any term* t *and any sort* s_t, $T \models (\forall X)\, t{:}s_t$ *implies that* t *is* T-strict. *We let* STRICT *denote the full subcategory of* $\mathbf{Th}_{\mathrm{MEL}}^{\top}$ *of strict theories.*

Proving the envelope of a sentence in the context of a strict theory is easier:

Proposition 2. *If* $T = (\Omega, \Gamma)$ *is strict then for any* $t, t' \in T_{\Sigma,k}(X)$,

$$T \models env_{\Omega}^{\top}((\forall X)\, t{:}s) \quad \textit{iff } T \models (\forall X)\, t{:}s, \textit{ and}$$
$$T \models env_{\Omega}^{\top}((\forall X)\, t = t') \textit{ iff } T \models (\forall X)\, t = t' \textit{ and } T \models (\forall X)\, t{:}\top_k.$$

Even if strictness is a necessary requirement for a total theory in order to be safely considered partial, as we shall see later in the paper, it is not sufficient. We call *duplex theories* those theories that, intuitively, can be viewed both as total and as partial. However, there are at least two important aspects of duplex theories, one proof theoretical and the other semantical.

Definition 2. *A theory* T *in* $\mathbf{Th}_{\mathrm{MEL}}^{\top}$ *is* **duplex proof-theoretic** *iff*

$$S(T) \models (\forall X)\, t{:}s \quad \textit{iff } T \models (\forall X)\, t{:}s, \textit{ and}$$
$$S(T) \models (\forall X)\, t = t' \textit{ iff } T \models (\forall X)\, t = t' \textit{ and } T \models (\forall X)\, t{:}\top_k.$$

T *is* **duplex semantic** *iff* $(\mathbf{Alg}(T))^{\circ} = \mathbf{PAlg}(S(T))$ *and is* **duplex** *iff it is both. Let* DUPLEX$_{\mathrm{PT}}$, DUPLEX$_{\mathrm{SEM}}$ *and* DUPLEX *denote the full subcategories of* $\mathbf{Th}_{\mathrm{MEL}}^{\top}$ *of duplex proof-theoretic, duplex semantic and duplex theories.*

Duplex proof-theoretic seems to be the right concept if one only considers the provability aspects of the partial and total theories. However, one may also be interested in models, in that a theory is expected to have the same models regardless of how it is viewed, as total or as partial, i.e., to be duplex semantic.

Theorem 1. *(for a detailed proof see [16]) The following hold:*

1. DUPLEX$_{\mathrm{PT}} \subsetneq$ STRICT;
2. DUPLEX$_{\mathrm{SEM}} \not\subseteq$ STRICT *and* STRICT $\not\subseteq$ DUPLEX$_{\mathrm{SEM}}$;
3. DUPLEX$_{\mathrm{PT}} \not\subseteq$ DUPLEX$_{\mathrm{SEM}}$ *and* DUPLEX$_{\mathrm{SEM}} \not\subseteq$ DUPLEX$_{\mathrm{PT}}$;
4. DUPLEX$_{\mathrm{SEM}} \cap$ STRICT $\subsetneq$ DUPLEX$_{\mathrm{PT}}$.

The four subcategories of topped total MEL theories can be represented as in the picture below, where each rectangle represents one subcategory and each inclusion of rectangles is proper:

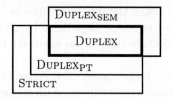

6   Envelope Invariance as a Central Concept

As argued above, in order for a partial theory to be safely regarded, via appropriate syntactic sugar conventions, as total it must be duplex. However, in algebraic specification practice we would like the strongest possible notion of "duplex" theory, assuring exact correspondence not only at the duplex semantic and proof-theoretic levels, but also ensuring exact correspondence of *initial algebra semantics* and of semantics of *free functors* associated to forgetful functors along a theory morphism (for *parameterized* specifications), so that proofs by *induction* at the total level can be "borrowed" at the partial level. As suggested by Proposition 1 and shown in full detail in [15], all these requirements hold when a partial theory P is translated to $\overline{env}(P)$ instead of $\overline{P}$, thus inspiring us to introduce the following key notion:

Definition 3. $T \in \mathbf{Th}_{\mathrm{MEL}}^{\top}$ *is* **envelope invariant**[2] *iff* $\mathbf{Alg}(T) = \mathbf{Alg}(env^{\top}(T))$. *We let* ENVINV *denote the full subcategory of envelope invariant theories.*

If P is a theory in PMEL such that $\overline{P}$ is envelope invariant then notice that $\mathbf{Alg}(\overline{P}) = \mathbf{Alg}(\overline{env}(P))$. Therefore, by Proposition 1 (also the comments after it) there is an exact correspondence between free models of P in PMEL and free models of $\overline{P}$ in MEL, which allows one to do inductive proofs in $\overline{P}$, using the well-understood machinery of total MEL efficiently implemented in Maude [5] and its inductive theorem prover [6], rather than in P at the partial level without tool support. The following result shows that envelope invariance also satisfies the duplex requirements, thus giving it a central role in our approach.

Theorem 2. *(for a detailed proof see [16])* ENVINV $\subsetneq$ DUPLEX.

7   Proving Envelope Invariance

Envelope invariance, besides being semantically and proof-theoretically the strongest and most satisfactory notion of "duplex" theory, is also a property

---

[2] What we here call an envelope invariant theory was called a *strict* theory in [15].

that holds for many theories of interest and can be established by practical proof methods.

First note that, since $T \in \mathbf{Th}^\top_{\mathrm{MEL}}$ is envelope invariant iff we have $\mathbf{Alg}(T) = \mathbf{Alg}(env^\top(T))$, by the completeness theorem of MEL [15], this holds iff $T \Leftrightarrow env^\top(T)$ (see Section 2). Using the inference rules in [15], this gives us a general *semidecision procedure* for envelope invariance by trying to prove both implications in parallel. We conjecture that for an arbitrary theory T the property may be undecidable.

Nevertheless, there are large classes of theories of interest in algebraic specification practice, for which either envelope invariance holds, or practical algorithms exist that will either decide the property or output useful proof obligations. For example, *sortable* theories are envelope invariant, where $T \in \mathbf{Th}^\top_{\mathrm{MEL}}$ is sortable iff its membership axioms consisting of the subsort inclusions and the operator declarations can prove $(\forall X)\ u : \top_k$, for any subterm u in any axiom in T. Such proofs are decidable, because *order-sorted parsing* of terms is decidable [10]; yet $S(T)$ may still have models in which some operations are partial.

A broad class of practical interest is the class of those $T \in \mathbf{Th}^\top_{\mathrm{MEL}}$ that are *Church-Rosser* and *terminating* in the precise sense of [2]. This is a key class of *executable specifications*, for which we can use tools like Maude [5]. In such a class we can try to decide the equivalence $T \Leftrightarrow env^\top(T)$ by rewriting. That is, we can use T and the implication introduction rule in [15] to try to prove $T \Rightarrow env^\top(T)$ by rewriting in an automated way; and in the exact same way we can use $env^\top(T)$ to try to prove $T \Leftarrow env^\top(T)$. The details of how this proof method establishes the envelope invariance of the category theory specification in Section 3 can be found in `http://ase.arc.nasa.gov/grosu/download/categ.maude`. For theories $T \in \mathbf{Th}^\top_{\mathrm{MEL}}$ that are Church-Rosser, terminating and *unconditional* (in their sugared form), checking envelope invariance by the above method becomes *decidable*, because in such a case we have $T \subseteq env^\top(T)$, so that we only need to prove $T \Rightarrow env^\top(T)$, which, with T Church-Rosser and terminating and with $env^\top(T)$ unconditional, is decidable by rewriting in T.

It does not seem obvious that $T \in \mathbf{Th}^\top_{\mathrm{MEL}}$ Church-Rosser and terminating implies that $env^\top(T)$ is so too. For this reason we are also experimenting with a more conservative *incremental* automated proof procedure. This method generalizes to conditional theories the method described above for unconditional ones: it only uses rewriting in (increasing subsets of) T to establish the equivalence $T \Leftrightarrow env^\top(T)$. Details of how this method proves envelope invariance for the theories of categories and of categories with pullbacks can be found on the web at the address `http://ase.arc.nasa.gov/grosu/download/categ-tower.maude`. More experimentation is needed to ascertain which of these two approaches, or some combination, is the best candidate for implementation, but it seems clear that an automated procedure along these lines will be able to decide envelope invariance of Church-Rosser and terminating specifications in many cases, or otherwise will output useful proof obligations to the user. Such proof obligations may also help uncover subtle flaws in a specification.

8   Conclusion and Future Work

We have proposed a total approach to partial equational specification that, under reasonable conditions, has almost-zero representational distance, in the sense that the partial theory and its total translation are identical up to syntactic sugar. For this purpose, we have studied and compared in full detail several natural notions of "strict" and "duplex" theory that can be viewed as partial or total without changing the axioms. The main conclusion of our study is that *envelope invariance* is the *semantically stronger* notion of duplex theory, including preservation *in both directions* of initial, free, and free extension algebras, and *allowing doing both execution and inductive theorem proving* for partial specifications at the total level in a very direct way.

We have also begun the study of and experimentation with practical *proof methods* that can be used to check envelope invariance and can decide the property in some cases. Our initial experiments suggest that many theories of practical interest will pass the check, and that the check is useful to *detect semantical errors* in large partial specifications, such as those in [14].

With the exception of [12], at present there seems to be limited direct theoretical support for *rewriting techniques* for explicitly partial equational specifications, and no automated deduction techniques directly supporting inductive theorem proving for such specifications, except for the indirect support via a translation into higher-order logic of the HOL-CASL tool [18]. In particular, we know of no tools supporting execution of explicitly partial equational specifications. One important practical application of our approach is that our methods, combined with existing tools for execution and theorem proving of MEL specifications such as the Maude interpreter [5], the Maude inductive theorem prover, and the rest of Maude's formal tool environment [6] offer a promising *short- and mid-term alternative* for very direct execution and formal analysis of explicitly partial equational specifications. In particular, our approach could be used to execute and prove properties of partial specifications in the equational subset of CASL [7].

However, more research is needed to further advance these ideas and to develop tools supporting them, including research in:

1. experimentation with *case studies* and systematic comparison of the relative advantages of different *proof procedures* for checking envelope invariance;
2. development of *checking tools* automating such proof procedures and integration of such tools in the Maude system;
3. study of *initial and free notions of envelope invariance*; that is, conditions under which theories with initiality and freeness constraints (as opposed to theories with "loose semantics") are envelope invariant;
4. more generally, study of *modularity issues*, including study of conditions under which module operations themselves are "duplex." We already know that PMEL and MEL are related by an "extension map of institutions" preserving many important semantic properties such as initial and free models [15], but, as shown for example in [19], modularity issues are quite delicate and should be studied carefully.

References

1. M. Bidoit, H.-J. Kreowski, P. Lescanne, F. Orejas, and D. Sannella, editors. *Algebraic System Specification and Development. A Survey and Annotated Bibliography*, volume 501 of *LNCS*. Springer, 1991.
2. A. Bouhoula, J.-P. Jouannaud, and J. Meseguer. Specification and proof in membership equational logic. *Theoretical Computer Science*, 236:35–132, 2000.
3. M. Broy and M. Wirsing. Partial abstract types. *Acta Informatica*, 18:47–64, 1982.
4. M. Cerioli and J. Meseguer. May I borrow your logic? (Transporting logical structures along maps). *Theoretical Computer Science*, 173(2):311–347, 1997.
5. M. Clavel, F. Durán, S. Eker, P. Lincoln, N. Martí-Oliet, J. Meseguer, and J. Quesada. Maude: specification and programming in rewriting logic. SRI International, January 1999, http://maude.csl.sri.com.
6. M. Clavel, F. Durán, S. Eker, and J. Meseguer. Building equational proving tools by reflection in rewriting logic. In *CAFE: An Industrial-Strength Algebraic Formal Method*. Elsevier, 2000. http://maude.csl.sri.com.
7. CoFI task group on semantics, CASL — The common algebraic specification language, Semantics. www.brics.dk/Projects/CoFI/Documents/CASL, July 1999.
8. R. Diaconescu and K. Futatsugi. *CafeOBJ Report: The Language, Proof Techniques, and Methodologies for Object-Oriented Algebraic Specification*. World Scientific, 1998. AMAST Series in Computing, volume 6.
9. M. Gogolla and M. Cerioli. What is an Abstract Data Type after all? Technical report, DISI – University of Genova, 1994.
10. J. Goguen and J. Meseguer. Order-sorted algebra I: Equational deduction for multiple inheritance, overloading, exceptions and partial operations. *Theoretical Computer Science*, 105(2):217–273, 1992.
11. J. Goguen, T. Winkler, J. Meseguer, K. Futatsugi, and J.-P. Jouannaud. Introducing OBJ. In *Software Engineering with OBJ: algebraic specification in action*, pages 3–167. Kluwer, 2000.
12. C. Hintermeier, C. Kirchner, and H. Kirchner. Dynamically-typed computations for order-sorted equational presentations. *Journal of Symbolic Computation*, 25(4):455–526, April 1998.
13. H. Kaphengst and H. Reichel. Initial algebraic semantics for non-context-free languages. In M. Karpinski, editor, *Fundamentals of Computation Theory*, volume 56 of *LNCS*, pages 120–126. Springer, 1977.
14. M. Lowry, T. Pressburger, and G. Roşu. Certifying domain-specific policies. In *Proceedings, International Conference on Automated Software Engineering (ASE'01)*, pages 81–90. IEEE, 2001. San Diego, California.
15. J. Meseguer. Membership algebra as a logical framework for equational specification. In *Proceedings, WADT'97*, volume 1376 of *LNCS*, pages 18–61, 1998.
16. J. Meseguer and G. Roşu. A total approach to partial algebraic specification. Extended version at ase.arc.nasa.gov/grosu/tapase.html, 2002.
17. T. Mossakowski. Equivalences among various logical frameworks of partial algebras. In *Computer Science Logic*, volume 1092 of *LNCS*, pages 403–433, 1996.
18. T. Mossakowski. Introduction into HOL-CASL. www.tzi.de/cofi/, 2001.
19. T. Mossakowski. Relating CASL with other specification languages: the institution level. *Theoretical Computer Science*, To appear. www.tzi.de/~{}till.
20. H. Reichel. *Initial Computability, Algebraic Specifications, and Partial Algebras*. Oxford University Press, 1987.

Axiomatising Divergence[*]

Markus Lohrey[1], Pedro R. D'Argenio[2], and Holger Hermanns[3]

[1] Institut für Informatik, Universität Stuttgart,
Breitwiesenstr. 20-22, 70565 Stuttgart, Germany
[2] FaMAF, Universidad Nacional de Córdoba,
Ciudad Universitaria, 5000 Córdoba, Argentina
[3] Formal Methods and Tools Group, Faculty of Computer Science,
University of Twente, P.O. Box 217, 7500 AE Enschede, The Netherlands

Abstract. This paper develops sound and complete axiomatisations for the divergence sensitive spectrum of weak bisimulation equivalence. The axiomatisations can be extended to a considerable fragment of the linear time – branching time spectrum with silent moves, partially solving an open problem posed in [5].

1 Motivation

The study of comparative concurrency semantics is concerned with a uniform classification of process behaviour, and has cumulated in Rob van Glabbeek's seminal papers on the *linear time-branching time spectrum* [4,5]. The main ('vertical') dimension of the spectrum with silent moves [5] spans between trace equivalence (TE) and branching bisimulation (BB), and identifies different ways to discriminate processes according to their branching structure, where BB induces the finest, and TE the coarsest reasonable semantics. Due to the presence of silent moves, this spectrum is spread in another ('horizontal') dimension, determined by the semantics of *divergence*. In the fragment spanning from weak bisimulation (WB) to BB, seven different criteria to distinguish divergence induce a 'horizontal' lattice, and this lattice appears for all the bisimulation relations.

To illustrate the spectrum, van Glabbeek lists a number of examples and counterexamples showing the differences among the various semantics [5]. *Process algebra* provides a different – and to our opinion more elegant – way to compare semantic issues, by providing distinguishing axioms that capture the essence of an equivalence (or preorder). For the 'vertical' dimension of the spectrum, these distinguishing axioms are well-known (see e.g. [4,7,2]). However, the 'horizontal' dimension has resisted an axiomatic treatment so far. We believe that this is mainly due to the fact that divergence only makes sense in the presence of recursion, and that recursion is hard to tackle axiomatically. Isolated points in the 'horizontal' dimension have however been axiomatised, most

---

[*] This work was done while the first author was on leave at IRISA, Campus de Beaulieu, 35042 Rennes Cedex, France and supported by the INRIA cooperative research action FISC.

P. Widmayer et al. (Eds.): ICALP 2002, LNCS 2380, pp. 585–596, 2002.

notably Milner's weak bisimulation (WB) congruence [10], and also convergent WB preorder [11], as well as divergence insensitive BB congruence [6] and stable WB congruence [8]. It is also worth to mention the works of [3] and [1], which axiomatised divergence sensitive WB congruence and convergent WB preorder, respectively, but without showing completeness in the presence of recursion.

This paper develops complete axiomatisations for the 'horizontal' dimension of weak bisimulation equivalence. A lattice of distinguishing axioms is shown to characterise the distinct semantics of divergence, and to precisely reflect the 'horizontal' lattice structure of the spectrum. We are confident that these axioms form the basis of complete axiomatisation for the bisimulation spectrum spanning from WB to BB.

The paper is organised as follows. Section 2 introduces the necessary notation and definitions, while Section 3 recalls the weak bisimulation equivalences and Section 4 introduces the axiom systems. Section 5 is devoted to soundness of the axioms and sets the ground for the completeness proof. Section 6 is devoted to the main step of the proof, only focusing on closed expressions, while Section 7 covers open expressions. Section 8 concludes the paper. Proofs that are omitted in this extended abstract will appear in the extended version of this paper.

2   Preliminaries

We assume a set of variables $\mathbb{V}$, and a set of actions $\mathbb{A}$, containing the silent action τ. We consider the set of open finite state agents with silent moves and explicit divergence, given as the set $\mathbb{E}$ of expressions generated by the grammar

$$\mathcal{E} \quad ::= \quad a.\mathcal{E} \quad | \quad \mathcal{E} + \mathcal{E} \quad | \quad recX.\mathcal{E} \quad | \quad X \quad | \quad \Delta(\mathcal{E})$$

where $X \in \mathbb{V}$ and $a \in \mathbb{A}$. $\Delta(E)$ is an expression that adds divergence explicitly to the root of E. It can be considered as a syntactic shorthand for $recX.(\tau.X+E)$ provided X does not occur in E. The explicit representation of divergence by means of Δ will prove handy in the sequel.

The syntactic equality on $\mathbb{E}$ is denoted by $\equiv$. With $\mathbb{V}(E)$ we denote the set of all variables that are free in $E \in \mathbb{E}$, i.e., not bounded by a $recX$-operator. We define $\mathbb{P} = \{E \in \mathbb{E} \mid \mathbb{V}(E) = \emptyset\}$. We use $E, F, G, H, \dots$ (resp. $P, Q, R, \dots$) to range over expressions from $\mathbb{E}$ (resp. $\mathbb{P}$). If $\boldsymbol{F} = F_1, \dots, F_n$ is a sequence of expressions, $\boldsymbol{X} = X_1, \dots, X_n$ is a sequence of variables, and $E \in \mathbb{E}$ then $E\{\boldsymbol{F}/\boldsymbol{X}\}$ denotes the expression that results from E by simultaneously replacing all free occurrences of X_i in E by F_i $(1 \leq i \leq n)$. The variable X is *guarded* in E, if every free occurrence of X in E lies within a subexpression of the form $a.F$ with $a \in \mathbb{A} \setminus \{\tau\}$, otherwise X is called unguarded in E. E is guarded if for every subexpression $recY.F$ of E the variable Y is guarded in F.

The semantics of $\mathbb{E}$ is given as the least transition relation satisfying the following rules, which are standard (except that, as indicated before, $\Delta(E)$ can diverge, in addition to exhibiting all the behaviour of E).

$$\frac{}{a.E \xrightarrow{a} E} \qquad \frac{E \xrightarrow{a} E'}{E + F \xrightarrow{a} E'} \qquad \frac{E \xrightarrow{a} E'}{F + E \xrightarrow{a} E'}$$

$$\frac{E\{recX.E/X\} \xrightarrow{a} E'}{recX.E \xrightarrow{a} E'} \qquad \frac{E \xrightarrow{a} E'}{\Delta(E) \xrightarrow{a} E'} \qquad \frac{}{\Delta(E) \xrightarrow{\tau} \Delta(E)}$$

3   The Bisimulations

Since we are working in the context of silent steps, we define a few standard abbreviations: $E \Longrightarrow F$ if $E \xrightarrow{\tau}{}^* F$; $E \xLongrightarrow{a} F$ if $E \Longrightarrow \xrightarrow{a} \Longrightarrow F$; $E \xLongrightarrow{\hat{a}} F$ if $(E \xLongrightarrow{a} F$ and $a \neq \tau)$ or $(E \Longrightarrow F$ and $a = \tau)$. We write $E \xrightarrow{a}$ (resp. $E \longrightarrow$) if $E \xrightarrow{a} F$ for some $F \in \mathbb{E}$ (resp. $E \xrightarrow{a} F$ for some $a \in \mathbb{A}$, $F \in \mathbb{E}$). With $E \xnrightarrow{a}$ and $E \nrightarrow$ we denote the corresponding negated conditions. We let $E \Uparrow$ denote $E \xrightarrow{\tau}{}^\omega$, i.e., E has the possibility to diverge. Finally, $E \Uparrow$ denotes that there is some F such that $E \Longrightarrow F$ and either $F \Uparrow$ or $F \nrightarrow$ (or equivalently, $E \Uparrow$ or $E \Longrightarrow F \nrightarrow$ for some F), i.e., E may either diverge, or silently decide to terminate. For a relation $\mathcal{R} \subseteq \mathbb{P} \times \mathbb{P}$ define the following conditions (in all conditions $P, Q, P' \in \mathbb{P}$ and $a \in \mathbb{A}$ are implicitly $\forall$-quantified):

(WB) if $(P,Q) \in \mathcal{R} \land P \xrightarrow{a} P'$ then $Q \xLongrightarrow{\hat{a}} Q' \land (P',Q') \in \mathcal{R}$ for some Q',

(S) if $(P,Q) \in \mathcal{R} \land P \xnrightarrow{\tau}$ then $Q \Longrightarrow Q' \xnrightarrow{\tau}$ for some Q',

(0) if $(P,Q) \in \mathcal{R} \land P \nrightarrow$ then $Q \Longrightarrow Q' \nrightarrow$ for some Q',

(Δ) if $(P,Q) \in \mathcal{R} \land P \Uparrow$ then $Q \Uparrow$,

(λ) if $(P,Q) \in \mathcal{R} \land P \Uparrow$ then $Q \Uparrow$.

Let $\mathcal{R} \subseteq \mathbb{P} \times \mathbb{P}$ be a symmetric relation. We say that $\mathcal{R}$ is a

- *weak bisimulation* (WB^ϵ or simply WB) if $\mathcal{R}$ satisfies (WB).
- *stable weak bisimulation* (WB^S) if $\mathcal{R}$ satisfies (WB) and (S).
- *completed weak bisimulation* (WB^0) if $\mathcal{R}$ satisfies (WB) and (0).
- *divergent weak bisimulation* (WB^λ) if $\mathcal{R}$ satisfies (WB) and (λ).
- *divergent stable weak bisimulation* (WB^Δ) if $\mathcal{R}$ satisfies (WB) and (Δ).

In the sequel, we let $*$ range over the set $\{\Delta, \lambda, S, 0, \epsilon\}$. The relation $\sim^* \subseteq \mathbb{P} \times \mathbb{P}$ is defined as the union of all WB^*, it is easily seen to be itself a WB^* as well as an equivalence relation.

Theorem 1. [5] *The equivalences $\sim^*$ are ordered by inclusion according to the lattice in Figure 1. The upper relation contains the lower if and only if both are connected by a line.*

Examples that distinguish these equivalences can be found in [5]. It is a well known deficiency that $\sim^*$ is not a congruence w.r.t. '+', moreover, for $* \in \{\Delta, \lambda, S, 0\}$ it is not a congruence w.r.t. $\Delta(.)$. For instance $\tau.0 \sim^\Delta 0$, but $\Delta(\tau.0) \not\sim^\Delta \Delta(0)$. To obtain the coarsest congruences in $\sim^*$ on $\mathbb{P}$, we define each $\simeq^*$ to be the relation that contains exactly the pairs $(P,Q) \in \mathbb{P} \times \mathbb{P}$ that satisfy the following *root conditions*:

- if $P \xrightarrow{a} P'$ then $Q \xLongrightarrow{a} Q'$ and $P' \sim^* Q'$ for some Q'
- if $Q \xrightarrow{a} Q'$ then $P \xLongrightarrow{a} P'$ and $P' \sim^* Q'$ for some P'

Fig. 1. Inclusions between the relations $\sim^*$

Fig. 2. Implications between the distinguishing axioms

We lift these relations from $\mathbb{P}$ to $\mathbb{E}$ as usual: let $E, F \in \mathbb{E}$, and let $\boldsymbol{X} = X_1, \ldots, X_n$ be a sequence of variables that contains all variables in $\mathbb{V}(E) \cup \mathbb{V}(F)$. Then $E \simeq^* F$ if $E\{\boldsymbol{P}/\boldsymbol{X}\} \simeq^* F\{\boldsymbol{P}/\boldsymbol{X}\}$ for all $\boldsymbol{P} = P_1, \ldots, P_n$ with $P_i \in \mathbb{P}$ (analogously for $\sim^*$).

Theorem 2. *The relation $\simeq^*$ is the coarsest congruence contained in $\sim^*$ w.r.t. the operators of $\mathbb{E}$. All inclusions from Figure 1 carry over from $\sim^*$ to $\simeq^*$.*

4   Axioms

This section introduces a lattice of axioms characterising the above weak bisimulations. For $* \in \{\Delta, S, 0, \epsilon\}$, the axioms for $\simeq^*$ are given in Table 1, plus the axiom $(*)$ from Table 2. The axioms for $\simeq^\lambda$ are given in Table 1, plus the axioms (Δ) and (λ) from Table 2. We write $E =^* F$ if $E = F$ can be derived by application of the axioms for $\simeq^*$.

The axioms from Table 1 are standard [10] except of $(rec5)$ and $(rec6)$. Axiom $(rec5)$ makes divergence explicit if introduced due to silent recursion; it defines the nature of the Δ-operator. Axiom $(rec6)$ states the redundancy of recursion on an unguarded variable in the context of divergence.

We discuss the distinguishing axioms in reverse order relative to how they are listed in Table 2. Axiom (λ) characterises the property of WB^λ that divergence cannot be distinguished when terminating. Axiom (ϵ) represents Milner's 'fair' setting, where divergence is never distinguished. The remaining three axioms state that divergence cannot be distinguished if the process can still perform an action to escape the divergence (0), that it cannot be distinguished if the process can perform a silent step to escape divergence (S), and that two consecutive divergences cannot be properly distinguished (Δ). It is a simple exercise to verify the implications between the distinguishing axioms as summarized in the lattice in Figure 2. It nicely reflects the inclusions between the respective congruences. The upper axioms turn into derivable laws given the lower ones (plus the core axioms from Table 1) as axioms.

The following two Δ-unfolding laws can be derived from the axioms for $\simeq^\Delta$ (and thus for all $\simeq^*$), they will be useful in Section 6.

$$(\tau\Delta)\ \ \Delta(E) =^\Delta \tau.\Delta(E) + E \qquad\qquad (\tau\Delta')\ \ \Delta(E) =^\Delta \tau.\Delta(E)$$

5   Soundness and Completeness

Checking soundness of the axioms is tedious but follows standard techniques.

Table 1. Core axioms

$(S1)\ E + F = F + E$	$(\tau 1)\ a.\tau.E = a.E$
$(S2)\ E + (F + G) = (E + F) + G$	$(\tau 2)\ \tau.E + E = \tau.E$
$(S3)\ E + E = E$	$(\tau 3)\ a.(E + \tau.F) = a.(E + \tau.F) + a.F$
$(S4)\ E + 0 = E$	

$(rec1)$ if Y is not free in $recX.E$ then $recX.E = recY.(E\{Y/X\})$

$(rec2)\ recX.E = E\{recX.E/X\}$

$(rec3)$ if X is guarded in E and $F = E\{F/X\}$ then $F = recX.E$

$(rec4)\ recX.(X + E) = recX.E$

$(rec5)\ recX.(\tau.(X + E) + F) = recX.\Delta(E + F)$

$(rec6)\ recX.(\Delta(X + E) + F) = recX.\Delta(E + F)$

Table 2. Distinguishing axioms

(Δ)	$\Delta(\Delta(E) + F) = \tau.(\Delta(E) + F)$
(S)	$\Delta(\tau.E + F) = \tau.(\tau.E + F)$
(0)	$\Delta(a.E + F) = \tau.(a.E + F)$
(ϵ)	$\Delta(E) = \tau.E$
(λ)	$\Delta(0) = \tau.0$

Theorem 3 (soundness). *If $E, F \in \mathbb{E}$ and $E =^* F$ then $E \simeq^* F$.*

In order to show completeness, i.e., that $E \simeq^* F$ implies $E =^* F$, we proceed along the lines of [10], except for the treatment of expressions from $\mathbb{E} \setminus \mathbb{P}$. We will work as much as possible in the setting of WB^Δ, the finest setting. As in [10] the first step consists in transforming every expression into a guarded one:

Theorem 4. *Let $E \in \mathbb{E}$. There exists a guarded F with $E =^\Delta F$.*

We do not consider $* = \epsilon$ in the sequel because by using axiom (ϵ), for every $E \in \mathbb{E}$ we find an E' such that E' does not contain the Δ-operator and $E =^\epsilon E'$. This allows to apply Milner's result [10] that in the absence of the Δ-operator the axioms from Table 1 with $(rec5)$ and $(rec6)$ replaced by Milner's rec-laws $(recX(\tau.X + E) = recX(\tau.E)$ and $recX.(\tau.(X + E) + F) = recX.(\tau.X + E + F)$, both can be easily derived from $(rec5)$ and $(\epsilon))$ are complete for $\simeq^\epsilon$.

The basic ingredients of the completeness proof are equation systems, and the manner in which these systems are set up constitutes the crucial deviation from the proof of Milner. Before we give detailed account of the proof, we illustrate the strategy by a small, informal example.

Consider an equation such as $X = a.X$. This equation is said to have a unique solution modulo Milner's observational congruence $\simeq^\epsilon$, since all expressions of $\mathbb{E}$ that satisfy this equation (such as $recX(a.\tau.X)$ for instance) are related by $\simeq^\epsilon$.

But, if we consider $a = \tau$, various inequivalent expressions satisfy the equation $X = \tau.X$ (such as $\tau.0$ and $\tau.b.0$, for $b \neq \tau$). So, this equation is said to *not* have a unique solution modulo $\simeq^\epsilon$, and therefore Milner resorts to 'guarded' equations only (and treats unguarded expressions in a preprocessing step analogously to Theorem 4). In principle, the situation is not much different for $\simeq^\Delta$, where the equation $X = \tau.X$ does neither possess a unique solution, since $\Delta(E)$ satisfies it, for arbitrary $E \in \mathbb{E}$ (in other words and as in [10], the axiom (*rec3*) is only sound if restricted to guarded expressions). However, we cannot erase all divergence in a preprocessing step, simply because the relations considered are divergence sensitive, and thus divergence must somehow be kept during the entire completeness proof. To solve this problem we use Δ as a placeholder. It 'swallows' divergence whenever it arises during the transformations, and hence to ensure 'guardedness' even in the presence of divergence. Concretely, we use equation systems that treat Δ as a 'first class' citizen: For each variable X occuring in an equation, we provide a 'divergent copy'X^Δ together with the equation $X^\Delta = \Delta(X)$. With this twist, it is still a matter of precise bookkeeping to establish the proof.

Let $V \subseteq \mathbb{V}$ be a set of variables and let $\boldsymbol{X} = X_1, \ldots, X_n$ be an ordered sequence of variables, where $X_i \notin V$. An *equation system over the free variables V and the formal variables $\boldsymbol{X}$* is a set of equations $\mathcal{E} = \{X_i = E_i \mid 1 \leq i \leq n\}$ such that $E_i \in \mathbb{E}$ and $\mathbb{V}(E_i) \subseteq \{X_1, \ldots, X_n\} \cup V$ for $1 \leq i \leq n$. Let $\boldsymbol{F} = F_1, \ldots, F_n$ be an ordered sequence of expressions. Then $\boldsymbol{F}$ $*$-provably satisfies the equation system $\mathcal{E}$ if $F_i =^* E_i\{\boldsymbol{F}/\boldsymbol{X}\}$ for all $1 \leq i \leq n$. An expression F $*$-provably satisfies $\mathcal{E}$ if there exists a sequence of expressions $F_1, \ldots, F_n$, which $*$-provably satisfies $\mathcal{E}$ and such that $F \equiv F_1$. We say that $\mathcal{E}$ is *guarded* if there exists a linear order $\prec$ on the variables $\{X_1, \ldots, X_n\}$ such that whenever the variable X_j is unguarded in the expression E_i then $X_j \prec X_i$.

For the next definition we take for each formal variable X_i ($1 \leq i \leq n$) a corresponding formal variable X_i^Δ such that $X_i^\Delta \notin \{X_1, \ldots, X_n\} \cup V$. The symbols $\alpha, \beta, \gamma, \ldots$ denote either Δ or $\_$. If e.g. $\alpha = \_$ then $X_i^\alpha \equiv X_i$ and $\alpha(E) \equiv E$. A *standard equation system (SES)* $\mathcal{E}$ over the free variables V and the formal variables $X_1, X_1^\Delta, \ldots, X_n, X_n^\Delta$ is an equation system of the form

$$\mathcal{E} = \{X_i = E_i \mid 1 \leq i \leq n\} \cup \{X_i^\Delta = \Delta(X_i) \mid 1 \leq i \leq n\}$$

where E_i is a sum of expressions $a.X_j$ ($a \in \mathbb{A}$, $1 \leq j \leq n$), $\tau.X_j^\Delta$ ($1 \leq j \leq n$), and variables $Y \in V$. We also say briefly that $\mathcal{E}$ is an SES over the free variables V and the formal variables $\boldsymbol{X} = X_1, \ldots, X_n$. If the sequence $F_1, \Delta(F_1), \ldots, F_n, \Delta(F_n)$ $*$-provably satisfies the SES $\mathcal{E}$ then we say briefly that $\boldsymbol{F} = F_1, \ldots, F_n$ $*$-provably satisfies $\mathcal{E}$. Furthermore $E_i\{\boldsymbol{F}/\boldsymbol{X}\}$ denotes the expression that results from substituting in E_i the variable X_i^α by $\alpha(F_i)$, where $1 \leq i \leq n$ and $\alpha \in \{\_, \Delta\}$. We write $X_i^\alpha \xrightarrow{a}_\mathcal{E} X_j^\beta$ if E_i contains the summand $a.X_j^\beta$. Note that $X_i \xrightarrow{a}_\mathcal{E} X_j^\beta$ if and only if $X_i^\Delta \xrightarrow{a}_\mathcal{E} X_j^\beta$. The notions $X_i^\alpha \Longrightarrow_\mathcal{E} X_j^\beta$, $X_i^\alpha \overset{a}{\Longrightarrow}_\mathcal{E} X_j^\beta$, $X_i \nrightarrow_\mathcal{E}, \ldots$ are derived from the relations $\xrightarrow{a}_\mathcal{E}$ analogously to the corresponding notions in Section 3. If the SES $\mathcal{E}$ is clear from the context then we will omit the subscript $\mathcal{E}$ in the following. Note that $\mathcal{E}$ is guarded if and only if the relation $\xrightarrow{\tau}_\mathcal{E}$ is acyclic.

Finally, the SES $\mathcal{E}$ is *saturated* if for all $1 \leq i, j \leq n$ and α, β, if $X_i \stackrel{a}{\Longrightarrow} X_j^\alpha$ then also $X_i \stackrel{a}{\longrightarrow} X_j^\alpha$ (since we use this notion only for systems without free variables, we do not need Milner's saturation condition for free variables). The introduction of the new variables X_i^Δ and the special form of an SES is crucial in order to carry over Milner's saturation property in the presence of the Δ-operator:

Theorem 5. *Every guarded expression E *-provably satisfies a guarded and saturated SES over the free variables $\mathbb{V}(E)$.*

Using axiom $(rec3)$, the following theorem can be shown analogously to [10].

Theorem 6. *Let $E, F \in \mathbb{E}$ and let $\mathcal{E}$ be a guarded equation system (not necessarily an SES) such that both E and F *-provably satisfy $\mathcal{E}$. Then $E =^* F$.*

6   Joining Two Equation Systems

In this section we restrict to expressions from $\mathbb{P}$. Our main technical result is

Theorem 7. *Let $P, Q \in \mathbb{P}$ such that $P \simeq^* Q$. Furthermore P (resp. Q) *-provably satisfies the guarded and saturated SES $\mathcal{E}_1 = \{X_i = E_i \mid 1 \leq i \leq m\}$ (resp. $\mathcal{E}_2 = \{Y_j = F_j \mid 1 \leq j \leq n\}$). Then there exists a guarded equation system $\mathcal{E}$ (not necessarily an SES) such that both P and Q *-provably satisfy $\mathcal{E}$.*

Let us postpone the proof of Theorem 7 for a moment and first see how completeness for $\mathbb{P}$ can be deduced:

Theorem 8 (completeness for $\mathbb{P}$). *If $P, Q \in \mathbb{P}$ and $P \simeq^* Q$ then $P =^* Q$.*

Proof. By Theorem 4 there exist guarded expressions P', Q' with $P' =^\Delta P$ and $Q' =^\Delta Q$. In particular, also $P', Q' \in \mathbb{P}$ and $P' \simeq^* Q'$ (due to soundness). By Theorem 5, P' (resp. Q') *-provably satisfies a guarded and saturated SES $\mathcal{E}_1$ (resp. $\mathcal{E}_2$) without free variables. By Theorem 7 there is some guarded equation system $\mathcal{E}$ which is *-provably satisfied by P' and Q'. Theorem 6 gives $P' =^* Q'$, and hence $P =^* Q$, concluding the proof.   □

In order to prove Theorem 7, we need the following two lemmas.

Lemma 1. *Let $\mathcal{E}$ be a guarded SES over the formal variables $X_1, \ldots, X_n$, and let X_i be such that there do not exist k, α, and $a \in \mathbb{A} \setminus \{\tau\}$ with $X_i \stackrel{a}{\Longrightarrow}_\mathcal{E} X_k^\alpha$. Then there exist j, β with $X_i \Longrightarrow_\mathcal{E} X_j^\beta \not\longrightarrow$.*

Proof. Induction along $\Longrightarrow_\mathcal{E}$, which is a partial order for a guarded SES.   □

For the further consideration it is useful to define a macro $\mathcal{M}^*(P)$ for $P \in \mathbb{P}$ by

$$\mathcal{M}^*(P) = \begin{cases} P \Uparrow & \text{if } * = \Delta, \\ P \stackrel{\tau}{\longrightarrow} & \text{if } * = S, \\ P \longrightarrow & \text{if } * = 0, \\ P \Uparrow\!\!\!\Uparrow & \text{if } * = \lambda. \end{cases}$$

Lemma 2. *If $\Delta(P) \sim^* \Delta(Q)$ then one of the following three cases holds:*

1. $\mathcal{M}^*(P)$ *and* $P \sim^* \Delta(Q)$
2. $\mathcal{M}^*(Q)$ *and* $\Delta(P) \sim^* Q$
3. *Neither* $\mathcal{M}^*(P)$ *nor* $\mathcal{M}^*(Q)$, *and* $P \sim^* Q$

Now we are able to prove Theorem 7.

Proof (Theorem 7). Assume that $\mathcal{E}_1$ is *-provably satisfied by the expressions $P_1, \ldots, P_m \in \mathbb{P}$, where $P \equiv P_1$, and that $\mathcal{E}_2$ is *-provably satisfied by the expressions $Q_1, \ldots, Q_n \in \mathbb{P}$, where $Q \equiv Q_1$. Thus $P_i =^* E_i\{\mathbf{P}/\mathbf{X}\}$ and $Q_j =^* F_j\{\mathbf{Q}/\mathbf{Y}\}$, and hence also $P_i \simeq^* E_i\{\mathbf{P}/\mathbf{X}\}$ and $Q_j \simeq^* F_j\{\mathbf{Q}/\mathbf{Y}\}$. Since $P, Q \in \mathbb{P}$, both $\mathcal{E}_1$ and $\mathcal{E}_2$ do not have free variables. The proof of the following two claims is tedious but straight-forward by using saturation of $\mathcal{E}_1$ and $\mathcal{E}_2$.

Claim 1 *If $\alpha(P_i) \sim^* \beta(Q_j)$ then the following implications hold:*

1. *If $X_i \xrightarrow{a} X_k^\gamma$ then either ($a = \tau$ and $\gamma(P_k) \sim^* \beta(Q_j)$) or there exist ℓ, δ such that $Y_j \xrightarrow{a} Y_\ell^\delta$ and $\gamma(P_k) \sim^* \delta(Q_\ell)$.*
2. *If $Y_j \xrightarrow{a} Y_\ell^\delta$ then either ($a = \tau$ and $\alpha(P_i) \sim^* \delta(Q_\ell)$) or there exist k, γ such that $X_i^\alpha \xrightarrow{a} X_k^\gamma$ and $\gamma(P_k) \sim^* \delta(Q_\ell)$.*
3. *Let $* = \Delta$. If $\alpha = \Delta$ then either $\beta = \Delta$ or $Y_j \xrightarrow{\tau} Y_\ell^\Delta$ for some ℓ.*
4. *Let $* = \Delta$. If $\beta = \Delta$ then either $\alpha = \Delta$ or $X_i \xrightarrow{\tau} X_k^\Delta$ for some k.*
5. *Let $* = \lambda$. If $\alpha = \Delta$ or ($\alpha = \_$ and $X_i \nrightarrow$) then either $\beta = \Delta$, or ($\beta = \_$ and $Y_j \nrightarrow$), or $Y_j \xrightarrow{\tau} Y_\ell^\Delta$ for some ℓ, or $Y_j \xrightarrow{\tau} Y_\ell \nrightarrow$ for some ℓ.*
6. *Let $* = \lambda$. If $\beta = \Delta$ or ($\beta = \_$ and $Y_j \nrightarrow$) then either $\alpha = \Delta$, or ($\alpha = \_$ and $X_i \nrightarrow$), or $X_i \xrightarrow{\tau} X_k^\Delta$ for some k, or $X_i \xrightarrow{\tau} X_k \nrightarrow$ for some k.*

Claim 2 *If $P_i \simeq^* Q_j$ then the following implications hold:*

1. *If $X_i \xrightarrow{a} X_k^\alpha$ then there exist ℓ, β such that $Y_j \xrightarrow{a} Y_\ell^\beta$ and $\alpha(P_k) \sim^* \beta(Q_\ell)$.*
2. *If $Y_j \xrightarrow{a} Y_\ell^\beta$ then there exist k, α such that $X_i \xrightarrow{a} X_k^\alpha$ and $\alpha(P_k) \sim^* \beta(Q_\ell)$.*

Now take for all $1 \le i \le m$, $1 \le j \le n$, and α, β with $\alpha(P_i) \sim^* \beta(Q_j)$ a variable $Z_{i,j}^{\alpha,\beta}$, and let $\mathbf{Z} = Z_{1,1}^{-,-}, \ldots$ be a sequence consisting of these variables (since $P_1 \simeq^* Q_1$, $Z_{1,1}^{-,-}$ is defined). Moreover, if $\alpha(P_i) \sim^* \beta(Q_j)$ and either $\alpha = \_$ or $\beta = \_$ then we define $G_{i,j}^{\alpha,\beta}$ as the sum, which contains the summand

$$a.Z_{k,\ell}^{\gamma,\delta} \quad \text{if } X_i \xrightarrow{a} X_k^\gamma, Y_j \xrightarrow{a} Y_\ell^\delta, \text{ and } \gamma(P_k) \sim^* \delta(Q_\ell),$$

$$\tau.Z_{k,j}^{\gamma,\beta} \quad \text{if } X_i \xrightarrow{\tau} X_k^\gamma \text{ but } \neg \exists \ell, \delta : Y_j \xrightarrow{\tau} Y_\ell^\delta \wedge \gamma(P_k) \sim^* \delta(Q_\ell)$$
$$\text{(this implies by Claim 1(1) that } \gamma(P_k) \sim^* \beta(Q_j)),$$

$$\tau.Z_{i,\ell}^{\alpha,\delta} \quad \text{if } Y_j \xrightarrow{\tau} Y_\ell^\delta \text{ but } \neg \exists k, \gamma : X_i \xrightarrow{\tau} X_k^\gamma \wedge \gamma(P_k) \sim^* \delta(Q_\ell)$$
$$\text{(this implies by Claim 1(2) that } \alpha(P_i) \sim^* \delta(Q_\ell)).$$

Furthermore $G_{i,j}^{\alpha,\beta}$ does not contain any other summands. Now the equation system $\mathcal{E}$ over the formal variables $\mathbf{Z}$ contains for each variable $Z_{i,j}^{\alpha,\beta}$ in $\mathbf{Z}$ the corresponding equation below, where for equation (E1) by Lemma 2 one of the three cases listed in (E1) holds (if the first and the second case hold, then we choose arbitrarily one of the two corresponding equations for (E1)).

$$\text{(E1)} \quad Z_{i,j}^{\Delta,\Delta} = \begin{cases} Z_{i,j}^{-,\Delta} & \text{if } \mathcal{M}^*(P_i) \text{ and } P_i \sim^* \Delta(Q_j) \\ Z_{i,j}^{\Delta,-} & \text{if } \mathcal{M}^*(Q_j) \text{ and } \Delta(P_i) \sim^* Q_j \\ \Delta(Z_{i,j}^{-,-}) & \text{if neither } \mathcal{M}^*(P_i) \text{ nor } \mathcal{M}^*(Q_j), \text{ and } P_i \sim^* Q_j \end{cases}$$

$$\text{(E2)} \quad Z_{i,j}^{\alpha,\beta} = \tau.G_{i,j}^{\alpha,\beta} \quad \text{if } \alpha = \Delta \neq \beta \text{ or } \alpha \neq \Delta = \beta$$

$$\text{(E3)} \quad Z_{i,j}^{-,-} = G_{i,j}^{-,-}$$

In general, $\mathcal{E}$ is not an SES, but from the guardedness of $\mathcal{E}_1$ and $\mathcal{E}_2$ it follows easily that also $\mathcal{E}$ is guarded. We will show that P *-provably satisfies $\mathcal{E}$, that also Q *-provably satisfies $\mathcal{E}$ can be shown analogously. For this we define for each variable $Z_{i,j}^{\alpha,\beta}$ in $\mathbf{Z}$ the corresponding expression $R_{i,j}^{\alpha,\beta}$ by

$$R_{i,j}^{\Delta,\Delta} \equiv R_{i,j}^{\Delta,-} \equiv \Delta(P_i), \qquad R_{i,j}^{-,\Delta} \equiv \tau.P_i, \text{ and}$$

$$R_{i,j}^{-,-} \equiv \begin{cases} P_i & \text{if } \forall \ell, \delta, a \left\{ Y_j \xrightarrow{a} Y_\ell^\delta \Rightarrow \exists k, \gamma \left\{ \begin{matrix} X_i \xrightarrow{a} X_k^\gamma \wedge \\ \gamma(P_k) \sim^* \delta(Q_\ell) \end{matrix} \right\} \right\} \\ \tau.P_i & \text{if } \exists \ell, \delta \left\{ Y_j \xrightarrow{\tau} Y_\ell^\delta \wedge \neg \exists k, \gamma \left\{ \begin{matrix} X_i \xrightarrow{\tau} X_k^\gamma \wedge \\ \gamma(P_k) \sim^* \delta(Q_\ell) \end{matrix} \right\} \right\} \end{cases}$$

Let $\mathbf{R} = R_{1,1}^{-,-}, \ldots$ be the sequence corresponding to the sequence $\mathbf{Z}$. First note that $R_{1,1}^{-,-} \equiv P_1 \equiv P$ by $P_1 \simeq^* Q_1$ and Claim 2(2). It remains to check that all equations are *-provably satisfied when every variable $Z_{i,j}^{\alpha,\beta}$ is replaced by $R_{i,j}^{\alpha,\beta}$. We start with equation (E1) defining $Z_{i,j}^{\Delta,\Delta}$. The case that $Z_{i,j}^{\Delta,\Delta}$ is defined by $Z_{i,j}^{\Delta,\Delta} = Z_{i,j}^{\Delta,-}$ is trivial, since $R_{i,j}^{\Delta,\Delta} \equiv R_{i,j}^{\Delta,-} \equiv \Delta(P_i)$. Thus, the following two cases 1 and 2 remain.

Case 1. Equation $Z_{i,j}^{\Delta,\Delta} = Z_{i,j}^{-,\Delta}$ belongs to $\mathcal{E}$: thus $\mathcal{M}^*(P_i)$ and $P_i \sim^* \Delta(Q_j)$. Since $R_{i,j}^{-,\Delta} \equiv \tau.P_i$ and $R_{i,j}^{\Delta,\Delta} \equiv \Delta(P_i)$ we have to prove that $\tau.P_i =^* \Delta(P_i)$. We distinguish on the value of *.

Case 1.1. $* = \Delta$: then $P_i \sim^\Delta \Delta(Q_j)$ and hence $X_i \xrightarrow{\tau} X_k^\Delta$ for some k by Claim 1(4). Thus there exists an expression R with (we use the derived law $(\tau\Delta')$ from Section 4) $\tau.P_i =^\Delta \tau.E_i\{\mathbf{P}/\mathbf{X}\} =^\Delta \tau.(R+\tau.\Delta(P_k)) =^\Delta \tau.(R+\Delta(P_k)) =^\Delta \Delta(R + \Delta(P_k)) =^\Delta \cdots =^\Delta \Delta(P_i)$.

Case 1.2. $* = S$: then $P_i \sim^S \Delta(Q_j)$ and $\mathcal{M}^S(P_i)$, i.e., $P_i \xrightarrow{\tau}$. Since $P_i \simeq^S E_i\{\mathbf{P}/\mathbf{X}\}$, also $E_i\{\mathbf{P}/\mathbf{X}\} \xrightarrow{\tau}$, i.e, $X_i \xrightarrow{\tau}$, and there exist expressions R, P_k with $\tau.P_i =^S \tau.E_i\{\mathbf{P}/\mathbf{X}\} =^\Delta \tau.(R+\tau.P_k) =^S \Delta(R+\tau.P_k) =^S \cdots =^S \Delta(P_i)$.

Case 1.3. $* = 0$: analogously to Case 1.2 with axiom (0) used instead of (S).

Case 1.4. $* = \lambda$: then $P_i \sim^\lambda \Delta(Q_j)$, and Claim 1(6) implies either $X_i \nrightarrow$, or $X_i \xrightarrow{\tau} X_k^\Delta$ for some k, or $X_i \xrightarrow{\tau} X_k \nrightarrow$ for some k.

Case 1.4.1. $X_i \nrightarrow$, i.e., $E_i \equiv 0$: [1] we obtain $\tau.P_i =^\lambda \tau.E_i\{P/X\} \equiv \tau.0 =^\lambda \Delta(0) =^\lambda \cdots =^\lambda \Delta(P_i)$.

Case 1.4.2. $X_i \xrightarrow{\tau} X_k^\Delta$: we can conclude as in Case 1.1.

Case 1.4.3. $X_i \xrightarrow{\tau} X_k \nrightarrow$: thus there exists R with $\tau.P_i =^\lambda \tau.E_i\{P/X\} =^\Delta \tau.(R + \tau.0) =^\lambda \tau.(R + \Delta(0)) =^\Delta \Delta(R + \Delta(0)) =^\lambda \cdots =^\lambda \Delta(P_i)$.

Case 2. Equation $Z_{i,j}^{\Delta,\Delta} = \Delta(Z_{i,j}^{-,-})$ belongs to $\mathcal{E}$: thus $P_i \sim^* Q_j$ and neither $\mathcal{M}^*(P_i)$ nor $\mathcal{M}^*(Q_j)$ holds. We have either $R_{i,j}^{-,-} \equiv P_i$ or $R_{i,j}^{-,-} \equiv \tau.P_i$. The case that $R_{i,j}^{-,-} \equiv P_i$ is trivial, thus let us assume that $R_{i,j}^{-,-} \equiv \tau.P_i$. Then there exist ℓ, δ such that $Y_j \xrightarrow{\tau} Y_\ell^\delta$ but there do not exist k, γ with $X_i \xrightarrow{\tau} X_k^\gamma$ and $\gamma(P_k) \sim^* \delta(Q_\ell)$. Since $\Delta(P_i) \sim^* \Delta(Q_j)$ (recall that variable $Z_{i,j}^{\Delta,\Delta}$ is defined), it follows $\Delta(P_i) \sim^* \delta(Q_\ell)$ by Claim 1(2). Using Claim 1 we can deduce for each value of $*$ a contradiction to $\neg\mathcal{M}^*(Q_j)$.

It remains to check the equations (E2) and (E3). Fix α, β such that $\alpha(P_i) \sim^* \beta(Q_j)$ and either $\alpha = \_$ or $\beta = \_$. We will distinguish two main cases 3 and 4:

Case 3. $\forall \ell, \delta, a \; (Y_j \xrightarrow{a} Y_\ell^\delta \Rightarrow \exists k, \gamma : X_i \xrightarrow{a} X_k^\gamma \wedge \gamma(P_k) \sim^* \delta(Q_\ell))$ (†)

With axiom $(\tau 1)$ and (S1)-(S3) we obtain $G_{i,j}^{\alpha,\beta}\{R/Z\} =^\Delta E_i\{P/X\} =^* P_i$ (this step is analogous to [10]). In case $\alpha = \_ = \beta$ (resp. $\alpha = \_, \beta = \Delta$), it is straight-forward to show that equation (E3) (resp. (E2)) is satisfied. So assume that $\alpha = \Delta, \beta = \_$. Thus $\Delta(P_i) \sim^* Q_j$. By inspecting equation (E2) and using the fact that $R_{i,j}^{\Delta,-} \equiv \Delta(P_i)$ and $G_{i,j}^{\alpha,\beta}\{R/Z\} =^* P_i$, we see that it remains to show $\Delta(P_i) =^* \tau.P_i$. We distinguish on the value of $*$.

Case 3.1. $* = \Delta$: thus $\Delta(P_i) \sim^\Delta Q_j$ and $Y_j \xrightarrow{\tau} Y_\ell^\Delta$ for some ℓ by Claim 1(3). Hence by (†) there exist k, γ with $X_i \xrightarrow{\tau} X_k^\gamma$ and $\gamma(P_k) \sim^\Delta \Delta(Q_\ell)$. By Claim 1(4) either $\gamma = \Delta$ or $X_k \xrightarrow{\tau} X_p^\Delta$ for some p. Saturation of $\mathcal{E}_1$ implies in both cases $X_i \xrightarrow{\tau} X_p^\Delta$ for some p, which allows to conclude as in Case 1.1.

Case 3.2. $* = S$: we have $\Delta(P_i) \sim^S Q_j$. If $Q_j \nrightarrow$ then $P_i \xrightarrow{\tau}$, and we can refer to Case 1.2. On the other hand, if $Q_j \xrightarrow{\tau}$ then $F_j\{Q/Y\} \xrightarrow{\tau}$, i.e, $Y_j \xrightarrow{\tau}$. Thus $X_i \xrightarrow{\tau}$ by (†), which allows again to refer to Case 1.2.

Case 3.3. $* = 0$: analogous to Case 3.2.

Case 3.4. $* = \lambda$: since $\Delta(P_i) \sim^\lambda Q_j$, Claim 1(5) implies either $Y_j \nrightarrow$, or $Y_j \xrightarrow{\tau} Y_\ell^\Delta$, or $Y_j \xrightarrow{\tau} Y_\ell \nrightarrow$ for some ℓ.

Case 3.4.1. $Y_j \nrightarrow$: by Claim 1(1) there cannot exist $a \in \mathbb{A} \setminus \{\tau\}$ with $X_i \xrightarrow{a}$. Lemma 1 and the saturation of $\mathcal{E}_1$ imply either $X_i \nrightarrow$, or $X_i \xrightarrow{\tau} X_k^\Delta$ for some k, or $X_i \xrightarrow{\tau} X_k \nrightarrow$ for some k. We can proceed as in Case 1.4.

---

[1] Note that if we would deal with equation systems containing free variables, then we could only conclude here that E_i must be a sum of free variables. This is the reason why Theorem 7 requires that $P, Q \in \mathbb{P}$, i.e., that $\mathbb{V}(P) = \mathbb{V}(Q) = \emptyset$.

Case 3.4.2. $Y_j \xrightarrow{\tau} Y_\ell^\Delta$ for some ℓ: by (†) there exist k, γ with $X_i \xrightarrow{\tau} X_k^\gamma$ and $\gamma(P_k) \sim^\lambda \Delta(Q_\ell)$. By Claim 1(6) either $\gamma = \Delta$, or ($\gamma = \_$ and $X_k \not\rightarrow$), or $X_k \xrightarrow{\tau} X_p^\Delta$ for some p, or $X_k \xrightarrow{\tau} X_p \not\rightarrow$ for some p. By saturation we obtain either $X_i \xrightarrow{\tau} X_p^\Delta$ for some p (see Case 1.4.2), or $X_i \xrightarrow{\tau} X_p \not\rightarrow$ for some p (see Case 1.4.3).

Case 3.4.3. $Y_j \xrightarrow{\tau} Y_\ell \not\rightarrow$ for some ℓ: by (†) there exist k, γ with $X_i \xrightarrow{\tau} X_k^\gamma$ and $\gamma(P_k) \sim^\lambda Q_\ell$. Using Claim 1(6) we can conclude as in Case 3.4.2.

Case 4. $\exists \ell, \delta \; (Y_j \xrightarrow{\tau} Y_\ell^\delta \wedge \neg \exists k, \gamma : X_i \xrightarrow{\tau} X_k^\gamma \wedge \gamma(P_k) \sim^* \delta(Q_\ell))$

We get $G_{i,j}^{\alpha,\beta} \{R/Z\} =^\Delta E_i \{P/X\} + \tau.\alpha(P_i) =^* P_i + \tau.\alpha(P_i)$ (as in Case 3, this step is analogous to [10]).

Case 4.1. $\alpha = \beta = \_$: we have $R_{i,j}^{-;-} \equiv \tau.P_i =^\Delta P_i + \tau.P_i =^* G_{i,j}^{-;-} \{R/Z\}$, thus (E3) is satisfied.

Case 4.2. $\alpha = \_, \beta = \Delta$: we obtain $R_{i,j}^{-;\Delta} \equiv \tau.P_i =^\Delta \tau.\tau.P_i =^\Delta \tau.(P_i + \tau.P_i) =^* \tau.G_{i,j}^{-;\Delta} \{R/Z\}$, thus (E2) is satisfied.

Case 4.3. $\alpha = \Delta, \beta = \_$: with $(\tau\Delta')$ and $(\tau\Delta)$ from Section 4 we get $R_{i,j}^{\Delta;-} \equiv \Delta(P_i) =^\Delta \tau.\Delta(P_i) =^\Delta \tau.(P_i + \tau.\Delta(P_i)) =^* \tau.G_{i,j}^{\Delta;-} \{R/Z\}$, thus (E2) is again satisfied. This concludes the proof of Theorem 7 and hence of Theorem 8.   □

7   Completeness for Open Expressions

In order to prove completeness for the whole set $\mathbb{E}$ we will argue in a purely syntactical way by investigating our axioms. The following observation is crucial:

Lemma 3. *Let* $* \neq 0$ *and* $E, F \in \mathbb{E}$. *If* $a \in \mathbb{A} \setminus \{\tau\}$ *does neither occur in* E *nor in* F *then* $E\{a.0/X\} =^* F\{a.0/X\}$ *implies* $E =^* F$.

Note that Lemma 3 is false for $* = 0$. We have $\tau.a.0 =^0 \Delta(a.0)$ but $\tau.X \neq^0 \Delta(X)$ (since $\tau.0 \neq^0 \Delta(0)$). Hence, in the following theorem we have to exclude $* = 0$.

Theorem 9. *Let* $* \neq 0$ *and* $E, F \in \mathbb{E}$. *If* $E \simeq^* F$ *then* $E =^* F$.

Proof. Let $E \simeq^* F$. We prove by induction on $|\mathbb{V}(E) \cup \mathbb{V}(F)|$ that $E =^* F$. If $\mathbb{V}(E) \cup \mathbb{V}(F) = \emptyset$ then in fact $E, F \in \mathbb{P}$ and $E =^* F$ by Theorem 8. Thus let $X \in \mathbb{V}(E) \cup \mathbb{V}(F)$. Since $E \simeq^* F$, we have $E\{a.0/X\} \simeq^* F\{a.0/X\}$. Thus by induction $E\{a.0/X\} =^* F\{a.0/X\}$ and hence $E =^* F$ by Lemma 3.   □

In order to obtain completeness for $\simeq^0$ on open expressions, we have to introduce the following additional axiom ($\mathbb{E}$), which can be shown to be sound for $\simeq^0$.

$(\mathbb{E})$ If $E\{0/X\} = F\{0/X\}$ and $E\{a.0/X\} = F\{a.0/X\}$ where $a \in \mathbb{A} \setminus \{\tau\}$ does neither occur in E nor in F then $E = F$.

If we add this axiom to the standard axioms for $\simeq^0$ then we can prove completeness in the same way as in the proof of Theorem 9.

Theorem 10. *Let* $E, F \in \mathbb{E}$. *If* $E \simeq^0 F$ *then* $E =^0 F$ *can be derived by the standard axioms for* $\simeq^0$ *plus the axiom* $(\mathbb{E})$.

8   Conclusion

This paper has developed sound and complete axiomatisations for the divergence sensitive spectrum of weak bisimulation equivalences. We have not covered the weak bisimulation preorders $WB^{\downarrow}$ and $WB^{\Downarrow}$ considered in [5]. We claim however that adding the axiom $\Delta(E) \leq E + F$ to the axioms of WB^{λ} (respectively WB^{Δ}) is enough to obtain completeness of $WB^{\Downarrow}$ ($WB^{\downarrow}$). Note that $WB^{\downarrow}$ is axiomatised in [11], so only $WB^{\Downarrow}$ needs further work.

We are confident that our axiomatisation form the basis of a complete equational characterisation of the bisimulation fragment of the linear time – branching time spectrum with silent moves. On the technical side, we are currently investigating whether the somewhat unsatisfactory auxiliary axiom ($\mathbb{E}$) is indeed necessary for achieving completeness of open expressions for $\simeq^0$.

References

1. L. Aceto and M. Hennessy. Termination, deadlock, and divergence. *J. ACM*, 39(1):147–187, 1992.
2. J.C.M. Baeten and W.P. Weijland. *Process Algebra*. Cambridge Univ. Press, 1990.
3. J.A. Bergstra, J.W. Klop, and E.-R. Olderog. Failure semantics with fair abstraction. Report CS-R8609, CWI, Amsterdam, 1986.
4. R.J. van Glabbeek. The Linear Time – Branching Time Spectrum I. The semantics of concrete, sequential processes. Chapter 1 in *Handbook of Process Algebra*, pages 3–99, Elsevier, 2001.
5. R.J. van Glabbeek. The Linear Time – Branching Time Spectrum II. The semantics of sequential systems with silent moves (Extended Abstract). In *Proc. CONCUR'93*, LNCS 715, pages 66–81. Springer, 1993.
6. R.J. van Glabbeek. A Complete Axiomatization for Branching Bisimulation Congruence of Finite-State Behaviours. In *Proc. MFCS'93*, LNCS 711, pages 473–484. Springer, 1993.
7. R.J. van Glabbeek and W.P. Weijland. Branching time and abstraction in bisimulation semantics. *J. ACM*, 43(3):555–600, 1996.
8. H. Hermanns and M. Lohrey. Observational Congruence in a Stochastic Timed Calculus with Maximal Progress. Technical Report IMMDVII-7/97, University of Erlangen-Nürnberg, IMMD7, 1997.
9. R. Milner. A Complete Inference System for a Class of Regular Behaviours. *J. Comput. System Sci.*, 28:439–466, 1984.
10. R. Milner. A Complete Axiomatisation for Observational Congruence of Finite-State Behaviours. *Inf. Comp.*, 91(227-247), 1989.
11. D.J. Walker. Bisimulation and divergence. *Inf. Comp.*, 85:202–241, 1990.

A Spatial Logic for Querying Graphs

Luca Cardelli[1], Philippa Gardner[*][2], and Giorgio Ghelli[3]

[1] Microsoft Research, Cambridge.
[2] Imperial College of Science, Technology and Medicine, London.
[3] University of Pisa, Pisa.

Abstract. We study a spatial logic for reasoning about labelled directed graphs, and the application of this logic to provide a query language for analysing and manipulating such graphs. We give a graph description using constructs from process algebra. We introduce a spatial logic in order to reason locally about disjoint subgraphs. We extend our logic to provide a query language which preserves the multiset semantics of our graph model. Our approach contrasts with the more traditional set-based semantics found in query languages such as TQL, Strudel and GraphLog.

1 Introduction

Semi-structured data plays an important role in the exchange of information between globally distributed applications: examples include BibTex files and XML documents. Whilst the research community mostly agree on defining semi-structured data using labelled directed graphs or trees with 'graphical' links, the study of how to query, modify and manipulate such data is still very active.

Motivating Examples. A standard example used by the semi-structured data community [ABS00] is a bibtex file with an article entry of the form:

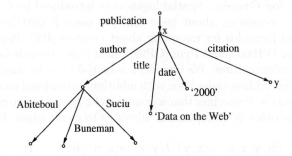

The global name (object identifier) x denotes the citation name of the publication, which is used to refer to the particular bibtex entry. The citation entry might be a simple text entry, or might point to another entry in the bibtex file.

---

[*] Supported by an EPSRC Advanced Fellowship.

P. Widmayer et al. (Eds.): ICALP 2002, LNCS 2380, pp. 597–610, 2002.

Another example with a more graphical emphasis is the correspondence between counties and towns, where counties contain towns and towns are in counties. A more complicated example is given by links between web pages, where names correspond to URLs. Such links display all manner of graphical linking. These simple examples illustrate that the typical data models for semi-structured data are either labelled directed graphs, or labelled trees with 'graphical links'. In this paper, we focus on labelled directed graphs.

Graph Model. We use a well-known graph description based on constructs from process algebra [CMR94]. The models consist of labelled edges and two kinds of nodes: the *global* nodes identified with unique names x, y, z and the *local* nodes whose identifiers are not known. In our bibtex example, the citation x corresponds to a global node labelled x, whereas the author field has no explicit citation. Similarly, the Internet's Domain Name Service globally registers IP addresses, but not all IP addresses are global. Our notation for describing graph (a) is $a(x, y) \,|\, b(y, x)$, where $a(x, y)$ denotes an edge and $\_|\_$ is the usual composition operator for processes used in this case to describe multisets of edges.

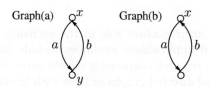

Graph (b) is given by $(\text{local } y)(a(x, y) \,|\, b(y, x))$. The local operator is analogous to restriction in the π-calculus. It means that the previously identified node cannot now have any more edges attached to it.

Spatial Logic for Graphs. Spatial logics were introduced by Caires, Cardelli and Gordon for reasoning about trees and processes [CG00,Cai99], and also by O'Hearn and Reynolds for reasoning about pointers [IO01,Rey00] using the bunched logic of O'Hearn and Pym [OP99]. Such logics provide local reasoning about disjoint substructures. We introduce a spatial logic for analysing graphs. It combines standard first-order logic with additional structural connectives. The structural formula $\phi \,|\, \psi$ specifies that a graph can be split into two parts: one part satisfying ϕ, the other ψ. Composition allows us to count edges. For example,

$$\exists \mathbf{x}, \mathbf{y}, \mathbf{z}, \mathbf{u}. \ a(\mathbf{x}, \mathbf{y}) \,|\, b(\mathbf{y}, \mathbf{z}) \,|\, a(\mathbf{z}, \mathbf{u}) \,|\, \text{true} \qquad (\dagger)$$

specifies that there are at least three different edges in the graph, with a following b following a. In contrast, conjunction allows us to describe paths with

$$\exists \mathbf{x}, \mathbf{y}, \mathbf{z}, \mathbf{u}. \ (a(\mathbf{x}, \mathbf{y}) \,|\, \text{true}) \wedge (b(\mathbf{y}, \mathbf{z}) \,|\, \text{true}) \wedge (a(\mathbf{z}, \mathbf{u}) \,|\, \text{true})$$

describing the existence of a *path* a followed by b followed by a. The path formula is satisfied by graph (a), but the composition formula (†) is not.

Our graph logic (without recursion) sits naturally between first-order logic FOL and monadic second-order logic MSOL: in FOL we can only quantify over single edges; in our logic, the formula $\phi \,|\, \mathsf{true}$ existentially quantifies a property ϕ over all subgraphs; in MSOL we can arbitrarily nest quantifications over sets of edges. Our logic can be viewed as a sublogic of MSOL. However, we can reason locally about disjoint subgraphs. FOL and MSOL require complex disjointness conditions to reason about such subgraphs: for example, the composition formula (†) requires such conditions to specify that the three edges are disjoint. Dawar, Gardner and Ghelli are studying expressivity results for the graph logic. Our current results are reported in [CGG01].

Query Language. We define a query language based on pattern matching and recursion. Our approach integrates well with our graph description, and contrasts with the standard set-based approach found in Cardelli and Ghelli's TQL, a query language based on the ambient logic [CG01a], and the graphical query languages StruQL [FFK+97] and GraphLog [CM90] based on first-order logic.

To illustrate the standard approach, consider a simple query $\mathsf{input\_graph} \models^?$ $\mathsf{a}(\mathbf{x}, \mathbf{y}) \,|\, \mathsf{true}$. This query asks for a substitution σ such that the satisfaction relation $\mathsf{input\_graph} \models^\sigma \mathsf{a}(\mathbf{x}, \mathbf{y}) \,|\, \mathsf{true}$ holds in our logic. For example, if the input graph is $a(x, y) \,|\, b(y, x)$, then there are two solutions:

$$(\mathbf{a} \mapsto a, \ \mathbf{x} \mapsto x, \ \mathbf{y} \mapsto y) \qquad \text{or} \qquad (\mathbf{a} \mapsto b, \ \mathbf{x} \mapsto y, \ \mathbf{y} \mapsto x)$$

The *from/select* expressions take such solutions and build new graphs. For example, the expression

$$\textbf{from } \mathsf{input\_graph} \models^? \mathsf{a}(\mathbf{x}, \mathbf{y}) \,|\, \mathsf{true} \textbf{ select } \mathsf{a}(\mathbf{y}, \mathbf{x}) \qquad (*)$$

takes *every* substitution σ which satisfies the query, and creates a new graph consisting of the composition of the edges $\mathbf{a}\sigma(\mathbf{y}\sigma, \mathbf{x}\sigma)$. In our example, the resulting new graph is $a(y, x) \,|\, b(x, y)$. Given the input graph $a(x, y) \,|\, a(x, y)$ instead, there is *one* substitution $\sigma : \mathbf{x} \mapsto x, \mathbf{y} \mapsto y$ which satisfies the query. The resulting graph is just $a(y, x)$. This collapse of information can be an advantage. It does mean however that we cannot accurately take a copy of a graph.

Instead we define a query language based on *queries* and *transducers*. Queries build new graphs from old. Transducers relate input graphs with output graphs. A basic transducer $\phi \Rightarrow Q$ relates any input graph satisfying ϕ with the query Q which might depend on witnesses from ϕ. For example, the transducer

$$\exists \mathbf{a}, \mathbf{x}, \mathbf{y}. \ (\mathbf{a}(\mathbf{x}, \mathbf{y}) \,|\, \mathsf{true} \Rightarrow \mathbf{a}(\mathbf{y}, \mathbf{x}))$$

relates an input graph with edge $\mathbf{a}\sigma(\mathbf{x}\sigma, \mathbf{y}\sigma)$ with the output graph $\mathbf{a}\sigma(\mathbf{y}\sigma, \mathbf{x}\sigma)$. Given the input graph $a(x, y) \,|\, b(y, x)$, there are two possible output graphs, either $a(y, x)$ or $b(x, y)$. This example does the pattern-matching part of the from/select expression (*). It does not combine the inverted edges. Instead this role is played by recursion. Consider the transducer

$$\mathbf{R} \stackrel{\text{def}}{=} (\text{nil} \Rightarrow \text{nil}) \vee (\exists \mathbf{a}, \mathbf{x}, \mathbf{y}. \ (\mathbf{a}(\mathbf{x}, \mathbf{y}) \Rightarrow \mathbf{a}(\mathbf{y}, \mathbf{x})) \,|\, \mathbf{R})$$

Either the input graph is empty and relates to the empty output graph. Or the input graph can be split into an edge and the rest of the graph. The output graph consists of the inverted edge composed with the output associated with the remaining graph. Given input graph $a(x, y) \,|\, a(x, y)$ for example, the output graph is the exact inverted copy $a(y, x) \,|\, a(y, x)$.

We study two query languages: a basic language which can express our motivating examples, and a general language which has a simple formalism but is too expressive to implement. We were surprised to observe that the from/select expressions can be embedded in our general language.

2   Labelled Directed Graphs

We use a simple graph algebra [CMR94] to describe labelled directed graphs. Assume an infinite set $\mathcal{X}$ of names ranged over by $u, \dots, z$, and an infinite set of edge labels $\mathcal{A}$ ranged over by a, b, c. We also use the notation $\tilde{z}$ to denote a sequence of names, and $|\tilde{z}|$ to denote the length of the sequence.

DEFINITION 1

The set $\mathcal{G}(\mathcal{X}, \mathcal{A})$ of graph terms generated by $\mathcal{X}$ and $\mathcal{A}$ is given by the grammar

$$
\begin{array}{lll}
G ::= & \text{nil} & \text{empty} \\
& a(x, y) & \text{edge} \\
& G \,|\, G & \text{composition} \\
& (\text{local } x)\, G & \text{hiding}
\end{array}
$$

We sometimes write $\mathcal{G}$ instead of $\mathcal{G}(\mathcal{X}, \mathcal{A})$. The definitions of *free* and *bound* names are standard: the hiding operator $(\text{local } x)\, G$ binds x in G; x is free in process $a(x, y)$. We write $\text{fn}(G)$ to denote the set of free names in G. We use the capture-avoiding substitution, denoted by $G\{y/x\}$.

Our graph model is based on a multiset semantics, with the graph term $a(x, y) \,|\, a(x, y)$ denoting a graph with two edges. We give a natural structural congruence on graph terms (definition 2) which corresponds to the usual notion of graph isomorphism [CMR94]. Our choice contrasts with the approach taken in the query language StruQL, which has a set-based semantics with $a(x, y) \,|\, a(x, y)$ corresponding to $a(x, y)$. It also contrasts with the language UnQL [BDHS96], which is based on graph bisimulation rather than graph isomorphism.

DEFINITION 2

The structural congruence between graph terms, written $\equiv$, is the smallest congruence closed with respect to $\_\,|\,\_$ and $(\text{local } x)\_$, and satisfying the axioms:

$$
\begin{array}{ll}
G \,|\, \text{nil} \equiv G & (\text{local } x)(\text{local } y)G \equiv (\text{local } y)(\text{local } x)G \\
(G_1 \,|\, G_2) \,|\, G_3 \equiv G_1 \,|\, (G_2 \,|\, G_3) & (\text{local } x)(G_1 \,|\, G_2) \equiv (\text{local } x)G_1 \,|\, G_2, \quad x \notin \text{fn}(G_2) \\
G_1 \,|\, G_2 \equiv G_2 \,|\, G_1 & (\text{local } x)\text{nil} \equiv \text{nil} \\
& (\text{local } x)G \equiv (\text{local } y)G\{y/x\}, \quad y \notin \text{fn}(G)
\end{array}
$$

2.1   Comparison with Courcelle

We give a set-theoretic description of graphs in the spirit of Courcelle [Cou97], which is equivalent to our graph description. We have made some different choices to Courcelle, which we will discuss after the definition. We assume disjoint infinite sets of vertices $\mathcal{V}$, edge identifiers $\mathcal{E}$, edge labels $\mathcal{A}$, and names $\mathcal{X}$.

DEFINITION 3
The graph structure $G_S = \langle V \cup E \cup A, \{edge \subseteq E \times A \times V \times V\}, src : X \to V \rangle$ is defined by

1. $V \subseteq \mathcal{V}$, $E \subseteq \mathcal{E}$, $A \subseteq \mathcal{A}$, $X \subseteq \mathcal{X}$ are finite sets;

2. each edge identifier has a unique label, domain node and codomain node:
   $\forall e, a_i, v_i, w_i.edge(e, a_1, v_1, w_1) \wedge edge(e, a_2, v_2, w_2) \Rightarrow a_1 = a_2 \wedge v_1 = v_2 \wedge w_1 = w_2;$

3. the edge identifiers, edge labels and vertices are related using edge:

   $\forall d \, \exists d_1, d_2, d_3.$
   $edge(d, d_1, d_2, d_3) \vee edge(d_1, d, d_2, d_3) \vee edge(d_1, d_2, d, d_3) \vee edge(d_1, d_2, d_3, d)$

4. src is an injective function.

This definition differs from Courcelle's approach in several ways. Courcelle permits nodes to be unattached to edges. He considers both finite and infinite graphs, whereas we use the finite case since it is enough for this paper. He also does not treat A as part of the domain. Instead, he defines a family of relations $edge_a \subseteq E \times V \times V$. This last point is significant when comparing our different logics for reasoning about graphs. Courcelle considers two systems, one where src is injective and one where it is not. The graphs presented here correspond to the injective case; the non-injective case corresponds to adding *name fusions* $x = y$ to our graphical description, as introduced by Gardner and Wischik [GW00].

In [Cou97], Courcelle studies a graph grammar which is similar to ours. Courcelle's motivation is to explore the expressive power of MSOL. In contrast, our motivation is to use our graphs to model semi-structured data, and to introduce a spatial logic for locally reasoning about such data.

3   The Graph Logic

We will only consider the simple case of graphs without hiding. It is possible to incorporate a quantifier for reasoning about hidden nodes [CC01, CG01b], and we believe that our query language will extend. For the rest of this paper, G ranges over the terms generated by the simple grammar: $G ::=$ nil $\mid$ $a(x, y)$ $\mid$ $G \mid G$. The set $\mathcal{G}(\mathcal{X}, \mathcal{A})$ denotes the set of all such terms.

3.1   Logical Formulae

Formulae are constructed from a name set $\mathcal{X}$ and label set $\mathcal{A}$. They also depend on the disjoint sets of name variables $V_{\mathcal{X}}$, label variables V_A and parametrised recursion variables $V_{\mathcal{R}}$. A recursion variable $\mathbf{R}$ comes with a fixed *arity* $|\mathbf{R}|$.

DEFINITION 4 (LOGICAL FORMULAE)
The set of pre-formulae $\mathcal{F}_{pre}(\mathcal{X}, \mathcal{A})$ is given by the grammars

name expressions	$\xi ::=$	x	name, $\quad x \in \mathcal{X}$
		$\mathbf{x}$	name variable
label expressions	$\alpha ::=$	a	label, $\quad a \in \mathcal{A}$
		$\mathbf{a}$	label variable
formulae	$\phi, \psi ::=$	*nil*	*empty*
		$\alpha(\xi_1, \xi_2)$	*edge*
		$\phi \mid \psi$	*composition*
		true	*true*
		$\phi \wedge \psi$	*conjunction*
		$\neg \phi$	*classical negation*
quantifiers		$\exists \mathbf{x}.\phi$	*exist. quant. over names*
		$\exists \mathbf{a}.\phi$	*exist. quant. over labels*
recursion		$\mathbf{R}(\tilde{\xi})$	$|\mathbf{R}| = |\tilde{\xi}|$
		$(\mu \mathbf{R}(\tilde{\mathbf{x}}).\ \phi)\tilde{\xi}$	*least fix-pt;* $|\tilde{\xi}| = |\tilde{\mathbf{x}}| = |\mathbf{R}|$, $\mathbf{R}(\tilde{\xi})$ *occurs positively,*
equality tests		$\xi_1 = \xi_2, \alpha_1 = \alpha_2$	*equalities*

The sets of free variables are standard. The set of formulae $\mathcal{F}(\mathcal{X}, \mathcal{A})$ are those pre-formulae with no free recursion variables. The order of binding precedence is $\_ = \_,\ \neg \_,\ \_ \mid \_,\ \_ \wedge \_$, with negation binding strongest. We write $x \neq y$ for $\neg(x = y)$. The scope of $\exists \mathbf{x}.\_$ and $\mu \mathbf{R}(\tilde{\mathbf{x}}).\_$ is always the maximum possible.

The nil formula specifies the empty graph. The edge formula $\alpha(\xi_1, \xi_2)$ specifies that a graph is just one edge. The composition formula $\phi \mid \psi$ specifies that a graph can be split into two parts with one part satisfying ϕ and the other ψ. The other formulae should be familiar. It is also logically natural to add other connectives such as a spatial negation and implication [OP99,CG00].

3.2   Satisfaction Relation

The satisfaction relation determines which graphs satisfy which formulae. It is defined by an interpretation function which maps pre-formulae to sets of graphs.

DEFINITION 5 (SATISFACTION)
We assume name set $\mathcal{X}$ and edge set $\mathcal{A}$. Let $\sigma : \mathcal{V}_\mathcal{X} \to \mathcal{X}$ denote a substitution from name and label variables to names and labels respectively, and let ρ send recursion variables of arity n to elements of the set of functions $(\mathcal{X}^n \to \mathcal{P}(\mathcal{G}))$.

The satisfaction interpretation $[\![\_]\!]_{\sigma;\rho} : \mathcal{F}_{pre} \to \mathcal{P}(\mathcal{G})$ is defined inductively by:

$$[\![nil]\!]_{\sigma;\rho} = \{G : G \equiv nil\}$$

$$[\![\alpha(\xi_1, \xi_2)]\!]_{\sigma;\rho} = \{G : G \equiv \alpha\sigma(\xi_1\sigma, \xi_2\sigma)\}$$

$$[\![\phi \,|\, \psi]\!]_{\sigma;\rho} = \{G : G \equiv G_1 \,|\, G_2 \wedge G_1 \in [\![\phi]\!]_{\sigma;\rho} \wedge G_2 \in [\![\psi]\!]_{\sigma;\rho}\}$$

$$[\![true]\!]_{\sigma;\rho} = \mathcal{G}$$

$$[\![\phi \wedge \psi]\!]_{\sigma;\rho} = [\![\phi]\!]_{\sigma;\rho} \cap [\![\psi]\!]_{\sigma;\rho}$$

$$[\![\neg\phi]\!]_{\sigma;\rho} = \mathcal{G}/[\![\phi]\!]_{\sigma;\rho}$$

$$[\![\exists \mathbf{x}.\ \phi]\!]_{\sigma;\rho} = \bigcup_{x \in \mathcal{X}} [\![\phi]\!]_{\sigma, \mathbf{x} \mapsto x; \rho}$$

$$[\![\exists \mathbf{a}.\ \phi]\!]_{\sigma;\rho} = \bigcup_{a \in \mathcal{A}} [\![\phi]\!]_{\sigma, \mathbf{a} \mapsto a; \rho}$$

$$[\![\mathbf{R}(\tilde{\xi})]\!]_{\sigma;\rho} = \mathbf{R}\rho(\tilde{\xi}\sigma), \quad |\tilde{\xi}| = n, \quad \mathbf{R}\rho : \mathcal{X}^n \to \mathcal{P}(\mathcal{G})$$

$$[\![(\mu\mathbf{R}(\tilde{\mathbf{x}}).\phi)\tilde{\xi}]\!]_{\sigma;\rho} = (\textstyle\bigcap\{S \in (\mathcal{X}^{|\tilde{\mathbf{x}}|} \to \mathcal{P}(\mathcal{G})) : (\lambda\tilde{y}.\ [\![\phi]\!]_{\sigma, \tilde{\mathbf{x}} \mapsto \tilde{y}; \rho, R \mapsto S}) \sqsubseteq S\})(\tilde{\xi}\sigma)$$

$$\text{where } S \sqsubseteq S' \text{ iff } \forall \tilde{y} \in \mathcal{X}^{|\tilde{\mathbf{x}}|}.\ S(\tilde{y}) \subseteq S'(\tilde{y})$$

$$[\![\xi_1 = \xi_2]\!]_{\sigma;\rho} = \mathcal{G}, \text{ if } \xi_1\sigma = \xi_2\sigma; \ \emptyset \text{ otherwise}$$

$$[\![\alpha_1 = \alpha_2]\!]_{\sigma;\rho} = \mathcal{G}, \text{ if } \alpha_1\sigma = \alpha_2\sigma; \ \emptyset \text{ otherwise}$$

Definition 5 is shown to be well-defined by structural induction on formulae. For the recursive case, observe that the set of all pointwise-ordered total functions of type $\mathcal{X}^{|\tilde{\mathbf{x}}|} \to \mathcal{P}(\mathcal{G})$ is a complete lattice. Define the *satisfaction relation* $G \vDash^\sigma \phi$ for formula ϕ if and only if $G \in [\![\phi]\!]_{\sigma;\_}$, where $\_$ denotes an arbitrary ρ.

PROPOSITION 6 (SATISFACTION PROPERTIES)
The satisfaction relation satisfies the following standard properties:

$$G \vDash^\sigma nil \Leftrightarrow G \equiv nil$$

$$G \vDash^\sigma \alpha(\xi_1, \xi_2) \Leftrightarrow G \equiv \alpha\sigma(\xi_1\sigma, \xi_2\sigma)$$

$$G \vDash^\sigma \phi \,|\, \psi \Leftrightarrow \exists G_1, G_2 \in \mathcal{G}.\ (G \equiv G_1 \,|\, G_2 \wedge G_1 \vDash^\sigma \phi \wedge G_2 \vDash^\sigma \psi)$$

$$G \vDash^\sigma true \Leftrightarrow G \in \mathcal{G}$$

$$G \vDash^\sigma \phi \wedge \psi \Leftrightarrow G \vDash^\sigma \phi \wedge G \vDash^\sigma \psi$$

$$G \vDash^\sigma \neg\phi \Leftrightarrow \neg(G \vDash^\sigma \phi)$$

$$G \vDash^\sigma \exists \mathbf{x}.\phi \Leftrightarrow \exists x \in \mathcal{X}.\ G \vDash^\sigma \phi\{x/\mathbf{x}\}$$

$$G \vDash^\sigma \exists \mathbf{a}.\phi \Leftrightarrow \exists a \in \mathcal{A}.\ G \vDash^\sigma \phi\{a/\mathbf{a}\}$$

$$G \vDash^\sigma (\mu\mathbf{R}(\tilde{\mathbf{x}}).\ \phi)(\tilde{\xi}) \Leftrightarrow G \vDash^\sigma \phi\{\tilde{\xi}/\tilde{\mathbf{x}}\}[(\mu\mathbf{R}(\tilde{\mathbf{x}}).\phi)/\mathbf{R}]$$

$$G \vDash^\sigma \xi_1 = \xi_2 \Leftrightarrow \xi_1\sigma = \xi_2\sigma$$

$$G \vDash^\sigma \alpha_1 = \alpha_2 \Leftrightarrow \alpha_1\sigma = \alpha_2\sigma$$

The recursion case requires a substitution and monotonicity lemma showing that the function $\lambda\tilde{y}. \ [\![\phi]\!]_{\sigma,\tilde{x}\mapsto\tilde{y};\rho,R\mapsto S}$ is monotone in S. Then we apply the fix-point theorem.

DEFINITION 7 (DERIVED FORMULAE)
We give some derived formulae which are used throughout the paper:

$$false \stackrel{\text{def}}{=} \neg true \qquad\qquad \phi \,||\, \psi \stackrel{\text{def}}{=} \neg(\neg\phi \,|\, \neg\psi)$$

$$\phi \vee \psi \stackrel{\text{def}}{=} \neg(\neg\phi \wedge \neg\psi) \qquad\qquad subgraph_\exists(\phi) \stackrel{\text{def}}{=} \phi \,|\, true$$

$$\phi \Rightarrow \psi \stackrel{\text{def}}{=} \neg\phi \vee \psi \qquad\qquad subgraph_\forall(\phi) \stackrel{\text{def}}{=} \phi \,||\, false$$

$$\forall \mathbf{x}.\ \phi \stackrel{\text{def}}{=} \neg\exists \mathbf{x}.\neg\phi$$

The connective $\_||\_$ is the de Morgan dual of $\_|\_$. The binding precedence is $\_\wedge\_$, $\_\vee\_,\_\Rightarrow\_$, with conjunction binding strongest. The scope of $\forall\mathbf{x}.\_$ is the maximum possible.

Example. We revisit the two examples discussed in the introduction:

$$\exists \mathbf{x}, \mathbf{y}, \mathbf{z}, \mathbf{u}.\quad a(\mathbf{x}, \mathbf{y}) \,|\, b(\mathbf{y}, \mathbf{z}) \,|\, a(\mathbf{z}, \mathbf{u}) \,|\, true$$

$$\exists \mathbf{x}, \mathbf{y}, \mathbf{z}, \mathbf{u}.\quad (a(\mathbf{x}, \mathbf{y}) \,|\, true) \wedge (b(\mathbf{y}, \mathbf{z}) \,|\, true) \wedge (a(\mathbf{z}, \mathbf{u}) \,|\, true)$$

Recall that the first formula specifies that a graph has at least three different edges; the second that a graph has a path of three edges.

Example. We specify the property that there exists a path from x to y in our logic without recursion. This is interesting since it is not expressible in first-order logic without recursion. First we give some preliminary derived formulae:

no edge into x $\qquad in_0(x) \stackrel{\text{def}}{=} \neg\exists \mathbf{y}, \mathbf{a}.\ \mathbf{a}(\mathbf{y}, x) \,|\, true$

$n+1$ edges into x $\qquad in_{n+1}(x) \stackrel{\text{def}}{=} \exists \mathbf{y}, \mathbf{a}.\ \mathbf{a}(\mathbf{y}, x) \,|\, in_n(x)$

a minimal graph satisfying ϕ $\qquad min(\phi) \stackrel{\text{def}}{=} \phi \wedge \neg(\phi \,|\, \neg nil)$

x is a node in the graph $\qquad in\_graph(x) \stackrel{\text{def}}{=} \exists \mathbf{y}, \mathbf{a}.\ (\mathbf{a}(x, \mathbf{y}) \vee \mathbf{a}(\mathbf{y}, x)) \,|\, true$

The formulae $out_n(x)$ are defined similarly to $in_n(x)$. We now give a formula which specifies that a graph is just a straight path from x to y and does not contain a cycle (when $x = y$ the formula is satisfied by the empty graph):

$$straight\_path(\mathbf{x}, \mathbf{y}) \stackrel{\text{def}}{=} min[\mathbf{x} = \mathbf{y} \vee (in_0(\mathbf{x}) \wedge out_1(\mathbf{x}) \wedge in_1(\mathbf{y}) \wedge out_0(\mathbf{y}) \wedge$$
$$\forall \mathbf{z}.\ \mathbf{z} \neq \mathbf{x} \wedge \mathbf{z} \neq \mathbf{y} \wedge in\_graph(\mathbf{z}) \Rightarrow out_1(\mathbf{z}) \wedge in_1(\mathbf{z}))]$$

This formula specifies that the graph contains one start node $\mathbf{x}$, one end node $\mathbf{y}$ and all the other nodes must have one incoming and one outgoing edge (hence no cycles). Minimality ensures that there are no disconnected cycles. The property that there exists a path from x to y is now specified by the formula $exists\_path(\mathbf{x}, \mathbf{y}) \stackrel{\text{def}}{=} subgraph_\exists(straight\_path(\mathbf{x}, \mathbf{y}))$.

Example. We give an equivalent formula to $exists\_path(x, y)$ using recursion—we use the notation $\mathbf{R}(\tilde{\mathbf{x}}) \stackrel{\text{def}}{=} \phi$, as an abbreviation for $\mathbf{R}(\tilde{\xi}) \stackrel{\text{def}}{=} (\mu \mathbf{R}(\tilde{\mathbf{x}}).\ \phi)(\tilde{\xi})$:

$$\mathbf{exists\_path}(\mathbf{x}, \mathbf{y}) \stackrel{\text{def}}{=} \mathbf{x} = \mathbf{y} \vee (\exists \mathbf{z}, \mathbf{a}.\ \mathbf{a}(\mathbf{x}, \mathbf{z}) \,|\, \mathbf{exists\_path}(\mathbf{z}, \mathbf{y})).$$

This combination of composition and recursion can be regarded as an induction on the graph structure. Consider the graph $a(x, z) \mid b(z, z) \mid c(z, y)$. There are just two ways to check that this graph satisfies the formula: either by checking that edge a is followed by c; or that a is followed by b is followed by c.

Example. A classic property associated with compiler optimisation is 'a node z *dominates* node y iff every path from some declared initial node x to y passes through z'. First we specify the property that a graph *is* a path from x to y:

$$\mathbf{path(x, y)} \stackrel{\text{def}}{=} (\mathbf{x = y} \wedge \mathsf{nil}) \vee (\exists \mathbf{z}, \mathbf{a}. \, \mathbf{a(x, z)} \mid \mathbf{path(z, y)})$$

The addition of nil ensures that *all* the edges are checked. For example, in graph $a(x, z) \mid b(z, z) \mid c(z, y)$ the only way that $\mathbf{path(x, y)}$ is satisfied is by checking that a follows b follows c. It is now simple to specify the property we seek:

$$\mathsf{dominates}(x, y, z) \stackrel{\text{def}}{=} \mathsf{subgraph}_\forall(\mathbf{path}(x, y) \Rightarrow \mathsf{in\_graph}(z)).$$

4   A Query Language

Our basic language consists of *queries* and *transducers*. Queries build new graphs from old. Transducers associate input graphs with output graphs. These concepts are related. The basic transducer $\phi \Rightarrow Q$ relates input graphs satisfying ϕ with output graphs given by Q. The query (**apply** τ **to** Q) applies the transducer τ to the input graphs given by Q, to yield the corresponding set of output graphs.

DEFINITION 8 (QUERY LANGUAGE)
The sets of pre-queries *and* pre-transducers, *denoted* $\mathcal{Q}_{pre}(\mathcal{X}, \mathcal{A})$ *and* $\mathcal{T}_{pre}(\mathcal{X}, \mathcal{A})$ *respectively, are given by the grammars from definition 4 and the grammars:*

$Q ::=$	queries	$\tau ::=$	transducers
$\mathbf{G}$	graph variable	$\phi \Rightarrow Q$	basic transducer
nil	empty graph	$\lambda \mathbf{G}.Q$	abstraction
$\alpha(\xi_1, \xi_2)$	edge graph	$\tau \mid \tau$	transducer composition
$Q \mid Q$	composition	$\tau \vee \tau$	disjunction
apply τ **to** Q	application	$\exists \mathbf{x}.\tau$	exist. quant. of names
		$\exists \mathbf{a}.\tau$	exist. quant. of labels
		$\mathbf{R_T}$	recursion
		$\mu \mathbf{R_T}.\tau$	least fix-pt, $\mathbf{R_T}$ positive

The sets of *queries* and *transducers*, denoted by $\mathcal{Q}(\mathcal{X}, \mathcal{A})$ and $\mathcal{T}(\mathcal{X}, \mathcal{A})$, contain those pre-queries and pre-transducers with no free recursion variables. We use $\mathbf{R_T} \stackrel{\text{def}}{=} \tau$ to denote $\mathbf{R_T} \stackrel{\text{def}}{=} \mu \mathbf{R_T}. \tau$. We overload notation: $\_\mid\_$ denotes the composition of formulae, queries and transducers. The connective $\_ \Rightarrow \_$ has the weakest binding strength; the other connectives are as before. A glaring omission is the absence of a renaming technique for node identifiers, such as Skolemization. Our approach is enough for this paper. Other transducer connectives are feasible. Our choice was determined by our aim to have a simple language in which to express our motivating examples. We describe a more general approach in section 4.1.

DEFINITION 9 (QUERY INTERPRETATION)
Assume name set $\mathcal{X}$ and label set $\mathcal{A}$. Let σ denote a substitution from name and label variables to names and labels respectively, let δ denote a substitution from graph variables to elements of $\mathcal{G}$, and let function ρ map transducer recursion variables to the set $\mathcal{P}(\mathcal{G} \times \mathcal{G})$. The query interpretation $[\![\_]\!]_{\sigma;\tau;\rho} : \mathcal{Q}_{pre} \to \mathcal{P}(\mathcal{G})$ and the transducer interpretation $[\![\_]\!]_{\sigma;\rho;\tau} : \mathcal{T}_{pre} \to \mathcal{P}(\mathcal{G} \times \mathcal{G})$, are defined by a simultaneous induction on the structure of pre-queries and pre-transducers:

$$[\![\mathbf{G}]\!]_{\sigma;\delta;\rho} = \{G : G \equiv \mathbf{G}\delta\}$$
$$[\![nil]\!]_{\sigma;\delta;\rho} = \{G : G \equiv nil\}$$
$$[\![\alpha(\xi_1,\xi_2)]\!]_{\sigma;\delta;\rho} = \{G : G \equiv \alpha\sigma(\xi_1\sigma,\xi_2\sigma)\}$$
$$[\![Q_1|Q_2]\!]_{\sigma;\delta;\rho} = \{G : G \equiv G_1 \,|\, G_2 \wedge G_1 \in [\![Q_1]\!]_{\sigma;\delta;\rho} \wedge G_2 \in [\![Q_2]\!]_{\sigma;\delta;\rho}\}$$
$$[\![\mathbf{apply}\ \tau\ \mathbf{to}\ Q]\!]_{\sigma;\delta;\rho} = \{G' : \exists G.\ (G,G') \in [\![\tau]\!]_{\sigma;\delta;\rho} \wedge G \in [\![Q]\!]_{\sigma;\delta;\rho}\}$$
$$[\![\phi \Rightarrow Q]\!]_{\sigma;\delta;\rho} = \{(G,G') : G \in [\![\phi]\!]_{\sigma;\_} \wedge G' \in [\![Q]\!]_{\sigma;\delta;\rho}\}$$
$$[\![\lambda\mathbf{G}.Q]\!]_{\sigma;\delta;\rho} = \{(G,G') : G' \in [\![Q]\!]_{\sigma;\delta,\mathbf{G} \mapsto G;\rho}\}$$
$$[\![\tau_1 \,|\, \tau_2]\!]_{\sigma;\delta;\rho} =$$
$$\{(G,G') : G \equiv G_1|G_2 \wedge G' \equiv G'_1|G'_2 \wedge (G_1,G'_1) \in [\![\tau_1]\!]_{\sigma;\delta;\rho} \wedge (G_2,G'_2) \in [\![\tau_2]\!]_{\sigma;\delta;\rho}\}$$
$$[\![\tau_1 \vee \tau_2]\!]_{\sigma;\delta;\rho} = [\![\tau_1]\!]_{\sigma;\delta;\rho} \cup [\![\tau_2]\!]_{\sigma;\delta;\rho}$$
$$[\![\exists\mathbf{x}.\tau]\!]_{\sigma;\delta;\rho} = \bigcup_{x\in\mathcal{X}}[\![\tau]\!]_{\sigma,\mathbf{x}\mapsto x;\delta;\rho}$$
$$[\![\exists\mathbf{a}.\tau]\!]_{\sigma;\delta;\rho} = \bigcup_{a\in\mathcal{A}}[\![\tau]\!]_{\sigma,\mathbf{a}\mapsto a;\delta;\rho}$$
$$[\![\mathbf{R_T}]\!]_{\sigma;\delta;\rho} = \mathbf{R_T}\rho$$
$$[\![\mu\mathbf{R_T}.\phi]\!]_{\sigma;\delta;\rho} = \bigcap\{S \in \mathcal{P}(\mathcal{G} \times \mathcal{G}) : [\![\phi]\!]_{\sigma;\delta;\rho,\mathbf{R_T}\mapsto S} \subseteq S\}$$

Example: inverting edges. Consider the transducer

$$\exists \mathbf{a}, \mathbf{x}, \mathbf{y}.\ \mathbf{a}(\mathbf{x},\mathbf{y}) \,|\, \mathsf{true} \Rightarrow \mathbf{a}(\mathbf{y},\mathbf{x})$$

It returns one inverted edge of any non-empty input graph. The transducer is *non-deterministic*: given input graph $a(x,y) \,|\, b(y,x)$, the set of possible output graphs is $\{a(y,x), b(x,y)\}$. Now consider the query

$$\mathbf{apply}\ (\exists \mathbf{a}, \mathbf{x}, \mathbf{y}.\ \mathbf{a}(\mathbf{x},\mathbf{y}) \,|\, \mathsf{true} \Rightarrow \mathbf{a}(\mathbf{y},\mathbf{x}))\ \mathbf{to}\ \mathsf{input\_graph}$$

When the input graph is $a(x,y) \,|\, b(y,x)$ the resulting output is either $a(y,x)$ or $b(x,y)$; when the input graph is $a(x,y) \,|\, a(x,y)$ the result can only be $a(y,x)$.

Example: case analysis. The connective $\_ \vee \_$ can be used for case analysis:

$$(nil \Rightarrow nil) \vee (\exists \mathbf{a}, \mathbf{x}, \mathbf{y}.\ \mathbf{a}(\mathbf{x},\mathbf{y}) \,|\, \mathsf{true} \Rightarrow \mathbf{a}(\mathbf{y},\mathbf{x}))$$

Either the input graph is empty and we return the empty output graph. Or the input graph is non-empty and we return an inverted edge.

Example: exact inverted copy. We can execute a query against every edge. For example, the transducer relating an input graph with its inverted copy is

$$\mathbf{R_T} \stackrel{\text{def}}{=} (\text{nil} \Rightarrow \text{nil}) \vee (\exists \mathbf{a}, \mathbf{x}, \mathbf{y}.\ \mathbf{a}(\mathbf{x}, \mathbf{y}) \Rightarrow \mathbf{a}(\mathbf{y}, \mathbf{x})) \,|\, \mathbf{R_T}$$

Either the input graph is empty and we return the empty graph. Or the graph can be split into an edge and the rest of the graph. We return the inverted edge and execute the transducer on the smaller graph. Given the input graph $a(x, y) \,|\, a(x, y)$, we return the exact inverted copy.

We can adapt this example to execute a query against every edge provided it satisfies a certain logical formula. For example, consider the transducer

$$\mathbf{R_T} \stackrel{\text{def}}{=} (\text{nil} \Rightarrow \text{nil}) \vee$$
$$(\exists \mathbf{a}, \mathbf{x}, \mathbf{y}.\ ((\mathbf{a}(\mathbf{x}, \mathbf{y}) \wedge \mathbf{x} \neq \mathbf{y} \Rightarrow \mathbf{a}(\mathbf{y}, \mathbf{x})) \vee (\mathbf{a}(\mathbf{x}, \mathbf{y}) \wedge \mathbf{x} = \mathbf{y} \Rightarrow \text{nil})) \,|\, \mathbf{R_T})$$

Either the input graph is empty and we return the empty graph. Or the input graph is non-empty and we pick an edge. If the domain and codomain of the edge are different then return the inverted edge; if they are the same then return the empty graph. Apply the transducer to the remaining smaller graph.

Example: transitive closure. A standard example is the *transitive closure* of a graph. It illustrates the power of mixing abstraction with recursion. For this example only, we assume the edge labelled set $A = \{a\}$. The following transducer, when applied to graph G, returns the minimum graph TC which contains G and satisfies the property: if $a(x, y)$ and $a(y, z)$ are in TC then so is $a(x, z)$:

$$\mathbf{R_T} \stackrel{\text{def}}{=} \lambda \mathbf{G}.\ (\neg \exists \mathbf{x}, \mathbf{y}, \mathbf{z}.\ (a(\mathbf{x}, \mathbf{y}) \,|\, \text{true} \wedge a(\mathbf{y}, \mathbf{z}) \,|\, \text{true} \wedge \neg(a(\mathbf{x}, \mathbf{y}) \,|\, \text{true})) \Rightarrow \mathbf{G}) \vee$$
$$\exists \mathbf{x}\, \mathbf{y}, \mathbf{z}.\ a(\mathbf{x}, \mathbf{y}) \,|\, \text{true} \wedge a(\mathbf{y}, \mathbf{z}) \,|\, \text{true} \wedge \neg(a(\mathbf{x}, \mathbf{z}) \,|\, \text{true}) \Rightarrow \mathbf{apply}\ \mathbf{R_T}\ \mathbf{to}\ (\mathbf{G} \,|\, a(\mathbf{x}, \mathbf{z}))$$

4.1   Generalised Transducers

We generalise the definition of transducers (definition 8). Our approach is simple, but too expressive to implement. The semantic interpretation (definition 11) gives us the flexibility to adapt our choice of basic language if we wish.

DEFINITION 10 (GENERALISED TRANSDUCERS)
Assume name set $\mathcal{X}$ and label set $\mathcal{A}$. The set of generalised pre-transducers, denoted $\mathcal{GT}_{pre}(\mathcal{X}, \mathcal{A})$, is given by the grammar:

$\tau ::=$ *id*	*identity*	*nil*	*empty input graph*
$\tau_1; \tau_2$	*composition*	$\ldots$	*analogous cases from definition 4*
G	*graph variable*	$\exists \mathbf{G}.\tau$	*existential quantification over graphs*

Generalised transducers relate input and output graphs. A logical formula ϕ regarded as a generalised transducer relates input graphs satisfying ϕ to arbitrary output graphs. The identity transducer relates structurally congruent graphs. The transducer composition $\tau_1; \tau_2$ is relational composition. Identity and composition allows us to specify properties of the output graphs. For example, the transducer $\mathbf{true}; (\phi \wedge \mathsf{id})$ relates arbitrary input graphs with output graphs satisfying ϕ. Queries correspond to such generalised transducers.

DEFINITION 11 (INTERPRETATION OF GENERALISED TRANSDUCERS)
Assume name set $\mathcal{X}$ and label set $\mathcal{A}$. The query interpretation $[\![\_]\!]_{\sigma;\delta;\rho} : \mathcal{GT} \to \mathcal{P}(\mathcal{G} \times \mathcal{G})$, where σ denotes a substitution from name and label variables to names and labels respectively, δ maps graph variables to graphs, and function ρ maps recursion variables of arity n to functions $\mathcal{X}^n \to \mathcal{R}(\mathcal{G} \times \mathcal{G})$, is defined by induction on the structure of the extended formulae:

$$[\![id]\!]_{\sigma;\delta;\rho} = \{(G, G') : G \equiv G'\}$$
$$[\![\tau_1; \tau_2]\!]_{\sigma;\delta;\rho} = \{(G, G') : \exists G_1. (G, G_1) \in [\![\tau_1]\!]_{\sigma;\delta;\rho} \wedge (G_1, G') \in [\![\tau_2]\!]_{\sigma;\delta;\rho}\}$$
$$[\![\mathbf{G}]\!]_{\sigma;\delta;\rho} = \{G : \mathbf{G}\delta = G\} \times \mathcal{G}$$
$$[\![nil]\!]_{\sigma;\delta;\rho} = \{G : G \equiv nil\} \times \mathcal{G}$$
$$[\![\alpha(\xi_1, \xi_2)]\!]_{\sigma;\delta;\rho} = \{G : G \equiv \alpha\sigma(\xi_1\sigma, \xi_2\sigma)\} \times \mathcal{G}$$
$$[\![\tau_1 \mid \tau_2]\!]_{\sigma;\delta;\rho} =$$
$$\{(G, G') : G \equiv G_1|G_2 \wedge G' \equiv G'_1 \mid G'_2 \wedge (G_1, G'_1) \in [\![\tau_1]\!]_{\sigma;\delta;\rho} \wedge (G_2, G'_2) \in [\![\tau_2]\!]_{\sigma;\delta;\rho}\}$$
$$[\![\mathbf{true}]\!]_{\sigma;\delta;\rho} = \mathcal{G} \times \mathcal{G}$$
$$[\![\tau_1 \wedge \tau_2]\!]_{\sigma;\delta;\rho} = [\![\tau_1]\!]_{\sigma;\delta;\rho} \cap [\![\tau_2]\!]_{\sigma;\delta;\rho}$$
$$[\![\neg\tau]\!]_{\sigma;\delta;\rho} = (\mathcal{G} \times \mathcal{G}) \setminus [\![\tau]\!]_{\sigma;\delta;\rho}$$
$$[\![\exists\mathbf{x}.\tau]\!]_{\sigma;\delta;\rho} = \bigcup_{x \in \mathcal{X}} [\![\tau]\!]_{\sigma,\mathbf{x} \mapsto x;\delta;\rho}$$
$$[\![\exists\mathbf{a}.\tau]\!]_{\sigma;\delta;\rho} = \bigcup_{a \in \mathcal{A}} [\![\tau]\!]_{\sigma,\mathbf{a} \mapsto a;\delta;\rho}$$
$$[\![\exists\mathbf{G}.\tau]\!]_{\sigma;\delta;\rho} = \bigcup_{G \in \mathcal{G}} [\![\tau]\!]_{\sigma,\delta,\mathbf{G} \mapsto G;\rho}$$
$$[\![\mathbf{R}(\tilde{\xi})]\!]_{\sigma;\delta;\rho} = \mathbf{R}\rho(\tilde{\xi}\sigma)$$
$$[\![(\mu\mathbf{R}(\tilde{\mathbf{x}}).\tau)(\tilde{\xi})]\!]_{\sigma;\delta;\rho} = (\sqcap\{S \in \mathcal{X}^{|\tilde{x}|} \to \mathcal{P}(\mathcal{G} \times \mathcal{G}) : \lambda\tilde{y}.[\![\tau]\!]_{\sigma,\tilde{\mathbf{x}} \mapsto \tilde{y};\delta;\rho,\mathbf{R} \mapsto S} \sqsubseteq S\})(\tilde{\xi}\sigma)$$
$$\text{where } S \sqsubseteq S' \text{ iff } \forall \tilde{y} \in \mathcal{X}^{|\tilde{x}|}. S(\tilde{y}) \subseteq S'(\tilde{y})$$
$$[\![\xi_1 = \xi_2]\!]_{\sigma;\rho} = \mathcal{G} \times \mathcal{G} \text{ if } \xi_1\sigma = \xi_2\sigma; \quad \emptyset \text{ otherwise}$$
$$[\![\alpha_1 = \alpha_2]\!]_{\sigma;\rho} = \mathcal{G} \times \mathcal{G} \text{ if } \alpha_1\sigma = \alpha_2\sigma; \quad \emptyset \text{ otherwise}$$

PROPOSITION 12
There exists embeddings $(\_)^\circ : \mathcal{Q}_{pre} \to \mathcal{GT}_{pre}$, $(\_)^\circ : \mathcal{F} \to \mathcal{GT}_{pre}$ and $(\_)^\circ : \mathcal{T}_{pre} \to \mathcal{GT}_{pre}$ such that

1. for all queries Q, $[\![Q^\circ]\!]_{\sigma;\delta;\rho} = \mathcal{G} \times [\![Q]\!]_{\sigma;\delta;\rho}$;
2. for all logical formulae ϕ, $[\![\phi^\circ]\!]_{\sigma;\delta;\rho} = [\![\phi]\!]_{\sigma;\_} \times \mathcal{G}$;
3. for all basic transducers τ, $[\![\tau^\circ]\!]_{\sigma;\delta;\rho} = [\![\tau]\!]_{\sigma;\delta;\rho}$.

Proof. The embeddings are give in [CGG01]. The query (**apply** τ **to** Q) is interpreted by the sequential composition. The basic transducer $\phi \Rightarrow Q$ is interpreted by conjunction. The abstraction $\lambda\mathbf{G}. Q$ by the existential quantification on $\mathbf{G}$.

Example. Consider the derived transducers:

$$\text{subgraph} \stackrel{\text{def}}{=} \text{id} \mid (\text{nil} \Rightarrow \text{true}) \qquad\qquad \text{strict\_subgraph} \stackrel{\text{def}}{=} \text{id} \mid (\text{nil} \Rightarrow \neg \text{nil})$$

$$\tau_1;;\tau_2 \stackrel{\text{def}}{=} \neg(\tau_1;\neg\tau_2) \qquad\qquad \text{min\_out}\,(\tau) \stackrel{\text{def}}{=} \tau \wedge \neg(\tau;\text{strict\_subgraph})$$

$$\text{finite\_lub}\,(\tau) \stackrel{\text{def}}{=} \text{min\_out}(\tau;;\text{subgraph})$$

The transducer subgraph relates G_1 to G_2 if and only if $G_1 \subseteq G_2$: that is, $G_1 \mid H \equiv G_2$ for some H. The strict_subgraph is the strict version. The connective ;; is the de Morgan dual of ;. Unravelling the definition, it states that

$$(G, G') \in [\![\tau_1;;\tau_2]\!]_{\sigma;\delta;\rho} \Leftrightarrow (\forall G_1.\,(G, G_1) \in [\![\tau_1]\!]_{\sigma;\delta;\rho} \Rightarrow (G_1, G') \in [\![\tau_2]\!]_{\sigma;\delta;\rho})$$

This operator allows us to work with *all* output graphs associated with a given input. For example, the transducer $\tau;;$ subgraph relates a graph G with all the finite upper bounds of $[\![\tau]\!](G)$ (where $[\![\tau]\!](G)$ is the set of all graphs G' such that $(G, G') \in [\![\tau]\!]$). These finite upper bounds do not necessarily exist, in which case $[\![\tau;;\text{subgraph}]\!](G)$ is the empty set. We may adapt our finite semantics to the infinite case, by using the infinite version of the set-theoretic presentation given in section 2.1. The $\text{min\_out}(\tau)$ transducer relates a graph G with the minimal graphs in $[\![\tau]\!](G)$. The transducer finite_lub (τ) relates a graph G with the minimal finite upper bound of $[\![\tau]\!](G)$, when it exists. The infinite semantics would give rise to a least upper bound. In the introduction, we discuss a standard set-theoretic language based on from/select expressions. These expressions are embeddable in our general language using this finite-lub construction [CGG01].

We must give an in-depth comparison between our query language and other query languages based on graphs [FFK+97,CM90,BDHS96]. Our language is closely related to XDuce [HP01], a processing language for XML documents based on pattern-matching and a simple typing scheme analogous to the structural component of our spatial logic. Our ambitious aim is to achieve a level of understanding of query languages for semi-structured data which rivals that of languages associated with the relational model.

References

[ABS00]   S. Abiteboul, P. Buneman, and D. Suciu. *Data on the Web*. Morgan Kaufmann, 2000.

[BDHS96]   P. Buneman, S. Davidson, G. Hillebrand, and D. Suciu. A query language and optimization techniques for unstructured data. In *SIGMOD*, LNCS 2044, pages 505–515, 1996.

[Cai99]   L. Caires. *A Model for Declarative Programming and Specification with Concurrency and Mobility*. PhD thesis, University of Lisbon, 1999.

[CC01]   L. Caires and L. Cardelli. A spatial logic for concurrency (part 1). In *TACS*, LNCS 2215. Springer, 2001. Journal paper to be in Information and Comp.

[CG00]   L. Cardelli and A. Gordon. Anytime, anywhere: Modal logics for mobile ambients. In *POPL*. ACM, 2000.

[CG01a]   L. Cardelli and G. Ghelli. A query language based on the ambient logic. In *ESOP/ETAPS*, LNCS 2028. Springer, 2001.

[CG01b]   L. Cardelli and A. Gordon. Logical properties of name restriction. In *TLCA*, LNCS 2044. Springer, 2001.

[CGG01]   L. Cardelli, P. Gardner, and G. Ghelli. A spatial logic for querying graphs. Fuller version found at http://www.doc.ic.ac.uk/˜pg, 2001.

[CM90]   M. Consens and A. Mendelzon. Graphlog: a visual formalism for real life recursion. In *Principles of Database Systems*, pages 404–416. ACM, 1990.

[CMR94]   A. Corradini, U. Montanari, and F. Rossi. An abstract machine for concurrent modular systems: Charm. *TCS*, 122:165–200, 1994.

[Cou97]   Bruno Courcelle. The expression of graph properties and graph transformations in monadic second-order logic. *Graph grammars and computing by graph transformations*, 1:313–400, 1997.

[FFK+97]   M. Fernandez, D. Florescu, J. Kang, A. Levy, and D. Suciu. Strudel: A web-site management system. In *SIGMOD Management of Data*, 1997.

[GW00]   P. Gardner and L. Wischik. Explicit fusions. *MFCS*, LNCS 1893, 2000. Journal version submitted to Theoretical Computer Science.

[HP01]   H. Hosoya and B. Pierce. Regular expression pattern matching for xml. In *POPL*. ACM, 2001.

[IO01]   S. Ishtiaq and P. O'Hearn. Bi as an assertion language for mutable data structures. In *POPL*, 664. ACM, 2001.

[OP99]   P. O'Hearn and D. Pym. The logic of bunched implications. *Bulletin of Symbolic Logic*, 5(2):215–244, 1999.

[Rey00]   J.C. Reynolds. Intuitionistic reasoning about shared mutable data structure. *Millenial Perspectives in Computer Science*, Palgrove, 2000.

Improving Time Bounds on Maximum Generalised Flow Computations by Contracting the Network

Tomasz Radzik*

Department of Computer Science
King's College London
London WC2R 2LS, United Kingdom
radzik@dcs.kcl.ac.uk

Abstract. We consider the maximum generalised network flow problem and a supply-scaling algorithmic framework for this problem. We present three network-modification operations, which may significantly decrease the size of the network when the remaining node supplies become small. We use these three operations in Goldfarb, Jin and Orlin's supply-scaling algorithm and prove a $\tilde{O}(m^2 n \log B)$ bound on the running time of the resulting algorithm. The previous best time bounds on computing maximum generalised flows were the $O(m^{1.5} n^2 \log B)$ bound of Kapoor and Vaidya's algorithm based on the interior-point method, and the $\tilde{O}(m^3 \log B)$ bound of Goldfarb, Jin and Orlin's algorithm.

1 Introduction

In a generalised flow network, each arc e has a gain factor $\gamma(e)$ associated with it, and if x units of flow enter e, then $\gamma(e)x$ units arrive at the other end. Each node has specified *supply* of one common commodity. The objective of the *maximum generalised flow problem* is to design flow which carries these node supplies through the network to one distinguished node, the *sink*. The designed flow must maximise the amount of commodity arriving at the sink and cannot violate the capacities of arcs. This problem models some optimisation problems arising in manufacturing, transportation and financial analysis [1,2,3].

The maximum generalised flow problem is a special case of *linear programming*, so it can be solved by any general-purpose linear programming method. The best asymptotic time bound on computing maximum generalised flows using this approach is the $O(m^{1.5} n^2 \log B)$ bound of Kapoor and Vaidya's algorithm [4, 5] based on Karmarkar's interior-point method. Here n is the number of nodes, m is the number of arcs, and B is the largest integer in the representations of the capacities and gain factors of arcs and the supplies at nodes, assuming that these numbers are given as ratios of two integers.

---

* This work was supported by the UK EPSRC grant GR/L81468.

P. Widmayer et al. (Eds.): ICALP 2002, LNCS 2380, pp. 611–622, 2002.

The other line of research in designing generalised-flow algorithms follows the *combinatorial approach* to network flow problems originated by Ford and Fulkerson [6]. A combinatorial algorithm for the maximum generalised flow problem exploits the combinatorial structures of the underlying network and of the flows in this network, and often uses as subroutines combinatorial algorithms for simpler network problems, such as the shortest paths problem and the maximum (non-generalised) flow problem. The first polynomial-time bound on computing maximum generalised flows by combinatorial algorithms was shown by Goldberg, Plotkin, and Tardos [7], and the best, prior to our paper, bound of this type is $\tilde{O}(m^3 \log B)$,[1] due to Goldfarb, Jin and Orlin [8].

Kapoor and Vaidya's algorithm and Goldfarb, Jin and Orlin's algorithm give the previous best asymptotic time bounds on computing maximum generalised flows. The main conclusion of our paper is a combinatorial algorithm which computes maximum generalised flows in $\tilde{O}(m^2 n \log B)$ time. This bound improves the previous bounds if $m = \Omega(n^{1+\epsilon})$ and $m = O(n^{2-\epsilon})$, for any constant $\epsilon > 0$.

Goldfarb, Jin and Orlin [8] two algorithms which have the following *supply-scaling* structure. The computation consists of scaling phases. During the current phase, the remaining node supplies are sent in chunks of Δ units towards the sink along the highest gain paths. The scaling parameter Δ decreases at least by half at the end of each phase. Each phase has $O(m)$ iterations and each iteration is dominated by one single-source shortest-path computation. We present in this paper three network-modification operations, which are intended to decrease the size of the network during the computation of a supply-scaling algorithm. If the network does become smaller, then the subsequent phases may run faster. The first operation is a standard operation of contracting two nodes, if there is enough arc capacity between them (in both direction) to accommodate all remaining node supplies. The other two operations are an operation of by-passing (and removing) some nodes and an operation of shortcutting some paths. We believe that in practice these operations may significantly decrease the size of the network and speed-up the computation, but in this paper we focus on the question if they can lead to improved asymptotic time bounds.

The computation of Goldfarb, Jin and Orlin's algorithms [8] terminates when the remaining node supplies total to less than B^{-m}, and an optimal flow is obtained by simple post-processing. The node supplies drop below B^{-m} in $O(m \log B)$ phases, so the total running time is $\tilde{O}(m^3 \log B)$. We show that the size of the network must decrease when the remaining node supplies become $B^{-\Omega(n)}$. More precisely, but abstracting from technical details, we show that if the node supplies are $B^{-\Omega(kn)}$, then our network-modification operations reduce the number of arcs to $O(m/k^2)$. This decreases the bound on the total running time of all $O(m \log B)$ phases by factor m/n to our new bound of $\tilde{O}(m^2 n \log B)$.

An essential tool in our analysis is a simple fact that the value of an expression composed of additions and multiplications of k fractional numbers with denominators bounded by B is a fractional number with the denominator bounded by B^k. The absolute value of such a number is either 0 or greater than B^{-k}. We

---

[1] Notation $\tilde{O}()$ hides a factor polylogarithmic in n.

use this fact in the following way. If the remaining supply at some node is still positive but less than B^{-k}, then $\Omega(k)$ arcs must "contribute" to the value of this remaining supply. We show a relation between the number of these contributing arcs and the decrement of the size of the network.

2 Definitions

A *generalised flow network* $G = (V, E, t, \gamma, u, \delta)$ consists of: a set of nodes V; a set of (directed) arcs E; a *sink* node $t \in V$; a *gain function* $\gamma : E \longrightarrow (0, \infty)$ ($\gamma(e)$ is the *gain factor* of arc e); a *capacity function* $u : E \longrightarrow [0, \infty]$; and a node *supply function* $\delta : V \setminus \{t\} \longrightarrow [0, \infty)$ ($\delta(v)$ is the *initial supply* at node v).

There may be multiple arcs in G, and we will normally denote an arc from a node v to a node w by $e_{v,w}$. We assume, without loss of generality, that all arcs are matched into pairs of reverse arcs. If an arc e' goes from a node $v \in V$ to a node $w \in V$, then its reverse arc e'' goes from w to v and $\gamma(e'') = 1/\gamma(e')$. Let E_v^- and E_v^+ denote the sets of arcs outgoing from v and incoming to v, respectively. If there is an arc from a node v to a node w, then we call the pair $\{v, w\}$ an *edge*. We denote the set of all edges in G by $\overline{E}$. The gain factors are sometimes called loss/gain factors. For a path or a cycle P in G, the gain factor of P is equal to $\gamma(P) = \Pi\{\gamma(e) : e \in P\}$. A cycle P of positive-capacity arcs is called a *flow-generating cycle* or a *flow-absorbing cycle* or a *1-gain cycle*, if $\gamma(P)$ is greater, less, or equal to 1, respectively. If network G does not contain a flow generating cycle, then we call it a *non-gain network*.

Let n and m denote the number of nodes and the number of arcs in network G. The arc gain factors and capacities and the node supplies are given as ratios of integers. We denote the set of these fractional input numbers by D_{frac}, and the largest integer among the enumerators and denominators in D_{frac} by B. To simplify asymptotic time bounds, we assume that $B \geq n$. We also assume in this extended abstract that, despite allowing multiple arcs, $m \leq n^2$. See [9] for the technical details covering the case when $m > n^2$.

A *(generalised) flow* $f : E \longrightarrow (-\infty, +\infty)$ satisfies the following conditions.

1. *Skew symmetry:* for each pair of reverse arcs e' and e'', $f(e'') = -\gamma(e')f(e')$.
2. *Capacity constraint:* for each arc $e \in E$, $f(e) \leq u(e)$.
3. *Flow conservation:* for each node $v \in V \setminus \{t\}$, $\sum_{e \in E_v^-} f(e) \leq \delta(v)$.

If $f(e_{v,w})$ units of flow enter arc $e_{v,w}$ at node v, then $\gamma(e_{v,w})f(e_{v,w})$ units arrive at node w. The actual flow is defined by those flow values $f(e)$ which are positive, while the negative flow values on the reverse arcs (Condition 1) are only for notational convenience. The sum in Condition 3 is the *net-flow outgoing* from node v. The *value of a flow* f is the net-flow into the sink t. The *maximum generalised network flow problem* is to compute a flow in a given network G with the maximum possible value. Such a flow is a *maximum flow* or an *optimal flow*.

For a flow f, the *residual capacity* of an arc e is $u_f(e) = u(e) - f(e)$, and the *residual supply* at a node v is $\delta_f(v) = \delta(v) - \sum_{e \in E_v^-} f(e)$. The *residual network* of G with respect to a flow f is the network $G_f = (V, E, t, \gamma, u_f, \delta_f)$. An optimal flow f'_{opt} in the residual network G_f gives an optimal flow $f''_{\text{opt}} = f + f'_{\text{opt}}$ in G.

A *labeling* in G is a function $\mu : V \longrightarrow (0, \infty)$, $\mu(t) = 1$. The *re-labeled network* $G_\mu = (V, E, t, \gamma_\mu, u_\mu, \delta_\mu)$ is network G "normalised" with labeling μ:

$$\delta_\mu(v) = \delta(v)\mu(v), \quad u_\mu(e_{v,w}) = u(e_{v,w})\mu(v), \quad \gamma_\mu(e_{v,w}) = \gamma(e_{v,w})\mu(w)/\mu(v).$$

If f is a flow in G, then the same flow expressed in the re-labeled network G_μ is denoted by f_μ, and $f_\mu(e_{v,w}) = f(e_{v,w})\mu(v)$, for each $e_{v,w} \in E$. We assume that for any flow f in G, there are paths of positive residual-capacity arcs to the sink t from all other nodes. (One can modify network G into an equivalent network G' satisfying this condition, and $n' = O(n)$, $m' = O(m+n)$, $B' = B$.) If a residual network G_f is a non-gain network, then the *canonical labeling* μ of G_f is defined by the maximum gains of paths to the sink. That is, $\mu(v)$ is equal to the maximum $\gamma(P)$ over all positive residual-capacity paths P from v to t. For this labeling μ, $\gamma_\mu(e) \le 1$, for each positive residual-capacity arc $e \in E$, and we call network $G_{f,\mu}$ a *canonical residual network*. If the gain of each positive residual-capacity arc is at most 1 (a common invariant in generalised flow algorithms), then the canonical labeling of G_f can be computed in $\tilde{O}(m)$ time using Dijkstra's shortest-paths algorithm. If f and μ are a flow and a labeling in network G, then the *total re-labeled residual supply* is defined as $\Delta_{f,\mu} = \sum_{v \in V \setminus \{t\}} \delta_{f,\mu}(v)$.

If the capacities of a pair of reverse arcs $e_{v,w}$ and $e_{w,v}$ are both positive, then we call such arcs *active arcs* and the edge $\{v, w\}$ an *active edge*. We call G a *basic network*, if it does not contain a cycle of active edges and does not contain a path of active edges between two nodes with positive supplies or between a node with positive supply and the sink t. We call a flow f *basic flow*, if the residual network G_f is basic. We call a flow f a *maximal flow*, if the residual network G_f is a non-gain network. For an arbitrary generalised flow network G, one can compute in $\tilde{O}(mn^2 \log B)$ time a maximal flow f' [7]. Having a maximal flow f' and the canonical labeling μ of $G_{f'}$, one can compute in $O(nm)$ time a basic maximal flow f'' such that: the value of flow f'' is not less than the value of flow f'; and the new total re-labeled residual supply $\Delta_{f'',\mu}$ is not greater than the previous total re-labeled residual supply $\Delta_{f',\mu'}$. Therefore we can assume that the computation always begins with a basic non-gain residual network.

The following theorem is proven in [7] (see also [8]).

Theorem 1. *Let f be a maximal flow in network G, and let μ be the canonical labeling of G_f. If $\Delta_{f,\mu} < B^{-m}$, then a maximum generalised flow in network G can be obtained by one maximum non-generalised flow computation in the network $G_{f,\mu}$ restricted to the arcs with re-label gain γ_μ equal to 1.*

If f is a maximal flow in G and $\Delta_{f,\mu} < B^{-m}$, where μ is the canonical labeling of G_f, then we call f a *near-optimal flow*. Theorem 1 implies that having a near-optimal flow, we can compute an optimal one in $\tilde{O}(nm)$ additional time, so from now on we assume that the objective is to compute a near-optimal flow.

3   An Underlying Approximation Algorithm

Let $\mathcal{A}$ be an iterative approximation algorithm for the maximum generalised flow problem which has the following properties. The input is a canonical non-gain residual network G_{f_0,μ_0} and the computation consists of a sequence of *phases*. The input to a phase is a canonical non-gain residual network $G_{f',\mu'}$ and the output (the input for the next phase) is another canonical non-gain residual network $G_{f'',\mu''}$. Each phase reduces the total residual supply at least by half: $\Delta_{f'',\mu''} \leq \Delta_{f',\mu'}/2$. The computation ends when the total residual supply falls below the desired approximation threshold. The running time of one phase is at most $T(n,m)$, where the bounding function is such that $T(n_1, cm) \leq cT(n_2, m)$, if $n_1 \leq n_2$ and $c \leq 1$. The total supply Δ_{f_0,μ_0} in G_{f_0,μ_0} is at most nB^n, so algorithm $\mathcal{A}$ computes a near-optimal flow in $O(m \log B)$ phases, or in $O(T(n,m)m \log B)$ total time. We prove in this paper the following main theorem.

Theorem 2. *A maximum generalised flow algorithm $\mathcal{A}$ which has the properties described above can be modified to an algorithm $\mathcal{A}'$ which computes a near-optimal flow in $O(T(n,m)n \log B) + \tilde{O}(m^2)$ time.*

Goldfarb, Jin and Orlin [8] algorithms have the properties of a maximum generalised flow algorithm $\mathcal{A}$ stated above, with $T(n,m) = \tilde{O}(m^2)$, so they compute near-optimal flows in $\tilde{O}(m^3 \log B)$ time. Theorem 2 implies that these algorithms can be modified to compute near-optimal flows in $\tilde{O}(m^2 n \log B)$ time, so we have the following main conclusion of our paper.

Theorem 3. *A maximum generalised flow can be computed in $\tilde{O}(m^2 n \log B)$ time.*

4   Reducing the Size of Network

Let G be a basic non-gain network and let μ be the canonical labeling of G. We say that an arc $e \in E$ has *large capacity*, if $u_\mu(e) > \Delta_\mu$; otherwise the arc has *small capacity*. If an arc $e \in E$ has large capacity, then for any basic maximal flow f in G which does not send flow along 1-gain cycles, $f(e) < u(e)$ (since such a flow can be decomposed into flows along simple paths). Thus if we change the capacity of arc e to infinity and compute a basic maximal flow f' in the modified network G', then this flow, after removal of flow from 1-gain cycles, is a basic maximal flow in network G. We assume that whenever large-capacity multiple arcs from a node v to a node w appear during the computation, we remove all of them except the one with the largest gain factor (and we remove all arcs reverse to the removed large-capacity arcs; their capacities must be zero). If the capacities of a pair of reverse arcs $e_{v,w}$ and $e_{w,v}$ are both large, then we call edge $\{v,w\}$ a *large-capacity edge* or a *contractable edge*. Observe that in such a case $\gamma_\mu(e_{v,w}) = \gamma_\mu(e_{w,v}) = 1$.

Contracting large-capacity edges. To contract a large-capacity edge $\{v,w\}$, where $w \neq t$ and $\delta(w) = 0$, means to create a new network H_μ by modifying network G_μ in the following way. For each arc adjacent to node w, replace the end node w with node v (the re-labeled gain γ_μ and capacity u_μ of this arc remain unchanged). Then after this changes remove node w.

For a basic non-gain network G, let $\overline{F}^a \subseteq \overline{E}$ and $\overline{F}^c \subseteq \overline{F}^a$ be the forest of active edges and the forest of large-capacity edges, respectively. A *component* $C \subseteq V$ of network G is a connected component of $(V, \overline{F}^a)$, and a *strong component* $C \subseteq V$ is a connected component of $(V, \overline{F}^c)$. In each tree of the forest $(V, \overline{F}^a)$, there is at most one node which has positive supply or is the sink t (see the definition of a basic network). Let H be the basic non-gain network obtained from network G by contracting all contractable edges. Network H can be computed in $O(m)$ time. The strong components of network G correspond to the nodes of network H. Every flow in network G_μ has the natural corresponding flow in network H_μ. Conversely, a basic maximal flow in network H_μ can be expanded in $O(m)$ time into a basic maximal flow in network G_μ which has the same value and the same residual node supplies.

Shortcutting small reverse-flow paths. Let G be a non-gain network and let μ be the canonical labeling of G. Let P be a (directed) path from a node v to a node w such that the sink t is not an intermediate node, the (re-labeled) capacities u_μ of all arcs on the path are the same, the capacities of all arcs reverse to the arcs on P are large, and there are no positive supplies at the intermediate nodes (see Figure 1). We call such a path P a *small reverse-flow path* because it would occur as the reverse residual path of a small flow from w to v. Modify network G by adding a new arc $e_{v,w}$ with capacity u_μ equal to the capacity u_μ of P and with gain γ_μ equal to 1 (the same as the gain factors γ_μ of the arcs on P). Add also a zero capacity reverse arc $e_{w,v}$. Finally, set the capacities of the arcs on P to zero. Observe, that if all positive-capacity arcs adjacent to the intermediate nodes on P other than the arcs of P have large capacities, then this shortcutting procedure makes all intermediate nodes on P free.

A flow in the modified non-gain network H can be easily converted into a flow of the same value and with the same residual node supplies in network G. Conversely, if we have a flow f in G, then we can obtain a flow h of the same value and with the same residual node supplies in network H in the following way. Let $M = \max\{0, \max\{f_\mu(e) : e \in P\}\}$, decrease the flow along P by M and send these M units of flow along the added arc $e_{v,w}$. That is, set $h_\mu(e_{v,w}) = M$, $h_\mu(e_{w,v}) = -M$, and $h_\mu(e) = f_\mu(e) - M$ and $h_\mu(e') = f_\mu(e') + M$, for every arc $e \in P$ and arc e' reverse to arc e.

By-passing and removing free nodes. Let G be a basic non-gain network. We say that a node $v \in V \setminus \{t\}$ is *free*, if it does not have positive supply and for each pair of reverse arcs adjacent to v, the capacity of one of them is large and the capacity of the other is zero. Let $V_{\text{free}} \cup V_{\text{non-free}} = V$ be the partitioning of the set of nodes into the free nodes and non-free nodes. We can remove a free node $v \in V_{\text{free}}$ from the network in the following way. For each pair of positive (hence large) capacity arcs $e_{x,v}$ and $e_{v,z}$, add a new arc $e_{x,z}$ with infinite capacity and gain equal to $\gamma(e_{x,v})\gamma(e_{v,z})$ (and the zero-capacity arc reverse to $e_{x,z}$). Then

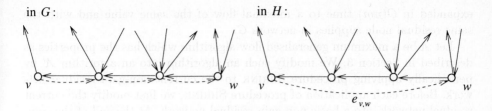

in G: in H:

Fig. 1. Shortcutting a small reverse-flow path from v to w. Only the positive-capacity arcs adjacent to the nodes on the path are shown. The dash arcs have all the same, small capacity u_μ. The other arcs have large capacities. After the shortcutting, all intermediate nodes on the path become free.

remove node v and all arcs adjacent to it. In this way we obtain a basic non-gain network H which, for our purpose, is equivalent to the original network G.

If we remove one or more free nodes, then up to 4 arcs may be added between any pair of the remaining nodes x and z: one infinite-capacity arc from x to z, one infinite-capacity arc from z to x, and the arcs reverse to these two arcs. (Remember that we always remove all multiple large-capacity arcs but one.) Let $\bar n = |V_{\text{non-free}}|$ and let m' denote the number of arcs which originate and end in non-free nodes. If we remove all free nodes from network G, then we get a network with $\bar n$ nodes and $\bar m \le m' + 2\bar n(\bar n - 1)$ arcs. For example, if the input network does not have multiple arcs, then $m' \le \bar n(\bar n - 1)$, so $\bar m \le 3\bar n(\bar n - 1)$. In such a case, if $\bar n = o(\sqrt{m})$, then $\bar m = o(m)$.

Simultaneous removal of all free nodes amounts to adding a new infinite-capacity arc $e_{x,y}$ for each pair of non-free nodes x and y. The gain factor of this arc is equal to the largest gain of a path from x to y which passes only through free nodes. We can compute the gain factors of all these new arcs in $\tilde O(\bar n m)$ time by applying Dijkstra's single-source shortest-paths algorithm to each non-free node as the source. Let H denote the basic non-gain network obtained from network G by removal of all free nodes. If we have a basic maximal flow in network H, and have recorded the $\bar n$ shortest-path trees computed to set the new edges in H as mentioned above, then we can obtain a maximal flow of the same value and with the same residual node supplies in network G in $O(\bar n n)$ time. (For each node x in H, shift the flow from the added arcs $e_{x,y}$ onto the arcs of the recorded tree; update the flow on the arcs of the tree starting from the leaves for the running time of $O(n)$ per one tree.)

Procedure SHRINK(G) modifies a basic non-gain network in the following way.

1. Contract all contractable edges in G. Let $G^{(1)}$ denote the obtained network.
2. Shortcut every small reverse-flow path in $G^{(1)}$ to obtain network $G^{(2)}$.
3. Remove all free nodes from $G^{(2)}$ to obtain network $G^{(3)}$.
4. Return network $G^{(3)}$ or the initial network G, whichever has fewer arcs.

The running time of procedure SHRINK(G) is dominated by the $\tilde O(nm)$ time required by step 3. A basic maximal flow in the shrunk network $G^{(3)}$ can be expanded in $O(nm)$ time to a maximal flow of the same value and with the same residual node supplies in network G.

Let $\mathcal{A}$ be a maximum generalised flow algorithm which has the properties as described in Section 3. We modify such an algorithm into an algorithm $\mathcal{A}'$ by periodically applying procedure SHRINK to the current non-gain residual network. Before each application of procedure SHRINK, we first modify the current residual network into a basic non-gain residual network. At the end of the computation of algorithm $\mathcal{A}'$, the final flow in the final shrunk network is expanded to a flow of the same value in the initial network G. We prove now our main Theorem 2, using the following theorem (proven in Section 5).

Theorem 4. *Let G_f be a basic non-gain residual network, let μ be a canonical labeling of G_f, and let $1 \le k \le m/n$. There exists a constant integer $\alpha > 0$ such that if $\Delta_{f,\mu} \le B^{-3kn}$, then procedure SHRINK$(G_f)$ returns a network with the number of nodes $n' \le n$ and the number of arcs $m' \le \min\{m, [(\alpha m)/(kn)]^2\}$.*

Proof of Theorem 2. We execute procedure SHRINK every $cn \log B$ phases of algorithm $\mathcal{A}$, for a suitably large constant c. We view the computation of this modified algorithm $\mathcal{A}'$ as a sequence of *stages*. Stage 0 consists of the initial $cn \log B$ phases, and each subsequent stage consists of the execution of procedure SHRINK followed by further $cn \log B$ phases of algorithm $\mathcal{A}$. Since there are $O(m \log B)$ phases in algorithm $\mathcal{A}$, there are $K = O(m/n)$ stages in algorithm $\mathcal{A}'$. To simplify the argument, we assume that procedure SHRINK is not applied to the current, possibly already shrunk network to shrink it even further, but to a residual network of the initial input network G.

Let f_k be the basic maximal flow in G before the k-th application of SHRINK, and let μ_k be the canonical labeling of G_{f_k}. The geometric decrease of the residual supply during the computation of $\mathcal{A}$ implies that the total residual supply in network G_{f_k, μ_k} is $\Delta_{f_k, \mu_k} \le nB^n/2^{(ckn \log B)} \le B^{-3kn}$. Let n_k and m_k be the numbers of nodes and arcs in the network computed by SHRINK(G_{f_k}), that is, the size of the network used during stage k. Theorem 4 implies that $n_k \le n$ and $m_k \le [(\alpha m)/(kn)]^2$. Hence $m_k \le (\alpha/k)^2 m$ for $k \ge \alpha$, assuming that $m \le n^2$. The running time of algorithm $\mathcal{A}'$ is $T(n, m)O(n \log B) + \tilde{O}(Knm)$ for the first α stages and all computations of procedure SHRINK, plus

$$\sum_{k=\alpha}^{K} O(T(n_k, m_k)n \log B) \le \sum_{k=\alpha}^{K} \frac{\alpha^2}{k^2}O(T(n, m)n \log B) \le O(T(n, m)n \log B).$$

The first inequality follows from the property of the bounding function $T(n, m)$ stated in Section 3. The second inequality holds because $\sum_{i=1}^{\infty} 1/i^2 = \Theta(1)$.

Remark 1. The above proof uses the assumption that $m \le n^2$, but can be extended to the case when $m > n^2$ by bounding separately the running time of the first $\Theta(m/n^2)$ stages and the running time of the remaining stages (see [9]).

5 Size of a Strong Component When Supply Is Very Small

In this section we prove Theorem 4. Define the degree $\deg(C)$ of a set of nodes $C \subseteq V$ as the number of arcs with at least one end in C. The proof is based on Lemmas 2, 3 and 4, which say that when the total residual supply is exponentially small, then a strong component must have large degree. These three lemmas considers different types of strong components. Their proofs are similar, so we include only the proof of Lemma 2. The proofs are based on Lemma 1, which gives an expression on the balance of flow at a subset of nodes, and on the following observation. To obtain an exponentially small value from the fractional input numbers D_{frac} using additions and multiplications, we have to use quite a few of them. We also need Lemma 5, which says that when the total residual supply is very small, at least one arc in each pair of reverse arcs must have large capacity.

A *bi-directional tree* $T \subseteq E$ spanning a set of nodes $C \subseteq V$, $C \neq \emptyset$, consists of $|C| - 1$ pairs of reverse arcs spanning C. The next lemma states the flow conservation property with respect to a tree.

Lemma 1. *Let $T \subseteq E$ be a bi-directional tree spanning a non-empty set of nodes $C \subseteq V \setminus \{t\}$. Let r be an arbitrary node in C, and for a node $v \in C$, let $P_v \subseteq T$ denote the tree path from node v to the root r. If f is a flow in network G, then the following flow-balance relation holds:*

$$\sum_{v \in C} \delta_f(v)\,\gamma(P_v) \;-\; \sum_{v \in C} \delta(v)\,\gamma(P_v) \;+\; \sum_{v \in C}\sum_{e \in E_v^- \setminus T} f(e)\,\gamma(P_v) \;=\; 0. \qquad (1)$$

Proof (Idea). From the definition of the residual supplies, we have

$$\delta_f(v) - \delta(v) + \sum_{e \in E_v^-} f(e) = 0, \quad \text{for each } v \in C. \qquad (2)$$

Eliminate in system (2) terms $f(e)$, for all $e \in T$, to obtain Equation (1).

Lemma 2. *Let G_f be a basic non-gain residual network, let μ be the canonical labeling of G_f, and let C be a strong component of G_f with the following properties. There are no active non-contractable edges adjacent to C and there is a positive residual supply at one node in C. Under these conditions, if a number d is such that $d \geq n$ and $\Delta_{f,\mu} \leq B^{-3d}$, then the degree of set C is at least d.*

Proof. If r is the node in C with positive supply (a component of a basic non-gain network has at most one node with positive supply) and T is the bi-directional tree of active arcs spanning C (see Fig. 2), then Equation (1) becomes

$$\delta_f(r) \;-\; \sum_{v \in C} \delta(v)\,\gamma(P_v) \;+\; \sum_{v \in C}\sum_{e \in E_v^- \setminus T} f(e)\,\gamma(P_v) \;=\; 0. \qquad (3)$$

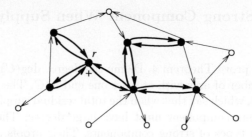

Fig. 2. A strong component of Lemma 2. Black nodes are in C, white nodes are adjacent to C. Only positive residual-capacity arcs are shown; the tree arcs in bold.

Assume that the degree of C is less than d. All arcs in the last sum in (3) are non-active because C is a component. Hence for each arc e in this sum, the flow $f(e)$ on this arc is either at its upper, $u(e)$, or at its lower, $-u(e')\gamma(e')$, bound, where e' is the reverse arc to arc e. Thus the left-hand side of (3) excluding $\delta_f(r)$ involves only numbers from set D_{frac}: $\delta(v)$, for each $v \in C$ ($|C|$ numbers); either $u(e)$, or $u(e')$ and $\gamma(e')$, for each non-tree arc e with tail in C, where e' is the reverse arc of arc e (at most $2(\deg(C) - 2(|C|-1))$ numbers); and $\gamma(e)$, for each arc $e \in T$ which is in the direction towards the root node r ($|C| - 1$ numbers). Thus at most $2\deg(C) + 1 \le 2d - 1$ numbers from D_{frac} are involved in (3), so the absolute value of the left-hand side of (3) excluding $\delta_f(r)$ is a fractional number with denominator less than B^{2d}. We get contradiction since

$$0 < \delta_f(r) = \delta_{f,\mu}(r)/\mu(r) \le \Delta_{f,\mu}B^{n-1} < B^{-3d+n} \le B^{-2d}.$$

Lemma 3. *Let G_f be a basic non-gain residual network, let μ be the canonical labeling of G_f, and let $C \in V \setminus \{t\}$ be a strong component of network G_f with the following properties. There is exactly one active but non-contractable edge adjacent to C, and there is no residual supply in C. Under these conditions, if $d \ge n$ and $\Delta_{f,\mu} \le B^{-3d}$, then the degree of set C is at least d.*

Lemma 4. *Let G_f be a basic non-gain residual network, let μ be the canonical labeling of G_f, and let $C \subseteq V \setminus \{t\}$ be a strong component of G_f with the following properties. There is no residual supply in C, and exactly two active but non-contractable edges $\{x, p\}$ and $\{y, q\}$ are adjacent to C, where $p, q \in C$. Let $e_{x,p}$ and $e_{p,x}$, and $e_{y,q}$ and $e_{q,y}$ be the two pairs of active reverse arcs. If $d \ge n$ and $\Delta_{f,\mu} \le B^{-3d}$, then at least one of the following two conditions must hold:*

(a) *the degree of set C is at least d,*
(b) $u_{f,\mu}(e_{p,x}) = u_{f,\mu}(e_{y,q}) \le \Delta_{f,\mu}$ *or* $u_{f,\mu}(e_{x,p}) = u_{f,\mu}(e_{q,y}) \le \Delta_{f,\mu}$.

Lemma 5. *Let G_f be a basic non-gain residual network and let μ be the canonical labeling of G_f. If $\Delta_{f,\mu} \le B^{-(n+1)}$, then at least one arc in every pair of reverse arcs in $G_{f,\mu}$ has large residual capacity.*

Proof. Each arc with non-positive flow f has large capacity.

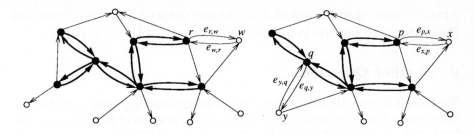

Fig. 3. Strong components of Lemmas 3 and 4.

Proof of Theorem 4. We apply procedure SHRINK to the basic non-gain resid-
ual network G_f given in the statement of the theorem. Let $G_f^{(1)}$, $G_f^{(2)}$ and $G_f^{(3)}$
denote the networks at the end of steps 1, 2 and 3 of procedure SHRINK, respectively. Call a strong component of network G_f large, if its degree is at least kn,
and small otherwise. The nodes of $G_f^{(3)}$ correspond to some strong components
of G_f. Call a node of $G_f^{(3)}$ large or small, depending whether it corresponds
to a large or small strong component of G_f. The core of the proof is to show
that $G_f^{(3)}$ has $O(m/(kn))$ nodes. We show this by showing that $G_f^{(3)}$ cannot have
more small nodes than the large ones. Let C be a component of G_f and estimate
how many small and large nodes it contributes to the final network $G_f^{(3)}$.

If C consists of only one strong component and contains the sink t, then it
becomes the sink node in $G_f^{(3)}$, and it may be either large or small. If C consists
of only one strong component, does not contain the sink t, and does not have
any residual supply, then it is contracted into a single free node in $G_f^{(1)}$ and then
removed. If C consists of only one strong component but does have residual
supply, then Lemma 2 implies that C is a large component, so it contributes
to $G_f^{(3)}$ only one node and this node is large. For the remaining case, when C
consists of at least two strong components, let W be the set of nodes of $G_f^{(3)}$
contributed by C, and let T be the tree in $G_f^{(3)}$ of the active edges spanning W.
One node in W may either be the sink or have positive residual supply; let
$W' \subseteq W$ denote the set of the other nodes in W. Lemma 3 implies that a node
in W' which is a leaf of tree T must be large. Lemma 4 implies that a node in
W' of degree 2 in T must be also large (otherwise it is shortcut and removed).
Thus a small node in W' has at least degree 3 in T. There are at most $|W|/2-1$
nodes in T of degree at least 3, so there are at most $|W|/2$ small nodes in W.

Let $\bar{n}$ denote the number of nodes in the final network $G_f^{(3)}$. We have shown
that at least $\bar{n}/2 - 1$ nodes in $G_f^{(3)}$ are large, so at least that many strong
components of G_f are large. Thus there must be at least $kn(\bar{n}/2 - 1)/2$ arcs in
network G_f which are adjacent to strong components. Hence $m \geq kn(\bar{n}/2-1)/2$
and $\bar{n} \leq 5m/(kn)$ (assuming $\bar{n} \geq 10$). Lemma 5 implies that there are at most 4
arcs between any two nodes in network $G_f^{(3)}$: a small-capacity arc in one direction,

a parallel large-capacity arc (with a smaller gain factor), and the two reverse arcs. Hence the number of arcs in network $G_f^{(3)}$ is at most $2(\bar{n})^2 \le O(m^2/(kn)^2)$.

6   Further Questions

Tardos and Wayne [10] proposed a simple combinatorial algorithms for the maximum generalised flow problem, including a generalisation of Goldberg and Tarjan's push-relabel algorithm for the min-cost flow problem [11]. Can our network modifications improve the time bounds of those algorithms by factor m/n?

Wayne [12] showed a polynomial-time combinatorial algorithm for the min-cost generalised flow problem. As in the maximum generalised flow case, the computation of this algorithm ends when the value of flow is within $B^{-\Omega(m)}$ from optimal. Can some network-modification operations work in this algorithm?

And finally, do the network-modification operations presented in this paper bring us any closer to settling the big open question of computing maximum generalised flows in strongly-polynomial time?

References

1. Ahuja, R.K., Magnanti, T.L., Orlin, J.B.: Network flows: Theory, algorithms, and applications. Prentice Hall (1993)
2. Glover, F., Hultz, J., Klingman, D., Stutz, J.: Generalized networks: A fundamental computer-based planning tool. Management Science **24** (1978)
3. Murray, S.M.: An interior point approach to the generalized flow problem with costs and related problems. PhD thesis, Stanford Univ. (1992)
4. Kapoor, S., Vaidya, P.M.: Speeding up Karmarkar's algorithm for multicommodity flows. Math. Programming **73** (1996) 111–127
5. Vaidya, P.M.: Speeding up Linear Programming Using Fast Matrix Multiplication. In: Proc. 30th IEEE Annual Symposium on Foundations of Computer Science. (1989) 332–337
6. Ford, Jr., L.R., Fulkerson, D.R.: Flows in Networks. Princeton Univ. Press, Princeton, NJ (1962)
7. Goldberg, A.V., Plotkin, S.A., Tardos, E.: Combinatorial Algorithms for the Generalized Circulation Problem. Math. Oper. Res. **16** (1991) 351–381
8. Goldfarb, D., Jin, Z., Orlin, J.: Polynomial-time highest-gain augmenting path algorithms for the generalized circulation problem. Math. Oper. Res. **22** (1997) 793–802
9. Radzik, T.: Contracting the network during maximum generalised network flow computation. Technical Report TR-01-08, Department of Computer Science, King's College London (2001) (available at http://www.dcs.kcl.ac.uk).
10. Tardos, E., Wayne, K.: Simple generalized maximum flow algorithms. In: Proc. 6th International Conference on Integer Programming and Combinatorial Optimization. (1998) 310–324
11. Goldberg, A.V., Tarjan, R.E.: Finding minimum-cost circulations by successive approximation. Math. Oper. Res. **15** (1990) 430–466
12. Wayne, K.: A polynomial combinatorial algorithm for generalized minimum cost flow. In: Proc. 31st Annual ACM Symposium on Theory of Computing. (1999)

Approximation Hardness of Bounded Degree MIN-CSP and MIN-BISECTION

Piotr Berman* and Marek Karpinski**

Dept. of Computer Science, University of Bonn, 53117 Bonn
berman@cse.psu.edu, marek@cs.uni-bonn.de

Abstract. We consider bounded occurrence (degree) instances of a minimum constraint satisfaction problem MIN-LIN2 and a MIN-BISECTION problem for graphs. MIN-LIN2 is an optimization problem for a given system of linear equations mod 2 to construct a solution that satisfies the minimum number of them. E3-OCC-MIN-E3-LIN2 is the bounded occurrence (degree) problem restricted as follows: each equation has exactly 3 variables and each variable occurs in exactly 3 equations. Clearly, MIN-LIN2 is equivalent to another well known problem, the Nearest Codeword problem, and E3-OCC-MIN-E3-LIN2 to its bounded occurrence version. MIN-BISECTION is a problem of finding a minimum bisection of a graph, while 3-MIN-BISECTION is the MIN-BISECTION problem restricted to 3-regular graphs only. We show that, somewhat surprisingly, these two restricted problems are exactly as hard to approximate as their general versions. In particular, an approximation ratio lower bound for E3-OCC-MIN-E3-LIN2 (bounded 3-occurrence 3-ary Nearest Codeword problem) is equal to MIN-LIN2 (Nearest Codeword problem) lower bound $n^{\Omega(1)/\log\log n}$. Moreover, an existence of a constant factor approximation ratio (or a PTAS) for 3-MIN-BISECTION entails existence of a constant approximation ratio (or a PTAS) for the general MIN-BISECTION.

1 Introduction

In this paper we study the approximation hardness of bounded occurence minimum constraint satisfaction problem E3-OCC-MIN-E3-LIN2 and 3-MIN-BISECTION. MIN-LIN2 problem has as its input a system of linear equations over $GF[2]$ and the minimized objective function is the number of equations satisfied by a solution; -Ek- denotes its restriction to equations with exactly k variables, and Ek-OCC- denotes another restriction, namely that each variable occurs exactly k times in a system. (Infix and prefix k denote more loose restrictions, to equations with at most k variables and to at most k occurrences of each

* Visiting from Pensylvania State University. Partially supported by DFG grant Bo 56/157-1 and NSF grant CCR-9700053.
** Research partially done while visiting Dept. of Computer Science, Yale University. Supported in part by DFG grants KA 673/4-1 and Bo 56/157, DIMACS, and IST grant 14036 (RAND-APX).

P. Widmayer et al. (Eds.): ICALP 2002, LNCS 2380, pp. 623–632, 2002.
© Springer-Verlag Berlin Heidelberg 2002

variable respectively.) The MIN-LIN2 problem or equivalently Nearest Codeword problem is known to be exceedingly hard to approximate. It is known to be NP-hard to approximate to within a factor $n^{\Omega(1)/\log\log n}$ (cf. [DKS98], [DKRS00]). Recently, also the first nonlinear approximation ratio $O(n/\log n)$ algorithm has been designed for that problem [BK01], see also [BFK00]. MIN-BISECTION problem has as its input an undirected graph, say with $2n$ nodes, a solution is a subset S of nodes with n elements, and the size of a cut $|Cut(S)|$ (i.e. the number of edges from S to its complement) is its minimizing objective function; prefix k- denotes its restriction to k-regular graphs. The general MIN-BISECTION problem belongs to the most intriguing problems in combinatorial optimization with no approximation hardness results known so far. Only recently Feige and Krauthgamer [FK00] have constructed a polylogarithmic factor approximation algorithm for that problem.

Our results are tight in the following sense: 2-OCC-MIN-LIN2 can be solved in polynomial time, and MIN-2-LIN2 is much easier to approximate than MIN-LIN2 (cf. [KST97], [DKS98], [DKRS00]), while E3-OCC-MIN-E3-LIN2 is as hard to approximate as MIN-LIN-2 itself, i.e. it is NP-hard to approximate to within $n^{\Omega(1)/\log\log n}$. Similarly, 2-MIN-BISECTION can be solved in polynomial time while 3-MIN-BISECTION is as hard to approximate as the general MIN-BISECTION.

2   MIN-LIN2 Problem

2.1   Our Terminology

For a system of linear equations S over $GF[2]$, we denote the set of variables with $V(S)$, the total number of occurrences of variables in S (equivalently, the number of non-zero coefficients) with $size(S)$, the set of assignments of values to variables of S with $AV(S)$. For $a \in AV(S)$ the number of equations of S that a satisfies is denoted with $sat(a, S)$. As mentioned before, $sat(a, S)$ is the minimized objective function.

A reduction of MIN-LIN-2 to 3-OCC-MIN-E3-LIN-2 will be described via the following triple of functions:

- an instance transformation f such that $f(S)$ is a system in which each variable occurs exactly 3 times and $size(f(S)) = O(size(S))$;
- a solution normalization g such that for $a \in AV(S)$ we have $g(a) \in AV(S)$ and $|sat(g(a), f(S))| \leq |sat(a, f(S))|$,
- a bijection h between $AV(S)$ and $g(AV(f(S)))$ such that $|sat(a, S)| = |sat(h(a), f(S))|$.

The above description of a reduction directly relates to the standard definition of aproximation preserving reductions. The implied solution transformation is $h^{-1} \circ g$, its desired properties follow immediately from the properties of a normalization and "equivalence bijection".

2.2   Consistency Gadgets

In approximation preserving reductions there exist a number of ways to replace a variable with a set of variables and auxiliary equations so that each element of the set occurs exactly three times. We have explored one such construction in [BK99] in a context of the maximization problems. In the reductions described in that paper the objective function is diluted, because part of the score assures that we can normalize each solution in such a way that the new set of variables correctly replaces a single variable. We refer to [BK99] for details. The same construction yields no such dilution in our minimization problems, because a normalized solution satisfies none of the auxiliary equations.

However, we describe here a different construction, which can be computed determinististically and the size of the new formula is a linear function of the size of the original formula. The construction of our reduction depends on the existence of special graphs which we call consistency gadgets.

In an undirected graph (V, E), we define $Cut(U) = \{e \in E : e \cap U \neq \emptyset$ and $e \cap (V - U) \neq \emptyset\}$. A consistency gadget $CG(K)$ is a graph with the following properties:

- $K \subset V_K$ where V_K is the set of nodes of $CG(K)$;
- if $A \subset V_K$ and $|A \cap K| \leq |K|/2$, then $|Cut(A)| \geq |A \cap K|$;
- each node in K has 2 neighbors and each node in $V_K - K$ has 3 neighbors. (Denote $k = |K|$.)

As described in Arora and Lund [AL95] (cf. also [PY91]), we could use a family of graphs that we may call *strong expanders*. The graph (V, E) is a strong expander if $|U| \leq |V|/2$ implies $|Cut(U)| \geq |U|$ for each $U \subset V$. Lubotzky *et al.* [LPS88] showed that a family of 14-regular strong expanders is constructible in polynomial time.

Arora and Lund suggest that one can obtain $CG(K)$ by constructing a 14-regular strong expander with node set K, and then by replacing each node with a cycle of 15 nodes: one node being the element of K and the other nodes being the terminals of the 14 connections to the other cycles.

One can see that this construction is not quite sufficient for our purposes, because it is conceivable that the new graph contains a set that contradicts the $CG(K)$ property, and which properly intersects some of the cycles. Aurora and Lund use a weaker notion of consistency, but in this paper we wish to establish an exact relationship between the tresholds of approximability.

To remedy this problem, we can modify the construction slightly. In particular, let S_{15} be a set of 15 elements, we will replace each node of the 14-regular strong expander with a copy of a $CG(S_{15})$ which is shown in Fig. 1.

2.3   Reduction to E3-OCC-MIN-3-LIN2

Recall that in E3-OCC-MIN-3-LIN2 we also allow equations in which exactly 2 variables appear.

Fig. 1. An example of $CG(S_{15})$; element of S_{15} are black.

Consider a system of equations S. Without loss of generality, we assume that each variable of S occurs at least 2 times, otherwise we reduce the problem by removing the equations in which a variable has its sole occurence.

Suppose that S contains an equation with more than 3 variables, $\xi + \zeta = b$, where ξ is a sum of 2 variables and ζ is a sum of at least 2 variables. We obtain $f(S)$ by replacing in S this equation with two equations, $\xi + x = 1$ and $x + \zeta = b$, where x is a new variable. We define the normalizing transformation $g(a)$ as follows: $g(a)(y) = a(y)$ for every variable y other than x, and $g(a)(x) = a(\xi)$ where $a(\xi)$ is the value of the linear expression ξ under assignment of values a. This assures that $g(a)$ does not satisfy equation $\xi + x = 1$. In turn, $g(a)$ satisfies $\xi + x = b$ if and only if a satisfies $\xi + \zeta = b$. The bijection h is defined by the same formula as g.

Now we assume that each equation consists of at most 3 variables. Consider a variable x that occurs in $k > 3$ equations. Let K be the set of these equations. We form the graph $CG_3(K)$ with node set V_K, for each node in V_K we introduce a new variable with the same name (so now V_K becomes a set of variables) and for each edge $\{u, g\}$ we introduce an equation $u + v = 1$. Then we replace an occurrence of x in equation e with the variable that replaced e (as a node of V_K). This is the instance transformation (obviously, we may need a sequence of such transformations to achieve our goal).

To define $h(a)$ for an assignment of values a of the new instance, we find value b that is assumed by at least half of the variables in K; then for every $x \in V_K$ we set $h(a)(x) = b$. Suppose that this normalization changed the values of l variables from K . Then up to l "old" equations may become satisfied. However, none of the equations that replaced the edges of $CG_3(K)$ is satisfied now. Because $2l < |K|$, we have l edge disjoint paths from the l variable/nodes that change the value to nodes that did not, in turn, on each path we have at least one edge corresponding to an equation that ceased to be satisfied, thus at least l "new" equations ceased to be satisfied. consequently, $sat(a, f(S))$ did not increase.

The bijection transformation simply $h(a)$ assigns $a(x)$ to each variable in V_K.

To summarize the reasoning of this section we introduce the following definition.

We call an approximation algorithm A for an optimization problem P, an $(r(n), t(n))$-approximation algorithm if A approximates P within an approximation ratio $r(n)$ and A works in $O(t(n))$ time with n the size of the problem instance.

We can formulate the following lemma.

Lemma 1. *There exists a constant c such that if there exists an $(r(n), t(n))$-approximation algorithm for E3-OCC-MIN-3-LIN2 then there exists an $(r(cn), t(cn))$-approximation algorithm for MIN-LIN2.*

2.4   Reducing MIN-LIN2 to E3-OCC-MIN-E3-LIN2

An existing method of converting equations with 2 variables into equations with 3 variables, described by Khanna, Sudan and Trevisan [KST97] cannot be applied here because it increases the number of occurences of variables. Therefore we need to provide a different technique.

Instead, we will modify the reduction of the previous sections. First, we can make sure that the variables in a resulting instance of E3-OCC-MIN-3-LIN2 can be colored with two colors blue and red, so that the following holds true: in an equation of three variables all three must be blue, in an equation with two variables, one must be blue and the other must be red. If we accomplish that, we can replace each red variable with a sum of two new variables.

This simple idea has some minor complications, so we describe it in more detail. The reduction again consists of functions f, g and h — transformation, normalization and bijection — except that now $|sat(a, S)| = 3|sat(h(a), f(S)|$, *i.e.* the normalized solutions for a transformed instance will satisfy 3 times as many equations as the equivalent solutions of the original instance.

We first color occurences of variables as described above: in an equation with 3 variables all occurences are blue, in an equation with 2 variables, one is blue, the other is red. Then we replicate the occurences 3 times, and each replicated occurence is a new variable. Each original equation is replaced with 3 new equations in an obvious manner.

Now, let K be the set of occurences of a variable. Note that the numbers of blue and red elements of K are divisible by 3. We create a gadget $CG(K)$ very similarly as before, so we will only mention the differences in a construction:

- we double the lenth of all cycles that replace nodes of a strong expander, the colors on the cycle alternate, if a contact belongs to such a cycle, it keeps its original color;
- an edge of the strong expander is replaced by two edges between the respective cycles, one being {blue, red} and the other {red, blue};
- at the moment, each cycle that contains a node that is connected only to its neighbors on the cycle and that has a color different than the color of the K element of this cycle, within $CG(K)$ we contact, we connect each 3 such blue nodes with a new red node, and each 3 reds with a new blue.

Now, each equation with 2 variables has one blue and one red variable, so we can replace each red variable with a sum of two new variables, and as a result each equation has 3 variables.

Because we increase the size of the transformed instance only by a constant factor, we can restate our lemma as the following theorem:

Theorem 1. *There exists a constant c such that if there exists an $(r(n), t(n))$-approximation algorithm for E3-OCC-MIN-E3-LIN2 then there exists an $(r(cn), t(cn))$-approximation algorithm for MIN-LIN2.*

3   MIN-BISECTION

We turn now to the problem of MIN-BISECTION and prove the approximation hardness of 3-MIN-BISECTION for that problem. We note that Bui and Jones [BJ92] have established a different kind of approximation hardness result for the bisectional vertex separators on graphs of maximum degree 3.

3.1   Notation

For a graph G, $V(G)$ is a set of nodes, $E(G)$ is a set of edges, $\overline{S} = V - S$, $B(G)$ is the set of *bisections*, e.g. sets $S \subset V(G)$ such that $|S| = |\overline{S}|$ and $Cut(S) = \{e \in E(V) : e \cap S \neq \varnothing$ and $e \cap \overline{S} \neq \varnothing\}$ for $S \subset V(G)$.

The MIN-BISECTION problem is to find $S \in B(G)$ such that $|Cut(S)|$ is minimum.

Our reduction is described using the following three functions:

- an instance transformation f such that if G is a graph with n nodes then $f(G)$ is a graph of degree 3 with $O(n^3)$ nodes;
- a solution normalization g such that for $S \in B(f(G))$ we have $g(S) \in B(f(G))$ and $|Cut(g(S))| \leq |Cut(S)|$;
- a bijection $h : B(G) \to g(B(f(G)))$ such that $|Cut(S)| = |Cut(h(S))|$.

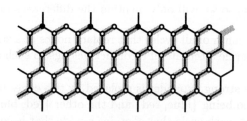

Fig. 2. Gadget for $d = 8$; black circles are the contact nodes.

3.2   The Reduction

We assume that the maximum node degree in G is bounded by some even d. Our instance transformation replaces every node v with a gadget A_v in which d nodes are the contacts (black circles in the Fig. 2). The diagram depicts a gadget for $d = 8$. The gadget is a cylindrical hexagonal mesh that can be alternatively decomposed into $d/2$ *horizontal cycles* and into d *diagonal paths*. Each diagonal path contains d nodes, for the total of d^2 nodes. Every other diagonal path has two *contact nodes* at its ends.

An edge $\{u, v\}$ is replaced with an edge between the contacts of A_u and A_v, these replacement edges are disjoint.

Solution normalization will assure that for each u either $A_u \subset g(S)$ or $A_u \subset \overline{g(S)}$. We will compute $g(S)$ in two stages. In the first stage we decide whether to place A_u within $g(S)$ or within $\overline{g(S)}$ based solely on $S \cap A_u$. As a result, k_u nodes that belong to $\overline{S}$ will become members of $g(S)$, or $-k_u$ nodes that belong to S will become members of $\overline{g(S)}$.

After the first stage, $|g(S)| - |S| = \sum_{u \in V(f(G))} k_u = s$. If $s \neq 0$, $g(S)$ is not a bisection. Therefore in the second stage we remove s/d^2 gadgets from $g(S)$ (if $s < 0$, we insert $-s/d^2$ gadgets).

The normalization causes some edges that did not belong to $Cut(S)$ to become members of $Cut(g(S))$. In the first stage we move some contacts from S to $\overline{g(S)}$, or from $\overline{S}$ to $g(S)$; for each such contact the incident replacement edge may become a member of $Cut(g(S))$. In the second stage, we move s/d^2 gadgets, thus s/d contacts, and again, the replacement edges incident to the contacts that moved may become members of $Cut(g(S))$. This allows us to estimate the consequences of the decision made during the first stage about each gadget A_u.

Let E_u be the set of edges inside A_u. A decision moves $k = |k_u|$ nodes, among them, i contacts. To offset the increases of $Cut(g(S))$ it suffices to gain $i + k/d$ edges, and this happens if $|Cut(S) \cap E_u| \geq i + k/d$.

Suppose first that $|Cut(S) \cap E_u| \geq d$. Let j be the number of contacts in $S \cap A_u$ and $k = |S \cap A_u|$. If $j + k/d \leq d$, we can place A_u in $g(S)$; otherwise $(d - j) + (d^2 - k)/d \leq d$ and we can place A_u in $\overline{g(S)}$.

Now we can assume that $|Cut(S) \cap E_u| < d$. We will make an observation. Our gadget contains $d/2$ horizontal cycles, if one of them contains an edge of $Cut(S)$, it must contain two such edges; we can conclude that at least one of the horizontal cycles has no edges in $Cut(S)$, and thus is contained in S or in $\overline{S}$. For similar reasons, one of the diagonal paths must be contained in S or in $\overline{S}$. Because every horizontal cycle overlaps every diagonal path, it is true for exactly one $\mathbf{S} \in \{S, \overline{S}\}$ that $\mathbf{S}$ contains a horizontal cycle and a diagonal path. We place A_u is $\mathbf{S}$.

Let U be a connected component of $A_u - \mathbf{S}$, and let $C(U) = Cut(U) \cap E_v$. Assume that U contains i_U contacts. To finish the proof of correctness of our normalization, it suffices to show that $i_U + |U|/d \leq |C(U)|$, or, equivalently, that $\beta(U) = |C(U)| - i_U - |U|/d \geq 0$.

Suppose, by the way of contradiction, that for some U that is disjoint with a horizontal cycle and with a diagonal path we have $\beta(U) < 0$. We choose such U

with the smallest possible $|C(U)|$, and, with this constraint, the smallest possible $\beta(U)$.

Observation 1. A node $v \in U$ cannot have two neighbors in $A_u - U$. Otherwise if v has no neighbors in U, then $U = \{v\}$ and $\beta(U) \geq 2 - 1 - 1/d$. If v has a neighbor in U, we could remove v from U and decrease both $C(U)$ and $\beta(U)$.

Observation 2. A node $v \in A_u - U$ must have at least two neighbors in $A_u - U$. otherwise we could insert v to U and decrease both $|C(U)|$ and $\beta(U)$.

Our gadget A_u can be covered with a collection of cycles of length 6 that we will call *hexagons*.

Observation 3. Set U cannot contain exactly 1,4, or 5 nodes of a hexagon. If U contains 1 or 5 nodes, it contradicts Observation 1 or Observation 2. If U contains 4 nodes, then the other two nodes, if they do not contradict Observation 2, form an edge with exactly two neighbors in U and two neighbors in $A_u - U$, thus we can insert this pair to U without increasing $|C|$ while decreasing $\beta(U)$.

Observation 4. Assume that D_0, D_1 are two adjacent diagonal paths, $U \cap D_0 = \emptyset$ and $B = U \cap D_1 \neq \emptyset$. Then B forms a path that has a contact node at one of its ends.

To show it, consider D_2, the other diagonal path adjacent to D_1, and A', a connected component of B. A brief and easy case analysis shows that A' has more neighbors in $D_0 \cup (D_1 - U)$ than in D_2. Because at least one horizontal cycle is disjoint with U, $|B| < d$. Therefore removing A' from U decreases $|C|$ and $|C(U)| - |U|/d$. Because $\beta(U)$ cannot be decreased in this fashion, A' must contain a contact node. Because U is disjoint with a horizontal cycle, there is only one contact node for which it is possible.

Another brief and easy case analysis shows that U must form a "trapesoid", with one basis being a fragment of a horizontal cycle that forms a path between two contact nodes, the other basis being a path that is a fragment of another horizontal cycle and the sides are the initial fragments of two diagonal paths.

Assume that such a set U contains a contact nodes and overlaps b horizontal cycles. U contains $4a - 3$ nodes in the basis that contains the contact nodes, then the consecutive horizontal cycles it contains $4a - 5$, $4a - 7, \ldots$ down to $4a - 3 - 2(b - 1) = 4a - 2b - 1$ nodes. Thus

$$|U| \;=\; b\,\frac{4a - 3 + 4a - 2b - 1}{2} \;=\; b\,(4a - b - 2).$$

In turn, $C(U)$ contains 2 edges in each of the b horizontal cycles, plus the edges extending to the horizontal cycle that is disjoint with U and adjacent to its smaller basis. This basis has $4a - 2b - 1$ nodes and the nodes with such edges alternate with the nodes without them, so we have

$$|C(U)| = 2b + \lfloor (4a - 2b - 1)/2 \rfloor = 2b + 2a - b - 1 = 2a + b - 1.$$

Thus $\beta(U) = (2a + b - 1) - a - b(4a - b - 2)/d = a + b - 1 - b(4a - b - 2)/d$, while we have the following constraint: $|C| < d$, *i.e.* $2a + b - 1 < d$. If we decrease d, $\beta(U)$ will also decrease, so we assume $d = 2a + b$ and thus $b = d - 2a$. Then we have

$$d\beta(U) = d(a + d - 2a - 1) - (d - 2a)(4a - d + 2a - 2) =$$
$$d(d - a - 1) - (d - 2a)(6a - d - 2) =$$
$$d^2 - da - d - 6da + d^2 + 2d + 12a^2 - 2da - 4a =$$
$$2d^2 + 12a^2 - 9da + d - 2a =$$
$$2(d - 2.25a)^2 + 1.875a^2 + d - 2a.$$

This concludes the proof of the following

Theorem 2. *If there exists an* $(r(n), t(n))$*-approximation algorithm for 3-MIN-BISECTION then there exists an* $(r(n^3), t(n^3))$*-approximation algorithm for MIN-BISECTION.*

□

The best currently known approximation ratio for MIN-BISECTION is $O(\log^2 n)$ due to Feige and Krauthgamer [FK00]. Theorem 2 implies that any asymptotic improvement on this ratio for the MIN-BISECTION problem restricted to 3-regular graphs will entail an improvement for the general MIN-BISECTION problem.

Acknowledgments. We thank Lars Engebretsen, Ravi Kannan, Mario Szegedy, and Ran Raz for stimulating discussions.

References

[AL95]   S. Arora, C. Lund, *Hardness of Approximations*, in *Approximation Algorithms for NP-Hard Problems*, D. S. Hochbaum (ed.), PWS Publishing, Boston 1995, 399-446.

[BFK00]   C. Bazgan, W. Fernandez de la Vega and M. Karpinski, Approximability of dense instances of NEAREST CODEWORD problem, ECCC Tech. Report TR00-091, 2000, to appear in Proc. 8th SWAT (2002).

[BK99]   P. Berman and M. Karpinski, *On some tighter inapproximability results*, Proc. of 26th ICALP, LNCS 1644, Springer-Verlag, Berlin, 1999, 200-209.

[BK01]   P. Berman and M. Karpinski, *Approximating Minimum Unsatisfiability of Linear Equations*, ECCC Technical Report, TR01-025, 2001, also in Proc. 13th ACM-SIAM SODA (2002), 514-516.

[BJ92]   T. N. Bui and C. Jones, *Finding Good Approximate Vertex and Edge Partitions is Hard*, Inform. Process. Letters 42 (1992), 153-159.

[GG81]   O. Gabber and Z. Galil, *Explicit construction of linear size superconcentrators,* JCSS 22(1981), 407-420.

[DKS98]   I. Dinur, G. Kindler and S. Safra, *Approximating CVP to within almost polynomial factors is NP-hard*, Proc. of 39th IEEE FOCS, 1998, 99-109.

[DKRS00] I. Dinur, G. Kindler, R. Raz and S. Safra, *An improved lower bound for approximating CVP*, 2000, submitted.

[FK00]   U. Feige and R. Krauthgamer, *A Polylogarithmic Approximation of the Minimum Bisection*, Proc. 41st IEEE FOCS (2000), pp. 105–115.

[KST97]   S. Khanna, M. Sudan and L. Trevisan, *Constraint Satisfaction: the approximability of minimization problems*, Proc. of 12th IEEE Computational Complexity 1997, 282-296.

[LPS88]   A. Lubotzky, R. Phillips and P. Sarnak, *Ramanujan graphs*, Combinatorica, 8 (1988), 261–277.

[PY91]   C. Papadimitriou and M. Yannakakis, *Optimization, approximation and complexity classes*, JCSS **43**, 1991, pp. 425–440.

Improved Bounds and New Trade-Offs for Dynamic All Pairs Shortest Paths[*]

Camil Demetrescu[1] and Giuseppe F. Italiano[2]

[1] Dipartimento di Informatica e Sistemistica,
Università di Roma "La Sapienza",
Via Salaria 113, 00198 Roma, Italy. demetres@dis.uniroma1.it,
http://www.dis.uniroma1.it/~demetres/
[2] Dipartimento di Informatica, Sistemi e Produzione,
Università di Roma "Tor Vergata",
Via di Tor Vergata 110, 00133 Roma, Italy. italiano@disp.uniroma2.it,
http://www.info.uniroma2.it/~italiano/

Abstract. Let G be a directed graph with n vertices, subject to dynamic updates, and such that each edge weight can assume at most S different arbitrary real values throughout the sequence of updates. We present a new algorithm for maintaining all pairs shortest paths in G in $O(S^{0.5} \cdot n^{2.5} \log^{1.5} n)$ amortized time per update and in $O(1)$ worst-case time per distance query. This improves over previous bounds. We also show how to obtain query/update trade-offs for this problem, by introducing two new families of algorithms. Algorithms in the first family achieve an update bound of $\widetilde{O}(S \cdot k \cdot n^2)$ [1] and a query bound of $\widetilde{O}(n/k)$, and improve over the best known update bounds for k in the range $(n/S)^{1/3} \leq k < (n/S)^{1/2}$. Algorithms in the second family achieve an update bound of $\widetilde{O}(S \cdot k \cdot n^2)$ and a query bound of $\widetilde{O}(n^2/k^2)$, and are competitive with the best known update bounds (first family included) for k in the range $(n/S)^{1/6} \leq k < (n/S)^{1/3}$.

1 Introduction

All Pairs Shortest Path (APSP) is perhaps one of the most fundamental algorithmic graph problems. The fastest static algorithm for APSP on graphs with arbitrary real weights is achieved with the Fibonacci heaps of Fredman and Tarjan [10] and has a running time of $O(mn + n^2 \log n)$, where m is the number of edges and n is the number of vertices in the graph. This is $\Omega(n^3)$ in the worst case. Fredman [9] and later Takaoka [29] showed how to break this cubic barrier: the best asymptotic bound is by Takaoka, who showed how to solve APSP in $O(n^3 \sqrt{\log \log n / \log n})$ time.

---

[*] Work partially supported by the IST Programme of the EU under contract n. IST-1999-14.186 (ALCOM-FT), by the Human Potential Programme of the European Union under contract no. HPRN-CT-1999-00104 (AMORE), and by CNR, the Italian National Research Council, under contract n. 01.00690.CT26.

[1] Throughout the paper, we use $\widetilde{O}(f(n))$ to denote $O(f(n)\text{polylog}(n))$.

P. Widmayer et al. (Eds.): ICALP 2002, LNCS 2380, pp. 633–643, 2002.
© Springer-Verlag Berlin Heidelberg 2002

The quest for faster subcubic algorithms has led to many new results for APSP. In particular, Alon *et al.* [1] were the first to show that fast matrix multiplication can be effectively used for solving APSP on directed graphs with small integer weights: they gave an algorithm whose running time is $\tilde{O}(n^{(3+\omega)/2})$ for APSP on directed graphs whose weights are integers in the set $\{-1, 0, 1\}$, where ω is the best known exponent for matrix multiplication: currently, $\omega < 2.376$ [3]. Galil and Margalit [13,14] and Seidel [27] gave $\tilde{O}(n^\omega)$ algorithms for solving APSP for unweighted undirected graphs whose weights are integers in the set $\{-1, 0, 1\}$. Shoshan and Zwick [28] and Zwick [31] achieve the best known bound for APSP with positive integer edge weights less than C: $\tilde{O}(Cn^\omega)$ for undirected graphs [28] and $O(C^{0.681}n^{2.575})$ for directed graphs [31]. All these subcubic algorithms are based on clever reductions to fast matrix multiplication.

In this paper we consider fully dynamic algorithms for maintaining all pairs shortest paths on directed graphs with *real-valued* edge weights. A dynamic graph algorithm maintains a given property $\mathcal{P}$ on a graph subject to dynamic changes, such as edge insertions, edge deletions and edge weight updates. A dynamic graph algorithm should process queries on property $\mathcal{P}$, and must perform update operations faster than recomputing from scratch, as carried out by the fastest static algorithm. We say that an algorithm is *fully dynamic* if it can handle both edge insertions and edge deletions. A *partially dynamic* algorithm can handle either edge insertions or edge deletions, but not both: we say that it is *incremental* if it supports insertions only, and *decremental* if it supports deletions only.

In the fully dynamic APSP problem we wish to maintain a directed graph $G = (V, E)$ with real-valued edge weights under an intermixed sequence of the following operations:

Increase(w):   increase the edge weights as specified by the cost function $w : E \to \mathcal{R} \cup \{+\infty\}$. Note that the same operation can increase the cost of many edges simultaneously.

Decrease(v, w): decrease the weights of edges incident to vertex v as specified by the cost function $w : E \to \mathcal{R} \cup \{+\infty\}$. Note that the same operation can decrease the cost of many edges incident to v simultaneously.

Distance(x, y): output the shortest distance from x to y.

Path(x, y):   report a shortest path from x to y, if any.

We note that edge deletions are a special case of Increase, whenever the new weights are set to $+\infty$. Similarly, edge insertions are a special case of Decrease. Throughout the paper, we denote by n the number of vertices in G.

Previous Work. The dynamic maintenance of shortest paths has been investigated for over three decades, as the first papers date back to 1967 [20,22,25]. In 1985 Even and Gazit [7] and Rohnert [26] presented algorithms for maintaining shortest paths on directed graphs with arbitrary real weights. Their algorithms required $O(n^2)$ per edge insertion; however, the worst-case bounds for edge deletions were comparable to recomputing APSP from scratch. Also Ramalingam and Reps [23,24] considered dynamic shortest path algorithms with arbitrary real weights, but in a different model. Namely, the running time of their algorithm is analyzed in terms of the output change rather than the input size (*output bounded complexity*). Frigioni *et al.* [11,12] designed fast algorithms for graphs with bounded genus, bounded degree graphs, and bounded treewidth graphs in the same model. Again, in the worst case the running times of output-bounded dynamic algorithms are comparable to recomputing APSP from scratch.

Recently several dynamic shortest path algorithms, which are provably faster than recomputing APSP from scratch, were proposed. In particular, Henzinger *et al.* [17] designed a fully dynamic algorithm for APSP on planar graphs with integer weights, with a running time of $O(n^{9/7} \log(nC))$ per edge insertion or deletion. This bound has been improved by Fakcharoemphol and Rao in [8], who designed a fully dynamic algorithm for single-source shortest paths in planar directed graphs that supports both queries and edge weight updates in $O(n^{4/5} \log^{13/5} n)$ amortized time per edge operation. King [18] presented a fully dynamic algorithm for maintaining all pairs shortest paths in directed graphs with positive integer weights less than C: the running time of her algorithm is $O(C^{0.5} \cdot n^{2.5} \log^{0.5} n)$ per update. The space required is $O(n^3)$: a simple method for reducing space to $O(n^?)$ is shown in [19]. Finally, Demetrescu and Italiano [5] showed how to maintain APSP on directed graphs with arbitrary real weights. In particular, given a direct graph G, subject to dynamic operations, and such that each edge weight can assume at most S different *real* values, they give a deterministic algorithm that supports each update in $O(S \cdot n^{2.5} \log^3 n)$ amortized time and each query in $O(1)$ worst-case time.

The best known bounds for dynamic all pairs shortest paths on general graphs are thus $O(1)$ worst-case time per distance query and $\widetilde{O}(C^{0.5} \cdot n^{2.5})$ amortized time per update in case of integer weights in the range $[0, C]$ [18], or $\widetilde{O}(S \cdot n^{2.5})$ amortized time per update in case of arbitrary real-valued weights (with at most S different values per edge) [5]. Note that update bounds of [5] are slightly worse than the update bounds of [18] when restricted to positive integer edge weights, and that in both cases there is a large gap in the running times of distance queries and updates. In this scenario, it seems quite natural to ask whether the bounds in [5] can be improved so as to match the bounds in [18] and whether it is possible to reduce the (high) update bounds at the expenses of increasing the distance query times. The latter question seems particularly important in many applications, in which after a distance query one is typically required to exhibit an actual shortest path realizing that distance. Since reporting a shortest path might require as much as $\Omega(n)$ time, then one might as well afford to slow down the distance queries in order to speed the updates up. We remark that, despite many decades of research in the area of dynamic graph problems, there seem to be very few query/update trade-offs available in the literature (see e.g., [4,16]).

Our Results. In this paper, we present a new fully dynamic shortest path algorithm for directed graphs and arbitrary real weights (with at most S different values per edge weight), which achieves $O(S^{0.5} \cdot n^{2.5} \log^{1.5} n)$ amortized time per update and $O(1)$ worst-case time per query. This improves over [5] and matches, up to logarithmic factors, the bounds of [18] in the special case of integer weights. Furthermore, we make a first step towards the study of effective query/update trade-offs in dynamic shortest paths problems. In particular, we introduce two new families of fully dynamic algorithms for maintaining APSP on directed graphs, which are able to obtain the following bounds:

1 Algorithms in the first family achieve an update bound of $\widetilde{O}(S \cdot k \cdot n^2)$ and a query bound of $\widetilde{O}(n/k)$, for any integer k in the range $(n/S)^{1/3} \leq k \leq n$. Note that this improves the best known update bounds (above algorithm included) whenever k is in the range $(n/S)^{1/3} \leq k < (n/S)^{1/2}$. The fastest update time for this family of algorithms can be obtained by choosing $k = (n/S)^{1/3}$: this yields $\widetilde{O}(S^{2/3} \cdot n^{7/3}) = \widetilde{O}(S^{0.667} \cdot n^{2.333})$ amortized time per update and $\widetilde{O}(S^{1/3} \cdot n^{2/3}) = \widetilde{O}(S^{0.333} \cdot n^{0.667})$ worst-case time per query.

2 Algorithms in the second family are able to achieve an update bound of $\widetilde{O}(S \cdot k \cdot n^2)$ and a query bound of $\widetilde{O}(n^2/k^2)$, for any parameter k in the range $(n/S)^{1/6} \leq k \leq n$. This is competitive with the best known update bounds (first family included) whenever k is in the range $(n/S)^{1/6} \leq k < (n/S)^{1/3}$. The fastest update time for this family of algorithms is obtained for $k = (n/S)^{1/6}$: in this case, we get $\widetilde{O}(S^{5/6} \cdot n^{13/6}) = \widetilde{O}(S^{0.833} \cdot n^{2.167})$ amortized time per update and $\widetilde{O}(S^{1/3} \cdot n^{5/3}) = \widetilde{O}(S^{0.333} \cdot n^{1.667})$ worst-case time per query.

All our algorithms are able to report a shortest path with ℓ edges in time $O(\ell)$ plus the time for the corresponding distance query. Like other dynamic shortest paths algorithms, their space usage is $O(n^3)$. To achieve our results, we find a way to combine in a novel fashion the matrix techniques of [5] and the path decomposition techniques of [18]. We remark that our algorithms do not need fast matrix multiplication, use simple data structures, and thus seem amenable to efficient implementations.

2   Short Paths and Long Paths

All our algorithms maintain track of paths according to their length. In particular, they keep explicit information about shortest paths with at most k edges. This can be done in $O(S \cdot k \cdot n^2 \log^2 n)$ amortized time per update, where S is the number of different real values that each edge can assume, and in $O(1)$ worst-case time per distance query.

We now sketch few details about how these bounds can be achieved. We denote the operations as k-Decrease, k-Increase, and k-Query. We use a standard logarithmic path decomposition by maintaining $\log k$ levels: level i includes information about all (but not only) the shortest paths with at most 2^i edges.

The main idea is to keep $\log k$ polynomials $P_i(Y) = Y^3$, $1 \leq i \leq \log k$, over the $\{\min, +\}$ semiring with instances of the data structure presented in [5]. The degree 3 for the polynomial $P_i(Y) = Y^3$ is used in a dynamic setting exactly as described in [4] for the simpler problem of fully dynamic transitive closure. The proof of the following theorem appears in the full paper [6]:

Theorem 1. *Any* k-Decrease *and* k-Increase *operation requires* $O(S \cdot k \cdot n^2 \log^2 n)$ *amortized time and any* k-Query *is answered in* $O(1)$ *worst-case time.*

For paths with more than k edges, we exploit the following combinatorial property: if a subset H of vertices is picked at random from a graph G, then a sufficiently long path will intersect H with high probability. This property appears already in [15], and later on it has been used may times in designing efficient algorithms for transitive closure and shortest paths (see e.g., [5,16,18, 30,31]). The following theorem is from [30].

Theorem 2. *Let* $H \subseteq V$ *be a set of vertices chosen uniformly at random. Then the probability that a given simple path has a sequence of more than* $\frac{cn}{|H|} \log n$ *vertices, none of which are from* H, *for any* $c > 0$, *is, for sufficiently large* n, *bounded by* $2^{-\alpha c}$ *for some positive* α.

As shown in [31], it is possible to choose set H deterministically by a reduction to a hitting set problem [2,21]. A similar technique has also been used in [18].

Most of the fully dynamic bounds presented in this paper can be made deterministic by simply choosing set H deterministically instead of randomly. For the sake of brevity, we will focus our attention on the randomized versions of our bounds.

3   Dynamic Shortest Paths with $O(1)$ Queries

In this section we show how to improve the bounds of [5] for dynamic all pairs shortest paths in directed graphs where each edge can assume at most S different real values. In particular, we achieve an amortized update time of $O(S^{0.5} \cdot n^{2.5} \log^{1.5} n)$ and a query time of $O(1)$ in the worst case. If the shortest path has at most $(n/(S \log n))^{1/2}$ edges, then the query is answered correctly; otherwise it is answered correctly with probability $1 - 1/n^c$, for any constant $c > 0$. These bounds can be made deterministic.

Data Structure. Given a weighted directed graph $G = (V, E, w)$, we denote by X the weight matrix of G. We recall that the Kleene closure X^* of X is the distance matrix of G. We maintain the following elementary data structures.

1. A set $H \subseteq V$ of vertices chosen uniformly at random, with $|H| = \frac{cn}{k} \log n$, for any constant $c > 0$, with $c \log n \leq k \leq n$.

2. An $n \times n$ matrix D such that $D[x, y]$ is the shortest distance from $u \in V$ to $v \in V$ in G using at most k edges. We maintain D using an instance of the data structure given in Section 2. We denote by A the $n \times |H|$ matrix obtained from D by considering only the columns corresponding to vertices in H. We denote by B the $|H| \times |H|$ matrix obtained from D by considering only the rows and the columns corresponding to vertices in H. Finally, we denote by C the $|H| \times n$ matrix obtained from D by considering only the rows corresponding to vertices in H.

3. An $n \times |H|$ matrix $AB^* = A \cdot B^*$.

4. An $n \times n$ matrix $AB^*C = A \cdot B^* \cdot C$.

Our data structure uses the same path decomposition as in [18], i.e., $X^* = AB^*C \oplus D$, where $\oplus$ is defined as follows: $(X \oplus Y)[u, v] = \min_h \{X[u, h] + Y[h, v]\}$. The definitions of matrices A, B, C, and D imply that: $A[u, h]$ is the shortest distance from $u \in V$ to $h \in H$ in G using at most k edges, $B^*[h, h']$ is the shortest distance from $h \in H$ to $h' \in H$ in G, and $C[h', v]$ is the shortest distance from $h' \in H$ to $v \in V$ in G using at most k edges. By Theorem 2, $AB^*[u, h']$ gives the shortest distance from $u \in V$ to $h' \in H$, and $(AB^*C \oplus D)[u, v]$ gives the shortest distance from u to v in G: both these distances are correct with high probability.

Implementation of Operations. Matrix D is maintained dynamically using the operations described in Section 2. Matrices B^*, AB^*, and AB^*C are recomputed from scratch, while queries about entries of X^* are computed by accessing AB^*C and D. Operations can be supported as follows.

- Decrease(v, w): update D by calling k-Decrease(v, w). Next, recompute B^*, AB^*, and AB^*C from scratch.
- Increase(w): update D by calling k-Increase(w). Next, recompute B^*, AB^*, and AB^*C from scratch.
- Query(x, y): return $\min \{D[x, y], (AB^*C)[x, y]\}$.

Analysis. We now discuss the running time of our algorithm.

Theorem 3. *Any* Decrease *and* Increase *operation can be supported in* $O(S^{0.5} \cdot n^{2.5} \log^{1.5} n)$ *amortized time, while any* Query *operation can be answered in* $O(1)$ *worst-case time.*

Proof. Recall that $|H| = \frac{cn}{k} \log n$. The bound for updates is given by summing up the times required for updating D, and for recomputing B^*, AB^*, and AB^*C from scratch.

1) D: By Theorem 1, the cost of updating D is $O(S \cdot k \cdot n^2 \cdot \log^2 n)$.
2) B^*:   Recomputing B^* from scratch can be done trivially in $O(|H|^3) = O((n^3/k^3) \cdot \log^3 n)$ using any cubic static algorithm for APSP.
3) AB^*: Recomputing $A \cdot B^*$ from scratch takes time $O(n \cdot |H|^2) = O((n^3/k^2) \cdot \log^2 n)$.

4) AB^*C: Recomputing $AB^* \cdot C$ from scratch takes time $O(n^2 \cdot |H|) = O((n^3/k) \cdot \log n)$.

Choosing $k = (\frac{n}{S \cdot \log n})^{1/2}$ yields the claimed bounds.

4   Trading off Updates and Queries

In this section we show how to achieve new trade-offs in dynamic shortest paths problems. In particular, we present two families of fully dynamic algorithms for maintaining all pairs shortest paths in directed graphs where each edge can assume at most S different real values.

4.1   Family $\mathcal{F}_1$

For the first family, we achieve an update time of $\widetilde{O}(S \cdot k \cdot n^2)$ and a query time of $\widetilde{O}(n/k)$, for any $(n/S)^{1/3} \leq k \leq n$. If the shortest path has at most k edges, then the query is answered correctly; otherwise it is answered correctly with probability $1 - 1/n^c$, for any constant $c > 0$.

Data Structure. We maintain the same data structures as in Section 3, with the only difference that now we do not maintain explicitly the matrix AB^*C:

1. A set $H \subseteq V$ of vertices chosen uniformly at random, with $|H| = \frac{cn}{k} \log n$, for any constant $c > 0$, with $c \log n \leq k \leq n$.

2. An $n \times n$ matrix D such that $D[x, y]$ is the shortest distance from $u \in V$ to $v \in V$ in G using at most k edges. We maintain D using an instance of the data structure given in Section 2. We denote by A the $n \times |H|$ matrix obtained from D by considering only the columns corresponding to vertices in H. We denote by B the $|H| \times |H|$ matrix obtained from D by considering only the rows and the columns corresponding to vertices in H. Finally, we denote by C the $|H| \times n$ matrix obtained from D by considering only the rows corresponding to vertices in H.

3. An $n \times |H|$ matrix $AB^* = A \cdot B^*$.

Implementation of Operations. Once again, information about paths with more than k edges is encoded implicitly in the product of matrices $AB^* \cdot C$, while distances realized with at most k edges are maintained explicitly in D. As in Section 3, matrix D is maintained dynamically using the operations described in Section 2, while matrix AB^* is recomputed from scratch. The main difference is that matrix B^* is now updated more efficiently by plugging in the algorithm of Section 3. Since the product of matrices $AB^* \cdot C$ is not recomputed explicitly, queries about entries of X^* are now answered by accessing on demand matrices AB^* and C. In more details, operations are supported as follows.

- **Decrease**(v, w): update D by calling k-Decrease(v, w). Next, perform a v-centered decrease on B^* using the algorithm of Section 3. Finally, recompute AB^* from scratch.

  Note that when $v \notin H$, we need to maintain H dynamically. This can be done similarly to [5]. Namely, add v to H; to keep H random, add another random vertex to H; to prevent H from getting bigger, remove the two oldest vertices from H, and update B^* accordingly.

- **Increase**(w): update D by calling k-Increase(w), perform an increase operation on B^* using the algorithm of Section 3, and then recompute AB^* from scratch.

- **Query**(x, y): return

$$\min \left\{ D[x, y], \min_{h \in H} \{ AB^*[x, h] + C[h, y] \} \right\}.$$

Analysis. We now discuss the running time of our algorithm.

Theorem 4. *Let* $(n/S)^{1/3} \leq k \leq n$. *Any* Decrease *and* Increase *operation can be supported in* $\widetilde{O}(S \cdot k \cdot n^2)$ *amortized time, while any* Query *operation can be answered in* $\widetilde{O}(\frac{n}{k})$ *worst-case time.*

Proof. We first observe that $|H| = \frac{cn}{k} \log n$ and that entries of D (and thus those of B) cannot assume more than $S \cdot k$ different real values. The bound for updates is given by summing up the times required for updating D, updating B^*, and recomputing AB^*.

1. By Theorem 1, the cost of updating D is $\widetilde{O}(S \cdot k \cdot n^2)$.
2. As shown in Section 3, maintaining all pairs shortest paths in a graph with N vertices and edges taking at most s different values requires $\widetilde{O}(s^{0.5} \cdot N^{2.5})$ amortized time. Since entries of B cannot assume more than $s = S \cdot k$ different values and B has size $|H| \times |H|$, then the time required for updating B^* is $\widetilde{O}(S^{0.5} \cdot k^{0.5} \cdot n^{2.5}/k^{2.5}) = \widetilde{O}(S^{0.5} \cdot n^{2.5}/k^2)$, which is $\widetilde{O}(S \cdot k \cdot n^2)$ for $k \geq (n/S)^{1/6}$.
3. Recomputing AB^* from scratch takes time $O(n \cdot |H|^2) = \widetilde{O}(n^3/k^2)$, which is $\widetilde{O}(S \cdot k \cdot n^2)$ for $k \geq (n/S)^{1/3}$.

Thus for any $k \geq (n/S)^{1/3}$ the running time of Decrease and Increase operations is dominated by updates in D, which is $\widetilde{O}(S \cdot k \cdot n^2)$ amortized. To conclude the proof, we observe that queries require $O(|H|) = \widetilde{O}\left(\frac{n}{k}\right)$ worst-case time per operation. ∎

Choosing $k = (n/S)^{1/3}$ yields the following bounds.

Corollary 1. *Any* Decrease *and* Increase *operation can be supported in* $\widetilde{O}(S^{2/3} \cdot n^{7/3})$ *amortized time, while any* Query *operation can be answered in* $\widetilde{O}(S^{1/3} \cdot n^{2/3})$ *worst-case time.*

4.2 Family $\mathcal{F}_2$

For the second family, we extend the update time of $\widetilde{O}\left(S \cdot k \cdot n^2\right)$ to the larger range $(n/S)^{1/6} \le k \le n$, at the price of increasing the query time to $\widetilde{O}(n^2/k^2)$. Once again, if the shortest path has at most k edges, then the query is answered correctly; otherwise it is answered correctly with probability $1 - 1/n^c$, for any constant $c > 0$.

The main difference with the algorithms in family $\mathcal{F}_1$ is that we avoid the recomputation of AB^* during the updates. Namely, we maintain matrices D and B^* exactly as in Section 4.1, while imposing more burden on queries about entries of X^*, which now have to be computed as follows:

- Query(x, y): return

$$\min \left\{ D[x, y], \min_{h, h' \in H} \{A[x, h] + B^*[h, h'] + C[h', y]\} \right\}.$$

Theorem 5. *Let* $n^{1/6} \le k \le n$. *Any* Decrease *and* Increase *operation can be supported in* $\widetilde{O}\left(S \cdot k \cdot n^2\right)$ *amortized time, while any* Query *operation can be answered in* $\widetilde{O}\left(n^2/k^2\right)$ *worst-case time.*

Proof. Following the same lines as in the proof of Theorem 4, we can show that the running time of an update is given by the cost of updating D, which is $\widetilde{O}(S \cdot k \cdot n^2)$, plus the cost of updating B^*, which is $\widetilde{O}(S^{1/2} \cdot n^{2.5}/k^2)$. For $k \ge (n/S)^{1/6}$, the update in D is dominating, and thus the overall update bound is $\widetilde{O}(S \cdot k \cdot n^2)$. Queries require $O(|H|^2) = \widetilde{O}\left(n^2/k^2\right)$ worst-case time per operation.

Choosing $k = (n/S)^{1/6}$ yields the following bounds.

Corollary 2. *Any* Decrease *and* Increase *operation can be supported in* $\widetilde{O}(S^{5/6} \cdot n^{13/6})$ *amortized time, while any* Query *operation can be answered in* $\widetilde{O}(S^{1/3} \cdot n^{5/3})$ *worst-case time.*

5 Conclusions

In this paper, we have presented a new fully dynamic shortest path algorithm for directed graphs and arbitrary real weights (with at most S different values per edge weight), which supports updates in $O(S^{0.5} \cdot n^{2.5} \log^{1.5} n)$ amortized time and distance queries in $O(1)$ worst-case time. We have also introduced two new families of algorithms that obtain effective query/update trade-offs for the same problem. Besides improving any of our bounds, there are several issues that seem worth further investigation. First, is it possible to achieve quadratic update time while maintaing costant query time? Note that constant update time and quadratic query time is trivially obtainable by just doing nothing at each update and by performing a single-source shortest path computation at each query. Second, is there any dynamic all pairs shortest path algorithm for general graphs that is able to achieve subquadratic times for both updates and queries? We remark that this is possible for the simpler problem of fully dynamic transitive closure on directed acyclic graphs [4].

References

1. N. Alon, Z. Galil, and O. Margalit. On the exponent of the all pairs shortest path problem. *Journal of Computer and System Sciences*, 54(2):255–262, April 1997.
2. V. Chvátal. A greedy heuristic for the set-covering problem. *Mathematics of Operations Research*, 4(3):233–235, 1979.
3. D. Coppersmith and S. Winograd. Matrix multiplication via arithmetic progressions. *Journal of Symbolic Computation*, 9:251–280, 1990.
4. C. Demetrescu and G.F. Italiano. Fully dynamic transitive closure: Breaking through the $O(n^2)$ barrier. In *Proc. of the 41st IEEE Annual Symposium on Foundations of Computer Science (FOCS'00)*, pages 381–389, 2000. Full paper available at the URL: http://arXiv.org/abs/cs.DS/0104001.
5. C. Demetrescu and G.F. Italiano. Fully dynamic all pairs shortest paths with real edge weights. In *Proc. of the 42nd IEEE Annual Symposium on Foundations of Computer Science (FOCS'01), Las Vegas, Nevada*, pages 260–267, 2001.
6. C. Demetrescu and G.F. Italiano. Fully dynamic all pairs shortest paths with real edge weights. Full paper available at: http://www.dis.uniroma1.it/pub/demetres/papers/JCSS.pdf, 2002.
7. S. Even and H. Gazit. Updating distances in dynamic graphs. *Methods of Operations Research*, 49:371–387, 1985.
8. J. Fakcharoemphol and S. Rao. Planar graphs, negative weight edges, shortest paths, and near linear time. In *Proc. of the 42nd IEEE Annual Symposium on Foundations of Computer Science (FOCS'01), Las Vegas, Nevada*, pages 232–241, 2001.
9. M. L. Fredman. New bounds on the complexity of the shortest path problems. *SIAM Journal on Computing*, pages 87–89, 1976.
10. M.L. Fredman and R.E. Tarjan. Fibonacci heaps and their use in improved network optimization algorithms. *Journal of the ACM*, 34:596–615, 1987.
11. D. Frigioni, A. Marchetti-Spaccamela, and U. Nanni. Semi-dynamic algorithms for maintaining single source shortest paths trees. *Algorithmica*, 22(3):250–274, 1998.
12. D. Frigioni, A. Marchetti-Spaccamela, and U. Nanni. Fully dynamic algorithms for maintaining shortest paths trees. *Journal of Algorithms*, 34:351–381, 2000.
13. Z. Galil and O. Margalit. All pairs shortest distances for graphs with small integer length edges. *Information and Computation*, 134(2):103–139, 1 May 1997.
14. Z. Galil and O. Margalit. All pairs shortest paths for graphs with small integer length edges. *Journal of Computer and System Sciences*, 54(2):243–254, April 1997.
15. D. H. Greene and D.E. Knuth. *Mathematics for the analysis of algorithms*. Birkhäuser, 1982.
16. M. Henzinger and V. King. Fully dynamic biconnectivity and transitive closure. In *Proc. 36th IEEE Symposium on Foundations of Computer Science (FOCS'95)*, pages 664–672, 1995.
17. M.R. Henzinger, P. Klein, S. Rao, and S. Subramanian. Faster shortest-path algorithms for planar graphs. *Journal of Computer and System Sciences*, 55(1):3–23, August 1997.

18. V. King. Fully dynamic algorithms for maintaining all-pairs shortest paths and transitive closure in digraphs. In *Proc. 40th IEEE Symposium on Foundations of Computer Science (FOCS'99)*, pages 81–99, 1999.

19. V. King and M. Thorup. A space saving trick for directed dynamic transitive closure and shortest path algorithms. In *Proceedings of the 7th Annual International Computing and Combinatorics Conference (COCOON), LNCS 2108*, pages 268–277, 2001.

20. P. Loubal. A network evaluation procedure. *Highway Research Record 205*, pages 96–109, 1967.

21. L. Lovász. On the ratio of optimal integral and fractional covers. *Discrete Mathematics*, 13:383–390, 1975.

22. J. Murchland. The effect of increasing or decreasing the length of a single arc on all shortest distances in a graph. Technical report, LBS-TNT-26, London Business School, Transport Network Theory Unit, London, UK, 1967.

23. G. Ramalingam and T. Reps. An incremental algorithm for a generalization of the shortest path problem. *Journal of Algorithms*, 21:267–305, 1996.

24. G. Ramalingam and T. Reps. On the computational complexity of dynamic graph problems. *Theoretical Computer Science*, 158:233–277, 1996.

25. V. Rodionov. The parametric problem of shortest distances. *U.S.S.R. Computational Math. and Math. Phys.*, 8(5):336–343, 1968.

26. H. Rohnert. A dynamization of the all-pairs least cost problem. In *Proc. 2nd Annual Symposium on Theoretical Aspects of Computer Science, (STACS'85), LNCS 182*, pages 279–286, 1985.

27. R. Seidel. On the all-pairs-shortest-path problem in unweighted undirected graphs. *Journal of Computer and System Sciences*, 51(3):400–403, December 1995.

28. A. Shoshan and U. Zwick. All pairs shortest paths in undirected graphs with integer weights. In *40th Annual Symposium on Foundations of Computer Science: October 17–19, 1999, New York City, New York,*, pages 605–614, 1999.

29. T. Takaoka. A new upper bound on the complexity of the all pairs shortest path problem. *Information Processing Letters*, 43(4):195–199, September 1992.

30. J.D. Ullman and M. Yannakakis. High-probability parallel transitive-closure algorithms. *SIAM Journal on Computing*, 20(1):100–125, 1991.

31. U. Zwick. All pairs shortest paths in weighted directed graphs - exact and almost exact algorithms. In *Proc. of the 39th IEEE Annual Symposium on Foundations of Computer Science (FOCS'98)*, pages 310–319, Los Alamitos, CA, November 8–11 1998.

Synthesis of Uninitialized Systems*

Thomas A. Henzinger[1], Sriram C. Krishnan[2], Orna Kupferman[3], and
Freddy Y.C. Mang[4]

[1] Electrical Engineering and Computer Sciences, University of California at Berkeley.
tah@eecs.berkeley.edu
[2] Cisco Systems, Inc.
srikrish@cisco.com
[3] School of Computer Science and Engineering, Hebrew University.
orna@cs.huji.ac.il
[4] Advanced Technology Group, Synopsys, Inc.
fmang@synopsys.com

Abstract. The sequential synthesis problem, which is closely related to
Church's solvability problem, asks, given a specification in the form of a
binary relation between input and output streams, for the construction
of a finite-state stream transducer that converts inputs to appropriate
outputs. For efficiency reasons, practical sequential hardware is often
designed to operate without prior initialization. Such hardware designs
can be modeled by uninitialized state machines, which are required to
satisfy their specification if started from any state. In this paper we
solve the sequential synthesis problem for uninitialized systems, that is,
we construct uninitialized finite-state stream transducers. We consider
specifications given by LTL formulas, deterministic, nondeterministic,
universal, and alternating Büchi automata. We solve this *uninitialized
synthesis problem* by reducing it to the well-understood initialized synthesis problem. While our solution is straightforward, it leads, for some
specification formalisms, to upper bounds that are exponentially worse
than the complexity of the corresponding initialized problems. However,
we prove lower bounds to show that our simple solutions are optimal
for all considered specification formalisms. We also study the problem
of deciding whether a given specification is uninitialized, that is, if its
uninitialized and initialized synthesis problems coincide. We show that
this problem has, for each specification formalism, the same complexity
as the equivalence problem.

1 Introduction

In *sequential synthesis*, we transform a temporal specification into a reactive
system that is guaranteed to satisfy the specification. A *closed system* that meets
the specification can be extracted from a model that satisfies the specification,
that is, the synthesis of closed systems amounts to solving a satisfiability ($\exists$)

---

* This research was supported in part by the SRC contract 99-TJ-683.003, the DARPA
contract NAG2-1214, and the NSF grant CCR-9988172.

P. Widmayer et al. (Eds.): ICALP 2002, LNCS 2380, pp. 644–656, 2002.

problem [EC82]. However, as argued for transformational systems in [MW80], and for reactive systems in [ALW89,Dil89,PR89], the synthesis of *open systems*, which interact with an unknown environment, requires the solution of a $\forall\exists$ problem: for all sequences of inputs, there exists a sequence of outputs that satisfies the specification. Consider, for example, a scheduler for a printer that serves two users. The scheduler is an open system. Each time unit it reads the input signals $J1$ and $J2$ (a job sent from the first or second user, respectively), and writes the output signals $P1$ and $P2$ (print a job of the first or second user, respectively). The scheduler should be designed so that jobs of the two users are not printed simultaneously, and whenever a user sends a job, the job is printed eventually. Of course, this should hold no matter how the users send jobs. We can specify the requirement for the scheduler in terms of a *linear temporal logic* (LTL) formula ψ [Pnu81], such as

$$\Box(J1 \Rightarrow \bigcirc(\neg J1\,\mathcal{U}\,P1)) \wedge \Box(J2 \Rightarrow \bigcirc(\neg J2\,\mathcal{U}\,P2)) \wedge \Box\neg(J1 \wedge J2).$$

Evidence of ψ's satisfiablity (note that ψ is satisfied in a structure in which the four signals never occur) is not of much help in extracting a correct scheduler: while such evidence only suggests a scheduler that is guaranteed to satisfy ψ for *some* input sequence, we want a scheduler that satisfies ψ for *all* possible scripts of jobs sent to the printer.

We now make this intuition formal. A *stream transducer* is a function that, given an infinite sequence of inputs, produces an infinite sequence of outputs. In particular, for the set I of inputs signals and the set O of output signals, a stream transducer is a function from $(2^I)^\omega$ to $(2^O)^\omega$. A *stream requirement* is a binary relation between input streams and output streams; that is, a stream requirement is a subset of $(2^I)^\omega \times (2^O)^\omega$ or, equivalently, a set of infinite words in $(2^{I\cup O})^\omega$. The stream transducer T *realizes* the stream requirement R if for every input stream $\tau \in (2^I)^\omega$, we have $R(\tau, T(\tau))$. Stream requirements can be specified by LTL formulas over the set $I \cup O$ of atomic propositions, or by *automata on infinite words* over the alphabet $2^{I\cup O}$. Stream transducers can be implemented by state machines that proceed ad infinitum. The *finite-state implementation* of a stream transducer is a deterministic finite-state machine that, from a given state on a given set of input signals, generates a set of output signals and moves to a successor state. The *realizability problem* (RP) asks, given a stream requirement R, if there is a finite-state implementation of a stream transducer that realizes R. The *sequential synthesis problem*, then, is to find a finite-state implementation (if one exists). The RP was first stated by Church [Chu62] for stream requirements specified in the *sequential calculus*. Since then, several solutions for the RP have been studied: [BL69,Rab72] showed that the RP is quadratic (exponential) if the specification is a deterministic (nondeterministic) Büchi automaton; [PR89] showed that the RP is doubly exponential if the specification is an LTL formula (researchers from control theory also studied the RP in the context of supervisory control for discrete-event systems [RW89]). The solutions to the RP can be extended, within the same complexity bounds, to construct finite-state implementations, so that a solution to the RP immediately provides a solution also to the sequential synthesis problem [Rab70,MS95,KV99].

In practice, sequential hardware is often designed to operate without prior initialization; that is, it is supposed to satisfy its input-output requirements if started from any state. *Uninitialized state machines*, which model such hardware designs, require no reset circuitry and therefore have an advantage of smaller area. A well-known example of an uninitialized state machine is the IEEE 1149.1 standard for boundary-scan test [IEEE93]. Uninitialized state machines are also necessary for the *safe replaceability* of sequential circuits [SP94], where a state machine is replaced by another one in such a way that the surrounding environment is unable to detect the changes. The replacing state machine may power-up in an arbitrary state, and is therefore uninitialized. The verification problem of deciding whether an uninitialized state machine safely replaces another machine, is studied in [SP94]. The optimization problem for uninitialized state machines is studied in [QBSP96]. In this paper, we study the synthesis problem for uninitialized state machines.

Given a stream requirement R, the *uninitialized realizability problem* (URP) asks if there is a finite-state implementation M that realizes R no matter what the initial state of M is. The *uninitialized synthesis problem*, then, is to find such an M (if one exists). We study the URP for stream requirements that are specified by LTL formulas or Büchi automata. We consider deterministic and nondeterministic Büchi automata, as well as universal Büchi automata, which accept a word iff all runs are accepting, and alternating Büchi automata, which allow both nondeterministic and universal branching modes. The solution of the URP is quite straightforward, and is done by a reduction to the RP: if the stream requirement R is specified by an LTL formula, then the URP for R can be reduced to the RP for the LTL formula $always(R) = \Box R$; if R is specified by a Büchi automaton, then the URP for R can be reduced to the RP for the automaton $always(R)$, which is obtained from R by adding a universal self-loop at the initial state. It is not hard to see that an infinite word $w \in (2^{I \cup O})^\omega$ satisfies the specification $always(R)$ iff w and all its suffixes satisfy R. This implies that R is realizable by an uninitialized implementation iff $always(R)$ is realizable. As in the initialized case, a solution to the uninitialized synthesis problem follows immediately from a solution to the URP.

While the above solution is straightforward, it may lead to upper bounds that are exponentially worse than the complexity of the RP for the corresponding specification formalism. For example, while for LTL specifications both RP and URP are doubly exponential, for deterministic Büchi automata, where the RP is quadratic, the presented solution of the URP is exponential. The reason is that the automaton $always(R)$ has a universal branching mode, which R may not have, and this makes the URP exponentially harder. In particular, if R is a deterministic automaton, then $always(R)$ is universal, and if R is nondeterministic, then $always(R)$ is alternating. Can the exponential blow-up be avoided by a more sophisticated solution? We answer this question in the negative by proving corresponding lower bounds for the URP of all discussed formalisms. Unlike the upper bounds, the lower-bound proofs are not immediate, and are the main technical contributions of this paper. Our results imply that specification formalisms that support an easy implementation of the *always* operator, such as LTL and alternating automata, have, unlike deterministic and nondeterministic automata, already "built-in" the complexity of uninitialized synthesis.

We say that a stream requirement R is *uninitialized* if it is suffix-closed; that is, for all infinite words $w \in (2^{I \cup O})^\omega$, if $w \in R$, then $w' \in R$ for all suffixes w' of w. For example, the LTL specification $\Box p$ for an output signal p is uninitialized, as w satisfies $\Box p$ iff all suffixes of w satisfy $\Box p$. The *uninitialized specification problem* (USP) asks if a given stream requirement R is uninitialized. This is the same as asking if the two formulas, or automata, that specify R and $always(R)$ are equivalent. In the final section, we show that the USP has, for all considered specification formalisms, the same complexity as the equivalence problem. For uninitialized stream requirements, the URP coincides with the RP. As the equivalence problem is easier than the corresponding URP in all cases, it follows that for specification formalisms whose URP is harder than RP, there is an advantage to first checking if the specification is uninitialized.

2   Preliminary Definitions

Trees. Given a finite set D of directions, a D-tree is a set $T \subseteq D^*$ such that if $x \cdot d \in T$, where $x \in D^*$ and $d \in D$, then also $x \in T$. The elements of T are called *nodes*, and the empty word ϵ is the *root* of T. For every $x \in T$, the nodes $x \cdot d \in T$, for $d \in D$, are the *successors* of x. Each node $x \in T$ has a direction $dir(x)$ in D, namely, $dir(\epsilon) = d^0$ for some designated $d^0 \in D$, and $dir(x \cdot d) = d$. A path π of the tree T is a set $\pi \subseteq T$ such that $\epsilon \in \pi$, and for every $x \in \pi$, exactly one successor of x is in π. Given two finite sets D and Σ, a Σ-labeled D-tree is a pair (T, V), where T is a D-tree, and $V : T \to \Sigma$ maps each node of T to a letter in Σ. We extend V to paths in the straightforward way: for a path $\pi = \{\epsilon, w_0, w_0 w_1, \ldots\}$, we have $V(\pi) = V(\epsilon) V(w_0) V(w_0 w_1) \ldots$ We say that a $(D \times \Sigma)$-labeled D-tree (T, V) is D-*exhaustive* if $T = D^*$, and for every node $w \in D^*$, we have $V(w) = (dir(w), \sigma)$ for some $\sigma \in \Sigma$.

Alternating Büchi automata. For a given finite set X, let $\mathcal{B}^+(X)$ be the set of positive boolean formulas over X. A subset $Y \subseteq X$ satisfies a formula $\theta \in \mathcal{B}^+(X)$ if the truth assignment that assigns *true* to the members of Y and assigns *false* to the members of $X \setminus Y$ satisfies θ. An *alternating Büchi automaton* $\mathcal{U} = (\Sigma, U, u_0, \delta, F)$ consists of a finite alphabet Σ, a finite set U of states, an initial state $u_0 \in U$, a transition function $\delta : U \times \Sigma \to \mathcal{B}^+(U)$, and a set $F \subseteq U$ of accepting states. The automaton $\mathcal{U}$ is *universal* if $\delta(u, \sigma)$, for all $u \in U$ and $\sigma \in \Sigma$, is a conjunction of states from U; *nondeterministic*, if $\delta(u, \sigma)$ is a disjunction of states from U; and *deterministic*, if $\delta(u, \sigma)$ is a single state from U. A *run* of $\mathcal{U}$ on an infinite word $w = w_0 w_1 \ldots$ in Σ^ω is an infinite U-labeled D-tree (T, r), where $D = \{1, \ldots, |U|\}$, such that $r(\epsilon) = u_0$ and the following holds: for all nodes $x \in T$, if $|x| = i$ and $r(x) = u$ and $\delta(u, w_i) = \theta$, then x has k successors $x_1, \ldots, x_k$, for some $k \leq |U|$, and $\{r(x_1), \ldots, r(x_k)\}$ satisfies θ. A run (T, r) is *accepting* if every infinite path of (T, r) visits the accepting set F infinitely often. An infinite word w is accepted by $\mathcal{U}$ if there exists a run (T, r) on w such that

(T, r) is accepting. The *language* $L(\mathcal{U})$ is the set of infinite words accepted by $\mathcal{U}$.

Finite-state machines. A *finite-state machine* (FSM) $M = (I, O, Q, q_{in}, \rho, \lambda)$ consists of a finite set I of input signals, a finite set O of output signals, a finite state set Q, an initial state $q_{in} \in Q$, a transition function $\rho : Q \times 2^I \to Q$, and an output function $\lambda : Q \to 2^O$. We assume that there is a special output signal *init* such that *init* $\in \lambda(q)$ iff $q = q_{in}$. We also assume that there is a nonempty set $In(q_{in}) \subseteq 2^I$ such that for each $i_0 \in In(q_{in})$, there is a $q \in Q$ such that $\rho(q, i_0) = q_{in}$; that is, the state q_{in} is reachable via some input (if this is not the case, we can add a new state from which q_{in} is reachable). An FSM M interacts with its environment through its input and output signals. Initially, M is at the initial state $q_0 = q_{in}$. The environment initiates the interaction by inputting some $i_0 \in In(q_{in})$. Then, M starts operating by outputting $\lambda(q_0)$, to which the the environment replies with some input $i_1 \in 2^I$. The FSM M replies by moving to the state $q_1 = \rho(q_0, i_1)$ and outputting $\lambda(q_1)$. Interaction then continues ad infinitum.

Hence, the FSM M can be viewed as a *strategy* $S_M : (2^I)^* \to 2^O$ that maps every finite sequence of inputs to an output. To define S_M formally, we first define the function $C_M : (2^I)^* \to Q$ that maps each finite input sequence to the state visited after the sequence has been read: $C_M(\epsilon) = q_{in}$, and $C_M(i_1 \ldots i_n) = \rho(C_M(i_1 \ldots i_{n-1}), i_n)$. The strategy S_M induced by M is then defined for every $w \in (2^I)^*$ by $S_M(w) = \lambda(C_M(w))$. Note that the first input i_0 merely initiates the interaction and does not have any effect on the behavior of M; it is disregarded in the definition of C_M. Each infinite sequence $i_0 i_1 \ldots \in In(q_{in}) \cdot (2^I)^\omega$ induces a *computation* $(i_0, S_M(\epsilon))(i_1, S_M(i_1))(i_2, S_M(i_1 i_2)) \ldots \in (2^I \times 2^O)^\omega$ of M. The *language* $L(M)$ is the set of all computations of M. We refer to the language also as a set of infinite words in $(2^{I \cup O})^\omega$, where $i_0 i_1 \ldots$ induces the computation $(i_0 \cup S_M(\epsilon)) \cdot (i_1 \cup S_M(i_1)) \cdot (i_2 \cup S_M(i_1 i_2)) \ldots \in (2^{I \cup O})^\omega$. The strategy S_M induces, for a given first input $i_0 \in In(q_{in})$, a *computation tree* whose branches correspond to external nondeterminism caused by different inputs, namely, the 2^I-exhaustive $(2^I \times 2^O)$-labeled 2^I-tree $((2^I)^*, V)$ such that each node $w \in (2^I \times 2^O)^*$ is labeled by $V(w) = (dir(w), S_M(w))$, where $dir(\epsilon) = i_0$. Note that all computation trees of M differ only in the first input.

3   The Uninitialized Realizability Problem

In this section we define and solve the uninitialized realizability problem. We first start with the (initialized) realizability problem. Given a specification R over the input signals I and output signals O, the *realizability problem* (RP) for R asks if there is an FSM M such that for all words $w \in L(M)$, we have $w \models R$. If so, we say that R is *realizable* by M. The *specification* R can be an LTL formula, or an alternating Büchi automaton. If R is an LTL formula, then the atomic propositions of R are $I \cup O$, and the relation $\models$ is the usual satisfaction relation. If R is an automaton, then the alphabet of R is $2^I \times 2^O$, and the relation $\models$ is the language membership relation, that is, $w \models R$ iff $w \in L(R)$. The realizability problem is closely related to *Church's solvability problem* [Chu62], and it has been shown that the problem is solvable in quadratic (exponential) time if R is a deterministic (nondeterministic) Büchi automaton [BL69,Rab72,Saf88,PR89], and in doubly exponential time if R is an LTL formula [PR89,KV99].

An *uninitialized FSM* $M = (I, O, Q, \rho, \lambda)$ is similar to an FSM except that there is no initial state. The language $L(M) = \bigcup_{q \in Q} L(M_q)$ of M is simply the union of the languages $L(M_q)$, where $M_q = (I, O, Q, q, \rho, \lambda)$ is the FSM obtained from M by regarding the state $q \in Q$ as the initial state. Given a specification R over the input signals I and output signals O, the *uninitialized realizability problem* (URP) for R asks if there is an uninitialized FSM M such that $w \models R$ for all words $w \in L(M)$. If the answer is yes, we say that R is *uninitialized realizable* by M.

3.1   Reducing URP to RP

We solve the URP by reducing it to the RP. For that, we define, given a specification R over the input signals I and output signals O, the specification $always(R)$ over I and O such that, for all words $w \in (2^I \times 2^O)^\omega$, we have w satisfies $always(R)$ iff all suffixes w' of w satisfy R. It is not hard to see that the URP for R can be reduced to the RP for $always(R)$, as stated in the following theorem.

Theorem 1. *Let I and O be finite sets of input and output signals, respectively, and let R be a specification over I and O. Then R is uninitialized realizable iff $always(R)$ is realizable.*

Given a specification R, we construct the specification $always(R)$ as follows. First, if R is an LTL formula, it is not hard to see that $always(R) = \Box R$. Now, if R is an alternating Büchi automaton $\mathcal{U} = (\Sigma, U, u_0, \delta, F)$, then $always(\mathcal{U}) = (\Sigma, U \cup \{u_0'\}, u_0', \delta', F \cup \{u_0'\})$, where u_0' is a new state, and for all $\sigma \in \Sigma$, we have $\delta'(u_0', \sigma) = \delta(u_0, \sigma) \wedge u_0'$; and for all $u \in U$, we have $\delta'(u, \sigma) = \delta(u, \sigma)$. Intuitively, the automaton $always(\mathcal{U})$ behaves like $\mathcal{U}$ except that $always(\mathcal{U})$ always sends a copy of itself to the suffix of the input word whenever it makes a transition. It follows that for every word w, not only w has to be accepted by $\mathcal{U}$, but so do all its suffixes. Note that one copy of $always(\mathcal{U})$ keeps visiting u_0' forever, which is why we have to duplicate the original initial state. Formally, we have the following.

Proposition 1. *Let $\mathcal{U} = (\Sigma, U, u_0, \delta, F)$ be an alternating Büchi automaton. For all words $w \in \Sigma^\omega$, the automaton $always(\mathcal{U})$ accepts w iff $\mathcal{U}$ accepts all suffixes of w.*

3.2   URP Complexity

We can solve the URP for R by solving the RP for $always(R)$. The complexity of this simple solution depends on the type of the specification $always(R)$. In Table 1 below we describe the type of $always(R)$ given the type of R. It follows

Table 1. The cost of moving from R to $always(R)$.

R	$always(R)$
an LTL formula	an LTL formula
a deterministic or universal Büchi automaton	a universal Büchi automaton
a nondeterministic or alternating Büchi automaton	an alternating Büchi automaton

that if R is a deterministic or a nondeterministic Büchi automaton, then the type of $always(R)$ is richer than that of R, which in turn implies that the presented reduction from URP to RP incurs a cost. We now analyze the complexity of the URP and show that this cost is unavoidable, that is, the simple solution to the URP is optimal. First we consider LTL specifications.

Theorem 2. *The URP for LTL is 2EXPTIME-complete.*

Proof. The upper bound follows from the fact that the RP for LTL is in 2EXP-TIME [PR89], and $always(R)$, for R in LTL, is also in LTL. For the lower bound, we show that the URP is at least as hard as the RP, which is 2EXPTIME-hard [Ros92]. Indeed, the RP for an LTL formula φ can be reduced to the URP for $init \Rightarrow \varphi$. ∎

We now turn to the various types of Büchi automata. While the upper bounds are easy, the lower bounds require complicated generic reductions. To illustrate the proof ideas, we begin by considering the *closed* RP and URP, where the set I of input signals is empty. In this case, the behavior of the desired FSM is independent of the environment, and RP coincides with the satisfiability problem. In particular, for deterministic Büchi automata, the closed RP is NLOGSPACE-complete. We show that the transition to *uninitialized* FSMs makes the problem exponentially harder.

Proposition 2. *The closed URP for deterministic Büchi automata is PSPACE-complete.*

An automata-theoretic problem that is well-known to be PSPACE-complete is the universality problem for *nondeterministic* automata [HU79,Wol82]. On the other hand, the universality problem for *deterministic* automata is NLOGSPACE-complete. The standard PSPACE lower bound proof is by reduction from the membership problem for polynomial space Turing machines: given a polynomial-space Turing machine T, one constructs a nondeterministic Büchi automaton $\mathcal{U}$ such that $\mathcal{U}$ accepts invalid or rejecting computations of T. The closed URP also has some flavor of "universality." It comes from the requirement that all suffixes of an infinite word need to satisfy the specification. However, the lower-bound proof is different; we construct a *deterministic* Büchi automaton that accepts an infinite word w as well as all its suffixes iff w is a *valid* and *accepting* computation of T.

Proof of Proposition 2. Consider a deterministic Büchi automaton R. When $I = \emptyset$, the set 2^I is a singleton and the universal Büchi automaton $always(R)$ is realizable iff its language is not empty. Since the latter can be checked in PSPACE [MH84,VW94], the upper bound follows.

For the lower bound, we do a reduction from the membership problem of a polynomial space Turing machine. Let $T = (\Gamma, Q, \mapsto, q_0, F_{acc}, F_{rej})$ be a polynomial space Turing machine, where Q is the set of states, $q_0 \in Q$ is the initial state, $F_{acc} \subseteq Q$ and $F_{rej} \subseteq Q$ are respectively accepting and rejecting states, and $\mapsto: Q \times \Gamma \to Q \times \Gamma \times \{L, R\}$ is the transition function. Assume T starts with the initial configuration, i.e., T at state q_0 with its reading head pointing at the leftmost cell of the empty tape. We also assume that once T reaches an accepting configuration, i.e., $q \in F_{acc}$, it "cleans" the tape content and restarts from the initial configuration. The machine T accepts the empty tape iff T has an infinite computation visiting the initial and accepting configurations infinitely often.

Assume T uses $s(n)$ tape cells to process an input of length n. We encode a configuration of T by a string $\sharp\gamma_1\gamma_2\ldots(q,\gamma_i)\ldots\gamma_{s(n)}$ in $(2^O)^*$, where subsets of the output signals O are selected to encode the the alphabets $\Gamma \cup (Q \times \Gamma) \cup \{\sharp\}$, i.e., $2^O = \Gamma \cup (Q \times \Gamma) \cup \{\sharp\}$. That is, a configuration starts with the letter $\sharp$, followed by a string of letters $\gamma_j \in \Gamma$, except for one in $Q \times \Gamma$. The meaning of the string is that γ_j is the letter on the j-th tape cell, while the letter (q, γ_i) indicates in addition that T is at state q with its reading head pointing at the i-th tape cell. Let $c = \sharp\sigma_1\sigma_2\ldots\sigma_{s(n)}$, and $c' = \sharp\sigma'_1\sigma'_2\ldots\sigma'_{s(n)}$ be two configurations. If c' is the successor of c, then we know by the transition function of T what σ'_i for $1 \leq i \leq s(n)$ should be. We let $next(\sigma_{i-1}, \sigma_i, \sigma_{i+1})$ denote our expectations for σ'_i.

We define a deterministic Büchi automaton $\mathcal{U} = (2^O, U, u_0, \delta, F)$ such that $\mathcal{U}$ accepts an input word $w = w_0w_1\ldots$ iff w satisfies the following conditions: (1) The $next$ relation of T is satisfied for the first three letters in w, i.e., $w_{s(n)+2} = next(w_0, w_1, w_2)$; and (2) the initial and the accepting configurations are eventually reached. It follows that $\mathcal{U}$ accepts an infinite word w as well as all its suffixes iff T has an infinite computation visiting the initial and accepting configuration infinitely often. Both conditions can be specified by a deterministic Büchi automaton of size polynomial in T.

Thus, if there exists a word w such that w and all its suffixes are accepted by $\mathcal{U}$, then there exists a suffix w' of w such that w' encodes an accepting run of T. On the other hand, if T has an accepting run, then it can be encoded as an infinite string $w \in \Sigma_0^*$ all of whose suffixes (including w itself) are accepted by $\mathcal{U}$. ∎

We now consider the general, *open* URP, where $I \neq \emptyset$. For deterministic Büchi automata, where the RP is quadratic, the URP is harder than both the RP and the closed URP.

Theorem 3. *The URP for deterministic or universal Büchi automata is EXPTIME-complete.*

Proof. Consider a deterministic or a universal Büchi automaton $\mathcal{U}$. The automaton $always(\mathcal{U})$ is a universal automaton, whose RP can be solved in EXP-TIME [Rab72,MS95].

For the lower bound, we use the input signals in I in order to encode branches and extend the proof of Theorem 2 to apply to alternating Turing machines. Consider an alternating linear-space Turing machine $T = (\Gamma, Q_u, Q_e, \mapsto , q_0, F_{acc}, F_{rej})$, where the disjoint sets of states Q_u and Q_e are respectively the universal and existential states, while the disjoint sets of states $F_{acc} \subseteq Q_e$ and $F_{rej} \subseteq Q_e$ are respectively the accepting and rejecting states. Their union is denoted by Q. Our model of alternation prescribes that $\mapsto \subseteq Q \times \Gamma \times Q \times \Gamma \times \{L, R\}$ has a binary branching degree. When a universal or an existential state of T branches into two states, we distinguish between the left and the right branches. Accordingly, we use $(q, a) \mapsto^l (q_l, b_l, \Delta_l)$ and $(q, a) \mapsto^r (q_r, b_r, \Delta_r)$ to indicate that when T is in state $q \in Q_u \cup Q_e$ reading input symbol a, it branches to the left with (q_l, b_l, Δ_l) and to the right with (q_r, b_r, Δ_r). We also assume that once T reaches an accepting configuration, it "cleans" the tape content and restarts from the initial configuration (i.e., empty tape and initial state at the left end of the tape).

Assume T uses $s(n)$ cells in its working tape in order to process an input of length n. A configuration of T is encoded in a similar way to how a configuration of a polynomial space Turing machine is encoded, except that a configuration starts with either $\sharp_l$ or $\sharp_r$. The letter $\sharp \in \{\sharp_l, \sharp_r\}$ marks the beginning of a configuration; moreover, since T has an existential mode, i.e., when the state q of T is in Q_e, the letter $\sharp$ also indicates a guess (left or right) for the accepting successor. A computation of T can then be encoded by a computation tree whose branches describe sequences of configurations of T. Note that the computation tree is unique if we ignore the distinction between $\sharp_l$ and $\sharp_r$. A run of T is a pruning of a computation tree in which all the universal configurations have both successors and all the existential configurations $c = \sharp\gamma_1\gamma_2 \ldots (q, \gamma_i) \ldots \gamma_{s(n)}$ have only the left (resp. right) successor if $\sharp = \sharp_l$ (resp. $\sharp_r$). The run is accepting if all branches in the pruned tree visit the initial and accepting configuration infinitely often.

Given an alternating linear-space Turing machine T as above, we construct a deterministic Büchi word automaton $\mathcal{U}$ such that $\mathcal{U}$ is uninitialized realizable iff T has an accepting run on the empty tape (clearly, proving a lower bound for deterministic automata, implies a bound also for universal ones). The automaton $\mathcal{U}$ has input signals I such that the subsets of I encode the set $\{l, r\}$, i.e., $2^I = \{l, r\}$. It also has output signals O such that $2^O = \{\sharp_l, \sharp_r\} \cup \Gamma \cup (Q \times \Gamma)$. Let $c = \sharp\sigma_1\sigma_2 \ldots \sigma_{s(n)}$ and $c' = \sharp'\sigma_1'\sigma_2' \ldots \sigma_{s(n)}'$ be two configurations, and let $(d_0, \sharp)(d_1, \sigma_1) \ldots (d_{s(n)}, \sigma_{s(n)})(d_0', \sharp')(d_1', \sigma_1') \ldots (d_{s(n)}', \sigma_{s(n)}')$ be a word in $(2^I \times 2^O)^*$. The letter d_0' indicates the direction of c' with respect to c: if $d_0' = l$, then c' is the left successor of c, and if $d_0' = r$, then c' is the right successor of c. Note that the direction of c' is given by d_0', not by the letter $\sharp$ or $\sharp'$; the letter $\sharp$ is only the guess that T makes at c if c is an existential configuration. If $\sharp = \sharp_l$ (resp. $\sharp_r$), then T guesses that the left (resp. right) successor of c leads to an accepting run:

if the guess of T is different from the successor information given by the input, we say that there is a mismatch between the input and the guess of T at the configuration. That is, a mismatch happens at c with $\sharp = \sharp_r$ and $d_0' = l$, as well as with $\sharp = \sharp_l$ and $d_0' = r$. Recall that we require every path of the computation tree of T to be legal and accepting. On the other hand, since T is alternating, only the paths in the computation tree that are guessed in existential configurations need to be accepting. We use $\sharp_l$ and $\sharp_r$ in order to detect mismatches, where paths that contain a mismatch are considered accepting.

If the configuration c' is a successor of the configuration c, we know by the transition relation of T what the "next" relation is. Now we have two "next" relations, one for left branching and one for right branching. Let $next^l$ and $next^r$ be the "next" relations for the left branch and right branch respectively. The definition of $next^l$ (resp. $next^r$) is similar to that of the $next$ relation in the polynomial space Turing machine case, except that only the transition function $\mapsto^l$ (resp. $\mapsto^r$) is considered, the letter $\sharp$ is in $\{\sharp_l, \sharp_r\}$, and $next^l(\sigma_{s(n)}, \sharp, \sigma_1') \in \{\sharp_l, \sharp_r\}$.

The automaton $\mathcal{U}$ can be constructed as follows. On input of a word $w = (d_0, \sigma_0)(d_1, \sigma_1)\ldots$, $\mathcal{U}$ checks the following:

1. The "next" transition relations of T are satisfied, i.e., $\sigma_{s(n)+2} = next^l(\sigma_0, \sigma_1, \sigma_2)$ if $d_0' = l$, and $\sigma_{s(n)+2} = next^r(\sigma_0, \sigma_1, \sigma_2)$ if $d_0' = r$; and
2. either of the following is true:
   a) Eventually there is a mismatch in the direction specified by the input and T at an existential configuration, i.e., w contains the string $(d_0, \sharp_0)\ldots(d_j, (q, \gamma_j))\ldots(d_{s(n)}, \sigma_{s(n)})(d_0', \sharp_0')$, where $q \in Q_e$ and either $\sharp_0 = \sharp_r$ and $d_0' = l$, or $\sharp_0 = \sharp_l$ and $d_0' = r$.
   b) The initial configuration is eventually reached, and thereafter the accepting configuration is also eventually reached.

All the above conditions can be specified by a deterministic Büchi word automaton linear in the size of T.

Given a path w of a (2^I)-exhaustive $(2^I \times 2^O)$-labeled 2^I-tree $((2^I)^*, \tau)$, if $\mathcal{U}$ accepts w and all the suffixes of w, then by condition (2), w is a valid branch of the computation tree of T; moreover, by condition (3), if w is a branch guessed by T, then infinitely often the initial and accepting configurations are reached. Thus, $\mathcal{U}$ accepts all suffixes of all branches of $((2^I)^*, \tau)$ iff T has an accepting run. ■

The RP for nondeterministic Büchi automata can be solved in exponential time, while the RP for alternating Büchi automata requires doubly exponential time. The following theorem shows that the URP for nondeterministic Büchi automata is exponentially harder than the RP, while for alternating Büchi automata, there is no additional cost.

Theorem 4. *The URP for nondeterministic or alternating Büchi automata is 2EXPTIME-complete.*

Proof. Consider a nondeterministic or an alternating Büchi automaton $\mathcal{U}$. The automaton $always(\mathcal{U})$ is an alternating automaton, whose RP can be solved in 2EXPTIME [Rab72,MS95]. The lower bound proof is similar to that for the URP for deterministic Büchi automata in Theorem 3, except that now we reduce from the membership problem of an alternating exponential space Turing machine. ∎

4   Uninitialized Specifications

For a specification R over the input signals I and output signals O, we say that R is *uninitialized* if for every infinite word $w \in (2^I \times 2^O)^\omega$, we have w satisfies R iff all suffixes of w satisfy R. It is easy to see that if R is uninitialized, then every FSM M that realizes R induces an uninitialized FSM M' (obtained from M by dropping its initial state) that uninitialized realizes R. Hence, a solution of the URP for R can be obtained from a solution of the RP for R. The *uninitialized specification problem* (USP) for a specification R asks whether or not R is uninitialized.

Solving the USP for the specification R amounts to checking if R is equivalent to $always(R)$. Clearly, $always(R)$ implies R, thus we only need to check whether R implies $always(R)$. For LTL formulas, this can be done by checking the validity of $R \Rightarrow always(R)$, and for alternating Büchi automata we need to solve the language-containment problem $L(R) \subseteq L(always(R))$. We show that this simple approach, like the simple solution for URP, is also optimal. The lower bounds can be obtained by reductions from both the satisfiability and validity problems for the various specification formalisms (for an alternating Büchi automaton $\mathcal{U}$ with alphabet Σ, we say that $\mathcal{U}$ is *satisfiable* if $L(\mathcal{U}) \neq \emptyset$, and $\mathcal{U}$ is *valid* if $L(\mathcal{U}) = \Sigma^\omega$).

Lemma 1. *The USP for LTL, deterministic, nondeterministic, universal, or alternating Büchi automata, is (1) at least as hard as the corresponding satisfiability problem, and (2) at least as hard as the corresponding validity problem.*

We can now obtain complexity bounds for the USP. Our results are summarized in Table 2. All bounds in the table are tight.

Theorem 5. *The USP for LTL, universal, nondeterministic, or alternating Büchi automata is PSPACE-complete.*

Proof. For the upper bound, recall that R is uninitialized iff R implies $always(R)$. If the specification R is an LTL formula, then checking validity of $R \Rightarrow always(R)$ is in PSPACE [SC85]. If R is an alternating (or universal) automaton, we have to check the language-containment problem $L(R) \subseteq L(always(R))$. For that, we can first construct a nondeterministic Büchi automaton R_n such that the size of R_n is exponential in the size of R and $L(R_n) = L(R)$ [MH84], and we construct a nondeterministic Rabin automaton R_c such that the size of R_c is exponential in the size of $always(R)$ and R_c complements $always(R)$ (that is, $L(R_c) = \Sigma^\omega \setminus L(always(R))$.) [MS95]. Now, the product of R_n and R_c

is a nondeterministic Rabin automaton whose emptiness can be checked in nondeterministic logarithmic space, implying a PSPACE upper bound for the USP.

The lower bound follows from Lemma 1, and from the fact that the satisfiability problem for LTL and universal (or alternating) Büchi automata, as well as the validity problem for nondeterministic Büchi automata are PSPACE-hard [SC85,HU79,Wol82]. ∎

Theorem 6. *The USP for deterministic Büchi automata is NLOGSPACE-complete.*

Proof. For a deterministic Büchi automaton R, the automaton $always(\mathcal{U})$ is universal, thus its complement R_c is a nondeterministic co-Büchi automaton. The product of R and R_c can be defined as a nondeterministic Rabin automaton, whose emptiness problem can be solved in nondeterministic logarithmic space, implying an NLOGSPACE upper bound for the USP.

The lower bound follows from Lemma 1, and from the fact that the satisfiability problem for deterministic Büchi automata is NLOGSPACE-hard. ∎

Table 2. The complexity of the RP, URP, and USP.

	URP	RP	USP = Equivalence
LTL formulas	2EXPTIME	2EXPTIME	PSPACE
deterministic Büchi automata	EXPTIME	quadratic	NLOGSPACE
nondeterministic Büchi automata	2EXPTIME	EXPTIME	PSPACE
universal Büchi automata	EXPTIME	EXPTIME	PSPACE
alternating Büchi automata	2EXPTIME	2EXPTIME	PSPACE

References

[ALW89]   M. Abadi, L. Lamport, and P. Wolper. Realizable and unrealizable concurrent program specifications. In *Proc. 16th Intl. Colloquium on Automata, Languages, and Programming*, LNCS 372, pages 1–17. Springer-Verlag, 1989

[BL69]   J.R. Büchi and L.H. Landweber. Solving sequential conditions by finite-state strategies. *Transactions of the American Mathematical Society*, 138:295–311, 1969.

[Chu62]   A. Church. Logic, arithmetic, and automata. In *Proc. Intl. Congress of Mathematicians*, pages 23–35. Institut Mittag-Leffler, 1962.

[Dil89]   D.L. Dill. *Trace Theory for Automatic Hierarchical Verification of Speed Independent Circuits*. MIT Press, 1989.

[EC82]   E.A. Emerson and E.M. Clarke. Using branching time logic to synthesize synchronization skeletons. *Science of Computer Programming*, 2:241–266, 1982.

[HU79]   J. E. Hopcroft and J. D. Ullman, *Introduction to Automata Theory, Languages, and Computation*. Addison-Wesley, 1987.

[IEEE93]   IEEE Standard 1149.1-1993. *IEEE Standard Test Access Port and Boundary Scan Architecture*. IEEE, 1993.

[KV99]   O. Kupferman and M.Y. Vardi. Church's problem revisited. *The Bulletin of Symbolic Logic*, 5:245–263, 1999.

[MH84]   S. Miyano and T. Hayashi. Alternating finite automata on ω-words. *Theoretical Computer Science*, 32:321–230, 1984.

[MS95]   D.E. Muller and P.E. Schupp. Simulating aternating tree automata by nondeterministic automata: New results and new proofs of theorems of Rabin, McNaughton, and Safra. *Theoretical Computer Science*, 141:69–107, 1995.

[MW80]   Z. Manna and R. Waldinger. A deductive approach to program synthesis. *ACM Transactions on Programming Languages and Systems*, 2:90–121, 1980.

[Pnu81]   A. Pnueli. The temporal semantics of concurrent programs. *Theoretical Computer Science*, 13:45–60, 1981.

[PR89]   A. Pnueli and R. Rosner. On the synthesis of a reactive module. In *Proc. 16th Symposium on Principles of Programming Languages*, pages 179–190. ACM Press, 1989.

[QBSP96]   S. Qadeer, R. K. Brayton, V. Singhal, and C. Pixley. Latch redundancy removal without global reset. In *Proc. Intl. Conference on Computer Design*, pages 432–439. IEEE Computer Society, 1996.

[Rab70]   M.O. Rabin. Weakly definable relations and special automata. *Mathematical Logic and Foundations of Set theory*, 1970.

[Rab72]   M.O. Rabin. *Automata on Infinite Objects and Church's Problem*. Number 13 in Regional Conference Series in Mathematics. American Mathematical Society, 1972.

[RW89]   P.J.G. Ramadge and W.M. Wonham. The control of discrete event systems. *IEEE Transactions on Control Theory*, 77:81–98, 1989.

[Ros92]   R. Rosner. *Modular Synthesis of Reactive Systems*. PhD thesis, Weizmann Institute of Science, 1992.

[Saf88]   S. Safra. On the complexity of omega-automata. In *Proc. 29th Symposium on Foundations of Computer Science*, pages 319–327. IEEE Computer Society, 1988.

[SC85]   A.P. Sistla and E.M. Clarke. The complexity of propositional linear temporal logic, *Journal of the ACM*, 32:733–749, 1985.

[SP94]   V. Singhal and C. Pixley. The verification problem for safe replaceability. In *Proc. Conference on Computer-Aided Verification*, LNCS 818, pages 311–323. Springer-Verlag, 1994.

[VW94]   M.Y. Vardi and P. Wolper. Reasoning about infinite computations. *Information and Computation*, 115:1–37, 1994.

[Wol82]   P. Wolper. *Synthesis of Communicating Processes from Temporal Logic Specifications*. PhD thesis, Stanford University, 1982.

Infinite-State High-Level MSCs: Model-Checking and Realizability

(Extended Abstract)

Blaise Genest[1*], Anca Muscholl[1*], Helmut Seidl[2], and Marc Zeitoun[1*]

[1] LIAFA, Université Paris VII
2, pl. Jussieu, case 7014, 75251 Paris cedex 05, France
[2] FB IV, Universität Trier, 54286 Trier, Germany

Abstract. We consider three natural classes of infinite-state HMSCs: globally-cooperative, locally-cooperative and local-choice HMSCs. We show first that model-checking for globally-cooperative and locally-cooperative HMSCs has the same complexity as for the class of finite-state (bounded) HMSCs. Surprisingly, model-checking local-choice HMSCs turns out to be exponentially more efficient in space than for locally-cooperative HMSCs. We also show that locally-cooperative and local-choice HMSCs can be always implemented by communicating finite states machines, provided we allow some additional (bounded) message data. Moreover, the implementation of local-choice HMSCs is deadlock-free and of linear-size.

1   Introduction

Message sequence charts (MSC) is a visual notation for asynchronously communicating processes and a standard of the ITU [1]. The usual application of MSCs in telecommunication is for capturing requirements of communication protocols in form of scenarios at early design stages. MSCs usually represent incomplete specifications, obtained from a preliminary view of the system that abstracts away several details such as variables or message contents. High-level MSCs (HMSCs) combine basic MSCs using choice and iteration, thus describing possibly infinite collections of scenarios. From the viewpoint of automatic verification, high-level MSCs are infinite-state systems. Moreover, certain basic questions as model-checking against HMSC properties [5,14] are undecidable.

A preliminary specification of a communication protocol can suffer from several deficiencies, either related to the partial order of events (e.g. race conditions [4,14]) or to the violation of user-defined properties specified in some logics or HMSCs (model-checking [5]). The detection of possible failures in early design stages is of critical importance, and the utility of HMSCs can be greatly enhanced by automatic validation methods. A natural question for HMSC specifications is to test whether the specification is implementable (or realizable) [2]

---

* Work partly supported by the European research project IST-1999-29082 ADVANCE and by the INRIA/IRISA ARC project FISC.

P. Widmayer et al. (Eds.): ICALP 2002, LNCS 2380, pp. 657–668, 2002.

by a communication protocol and to construct such an implementation. Since an abstract communication protocol is usually described by communicating finite-state machines (CFM), we test implementability against CFMs. Opposed to HMSC specifications, no global control is available in CFMs. In order to install a distributed control the CFM realization therefore may have to add further data to messages or even exchange additional (synchronization) messages. Our goal is to exhibit general techniques for synthesizing such distributed control. Once an implementation is available, one can easily simulate executions of the HMSC using the ITU-standard model SDL (Specification and Description Language) and use SDL tools for model-checking the HMSC specification.

We consider in this paper three natural classes of infinite-state HMSCs: globally-cooperative, locally-cooperative and local-choice HMSCs. Locally-cooperative and local-choice HMSCs are (syntactically incomparable) subclasses of globally-cooperative HMSCs. The crucial property of globally-cooperative HMSCs is there exists a *regular* set of representative behaviors, however buffers are unbounded (making the system infinite-state). For telecommunication applications it is of course essential to cope with unbounded communication buffers. Globally-cooperative HMSCs have been introduced independently in [12], whereas locally-cooperative HMSCs are defined in this paper. The local-choice property we use here has been considered in [8].

In the first part of the paper (Section 3) we consider the model-checking problem and we show that it is decidable for globally-cooperative HMSCs. The model-checking problem is stated as intersection (negative property) or inclusion (positive property) of HMSCs, that is, the property to be tested is also described by an HMSC. Recall that both questions are undecidable for unrestricted HMSCs, which has been the motivation for considering regular (finite-state) HMSCs [5,14,10]. Here, we show that negative and positive model-checking are PSPACE- and EXPSPACE-complete, respectively, for both globally-cooperative and locally-cooperative HMSCs — thus generalizing the complexity bounds for regular HMSCs to infinite-state HMSCs. For local-choice HMSCs we obtain better algorithms, specifically negative model-checking is quadratic time whereas positive model-checking is PSPACE-complete.

In the second part of the paper we consider the synthesis of communicating finite-state machines from locally-cooperative, resp. local-choice HMSCs (Sections 4 and 4.3). We adopt a moderate view by allowing additional message contents, while ruling out extra control messages. The reason is that additional messages mean additional process synchronization. This is not desirable, or even not realizable, in a given environment. Still, our implementation semantics by CFMs is more general than the one introduced in [2] and used in [3,12] where a parallel product of finite-state automata communicating over FIFO channels is employed to realize the (linear) behavior of each process of the given HMSC. Moreover, implementability in this framework is undecidable even for bounded HMSCs (however, deadlock-free implementability is in EXPSPACE). Other notions of implementation are proposed in [7],[8],[13]. We show that both HMSCs classes are *always implementable* by CFMs. Moreover, the CFM implementation of local-choice HMSCs is deadlock-free and of linear-size.

The proofs use standard techniques from Mazurkiewicz trace theory as well as specific partial orders methods. Proofs are in most cases omitted in this abstract, due to lack of space.

2   Preliminaries

In this section we recall the specification formalism of message sequence charts (MSC) and high-level message sequence charts (HMSC) based on the ITU standard Z.120 [1]. An MSC describes a scenario or an execution of a communication protocol in which processes communicate with each other over point-to-point channels. Such a scenario is given by a description of the messages sent and received, the local events, and the ordering between them. The event ordering is based on a process ordering and a message ordering. In the visual description of MSCs, each process is represented by a vertical line, which shows the total order on the events belonging to that process. Messages are represented by horizontal or slanted arrows from the sending process to the receiving one.

MSC. An MSC over process set $\mathcal{P}$ is a tuple $M = \langle E, <, \mathcal{P}, t, \mathcal{C}, m \rangle$ where:

- $E = \bigcup_{p \in \mathcal{P}} E_p$ is a finite set of events, each located on some process from the set $\mathcal{P}$, with E_p denoting the set of events of process p. We denote by $P(e) \in \mathcal{P}$ the process to which event e belongs.
- Every event is either a communication event (send or receive) or a local event. We write $E = S \uplus R \uplus L$ as a disjoint union, with S denoting the sends, R the receives and L the local events.
- $\mathcal{C}$ is a finite set of message contents (names) and local action names.
- $t : E \to A = \{p!q(a), p?q(a), l_p(a) \mid p, q \in \mathcal{P}, p \neq q, a \in \mathcal{C}\}$ labels each event by its *type* $t(e)$, with $t(e) = p!q(a)$ if $e \in E_p \cap S$ is a send event of message a from p to q, $t(e) = p?q(a)$ if $e \in E_p \cap R$ is a receive event of message a by p from q and $t(e) = l_p(a)$ if $e \in E_p \cap L$ is a p-local event describing the local action a.
- $m : S \longrightarrow R$ is a bijection that pairs up send and receive events (*matching* function). We have that $m(e) = f$ only if $t(e) = p!q(a)$, $t(f) = q?p(a)$ for some $p, q \in \mathcal{P}$ and $a \in \mathcal{C}$.
- $< \subseteq E \times E$ is the least acyclic relation satisfying the following requirements:
  - The restriction of $<$ to E_p is a total order, for every process $p \in \mathcal{P}$.
  - For all $e, f \in E$, $m(e) = f$ implies $e < f$.

A *message* (e, f) is a pair of matching send and receive events, i.e., $m(e) = f$. Often one assumes that channels are FIFO, that is, there is no overtaking of messages in the channel from p to q, for every $p \neq q$. The results of this paper are depending on this assumption. For non-FIFO channels we just have to add some information in the type $t(e)$ of an event e. Formally, we extend the type of each event by an integer. We require that $m(e) = f$ only if $t(e) = (p!q(a), k)$ and $t(f) = (q?p(a), k)$ for some p, q, a and $k \in \mathbb{N}$.

Since $<$ is required to be acyclic, its reflexive-transitive closure $<^*$ is a partial order on E. For sake of simplicity we will use the same notation $\leq$ for the partial order $<^*$. A *linearization* of $<$ is defined as usual, as a total order $\preceq$ extending $\leq$, i.e., $\leq \subseteq \preceq$. For any MSC M we denote by $\text{Lin}(M)$ the set of labeled linearizations of M: $\text{Lin}(M) = \{t(e_1) \cdots t(e_k) \mid e_1 \cdots e_k \text{ is a linearization of } M\}$. Note that any $x \in \text{Lin}(M)$ suffices to reconstruct the MSC M, since the type mapping $t : E \to A$ encodes all the relevant information about M. If the matching m is a partial function then we speak about a *partial MSC*. For every $x \in A^*$ we denote by $\text{msc}(x)$ the (partial) MSC defined by x, if it exists. The *size* of an MSC is the number of events it contains.

Since the specification of a communication protocol includes many scenarios, a high-level description is needed for combining them together and defining infinite sets of (finite or infinite) scenarios. The standard description of the norm Z.120 uses non-deterministic branching, concatenation and iteration for defining finite or infinite sets of MSCs. Formally, a *high-level MSC* $G = \langle V, R, v^0, v^f, \lambda \rangle$ (HMSC for short) is a finite transition system (V, R, v^0, v^f) with transition set $R \subseteq V \times V$, initial node v^0 (with no ingoing edge) and terminal node v^f (with no outgoing edge). Each node v is labeled by the finite MSC $\lambda(v)$. We let $P(v) = P(\lambda(v))$, and we assume that $P(v) \neq \emptyset$, except possibly for $v = v^f$. We also assume that every node is accessible from v^0 and from each node there is a path to v^f. An *execution* of G is the labeling $\lambda(v_0)\lambda(v_1) \cdots \lambda(v_k)$ of some path $v^0 = v_0, v_1, \ldots, v_k = v^f$ in G, i.e., $(v_i, v_{i+1}) \in R$ for every $0 \leq i < k$. The set of executions of G is denoted by $\mathcal{L}(G)$, the set of linearizations of executions of G is denoted by $\text{Lin}(G)$. The *size* of an HMSC is the sum of the sizes of its nodes.

The semantics of HMSCs depends on the definition of the MSC product. We consider the usual *weak product* of MSCs, as defined in the following. Let $M_1 = \langle E_1, <_1, \mathcal{P}, t_1, \mathcal{C}_1, m_1 \rangle$ and $M_2 = \langle E_2, <_2, \mathcal{P}, t_2, \mathcal{C}_2, m_2 \rangle$ be MSCs over the same set of processes $\mathcal{P}$. Their product $M_1 M_2$ is defined as the MSC $\langle E_1 \cup E_2, <, \mathcal{P}, t_1 \cup t_2, \mathcal{C}_1 \cup \mathcal{C}_2, m_1 \cup m_2 \rangle$ over the disjoint union of events $E_1 \uplus E_2$, with the visual order given by:

$$< = <_1 \cup <_2 \cup \{(e, f) \in E_1 \times E_2 \mid P(e) = P(f)\}.$$

That is, events of M_1 precede the events of M_2 for each process, respectively. Note that there is no synchronization between different processes when moving from one node to the next one (this is called *weak sequencing*). Hence, it is possible that one process is still involved in some actions of M_1, while another process has advanced to an event of M_2. We also say that M_1 is a *prefix* of $M_1 M_2$.

3   Model-Checking Cooperative HMSCs

In this section we introduce globally-cooperative HMSCs and the subclass of locally-cooperative HMSCs and we show that the model-checking problem for these infinite-state HMSCs is decidable.

For MSCs M_1, M_2 with $P(M_1) \cap P(M_2) = \emptyset$, we write $M_1 \parallel M_2$ and we say that M_1, M_2 are *independent*. Observe that $M_1 \parallel M_2$ implies $M_1 M_2 = M_2 M_1$.

An MSC M is called *linked* if it cannot be written as $M = M_1 M_2$ with $M_1 \parallel M_2$ both non empty MSCs. If v_1, v_2 are nodes of an HMSC $G = \langle V, R, v^0, v^f, \lambda \rangle$, we write $v_1 \parallel v_2$ if $\lambda(v_1) \parallel \lambda(v_2)$. The relation Sync $= V \times V \setminus \parallel$ is called the *synchronization relation*.

Globally-cooperative and locally-cooperative HMSCs

Let $G = \langle V, R, v^0, v^f, \lambda \rangle$ be an HMSC with nodes labeled by linked MSCs.

1. G is called *globally-cooperative* if no strongly connected R-component $U \subseteq V$ can be partitioned as $U = U_1 \cup U_2$ such that $U_1, U_2 \neq \emptyset$ and $v_1 \parallel v_2$ for all $v_1 \in U_1, v_2 \in U_2$.
2. G is called *locally-cooperative* if $R \subseteq$ Sync.

In particular, any MSC labeling a loop of a globally-cooperative HMSC is linked. We will assume in the sequel that all nodes of an HMSC are labeled by linked MSCs. This is not a restriction, since any non-linked node can be split in a sequence of linked nodes. Note that this transformation preserves both classes of regular and globally-cooperative HMSCs.

The motivation behind the definition of globally-cooperative HMSCs comes from Mazurkiewicz trace theory. Let $\parallel \subseteq V \times V$ be a symmetric, irreflexive independence relation on V. A set $K \subseteq V^*$ is called $\parallel$-*closed* when $\sigma uv\sigma' \in K \Leftrightarrow \sigma vu\sigma' \in K$ for all $\sigma, \sigma' \in V^*$ and $u \parallel v$. The $\parallel$-*closure* of $K \subseteq V^*$ is the smallest $\parallel$-closed set containing K and is denoted $\mathrm{clos}_\parallel(K)$. In general, the closure does not preserve regularity. A sufficient condition for $\mathrm{clos}_\parallel(\mathcal{L}(A))$ being regular is that the set of edge labels $U \subseteq V$ occurring in any strongly connected component of $\mathcal{A}$ induces a connected subgraph of $(V, (V \times V) \setminus \parallel)$ [11, 15]. Moreover, the size of a non-deterministic automaton recognizing $\mathrm{clos}_\parallel(\mathcal{L}(A))$ is at most $2^{O(n \cdot \wp)}$ [5,14], where $n = |\mathcal{A}|$ and $\wp$ is the minimal number of cliques covering the graph $(V, (V \times V) \setminus \parallel)$. Testing that an automaton $\mathcal{A}$ satisfies the above condition is co-NP-complete, [5,14]. This yields:

Proposition 1. *Checking whether an HMSC is globally-cooperative is co-NP-complete, whereas checking whether it is locally-cooperative is in P.*

Notice that globally-cooperative and locally-cooperative HMSCs have in general an infinite state space, thus they are not *bounded* in the sense of [5,14]. Formally, the set of linearizations of executions cannot be described by finite automata. However, although globally-cooperative HMSCs are more general than bounded HMSCs, we are still able to do automata-based model-checking. The underlying idea is that the executions of a globally-cooperative HMSC can be captured by a regular set of representatives. As an example consider the HMSC G of Figure 1. The set Lin(G) of linearizations of executions of G is obviously non-regular (consider e.g. linearizations in the set $A_p^*(A_r + A_s)^* A_q^*$, where $A_p = \{t(e) \mid P(e) = p\}$ is the set of types of events located on p). A regular set of representatives is given by the set Lin$(G) \cap (A_p(A_r + A_s)A_q)^*$.

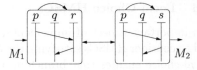

Fig. 1. A locally-cooperative HMSC

A non empty MSC is called *atomic* (atom, for short) if it cannot be written as $M = M_1 M_2$ for non empty MSCs M_1, M_2. It is not hard to see that any MSC M can be written as $M = M_1 \cdots M_k$ where each M_i is a non empty atom, and that this factorization is unique up to commutation of adjacent independent atoms. In [9], it is shown how to compute this factorization in linear time. Note that any atomic MSC is linked. Note that replacing any node by a sequence of atoms preserves globally-cooperative, but not locally-cooperative HMSCs, as shown through Proposition 2.

Let $G = \langle V, R, v^0, v^f, \lambda \rangle$ be an HMSC. We define Atom(G) as the set of atoms occurring in the decomposition of MSCs from $\lambda(V)$. Let $\mathrm{Lin}^a(G) = \mathrm{Lin}(G) \cap \mathrm{Lin}(\mathrm{Atom}(G))^*$, where $\mathrm{Lin}(\mathrm{Atom}(G)) = \bigcup_{M \in \mathrm{Atom}(G)} \mathrm{Lin}(M)$.

We show now that $\mathrm{Lin}^a(G)$ is a regular set of representatives for $\mathrm{Lin}(G)$.

Theorem 1. *Let G be a globally-cooperative HMSC. Then $msc(\mathrm{Lin}^a(G)) = \mathcal{L}(G)$. Let s denote the size of G, $\wp$ the number of processes, and let μ be the maximal number of events on the same process in an MSC from Atom(G). Then:*

1. *$\mathrm{Lin}^a(G)$ is recognized by a non-deterministic finite automaton of size in $2^{O(\wp s)}(\mu + 1)^{\wp}$.*
2. *Moreover, if G is locally-cooperative, then the size of the automaton is in $s^{O(\wp)}(\mu + 1)^{\wp}(\wp + 1)^{\wp}$.*

The lower bounds below follow constructions similar to [14]:

Corollary 1. *Model-checking globally-cooperative HMSCs is decidable. Precisely, let G_1, G_2 be globally-cooperative HMSCs, then we have:*

1. *Deciding whether $\mathcal{L}(G_1) \cap \mathcal{L}(G_2) \neq \emptyset$ is a PSPACE-complete problem.*
2. *Deciding whether $\mathcal{L}(G_1) \subseteq \mathcal{L}(G_2)$ is an EXPSPACE-complete problem.*

Both lower bounds hold also when G_1, G_2 are locally-cooperative.

We obtain better complexity bounds for the model-checking problems for locally-cooperative HMSCs if the nodes are labeled by *atomic* MSCs. For this we use the unicity of the decomposition of an MSC into atoms:

Proposition 2. *Let G_1, G_2 be locally-cooperative HMSCs such that each node is labeled by an atomic MSC. Then:*

1. *Deciding whether $\mathcal{L}(G_1) \cap \mathcal{L}(G_2) \neq \emptyset$ is a NLOGSPACE-complete problem.*
2. *Deciding whether $\mathcal{L}(G_1) \subseteq \mathcal{L}(G_2)$ is an PSPACE-complete problem.*

3.1   Local-Choice HMSCs

An important aspect of implementation is the absence of deadlocks. So we are led to consider HMSCs satisfying the *local-choice* property [8]. Roughly speaking, local-choice ensures that branching between executions is always controlled by a unique process.

Local-choice HMSCs. An HMSC $N = \langle V, R, v^0, v^f, \lambda \rangle$ is called *local-choice* if the following conditions are satisfied:

1. Every <u>path</u> starting in v^0 has a unique minimal event.
2. For each node $v \in V$ having at least two outgoing edges, there is a process $root(v)$ such that every <u>path</u> $w_1 w_2 \cdots$, starting in a node w_1 successor of v, has a unique minimal event located on $root(v)$.

It is easy to see that locally-cooperative and local-choice HMSCs are syntactically incomparable. However, local-choice HMSCs are globally-cooperative. Actually we can transform local-choice HMSCs into locally-cooperative, local-choice HMSCs of quadratic size, as shown in Proposition 3 below.

We call a node with at least two outgoing edges a *branching node*. Notice that every path $v_0 \cdots v_l$ in N where all of $v_1, \ldots, v_n$ are non-branching, is of length $l + 1 \le |V|$. Such a path will be called a *non-branching path*. Moreover, if the non-branching path $\sigma = v_0 \cdots v_l$ is maximal, then it is determined by v_0 and v_l is either branching or the terminal node v^f. We denote the maximal non-branching path starting in v by $\mathrm{NPath}(v)$. Consider now an accepting path σ of N. We decompose σ as $\sigma = \sigma_0 \sigma_1 \cdots \sigma_{k+1}$ where each σ_i is a maximal non-branching path (note that this decomposition is unique). Let v_i be the last node of σ_{i-1} (see Figure 2, where the triangles illustrate the partial order graphs of the subpaths σ_i). Note that v_i is branching for all $i \le k$. Let also w_i be the first node of σ_i, hence $\sigma_i = \mathrm{NPath}(w_i)$. By definition, $p_i = root(v_i) \in P(w_i)$ is the process on which the minimal event of σ_i is located (which is the unique minimal event of the path $\sigma_i \cdots \sigma_k$). Moreover, the local choice condition applied to the branching node v_{i-1} ensures that p_i also belongs to $P(\sigma_{i-1})$. If $p_i \ne p_{i-1}$ then the first action of p_i in σ_{i-1} is a receive action.

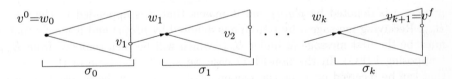

Fig. 2. Path decomposition in a local-choice HMSC.

The above decomposition of paths in a local-choice HMSC will be used by the implementation algorithm (Section 4.3).

Proposition 3. *For every local-choice HMSC we can construct an equivalent locally-cooperative, local-choice HMSC of quadratic size.*

The main technical argument establishing the model-checking algorithm for local-choice HMSCs is based on the following property. Consider an equality $M_1 \cdots M_k = M_1' \cdots M_l'$, where every $M_i \cdots M_k$, resp. $M_j' \cdots M_l'$ has a unique minimal event, for all i, j. Then we can show that either M_1 is a prefix of M_1' (or vice-versa) or $M_1 = X M_2' \cdots M_l'$, $M_1' = X M_2 \cdots M_k$ hold for some MSC X.

This observation allows to consider only configurations with a unique minimal event (instead of arbitrary configurations, which would require polynomial space).

Theorem 2. *Let G and G' be two local-choice HMSCs. Then we have:*

1. *Deciding whether $\mathcal{L}(G) \cap \mathcal{L}(G') \neq \emptyset$ is NLOGSPACE-complete. Moreover, this question can be decided in time $O(|G| \cdot |G'|)$.*
2. *Deciding whether $\mathcal{L}(G) \subseteq \mathcal{L}(G')$ is PSPACE-complete.*

4   Implementing HMSCs by Communicating Finite-State Machines

4.1   Communicating Finite-State Machines

The most natural implementation model for HMSCs are communicating finite state machines (CFM), as used for instance in the ITU standard specification language SDL.

A CFM $\mathcal{A}$ consists of a network of finite state machines $\mathcal{A} = (\mathcal{A}_p)_{p \in \mathcal{P}}$ that communicate over unbounded, error-free buffers. In general we assume that buffers are FIFO (if for instance the given HMSC is FIFO), but we can modify the semantics of receives if the MSCs contain overtaking of messages. The content of a buffer is a word over a finite alphabet $\mathcal{C}$. With each pair $(p, q) \in \mathcal{P}^2$ of distinct processes we associate a buffer $B_{p,q}$. Each finite state machine $\mathcal{A}_p$ is described by a tuple $\mathcal{A}_p = (S_p, A_p, \rightarrow_p, F_p)$ consisting of a set of local states S_p, a set of actions A_p, a set of local final states F_p and a transition relation $\rightarrow_p \subseteq S_p \times A_p \times S_p$. The computation begins in an initial state $s^0 \in \prod_{p \in \mathcal{P}} S_p$. The actions of $\mathcal{A}_p$ are either local actions or sending/receiving a message. We use the same notations as for MSCs. Sending message $a \in \mathcal{C}$ from process p to process q is denoted by $p!q(a)$ and it means that a is appended to the buffer $B_{p,q}$. Receiving message a by p from q is denoted by $p?q(a)$ and it means that a must be the first message in buffer $B_{q,p}$, which will be then removed from $B_{q,p}$ (supposing FIFO). In the non-FIFO case we specify the type of the message that can be received next (cf. the semantics of a receive in the *message queue* of UNIX system V). A local action m is denoted by $l_p(a)$. We denote a run of the CFM as *successful*, if each process p finishes the execution in some final state F_p and all buffers are empty. The set of successful runs (i.e., MSCs) of $\mathcal{A}$ is denoted $\mathcal{L}(\mathcal{A})$. The *size* of $\mathcal{A}$ is $\sum_p |A_p|$.

A CFM implementation of an HMSC N will add in general data to the message contents of N. We will call $\mathcal{A}$ a *CFM implementation* of N if the MSCs defined by the successful runs of $\mathcal{A}$ with the additional data removed, correspond precisely to the executions of N.

4.2   Implementing Locally-Cooperative HMSCs

The simplest realization of an HMSC N by a CFM is the one where the automaton $\mathcal{A}_p$ corresponding to process p generates the projection of $\mathcal{L}(N)$ on p. This approach is used in [3,12]. Consider again the HMSC G_1 of Figure 1 (page 661), and let M be the MSC given by the projections $\pi_p(M) = p!r\ p!r\ p!s\ p!s$, $\pi_q(M) = q?r\ q?s\ q?r\ q?s$, $\pi_r(M) = r?p\ r!q\ r?p\ r!q$ and $\pi_s(M) = s?p\ s!q\ s?p\ s!q$. Then M

does not belong to $\mathcal{L}(G_1)$. However, it is easy to verify that $\pi_t(M) \in \pi_t(\mathcal{L}(G_1))$ for all $t \in \{p,q,r,s\}$. Hence G_1 is not realizable according to [3].

We describe our implementation of locally-cooperative HMSCs first on our example G_1. One can observe that G_1 can be implemented if process p anticipates the next choice and sends the prediction with the current message. Processes r and s then forward the prediction to q. In this way, process q knows whether the next message should be received from r or from s. The general solution will involve a leader process (p in the example) for each transition, i.e., a process that occurs in both nodes and decides about certain nodes in the future (prediction), as described below.

For a node $v \in V$ of $N = \langle V, R, v^0, v^f, \lambda \rangle$, let $P(v)$ denote the processes occurring in $\lambda(v)$. For a path $\sigma = v_0 v_1 \cdots v_k$ of N let $P(\sigma) = \bigcup_i P(v_i)$ be the processes occurring in σ. Moreover, we define $\text{first}(\sigma, p)$ for all $p \in \mathcal{P}$ as the first node containing p in σ: $\text{first}(\sigma, p) = \bot$ if $p \notin P(\sigma)$, and $\text{first}(\sigma, p) = v_j$, where $j = \min\{k \geq 0 \mid p \in P(v_k)\}$ otherwise. Similarly, if σ has at least $i + 1$ nodes, let $\text{last}(\sigma, i, p)$ be the last node among the first $(i+1)$ nodes of σ containing process p (resp., $\text{first}(\sigma, p) = \bot$ if $p \notin P(v_0 \cdots v_i)$).

Let $N = \langle V, R, v^0, v^f, \lambda \rangle$ be a locally-cooperative HMSC. A triple $(v, \nu, l) \in V \times (V \cup \{\bot\})^{\mathcal{P}} \times \mathcal{P}$ is a *realizable prediction* if either $\nu(p) = \bot$ for all $p \in \mathcal{P}$ and $v = v^f$, or if all conditions below hold:

1. There exists a path $\sigma = v_0 v_1 \cdots$ in N such that $v_0 = \nu(l)$ and $\nu(p) = \text{first}(\sigma, p)$ for each process p;
2. $(v, \nu(l)) \in R$;
3. $l \in P(v) \cap P(\nu(l))$.

The process l is called the *leader* of the transition $(v, \nu(l))$ with respect to (v, ν, l).

From a locally-cooperative HMSC $N = \langle V, R, v^0, v^f, \lambda \rangle$, we build a communicating automaton $\mathcal{A}_N$ as follows. Each process is initialized with the same input $i_0 = (v_0, \nu_0, l_0)$ which is some realizable prediction with $v_0 = v^0$. The algorithm for process p is described below:

```
(v,ν,ℓ) =  (v₀,ν₀,ℓ₀);
while (true)
{ m = (v,ν,ℓ);
if (p ∈ P(v))     // test useful only for the first node of p
execute(v,m);
v' = ν(p);
if (v' == ⊥)
halt();
if (v' == ν(ℓ))          // v' is the successor of v
(ν',ℓ') = guess_next(v',ν);
else (ν',ℓ') = guess(v');
v = v'; ν = ν'; ℓ = ℓ';}
```

The call guess_next(v', ν) guesses nondeterministically a prediction and a leader (ν', ℓ') for the next node v', such that (v', ν', ℓ') is realizable and the new prediction ν' is *compatible* with the old prediction ν for processes not occurring in v', that is, $\nu_{|\mathcal{P}\backslash P(v')} = \nu'_{|\mathcal{P}\backslash P(v')}$. The call guess(v') guesses nondeterministically a pair (ν', ℓ') such that (v', ν', ℓ') is realizable. In this case, process p makes a prediction about a node p' that is not a direct R-successor of v. This prediction is needed since all processes of a node must agree on some future information. The call halt() terminates the execution of p in an accepting state. Finally, the call execute(v,m) consists in executing the actions of p of the MSC labeling v, but overloading the messages to be sent or received with m. Note that if two communicating processes do not choose the same value for m, then a deadlock occurs. By transitivity and weak connectivity of each MSC, the deadlock-free execution of a node means that all processes in the node have chosen the same value for m.

Proposition 4. *Let N be a locally-cooperative HMSC. Then $\mathcal{A}_N$ is a CFM implementation of $\mathcal{L}(N)$ of size $n^{O(\wp)}$, where n is the number of nodes of N and $\wp$ is the number of processes.*

Remark 1. Note that we can fix a leader for each transition of the HMSC beforehand. This would decrease the degree of non-determinism and deadlocks.

4.3 Implementing Local-Choice HMSCs

The implementation algorithm described in the previous section cannot avoid deadlocks for the resulting CFM, since the future predictions are chosen by each process separately. One reason is that branching in an HMSC is not controlled by a single process, as it is the case for local-choice HMSCs. The results of [8] give a sufficient condition (called *reconstructibility*) for a local-choice HMSC to be implementable with no addition of extra message data.

Recall from Section 3.1 that any accepting path σ of a local-choice HMSC has a canonical decomposition $\sigma = \sigma_0 \cdots \sigma_k$ in subpaths σ_i, such that the MSC execution of σ_i has an unique minimal process located in the first node w_i of σ_i (recall Figure 2 in Section 3.1). We can use the minimal nodes w_i as future predictions in the CFM implementation. Each process $q \in P(\sigma_i)$ transmits w_i with each send action. Recall that p_{i+1} is the minimal process of w_{i+1}. When process $p_{i+1} \in P(\sigma_i)$ finishes σ_i then it has to choose the starting node w_{i+1} of the next non-branching path σ_{i+1} such that $(v_{i+1}, w_{i+1}) \in R$ (if $v_{i+1} \neq v^f$). Process p_{i+1} will be the only process in σ_i that knows the next node (w_{i+1}) to be executed. Every other process $q \in P(\sigma_i)$ will execute $\pi_q(\sigma_i)$ and then get into a polling state in which it accepts any incoming message. The first message received by q will inform it about a node w_j, $j > i$. Knowing N and w_j, process q determines the path $\sigma_j = \text{NPath}(w_j)$ that it is executed.

The algorithm $\mathcal{A}'_p$ for process p is given below. The call execute_path(w) means that p has to execute $\pi_p(\sigma)$, where $\sigma = \text{NPath}(w)$. Note that p must only remember w and its current node in σ. The call execute1_path(w) is similar, except for the fact that p executes $\pi_p(\sigma)$ without its first action. The call guess(w) guesses w' such that w' is a successor node of the last node of $\text{NPath}(w)$. The call poll() means that process p is waiting for an incoming message from an arbitrary process. By receiving a message, process p gets the current node w' and also executes its first action in $\sigma' = \text{NPath}(w')$ (a receive action).

```
w = v^0; not_polling = true;
  while (true)
  { if (p ∈ P(NPath(w)))  // test useful only when starting
  if (not_polling)  execute_path(w);
  else execute1_path(w);
  v = last node of NPath(w);
  if (p = p_min(v))
  { w' = guess(w); not_polling = true; }
  else { w' = poll();  not_polling = false; }
  w = w';}
```

Proposition 5. *Let N be a local-choice HMSC. Then $\mathcal{A}'_N$ is a deadlock-free CFM implementation of N. Moreover, the size of $\mathcal{A}'_N$ is linear in the size of N.*

5   Channel-Bounded CFM and Deadlock Detection

In this section we consider a subclass of communicating finite state machines, called *channel-bounded CFM*. Intuitively, a CFM is bounded if every execution can be simulated by an execution using bounded buffers. Since implementations of HMSCs yield channel-bounded CFMs, it is natural to ask whether a channel-bounded CFM is deadlock-free.

A configuration $C = (q, B)$ of a CFM $\mathcal{A} = (\mathcal{A}_p)_{p \in \mathcal{P}}$ is described by a global state q of $S = \prod_{p \in \mathcal{P}} S_p$ and the contents $B \in (\mathcal{C}^*)^{\mathcal{P} \times \mathcal{P}}$ of all buffers. The transition relation of the CFM is denoted by $\to$, its transitive-reflexive closure is denoted as usual by $\overset{*}{\to}$. The configuration with global state s^0 and empty buffers is the initial configuration. An *execution* $\sigma = C_1 \overset{a_1}{\to} C_2 \overset{a_2}{\to} \cdots \overset{a_{m-1}}{\to} C_m$ of $\mathcal{A}$ is a finite $\to$-path. The labeling of the execution σ is the sequence $a_1 \cdots a_{m-1}$. Note that the labeling of an execution σ defines in a natural way a partial MSC $\text{msc}(\sigma)$. Recall that an execution is *successful* if it ends with empty buffers and each process reaches some local final state. A configuration C is a *deadlock* if there is no successful execution starting from C. Let $A = \bigcup_{p \in \mathcal{P}} A_p$ be the set of possible actions of a CFM over process set $\mathcal{P}$. Two executions σ, σ' are *equivalent* (and we write $\sigma \sim \sigma'$) if $\text{msc}(\sigma) = \text{msc}(\sigma')$ and σ, σ' start in the same configuration.

An execution σ of a CFM is called *b-bounded*, if every configuration of σ is such that the size of every buffer is bounded by b. If $C \overset{*}{\to} C'$ is b-bounded, then we say that C' is *b-reachable* from C.

A CFM is *b-bounded* if every successful execution σ starting in the initial configuration admits some successful, b-bounded equivalent execution $\sigma' \sim \sigma$. Let $\mathcal{A}$ be a b-bounded CFM and let C be b-reachable from the initial configuration of $\mathcal{A}$. Then C is not a deadlock if and only if there is some b-bounded, successful execution starting from C. This allows to show the following proposition.

Proposition 6. *Reachability and deadlock detection for b-bounded CFMs are both PSPACE-complete (with b in unary representation).*

The last proposition allows to connect the model-checking problem and the implementation by CFMs.

Theorem 3. *Let C be a class of HMSCs that are CFM implementable. Then the model-checking problem for C (intersection and inclusion, resp.) is decidable.*

Conclusion. We have shown that model-checking a natural class of infinite-state HMSCs, globally-cooperative HMSCs, is decidable and of the same complexity as for regular HMSCs. For a natural subclass (local-choice HMSCs) the complexity of model-checking is the same as in the sequential case. The implementation of locally-cooperative HMSCs raises the question whether we can decide for a given HMSC if it can be implemented without deadlocks in our framework (with finite additional message contents).

References

1. ITU-TS recommendation Z.120, 1996.
2. R. Alur, K. Etessami, and M. Yannakakis. Inference of message sequence charts. In *22nd Int. Conf. on Software Engineering*, pages 304–313. ACM, 2000.
3. R. Alur, K. Etessami, and M. Yannakakis. Realizability and verification of MSC graphs. In *ICALP'01*, LNCS 2076, pages 797–808, 2001.
4. R. Alur, G. H. Holzmann, and D. A. Peled. An analyzer for message sequence charts. *Software Concepts and Tools*, 17(2):70–77, 1996.
5. R. Alur and M. Yannakakis. Model checking of message sequence charts. In *CONCUR'99*, LNCS 1664, pages 114–129, 1999.
6. D. Brand and P. Zafiropulo. On communicating finite-state machines. *Journal of the ACM*, 30(2):323–342, 1983.
7. B. Caillaud, P. Darondeau, L. Hélouët, and G. Lesventes. HMSCs as partial specifications... with PNs as completions. In *MOVEP*, 2000.
8. L. Hélouët and C. Jard. Conditions for synthesis of communicating automata from HMSCs. In *5th Int. Workshop on Formal Methods for Ind. Crit. Systems*, 2000.
9. L. Hélouët and P. Le Maigat. Decomposition of Message Sequence Charts. In *SAM2000*, pages 46–60, 2000.
10. J. G. Henriksen, M. Mukund, K. Narayan Kumar, and P. Thiagarajan. On message sequence graphs and finitely generated regular msc languages. In *ICALP'00*, LNCS 1853, pages 675–686, 2000.
11. Y. Métivier. On recognizable subsets of free partially commutative monoids. *Theoretical Computer Science*, 58:201–208, 1988.
12. R. Morin. Recognizable Sets of Message Sequence Charts. In *STACS'02*, LNCS 2285, pages 523–534, 2002.
13. M. Mukund, K. Narayan Kumar, and M. Sohoni. Synthesizing distributed finite-state systems from MSCs. In *CONCUR'00*, LNCS 1877, pages 521–535, 2000.
14. A. Muscholl and D. Peled. Message sequence graphs and decision problems on Mazurkiewicz traces. In *MFCS'99*, LNCS 1672, pages 81–91, 1999.
15. E. Ochmański. Recognizable trace languages. In *The Book of Traces*, chapter 6, pages 167–204. World Scientific, Singapore, 1995.

Universal Inherence of Cycle-Free Context-Free Ambiguity Functions

Klaus Wich

Institut für Informatik, Universität Stuttgart,
Breitwiesenstr. 20-22, 70565 Stuttgart `wich@informatik.uni-stuttgart.de`

Abstract. It is shown that the set of inherent ambiguity functions for context-free languages and the set of ambiguity functions for cycle-free context-free grammars coincide. Moreover for each census function γ of an unambiguous context-free language the least monotone function larger than or equal to γ is an inherent ambiguity function. Both results are based on a more general theorem. Informally it states that the loss of information induced by a length preserving homomorphism on an unambiguous context-free language can be turned into inherent ambiguity.

1 Introduction

A context-free (for short cf) grammar G is ambiguous if there is some word w which can be derived by G with at least two different derivation trees. Otherwise G is unambiguous. A cf language is (inherently) ambiguous if it cannot be generated by an unambiguous cf grammar. The existence of ambiguous cf languages is shown in [5]. Ambiguous cf grammars and languages can be distinguished by their degree of ambiguity, that is, the least upper bound for the number of derivation trees which a word may have. There are examples for k ambiguous languages for all $k \in \mathbb{N}$ [3]. But even languages with infinite degree of ambiguity exist [2]. They can be distinguished by the asymptotic behaviour of their ambiguity with respect to the length of the words. In [6] it has been shown that each cf grammar is either $\Omega(2^n)$ or $\mathcal{O}(n^k)$ ambiguous. An appropriate k can be constructed effectively. Languages with inherent ambiguity $2^{\Theta(n)}$ and $\Theta(n^k)$ for all $k \in \mathbb{N}$ are presented in [4]. Languages with sublinear ambiguity are presented in [7].

So far the questions whether for a given function f there is an f-ambiguous cf grammar or whether there is an f-ambiguous cf language have been studied separately. The latter question is considered to be much harder then the first. In this paper we reduce the problem for cf languages to the corresponding problem for cycle-free cf grammars.

In Section 3 for each pair consisting of an unambiguous cf language L and a length preserving homomorphism h we show a strong connection between the loss of information induced by h on L, and the ambiguity of the constructed cf language. In Section 4 we apply the results of Section 3 and obtain:

P. Widmayer et al. (Eds.): ICALP 2002, LNCS 2380, pp. 669–680, 2002.
© Springer-Verlag Berlin Heidelberg 2002

- The least monotone function $\hat{\gamma}$ larger than or equal to the census function γ of an arbitrary unambiguous cf language is an inherent ambiguity function. (See Definition 4.1.)
- The set of ambiguity functions for cycle-free context-free grammars and the set of inherent ambiguity functions for cf languages coincide.

2   Preliminaries

Let A be a set. Then $|A|$ denotes the cardinality of A and 2^A the power set of A. For arbitrary $i, j \in \mathbb{N}$ the *interval* from i to j is $[i, j] := \{k \in \mathbb{N} \mid i \le k \le j\}$. We generally assume alphabets to be finite. Let A be an alphabet. Let $u := a_1 \cdots a_n \in A^*$ be a word, where $a_i \in A$ for all $i \in [1, n]$. The symbol at *position* i is $u[i] := a_i$. The *length* of u is $|u| = n$. The words over A of length at most n are denoted by $A^{\le n} := \{w \in A^* \mid |w| \le n\}$. The *empty word* ε is the unique word with length 0. For all $i \in [1, n+1]$ and $j \in [0, n]$ we define the *factor* of u from position i to position j as $u[i, j] := a_i \cdots a_j$. If $j < i$ then $u[i, j] = \varepsilon$. The word $u[1, j]$ is a *prefix* of u, it is a *proper* prefix if $j < n$. The word $u[i, n]$ is a *suffix* of u. A homomorphism $h : A^* \to \Gamma^*$ is *length preserving* if $|h(X)| = 1$ for all $X \in A$. The *projection* on a subalphabet $\Gamma \subseteq A$ is the homomorphism $\pi_\Gamma : A^* \to \Gamma^*$ given by $\pi_\Gamma(X) = X$ for $X \in \Gamma$ and $\pi_\Gamma(X) = \varepsilon$ for $X \in A \setminus \Gamma$. If A and Γ are two alphabets then we call a homomorphism $h : A^* \to 2^{\Gamma^*}$ a *substitution*, where the operation on 2^{Γ^*} is the concatenation of languages defined by $L_1 \cdot L_2 := \{uv \mid u \in L_1 \text{ and } v \in L_2\}$.

A context free grammar is a quadruple $G = (N, A, P, S)$ where N and A are two disjoint alphabets of *nonterminals* and *terminals*, respectively, $P \subseteq N \times (N \cup A)^*$ is a finite set of *productions*, and $S \in N$ is the *start symbol*.

The usual way to continue at this point is to introduce a derivation relation and sentential forms. A sentential form can be considered as the sequence of leaves obtained from the preorder traversal of a derivation tree. But for ambiguity considerations we need a formalism which describes derivation trees completely. The well-known left parse of a derivation tree can be used for this purpose if we restrict ourselves to trees without nonterminal leaves. The left parse of a tree can be seen as the result of a preorder traversal of a derivation tree, where the internal nodes are represented by the productions applied to them, while the leaves are omitted. Thus sentential forms and left parses are complementary parts of derivation trees. It is useful to shuffle both according to the preorder thus forming a single more general tree formalism.

The tree alphabet of a cf grammar $G = (N, A, P, S)$ is $T_G := N \cup A \cup P$. A word $\rho \in T_G^*$ is a *partial derivation tree of G* if

(1) $\rho \in N \cup A$ or
(2) if $\rho = \tau_1(X, \alpha)\alpha\tau_2$ for some partial derivation tree $\tau_1 X \tau_2$ and $(X, \alpha) \in P$.

Note that (X, α) is a single letter of the tree alphabet. The set of G's partial derivation trees is denoted Λ_G. It is easily seen that $\Lambda_G \subseteq N \cup A \cup PT_G^*$. The root of a partial derivation tree ρ is $\uparrow_G(\rho) := \rho$ if $\rho \in N \cup A$ and $\uparrow_G(\rho) := X$ if $\rho =$

$(X, \alpha)\tau$ for some $(X, \alpha) \in P$ and $\tau \in T_G^*$. The *frontier* of ρ is $\downarrow_G(\rho) := \pi_{N \cup A}(\rho)$. If $\uparrow_G(\rho) = S$ then the frontier of ρ is a *sentential form*. We drop the subscripts if G is clear from the context. The arrows for the root and the frontier of partial derivation trees point into the direction where they are usually displayed in a diagram. A *node* of ρ is an element of $[1, |\rho|]$. A node i is a *leaf* if $\rho[i] \in N \cup A$, it is an *internal node* if $\rho[i] \in P$. The *label* of a node i is $\rho[i]$ if i is a leaf and it is the left-hand side of the production $\rho[i]$ if i is an internal node, i.e., it is $X \in N$ if $\rho[i] \in \{X\} \times (N \cup A)^*$. The set of G's derivation trees is defined by $\Delta_G := \{\rho \in \Lambda_G \mid \uparrow(\rho) = S \wedge \downarrow(\rho) \in A^*\}$. The context-free language *generated by* G is $L(G) = \{\downarrow(\rho) \mid \rho \in \Delta_G\}$. The grammar G is *cycle-free* if for all $\rho \in \Lambda_G$ the equation $\uparrow(\rho) = \downarrow(\rho)$ implies $\rho \in N \cup A$, otherwise it is called *cyclic*. It is *reduced* if for each $X \in N$ there are $\tau_1, \tau_2 \in T_G^*$ and $p \in \{X\} \times (N \cup A)^*$ such that $\tau_1 p \tau_2 \in \Delta_G$. Finally it is *$\varepsilon$-free* if $P \subseteq N \times (N \cup A)^+$.

The set of mappings from a monoid M into a semiring S is denoted $S\langle\langle M \rangle\rangle$. An element $r \in S\langle\langle M \rangle\rangle$ is a *formal power series*. For $m \in M$ the value $r(m)$ is called *coefficient* of m. A formal power series can be represented by a formal sum $r := \sum_{m \in M} r(m)m$. For $r \in \mathbb{N}\langle\langle A^* \rangle\rangle$ we define $\hat{r} : \mathbb{N} \to \mathbb{N}$ by $\hat{r}(n) := \max\{r(w) \mid w \in A^{\leq n}\}$. The *characteristic ambiguity power series* of G is the formal power series $d_G := \sum_{w \in A^*} |\Delta_G(w)| \cdot w$ where $\Delta_G(w) := \{\rho \in \Delta_G \mid \downarrow(\rho) = w\}$ for each $w \in A^*$, i.e., the coefficients of d_G are the numbers of derivation trees for the corresponding words. The function $\hat{d}_G$ is called the *ambiguity function* of G. It maps each $n \in \mathbb{N}$ to the ambiguity of the most ambiguous word of length up to n. The grammar G is *$\hat{d}_G$-ambiguous*. We call G *k-ambiguous* for a $k \in \mathbb{N}$ if $\hat{d}_G$ is bounded by k but not by $k - 1$. It is *unambiguous* if it is 1-ambiguous or 0-ambiguous. Note that a cf grammar is 0-ambiguous if and only if $L(G) = \emptyset$.

Let $f : \mathbb{N} \to \mathbb{N}$ be a monotone function. A cf language L is *$\mathcal{O}(f)$-ambiguous* if it is generated by a cf grammar G such that $\hat{d}_G \in \mathcal{O}(f)$, it is *$\Omega(f)$-ambiguous* if it is only generated by cf grammars G' such that $\hat{d}_{G'} \in \Omega(f)$, and it is *$\Theta(f)$-ambiguous* if it is $\mathcal{O}(f)$- and $\Omega(f)$-ambiguous. But the $\mathcal{O}$ and Ω notations are very rough for low ambiguities and at the same time too precise for exponential ambiguity. For example with this notation all constant degrees of ambiguity would be indistinguishable subsumed to $\Theta(1)$-ambiguity. On the other hand exponentially ambiguous languages are $2^{\Omega(n)}$-ambiguous but not $\Omega(2^{cn})$-ambiguous for any $c \in \mathbb{R}_+$. While the $\mathcal{O}$ and Ω notations specify the value of a function up to a constant factor for a fixed argument, in our setting it is more appropriate to specify the length of a word (argument) up to a constant factor for a fixed ambiguity (value). This leads us to the following definition:

Definition 2.1. *Let L be a cf language and $f : \mathbb{N} \to \mathbb{N}$ a function. The language L is f-ambiguous if*

(1) there is a cf grammar G such that $L = L(G)$ and $f = \hat{d}_G$ and
(2) for each cf grammar G' such that $L = L(G')$ there exists a $c \in \mathbb{N}$ such that $f(n) \leq \hat{d}_{G'}(c \cdot n)$ for all $n \in \mathbb{N} \setminus \{0\}$.

We implicitly identify the constant $k \in \mathbb{N}$ with the corresponding constant function. A language is unambiguous if it is 1-ambiguous or 0-ambiguous.

A function $f : \mathbb{N} \to \mathbb{N}$ is an *inherent ambiguity function* if there is a cf language L such that L is f-ambiguous. If L is f-ambiguous then L is f'-ambiguous for all monotone functions f' such that f' agrees with f for all but a finite number of arguments. Note that it is not clear whether each context-free language has an inherent ambiguity function.

It is easily seen that derivation trees must not overlap, i.e., no non empty suffix of a partial derivation tree is a proper prefix of a partial derivation tree. Moreover each position in a partial derivation tree $\rho \in \varLambda_G$ is the beginning of a uniquely determined partial derivation tree. That is, for each $i \in [1, |\rho|]$ there is a uniquely defined $j \in [i, |\rho|]$ such that $\rho[i, j] \in \varLambda_G$. If $\rho[i, j] \in \varLambda_G$ we call $\rho[i, j]$ a *subtree* of ρ and the interval $[i, j]$ a *phrase* of ρ. Then the word $\rho[1, i-1] \cdot \Upsilon(\rho[i, j]) \cdot \rho[j + 1, |\rho|] \in \varLambda_G$ is called the *remainder tree* obtained by *truncation* of the phrase $[i, j]$. Obviously we can append a partial derivation tree ρ' with root X to a leaf i of a partial derivation tree ρ, labeled with X, by a replacement of $\rho[i]$ with ρ'. Let $G = (N, A, P, S)$ be a reduced cf grammar. The terminals of a partial derivation tree $\rho \in \varLambda_G \setminus A$ can be retrieved from the remaining symbols, i.e., the restriction of the projection $\pi_{P \cup N}$ to $\varLambda_G \setminus A$ is injective. Therefore we define the *parse* of a partial derivation tree $\rho \in \varLambda_G \setminus A$, as a more compact tree representation, by $parse_G(\rho) := \pi_{P \cup N}(\rho)$. The reader familiar with the notion of left parses may note that $parse(\rho)$ and the left parse of ρ coincides for all derivation trees. But in contrast to the left parse, which is only defined for partial derivation trees of the form $(P \cup A)^* (N \cup A)^*$, our parse notion is a unique representation for all partial derivation trees, but those in A. We extend the parse notion in the natural way to sets and observe:

Lemma 2.2. *For each context-free grammar G the sets $\varDelta_G$, $\varLambda_G$, $parse_G(\varDelta_G)$, and $parse_G(\varLambda_G)$ are unambiguous context-free languages.*

We take Ogden's iteration Lemma for cf grammars and for cf languages presented in [1, Lemma 2.3 and 2.5] and combine them to:

Lemma 2.3. *For each context-free grammar $G = (N, A, P, S)$ there is an integer $n \in \mathbb{N}$ such that for each $\rho \in \varDelta(G)$ and any choice of at least n marked positions in ρ there are $\alpha, \beta, \gamma, \delta, \eta \in T_G^*$ and a nonterminal $X \in N$ such that:*

(1) $\rho = \alpha\beta\gamma\delta\eta$.
(2) (α and β and γ) or (γ and δ and η) contain at least one marked position.
(3) $\beta\delta$ contains at most n marked positions.
(4) $\alpha\beta^i\gamma\delta^i\eta \in \varDelta_G$ and $\alpha\beta^i X\delta^i\eta$, $\beta^i\gamma\delta^i$, $\beta^i X\delta^i \in \varLambda_G$ for all $i \in \mathbb{N}$.

A tuple $\vartheta = (|\alpha| + 1, |\alpha\beta|, |\alpha\beta\gamma| + 1, |\alpha\beta\gamma\delta|)$ satisfying the conditions above is called a *pumping phrase* and $\beta\gamma\delta$ the *subtree corresponding to ϑ*. Note that $\rho \in \varDelta_G$ implies $\downarrow(\rho) \in L(G)$. Therefore, if we only mark leaves in ρ we obtain Ogden's iteration Lemma for cf languages. The advantage of pumping derivation trees instead of their frontiers is that they have a unique phrase structure even if the generated words are ambiguous. As we will see this additional information can be useful if we apply Ogden's Lemma several times to a tree with intermediate shifts of the marked positions.

3   The Hiding Theorem

In this section it is shown how the loss of information induced by a length preserving homomorphism can be turned into inherent ambiguity.

Definition 3.1. *For the remainder of the section we define the pairwise disjoint alphabets Γ, $\{0,1\}$, and $A := \{a_1, \ldots, a_k\}$. Further $L \subseteq A^*$ is an unambiguous context-free language, $h : A^* \to \Gamma^*$ a length preserving homomorphism, $p \in \mathbb{N}$ a positive integer, and $q := p! + p$.*

First we define a system of languages which has an "inherent capacity" to hide information:

Definition 3.2. *For arbitrary $j \in \mathbb{N}$ we write $[j] := 0^j 1$. For $i \in [1, k]$ we define:*

$$L_i := \{\varepsilon\} \cup \{[j_0] \cdots [j_k] \mid j_0, \ldots, j_k \in \mathbb{N} \text{ and } j_0 = j_i\}.$$

All the languages defined in the previous definition are unambiguous.

Definition 3.3. *We define:*

- *The formal power series $r_{h,L} := \sum_{w \in \Gamma^*} |h^{-1}(w) \cap L| \cdot w$.*
- *The substitution $\sigma_h : A^* \to 2^{(\Gamma \cup \{0,1\})^*}$ given by*
  $\sigma_h(a_i) := \{h(a_i)\} L_i$ for all $i \in [1, k]$.
- *The homomorphism $\mathit{fill}_p : \Gamma^* \to (\Gamma \cup \{0,1\})^*$ defined by*
  $\mathit{fill}_p(X) := X[q]^{k+1}$ for all $X \in \Gamma$.
- *The homomorphism $\mathit{code}_{h,p} : A^* \to (\Gamma \cup \{0,1\})^*$ defined by*
  $\mathit{code}_{h,p}(a_i) := h(a_i)[p][q]^{i-1}[p][q]^{k-i}$ for all $i \in [1, k]$.

Words in $\sigma_h(L)$ can be broken into blocks and subblocks. A block is an element of $\sigma_h(a_i)$ for some $i \in [1, k]$. They are numbered from left to right beginning with 1. The blocks are uniquely determined since they have the form $\Gamma\{0,1\}^*$ and do not end before the end of the word or the beginning of the next block. A subblock is a word of 0^*1 not immediately preceded by a 0-symbol. The subblocks are numbered from left to right beginning with 0.

The main idea of this work is outlined as follows: For each $v \in \Gamma^*$ the coefficient $r_{h,L}(v)$ is the number of words in L which are mapped by h onto v. Thus it can be seen as the degree of information hiding induced by h on L. The mapping $\mathit{code}_{h,p}$ is injective since the mapped symbol is coded in the blocks of 0- and 1-symbols trailing the image under h. Now for each $w \in A^*$ both $\mathit{code}_{h,p}(w)$ and $(\mathit{fill}_p \circ h)(w)$ are elements of $\sigma_h(w)$. Thus $\sigma_h(w)$ contains at the same time words which allow to retrieve w and words which hide all information about w but $h(w)$. Let G be an arbitrary cf grammar generating $\sigma_h(L)$ and let $w \in L$. We will see that for large enough $p \in \mathbb{N}$ the set Δ_G contains a derivation tree with frontier $(\mathit{fill}_p \circ h)(w)$ obtained by pumping a derivation tree with frontier $\mathit{code}_{h,p}(w)$. Assume two different words $w_1, w_2 \in L$ have the same image under h. Then there are derivation trees ω_1 and ω_2, both having the frontier $(\mathit{fill}_p \circ h)(w_1)$ obtained by pumping up trees with frontiers $\mathit{code}_{h,p}(w_1)$

and $code_{h,p}(w_2)$, respectively. The main point of Theorem 3.6 is to show that these trees do not coincide. Thus the "information hiding" which h induces from the "outside" of L is an inherent feature of $\sigma_h(L)$ "carried out" by the "internal pumping structure" of $\sigma_h(L)$.

As an immediate consequence of the definition we observe:

Lemma 3.4.

(1) $\forall u \in A^* : \sigma_h(u) \cap \Gamma^* = \{h(u)\}$ and
(2) $\{h(a_i)\}L_i = \sigma_h(a_i)$ is an unambiguous cf language for all $i \in [1,k]$.

Definition 3.5. For each $i \in [1,k]$ let $G_i = (N_i, \Gamma \cup \{0,1\}, P_i, a_i)$ be an unambiguous cf grammar generating $\sigma_h(a_i)$. Further let $G_L = (N_L, A, P_L, S)$ be an unambiguous cf grammar generating L, such that $N_1, \dots, N_k$, and N_L are pairwise disjoint. We compose the cf grammar:

$$G(h,L) := (N_L \cup (\cup_{i \in [1,k]} N_i), \Gamma \cup \{0,1\}, P_L \cup (\cup_{i \in [1,k]} P_i), S).$$

Note that $L(G(h,L)) = \sigma_h(L)$.

Theorem 3.6. The substitution σ_h has the following properties:

(1) For all $v \in (\Gamma \cup \{0,1\})^*$ we have
$r_{h,L}(\pi_\Gamma(v)) = d_{G(h,L)}(\pi_\Gamma(v)) \geq d_{G(h,L)}(v)$.
(2) For each cf grammar G' such that $\sigma_h(L) = L(G')$ there is a $p \in \mathbb{N}$ such that
$r_{h,L}(w) \leq d_{G'}(fill_p(w))$ holds for all $w \in \Gamma^*$.

The proof of Theorem 3.6 continues until Theorem 3.15. The definitions and lemmas in between belong to the proof of Theorem 3.6.

Proof of Theorem 3.6 part (1): Each derivation tree $\rho \in \Delta_{G(h,L)}$ consists of a partial derivation tree $\rho' \in \Delta_{G_L} \subset \Lambda_{G(h,L)}$ appended with subtrees belonging to $\Delta_{G_i} \subset \Lambda_{G(h,L)}$ for some $i \in [1,k]$. We often need to refer to the remainder tree ρ' in the sequel. Therefore we define:

Definition 3.7. The G_L remainder of a derivation tree $\rho \in \Delta_{G(h,L)}$, denoted by $rem(\rho)$, is the uniquely defined derivation tree in Δ_{G_L} obtained from ρ by truncation of all phrases $[j,j']$ for which $\rho[j,j'] \in \cup_{i \in [1,k]} \Delta_{G_i}$.

Lemma 3.8. For all $\rho \in \Delta_{G(h,L)}$ the statement $\downarrow(\rho) \in \sigma_h((\downarrow \circ rem)(\rho))$ is true.

Proof. The expression $\sigma_h((\downarrow \circ rem)(\rho))$ describes the set of words in $\sigma_h(L)$ which are frontiers of those derivation trees in $\Delta_{G(h,L)}$ having the G_L remainder $rem(\rho)$. Obviously ρ is such a tree. Therefore $\downarrow(\rho) \in \sigma_h((\downarrow \circ rem)(\rho))$. □

Lemma 3.9. For $w \in \Gamma^*$ and $\rho \in \Delta_{G(h,L)}(w)$ we have $(\downarrow \circ rem)(\rho) \in h^{-1}(w) \cap L$.

Proof. Let $w \in \Gamma^*$ and $\rho \in \Delta_{G(h,L)}(w)$. By definition $rem(\rho) \in \Delta_{G_L}$. Thus $(\downarrow \circ rem)(\rho) \in L$. It remains to show that $(\downarrow \circ rem)(\rho) \in h^{-1}(w)$. By Lemma 3.8 we obtain $w = \downarrow(\rho) \in \sigma_h((\downarrow \circ rem)(\rho))$. Since $w \in \Gamma^*$ we obtain $w \in \sigma_h(u) \cap \Gamma^*$ for $u := (\downarrow \circ rem)(\rho) \in A^*$. Then $w = h((\downarrow \circ rem)(\rho))$ follows by Lemma 3.4. This implies $h^{-1}(w) = (h^{-1} \circ h)((\downarrow \circ rem)(\rho)) \ni (\downarrow \circ rem)(\rho)$. □

Lemma 3.10. *For arbitrary* $v \in (\Gamma \cup \{0,1\})^*$ *the restriction of rem to the set* $\Delta_{G(h,L)}(v)$ *is injective.*

Proof. Let $rem(\rho) = rem(\rho')$ for some $\rho, \rho' \in \Delta_{G(h,L)}(v)$ and let $n = |rem(\rho)|$. We can retrieve $rem(\rho)$ from ρ and ρ' by truncation of all those phrases which correspond to subtrees with roots in A. Let $\rho_1, \ldots, \rho_n \in \cup_{i \in [1,k]} \Delta_{G_i}$ and $\rho'_1, \ldots, \rho'_n$ $\in \cup_{i \in [1,k]} \Delta_{G_i}$ be these subtrees for ρ and ρ' in a left to right order, respectively. For all $i \in [1,n]$ we observe $\Uparrow(\rho_i) = \Uparrow(\rho'_i)$. Thus ρ_i and ρ'_i both must be generated by the same grammar G_{j_i} for some $j_i \in [1,k]$. Since ρ_i and ρ'_i generates the i-th block of v we have $\downarrow(\rho_i) = \downarrow(\rho'_i)$ as well. But G_{j_i} is unambiguous for all $i \in [1,n]$. Hence $\rho_i = \rho'_i$ and we finally obtain $\rho = \rho'$. □

Definition 3.11. *For all* $i \in [1,k]$ *let* $\omega_i \in \Delta_{G_i}$ *be a derivation tree such that* $\Uparrow(\omega_i) = a_i$ *and* $\downarrow(\omega_i) = h(a_i)$. *Trees with these properties must exist, since* $h(a_i) \in$ $\sigma_h(a_i) = L(G_i)$. *The homomorphism append* : $\Delta_{G_L} \to \Delta_{G(h,L)}$ *is defined by* $append(p) = p$ *if* $p \in P_L$ *and* $append(a_i) = \omega_i$ *for all* $i \in [1,k]$.

Note that for all $\rho \in \Delta_{G_L}$ we have $(\downarrow \circ append)(\rho) = (h \circ \downarrow)(\rho)$. Since $\rho = rem(\rho)$ and the G_L remainder is invariant under appending trees we observe that *append* is injective.

Lemma 3.12. *For all* $w \in \Gamma^*$ *the restriction of* $(\downarrow \circ rem)$ *to* $\Delta_{G(h,L)}(w)$ *is onto* $h^{-1}(w) \cap L$.

Proof. Let $v \in h^{-1}(w) \cap L$. Since $v \in L$ there is a $\rho \in \Delta_{G_L}$ with $v = \downarrow(\rho)$. Since all the symbols and productions of G_L are contained in $G(h,L)$ we obtain $\rho \in \Lambda_{G(h,L)}$. Let $\rho' := append(\rho) \in \Delta_{G(h,L)}$. Obviously $\rho = rem(\rho')$. Therefore $v = \downarrow(\rho) = \downarrow(rem(\rho')) = (\downarrow \circ rem)(\rho')$. It remains to show that $\downarrow(\rho') = w$. Since $v \in h^{-1}(w)$ we have $h^{-1}(w) \neq \emptyset$. Therefore $h(v) \in h(h^{-1}(w)) = \{w\}$. Finally $\downarrow(\rho') = \downarrow(append(\rho)) = (\downarrow \circ append)(\rho) = (h \circ \downarrow)(\rho) = h(\downarrow(\rho)) = h(v) = w$. □

Lemma 3.13. *The equation* $r_{h,L}(w) = d_{G(h,L)}(w)$ *holds for all* $w \in \Gamma^*$.

Proof. Since $r_{h,L}(w) = |h^{-1}(w) \cap L|$ and $d_{G(h,L)}(w) = |\Delta_{G(h,L)}(w)|$ it is sufficient to show that the restriction of $(\downarrow \circ rem)$ to $\Delta_{G(h,L)}(w)$ is a bijection onto $h^{-1}(w) \cap L$. Let $(\downarrow \circ rem)(\rho) = (\downarrow \circ rem)(\rho')$ for some $\rho, \rho' \in \Delta_{G(h,L)}(w)$. Since $rem(\rho), rem(\rho') \in \Delta_{G_L}$ and G_L is unambiguous, $rem(\rho) = rem(\rho')$ follows. By Lemma 3.10 this implies $\rho = \rho'$. Hence the restriction of $(\downarrow \circ rem)$ to $\Delta_{G(h,L)}(w)$ is injective. Moreover by Lemma 3.9 it is a mapping into $h^{-1}(w) \cap L$, and by Lemma 3.12 it is a mapping onto $h^{-1}(w) \cap L$. □

Lemma 3.14. *The inequality* $d_{G(h,L)}(\pi_\Gamma(v)) \geq d_{G(h,L)}(v)$ *holds for all* $v \in$ $(\Gamma \cup \{0,1\})^*$.

Proof. If $v \in (\Gamma \cup \{0,1\})^* \setminus \sigma_h(L)$ the claim follows trivially by $d_{G(h,L)}(v) = |\Delta_{G(h,L)}(v)| = 0$. Now let us consider the case $v \in \sigma_h(L)$. Since $d_{G(h,L)}(v) = |\Delta_{G(h,L)}(v)|$ and $d_{G(h,L)}(\pi_\Gamma(v)) = |\Delta_{G(h,L)}(\pi_\Gamma(v))|$, it suffices to show that the restriction of $(append \circ rem)$ to the set $\Delta_{G(h,L)}(v)$ is an injection into the set $\Delta_{G(h,L)}(\pi_\Gamma(v))$. Let $\rho \in \Delta_{G(h,L)}(v)$. First we show that $(append \circ rem)$ is into

$\Delta_{G(h,L)}(\pi_\Gamma(v))$. Since $rem(\rho) \in \Delta_{G_L}$ we obtain $(\downarrow \circ rem)(\rho) \in L \subseteq A^*$. Thus we can write $(\downarrow \circ rem)(\rho) = a_{j_1} \cdots a_{j_n}$ for some $j_1, \ldots, j_n \in [1, k]$ and some $n \in \mathbb{N}$.

By definition $v = \downarrow(\rho)$. Therefore $\pi_\Gamma(v) = \pi_\Gamma(\downarrow(\rho))$ and by Lemma 3.8 we obtain the following:

$$
\begin{aligned}
\pi_\Gamma(\downarrow(\rho)) &\subseteq \pi_\Gamma(\sigma_h((\downarrow \circ rem)(\rho))) \subseteq \pi_\Gamma(\sigma_h(a_{j_1} \cdots a_{j_n})) \\
&\subseteq \quad \pi_\Gamma(h(a_{j_1})\{0,1\}^* \cdots h(a_{j_n})\{0,1\}^*) \quad = \{h(a_{j_1} \cdots a_{j_n})\}
\end{aligned}
$$

This implies:

$$
\begin{aligned}
\pi_\Gamma(v) &= h(a_{j_1} \cdots a_{j_n}) = h((\downarrow \circ rem)(\rho)) = (h \circ \downarrow)(rem(\rho)) \\
&= \quad (\downarrow \circ append)(rem(\rho)) = \downarrow((append \circ rem)(\rho)).
\end{aligned}
$$

Therefore $(append \circ rem)(\rho) \in \Delta_{G(h,L)}(\pi_\Gamma(v))$. It remains to show that the restriction of $(append \circ rem)$ to $\Delta_{G(h,L)}(v)$ is injective. This follows by Lemma 3.10 and the observation that $append$ is injective. $\qquad\square$

Lemma 3.13 and Lemma 3.14 immediately imply part *(1)* of Theorem 3.6. $\qquad\square$

Proof of Theorem 3.6 part *(2)*: Assume G' is a cf grammar such that $L(G') = \sigma_h(L)$, p is the maximum of 3 and the pumping constant of G', $q := p! + p$, and $w \in \Gamma^*$. Obviously $code_{h,p}(h^{-1}(w) \cap L) \subseteq \sigma_h(L)$. In case $h^{-1}(w) \cap L = \emptyset$ the inequality of Theorem 3.6 part *(2)* is trivially satisfied. Now assume $h^{-1}(w) \cap L \neq \emptyset$. Let $v \in h^{-1}(w) \cap L$ and $n := |w|$. Then for some $j_1, \ldots, j_n \in [1, k]$ we have:

$$
code_p(v) = h(a_{j_1})[p][q]^{j_1-1}[p][q]^{k-j_1} \cdots h(a_{j_n})[p][q]^{j_n-1}[p][q]^{k-j_n}.
$$

We say that an interval of a derivation tree lies *within* a subblock (block) if the corresponding nodes do not contain any leaf belonging to another subblock (block). We prove by induction that for each $i \in [0, n]$ there is a derivation tree $\rho_i \in \Delta_{G'}$ such that

$$
\downarrow(\rho_i) = (fill_p \circ h)(a_{j_1} \cdots a_{j_i}) \cdot code_{h,p}(a_{j_{i+1}} \cdots a_{j_n})
$$

and for each $m \in [1, i]$ the derivation tree ρ_i has a pumping phrase allowing to pump the same number of 0-symbols into the 0-th subblock and the j_m-th subblock of block m jointly. For $i = 0$ we only have to show that $code_p(v) = \downarrow(\rho_0)$ for some $\rho_0 \in \Delta_{G'}$. This follows by $code_p(v) \in \sigma_h(L) = L(G')$. Assume the statement is true for $i - 1$. Then there is a derivation tree $\rho_{i-1} \in \Delta_{G'}$ with the required phrase structure and the sentential form:

$$
\downarrow(\rho_{i-1}) = (fill_p \circ h)(a_{j_1} \cdots a_{j_{i-1}}) \cdot h(a_{j_i})[\underline{p}][q]^{j_i-1}[p][q]^{k-j_i} \cdot code(a_{j_{i+1}} \cdots a_{j_n}).
$$

The 0-th subblock of the i-th block in $\downarrow(\rho_{i-1})$ is underlined to indicate that the leaves of ρ_{i-1} forming the 0-symbols of this subblock are marked. According to Ogden's Lemma 2.3 the tree $\rho_{i-1} = \alpha\beta\gamma\delta\eta$ for some $\alpha, \beta, \gamma, \delta, \eta \in T_{G'}^*$ such that $\downarrow(\alpha\beta^l\gamma\delta^l\eta) \in \sigma_h(L)$ for each $l \in \mathbb{N}$. Moreover $\beta\delta$ must contain at least one marked position and at least one of the intervals $[|\alpha| + 1, |\alpha\beta|]$ and $[|\alpha\beta\gamma| +$

$1, |\alpha\beta\gamma\delta|]$ lies within the 0-th subblock of block i. Let $\tau := [|\alpha\beta\gamma| + 1, |\alpha\beta\gamma\delta|]$ if $[|\alpha| + 1, |\alpha\beta|]$ has this property and $\tau := [|\alpha| + 1, |\alpha\beta|]$ otherwise.

By the choice of p and q the insertion of at most p many 0-symbols into a subblock $[p]$ yields a subblock shorter than $[q]$. We will implicitly apply this argument in the sequel.

Assume τ is not within the i-th block, i.e., it is outside or it overlaps with block i and some neighbouring blocks. Let $i' = i + c$ where c is the number of Γ symbols in β if $\tau = [|\alpha| + 1, |\alpha\beta|]$ and $i' := i$ otherwise.

Then block i' of $\downarrow(\alpha\beta^2\gamma\delta^2\eta)$ equals block i of $\downarrow(\alpha\beta\gamma\delta\eta)$, except for a proper insertion of at most p many 0-symbols in the 0-th subblock of block i'. Therefore within block i' the 0-th subblock does not agree with any other subblock.

Now assume τ lies within block i. Then it cannot contain a 1-symbol because otherwise the i-th block of $\downarrow(\alpha\beta^2\gamma\delta^2\eta)$ would contain more than $k+1$ subblocks. Hence each of β and δ lie within one subblock of the i-th block. We can easily verify that $\downarrow(\alpha\beta^2\gamma\delta^2\eta)$ does not contain more than $2p$ occurrences of 0-symbols in the 0-th subblock of block i in this case. This implies that β lies within the 0-th subblock and δ within the j_i-th subblock of block i and $1 \le |\downarrow(\beta)| = |\downarrow(\delta)| \le p$. Thus for $l = p! \cdot |\downarrow(\beta)|^{-1} + 1$ the derivation tree $\rho_i := \alpha\beta^l\gamma\delta^l\eta$ has the property $\downarrow(\rho_i) = (fill_p \circ h)(a_{j_1} \cdots a_{j_i}) \cdot code(a_{j_i+1} \cdots a_{j_k})$. Now ρ_i contains a pumping phrase allowing to pump the same number of 0-symbols into the 0-th subblock and into the j_i-th subblock of block i jointly. Moreover the pumping phrases of ρ_i to the left of block i are the same as in ρ_{i-1}, which completes the induction.

Eventually for $i = n$ we obtain a derivation tree ρ_n with the frontier $\downarrow(\rho_n) = (fill_p \circ h)(w)$ and the claimed phrase structure starting from an arbitrary word of $code_{h,p}(h^{-1}(w) \cap L)$. It remains to show that two trees obtained in this way beginning with different words in $h^{-1}(w) \cap L$ must not coincide. Let $v_1, v_2 \in h^{-1}(w) \cap L$ be two different words and let ω_1 and ω_2 be the corresponding derivation trees obtained by the pumping sequence described above. Then ω_1 and ω_2 both generate $(fill_p \circ h)(v_1)$. Assume $\omega_1 = \omega_2$. Since h is length preserving we observe $|v_1| = |v_2|$. Therefore v_1 and v_2 differ in at least one position $i \in [1, |v_1|]$. Then $a_j = v_1[i] \ne v_2[i] = a_{j'}$ for some $j, j' \in [1, k]$. W.l.o.g. we assume $j > j'$. Then ω_1 contains a pumping phrase ϑ_1 allowing to pump the 0-th subblock and the j-th subblock of block i jointly and it contains a pumping phrase ϑ_2 allowing to pump the 0-th subblock and the j'-th subblock of block i jointly. Pumping once ω_1 corresponding to ϑ_1 we obtain a derivation tree ω' with a word $\beta \in (P \cup \{0\})^*$ inserted to the left of ϑ_2. Since β does only contain leaves labeled with 0-symbols ϑ_2 is shifted to the right, but remains within the same subblock as in ω_1. Thus in ω' it is still possible to pump the 0-th subblock and the j'-th subblock of block i jointly but now in $\downarrow(\omega')$. This pumping yields a derivation tree ω'' for which the 0-th subblock of block i does no longer agree with any other subblock of block i, which is a contradiction. Hence $\omega_1 \ne \omega_2$. This implies that $(fill_p \circ h)(w)$ can be generated by at least $|code_{h,p}(h^{-1}(w) \cap L)|$ many different derivation trees. Moreover since $code_{h,p}$ is injective we finally obtain $d_{G'}(fill(w)) \ge |code_{h,p}(h^{-1}(w) \cap L)| = |h^{-1}(w) \cap L| = r_{h,L}(w)$. $\qquad\square$

Theorem 3.15. *The context-free language $\sigma_h(L)$ is $\hat{r}_{h,L}$-ambiguous.*

Proof. Recall that $L(G(h, L)) = \sigma_h(L)$. Now Theorem 3.6 *(1)* implies:

$$\max\{d_{G(h,L)}(v) \mid v \in (\Gamma \cup \{0,1\})^{\leq n}\}$$
$$\overset{3.6}{\leq} \max\{d_{G(h,L)}(\pi_\Gamma(v)) \mid v \in (\Gamma \cup \{0,1\})^{\leq n}\}$$
$$= \max\{d_{G(h,L)}(w) \mid w \in \Gamma^{\leq n}\}$$
$$\leq \max\{d_{G(h,L)}(v) \mid v \in (\Gamma \cup \{0,1\})^{\leq n}\}$$

Hence all the expressions above are equal and again by Theorem 3.6 *(1)* we obtain:

$$\hat{r}_{h,L}(n) = \max\{r_{h,L}(w) \mid w \in \Gamma^{\leq n}\} \overset{3.6}{=} \max\{d_{G(h,L)}(w) \mid w \in \Gamma^{\leq n}\}$$
$$= \max\{d_{G(h,L)}(\pi_\Gamma(v)) \mid v \in (\Gamma \cup \{0,1\})^{\leq n}\} = \hat{d}_{G(h,L)}(n)$$

Thus $G(h, L)$ is appropriate to satisfy property *(1)* of Definition 2.1. By Theorem 3.6 *(2)* we obtain that for each cf grammar G' such that $L(G') = \sigma_h(L)$ and all words $w \in \Gamma^*$ we have $r_{h,L}(w) \leq d_{G'}(fill_p(w))$ for some $p \in \mathbb{N}$. This implies $\hat{r}_{h,L}(|w|) \leq \hat{d}_{G'}(|fill_p(w)|) = \hat{d}_{G'}(c \cdot |w|)$ where $c = 1 + (k+1)(p! + p + 1)$. Thus $\sigma_h(L)$ and $\hat{r}_{h,L}$ also satisfy the property *(2)* of Definition 2.1.   □

4 Applications

4.1 Census Functions

Definition 4.1. *Let $L \subseteq A^*$ be a formal language. The* census function $\gamma_L :$ $\mathbb{N} \to \mathbb{N}$ *is defined by $\gamma_L(n) := |A^n \cap L|$, and the function $\hat{\gamma}_L : \mathbb{N} \to \mathbb{N}$ is defined by $\hat{\gamma}_L(n) := \max\{\gamma_L(i) \mid i \leq n\}$. The* homomorphism hide $: A^* \to \{\$\}$ *is defined by $hide(X) := \$$ for all $X \in A$.*

Theorem 4.2. *Let $L \subseteq A^*$ be an unambiguous context-free language. Then*

$$\sigma_{hide}(L) \text{ is } \hat{\gamma}_L\text{-ambiguous.}$$

Proof. By Theorem 3.15 the language $\sigma_{hide}(L)$ is $\hat{r}_{hide,L}$-ambiguous. We get $\hat{r}_{hide,L}(n) = \max\{r_{hide,L}(w) \mid w \in A^{\leq n}\} = \max\{|hide^{-1}(w) \cap L| \mid w \in A^{\leq n}\} = \max\{|A^{|w|} \cap L| \mid w \in A^{\leq n}\} = \max\{|A^j \cap L| \mid j \leq n\} = \hat{\gamma}_L(n).$ □

Corollary 4.3. *There is an unambiguous context-free language L such that L^+ is 2^{n-1}-ambiguous and L^k is $\binom{n-1}{k-1}$-ambiguous (implying $\Theta(n^{k-1})$-ambiguity) for each $k \in \mathbb{N} \setminus \{0\}$.*

Proof. Let $\{a, b\}$ be an alphabet. We observe that $\hat{\gamma}_{(a+b)^*b}(n) = 2^{n-1}$ for $n > 0$ and $\hat{\gamma}_{(a^*b)^k}(n) = \binom{n-1}{k-1}$ for each $k \in \mathbb{N} \setminus \{0\}$. Using Theorem 4.2 we obtain that $\sigma_{hide}((a+b)^*b)$ is 2^{n-1}-ambiguous and $\sigma_{hide}((a^*b)^k)$ is $\binom{n-1}{k-1}$-ambiguous. Therefore $\sigma_{hide}((a^*b)^1)$ is unambiguous. Finally since σ_{hide} is a homomorphism we get $\sigma_{hide}((a+b)^*b) = \sigma_{hide}((a^*b)^+) = (\sigma_{hide}(a^*b))^+$ and $\sigma_{hide}((a^*b)^k) = (\sigma_{hide}(a^*b))^k$. Thus $L := \sigma_{hide}(a^*b)$ is a language with the required properties.   □

4.2   Cycle-Free Context-Free Grammars

In this part our aim is to find out when the ambiguity function of a cf grammar is inherent for some cf language.

Definition 4.4. *A grammar* $G = (N, A, P, S)$ *is in* Greibach-normal-form *if* $P \subseteq N \times AN^*$.

Symbols which cannot occur in any derivation tree do not contribute to the ambiguity function. Therefore it is sufficient to consider reduced cf grammars. Reduced cyclic cf grammars generate for some words infinitely many derivation trees. This situation does not occur in Greibach-normal-form grammars. Since each cf grammar can be transformed into an equivalent grammar in Greibach-normal-form the ambiguity function of a cyclic reduced cf grammar cannot be inherent for any cf language.

By the definition of the Greibach-normal-form we immediately obtain that the parse of a derivation tree has the same length as its frontier. Moreover the i-th symbol of the frontier is uniquely determined by the i-th symbol of the parse. Thus we obtain:

Lemma 4.5. *Let* $G = (N, A, P, S)$ *be a grammar in Greibach-normal-form and* $h_G : P^* \to A^*$ *the length preserving homomorphism defined by* $h_G(p) = X$ *where* X *is for each* $p \in P$ *the terminal at the beginning of* p's *right-hand side. Then* $\downarrow(\rho) = h_G(parse(\rho))$ *for all* $\rho \in \Delta_G$.

Theorem 4.6. *Let* $G = (N, A, P, S)$ *be a context-free grammar in Greibach-normal-form, and* h_G *defined as in Lemma 4.5. Then the context-free language* $\sigma_{h_G}(parse_G(\Delta_G))$ *is* $\hat{d}_G$-*ambiguous.*

Proof. Let $L := parse_G(\Delta_G)$ and $h := h_G$.

$$\hat{r}_{h,L}(n) = \max\left\{|h^{-1}(w) \cap L| \,\big|\, w \in A^{\leq n}\right\}$$
$$= \max\left\{\left|\{\rho \in \Delta_G \,|\, \downarrow(\rho) = w\}\right| \,\big|\, w \in A^{\leq n}\right\}$$
$$= \max\left\{d_G(w) \,|\, w \in A^{\leq n}\right\} \qquad = \hat{d}_G(n).$$

By Lemma 2.2 the cf language $parse_G(\Delta_G)$ is unambiguous. Moreover h_G is length preserving. Thus the claim follows by Theorem 3.15   □

By definition each inherent ambiguity function is an ambiguity function for some cf grammar G. By the considerations at the beginning of section 4.2 we can even find a reduced cycle-free cf grammar G with this property. On the other hand for each ambiguity function of a cycle-free cf grammar G we can find an ε-free cycle-free cf grammar G', not necessarily for the same language, such that $\hat{d}_G(n) = \hat{d}_{G'}(n)$ for all $n \in \mathbb{N} \setminus \{0\}$. It can be shown that G' can be transformed into a grammar G'' in Greibach-normal-form which has the same characteristic ambiguity power series as G'. Then by Theorem 4.6 the language $\sigma_{h_{G''}}(parse_{G''}(\Delta_{G''}))$ is $\hat{d}_{G'}$-ambiguous. By inspection of Definition 2.1 we can easily verify that this already implies $\hat{d}_G$-ambiguity for $\sigma_{h_{G''}}(parse_{G''}(\Delta_{G''}))$. This finally implies:

Theorem 4.7. *The set of ambiguity functions for cycle-free context-free grammars and the set of inherent ambiguity functions coincide.*

5 Conclusion

It has been shown that the loss of information induced by a length preserving homomorphism can be turned into inherent ambiguity. The result has been used to prove that:

- The least monotone function $\hat{\gamma}$ larger than or equal to the census function of an arbitrary unambiguous context-free language is an inherent ambiguity function.
- The set of ambiguity functions for cycle-free context-free grammars and the set of inherent ambiguity functions coincide.

The latter result is particularly useful for future research. As was mentioned in the introduction some examples of languages with sublinear ambiguity have recently been discovered. But a characterization of the obtainable ambiguities is still missing. The author conjectures that context-free languages with infinite sublogarithmic ambiguity can be found. To prove that a given function f is an inherent ambiguity function, with the result of this paper, it is sufficient to find an f-ambiguous cycle-free context-free grammar. Finally we can raise the question whether each context-free language is f-ambiguous for some function f.

Acknowledgments. I would like to thank Friedrich Otto for proofreading and valuable discussions.

References

1. J. Berstel. *Transductions and context-free languages.* Teubner Studienbücher, Stuttgart, 1979.
2. J. Crestin. Un langage non ambigu dont le carré est d'ambiguité non bornée. In M. Nivat, editor, *Automata, Languages and Programming*, pages 377–390. Amsterdam, North-Holland, 1973.
3. H. Maurer. The existence of context-free languages which are inherently ambiguous of any degree. Research series, Department of Mathematics, University of Calgary, 1968.
4. M. Naji. Grad der Mehrdeutigkeit kontextfreier Grammatiken und Sprachen, 1998. Diplomarbeit, FB Informatik, Johann–Wolfgang–Goethe–Universität Frankfurt/M.
5. R. J. Parikh. Language–generating devices. In *Quarterly Progress Report*, volume 60, pages 199–212. Research Laboratory of Electronics, M.I.T, 1961.
6. K. Wich. Exponential ambiguity of context-free grammars. In G. Rozenberg and W. Thomas, editors, *Proceedings of the 4th International Conference on Developments in Language Theory, July 1999*, pages 125–138. World Scientific, Singapore, 2000.
7. K. Wich. Sublinear ambiguity. In M. Nielsen and B. Rovan, editors, *Proceedings of the MFCS 2000*, number 1893 in Lecture Notes in Computer Science, pages 690–698, Berlin-Heidelberg-New York, 2000. Springer.

Histogramming Data Streams with Fast Per-Item Processing

Sudipto Guha[1], Piotr Indyk[2], S. Muthukrishnan[3], and Martin J. Strauss[3]

[1] Department of Computer and Information Science, University of Pennsylvania, Philadelphia, PA 19104, sudipto@cis.upenn.edu
[2] MIT Laboratory for Computer Science; 545 Technology Square, NE43-373; Cambridge, Massachusetts 02139-3594; indyk@theory.lcs.mit.edu
[3] AT&T Labs—Research, 180 Park Avenue, Florham Park, NJ 07932 USA, {muthu,mstrauss}@research.att.com

Abstract. A vector $\mathbf{A}$ of length N can be approximately represented by a histogram $\mathbf{H}$, by writing $[0, N)$ as the non-overlapping union of B intervals I_j, assigning a value b_j to I_j, and approximating $\mathbf{A}_i$ by $\mathbf{H}_i = b_j$ for $i \in I_j$. An optimal histogram representation $\mathbf{H}_{\mathrm{opt}}$ consists of the choices of I_j and b_j that minimize the sum-square-error $\|\mathbf{A} - \mathbf{H}\|_2^2 = \sum_i |\mathbf{A}_i - \mathbf{H}_i|^2$. Numerous applications in statistics, signal processing and databases rely on histograms; typically B is (significantly) smaller than N and, hence, representing $\mathbf{A}$ by $\mathbf{H}$ yields substantial compression.

We give a deterministic algorithm that approximates $\mathbf{H}_{\mathrm{opt}}$ and outputs a histogram $\mathbf{H}$ such that

$$\|\mathbf{A} - \mathbf{H}\|_2^2 \le (1 + \epsilon) \|\mathbf{A} - \mathbf{H}_{\mathrm{opt}}\|_2^2.$$

Our algorithm considers the data items $\mathbf{A}_0, \mathbf{A}_1, \ldots$ in order, *i.e.*, in one pass, spends processing time $O(1)$ per item, uses total space $B \ poly(\log(N), \log \|\mathbf{A}\|, \frac{1}{\epsilon})$, and determines the histogram in time $poly((B, \log(N), \log \|\mathbf{A}\|, \frac{1}{\epsilon})$. Our algorithm is suitable to emerging applications where signal is presented in a stream, size of the signal is very large, and one must construct the histogram using significantly smaller space than the signal size. In particular, our algorithm is suited to high performance needs where the per-item processing time must be minimized. Previous algorithms either used large space, i.e., $\Omega(N)$, or worked longer, *i.e.*, $N \log^{\Omega(1)}(N)$ total time over the N data items. Our algorithm is the first that simultaneously uses small space as well as runs fast, taking $O(1)$ worst case time for per-item processing. In addition, our algorithm is quite simple.

Keywords: Histograms, streaming algorithms

1 Introduction

We study the problem of representing signals succinctly using histograms. The *signal* is a vector $\mathbf{A}$ of length N. A *histogram* $\mathbf{H}$ on the signal is obtained by

P. Widmayer et al. (Eds.): ICALP 2002, LNCS 2380, pp. 681–692, 2002.
© Springer-Verlag Berlin Heidelberg 2002

writing $[0, N)$ as the non-overlapping union of B intervals I_j and assigning a value b_j to I_j. The histogram $\mathbf{H}$ can be used to approximately represent the signal $\mathbf{A}$ by approximating $\mathbf{A}_i$ as $\mathbf{H}_i = b_j$ for $i \in I_j$. Equivalently, a histogram is a piecewise constant approximation to the signal. Histogram $\mathbf{H}$ takes only $O(B)$ values to store, namely, the I_j's and b_j's, in contrast to the $O(N)$ values needed to store the signal $\mathbf{A}$. Typically $B \ll N$ in applications and thus $\mathbf{H}$ is a succinct representation for $\mathbf{A}$.

Histograms must nevertheless capture the trends in the signal. Numerous measures evaluate how well a histogram achieves this; the most common form is the sum-square-error, which measures the sum of the squares of the devation from the signal. An *optimal histogram* $\mathbf{H}_{\text{opt}}$, sometimes written $\mathbf{H}_{\text{opt}}^B$, is a B-bucket histogram, *i.e.*, choices of I_j and b_j, that minimize the sum-square-error $\|\mathbf{A} - \mathbf{H}\|_2^2 = \sum_i |\mathbf{A}_i - \mathbf{H}_i|^2$. The problem of interest therefore is to find $\mathbf{H}_{\text{opt}}$ or to approximate it. An approximation factor of α would indicate $\|\mathbf{A} - \mathbf{H}\|_2^2 \leq \alpha \|\mathbf{A} - \mathbf{H}_{\text{opt}}\|^2$.

Histogram representations are used extensively in signal processing, statistics and databases. In database systems for example, they are used to approximate sizes of database operations which in turn help determine efficient execution plans for complex queries (See [1] for an overview). Almost all commercial database systems use histograms; finding best histograms—and other succinct representations such as wavelets or discrete fourier coefficients of the signal—is a thriving area of research in the database community.

Our motivation lies in an emerging application scenario in large databases and in processing massive data in general. Signals (such as time series of network events, web accesses, IP traffic patterns) are huge and appear in a stream. In most cases, they are not captured in databases because they are far too voluminous. (For an overview, see [2] for sizes, description of data feeds, etc.) Nevertheless, there is a growing need (eg., for network management purposes) to summarize the signals succinctly using histograms. Even when signals are small enough to be stored in disks within databases, practitioners seek algorithms that read the signal in one pass and compute (or estimate) functions of interest. This is because multiple passes or random accesses to disk-resident data is expensive. Thus, the focus in dealing with massive data is to seek algorithms that process signals on a stream, be they available as a stream from the data source (as in IP router or server logs) or read in one pass over stored data (as in large databases). Systems researchers have extensively articulated the need for stream processing (See [3,4,5] for references).

In the past few years, models have been developed to design, analyze and study data stream algorithms. Specifically, algorithms are required to read each item in the stream in turn, work using some additional workspace, and compute functions of interest. No backtracking is allowed on the input data stream. There are three parameters of performance: **(1)** time to process each data stream item, **(2)** amount of workspace used, and **(3)** time to compute functions of interest. Clearly the model is of interest only when the workspace provided is smaller than the space needed to store the entire signal. Furthermore, with this restriction, almost no function can be computed exactly (for example, computing the median of signal items is hard [6]). Hence, the emphasis is on estimating functions accurately, rather than computing them exactly. Specifically, algorithms in data stream model are designed with polylogarithmic workspace and they attempt to optimize the other two parameters pertaining to speed. Data stream algorithms have been designed for estimating norms [7,8,9], clustering [10], wavelet and histogram estimations [11,12,14,13], etc.

Our work involves designing data stream algorithms. Our departure from previous work begins with observing that not all parameters above are equally important in applications. In particular, the per-item processing time is highly critical. This is because in data streaming instances such as IP routers that generate logs of packets they forward, flows and TCP connections they maintain etc, work at blistering speeds, and process millions of items per second. It is imperative that any per-item processing be very small in order to deal with this deluge. Likewise, disk systems scan large databases at very high speeds and to keep up with the pipeline, it is desirable that data stream algorithms minimize per-item processing. Our work here is inspired by this requirement.

1.1   Our Result and Previous Work

The problem of constructing (near) optimal histograms has been investigated theoretically. Most of the solutions (including ours) can be viewed as a *streaming algorithm*, that actually consists of two algorithms: a *sketching* algorithm that preprocess the input to some data structure and a *reconstruction* algorithm that performs some computation on the structure to output the final approximation. The total time is dominated by the preprocessing time to build (and maintain) the data structure. Thus we will quote results in per-item time in context of streaming algorithms. Since we can always store a set of elements and then perform the entire computation required, the per-item time is meaningful only if the space allowed is sublinear.

In [15], an $O(N^2 B)$ time and $O(NB)$ space dynamic programming based solution was presented for optimal offline histogram computation. They also presented an approximate algorithm, that, when adopted to the streaming model, use space $O(B \log \|\mathbf{A}\|)$ space and $O(\log \mathbf{A})$ per-item processing time to output a $\mathbf{H}$ of at most $3B$ intervals such that $\|\mathbf{A} - \mathbf{H}\|_2^2 \leq 3 \|\mathbf{A} - \mathbf{H}_{opt}^B\|_2^2$. Gilbert et al [14] presented an algorithm that uses $O(B \log N)$ space and $O(\log N)$ per-item processing time[1] algorithm to output $\mathbf{H}$ using $O(B \log N)$ intervals with $\|\mathbf{A} - \mathbf{H}\|_2^2 \leq \|\mathbf{A} - \mathbf{H}_{opt}\|_2^2$. The essential tradeoff however has been between accuracy of the approximation and the number of buckets.

[11] presented a substantially more accurate algorithm taking $(B^2 \log(N))/\epsilon$ per-item processing and histogram computation time as well as $(B^2 \log(N))/\epsilon$ space to output a $1 + \epsilon$ approximation; that is an $\mathbf{H}$ such that $\|\mathbf{A} - \mathbf{H}\|_2^2 \leq (1 + \epsilon) \|\mathbf{A} - \mathbf{H}_{opt}\|_2^2$. Subsequently in [12] it was modified to give a $\theta(N)$ space,

---

[1] This algorithm may be converted to use $O(1)$ time per-item time using techniques in this paper.

$O(N)$ total time algorithm. These results capture a tradeoff between space and the time required for histogram construction.

Our result here improves both the tradeoff directions above and achieves the best possible. We provide a $1 + \epsilon$ approximation in $O(1)$ per-item time. The work space we use is only $B(\log N \log \|\mathbf{A}\| / \epsilon)^{O(1)}$ which is of independent interest since previous algorithms used $\Omega(B^2)$ space, a limiting condition. The reconstruction time is polylogarithmic.

1.2   Overview of Our Techniques

Our approach uses signal processing techniques. Our overall algorithm divides neatly into two modules. The first module reads over the stream and outputs a certain number of wavelet coefficients of $\mathbf{A}$. Wavelet coefficients are inner products of the signal with dyadic basis vectors as defined in Section 2. This module is deterministic and exact, taking only $O(1)$ time for per-item processing. The second module processes these coefficients to construct $\mathbf{H}$. It has two major components. The first component involves constructing what we call a *robust* histogram of $\mathbf{A}$. A robust histogram has the property that refinement by further buckets does not decrease the overall error significantly. The notion of robust histogram was introduced in [13]; there the robust histogram had to be constructed directly from the signal via sophisticated randomized techniques. Here, we construct the robust histogram via wavelets, iteratively and deterministically, which we can do in our streaming model. The robust approximation is already a good approximation to the final histogram, but has a few too many intervals. The second major component involves culling the output histogram using dynamic programming from the robust histogram. The overall algorithm is quite simple. The crux throughout is the proof of various structural properties of intermediate histograms that yields the final histogram.

1.3   Map

In Section 2 we present formal definitions and background. In Section 3, we present the first module, namely, the streaming algorithm for computing various wavelet coefficients. In Section 4, we present the second module of constructing robust histogram and culling the final histogram from it. In Section 5, we present concluding remarks.

2   Preliminaries

We consider signals indexed on $\{0, 1, \ldots, N - 1\}$, where N is a power of 2. A dyadic interval is an interval of the form $[k2^j, (k + 1)2^j)$, where j and k are integers. The function that equals 1 on set S and zero elsewhere is denoted χ_S.

A (Haar) wavelet is a function ψ on $[0, N)$ of one of the following forms:

- $\frac{1}{\sqrt{N}} \chi_{[0,N)}$
- $2^{-j/2} \left(-\chi_{[k2^{j-1}, (k+1)2^{j-1})} + \chi_{[(k+2)2^{j-1}, (k+3)2^{j-1})} \right).$

Example wavelets of the second type are

$$2^{-1/2}(-1, 1, 0, 0, 0, 0, \dots,),$$
$$2^{-1/2}(0, 0, -1, 1, 0, 0, 0, 0, \dots,), \dots,$$
$$2^{-1}(-1, -1, 1, 1, 0, 0, 0, 0, \dots,) \dots$$

There are N wavelets altogether, and they form an orthonormal basis, i.e., $\langle \psi, \psi' \rangle$ is 1 if $\psi = \psi'$ and 0 otherwise.

Every signal can be reconstructed exactly from all its wavelet coefficients (its full *wavelet transform*, an orthonormal linear transformation), as $\mathbf{A} = \sum_j \langle \mathbf{A}, \psi_j \rangle \psi_j$, whence a formal linear combination of distinct wavelets is its own wavelet transform.

Parsefal's equality states that the L^2 norm of a signal is invariant under orthonormal change of basis:

$$\sum_i \mathbf{A}_i^2 = \sum_j \langle \mathbf{A}, \psi_j \rangle^2 .$$

It follows from Parsefal's equality that, for any set Λ of B wavelet terms, the error in using those terms to approximate $\mathbf{A}$ is the sum of the squares of the omitted terms. This is minimized when Λ contains the wavelet terms whose coefficients have the largest squares.

A simple classical wavelet algorithm computes the full wavelet transform in linear time and space. In the first of several passes, compute and output $N/2$ of the wavelets of the form $\mathbf{A}_{2i+1} - \mathbf{A}_{2i}$. Also compute $N/2$ quantities of the form $\mathbf{A}_{2i+1} + \mathbf{A}_{2i}$, and recursively compute a wavelet decomposition of that.

Note that each wavelet is a piecewise-constant function of 4 pieces (3 boundaries). Conversely, each χ_I can be written as $\chi_I = \sum_j \langle \chi_I, \psi_j \rangle \psi_j$, and there are only $2\log(N)$ wavelets for which the dot product is zero—those wavelets whose support intersects an endpoint of I, where the support of a vector is the set of positions where it is non-zero. Thus wavelet representations simulate histograms with at most a $O(\log(N))$ blowup in the number of buckets/terms. It follows that, using wavelets, one can easily find a $O(B \log(N))$-bucket histogram which approximates given signal as well as the best B-bucket histogram. Specifically, the $(6B \log(N) + 1)$-bucket histogram representation defined as the best $2B \log N$-term wavelet representation has this property.

In what follows, we first show how to find the top B' wavelet coefficients quickly. For $B' = 2B \log(N)$, this already gives an efficient construction of a $O(B \log(N))$-bucket histogram. Next, we show how to output instead a B-bucket histogram though the error of our histogram is worse than optimal by the factor $(1 + \epsilon)$.

3   Efficient Computation of Wavelet Coefficients from a Stream

We will be interested in finding the largest B' coefficients of a wavelet decomposition of the stream we receive.

It is easy to modify the $\log(N)$-pass classical algorithm from the previous section to work in one pass; the order in which the coefficients are output is altered only. We need to view the $\log N$ passes as happening concurrently. The result of the first pass is the input to the second pass in a stream fashion and so forth. Thus, with space $O(\log N)$, we can output a stream of wavelet coefficients of the original stream.

Lemma 1. *There is an algorithm that reads in a stream $\mathbf{A}_0, \mathbf{A}_1, \ldots$ and outputs the N wavelet coefficients (in arbitrary order), using per-item time $O(1)$ and space $O(\log(N))$.*

We next show how to find the B' largest items in a stream by an algorithm that uses per-item time $O(1)$ and space $O(B')$.

We maintain a list of size at most $2B'$. Initially we store the first $2B'$ elements we receive. After we have $2B'$ elements, we run a selection algorithm [6] which finds the median; we then retain only the elements larger than it. The process of finding the median of $2B'$ elements and discarding the bottom half takes $O(B')$ time. At this point we have only B' elements; we store the next B' elements without any computation, then perform the reduction as above. We perform $O(N/B')$ reductions on sets of size $O(B')$; altogether, this takes $O(N)$ time with worst-case per-item time $O(B')$. By buffering the input in a buffer of size $O(B')$ we can reduce the worst-case per-item time to $O(1)$. (For example, upon reading an item that grows our list to size $2B'$, perform $O(1)$ steps of the reduction algorithm for each input, while incoming input sits in a buffer of size B'. When that buffer fills, we'll have completed the reduction, leaving a set of size B', which we combine with the input buffer of B' items.)

Lemma 2. *There is an algorithm that takes B' as input, reads in a stream $\mathbf{A}_0, \mathbf{A}_1, \ldots$ and outputs the top B' wavelet coefficients, using per-item time $O(1)$ and space $O(B')$.*

4   Wavelets to Histograms via Robust Representations

In this section, we show how to find a nearly optimal B-bucket histogram representation, using the techniques of the previous section. There are two algorithmic parts to this.

- Construction of a "robust" approximation. Given B, ϵ, M, and N, we define a particular $B' \leq (B \log(N) \log(M)/\epsilon)^{O(1)}$. We then show, for each signal $\mathbf{A}$ with $\|\mathbf{A}\|^2 \leq M$, how to construct, greedily, an approximation $\mathbf{H}_r$ from the top B' wavelet coefficients of $\mathbf{A}$ such that, if $\mathbf{H}$ refines $\mathbf{H}_r$ by an additional $B - 1$ boundaries and $\mathbf{H}$ has optimal parameters, then $\|\mathbf{A} - \mathbf{H}_r\|^2 \leq (1 + \epsilon_r) \|\mathbf{A} - \mathbf{H}\|^2$. That is, $\mathbf{H}_r$ is not significantly improved by additional boundaries.
- Construction of our output, $\mathbf{H}$, which is defined to be the best B-term representation to $\mathbf{H}_r$. We first give an efficient construction of $\mathbf{H}$ and then we show that $\|\mathbf{A} - \mathbf{H}\|^2 \leq (1 + \epsilon) \|\mathbf{A} - \mathbf{H}_{\text{opt}}\|^2$, provided ϵ_r is chosen properly given ϵ. (They are polynomially related.)

4.1   Robust Histograms

Definition 1 (see [13]). *Fix a signal,* $\mathbf{A}$. *A representation* $\mathbf{H}_r$ *is called a* (B_r, ϵ_r)-*robust approximation to* $\mathbf{A}$ *if, for any representation* $\mathbf{H}$ *on the boundaries of* $\mathbf{H}_r$ *and any other* $B_r - 1$ *boundaries, with optimal parameters, we have*

$$(1 - \epsilon_r) \|\mathbf{A} - \mathbf{H}_r\|^2 \leq \|\mathbf{A} - \mathbf{H}\|^2 .$$

In [13], an algorithm for constructing a robust approximation was given. We obtain the same result here, but in a significantly simpler way, which is possible in our model. We include a sketch of the construction here. In this paper, we will need $B_r = B \log n$ and $\epsilon_r = \epsilon^4$ to get a B-bucket histogram with error $(1 + \epsilon)$ times the optimal.

Lemma 3. *Given* $B_r, N, \epsilon_r,$ *and* M, *for any integer-valued signal* $\mathbf{A}$ *with* $\|\mathbf{A}\| \leq M$, *there exists a* $B' \leq (B_r \log(N) \log(M)/\epsilon_r)^{O(1)}$ *and a* $(B')^{O(1)}$-*time algorithm to find a* (B_r, ϵ_r)-*robust approximation to* $\mathbf{A}$ *from among the top* B' *wavelet coefficients for* $\mathbf{A}$.

Proof. First note that the characteristic function χ_I can be written as the sum of $O(\log(N))$ wavelets—corresponding to those wavelets whose support intersects an endpoint of I. Thus a B-bucket histogram can be viewed as a $O(B \log(N))$-term wavelet representation, refinenment of a histogram by $B - 1$ boundaries can be simulated by refining a wavelet representation by $O(B \log(N))$ terms, and it suffices to find a wavelet representation that is not much improved by $O(B_r \log(N))$ additional wavelet terms.

The algorithm is as follows. Start with the zero representation $\mathbf{R}$. If $\mathbf{R}$ is not already robust, then some $B_r \log(N)$ wavelet terms improve it. The terms giving the best improvement are, by Parsefal, those with the largest coefficients not already in $\mathbf{R}$. It follows that the largest single coefficient gives some improvement, namely,

$$\|\mathbf{A} - (\mathbf{R} + \langle \mathbf{A}, \psi \rangle \psi)\|^2 \leq \left(1 + \Omega \left(\frac{\epsilon_r}{B_r \log(N)}\right)\right)^{-1} \|\mathbf{A} - \mathbf{R}\|^2 .$$

Replace $\mathbf{R} \leftarrow \mathbf{R} + \langle \mathbf{A}, \psi \rangle \psi$ and remove ψ from the list of available terms. Repeat this procedure until the representation is robust. Observe that after $O\left(\frac{B_r \log(N) \log(M)}{\epsilon_r}\right)$ iterations, the value of $\|\mathbf{A} - \mathbf{R}\|^2$ has been reduced from $\|\mathbf{A}\|^2$ by the factor $\frac{1}{4M^2}$. Since we assumed that $\mathbf{A}$ is integer-valued and $\|\mathbf{A}\| \leq M$, it follows that a rounding of $\mathbf{R}$ to integer coefficients (a legitimate wavelet representation) equals $\mathbf{A}$. Thus, at this iteration, one can choose $\mathbf{A}$ itself as a robust representation.

Thus we put $B' = \Theta\left(\frac{B_r \log(N) \log(M)}{\epsilon_r}\right)$. We construct a $(O(B_r \log(N)), \epsilon_r)$-robust wavelet represntation (generalizing the definition of robustness from histograms to wavelets in the obvious way), for which we need the top B' wavelet terms. The result can be viewed as a (B_r, ϵ_r)-robust histogram approximation with at most $O(B')$ buckets.

4.2   Approximating the Robust Histogram

In this section, we show how to find the best B-bucket approximation $\mathbf{H}$ to $\mathbf{H}_r$. (Later we will argue that $\mathbf{H}$ is, in fact, a good approximation to $\mathbf{A}$.) We will show that $\mathbf{H}$ only needs to use the boundaries in $\mathbf{H}_r$, of which there are at most $O(B')$. Thus a dynamic programming algorithm [15] will find $\mathbf{H}$ in time polynomial in B' using space $O(B')$.

Lemma 4. *Given a (sparsely presented) B'-bucket histogram $\mathbf{H}'$ on $N \geq B'$ numbers, the best B-bucket approximation $\mathbf{H}$ to $\mathbf{H}'$ uses only the boundaries of $\mathbf{H}'$.*

Proof. Suppose not. Let $r_1 < b < r_2$, where b is a boundary in $\mathbf{H}$ and r_1 and r_2 are consecutive boundaries in $\mathbf{H}'$. Let c_1 and c_2 be the coefficients in buckets to the left and right of b and let d be the coefficient in bucket $[r_1, r_2)$. (See Figure 1.)

Suppose $|c_1 - d| \leq |c_2 - d|$. Then the result of moving b to the right to r_2 is no worse, since, in $\mathbf{H}$, now more of $[r_1, r_2)$ gets the value c_1 and less gets c_2.

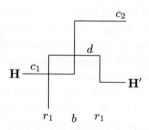

Fig. 1. Illustration of histograms in Lemma 4.

4.3   Correctness of the Output

We now show that if $\mathbf{H}_r$ is a (B, ϵ_r)-robust approximation to $\mathbf{A}$ for suitable B and ϵ_r, then the best B-bucket approximation to $\mathbf{H}_r$ is a nearly optimal representation for $\mathbf{A}$. We will have $\epsilon_r = \epsilon^4/8 < \epsilon/8$. Let $\mathbf{H}^*$ denote the best linear combination of $\mathbf{H}_r$ with $\mathbf{H}_{\mathrm{opt}}$. Since this is a refinement of $\mathbf{H}_r$ by $B - 1$ boundaries,

$$(1 - \epsilon_r) \|\mathbf{A} - \mathbf{H}_r\|^2 \leq \|\mathbf{A} - \mathbf{H}^*\|^2 \leq \|\mathbf{A} - \mathbf{H}_{\mathrm{opt}}\|^2 \tag{1}$$

Lemma 5. *One of the following condition holds:*

Either $\|\mathbf{H} - \mathbf{H}_r\| > \frac{\epsilon}{2} \|\mathbf{A} - \mathbf{H}_{\mathrm{opt}}\| \geq \epsilon_r \|\mathbf{A} - \mathbf{H}_{\mathrm{opt}}\|$

or $\|\mathbf{H} - \mathbf{A}\|^2 \leq (1 + O(\epsilon)) \|\mathbf{H}_{\mathrm{opt}} - \mathbf{A}\|^2$.

Proof. Assume $\|\mathbf{H} - \mathbf{H}_r\| \leq \frac{\epsilon}{2}\|\mathbf{A} - \mathbf{H}_{\mathrm{opt}}\|$. Then, by the triangle inequality,

$$
\begin{aligned}
\|\mathbf{H} - \mathbf{A}\| &\leq \|\mathbf{H} - \mathbf{H}_r\| + \|\mathbf{H}_r - \mathbf{A}\| \\
&\leq \frac{\epsilon}{2}\|\mathbf{A} - \mathbf{H}_{\mathrm{opt}}\| + \|\mathbf{H}_r - \mathbf{A}\| \\
&\leq \frac{\epsilon}{2}\|\mathbf{A} - \mathbf{H}_{\mathrm{opt}}\| + \frac{1}{\sqrt{1-\epsilon_r}}\|\mathbf{A} - \mathbf{H}_{\mathrm{opt}}\| \\
&\leq (1+\epsilon)\|\mathbf{A} - \mathbf{H}_{\mathrm{opt}}\|.
\end{aligned}
$$

Lemma 6. *Fix a signal* $\mathbf{A}$*, and let* $\mathbf{H}_r$ *be a* (B, ϵ_r)*-robust approximation to* $\mathbf{A}$*. Let* $\mathbf{H}$ *be the best B-bucket approximation to* $\mathbf{H}_r$*. Then*

$$
\|\mathbf{A} - \mathbf{H}\|^2 \leq (1 + O(\epsilon))\|\mathbf{A} - \mathbf{H}_{\mathrm{opt}}\|^2,
$$

where $\epsilon_r = \Theta(\epsilon^4)$.

Proof. Assume that the inequality $\frac{\epsilon}{2}\|\mathbf{A} - \mathbf{H}_{\mathrm{opt}}\| < \|\mathbf{H} - \mathbf{H}_r\|$ in Lemma 5 holds. Let $\widehat{\mathbf{H}}$ be the best linear combination of $\mathbf{H}$ and $\mathbf{H}_r$ (see Figure 2). Thus there are right angles at $\mathbf{H}^*$ and at $\widehat{\mathbf{H}}$. In Equations (2) to (5), we show that $\mathbf{H}^*, \widehat{\mathbf{H}}$, and $\mathbf{H}_r$ are all close together, so that, in that sense, the angles $\mathbf{A}$-$\mathbf{H}_r$-$\mathbf{H}$ and $\mathbf{A}$-$\mathbf{H}_r$-$\mathbf{H}$ are close to right angles. We then give the conclusion.

Note that each of $\mathbf{H}^*$ and $\widehat{\mathbf{H}}$ refines $\mathbf{H}_r$ by B buckets, so, it follows that

$$
\begin{aligned}
\left\|\widehat{\mathbf{H}} - \mathbf{H}_r\right\|^2 &= \|\mathbf{A} - \mathbf{H}_r\|^2 - \left\|\mathbf{A} - \widehat{\mathbf{H}}\right\|^2 \\
&\leq \epsilon_r \|\mathbf{A} - \mathbf{H}_r\|^2, \quad \text{by robustness} \tag{2}\\
&\leq \epsilon_r(1 + O(\epsilon_r))\|\mathbf{A} - \mathbf{H}_{\mathrm{opt}}\|^2, \quad \text{by (1)} \\
&\leq \frac{4\epsilon_r(1 + O(\epsilon_r))}{\epsilon^2}\|\mathbf{H} - \mathbf{H}_r\|^2 \\
&\leq \epsilon^2 \|\mathbf{H} - \mathbf{H}_r\|^2, \quad \text{by Lemma 5.} \tag{3}
\end{aligned}
$$

Similarly,

$$
\begin{aligned}
\|\mathbf{H}^* - \mathbf{H}_r\|^2 &= \|\mathbf{A} - \mathbf{H}_r\|^2 - \|\mathbf{A} - \mathbf{H}^*\|^2 \\
&\leq \epsilon^2 \|\mathbf{H} - \mathbf{H}_r\|^2 \\
&\leq \epsilon^2 \|\mathbf{H}_{\mathrm{opt}} - \mathbf{H}_r\|^2, \quad \text{by optimality of } \mathbf{H} \text{ for } \mathbf{H}_r \tag{4}
\end{aligned}
$$

Furthermore, it also follows that

$$
\begin{aligned}
\left\|\widehat{\mathbf{H}} - \mathbf{H}^*\right\| &\leq \left\|\widehat{\mathbf{H}} - \mathbf{H}_r\right\| + \|\mathbf{H}_r - \mathbf{H}^*\| \\
&\leq 2\sqrt{\epsilon_r}\|\mathbf{A} - \mathbf{H}_r\|, \quad \text{by (2)} \\
&\leq \frac{2\sqrt{\epsilon_r}}{\sqrt{1-\epsilon_r}}\|\mathbf{A} - \mathbf{H}^*\| \quad \text{by (1)} \\
&\leq \epsilon \|\mathbf{A} - \mathbf{H}^*\|. \tag{5}
\end{aligned}
$$

See Figure 2. Finally, we have:

$$\|\mathbf{H} - \mathbf{A}\|^2 = \left\|\mathbf{H} - \widehat{\mathbf{H}}\right\|^2 + \left\|\widehat{\mathbf{H}} - \mathbf{A}\right\|^2$$

$$\leq \left(\|\mathbf{H} - \mathbf{H}_r\| + \left\|\mathbf{H}_r - \widehat{\mathbf{H}}\right\|\right)^2 + \left\|\widehat{\mathbf{H}} - \mathbf{A}\right\|^2$$

$$\leq (1+\epsilon)^2 \|\mathbf{H} - \mathbf{H}_r\|^2 + \left\|\widehat{\mathbf{H}} - \mathbf{A}\right\|^2, \quad \text{by (3)}$$

$$\leq (1+\epsilon)^2 \|\mathbf{H}_{\text{opt}} - \mathbf{H}_r\|^2 + \left\|\widehat{\mathbf{H}} - \mathbf{A}\right\|^2, \quad \text{since } \mathbf{H} \text{ is optimal for } \mathbf{H}_r$$

$$\leq (1+\epsilon)^2 \left(\|\mathbf{H}_{\text{opt}} - \mathbf{H}^*\| + \|\mathbf{H}^* - \mathbf{H}_r\|\right)^2 + \left\|\widehat{\mathbf{H}} - \mathbf{A}\right\|^2$$

$$\leq (1+\epsilon)^4 \|\mathbf{H}_{\text{opt}} - \mathbf{H}^*\|^2 + \left\|\widehat{\mathbf{H}} - \mathbf{A}\right\|^2, \quad \text{by (4)}$$

$$\leq (1+O(\epsilon)) \|\mathbf{H}_{\text{opt}} - \mathbf{H}^*\|^2 + \left(\left\|\widehat{\mathbf{H}} - \mathbf{H}^*\right\| + \|\mathbf{H}^* - \mathbf{A}\|\right)^2$$

$$\leq (1+\epsilon)^4 \|\mathbf{H}_{\text{opt}} - \mathbf{H}^*\|^2 + (1+\epsilon)^2 \|\mathbf{H}^* - \mathbf{A}\|^2, \quad \text{by (5)}$$

$$\leq (1+\epsilon)^4 \|\mathbf{H}_{\text{opt}} - \mathbf{A}\|^2, \quad \text{by the Pythagorean theorem}$$

$$\leq (1+O(\epsilon)) \|\mathbf{H}_{\text{opt}} - \mathbf{A}\|^2.$$

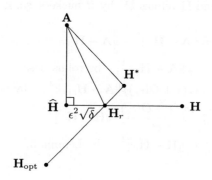

Fig. 2. Illustration of histograms, which are not necessarily coplanar. The histograms indicated as colinear are, in fact colinear; the order of three histograms in a line is not necessarily the order indicated. The points $\widehat{\mathbf{H}}, \mathbf{H}_r$, and $\mathbf{H}^*$ are all close can can be roughly collapsed. Then, since $\|\mathbf{H} - \mathbf{H}_r\| \leq \|\mathbf{H}_{\text{opt}} - \mathbf{H}_r\|$, it follows that $\|\mathbf{H} - \mathbf{A}\|^2 \leq (1+\epsilon) \|\mathbf{H}_{\text{opt}} - \mathbf{A}\|^2$.

4.4   Main Theorem

Combining the above results, we have

Theorem 1. *There exists an algorithm that, given B, N and ϵ, on input the N values of an integer-valued signal $\mathbf{A}$ with $\|\mathbf{A}\| \leq M$, outputs a B-bucket histogram $\mathbf{H}$ with*

$$\|\mathbf{A} - \mathbf{H}\|_2^2 \le (1 + \epsilon) \|\mathbf{A} - \mathbf{H}_{\mathrm{opt}}\|_2^2,$$

where $\mathbf{H}_{\mathrm{opt}}$ is the best possible B-bucket histogram representation to $\mathbf{A}$. The algorithm uses space $B(\log(N)\log(M)/\epsilon)^{O(1)}$. The algorithm requires two modules. The first (sketching) uses time $O(N)$ and the second (reconstruction) uses time in $(B\log(M)\log(N)/\epsilon)^{O(1)}$.

Proof. Use the classic wavelet algorithm to produce all coefficients, reading in a stream and spitting out a stream. Use the median algorithm to select the B' coefficients with largest square. Use the greedy algorithm to produce a $(O(B\log(N)), \Theta(\epsilon^4))$-robust wavelet approximation $\mathbf{H}_r$, which is a $(B, \Theta(\epsilon^4))$-robust histogram representation. Use dynamic programming to find the best B-bucket histogram $\mathbf{H}$ to $\mathbf{H}_r$, by observing that the boundaries of $\mathbf{H}$ are among the boundaries of $\mathbf{H}_r$. By Lemma 6, this output is correct.

5   Concluding Remarks

We have presented a simple algorithm that takes $O(1)$ time to process each item of a data stream and using $B(\log(N)\log(M)/\epsilon)^{O(1)}$ space and polylogarithmic reconstruction time, obtains an $1 + \epsilon$ approximation to the optimal histogram.

A more general streaming model is one in which we are allowed to observe the signal only through updates (not necessarily left-to-right pass). Under this model, the best known algorithm for histogram constructing takes $(B\log(N)\log(M)/\epsilon)^{O(1)}$ resources for all three parameters [13]. Some of the exponents involved are rather large, *e.g.*, at least 6. It is an open problem to see if per-item processing time can be reduced further in this model.

Our approach of first obtaining large wavelet coefficients, then obtaining a robust histogram from which the final histogram is culled, is of interest. For example, in practice, one may experimentally compare each of these intermediate and final representations in terms of their overall accuracy in capturing signal trends. Our algorithm is simple and implementable, so this is a feasible study for the future.

References

1. V. Poosala. Histogram techniques for databases. Ph. D Thesis, Univ. Wisconsin, Madison, 1996.
2. Anja Feldmann, Albert G. Greenberg, Carsten Lund, Nick Reingold, Jennifer Rexford, Fred True. Deriving traffic demands for operational IP networks: methodology and experience. SIGCOMM 2000: 257-270
3. http://www-db.stanford.edu/stream/
4. http://www.cs.cornell.edu/database/cougar/index.htm.
5. Fjording the Stream: An Architecture for Queries over Streaming Sensor Data. Sam Madden and Michael J. Franklin, ICDE Conference, February, 2002, San Jose.
6. Selection and sorting with limited storage. J. I. Munro and M. S. Paterson. Theoretical Computer Science, pages 315-323, 1980.

7. Stable Distributions, Pseudorandom Generators, Embeddings and Data Stream Computation. Piotr Indyk: FOCS 2000: 189-197
8. The Space Complexity of Approximating the Frequency Moments. Noga Alon, Yossi Matias, Mario Szegedy. STOC 1996: 20-29
9. An Approximate L1-Difference Algorithm for Massive Data Streams. Joan Feigenbaum, Sampath Kannan, Martin Strauss, Mahesh Viswanathan. FOCS 1999: 501-511
10. Clustering Data Streams. Sudipto Guha, Nina Mishra, Rajeev Motwani, Liadan O'Callaghan. FOCS 2000: 359-366
11. Data-streams and histograms. Sudipto Guha, Nick Koudas, Kyuseok Shim. STOC 2001: 471-475
12. Approximating a Data Stream for Querying and Estimation: Algorithms and Performance Evaluation. Sudipto Guha, Nick Koudas. ICDE 2002
13. Dynamic maintenance of histograms. A. Gilbert, S. Guha, P. Indyk, Y. Kotidis, S. Muthukrishnan and M. Strauss. To appear in STOC 2002.
14. Surfing Wavelets on Streams: One-Pass Summaries for Approximate Aggregate Queries. Anna C. Gilbert, Yannis Kotidis, S. Muthukrishnan, Martin Strauss. VLDB 2001: 79-88
15. H. V. Jagadish, Nick Koudas, S. Muthukrishnan, Viswanath Poosala, Kenneth C. Sevcik, Torsten Suel. Optimal Histograms with Quality Guarantees. VLDB 1998: 275-286.

Finding Frequent Items in Data Streams

Moses Charikar[⋆1], Kevin Chen[⋆⋆2], and Martin Farach-Colton[3]

1 Princeton University
moses@cs.princeton.edu
2 UC Berkeley
kevinc@cs.berkeley.edu
3 Rutgers University and Google Inc.
martin@google.com

Abstract. We present a 1-pass algorithm for estimating the most frequent items in a data stream using very limited storage space. Our method relies on a novel data structure called a COUNT SKETCH, which allows us to estimate the frequencies of all the items in the stream. Our algorithm achieves better space bounds than the previous best known algorithms for this problem for many natural distributions on the item frequencies. In addition, our algorithm leads directly to a 2-pass algorithm for the problem of estimating the items with the largest (absolute) change in frequency between two data streams. To our knowledge, this problem has not been previously studied in the literature.

1 Introduction

One of the most basic problems on a data stream [HRR98,AMS99] is that of finding the most frequently occurring items in the stream. We shall assume here that the stream is large enough that memory-intensive solutions such as sorting the stream or keeping a counter for each distinct element are infeasible, and that we can afford to make only one pass over the data. This problem comes up in the context of search engines, where the streams in question are streams of queries sent to the search engine and we are interested in finding the most frequent queries handled in some period of time.

A wide variety of heuristics for this problem have been proposed, all involving some combination of sampling, hashing, and counting (see [GM99] and Section 2 for a survey). However, none of these solutions have clean bounds on the amount of space necessary to produce good approximate lists of the most frequent items. In fact, the only algorithm for which theoretical guarantees are available is the straightforward SAMPLING algorithm, in which a uniform random sample of the data is kept. For this algorithm, the space bound depends on the distribution of the frequency of the items in the data stream. Our main contribution is a simple algorithm with good theoretical bounds on its space requirements that also beats the naive sampling approach for a wide class of common distributions.

⋆ This work was done while the author was at Google Inc.
⋆⋆ This work was done while the author was at Google Inc.

P. Widmayer et al. (Eds.): ICALP 2002, LNCS 2380, pp. 693–703, 2002.

Before we present the details of our result, however, we need to introduce some definitions. Let $S = q_1, q_2, \ldots, q_n$ be a data stream, where each $q_i \in \mathcal{O} = \{o_1, \ldots, o_m\}$. Let object o_i occur n_i times in S, and order the o_i so that $n_1 \geq n_2 \geq \cdots \geq n_m$. Finally, let $f_i = n_i/n$.

We consider two notions of approximating the frequent-elements problem:

FINDCANDIDATETOP(S, k, l)

- Given: An input stream S, and integers k and l.
- Output: A list of l elements from S such that the k most frequent elements occur in the list.

Note that for a general input distribution, FINDCANDIDATETOP(S, k, l) may be very hard to solve. Suppose, for example, that $n_k = n_{l+1} + 1$, that is, the kth most frequent element has almost the same frequency as the $l + 1$st most frequent element. Then it would be almost impossible to find only l elements that are likely to have the top k elements. We therefore define the following variant:

FINDAPPROXTOP(S, k, ϵ)

Given: An input stream S, integer k, and real ϵ.
Output: A list of k elements from S such that every element i in the list has $n_i > (1 - \epsilon)n_k$.

A somewhat stronger guarantee on the output is that every item o_i with $n_i > (1 + \epsilon)n_k$ will be in the output list, w.h.p.. Our algorithm will, in fact, achieve this stronger guarantee. Thus, it will only err on the boundary cases.

A summary of our final results are as follows: We introduce a simple data structure called a COUNT SKETCH, and give a 1-pass algorithm for computing the count sketch of a stream. We show that using a count sketch, we reliably estimate the frequencies of the most common items, which directly yields a 1-pass algorithm for solving FINDAPPROXTOP(S, k, ϵ). The Sampling algorithm does not give any bounds for this version of the problem. For the special case of Zipfian distributions, we also give bounds on using our algorithm to solve FINDCANDIDATETOP(S, k, ck) for some constant c, which beat the bounds given by the Sampling algorithm for reasonable values of n, m and k.

In addition, our count sketch data structure is additive, i.e. the sketches for two streams can be directly added or subtracted. Thus, given two streams, we can compute the difference of their sketches, which leads directly to a 2-pass algorithm for computing the items whose frequency changes the most between the streams. None of the previous algorithms can be adapted to find max-change items. This problem also has a practical motivation in the context of search engine query streams, since the queries whose frequency changes most between two consecutive time periods can indicate which topics people are currently most interested in [Goo].

We defer the actual space bounds of our algorithms to the Section 4. In Section 2, we survey previous approaches to our problem. We present our algorithm for constructing count sketches in Section 3, and in Section 4, we analyze the space requirements of the algorithm. In Section 4.2, we show how the algorithm can be adapted to find elements with the largest change in frequency. We conclude in Section 5 with a short discussion.

2   Background

The most straightforward solution to the FINDCANDIDATETOP$(\mathcal{S}, k, l)$ problem is to keep a uniform random sample of the elements stored as a list of items plus a counter for each item. If the same object is added more than once, we simply increment its counter, rather than adding a new object to the list. We refer to this algorithm as the SAMPLING algorithm. If x is the size of the sample (counting repetitions), to ensure that an element with frequency f_k appears in the sample, we need to set x/n, the probability of being included in the sample, to be $x/n > O(\log n/n_k)$, thus $x > O(\log n/f_k)$. This guarantees that all top k elements will be in the sample, and thus gives a solution to FINDCANDIDATETOP$(\mathcal{S}, k, O(\log n/f_k))$.

Two variants of the basic sampling algorithm were given by Gibbons and Matias [GM98]. The concise samples algorithm keeps a uniformly random sample of the data, but does not assume that we know the length of the data stream beforehand. Instead, it begins optimistically assuming that we can include elements in the sample with probability $\tau = 1$. As it runs out of space, it lowers τ until some element is evicted from the sample, and continues the process with this new, lower τ'. The invariant of the algorithm is that, at any point, each item is in the sample with the current threshold probability. The sequence can be chosen arbitrarily to adapt to the input stream as it is processed. At the end of the algorithm, there is some final threshold τ_f, and the algorithm gives the same output as the Sampling algorithm with this inclusion probability. However, the value of τ_f depends on the input stream in some complicated way, and no clean theoretical bound for this algorithm is available.

The counting samples algorithm adds one more optimization based on the observation that so long as we are setting aside space for a count of an item in the sample anyway, we may as well keep an exact count for the occurrences of the item after it has been added to the sample. This change improves the accuracy of the counts of items, but does not change who will actually get included in the sample.

Fang et al. [FSGM+96] consider the related problem of finding all items in a data stream which occur with frequency above some fixed threshold, which they call *iceberg queries*. They propose a number of different heuristics, most of which involve multiple passes over the data set. They also propose a heuristic 1-pass multiple-hash scheme which has a similar flavor to our algorithm.

Though not directly connected, our algorithm also draws on a quite substantial body of work in data stream algorithms   [FKSV99,FKSV00,GG+02]

[GMMO00,HRR98,Ind00]. In particular, Alon, Matias and Szegedy [AMS99] give an $\Omega(n)$ lower bound on the space complexity of any algorithm for estimating the frequency of the largest item given an arbitrary data stream. However, their lower bound is brittle in that it only applies to the FINDCANDIDATETOP$(\mathcal{S}, 1, 1)$ problem and not to the relaxed versions of the problem we consider, for which we achieve huge space reduction. In addition, they give an algorithm for estimating the second frequency moment, $F_2 = \Sigma_{i=1}^m n_i^2$, in which they use the idea of random ± 1 hash functions that we use in our algorithm (see also [Ach01]).

3   The COUNT SKETCH Algorithm

Before we give the algorithm itself, we begin with a brief discussion of the intuition behind it.

3.1   Intuition

Recall that we would like a data structure that maintains the approximate counts of the high frequency elements in a stream and is compact.

First, consider the following simple algorithm for finding estimates of all n_i. Let s be a hash function from objects to $\{+1, -1\}$ and let c be a counter. While processing the stream, each time we encounter an item q_i, update the counter $c \mathrel{+}= s[q_i]$. The counter then allows us to estimate the counts of all the items since $\mathbf{E}[c \cdot s[q_i]] = n_i$. However, it is obvious that there are a couple of problems with the scheme, namely that, the variance of every estimate is very large, and $O(m)$ elements have estimates that are wrong by more than the variance.

The natural first attempt to fix the algorithm is to select t hash functions $s_1, \ldots, s_t$ and keep t counters, $c_1, \ldots, c_t$. Then to process item q_i we need to set $c_j \mathrel{+}= s_j[q_i]$, for each j. Note that we still have that each $\mathbf{E}[c_i \cdot s_i[q_i]] = n_i$. We can then take the mean or median of these estimates to achieve an estimate with lower variance.

However, collisions with high frequency items, like o_1, can spoil most estimates of lower frequency elements, even important elements like o_k. Therefore rather than having each element update every counter, we replace each counter with a hash table of b counters and have the items update different subsets of counters, one per hash table. In this way, we will arrange matters so that every element will get enough high-confidence estimates – those untainted by collisions with high-frequency elements – to estimate its frequency with sufficient precision.

As before, $\mathbf{E}[h_i[q] \cdot s[q]] = n_q$. We will show that by making b large enough, we will decrease the variance to a tolerable level, and that by making t large enough – approximately logarithmic in n – we will make sure that each of the m estimates has the desired variance.

3.2   Our Algorithm

Let t and b be parameters with values to be determined later. Let $h_1, \ldots, h_t$ be hash functions from objects to $\{1, \ldots, b\}$ and $s_1, \ldots, s_t$ be hash functions from objects to $\{+1, -1\}$. The CountSketch data structure consists of these hash functions along with a $t \times b$ array of counters, which should be interpreted as an array of t hash tables, each containing b buckets.

The data structure supports two operations:

ADD(C, q): For $i \in [1, t], h_i[q] += s_i[q]$.
ESTIMATE(C, q): return median$_i\{h_i[q] \cdot s_i[q]\}$.

Why do we take the median instead of the mean? The answer is that even in the final scheme, we have not eliminated the problem of collisions with high-frequency elements, and these will still spoil some subset of the estimates. The mean is very sensitive to outliers, while the median is sufficiently robust, as we will show in the next section.

Once we have this data structure, our algorithm is straightforward and simple to implement. For each element, we use the CountSketch data structure to estimate its count, and keep a heap of the top k elements seen so far. More formally:
Given a data stream $q_1, \ldots, q_n$, for each $j = 1, \ldots, n$:

1. ADD(C, q_j)
2. If q_j is in the heap, increment its count. Else, add q_j to the heap if ESTIMATE(C, q_j) is greater than the smallest estimated count in the heap. In this case, the smallest estimated count should be evicted from the heap.

This algorithm solves FINDAPPROXTOP$(\mathcal{S}, k, \epsilon)$, where our choice of b will depend on ϵ. Also, notice that if two sketches share the same hash functions – and therefore the same b and t – that we can add and subtract them. The algorithm takes space $O(tb + k)$. In the next section we will bound t and b.

4   Analysis

To make the notation easier to read, we will sometimes drop the subscript of q_i and simply write q, when there is no ambiguity. We will further abuse the notation by conflating q with its index i.

We will assume that each hash function h_i and s_i is pairwise independent. Further, all functions h_i and s_i are independent of each other. Note that the amount of randomness needed to implement these hash functions is $O(t \log m)$. We will use $t = O\left(\log \frac{n}{\delta}\right)$, where the algorithm fails with probability at most δ. Hence the total randomness needed is $O\left(\log m \log \frac{n}{\delta}\right)$.

Consider the estimation of the frequency of an element at position ℓ in the input. Let $n_q(\ell)$ be the number of occurrences of element q up to position ℓ. Let $A_i[q]$ be the set of elements that hash onto the same bucket in the ith row as q does, i.e. $A_i[q] = \{q' : q' \neq q, h_i[q'] = h_i[q]\}$. Let $A_i^{>k}[q]$ be the elements of

$A_i[q]$ other than the k most frequent elements, i.e. $A_i^{>k}[q] = \{q' : q' \neq q, q' > k, h_i[q'] = h_i[q]\}$. Let $v_i[q] = \sum_{q' \in A_i[q]} n_{q'}^2$. We define $v_i^{>k}[q]$ analogously for $A_i^{>k}[q]$.

Lemma 1. *The variance of $h_i[q]s_i[q]$ is bounded by $v_i[q]$.*

Lemma 2. $E[v_i^{>k}[q]] = \dfrac{\sum_{q'=k+1}^{m} n_{q'}^2}{b}.$

Let SMALL-VARIANCE$_i[q]$ be the event that $v_i^{>k}[q] \leq \dfrac{8\sum_{q'=k+1}^{m} n_{q'}^2}{b}$. By the Markov inequality,

$$\mathbf{Pr}[\text{SMALL-VARIANCE}_i[q]] \geq 1 - \frac{1}{8} \tag{1}$$

Let NO-COLLISIONS$_i[q]$ be the event that $A_i[q]$ does not contain any of the top k elements.
If $b \geq 8k$,

$$\mathbf{Pr}[\text{NO-COLLISIONS}_i[q]] \geq 1 - \frac{1}{8} \tag{2}$$

Let SMALL-DEVIATION$_i[q](\ell)$ be the event that

$$|h_i[q]s_i[q] - n_q(\ell)|^2 \leq 8\,\mathbf{Var}[h_i[q]s_i[q]].$$

Then,

$$\mathbf{Pr}[\text{SMALL-DEVIATION}_i[q](\ell)] \geq 1 - \frac{1}{8}. \tag{3}$$

By the union bound,

$$\mathbf{Pr}[\text{NO-COLLISIONS}_i[q] \text{ and } \text{SMALL-VARIANCE}_i[q] \tag{4}$$
$$\text{and } \text{SMALL-DEVIATION}_i[q]] \geq \frac{5}{8}$$

We will express the error in our estimates in terms of a parameter γ, defined as follows:

$$\gamma = \sqrt{\frac{\sum_{q'=k+1}^{m} n_{q'}^2}{b}}. \tag{5}$$

Lemma 3. *With probability $(1 - \frac{\delta}{n})$,*

$$|median\{h_i[q]s_i[q]\} - n_q(\ell)| \leq 8\gamma \tag{6}$$

Proof. We will prove that, with high probability, for more than $\frac{t}{2}$ indices $i \in [1, t]$,

$$|h_i[q]s_i[q] - n_q(\ell)| \leq 8\gamma$$

This will imply that the median of $h_i[q]s_i[q]$ is within the error bound claimed by the lemma. First observe that for an index i, if all three events NO-COLLISIONS$_i[q]$, SMALL-VARIANCE$_i[q]$, and SMALL-DEVIATION$_i[q]]$ occur, then $|h_i[q]s_i[q] - n_q(\ell)| \leq 8\gamma$. Hence, for a fixed i,

$$\mathbf{Pr}[|h_i[q]s_i[q] - n_q(\ell)| \leq 8\gamma] \geq \frac{5}{8}.$$

The expected number of such indices i is at least $5t/8$. By Chernoff bounds, the number of such indices i is more than $t/2$ with probability at least $1 - e^{O(t)}$. Setting $t = \Omega(\log(n) + \log(\frac{1}{\delta}))$, the lemma follows.

Lemma 4. *With probability $1 - \delta$, for all $\ell \in [1, n]$,*

$$|median\{h_i[q]s_i[q]\} - n_q(\ell)| \leq 8\gamma \tag{7}$$

where q is the element that occurs in position ℓ.

Lemma 5. *If $b \geq 8 \max\left(k, \dfrac{32\sum_{q'=k+1}^m n_{q'}^2}{(\epsilon n_k)^2}\right)$, then the estimated top k elements occur at least $(1 - \epsilon)n_k$ times in the sequence; further all elements with frequencies at least $(1 + \epsilon)n_k$ occur amongst the estimated top k elements.*

Proof. By Lemma 4, the estimates for number of occurrences of all elements are within an additive factor of 8γ of the true number of occurrences. Thus for two elements whose true number of occurrences differ by more than 16γ, the estimates correctly identify the more frequent element. By setting $16\gamma \leq \epsilon n_k$, we ensure that the only elements that can replace the true most frequent elements in the estimated top k list are elements with true number of occurrences at least $(1 - \epsilon)n_k$.

$$16\gamma \leq \epsilon n_k$$

$$\Leftrightarrow 16\sqrt{\frac{\sum_{q'=k+1}^m n_{q'}^2}{b}} \leq \epsilon n_k$$

$$\Leftrightarrow b \geq \frac{256\sum_{q'=k+1}^m n_{q'}^2}{(\epsilon n_k)^2}$$

This combined with the condition $b \geq 8k$ used to prove (2), proves the lemma.

We conclude with the following summarization:

Theorem 1. *The* COUNT SKETCH *algorithm solves* FINDAPPROXTOP$(\mathcal{S}, k, \epsilon)$ *in space*

$$O\left(k \log\frac{n}{\delta} + \frac{\sum_{q'=k+1}^m n_{q'}^2}{(\epsilon n_k)^2} \log\frac{n}{\delta}\right)$$

4.1   Analysis for Zipfian Distributions

Note that in the algorithm's (ordered) list of estimated most frequent elements, the k most frequent elements can only by preceded by elements with number of occurrences at least $(1 - \epsilon)n_k$. Hence, by keeping track of $l \geq k$ estimated most frequent elements, the algorithm can ensure that the most frequent k elements are in the list. For this to happen l must be chosen so that $n_{l+1} < (1 - \epsilon)n_k$. When the distribution is Zipfian with parameter z, $l = O(k)$ (in fact $l = k/(1 - \epsilon)^{1/z}$). If the algorithm is allowed one more pass, the true frequencies of all the l elements in the algorithms list can be determined allowing the selection of the most frequent k elements.

In this section, we analyze the space complexity of our algorithm for Zipfian distributions. For a Zipfian distribution with parameter z, $n_q = \frac{c}{q^z}$ for some scaling factor c. (We omit c from the calculations)[1]. We will compare the space requirements of our algorithm with that of the sampling based algorithm for the problem $\text{FINDCANDIDATETOP}(\mathcal{S}, k, l)$. We will use the bound on b from Lemma 5, setting ϵ to be a constant so that, with high probability, our algorithms' list of $l = O(k)$ elements is guaranteed to contain the most frequent k elements. First note that

$$\sum_{q'=k+1}^{m} n_{q'}^2 = \sum_{q'=k+1}^{m} \frac{1}{(q')^{2z}} = \begin{cases} O(m^{1-2z}), & z < \frac{1}{2} \\ O(\log m), & z = \frac{1}{2} \\ O(k^{1-2z}), & z > \frac{1}{2} \end{cases}$$

Substituting this into the bound in Lemma 5 (and setting ϵ to be a constant), we get the following bounds on b (correct up to constant factors). The total space requirements are obtained by multiplying this by $O(\log \frac{n}{\delta})$.

Case 1: $z < \frac{1}{2}$.

$$b = m^{1-2z}k^{2z}$$

Case 2: $z = \frac{1}{2}$.

$$b = k \log m$$

Case 3: $z > \frac{1}{2}$.

$$b = k$$

We compare these bounds with the space requirements for the random sampling algorithm. The size of the random sample required to ensure that the k most frequent elements occur in the random sample with probability $1 - \delta$ is

$$\frac{n}{n_k} \log(k/\delta).$$

We measure the space requirement of the random sampling algorithm by the expected number of distinct elements in the random sample. (Note that the size

---

[1] While c need not be a constant, it turns out that all occurrences of c cancel in our calculations, and so, for ease of presentation, we omit them from the beginning.

of the random sample could be much larger than the number of distinct elements due to multiple copies of elements).

Furthermore, the sampling algorithm as stated, solves the FINDCANDIDATETOP($\mathcal{S}, k, x$), where x is the number of distinct elements in the sample. This does not constitute a solution of FINDCANDIDATETOP($\mathcal{S}, k, O(k)$), as does our algorithm. We will be reporting our bounds for the latter and the sampling bounds for the former. However, this only gives the sampling algorithm an advantage over ours.

We now analyze the space usage of the sampling algorithm for Zipfians. It turns out that for Zipf parameter $z \leq 1$, the expected number of distinct elements is within a constant factor of the sample size. We analyze the number of distinct elements for Zipf parameter $z > 1$.

Items are placed in the random sample S with probability $\frac{\log(k/\delta)}{n_k}$.

$$\mathbf{Pr}[q \in S] = 1 - \left(1 - \frac{\log(k/\delta)}{n_k}\right)^{n_q}$$

$$
\begin{aligned}
\mathbf{E}[\text{no. of distinct elements in } S] &= \sum_{q=1}^{m} \mathbf{Pr}[q \in S] = \sum_{q=1}^{m} 1 - \left(1 - \frac{\log(k/\delta)}{n_k}\right)^{n_q} \\
&= \sum_{q=1}^{m} O\left(\min\left(1, \frac{n_q \log(k/\delta)}{n_k}\right)\right) \\
&= \sum_{q=1}^{m} O\left(\min\left(1, \frac{k^z \log(k/\delta)}{q^z}\right)\right) \\
&= O\left(k\left(\log\frac{k}{\delta}\right)^{1/z}\right)
\end{aligned}
$$

The bounds for the two algorithms are compared in Table 1.

4.2   Finding Items with Largest Frequency Change

For object q and sequence S, let n_q^S be the number of occurrences of q in S. Given two streams S_1, S_2, we would like to find the items q such that the values of $|n_q^{S_2} - n_q^{S_1}|$ are the largest amongst all items q. We can adapt our algorithm for finding most frequent elements to this problem of finding elements whose frequencies change the most.

We make two passes over the data. In the first pass, we only update counters. In the second pass, we actually identify elements with the largest changes in number of occurrences.

We first make a pass over S_1, where we perform the following step:

For each q, for $i \in [1, t], h_i[q] \mathbin{-}= s_i[q]$.

Next, we make a pass over S_2, doing the following:

Table 1. Comparison of space requirements for random sampling vs. our algorithm

Zipf parameter	random sampling	Count Sketch Algorithm
$z < \dfrac{1}{2}$	$m\left(\dfrac{k}{m}\right)^z \log \dfrac{k}{\delta}$	$m^{1-2z}k^{2z} \log \dfrac{n}{\delta}$
$z = \dfrac{1}{2}$	$\sqrt{km} \log \dfrac{k}{\delta}$	$k \log m \log \dfrac{n}{\delta}$
$\dfrac{1}{2} < z < 1$	$m\left(\dfrac{k}{m}\right)^z \log \dfrac{k}{\delta}$	$k \log \dfrac{n}{\delta}$
$z = 1$	$k \log m \log \dfrac{k}{\delta}$	$k \log \dfrac{n}{\delta}$
$z > 1$	$k\left(\log \dfrac{k}{\delta}\right)^{\frac{1}{z}}$	$k \log \dfrac{n}{\delta}$

For each q, for $i \in [1, t], h_i[q] += s_i[q]$.

We make a second pass over S_1 and S_2:
For each q,

1. $\hat{n}_q = \text{median}\{h_i[q]s_i[q]\}$.
2. Maintain set A of the l objects encountered with the largest values of $|\hat{n}_q|$.
3. For every item $q \in A$ maintain an exact count of the number of occurrences in S_1 and S_2.

(Note that though A can change, items once removed are never added back. Thus accurate exact counts can be maintained for all q currently in A).

Finally, we report the k items with the largest values of $|n_q^{S_2} - n_q^{S_1}|$ amongst the items in A.

We can give a guarantee similar to Lemma 5 with n_q replaced by $\Delta_q = |n_q^{S_1} - n_q^{S_2}|$.

5   Conclusions

We make a final note comparing the Count Sketch algorithm with the Sampling algorithm. So far, we have neglected the space cost of actually storing the elements from the stream. This is because different encodings can yield very different space use. Both algorithms need counters that require $O(\log n)$ bits, however, we only keep k objects from the stream, while the Sampling algorithm keeps a potentially much larger set of items from the stream. For example, if the space used by an object is Ψ, and we have a Zipfian with $z = 1$, then the sampling algorithm uses $O(k \log m \log \frac{k}{\delta} \Psi)$ space while the Count Sketch algorithm uses $O(k \log \frac{n}{\delta} + k\Psi)$ space. If $\Psi >> \log n$, as it will often be in practice, this give the Count Sketch algorithm a large advantage over the Sampling algorithm.

As for the max-change problem, we note that there is still an open problem of finding the elements with the max-percent change, or other objective functions that somehow balance absolute and relative changes.

References

[Ach01]   Dimitris Achlioptas. Database-friendly random projections. In *Proc. 20th ACM Symposium on Principles of Database Systems*, pages 274–281, 2001.

[AMS99]   Noga Alon, Yossi Matias, and Mario Szegedy. The space complexity of approximating the frequency moments. *Journal of Computer and System Sciences*, 58(1):137–147, 1999.

[FKSV99]   Joan Feigenbaum, Sampath Kannan, Martin Strauss, and Mahesh Viswanathan. An approximate l_1-difference algotihm for massive data streams. In *Proc. 40th IEEE Symposium on Foundations of Computer Science*, pages 501–511, 1999.

[FKSV00]   Joan Feigenbaum, Sampath Kannan, Martin Strauss, and Mahesh Viswanathan. Testing and spot-checking of data streams. In *Proc. 11th ACM-SIAM Symposium on Discrete Algorithms*, pages 165–174, 2000.

[FSGM+96] Min Fang, Narayanan Shivakumar, Hector Garcia-Molina, Rajeev Motwani, and Jeffrey Ullman. Computing iceberg queries efficiently. In *Proc. 22nd International Conference on Very Large Data Bases*, pages 307–317, 1996.

[GG+02]   Anna Gilbert, Sudipto Guha, Piotr Indyk, Yannis Kotidis, S. Muthukrishnan, and Martin Strauss. Fast, small-space algorithms for approximate histogram maintenance. In *to appear in Proc. 34th ACM Symposium on Theory of Computing*, 2002.

[GM98]   Phillip Gibbons and Yossi Matias. New sampling-based summary statistics for improving approximate query answers. In *Proc. ACM SIGMOD International Conference on Management of Data*, pages 331–342, 1998.

[GM99]   Phillip Gibbons and Yossi Matias. Synopsis data structures for massive data sets. In *Proc. 10th Annual ACM-SIAM Symposium on Discrete Algorithms*, pages 909–910, 1999.

[GMMO00]   Sudipto Guha, Nina Mishra, Rajeev Motwani, and Liadan O'Callaghan. Clustering data streams. In *Proc. 41st IEEE Symposium on Foundations of Computer Science*, pages 359–366, 2000.

[Goo]   Google. Google zeitgeist - search patterns, trends, and surprises according to google. http://www.google.com/press/zeitgeist.html.

[HRR98]   Monika Henzinger, Prabhakar Raghavan, and Sridhar Rajagopalan. Computing on data streams. Technical Report SRC TR 1998-011, DEC, 1998.

[Ind00]   Piotr Indyk. Stable distributions, pseudorandom generators, embeddings and data stream computation. In *Proc. 41st IEEE Symposium on Foundations of Computer Science*, pages 148–155, 2000.

Symbolic Strategy Synthesis for Games on Pushdown Graphs

Thierry Cachat

Lehrstuhl für Informatik VII, RWTH, D-52056 Aachen
Fax: (49) 241-80-22215, `cachat@informatik.rwth-aachen.de`

Abstract. We consider infinite two-player games on pushdown graphs, the reachability game where the first player must reach a given set of vertices to win, and the Büchi game where he must reach this set infinitely often. We provide an automata theoretic approach to compute uniformly the winning region of a player and corresponding winning strategies, if the goal set is regular. Two kinds of strategies are computed: positional ones which however require linear execution time in each step, and strategies with pushdown memory where a step can be executed in constant time.

1 Introduction

Games are an important model of reactive computation and a versatile tool for the analysis of logics like the μ-calculus [5,6]. In recent years, games over infinite graphs have attracted attention as a framework for the verification and synthesis of infinite-state systems [8]. In the present paper we consider pushdown graphs (transition graphs of pushdown automata). It was shown in [12] that in two-player parity games played on pushdown graphs, winning strategies can be realized also by pushdown automata. The drawback of these results [8,12] are a dependency of the analysis on a given initial game position, and a lack of algorithmic description of the (computation of) winning strategies. Such an algorithmic (or "symbolic") solution must transform the finite presentation of the pushdown game into a finite description of the winning regions of the two players as well as of their strategies.

In this paper we develop such an algorithmic approach, also leading to uniform complexity bounds. The nodes of the game graph, *i.e.*, the game positions, are *unbounded* finite objects: stack contents of a given pushdown automaton. Our "symbolic approach" uses finite automata as defining devices for sets of game positions and for the description of strategies. This lifts the results of [1] and [2] from CTL and LTL model checking over pushdown systems to the level of program synthesis.

In this paper we only consider reachability games and Büchi games (where a winning play should reach a given regular set of nodes, respectively should pass infinitely often through this set). This restriction is justified by two aspects: First, reachability and Büchi games are the typical cases in the analysis of safety

P. Widmayer et al. (Eds.): ICALP 2002, LNCS 2380, pp. 704–715, 2002.
© Springer-Verlag Berlin Heidelberg 2002

and liveness conditions. Secondly, as our construction shows, these cases can be handled with set-oriented computations (in determining fixed-points) which fits well to the symbolic approach. So far, it seems open whether in the more general case of parity games the treatment of individual game positions can be avoided in order to have a symbolic computation.

Our paper is structured as follows: we first exhibit a computation of the winning region of a reachability game. In third section, we derive from it two kinds of winning strategies. Finally in fourth section this material is used to solve Büchi games on pushdown graphs. The proofs can be found on the Web at http://www-i7.informatik.rwth-aachen.de/~cachat/ .

2   Reachability Game, Computing the Attractor

2.1   Technical Preliminaries

A Pushdown Game System (PDS) $\mathcal{P}$ is a triple (P, Γ, Δ), where Γ is the finite stack alphabet, $P = P_0 \uplus P_1$ the partitioned finite set of control locations, where P_i indicates the game positions of Player i, and $\Delta \subseteq P \times \Gamma \times P \times \Gamma^*$ the finite set of (unlabelled) transition rules.

Each macro-state or *configuration* is a pair pw of a control location p and a stack content w. The set of nodes of the pushdown game graph $\mathcal{G} = (V, \hookrightarrow)$ is the set of *all* configurations: $V = P\Gamma^*$, and the arcs are exactly the pairs

$$p\gamma v \hookrightarrow qwv, \text{ for } (p, \gamma, q, w) \in \Delta, \text{ where } \gamma \in \Gamma \text{ and } v \in \Gamma^* .$$

Referring to the case $v = \epsilon$ we also write rules of Δ in the form $p\gamma \hookrightarrow qw$. In the following γ is always a single letter from Γ. If one needs a bottom stack symbol ($\bot$) one has to declare it explicitly in Γ and Δ. The set of nodes of Player 0 is $V_0 = P_0\Gamma^*$, that of Player 1 is $V_1 = P_1\Gamma^*$. Starting in a given configuration $\pi_0 \in V$, a play in $\mathcal{G}$ proceeds as follows: if $\pi_0 \in V_0$, Player 0 picks the first transition (move) to π_1, else Player 1 does, and so on from the new configuration π_1. A play is a (possibly infinite) maximal sequence $\pi_0\pi_1\cdots$

We describe sets of configurations (and thus also winning conditions in pushdown games) by finite automata. We are thus interested in regular sets of configurations. We define them from alternating $\mathcal{P}$-*automata*. They are alternating word automata with a special convention about initial states. A $\mathcal{P}$-automaton $\mathcal{A}$ is a tuple $(\Gamma, Q, \longrightarrow, P, F)$, where Q is a finite set of states, $\longrightarrow \subseteq Q \times \Gamma \times 2^Q$ a set of transitions, $P \subseteq Q$ a set of initial states (which are taken here as the control locations of $\mathcal{P}$), and $F \subseteq Q$ a set of final states. For each $p \in P$ and $w \in \Gamma^*$, the automaton $\mathcal{A}$ accepts a configuration pw iff there exists a successful $\mathcal{A}$-run on w *from* p. Formally a transition has the form $r \overset{\gamma}{\longrightarrow} \beta$, where β is a positive boolean formula over Q in Disjunctive Normal Form. To simplify the exposition we allow AND-transitions $r \overset{\gamma}{\longrightarrow} r_1 \wedge \cdots \wedge r_n$, written as $r \overset{\gamma}{\longrightarrow} \{r_1, \cdots, r_n\}$, and we capture disjunction by nondeterminism. So a transition like $r \overset{\gamma}{\longrightarrow} (r_1 \wedge r_2) \vee (r_3 \wedge r_4)$ is represented here by *two* transitions $r \overset{\gamma}{\longrightarrow} \{r_1, r_2\}$ and $r \overset{\gamma}{\longrightarrow} \{r_3, r_4\}$.

We define the global transition relation of $\mathcal{A}$, the reflexive and transitive closure of $\longrightarrow$, denoted $\longrightarrow^* \subseteq Q \times \Gamma^* \times 2^Q$, as follows:

- $r \xrightarrow{\epsilon}_* \{r\}$, ($\epsilon$ is the empty word),
- $r \xrightarrow{\gamma} \{r_1, \cdots, r_n\} \wedge \forall i, \ r_i \xrightarrow{w}_* S_i \ \Rightarrow \ r \xrightarrow{\gamma w}_* \bigcup_i S_i$.

The automaton $\mathscr{A}$ accepts the word pw iff there *exists* a run $p \xrightarrow{w}_* S$ with $S \subseteq F$, *i.e.*, all finally reached states are final.

In section 3 we will need the description of a run. The run trees of an alternating automaton $\mathscr{A}$ (where the branching captures the AND-transitions) can be transformed to "run DAGs" (Directed Acyclic Graphs, see [9,10]). In such a run DAG, the states occurring on each level of the tree are collected in a set, and a transition $r \longrightarrow \{r_1, \cdots, r_k\}$ connects state r of level i with states $\{r_1, \cdots, r_k\}$ of level $i+1$. Note that every transition of level i is labelled by the same i-th letter of the input word. Let Φ be the set of partial functions from Q to the transition relation $\longrightarrow$ of $\mathscr{A}$. A run DAG from state p labelled by $w = \gamma_0 \cdots \gamma_n$ is described by a sequence $\sigma_0, \cdots, \sigma_n$ of elements of Φ and a sequence $Q_0, Q_1, \cdots, Q_n$ of subsets of Q, such that $Q_i = Dom(\sigma_i)$, and from each $q \in Q_i$ the transition $\sigma_i(q)$ is used from q:

$$Q_0 = \{p\} \xrightarrow[\sigma_0]{\gamma_0} Q_1 \xrightarrow[\sigma_1]{\gamma_1} \cdots Q_n \xrightarrow[\sigma_n]{\gamma_n} S .$$

So σ_i describes the step $Q_i \xrightarrow{\gamma_i} Q_{i+1}$ by the transitions used. We write shortly $\{p\} \xrightarrow{w}_{\sigma}_* S$, assuming $\sigma = \sigma_0, \cdots, \sigma_n$, or just $\{p\} \xrightarrow{w}_* S$ to denote the run.

2.2 Reachability

We consider a regular *goal set* $R \subseteq P\Gamma^*$, defined by a $\mathscr{P}$-automaton $\mathscr{A}_R$. Player 0 wins a play iff it reaches a configuration of R. Our goal is to compute the winning region W_0 of this game: the set of nodes from which Player 0 can force the play to reach the set R or a deadlock for Player 1. The set W_0 is clearly the "0-attractor of R" (see [11]), denoted $Attr_0(R)$, and defined inductively by

$$Attr_0^0(R) = R ,$$
$$Attr_0^{i+1}(R) = Attr_0^i(R) \cup \{u \in V_0 \mid \exists v, u \hookrightarrow v, \ v \in Attr_0^i(R)\}$$
$$\cup \{u \in V_1 \mid \forall v, u \hookrightarrow v \Rightarrow v \in Attr_0^i(R)\} ,$$
$$Attr_0(R) = \bigcup_{i \in \mathbb{N}} Attr_0^i(R) .$$

As the degree of the game graph is finite, an induction on ω is sufficient. According to this definition, we adopt the convention that if the play is in a deadlock (before reaching R), the Player who should play has lost.

Our task is to transform a given automaton $\mathscr{A}_R$ recognizing R into an automaton $\mathscr{A}_{Att(R)}$ recognizing $Attr_0(R)$. Without loss of generality, we can assume that there is no transition in $\mathscr{A}_R$ leading to an initial state (a state of P).

Algorithm 1 (saturation procedure)
Input: *a PDS $\mathscr{P}$, a $\mathscr{P}$-automaton $\mathscr{A}_R$ that recognizes the goal set R, without transition to the initial states.*
Output: *a $\mathscr{P}$-automaton $\mathscr{A}_{Att(R)}$ that recognizes $Attr_0(R)$.*

Let $\mathscr{A}_{Att(R)} := \mathscr{A}_R$. Transitions are added to $\mathscr{A}_{Att(R)}$ according to the following saturation procedure.
repeat

(Player 0) *if* $p \in P_0$, $p\gamma \hookrightarrow qv$ *and* $q \xrightarrow{v}_* S$ *in* $\mathcal{A}_{Att(R)}$, *then add a new transition* $p \xrightarrow{\gamma} S$.

(Player 1) *if* $p \in P_1$, $\begin{cases} p\gamma \hookrightarrow q_1 v_1 \\ \vdots \quad \vdots \\ p\gamma \hookrightarrow q_n v_n \end{cases}$ *are all the moves (rules) starting from*

$p\gamma$ *and* $\begin{cases} q_1 \xrightarrow{v_1}_* S_1 \\ \vdots \quad \vdots \\ q_n \xrightarrow{v_n}_* S_n \end{cases}$ *in* $\mathcal{A}_{Att(R)}$, *then add a new transition* $p \xrightarrow{\gamma} \bigcup_i S_i$.

until *no new transition can be added.*

Note that $\mathcal{A}_{Att(R)}$ has exactly the same state space as $\mathcal{A}_R$. The algorithm eventually stops because there are only finitely many possible new transitions, and the "saturation" consists in adding as many transitions as possible. The idea of adding a new transition $p \xrightarrow{\gamma} S$ for $p \in P_0$ is that, if $qv \in Attr_0(R)$, and $p\gamma \hookrightarrow qv$, then $p\gamma \in Attr_0(R)$ too, and then $p\gamma$ *should* have the same behavior as qv in the automaton. For $p \in P_1$, $Attr_0(R)$ is defined by a conjunction, expressed in $\mathcal{A}_{Att(R)}$ by the AND-transition. The algorithm and the proof is a generalization of [2,7] from nondeterministic automata (for simple reachability) to alternating automata (for game reachability). In [2], one deals with the case $P = P_0$, and the "winning region" is the set of "predecessors" of R, denoted $pre^*(R)$. In [1], alternating (pushdown) automata were already considered, but they were not used to solve a game, and winning strategies were not treated.

Theorem 2 *The automaton $\mathcal{A}_{Att(R)}$ constructed by Algorithm 1 recognizes the set $Attr_0(R)$, if $\mathcal{A}_R$ recognizes R.*

The algorithm runs in time $\mathcal{O}(|\Delta| \, 2^{c|Q|^2})$, where $|\Delta|$ is the sum of the lengths of the rules in Δ. An implementation of Algorithm 1 was developed in [4].

We have chosen a *regular* goal set R, and proved that $Attr_0(R)$ is also regular. If we consider a context free goal set R, the situation diverges for the cases of simple reachability and game reachability:

Proposition 3 *If the goal set R is a context free language, then $pre^*(R)$ is also context free, but $Attr_0(R)$ is not necessarily context free.*

The first part can be deduced from [3]. For the proof of the second part, we just remark that the intersection of two context free languages may not be context free. If R_1 and R_2 are two context free languages over $\{a, b, c\}$ and the first move of Player 1 goes from pu to $q_1 u$ or $q_2 u$, $u \in \{a, b, c\}^*$, and if the goal set is $q_1 R_1 \cup q_2 R_2$, then the winning region is $p(R_1 \cap R_2) \cup q_1 R_1 \cup q_2 R_2$.

2.3   Determining Membership in the Attractor

In [8] and [12], for a given initial position of the game, an EXPTIME procedure determines if it is in the winning region W_0 of Player 0. In contrast our solution is uniform: after a single EXPTIME procedure, we can determine in linear time if any given configuration is in the winning region W_0.

To determine whether a given configuration belongs to $Attr_0(R)$, we can use a polynomial time algorithm, that searches backwards all the accepting runs of the automaton $\mathcal{A}_{Att(R)}$ (from now on, we skip corresponding claims for Player 1). We repeat here the classical algorithm because variants of it will be used in the next section. The correctness proof is easy and omitted here.

Algorithm 4 (Membership)
Input: *an alternating $\mathcal{P}$-automaton $\mathcal{B} = (\Gamma, Q, \longrightarrow, P, F)$ recognizing* $Attr_0(R) = L(\mathcal{B})$, *a configuration $pw \in P\Gamma^*$, $w = a_1 \ldots a_n$.*
Output: *Answer whether $pw \in L(\mathcal{B})$ or $pw \notin L(\mathcal{B})$.*

Let $S := F$;
for $i := n$ down to 1 **do** $S := \{s \in Q \mid \exists\, (s \xrightarrow{a_i} X) \text{ in } \mathcal{B}, X \subseteq S\}$ **end for**
If $p \in S$, *answer "$pw \in L(\mathcal{B})$" else answer "$pw \notin L(\mathcal{B})$"*

The space complexity of Algorithm 4 is $\mathcal{O}(|Q|)$, the time complexity is $\mathcal{O}(nm|Q|)$ where m is the number of transitions of $\mathcal{B}$, and n is the length of the input configuration.

3   Winning Strategy for Player 0

A *strategy* for Player 0 is a function which associates to a prefix $\pi_0 \pi_1 \cdots \pi_n \in V^* V_0$ of a play a "next move" π_{n+1} such that $\pi_n \hookrightarrow \pi_{n+1}$.

3.1   Preparation

A move of Player 0 consists in a choice of a PDS-Rule. Given a configuration $pw \in Attr_0(R)$, our aim is to extract such a choice from an accepting run of $\mathcal{A}_{Att(R)}$ on pw. In Algorithm 1, a new transition $p \xrightarrow{\gamma} S$ of $\mathcal{A}_{Att(R)}$ is generated by a (unique) rule $p\gamma \hookrightarrow qv$ of the PDS under consideration, if $p \in P_0$. We extend now the algorithm so that it computes the partial function $Rule$ from $\longrightarrow$ to Δ. This function remembers the link between a new transition of the finite automaton for $Attr_0(R)$ and the rule of the Pushdown Graph that was used to construct it. We shall write in the algorithm $Rule(p \xrightarrow{\gamma} S) := p\gamma \hookrightarrow qv$. For transitions $p \xrightarrow{\gamma} S$ of the original automaton $\mathcal{A}$, $Rule(p \xrightarrow{\gamma} S)$ is undefined.

Now, given a configuration $p\gamma w \in V_0$ accepted by $\mathcal{A}_{Att(R)}$, with a run $\{p\} \xrightarrow{\gamma} S \xrightarrow{w}_* T$ (and if $p\gamma w \notin R$), a first idea would be to choose the move $Rule(p \xrightarrow{\gamma} S) = p\gamma \hookrightarrow qv$, *hoping to get closer to R.* Unfortunately this does not, in general, define a winning strategy. Still it ensures that we remain in the winning region. The following example illustrates this situation.

Example 5 *Let $\Gamma = \{a\}$, $P = P_0 = \{p\}$, $P_1 = \emptyset$ and $\Delta = \{pa \hookrightarrow p, pa \hookrightarrow paa\}$.*

It is clear that Player 0 can add and remove as many a's as he wants. Let $R = \{pa^3\}$ (R is regular), then the winning region is $Attr_0(R) = pa^+$, as shown

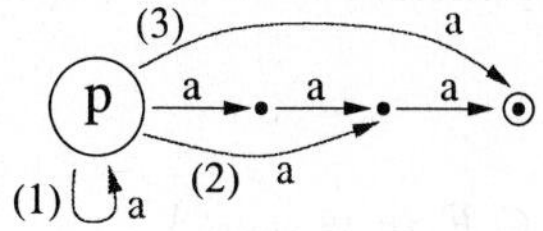

$Rule((1)) = pa \hookrightarrow p$
$Rule((2)) = pa \hookrightarrow paa$
$Rule((3)) = pa \hookrightarrow paa$ $p \hookleftarrow pa \overset{\leftrightarrow}{\hookrightarrow} paa \overset{\leftrightarrow}{\hookrightarrow} [paaa] \cdots$

Fig. 1. Automaton from Algorithm 1 and Example 5, game graph

by the automaton in Figure 1, from Algorithm 1. There are two different runs of $\mathscr{A}_{Att(R)}$ that accept paa: through transitions (1)(3), or through (2). The strategy associated to (1)(3) plays to pa, and therefore is not successful. To define a winning strategy using finite automaton $\mathscr{A}_{Att(R)}$, we need to select the most suitable run on a given configuration, or to remember information about an accepting run, to play coherently the following moves. We give two solutions: the first one, a positional strategy, associates a cost to each transition added while constructing $\mathscr{A}_{Att(R)}$, in order to compute the distance to R. The second one, a pushdown strategy, uses a stack to remember how $\mathscr{A}_{Att(R)}$ accepts the current configuration.

3.2 Positional Min-Rank Strategy

The *rank* of a configuration pw is the smallest i such that $pw \in Attr_0^i(R)$ (it is ∞ if $pw \notin Attr_0(R) = W_0$). It is the "distance" of the configuration pw to R. In the following we consider only configurations in W_0. Then Player 0 will be able, from a configuration in $Attr_0^i(R)$, to move to $Attr_0^{i-1}(R)$, and Player 1 does this with each possible move. In order to implement this, during the construction of $\mathscr{A}_{Att(R)}$ we will attribute to each $\mathscr{A}_{Att(R)}$-transition τ a cost $Cost(\tau)$. Initially, each transition of $\mathscr{A}_R$ has the cost 0 (with these transitions $\mathscr{A}_{Att(R)}$ recognizes configurations that are already in R).

The function $Cost$ from the transition relation $\longrightarrow$ of $\mathscr{A}_{Att(R)}$ to $\mathbb{N}$ is extended to a function $Cost^*$ from the run DAGs to $\mathbb{N}$. Given a fixed run $\{q\} \overset{w}{\longrightarrow}_* S$ of the automaton $\mathscr{A}_i$ (obtained at step i in the construction of $\mathscr{A}_{Att(R)}$), its cost $Cost^*(\{q\} \overset{w}{\longrightarrow}_* S)$ is the maximal sum of the costs of the transitions along a single path (branch) of the run DAG $\{q\} \overset{w}{\longrightarrow}_* S$. Inductively $Cost^*$ is defined by the following clauses:
$Cost^*(\{q\} \overset{\epsilon}{\longrightarrow}_* \{q\}) = 0$
$Cost^*(\{q\} \overset{\gamma}{\longrightarrow} \{q_1, \cdots, q_n\} \overset{u}{\longrightarrow}_* \bigcup_i S_i) =$

$$Cost(q \overset{\gamma}{\longrightarrow} \{q_1, \cdots, q_n\}) + \max_{1 \leqslant i \leqslant n} (Cost^*(\{q_i\} \overset{u}{\longrightarrow}_* S_i)) .$$

When adding a new transition $p \overset{\gamma}{\longrightarrow} S$ to $\mathscr{A}_i$, to obtain $\mathscr{A}_{i+1}$, its cost is computed by an extension of Algorithm 1, using the costs of the existing transitions. In the main loop of Algorithm 1, we add the following assignments:
- if $p \in P_0 \ldots$, let $Cost(p \overset{\gamma}{\longrightarrow} S) := 1 + Cost^*(\{q\} \overset{v}{\longrightarrow}_* S)$,
- if $p \in P_1 \ldots$, let $Cost(p \overset{\gamma}{\longrightarrow} \bigcup_i S_i) := 1 + \max_j (Cost^*(\{q_j\} \overset{v_j}{\longrightarrow}_* S_j))$.

The significance of $Cost^*$ follows clearly from next proposition:

Proposition 6 *For any configuration $pw \in Attr_0(R)$,*

$$rank(pw) = \min\{Cost^*(\{p\} \xrightarrow{w}_* S) \mid \{p\} \xrightarrow{w}_* S \subseteq F \text{ in } \mathscr{A}_{Att(R)}\} .$$

In the Example 5, one gets $Cost((1)) = 1$, $Cost((2)) = 1$, $Cost((3)) = 2$. So using the transitions (1) and (3) is not the best way to accept paa, and transition (2) is taken. We are now able to define the desired strategy.

Min-rank Strategy for Player 0
Input: alternating automaton $\mathscr{A}_{Att(R)}$ for $Attr_0(R)$, functions *Rule* and *Cost* (as computed from Algorithm 1 from PDS $\mathscr{P}$), configuration $pw \in Attr_0(R)$, $p \in P_0$.
Output: "next move" from configuration pw.

Find an accepting run $\{p\} \xrightarrow{w}_* S \subseteq F$ of $\mathscr{A}_{Att(R)}$ with minimal cost $Cost(\{p\} \xrightarrow{w}_* S)$.
If the cost is 0, $pw \in R$ and the play is won, else decompose this run: $w = \gamma w'$, $\{p\} \xrightarrow{\gamma} T \xrightarrow{w'}_* S$, and choose the rule $Rule(p \xrightarrow{\gamma} T)$.

Theorem 7 *The min-rank strategy is positional, winning from all configurations of the winning region W_0 of Player 0. It can be computed in time $\mathcal{O}(n)$ in the length n of the input configuration.*

Algorithm 4 can be easily extended to compute the distance to R and the strategy. By Proposition 6, the min-rank strategy is optimal in the sense that it finds a shortest path to R. It reevaluates its choices at each step of the game (particularly if Player 1 goes much "closer" to R than needed). We will present in the next subsection a strategy that is not necessarily optimal but easier to compute.

3.3   Pushdown Strategy

A *pushdown strategy*, as defined in [12] is a deterministic pushdown automaton with input and output. It "reads" the moves of Player 1 and outputs the moves (choices) of Player 0, like a pushdown transducer. For simplicity, we will restrict our presentation to the following form of pushdown strategy:

Definition 8 *Given a PDS (P, Γ, Δ), $P = P_0 \uplus P_1$, where Δ_i is the set of transition rules in Δ departing from Player i configurations, a pushdown strategy for Player 0 in this game is a deterministic pushdown automaton $\mathscr{S} = (P, A, \Pi)$, where $A = \Gamma \times \Sigma$, Σ is any alphabet, $\Pi \subseteq ((P_1 \times A \times \Delta_1) \times (P \times A^*)) \cup ((P_0 \times A) \times (P \times A^* \times \Delta_0))$ is a finite set of transition rules.*

A transition of $\mathscr{S}$ either reads a move of Player 1 or outputs a move for Player 0, in both cases updating its stack. We will now define a pushdown strategy, starting from the automaton $\mathcal{A}_{Att(R)}$. Given a configuration $pw \in Attr_0(R)$, there is an accepting run $\{p\} \xrightarrow{w}_* S$ of $\mathcal{A}_{Att(R)}$:

$$Q_0 = \{p\} \xrightarrow[\sigma_0]{\gamma_0} Q_1 \xrightarrow[\sigma_1]{\gamma_1} \cdots Q_n \xrightarrow[\sigma_n]{\gamma_n} S, \qquad w = \gamma_0 \gamma_1 \cdots \gamma_n .$$

Our aim is to store in the stack of the strategy the description of this run. The corresponding configuration of $\mathscr{S}$ is $p(\gamma_0, \sigma_0) \cdots (\gamma_n, \sigma_n)$. We fix for the alphabet Σ the set Φ (see Section 2.1).

At the beginning of the play, if the initial configuration pw is in $Attr_0(R)$, we have to initialize the stack of $\mathscr{S}$ with the description of an accepting run of $\mathcal{A}_{Att(R)}$ (not necessarily the cheapest according to the costs defined above). Algorithm 4 can initialize the stack at the same time when searching an accepting run (in linear time). We define now the unique transition rule of Π from $(p, (\gamma_0, \sigma_0))$ or $(p, (\gamma_0, \sigma_0), \delta_1)$ (with $\delta_1 \in \Delta_1$). By construction $\sigma_0(p)$ is the "good" transition $\tau = p \xrightarrow{\gamma_0} Q_1$ used in the run of $\mathcal{A}_{Att(R)}$.

- If $p \in P_0$ then output the move $Rule(\sigma_0(p)) = p\gamma_0 \hookrightarrow qv$ that corresponds to τ. Remove the first letter of the stack. Push on the stack the description of the run $\{q\} \xrightarrow{v}_* Q_1$ used in Algorithm 1) to generate τ. Go to control state q.
- If $p \in P_1$ and Player 1 chooses the transition $\delta_1 = p\gamma_0 \hookrightarrow qv$ in Δ_1, by construction of the automaton $\mathcal{A}_{Att(R)}$, $q \xrightarrow{v}_* S$, and S is a subset Q_1. Go to control state q, remove the first letter of the stack, push the description of the run $\{q\} \xrightarrow{v}_* S$ used in Algorithm 1) to generate τ.

For Example 5 (Figure 1), we can see that the pushdown strategy is winning even if the initialization is not optimal. The configuration paa is in the winning region and an accepting run is coded on the stack of the strategy:

$$p(a, \{(p, 1)\})(a, \{(p, 3)\}) .$$

According to the strategy, the following play is generated (the symbol "$-$" denotes a value that is not relevant):

(by $Rule((1)) = pa \hookrightarrow p$)	proceed to	$p(a, \{(p, 3)\})$
(by $Rule((3)) = pa \hookrightarrow paa$)	proceed to	$p(a, \{(p, 2)\})(a, -)$
(by $Rule((2)) = pa \hookrightarrow paa$)	proceed to	$p(a, -)(a, -)(a, -)$.

Theorem 9 *One can construct effectively a pushdown strategy that is winning from each node of the winning region of a pushdown reachability game. Its transition function is defined uniformly for the whole winning region. The initialization of the stack is possible in linear time in the length of the initial game position, and the computation of the "next move" is in constant time (for fixed Δ).*

Although there is no need to compute costs to define this strategy, it is useful to refer to the costs of the previous subsection for the correctness proof. The strategy for Player 1 in this "safety game" is much easier to define and compute: he just has to stay in $V \backslash Attr_0(R)$.

3.4   Discussion

The stack of the pushdown strategy needs to be initialized at the beginning of a play (in linear time in the length of the configuration); then the computation of the "next move" is done in constant time (execution of one transition of the strategy). In contrast, the min-rank strategy needs for each move a computation in linear time in the length of the configuration. So we can say that in the case of pushdown graphs a positional strategy can be more expensive than a strategy with memory. This effect does not appear over finite-state game graphs.

4   Büchi Condition

Given $\mathcal{P}$ and R as in the preceding sections, the (Büchi) winning condition is now the following:

> Player 0 wins a play iff it meets infinitely often the goal set R, or ends in a deadlock for Player 1.

To determine the winning region of this game one defines $Attr_0^+$ inductively, similarly to $Attr_0$: for $T \subseteq V$, let

$$
\begin{aligned}
X_0(T) &= \emptyset , \\
X_{i+1}(T) &= X_i(T) \cup \{u \in V_0 \mid \exists v, u \hookrightarrow v, \ v \in T \cup X_i(T)\} \\
&\quad \cup \{u \in V_1 \mid \forall v, u \hookrightarrow v \Rightarrow v \in T \cup X_i(T)\} , \\
Attr_0^+(T) &= \bigcup_{i \geqslant 0} X_i(T) \qquad \text{(the degree of the game graph is bounded).}
\end{aligned}
$$

Following the approach as in [11] (where the definition of the X_i's needs adjustment) the following claims are easy:

Proposition 10 $Attr_0^+(T)$ *is the set of nodes from which Player 0 can join T in at least one move, whatever Player 1 does.*

Proposition 11 *Let $Y_0 = V$, and $\forall i \geqslant 0, Y_{i+1} = Attr_0^+(Y_i \cap R)$, then the fixed point $Y^\infty = \bigcap_{i \geqslant 0} Y_i$ is the winning region of the Büchi game with goal set R.*

In the case of finite graphs, this induction can be effectively carried out: the sequence $(Y_i)_{i>0}$ is strictly decreasing until it reaches a fixed point. For pushdown graphs, the regular languages Y_i are also smaller and smaller, but the question of convergence is nontrivial. Following the symbolic approach, we will construct finite automata that are strictly "decreasing" until they reach a fixed point.

In the following we will proceed in two steps: firstly Algorithm 12 computes an automaton that recognizes Y_i for a given i, secondly Algorithm 14 computes directly an automaton for Y^∞. The first algorithm helps to understand the second, but is not a pre-computation.

Remark 1. In this section we will consider a simple goal set of the form $R = F\Gamma^*$ for some $F \subseteq P$ (it only depends on the control state). This is not an essential restriction: given a PDS $\mathcal{P} = (P, \Gamma, \Delta)$ that defines the original game and a regular set R of configurations, we can reduce to the case of simple goal sets by proceeding to a new PDS $\mathcal{P} \times \mathcal{D}$, where $\mathcal{D}$ is a finite automaton recognizing R.

So from now on we consider a simple goal set $R = F\Gamma^*$ for some $F \subseteq P$. Of course it is regular. Algorithm 1 is modified (by new steps involving ϵ-transitions) to compute $Attr_0^+(R)$. The intersection with R is easy to compute, and Algorithm 12 determines successively $Y_1, Y_2, Y_3 \cdots$

Algorithm 12 (computation of Y_j)
Input: *PDS* $\mathscr{P} = (P, \Gamma, \Delta)$, $j \geqslant 1$ *and* $F \subseteq P$ *that defines the goal set* $R = F\Gamma^*$.
Output: *a $\mathscr{P}$-automaton $\mathscr{B}_j$ that recognizes Y_j.*

Initialization: the state space of $\mathscr{B}_j$ is $\{f\} \cup (P \times [1, j])$, and f is the unique final state (we write (p, i) as p^i). For all $\gamma \in \Gamma, f \xrightarrow{\gamma} f$. For all $p \in P$, p^0 is set to be f.

for $i := 1$ **to** j **do**
   (compute generation i, consider now the p^i's as initial states.)
   Add an ϵ-transition from p^i to p^{i-1} for each $p \in F$ (only). (If $i = 1$, p^0 is f.)
   Add new transitions to $\mathscr{B}_j$ according to Algorithm 1:
   repeat
      (Player 0) if $p \in P_0$, $p\gamma \hookrightarrow qw$ and $q^i \xrightarrow{w}_* S$ in the current automaton, then *add a new transition* $p^i \xrightarrow{\gamma} S$.
      (Player 1) if $p \in P_1$, $\{p\gamma \hookrightarrow q_1w_1, \cdots, p\gamma \hookrightarrow q_nw_n\}$ are all the moves (rules) starting from $p\gamma$ and $\forall k$, $q_k^i \xrightarrow{w_k}_* S_k$ in the current automaton, then *add a new transition* $p^i \xrightarrow{\gamma} \bigcup_k S_k$.
   until *no new transition can be added*
   remove the ϵ-transitions. (generation number i is done)
end for

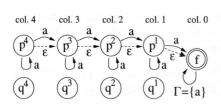

col. 4   col. 3   col. 2   col. 1   col. 0

$\Gamma = \{a\}$

Example of execution where $j = 4$. We consider $\Delta = \{pa \hookrightarrow p\}$, $R = p\Gamma^*$ ($F = \{p\}$), $P = P_0 = \{p, q\}$, $P_1 = \emptyset$. The i-th generation is given by the states of column i. The a-edges from p^i are added in the construction of generation i, since $pa \hookrightarrow p$ and $p^i \xrightarrow{\epsilon}_* p^i$, $p^i \xrightarrow{\epsilon}_* p^{i-1}$.

Fig. 2. Automaton from Algorithm 12

Proposition 13 *Algorithm 12 constructs a $\mathscr{P}$-automaton $\mathscr{B}_j$ that recognizes exactly Y_j (using the states p^j as initial states).*

The sequence (Y_i) might be strictly decreasing, so that the previous algorithm can not reach a fixed point: for the example above one gets $Y_i = pa^i\Gamma^*$, $\forall i > 0$. But on the symbolic level we can modify Algorithm 12 to obtain an automaton for Y^∞ directly. As a preparation one defines a function ϕ which is the translation one column to the right of the elements of S in the figure above, with the convention that f stays f. For all finite sets $S \subseteq P \times \mathbb{N}$ let

$$\phi(S) = \begin{cases} \{q^i \mid q^{i+1} \in S\} \cup \{f\} & \text{if } f \in S \text{ or } \exists q^1 \in S \\ \{q^i \mid q^{i+1} \in S\} & \text{else} \end{cases}$$

We also use the projection $\pi^i(S)$ of the states in $P \times [1,i]$ to the i-th column (except for f). For all $i > 0$ and sets $S \subseteq P \times [1,i] \cup \{f\}$, let

$$\pi^i(S) = \{q^i \mid \exists i \geqslant k > 0, q^k \in S\} \cup \{f \mid f \in S\} \ .$$

Algorithm 14 (computation of Y^∞)
Input: *PDS $\mathcal{P}$, $F \subseteq P$ that defines the goal set $R = F\Gamma^*$*
Output: *a $\mathcal{P}$-automaton $\mathcal{C}$ that recognizes Y^∞*

Initialization: *the state space of $\mathcal{C}$ is a subset of $(P \times \mathbb{N}) \cup \{f\}$, where (p,i) is denoted by p^i and f is the unique final state. For all $\gamma \in \Gamma, f \xrightarrow{\gamma} f$ in $\mathcal{C}$. By convention p^0 is f for all $p \in F$.*
$i := 0$.

repeat
   $i := i + 1$ *(consider now the p^i's as initial states.)*
   Add an ϵ-transition from p^i to p^{i-1} for each $p \in F$ (only)
   Add new transitions to $\mathcal{C}$ according to Algorithm 1 (see inner loop of Algorithm 12). Remove the ϵ-transitions.
   Replace each transition $p^i \xrightarrow{\gamma} S$ by $p^i \xrightarrow{\gamma} \pi^i(S)$
until $i > 1$ *and the outgoing transitions from the p^i's are "the same" as from the p^{i-1}'s:*

$$p^i \xrightarrow{\gamma} S \Leftrightarrow p^{i-1} \xrightarrow{\gamma} \phi(S) \ .$$

Applying the algorithm to the simple example above, we see that the ϵ-transitions and the a-transitions from p^2 to p^1 and from p^3 to p^2 are deleted (by the projection), and that only three generations are created.

Theorem 15 *The automaton constructed by Algorithm 14 recognizes Y^∞, the winning region of the Büchi game given by the PDS $\mathcal{P}$ and goal set R.*

The theorem implies that Y^∞ is regular. Similarly to Section 2 each execution of the inner loop (saturation procedure) is done in time $|\Delta| \, 2^{\mathcal{O}(|Q|^2)}$, where $|Q| = 2|P| + 1$. At each step of the outer loop, we "lose" at least one transition. Since there are at most $|\Gamma| \, |P| \, 2^{|Q|}$ of them, the global time complexity of the algorithm is $\mathcal{O}(|\Gamma| \, |\Delta| \, 2^{c|P|^2})$. With an extension of the arguments of the previous section we obtain corresponding winning strategies:

Theorem 16 *For a Büchi game given by a PDS $\mathcal{P}$ and goal set R, one can compute from the automaton constructed by Algorithm 14 a min-rank (positional) winning strategy and a pushdown winning strategy, uniformly for the winning region of Player 0.*

5   Conclusion

We developed a symbolic approach to solve pushdown games with reachability and Büchi winning conditions. It allows to handle uniformly the whole winning region (of a given player) and to define uniformly two types of winning strategies, a positional one and a pushdown strategy. As an extension to the results presented here, one could consider parity games on pushdown graphs, or more general winning conditions. In this context it would be interesting to connect the present symbolic approach in a tighter way with the work of Walukiewicz [12]. Another generalization is to study games on prefix recognizable graphs.

Acknowledgement. Great thanks to Wolfgang Thomas, and also to Christophe Morvan, Christof Löding and Tanguy Urvoy for their helpful advice, to Olivier Corolleur for implementing the algorithm of Section 2, and to the referees for their comments.

References

1. A. BOUAJJANI, J. ESPARZA, and O. MALER, *Reachability analysis of pushdown automata: Application to model-checking*, CONCUR '97, LNCS 1243, pp 135-150, 1997.
2. A. BOUAJJANI, J. ESPARZA, A. FINKEL, O. MALER, P. ROSSMANITH, B. WILLEMS, and P. WOLPER, *An efficient automata approach to some problems on context-free grammars*, Information Processing Letters, Vol 74, 2000.
3. D. CAUCAL, *On the regular structure of prefix rewriting*, CAAP '90, LNCS 431, pp. 87-102, 1990.
4. O. COROLLEUR, *Etude de jeux sur les graphes de transitions des automates à pile*, Rapport de stage d'option informatique, Ecole Polytechnique, 2001.
5. E. A. EMERSON and C. S. JUTLA, *Tree automata, mu-calculus and determinacy*, FoCS '91, IEEE Computer Society Press, pp. 368–377, 1991.
6. E. A. EMERSON, C. S. JUTLA, and A. P. SISTLA, *On model-checking for fragments of μ-calculus*, CAV '93, LNCS 697, pp. 385–396, 1993.
7. J. ESPARZA, D. HANSEL, P. ROSSMANITH, and S. SCHWOON, *Efficient Algorithm for Model Checking Pushdown Systems*, Technische Universität München, 2000.
8. O. KUPFERMAN and M. Y. VARDI, *An Automata-Theoretic Approach to Reasoning about Infinite-State Systems*, CAV 2000, LNCS 1855, 2000.
9. O. KUPFERMAN and M. Y. VARDI, *Weak Alternating Automata Are Not That Weak*, ISTCS'97, IEEE Computer Society Press, 1997.
10. C. LÖDING and W. THOMAS, *Alternating Automata and Logics over Infinite Words*, IFIP TCS '00, LNCS 1872, pp. 521-535, 2000.
11. W. THOMAS, *On the synthesis of strategies in infinite games*, STACS '95, LNCS 900, pp. 1-13, 1995.
12. I. WALUKIEWICZ, *Pushdown processes: games and model checking*, CAV '96, LNCS 1102, pp 62-74, 1996. Full version in Information and Computation 157, 2000.

Strong Bisimilarity and Regularity of Basic Process Algebra Is PSPACE-Hard

Jiří Srba*

BRICS**
Department of Computer Science, University of Aarhus,
Ny Munkegade bld. 540, 8000 Aarhus C, Denmark
srba@brics.dk

Abstract. Strong bisimilarity and regularity checking problems of Basic Process Algebra (BPA) are decidable, with the complexity upper bounds 2-EXPTIME. On the other hand, no lower bounds were known. In this paper we demonstrate PSPACE-hardness of these problems.

1 Introduction

Despite several positive decidability results for process algebras which generate infinite-state transition systems (see [3]), the complexity issues often remain open. Two main problems studied are *equivalence checking* and *model checking*. In this paper we investigate the first sort of problems, with a special focus on *strong bisimulation equivalence* of Basic Process Algebra (BPA), sometimes called also context-free processes.

BPA represents the class of processes introduced by Bergstra and Klop (see [2]). This class corresponds to the transition systems associated with context-free grammars in Greibach normal form (GNF), in which only left-most derivations are allowed. While it is a well known fact that language equivalence is undecidable for context-free grammars (BPA processes), strong bisimilarity is decidable [7] and the best known upper bound is 2-EXPTIME [4] so far. Language equivalence of context-free grammars with no redundant nonterminals (unnormed variables in the terminology of process algebra) is still undecidable, whereas strong bisimilarity is decidable even in polynomial time [8]. In fact the strong bisimilarity checking problem for this subclass, called *normed* BPA, is P-complete (for P-hardness see [1]).

We prove in this paper that the complexity of strong bisimilarity checking of BPA is indeed different (unless P=PSPACE) from the case of normed BPA by giving the first nontrivial lower bound for unnormed BPA. We show that strong bisimilarity of BPA is PSPACE-hard by describing a polynomial time reduction from the problem of quantified boolean formula.

---

* The author is supported in part by the GACR, grant No. 201/00/0400.
** **B**asic **R**esearch **i**n **C**omputer **S**cience,
Centre of the Danish National Research Foundation.

Another interesting problem that has attracted much attention is that of *regularity checking*. The question is whether a given BPA process is strongly bisimilar to some finite-state process. Strong regularity checking is known to be decidable in 2-EXPTIME for BPA [5,4] and in polynomial time for normed BPA [10]. As far as we know, no lower bound was given for strong regularity of BPA. We describe a polynomial time reduction from bisimilarity of strongly regular BPA processes to the strong regularity problem of BPA. This, using the fact that the processes in the PSPACE-hardness proof of strong bisimilarity of BPA are strongly regular, implies a PSPACE lower bound for strong regularity checking of BPA processes.

Finally, it is worth mentioning what is known about strong bisimilarity and regularity problems of Basic Parallel Processes (BPP). BPP are a parallel analogue of BPA, where the associative operator of sequential composition is replaced with an associative and commutative one. The strong bisimilarity checking problem for BPP is known to be decidable [6], but no primitive recursive upper bound has been given so far. Only recently Mayr proved the problem to be co-NP-hard [11]. We improved the result to PSPACE [15] by a reduction from quantified boolean formula, using similar ideas as in this paper. However, in the case of BPP it is easier to check which clauses of a given boolean formula are satisfied, because we have a parallel access to all process constants contained in the current state. This technique had to be substantially modified to work for BPA as well, since we have only sequential access to the process constants contained in a state, and there is no possibility of remembering any information in e.g. a finite-state control unit as in pushdown systems. Hence we have to encode the information about satisfied clauses in a unary way to achieve our result.

Similarly, the strong regularity checking problem for BPP is decidable [9], but no primitive recursive upper bound is known. Mayr proved the problem to be co-NP-hard [11] and we recently improved the result to PSPACE [15].

Note: full and extended version of this paper appears as [16].

2   Basic Definitions

A *labelled transition system* is a triple $(S, \mathcal{A}ct, \longrightarrow)$ where S is a set of *states* (or *processes*), $\mathcal{A}ct$ is a set of *labels* (or *actions*), and $\longrightarrow \subseteq S \times \mathcal{A}ct \times S$ is a *transition relation*, written $\alpha \xrightarrow{a} \beta$, for $(\alpha, a, \beta) \in \longrightarrow$.

Let $\mathcal{A}ct$ and $\mathcal{C}onst$ be countable sets of *actions* and *process constants*, respectively, such that $\mathcal{A}ct \cap \mathcal{C}onst = \emptyset$. We define the class of BPA *process expressions* $\mathcal{E}^{\mathcal{C}onst}$ over $\mathcal{C}onst$ by $E ::= \epsilon \mid X \mid E.E$ where 'ϵ' is the *empty process* and X ranges over $\mathcal{C}onst$. The operator '.' stands for a *sequential composition*. We do not distinguish between process expressions related by a *structural congruence*, which is the smallest congruence over process expressions such that '.' is associative, and 'ϵ' is a unit for '.'.

A BPA *process rewrite system* (or a $(1,S)$-PRS in the terminology of [12]) is a finite set $\Delta \subseteq \mathcal{C}onst \times \mathcal{A}ct \times \mathcal{E}^{\mathcal{C}onst}$ of *rewrite rules*, written $X \xrightarrow{a} E$ for $(X, a, E) \in \Delta$. Let us denote the set of actions and process constants that appear in Δ by $\mathcal{A}ct(\Delta)$ resp. $\mathcal{C}onst(\Delta)$. Note that $\mathcal{A}ct(\Delta)$ and $\mathcal{C}onst(\Delta)$ are finite sets.

A process rewrite system Δ determines a labelled transition system where *states* are process expressions over $Const(\Delta)$, $Act(\Delta)$ is the set of *labels*, and *transition relation* is the least relation satisfying the following SOS rules.

$$\frac{(X \xrightarrow{a} E) \in \Delta}{X \xrightarrow{a} E} \qquad \frac{E \xrightarrow{a} E'}{E.F \xrightarrow{a} E'.F}$$

As usual we extend the transition relation to the elements of Act^*. We also write $E \longrightarrow^* E'$, whenever $E \xrightarrow{w} E'$ for some $w \in Act^*$. A state E' is *reachable from a state* E iff $E \longrightarrow^* E'$. We write $E \xrightarrow{a}\!\!\!\!\!/\,$ whenever there is no F such that $E \xrightarrow{a} F$, and $E \not\longrightarrow$ whenever $E \xrightarrow{a}\!\!\!\!\!/\,$ for all $a \in Act$.

A BPA *process* is a pair (P, Δ), where Δ is a BPA process rewrite system and $P \in \mathcal{E}^{Const(\Delta)}$ is a BPA process expression. *States* of (P, Δ) are the states of the corresponding transition system. We say that a state E is *reachable* in (P, Δ) iff $P \longrightarrow^* E$. Whenever (P, Δ) has only finitely many reachable states, we call it a *finite-state process*. A process (P, Δ) is *normed* iff from every reachable state E in (P, Δ) there is a terminating computation, i.e., $E \longrightarrow^* \epsilon$.

Let Δ be a process rewrite system. A binary relation $R \subseteq \mathcal{E}^{Const(\Delta)} \times \mathcal{E}^{Const(\Delta)}$ over process expressions is a *strong bisimulation* iff whenever $(E, F) \in R$ then for each $a \in Act(\Delta)$: if $E \xrightarrow{a} E'$ then $F \xrightarrow{a} F'$ for some F' such that $(E', F') \in R$; and if $F \xrightarrow{a} F'$ then $E \xrightarrow{a} E'$ for some E' such that $(E', F') \in R$.

Processes (P_1, Δ) and (P_2, Δ) are *strongly bisimilar*, and we write $(P_1, \Delta) \sim (P_2, \Delta)$, iff there is a strong bisimulation R such that $(P_1, P_2) \in R$. Given a pair of processes (P_1, Δ_1) and (P_2, Δ_2) such that $\Delta_1 \neq \Delta_2$, we write $(P_1, \Delta_1) \sim (P_2, \Delta_2)$ iff $(P_1, \Delta) \sim (P_2, \Delta)$ where Δ is a disjoint union of Δ_1 and Δ_2.

We say that a process (P, Δ) is *strongly regular* iff there exists some finite-state process bisimilar to (P, Δ).

Bisimulation equivalence has an elegant characterisation in terms of *bisimulation games* [19,17]. A bisimulation game on a pair of processes (P_1, Δ) and (P_2, Δ) is a two-player game of an 'attacker' and a 'defender'. The game is played in rounds. In each round the attacker chooses one of the processes and makes an $\xrightarrow{a}$-move for some $a \in Act(\Delta)$, and the defender must respond by making an $\xrightarrow{a}$-move in the other process under the same action a. Now the game repeats, starting from the new processes. If one player cannot move, the other player wins. If the game is infinite, the defender wins. Processes (P_1, Δ) and (P_2, Δ) are strongly bisimilar iff the defender has a winning strategy (and non-bisimilar iff the attacker has a winning strategy).

3   The Main Idea

We try to explain here the main idea of the PSPACE-hardness proof given in Section 4. Our aim is to make the rewrite rules defined in Section 4 more readable by demonstrating a general pattern used heavily (with small modifications) later on. Let us consider the following BPA system Δ where α_{one} and α_{two} are some sequences of process constants.

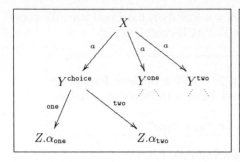 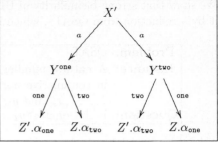

Fig. 1. Processes (X, Δ) and (X', Δ)

$$X \xrightarrow{a} Y^{\text{choice}}$$
$$X \xrightarrow{a} Y^{\text{one}} \qquad\qquad X' \xrightarrow{a} Y^{\text{one}}$$
$$X \xrightarrow{a} Y^{\text{two}} \qquad\qquad X' \xrightarrow{a} Y^{\text{two}}$$

$$Y^{\text{choice}} \xrightarrow{\text{one}} Z.\alpha_{\text{one}} \qquad Y^{\text{one}} \xrightarrow{\text{one}} Z'.\alpha_{\text{one}}$$
$$Y^{\text{choice}} \xrightarrow{\text{two}} Z.\alpha_{\text{two}} \qquad Y^{\text{two}} \xrightarrow{\text{two}} Z'.\alpha_{\text{two}}$$
$$Y^{\text{one}} \xrightarrow{\text{two}} Z.\alpha_{\text{two}} \qquad Y^{\text{two}} \xrightarrow{\text{one}} Z.\alpha_{\text{one}}$$

Transition systems generated by processes (X, Δ) and (X', Δ) are depicted in Figure 1. The intuition behind the construction can be nicely explained in terms of bisimulation games. Consider a bisimulation game starting from X and X'.

The attacker is forced to make the first move by playing $X \xrightarrow{a} Y^{\text{choice}}$ because in all other possible moves, either from X or X', the defender can make the resulting processes syntactically equal and hence bisimilar. The defender's answer to the move $X \xrightarrow{a} Y^{\text{choice}}$ is either (i) $X' \xrightarrow{a} Y^{\text{one}}$ or (ii) $X' \xrightarrow{a} Y^{\text{two}}$.

In the next round starting from (i) Y^{choice} and Y^{one} or (ii) Y^{choice} and Y^{two}, the attacker can use either the action **one** or **two** — obviously it is irrelevant whether the chosen action is performed in the first or in the second process. In case (i), if the attacker chooses the action **one** then the players reach the pair $Z.\alpha_{\text{one}}$ and $Z'.\alpha_{\text{one}}$. If he chooses the action **two** then the players reach a pair of syntactically equal states, namely $Z.\alpha_{\text{two}}$ and $Z.\alpha_{\text{two}}$, from which the defender has an obvious winning strategy. In case (ii), if the attacker chooses the action **two** then the players reach the pair $Z.\alpha_{\text{two}}$ and $Z'.\alpha_{\text{two}}$. If he chooses the action **one** then he loses as in case (i). Now, either the defender won by reaching syntactically equal states, or the resulting processes after two rounds are (i) $Z.\alpha_{\text{one}}$ and $Z'.\alpha_{\text{one}}$ or (ii) $Z.\alpha_{\text{two}}$ and $Z'.\alpha_{\text{two}}$. Note that it was the defender who had the possibility to decide between adding α_{one} or α_{two}.

We can repeat this construction several times in a row as it is explained later.

4   Hardness of Strong Bisimilarity

Problem:	Strong bisimilarity of BPA
Instance:	Two BPA processes (P_1, Δ) and (P_2, Δ).
Question:	$(P_1, \Delta) \sim (P_2, \Delta)$?

We show that strong bisimilarity of BPA is a PSPACE-hard problem. We prove it by a reduction from QSAT[1], which is PSPACE-complete [13].

Problem:	QSAT
Instance:	A natural number $n > 0$ and a Boolean formula ϕ in conjunctive normal form with Boolean variables $x_1, \dots, x_n$ and $y_1, \dots, y_n$.
Question:	Is $\exists x_1 \forall y_1 \exists x_2 \forall y_2 \dots \exists x_n \forall y_n. \phi$ true?

A *literal* is a variable or the negation of a variable. Let

$$C \equiv \exists x_1 \forall y_1 \exists x_2 \forall y_2 \dots \exists x_n \forall y_n.\ C_1 \wedge C_2 \wedge \dots \wedge C_k$$

be an instance of QSAT, where each *clause* $C_j,\ 1 \leq j \leq k$, is a disjunction of literals. For each i, $1 \leq i \leq n$, let

α_i be a sequential composition $Q_{i_1}.Q_{i_2}.\cdots.Q_{i_\ell}$ s.t. $1 \leq i_1 < i_2 < \cdots < i_\ell \leq k$ and $C_{i_1}, C_{i_2}, \dots, C_{i_\ell}$ are all the clauses where x_i occurs positively,

$\overline{\alpha_i}$ be a sequential composition $Q_{i_1}.Q_{i_2}.\cdots.Q_{i_\ell}$ s.t. $1 \leq i_1 < i_2 < \cdots < i_\ell \leq k$ and $C_{i_1}, C_{i_2}, \dots, C_{i_\ell}$ are all the clauses where x_i occurs negatively,

β_i be a sequential composition $Q_{i_1}.Q_{i_2}.\cdots.Q_{i_\ell}$ s.t. $1 \leq i_1 < i_2 < \cdots < i_\ell \leq k$ and $C_{i_1}, C_{i_2}, \dots, C_{i_\ell}$ are all the clauses where y_i occurs positively,

$\overline{\beta_i}$ be a sequential composition $Q_{i_1}.Q_{i_2}.\cdots.Q_{i_\ell}$ s.t. $1 \leq i_1 < i_2 < \cdots < i_\ell \leq k$ and $C_{i_1}, C_{i_2}, \dots, C_{i_\ell}$ are all the clauses where y_i occurs negatively.

Let $\mathcal{S}(\gamma)$ be the set of all suffixes of γ for $\gamma \in \mathcal{E}^{\{Q_1,\dots,Q_k\}}$, i.e., $\mathcal{S}(\gamma) \stackrel{\text{def}}{=} \{\gamma' \in \mathcal{E}^{\{Q_1,\dots,Q_k\}} \mid \exists \gamma'' \in \mathcal{E}^{\{Q_1,\dots,Q_k\}}$ such that $\gamma''.\gamma' = \gamma\}$. Let M be the least natural number such that $M \geq 2n + 1$ and $M = 2^K$ for some natural number $K > 0$. Of course, $M > 1$.

We define BPA processes $(X_1.S, \Delta)$ and $(X_1'.S, \Delta)$, where $Const(\Delta) \stackrel{\text{def}}{=}$

$$\begin{aligned}
&\{A_0, A_1, A_2, \dots, A_{kK-1}\} \cup \{Q_1, \dots, Q_k\} \cup \\
&\{V_i^\gamma, V_i^{\gamma,\text{choice}}, V_i^{'\gamma}, V_i^{\gamma,\text{yes}}, V_i^{\gamma,\text{no}} \mid 1 \leq i \leq n \wedge \gamma \in \mathcal{S}(\alpha_i) \cup \mathcal{S}(\overline{\alpha_i})\} \cup \\
&\{W_i^\gamma, W_i^{\gamma,\text{choice}}, W_i^{'\gamma}, W_i^{\gamma,\text{yes}}, W_i^{\gamma,\text{no}} \mid 1 \leq i \leq n \wedge \gamma \in \mathcal{S}(\beta_i) \cup \mathcal{S}(\overline{\beta_i})\} \cup \\
&\{X_i, Y_i^{\text{choice}}, Z_i, X_i', Y_i^{\text{tt}}, Y_i^{\text{ff}}, Z_i' \mid 1 \leq i \leq n\} \cup \{X_{n+1}, X_{n+1}', S\}
\end{aligned}$$

and $Act(\Delta) \stackrel{\text{def}}{=} \{a, c, \text{tt}, \text{ff}, \text{yes}, \text{no}, s\}$. The first part of the rewrite system Δ consists of the rewrite rules:

$$A_0 \stackrel{c}{\longrightarrow} \epsilon$$
$$A_\ell \stackrel{c}{\longrightarrow} A_{\ell-1}.A_{\ell-2}.\cdots.A_1.A_0 \qquad \text{for all } \ell,\ 1 \leq \ell \leq kK - 1$$

$$Q_j \stackrel{c}{\longrightarrow} A_{jK-1}.A_{jK-2}.\cdots.A_1.A_0 \qquad \text{for all } j,\ 1 \leq j \leq k.$$

---

[1] This problem is known also as QBF, for *Quantified Boolean formula*.

Remark 1. Notice that the size of the previously introduced rewrite rules is polynomial w.r.t. the size of the formula C. Moreover $A_\ell \xrightarrow{c^{2^\ell}} \epsilon$ for all ℓ, $0 \leq \ell \leq kK - 1$, which implies by using the equation $2^{jK} = M^j$ that $Q_j \xrightarrow{c^{M^j}} \epsilon$ for all j, $1 \leq j \leq k$. Hence Q_j can perform exactly M^j transitions labelled by the "counting" action c and then disappears.

The intuition is that each clause C_j, $1 \leq j \leq k$, is coded by the process constant Q_j, which enables to perform exactly M^j of c actions. The key idea of our proof is then that the defender and the attacker will choose truth values for the variables $x_1, \dots, x_n$ and $y_1, \dots, y_n$, respectively. During this process some of the clauses $C_1, \dots, C_k$ become satisfied, and the defender will have the possibility to add the corresponding process constants $Q_1, \dots, Q_k$ to the current state.

Moreover, the defender will be able to select which of the process constants (corresponding to the satisfied clauses) appear in the current state in such a way that each of them appears there exactly once.

The following lemma shall be essential for proving our reduction correct.

Lemma 1. *Assume that M and k are constants introduced above, i.e., $M > 1$ and $k > 0$. Let a_j, $1 \leq j \leq k$, be natural numbers such that $0 \leq a_j \leq M - 1$ for all j. The following two statements are equivalent:*

$$(i) \quad \sum_{j=1}^{k} a_j M^j = \sum_{j=1}^{k} M^j \qquad (ii) \quad a_j = 1 \text{ for all } j, 1 \leq j \leq k.$$

Proof. By uniqueness of M-ary representations. Details are in [16]. □

We continue with the definition of the set of rewrite rules Δ. For all i, $1 \leq i \leq n$, and $Q.\gamma \in \mathcal{S}(\alpha_i) \cup \mathcal{S}(\overline{\alpha_i})$ where $Q \in \{Q_1, \dots, Q_k\}$ and $\gamma \in \mathcal{E}^{\{Q_1, \dots, Q_k\}}$, Δ contains the rules:

$$V_i^{Q.\gamma} \xrightarrow{a} V_i^{Q.\gamma,\text{choice}}$$
$$V_i^{Q.\gamma} \xrightarrow{a} V_i^{Q.\gamma,\text{yes}} \qquad V_i'^{Q.\gamma} \xrightarrow{a} V_i^{Q.\gamma,\text{yes}}$$
$$V_i^{Q.\gamma} \xrightarrow{a} V_i^{Q.\gamma,\text{no}} \qquad V_i'^{Q.\gamma} \xrightarrow{a} V_i^{Q.\gamma,\text{no}}$$

$$V_i^{Q.\gamma,\text{choice}} \xrightarrow{\text{yes}} V_i^\gamma.Q \qquad V_i^{Q.\gamma,\text{yes}} \xrightarrow{\text{yes}} V_i'^\gamma.Q$$
$$V_i^{Q.\gamma,\text{choice}} \xrightarrow{\text{no}} V_i^\gamma \qquad V_i^{Q.\gamma,\text{no}} \xrightarrow{\text{no}} V_i'^\gamma$$
$$V_i^{Q.\gamma,\text{yes}} \xrightarrow{\text{no}} V_i^\gamma \qquad V_i^{Q.\gamma,\text{no}} \xrightarrow{\text{yes}} V_i'^\gamma.Q$$

Similarly, for all i, $1 \leq i \leq n$, and $Q.\gamma \in \mathcal{S}(\beta_i) \cup \mathcal{S}(\overline{\beta_i})$ where $Q \in \{Q_1, \dots, Q_k\}$ and $\gamma \in \mathcal{E}^{\{Q_1, \dots, Q_k\}}$, Δ contains the rules:

$$W_i^{Q.\gamma} \xrightarrow{a} W_i^{Q.\gamma,\text{choice}}$$
$$W_i^{Q.\gamma} \xrightarrow{a} W_i^{Q.\gamma,\text{yes}} \qquad W_i'^{Q.\gamma} \xrightarrow{a} W_i^{Q.\gamma,\text{yes}}$$
$$W_i^{Q.\gamma} \xrightarrow{a} W_i^{Q.\gamma,\text{no}} \qquad W_i'^{Q.\gamma} \xrightarrow{a} W_i^{Q.\gamma,\text{no}}$$

$$W_i^{Q.\gamma,\text{choice}} \xrightarrow{\text{yes}} W_i^{\gamma}.Q \quad W_i^{Q.\gamma,\text{yes}} \xrightarrow{\text{yes}} W_i^{'\gamma}.Q$$
$$W_i^{Q.\gamma,\text{choice}} \xrightarrow{\text{no}} W_i^{\gamma} \quad W_i^{Q.\gamma,\text{no}} \xrightarrow{\text{no}} W_i^{'\gamma}$$
$$W_i^{Q.\gamma,\text{yes}} \xrightarrow{\text{no}} W_i^{\gamma} \quad W_i^{Q.\gamma,\text{no}} \xrightarrow{\text{yes}} W_i^{\gamma}.Q$$

Assume now a bisimulation game starting from $(V_i^{\alpha_i}, \Delta)$ and $(V_i^{'\alpha_i}, \Delta)$. As shown in Section 3, either in some round the states become syntactically equal, or the defender has the possibility to choose in the first round the next states (i) $V_i^{\alpha_i,\text{choice}}$ and $V_i^{\alpha_i,\text{yes}}$ or (ii) $V_i^{\alpha_i,\text{choice}}$ and $V_i^{\alpha_i,\text{no}}$. This means that in the next round a process constant Q such that $\alpha_i = Q.\alpha_i'$ for some α_i' is either (i) added to a current state or (ii) left out. Now the game continues either from (i) $V_i^{\alpha_i'}.Q$ and $V_i^{'\alpha_i'}.Q$ or from (ii) $V_i^{\alpha_i'}$ and $V_i^{'\alpha_i'}$. This repeats in similar fashion until the states $V_i^{\epsilon}.\gamma_i$ and $V_i^{'\epsilon}.\gamma_i$ are reached, such that γ_i is some subsequence of α_i (in a reverse order) and it was the defender who had the possibility to decide which of the process constants contained in α_i appear also in γ_i.

The same happens if we start playing the bisimulation game from the pairs $(V_i^{\overline{\alpha_i}}, \Delta)$ and $(V_i^{'\overline{\alpha_i}}, \Delta)$, or $(W_i^{\beta_i}, \Delta)$ and $(W_i^{'\beta_i}, \Delta)$, or $(W_i^{\overline{\beta_i}}, \Delta)$ and $(W_i^{'\overline{\beta_i}}, \Delta)$.

We finish the definition of Δ by adding the rules:

– for all i, $1 \le i \le n$,

$$X_i \xrightarrow{a} Y_i^{\text{choice}}$$
$$X_i \xrightarrow{a} Y_i^{\text{tt}} \qquad X_i' \xrightarrow{a} Y_i^{\text{tt}}$$
$$X_i \xrightarrow{a} Y_i^{\text{ff}} \qquad X_i' \xrightarrow{a} Y_i^{\text{ff}}$$

$$Y_i^{\text{choice}} \xrightarrow{\text{tt}} V_i^{\alpha_i} \qquad Y_i^{\text{tt}} \xrightarrow{\text{tt}} V_i^{'\alpha_i}$$
$$Y_i^{\text{choice}} \xrightarrow{\text{ff}} V_i^{\overline{\alpha_i}} \qquad Y_i^{\text{ff}} \xrightarrow{\text{ff}} V_i^{'\overline{\alpha_i}}$$
$$Y_i^{\text{tt}} \xrightarrow{\text{ff}} V_i^{\overline{\alpha_i}} \qquad Y_i^{\text{ff}} \xrightarrow{\text{tt}} V_i^{\alpha_i}$$

$$V_i^{\epsilon} \xrightarrow{a} Z_i \qquad V_i^{'\epsilon} \xrightarrow{a} Z_i'$$

$$Z_i \xrightarrow{\text{tt}} W_i^{\beta_i} \qquad Z_i' \xrightarrow{\text{tt}} W_i^{'\beta_i}$$
$$Z_i \xrightarrow{\text{ff}} W_i^{\overline{\beta_i}} \qquad Z_i' \xrightarrow{\text{ff}} W_i^{'\overline{\beta_i}}$$

$$W_i^{\epsilon} \xrightarrow{a} X_{i+1} \qquad W_i^{'\epsilon} \xrightarrow{a} X_{i+1}'$$

– and $\quad X_{n+1} \xrightarrow{a} Q_1.Q_2.\ldots.Q_{k-1}.Q_k.S \qquad X_{n+1}' \xrightarrow{a} \epsilon \qquad S \xrightarrow{s} S.$

Lemma 2. *If* $(X_1.S, \Delta) \sim (X_1'.S, \Delta)$ *then the quantified formula* C *is true.*

Proof. We show that $(X_1.S, \Delta) \not\sim (X_1'.S, \Delta)$ under the assumption that C is false. If C is false then C' defined by

$$C' \stackrel{\text{def}}{=} \forall x_1 \exists y_1 \forall x_2 \exists y_2 \ldots \forall x_n \exists y_n. \ \neg(C_1 \wedge C_2 \wedge \ldots \wedge C_k)$$

is true and we claim that the attacker has a winning strategy in the bisimulation game starting from $(X_1.S, \Delta)$ and $(X'_1.S, \Delta)$. As mentioned in Section 3, in the first round the attacker is forced to perform the move $X_1.S \xrightarrow{a} Y_1^{\text{choice}}.S$. The defender can respond either by (i) $X'_1.S \xrightarrow{a} Y_1^{\text{tt}}.S$ (which corresponds to setting the variable x_1 to true) or by (ii) $X'_1.S \xrightarrow{a} Y_1^{\text{ff}}.S$ (which corresponds to setting the variable x_1 to false). In the second round the attacker performs the action (i) tt or (ii) ff, and the defender must answer by the same action in the other process. Now the game continues from (i) $V_1^{\alpha_1}.S$ and $V_1^{'\alpha_1}.S$ or (ii) $V_1^{\overline{\alpha_1}}.S$ and $V_1^{'\overline{\alpha_1}}.S$. Within the next (i) $2 \cdot |\alpha_1|$ or (ii) $2 \cdot |\overline{\alpha_1}|$ rounds (where $|w|$ is the length of w) the defender has the possibility to choose some subsequence of (i) α_1 or (ii) $\overline{\alpha_1}$ and add it in a reverse order to the current state. Then the game continues either from (i) $V_1^{\epsilon}.\gamma_1.S$ and $V_1^{'\epsilon}.\gamma_1.S$ or (ii) $V_1^{\epsilon}.\overline{\gamma_1}.S$ and $V_1^{'\epsilon}.\overline{\gamma_1}.S$, such that (i) γ_1 is a subsequence (in a reverse order and chosen by the defender) of α_1 or (ii) $\overline{\gamma_1}$ is a subsequence (in a reverse order and chosen by the defender) of $\overline{\alpha_1}$. The players have only one possible continuation of the game by using the rewrite rules $V_1^{\epsilon} \xrightarrow{a} Z_1$ and $V_1^{'\epsilon} \xrightarrow{a} Z'_1$, thus reaching the states (i) $Z_1.\gamma_1.S$ and $Z'_1.\gamma_1.S$ or (ii) $Z_1.\overline{\gamma_1}.S$ and $Z'_1.\overline{\gamma_1}.S$.

Now, it is the attacker who has the possibility of making a choice between the rewrite rules $Z_1 \xrightarrow{\text{tt}} W_1^{\beta_1}$ or $Z_1 \xrightarrow{\text{ff}} W_1^{\overline{\beta_1}}$ in the first process. This corresponds to setting the variable y_1 to true or false. The defender can only imitate the same action by using the rules $Z'_1 \xrightarrow{\text{tt}} W_1^{'\beta_1}$ or $Z'_1 \xrightarrow{\text{ff}} W_1^{'\overline{\beta_1}}$ in the other process. From the current states starting with $W_1^{\beta_1}$ and $W_1^{'\beta_1}$, or $W_1^{\overline{\beta_1}}$ and $W_1^{'\overline{\beta_1}}$, the same happens as before: the defender has the possibility of choosing a subsequence δ_1 (in a reverse order) of β_1 or a subsequence $\overline{\delta_1}$ (in a reverse order) of $\overline{\beta_1}$. So precisely after $2 \cdot |\beta_1|$ or $2 \cdot |\overline{\beta_1}|$ rounds the following four possible pairs of states can be reached: (1) $W_1^{\epsilon}.\delta_1.\gamma_1.S$ and $W_1^{'\epsilon}.\delta_1.\gamma_1.S$, or (2) $W_1^{\epsilon}.\delta_1.\overline{\gamma_1}.S$ and $W_1^{'\epsilon}.\delta_1.\overline{\gamma_1}.S$, or (3) $W_1^{\epsilon}.\overline{\delta_1}.\gamma_1.S$ and $W_1^{'\epsilon}.\overline{\delta_1}.\gamma_1.S$, or (4) $W_1^{\epsilon}.\overline{\delta_1}.\overline{\gamma_1}.S$ and $W_1^{'\epsilon}.\overline{\delta_1}.\overline{\gamma_1}.S$. We have now only one possible continuation of the game in the next round, reaching the states (1) $X_2.\delta_1.\gamma_1.S$ and $X'_2.\delta_1.\gamma_1.S$, or (2) $X_2.\delta_1.\overline{\gamma_1}.S$ and $X'_2.\delta_1.\overline{\gamma_1}.S$, or (3) $X_2.\overline{\delta_1}.\gamma_1.S$ and $X'_2.\overline{\delta_1}.\gamma_1.S$, or (4) $X_2.\overline{\delta_1}.\overline{\gamma_1}.S$ and $X'_2.\overline{\delta_1}.\overline{\gamma_1}.S$.

We remind the reader of the fact that the defender had the possibility to set the variable x_1 to true or false, and the attacker decided on the truth value for the variable y_1. In the meantime, all the process constants from $\{Q_1, \ldots, Q_k\}$ corresponding to the clauses that became satisfied by this assignment could have been potentially added to the current state, but it was the defender who had the possibility to filter some of them out.

In the next rounds the same schema of the game repeats, until we reach the states $X_{n+1}.\omega.S$ and $X'_{n+1}.\omega.S$. The defender decides on the truth values for each of the variables $x_2, \ldots, x_n$, and the attacker has the possibility to respond by choosing the truth values for the variables $y_2, \ldots, y_n$. During this some of the clauses appear to be satisfied and ω consists of a selection (made by the defender) of process constants corresponding to these clauses.

Since we assume that the formula C' is true, the attacker can decide on the truth values for $y_1, \ldots, y_n$ in such a way that at least one of the clauses

$C_1, \ldots, C_k$ is not satisfied. Let us suppose that it is C_m for some m, $1 \leq m \leq k$, that is not satisfied. Hence Q_m cannot appear in ω and the attacker has the following winning strategy. He plays $X_{n+1}.\omega.S \xrightarrow{a} Q_1.Q_2.\ldots.Q_{k-1}.Q_k.S.\omega.S$, to which the defender can only answer by $X'_{n+1}.\omega.S \xrightarrow{a} \omega.S$.

The state $Q_1.Q_2.\ldots.Q_{k-1}.Q_k.S.\omega.S$ can perform exactly $\sum_{j=1}^{k} M^j$ of actions c (Remark 1) followed by an infinite sequence of actions s. On the other hand, $\omega.S$ can never perform exactly $\sum_{j=1}^{k} M^j$ of actions c and then the infinite sequence of actions s. This follows from the fact that Q_m does not appear in ω and from Lemma 1 — obviously, any process constant from $\{Q_1, \ldots, Q_k\}$ can occur at most $2n$ times in ω ($2n \leq M - 1$), which justifies the assumption of Lemma 1. Hence the attacker has a winning strategy and $(X_1.S, \Delta) \not\sim (X'_1.S, \Delta)$. □

Lemma 3. *If the quantified formula C is true then $(X_1.S, \Delta) \sim (X'_1.S, \Delta)$.*

Proof. Assume a bisimulation game starting from $(X_1.S, \Delta)$ and $(X'_1.S, \Delta)$. We show that the defender has a winning strategy. As mentioned in Section 3 and in the proof above, the attacker is forced to play according to a strictly defined strategy, otherwise the defender can make the resulting processes immediately syntactically equal and hence bisimilar. As shown before the defender can make the choices between setting the variables $x_1, \ldots, x_n$ to true or false, whereas the attacker can decide on truth values for $y_1, \ldots, y_n$. Thus the defender can play the bisimulation game such that finally every clause $C_1, \ldots, C_k$ in C is satisfied. The defender has the possibility to add the corresponding process constants $Q_1, \ldots, Q_k$ to the current state in such a way that when reaching the states $X_{n+1}.\omega.S$ and $X'_{n+1}.\omega.S$, the sequential composition ω contains every Q_j exactly once for each j, $1 \leq j \leq k$. This can be easily achieved by following the strategy: "add Q_j to a current state if and only if it is not already present there". After performing the moves $X_{n+1}.\omega.S \xrightarrow{a} Q_1.Q_2.\ldots.Q_{k-1}.Q_k.S.\omega.S$ and $X'_{n+1}.\omega.S \xrightarrow{a} \omega.S$, the defender wins since $S.\omega.S \sim S$ and both $Q_1.Q_2.\ldots.Q_{k-1}.Q_k$ and ω can perform the same number of actions c. Hence $(X_1.S, \Delta) \sim (X'_1.S, \Delta)$. □

Theorem 1. *Strong bisimilarity of BPA is PSPACE-hard.*

Proof. By Lemma 2 and Lemma 3. □

Remark 2. Notice that there are only finitely many reachable states from both $(X_1.S, \Delta)$ and $(X'_1.S, \Delta)$. Hence $(X_1.S, \Delta)$ and $(X'_1.S, \Delta)$ are strongly regular.

5   Hardness of Strong Regularity

Problem:	Strong regularity of BPA
Instance:	A BPA process (P, Δ).
Question:	Is there a finite-state process (F, Δ') such that $(P, \Delta) \sim (F, \Delta')$?

The idea to reduce bisimilarity to regularity first appeared in the literature due to Mayr [11]. He showed a technique for reducing weak bisimilarity of regular BPP to weak regularity of BPP. However, in his reduction τ actions are used. Developing this approach, we provide a polynomial time reduction from strong bisimilarity of regular BPA to strong regularity of BPA.

Theorem 2 (Reduction from bisimilarity to regularity).
Let (P_1, Δ) and (P_2, Δ) be strongly regular BPA processes. We can construct in polynomial time a BPA process (P, Δ') such that

$$(P_1, \Delta) \sim (P_2, \Delta) \quad \text{if and only if} \quad (P, \Delta') \text{ is strongly regular.}$$

Proof. Assume that (P_1, Δ) and (P_2, Δ) are strongly regular processes. We construct a BPA process (P, Δ') with $Const(\Delta') \overset{\text{def}}{=} Const(\Delta) \cup \{X, A, C, S, P_1', P_2'\}$ and $Act(\Delta') \overset{\text{def}}{=} Act(\Delta) \cup \{a, s\}$ where X, A, C, S, P_1', P_2' are new process constants and a, s are new actions. Let Δ' contain all the rules from Δ plus

$$X \xrightarrow{a} X.A \qquad X \xrightarrow{a} \epsilon \qquad A \xrightarrow{a} \epsilon \qquad S \xrightarrow{s} S$$

$$
\begin{array}{llll}
A \xrightarrow{a} P_1'.S & A \xrightarrow{a} P_1.S & P_1' \xrightarrow{a} P_1' & P_1' \xrightarrow{a} P_1 \\
X \xrightarrow{a} P_1'.S & X \xrightarrow{a} P_1.S & & \\
C \xrightarrow{a} P_1'.S & C \xrightarrow{a} P_1.S & & \\
A \xrightarrow{a} P_2'.S & A \xrightarrow{a} P_2.S & P_2' \xrightarrow{a} P_2' & P_2' \xrightarrow{a} P_2 \\
X \xrightarrow{a} P_2'.S & X \xrightarrow{a} P_2.S & & \\
C \xrightarrow{a} P_2'.S & C \xrightarrow{a} P_2.S. & &
\end{array}
$$

Let $P \overset{\text{def}}{=} X.C$. It remains to establish that (P, Δ') is strongly regular if and only if $(P_1, \Delta) \sim (P_2, \Delta)$. Details can be found in [16]. □

Theorem 3. *Strong regularity of BPA is PSPACE-hard.*

Proof. By Theorem 1, Remark 2 and Theorem 2. □

6 Conclusion

The main results of this paper are the first nontrivial lower bounds for strong bisimilarity and strong regularity of BPA. We proved both problems to be PSPACE-hard. Another contribution of this paper is with regard to the normedness notion. Our results show a substantial difference (unless P=PSPACE) between *normed* and *unnormed* processes — strong bisimilarity and regularity checking is PSPACE-hard for unnormed BPA, whereas it is decidable in polynomial time for normed BPA [8,10].

An interesting observation is that only one unnormed process constant (namely S) is used in the hardness proofs for BPA. In contrast, the hardness proofs for strong bisimilarity of BPP (see [11] and [15]) require a polynomial number of unnormed process constants.

Recently some lower bounds appeared for *weak* bisimilarity of BPA and BPP [18,11,14], even though the problems are not known to be decidable. In the following tables we compare the results for strong/weak bisimilarity and regularity. New results achieved in this paper are in boldface. Obviously, the lower bounds for strong bisimilarity and regularity checking apply also to weak bisimilarity and regularity, thus improving the DP lower bound for weak regularity of BPA from [14] to PSPACE.

	strong bisimilarity	weak bisimilarity
BPA	decidable in 2-EXPTIME [4] **PSPACE-hard**	? PSPACE-hard [18]
normed BPA	decidable in P [8] P-hard [1]	? NP-hard [18]

	strong regularity	weak regularity
BPA	decidable in 2-EXPTIME [5,4] **PSPACE-hard**	? **PSPACE-hard**
normed BPA	decidable in NL [10] **NL-hard**	? NP-hard [14]

Remark 3. Complexity of strong regularity of normed BPA needs more explanation. Kucera in [10] argues that the problem is decidable in polynomial time but it is easy to see that a test whether a BPA process contains an *accessible* and *growing* process constant (a condition equivalent to regularity) can be performed even in nondeterministic logarithmic space (NL).

In order to prove NL-hardness, we reduce the reachability problem for acyclic directed graphs (NL-complete problem, see [13]) to strong regularity checking of normed BPA. Given an acyclic directed graph G with a pair of nodes v_1 and v_2 (w.l.o.g. assume that both v_1 and v_2 have out-degree at least one), we naturally construct a BPA system Δ by introducing a new process constant for each node of G with out-degree at least one, and with a-labelled transitions respecting the edges of G. All nodes with out-degree 0 are represented by the empty process ϵ to ensure that the system is deadlock-free. Moreover, a process constant representing the node v_2 has a transition to a new process constant A such that Δ contains also the rewrite rules $A \xrightarrow{a} A.A$ and $A \xrightarrow{b} \epsilon$. It is easy to see that (A, Δ) is a normed and non-regular process. Let X be a process constant representing the node v_1. Since G is acyclic, (X, Δ) is a normed BPA process. Obviously, there is a directed path from v_1 to v_2 in G if and only if (X, Δ) is not a strongly regular process. Recall that NL = co-NL (see e.g. [13]). Hence strong regularity of normed BPA is NL-complete.

Acknowledgement. I would like to thank my advisor Mogens Nielsen for his kind supervision. I also thank Pawel Sobocinski and the anonymous referees for useful comments and suggestions.

References

[1] J. Balcazar, J. Gabarro, and M. Santha. Deciding bisimilarity is P-complete. *Formal Aspects of Computing*, 4(6A):638–648, 1992.

[2] J.A. Bergstra and J.W. Klop. Algebra of communicating processes with abstraction. *Theoretical Computer Science*, 37:77–121, 1985.

[3] O. Burkart, D. Caucal, F. Moller, and B. Steffen. Verification on infinite structures. In J. Bergstra, A. Ponse, and S. Smolka, editors, *Handbook of Process Algebra*, chapter 9, pages 545–623. Elsevier Science, 2001.

[4] O. Burkart, D. Caucal, and B. Steffen. An elementary decision procedure for arbitrary context-free processes. In *Proceedings of MFCS'95*, volume 969 of *LNCS*, pages 423–433. Springer-Verlag, 1995.

[5] O. Burkart, D. Caucal, and B. Steffen. Bisimulation collapse and the process taxonomy. In *Proceedings of CONCUR'96*, volume 1119 of *LNCS*, pages 247–262. Springer-Verlag, 1996.

[6] S. Christensen, Y. Hirshfeld, and F. Moller. Bisimulation is decidable for basic parallel processes. In *Proceedings of CONCUR'93*, volume 715 of *LNCS*, pages 143–157. Springer-Verlag, 1993.

[7] S. Christensen, H. Hüttel, and C. Stirling. Bisimulation equivalence is decidable for all context-free processes. *Information and Computation*, 121:143–148, 1995.

[8] Y. Hirshfeld, M. Jerrum, and F. Moller. A polynomial algorithm for deciding bisimilarity of normed context-free processes. *Theoretical Computer Science*, 158(1–2):143–159, 1996.

[9] P. Jancar and J. Esparza. Deciding finiteness of Petri nets up to bisimulation. In *Proceedings of ICALP'96*, volume 1099 of *LNCS*, pages 478–489. Springer-Verlag, 1996.

[10] A. Kucera. Regularity is decidable for normed BPA and normed BPP processes in polynomial time. In *Proceedings of SOFSEM'96*, volume 1175 of *LNCS*, pages 377–384. Springer-Verlag, 1996.

[11] R. Mayr. On the complexity of bisimulation problems for basic parallel processes. In *Proceedings of ICALP'00*, volume 1853 of *LNCS*, pages 329–341. Springer-Verlag, 2000.

[12] R. Mayr. Process rewrite systems. *Information and Computation*, 156(1):264–286, 2000.

[13] Ch.H. Papadimitriou. *Computational Complexity*. Addison-Wesley, 1994.

[14] J. Srba. Complexity of weak bisimilarity and regularity for BPA and BPP. In *Proceedings of EXPRESS'00*, volume 39 of *ENTCS*. Elsevier Science Publishers, 2000. To appear.

[15] J. Srba. Strong bisimilarity and regularity of basic parallel processes is PSPACE-hard. In *Proceedings of STACS'02*, volume 2285 of *LNCS*, pages 535–546. Springer-Verlag, 2002.

[16] J. Srba. Strong bisimilarity of simple process algebras: Complexity lower bounds. Technical Report RS-02-16, BRICS Research Series, 2002.

[17] C. Stirling. Local model checking games. In *Proceedings of CONCUR'95*, volume 962 of *LNCS*, pages 1–11. Springer-Verlag, 1995.

[18] J. Stribrna. Hardness results for weak bisimilarity of simple process algebras. In *Proceedings of the MFCS'98 Workshop on Concurrency*, volume 18 of *ENTCS*. Springer-Verlag, 1998.

[19] W. Thomas. On the Ehrenfeucht-Fraïssé game in theoretical computer science (extended abstract). In *Proceedings of TAPSOFT'93*, volume 668 of *LNCS*, pages 559–568. Springer-Verlag, 1993.

Solving the String Statistics Problem in Time $\mathcal{O}(n \log n)$

Gerth Stølting Brodal[1][*][**], Rune B. Lyngsø[3],
Anna Östlin[1][**], and Christian N.S. Pedersen[1,2][**]

[1] BRICS[†], Department of Computer Science, University of Aarhus, Ny Munkegade,
DK-8000 Århus C, Denmark. {gerth,annao,cstorm}@brics.dk
[2] BiRC[‡], University of Aarhus, Ny Munkegade, DK-8000 Århus C, Denmark.
[3] Department of Statistics, Oxford University, Oxford OX1 3TG, UK.
lyngsoe@stats.ox.ac.uk

Abstract. The string statistics problem consists of preprocessing a
string of length n such that given a query pattern of length m, the
maximum number of non-overlapping occurrences of the query pattern
in the string can be reported efficiently. Apostolico and Preparata introduced the minimal augmented suffix tree (MAST) as a data structure for
the string statistics problem, and showed how to construct the MAST in
time $\mathcal{O}(n \log^2 n)$ and how it supports queries in time $\mathcal{O}(m)$ for constant
sized alphabets. A subsequent theorem by Fraenkel and Simpson stating
that a string has at most a linear number of distinct squares implies
that the MAST requires space $\mathcal{O}(n)$. In this paper we improve the construction time for the MAST to $\mathcal{O}(n \log n)$ by extending the algorithm
of Apostolico and Preparata to exploit properties of efficient joining and
splitting of search trees together with a refined analysis.

1 Introduction

The *string statistics problem* consists of preprocessing a string S of length n
such that given a query pattern α of length m, the maximum number of non-
overlapping occurrences of α in S can be reported efficiently. Without preprocessing the maximum number of non-overlapping occurrences of α in S can be
found in time $\mathcal{O}(n)$, by using a linear time string matching algorithm to find
all occurrences of α in S, e.g. the algorithm by Knuth, Morris, and Pratt [14],
and then in a greedy fashion from left-to-right compute the maximal number of
non-overlapping occurrences.

---

[*] Partially supported by the Future and Emerging Technologies programme of the EU
under contract number IST-1999-14186 (ALCOM-FT).
[**] Supported by the Carlsberg Foundation (contract number ANS-0257/20).
[†] Basic Research in Computer Science (BRICS), www.brics.dk, funded by the Danish
National Research Foundation.
[‡] Bioinformatics Research Center (BiRC), www.birc.dk, funded by Aarhus University
Research Fundation.

P. Widmayer et al. (Eds.): ICALP 2002, LNCS 2380, pp. 728–739, 2002.
© Springer-Verlag Berlin Heidelberg 2002

Apostolico and Preparata in [3] described a data structure for the string statistics problem, *the minimal augmented suffix tree* MAST(S), with preprocessing time $\mathcal{O}(n \log^2 n)$ and with query time $\mathcal{O}(m)$ for constant sized alphabets. In this paper we present an improved algorithm for constructing MAST(S) with preprocessing time $\mathcal{O}(n \log n)$, and prove that MAST(S) requires space $\mathcal{O}(n)$, which follows from a recent theorem of Fraenkel and Simpson [9].

The basic idea of the algorithm of Apostolico and Preparata and our algorithm for constructing MAST(S), is to perform a traversal of the suffix tree of S while maintaining the leaf-lists of the nodes visited in appropriate data structures (see Section 1.1 for definition details). Traversing the suffix tree of a string to construct and examine the leaf-lists at each node is a general technique for finding regularities in a string, e.g. for finding squares in a string (or tandem repeats) [2,17], for finding maximal quasi-periodic substrings, i.e. substrings that can be covered by a shorter substring, [1,6], and for finding maximal pairs with bounded gap [4]. All these problems can be solved using this technique in time $\mathcal{O}(n \log n)$. Other applications are listed by Gusfield in [10, Chapter 7].

A crucial component of our algorithm is the representation of a leaf list by a collection of search trees, such that the leaf-list of a node in the suffix tree of S can be constructed from the leaf-lists of the children by efficient merging. Hwang and Lin [13] described how to optimally merge two sorted lists of length n_1 and n_2, where $n_1 \leq n_2$, with $\mathcal{O}(n_1 \log \frac{n_1+n_2}{n_1})$ comparisons. Brown and Tarjan [7] described how to achieve the same number of comparisons for merging two AVL-trees in time $\mathcal{O}(n_1 \log \frac{n_1+n_2}{n_1})$, and Huddleston and Mehlhorn [12] showed a similar result for level-linked (2,4)-trees. In our algorithm we will use a slightly extended version of level-linked (2,4)-trees where each element has an associated weight. Due to lack of space proofs have been omitted. The omitted details can be found in [5].

1.1   Preliminaries

Some of the terminology and notation used in the following originates from [3], but with minor modifications. We let Σ denote a finite alphabet, and for a string $S \in \Sigma^*$ we let $|S|$ denote the length of S, $S[i]$ the ith character in S, for $1 \leq i \leq |S|$, and $S[i..j] = S[i]S[i+1] \cdots S[j]$ the substring of S from the ith to the jth character, for $1 \leq i \leq j \leq |S|$. The suffix $S[i..|S|]$ of S starting at position i will be denoted $S[i..]$.

An integer p, for $1 \leq p \leq |S|$, is denoted *a period* of S if and only if the suffix $S[p+1..]$ of S is also a prefix of S, i.e. $S[p+1..] = S[1..|S|-p]$. The shortest period p of S is denoted *the period* of S, and the string S is said to be *periodic* if and only if $p \leq |S|/2$. A nonempty string S is a *square*, if $S = \alpha\alpha$ for some string α.

In the rest of this paper S denotes the input string with length n and α a substring of S. A non-empty string α is said to *occur* in S at position i if $\alpha = S[i..i+|\alpha|-1]$ and $1 \leq i \leq n-|\alpha|+1$. E.g. in the string $b\,a\,b\,a\,a\,a\,a\,b\,a\,b\,a\,a\,b$ the substring $b\,a\,b$ occurs at positions 1 and 8. The *maximum number of non-overlapping occurrences* of a string α in a string S, is the maximum number of occurrences of α where no two occurrences overlap. E.g. the maximum number of non-overlapping occurrences of $b\,a\,b$ in $\underline{ba}ba\underline{ba}ba\underline{ba}b$ is three, since the occurrences at positions 1, 5 and 9 do not overlap.

The *suffix tree* $\mathsf{ST}(S)$ of the string S is the compressed trie storing all suffixes of the string $S\$$ where $\$ \notin \Sigma$. Each leaf in $\mathsf{ST}(S)$ represents a suffix $S[i\mathinner{..}]\$$ of $S\$$ and is annotated with the index i. Each edge in $\mathsf{ST}(S)$ is labeled with a nonempty substring of $S\$$, represented by the start and end positions in S, such that the path from the root to the leaf annotated with index i spells the suffix $S[i\mathinner{..}]\$$. We refer to the substring of S spelled by the path from the root to a node v as the *path-label* of v and denote it $\mathsf{L}(v)$. We refer to the set of indices stored at the leaves of the subtree rooted at v as the *leaf-list* of v and denote it $\mathsf{LL}(v)$. Since $\mathsf{LL}(v)$ is exactly the set of start positions i where $\mathsf{L}(v)$ is a prefix of the suffix $S[i\mathinner{..}]\$$, we have Fact 1 below.

Fact 1 *If v is an internal node of $\mathsf{ST}(S)$, then $\mathsf{LL}(v) = \bigcup_{c \text{ child of } v} \mathsf{LL}(c)$, and $i \in \mathsf{LL}(v)$ if and only if $\mathsf{L}(v)$ occurs at position i in S.*

The problem of constructing $\mathsf{ST}(S)$ has been studied intensively and several algorithms have been developed which for constant sized alphabets can construct $\mathsf{ST}(S)$ in time and space $\mathcal{O}(|S|)$ [8,15,18,19]. For non-constant alphabet sizes the running time of the algorithms become $\mathcal{O}(|S| \log |\Sigma|)$.

In the following we let the height of a tree T be denoted $h(T)$ and be defined as the maximum number of edges in a root-to-leaf path in T, and let the size of T be denoted $|T|$ and be defined as the number of leaves of T. For a node v in T we let T_v denote the subtree of T rooted at node v, and let $|v| = |T_v|$ and $h(v) = h(T_v)$. Finally, for a node v in a binary tree we let $\mathrm{small}(v)$ denote the child of v with smaller size (ties are broken arbitrarily).

The basic idea of our algorithm in Section 5 is to process the suffix tree of the input string bottom-up, such that we at each node v spend amortized time $\mathcal{O}(|\mathrm{small}(v)| \cdot \log(|v|/|\mathrm{small}(v)|))$. Lemma 1 then states that the total time becomes $\mathcal{O}(n \log n)$ [16, Exercise 35].

Lemma 1. *Let T be a binary tree with n leaves. If for every internal node v, $c_v = |\mathrm{small}(v)| \cdot \log(|v|/|\mathrm{small}(v)|)$, and for every leaf v, $c_v = 0$, then $\sum_{v \in T} c_v \le n \log n$.*

2   The String Statistics Problem

Given a string S of length n and a pattern α of length m the following greedy algorithm will compute the maximum number of non-overlapping occurrences of α in S. Find all occurrences of α in S by using an exact string matching algorithm. Choose the leftmost occurrence. Continue to choose greedily the leftmost occurrence not overlapping with any so far chosen occurrence. This greedy algorithm will compute the maximum number of occurrences of α in S in time $\mathcal{O}(n)$, since all matchings can be found in time $\mathcal{O}(n)$, e.g. by the algorithm by Knuth, Morris, and Pratt [14].

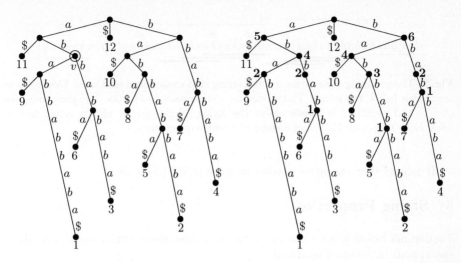

Fig. 1. To the left is the suffix tree $\mathsf{ST}(S)$ of the string $S = ababbabbaba$. The node v has path-label $\mathsf{L}(v) = ab$ and leaf-list $\mathsf{LL}(v) = \{1, 3, 6, 9\}$. To the right is the minimal augmented suffix tree $\mathsf{MAST}(S)$ for the string $S = ababbabbaba$. Numbers in the internal nodes are the c-values.

In the *string statistics problem* we want to preprocess a string S such that queries of the following form are supported efficiently: Given a query string α, what is the maximum number of non-overlapping occurrences of α in S? The maximum number of non-overlapping occurrences of α is called the *c-value* of α, denoted $c(\alpha)$. The preprocessing will be to compute the *minimal augmented suffix tree* described below. Given the minimal augmented suffix tree, string statistics queries can be answered in time $\mathcal{O}(m)$.

For any substring, α, of S there is exactly one path from the root of $\mathsf{ST}(S)$ ending in a node or on an edge of $\mathsf{ST}(S)$ spelling out the string α. This node or edge is called the *locus* of α. In a suffix tree $\mathsf{ST}(S)$ the number of leaves in the subtree below the locus of α in $\mathsf{ST}(S)$ tells us the number of occurrences of α in S. These occurrences may overlap, hence the suffix tree is not immediately suitable for the string statistics problem. The minimal augmented suffix tree for S, denoted $\mathsf{MAST}(S)$ can be constructed from the suffix tree $\mathsf{ST}(S)$ as follows. A minimum number of new auxiliary nodes are inserted into $\mathsf{ST}(S)$ in such a way that the c-value for all substrings with locus on an edge (u, v), where u is the parent of v, have c-value equal to $c(\mathsf{L}(v))$, i.e. the c-value only changes at internal nodes along a path from a leaf to the root. Each internal node v in the augmented tree is then labeled by $c(\mathsf{L}(v))$ to get the minimal augmented suffix tree. Figure 1 shows the suffix tree and the minimal augmented suffix tree for the string *ababbabbaba*.

Fraenkel and Simpson in [9] prove that a string S contains less than $2|S|$ distinct squares, which implies the following lemma.

Lemma 2. *The minimal augmented suffix tree for a string S has at most $3|S|$ internal nodes.*

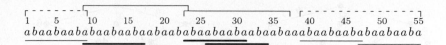

Fig. 2. The grouping of occurrences in a string into chunks and necklaces. Occurrences are shown below the string. Thick lines are occurrences in chunks. The grouping into chunks and necklaces is shown above the string. Necklaces are shown using dashed lines. Note that a necklace can consist of a single occurrence.

It follows that the space needed to store $\mathsf{MAST}(S)$ is $\mathcal{O}(n)$.

3   String Properties

The lemma below gives a characterization of how the occurrences of a string α can appear in S (proof omitted).

Lemma 3. *Let S be a string and α a substring of S. If the occurrences of α in S are at positions $i_1 < \cdots < i_k$, then for all $1 \le j < k$ either $i_{j+1} - i_j = p$ or $i_{j+1} - i_j > \max\{|\alpha| - p, p\}$, where p denotes the period of α.*

A consequence of Lemma 3 is that if $p \ge |\alpha|/2$, then an occurrence of α in S at position i_j can only overlap with the occurrences at positions i_{j-1} and i_{j+1}. If $p < |\alpha|/2$, then two consecutive occurrences i_j and i_{j+1}, either satisfy $i_{j+1} - i_j = p$ or $i_{j+1} - i_j > |\alpha| - p$.

Corollary 1. *If $i_{j+1} - i_j \le |\alpha|/2$, then $i_{j+1} - i_j = p$ where p is the period of α.*

Motivated by the above observations we group the occurrences of α in S into *chunks* and *necklaces*. Let p denote the period of α. Chunks can only appear if $p < |\alpha|/2$. A chunk is a maximal sequence of occurrences containing at least two occurrences and where all consecutive occurrences have distance p. The remaining occurrences are grouped into necklaces. A necklace is a maximal sequence of overlapping occurrences, i.e. only two consecutive occurrences overlap at a given position and the overlap of two occurrences is between one and $p - 1$ positions long. Figure 2 shows the occurrences of the string *abaabaaba* in a string of length 55 grouped into chunks and necklaces. By definition two necklaces cannot overlap, but a chunk can overlap with another chunk or a necklace at both ends. By Lemma 3 the overlap is at most $p - 1$ positions.

We now turn to the contribution of chunks and necklaces to the c-values. We first consider the case where chunks and necklaces do not overlap. An *isolated* necklace or chunk is a necklace or chunk that does not overlap with other necklaces and chunks. Figure 3 gives an example of the contribution to the c-values by an isolated necklace and chunk. More formally, we have the following lemma, which we state without proof.

Lemma 4. *An isolated necklace of k occurrences of α contributes to the c-value of α with $\lceil k/2 \rceil$. An isolated chunk of k occurrences of α contributes to the c-value of α with $\lceil k/\lceil |\alpha|/p \rceil \rceil$, where p is the period of α.*

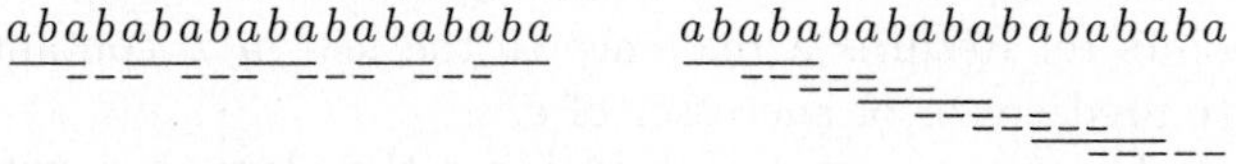

$$ababababababababababa$$

$$ababababababababababa$$

Fig. 3. Examples of the contribution to the c-values by an isolated necklace (left; $\alpha = aba$ and the contribution is $5 = \lceil 9/2 \rceil$) and an isolated chunk (right; $\alpha = ababa$, $p = 2$, and the contribution is $3 = \lceil 8/\lceil 5/2 \rceil \rceil$)

Motivated by Lemma 4, we define the *nominal contribution* of a necklace of k occurrences of α to be $\lceil k/2 \rceil$ and the nominal contribution of a chunk of k occurrences of α to be $\lceil k/\lceil |\alpha|/p \rceil \rceil$. The nominal contribution of a necklace or chunk of α's is the contribution to the c-value of α if the necklace or chunk appeares isolated. If the necklace of chunk does not appear isolated, i.e. it overlaps with a neighboring necklace or chunk, then its actual contribution to the c-value of α is at most be one less than its nominal contribution to the c-value of α.

We define the *excess* of a necklace of k occurrences to be $(k-1) \bmod 2$, and the excess of a chunk of k occurrences to be $(k-1) \bmod \lceil |\alpha|/p \rceil$. The excess describes the number of occurrences of $\alpha[1 .. p]$ which are covered by the necklace or chunk, but not covered by the maximal sequence of non-overlapping occurences.

We group the chunks and necklaces into a collection of *chains* $\mathcal{C}$ by the following two rules:

1. A chunk with excess at least two is a chain by itself.
2. A maximal sequence of overlapping necklaces and chunks with excess zero or one is a chain.

For a chain $c \in \mathcal{C}$ we define $\#_0(c)$ to be the number of chunks and necklaces with excess zero in the chain.

We are now ready to state our main lemma enabling the efficient computation of the c-values. The lemma gives an alternative to the characterization in [3, Proposition 2] (proof omitted).

Lemma 5. *The maximum number of non-overlapping occurrences of α in S equals the sum of the nominal contributions of all necklaces and chunks minus* $\sum_{c \in \mathcal{C}} \lfloor \#_0(c)/2 \rfloor$.

4   Level-Linked (2,4)-Trees

In this section we consider how to maintain a set of sorted lists of elements as a collection of level-linked (2,4)-trees where the elements are stored at the leaves in sorted order from left-to-right, and each element can have an associated real valued weight. For a detailed treatment of level-linked (2,4)-trees see [12] and [16, Section III.5]. The operations we consider supported are:

NewTree(e, w): Creates a new tree T containing the element e with associated weight w.

Search(p, e): Search for the element e starting the search at the leaf of a tree T
  that p points to. Returns a reference to the leaf in T containing e or the
  immediate predecessor or successor of e.

Insert(p, e, w): Creates a new leaf containing the element e with associated
  weight w and inserts the new leaf immediate next to the leaf pointed to
  by p in a tree T, provided that the sorted order is maintained.

Delete(p): Deletes the leaf and element that p is a pointer to in a tree T.

Join(T_1, T_2): Concatenates two trees T_1 and T_2 and returns a reference to the
  resulting tree. It is required that all elements in T_1 are smaller than the
  elements in T_2 w.r.t. the total order.

Split(T, e): Splits the tree T into two trees T_1 and T_2, such that e is larger than
  all elements in T_1 and smaller than or equal to all elements in T_2. Returns
  references to the two trees T_1 and T_2.

Weight(T): Returns the sum of the weights of the elements in the tree T.

Theorem 1 (Hoffmann et al. [11, Section 3]). *Level-linked (2,4)-trees sup-*
port NewTree, Insert *and* Delete *in amortized constant time,* Search *in time*
$\mathcal{O}(\log d)$ *where d is the number of elements in T between e and p, and* Join
and Split *in amortized time* $\mathcal{O}(\log \min\{|T_1|, |T_2|\})$.

To allow each element to have an associated weight we extend the construction from [11, Section 3] such that we for all nodes v in a tree store the sum of
the weights of the leaves in the subtree T_v, except for the nodes on the paths to
the leftmost and rightmost leaves. These sums are straightforward to maintain
while rebalancing a (2,4)-tree under node splittings and fusions, since the sum
at a node is the sum of the weights at the children of the node. For each tree we
also store the total weight of the tree.

Theorem 2. *Weighted level-linked (2,4)-trees support* NewTree *and* Weight *in*
amortized constant time, Insert *and* Delete *in amortized time* $\mathcal{O}(\log |T|)$, Search
in time $\mathcal{O}(\log d)$ *where d is the number of elements in T between e and p, and*
Join *and* Split *in amortized time* $\mathcal{O}(\log \min\{|T_1|, |T_2|\})$.

5 The Algorithm

In this section we describe the algorithm for constructing the minimal augmented
suffix tree for a string S of length n.

Algorithm idea: The algorithm starts by constructing the suffix tree, ST(S),
for S. The suffix tree is then augmented with extra nodes and c-values for all
nodes to get the minimal augmented suffix tree, MAST(S), for S. The augmentation of ST(S) to MAST(S) starts at the leaves and the tree is processed in
a bottom-up fashion. At each node v encountered on the way up the tree the
c-value for the path-label L(v) is added to the tree, and at each edge new nodes
and their c-values are added if there is a change in the c-value along the edge.
To be able to efficiently compute the c-values and decide if new nodes should
be added along edges the indices in the leaf-list of v, LL(v), are stored in a
data structure that keeps track of necklaces, chunks, and chains, as defined in
Section 3.

Data structure: Let α be a substring of S. The data structure $\mathsf{D}(\alpha)$ is a search tree for the indices of the occurrences of α in S. The leaves in $\mathsf{D}(\alpha)$ are the leaves in $\mathsf{LL}(v)$, where v is the node in $\mathsf{ST}(S)$ such that the locus of α is the edge directly above v or the node v. The search tree, $\mathsf{D}(\alpha)$, will be organized into three levels to keep track of chains, chunks, and necklaces. The top level in the search tree stores chains, the middle level chunks and necklaces, and the bottom level occurrences.

Top level: Unweighted (2,4)-tree (cf. Theorem 1) with the chains as leaves. The leftmost indices in each chain are the keys.

Middle level: One weighted (2,4)-tree (cf. Theorem 2) for each chain, with the chunks and necklaces as leaves. The leftmost indices in each chunk or necklace are the keys. The weight of a leaf is 1 if the excess of the chunk or necklace is zero, otherwise the weight is 0. The total weight of a tree on the middle level is $\#_0(c)$, where c denotes the chain represented by the tree.

Bottom level: One weighted (2,4)-tree for each chunk and necklace, with the occurrences in the chunk or necklace as the leaves. The weight of a leaf is one. The total weight of a tree is the number of occurrences in the chunk or the necklace.

Together with each of the 3-level search trees, $\mathsf{D}(\alpha)$, some variables are stored. $\mathsf{NCS}(\alpha)$ stores the sum of the nominal contribution for all chunks and necklaces, $\mathsf{ZS}(\alpha)$ stores the sum $\sum_{c \in \mathcal{C}} \lceil \#_0(c)/2 \rceil$, where $\mathcal{C}$ is the set of chains. By Lemma 5 the maximum number of non-overlapping occurrences of α is $\mathsf{NCS}(\alpha) - \mathsf{ZS}(\alpha)$. We also store the total number of indices in $\mathsf{D}(\alpha)$ and a list of all chunks denoted $\mathsf{CHUNKLIST}(\alpha)$. Finally we store, $p(\alpha)$, which is the smallest difference between the indices of two consecutive occurrences in $\mathsf{D}(\alpha)$. Note that, by Corollary 1, $p(\alpha)$ is the period of α if there is at least one chunk. To make our presentation more readable we will sometimes refer to the tree for a chain, chunk, or necklace just as the chain, chunk, or necklace.

For the top level tree in $\mathsf{D}(\alpha)$ we will use level-linked (2,4)-trees, according to Theorem 1, and for the middle and bottom level trees in $\mathsf{D}(\alpha)$ we will use weighted level-linked (2,4)-trees, according to Theorem 2. In these trees predecessor and successor queries are supported in constant time. We denote by $\ell(e)$ and $r(e)$ the indices to the left and right of index e. To be able to check fast if there are overlaps between two consecutive trees on the middle and bottom levels we store the first and last index in each tree in the root of the tree. This can easily be kept updated when the trees are joined and split.

We will now describe how the suffix tree is processed and how the data structures are maintained during this process.

Processing events: We want to process edges in the tree bottom-up, i.e. for decreasing length of α, so that new nodes are inserted if the c-value changes along the edge, the c-values for nodes are added to the tree, and the data structure is kept updated. The following events can cause changes in the c-value and the chain, chunk, and necklace structure.

1. Excess change: When $|\alpha|$ becomes $i \cdot p(\alpha)$, for $i = 2, 3, 4, \ldots$ the excess and nominal contribution of chunks changes and we have to update the data structure and possibly add a node to the suffix tree.
2. Chunks become necklaces: When $|\alpha|$ decreases and becomes $2p$ a chunk degenerates into a necklace. At this point we join all overlapping chunks and necklaces into one necklace and possibly add a node to the suffix tree.
3. Necklace and chain break-up: When $|\alpha|$ decreases two consecutive occurrences at some point no longer overlap. The result is that a necklace or a chain may split, and we have to update the necklace and chain structure and possibly add a node to the suffix tree.
4. Merging at internal nodes: At internal nodes in the tree the data structures for the subtrees below the node are merged into one data structure and the c-value for the node is added to the tree.

To keep track of the events we use an event queue, denoted EQ, that is a common priority queue of events for the whole suffix tree. The priority of an event in EQ is equal to the length of the string α when the event has to be processed. Events of type 1 and 2 store a pointer to any leaf in $D(\alpha)$. Events of type 3, i.e. that two consecutive overlapping occurrences with index e_1 and e_2, $e_1 < e_2$, terminate to overlap, store a pointer to the leaf e_1 in the suffix tree. For the leaf e_1 in the suffix tree also a pointer to the event in EQ is stored. Events of type 4 stores a pointer to the internal node in the suffix tree involved in the event. When the suffix tree is constructed all events of type 4 are inserted into EQ. For a node v in $ST(S)$ the event has priority $|L(v)|$ and stores a pointer to v. The pointers are used to be able to decide which data structure to update. The priority queue EQ is implemented as a table with entries $EQ[1] \ldots EQ[|S|]$. All events with priority x are stored in a linked list in entry $EQ[x]$. Since the priorities of the events considered are monotonic decreasing, it is sufficient to consider the entries of EQ in a single scan starting at $EQ[|S|]$.

The events are processed in order of the priority and for events with the same priority they are processed in the order as above. Events of the same type and with the same priority are processed in arbitrary order. In the following we only look at one edge at the time when events of type 1, 2, and 3 are taken care of. Due to space limitations many algorithmic details are left out in the following. See [5] for a detailed description of the algorithm.

1. Excess change. The excess changes for all chunks at the same time, namely when $|\alpha| = i \cdot p(\alpha)$ for $i = 2, 3, 4, \ldots$. For each chunk in CHUNKLIST(α) we will remove the chunk from $D(\alpha)$, recompute the excess and nominal contribution based on the number of occurrences in the chunk, update NCS(α), reinsert the chunk with the new excess and finally update ZS(α). This is done as follows:

First decide which chain each chunk belongs to by searching the tree. Remove each chunk from its chain by splitting the tree for the chain. Recompute the excess for each chunk and reconstruct the tree. In the new tree the chain structure may have changed. Chunks for which the excess increases to two will be separate chains, while chunks where the excess become less than two may join two or three chains into one chain. NCS(α) and ZS(α) are always kept updated during the processing of the event.

If $|\alpha| = 2p(\alpha)$ then insert an event of type 2 with priority $2p(\alpha)$ into EQ, with a pointer to any leaf in D(α). If $|\alpha| = ip(\alpha) > 2p(\alpha)$, then insert an event of type 1 with priority $(i-1)p(\alpha)$ into EQ, with a pointer to any leaf in D(α).

2. Chunks become necklaces. When $|\alpha|$ decreases to $2p$ all chunks become necklaces at the same time. At this point all chunks and necklaces that overlap shall be joined into one necklace. Note that all chunks have excess 0 or 1 when $|\alpha| = 2p$ and since we first recompute the excess all overlapping chunks and necklaces are in the same chain. Hence, what we have to do is to join all chunks and necklaces from left to right, in each chain.

This is done by first deciding for each chunk which chain it belongs to. Next, for each chain containing at least one chunk, join all chunks and necklaces from left to right. Update NCS(α) and ZS(α).

3. Necklace and chain break-up. When two consecutive occurrences of α with indices e_1 and e_2 terminate to overlap this may cause a necklace or a chain to break up into two necklaces or chains.

If e_1 and e_2 belong to the same chain then the chain breaks up in two chains. If e_1 and e_2 belongs to the same necklace then split both the necklace and the chain between e_1 and e_2. If e_1 and e_2 belong to different necklaces or chunks in the chain then split the chain between the two subtrees including e_1 and e_2 respectively. Update NCS(α) and ZS(α).

4. Merging at internal nodes. Let α be a substring such that the locus of α is a node v in the suffix tree. Then the leaf-list, LL(v) for v is the union of the leaf-lists for the subtrees below v, hence at the nodes in the suffix tree the data structures for the subtrees should be merged into one. We assume that the edges below v are processed for α as described above.

Let $T_1, \ldots, T_t$ be the subtrees below v in the suffix tree. We never merge more than two data structures at the time. If there are more than two subtrees the merging is done in the following order: $T = \text{Merge}(T, T_i)$, for $i = 2, \ldots, t$, where $T = T_1$ to start with. This can also be viewed as if the suffix tree is made binary by replacing all nodes of degree larger than 2 by a binary tree with edges without labels. From now on we will describe how to merge the data structures for two subtrees.

The merging will be done by inserting all indices from the smaller of the two leaf-lists into the data structure for the larger one. Let T denote the 3-level search tree to insert new indices in and denote by $e_1, \ldots, e_m$ the indices to insert, where $e_i < e_{i+1}$. The insertion is done by first splitting the tree T at all positions e_i for $i = 1, \ldots, m$. The tree is then reconstructed from left to right at the same time as the new indices are inserted in increasing order. Assume that the tree is reconstructed for all indices, in both trees, smaller than e_i. The next step is to insert e_i and all indices between e_i and e_{i+1}. This is done as follows:

Check if the occurrence with index e_i overlaps any occurrences to the left, i.e. an occurrence in the tree reconstructed so far. Insert e_i into the tree. If e_i

overlaps with an occurrence already in the tree then check in what way this affects the chain, chunk, and necklace structure and do the appropriate updates. Do the corresponding check and updates when the tree to the right of e_i (the tree for indices between e_i and e_{i+1}) is incorporated, i.e. check if e_i will cause any further changes in the chain, chunk, and necklace structure due to overlaps to the right. Update $NCS(\alpha)$ and $ZS(\alpha)$.

Every time, during the above described procedure, when two overlapping occurrences with indices e_i and e_j, $e_i < e_j$, from different subtrees are encountered the event (e_i, e_j) with priority $e_j - e_i$ is inserted into the event queue EQ and the previous event, if any, with a pointer to e_i is removed from EQ. Update $p(\alpha)$ to $e_j - e_i$ if this is smaller than the current $p(\alpha)$ value. If $|\alpha| > 2p(\alpha)$ then insert an event of type 1 with priority $\lfloor |\alpha|/p(\alpha) \rfloor p(\alpha)$ into EQ, with a pointer to any leaf in $D(\alpha)$.

6   Analysis

Theorem 3. *The minimal augmented suffix tree, $MAST(S)$, for a string S of length n can be constructed in time $\mathcal{O}(n \log n)$ and space $\mathcal{O}(n)$.*

In the full version of the paper [5] we show that the running time of the algorithm in Section 5 is $\mathcal{O}(n \log n)$. Here we only state the main steps of the proof. The proof uses an amortization argument, allowing each edge to be processed in amortized constant time, and each binary merge at a node (in the binary version) of $ST(S)$ of two leaf-lists of sizes n_1 and n_2, with $n_1 \geq n_2$, in amortized time $\mathcal{O}(n_2 \log \frac{n_1 + n_2}{n_2})$. From Lemma 1 it then follows that the total time for processing the internal nodes and edges of $ST(S)$ is $\mathcal{O}(n \log n)$.

Using Theorem 1 and 2 we can prove that: Processing events of types 1 and 2 take time $\mathcal{O}(m \log \frac{|LL(v)|}{m})$, where $m = |CHUNKLIST(\alpha)|$. Processing an event of type 3 takes time $\mathcal{O}(\log |c|)$, where c is the chain being split. An event of type 4 has processing time $\mathcal{O}(n_1 \log \frac{n_1 + n_2}{n_1})$.

Let v be a node in the suffix tree and let α be a string with locus v or locus on the edge immediately above v. For the data structure $D(\alpha)$ we define a potential $\Phi(D(\alpha))$. Let $\mathcal{C}$ be the set of chains stored in $D(\alpha)$, and for a chain c let $|c|$ denote the number of occurrences of α in c. We define the potential of $D(\alpha)$ by $\Phi(D(\alpha)) = \Phi_1(\alpha) + \Phi_2(\alpha) + \sum_{c \in \mathcal{C}} \Phi_3(c)$, where the rôle of Φ_1, Φ_2, and Φ_3 is to account for the potential required to be able to process events of type 1, 2, and 3 respectively. For a chunk, with leftmost occurrence of α at position i, consider the substring $S[i..j]$ with maximal j and $S[i..j]$ having period p, where $p = p(\alpha)$ is the period of α. We denote the chunk *green* if and only if $|\alpha| \bmod p \leq j - i + 1 \bmod p$. Otherwise the chunk is *red*. Let k denote the number of chunks in $D(\alpha)$ and let g denote the number of green chunks in $D(\alpha)$. We define $\Phi_1(\alpha) = 7g \log \frac{|v| \cdot e}{g}$, $\Phi_2(\alpha) = k \log \frac{|v| \cdot e}{k}$, and $\Phi_3(c) = 2|c| - \log |c| - 2$, with the exceptions that $\Phi_1(\alpha) = 0$ if $g = 0$, and $\Phi_2(\alpha) = 0$ if $k = 0$.

We can prove that processing events of type 1, 2, and 3 release sufficient potential to pay for the processing, while processing an event of type 4 increases the potential by $\mathcal{O}(n_1\log\frac{n_1+n_2}{n_1})$. By Lemma 1 the total amortized time for handling all events is $\mathcal{O}(n\log n)$.

References

1. A. Apostolico and A. Ehrenfeucht. Efficient detection of quasiperiodicities in strings. *Theoretical Computer Science*, 119:247–265, 1993.
2. A. Apostolico and F. P. Preparata. Optimal off-line detection of repetitions in a string. *Theoretical Computer Science*, 22:297–315, 1983.
3. A. Apostolico and F. P. Preparata. Data structures and algorithms for the string statistics problem. *Algorithmica*, 15:481–494, 1996.
4. G. S. Brodal, R. Lyngsø, C. N. S. Pedersen, and J. Stoye. Finding maximal pairs with bounded gap. *Journal of Discrete Algorithms, Special Issue of Matching Patterns*, 1(1):77–104, 2000.
5. G. S. Brodal, R. B. Lyngsø, A. Östlin, and C. N. S. Pedersen. Solving the string statistics problem in time $O(n\log n)$. Technical Report RS-02-13, BRICS, Department of Computer Science, University of Aarhus, 2002.
6. G. S. Brodal and C. N. S. Pedersen. Finding maximal quasiperiodicities in strings. In *Proc. 11th Combinatorial Pattern Matching*, volume 1848 of *Lecture Notes in Computer Science*, pages 397–411. Springer Verlag, Berlin, 2000.
7. M. R. Brown and R. E. Tarjan. A fast merging algorithm. *Journal of the ACM*, 26(2):211–226, 1979.
8. M. Farach. Optimal suffix tree construction with large alphabets. In *Proc. 38th Ann. Symp. on Foundations of Computer Science (FOCS)*, pages 137–143, 1997.
9. A. S. Fraenkel and J. Simpson. How many squares can a string contain? *Journal of Combinatorial Theory, Series A*, 82(1):112–120, 1998.
10. D. Gusfield. *Algorithms on Strings, Trees and Sequences: Computer Science and Computational Biology*. Cambridge University Press, 1997.
11. K. Hoffmann, K. Mehlhorn, P. Rosenstiehl, and R. E. Tarjan. Sorting Jordan sequences in linear time using level-linked search trees. *Information and Control*, 86(1-3):170–184, 1986.
12. S. Huddleston and K. Mehlhorn. A new data structure for representing sorted lists. *Acta Informatica*, 17:157–184, 1982.
13. F. K. Hwang and S. Lin. A simple algorithm for merging two disjoint linearly ordered sets. *SIAM Journal of Computing*, 1(1):31–39, 1972.
14. D. E. Knuth, J. H. Morris, and V. R. Pratt. Fast pattern matching in strings. *SIAM Journal of Computing*, 6:323–350, 1977.
15. E. M. McCreight. A space-economical suffix tree construction algorithm. *Journal of the ACM*, 23(2):262–272, 1976.
16. K. Mehlhorn. *Sorting and Searching*, volume 1 of *Data Structures and Algorithms*. Springer Verlag, Berlin, 1984.
17. J. Stoye and D. Gusfield. Simple and flexible detection of contiguous repeats using a suffix tree. *Theoretical Computer Science*, 270:843–856, 2002.
18. E. Ukkonen. On-line construction of suffix trees. *Algorithmica*, 14:249–260, 1995.
19. P. Weiner. Linear pattern matching algorithms. In *Proc. 14th Symposium on Switching and Automata Theory*, pages 1–11, 1973.

A PTAS for Distinguishing (Sub)string Selection

Xiaotie Deng[1*], Guojun Li[2], Zimao Li[1], Bin Ma[3], and Lusheng Wang[1**]

[1] Department of Computer Science, City University of Hong Kong, Kowloon, Hong Kong. deng,lizm,lwang@cs.cityu.edu.hk
[2] Department of Mathematics, Shandong University, Jinan 250100, P. R. China.
[3] Department of Computer Science, University of Western Ontario, London, Ont. N6A 5B7, Canada. bma@csd.uwo.ca

Abstract. Consider two sets of strings, $\mathcal{B}$ (bad genes) and $\mathcal{G}$ (good genes), as well as two integers d_b and d_g ($d_b \leq d_g$). A frequently occurring problem in computational biology (and other fields) is to find a (distinguishing) substring s of length L that distinguishes the bad strings from good strings, i.e., for each string $s_i \in \mathcal{B}$ there exists a length-L substring t_i of s_i with $d(s, t_i) \leq d_b$ (close to bad strings) and for every substring u_i of length L of every string $g_i \in \mathcal{G}$, $d(s, u_i) \geq d_g$ (far from good strings). We present a polynomial time approximation scheme to settle the problem, i.e., for any constant $\epsilon > 0$, the algorithm finds a string s of length L such that for every $s_i \in \mathcal{B}$, there is a length-L substring t_i of s_i with $d(t_i, s) \leq (1+\epsilon)d_b$ and for every substring u_i of length L of every $g_i \in \mathcal{G}$, $d(u_i, s) \geq (1 - \epsilon)d_g$, if a solution to the original pair ($d_b \leq d_g$) exists.

1 Introduction

Research effort in molecular biology, such as the human genome project, has been revealing the secret of our genetical composition, the long DNA sequences that determine almost every aspect of life. Applications that use this information have posed new challenge to design and analysis of efficient computational methods.

A frequently surfacing problem in many biological applications is to find one substring of length L that appears (with a few substitutions) at least once in each of bad strings (such as bacterial sequences) and is not "close" to any substring of length L in *each* of good strings (such as Human and live stock sequences). The problem and its formulations are originally proposed in [7]. The problem has various applications in molecular biology such as universal PCR primer design, genetic drug target identification and genetic probes design [6,2,12,7]. In particular, the genetic drug target identification problem searches for a sequence of genes that is close to bad genes (the target) but far from all good genes (to

---

* Fully supported by a grant from the Natural Science Foundation of China and Research Grants Council of the HKSAR Joint Research Scheme [Project No: N_CityU 102/01].
** Fully supported by a grant from the Research Grants Council of the Hong Knog SAR, China [Project No: CityU 1130/99E].

P. Widmayer et al. (Eds.): ICALP 2002, LNCS 2380, pp. 740–751, 2002.
© Springer-Verlag Berlin Heidelberg 2002

avoid side-effect). Our study develops a polynomial time approximation scheme in both measures simultaneously.

1.1   The Mathematical Model

Formally, the *distinguishing substring selection* problem is as follows: given a set $\mathcal{B} = \{s_1, s_2, \ldots, s_{n_1}\}$ of n_1 (bad) strings of length at least L, a set $\mathcal{G} = \{g_1, g_2, \ldots g_{n_2}\}$ of n_2 (good) strings of length exactly L, and two integers d_b and d_g ($d_b \leq d_g$), the distinguishing substring selection problem ($DSSP$) is to find a string s such that for each string $s_i \in \mathcal{B}$ there exists a length-L substring t_i of s_i with $d(s, t_i) \leq d_b$ and for any string $g_i \in \mathcal{G}$, $d(s, g_i) \geq d_g$. Here $d(,)$ represents the Hamming distance between two strings. Note that every string in $\mathcal{G}$ has the same length L since $\mathcal{G}$ can be reconstructed to contain all substrings of length L in each of all the good strings.

1.2   Previous Results

The best previous known approximation ratio is two [7]. Many simplified versions were proposed to make the task easier in terms of computation/approximation at the sacrifice of general applicability.

The problem is NP-hard, even when only one objective is to be met [1,7]. To approximate the center string that is far away from each of the good strings is not difficult, and is shown to have a polynomial time approximation scheme by Lanctot, et al. [7]. The problem to find a center string s that is close to all the bad strings is non-trivial, and has demanded intensive investigation of several research groups [7,4] to finally have a polynomial time approximation scheme [8].

There are also many other related modifications and variations, e.g., the *d-characteristic string* [5], the *far from most string* [7], and the formulation of Ben-Dor *et al.* [1].

To meet the requirements of many application problems, however, we would need to have a solution that is close to one group of given strings and far away from the other. The DSSP problem that contains two objective functions has remained evasive despite of the above mentioned recent intensive research effort.

1.3   Our Contribution

In this paper, we settle this difficult problem by presenting a solution that is a polynomial time approximation scheme in both requirements. If a solution to the original problem exists, for any constant $\epsilon > 0$, our algorithm computes a string s of length L such that for every $s_i \in \mathcal{S}$, there is a length-L substring t_i of s_i with $d(t_i, s) \leq (1 + \epsilon)d_b$ and for every $g_i \in \mathcal{G}$, $d(g_i, s) \geq (1 - \epsilon)d_g$. Here d_b and d_g are two important parameters supplied by users to specify the distances from the distinguishing string s to the set of bad strings and the set of good strings. Our algorithm gives a solution that roughly meets the requirements, i.e., $(1 + \epsilon)d_b$ and $(1 - \epsilon)d_g$, if a solution to the original requirements, i.e., d_b and d_g, exists.

1.4   Sketch of Our Approach

Design and analysis of good algorithms for approximating multiple objective functions are not simple in general (see [11] for a general approach and related works).

The standard techniques for related problems follow a linear program approach combined with randomized rounding. That works when the parameters are sufficiently large. The main difficulty for our problem lies in the case when objective function value is relatively small but still too large for enumeration methods to work.

To overcome this difficulty, we sample a constant number of strings and find the positions where the sample strings have the same letters. We denote the set of such positions by Q and the set of the remaining positions by P. Our breakthrough starts with a key lemma that shows that there is a set of constant number of sample strings, for which there are y positions in Q such that when we change the letters at the y positions in Q, keep the letters at the rest of $|Q| - y$ positions the same as in the sample strings, and choose letters at positions in P by the linear programming approach, then we can obtain the right approximate solution.

An interesting case is that $y < O(\log n)$, i.e., y is small, but not small enough for a brute-force enumeration method to go through directly. A new method is designed to handle this case. Since similar situations occur in many combinatorial optimization problems, we expect that this idea may have wider application values.

1.5   Organization of the Paper

The rest of the paper is organized as follows: Section 2 introduces the related notations and the standard integer programming formulation which works well when both d_b and d_g are large, i.e., at least $\Omega(L)$. Section 3 gives the key lemma. We discuss the methods to find a good approximation of the set of y positions in Q in section 4. In Section 5, we show that the algorithm can be extended to work for the general case, the distinguishing substring selection problem.

2   Preliminaries

We consider two sets of strings: $\mathcal{G}$ (good strings) and $\mathcal{B}$ (bad strings). We call d_b the *upper* radius for bad strings ($\mathcal{B}$) and d_g the *lower* radius for good strings ($\mathcal{G}$). Let $n = n_1 + n_2$ be the total number of good and bad strings in $\mathcal{G} \cup \mathcal{B}$.

For the distinguishing string selection problem (DSP), every good or bad string is of the same length L. The distance $d(x, y)$ of two strings x and y is their Hamming distance, i.e., the number of positions where they differ from each other. We are to find the string x of length L such that

$$\begin{cases} d(s_i, x) \leq d_b, & s_i \in \mathcal{B}; i = 1, \cdots, n_1; \\ d(g_j, x) \geq d_g, & g_j \in \mathcal{G}; j = 1, \cdots, n_2. \end{cases} \tag{1}$$

In this section, we present an approximation algorithm that works well for a special case of the DSP problem. The restriction is that L, d_b and d_g are large, more specifically, $L \geq (4\log(n_1 + n_2))/\epsilon^2$, $d_g = \Omega(L)$, and $d_b = \Omega(L)$, where ϵ is the parameter to control the performance ratio, and n_1 and n_2 are the numbers of bad and good strings, respectively.

This is achieved via a standard method using integer linear programming. Define $\chi(s, i, k, a) = 0$ if $s_i[k] = a$ and $\chi(s, i, k, a) = 1$ if $s_i[k] \neq a$. Similarly, define $\chi(g, j, k, a) = 0$ if $g_j[k] = a$ and $\chi(g, j, k, a) = 1$ if $g_j[k] \neq a$. The problem becomes the following integer linear programming problem:

$$\begin{cases} \sum_{a \in \Sigma} x_{k,a} = 1, & k = 1, 2, \ldots, L; \\ \sum_{1 \leq k \leq L} \sum_{a \in \Sigma} \chi(s, i, k, a) x_{k,a} \leq d_b, & i = 1, 2, \ldots, n_1; \\ \sum_{1 \leq k \leq L} \sum_{a \in \Sigma} \chi(g, j, k, a) x_{k,a} \geq d_g, & j = 1, 2, \ldots, n_2; \\ x_{k,a} \in \{0, 1\}, & a \in \Sigma, \ k = 1, 2, \ldots, L. \end{cases} \tag{2}$$

Let $\bar{x}_{k,a}$ be a solution for LP relaxation of (2). For each $0 \leq k \leq L$, with probability $\bar{x}_{k,a}$, we set $x_{k,a} = 1$ and set $x_{k,a'} = 0$ for any $a' \neq a$. We choose the random variables independently for different k. This results in an integer solution for (2) (and hence a solution for (1)) if d_b is replaced by $(1 + \epsilon)d_b$ and d_g by $(1 - \epsilon)d_g$, as shown in Theorem 1. Standard derandomization method [10] transfers it to a deterministic algorithm in Theorem 2.

Theorem 1. *Let $\delta > 0$ be any constant. Suppose that $L \geq (4\log(n_1 + n_2))/\delta^2$, $d_g \geq c_g L$ and $d_b \geq c_b L$. There is a randomized algorithm that finds a solution i.e., a string x of length L, with high probability such that for each string s_i in $\mathcal{B}$, $d(s_i, x) < (1 + \delta/c_b)d_b$, and for any string g_j in $\mathcal{G}$, $d(g_j, x) > (1 - \delta/c_g)d_g$,*

Theorem 2. *There exists a PTAS for DSP problem when $d_g = \Omega(L)$, and $d_b = \Omega(L)$.*

3   A Key Lemma

To obtain our PTAS algorithm, we need to introduce two parameters: an integer r and a positive number δ. The constant r is introduced in this section and r depends purely on ϵ. The constant $\delta > 0$ will be introduced in the next section and $\delta > 0$ depends on both $\epsilon > 0$ and r.

For any $r > 2$, let $1 \leq i_1, i_2, \ldots, i_r \leq n_1$ be r distinct numbers. Let $Q_{i_1, i_2, \ldots, i_r}$ be the set of positions where s_{i_1}, s_{i_2}, $\ldots$, s_{i_r} agree. Let s be a feasible solution of length L for the distinguishing string selection problem with the upper radius d_b for bad strings ($\mathcal{B}$) and lower radius d_g for good strings ($\mathcal{G}$). $P_{i_1, i_2, \ldots, i_r} = \{1, 2, \ldots, L\} - Q_{i_1, i_2, \ldots, i_r}$ is the set of positions where at least one pair of strings s_{i_j} and $s_{i_{j'}}$ differ. Then, at such positions, s differs from at least one of s_{i_j}. The number of positions that s differs from one of s_{i_j} is no more than d_b. Therefore, the total number of positions that s differs from at least one of s_{i_j} ($j = 1, 2, \cdots, r$) is no more than rd_b. It follows that $rd_b \geq |P_{i_1, i_2, \ldots, i_r}| = L - |Q_{i_1, i_2, \ldots, i_r}|$. That is, $|Q_{i_1, i_2, \ldots, i_r}| \geq L - rd_b$. Thus, we have the following result which is often applied late in the proof of our main result.

Proposition 1. *Assuming that $d_g \geq d_b$, we have $rd_g \geq rd_b \geq L - |Q_{i_1,i_2,\ldots,i_r}| = |P_{i_1,i_2,\ldots,i_r}|$.*

Let $\mathcal{B} = \{s_1, s_2, \ldots, s_{n_1}\}$. Let $p_{i_1,i_2,\ldots,i_k}$ be the number of mismatches between s_{i_1} and s at the positions in $Q_{i_1,i_2,\ldots,i_k}$. Let $\rho_k = \min_{1 \leq i_1,i_2,\ldots,i_k \leq n_1} p_{i_1,i_2,\ldots,i_k}/d_b$. The following proposition is a variant of a claim in [8] but the proof is identical.

Proposition 2. *For any k such that $2 \leq k \leq r$, where r is a constant, there are indices $1 \leq i_1, i_2, \ldots, i_r \leq n_1$ such that for any $1 \leq l \leq n_1$*

$$|\{j \in Q_{i_1,i_2,\ldots,i_r} \mid s_{i_1}[j] \neq s_l[j] \text{ and } s_{i_1}[j] \neq s[j]\}| \leq (\rho_k - \rho_{k+1}) d_b.$$

From Proposition 2, we can immediately obtain the following lemma.

Lemma 1. *For any constant r, there are indices $1 \leq i_1, i_2, \ldots, i_r \leq n_1$ such that for any $s_l \in \mathcal{B}$,*

$$|\{j \in Q_{i_1,i_2,\ldots,i_r} \mid s_{i_1}[j] \neq s_l[j] \text{ and } s_{i_1}[j] \neq s[j]\}| \leq \frac{1}{r-1} d_b.$$

The following corollary will also be useful later.

Corollary 1. *For any $Q \subseteq Q_{i_1,i_2,\ldots,i_r}$ (as in Lemma 1), we have that, for any $s_l \in \mathcal{B}$,*

$$|\{j \in Q \mid s_{i_1}[j] \neq s_l[j] \text{ and } s_{i_1}[j] \neq s[j]\}| \leq \frac{1}{r-1} d_b.$$

Moreover, let $P = \{1, 2, \ldots, L\} - Q$. Then, for any $s_l \in \mathcal{B}$,

$$d(s_l, s_{i_1}|Q) + d(s_l, s|P) \leq \frac{r}{r-1} d_b.$$

Informally, Lemma 1 implies that s_{i_1} is a good approximation of s at positions in $Q_{i_1,i_2,\ldots,i_r}$ for the bad strings, i.e., for any $Q \subseteq Q_{i_1,i_2,\ldots,i_r}$,

$$d(s_l, s_{i_1}|Q) \leq d(s_l, s|Q) + \frac{1}{r-1} d_b.$$

Before we present our key lemma, we need a boosting proposition that, when applied together with Corollary 1, it obtains a better and better solution.

Proposition 3. *Let $Q \subset Q_{i_1,i_2,\ldots,i_r}$ (as in Lemma 1). Consider the index k : $1 \leq k \leq n_1$ and the number $y \geq 0$ such that, for any $s_l \in \mathcal{B}$, $d(s_l, s_{i_1}|Q) + d(s_l, s|P) \leq d(s_k, s_{i_1}|Q) + d(s_k, s|P) = \frac{r}{r-1} d_b - y$, (where $P = \{1, 2, \ldots, L\} - Q$). Then we have*

$$|\{j \in Q^k | s_{i_1}[j] \neq s[j]\}| \leq y,$$

where $Q^k = \{j \in Q : s_{i_1}[j] = s_k[j]\}$.

Proof. Dividing $d(s_k, s_{i_1}|Q)$ into two parts, we have

$$d(s_k, s_{i_1}|Q - Q^k) + d(s_k, s_{i_1}|Q^k) + d(s_k, s|P) = \frac{r}{r-1}d_b - y. \qquad (3)$$

By Corollary 1 and with the further restriction that $s_l[j] = s[j]$, we have that, for any string $s_l \in \mathcal{B}$,

$$|\{j \in Q \mid s_{i_1}[j] \neq s_l[j] \text{ and } s_{i_1}[j] \neq s[j] \text{ and } s_l[j] = s[j]\}| \leq \frac{1}{r-1}d_b.$$

That is, for any string $s_l \in \mathcal{B}$

$$|\{j \in (Q - Q^l)|s_l[j] = s[j]\}| \leq \frac{1}{r-1}d_b. \qquad (4)$$

From (4), there exists β with $\beta \leq 1$ such that for any $s_l \in \mathcal{B}$, $|\{j \in (Q - Q^l) \mid s_l[j] = s[j]\}| \leq |\{j \in (Q - Q^k) \mid s_k[j] = s[j]\}| = \beta\frac{1}{r-1}d_b$. On the other hand, $|\{j \in (Q - Q^k) \mid s_k[j] \neq s_{i_1}[j]\}|$ is no more than $|\{j \in (Q - Q^k) \mid \}|$ which is the sum of $|\{j \in (Q - Q^k) \mid s_k[j] = s[j]\}|$ and $|\{j \in (Q - Q^k) \mid s_k[j] \neq s[j]\}|$. Combining those two formulae, we have

$$d(s_k, s|Q - Q^k) \geq d(s_k, s_{i_1}|Q - Q^k) - \beta\frac{1}{r-1}d_b. \qquad (5)$$

Moreover, combining (3) and (5), we have

$$d(s_k, s|Q - Q^k) + \frac{\beta}{r-1}d_b + d(s_k, s_{i_1}|Q^k) + d(s_k, s|P) \geq \frac{r}{r-1}d_b - y. \qquad (6)$$

Since

$$d(s_k, s|Q - Q^k) + d(s_k, s|Q^k) + d(s_k, s|P) \leq d_b, \qquad (7)$$

(6) implies

$$d(s_k, s|Q^k) \leq d(s_k, s_{i_1}|Q^k) + \frac{\beta - 1}{r-1}d_b + y = \frac{\beta - 1}{r-1}d_b + y \leq y. \qquad (8)$$

(8) is from the facts that $d(s_k, s_{i_1}|Q^k) = 0$ and $\beta \leq 1$. Therefore,

$$|\{j \in Q^k|s_{i_1}[j] \neq s[j]\}| \leq y. \qquad (9)$$

This completes the proof. $\qquad \square$

Here is our main lemma.

Lemma 2. *For any constant r, there exist indices $1 \leq i_1, i_2, \ldots, i_t \leq n_1$, $r \leq t \leq 2r$, and a number $0 \leq y \leq d_b$ such that $|\{j \in Q_{i_1, i_2, \ldots, i_t}|s_{i_1}[j] \neq s[j]\}| \leq y$ (Note that it implies $d(g, s_{i_1}|Q_{i_1, i_2, \ldots, i_t}) \geq d(g, s|Q_{i_1, i_2, \ldots, i_t}) - y$). Moreover, for any $s_l \in \mathcal{B}$, $d(s_l, s_{i_1}|Q_{i_1, i_2, \ldots, i_t}) + d(s_l, s|P_{i_1, i_2, \ldots, i_t}) \leq \frac{r+1}{r-1}d_b - y$, where $P_{i_1, i_2, \ldots, i_t} = \{1, 2, \ldots, L\} - Q_{i_1, i_2, \ldots, i_t}$.*

Proof. We shall repeatedly apply Proposition 3 upto r times for the proof.

We start with $Q = Q_{i_1,i_2,\ldots,i_r}$. By Corollary 1, for any $s_l \in \mathcal{B}$, $d(s_l, s_{i_1}|Q) - |\{j \in Q \,|\, s_{i_1}[j] \neq s_l[j]$ and $s_{i_1}[j] = s[j]\}| \leq \frac{1}{r-1}d_b$. Notice that $|\{j \in Q \,|\, s_{i_1}[j] \neq s_l[j]$ and $s_{i_1}[j] = s[j]\}| \leq d(s_l, s \,|\, Q)$. We get

$$d(s_l, s_{i_1}|Q) + d(s_l, s|P) \leq \frac{1}{r-1}d_b + d(s_l, s) \leq \frac{r}{r-1}d_b.$$

Therefore, there are index k (and denote it by i_{r+1}) and number z^0 ($\frac{r}{r-1}d_b \geq z^0 \geq 0$) such that for any $s_l \in \mathcal{B}$,

$$d(s_l, s_{i_1}|Q) + d(s_l, s|P) \leq d(s_{i_{r+1}}, s_{i_1}|Q) + d(s_{i_{r+1}}, s|P) = \frac{r}{r-1}d_b - z^0.$$

By Proposition 3, $|\{j \in Q_{i_1,i_2,\ldots,i_r,i_{r+1}}|s_{i_1}[j] \neq s[j]\}| \leq z^0$. Now apply Corollary 1 again to the index set $Q' = Q_{i_1,i_2,\ldots,i_{r+1}}$. Then, for any $s_l \in \mathcal{B}$, $d(s_l, s_{i_1}|Q_{i_1,i_2,\ldots,i_{r+1}}) + d(s_l, s|\{1, \ldots, L\} - Q_{i_1,i_2,\ldots,i_{r+1}}) \leq \frac{r}{r-1}d_b$. Again, choose k' (and denote it by i_{r+2} later) and $y' \geq 0$ such that

$$d(s_l, s_{i_1}|Q_{i_1,i_2,\ldots,i_{r+1}}) + d(s_l, s|\{1, \ldots, L\} - Q_{i_1,i_2,\ldots,i_{r+1}}) \leq \frac{r}{r-1}d_b - y'$$

and such that equality holds for $l = k'$.

If $y' \geq z^0 - \frac{1}{r-1}d_b$, then our lemma follows with $t = r + 1$ and $y = z^0$. Therefore, we only need to consider the case $y' \leq z^0 - \frac{1}{r-1}d_b$. Notice that if $z^0 < \frac{1}{r-1}d_b$ we will reach a contradiction here. We should also have $z^0 \geq \frac{1}{r-1}d_b$, or the lemma holds.

Denote by $z^1 = y'$, the conditions of Proposition 3 hold with $y = z^1$ and $Q = Q_{i_1,i_2,\ldots,i_{r+1}}$. Therefore, we are able to repeat the above process. Notice that $z^1 \leq z^0 - \frac{1}{r-1}d_b$ and it is no smaller than $\frac{1}{r-1}d_b$. The same holds each time we repeat the process. Therefore, the process can only be repeated upto r times before we obtain the result as claimed. □

In Lemma 2, we decompose the L positions into two parts, $Q_{i_1,i_2,\ldots i_t}$ and $P_{i_1,i_2,\ldots i_t}$. In $P_{i_1,i_2,\ldots i_t}$, either the linear programming approach in Section 2 or brute-force enumeration method can go through since $|P_{i_1,i_2,\ldots i_t}| \leq td_b$. Details will be given in the next section. Thus, we can get a good approximation solution at positions in $P_{i_1,i_2,\ldots i_t}$.

By Lemma 2, at positions in $Q_{i_1,i_2,\ldots i_t}$, there are at most y positions where the letters for a feasible (optimal) solution are different from the letters in s_{i_1}, this implies that $d(g, s_{i_1}|Q_{i_1,i_2,\ldots,i_t}) \geq d(g, s|Q_{i_1,i_2,\ldots,i_t}) - y$. We may need to carefully choose y positions in $Q_{i_1,i_2,\ldots,i_t}$ in s_{i_1} to change to s'_{i_1} so that the condition $d(g, s'_{i_1}|Q_{i_1,i_2,\ldots,i_t}) \geq d(g, s|Q_{i_1,i_2,\ldots,i_t})$ is satisfied (with negligible error) for each $g \in \mathcal{G}$.

On the other hand, by Lemma 2 again, if any y positions chosen in $Q_{i_1,i_2,\ldots i_t}$ are changed to obtain a string s'_{i_1}, i.e, $d(s_{i_1}, s'_{i_1}) = y$, then $d(s_l, s|P_{i_1,i_2,\ldots,i_t}) + d(s_l, s'_{i_1}|Q_{i_1,i_2,\ldots,i_t}) \leq (1 + \frac{2}{r-1})d_b$. That is, the total error produced for every string in $\mathcal{B}$ will be at most $\frac{2}{r-1}d_b$ no matter which y positions in $Q_{i_1,i_2,\ldots i_t}$ are chosen. Therefore, to choose y positions in $Q_{i_1,i_2,\ldots,i_t}$, we can simply consider the good strings in $\mathcal{G}$ and ignore the strings in $\mathcal{B}$. This dramatically reduces the difficulties of the problem.

The main contribution of Lemma 2 is to transfer the original DSP problem into the problem of how to choose y positions in $Q_{i_1,i_2,\ldots i_t}$ to modify the string s_{i_1} into a string s'_{i_1} such that $d(g, s'_{i_1}|Q_{i_1,i_2,\ldots,i_t}) \geq d(g, s|Q_{i_1,i_2,\ldots,i_t})$. Though we do not know the value of y, we can enumerate all y: $y = 0, 1, \ldots, d_b$ in the algorithm. We will elaborate on this in the next section.

4   Choice of a Distinguishing String

In general, we aim at constructing an approximate solution s'_{i_1} that differs from s_{i_1} at y positions (upto negligible error) in $Q_{i_1,i_2,\ldots,i_t}$. Though we do not know the value of y, we can enumerate all y: $y = 0, 1, \ldots, d_b$ in the algorithm. However, if $d_g \geq (r-1)y$, we can simply use s_{i_1} at all positions in $Q_{i_1,i_2,\ldots,i_t}$. For good strings $g \in \mathcal{G}$, the error at positions in $Q_{i_1,i_2,\ldots,i_t}$ will be $\frac{1}{r-1}d_g$, which will be good enough for a PTAS. For bad strings $s_l \in \mathcal{B}$, the error created by making s_{i_1} as the approximate solution at positions in $Q_{i_1,i_2,\ldots,i_t}$ will be at most $\frac{2}{r-1}d_b$ (again good enough for a PTAS) by Lemma 2. Therefore, we can assume that $d_g < (r-1)y$.

For any string $g_i \in \mathcal{G}$, if $d(g_i, s_{i_1}|Q_{i_1,i_2,\ldots,i_t}) \geq (r-1)y$, then the selected y positions in $Q_{i_1,i_2,\ldots,i_t}$ will cause error by at most $\frac{1}{r-1}d_g$. In fact, let s'_{i_1} be the string obtained from s_{i_1} by changing y positions in $Q_{i_1,i_2,\ldots,i_t}$. Then $d(g_i, s'_{i_1}|Q_{i_1,i_2,\ldots,i_t}) \geq (r-2)y$. Recalling the assumption that $d_g < (r-1)y$, we get for any $g_i \in \mathcal{G}$,

$$d(g_i, s'_{i_1}|Q_{i_1,i_2,\ldots,i_t}) \geq \frac{r-2}{r-1}d_g = (1 - \frac{1}{r-1})d_g.$$

Therefore, we only have to consider the strings $g_i \in \mathcal{G}$ with the restriction that $d(g_i, s_{i_1}|Q_{i_1,i_2,\ldots,i_t}) < (r-1)y$. (This condition can be verified in polynomial time for each $g \in \mathcal{G}$.) In summary, without loss of generality, we have

Assumption 1: For every string $g_i \in \mathcal{G}$, $d(g_i, s_{i_1}|Q_{i_1,i_2,\ldots,i_t}) < (r-1)y$ and $d_g \leq (r-1)y$.

In the rest part of the section, several different methods will be used to carefully select y positions in $Q_{i_1,i_2,\ldots,i_t}$, each dealing with one of the three cases. Let $L' = |Q_{i_1,i_2,\ldots,i_t}|$.

In addition to the constant integer r, we need to introduce another positive number $\delta > 0$. We should first choose r to be sufficiently large but remain a constant. Then we should choose δ to be sufficiently small to achieve the required bound $(1+\epsilon)$ (and $1-\epsilon$) for PTAS. Notice that since δ is chosen after r is fixed, it may be a function of r. However, since r is a constant, we can make δ sufficiently small (as a function of r) and at the same time δ remains a constant. Therefore, in some of the asymptotic notations (Ω and big-O) used in this section, the constant may be a function of r.

Case 1: $L' \leq (r-1)^2 y$ and $y \geq (8\log(n_1 + n_2))/\delta^2$.

In this case, it follows that $d_b \geq y \geq (8\log(n_1 + n_2))/\delta^2$. Moreover, we should prove that $d_b \geq c_b L$ and $d_g \geq c_g L$ where $c_b, c_g = \max\{\frac{1}{4r}, \frac{1}{2(r-1)^2}\}$.

Because $L' \geq y$ and $L' \leq (r-1)^2 y$, it follows that $y = eL'$ for some e : $\frac{1}{(r-1)^2} \leq e \leq 1$. For $d_b \geq y$ and $d_b \leq d_g \leq (r-1)y$ (from Assumption 1), we have $d_b = xy$ where $1 \leq x \leq r-1$. So $d_b = xy = xeL'$ and $\frac{1}{(r-1)^2} \leq xe \leq r-1$. Note that $L = L' + |P|$ and $|P| \leq td_b$ by Proposition 1. We consider two subcases: (a) $|P| \leq 0.5L$: Then $L' \geq 0.5L$. Thus $d_b = x_b eL' \geq 0.5 x_b eL$. Since $d_b \leq L$, we have $d_b = c'L$ for some c' : $\frac{1}{2(r-1)^2} \leq c' \leq 0.5(r-1)$. (b) $|P| > 0.5L$. Then by Proposition 1, we have $d_b \geq |P|/t \geq 0.5L/t \geq L/4r$.

Recall that we also assume that $d_g \geq d_b$ in the problem setting. It follows that $d_g \geq d_b \geq c_g L$. Thus, we can directly apply the linear programming approach in Section 2 to all the L positions.

Case 2: $L' > (r-1)^2 y$ and $y \geq (8\log(n_1 + n_2))/\delta^2$.

We can apply the following simple randomized algorithm:

1. Arbitrarily divide the L' positions into y sets of positions $Y_1, Y_2, \ldots, Y_y$ such that $|Y_1| = |Y_2| = \ldots |Y_{y-1}| = \lfloor \frac{L'}{y} \rfloor$.
2. For each Y_i, independently, randomly select a position in Y_i.
3. Let $Y = \{j_1, j_2, \ldots, j_y\}$ be the y selected positions in Step 2, and y_{i_1} be a string of length $|Q_{i_1, i_2, \ldots, i_t}|$ obtained from s_{i_1} by changing the letters at these positions in Y (to any different letters) such that $d(y_{i_1}, s_{i_1}|Q_{i_1, i_2, \ldots, i_t}) = y$.

This algorithm gives us the needed solution for Case 2. We have the following proposition.

Proposition 4. *The randomized algorithm for Case 2 produces a center string y_{i_1} of length $|Q_{i_1, i_2, \ldots, i_t}|$ with high probability such that*
(i) for any $s_i \in \mathcal{B}$, $d(y_{i_1}, s_i|Q_{i_1, i_2, \ldots, i_t}) + d(s, s_i|P_{i_1, i_2, \ldots, i_t}) \leq \frac{r+1}{r-1}d_b$, and
(ii) for any $g_j \in \mathcal{G}$, $d(y_{i_1}, g_j|Q_{i_1, i_2, \ldots, i_t}) + d(s, g_j|P_{i_1, i_2, \ldots, i_t}) \geq (1 - \frac{2}{r-1} - \delta)d_g$,
where s is a feasible (optimal) distinguishing string.

Case 3: $y < (8\log(n_1 + n_2))/\delta^2$.

For any q good strings $g_{j_1}, g_{j_2}, \ldots, g_{j_q}$, we define

$$R_{j_1, j_2, \ldots, j_q} = \{j \in Q_{i_1, i_2, \ldots, i_t} \mid s_{i_1}[j] \neq g_{j_l}[j] \text{ for at least one } l \in \{1, 2, \ldots, q\}\},$$

and $\bar{R}_{j_1, j_2, \ldots, j_q} = Q_{i_1, i_2, \ldots, i_t} - R_{j_1, j_2, \ldots, j_q}$. From the definition of $R_{j_1, j_2, \ldots, j_q}$ and $\bar{R}_{j_1, j_2, \ldots, j_q}$, we have that for any integers p, q $(p < q)$, $R_{j_1, j_2, \ldots, j_p} \subseteq R_{j_1, j_2, \ldots, j_q}$, $\bar{R}_{j_1, j_2, \ldots, j_q} \subseteq \bar{R}_{j_1, j_2, \ldots, j_p}$. Let $y_1 = d(s_{i_1}, s|R_{j_1, j_2, \ldots, j_r})$, $y_2 = d(s_{i_1}, s|\bar{R}_{j_1, j_2, \ldots, j_r})$. Then $y_1 + y_2 \leq y$.

The following lemma is important in dealing with Case 3.

Lemma 3. *There exist* r *indices* $1 \leq j_1, j_2, \ldots, j_r \leq n_2$ *such that for any* $g \in \mathcal{G}$,

$$|\{j \in \bar{R}_{j_1, j_2, \ldots, j_r} \mid s_{i_1}[j] \neq g[j] \text{ and } s_{i_1}[j] \neq s[j]\}| \leq \frac{y}{r}.$$

Now, we consider the two parts $R_{j_1, j_2, \ldots, j_r}$ and $\bar{R}_{j_1, j_2, \ldots, j_r}$ respectively. Assumption 1 and the definition of $R_{j_1, j_2, \ldots, j_r}$ ensure that

$$\begin{aligned}
|R_{j_1, j_2, \ldots, j_r}| &= |\{j \in Q_{i_1, i_2, \ldots, i_t} | s_{i_1}[j] \neq g_{j_l}[j] \text{ for at least one } l \in \{1, 2, \ldots, r\}\}| \\
&\leq \textstyle\sum_{l=1}^{r} |\{j \in Q_{i_1, i_2, \ldots, i_t} \mid s_{i_1}[j] \neq g_{j_l}[j]\}| \\
&= \textstyle\sum_{l=1}^{r} d(s_{i_1}, g_{j_l}) |Q_{i_1, i_2, \ldots, i_t}) \leq r(r-1)y < r^2 y.
\end{aligned}$$

Keep in mind that we are dealing with the case $y \leq (8\log(n_1+n_2))/\delta^2$. Thus, we have $|R_{j_1, j_2, \ldots, j_r}| \leq O(\log n)$. We can try all possible ways in polynomial time to achieve the optimal distinguishing string s at the positions in $R_{j_1, j_2, \ldots, j_r}$. So we can assume that we know the optimal distinguishing string s at the positions in $R_{j_1, j_2, \ldots, j_r}$.

Now, let us focus on the positions in $\bar{R}_{j_1, j_2, \ldots, j_r}$. For any $g \in \mathcal{G}$, define $Set(g, s_{i_1})$ be the set of positions in $\bar{R}_{j_1, j_2, \ldots, j_r}$, where g and s_{i_1} do not agree. From Lemma 3, we know that there exists a set of y_2 positions $K_{y_2} = \{k_j \in \bar{R}_{j_1, j_2, \ldots, j_r} \mid s_{i_1}[k_j] \neq s[k_j]\}$ such that for any $g \in \mathcal{G}$, $|K_{y_2} \cap Set(g, s_{i_1})| \leq \frac{y}{r}$. Though we do not know the exact value of y_2, again we can guess it in $O(y)$ time. Thus, we can assume that y_2 is known.

Let $s_{i_1}(K_{y_2})$ be the string obtained from s_{i_1} by changing the letters at the positions in K_{y_2} to any different letters. From Lemma 3, we can see that for any $g \in \mathcal{G}$,

$$d(g, s_{i_1}(K_{y_2}) | \bar{R}_{j_1, j_2, \ldots, j_r}) \geq d(g, s | \bar{R}_{j_1, j_2, \ldots, j_r}) - \frac{1}{r} y. \tag{10}$$

Now, the only task left is to select y_2 positions in $\bar{R}_{j_1, j_2, \ldots, j_r}$ to approximate the set K_{y_2}. Recall that $|\{j \in \bar{R}_{j_1, j_2, \ldots, j_r} \mid s_{i_1}[j] \neq s[j]\}| = y_2$. From which we have that for every good string $g \in \mathcal{G}$,

$$d(g, s_{i_1} | \bar{R}_{j_1, j_2, \ldots, j_r}) - d(g, s | \bar{R}_{j_1, j_2, \ldots, j_r}) \geq -y_2. \tag{11}$$

Next we are going to show that we can find a set K'_{y_2} of y_2 positions in $\bar{R}_{j_1, j_2, \ldots, j_r}$ in polynomial time such that for any $g \in \mathcal{G}$, $|K'_{y_2} \cap Set(g, s_{i_1})| \leq \frac{y}{r}$. If this is true, by changing the letters of s_{i_1} on the positions in K'_{y_2}, the distance between $s_{i_1}(K'_{y_2})$ and g has to increase at least $y_2 - \frac{y}{r}$. To do so, let $m \leq |\bar{R}_{j_1, j_2, \ldots, j_r}|$ be an integer, and $\bar{R}_{j_1, j_2, \ldots, j_r}(m) \subseteq \bar{R}_{j_1, j_2, \ldots, j_r}$ be any subset of $\bar{R}_{j_1, j_2, \ldots, j_r}$ with m elements.

The following two lemmas are the keys to solve the problem.

Lemma 4. *If* $m \geq 2^{r^2} \times y \times (n+1)^{\frac{r}{y}} + \frac{y}{r}$, *then there exist a set* $K'_{y_2} \subseteq \bar{R}_{j_1, j_2, \ldots, j_r}(m)$ *of* y_2 *positions such that for any* $g \in \mathcal{G}$

$$|K'_{y_2} \cap Set(g, s_{i_1})| \leq \frac{y}{r}.$$

Lemma 4 says that a good solution exists at the positions in $\bar{R}_{j_1,j_2,\ldots,j_r}(m)$ if $m = 2^{r^2} \times y \times (n+1)^{\frac{r}{y}} + \frac{y}{r}$. On the other hand, the following lemma shows that $m = 2^{r^2} \times y \times (n+1)^{\frac{r}{y}} + \frac{y}{r}$ is not too big so that we can find such a good solution in polynomial time by looking at all $C_m^{y_2}$ possible choices.

Lemma 5. *If* $m = 2^{r^2} \times y \times (n+1)^{\frac{r}{y}} + \frac{y}{r}$ *and* $y \leq (8\log(n_1 + n_2))/\delta^2$, *then* C_m^y *is polynomial in* n.

Algorithm gdistString

Input $\mathcal{B} = \{s_1, s_2, \ldots, s_{n_1}\}$ and $\mathcal{G} = \{g_1, g_2, \ldots, g_{n_2}\}$.
Output   a center string s of length L. .
1.   **for** each t-element subset $\{s_{i_1}, s_{i_2}, \ldots, s_{i_t}\}$ of the n_1 input strings in $\mathcal{B}$
      where $r \leq t \leq 2r$ **do**
      (a) $Q = \{1 \leq j \leq L \mid s_{i_1}[j] = s_{i_2}[j] = \ldots = s_{i_t}[j]\}$, $P = \{1, 2, \ldots, L\} - Q$.
      (b) For the positions in $Q_{i_1,i_2,\ldots,i_t}$, use the methods described in Case 1, Case 2, and Case 3 to obtain a string of length $|Q_{i_1,i_2,\ldots,i_t}|$. Notice that the algorithm described in Case 1 produces directly a string s' of length L which is a PTAS.
      (c) Solve the optimization problem at the positions in $|P|$ using either linear programming formulation (randomized rounding approach) (when $|P| \geq O(\log n)$) or exhaustive search method (when $|P| \leq O(\log n)$) to get an approximate solution x of length $|P|$.
2.   Output the best solution obtained in Step 1.

Fig. 1. Algorithm for DSP.

The complete algorithm is given in Fig 1. Note that all the randomized algorithms we have used can be derandomized by the standard methods in [10]. Therefore, we can conclude that

Theorem 3. *There is a PTAS for the DSP.*

5   The Distinguishing Substring Selection Problem

In this section, we present the algorithm for the distinguishing substring selection problem. The idea is to combine the sampling technique in [9] with the algorithm for DSP. The difficulty here is that for each $s_i \in \mathcal{B}$, we do not know the substring t_i of s_i. The sampling approach in choices of r substrings of length L from $\mathcal{B}$, we can assume that $t_{i_1}, t_{i_2}, \ldots, t_{i_r}$ are the r substrings of length L that satisfy Lemma 1 by replacing s_l with t_l and s_{i_j} with t_{i_j}. Let Q be the set of positions where $t_{i_1}, t_{i_2}, \ldots, t_{i_r}$ agree and $P = \{1, 2, \ldots, L\} - Q$. By Lemma 1, $t_{i_1}|Q$ is a good approximation to $s|Q$. However, we do not know the letters at positions in P. So, we randomly pick $O(\log(mn))$ positions from P where m is the length of bad strings. Suppose the multiset of these random positions is R. By trying all length $|R|$ strings, we can assume that we know $s|R$. Then for each $1 \le i \le n_1$, we find the substring t_i' from s_i such that $f(t_i') = d(s, t_i'|R) \times \frac{|P|}{|R|} + d(t_{i_1}, t_i'|Q)$ is minimized. Let s be the optimal distinguishing string. t_i denotes the substring of s_i that is closest to s. Let s^* be a string such that $s^*|P = s|P$ and $s^*|Q = t_{i_1}|Q$. [9] shows that

Fact 1 With probability $1 - ((nm)^{-2} + (nm)^{-\frac{4}{3}})$, $d(s^*, t_i') \le d(s^*, t_i) + 2\epsilon|P|$ for all $1 \le i \le n$.

After obtaining a t_i' of s_i for every $s_i \in \mathcal{B}$, we have the DSP problem which can be solved by using the algorithms developed in Section 3.

Theorem 4. *There is a polynomial time approximation scheme that takes $\epsilon > 0$ as part of the input and computes a center string s of length L such that for every $s_i \in \mathcal{B}$, there is a length-L substring t_i of s_i with $d(s_i, s) \le (1 + \epsilon)d_b$ and for every $g_i \in \mathcal{G}$, $d(g_i, s) \ge (1 - \epsilon)d_g$, if a solution exists.*

References

1. A. Ben-Dor, G. Lancia, J. Perone, and R. Ravi, Banishing bias from consensus sequences, *Proc. 8th Ann. Combinatorial Pattern Matching Conf.*, pp. 247-261, 1997.
2. J. Dopazo, A. Rodríguez, J. C. Sáiz, and F. Sobrino, Design of primers for PCR amplification of highly variable genomes, *CABIOS*, 9(1993), 123-125.
3. M. Frances, A. Litman, On covering problems of codes, *Theor. Comput. Syst.*, 30(1997), 113-119.
4. L. Gąsieniec, J. Jansson, and A. Lingas, Efficient approximation algorithms for the Hamming center problem, *Proc. 10th ACM-SIAM Symp. on Discrete Algorithms*, pp. S905-S906, 1999.
5. M. Ito, K. Shimizu, M. Nakanishi, and A. Hashimoto, Polynominal-time algorithms for computing characteristic strings, *Proc. 5th Annual Symposium on Combinatorial Pattern Matching*, pp. 274-288, (1994).
6. K. Lucas, M. Busch, S. Mössinger and J.A. Thompson, An improved microcomputer program for finding gene- or gene family-specific oligonucleotides suitable as primers for polymerase chain reactions or as probes, *CABIOS*, 7(1991), 525-529.
7. K. Lanctot, M. Li, B. Ma, S. Wang, and L. Zhang, Distinguishing string selection problems, *SODA '99*, pp. 633-642..
8. Ming Li, Bin Ma,and Lusheng Wang, "Finding similar regions in many strings", the 31th ACM Symp. on Theory of Computing, pp. 473-482, 1999.
9. B. Ma, A polynomial time approximation scheme for the closest substring problem, *Proc. 11th Annual Symposium on Combinatorial Pattern Matching*, pp. 99-107, Montreal, (2000).
10. R. Motwani and P. Raghavan, *Randomized Algorithms*, Cambridge Univ. Press, 1995.
11. C.H. Papadimitriou and M. Yannakakis, On the approximability of trade-offs and optimal access of web sources, FOCS00 , pp. 86-92, 2000.
12. V. Proutski and E. C. Holme, Primer Master: a new program for the design and analysis of PCR primers, *CABIOS*, 12(1996), 253-255.

On the Theory of One-Step Rewriting in Trace Monoids[*]

Dietrich Kuske[1] and Markus Lohrey[2]

[1] Department of Mathematics and Computer Science
University of Leicester, LEICESTER, LE1 7RH, UK
[2] Universität Stuttgart, Institut für Informatik,
Breitwiesenstr. 20-22, 70565 Stuttgart, Germany
D.Kuske@mcs.le.ac.uk, lohrey@informatik.uni-stuttgart.de

Abstract. We prove that the first-order theory of the one-step rewriting relation associated with a trace rewriting system is decidable and give a nonelementary lower bound for the complexity. The decidability extends known results on semi-Thue systems but our proofs use new methods; these new methods yield the decidability of local properties expressed in first-order logic augmented by modulo-counting quantifiers. Using the main decidability result, we describe a class of trace rewriting systems for which the confluence problem is decidable. The complete proofs can be found in the Technical Report [14].

1 Introduction

Rewriting systems received a lot of attention in mathematics and theoretical computer science and are still an active field of research. Historically, rewriting systems were introduced to solve word problems in certain structures [28]. By the work of Markov [18] and Post [24], this hope vanished as they showed that there exist fixed semi-Thue systems with an undecidable word problem. Despite this result, there are plenty of rewriting systems with a decidable word problem, the most famous class being that of confluent and terminating systems. By Newman's Lemma, confluence can be decided for terminating semi-Thue systems as well as for terminating term rewriting systems. In general, both confluence and termination are undecidable properties of a semi-Thue system. A large deal of research tries to identify sufficient conditions for confluence/termination of rewriting systems (cf. [26]), or to describe classes of rewriting systems where confluence/termination is decidable.

These two properties which are in the heart of research in this area are typical second-order properties of the rewrite graph: its nodes are the structures that are rewritten (e.g., words in case of a semi-Thue system or terms in case of a term rewriting system), and directed edges indicate that one such structure can be rewritten into the other in one step. In order to define confluence and termination, one needs to quantify over paths in this graph. Hence the monadic second-order theory of rewrite graphs is in general undecidable. The situation changes for semi-Thue systems when one considers the first-order theory: the edges of the rewrite graph of a semi-Thue system can be described by

[*] This work was partly done while the second author was on leave at IRISA, Campus de Beaulieu, 35042 Rennes Cedex, France and supported by the INRIA cooperative research action FISC.

P. Widmayer et al. (Eds.): ICALP 2002, LNCS 2380, pp. 752–763, 2002.
© Springer-Verlag Berlin Heidelberg 2002

two-tape automata that move their heads synchronously on both tapes.[1] Using the well known closure properties of regular sets, the decidability of the first-order theory of these graphs follows [5,13]. This result also holds for rewrite graphs of ground term rewriting systems [5], but not for term rewriting systems in general [29]. Another result in this direction is the decidability of the monadic second-order theory of the rewrite graph of a prefix semi-Thue system [2] (a prefix semi-Thue system is a semi-Thue system where only prefixes can be rewritten). In particular confluence and termination are decidable for prefix semi-Thue systems.

This paper investigates the first-order theory of the rewrite graph of a trace rewriting system. Cartier and Foata [1] investigated the combinatorics of free partially commutative monoids that became later known as trace monoids. Mazurkiewicz [20] introduced them into computer science. They form a mathematically sound model for the concurrent behaviour of systems of high abstraction. Since trace monoids are a generalization of free monoids, it was tempting to extend the investigation of free monoids to free partially commutative monoids. This resulted, e.g., in the extensive consideration of recognizable and rational trace languages (cf. [9] for a collection of surveys on this field), trace equations [10,19,8], and trace rewriting systems [6,7,16,17].

Our main result states that for any finite trace rewriting system, the first-order theory of the associated rewrite graph is decidable. Because of the non-local effects of trace rewriting,[2] the automata-theoretic techniques from Dauchet and Tison [5] and Jacquemard [13] are not applicable here and we had to search for other ideas. The first is an application of Gaifman's locality theorem: the validity of a first-order sentence in a structure S depends on first-order properties of spheres around elements of S. Since this theorem is effective, we were left with the question how to describe the set of traces that are centers of an r-sphere satisfying a given first-order formula. Our second idea is that the r-sphere around a trace can be described in the dependence graph of this trace by a sentence of monadic second-order logic. Note that this logic does not speak about the infinite rewrite graph, but about a single finite dependence graph. We show that this is indeed effectively possible. Hence, by a result of Thomas [27], this implies the recognizability of the set of traces that are centers of an r-sphere satisfying a given first-order formula. Taking these two ideas together, we obtain that the first-order theory of the graph of any trace rewriting system is decidable.

We actually show a more general result since we do not only consider trace rewriting systems, but scattered rewriting systems. The idea is that of a parallel rewrite step where the intermediate factors of a trace have to satisfy some recognizable constraints and can be permuted as long as they are independent in the trace monoid.

As mentioned above, the first step in our decidability proof is an application of Gaifman's Theorem. To the knowledge of the authors, all known translations of a first-order sentence into a Boolean combination of local sentences are not elementary, thus our decision procedure is far from efficient. We also show that one cannot avoid this nonelementary complexity. To this aim, we construct a trace rewrite graph whose first-order theory is not elementary. Thus, our use of Gaifman's translation does not lead to

---

[1] As opposed to rational graphs where the movement is asynchronous.

[2] With a and c the only independent letters, one can, e.g., rewrite $a^n b c^m$ into $c^m b a^n$ in just two steps using the rules $abc \to ac$ and $ca \to cba$.

an unreasonable inefficiency. We actually show a slightly stronger result, namely that the set of valid local sentences for a fixed radius is not elementary. In other words, the complexity of the decision question is already present when restricting to local sentences. This nonelementary lower bound is shown for a nontrivial independence alphabet and the proof does not carry over to semi-Thue systems. We show a lower bound of doubly exponential nondeterministic time for this problem. Again this lower bound holds for local sentences for a fixed radius.

In the last section, we return to the confluence problem for trace rewriting systems. For terminating rewriting systems, confluence and local confluence are equivalent. The problem with trace rewriting systems is that there can be infinitely many critical pairs which makes it impossible to check all of them in turn [6,7]. Even worse, by [22], it is undecidable whether a length-reducing trace rewriting system is confluent. We describe classes of terminating trace rewriting systems for which confluence is decidable. The classes of trace rewriting systems we consider in this last section ensure that local confluence is effectively expressible by a sentence of first-order logic (which is not the case in general). This then allows to apply our main result on the decidability of these first-order properties and therefore the decidability of confluence for these classes when restricted to terminating systems.

2 Rewriting in Trace Monoids

2.1 Trace Monoids and Recognizable Trace Languages

In the following we introduce some notions from trace theory, see [9] for more details. An *independence relation* on an alphabet Σ is an irreflexive and symmetric relation $I \subseteq \Sigma \times \Sigma$, the complementary relation $D = (\Sigma \times \Sigma) \backslash I$ is called a *dependence relation*. The pair (Σ, I) (resp. (Σ, D)) is called an *independence alphabet* (resp. a *dependence alphabet*). A *dependence graph* or *trace* is a triple (V, E, λ) where (V, E) is a directed acyclic and finite graph (possibly empty) and $\lambda : V \to \Sigma$ is a labeling function such that, for all $p, q \in V$ with $p \neq q$, we have

$$(\lambda(p), \lambda(q)) \in D \text{ if and only if } (p, q) \in E \text{ or } (q, p) \in E.$$

We will identify traces that are isomorphic as labeled graphs. The set of all (isomorphism classes of) traces is denoted by $\mathbb{M} = \mathbb{M}(\Sigma, I)$. For a trace $t = (V, E, \lambda)$, let $\mathrm{alph}(t) = \lambda(V)$. The independence relation I can be lifted to $\mathbb{M}$ by setting $(u, v) \in I$ if $\mathrm{alph}(u) \times \mathrm{alph}(v) \subseteq I$. On the set $\mathbb{M}$, one defines a binary operation $\circ$ by

$$(V_1, E_1, \lambda_1) \circ (V_2, E_2, \lambda_2) = (V_1 \dot{\cup} V_2, E_1 \cup E \cup E_2, \lambda_1 \cup \lambda_2)$$

where $E = \{(p_1, p_2) \in V_1 \times V_2 \mid (\lambda(p_1), \lambda(p_2)) \in D\}$. Then $(\mathbb{M}, \circ)$ becomes a monoid, its neutral element is the empty trace 1. If $I = \emptyset$ then $\mathbb{M}$ is isomorphic to the free monoid Σ^*. On the other extreme if $D = \mathrm{Id}_\Sigma$, then $\mathbb{M}$ is isomorphic to the free commutative monoid $\mathbb{N}^{|\Sigma|}$. We will identify the letter $a \in \Sigma$ with the singleton trace whose node is labeled by a. In this sense, a word $w = a_1 a_2 \ldots a_n \in \Sigma^*$ defines the trace $[w]_I = a_1 \circ a_2 \circ \cdots \circ a_n$. We write $u \equiv_I v$ for two words u and v if $[u]_I = [v]_I$.

This relation is the congruence on the free monoid Σ^* generated by all pairs $ab \equiv_I ba$ for $(a, b) \in I$. In the following for $u, v \in \mathbb{M}$ we will also write uv instead of $u \circ v$.

Let $t = (V, E, \lambda)$ be a trace. Then the transitive reflexive closure E^* of E is a partial order. Let $U \subseteq V$ such that, for $p_1, p_2 \in U$ and $q \in V$ with $(p_1, q), (q, p_2) \in E^*$ it holds $q \in U$ (i.e., U is convex w.r.t. E^*). Then $u = t\lceil_U \in \mathbb{M}$ is a trace and, furthermore, there exist $t_1, t_2 \in \mathbb{M}$ with $t = t_1 u t_2$. Vice versa if t can be factorized as $t = t_1 u t_2$ then there exists a convex $U \subseteq V$ such that $t\lceil_U = u$.

A set $L \subseteq \mathbb{M}$ is called *recognizable* if there exists a morphism $h : \mathbb{M} \to Q$ from $(\mathbb{M}, \circ)$ into a finite monoid Q and a subset $F \subseteq Q$ such that $L = h^{-1}(F)$. The set of all recognizable subsets of $\mathbb{M}$ is denoted by $\mathrm{REC}(\mathbb{M})$. It is well-known that $\mathrm{REC}(\mathbb{M})$ is effectively closed under Boolean operations and concatenation of languages.[3] Furthermore emptiness and finiteness are decidable for recognizable trace languages, and if $L \in \mathrm{REC}(\mathbb{M})$ is finite then its elements can be calculated effectively.

2.2   Scattered Trace Rewriting

Let us fix a countable infinite set Ω of (first-order) variables ranging over $\mathbb{M}$ for the rest of this paper. In order to make notations more succinct, we associate with every first-order variable $x \in \Omega$ a recognizable trace language $L(x)$. We assume that for every $L \in \mathrm{REC}(\mathbb{M})$ there is a countably infinite supply of variables $x \in \Omega$ with $L(x) = L$. The mapping $x \mapsto L(x)$ will be fixed for the rest of this paper. The intuition of this mapping is that the variable $x \in \Omega$ will be restricted to its associated set $L(x)$. On the set Ω we define an independence relation J by

$$J = \{(x, y) \mid \forall t \in L(x)\, \forall u \in L(y) : (t, u) \in I\} \setminus \mathrm{Id}_\Omega.$$

Let $x_1, \ldots, x_m$ be pairwise different variables from Ω. A *pattern* S over $\mathbb{M}$ and the variables $x_1, \ldots, x_m$ is a sequence $x_{\pi(1)}\, t_1\, x_{\pi(2)}\, t_2 \cdots x_{\pi(m)}$ where $t_1, \ldots, t_{m-1} \in \mathbb{M}$ and π is a permutation of $\{1, 2, \ldots, m\}$. We define $\mathrm{type}(S) = x_{\pi(1)} x_{\pi(2)} \cdots x_{\pi(m)}$. A pattern S over the variables $x_1, \ldots, x_m$ is also denoted by $S(x_1, \ldots, x_m)$. Note that in a pattern a variable occurs precisely once, but the variables may occur in an arbitrary order. If the variable x_i evaluates to $u_i \in \mathbb{M}$, $1 \leq i \leq m$, then the trace $S(u_1, \ldots, u_m) \in \mathbb{M}$ is defined in the obvious way. A *scattered rewrite rule* over $\mathbb{M}$ and the variables $x_1, \ldots, x_m$ is a pair $(S(x_1, \ldots, x_m), T(x_1, \ldots, x_m))$ of patterns over $\mathbb{M}$ such that $\mathrm{type}(S) \equiv_J \mathrm{type}(T)$. The set of all scattered rewrite rules over $\mathbb{M}$ is denoted by $\mathbb{S}$. A *scattered rewriting system* over $\mathbb{M}$ is a finite subset of $\mathbb{S}$. For a scattered rewrite rule $\rho = (S(x_1, \ldots, x_m), T(x_1, \ldots, x_m))$ and $s, t \in \mathbb{M}$ we write $s \to_\rho t$ if there exist traces $u_i \in L(x_i)$ such that $s = S(u_1, \ldots, u_m)$ and $t = T(u_1, \ldots, u_m)$. For a scattered rewriting system $\mathcal{R}$ we write $s \to_\mathcal{R} t$ if $s \to_\rho t$ for some $\rho \in \mathcal{R}$.

An important special case of scattered rewriting systems are *trace rewriting* systems [6,7], i.e., scattered rewriting systems whose rules are all of the form $(x\ell y, xry)$ for $x, y \in \Omega$, $\ell, r \in \mathbb{M}$ such that $L(x) = L(y) = \mathbb{M}$. If $I = \emptyset$, i.e., $\mathbb{M} \simeq \Sigma^*$, then a trace rewriting system over $\mathbb{M}$ is also called a *semi-Thue system* over Σ^*. On the other hand if $I = (\Sigma \times \Sigma) \setminus \mathrm{Id}_\Sigma$, i.e., $\mathbb{M} \simeq \mathbb{N}^{|\Sigma|}$, then a trace rewriting system over $\mathbb{M}$ is also called

---

[3] In these effectiveness statements, a recognizable language is given as a triple (Q, F, h).

a *vector replacement system* over $\mathbb{N}^{|\Sigma|}$. A rule $(x\ell y, xry)$ of a trace rewriting system will be briefly denoted by (ℓ, r).

In this paper we will be concerned with the first-order theory of the structure

$$\mathcal{M} = (\mathbb{M}, (L)_{L \in \mathrm{REC}(\mathbb{M})}, (\to_\rho)_{\rho \in \mathbb{S}})$$

and its reducts $\mathcal{M}_{\mathcal{R}} = (\mathbb{M}, (L)_{L \in \mathrm{REC}(\mathbb{M})}, (\to_\rho)_{\rho \in \mathcal{R}})$, where $\mathcal{R}$ is a scattered rewriting system. Each recognizable language $L \in \mathrm{REC}(\mathbb{M})$ is put into $\mathcal{M}$ as a unary predicate. Furthermore, $\mathcal{M}$ contains all binary relations $\to_\rho \subseteq \mathbb{M} \times \mathbb{M}$ for $\rho \in \mathbb{S}$, while $\mathcal{M}_{\mathcal{R}}$ contains only those relations $\to_\rho$ for $\rho \in \mathcal{R}$.

Formally, trace rewriting systems are more general than semi-Thue and vector replacement systems since they work modulo a partial commutation. Even more, there are trace rewriting systems $\mathcal{R}$ such that the graph $(\mathbb{M}, \to_{\mathcal{R}})$ is not isomorphic to the graph $(\Sigma^*, \to_{\mathcal{S}})$ for any semi-Thue system $\mathcal{S}$. To see this let us introduce some notions concerning confluence. We say that the trace rewriting system $\mathcal{R}$ is *confluent* (resp. *locally confluent*) if for all $t, t_1, t_2 \in \mathbb{M}$ with $t \xrightarrow{*}_{\mathcal{R}} t_1$ and $t \xrightarrow{*}_{\mathcal{R}} t_2$ (resp. $t \to_{\mathcal{R}} t_1$ and $t \to_{\mathcal{R}} t_2$) there exists $u \in \mathbb{M}$ with $t_1 \xrightarrow{*}_{\mathcal{R}} u$ and $t_2 \xrightarrow{*}_{\mathcal{R}} u$. These two notions are standard. The following notion seems to be new: We say that $\mathcal{R}$ is α-*confluent*, where $\alpha \in \mathbb{N}$, if for all $t, t_1, t_2 \in \mathbb{M}$ with $t \to_{\mathcal{R}} t_1$ and $t \to_{\mathcal{R}} t_2$ there exists $u \in \mathbb{M}$ with $t_1 \xrightarrow{\leq \alpha}_{\mathcal{R}} u$ and $t_2 \xrightarrow{\leq \alpha}_{\mathcal{R}} u$ (where $t_i \xrightarrow{\leq \alpha}_{\mathcal{R}} u$ denotes that u can be obtained from t_i in at most α steps). Using critical pairs one can show that any locally confluent semi-Thue system is α-confluent for some $\alpha \in \mathbb{N}$. On contrast, the trace rewriting system $\{(ab, 1), (ba, 1), (c, 1)\}$ over the trace monoid $\mathbb{M}(\{a, b, c\}, \{(a, c), (c, a)\})$ is locally confluent [17] but not α-confluent (consider $c^n b \leftarrow bac^n b \equiv_I bc^n ab \to bc^n$ for $n > \alpha$).

2.3   The Main Result

The main result of this paper states that the first-order theory of any structure $\mathcal{M}$ is decidable. Since our decision procedure is uniform in the underlying alphabet, we obtain

Theorem 2.1. *There exists an algorithm that, on input of an independence alphabet (Σ, I) and a first-order sentence φ over the signature of the structure $\mathcal{M}$, decides whether $\mathcal{M} \models \varphi$.*

Note that $\mathcal{M} \models \varphi$ if and only if $\mathcal{M}_{\mathcal{R}} \models \varphi$ where $\mathcal{R} \subseteq \mathbb{S}$ is finite and contains the set of rewrite rules mentioned in φ. Thus, in order to prove Theorem 2.1, it suffices to prove the decidability of the first-order theory of $\mathcal{M}_{\mathcal{R}}$ for any scattered rewriting system $\mathcal{R}$.

An immediate corollary of Theorem 2.1 is that the rewrite graph of a trace rewriting system has a decidable first-order theory. This generalizes the corresponding result for semi-Thue systems in [5,13], furthermore this generalization is a strict one by the observation at the end of the previous section. The basic fact used in [5,13] is that for a semi-Thue system $\mathcal{R}$ the relation $\to_{\mathcal{R}}$ is a synchronized rational transduction [11]. While these methods can be generalized to work for the case that $\mathbb{M}$ is a direct product of free monoids, there seems to be no way to generalize them to arbitrary trace monoids. Hence in our proof of Theorem 2.1 we will follow a completely different and new strategy.

Let us close this section with some remarks on the limitations of our results. First, if one omits the restriction $\mathrm{type}(S) \equiv_J \mathrm{type}(T)$ for scattered rewrite rules (S, T), the theory of $\mathcal{M}$ becomes undecidable [14, Thm. 3.6]. A prefix rewriting system over $\mathbb{M}$ is a scattered rewriting system $\mathcal{R}$ where all rules have the form $(x\ell y, xry)$ with $L(x) = \{1\}$ and $L(y) = \mathbb{M}$ (we abbreviate this rule by (ℓ, r)). Based on results from [4], Caucal has shown in [2] that for a prefix rewriting system $\mathcal{R}$ over a free monoid Σ^* the monadic second-order theory of the graph $(\Sigma^*, \rightarrow_{\mathcal{R}})$ is decidable (this does not hold for semi-Thue systems). In contrast to this, let $\mathcal{R}$ be the prefix rewriting system $\{(1, a), (1, b)\}$ over the free commutative monoid $\mathbb{M}(\{a, b\}, \{(a, b), (b, a)\})$. The graph $(\mathbb{M}, \rightarrow_{\mathcal{R}})$ is a two-dimensional grid which has an undecidable monadic second-order theory. Hence, in general, the monadic second-order theory of the relation $\rightarrow_{\mathcal{R}}$ for a prefix rewriting system $\mathcal{R}$ is undecidable.

3 Decidability of Scattered Rewriting

In this section we will prove Theorem 2.1. It is important to note that all statements are effective although we do not state this fact explicitly in order to smoothen the formulations. Let $\mathcal{R}$ be a fixed scattered rewriting system over the trace monoid $\mathbb{M}$.

3.1 Reduction to Local Properties

The main tool in this section is Gaifman's locality theorem for first-order logic [12]. For two traces $s, t \in \mathbb{M}$, let $d_{\mathcal{R}}(s, t)$ denote the length of a shortest undirected path from s to t in the graph $(\mathbb{M}, \rightarrow_{\mathcal{R}})$. For $r \geq 0$ and $t \in \mathbb{M}$, the r-*sphere* around t is $S(r, t) = \{s \in \mathbb{M} \mid d_{\mathcal{R}}(s, t) \leq r\}$. The r-sphere around t is definable in $\mathcal{M}_{\mathcal{R}}$, i.e., there exists a first-order formula with two free variables expressing $d_{\mathcal{R}}(x, y) \leq r$. Now let φ be a first-order formula in the signature of $\mathcal{M}_{\mathcal{R}}$. Then the first-order formula $\varphi^{S(r,x)}$ results from φ by relativizing all quantifiers to $S(r, x)$. It can be defined inductively, in particular $(\exists y \phi)^{S(r,x)} \equiv \exists y \{d_{\mathcal{R}}(x, y) \leq r \ \wedge \ \phi^{S(r,x)}\}$. Now Gaifman's theorem applied to the structure $\mathcal{M}_{\mathcal{R}}$ states the following:

Theorem 3.1. *For a given first-order sentence ϕ over the signature of $\mathcal{M}_{\mathcal{R}}$ one can effectively compute a natural number $r \geq 1$ and a Boolean combination $\widehat{\phi}$ of sentences of the form*

$$\exists x_1 \cdots \exists x_m \left\{ \bigwedge_{1 \leq i < j \leq m} d_{\mathcal{R}}(x_i, x_j) > 2r \ \wedge \ \bigwedge_{1 \leq i \leq m} \psi^{S(r, x_i)}(x_i) \right\}$$

where ψ is a first-order sentence over the signature of $\mathcal{M}_{\mathcal{R}}$ such that $\mathcal{M}_{\mathcal{R}} \models \phi$ if and only if $\mathcal{M}_{\mathcal{R}} \models \widehat{\phi}$.

In order to use Gaifman's locality theorem for decidability purposes, we will need a "useful" description of the set of all traces $t \in \mathbb{M}$ with $\mathcal{M}_{\mathcal{R}} \models \varphi^{S(r,x)}(t)$. We will show that this set is recognizable and that it is indeed a "useful" description.

3.2 Reduction to 1-Spheres

The aim of this section is to show that by enlarging the set $\mathcal{R}$ it suffices to restrict to the case $r = 1$ in Theorem 3.1. The basic idea is the following: let s, t, v be traces and (S, T), (U, V) be scattered rewrite rules such that $s \to_{(S,T)} t \to_{(U,V)} v$. Then the trace t can be factorized in two ways, one according to the pattern T and one according to the pattern U. Using Levi's Lemma for traces (see e.g. [9]), one can then refine the scattered rewrite rules (S, T) and (U, V) to (S', T') and (U', V') such that $s \to_{(S',T')} t \to_{(U',V')} v$, and the two factorizations of t according to T' and to U' are actually the same. Then also (S', V') is a scattered rewrite rule and $s \to_{(S',V')} v$. Any pair of rewrite rules (S, T) and (U, V) from $\mathcal{R}$ gives rise to a finite set of refinements. Using this process of refinement inductively, one obtains

Lemma 3.2. *For $r \geq 0$, there exists a scattered rewriting system $\mathcal{R}_r$ over $\mathbb{M}$ such that $\mathcal{R} \subseteq \mathcal{R}_r$ and for all $s, t \in \mathbb{M}$ it holds $d_{\mathcal{R}}(s, t) \leq r$ if and only if $s \to_{\mathcal{R}_r} t$.*

It should be noted that if $\mathcal{R}$ is a trace rewriting system then the system $\mathcal{R}_r$ is in general not a trace rewriting systems. This was one of the reasons for generalizing trace rewriting systems to scattered rewriting systems.

3.3 Internalizing the 1-Sphere

As a major tool in the further consideration we will use monadic second-order logic (MSO logic) over dependence graphs. Formulae in this logic are interpreted over dependence graphs (V, E, λ). There exist first-order variables $x, y, z, \ldots$ ranging over elements of V and second-order variables $X, Y, Z, \ldots$ ranging over subsets of V. Atomic formulae are of the form $Q_a(x)$, $x \preceq y$, and $x \in X$ where x and y are first-order variables, X is a second-order variable, and Q_a is a unary relation symbol for every $a \in \Sigma$. The interpretation of $Q_a(x)$ is $\lambda(x) = a$ whereas $x \preceq y$ is interpreted as $(x, y) \in E^*$. From atomic formulae, MSO-formulae are constructed using Boolean connectives and quantification over first-order and second-order variables.

Note that this logic is *not* an extension of first-order logic as considered so far in this paper. The reason is simply that it speaks on finite dependence graphs (V, E, λ) while the first-order logic we are interested in speaks on the infinite structure $\mathcal{M}$. Since the elements of this latter structure are traces, we will use the following terminology: formulae of the first-order logic considered so far are called *external first-order formulae* and formulae of the MSO-logic on dependence graphs are called *internal MSO-formulae*. Similarly, we will speak of *external first-order variables* that range over traces and *internal second-order variables* (resp. *internal first-order variables*) that range over subsets (resp. elements) of a dependence graph (V, E, λ).

Theorem 3.3. *Let $\varphi(x)$ be an external first-order formula. There exists an internal MSO-sentence $\mathrm{int}(\varphi)$ such that we have for all dependence graphs $s = (V, E, \lambda)$*

$$s \models \mathrm{int}(\varphi) \text{ if and only if } \mathcal{M}_{\mathcal{R}_r} \models \varphi^{S(1,x)}(s).$$

Proof (sketch). The underlying idea is as follows: suppose $s, t \in \mathbb{M}$ are traces such that $s \rightarrow_{(S,T)} t$ for some scattered rewrite rule (S, T). Then the dependence graphs $s = (V_s, E_s, \lambda_s)$ and $t = (V_t, E_t, \lambda_t)$ coincide on large parts. In order to make this more precise, let $S = x_1\, s_1\, x_2 \ldots s_{m-1}\, x_m$ and $T = x_{\pi(1)}\, t_1\, x_{\pi(2)} \ldots t_{m-1}\, x_{\pi(m)}$ where π is a permutation of $\{1, 2, \ldots m\}$. There are traces $u_1, u_2, \ldots, u_m$ such that $s = u_1 \circ s_1 \circ u_2 \circ \ldots s_{m-1} \circ u_m$ and $t = u_{\pi(1)} \circ t_1 \circ u_{\pi(2)} \circ \ldots t_{m-1} \circ u_{\pi(m)}$. Hence the trace t can be represented by a tuple $u_1, u_2, \ldots u_m$ of factors of s (i.e., convex subsets of V) and the scattered rewrite rule (S, T). It is therefore possible to replace the external quantification over neighbors of s in $\mathcal{M}_{\mathcal{R}_r}$ by a finite disjunction over all rules from $\mathcal{R}_r$ and an internal quantification over m-tuples of factors of s. For $i = 1, 2$, let t_i be a neighbor of s represented by the rule (S_i, T_i) and the tuple $(u_1^i, u_2^i, \ldots u_m^i)$. One than has to express internally that $t_1 = t_2$ as well as $t_1 \rightarrow_\rho t_2$. This is achieved using a thorough analysis of the interplay of the scattered rewrite rules, the independence relation I, and the recognizable constraints $L(x)$ on the external first order variables $x \in \Omega$. Only here the restriction that $\text{type}(S) \equiv_J \text{type}(T)$ for $(S, T) \in \mathbb{S}$ becomes important. The proof can be found in [14]. □

Now we can prove Theorem 2.1, our main result:

Proof sketch of Theorem 2.1. By Gaifman's Theorem and Lemma 3.2, it suffices to check whether a sentence of the form

$$\exists x_1 \cdots \exists x_m \left\{ \bigwedge_{1 \leq i < j \leq m} d_{\mathcal{R}_r}(x_i, x_j) > 2 \ \wedge \ \bigwedge_{1 \leq i \leq m} \varphi^{S(1,x_i)}(x_i) \right\} \qquad (1)$$

holds in $\mathcal{M}_{\mathcal{R}_r}$. By Theorem 3.3, $\mathcal{M}_{\mathcal{R}_r} \models \varphi^{S(1,x)}(t)$ if and only if $t \models \text{int}(\varphi)$. Hence the set $L = \{t \in \mathbb{M} \mid \mathcal{M}_{\mathcal{R}_r} \models \varphi^{S(1,x)}(t)\}$ is recognizable by [27]. Thus we can check whether L is infinite. If this is the case than L contains traces of arbitrary size; there are in particular infinitely many traces $t_i \in L$, $i \in \mathbb{N}$, such that $d_{\mathcal{R}_r}(t_i, t_j) > 2$ for $i < j$. Hence (1) is true. On the other hand if L is finite, then we can enumerate all elements of L and calculate their 1-spheres with respect to $\mathcal{R}_r$. In this way we can check whether there are at least m traces $t_1, \ldots, t_m \in L$ such that $d_{\mathcal{R}_r}(t_i, t_j) > 2$ for $i < j$. Hence the decidability follows. □

First-order logic can be extended by modulo counting quantifiers [25]; the resulting logic is called FO+MOD. The only difference between FO+MOD and FO is that we now have a second type of quantifiers: if φ is a formula of FO+MOD, then $\exists^{(p,q)} x \varphi$ is a formula as well for $p, q \in \mathbb{N}$ and $p < q$. Then $\mathcal{M} \models \exists^{(p,q)} x \varphi$ if the number of traces $t \in \mathbb{M}$ with $\mathcal{M} \models \varphi(t)$ is finite and congruent p modulo q.

Since there is no locality theorem known for this logic,[4] our decidability proof for the first-order theory of $\mathcal{M}$ does not work for this more expressive logic; but the second step of our proof, i.e., the recognizability of the set of traces satisfying some local formula in FO extends to the logic FO+MOD. Thus, we obtain the decidability of local properties expressed in the logic FO+MOD. It seems that this result is new even for semi-Thue

---

[4] Libkin [15] and Nurmonen [23] proved locality theorems for counting logics including modulo counting, but not in the form of Theorem 3.1. We could not make them work in our situation.

systems and, as far as we see, cannot be shown using the automata theoretic methods from [5,13].

Theorem 3.4. *There is an algorithm that, on input of an independence alphabet (Σ, I), a natural number $r \geq 0$, and a sentence φ of FO+MOD in the language of the model $\mathcal{M}$, decides whether there exists $t \in \mathbb{M}$ with $\mathcal{M} \models \varphi^{S(r,x)}(t)$.*

4   Complexity Issues

We prove a nonelementary lower bound for the first-order theory of $\mathcal{M}$ by reducing the first-order theory of finite labeled linear orders. In order to formulate this, we take the MSO-logic over dependence graphs from Section 3.3 but forbid the use of second-order variables. The resulting formulae are called first-order formulae over dependence graphs. For the further consideration we will use this logic only for dependence graphs t where t is in fact a word $t \in \Sigma^*$. In this case the relation symbol $\preceq$ is interpreted by the usual order on the set $\{1, \dots, |t|\}$, and we speak of first-order formulae over words. Throughout this section, let $\Gamma = \{\alpha, \beta\}$ be an alphabet with two elements. The *first-order theory of Γ^** is the set of all first-order sentences over words φ such that $w \models \varphi$ for all $w \in \Gamma^*$. It is known that the first-order theory of Γ^* is not elementary decidable. This lower bound was announced in [21] where it is attributed to Stockmeyer. Stockmeyer's proof can only be found in his thesis and the same holds for the sharpening by Führer while Robertson's independent proof appeared as an extended abstract, only. The only proof that has been published seems to be [3, Example 8.1].

Let $\Sigma_1 = \Gamma \times \{0, 1, 2, 3\}$. On this set, we define a dependence relation D_1 as follows: $((a, i), (b, j)) \in D_1$ if and only if $i = j$ or $i, j \leq 2$ or $(a = b$ and $\{i, j\} \subseteq \{1, 2, 3\})$. The complementary relation is denoted by I_1. Next, we will consider the (preliminary) trace rewriting system $\mathcal{R}_1$ over $\mathbb{M}(\Sigma_1, I_1)$ defined by

$$\mathcal{R}_1 = \{(a, 0) \to (a, 3)(a, 1)(a, 3), \ (a, 1) \to (a, 2) \mid a \in \Gamma\}.$$

We first reduce the first-order theory of Γ^* to the first-order theory of the structure $(\mathbb{M}(\Sigma_1, I_1), \to_{\mathcal{R}_1}, (F_t)_{t \in F})$ where $F \subseteq \mathbb{M}(\Sigma_1, I_1)$ is finite and $F_t = \mathbb{M} \circ t \circ \mathbb{M}$ is the set of all traces that contain the factor t:

Lemma 4.1. *The first-order theory of Γ^* can be reduced in polynomial time to the first-order theory of $(\mathbb{M}(\Sigma_1, I_1), \to_{\mathcal{R}_1}, (F_t)_{t \in F})$ for some finite set $F \subseteq \mathbb{M}(\Sigma_1, I_1)$.*

Proof (sketch). A trace in $\mathbb{M}$ contains only letters from $\Gamma \times \{0\}$ if and only if it does not contain any factor of the form (a, i) for $1 \leq i \leq 3$. Hence (with $\Sigma_1 \times \{1, 2, 3\} \subseteq F$) the set of words over $\Gamma \times \{0\}$ (which will be identified with the words over Γ) can be defined in $(\mathbb{M}(\Sigma_1, I_1), \to_{\mathcal{R}_1}, (F_t)_{t \in F})$. The successors of such a word w are in one-to-one correspondence with the positions in w, hence the internal quantification over positions in w gets replaced by external quantifications over neighbors of w. The label of a position can be recovered using the predicates $F_{(a,1)}$ for $a \in \Gamma$. The order between positions requires the use of predicates F_t with $t = (a, 2)(a, 3)(b, 3)$ and $t = (a, 3)(b, 3)(b, 1)$.   □

In order to get rid of the predicates F_t in the lemma above, one extends the alphabet Σ_1 and the trace rewriting system $\mathcal{R}_1$ in such a way that loops of characteristic lengths are attached to traces from F_t. This allows to reduce the first-order theory of Γ^* to the set of valid local sentences of the resulting structure $(\mathbb{M}\,(\Sigma_2, I_2), \to_{\mathcal{R}_2})$. Hence one obtains

Theorem 4.2. *There exists an independence alphabet (Σ_2, I_2) and a trace rewriting system $\mathcal{R}_2$ over $\mathbb{M}\,(\Sigma_2, I_2)$ such that the first-order theory of $(\mathbb{M}\,(\Sigma_2, I_2), \to_{\mathcal{R}_2})$ is not elementary decidable.*

For semi-Thue systems, we can only show a weaker lower bound:

Theorem 4.3. *There exists an alphabet Σ_3 and a semi-Thue system $\mathcal{R}_3$ over Σ_3^* such that any decision procedure for the first-order theory of $(\Sigma_3^*, \to_{\mathcal{R}_3})$ requires at least doubly exponential nondeterministic time.*

5   Applications to the Confluence Problem

In this section we present applications of Theorem 2.1 to the confluence problem for trace rewriting systems. For terminating semi-Thue systems, i.e., systems without infinite derivations, confluence is decidable by Newman's Lemma and the use of critical pairs. For trace rewriting systems, the situation becomes more complicated since even finite length-reducing trace rewriting systems can have infinitely many critical pairs [6,7]. Generalizing a result from [22], it is shown in [17] that confluence of length-reducing trace rewriting systems is decidable if and only if $I = \emptyset$ or $I = (\Sigma \times \Sigma) \backslash \mathrm{Id}_\Sigma$, i.e., undecidable in most cases. In this section we describe specific classes of trace rewriting systems with a decidable confluence problem, see [7,17] for related results.

First we have to introduce some notation. For $t \in \mathbb{M}$ define $\mathrm{D}(t) = \{a \in \Sigma \mid (a, t) \notin I\}$. For a subalphabet $\Gamma \subseteq \Sigma$ and a trace rewriting system $\mathcal{R}$ we define the trace rewriting system $\pi_\Gamma(\mathcal{R})$ by $\pi_\Gamma(\mathcal{R}) = \{(\pi_\Gamma(\ell), \pi_\Gamma(r)) \mid (\ell, r) \in \mathcal{R}\}$, where $\pi_\Gamma(t)$ denotes the projection of the trace t to the alphabet Γ. A *clique covering* of a dependence alphabet (Σ, D) is a sequence $(\Gamma_1, \dots, \Gamma_n)$ with $\Gamma_i \subseteq \Sigma$ such that $\Sigma = \bigcup_{i=1}^n \Gamma_i$ and $D = \bigcup_{i=1}^n \Gamma_i \times \Gamma_i$. Finally a trace rewriting system $\mathcal{R}$ is *terminating on a trace t* if there does not start an infinite $\to_{\mathcal{R}}$ path in t.

Theorem 5.1. *Confluence is decidable for the class of terminating trace rewriting systems $\mathcal{R}$ over $\mathbb{M}(\Sigma, I)$ satisfying the following conditions:*

(1) For all $(\ell, r) \in \mathcal{R}$ and all $a \in \Sigma$ with $(a, \ell) \in I$ it holds $ar = ra$.

(2) For all $p_0, p_1, q_0, q_1, \in \mathbb{M} \backslash \{1\}$, $r_0, r_1 \in \mathbb{M}$ with $(p_0 q_0, r_0), (p_1 q_1, r_1) \in \mathcal{R}$ and $(p_0, p_1), (q_0, q_1) \in I$ there exist $s_0, s_1, t_0, t_1 \in \mathbb{M}$ such that $r_i = s_i t_i$, $\mathrm{D}(s_i) \subseteq \mathrm{D}(p_i)$, and $\mathrm{D}(t_i) \subseteq \mathrm{D}(q_i)$ for $i = 0, 1$.

(3) For all $p_0, p_1, q_0, q_1, r_0, r_1 \in \mathbb{M}$, $s \in \mathbb{M} \backslash \{1\}$ with $(p_0 s q_0, r_0), (p_1 s q_1, r_1) \in \mathcal{R}$ and $(p_0, p_1), (q_0, q_1) \in I$, the trace rewriting system $\pi_\Gamma(\mathcal{R})$ is terminating on the traces $\pi_\Gamma(p_1 r_0 q_1)$ and $\pi_\Gamma(p_0 r_1 q_0)$, where $\Gamma = \mathrm{D}(p_0 s q_1) \cap \mathrm{D}(p_1 s q_0)$.

Proof (sketch). One shows that in this case $\alpha \in \mathbb{N}$ can be computed effectively such that confluence and α-confluence (see Section 2.3) are equivalent. Since α-confluence is first-order expressible, it is decidable by Theorem 2.1.   □

From this very technical decidability criterion, one can infer [16, Thm. 2] and the following new special case:

Corollary 5.2. *Confluence is decidable for the class of trace rewriting systems $\mathcal{R}$ over $\mathbb{M}(\Sigma, I)$ such that*

(1) for all $(\ell, r) \in \mathcal{R}$, the graph $(\mathrm{alph}(\ell), D)$ is connected, and
(2) there exists a clique covering $(\Gamma_1, \dots, \Gamma_n)$ of $(\Sigma, (\Sigma \times \Sigma) \backslash I)$ such that for all $i \in \{1, \dots, n\}$ the semi-Thue system $\pi_i(\mathcal{R})$ is terminating.

6   Open Questions

In Section 4, we gave a lower bound for the complexity of the first-order theory of the one-step rewriting by a semi-Thue system. There is a huge gap between this doubly exponential lower and the nonelementary upper bound that follows immediately from the proofs in [5,13].

Although our decidability result is very similar to corresponding results in [5,13], our technique is new. It could provide a means to identify term rewriting systems whose rewrite graph has a decidable first-order theory. Several classes of term rewriting systems with this property have been identified, like for instance ground term rewriting systems [5], but in general the problem is undecidable [29].

Semi-Thue systems can be seen as term rewriting systems modulo associativity (it is a very simple case since there are no further symbols). Similarly, trace rewriting is term rewriting modulo associativity and partial commutativity. Is it possible to use the technique developed in this paper to handle other "term rewriting modulo ..." theories?

References

1. P. Cartier and D. Foata. *Problèmes combinatoires de commutation et réarrangements.* Lecture Notes in Mathematics vol. 85. Springer, Berlin - Heidelberg - New York, 1969.
2. D. Caucal. On the regular structure of prefix rewriting. *Theoretical Computer Science,* 106:61–86, 1992.
3. K. Compton and C. Henson. A uniform method for proving lower bounds on the computational complexity of logical theories. *Annals of Pure and Applied Logic,* 48:1–79, 1990.
4. B. Courcelle. The monadic second-order logic of graphs, II: Infinite graphs of bounded width. *Mathematical Systems Theory,* 21:187–221, 1989.
5. M. Dauchet and S. Tison. The theory of ground rewrite systems is decidable. In *Proceedings of the 5th Annual IEEE Symposium on Logic in Computer Science (LICS '90),* pages 242–256. IEEE Computer Society Press, 1990.
6. V. Diekert. On the Knuth-Bendix completion for concurrent processes. In Th. Ottmann, editor, *Proceedings of the 14th International Colloquium on Automata, Languages and Programming (ICALP 87), Karlsruhe (Germany),* number 267 in Lecture Notes in Computer Science, pages 42–53. Springer, 1987.
7. V. Diekert. *Combinatorics on Traces.* Number 454 in Lecture Notes in Computer Science. Springer, 1990.
8. V. Diekert, Y. Matiyasevich, and A. Muscholl. Solving word equations modulo partial commutations. *Theoretical Computer Science,* 224(1–2):215–235, 1999.

9. V. Diekert and G. Rozenberg, editors. *The Book of Traces*. World Scientific, Singapore, 1995.

10. C. Duboc. On some equations in free partially commutative monoids. *Theoretical Computer Science*, 46:159–174, 1986.

11. C. Frougny and J. Sakarovitch. Synchronized rational relations of finite and infinite words. *Theoretical Computer Science*, 108(1):45–82, 1993.

12. H. Gaifman. On local and nonlocal properties. In J. Stern, editor, *Logic Colloquium '81*, pages 105-135, 1982, North Holland.

13. F. Jacquemard. *Automates d'arbres et Réécriture de termes*. PhD thesis, Université de Paris-Sud, 1996.

14. D. Kuske and M. Lohrey. On the theory of one-step rewriting in trace monoids. Technical Report 2002-01, Department of Mathematics and Computer Science, University of Leicester. Available at www.mcs.le.ac.uk/~dkuske/pub-rest.html#UNP9.

15. L. Libkin. Logics capturing local properties. *ACM Transactions on Computational Logic*. To appear.

16. M. Lohrey. On the confluence of trace rewriting systems. In V. Arvind and R. Ramanujam, editors, *Proceedings of the 18th Conference on Foundations of Software Technology and Theoretical Computer Science, (FSTTCS'98), Chennai (India)*, number 1530 in Lecture Notes in Computer Science, pages 319–330. Springer, 1998.

17. M. Lohrey. Confluence problems for trace rewriting systems. *Information and Computation*. 170:1–25, 2001.

18. A. Markov. On the impossibility of certain algorithms in the theory of associative systems. *Doklady Akademii Nauk SSSR*, 55, 58:587–590, 353–356, 1947.

19. Y. Matiyasevich. Some decision problems for traces. In S. Adian and A. Nerode, editors, *Proceedings of the 4th International Symposium on Logical Foundations of Computer Science (LFCS'97), Yaroslavl (Russia)*, number 1234 in Lecture Notes in Computer Science, pages 248–257. Springer, 1997.

20. A. Mazurkiewicz. Concurrent program schemes and their interpretation. Technical report, DAIMI Report PB-78, Aarhus University, 1977.

21. A. Meyer. Weak monadic second order theory of one successor is not elementary recursive. In *Proc. Logic Colloquium*, Lecture Notes in Mathematics vol. 453, pages 132–154. Springer, 1975.

22. P. Narendran and F. Otto. Preperfectness is undecidable for Thue systems containing only length-reducing rules and a single commutation rule. *Information Processing Letters*, 29:125–130, 1988.

23. J. Nurmonen. Counting modulo quantifiers on finite structures. *Information and Computation*, 160:62–87, 2000. LICS 1996, Part I (New Brunswick, NJ).

24. E. Post. Recursive unsolvability of a problem of Thue. *Journal of Symbolic Logic*, 12(1):1–11, 1947.

25. H. Straubing, D. Thérien, and W. Thomas. Regular languages defined with generalized quantifiers. *Information and Computation*, 118:289–301, 1995.

26. Terese. Term Rewriting Systems. To appear with Cambridge University Press, 2001.

27. W. Thomas. On logical definability of trace languages. In V. Diekert, editor, *Proceedings of a workshop of the ESPRIT Basic Research Action No 3166: Algebraic and Syntactic Methods in Computer Science (ASMICS), Kochel am See (Germany)*, Report TUM-I9002, Technical University of Munich, pages 172–182, 1990.

28. A. Thue. Probleme über die Veränderungen von Zeichenreihen nach gegebenen Regeln. *Skr. Vid. Kristiania, I Math. Natuv. Klasse*, No. 10, 34 S., 1914.

29. R. Treinen. The first-order theory of linear one-step rewriting is undecidable. *Theoretical Computer Science*, 208(1-2):149–177, 1998.

Navigating with a Browser[*]

Michał Bielecki,[1][**] Jan Hidders[2], Jan Paredaens[2], Jerzy Tyszkiewicz[1], and
Jan Van den Bussche[3]

[1] Warsaw University, Poland
[2] University of Antwerp, Belgium
[3] University of Limburg, Belgium

NAVIGARE NECESSE EST, VIVERE NON
EST NECESSE.   POMPEIUS

Abstract. We consider the navigation power of Web browsers, such as
Netscape Navigator, Internet Explorer or Opera. To this end, we formally
introduce the notion of a navigational problem. We investigate various
characteristics of such problems which make them hard to visit with
small number of clicks.

The Web browser is an indispensable piece of application software for the modern computer user. All the popular browsers essentially implement a very basic machinery for navigating the Web: a user can enter a specific URL as some kind of "source node" to start his navigation; he can then further click on links to visit other Web nodes; and he can go "back" and "forward" along a stack of already visited nodes.

A lot of application software, such as word processors or spreadsheets, allows in addition to the standard "manual" use of the software, also some kind of "programmed" use, by allowing the user to write macros which are then executed by the software tool. Such macros are typically simple programs, which offer the standard test and jump control constructs; some variables to store temporary information; and for the rest are based on the basic features offered by the application.

In this paper, we study such a macro mechanism for Web browsers. Thereto, we introduce the *browser stack machine*. This is a finite-memory automaton as introduced by Kaminski and Francez [4], i.e., an automaton with finite control and a finite number of registers which can store Web nodes, which is extended with the basic features offered by a Web browser and already summarized above: clicking on a link; going "back"; and going "forward". The browser stack machine is a restriction of the *browser machine* introduced by Abiteboul and Vianu [2], which has an unlimited Turing tape for storing Web nodes, rather than just the finite memory plus the stack which we have here.

---

[*] Research supported in parts by Polish KBN grant 7T11C 007 21 (M.B. and J.T.)
and by FWO grant G.0246.99 (J.H.)

[**] Contact author: Institute of Informatics, Warsaw University, Banacha 2, PL-02-097
Warszawa, Poland, mab@mimuw.edu.pl.

P. Widmayer et al. (Eds.): ICALP 2002, LNCS 2380, pp. 764–775, 2002.
© Springer-Verlag Berlin Heidelberg 2002

Just like Alan Turing was interested in understanding the problems solvable by a clerk following a formal algorithm, using only pencil and sufficient supply of paper, we are here interested in the problems solvable by such a browser stack machine. However, while Turing could easily define a "problem" as function on the natural numbers, what kind of "problems" over the Web can we consider in our setting? Abiteboul and Vianu considered the Web as a database and studied the power of browser machines in answering *queries* to this database. Browser stack machines from such a database querying perspective were studied in another paper [6]. We will take a different angle here, and want to focus on *navigational problems*. A navigational problem asks the browser to visit a certain specified set of Web nodes, and no others. It thus corresponds to avoiding "getting lost in hyperspace" and getting your job done. Specifically, we focus on *structural* navigational problems only, where the browser has to solve the problem purely on structural information of the Web graph alone. More advanced models could also introduce various predicates on Web nodes so that the browser can detect various properties of the nodes, but we feel the basic "uncolored" model remains fundamental.

Concretely, we offer the following contributions:

1. We show that browser stack machines, simple as they may appear, can simulate arbitrary Turing machines.
2. We introduce a notion of "data transfer-optimal" browser programs which never download a node more than once. We show that this is a real restriction, by exhibiting various natural navigational problems that cannot be solved in such an optimal manner, and by providing a rather general necessary condition on the structure of such problems.
3. We provide concrete lower bounds on the number of data transfers a browser program has to make to solve certain simple problems. Interestingly, our proof employs a basic result from communication complexity.
4. Finally, we propose a new feature for Web browsers: switching the contents of the "back" and "forward" stacks. We show that this feature allows problems to be solvable with provably less data transfer, and that it allows solving navigational problems unsolvable without it.

1   Web Instances and Browser Stack Machines

1.1   Web Instances

Definition 1. *A* Web graph *is a finite, locally ordered directed graph* $\mathbb{V} = \langle V, l, < \rangle$. *V is the finite set of vertices of $\mathbb{V}$ (we always use the matching Roman letter for the set of vertices of any Web graph denoted by a blackboard-font letter), l is the edge relation (we call it also the link relation), and < is a ternary relation giving the local ordering of the vertices reachable by edges outgoing from the current node.*

766 M. Bielecki et al.

Definition 2. *A* page *of a Web graph* $\mathbb{V}$ *is the following structure: it is a node* $v \in V$ *together with the ordered list* $t_1, \ldots, t_k$ *of all the vertices* $t_i \in V$ *such that* $l(v, t_i)$, *the list being given in the local order of all the vertices reachable by one edge from* v. *We will often depict such a page as follows:* $\boxed{\begin{array}{c} v \\ \hline t_1,\ldots,t_k \end{array}}$. *A* Web graph *can be equivalently represented by the set of its pages.*

Definition 3. *A* Web instance $(\mathbb{V}, s)$ *is a Web graph with a distinguished node* s, *and such that all vertices of* $\mathbb{V}$ *are reachable by links from* s, *which is henceforth called the* source.

The source node is where browsing starts in the Web graph. Obviously, nodes not reachable from the source are irrelevant to browsing, hence the reachability requirement. This formalization of Web instance is similar to earlier formal models of the Web, e.g., that by Abiteboul and Vianu [2]

1.2 Browser Stack Machine

We next define *browser stack machines*, which we abbreviate later on as BSM, for the automatic navigation of Web instances.

Definition 4. *A* browser stack machine *is a finite state computing device* B *equipped with the following ingredients:*

1. *The components of* B *are:*
   a) *A finite state control.*
   b) *A read-only tape, which stores a sequence of vertices of a Web graph. The tape can be accessed by a head, which can move backwards and forwards, and can sense the beginning and the end of the tape.*
   c) *A finite number of registers* $r_1, \ldots, r_k$, *each one capable of storing a single node of a Web graph.*
   d) *Two stacks, called* $\Rightarrow$ *and* $\Leftarrow$, *on which* B *can store always the entire content of its tape (as a single stack item), together with the current location of the head.*
2. *The following actions can be undertaken by* B, *as ordered by its finite control:*
   a) B *can change its control state.*
   b) B *can move the head forward or backward on the tape.*
   c) B *can store in any of the registers the identity of the node currently seen by the head on the tape.*
   d) B *can make a forward move, which consists of storing the current content of the tape, together with the head position, on the* $\Leftarrow$ *stack, removing the top of the* $\Rightarrow$ *stack and restoring it as the current tape and setting the head in the recorded position. This move is impossible if the* $\Rightarrow$ *stack is empty.*
   e) B *can make a backward move, which consists of storing the current content of the tape, together with the head position, on the* $\Rightarrow$ *stack, removing the top of the* $\Leftarrow$ *stack and restoring it as the current tape and setting the head in the recorded position. This move is impossible if the* $\Leftarrow$ *stack is empty.*

f) B can click, *which causes the current content of the tape together with the current head position to be stored at the top of the ⇐ stack, removes the* entire content *of the* ⇒ *stack, and fills the tape with the list of the vertices, which are accessed by the edges outgoing from the node seen by the head on the tape at the moment of clicking. The head is set at the first cell of the tape. If the link points to the current page itself, nothing happens.*[1]

g) B can halt.

3. *The following information determines the next state and the next move of the machine:*

   a) *The current control state.*

   b) *Equalities and non-equalities between values of registers and/or the node currently under the head.*

   c) *Information whether the head scans the leftmost or rightmost cell of the tape. (Recall that, according to the mechanism of changing the tape content, the length of the tape can vary.)*

   d) *Information whether any of the stacks is empty.*

4. *The initial configuration of the machine in a Web instance* $(\mathbb{V}, s)$ *is as follows:*

   a) s *is the value of all the registers of B.*

   b) s *is the only node on the tape.*

   c) *Both stacks are empty.*

A formal definition of the *computation* of a machine on an instance is easily produced and omitted from this extended abstract.

2 Navigational Problems

Our central idea is to study the power of BSMs in solving *navigational problems*:

Definition 5. *A navigational problem is a partial computable function P from Web instances to finite sets of nodes, such that*

- *whenever* $P(\mathbb{V}, s)$ *is defined, it is a subset of* V; *and*
- *if* $\alpha : (\mathbb{V}, s) \to (\mathbb{V}', s')$ *is an isomorphism, then* $P(\mathbb{V}', s') = \alpha(P(\mathbb{V}, s))$.

The second condition is a common "consistency criterion" found in database querying [3,1] and corresponds to the conceptual practice not to distinguish between isomorphic logical structures.

Definition 6. *Recall that the set of nodes on which a BSM B clicks during its computation on* $(\mathbb{V}, s)$ *is denoted* $B(\mathbb{V}, s)$. *B solves a navigational problem P if for every instance* $(\mathbb{V}, s)$ *on which P is defined* $B(\mathbb{V}, s) = P(\mathbb{V}, s)$.

Definition 7. *A navigational problem P is called "visitable" if it can be solved by a BSM.*

---

[1] The same behavior is shown by real life browsers.

We begin with the most crude approximation of the computational power of BSMs.

Theorem 1. *There is a translation of Turing machines M into BSMs B_M and of input words w into Web instances $(\mathbb{V}_w, s)$ such that the computation of B_M in $(\mathbb{V}_w, s)$ simulates the computation of M on w.*

Proof. We give a direct encoding of the computations of any single-tape Turing machine in the model of BSM.

Turning to the construction, let M be a Turing machine with a a single input-work tape with tape alphabet consisting of symbols $\{0, 1\}$. Blank is allowed too, but cannot be written by the machine on the tape. Let $w = w_1 \ldots w_n \in \{0, 1\}^*$ be an input word for M. We convert w into a Web instance $(\mathbb{V}_w, s)$ as follows:

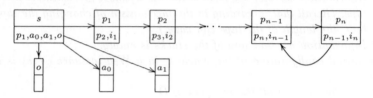

and where additionally each link i_j points on the node a_l iff $w_j = l$. Note the loop from p_n to p_{n-1}.

Now we construct a BSM B_M able to mimic in $(\mathbb{V}_w, s)$ the computation of M on w as follows:

1. Starting from s, B stores the addresses of a_0 and a_1 in its registers. (Here and in the following, B identifies links by their order of appearance on the pages. There are always at most 4 of them, so all of the necessary information can be encoded in the finite control of B.)
2. Next it follows the first link on each page it arrives at, until it finds a two-page-large loop, which can be easily detected by comparing the link with the address of the previous page (stored on the $\Leftarrow$ stack). When this loop is found, B must be in p_n.
3. B memorizes p_n in a register.
4. On page p_i B repeats the actions
   a) Compare the second link on the page with the stored addresses a_0, a_1.
   b) If it is a_j, then move the head to the j-th link on that page.
   c) Go back one step.
   until it comes back to s.
5. Presently, the position of the head on each of the pages $p_1, \ldots, p_n$ which are on the $\Rightarrow$ stack indicates the corresponding symbol of M's input word w.
6. B starts simulating M, which is done by walking backward and forward on the stacks exactly as the head of M does, updating always the head position on the actually visited page to be over the j-th link, with j the symbol M writes to the current tape cell. In this simulation, the page currently visited by B corresponds to the tape cell with the head of M, the $\Leftarrow$ stack

corresponds to the portion of M's tape to the left of the head, and the $\Rightarrow$ stack corresponds to the portion of M's tape to the right of the head, except that the not-yet-used portion of the tape will be only created by B when it is necessary.

7. If B senses that the $\Rightarrow$ stack is empty and M wants to go right, then B memorizes in a register the current location of the head on the present page (which is always either p_n or p_{n-1} in such cases), clicks on the first link (so that it goes to either p_{n-1} or p_n, respectively), makes one move backward, restores the head position to the memorized position, makes a move forward, and prints there the intended symbol, assuming that what is saw there was a blank symbol.

8. If M accepts and halts, B goes all the way back to s and clicks on link o.

Note that in longer computations the vast majority of the pages stored on the stacks are copies of p_{n-1} and p_n. However, the position of the head is always the page currently visited[2], so the BSM never gets lost in the simulation. As it is readily seen, M accepts w iff $o \in B(\mathbb{V}_w, s)$. $\square$

Remark 1. Now, if we take a universal Turing machine as M, a standard argument shows that the navigational problem solved by B is RE-complete.

It is worth noting that the above construction depends crucially on the Web instance possessing a cycle. Hence, it is a natural question to ask what happens if we restrict attention to acyclic Web graphs. The following result answers this question.

Theorem 2. *Every visitable navigational problem, when restricted to acyclic instances, is in PSPACE, and there exist such problems that are PSPACE-complete.*

Proof. The first part follows by a straightforward simulation of B by a Turing machine, which stores the stacks of B on its work tape. In the absence of cycles, these stacks never exceed height equal to the cardinality of the input Web instance, and the pages themselves are of size linearly dependent on the number of pages of the input. This gives altogether a polynomial amount of space, necessary to perform the simulation.

After minor technical modifications, the proof method of Theorem 1 yields the PSPACE-hardness part. $\square$

Let $Reach_k$ be the navigational problem $(\mathbb{V}, s) \mapsto$ "the set of vertices reachable from s in at most k steps" (we permit $k \in \{1, 2, \ldots\} \cup \{\infty\}$).

Theorem 3. *For each $k > 0$ $Reach_k$ is visitable.*

Proof. The machine does depth-first-search, limited to k steps. To avoid loops and to assure that the BSM always halts, it clicks only pages that are not already on the $\Leftarrow$ stack. Before clicking a page it stores it in a register then goes back to the source comparing the page to all pages on the $\Leftarrow$ stack. $\square$

---

[2] With the small and only temporary exception when the BSM clicks to extend the stacks by a new page.

2.1   Navigational Problems and Data Transfer

The above theorems do not exhaust all the interesting questions connected with the navigational abilities of BSMs.

Clicking on a link causes data transfer to happen, even if the page has been already seen by the browser. The pages stored on the stacks are cached locally, so visiting them does not cause any data transfer. Thus, by measuring how many times a BSM clicks during its computation, we indeed measure the data transfer it generates.

Definition 8. *A BSM B is called a "1-visitor" if in its computation on any Web instance, B clicks every node at most once.*
A navigational problem is called "1-visitable" if it is solvable by a 1-visitor.

1-visitability is an important feature of problems. Such problems can be solved without making any unnecessary data transfer. In certain sense, for 1-visitable problem there exist BSM which achieve the optimal possible communication cost.

We next present an intrinsic necessary condition on navigational problems to be 1-visitable:

Definition 9. *Given a Web instance $(\mathbb{V}, s)$ and a subset V' of the vertices of $\mathbb{V}$, a node n of $\mathbb{V}$ is said to be* distant from V' *if $n \notin V'$ and all paths in $\mathbb{V}$ from a node in V' to n contain more than one edge.*

Definition 10. *Given a Web instance $(\mathbb{V}, s)$, a subset V' of V is called a "PD[k] set" (Path with k Distant nodes) if there is a subset $S \subseteq V'$ such that*

- $s \in S$,
- *if $|S| > 1$ then there is a simple path in $\mathbb{V}$ that starts in s and contains exactly the vertices in S, and*
- *there are at most k vertices in V' that are distant from S.*

Theorem 4. *A navigational problem P can be 1-visitable only if there is a natural number k such that for every Web instance $(\mathbb{V}, s)$, $P(\mathbb{V}, s)$ is a PD[k] set.*

Proof. (Sketch) Let B be the BSM solving P. If B has no registers then it cannot compare pages and can therefore only visit the source s without the risk of visiting a page twice. If, on the other hand, B has registers then it can compare pages and check if a certain page that is to be visited was not visited before.

Let us call the path that is formed by the $\Leftarrow$ stack plus the current page the Current path. It either holds that B has directly after every click all the visited pages that are distant from the Current path in its registers, or it does not. Let us first assume that it does. Since this will then also hold after the last click, the visited pages will be an $PD[k]$ set with S the Current path just after the last click. Let us assume that directly after a click to a certain page the BSM holds no longer all the visited pages that are distant from the Current path in its registers. Let p be the first page for which this holds, and S the corresponding

Current path. Let p' be the page that B clicked to just before p and let S' be the corresponding Current path. After the click to page p the BSM can only visit pages that are children of pages in S' or were in the registers of B just after the click to p'. Since after this click all visited pages distant from S' are also in the registers it follows that at that moment all the pages distant from S' that were and will be visited by B are then in its k registers. It follows that all the pages visited by B form an $PD[k]$ set. □

It is obvious that the above condition is not sufficient for a problem to be visitable at all, let alone 1-visitable. The reason is that, according to Theorem 2, the problem must obey certain complexity requirements to have at least a chance to be visitable.

A counterpart of Theorem 3 is the following one.

Theorem 5. *Reach$_1$ is 1-visitable. For $k > 1$ the problem Reach$_k$ is not 1-visitable.*

Proof. The proof follows from Theorem 4 or can be done in a similar way to the Proof 2 of Theorem 7. □

An obvious approach for showing certain navigational problems to be 1-visitable is to start from a BSM B, not necessarily a 1-visitor, that solves the problem, and then to try to "optimize" B's program, making it more "careful" not to click on a page that has been clicked before. Putting this idea in practice in full generality is impossible (see Theorem 7), but we can at least show that there are cases where such an approach is successful.

Theorem 6. *If a navigational problem P is visitable by a BSM B and, for every Web instance $(\mathbb{V}, s)$, the trace of the computation of B on $(\mathbb{V}, s)$ is a simple path (perhaps with a link from the last node to some earlier one on the path), then P is indeed 1-visitable.*

Proof. Omitted. □

3   Click Complexity

We now know that many navigational problems are not 1-visitable: they cannot be solved by a BSM without clicking more often than the actual number of nodes that are visited by the BSM. Hence, it is natural to wonder about the actual "click complexity" of a visitable navigational problem: how many clicks must any BSM make to solve them? We next prove a quite negative result, to the effect that the number of needed clicks in general *cannot be bounded* as a function of the number of visited nodes.

Definition 11. *For a BSM B and a Web instance $(\mathbb{V}, s)$ let $C_B(\mathbb{V}, s)$ be the number of clicks during computation of B on $(\mathbb{V}, x)$.*

Theorem 7. *There exists no fixed recursive function $f : \mathbb{N} \to \mathbb{N}$ such that for every visitable navigational problem P on Web instances, there exists a BSM B solving P for all Web instances $(\mathbb{V}, s)$ with $C_B(\mathbb{V}, x) \le f(|B(\mathbb{V}, x)|)$.*

Proof. We give two fundamentally different proofs of the theorem. Each of them highlights a quite different reason for which a BSM might have to click many times on the same page.

Proof 1. Suppose such a function f exists.
We proceed as in the proof of Theorem 1. By appropriately choosing a Turing machine M to be simulated by a BSM B, we can construct a visitable navigational problem such that the decision problem $o \overset{?}{\in} B(\mathbb{V}, s)$ is not in $\text{DSPACE}(n^3 \cdot f(n))$. This follows from the well-known hierarchy theorem for deterministic space complexity.
Now suppose there exists another BSM B' such that $B'(\mathbb{V}, s) = B(\mathbb{V}, s)$ for all Web instances $(\mathbb{V}, s)$ and $C_{B'}(\mathbb{V}, s) \leq f(|B(\mathbb{V}, s)|)$. By a straightforward modification of the argument given in the proof of Theorem 2, we can simulate B' by a Turing machine using at most $(n^3 \cdot f(n))$ space for input Web instances of size n. By this simulation, $o \overset{?}{\in} B(\mathbb{V}, s)$ is in $\text{DSPACE}(n^3 \cdot f(n))$, a contradiction.

Proof 2. Consider the following Web instance $(\mathbb{V}, s)$.

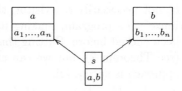

where the pages of $a_1, \ldots, a_n, b_1, \ldots, b_n$ contain no further links. Moreover, some of the a_i'a may equal to some of the b_j's. The task of the BSM is to verify if the sequences $a_1, \ldots, a_n$ and $b_1, \ldots, b_n$ are identical, and if so, to click on b_n.
The following BSM B does that, clicking altogether on a number of vertices linear in n.
After clicking on s, B clicks on a, then memorizes a_1 in a register x_a, goes back to s, clicks on b, memorizes b_1 in a register x_b and compares x_a with x_b. If they are distinct then B immediately halts.
Otherwise B repeats the following action until x_a and x_b are the last links on the pages a and b: click on a, then move the head to the link next to the value of x_a, memorize its value in x_a, go back to s, click on b, move the head to the link next to the value of x_b, memorize its value in x_b, and compare x_a with x_b. If they are distinct then immediately halt.
If B finds a_n and b_n equal (i.e., the above loops terminates without discovering any difference), click on b_n.
Now we prove that any BSM B solving the same navigational problem must indeed use a large number of clicks, even though $|B(\mathbb{V}, s)| \leq 3$.
We use an argument from communication complexity. Let there be two players, Alice and Bob, each of whom has an ordered subset (unknown to the other person) of a (known to both Bob and Alice) set $\Sigma = \{\sigma_1, \ldots, \sigma_t\}$. Their task is to decide if the sequences are equal, by sending each other messages: words

over the same alphabet Σ. They do so according to a predefined communication protocol. This protocol specifies whose turn is to send the next message, based on the full history of prior messages. At the end of the protocol, the players must be able to decide if the words are equal. A theorem of communication complexity asserts that for any such protocol, which gives the correct answer for every pair of words of length n, there exists a pair of words for which the total length of exchanged messages is at least $\log_t(\binom{t}{n}n!)$, see [5, Section 1.3].

We show that if there exists a BSM B with k registers and $r < |\Sigma|$ control states, solving our initial navigational problem with c clicks, then Alice and Bob can use it to construct a protocol to decide the equality of any two ordered subsets of Σ of size n, with communication of $(c-1)/(k+1)$ elements. This protocol is valid for all Σ of sufficiently high cardinality.

Indeed, suppose such a B exists and that it has k registers and r control states. Let the ordered subsets (i.e., words) Alice and Bob have be $a_1 \ldots a_n$ and $b_1 \ldots b_n$, respectively. Each of the players converts his/her sequence into a Web page

a
$a_1, \ldots, a_n$

and

b
$b_1, \ldots, b_n$

, respectively. Now each of them simulates the BSM B in the so formed instance of the above figure. Whenever B enters a (b, respectively) then only Alice (Bob, respectively) can continue the simulation, and does so until B enters the other node from among a and b. At this moment that player sends to the other one a message "B clicks on the link to your page with $x_1, \ldots, x_k$ as the values of the registers and q as the control state". Note that indeed it is enough to send $x_1, \ldots, x_k, \sigma_q$, where σ_q is used to encode the control state. Then the other player takes over the simulation, and so on. Note that he/she can do it, since at this moment the content of the $\Leftarrow$ stack is

s
a, b

with head pointing at

the entered page, the $\Rightarrow$ stack is empty, and all the remaining information B has are the values of the registers and the control state. If B either halts or clicks on b_n, Bob notifies Alice about the decision of B. Now, besides the last message of length 1, each message exchanged has length $r+1$ and the number of such messages is equal to the number c of clicks B has done minus one (the initial click on s). The total amount of communication is then $(c-1)(r+1) \geq \log_t(\binom{t}{n}n!) \geq \log_t((t-n)^n)$, by the communication complexity theorem. This estimation is valid for any sufficiently large t. Hence the number c of clicks satisfies

$$c \geq 1 + \frac{\lim_{t \to \infty} \log_t((t-n)^n)}{r+1} = 1 + \frac{n}{r+1}.$$

$\square$

Remark 2. We haven't made one specific feature of real-life browsers a part of our BSM. Namely, they are capable of remembering which pages they visited, and represent this information by displaying the already visited links in a specific color. This feature could be, without much difficulty, incorporated into our model. Now the remark is that even in presence of this feature, the last theorem remains true, and, indeed, both proofs remain valid. In particular, there are visitable navigational problems which are not 1-visitable, even by BSM with memory.

4 Enhanced Browsing

If we have a look at the navigation mechanism of browsers, it appears that they behave very much like Theseus in the maze, in the old Greek myth. Theseus had been equipped with a roll of veil by Ariadne. He set one end of the veil at the entrance to the maze, and following it, he could find the way back from the maze after killing Minotaur. Modern browsers set one end of the veil at the entrance to the WWW and, at any time, they can leave the roll at any place and walk back and forth along the veil. If they decide to do so, they can take the roll again and relocate it to any other place in the maze.

We propose another, more powerful navigation mechanism. We suggest that, in addition to what they already can do, browsers should be able to relocate the *beginning* of the veil, too. This is about the way how professional climbers use their ropes, reusing them over and over again on their way up. On the level of user interface, it would amount to giving the user the choice, which of the two stacks should be reset to empty upon a click: $\Rightarrow$ or $\Leftarrow$. This can be achieved quite easily, by adding a button to exchange the contents of the two stacks, leaving unchanged the rule that the forward stack is always discarded. It is a conservative enhancement, i.e., those not interested can still use their old way of navigating.

The new style of navigation is indeed provably more efficient than the old one.

Theorem 8. *The navigational problem defined in the second proof of Theorem 7 requires 4 clicks altogether to compute by an enhanced BSM.*

Proof. First we link a and b by a stack, as shown below. Next, we can walk between a and b on the stack, comparing their sons, without any more clicks.

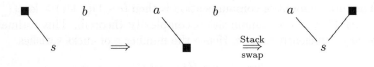

$\square$

The class of 1-visitable problems increases even more substantially, when we move from normal BSMs to the enhanced ones. Formally, we show that Theorem 4 is not valid for enhanced BSMs. Note that the problem we have shown above to be 1-visitable is not 1-visitable for standard BSMs, even with memory of visited pages, according to the Remark 2.

Let Q be the following navigational problem: $Q(\mathbb{V}, s) = $ the set of all nodes reachable from s by following always the leftmost link union the set of all nodes reachable from s by following always the rightmost link to a page not yet visited. Q has been shown in Theorem 4 not to be 1-visitable.

Theorem 9. *The necessary condition on 1-visitability expressed by Theorem 4 fails for enhanced BSMs.*

Proof. Since $Q(\mathbb{V}, s)$ doesn't satisfy the condition it suffices to show that $Q(\mathbb{V}, s)$ is 1-visitable by an enhanced BSM. The BSM follows the path formed by choosing always the first link on each page whenever this leads to a new page. This can be checked by memorizing the page the new link points to in a register and going down the $\Leftarrow$ stack to see if this page is a new one. When this path ends (because either there are no more links to follow or the leftmost link points to a page already visited), the BSM goes back to s, swaps the stacks (so that now the $\Leftarrow$ stack contains the whole so far visited path), finds the rightmost link on s which leads to a new node, and follows this procedure as long as on the current page there is at least one link not yet on the $\Leftarrow$ stack, choosing always the rightmost one. □

However, the first proof of Theorem 7 is easily seen to carry over to enhanced browser stack machines, so

Theorem 10. *There exists no recursive function $f : \mathbb{N} \to \mathbb{N}$ such that for every visitable navigational problem P on Web instances, there exists an enhanced BSM B solving P for all Web instances $(\mathbb{V}, s)$ with $C_B(\mathbb{V}, x) \leq f(|B(\mathbb{V}, x)|)$.*

Enhanced browsers are not only more efficient but also more powerful than the standard ones.

Theorem 11. *There exists a navigational problem which is solvable by an enhanced BSM, but not by a standard one.*

Proof. Omitted due to space limitations. □

References

1. S. Abiteboul and V. Vianu. *Foundations of Databases.* Addison-Wesley, 1995.
2. S. Abiteboul and V. Vianu. Queries and computation on the Web. *Theoretical Computer Science*, 239(2):231–255, 2000.
3. A. Chandra and D. Harel. Computable queries for relational data bases. *Journal of Computer and System Sciences*, 21(2):156–178, 1980.
4. M. Kaminski and N. Francez. Finite-memory automata. *Theoretical Computer Science*, 134(2):329–363, 1994.
5. E. Kushilevitz and N. Nisan. *Communication complexity.* Cambridge University Press, 1997.
6. M. Spielmann, J. Tyszkiewicz, and J. Van den Bussche. Distributed computation of Web queries using automata. In *Proceedings 21st ACM Symposium on Principles of Database Systems*. ACM Press, 2002.

Improved Results for Stackelberg Scheduling Strategies[*]

V.S. Anil Kumar and Madhav V. Marathe

Basic and Applied Simulation Science (D-2), P. O. Box 1663, MS M997
Los Alamos National Laboratory, Los Alamos NM 87545.
{anil,marathe}@lanl.gov

Abstract. We continue the study initiated in [13] on *Stackelberg Scheduling Strategies*. We are given a set of m independent parallel machines or equivalently a set of m parallel edges, each with a load dependent latency function. The setting is that of a non-cooperative game: players route their flow so as minimize their individual latencies. Additionally, there is a single player (the *leader*), who controls an α fraction of the total flow. The goal is to find a strategy for the leader (i.e. an assignment of flow to individual links) such that the selfish users react so as to minimize the total latency of the system. Building on the recent results in [13,14], we devise a *fully polynomial approximate Stackelberg scheme* that runs in time $poly(m, 1/\epsilon)$ and results in an assignment whose cost is within a $(1 + \epsilon)$ factor of the optimum Stackelberg strategy. We also study the generalization to multiple rounds. It is easy to see that more than two rounds do not help. We show that the two round Stackelberg strategy (denoted 2SS) always dominates the one round scheme. We also consider extensions of the above results to special graphs, and special kind of latency functions.

1 Introduction and Motivation

The dynamic behavior of large scale networks can often be modeled by non-cooperative games, with agents acting in a selfish manner. The fixed points of such dynamical systems often correspond to *Nash equilibrium* of the corresponding non-cooperative game. Although Nash equilibria are adequate from the standpoint of user optimum, these operating points are usually inefficient as measured by the way system resources are used (a.k.a. system/social optimum) [13,8,10]. The inefficient use of a system can be overcome by a number of possible strategies that aim to bring the operating point of the system closer to a social or a system optimum. Examples of this include: (i) **Pricing:** Use pricing mechanisms that lead to strategies by players with equilibria that are more efficient [4,15], (ii) **Algorithmic Mechanisms:** Network wide rules on how commodities are stored, routed and scheduled [11,5], (iii) **Network Design:** Designing networks in which Nash equilibria are close to global optimum [10]. The above approaches demand either the addition of a new component to the networking

---

[*] Work supported by the Department of Energy under Contract W-7405-ENG-36.

P. Widmayer et al. (Eds.): ICALP 2002, LNCS 2380, pp. 776–787, 2002.

structure, such as price or apriori design decisions regarding the network topology or policies used. Here we consider an alternative approach motivated by the earlier work of [8,13]. In this setting, we have two types of players: set of selfish players who wish to minimize the latency they experience and a leader whose aim is to optimize the overall system. The leader controls an α fraction of the total flow, and by routing it suitably, can help move the final equilibrium point more closer to the system optimum (in terms of the global objective function under consideration). Such games are referred to as *Stackelberg games* and also arise in the design and development of large scale socio-technical simulations. See [16] for more details on these projects.

Here we consider a particular Stackelberg game (referred to as *Stackelberg Flow Routing Game*), as studied by [8,13]. We have a single source destination pair joined by m parallel links from the source to the sink. Latency functions are specified for the links, and they are required to be non-decreasing. This can also be viewed as a machine scheduling problem. Each agent is assumed to constitute an infinitesimal fraction of the flow, and the total flow to be set up is denoted by r. In addition, there is one distinguished player called the leader, who controls an α fraction of the flow r. The protocol of the game is as follows: First, the leader chooses an assignment $\mathbf{s} = (s_1, \ldots, s_m)$ of flows on the links, taking into account that remaining players are going to play selfishly. Next, all the selfish players route their flows so that the system reaches a Nash equilibrium, t. The assignment chosen by the leader is called a Stackelberg strategy and it satisfies $\sum_i s_i = \alpha r$. The goal is to minimize the cost of the flow $s + t$. In this paper, the time it takes to reach the unique Nash equilibrium is not of interest. Instead, we are interested in the computational cost of finding a Stackelberg strategy. We will say more about this later.

As argued in [8,13], despite its simplicity, the above setting models a number of practical situations that arise in the design of communication networks and machine scheduling. For example, as noted in [8], in broadband networks, bandwidth is separated among different virtual paths resulting effectively, in a system of parallel and non-interfering links. Moreover, recent IP specifications provides the option of choosing a particular paths to route their packets [3,6]. Similarly, as noted in [2], many ISPs have chosen to increase their network capacity by placing a set of parallel fiber optic links between consecutive switching centers. In this setting, the ISPs as owners of the infrastructure can reserve certain amount of bandwidth for itself and allow the remainder of the bandwidth to be used by the customers.

2   Our Contributions and Related Work

We continue the study initiated in [13] on finding polynomial time computable Stackelberg strategies that improve upon the cost of Nash equilibria obtained without the presence of any leader.

1. Given a set of m parallel links with latency functions represented as polynomials with non-negative coefficients, we devise a family of Stackelberg algorithms

1SS$_\epsilon$ that for each $\epsilon > 0$, yield an assignment of flows to the links with the following property: The cost of the solution induced by **1SS$_\epsilon$** is no more than $(1+\epsilon)$ times the cost of the solution induced by an Optimal Stackelberg strategy. The algorithms run in time $poly(m, 1/\epsilon)$, and thus constitute a *fully polynomial time approximate Stackelberg scheme* (FPTAS) for the Stackelberg flow routing game on parallel links. Note that as shown in [13], the problem of computing the optimal Stackelberg strategies is **Weakly NP**-hard even for instances consisting of m parallel links between a given source destination pair, even when restricted to linear latency functions on each edge. Roughgarden's [13] results imply a $\frac{1}{\alpha}$ approximation algorithm for the case where the latency functions are nonnegative, continuous and nondecreasing, and a $\frac{4}{3+\alpha}$ approximation algorithm for the case of linear latency functions. In [13], Roughgarden left open the question of designing approximation algorithm with a better performance guarantee. Thus our results answer the above question affirmatively.

As a first step towards understanding this problem in general graphs, we consider layered graphs with bounded width, and show that there is an approximation scheme for the optimum stackelberg strategy.

2.   We then consider two natural variants of the basic Stackelberg Strategy. Given that a move by the leader followed by a round of moves by all other selfish followers improves the solution, it is natural to ask what happens if this process is repeated. It is easy to see that any further pair of leader-follower moves will yield the same solution as one round of leader-follower moves. What will make a difference is if we allow only the leader to make an extra move: in round 1, the leader assigns certain flow s to each of the links. In round 2, the selfish players then assign the remaining $(1 - \alpha)r$ flow (denoted by t) such that the flow $(s + t)$ is a Nash equilibrium. Finally, in round 3, the leader is allowed to reroute some of the αr flow it controls. Call this assignment s'. Thus the resulting assignment is $s' + t$. This kind of game is denoted by 2SS (two -round stackelberg). We show that 2SS indeed strictly dominates the 1 round Stackelberg Strategy, i.e. the cost of assignment is no more than the cost of 1 round Stackelberg. For some special classes of latency functions, we obtain better factors. An interesting aspect of the problem is that whenever one and two round Stackelberg strategies guarantee only a factor $\frac{1}{\alpha}$ of the system optimal, a simple Nash equilibrium can also guarantee the same factor. Finally, we also study the variant where the selfish followers move first, and then the leader moves. It turns out that this variant is also better than the one round stackelberg strategy.

Organization: The rest of the paper is organized as follows. Section 3 defines the basic model. In Section 4, we discuss the PTAS for the problem on parallel links. Section 5 discusses the two round stackelberg game, and Section 6 has some conclusions and open questions. Many proofs are ommitted due to lack of space, and will appear in a full version.

3   Basic Model and Preliminaries

For sake of consistency, to the extent possible, we use the notation used in [13]. In general, we have a directed network $G(V, E)$, with latency functions $\ell_e()$ specified on each edge e. A vector $\bar{r}$ specifies the flow requirement between different pairs of nodes in G. For a function f, we use $f'(x)$ to denote the derivative of f at x. Here we will assume that the latency functions $\ell_i()$ are specified as polynomials with nonnegative coefficients that have $poly(m)$ sized description [1]. For the most part, this paper deals with networks consisting of two nodes, i.e., $V = \{v_s, v_t\}$, with a set $M = \{e_1, \ldots, e_m\}$ of m parallel links between v_s and v_t with r units of flow to be sent from v_s to v_t. Throughout this paper we will use $\mathbf{z}$ to denote a vector of flow values assigned to edges and use z_i to denote the flow on edge i. The cost, $C(\mathbf{z})$, of a flow assignment $\mathbf{z} = (z_1, \ldots, z_m)$ is defined as $C(\mathbf{z}) = \sum_i z_i \ell_i(z_i)$.

By $\mathbf{x} = OPT(G, r)$ and $\mathbf{y} = Nash(G, r)$, we denote the optimum flow assignment (i.e., one that minimizes $\sum_i x_i \ell_i(x_i)$) and the Nash flow assignment vectors, respectively, when the flow requirements are specified by r. Order the links so that $\ell_i(x_i) \leq \ell_j(x_j), \forall i < j$. Sometimes, we will need to consider a subset X of links rather than all the links, and given an assignment $\mathbf{u}$, we use:

- $\mathbf{u}_X$ to denote the projection of $\mathbf{u}$ on X,
- $C(\mathbf{u}_X) = \sum_{i \in X} u_i \ell_i(u_i)$ to denote the cost of the assignment restricted to X and
- $\mathbf{u}(X) = \sum_{i \in X} u_i$ to denote the sum of flows on links restricted to X.

In general, we will use $\mathbf{u}$ and $\mathbf{u}_E$ interchangeably.

Definition 1. *A **Stackelberg Strategy** is an assignment vector $\mathbf{s}$ such that $\sum_i s_i = \alpha r$ and the Nash equilibrium $\mathbf{t}$ [2] induced by $\mathbf{s}$ is a vector satisfying the following properties.*

1. $\sum_i t_i = (1 - \alpha)r$
2. $\ell_i(s_i + t_i) \leq \ell_j(s_j + t_j)$ *for all i, j such that $t_i > 0$.*

From the definition above, given the Stackelberg assignment $\mathbf{s}$, the induced Nash assignment $\mathbf{t}$ is well defined, and the cost induced by $\mathbf{s}$ is defined as $C(\mathbf{s}) = C(\mathbf{s} + \mathbf{t}) = \sum_i (s_i + t_i) \ell_i(s_i + t_i)$.

An instance of the Stackelberg Routing problem is given by (G, α, r). Here G is the graph consisting of parallel links, α is the fraction of the flow can be chosen by the leader and r is the total flow to be routed. Thus $(1 - \alpha)r$ units of flow are routed by selfish players and each controls an insignificantly small quantity of the this flow. The game is played in two steps:

1. In Step 1, the Stackelberg player (leader) chooses a flow vector $\mathbf{s}$ such that $\sum_i s_i = \alpha r$.

---

[1] The conditions assumed in [14] were that the each latency function $\ell_i()$ is continuous, differentiable and non-decreasing, and $x\ell_i(x)$ is convex.

[2] Technically t should be indexed by $\mathbf{s}$; but in the current setting this will be clear from context and will thus be omitted.

2. In Step 2, the selfish users route the remainder of flow i.e. choose an assignment **t** of $(1 - \alpha)r$ units of flow to the links to reach a Nash equilibrium induced by **s**.

The cost of the game is $C(\mathbf{s}) = C(\mathbf{s} + \mathbf{t}) = \sum_i (s_i + t_i)\ell_i(s_i + t_i)$. Let $\mathbf{s}^*$ be the optimal Stackelberg strategy, and $\mathbf{t}^*$ the (unique) Nash equilibrium induced by $\mathbf{s}^*$. Thus $\mathbf{s}^* = \arg\min\{C(\mathbf{s}) : \mathbf{s} \text{ is a Stackelberg Strategy }\}$.

Definition 2. *An ρ-**approximate Stackelberg strategy (algorithm)** for the Stackelberg Flow Routing problem is a polynomial time algorithm that outputs an assignment s of flows such that its induced cost $C(s)$ is no more than a multiplicative factor ρ more than the cost of the assignment induced by the optimal Stackelberg Strategy s^*, i.e. $C(s) \leq \rho C(s^*)$. A **fully polynomial time approximate Stackelberg scheme** for the Stackelberg Flow Routing problem is a family of algorithms that for each a given performance requirement $\epsilon > 0$, run in time polynomial in $1/\epsilon$ and the problem specification and output an assignment vector s_ϵ such that $C(s_\epsilon) \leq (1 + \epsilon)C(s^*)$.*

Finally, we recall the results in [13,14] that will be used in the rest of the paper.

Lemma 1. *([13,14])   Suppose M is a set of machines (parallel links) with continuous, nondecreasing latency functions. Then the following hold.*

1. *For any flow requirement $r > 0$, a Nash equilibrium exists.*
2. *If x and x' are assignments at Nash equilibrium for (M, r), then $\forall i \in M$, $l_i(x_i) = l_i(x_i')$.*
3. *Suppose $x_i l_i(x_i)$ is a convex function for each machine i. Then an assignment x to the machines M is optimal iff $\forall i, j \in M$, if $x_i > 0$, then $l_i(x_i) + x_i l_i'(x_i) \leq l_i(x_j) + x_j l_j'(x_i)$. In other words, all machines with positive flow assignment have the same marginal cost function. Moreover, the optimal assignment can be computed in polynomial time.*

4   A FPTAS for Stackelberg Strategies

4.1   Properties of s*

We first isolate certain invariants of the optimal strategy, and show that the knowledge of these invariants reduces the problem of finding s^* to solving a mulidimensional knapsack instance. To get a $(1 + \epsilon)$-approximate solution, it is sufficient to guess these invariants, and this is demonstrated in the next section. Recall that $\mathbf{s}^*$ denotes the optimal strategy and $\mathbf{t}^*$ denotes the optimal Nash equilibrium induced by $\mathbf{s}^*$. Let $M_{=0} = \{i : t_i^* = 0\}$ and $M_{>0} = \{i : t_i^* > 0\}$; thus $E = M_{=0} \cup M_{>0}$ and $M_{=0} \cap M_{>0} = \phi$ Thus $M_{=0}$ consists of those machines which are no assigned flow by the selfish users and $M_{>0}$ is the set of machines that are assigned a positive flow by the selfish users. The cost induced by $\mathbf{s}^* + \mathbf{t}^*$ is the sum of the cost of assignments on $M_{=0}$ and on $M_{>0}$. Then:

- Since $\mathbf{t}^*$ is a Nash equilibrium, by Lemma 1, the latency on all $i \in M_{>0}$ is the same. Let us denote this latency by L^*.
- Second, since $\mathbf{s}^*$ is an optimal Stackelberg strategy, by Lemma 1, the marginal costs of increasing cost on any $i \in M_{=0}$ are the same. We will denote this by D^*.
- Finally, since $\forall i \in M_{=0}, \; t_i^* = 0$ it must follow that $\forall i \in M_{=0}, \; \ell_i(s_i^*) \geq L^*$ (otherwise, the Nash assignment would choose to add some flow on link i).

The following observation shows that the assignment of $\mathbf{s}^*$ to $M_{>0}$ is not unique.

Observation 1 *Let $\hat{\mathbf{s}}$ be any assignment such that $\hat{s}_i = s_i^*, \forall i \in M_{=0}$ and $\hat{s}_i \leq s_i^* + t_i^*, \forall i \in M_{>0}$, while satisfying $\sum_i \hat{s}_i = \sum_i s_i^*$. Let $\hat{t}_i = s_i^* + t_i^* - \hat{s}_i, \forall i$. Then, $\hat{t}$ is a Nash equilibrium induced by the Stackelberg strategy $\hat{\mathbf{s}}$ and $C(\hat{\mathbf{s}} + \hat{\mathbf{t}}) = C(\mathbf{s}^* + \mathbf{t}^*)$.*

4.2   Reduction to Multidimensional Knapsack

Assume now that we know L^* and D^*, and $S_0^* = s^*(M_{=0})$. Then $U_{>0}^* = r - S_0^*$ is the total assignment on $M_{>0}$ by $s^* + t^*$. Also assume that we can solve for the roots of the latency functions exactly. All these assumptions will be relaxed within a $1 + \epsilon$ factor when we look for an approximate solution in the next section.

For each link i, the basic difficulty is deciding whether it must belong to $M_{=0}$ or to $M_{>0}$. Once this decision is made, the assignment on it is easily fixed: if $i \in M_{>0}$, solve for u_i in $\ell_i(u_i) = L^*$; else (i.e. $i \in M_{=0}$,) solve for s_i in $(s_i \ell_i(s_i))' = D^*$, where the prime as stated denotes the derivative. The assumptions that the latency functions are polynomial and non decreasing imply that the roots are unique.

Since, we do not know if a link belongs to $M_{=0}$ or to $M_{>0}$, we compute for each link i, a tuple (s_i, u_i) where $\ell_i(u_i) = L^*$ and s_i is defined as follows: let y be the solution to $(x\ell_i(x))' = D^*$. If $\ell_i(y) \geq L^*$, define $s_i = y$, otherwise $s_i = \infty$. The reason is that if y represents the flow on the machine on which the selfish users do not assign any flows, then the latency on this machine should be at least L^*. Let $U^* = \sum_i u_i$ and $U_{>0}^* = \sum_{i \in M_{>0}} u_i$. The tuple tells us the assignment of flows on the link once the decision about the link being in the set $M_{=0}$ or to $M_{>0}$ is made.

Lemma 2. *Let X be a subset of links that minimizes $\sum_{i \in X} s_i \ell_i(s_i)$, while satisfying $\sum_{i \in X} s_i = S_0^*$ and $\sum_{i \in X} u_i = U^* - U_{>0}^*$. Consider the Stackelberg strategy $\mathbf{w}'$ defined as $w_i' = s_i, \forall i \in X$ and $w_i' = 0, \forall i \notin X$. Then $C(\mathbf{w}') = C(\mathbf{s}^*)$.*

Lemma 2 provides a method for choosing the membership of each link to one of the sets $M_{=0}$ or to $M_{>0}$ and also specifies a method for assigning the flow values appropriately. Note that the Stackelberg assignment does not need to assign anything on $\bar{X}$.

Given Lemma 2, the problem now reduces to of finding such an X. As mentioned before, each link $i, i = 1, \ldots, m$, is associated with a pair (s_i, u_i) and cost $c_i = s_i\ell_i(s_i)$. Suppose we are given S_0^*, U_0^* and L^*. We need to compute the cheapest subset, X, satisfying $\mathbf{s}(X) = S_0^*$ and $\mathbf{u}(X) = U^* - U_{>0}^*$. It is easy to see that the normal dynamic programming for the knapsack problem can be easily adapted here. When we try to find an approximate solution in the next section, We will actually need a solution to a slightly more general problem: given bounds A_1, A_2, B_1, B_2, determine the cheapest subset X satisfying $\mathbf{s}(X) \in [A_1, A_2], \mathbf{u}(X) \in [B_1, B_2]$.

4.3   Finding an Approximate Solution

In the previous section, we showed that if we knew the invariants L^*, D^*, S_0^* exactly, we could compute the optimum Stackelberg strategy. We cannot expect to know these quantities exactly, but can guess them within a factor of $1 + \delta$, simply by trying all possible powers of $1+\delta$. If these quantities are polynomially bounded, the number of trials is bounded by a polynomial in $\frac{\log n}{\log (1+\delta)}$. We show now that with this slack, we can still obtain an approximate solution.

We assume here that all the latency functions are rational functions of polynomials with polynomially bounded integral coefficients and exponents. This allows us to estimate the assignments on the links, given the latencies on them (which we guess, as mentioned above) and also ensures that when the assignment on a link is increased by a small factor, the latency does not blow up. We will have a fixed parameter δ, which depends on ϵ, and another parameter, δ_1, is chosen so that $\ell_i((1 + 2\delta)x) \le (1 + \delta_1)\ell_i(x)$ for any i, x. For our purposes, δ will be chosen to be inverse polynomial.

Following the discussion above, assume that we have guessed $L, D, S_0, U_{>0}$ so that $L \in [L^*, (1 + \delta_2)L^*], D \in [D^*, (1 + \delta_2)D^*], S_0 \in [S_0^*, (1 + \delta_2)S_0^*]$ and $U_{>0} \in [U_{>0}^*, (1 + \delta_2)U_{>0}^*]$ for a parameter δ_2 to be specified below. For each link i, s_i^*, u_i^* are defined as in the previous section. For each link i, solve for $\ell_i(x_i) = L$ and $(y_i\ell_i(y_i))' = D$ so that the estimates are at least as large as the exact roots of these equations, but not exceeding by a factor of $1+\delta_2$. By choosing $\delta_2 < \delta$ appropriately, we can ensure that $u_i = x_i$ satisfies $u_i \in [u_i^*, (1 + \delta)u_i^*]$. If $\ell_i(y_i) \ge (1-\delta)L$, define $s_i = y_i{}^3$, otherwise $s_i = \infty$. This gives us a tuple (s_i, u_i) for each link i.

As before, s_i is intended to be the assignment to link i if it is in $M_{=0}$ and u_i is the assignment to link i if it is in $M_{>0}$. The extra complication we will face is that even if $i \in M_{=0}$, we may have $t_i > 0$ in the approximate Stackelberg solution we find.

The next lemma – a refinement of Lemma 2, shows how the problem of approximating the Stackelberg strategy can be viewed as an approximation to the knapsack problem. The proof of the Lemma is based on Proposition 1.

Lemma 3. *Let $X \subset E$ be a subset satisfying the following conditions: (i) $\sum_{i \in X} s_i \in [(1-\delta)S_0^*, (1+\delta)S_0^*]$,   (ii) $\sum_{i \notin X} u_i \in [(1-\delta)(r-S_0^*), (1+\delta)(r-S_0^*)]$ and (iii) X minimizes the cost $\sum_{i \in X} s_i\ell_i(s_i)$.*

---

[3] This ensures that if s_i^* is finite and $\ell_i(s_i^*) \ge L^*$, $\ell_i(s_i) \ge (1 - \delta)L^*$

Consider the following Stackelberg strategy $\boldsymbol{w'}$ induced by X: if $s(X) \leq \alpha r/(1+2\delta)$, $w'_i = (1+2\delta)s_i, \forall i \in X$ and if $s(X) > \alpha r/(1+2\delta)$, $w'_i = \frac{\alpha r}{s(X)} s_i$. Then, $C(\boldsymbol{w'}) \leq (1+\epsilon)C(\boldsymbol{s^*})$

The following proposition is needed in the proof of Lemma 3.

Proposition 1. *Let $\boldsymbol{z'}$ be the Nash assignment induced by $\boldsymbol{w'}$ and $u' = w' + z'$. Let L' be the common Nash latency on all edges i such that $z'_i > 0$. Then the following hold: (i) $\forall i \in X$, $w'_i \leq (1+2\delta)s_i$. (ii) $\boldsymbol{w'}(X) \geq S^*_0$, and (iii) $L' \leq (1+\delta_1)L$.*

Proof Sketch of Lemma 3: Let $\boldsymbol{z'}$ be the Nash assignment induced by $\boldsymbol{w'}$ and $\mathbf{u'} = \mathbf{w'} + \mathbf{z'}$. Let L' be the common Nash latency on all edges i such that $z'_i > 0$.

1. The cost of $\mathbf{w'} + \mathbf{z'}$ restricted to $\bar{X}$ is $C(\mathbf{w'}_{\bar{X}}) = \sum_{i \in \bar{X}} u'_i L' = u'(\bar{X})L'$. In the remaining steps, we bound the cost restricted to set X.

2. By part 3 of Proposition 1, if $z'_i > 0$ for some $i \in X$, $\ell_i(w'_i + z'_i) = L' \leq (1+\delta_1)L$.

3. Bound on the sum $\sum_i w'_i \ell_i(w'_i + z'_i)$: By construction, $\forall i \in X, \ell_i(s_i) \geq (1 - \delta)L \geq (1-\delta)\ell_i(w'_i + z'_i)/(1+\delta_1)$, and therefore, by part 1 of Proposition 1, we have $\sum_{i \in X} w'_i \ell_i(w'_i + z'_i) \leq (1+2\delta)^2(1+\delta_1)\sum_{i \in X} s_i \ell_i(s_i) \leq (1+2\delta)^3(1+\delta_1)^2 C(s^*_{M=0})$, where the last inequality uses the property that X is the cheapest set satisfying the feasibility conditions. In the remaining steps, we bound the quantity $\sum_i z'_i \ell_i(w'_i + z'_i) = z'(X)L'$.

4. $\mathbf{z'}(X) = r - \mathbf{w'}(X) - \mathbf{u'}(\bar{X}) \leq 2\delta\mathbf{w'}(X) + (1+2\delta)\mathbf{u}(X) - \mathbf{u'}(X)$ because $\mathbf{w'}(X) + \mathbf{u}(X) \geq (1-\delta)r$.

5. $\mathbf{w'}(X)L' \leq (1+2\delta)(1+\delta_1)\mathbf{s}(X)(1-\delta)L \leq (1+2\delta)(1+\delta_1)\sum_{i \in X} s_i \ell_i(s_i) \leq (1+2\delta)^2(1+\delta_1)^2 C(s^*_{M=0})$, using $\mathbf{w'}(X) \leq \mathbf{s}(X)(1+2\delta)$, $L' \leq (1+\delta_1)L$ and $\ell_i(s_i) \geq (1-\delta)L, \forall i \in X$.

6. Putting all the above together yields $C(\mathbf{w'}) = \sum_{i \in X}(w'_i + z'_i)\ell_i(w'_i + z'_i) + \sum_{i \in \bar{X}} u'_i \ell_i(u'_i) \leq (1+4\delta)^3(1+\delta_1)^2 C(s^*_{M=0}) + (1+2\delta)(1+\delta_1)\mathbf{u}(\bar{X})L \leq (1+\epsilon)C(s^*)$, where ϵ is chosen so that $1+\epsilon = (1+4\delta)^3(1+2\delta_1)^2$.

□

Recall that we have estimates $S_0 \in [S^*_0, (1+\delta_2)S^*_0]$ and $U_{>0} \in [U^*_{>0}, (1+\delta_2)U^*_{>0}]$ for appropriate $\delta_2 < \delta$. Since we do not know $S^*_0, U^*_{>0}$ exactly, we will actually find the cheapest subset X such that $s(X) \in [(1-\delta_2)S_0, (1+\delta_2)S_0] \subset [(1-\delta)S^*_0, (1+\delta)S^*_0]$ and $\mathbf{u}(X) \in [U - (1+\delta_2)U_{>0}, U - (1-\delta_2)U_{>0}] \subset [U - (1+\delta)U^*_{>0}, U - (1-\delta)U^*_{>0}]$ (which automatically ensures that $\mathbf{u}(\bar{X}) \in [(1-\delta)U^*_{>0}, (1+\delta)U^*_{>0}]$). This leaves us with the problem of finding an approximate solution and this is solved in the following steps given in Figure 1.

Proposition 2. *Let X be as constructed in Step 3 of Algorithm described in Figure 1. Construct a Stackelberg strategy $\boldsymbol{w'}$ from X, as in Lemma 3. Then $C(\boldsymbol{w'}) \leq (1+\epsilon)C(\boldsymbol{s^*})$.*

5   Two-Round Stackelberg Flow Routing Game

We consider below a two round modification of the game. The corresponding Stackelberg strategy, called 2SS is denoted by $(\mathbf{s}, \mathbf{s'})$.

1. Guess the values of $L, D, S_0, U_{>0}$ as described before, which means we try out every power of $1 + \delta_2$ as a potential candidate for these values. Once this is done, solve for s_i, u_i and then perform the steps below for each set of candidates to get a solution corresponding to these values if it is feasible. Finally, choose the set of candidates that gives the cheapest solution.

2. **Scaling** Let $m_s = \max_i\{s_i : s_i < \infty\}$ and $m_u = \max_i\{u_i\}$. Define $\hat{s}_i = \lfloor \frac{s_i m}{\gamma m_s} \rfloor$, $\hat{u}_i = \lfloor \frac{u_i m}{\gamma m_u} \rfloor$, $\hat{S}_0 = \lfloor \frac{S_0 m}{\gamma m_s} \rfloor$ and $\hat{U}_{>0} = \lfloor \frac{U_{>0} m}{\gamma m_u} \rfloor$. Let $\hat{U} = \sum_i \hat{u}_i$. If $s_i > S_0, s_i < \infty$ for some i, it is clear that $i \in M_{>0}$, and we can remove link i from consideration. Therefore, wlog we can assume that $m_s \le S_0$. Similarly, we can assume that $m_u \le U_{>0}$.

3. **The Dynamic Program** Run the same dynamic program described in the preceding section: compute the cheapest set $\mathcal{S}(m, (1 - \delta_3)\hat{S}_0, (1 + \delta_3)\hat{S}_0, \hat{U} - (1+\delta_3)\hat{U}_{>0}, \hat{U} - (1-\delta_3)\hat{U}_{>0})$, in the notation of the previous section, where δ_3 is a small enough parameter to be fixed later. This gives us a set X such that $\hat{s}(X) \in [(1 - \delta_3)\hat{S}_0, (1 + \delta_3)\hat{S}_0]$, $\hat{u}(X) \in [\hat{U} - (1 + \delta_3)\hat{U}_{>0}, \hat{U} - (1 - \delta_3)\hat{U}_{>0})]$ and the cost of X is minimized. The running time of this step is $O(m^5/\gamma)$.

Fig. 1. Overall Description of The Stackelberg Strategy for the Parallel Links Graph.

1. Choose a Stackelberg strategy $\mathbf{s} = (s_1, \ldots, s_m)$ satisfying $\sum_i s_i \le \alpha r$
2. Let $\mathbf{t}$ be the Nash-equilibrium induced by $\mathbf{s}$.
3. Keep $\mathbf{t}$ fixed and change $\mathbf{s}$ to vector $\mathbf{s}'$

The goal of two-round Stackelberg strategy $(\mathbf{s}, \mathbf{s}')$ is to choose $\mathbf{s}$ and $\mathbf{s}'$ so that $C(\mathbf{s}' + \mathbf{t})$ (also denoted by $C(\mathbf{s}, \mathbf{s}')$) is minimized. Roughgarden [13] showed that the one-round Stackelberg strategy leads to an assignment with cost at most $\frac{1}{\alpha}$ times the social (system) optimum, and so the question is whether a two-round strategy leads to a constant factor improvement. It is easy to see that further rounds do not help. If we have k alternating Stackelberg/Nash strategy, the final solution just depends on the final round. Therefore, the only meaningful setting is when the leader plays once more in the end.

The quality of 2SS In general, 2SS might not guarantee a factor better than 1SS but we point out instances where it does better. Also, it is interesting to note that *whenever 2SS does not do better than $\frac{1}{\alpha}$ of the optimum, just the Nash assignment is within $\frac{1}{\alpha}$ of the optimum, and so in this case, the Nash solution is as good as the 1SS solution.* This is summarized in the following lemma.

Lemma 4. *Let $\boldsymbol{x} = OPT(r)$ and $\boldsymbol{y} = Nash(r)$. Let $A = \{i : x_i \ge y_i\}$. If $x(A) - y(A) \ge \alpha r$, $C(\boldsymbol{y}) \le \frac{1}{\alpha}C(\boldsymbol{x})$. If $x(A) - y(A) < \alpha$, 2SS leads to the optimum solution.*

Proof Let $L(y)$ be the common Nash latency of $\mathbf{y}$. Assume first that $x(A) - y(A) \ge \alpha r$. Then, $C(x_A) \ge (\alpha r + y(A))L(y)$. Therefore, $C(y) = rL(y) \le \frac{1}{\alpha + y(A)/r}C(x)$. Next, consider the case $x(A) - y(A) < \alpha r$. Choose the vector s to be $s_i = y_i - x_i, i \notin A$ and $s_i = 0, i \in A$. Then $\sum_i s_i \le \alpha r$. The induced Nash equilibrium will then be $t_i = y_i - s_i, \forall i$. In the second round, choose

$s_i' = x_i - y_i, i \in A$ and $s_i' = 0, i \notin A$. Then $\sum_i s_i' = \sum_i s_i$ and $s' + t$ gives exactly the optimum solution x.   □

Note that the above scheme for 2SS actually results in a factor of at most $\frac{1}{\alpha + y(A)/r}$. The LLF strategy for 1SS only guarantees a factor of $\frac{1}{\alpha}$ and it can be shown that no strategy for 1SS can actually do better.

The following Lemma shows that the worst case bounds can be improved when the latency functions are restricted. If $\ell_i(z(1 + \delta)) \geq \phi(\delta)\ell_i(z)$ for each i, the guarantee achieved by 2SS can be improved. Such an assumption is not too unrealistic, since functions growing as fast as a polynomial have this property at least asymptotically.

Lemma 5. *Let* $\forall i, u \quad \ell_i(u(1 + \delta)) \geq \phi(\delta)\ell_i(u)$. *Then if* $\alpha < 1$ *and* $\phi(\frac{1}{1-\alpha}) > 1$, *then there is a polynomial time computable 2SS strategy* $(\mathbf{s}, \mathbf{s}')$, *such that*

$$C(\mathbf{s}, \mathbf{s}') \leq \frac{1 - \alpha + \alpha\phi(\frac{1}{1-\alpha})}{\alpha\phi(\frac{1}{1-\alpha})}C(\boldsymbol{x}) < \frac{1}{\alpha}C(\boldsymbol{x}).$$

5.1   Linear Latency Functions

Roughgarden [13] showed that the LLF strategy for 1SS yields an assignment of cost bounded by $\frac{4}{3+\alpha}$ times the optimal, when the latency functions are all linear. We show that 2SS gives a strictly better bound for this case. In this section, we assume that the latency function for link i has the form $\ell_i(u) = a_i u + b_i, i = 1, \ldots, m$ and $a_i, b_i \geq 0, \forall i$. As observed in [13], our ordering on the edges also corresponds to the order $b_1 \leq \ldots \leq b_m$.

The following lemma relates the costs of the optimum flow of r and the Nash flow of r'.

Lemma 6. *Suppose* $\mathbf{x} = OPT(S, r)$ *and* $\mathbf{y} = Nash(S, r')$. *Then,* $C(\mathbf{x}) = (r - \frac{r'}{4})L(\mathbf{y}) + \sum_i a_i(x_i - \frac{y_i}{2})^2 + \frac{b_i y_i}{4}$, *where* $L(\mathbf{y})$ *is the common Nash latency for* $\mathbf{y}$.

Lemma 7. *If the latency functions are all linear, then there exists a 2SS strategy* $(\mathbf{s}, \mathbf{s}')$ *such that*

$$C(\mathbf{s}, \mathbf{s}') \leq \max(\frac{4}{3 + \alpha + \alpha^3/32}, \frac{4}{3 + \alpha + \alpha(1 - \alpha)/8})C(\mathbf{x})$$

Proof Sketch: Let $A = \{i : x_i \geq y_i\}$. Let $x(A) = \beta r, y(A) = \beta' r$. As before, the initial Stackelberg assignment s is concentrated on $\bar{A}$, in such a way that $s + t = y$, where t is the Nash assignment induced by s. The best 2SS strategy would be to choose s so that after the first round, when s is transferred to elements of A, the remainder on $\bar{A}$ is assigned optimally. Because of the difficulty of analyzing this, we consider a different scheme below. Following earlier remarks, we will assume that $x(A) - y(A) \geq \alpha$.

We have three different cases, and the choice of s is different in each. The choice of $\mathbf{s}'$ is the same in all cases, and is described below. Let $A = \{a, \ldots, m\}$ and let $a' \in A$ be the smallest index such that (i) $\sum_{i \in A_1} x_i - y_i \leq \alpha$, where

$A_1 = \{a, \ldots, a'\}$ and $A_2 = \{a' + 1, \ldots, m\}$, (ii) $x_i \geq z_i, \forall i \in A_2$, where $\mathbf{z}_A$ is defined as $\mathbf{z}_{A_2} = Nash(A_2, \beta' + \alpha - \mathbf{x}(A_1))$, $\mathbf{z}_{A_1} = \mathbf{x}_{A_1}$, (iii) $\ell_{a'}(x_{a'}) \leq L(\mathbf{z}_{A_2})$, where $L(\mathbf{z}_{A_2}) = \ell_i(z_i), i \in A_2$ is the common Nash latency of $\mathbf{z}_{A_2}$. Then $s'_i = x_i - y_i, i = a, \ldots, a'$ and $s'_i = z_i - y_i, i = a' + 1, \ldots, m$. It can be shown that $C(\mathbf{x}_A) \geq \frac{\beta}{\beta' + \alpha} C(\mathbf{z}_A)$.

Case 1: $\exists i \in \bar{A}$ such that $y_i - x_i \geq \alpha r$ and $b_i \geq \delta L(y)$:   Define $\mathbf{z}_{\bar{A}} = Nash(\bar{A}, r(1 - \beta' - \alpha))$. Choose $\mathbf{s}$ as $s_j = y_j - z_j, \forall j \in \bar{A}$ and $s_j = 0, \forall j \in A$. By lemma 6, $C(\mathbf{x}_{\bar{A}}) \geq (1 - \beta - \frac{1 - \beta' - \alpha}{4}) r L(\mathbf{z}_{\bar{A}}) + b_i z_i / 4$. It can be shown that since $y_i - x_i \geq \alpha r$, which yields $C(\mathbf{x}_{\bar{A}}) \geq \frac{3 - 4\beta + \beta' + \alpha + \delta\alpha}{4} L(\mathbf{z})$. The lemma now follows by fixing $\delta = \alpha^2/4$ and combining the previous relations.

Case 2: $\exists i \in \bar{A}$ such that $y_i - x_i \geq \alpha r$ and $b_i \leq \delta L(y)$: Choose $s_i = \alpha r$ and $s_j = 0, j \neq i$. Define $\mathbf{z}_{\bar{A}}$ as $z_i = y_i - \alpha r$ and $z_j = y_j, \forall j \in \bar{A} \setminus \{i\}$. As in the previous case, $x_i \geq y_i - x_i \geq \alpha r$. Let $x_i = \gamma r \geq \alpha r$ and $y_i = \gamma' r \geq 2\alpha r$. By lemma 6, $C(\mathbf{x}_{\bar{A} \setminus \{i\}}) \geq (1 - \beta - \gamma - \frac{1 - \beta' - \gamma'}{4}) r L(y)$ Since $b_i \leq \delta L(y)$, it can be shown that $C(\mathbf{x}_{\{i\}}) = \gamma(a_i\gamma + b_i) \geq \frac{\gamma^2}{\gamma'} L(y)$ and putting all of these together, we get $C(\mathbf{x}) \geq \frac{3 - 4\gamma + \beta' + \gamma' + 4\epsilon + 4\gamma^2/\gamma'}{4} L(y)$ and $C(\mathbf{z}) \leq (1 + \epsilon - \alpha \frac{\gamma' - \alpha}{\gamma'} + \alpha\delta) L(y)$. The lemma now follows by setting $\delta = \alpha^2/4$ and by considering different cases for ϵ.

Case 3: $y_i - x_i < \alpha r, \forall i \in \bar{A}$: In this case, there always exists a set $B \subset \bar{A}$ such that $\alpha r/2 \leq y(B) - x(B) \leq \alpha r$. Let $x(B) = \gamma r, y(B) = \gamma' r$. Define $\mathbf{z}_{\bar{A} \setminus B} = Nash(\bar{A} \setminus B, r(1 - \beta' - \gamma - \alpha))$ and $\mathbf{z}_B = \mathbf{x}_B$. Choose $\mathbf{s}$ so that $s_j = y_j - x_j, \forall j \in B$ and $s_j = y_j - z_j, j \in \bar{A} \setminus B$. By Lemma 6 $C(\mathbf{x}_{\bar{A} \setminus B}) \geq \frac{3 - 4\beta - 3\gamma + \alpha}{4} L(\mathbf{z})$. Also, $C(\mathbf{z}_{\bar{A} \setminus B}) = (1 - \beta' - \gamma - \alpha) L(\mathbf{z})$. Adding the above inequalities, we get $C(\mathbf{x}_{M \setminus B}) \geq \frac{3 - 3\gamma + \alpha + 4\beta\epsilon/(\alpha + \beta')}{4} L(\mathbf{z})$. The lemma now follows by considering different cases for ϵ.   □

6   Conclusions

We gave polynomial time approximation scheme for approximating the optimal stackelberg strategy on graphs consisting of m parallel links. The immediate problem in attempting to generalize this approach to more general topologies is that the particular invariants we identified for designing the dynamic program do not hold.

Nevetherless, for layered graphs with bounded number of nodes at each layer, it is still possible to extend this approach: for each node v, we guess the flow v to the sink, the common latency of flow paths through v carrying flow corresponding to the selfish users and the marginal cost of paths through v that only carry the leader's flow. Once this is done, the flow assignment between two successive layers can be obtained by solving the same convex program given in [14,13]. Finally, it is easy to see that the number of possible guesses at all vertices is exponential in the width, and polynomial in the number of vertices and the error parameter.

An immediate open question is to investigate if similar results hold for general graph topologies. A more difficult conceptual question is to study the Stackelberg game when the flow is not infinitely divisible, for instance in the model of [9].

References

1. M. Beckman, C. Mcguire and C. Winstein. *Studies in the Economics of Transportation.* Yale University Press, 1956.
2. J. Bennett, C. Partridge and N. Shectman. "Packet reordering is not Pathological Network Behavior," *IEEE/ACM Transactions on Networking,* 7(6), Dec. 1999, pp. 789-798.
3. J. Carrahan, P. Russo, K. Kitami and R. Kung. Intelligent Network Overview. *IEEE Communications Magzine,* 31, pp. 30-36, 1993.
4. R. Cocchi, S. Shenker, D. Estrin and L. Zhang. Pricing in Computer Networks: Motivation, Formulation and Example. *IEEE/ACM Transactions on Networking.* 1(6), pp. 614-627, 1993.
5. T. O'Connell, and R. Stearns. Polynomial Time Mechanisms for Collective Decision Making. *Game Theory and Decision Theory in Agent-Based Systems.* S. Parsons, P. Gmytrasiewicz, P. and M. Wooldridge, (eds.), Kluwer Academic Publishers, 2000.
6. S. Deering and R. Hinden. *Internet Protocol Version 6 Specification.* Internet Draft IETF March 1995.
7. S. Dafermos and F. Sparrow. The Traffic Assignment Problem for a General Network. *J. Research of the National Bureau of Standards Series B.* 73B(2), pp. 91-118, 1969.
8. Y. Korillis, A. Lazar and A. Orda. Achieving Network Optima Using Stackelberg Routing Strategies. *IEEE/ACM Transactions on Networking.* 5(1), pp. 161-173, 1997.
9. E. Koutsoupias and C. Papadimitriou. Worst-Case Equilibria. *Proc. 16th Annual Symposium on Theoretical Aspects of Computer Science (STACS),* pp. 403-413, 1999.
10. Y. Korillis, A. Lazar and A. Orda. Capacity Allocation Under Non-cooperative Routing. *IEEE Transactions on Automatic Control.* 42(3), pp. 309-325, 1997.
11. N. Nisan and A. Ronen. Algorithmic Mechanism Design. *Proc. 31st Annual ACM Symposium on Theory of Computing (STOC).* pp. 129-140, 1999.
12. C. Papadimitriou. Algorithms, games and the Internet. *Proc. 33rd Annual ACM Symposium on Theory of Computing.* pp. 749-753, 2001.
13. T. Roughgarden. Stackelberg Scheduling Strategies. *Proc. 31st ACM Symposium on Theory of Computing (STOC).* pp. 2001.
14. T. Roughgarden and E. Tardos. How bad is Selfish Routing. *Proc. 41st Annual Symposium on Foundations of Computer Science.* pp. 93-102, 2000.
15. C. Saraydar, N. Mandayam and D. Goodman. Efficient Power Control via Pricing in Wireless Data Networks To Appear in *IEEE Trans. on Communications,* 2001.
16. http://www.lanl.gov/orgs/d/d5/projects/MESA/mesa.html
    http://www.lanl.gov/orgs/d/d2/projects.html.

Call Control in Rings[*]

Udo Adamy[1], Christoph Ambuehl[1], R. Sai Anand[2][**], and Thomas Erlebach[2]

[1] Institute for Theoretical Computer Science, ETH Zürich, 8092 Zürich, Switzerland.
{adamy|ambuehl}@inf.ethz.ch
[2] Computer Engineering and Networks Laboratory, ETH Zürich, 8092 Zürich, Switzerland.
{anand|erlebach}@tik.ee.ethz.ch

Abstract. The call control problem is an important optimization problem encountered in the design and operation of communication networks. The goal of the call control problem in rings is to compute, for a given ring network with edge capacities and a set of paths in the ring, a maximum cardinality subset of the paths such that no edge capacity is violated. We give a polynomial-time algorithm to solve the problem optimally. The algorithm is based on a decision procedure that checks whether a solution with at least k paths exists, which is in turn implemented by an iterative greedy approach operating in rounds. We show that the algorithm can be implemented efficiently and, as a by-product, obtain a linear-time algorithm to solve the call control problem in chains optimally.

1 Introduction

Due to the ever-increasing importance of communication networks for our society and economy, optimization problems concerning the efficient operation of such networks are receiving considerable attention in the research community. Many of these problems can be modeled as graph problems or path problems in graphs. A prominent example is the *call admission control problem*, where the task is to determine which of the requests in a given set of connection requests to accept or reject so as to optimize some objective, e.g., the number of accepted requests. In this paper, we consider a call admission control problem in ring networks and prove that it can be solved optimally by an efficient polynomial-time algorithm. The ring topology is a fundamental network topology that is frequently encountered in practice. As an additional application of our algorithm, we show that it can also be used to compute optimal solutions to periodic scheduling problems with rejection.

Problem Definition and Applications. The CALLCONTROL problem considered in this paper is defined as follows. An instance of the problem is given by an undirected graph (V, E) with edge capacities $c : E \rightarrow \mathbb{N}$ and a multi-set P of m paths in (V, E). The paths represent connection requests whose acceptance requires the reservation of one unit of bandwidth on all edges of the path. A feasible solution is a multi-set $Q \subseteq P$ such that for every edge $e \in E$, the number of paths in Q that contain e is at most $c(e)$. Such

---

[*] Research partially supported by the Swiss National Science Foundation.
[**] Supported by the joint Berlin/Zurich graduate program Combinatorics, Geometry, and Computation (CGC), financed by ETH Zurich and the German Science Foundation (DFG).

P. Widmayer et al. (Eds.): ICALP 2002, LNCS 2380, pp. 788–799, 2002.

a multi-set of paths is called a *feasible set* and the paths in it are called *accepted*. The objective is to maximize the number of accepted paths. Whenever we talk about a set of paths in the following, we allow that the set is actually a multi-set, i.e., that it can contain several instances of the same path. In this paper we deal with CALLCONTROL mainly in ring networks. A *ring network* with n nodes is an undirected graph (V, E) that is a cycle on the nodes $V = \{0, \dots, n-1\}$. We imagine the cycle drawn in the plane with the nodes labeled clockwise. The edge $e_i \in E$, $0 \le i < n$, connects the two neighboring nodes i and $(i + 1) \bmod n$ and has a non-negative integer capacity $c(e_i)$.

The problem of CALLCONTROL in ring networks as defined above applies to various types of existing communication networks with ring topology. For example, the problem applies to ring networks that support bandwidth reservation (e.g., ATM networks) and in which the route taken by a request is determined by some other mechanism and cannot be modified by the call control algorithm. Furthermore, it applies to bidirectional self-healing rings with full protection. In such rings, one direction of the ring (say, clockwise) is used to route all accepted requests during normal operation, and the other direction is used only in case of a link failure in order to reroute the active connections that are affected. In all-optical WDM ring networks with w wavelengths that have a wavelength converter in at least one node, any set of lightpaths with maximum link load w can be established simultaneously [13]. Thus, call admission control in such networks can be modeled as CALLCONTROL with all edge capacities equal to w.

Furthermore, it should be noted that problems related to call admission control are often encountered in an on-line setting, where the requests are presented to the algorithm one by one and the algorithm must accept or reject each request without knowledge of future requests. However, we think that it is meaningful to study the off-line version as well for several reasons. First, an off-line call control algorithm is needed in the network *design* phase, when a candidate topology with link capacities is considered and one wants to know how many of the forecasted traffic requirements can be satisfied by the network. Second, an off-line call control algorithm is useful in a scenario that supports advance reservation of connections, because then it is possible to collect a number of reservation requests before the admission control is carried out for a whole batch of requests. Finally, an optimal off-line call control algorithm is helpful as a benchmark for the evaluation of other off-line or on-line call control strategies.

We briefly discuss an application in periodic scheduling. Without loss of generality, we assume a time period of one day. There are k machines and a set of tasks with fixed start and end times. (For example, there could be a task from 10am to 5pm and another task from 3pm to 2am on the following day.) Each task must be accepted or rejected. If it is accepted, it must be executed every day from its start time to its end time on one of the k machines, and each machine can execute only one task at a time. The goal is to select as many tasks as possible while ensuring that at most k of the selected tasks are to be executed simultaneously at any point in time. By taking the start times and end times of all given tasks as nodes in a ring, we can view the tasks as calls and compute an optimal selection of accepted tasks by solving the corresponding CALLCONTROL problem with all edge capacities set to k. Even if the number of available machines changes throughout the day (and the changes are the same every day), the problem can still be handled as a CALLCONTROL problem with arbitrary edge capacities.

Related Work. As paths in a ring network can be viewed as arcs on a circle, path problems in rings are closely related to *circular-arc graphs*. A graph is a circular-arc graph if its vertices can be represented by arcs on a circle such that two vertices are joined by an edge if and only if the corresponding arcs intersect [9]. Circular-arc graphs can be recognized efficiently [5]. For a given circular-arc graph, a maximum clique or a maximum independent set can be computed in polynomial time [9]. Coloring a circular-arc graph with the minimum number of colors is NP-hard [8]. A coloring with at most $1.5\,\omega$ colors always exists and can be computed efficiently [10], where ω is the size of a maximum clique in the graph. Concerning our CALLCONTROL problem, we note that the special case where all edges have capacity 1 is equivalent to the maximum independent set problem in circular-arc graphs. We are interested in the case of arbitrary edge capacities, which has not been studied previously.

Many authors have investigated call control problems for various network topologies in the off-line and on-line setting. For topologies containing cycles, an important distinction for call control is whether the paths are specified as part of the input (like we assume in this paper) or can be determined by the algorithm. In the latter case, only the endpoints are specified in the input, and we refer to the problem as CALLCONTROLANDROUTING. The special case of CALLCONTROLANDROUTING where all edges have capacity 1 is called the *maximum edge-disjoint paths* problem (MEDP). We refer to [3, Chapter 13] and [11,12] for surveys on on-line algorithms for call control problems and mention only some of the known results here.

For chains, the off-line version of CALLCONTROL is closely related to the maximum k-colorable induced subgraph problem for interval graphs. The latter problem can be solved optimally in linear time by a clever implementation of a greedy algorithm provided that a sorted list of interval endpoints is given [4]. This immediately gives a linear-time algorithm for CALLCONTROL in chains where all edges have the same capacity. It is not difficult to adapt the approach to chains with arbitrary capacities incurring an increase in running-time. As a by-product of our algorithm for rings, we will obtain a linear-time algorithm for CALLCONTROL in chains with arbitrary capacities.

The on-line version of CALLCONTROL in chains with unit edge capacities was studied for the case with preemption (where interrupting and discarding a call that was accepted earlier is allowed) in [7], where competitive ratio $O(\log n)$ is achieved for a chain with n nodes by a deterministic algorithm. A randomized preemptive $O(1)$-competitive algorithm for CALLCONTROL in chains where all edges have the same capacity is given in [1]. It can be adapted to ring networks with equal edge capacities.

In [2], the preemptive on-line version of CALLCONTROL is studied with the number of *rejected* calls as the objective function. They obtain competitive ratio 2 for chains with arbitrary capacities, 2 for arbitrary graphs with unit capacities, and $O(\sqrt{m})$ for arbitrary graphs with m edges and arbitrary capacities. For the off-line version, they give an $O(\log m)$-approximation algorithm for arbitrary graphs and arbitrary capacities.

2   Preliminaries

Let $P = \{p_1, \ldots, p_m\}$ denote the given set of m paths, each connecting two nodes in the ring network. Every $p_i \in P$ is an ordered pair of nodes $p_i = (s_i, t_i)$ from V^2 with $s_i \neq t_i$.

The path p_i contains all edges from the *source node* s_i to the *target node* t_i in clockwise direction. For a subset $Q \subseteq P$, the *ringload* $L(Q, e_i)$ of an edge e_i with respect to Q is the number of paths in Q that contain the edge e_i, i.e. $L(Q, e_i) := |\{p \in Q : e_i \in p\}|$. A subset $Q \subseteq P$ is called *feasible* if the ringload does not exceed the capacity of any edge, i.e. $L(Q, e_i) \leq c(e_i)$ for all $e_i \in E$.

By opening the ring at node 0, we partition the set P of paths into two disjoint subsets P_1 and P_2, where P_1 is the set of paths that do not have node 0 as an internal node and P_2 are the remaining paths, i.e., the paths going through node 0. Each path in P_2 consists of two pieces: the *head* of the path extends from its source node to node 0, the *tail* from node 0 to its target node. To simplify the explanation we introduce a node n and identify it with node 0. From now on, the paths with target node 0 are treated as if they end at node n. Thus we have the characterization $P_1 = \{p_i \in P : s_i < t_i\}$ and $P_2 = \{p_i \in P : s_i > t_i\}$. Note that $P = P_1 \cup P_2$.

We define a linear ordering on the paths in P as follows. All paths in P_1 are strictly less than all paths in P_2. Within both subsets we order the paths by increasing target nodes. Paths with equal target nodes are ordered arbitrarily. We call this ordering *greedy*. In the example of Fig. 1(a) the paths $p_1, \ldots, p_6$ are in greedy order. The solid paths $p_1, \ldots, p_4$ are in P_1. P_2 consists of the dotted paths p_5 and p_6.

The algorithm considers a chain of $2n$ edges consisting of two copies of the ring glued together. The chain begins with the first copy of e_0 and ends with the second copy of e_{n-1}, see Fig. 1(b). The tails of the P_2-paths are in the second copy of the ring, while the P_1-paths and the heads of the P_2-paths are in the first copy. Note that the greedy order of the paths corresponds to an ordering by right endpoints in this chain.

For a given set Q of paths, we define $L_1(Q, e_i)$ and $L_2(Q, e_i)$ to be the load of the paths in Q on the first copy of e_i and their load on the second copy of e_i, respectively. Thus, the paths in P_1 and the heads of the paths in P_2 contribute to the load values $L_1(P, e_i)$ of the first copy of the ring. The tails of the P_2-paths determine the load values $L_2(P, e_i)$. The ringload L is simply the sum of L_1 and L_2. With this definition of L_2, we can introduce the central notion of profiles.

Definition 1 (Profiles). *Let Q be a set of paths. The profile π of Q is the non-increasing sequence of n load values L_2 for the edges $e_0, \ldots, e_{n-1}$ in the second copy of the ring,*

$$\pi_Q := L_2(Q, e_0) \ldots L_2(Q, e_{n-1}).$$

With $\pi_Q(e_i)$ we denote the profile values $L_2(Q, e_i)$ for all edges $e_i \in E$. The empty profile is zero everywhere. For profiles π and π' we have $\pi \leq \pi'$, iff $\pi(e_i) \leq \pi'(e_i)$ for all edges $e_i \in E$.

A set Q of paths is called *chain-feasible* if it does not exceed the capacity of any edge in this chain of length $2n$. In other words, Q is *chain-feasible*, if it does not exceed the capacities in both copies of the ring, i.e. $L_1(Q, e_i) \leq c(e_i)$ and $L_2(Q, e_i) \leq c(e_i)$ for all $e_i \in E$. It is called *chain-feasible for (starting) profile π* if it is chain-feasible and in the first copy of the ring the stronger inequalities $L_1(Q, e_i) + \pi(e_i) \leq c(e_i)$ hold for all $e_i \in E$. Observe that a set Q of paths is feasible (in the ring) if and only if it is chain-feasible for starting profile π_Q.

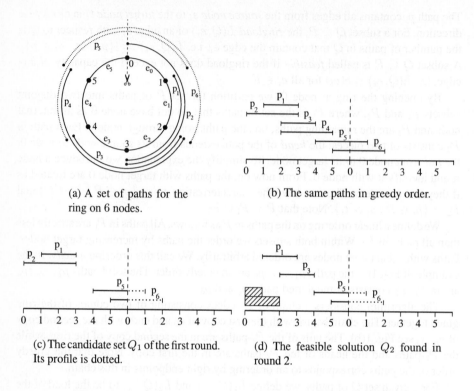

(a) A set of paths for the ring on 6 nodes.

(b) The same paths in greedy order.

(c) The candidate set Q_1 of the first round. Its profile is dotted.

(d) The feasible solution Q_2 found in round 2.

Fig. 1. The decision procedure. Is there a feasible solution with 4 paths?

3   The Algorithm

The goal of the algorithm is to find a maximum sized feasible subset of paths in P. The algorithm builds a chain of $2n$ edges consisting of two copies of the ring glued together. It sorts the paths in P according to the greedy order. The heart of our algorithm is a decision procedure that, given a parameter k, decides whether there exists a feasible solution $Q \subseteq P$ of size k, or not. Clearly, the maximum k can be found by several calls to this procedure. The decision procedure makes heavy use of the *greedy algorithm*, which processes the paths one by one in greedy order. If adding the current path does not exceed any capacity constraint on its edges, the path is accepted and the edge capacities are reduced accordingly; otherwise it is rejected.

We are now ready to describe the decision procedure. We start with the empty profile. The decision procedure works in rounds. In each round, it computes a greedy solution of k paths for a given profile as follows. It initializes both copies of the ring with the edge capacities $c(e_i)$ and subtracts the profile values from the initial capacities of the edges in the first copy, since these capacities are occupied by the profile. Then, it starts to place k paths using the greedy algorithm. If the procedure runs out of paths before it can select k of them, there is no feasible solution of size k. It answers "no" and stops. Otherwise, let Q_i denote the candidate set of k chosen paths in round i. By construction, the set Q_i is chain-feasible for the given starting profile, but not necessarily feasible in the ring, since the tails of the chosen P_2-paths together with the selected paths in P_1 and the heads of the chosen P_2-paths may violate some capacity constraints.

At the end of round i, the procedure compares the profile of Q_i with the profile of the previous round. If both are equal, the paths in Q_i form a feasible solution of size k. The procedure outputs Q_i, answers "yes", and stops. Otherwise, the procedure uses the profile of Q_i as the starting point for the round $i + 1$. As we will prove later, the profiles of such a greedily chosen Q_i serve as a lower bound for any feasible solution in the sense that there exists no feasible solution with a smaller profile.

We illustrate the decision procedure at the example in Fig. 1. Let the capacities be $c(e_i) = 2$ for every edge $e_0, \ldots, e_5$. We ask for a feasible solution consisting of $k = 4$ paths. The paths are always processed in the greedy order, which is shown in Fig. 1(b). In the first round the paths p_1 and p_2 are accepted. The paths p_3 and p_4 are rejected, because they violate the capacity constraint of the edge e_2 after the paths p_1 and p_2 have been accepted. The paths p_5 and p_6 are both accepted to form the candidate set $Q_1 = \{p_1, p_2, p_5, p_6\}$ of 4 paths shown in Fig. 1(c). The profile of Q_1 is 2 for the edge e_0, 1 for the edge e_1, and 0 elsewhere. Q_1 is not feasible because $L(Q_1, e_0) = 3$ exceeds the capacity $c(e_0) = 2$.

The procedure starts a second round, this time with the profile of Q_1 as the starting profile. In this round the procedure accepts the paths in $Q_2 = \{p_1, p_3, p_5, p_6\}$ illustrated in Fig. 1(d). The path p_2 is rejected this time, because both edges e_0 and e_1 are saturated by the profile of Q_1 and the path p_1. The path p_4 is rejected for the same reason as before. The profile of Q_2 is again 2 for the edge e_0, 1 for the edge e_1, and 0 elsewhere. Since the resulting profile π_{Q_2} is equal to the starting profile π_{Q_1}, Q_2 is a feasible solution of size 4. The procedure stops.

4   Correctness of the Algorithm

The decision procedure will generate a sequence of profiles and chain-feasible solutions

$$\pi_0 \quad Q_1 \quad \pi_1 \quad Q_2 \quad \pi_2 \quad \ldots,$$

where π_0 is the empty profile we start with, and Q_i denotes the chain-feasible solution computed in round i. We set the profile $\pi_i := \pi_{Q_i}$.

We represent a chain-feasible solution A by the indices of the chosen paths in greedy order. A chain-feasible set A of k paths corresponds to a k-vector $A = (a_1, a_2, \ldots, a_k)$, where a_i is the index of the ith path chosen by the greedy algorithm. If A and B are two chain-feasible solutions, we write $A \le B$, iff $a_i \le b_i$ for all $1 \le i \le k$.

Note that $A \le B$ implies $\pi_A \le \pi_B$. This can be seen by comparing the ith path in A with the ith path in B: Since their indices a_i and b_i satisfy the condition $a_i \le b_i$ for all i, the paths in A contribute no more to the profile values $\pi_A(e_j)$ than the paths in B add to their respective profile values $\pi_B(e_j)$ for all edges e_j. Thus, $\pi_A \le \pi_B$.

From $\pi \le \pi'$ it follows easily that any chain-feasible solution for profile π' is also chain-feasible for profile π. In the following, we call a solution A that is chain-feasible for profile π *minimal* if for any other solution B that is chain-feasible for π and has the same cardinality as A, we have $A \le B$.

Lemma 1 (Optimality of greedy algorithm). *Let* π *be some starting profile. If there exists a solution of size* k *that is chain-feasible for profile* π, *there is also a minimal such solution, and the greedy algorithm computes this minimal solution.*

Proof. Let Q be any chain-feasible solution for profile π of size k. We transform Q step by step into the greedy solution G by replacing paths in Q by paths in G with smaller index. This is done during the execution of the greedy algorithm as it processes the paths in greedy order. We maintain the invariant that Q is always a chain-feasible solution of size k and that Q is equal to G with respect to the paths that have been processed so far.

Initially, the invariant clearly holds. Suppose the invariant holds up to path p_{i-1}, and the greedy algorithm processes the path p_i.

If adding the path p_i violates some capacity constraint, p_i is not selected by the greedy algorithm. Because of the invariant, the path p_i is not in Q either. Otherwise, the path p_i is chosen by the greedy algorithm. We distinguish two cases:

Case 1: $p_i \in Q$. Since the path p_i is in both G and Q, no transformation is needed, and Q remains feasible.

Case 2: $p_i \notin Q$. From the set of paths in Q with indices larger than i, we select a path p_j with the smallest source node (starting leftmost). We transform Q by replacing p_j by p_i. Since $j > i$, the index j in Q is reduced to i. We have to check the invariant. If the path p_i is contained in p_j, the invariant clearly holds, since replacing p_j by p_i does not affect feasibility. Otherwise, look at the remaining capacities. The edges to the left of the path p_j do not matter, because p_j has the smallest source node among all paths in Q greater than p_i. On the edges in the intersection of the paths p_i and p_j, taking either path p_i or p_j does not affect the capacities. Finally, we even gain one unit of capacity on all edges between the target node of the path p_i and the target node of the path p_j, since $i < j$. Altogether, Q is again feasible. The invariant holds.

At the end of the greedy algorithm, Q equals G. During the transformation we always replaced paths $p_j \in Q$ by paths $p_i \in G$ with $i < j$. This implies that G is less than or equal to the initial chain-feasible solution Q, i.e. $G \leq Q$.   □

Lemma 2. *The sequence of profiles generated by the decision procedure is monotonically increasing, i.e., we have* $\pi_i \leq \pi_{i+1}$ *for all* i.

Proof. (by induction) For $i = 0$, the claim holds, since π_0 is the empty profile. Assume that the claim holds for $i-1$. The induction hypothesis $\pi_{i-1} \leq \pi_i$ implies that the greedy solution Q_{i+1}, which is chain-feasible for profile π_i, is also chain-feasible for the profile π_{i-1}. Because Q_i is the greedy solution for the profile π_{i-1}, we obtain $Q_i \leq Q_{i+1}$ by Lemma 1. Therefore, $\pi_i \leq \pi_{i+1}$.   □

Lemma 3. *If a feasible solution* Q^* *with* k *paths exists, then each profile in the sequence of profiles generated by the decision procedure is bounded by the profile of* Q^*, *i.e., we have* $\pi_i \leq \pi_{Q^*}$ *for all* i.

Proof. (by induction) Since π_0 is the empty profile, the case $i = 0$ holds trivially. Now suppose $\pi_i \leq \pi_{Q^*}$ holds for some i. Because Q^* is chain-feasible for π_{Q^*}, it is also chain-feasible for π_i. Then, the greedy solution Q_{i+1} satisfies $Q_{i+1} \leq Q^*$ by Lemma 1, which immediately implies $\pi_{i+1} \leq \pi_{Q^*}$.   □

Lemma 4. *The decision procedure gives correct results and terminates after at most* $n \cdot c(e_0)$ *rounds.*

Proof. Assume first that there exists a feasible solution Q^* with k paths. By Lemma 3, the profile of the chain-feasible solutions computed by the algorithm always stays below the profile of Q^*. By Lemma 2, in each round the profile either stays the same or grows. If the profile stays the same, a feasible solution has been found by the algorithm. If the profile grows, the algorithm will execute the next round, and after finitely many rounds, a feasible solution will be found.

Now assume that the answer is "no". Again, the profile grows in each round, so there can be only finitely many rounds until the algorithm does not find k paths anymore and says "no".

We have $\sum_{j=0}^{n-1} \pi_{Q_i}(e_j) \leq n \cdot \pi_{Q_i}(e_0) \leq n \cdot c(e_0)$ for every generated profile π_{Q_i}, since profiles are non-increasing sequences and each Q_i is chain-feasible. As the profile grows in each round, the number of rounds is bounded by $n \cdot c(e_0)$. $\square$

Theorem 1. *There is a polynomial-time algorithm that computes an optimal solution for* CALLCONTROL *in rings.*

Proof. By Lemma 4, we have a decision procedure with $n \cdot c(e_0)$ rounds to decide whether there exists a feasible solution with k paths. Each round is a pass through the m given paths in greedy order, which can obviously be implemented in polynomial time. The number of rounds is polynomial as well, since we can assume without loss of generality that $c(e_0) \leq m$.

Given the decision procedure, we can use binary search on k to determine the maximum value for which a feasible solution exists with $O(\log m)$ calls of the decision procedure. $\square$

5   Efficient Implementation

In this section, we discuss how the algorithm can be implemented efficiently and analyze the worst-case running-time. Let an instance of CALLCONTROL be given by m paths in a ring with n nodes. Each path is specified by its counterclockwise and clockwise endnode. We assume $n \leq 2m$ since every node that is not an endpoint of a path can be removed. A sorted list of all path endpoints can be computed in time $O(m + n)$ using bucketsort, and it suffices to do this once at the start of the algorithm. From this list it is easy to determine the greedy order of the paths in linear time.

First we consider the implementation of the greedy algorithm for CALLCONTROL in chain networks with arbitrary capacities that is executed in each round of the decision procedure. While an $O(mn)$ implementation of the greedy algorithm is straightforward, we show in the following that it can even be implemented in linear time $O(m)$.

5.1   Implementation of the Greedy Algorithm for Chains

The input of the greedy algorithm consists of a chain with $N = 2n + 1$ nodes and arbitrary edge capacities, a set of m paths in the chain, and a parameter k. The algorithm

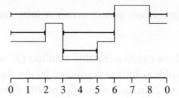

Fig. 2. The dummy paths for a given capacity function.

processes the paths in greedy order and accepts each path if it does not violate any edge capacity. It stops when either k paths are accepted or all paths have been processed.

Let $C = \max_{e \in E} c(e)$ denote the maximum edge capacity. Without loss of generality, we can assume $C \leq m$. In the following, we assume that we let the greedy algorithm run until all paths have been processed even if it accepts more than k paths. In this way the greedy algorithm actually computes a maximum cardinality subset of the paths that does not violate any edge capacity [4]. Stopping the greedy algorithm as soon as k paths are accepted is then a trivial modification.

For the case that all edges have the same capacity C, a linear-time implementation of the greedy algorithm was given in [4]. The main idea of their algorithm is to actually compute a C-coloring of the accepted paths and to maintain the *leader* for each color (the greatest path in greedy order colored with that color so far) in a data structure. When a path p is processed, the rightmost (greatest in greedy order) leader not intersecting p, denoted by $leader(p)$, is determined. If no such leader exists, p is rejected. Otherwise p is assigned the color of $leader(p)$ and becomes the new leader of that color.

The union-find data structure of [6] is used to compute leaders in amortized constant time. For this purpose, each path p has a preferred leader $adj(p)$, which is the greatest path in greedy order ending to the left of p. When p is processed and $adj(p)$ is really a leader, the correct leader for p is $adj(p)$. Otherwise, $adj(p)$ has either been rejected or is no longer a leader, and an operation find($adj(p)$) is used to determine the rightmost leader ending no later than $adj(p)$, which is the correct leader for p. If such a leader is found, p is colored with the color of that leader and the sets containing $leader(p)$ and $pred(leader(p))$ are merged, where $pred(q)$ denotes the last path before q in the greedy order. If no leader is found, p is rejected and the sets containing p and $pred(p)$ are merged. We refer to [4] for a detailed explanation why this yields a correct implementation of the greedy algorithm.

In order to adapt this approach to the case of arbitrary capacities, we add dummy paths to the instance to fill up the $C - c(e_i)$ units of extra capacity on every edge e_i as shown in Fig. 2 for an example. After setting all edge capacities equal to C, we compute an optimal solution containing all dummy paths. Removing them from the solution yields an optimal solution for the original problem. We will show later how to modify the algorithm of [4] to ensure that all dummy paths are colored. The dummy paths are computed by scanning the chain from left to right and deciding at each node how many dummy paths should start or end here: If the edges to the left and to the right of the current node are e_i and e_{i+1}, then $c(e_{i+1}) - c(e_i)$ dummy paths end at the node if $c(e_{i+1}) > c(e_i)$ and $c(e_i) - c(e_{i+1})$ dummy paths begin at the node otherwise.

In order to achieve a linear running-time, the number of dummy paths should be $O(m)$. However, there are capacity functions where $\Omega(mn)$ dummy paths are needed (e.g., capacities alternating between 1 and m). Therefore, we introduce the following preprocessing step in order to somewhat flatten the capacity function. We scan the chain of nodes from left to right. Let $n(i)$ denote the number of original paths that have node i as their left endpoint. For each edge e_i we set the new capacity $c'(e_i)$ for the edge e_i to the minimum of the original capacity $c(e_i)$ and $c'(e_{i-1}) + n(i)$. Hence, a decrease in the original capacity function is replicated by the new capacity function, while an increase is limited to the number of paths starting at the current node. We have $c'(e_i) \leq c(e_i)$ for all edges e_i and that any subset of paths that is feasible for capacity function c is also feasible for capacity function c'. To see the latter, note that the number of paths using edge e_i in any feasible solution for capacity function c is at most $c(e_{i-1}) + n(i)$. The new capacity function c' can clearly be computed in linear time.

Lemma 5. *With the new capacity function c', the number of dummy paths added by the algorithm is $O(m)$.*

Proof. Let us define the *total increase* of the capacity function c' to be the sum of the values $\max\{c'(e_i) - c'(e_{i-1}), 0\}$ for $i = 0, \ldots, N - 1$, where we take $c'(e_{-1}) = 0$. By definition of c', the total increase of c' is at most m, since every increase by 1 can be charged to a different path. Now consider the dummy paths added by the algorithm. Every dummy path ends because of an increase by 1 of c' or because the right end of the chain is reached. Therefore, there can be at most $m + C = O(m)$ dummy paths. □

After preprocessing the capacity function and adding the dummy paths, we compute a maximum C-colorable subset of paths in which all dummy paths are colored. It is clear that then the set of colored original paths forms an optimal solution in the original chain (with capacities $c(e_i)$ or $c'(e_i)$).

We must modify the algorithm of [4] to make sure that all dummy paths are accepted and colored. We assume that all paths including dummy paths are given as a sorted list of their endpoints such that for every node i, the right endpoints of paths ending at i come before the left endpoints of paths starting at i. The endpoints are processed in this order. Now the idea is to process original paths at their right endpoints and dummy paths at their left endpoints to make sure that all dummy paths are accepted and colored.

We give a rough sketch of the resulting algorithm, omitting some details such as the initialization of the union-find data structure (which is the same as in [4]). The algorithm maintains at any point the last path whose right endpoint has already been processed. This path is called *last* and is stored in a variable with the same name.

Let x be the path endpoint currently being processed and p the respective path. First, consider the case that x is the left endpoint of p. Then we set $adj(p)$ to be the path stored in *last*. If p is a dummy path, we want to color p immediately and perform a find operation on $adj(p)$ to find $q = leader(p)$. We color p with the color of q and perform a union operation to merge the set containing q with the set containing $pred(q)$. If p is not a dummy path, nothing needs to be done for p now, because p will be colored later when its right endpoint is processed.

Now, consider the case that x is the right endpoint of p. Then p is stored in *last*, since it is now the last path whose right endpoint has already been processed. If p is an original path, we want to color it now, if possible. Therefore, we perform a find operation on $adj(p)$ in order to find its leader. If such a leader q is found, the color of p is set to the color of q, and the set containing q is merged with the set containing $pred(q)$; otherwise, p is rejected and the set containing p is merged with the set containing $pred(p)$. If p is a dummy path, p has already been colored at its left endpoint, so nothing needs to be done for p anymore.

The union-find data structure of [6] is applicable, since the structure of the potential union operations is a tree (actually, even a chain). Therefore, the algorithm runs in time linear in the number of all paths including the dummy paths. The arguments for proving that this gives a correct implementation of the greedy algorithm are similar to the ones given in [4] and are omitted here. Furthermore, it can be shown similar to Lemma 1 that the computed solution is optimal.

Summing up, the algorithm does a linear-time preprocessing of the capacity function, then adds $O(m)$ dummy paths in linear time, and then uses an adapted version of the algorithm in [4] to run the greedy algorithm in time linear in the number of paths.

Theorem 2. *The greedy algorithm computes optimal solutions for* CALLCONTROL *in chains with arbitrary edge capacities and can be implemented to run in time* $O(n + m)$, *where* n *is the number of nodes in the chain and* m *is the number of given paths.*

5.2 Analysis of Total Running Time for Rings

An instance of CALLCONTROL in ring networks is given by a capacitated ring with n nodes and m paths in the ring. To implement the algorithm of Sect. 3, we use binary search on k to determine the maximum value for which a feasible solution exists. This amounts to $O(\log m)$ calls of the decision procedure. In each call of the decision procedure, the number of rounds is bounded by $n \cdot c(e_0)$ according to Lemma 4. This can be improved to $n \cdot c_{min}$ by labeling the nodes such that $c(e_0)$ equals the minimum edge capacity c_{min}. Each round consists of one execution of the greedy algorithm, which takes time $O(m)$ as shown in Sect. 5.1. Thus the total running-time of our algorithm is bounded by $O(mnc_{min} \log m)$.

Theorem 3. *There is an algorithm that solves* CALLCONTROL *in ring networks optimally in time* $O(mnc_{min} \log m)$, *where* n *is the number of nodes in the ring,* m *is the number of paths, and* c_{min} *is the minimum edge capacity.*

A minor improvement in the number of calls of the decision procedure may be obtained on certain instances as follows. We first run the greedy algorithm of Sect. 5.1 on the paths in P_1. This yields an optimal feasible subset Q of P_1. Let $t = |Q|$. Then we know that the size of an optimal feasible subset of P lies in the interval $[t, t + \min\{|P_2|, c(e_0), c(e_{n-1})\}]$. The number of calls of the decision procedure is reduced to $O(\log \min\{|P_2|, c(e_0), c(e_{n-1})\})$.

6   Conclusion and Open Problems

We have presented an algorithm for CALLCONTROL in ring networks that always computes an optimal solution in polynomial time. CALLCONTROL in rings is significantly more general than the maximum edge-disjoint paths problem for rings and appears to be close to the maximum k-colorable subgraph problem for circular-arc graphs, which is NP-hard. Therefore, we find it interesting to see that CALLCONTROL in rings is still on the "polynomial side" of the complexity barrier. Besides its applications in call admission control for communication networks, the algorithm can also be used to solve periodic scheduling problems with rejection. Furthermore, the algorithm can be implemented efficiently, and as a by-product we obtain a linear-time implementation of the greedy algorithm that solves CALLCONTROL in chains optimally.

These results lead to some open questions for future research. First, one could consider a weighted version of CALLCONTROL where each request has a certain profit and the goal is to maximize the total profit of the accepted requests. Second, one could try to tackle the version of CALLCONTROL where the paths for the accepted requests can be determined by the algorithm. For both problem variants, we do not yet know whether they can be solved optimally in polynomial time as well. We remark that the weighted version of CALLCONTROL in chains can be solved in polynomial time by adapting the approach based on min-cost network flow of [4].

References

1. R. Adler and Y. Azar. Beating the logarithmic lower bound: Randomized preemptive disjoint paths and call control algorithms. In *Proceedings of the 10th Annual ACM–SIAM Symposium on Discrete Algorithms SODA'99*, pages 1–10, 1999.
2. A. Blum, A. Kalai, and J. Kleinberg. Admission control to minimize rejections. In *Proceedings of the 7th International Workshop on Algorithms and Data Structures (WADS 2001)*, LNCS 2125, pages 155–164, 2001.
3. A. Borodin and R. El-Yaniv. *Online Computation and Competitive Analysis*. Cambridge University Press, 1998.
4. M. C. Carlisle and E. L. Lloyd. On the k-coloring of intervals. *Discrete Applied Mathematics*, 59:225–235, 1995.
5. E. M. Eschen and J. P. Spinrad. An $O(n^2)$ algorithm for circular-arc graph recognition. In *Proceedings of the Fourth Annual ACM-SIAM Symposium on Discrete Algorithms SODA'93*, pages 128–137, 1993.
6. H. Gabow and R. Tarjan. A linear-time algorithm for a special case of disjoint set union. *Journal of Computer and System Sciences*, 30(2):209–221, 1985.
7. J. A. Garay, I. S. Gopal, S. Kutten, Y. Mansour, and M. Yung. Efficient on-line call control algorithms. *Journal of Algorithms*, 23:180–194, 1997.
8. M. R. Garey, D. S. Johnson, G. L. Miller, and C. H. Papadimitriou. The complexity of coloring circular arcs and chords. *SIAM J. Algebraic Discrete Methods*, 1(2):216–227, 1980.
9. M. C. Golumbic. *Algorithmic Graph Theory and Perfect Graphs*. Academic Press, New York, 1980.
10. I. A. Karapetian. On the coloring of circular arc graphs. *Journal of the Armenian Academy of Sciences*, 70(5):306–311, 1980. (in Russian)
11. S. Leonardi. On-line network routing. In A. Fiat and G. J. Woeginger, editors, *Online Algorithms: The State of the Art*, LNCS 1442. Springer-Verlag, Berlin, 1998.
12. S. Plotkin. Competitive routing of virtual circuits in ATM networks. *IEEE Journal of Selected Areas in Communications*, 13(6):1128–1136, August 1995.
13. G. Wilfong and P. Winkler. Ring routing and wavelength translation. In *Proceedings of the Ninth Annual ACM-SIAM Symposium on Discrete Algorithms SODA'98*, pages 333–341, 1998.

Preemptive Scheduling in Overloaded Systems

Marek Chrobak[1], Leah Epstein[2], John Noga[3], Jiří Sgall[4], Rob van Stee[5],
Tomáš Tichý[4], and Nodari Vakhania[6]

[1] Department of Computer Science, University of California, Riverside, CA 92521,
U.S.A., marek@cs.ucr.edu
[2] School of Computer Science, The Interdisciplinary Center, P.O.B. 167,
46150 Herzliya, Israel, lea@idc.ac.il
[3] Department of Computer Science, California State University,
Northridge, CA 91330, U.S.A., jnoga@ecs.csun.edu
[4] Mathematical Institute, AS CR, Žitná 25, CZ-11567 Praha 1, Czech Republic,
sgall,tichy@math.cas.cz
[5] Institut für Informatik, Albert-Ludwigs-Universität, Georges-Köhler-Allee,
79110 Freiburg, Germany, vanstee@informatik.uni-freiburg.de
[6] Facultad de Ciencias, Universidad Autonoma del Estado de Morelos,
62251 Cuernavaca, Morelos, Mexico, nodari@servm.fc.uaem.mx

Abstract. The following scheduling problem is studied: We are given
a set of tasks with release times, deadlines, and profit rates. The objective is to determine a 1-processor preemptive schedule of the given
tasks that maximizes the overall profit. In the standard model, each
completed task brings profit, while non-completed tasks do not. In the
metered model, a task brings profit proportional to the execution time
even if not completed. For the metered task model, we present an efficient offline algorithm and improve both the lower and upper bounds on
the competitive ratio of online algorithms. Furthermore, we prove three
lower bound results concerning resource augmentation in both models.

1 Introduction

In most task scheduling problems the objective is to minimize some function
related to the completion time. This approach is not useful in overloaded systems,
where the number of tasks and their processing times exceed the capacity of the
processor and not all tasks can be completed. In such systems, the goal is usually
to maximize the number of executed tasks or, more generally, to maximize their
value or profit.

The problem can be formalized as follows: we have a set of n tasks, each task
j is specified by its release time r_j, deadline d_j, processing time p_j, and weight
w_j representing its profit rate. Preemption is allowed, i.e., each task can be
divided into any number of intervals, with arbitrary granularity. The objective
is to determine a 1-processor preemptive schedule that maximizes the overall
profit. The profit gained from processing task j can be defined in two ways. In
the *standard model*, each completed task j brings profit $w_j p_j$, but non-completed

P. Widmayer et al. (Eds.): ICALP 2002, LNCS 2380, pp. 800–811, 2002.

tasks do not bring any profit. In the *metered model*, a task w_j executed for time $t \leq p_j$ brings profit $w_j t$ even if it is not completed.

In many real-world applications, algorithms for task scheduling are required to be *online*, i.e., to choose the task to process based only on the specification of the tasks that have already been released. An algorithm that approximates the optimal solution within a factor R is called R-*competitive*. Online algorithms are also studied in the framework called *resource augmentation*. The idea is to allow an online algorithm to use more resources (a faster processor or more processors) and then to compare its performance to the optimum solution (with no additional resources). For the scheduling problems, we then ask what competitive ratio can be achieved for a given speed-up factor s, or what speed-up is necessary to achieve 1-competitiveness. See [9,2] for more information on competitive analysis.

The standard model. This problem has been extensively studied. Koren and Shasha [6] give a $(\sqrt{\xi}+1)^2$-competitive algorithm, where $\xi = \max_j w_j / \min_j w_j$ is called the *importance factor*. This ratio is in fact optimal [1,6]. Since no constant-competitive algorithms are possible in this model, it is natural to study this problem under the resource augmentation framework. Kalyanasundaram and Pruhs [4] present an online algorithm that uses a processor with speed 32 and achieves a constant competitive ratio. Lam and To [8] show an online algorithm with speed-up $O(\log \xi)$ and competitive ratio 1. One natural special case of this problem is when the tasks are *tight*, that is, for each j we have $d_j = r_j + p_j$. For this case, Koo *et al.* [5] give a 1-competitive algorithm with speed-up 14, and Lam *et al.* [7] show that in order to achieve 1-competitiveness the speed-up must be at least $\phi \approx 1.618$.

The metered model. This version was introduced (in a different terminology) by Chang and Yap [3] in the context of thinwire visualization, where the profit represents overall *quality of service*. Metered preemptive tasks also provide a natural model for various decision making processes where an entity with limited resources needs to choose between engaging in several profitable activities. Chang and Yap proved that two online algorithms called FirstFit and EndFit have competitive ratio 2. They also proved that no online algorithm can achieve a competitive ratio better than $2(2 - \sqrt{2}) \approx 1.17$.

Our results. We first focus on the metered profit model. In Section 3, we consider offline algorithms. We characterize the structure of optimal solutions and provide a polynomial time algorithm based on bipartite matching. This addresses a problem stated in [3].

The online metered case is studied in Section 4 to 7. In Section 4 we present a 1.8-competitive algorithm. In Section 5, we show that our analysis of this algorithm is tight and that algorithm FirstEndFit, conjectured in [3] to be 1.5-competitive, is only 2-competitive. In Section 6 we prove a lower lower bound of $\sqrt{5} - 1 \approx 1.236$ on the competitive ratio of algorithms for this problem. These results improve both the lower and upper bounds from [3].

In Section 7 we study the resource augmentation version of this problem, and prove that no online algorithm with constant speed-up can be 1-competitive, neither in the metered profit model, nor in the standard model. In fact, we prove that the minimal speed-up needed to achieve 1-competitiveness is $\Omega(\log\log\xi)$. Thus we disprove a conjecture from [5] by showing that the problem with general deadlines is provably harder than the special case of tight deadlines.

Furthermore, we prove some lower bounds for the restricted case of tight tasks in the standard model. We improve the lower bound from [7], by proving that, in order to achieve 1-competitiveness, an online algorithm needs speed-up at least 2. Our last result concerns the model where an online algorithm is allowed to use m processors of speed 1, rather than a single faster processor. For this case we prove that the competitive ratio is $\Omega(\sqrt[m]{\xi}/m)$, even if all tasks are restricted to be tight. For tight tasks constant speed-up is sufficient for 1-competitiveness, so the lower bound shows that increasing the speed of a single processor is more powerful than increasing the number of processors of speed 1.

2   Preliminaries

Let $J = \{1, 2, \ldots, n\}$ be the given set of tasks, with task j specified by the values (r_j, d_j, p_j, w_j), where r_j is its release time, d_j is the deadline, p_j is the processing time, and w_j is the weight of task j representing its *profit rate*. (In the literature, w_j is sometimes called the *value density*, and the product $w_j p_j$ is called the *value* of task j.) We assume $\min_j r_j = 0$ and we denote by $D = \max_j d_j$ the latest deadline. If $r_j \le t \le d_j$, then we say that task j is *feasible at time t*.

Schedules. We define a *schedule* for J to be a measurable function $S : \mathbb{R} \to J \cup \{\bot\}$ such that, for each j and t, $|S^{-1}(j)| \le p_j$ and $S(t) \ne j$ for $t \notin [r_j, d_j]$. In this definition, $S(t)$ denotes the task that is scheduled at time t, and $S(t) = \bot$ if no task is scheduled. For a set $X \subseteq \mathbb{R}$, $|X|$ denotes the size (measure) of X.

The *profit* of a schedule S depends on the model: In the *standard model*, the profit is the sum of the profits of the completed tasks, that is $profit_S(J) = \sum_j w_j p_j$, where the sum is taken over all j for which $|S^{-1}(j)| = p_j$. In the *metered model*, even partially executed tasks count, that is $profit_S(J) = \sum_j w_j |S^{-1}(j)|$. The optimal profit is $profit_{\text{OPT}}(J) = \sup_S profit_S(J)$. It is easy to see that this supremum is achieved. Moreover, each schedule can be transformed into a piecewise constant schedule without changing the total profit (see [3]). The profit of a schedule generated by an algorithm $\mathcal{A}$ on the instance J is denoted by $profit_{\mathcal{A}}(J)$.

For the metered model, it is important to keep in mind that the optimum profit is not changed if any task is divided into several tasks with the same release times, deadlines, and weights, and whose total processing time is equal to the processing time of the original task. (For this reason it is more natural to define the weight as the profit rate instead of the total profit.)

For a schedule S, let $done_{S,j}(t) = |S^{-1}(j) \cap [0, t]|$ be the amount of task j that has been processed in S by time t. We define a task j to be *active* in S at time t if $r_j \le t < d_j$ and $done_{S,j}(t) < p_j$. In other words, the active tasks are those that are feasible at time t and have not been yet completely processed.

We say that a schedule S is *canonical* if for any two times $t_1 < t_2$, if $j_2 = S(t_2) \ne \bot$, then either $r_{j_2} > t_1$, or $j_1 = S(t_1) \ne \bot$ and $d_{j_1} \le d_{j_2}$. One way to think about canonical schedules is this: at each time t, if j is the earliest-deadline task among the active tasks at time t, then we either process j at time t, or discard j irrevocably so that it will never be processed in the future. Any schedule S, including an optimal one, can be converted into a canonical schedule as follows. Consider the instance J' consisting of the portions of tasks that are processed in S. Reschedule the tasks in J' so that at each time we schedule the active task with the earliest deadline. Using a standard exchange argument, it is easy to verify that all tasks are fully processed.

Online algorithms. A scheduling algorithm $\mathcal{A}$ is *online* if, at any time t, its schedule depends only on the tasks that have been released before or at time t. An online algorithm $\mathcal{A}$ is called *R-competitive* if $profit_{\mathcal{A}}(J) \geq profit_{\text{OPT}}(J)/R$ for every instance J. The *competitive ratio* of $\mathcal{A}$ is the smallest R for which $\mathcal{A}$ is R-competitive.

In the online case, it is convenient to allow time-sharing of tasks, which means that several tasks may be processed simultaneously at appropriately reduced speeds. Formally, a *generalized schedule* is a function V that, for each task j and time $t \in [0, D]$, specifies the speed $V(j, t)$ at which we perform task j at time t. We impose the following restrictions on $V(j, t)$:

$$\sum_j V(j, t) \leq 1, \quad \int_0^\infty V(j, t)dt \leq p_j, \quad \text{and} \quad V(j, t) = 0 \text{ for } t \notin [r_j, d_j]$$

The profit of a generalized schedule V is

$$profit_V(J) = \sum_j w_j \int_0^\infty V(j, t)dt = \int_0^\infty \sum_j w_j V(j, t)dt.$$

Clearly, this definition generalizes the previous one. Both definitions are equivalent in the offline case. Even in the online case, any generalized schedule V can be transformed into a schedule S which simulates the time-sharing in V by alternating the tasks. It is easy to see that if the tasks are alternated with sufficiently high frequency (compared to the processing times), this transformation increases the competitive ratio only by an arbitrarily small $\varepsilon > 0$. So both definitions are equivalent in the online case as well, in the sense that the infima of achievable competitive ratios are the same. Throughout the paper we slightly abuse terminology and refer to the function V simply as a *schedule*.

3   An Offline Algorithm for Metered Tasks

The release times and deadlines partition the range $[0, D]$ into $2n - 1$ intervals that we call *stages*. We number the stages $1, 2, \ldots, 2n - 1$. If stage s is $[a, b]$, we say that task j is *feasible in stage s* if it is feasible at any time $t \in [a, b]$. It is straightforward to set up a linear program describing all feasible schedules, using variables $x_{j,s}$ denoting the amount of task j processed in stage s, and thus an optimal schedule can be computed in polynomial time. The goal of this section is to present a more efficient algorithm based on bipartite matchings and flows.

Before giving the algorithm, we prove the following property: all optimal schedules include the same portion of the tasks of any given weight. Thus, perhaps surprisingly, the set of optimal schedules depends only on the ordering of the weights but not on their values, and every optimal schedule contains an optimal schedule for any instance restricted to heavy tasks.

Order the tasks in an instance J so that $w_1 \geq w_2 \geq \cdots \geq w_n$. W.l.o.g., $w_n > 0$. For convenience, write $w_{n+1} = 0$. Let J_k denote the sub-instance consisting of tasks $1, \ldots, k$. Given a schedule S for J, let S_k be the restriction of S to J_k. In particular, $S_n = S$. Let $busy(S) = |S^{-1}(J)|$ be the total time when any task is scheduled in S.

Assume now that all release times and deadlines are integers and that we only have unit tasks (with $p_j = 1$). In this scenario preemptions are not necessary in the offline case. Construct a bipartite graph G with vertices $X = \{x_1, \ldots, x_n\}$ corresponding to tasks and $Y = \{y_1, \ldots, y_D\}$ corresponding to the unit time slots. If task j is feasible in time unit t then connect x_j and y_t with an edge of weight w_j. Let G_k denote the subgraph of G induced by $\{x_1, \ldots, x_k\} \cup Y$. Any schedule defines a matching in G and any matching is a schedule. So computing an optimal schedule is equivalent to computing a maximum-weight matching.

Lemma 1. (a) *There exists a maximum-weight matching M in G such that, for each $k = 1, \ldots, n$, M contains a maximum-cardinality matching of G_k.*
(b) *If M is any maximum-weight matching in G then, for each $k = 1, \ldots, n$ such that $w_k > w_{k+1}$, M contains a maximum-cardinality matching of G_k.*

Proof. Let M be a maximum-weight matching. Denote by M_k the sub-matching of M restricted to G_k, and suppose that $\hat{M}$ is a matching in G_k with cardinality larger than M_k. Consider the symmetric difference of M_k and $\hat{M}$. It consists of disjoint alternating cycles and paths. Moreover, it contains at least one odd-length alternating path P with more edges from $\hat{M}$ than from M_k. Let y be the endpoint of P in Y. Since the weights of the edges depend only on the endpoints in X, the consecutive pairs of adjacent edges (1st and 2nd, 3rd and 4th, etc) on P have equal weights, and the last edge has weight $w_p \geq w_k$. If y is matched in M then the matching edge containing y has weight w_l, for some $l > k$. Let M' be the matching obtained from M by unmatching y and swapping the edges on P. We have $w(M') = w(M) + w_p - w_l$. We cannot have $w_l < w_p$, since this would contradict the maximality of M. This proves part (b). So $w_l = w_p$, that is, M' has the same weight as M but it has one more edge from G_k. By repeating this process we get a matching that satisfies part (a).   □

For a general instance, round each processing time, release time, and deadline to the nearest multiple of $\varepsilon > 0$. Taking a limit $\varepsilon \to 0$, Lemma 1 implies the following theorem.

Theorem 1. (a) *There exists an optimal schedule S for J such that, for each $k = 1, \ldots, n$, S_k is optimal for J_k and $busy(S_k)$ is maximized.*
(b) *If S is any optimal schedule then, for each $k = 1, \ldots, n$ such that $w_k > w_{k+1}$, S_k is optimal for J_k and $busy(S_k)$ is maximized.*

Algorithm OPT.

(i) Construct a flow network H with source s, sink t, and vertices x_j for each task j and z_i for each stage i. The edges are: (s, x_j) for each task j, (x_j, z_i) for each stage i and each task j feasible in stage i, and (z_i, t) for each stage i. The capacity of each edge (x_j, z_i) is p_j and the capacity of each edge (z_i, t) is ℓ_i, the length of stage i. The capacities of edges (s, x_j) are initialized to 0.

(ii) For $j = 1, \ldots, n$, do the following: Set the capacity of (s, x_j) to p_j. Compute the maximal flow, denote it f_j. Let $b_j = |f_j| - |f_{j-1}|$ be the increase in the flow value. Set the capacity of (s, x_j) to b_j. Recompute the maximal flow, denote it f'_j.

The resulting maximum flow f'_n defines a schedule: we take $f_n(x_j, z_i)$ to be the amount of task j scheduled during stage i. Theorem 1 implies that $|f_j| = |f'_j|$ and after iteration j, the flow on (s, x_j) will remain b_j until the end. Thus Algorithm OPT computes step by step $busy(S_j) = |f_j|$, $j = 1, \ldots, n$, for an optimal schedule S. Therefore f'_n defines an optimal schedule. The running time of Algorithm OPT is no worse than $O(n)$ times the complexity of the maximum flow, which is not worse than $O(n^4)$.

4   A Competitive Online Algorithm for Metered Tasks

We now present our 1.8-competitive online algorithm. Algorithm FIRSTFIT [3] is a greedy algorithm that always processes the heaviest task. The drawback of FIRSTFIT is that it may schedule a task with a distant deadline, discarding an only slightly less profitable task with a tight deadline. To avoid this, our algorithm MIXED schedules concurrently two tasks: the heaviest task and a "relatively heavy" task with the earliest deadline. Note that MIXED keeps processing the same tasks in-between any of following at most $2n$ events: task arrivals, deadlines or task completions.

Algorithm MIXED. If no tasks are active, do nothing. Otherwise, let h be the active task with maximum w_h, and let e the active task with $w_e \geq \frac{2}{3}w_h$ and minimum d_e. Schedule h and e with equal speed. More precisely, if $e \neq h$ set $V(e, t) = V(h, t) = \frac{1}{2}$, otherwise set $V(e, t) = 1$. For $j \neq e, h$ set $V(j, t) = 0$.

Theorem 2. *Algorithm MIXED is 1.8-competitive.*

Proof. Let V be the schedule generated by MIXED and S be some canonical optimal schedule. We assume that the ties among the deadlines of tasks are resolved consistently, both in MIXED and in S.

We devise an appropriate charging scheme, described by a function $F : \mathbb{R} \to \mathbb{R}$ which maps each time in S to a time in V. The intention is that any profit achieved at time t in S is "charged" to the time $F(t)$ in V. We then argue that each time u in V is charged at most 1.8 times the profit in V at time u. Further, all profit from S is charged. These two facts imply 1.8-competitiveness. Since the "profit at time t" is infinitesimally small, in the formal proof we need to express our argument in terms of profit rates instead.

Consider a time t, and let $j = S(t)$. If $done_{V,j}(t) < done_{S,j}(t)$, define $F(t) = t$; the charged profit rate of j at time t is w_j. Otherwise, let $F(t) = u \leq t$, where u is the minimum time such that $done_{V,j}(u) = done_{S,j}(t)$; the charged profit rate of j at time u is $w_j V(j, u)$ (i.e., $\frac{1}{2} w_j$ or w_j, depending on whether j is scheduled with another task or not). It is easy to check that the total charged profit (i.e., the rate integrated over the whole schedule) equals the total profit of S.

Let h and e be the task(s) scheduled in V at time t, as chosen by MIXED. $F^{-1}(t)$ consists of at most three points, t and the minimal times t_h, t_e such that $done_{V,e}(t) = done_{S,e}(t_e)$ and $done_{V,h}(t) = done_{S,h}(t_h)$, if such t_e and t_h exist and satisfy $t_e, t_h > t$. The (combined) charged profit rate at t depends on which points are present. For example, if all three points are present, this rate is $w_j + \frac{1}{2}(w_e + w_h)$. We show by case analysis that this rate is at most 1.8 times the profit rate of V at time t, that is $1.8 \times \frac{1}{2}(w_e + w_h) = \frac{9}{10}(w_e + w_h)$. Integrating over all times we obtain that $profit_S(J) \leq 1.8 \cdot profit_V(J)$ and the theorem follows.

Case 1: $F(t) < t$. Then t is charged at most $\frac{1}{2}(w_e + w_h) \leq \frac{9}{10}(w_e + w_h)$.

Case 2: $F(t) = t$. By the definition of F, j is active at t in V. By the choices of h and e in MIXED, we have $w_j, w_e \leq w_h$ and $w_e \geq \frac{2}{3} w_h$. We have two subcases.

Case 2.1: $j \neq e$ and $F(t_e) = t$ (i.e., t_e exists and is charged to t). In this case, both tasks j and e are active at time t both in V and S. Since S is a canonical schedule, we have $d_j \leq d_e$. MIXED did not choose j as e despite the earlier deadline, thus it has to be the case that $w_j < \frac{2}{3} w_h \leq w_e$. Therefore the weight charged at t is at most $w_j + \frac{1}{2}(w_e + w_h) = \frac{2}{5} w_j + \frac{3}{5} w_j + \frac{1}{2}(w_e + w_h) < \frac{2}{5} w_e + \frac{2}{5} w_h + \frac{1}{2}(w_e + w_h) = \frac{9}{10}(w_e + w_h)$.

Case 2.2: Otherwise. If $j = e$ and $F(t) = t$ then $t_e > t$ cannot be charged to t, as $done_{V,e}(t) \leq done_{S,e}(t)$. If $j \neq e$ and Case 2.1 does not occur, t_e is not charged to t either. Thus, in all remaining scenarios, the weight charged to t is at most $w_j + \frac{1}{2} w_h \leq \frac{3}{2} w_h \leq \frac{9}{10}(w_e + w_h)$.   □

5   Hard Examples for Online Algorithms

We first show that our analysis of algorithm MIXED is tight, in the sense that we cannot achieve a ratio better than 1.8 by choosing different parameters. Then, we consider algorithm FIRSTENDFIT proposed in [3], and conjectured to be 1.5-competitive. We disprove this conjecture by showing that FIRSTENDFIT is no better than 2-competitive. In fact, our examples indicate that the idea of combining FIRSTFIT and ENDFIT is not likely to give a competitive ratio better than 2.

In all the examples we prescribe how the algorithm breaks ties using an arbitrarily small $\varepsilon > 0$ to modify the deadlines and/or processing times. As ε is arbitrarily small, it can be omitted in the calculations.

Algorithm GENMIXED. We generalize algorithm MIXED by replacing constants $\frac{2}{3}$ and $\frac{1}{2}$ by arbitrary $\beta, \gamma \in [0, 1]$, respectively. We now choose e to be the active task with $w_e \geq \beta w_h$ that minimizes d_e. The algorithm schedules task e at speed γ and h at speed $1 - \gamma$.

Theorem 3. *For any $\beta, \gamma \in [0,1]$, the competitive ratio of* GENMIXED *is at least 1.8.*

Proof. We distinguish two cases. If $\beta \leq 1/(1+\gamma)$ we give the algorithm an instance of three tasks $(0,1,1,1-\varepsilon)$, $(0,1-\varepsilon,\gamma,\beta)$, and $(0,2-\gamma,1-\gamma,1)$. The optimal solution schedules the first and third tasks, in this order, achieving profit $2-\gamma$. In the time interval $[0,1]$, MIXED chooses the second task as e and the third task as h, missing the first task, so its profit is $\beta\gamma+1-\gamma$. Thus, since $\beta \leq 1/(1+\gamma)$, its competitive ratio is at least

$$\frac{2-\gamma}{\beta\gamma+1-\gamma} \geq \frac{2-\gamma}{\frac{\gamma}{1+\gamma}+1-\gamma} = 1 + \frac{1}{1+\gamma-\gamma^2}.$$

If $\beta > 1/(1+\gamma)$ we give the algorithm an instance of three tasks $(0,1,1,\beta-\varepsilon)$, $(0,1+\gamma,\gamma,\beta)$, and $(0,2,1-\gamma,1)$. The optimal solution schedules all tasks in this order, achieving profit $\beta+\beta\gamma+1-\gamma$. In the time interval $[0,1]$, GENMIXED chooses the second task as e and the third task as h, missing the first task, so its profit is $\beta\gamma+1-\gamma$. Thus, since $\beta > 1/(1+\gamma)$, its competitive ratio is at least

$$1 + \frac{\beta}{\beta\gamma+1-\gamma} = 1 + \frac{1}{\frac{1-\gamma}{\beta}+\gamma} > 1 + \frac{1}{(1-\gamma)(1+\gamma)+\gamma} = 1 + \frac{1}{1+\gamma-\gamma^2}.$$

Since $\gamma - \gamma^2 \leq \frac{1}{4}$, the competitive ratio of GENMIXED is at least 1.8.   □

Algorithm FIRSTENDFIT. This algorithm chooses with probability $\frac{1}{2}$ either FIRSTFIT or ENDFIT and executes it. FIRSTFIT always schedules the heaviest task, while ENDFIT at each time computes an optimal schedule for the remaining portions of the active tasks, pushing the heavier tasks to the latest times possible, and follows this plan until a new task arrival (see [3]). We show that FIRSTENDFIT is not better than 2-competitive.

Our example consists of n tasks, where n is odd. For j odd, task j is $(j-1,j,1,1)$, and for j even, the task j is $(j-2,j+1,1,1+\varepsilon)$. The optimal schedule processes task j during time interval $[j-1,j]$, achieving profit n. Both FIRSTFIT and ENDFIT schedule all even tasks and only one odd task, achieving profit $(n+1)/2$. Thus the competitive ratio cannot be better than 2.

Another way to combine FIRSTFIT and ENDFIT is to schedule both the task chosen by FIRSTFIT and by ENDFIT, each at speed $\frac{1}{2}$. This algorithm is also only 2-competitive. To see this, consider tasks $(0,3,1,1+\varepsilon)$, $(0,2,1,1)$, and $(1,2,1,1)$. During the interval $[0,2]$, the first task is chosen by FIRSTFIT, thus it is finished by time 2; no task is chosen by ENDFIT until time 1, then either the second one or the third one is chosen during the time interval $[1,2]$. This achieves profit 1.5, while the optimum schedules all the tasks at profit 3.

6   A Lower Bound for Metered Tasks

Theorem 4. *The competitive ratio of any online algorithm for scheduling metered tasks is at least $\sqrt{5} - 1 \approx 1.236$.*

Proof. Fix an online algorithm $\mathcal{A}$ and $\varepsilon > 0$ arbitrarily small. We show that the competitive ratio of $\mathcal{A}$ is at least $\sqrt{5} - 1 - \varepsilon$, which proves the theorem.

Let $\sigma = \sqrt{5} - 2$ and let $\phi = (\sqrt{5} + 1)/2$ be the golden ratio. Define the sequence $\{v_i\}_{i=0}^{\infty}$ by $v_0 = 1$, $v_1 = \phi + \varepsilon$, and $v_{i+1} = (v_i - v_{i-1})/\sigma$ for $i > 1$. We solve the recurrence: $\phi + 1$ and ϕ are the roots of the characteristic equation $\sigma x^2 - x + 1 = 0$, and we have $v_i = (1 - \varepsilon)\phi^i + \varepsilon(\phi + 1)^i$.

The adversary strategy is this: Pick some large integer n. For each time $i = 0, 1, 2, \ldots$ the following two tasks arrive

$$\text{task } i: \quad (i, i+1, 1, v_i), \qquad \text{task } i': \quad (i, i+2, 1, v_{i+1}).$$

If there is an integer time $1 \le j < n$ when $\mathcal{A}$ has completed at least half of task $j - 1$, the adversary terminates the sequence (prior to releasing tasks j and j'). If this case occurs, $\mathcal{A}$ earns at most $profit_{\mathcal{A}}(J) \le \frac{1}{2}v_0 + v_1 + \ldots + v_j$, and the optimal profit is $profit_{\text{OPT}}(J) = v_1 + \ldots + v_{j-2} + 2v_{j-1} + v_j$. Using the recurrence and $v_{j-1} \ge 1$ (in the last inequality) we obtain

$$\frac{profit_{\text{OPT}}}{profit_{\mathcal{A}}(J)} \ge \frac{(1 + 2\sum_{i=1}^{j} v_i) + 2v_{j-1} - 1}{1 + 2\sum_{i=1}^{j} v_i} = 1 + \frac{2v_{j-1} - 1}{1 + 2v_1 + 2\sum_{i=2}^{j} \frac{v_{i-1} - v_{i-2}}{\sigma}}$$

$$= 1 + \frac{\sigma(2v_{j-1} - 1)}{\sigma + 2\sigma v_1 + 2v_{j-1} - 2} = 1 + \frac{\sigma(2v_{j-1} - 1)}{2\varepsilon\sigma + (2v_{j-1} - 1)} > 1 + \sigma - \varepsilon.$$

Otherwise, the adversary issues all tasks up to time $n - 1$, and at time n he releases task n only. Now $profit_{\mathcal{A}}(J) \le \frac{1}{2}v_0 + v_2 + \ldots + v_{n-1} + \frac{3}{2}v_n$ and $profit_{\text{OPT}}(J) = v_1 + \ldots + v_{n-1} + 2v_n$. Using the recurrence and letting $n \to \infty$

$$\frac{profit_{\text{OPT}}}{profit_{\mathcal{A}}(J)} = \frac{2v_n + 2\sum_{i=1}^{n} v_i}{1 + v_n + 2\sum_{i=2}^{n} v_i} = 1 + \frac{v_n - 1}{1 + v_n + 2\frac{v_{n-1} - 1}{\sigma}}$$

$$\longrightarrow 1 + \frac{\phi + 1}{(\phi + 1) + \frac{2}{\sigma}} = 1 + \sigma.$$

In both cases, the competitive ratio is at least $1 + \sigma - \varepsilon$, as claimed.   □

7   Lower Bounds for Resource Augmentation

Theorem 5. *Both in the metered and standard profit model, any online 1-competitive algorithm has speedup at least $\Omega(\log \log \xi)$, where ξ is the importance factor. In particular, there is no constant speed-up 1-competitive algorithm.*

Proof. Fix an integer m. We construct an instance such that any online 1-competitive algorithm needs speed-up $m/2$. All tasks ending at the same deadline t have the same profit rate $h(t) = (2^m)^t$. The tasks are grouped into $m+1$ classes numbered $k = 0, \ldots, m$. For $k = 0$, the tasks in class 0 are

$$(i, i+1, 1, h(i+1)), \quad i = 0, \ldots, 2^m - 1.$$

For $k = 1, \ldots, m$, the tasks in class k are

$$(i2^k, (i+1)2^k, 2^{k-1}, h((i+1)2^k)), \quad i = 0, \ldots, 2^{m-k} - 1.$$

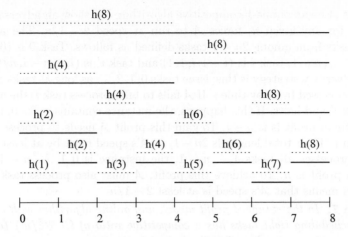

The figure shows this set of tasks for $m = 3$. Solid lines represent feasibility ranges, dotted lines represent processing times, and the profit rates $h(x)$ are shown above these lines.

For each time t consider the sub-instance consisting of the tasks that are released before t. We claim that the optimal solution of this sub-instance schedules exactly all the tasks with deadline t or later. To prove this claim, note that in the sub-instance, there is exactly one task in each class with deadline at least t; all these task can be scheduled from time 0 to 2^m, completely filling the capacity of the processor at any time. Since all the other tasks in the sub-instance have smaller profit rate, this gives the optimal solution (both for metered and standard tasks).

The weights increase so fast that the profit rate $h(t)$ is at least the total profit of all the tasks with deadlines before t: The total processing time of all tasks with deadline equal to $t - 1$ is at most 2^{m-1}, thus their total profit is at most $2^{m-1}h(t - 1)$. By induction, $h(t - 1)$ bounds the total profit of all tasks with deadline before $t - 1$. Thus the total profit of all tasks with deadline before t is at most $(2^{m-1} + 1)h(t - 1) \leq 2^m h(t - 1) = h(t)$.

The previous considerations show that to achieve optimal profit, the online algorithm has to completely execute all the tasks with deadline t, with the exception of tasks or their parts (in metered model) with processing time bounded by 1 (since only that much can be replaced by the hypotetic profit of tasks with earlier deadlines). This holds for any time t, thus all the tasks must be completed, with a possible exception of processing time 2^m. The total processing time of all tasks is $(m + 2)2^{m-1}$, thus $m2^{m-1}$ has to be executed by time 2^m, and $\mathcal{A}$ must run at speed at least $m/2$.

Since the deadlines range from 1 to 2^m, the importance ratio is $\xi = (2^m)^{2^m - 1}$ and the lower bound is $m/2 = \Omega(\log \log \xi)$. $\qquad\square$

Theorem 6. *In the standard profit model, there is no online 1-competitive algorithm with speed-up $s < 2$ for scheduling tight tasks.*

Proof. Let $\mathcal{A}$ be an online 1-competitive algorithm. We show an adversary strategy that, for any given n, forces $\mathcal{A}$ to run at speed $2 - 1/n$. The adversary chooses tasks from among $2n - 1$ tasks defined as follows. Task 0 is $(0, n, n, 1)$. For $i = 1, \ldots, n - 1$, task i is $(i - 1, i, 1, 1)$ and task i' is $(i, n, n - i, n/(n - i))$.

The adversary strategy is this: issue tasks $0, 1, 2, \ldots$, as long as tasks $1, 2, \ldots, i$ are fully processed by $\mathcal{A}$ by time i. If $\mathcal{A}$ fails to fully process task i, the adversary issues task i' and halts. If this happens, the instance contains tasks $0, 1, \ldots, i, i'$ whose optimal profit is is $n + i$. To gain this profit $\mathcal{A}$ needs to process all tasks other than i. Their total length is $2n - 1$, so $\mathcal{A}$'s speed must be at least $2 - 1/n$.

If $\mathcal{A}$ processes all tasks $1, \ldots, n - 1$, the instance is $0, 1, \ldots, n - 1$ and its maximum profit is n. To achieve this profit, $\mathcal{A}$ must also process task 0. Once again, this means that $\mathcal{A}$'s speed is at least $2 - 1/n$. $\square$

Theorem 7. *In the standard profit model, any online algorithm with m processors for scheduling tight tasks has a competitive ratio of $\Omega(\sqrt[m]{\xi}/m)$ (against a 1-processor optimum), where ξ is the importance ratio.*

Proof. Let M be large constant. Suppose $\mathcal{A}$ has m machines. The adversary chooses tasks from $m + 1$ task classes numbered $0, 1, \ldots, m$. The tasks in class i have all equal processing time $p_j = M^{2i}$, profit rate $w_j = M^{-i}$, and profit $w_j p_j = M^i$; their release times are aM^{2i}, for $a = 0, 1, \ldots, M^{2m-2i} - 1$. The importance ratio is $\xi = M^m$.

The adversary strategy is as follows. Since the tasks are tight and we consider the standard model, we can assume that once $\mathcal{A}$ fails to run a task, it never starts it again. If $\mathcal{A}$ stops executing a task j at time t (where t could be d_j) then from time $t + 1$ until the deadline of j no tasks from classes $0, 1, \ldots, j - 1$ are released. (In other words, a task arrives if at its release time all the active tasks are running; note that these tasks are only from higher classes.) It follows that at each time there exists at least one task that was released but is not being executed by $\mathcal{A}$. At time t, let j_t be such a task from the smallest class.

Let P be the total profit of all the dropped tasks, i.e., tasks not finished in $\mathcal{A}$. We prove that (i) the optimal solution schedules tasks with profit at least $P/(m+1)$, and (ii) the algorithm $\mathcal{A}$ schedules tasks with profit at most $2P/(M-1)$. The bound on the competitive ratio follows.

The proof of (i) is trivial: The dropped tasks in each class are disjoint, so the dropped tasks in one of the classes have weight at least $P/(m + 1)$.

Now we prove (ii). If a task j running in $\mathcal{A}$ at some time t is from a lower-numbered class than j_t, we assign it to j_t. A task executed by $\mathcal{A}$ can be assigned to none, one or even more dropped tasks (as j_t may change). Any running task not assigned at all is always from a higher class than the current j_t. At each time, the total profit rate of all such tasks is at most $1/(M - 1)$ fraction of the profit rate of j_t. Thus the overall profit of all unassigned completed tasks is at most $P/(M - 1)$. Now consider all the executed tasks assigned to a particular dropped task j from class i. From the definition of the sequence it follows that there is at most one such task from each class $i' < i$. Thus their total profit (not profit rate) is at most $1/(M - 1)$ fraction of the profit of j. Hence the overall profit of all assigned completed tasks is at most $P/(M - 1)$, and (ii) follows. $\square$

8   Final Comments

The main remaining open problem is to determine the best competitive ratio for the metered profit model. We showed that this ratio is between 1.236 and 1.8, but the gap between these two bounds is still very wide. Similarly, in the standard model, we know that the minimum speedup needed to obtain a 1-competitive algorithm is between $\Omega(\log\log\xi)$ and $O(\log\xi)$. It would be very interesting to determine the optimal speedup for this problem.

Acknowledgements. M. Chrobak was supported by NSF grant CCR-9988360. L. Epstein was supported by the Israel Science Foundation, grant No. 250/01-1. J. Sgall and T. Tichý were supported by Institute for Theoretical Computer Science, Prague (project LN00A056 of MŠMT ČR), grant 201/01/1195 of GA ČR, and grant A1019901 of GA AV ČR. M. Chrobak, J. Sgall and T. Tichý were supported by cooperative grant KONTAKT-ME476/CCR-9988360-001 from MŠMT ČR and NSF. R. van Stee was supported by the Deutsche Forschungsgemeinschaft, Project AL 464/3-1, and by the European Community, Projects APPOL and APPOL II. N. Vakhania was supported by NSF-CONACyT grant E120.1914.

References

1. Sanjoy Baruah, Gilad Koren, Decao Mao, Bud Mishra, Arvind Raghunathan, Louis Rosier, Dennis Shasha, and Fuxing Wang. On the competitiveness of on-line real-time task scheduling. *Real-Time Systems*, 4:125–144, 1992.
2. Allan Borodin and Ran El-Yaniv. *Online Computation and Competitive Analysis*. Cambridge University Press, 1998.
3. Ee-Chien Chang and Chee Yap. Competitive online scheduling with level of service. In *Proc. 7th Annual International Computing and Combinatorics Conference*, volume 2108 of *Lecture Notes in Computer Science*, pages 453–462. Springer, 2001.
4. Bala Kalyanasundaram and Kirk Pruhs. Speed is as powerful as clairvoyance. *Journal of the ACM*, 47(4):214–221, 2000.
5. Chiu-Yuen Koo, Tak-Wah Lam, Tsuen-Wan Ngan, and Kar-Keung To. On-line scheduling with tight deadlines. In *Proc. 26th Symp. on Mathematical Foundations of Computer Science*, volume 2136 of *Lecture Notes in Computer Science*, pages 464–473, 2001.
6. G. Koren and D. Shasha. d^{over}: an optimal on-line scheduling algorithm for overloaded uniprocessor real-time systems. *SIAM Journal on Computing*, 24:318–339, 1995.
7. Tak-Wah Lam, Tsuen-Wan Ngan, and Ker-Keung To. On the speed requirement for optimal deadline scheduling in overloaded systems. In *Proc. 15th International Parallel and Distributed Processing Symposium*, page 202, 2001.
8. Tak-Wah Lam and Ker-Keung To. Trade-offs between speed and processor in hard-deadline scheduling. In *Proc. 10th Symp. on Discrete Algorithms*, pages 755–764, 1999.
9. Jiří Sgall. Online scheduling. In *Online Algorithms: The State of Art*, pages 196–227. Springer-Verlag, 1998.

The Equivalence Problem of Finite Substitutions on ab^*c, with Applications[*]

J. Karhumäki[1] and L.P. Lisovik[2]

[1] Department of Mathematics and Turku Centre for Computer Science, University of Turku, FIN-20014 Turku, Finland karhumak@cs.utu.fi

[2] Department of Cybernetics, Kiev National University, Kiev, 252017, Ukraine office@nan.kiev.ua

Abstract. We show that it is undecidable whether or not two finite substitutions are equivalent on the fixed regular language ab^*c. This gives an unexpected answer to a question proposed in 1985 by Culik II and Karhumäki. At the same time it can be seen as the final result in a series of undecidability results for finite transducers initiated in 1968 by Griffiths. An application to systems of equations over finite languages is given.

1 Introduction

The undecidability of the equivalence problem for ε-free nondeterministic generalized sequential machines from the year 1968 is among the oldest undecidability results in automata theory, cf. [6]. It initiated a long lasting research on the equivalence problems of different types of finite transducers, cf. [2] for definitions. Remarkable extensions of the above results were achieved in [12] and [20] where it was shown that the problem remains undecidable even if the output alphabet is unary, and in [23] where it was shown that it remains undecidable also for so-called input deterministic transducers. Here the input determinism means that the transducer is, with respect to its input structure, a deterministic finite automaton.

On the other side the decidability of the equivalence problem for deterministic gsm's, even for single-valued finite transducers, seems to be folklore, cf. [2] and [26]. The other milestones on the decidable side are the papers [7] and [5] which show that the problem is decidable for finitely ambiguous and finite-valued finite transducers, respectively. The latter result was reproved and extended in [29], and generalized to the case of arbitrary output semigroups embeddable in finitely presented groups with a solvable word problem in [22]. Another remarkable result in that direction was that of [3] showing the decidability of deterministic two tape finite automata – a result which was extended to arbitrary number of tapes only in [10].

What remained open is as follows:

---

[*] Research was supported by the grant 44087 of the Academy of Finland

P. Widmayer et al. (Eds.): ICALP 2002, LNCS 2380, pp. 812–820, 2002.

Problem 1. Let $L = ab^*c$ for different letters a, b and c. Is it decidable whether two finite substitutions φ and ψ are equivalent on L, i.e. whether or not

$$\varphi(x) = \psi(x) \quad \text{for all} \quad x \in L?$$

This problem was stated implicitly in [4], more explicitly in [13] and recalled as one of the remarkable open problems in combinatorics on words in [16].

The problem can be viewed in at least two different ways. It is a special case of the equivalence problem for two-state 1-free finite transducers with a unary input alphabet, cf. Section 4. It is also a very natural question on finite sets of words, and thus potentially related to Conway's Problem, cf. below.

The problem allows a number of variants. For example we can have instead of L language $L' = a\{b,\ c\}^*d$ or instead of the equality requirement the inclusion requirement: "$\varphi(x) \subseteq \psi(x)$ for all x in L". The former variant was used in [23] to prove the above mentioned undecidability for input deterministic finite transducers, cf. also [9]. The latter problem – as a preliminary step of this paper – was shown to be undecidable in [18]. Actually, there exists quite a large literature considering these and related questions, cf. in addition to above e.g. [17], [24], [27] and [28].

The goal of this paper is to show that – on the contrary to general expectations – Problem 1 is undecidable. First hints of potential difficulties of Problem 1 were obtained in [19] where it was shown that no finite subset of L would be enough to be used as a test set to test whether arbitrary pairs of finite substitutions are equivalent on the whole L. In other words [19] showed that Ehrenfeucht Compactness Property does not hold in the monoid of finite languages, as it does in the case of word monoids, cf. [11]. Further evidence was obtained only recently: first in [21] where the problem was solved for the language $a\{b,\ c\}^*d$, and later even more in [18] when the corresponding inclusion problem was solved. In despite of these achievements it was expected that the Problem 1 should be decidable.

The undecidability of the problem, in turn, can be seen as an evidence of the difficulty of so-called *Conway's Problem*, cf. [14]. The problem asks whether the maximal set commuting with a given finite set X, referred to as the *centralizer* of X, is rational. Of course, it cannot be finite since X^* is included in the centralizer. Actually, originally the question was asked for rational X instead of finite one. This, however, does not seem to make any difference – according to the current knowledge. The problem has turned out extremely challenging: It is not even known whether the centralizer is recursive, although in all examples it is rather easily computable rational set! For a survey on Conway's Problem we refer to [14].

Finally, this paper is organized as follows. In Section 2 we fix the terminology, and in particular define the crucial notion of a *defence system*. Section 3 contains our theorem and its detailed proof. In Section 4 we point out some applications of our result.

2   Notation and Defence Systems

Let Σ be a finite alphabet, and Σ^* (resp. Σ^+) the free monoid (resp. semigroup) generated by Σ. We denote by 1 the unit element of Σ^*, so that $\Sigma^* = \Sigma^+ \setminus \{1\}$. For two finite alphabets Σ and Δ we consider *finite substitutions* $\varphi : \Sigma^* \to \Delta^*$ which are many-valued mappings and can be defined as morphisms from Σ^* into the monoid of finite subsets of Δ^*, i.e. into 2^{Δ^*}. More formally, for each $a \in \Sigma$, $\varphi(a)$ is a finite subset of Δ^*, and φ satisfies conditions $\varphi(1) = \{1\}$, and $\varphi(uv) = \varphi(u)\varphi(v)$ for all $u, v \in \Sigma^*$. If φ is single-valued then it is an ordinary morphism, and our problems become trivial.

Let φ, ψ be finite substitutions $\Sigma^* \to \Delta^*$ and $L \subseteq \Sigma^+$ a language. We say that φ and ψ *are equivalent* on L if and only if

$$\varphi(w) = \psi(w) \quad \text{for all } w \in L. \tag{1}$$

The *equivalence problem for finite substitutions on L* is a decision problem asking to decide whether two finite substitutions are equivalent on L. Problem 1 defined in Introduction is an instance of this problem where $L = ab^*c$. Similarly the *inclusion* problem can be defined by replacing the equality in (1) by the inclusion.

The above equivalence problem is closely related to that for finite transducers, cf. e.g. [17]. For definitions of finite transducers we refer to [2]. A related model of automata, so-called defence systems, is defined as follows. A *nondeterministic defence system*, ND-system for short, over the alphabet Δ is a triple $V = (Q, P, q_1)$, where Q is a finite set of states, q_1 is the unique initial state and P is a finite set of transitions of the form

$$(p, a, q, z),$$

where $p, q \in Q$, $a \in \Delta$ and $z \in \{-1, 0, 1\}$, that is

$$P \subseteq Q \times \Delta \times Q \times \{-1, 0, 1\}.$$

We say that the ND-system is *reliable* if and only if each input word $w = a_1 \ldots a_t$, with $a_i \in \Delta$ for $i = 1, \ldots, t$, possesses a *defending computation*, that is there exist states $q_1, \ldots, q_{t+1}$ and numbers $z_1, \ldots, z_t$ such that

$$(q_i, a_i, q_{i+1}, z_i) \in P \quad \text{for } i = 1, \ldots, t \tag{2}$$

and moreover,

$$\sum_{i=1}^{t} z_i = 0. \tag{3}$$

The requirement (2) defines a computation according to V in the standard way, and the condition (3) requires that the sum of the outputs, i.e. z_i's, is zero.

For our considerations the following result proved in [21] is crucial.

Lemma 1. *The reliability problem for nondeterministic defence systems is undecidable.*

As in many problems in formal language theory it is no restriction to assume here that the alphabet Δ is binary, say $\Delta = \{0, 1\}$.

3   The Result

In this section we prove our main result, which solves a longstanding open problem of [4]. Moreover, the answer is not what it was expected to be when the problem was stated. The proof resembles the one presented in [18] to solve a related problem, namely the inclusion problem for finite substitution on the same language ab^*c. However, it is essentially more complicated.

Theorem 1. *The equivalence problem for finite substitutions on the language ab^*c is undecidable.*

Ideas of Proof. The detailed proof is rather long and technical. It can be found in the final version of this paper. Here we give constructions needed and some intuitive explanations which make these to work.

We reduce the undecidability of the current problem to that of the reliability of ND-systems. Let $V = (Q, P, 1)$ be an ND-system over $\Sigma = \{0, 1\}$ and with $Q = \{1, \dots, s\}$. We associate V with a pair (φ, ψ) of finite substitutions

$$\varphi, \psi : \{a, b, c\}^* \to \{0, 1, 2, 3, 4, 5, 6\}^*$$

such that

$$V \text{ is reliable}$$

if and only if

$$\varphi(ab^ic) = \psi(ab^ic) \text{ for all } i \geq 0.$$

Construction of φ and ψ. First we fix some notation. We set

$$W = v_1 \dots v_{s+1} \text{ with } v_i = 0^i 1234 \text{ for } i = 1, \dots, s+1.$$

Consequently, $W \in \{0, 1, 2, 3, 4\}^+$. Further we set

$$w_{k,j} = v_k \dots v_j \text{ for } 1 \leq k \leq j \leq s+1.$$

Next for $k, j \in \{1, \dots, s\}$, $d \in \{0, 1\}$ and $y \in \{-1, 0, 2\}$ we define words

$$F(d, k, j, y) = w_{k+1,s+1}(S(d)S(d)W)^{y+1}S(d)S(d)w_{1,j}$$

and

$$B_d = S(d)S(d)WS(d)S(d)W,$$

where $S(d) = d + 5$. Hence $F(d, k, j, y), B_d \in \{0, 1, 2, 3, 4, 5, 6\}^*$. Using above F-words we define three new sets of words

(i) $T_1(d, k, j, -1) = F(d, k, j, 2),$

(ii) $I_2(d, k, j, 0) = F(d, k, j, 0)(34)^{-1}$
         $T_2(d, j, j, 0) = 34\, F(d, j, j, 0),$

         $I_3(d, k, j, 1) \; = F(d, k, j, -1)(234)^{-1}$

(iii) $M_3(d, j, j, 1) = 234\, F(d, j, j, 0)4^{-1}$
         $T_3(d, j, j, 1) \; = 4\, F(d, j, j, -1).$

Note that the fourth argument of these words inside any group (i), (ii) or (iii) is always a constant, namely -1, 0 or 1. The abbreviations I, M and T come from words initial, middle and terminal; notice that, in the group (i) we call the only word terminal, and not initial. From now on we can talk about I- or T_2-words, for example.

Now, we select some of the above I-, M-, and T-words, based on V, to constitute a language L_0. For a transition $(k,\ d,\ j,\ z)$ in P we choose

$$\left. \begin{array}{ll} T_1(d,\ k,\ j,\ -1) & \text{if } z = -1, \\ I_2(d,\ k,\ j,\ 0) \text{ and } T_2(d,\ j,\ j,\ 0) & \text{if } z = 0, \\ I_3(d,\ k,\ j,\ 1),\ M_3(d,\ j,\ j,\ 1) \text{ and } T_3(a,\ j,\ j,\ 1) \text{ if } z = 1. \end{array} \right\} \quad (4)$$

So

$$L_0 = \{w \mid w \text{ is a word defined in (4) when } P$$
$$\text{ranges over all transitions of } V\}.$$

Up to this point the construction is exactly as it was in [18] where the corresponding inclusion problem was solved. For the equivalence problem we continue by defining

$$L = L_0 + L_1 + L_2,$$

where the languages L_1 and L_2 are defined as follows:

Let T_j denote the T-words in L_0 having j as the value of the third parameter. Then

$$L_1 = \{ww_{j+1,s+1}w_{n,s+1} \mid w \in T_j \text{ for some } j \text{ and } 2 \le n \le s+1\}.$$

Further

$$L_2 = \{w_{m,s+1} \mid 2 \le m \le s+1\} \cdot (L_0 + L_1).$$

Now, finally we can define the substitutions φ and ψ. By denoting

$$D = B_0 B_0 + B_1 B_1$$

we set:

$$\psi : \begin{array}{l} a \longmapsto S(0)S(0)WS(0)S(0)w_{1,1} + \{B_0 w_{k,s+1} \mid 2 \le k \le s+1\} \\ b \longmapsto D + LL \\ c \longmapsto W + \{w_{k,s+1}W \mid 2 \le k \le s+1\} \end{array}$$

and

$$\varphi : \begin{array}{l} a \longmapsto \psi(a) + B_0 \\ b \longmapsto \psi(b) \\ c \longmapsto \psi(c). \end{array}$$

It follows that $\psi(x) \subseteq \varphi(x)$ for all x in $\{a,\ b,\ c\}$, and therefore the proof of the theorem is reduced to the proof of the following

Claim I. The defence system V is reliable if and only if φ and ψ satisfy:

$$\varphi(ab^i c) \subseteq \psi(ab^i c) \quad \text{for all} \ \ i \geq 0. \tag{5}$$

Instead of proving Claim I we explain the ideas behind the above constructions. Words

$$
\begin{array}{ll}
T_1 & (-1) \\
I_2 T_2 & (\ 0) \\
I_3 M_3 T_3 & (\ 1)
\end{array}
$$

play a central role in the proof. They have several important properties. They are the only products of all I-, M-, and T-words which satisfy

(i) they contain $(WS(d)S(d))^2 W$ as a factor for $d = 0$ or 1,
(ii) they are factors of $(WS(d)S(d))^4 W$ for $d = 0$ or 1.

Moreover, each of these words corresponds a computation step of the defence system V. Indeed, a prefix (resp. a suffix) of these words identifies the state and the value of d of the corresponding input symbol. Finally, words in different groups are associated with numbers $-1, 0, 1$, respectively. In terms of the defence system these correspond the movements of the head to the left, nowhere or to the right. In terms of finite substitutions they identify how many words of L_0 are used in the images $\varphi(b)$ and $\psi(b)$, or more precisely whether this number is 1 (read $2-1$) 2 (read $2-0$) or 3 (read $2+1$). This allows to get the correspondence between defending computations and the use of right number of b's.

More formally, based on above (and other parts of the construction) one can show on the one hand

that if a word $w = a_1 \dots a_n$ does not have a defending computation, then $\alpha = B_0 E_{a_1} \dots E_{a_n} W \in \varphi(ab^n c) \setminus \psi(ab^n c)$,

and on the other hand

that if $\beta = u_0 u_1 \dots u_{n+1}$ with $u_0 \in \varphi(a)$, $u_i \in \varphi(b)$, for $i = 1, \dots, n$ and $u_{n+1} \in \varphi(c)$, and if V is reliable then β can also be factorized as $\beta = v_0 v_1 \dots v_n v_{n+1}$ with $v_0 \in \psi(a)$, $v_i \in \psi(b)$ for $i = 1, \dots, n$ and $v_{n+1} = \psi(c)$. □

4   Consequences

We have solved a long standing open decision problem, and moreover settled it into unexpected direction. As consequences we obtain the following two undecidability results.

Corollary 1. *The equivalence problem for two state finite transducers with unary input alphabet is undecidable.*

Proof. Follows directly from Theorem 1 when we associate a finite substitution $\tau : \{a,\ b,\ c\}^* \to \Sigma^*$ with a transducer T_τ:

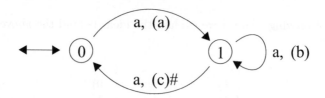

where $\# \notin \Sigma$. Indeed, the introduction of the endmaker $\#$ yields directly: For any finite substitution φ and ψ:

$$\varphi \text{ and } \psi \text{ are equivalent on } ab^*c$$

if and only if

$$T_\varphi \text{ and } T_\psi \text{ are equivalent.}$$

$\square$

The other corollary concerns equations over finite sets of words. Let $\varXi$ be a finite set of *variables* and Σ a finite set of *constants*. A *word equation* with $\varXi$ as variables and Σ as constants is a pair $(u,\ v) \in (\varXi \cup \Sigma)^2$, usually written as $u = v$. It has a *solution* in Σ^* if there exists a morphism $\varphi : (\varXi \cup \Sigma)^* \to \Sigma^*$ such that $\varphi(u) = \varphi(v)$ and $\varphi(a) = a$ for each constant $a \in \Sigma$. Two fundamental results on word equations are:

(i) *It is decidable whether a given equation or a finite system of equations possesses a solution, cf. [25].*
(ii) *Each system of equations is equivalent to some of its finite subsystems, cf. [1] or [8].*

Now, natural questions are whether these results extend to equations over finite sets of words. Then, in general, constants in equations $u = v$ are finite languages and solutions φ are finite substitutions such that $\varphi(u) = \varphi(v)$ and $\varphi(A) = A$ for each constant A in u or v.

In [19] it was shown that (ii) does not hold for finite languages even in the case of constant free equations. On the other hand, as noted in [15], (i) holds true if the constants are singletons or even prefix sets. In general, the validity of (i) for systems of finite equations over finite languages is open.

In order to formulate our contribution we recall that a system S of equations is *rational*, if it can be defined by a finite transducer, cf. [2]. Here the constants can be finite languages, the restriction being that there are only finitely many different ones in S.

We obtain:

Corollary 2. *It is undecidable whether a rational system of equations over finite sets of words has a solution.*

Proof. We associate with an instance $(\varphi,\ \psi)$ of Theorem 1 a system

$$S: \begin{cases} XY^iZ = UW^iV & \text{for } i \geq 0 \\ X = \varphi(a),\ Y = \varphi(b),\ Z = \varphi(c) \\ U = \psi(a),\ W = \psi(b),\ V = \psi(c). \end{cases}$$

Here X, Y, Z, U, W and V are variables, and $\varphi(a)$, $\varphi(b)$, $\varphi(c)$, $\psi(a)$, $\psi(b)$ and $\psi(c)$ are constants.

Clearly, φ and ψ are equivalent on ab^*c if and only if the system S has a solution. Hence, Corollary 2 follows.   □

Interestingly, the problem of Corollary 2 is decidable for word equations, cf. [11].

Acknowledgement. The authors are grateful to an anonymous referee of ICALP conference for very careful reading of the manuscript.

References

1. Albert, M. H. and Lawrence, J., A proof for Ehrenfeucht's Conjecture, Theoret. Comput. Sci. 41, 1985, 121–123.
2. Berstel, J., Transductions and Context-Free Languages, Teubner, 1979.
3. Bird, M., The equivalence problem for deterministic two-tape automata, J. Comput. Syst. Sciences 7, 1973, 218–236.
4. Culik II, K. and Karhumäki, J., Systems of equations and Ehrenfeucht's conjecture, Discr. Math. 43, 1983, 139–153.
5. Culik II, K. and Karhumäki, J., The equivalence problem of finite valued transducers (on *HDT0L* languages) is decidable, Theoret. Comput. Sci. 47, 1986, 71–84.
6. Griffiths, T. V., The unsolvability of the equivalence problem for λ-free nondeterministic generalized machines, J. Assoc. Comput. Mach. 15, 1968, 409–413.
7. Gurari, E. and Ibarra, O., A note on finite-valued and finitely ambiguous transducers, Math. Syst. Theory 16, 1983, 61–66.
8. Guba, V. S., The equivalence of infinite systems of equations in free groups and semigroups with finite systems (in Russian), Mat. Zametki, 40, 1986, 321–324.
9. Halava, V. and Harju, T., Undecidability of the equivalence of finite substitutions on regular language, Theoret. Informat. and Appl. 33, 1999, 117–124.
10. Harju, T. and Karhumäki, J., The equivalence problem of multitape finite automata, Theoret. Comput. Sci. 78, 1991, 345–353.
11. Harju, T. and Karhumäki, J., Morphisms, in: G. Rozenberg and A. Salomaa (eds), Handbook of Formal Languages, Vol. I, Springer, 1997, 439–510.
12. Ibarra, O., The unsolvability of the equivalence problem for ε-free NGSM's with unary input (output) alphabet and applications, SIAM J. Comput. 7, 1978, 524–532.
13. Karhumäki, J., Problem P 97, Bull. EATCS 25, 1985, 185.
14. Karhumäki, J., Challenges of commutation: an advertisement, Springer LNCS 2138, 2001, 15–23.

15. Karhumäki, J., Equations over finite sets of words and equivalence problems in automata theory, Theoret. Comput. Sci. 108, 1993, 103–118.
16. Karhumäki, J., Some open problems in combinatorics of words and related areas, RIMS Prodeedings 1166, Research Institute of Mathematical Sciences, Kyoto, 2000, 118–130.
17. Karhumäki, J. and Lisovik, L. P., On the equivalence of finite substitutions and transducers, in: J. Karhumäki, H. Maurer, Gh. Paun and G. Rozenberg (eds), Jewels are Forever, Springer, 1999, 97–108.
18. Karhumäki, J. and Lisovik, L. P., A simple undecidable problem: The inclusion problem for finite substitutions on ab^*c, Inf. and Comput. (to appear); preliminary version in Springer LNCS 2010, 2001, 388–395.
19. Lawrence, J., The nonexistence of finite test set for set-equivalence of finite substitutions, Bull. EATCS 28, 1986, 34–37.
20. Lisovik, L. P., The identity problem of regular events over cartesian product of free and cyclic semigroups, Doklady of Academy of Sciences of Ukraine 6, 1979, 410–413.
21. Lisovik, L. P., An undecidability problem for countable Markov chains, Kibernetika 2, 1991, 1–8.
22. Lisovik, L. P., The problem of inclusion and ambiguity for regular events in semigroups, Discritnaya Matematika (Moscow Nauka) 5, 1993, 54–74.
23. Lisovik, L. P., The equivalence problem for finite substitutions on regular languages, Doklady of Academy of Sciences of Russia 357, 1997, 299–301.
24. Lisovik, L. P., The equivalence problem for transducers with bounded number of states, Kibernetika and Sistemny Analiz 6, 1997, 109–114.
25. Makanin, G. S., The problem of solvability of equations in a free semigroups, Mat. Sb. 103, 1977, 147–236; Math. USSR Sb. 32, 1977, 129–198.
26. Schützenberger, M. P., Sur les relations rationelles entre monoides libres, Theoret. Comput. Sci. 3, 1976, 243–259.
27. Turakainen, P., On some transducer equivalence problems for families of languages, Intern. J. Comput. Math. 23, 1988, 99–124.
28. Turakainen, P., The undecidability of some equivalence problems concerning ngsm's and finite substitutions, Theoret. Comput. Sci. 174, 1997, 269–274.
29. Weber, A., Decomposing finite-valued transducers and deciding their equivalence, SIAM J. Comput. 22, 1993, 175–202.

Deciding DPDA Equivalence Is Primitive Recursive

Colin Stirling

Division of Informatics
University of Edinburgh
cps@dcs.ed.ac.uk

Abstract. Recently Sénizergues showed decidability of the equivalence problem for deterministic pushown automata. The proof of decidability is two semi-decision procedures that do not give a complexity upper bound for the problem. Here we show that there is a simpler deterministic decision procedure that has a primitive recursive upper bound.

1 Introduction

Recently Sénizergues showed decidability of the equivalence problem for deterministic pushown automata, that language equivalence is decidable for deterministic context-free languages, see [6,7,8,9]. These proofs of decidability involve two semi-decision procedures that do not give a complexity upper bound for the problem. Here we show that there is a simpler deterministic decision procedure (first described in [10]) that has a primitive recursive upper bound.

2 Preliminaries

A deterministic pushdown automaton, a DPDA, consists of finite sets of states P, stack symbols S, terminals A and basic transitions T. A basic transition is $pS \xrightarrow{a} q\alpha$ where p, q are states in P, $a \in A \cup \{\varepsilon\}$, S is a stack symbol in S and α is a sequence of stack symbols in S^*. Basic transitions are restricted.

if $pS \xrightarrow{a} q\alpha \in T$ and $pS \xrightarrow{a} r\beta \in T$ and $a \in A \cup \{\varepsilon\}$, then $q = r$ and $\alpha = \beta$

if $pS \xrightarrow{\epsilon} q\alpha \in T$ and $pS \xrightarrow{a} r\beta \in T$ then $a = \epsilon$

A configuration of a DPDA has the form $p\delta$ where $p \in P$ and $\delta \in S^*$. The transitions of a configuration are determined by the following prefix rule: if $pS \xrightarrow{a} q\alpha \in T$ then $pS\beta \xrightarrow{a} q\alpha\beta$. The transition relation $\xrightarrow{a}$, $a \in A \cup \{\varepsilon\}$, is extended to words $\xrightarrow{w}$, $w \in A^*$. First, $p\alpha \xrightarrow{\epsilon} p_n\alpha_n$, if $p_n = p$ and $\alpha_n = \alpha$ or there is a sequence of transitions $p\alpha \xrightarrow{\epsilon} p_1\alpha_1 \xrightarrow{\epsilon} \ldots \xrightarrow{\epsilon} p_n\alpha_n$. If $w = av \in A^+$, then $p\alpha \xrightarrow{w} q\beta$ if $p\alpha \xrightarrow{\epsilon} p'\alpha' \xrightarrow{a} q'\beta' \xrightarrow{v} q\beta$. The language accepted by a configuration $p\delta$, $L(p\delta)$, is the set of words $\{w \in A^* : \exists q \in P. p\delta \xrightarrow{w} q\epsilon\}$. Acceptance is by empty stack. The DPDA problem is whether $L(p\alpha) = L(q\beta)$.

P. Widmayer et al. (Eds.): ICALP 2002, LNCS 2380, pp. 821–832, 2002.
© Springer-Verlag Berlin Heidelberg 2002

Moreover, one can assume that the DPDA is in normal form: if $pS \xrightarrow{a} q\alpha \in \mathsf{T}$, then $|\alpha| \leq 2$ and if $pS \xrightarrow{\varepsilon} q\alpha \in \mathsf{T}$ then $\alpha = \varepsilon$.

An important step in the decidability proof is a syntactic representation of DPDA configurations that dispenses with ε-transitions. The key is *nondeterministic* pushdown automata with a single state and without ϵ-transitions, introduced by Harrison and Havel [3] as grammars, and further studied in [2,4]. Because the state is redundant, a configuration of a pushdown automaton with a single state is a sequence of stack symbols. Ingredients of such an automaton without ε-transitions, an SDA, are finite sets of stack symbols S, terminals A and basic transitions T of the form $S \xrightarrow{a} \alpha$ where $a \in \mathsf{A}$, $S \in \mathsf{S}$ and $\alpha \in \mathsf{S}^*$. A configuration of an SDA is a sequence of stack symbols whose transitions are determined by the prefix rule: if $S \xrightarrow{a} \alpha \in \mathsf{T}$, then $S\beta \xrightarrow{a} \alpha\beta$. The language $\mathsf{L}(\alpha)$ accepted, or generated, by a configuration α is the set $\{w \in \mathsf{A}^* : \alpha \xrightarrow{w} \epsilon\}$, so acceptance is again by empty stack. We assume the SDA is in normal form: if $S \xrightarrow{a} \alpha \in \mathsf{T}$, then $|\alpha| \leq 2$ and no element of S is redundant ($S \in \mathsf{S}$ is redundant, if $\mathsf{L}(S) = \emptyset$).

If the SDA is deterministic, then the decision problem is decidable, as proved by Korenjak and Hopcroft [5]. However, the languages generable by deterministic SDA are strictly contained in those generable by DPDA. Instead of assuming determinism, Harrison and Havel include an extra component, $\equiv$, in the definition of an SDA that is an equivalence relation on stack symbols that partitions S. The relation, $\equiv$, on S is extended to a relation on sequences of stack symbols, and the same relation, $\equiv$, is used for the extension: $\alpha \equiv \beta$ if, either $\alpha = \beta$, or $\alpha = \delta X \alpha'$ and $\beta = \delta Y \beta'$ and $X \equiv Y$ and $X \neq Y$. Some simple properties of $\equiv$ are: $\alpha\beta \equiv \alpha$ if, and only if, $\beta = \epsilon$; $\alpha \equiv \beta$ if, and only if, $\delta\alpha \equiv \delta\beta$; if $\alpha \equiv \beta$ and $\gamma \equiv \delta$, then $\alpha\gamma \equiv \beta\delta$; if $\alpha \equiv \beta$ and $\alpha \neq \beta$, then $\alpha\gamma \equiv \beta\delta$; if $\alpha\gamma \equiv \beta\delta$ and $|\alpha| = |\beta|$, then $\alpha \equiv \beta$.

Definition 1 The relation $\equiv$ on S is *strict* when the following two conditions hold. (1) If $X \equiv Y$ and $X \xrightarrow{a} \alpha$ and $Y \xrightarrow{a} \beta$, then $\alpha \equiv \beta$. (2) If $X \equiv Y$ and $X \xrightarrow{a} \alpha$ and $Y \xrightarrow{a} \alpha$, then $X = Y$. An SDA is *strict* (deterministic) if its partition is strict.

Proposition 1 *Assume a strict SDA.* (1) *If* $\alpha \xrightarrow{w} \alpha'$ *and* $\beta \xrightarrow{w} \beta'$ *and* $\alpha \equiv \beta$ *then* $\alpha' \equiv \beta'$. (2) *If* $\alpha \xrightarrow{w} \alpha'$ *and* $\beta \xrightarrow{w} \alpha'$ *and* $\alpha \equiv \beta$ *then* $\alpha = \beta$. (3) *If* $\alpha \equiv \beta$ *and* $w \in \mathsf{L}(\alpha)$, *then for all words* v, *and* $a \in \mathsf{A}$, $wav \notin \mathsf{L}(\beta)$. (4) *If* $\alpha \equiv \beta$ *and* $\alpha \neq \beta$, *then* $\mathsf{L}(\alpha) \cap \mathsf{L}(\beta) = \emptyset$.

The definition of a configuration of a strict SDA is extended to sets of sequences of stack symbols, $\{\alpha_1, \ldots, \alpha_n\}$, written in sum form $\alpha_1 + \ldots + \alpha_n$. Two sum configurations are equal, written using $=$, if they are the same set. A degenerate case is the empty sum, written $\emptyset$. The language of a sum configuration is defined using union: $\mathsf{L}(\alpha_1 + \ldots + \alpha_n) = \bigcup \{\mathsf{L}(\alpha_i) : 1 \leq i \leq n\}$.

Definition 2 A sum configuration $\beta_1 + \ldots + \beta_n$ is *admissible*, if $\beta_i \equiv \beta_j$ for each pair of components, and $\beta_i \neq \beta_j$ when $i \neq j$.

In [4] admissible configurations are called "associates". A simple corollary of Proposition 1 is that admissibility is preserved by word transitions: if $\{\beta_1, \ldots, \beta_n\}$ is admissible, then for any $w \in A^*$, $\{\beta' : \beta_i \xrightarrow{w} \beta', 1 \leq i \leq n\}$ is admissible.

A strict SDA can be determinised, by determinising the basic transitions T to T^d: for each stack symbol X and $a \in A$, the transitions $X \xrightarrow{a} \alpha_1, \ldots, X \xrightarrow{a} \alpha_n$ in T are replaced by the single transition $X \xrightarrow{a} \alpha_1 + \ldots + \alpha_n$ in T^d. The resulting sum configuration is admissible. For each stack symbol X and $a \in A$ there is a unique transition $X \xrightarrow{a} \sum \alpha_i \in \mathsf{T}^\mathsf{d}$ (where the empty sum is $\emptyset$). The prefix rule for generating transitions is extended to admissible configurations: if $X_1\beta_1 + \ldots + X_m\beta_m$ is admissible and $X_i \xrightarrow{a} \sum \alpha_{ij} \in \mathsf{T}^\mathsf{d}$ for each i, then $X_1\beta_1 + \ldots + X_m\beta_m \xrightarrow{a} \sum \alpha_{1j}\beta_1 + \ldots + \sum \alpha_{mj}\beta_m$. The resulting configuration is admissible.

Admissible configurations of strict SDAs generate the same languages as configurations of DPDA with empty stack acceptance [3]. The DPDA problem is the same problem as deciding language (or, bisimulation) equivalence between admissible configurations of a determinised strict SDA in normal form.

3   Heads, Tails, and Extensions

Assume a fixed determinised strict SDA in normal form with ingredients S, A, $\mathsf{T_d}$ and $\equiv$. We assume a total ordering on A, and we say that u is shorter than v if $|u| < |v|$ or $|u| = |v|$ and u is lexicographically smaller than v. Let $E, F, G, \ldots$ range over admissible configurations. The configuration E after the word u, $E \cdot u$, is the unique admissible configuration F such that $E \xrightarrow{u} F$. The language accepted by E, $\mathsf{L}(E)$, is $\{u : (E \cdot u) = \epsilon\}$. E and F are language equivalent, $E \sim F$, if they accept the same language, $\mathsf{L}(E) = \mathsf{L}(F)$. Language equivalence can also be approximated. If $n \geq 0$, then E and F are n-equivalent, $E \sim_n F$, provided that they reject the same words whose length is at most n: for all w such that $|w| \leq n$, $(E \cdot w) = \emptyset$ if, and only if, $(F \cdot w) = \emptyset$.

Proposition 1 (1) $E \sim F$ if, and only if, for all $n \geq 0$, $E \sim_n F$. (2) If $E \not\sim F$ and $E, F \neq \emptyset$, then there is an $n \geq 0$ such that $E \sim_n F$ and $E \not\sim_{n+1} F$. (3) $E \sim F$ if, and only if, for all $u \in A^*$, $(E \cdot u) \sim (F \cdot u)$. (4) $E \sim_n F$ if, and only if, for all $u \in A^*$ where $|u| \leq n$, $(E \cdot u) \sim_{n-|u|} (F \cdot u)$. (5) If $E \sim_n F$ and $0 \leq m < n$, then $E \sim_m F$. (6) If $E \sim_n F$ and $F \not\sim_n G$, then $E \not\sim_n G$.

Definition 1 (1) For each stack symbol X, the word $w(X)$ is the shortest word in the set $\{u : (X \cdot u) = \epsilon\}$. (2) The *norm* M of an SDA in normal form is $\max \{|w(X)| : X \in \mathsf{S}\}$.

The operation $+$, as used in sum form, can be extended: if E and F are admissible and $E \cup F$ is admissible and E, F are disjoint, $E \cap F = \emptyset$, then $E + F$ is the admissible configuration $E \cup F$. Sequential composition, written as juxtaposition, is also used: if E and F are admissible, then EF is the configuration $\{\beta\gamma : \beta \in E \text{ and } \gamma \in F\}$, that is admissible. Some properties are: if $E + F$ is admissible and $u \in \mathsf{L}(E)$, then $uv \notin \mathsf{L}(F)$; if $E + F$ is admissible, then $\mathsf{L}(E) \cap \mathsf{L}(F) = \emptyset$; $\mathsf{L}(EF) = \{uv : u \in \mathsf{L}(E)$ and $v \in \mathsf{L}(F)\}$. Also, the following identities hold: $E + \emptyset = E = \emptyset + E$, $E\emptyset = \emptyset = \emptyset E$, $E\varepsilon = E = \varepsilon E$, $(E + F)G = EG + FG$ and $G(E + F) = GE + GF$. Admissible configurations can have different "shapes", using $+$ and sequential composition.

Definition 2 Assume $k \geq 1$ and $E = \{\beta'_1\delta_1, \ldots, \beta'_m\delta_m\}$ and $|\beta'_i| = k$, or $|\beta'_i| < k$ and $\delta_i = \varepsilon$, for each $i : 1 \leq i \leq m$. If $\beta_1, \ldots, \beta_n$ are the distinct elements in $\{\beta'_1, \ldots, \beta'_m\}$ and $H_l = \{\delta_j : \beta'_j = \beta_l\}$ for each $l : 1 \leq l \leq n$, then $E = \beta_1 H_1 + \ldots + \beta_n H_n$ is in k-head form.

Fact 1 If $E = \beta_1 G_1 + \ldots + \beta_n G_n$ is in k-head form, then $\beta_1 + \ldots + \beta_n$ is admissible and each G_i is admissible and different from $\emptyset$.

Proposition 2 Assume $E = \beta_1 G_1 + \ldots + \beta_n G_n$ is in k-head form, and each $\beta_j = X_1^j \ldots X_{k_j}^j$. If $(\beta_i \cdot u) = X_m^i \ldots X_{k_i}^i$, then $(E \cdot u) = E_1 G_1 + \ldots + E_n G_n$ where $E_i = (\beta_i \cdot u)$ and for $j \neq i$, either $E_j = \emptyset$ or $E_j = X_m^j \ldots X_{k_j}^j$ and $X_1^j \ldots X_{m-1}^j$ is the same sequence as $X_1^i \ldots X_{m-1}^i$.

Definition 3 $E = E_1 G_1 + \ldots + E_n G_n$ is in *head/tail* form, if the head $E_1 + \ldots + E_n$ is admissible and at least one $E_i \neq \emptyset$, and each tail $G_i \neq \emptyset$.

If a configuration E is presented as $E_1 G_1 + \ldots + E_n G_n$, then we assume that it fulfills the conditions of Definition 3 of a head/tail form. If E is $E_1 G_1 + \ldots + E_n G_n$ and F is $F_1 G_1 + \ldots + F_n G_n$, then the imbalance between E and F, relative to this presentation, is $\max\{|E_i|, |F_i| : 1 \leq i \leq n\}$.

Proposition 3 Assume $E = E_1 G_1 + \ldots + E_n G_n$. (1) If $(E_i \cdot u) = \epsilon$, then for all $j \neq i$, $(E_j \cdot u) = \emptyset$ and $(E \cdot u) = G_i$. (2) If $(E_i \cdot u) \neq \emptyset$, then $(E \cdot u) = (E_1 \cdot u)G_1 + \ldots + (E_n \cdot u)G_n$. (3) If $H_i \neq \emptyset$ for all $i : 1 \leq i \leq n$, then $E_1 H_1 + \ldots + E_n H_n$ is a head/tail form. (4) If each $H_i \neq \emptyset$ and each $E_i \neq \varepsilon$ and for each j such that $E_j \neq \emptyset$, $H_j \sim_m G_j$, then $E \sim_{m+1} E_1 H_1 + \ldots + E_n H_n$. (5) If each $H_i \neq \emptyset$ and for each j such that $E_j \neq \emptyset$, $H_j \sim G_j$, then $E \sim E_1 H_1 + \ldots + E_n H_n$.

Proposition 4 Assume $E = E_1 G_1 + \ldots + E_n G_n$, $F = F_1 G_1 + \ldots + F_n G_n$, $E' = E_1 H_1 + \ldots + E_n H_n$ and $F' = F_1 H_1 + \ldots + F_n H_n$. If $E \sim_m F$ and $E' \not\sim_m F'$, then there is a word u, $|u| \leq m$, and an i such that either 1 or 2. (1)$(E' \cdot u) = H_i$ and $(F' \cdot u) = (F_1 \cdot u)H_1 + \ldots + (F_n \cdot u)H_n$ and $(F' \cdot u) \neq \emptyset$ and $(E' \cdot u) \not\sim_{m-|u|} (F' \cdot u)$. (2) $(F' \cdot u) = H_i$ and $(E' \cdot u) = (E_1 \cdot u)H_1 + \ldots + (E_n \cdot u)H_n$ and $(E' \cdot u) \neq \emptyset$ and $(E' \cdot u) \not\sim_{m-|u|} (F' \cdot u)$.

Definition 4 If $E = E_1 G_1 + \ldots + E_n G_n$ and $F = F_1 H_1 + \ldots + F_m H_m$, then F in its head/tail form is a *tail extension* of E in its head/tail form provided that each $H_i = K_1^i G_1 + \ldots + K_n^i G_n$, $1 \leq i \leq m$. When F is a tail extension of E, the associated *extension* e is the m-tuple $(K_1^1 + \ldots + K_n^1, \ldots, K_1^m + \ldots + K_n^m)$ without the G_is, and F is said to extend E by e.

Proposition 5 If $E = E_1 G_1 + \ldots + E_l G_l$ and $E' = E'_1 G'_1 + \ldots + E'_m G'_m$ and $E'' = E''_1 G''_1 + \ldots + E''_n G''_n$ and E' extends E by $e = (J_1^1 + \ldots + J_l^1, \ldots, J_1^m + \ldots + J_l^m)$ and E'' extends E' by $f = (K_1^1 + \ldots + K_m^1, \ldots, K_1^n + \ldots + K_m^n)$, then E'' extends

E by $ef = (H_1^1 + \ldots + H_l^1, \ldots, H_1^n + \ldots + H_l^n)$ where $H_j^i = K_1^i J_j^1 + \ldots + K_m^i J_j^m$ and $|ef| \leq |e| + |f|$.

Extensions are matrices, written in a linear notation. A special case is when the tails are the same: if $E = E_1 G_1 + \ldots + E_n G_n$ and $F = F_1 G_1 + \ldots + F_n G_n$, then F extends E by $e = (\varepsilon + \emptyset + \ldots + \emptyset, \ldots, \emptyset + \emptyset + \ldots + \varepsilon)$, and e is then abbreviated to the identity (ε).

4   The Decision Procedure

The procedure for deciding $E \sim F$ is to build a goal directed proof tree, a tableau, with initial goal $E \doteq F$, "is $E \sim F$?", using proof rules that reduce goals to subgoals. There are just three rules. The first is UNF, for "unfold".

$$\frac{E \doteq F}{(E \cdot a_1) \doteq (F \cdot a_1) \quad \ldots \quad (E \cdot a_k) \doteq (F \cdot a_k)} \quad A = \{a_1, \ldots, a_k\}$$

The goal is true, if, and only if, all the subgoals are true. A finer version of soundness uses approximants: if $E \not\sim_{m+1} F$, then at least one subgoal fails at level m, $(E \cdot a) \not\sim_m (F \cdot a)$. If $E' \doteq F'$ is a subgoal that is a result of m consecutive applications of UNF (and no other rule) to $E \doteq F$, then there is a word u such that $|u| = m$ and $E' = (E \cdot u)$ and $F' = (F \cdot u)$.

$$\text{BAL(R)} \qquad F \doteq X_1 H_1 + \ldots + X_k H_k$$

$$\vdots \qquad\qquad\qquad\qquad C$$

$$\frac{F' \doteq E_1 H_1 + \ldots + E_k H_k}{F' \doteq E_1 (F \cdot w(X_1)) + \ldots + E_k (F \cdot w(X_k))}$$

$$\text{BAL(L)} \qquad X_1 H_1 + \ldots + X_k H_k \qquad\qquad \doteq F$$

$$\vdots \qquad\qquad\qquad\qquad C$$

$$\frac{E_1 H_1 + \ldots + E_k H_k \qquad\qquad \doteq F'}{E_1 (F \cdot w(X_1)) + \ldots + E_k (F \cdot w(X_k)) \doteq F'}$$

where C is the condition

1. Each $E_i \neq \varepsilon$ and at least one $H_i \neq \varepsilon$.
2. There are precisely $\max\{ |w(X_i)| : E_i \neq \emptyset$ for $1 \leq i \leq k\}$ applications of UNF between the top goal and the bottom goal, and no application of any other rule.
3. If u is the word associated with the sequence of UNFs, then $E_i = (X_i \cdot u)$ for each $i : 1 \leq i \leq k$.

Fig. 1. Balance rules

The balance rules in Figure 1 involve two premises: the second premise goal reduces to the (balanced) subgoal beneath it provided that the first premise is above it (on the path back to the root goal). An application of BAL *uses* F, if F is the configuration in the initial goal of the rule, see Figure 1. The BAL rules are sound and complete.

Proposition 1 (1) *If* $X_1 H_1 + \ldots + X_k H_k \sim F$ *and* $E_1 H_1 + \ldots + E_k H_k \sim F'$, *then* $E_1 (F \cdot w(X_1)) + \ldots + E_k (F \cdot w(X_k)) \sim F'$. (2) *If* $X_1 H_1 + \ldots + X_k H_k \sim_{n+m} F$ *and* $E_1 H_1 + \ldots + E_k H_k \not\sim_{n+1} F'$ *and each* $E_i \neq \varepsilon$ *and* $m \geq \max\{|w(X_i)| : E_i \neq \emptyset\}$, *then* $E_1 (F \cdot w(X_1)) + \ldots + E_k (F \cdot w(X_k)) \not\sim_{n+1} F'$.

It is intended that there be a unique tableau associated with any initial goal. So there are restrictions on which rule is to be applied when. First, the initial premise of a BAL is the one that is "closest" to the goal and, therefore, the one that involves the least number of applications of UNF. To resolve which rule should be applied, the following priority order is assumed: (1) if BAL(L) is permitted, then apply BAL(L), (2) if BAL(R) is permitted, then apply BAL(R), (3) otherwise, apply UNF. However, whether an application of BAL is permitted involves more than fulfillment of the side condition. It also depends on the previous application of a BAL. The motivation for this restriction is Proposition 2, below. Initially, both BALs are permitted provided the side conditions hold. If an application of BAL uses F, then the resulting goal contains the configuration $E_1 (F \cdot w(X_1)) + \ldots + E_k (F \cdot w(X_k))$. E_i is a "top" of the application of BAL and $(F \cdot w(X_i))$ is a "bottom". Assume an application of BAL(L). A subsequent application of BAL(L) is permitted provided the side condition of the rule is fulfilled. However, BAL(R) is not permitted until a bottom of the previous application of BAL(L) is exposed and the side condition of the rule is true. Between the application of BAL(L) and the goal $G_1 \doteq H_1$, below,

$$F$$

$$\vdots \quad \text{BAL(L)}$$

$$E_1 (F \cdot w(X_1)) + \ldots + E_k (F \cdot w(X_k)) \doteq H$$

$$\vdots \quad \vdots \quad \text{UNFs}$$

$$(F \cdot w(X_i)) = G_1 \doteq H_1$$

there are no other applications of BAL(L), and G_1 is a bottom, $(F \cdot w(X_i))$, of the previous application of BAL(L). BAL(R) is now permitted provided it uses configuration G_1, or a configuration beneath it, when the side condition holds. BAL(R) is not permitted using a configuration from a goal above $G_1 \doteq H_1$, even when the side condition is true. The strategy is to apply a BAL rule whenever it is permitted, and if both BAL rules are permitted, then priority lies with BAL(L). If BAL(R) is applied, then the strategy is to repeatedly apply BAL(R), and to use UNF otherwise. BAL(L) is only permitted once a bottom of the previous application of BAL(R) becomes the right hand configuration of a goal and the side condition holds. One consequence is that when building a tableau proof tree, there is just one choice of which rule to apply next to any subgoal. A branch of a tableau from a subgoal $g(0)$ is a sequence of goals that start from $g(0)$.

Proposition 2 *If $F_0, F_1, \ldots, F_n$ are successive configurations used in applications of BAL in a branch, then there are $u_1, \ldots, u_n$ and $F_i = (F_0 \cdot u_1 \ldots u_i)$.*

To be a decision procedure, a notion of final goal is needed so that a tableau can be terminated. The tableau proof rules are locally complete, if a goal is true then so are the subgoals. If an obviously false subgoal is reached then the root goal is also false. The tableau proof rules are also locally sound, if all the subgoals are true then so is the goal. If an obviously true subgoal, $E \doteq E$, is reached then it is a successful final goal. However, the rules are sound in a finer version. In the case of UNF, if the goal is false at level $m+1$, then at least one subgoal fails at level m and applications of BAL preserve the falsity index. Consequently, if a subgoal $E \doteq F$ is repeated in a branch, and there is at least one application of UNF between the two occurrences, then the second occurrence of $E \doteq F$ is also a successful final goal.

A repeat is an instance of a more general situation where goals may be growing in size, formally captured below by the "extension theorem". Roughly speaking, in a branch if there are goals where the rates of change of tails are repeating, then there is a successful final goal. A repeat is an instance when the rate of change is zero. In a long enough branch with multiple applications of BAL, there must be goals within the branch that have the same heads. The idea is to discern patterns of relations between their tails. Definition 4 of the previous section is lifted to goals. Assume $E = E_1 H_1 + \ldots + E_n H_n$, $F = F_1 H_1 + \ldots + F_n H_n$, $E' = E'_1 G_1 + \ldots + E'_m G_m$ and $F' = F'_1 G_1 + \ldots + F'_m G_m$ and goal h is $E \doteq F$ and goal g is $E' \doteq F'$. Goal h extends g by extension e, if E extends E' by e (and F extends F' by e). The main insight, "the extension theorem", below, underpins when a subgoal counts as a successful final goal. It involves families of extensions with the same heads. In this theorem, a goal $E \doteq F$ is true at level m, if $E \sim_m F$. To illustrate it, consider $n = 3$. Assume goals $g(i), h(i), i : 1 \leq i \leq 8$ and each goal $g(i)$ has the form $E_1 G_1^i + \ldots + E_3 G_3^i \doteq F_1 G_1^i + \ldots + F_3 G_3^i$ and each goal $h(i)$ has the form $E_1 H_1^i + \ldots + E_3 G_3^i \doteq F_1 H_1^i + \ldots + F_3 H_3^i$. They all have the same heads. Assume extensions e_1, e_2, and e_3 as follows (where, for example, $g(2)$ extends $g(1)$ by e_1 and $g(5)$ extends $g(4)$ by e_3).

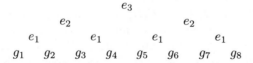

And assume the same extensions for the goals $h(i)$. The theorem says that if each $g(i), 1 \leq i \leq 8$, is true at level m and each $h(i), 1 \leq i < 8$, is true at level m, then $h(8)$ is also true at level m.

Definition 1 Assume a family of goals $E_1 G_1^s + \ldots + E_n G_n^s \doteq F_1 G_1^s + \ldots + F_n G_n^s$ with the same heads where $s \in I$. (1) A *known* at level m is either $G_i^s \sim_m G_j^s$ for some $j < i$ and for all $s \in I$, or $G_i^s \sim_m J_1 G_1^s + \ldots + J_n G_n^s$ for all $s \in I$ and each $J_k \neq \varepsilon$. (2) Two knowns $G_i^s \sim_m H_1^s$ and $G_j^s \sim_m H_2^s$ are distinct, if $i \neq j$.

Theorem 1 [The extension theorem] *Assume there are two families of goals* $g(i)$, $h(i)$, $1 \leq i \leq 2^{n-k}$, *and each goal* $g(i)$ *has the form* $E_1 G_1^i + \ldots + E_n G_n^i \doteq F_1 G_1^i + \ldots + F_n G_n^i$ *and each goal* $h(i)$ *has the form* $E_1 H_1^i + \ldots + E_n H_n^i \doteq F_1 H_1^i + \ldots + F_n H_n^i$, *and* k *is the number of distinct knowns for the family* $g(i) \cup h(i)$ *at level* m. *Assume extensions* $e_1, \ldots, e_{n-k}$ *such that for each* e_j *and* $i \geq 0$ $g(2^j i + 2^{j-1} + 1)$ *extends* $g(2^j i + 2^{j-1})$ *by* e_j *and* $h(2^j i + 2^{j-1} + 1)$ *extends* $h(2^j i + 2^{j-1})$ *by* e_j. *If each goal* $g(i)$ *is true at level* m, $i : 1 \leq i \leq 2^{n-k}$, *and each goal* $h(j)$, $j : 1 \leq j < 2^{n-k}$, *is true at level* m, *then* $h(2^{n-k})$ *is true at level* m.

Proof: The proof is by induction on $n - k$. For the base case assume $n - k = 0$. Assume that there are two goals $g(1)$, $E_1 G_1 + \ldots + E_n G_n \doteq F_1 G_1 + \ldots + F_n G_n$, abbreviated to $E \doteq F$, and $h(1)$, $E_1 H_1 + \ldots + E_n H_n \doteq F_1 H_1 + \ldots + F_n H_n$, abbreviated to $E' \doteq F'$, and n distinct knowns for $\{g(1), h(1)\}$ at level m. That is, for each $i : 1 \leq i \leq n$ either $G_i \sim_m G_j$ and $H_i \sim_m H_j$ and $j < i$, or $G_i \sim_m J_1 G_1 + \ldots + J_n G_n$ and $H_i \sim_m J_1 H_1 + \ldots + J_n H_n$ where $J_i \neq \varepsilon$. We prove that if $g(1)$ is true at level m, $E \sim_m F$, then $h(1)$ is also true at level m, $E' \sim_m F'$. Suppose not. Then, without loss of generality, by Proposition 4 of Section 3 there is a word u, $|u| \leq m$, and $(E' \cdot u) = H_i$ and $(F' \cdot u) = (F_1 \cdot u) H_1 + \ldots + (F_n \cdot u) H_n$ and $(F' \cdot u) \neq \emptyset$ and $(E' \cdot u) \not\sim_{m-|u|} (F' \cdot u)$. However, because $E \sim_m F$ it follows that $G_i \sim_{m-|u|} (F_1 \cdot u) G_1 + \ldots + (F_n \cdot u) G_n$. The first case is that $G_i \sim_m G_j$ and $H_i \sim_m H_j$ where $j < i$. Therefore, using Proposition 1 of Section 3, $G_j \sim_{m-|u|} (F_1 \cdot u) G_1 + \ldots + (F_n \cdot u) G_n$ and $H_j \not\sim_{m-|u|} (F_1 \cdot u) H_1 + \ldots + (F_n \cdot u) H_n$. This can only be repeated finitely many times, $G_j \sim_m G_{j'}$ and $H_j \sim_m H_{j'}$, and $j' < j$. Hence, the second case must then occur, $G_i \sim_m J_1 G_1 + \ldots + J_n G_n$ and $H_i \sim_m J_1 H_1 + \ldots + J_n H_n$ and no $J_i = \varepsilon$. Using Proposition 1 of Section 3, $J_1 G_1 + \ldots + J_n G_n \sim_{m-|u|} (F_1 \cdot u) G_1 + \ldots + (F_n \cdot u) G_n$ and $J_1 H_1 + \ldots + J_n H_n \not\sim_{m-|u|} (F_1 \cdot u) H_1 + \ldots + (F_n \cdot u) H_n$. Now Proposition 4 of Section 3, as above, is applied again. Therefore, by repeating this argument a contradiction will be obtained.

For the general step, assume that it holds for $n - k < t$ and consider $n - k = t$. So, there are two families of goals $g(i)$, $h(i)$, $1 \leq i \leq 2^{n-k}$, and each goal $g(i)$ has the form $E_1 G_1^i + \ldots + E_n G_n^i \doteq F_1 G_1^i + \ldots + F_n G_n^i$ and each goal $h(i)$ has the form $E_1 H_1^i + \ldots + E_n H_n^i \doteq F_1 H_1^i + \ldots + F_n H_n^i$, and k is the number of distinct knowns for the family $g(i) \cup h(i)$ at level m. Assume extensions $e_1, \ldots, e_{n-k}$ such that for each e_j and $i \geq 0$, $g(2^j i + 2^{j-1} + 1)$ extends $g(2^j i + 2^{j-1})$ by e_j and $h(2^j i + 2^{j-1} + 1)$ extends $h(2^j i + 2^{j-1})$ by e_j. Assume that each goal $g(i)$ is true at level m, $i : 1 \leq i \leq 2^{n-k}$, and each goal $h(j)$, $j : 1 \leq j < 2^{n-k}$, is true at level m. The aim is to show that $h(2^{n-k})$ is also true at level m. Suppose not. Let $q = 2^{n-k}$. Therefore, $E_1 H_1^q + \ldots + E_n H_n^q \not\sim_m F_1 H_1^q + \ldots + F_n H_n^q$ and $E_1 G_1^q + \ldots + E_n G_n^q \sim_m F_1 G_1^q + \ldots + F_n G_n^q$. Without loss of generality, by Proposition 4 of Section 3 there is a word u, $|u| \leq m$, and $(E_1 H_1^q + \ldots + E_n H_n^q \cdot u) = H_i^q$ and $(F_1 H_1^q + \ldots + F_n H_n^q \cdot u) = (F_1 \cdot u) H_1^q + \ldots + (F_n \cdot u) H_n^q$ and $H_i^q \not\sim_{m-|u|} (F_1 \cdot u) H_1^q + \ldots + (F_n \cdot u) H_n^q$ and for all $j : 1 \leq j \leq q$, $G_i^q \sim_{m-|u|} (F_1 \cdot u) G_1^j + \ldots + (F_n \cdot u) G_n^j$ and for all $j : 1 \leq j < q$, $H_i^j \sim_{m-|u|} (F_1 \cdot u) H_1^j + \ldots + (F_n \cdot u) H_n^j$. There are two cases. First is that each G_i^j and H_i^j, $1 \leq j \leq q$, is a known at level m. The proof proceeds as in the base case, but here with respect to the whole family of goals. Therefore, at some point the second case happens, that G_i^j and H_i^j are not knowns. However, each even goal g(2j) and $h(2j)$ extends the previous goal g(2j-1) and $h(2j-1)$ by e_1. Therefore, each $G_i^{2j} \sim_{m-|u|} (F_1 \cdot u)G_1^{2j} + \ldots + (F_n \cdot u)G_n^{2j}$ becomes $J_1 G_1^{2j-1} + \ldots + J_n G_n^{2j-1} \sim_{m-|u|} F_1' G_1^{2j-1} + \ldots + F_n' G_n^{2j-1}$ by substituting in the entries of e_1, and $J_1 H_1^{2j-1} + \ldots + J_n H_n^{2j-1} \sim_{m-|u|} F_1' H_1^{2j-1} + \ldots + F_n' H_n^{2j-1}$ when $j < (q-1)$ and $J_1 H_1^{q-1} + \ldots + J_n H_n^{q-1} \nsim_{m-|u|} F_1' H_1^{q-1} + \ldots + F_n' H_n^{q-1}$. Let these goals be $g'(i)$, $h'(i)$, $1 \leq i \leq 2^{n-(k+1)}$ and consider the extensions $e_1', \ldots e_{n-(k+1)}'$ where $e_i' = e_1 e_{i+1}$. Using Proposition 5 of Section 3, for each e_j' and $i \geq 0$, $g'(2^j i + 2^{j-1} + 1)$ extends $g'(2^j i + 2^{j-1})$ by e_j' and the same for the $h'(i)$s. Moreover, the number of knowns is now $(k+1)$ because the previous k knowns at level m remain knowns at level $m - |u|$ and there is the new known at level $m - |u|$ because for each j, $G_i^j \sim_{m-|u|} (F_1 \cdot u)G_1^j + \ldots + (F_n \cdot u)G_n^j$ and $H_i^j \sim_{m-|u|} (F_1 \cdot u)H_1^j + \ldots + (F_n \cdot u)H_n^j$. By the induction hypothesis if every $g'(i)$, $1 \leq i \leq 2^{n-(k+1)}$ is true at level $m - |u|$ and every $h'(i)$, $1 \leq i < 2^{n-(k+1)}$, is true at level $m - |u|$, then $h'(2^{n-(k+1)})$ is also true at level $m - |u|$, which contradicts that $J_1 H_1^{q-1} + \ldots + J_n H_n^{q-1} \nsim_{m-|u|} F_1' H_1^{q-1} + \ldots + F_n' H_n^{q-1}$.   □

Definition 2 Assume $d(0), \ldots, d(l)$ is a branch of goals and $d(l)$ is $E_1 H_1 + \ldots + E_n H_n \doteq F_1 H_1 + \ldots + F_n H_n$. Goal $d(l)$ *obeys the extension theorem* if the following hold: (1) There are families of goals $g(i)$, $h(i)$, $1 \leq i \leq 2^n$ belonging to $\{d(0), \ldots d(l)\}$, and each goal $g(i)$ has the form $E_1 G_1^i + \ldots + E_n G_n^i \doteq F_1 G_1^i + \ldots + F_n G_n^i$ and each goal $h(i)$ has the form $E_1 H_1^i + \ldots + E_n H_n^i \doteq F_1 H_1^i + \ldots + F_n H_n^i$. (2) The goal $h(2^n)$ is $d(l)$ and there is at least one application of UNF between goal $h(2^n - 1)$ and $d(l)$. (3) There are extensions $e_1, \ldots, e_n$ such that for each e_j and $i \geq 0$, $g(2^j i + 2^{j-1} + 1)$ extends $g(2^j i + 2^{j-1})$ by e_j and $h(2^j i + 2^{j-1} + 1)$ extends $h(2^j i + 2^{j-1})$ by e_j.

Definition 3 If $g(0), \ldots, g(n)$ where $g(0)$ is the root goal is a branch of goals, then $g(n)$ is a *final goal* in the following circumstances. (1) If $g(n)$ is an identity $E \doteq E$, then $g(n)$ is a successful final goal. (2) If $g(n)$ obeys the extension theorem, then $g(n)$ is a successful final goal. (3) If $g(n)$ has the form $E \doteq \emptyset$ or $\emptyset \doteq E$ and $E \neq \emptyset$, then $g(n)$ is an unsuccessful final goal.

The deterministic procedure that decides whether $E \sim F$ is defined iteratively.

1. Stage 0: start with the root goal $g(0)$, $E \doteq F$, that becomes a frontier node of the branch $g(0)$.
2. Stage $n + 1$: if a current frontier node $g(n)$ of branch $g(0), \ldots, g(n)$ is an unsuccessful final goal, then halt and return "unsuccessful tableau"; if each frontier node $g(n)$ of branch $g(0), \ldots, g(n)$ is a successful final goal, then return "successful tableau"; otherwise, for each frontier node $g(n)$ of branch $g(0), \ldots, g(n)$ that is not a final goal, apply the next rule to it, and the subgoals that result are the new frontier nodes of the extended branches.

5   Correctness and Complexity of the Decision Procedure

For the complexity upper bound, some basis functions are introduced assuming a fixed SDA with s stack symbols and largest partition p: a DPDA in normal form with s_1 stack symbols and p_1 states is transformed into an SDA with at most $s_1 \times p_1^2$ stack symbols and whose largest partition is no more than p_1. The size of an admissible configuration $E = \beta_1 + \ldots + \beta_n$, written $|E|$, is $\max\{|\beta_i| : 1 \le i \le n\}$. Let goal($h$) be the number of different goals $E \doteq F$ such that $|E|, |F| \le h$. Let width(h) be the maximum n of an admissible configuration $\beta_1 + \ldots + \beta_n$ such that $|\beta_i| \le h$ (so, width(h) $\le p^h$). The function ext(d, w) is the number of different extensions e such that $|e| \le d$ and e has width at most w (so, ext(d, w) $\le 2^{s^{dw}}$). The other measure used is the norm M that is bounded by 2^s. Some auxiliary functions are now defined from the basis functions.

Definition 1   (1) The function $f_1(k) = M(k + 1 + M^2 + 2M) + \text{goal}(M)$. (2) The function $f_2[d, h, w](n)$ is defined recursively: $f_2[d, h, w](0) = \text{goal}(h)$, $f_2[d, h, w](j+1) = \text{ext}(f_2[d, h, w](j) \times d, w) \times f_2[d, h, w](j)$. (3) Let $d = M^2 + 2M$, $h = M^2 + 5M + 2$, $w = \text{width}(h)$ and $b = f_2[d, h, w](w)$. Then, $f(n) = n \times d^b \times f_1(n + bd)$.

Theorem 1 *If $E \not\sim F$, then the decision procedure terminates with "unsuccessful tableau".*

In the following result, symmetric properties hold if BAL(R) replaces BAL(L) in parts 2 and 4.

Proposition 1 *Assume that $g(0), \ldots, g(n)$ is a sequence of goals in a branch of a tableau and $g(0)$ is $E \doteq F$. (1) If $g(0)$ is the root goal and $E = \beta_1 G_1 + \ldots + \beta_m G_m$ and $F = \delta_1 H_1 + \ldots + \delta_l H_l$ are in k-head form and $|E|, |F| > k$, and there is no application of BAL between $g(0)$ and $g(kM)$, then there are goals $g(i)$ and $g(j)$, $1 \le i, j \le kM$, such that $g(i)$ is $G_{i'} \doteq F'$ and $g(j)$ is $E' \doteq H_{j'}$ for some i' and j'. (2) If $g(0)$ is a result of BAL(L) using F' and there is no application of a BAL between $g(0)$ and $g(k)$ where $k = M^2 + M$ and $E = E_1(F' \cdot w(X_1)) + \ldots + E_k(F' \cdot w(X_k))$, then there is a goal $g(i)$, $i : 1 \le i \le k$, that has the form $(F' \cdot w(X_j)) \doteq F''$. (3) If $g(0)$ is a result of BAL using F' and the next application of BAL is $g(k)$ using F'', then $|F''| \le |F'| + M^2 + 2M$. (4) If $g(0)$ is a result of BAL(L) using F' and the next application of BAL is $g(k)$ using F'' and $k \ge M^3 + 3M^2 + 3M$, then $|F''| < |F'|$.*

Lemma 1 *Assume that $g(0), \ldots, g(n)$ is a sequence of goals in a branch and $g(0)$ is $E \doteq F$ and there are no applications of BAL in this branch. If $|E|, |F| \le k$, and $n \ge f_1(k)$, then the branch contains a successful final goal.*

Lemma 1 establishes termination of the decision procedure for a sufficiently long branch of goals that does not involve applications of BAL. An analysis is now developed for branches that involve repeated applications of BAL.

Definition 2   Assume $F_0, F_1, \ldots, F_t$ are successive configurations used in applications of BAL in a branch and assume words $u_1, \ldots, u_t$ such that $F_i =$

$(F_0 \cdot u_1 \dots u_i)$. Let $F_i = \beta_1 G_1 + \dots + \beta_n G_n$ be in m-head form, $m \geq 1$. F_i is m-low in the interval $F_j F_{j+1} \dots F_{j+k}$, $j \geq i$ and $j + k \leq t$, if there is a prefix v of $u_{j+1} \dots u_{j+k}$ such that $(F_i \cdot u_{i+1} \dots u_j v) = G_l$ for some l, and F_i is then said to be m-low with G_l using v in this interval. The definition is extended to 0-low: F_i is 0-low with F_i using ε in the interval $F_j F_{j+1} \dots F_{j+k}$ provided that $j = i$ and F_i is not m-low for any $m > 0$ in this interval.

Proposition 2 *Assume $F_0, F_1, \dots, F_t$ are successive configurations used in applications of BAL in a branch and assume words $u_1, \dots, u_t$ such that $F_i = (F_0 \cdot u_1 \dots u_i)$. Assume F_i is m_1-low in $F_j F_{j+1}$, $m_1 \geq 0$ and $j \geq i$ with G using v, and F_i is not m_1'-low for any $m_1' > m_1$ in $F_j F_{j+1} \dots F_t$. Assume F_{j+1} is m_2-low in $F_l F_{l+1}$, $j + 1 \leq l < t$, and $m_2 \geq 0$, with G' and F_{j+1} is not m_2'-low in $F_l F_{l+1}$ for any $m_2' > m_2$. (1) If $G = X_1 H_1 + \dots + X_k H_k$ is in 1-head form and $u_{j+1} = vv'$, then for all prefixes w of $v' u_{j+1} \dots u_t$, $(G \cdot w) = (X_1 \cdot w) H_1 + \dots + (X_k \cdot w) H_k$. (2) $|F_{j+1}| \leq |G| + (M^2 + 2M)$ and $m_2 \leq (M^2 + 2M)$. (3) If $G = \beta_1 G_1 + \dots + \beta_n G_n$ is in $M + 1$-head form, then the goal that is the result of BAL using F_{j+1} has the head/tail form $E_1 G_1 + \dots + E_n G_n \doteq F_1 G_1 + \dots + F_n G_n$, where $|E_i|, |F_i| \leq M^2 + 5M + 2$. (4) If $G = \beta_1 G_1 + \dots + \beta_n G_n$ and $G' = \delta_1 G_1' + \dots + \delta_{n'} G_{n'}'$ are in $M + 1$-head form, then in these forms G' is an extension of G with extension e and $|e| \leq M^2 + 2M$.*

Lemma 2 *Assume that there is a subsequence of goals in a branch $g(i)$, $E_1^i G_1^i + \dots + E_{n_i}^i G_{n_i}^i \doteq F_1^i G_1^i + \dots + F_{n_i}^i G_{n_i}^i$, $i : 0 \leq i \leq t$ and each $g(j + 1)$ extends $g(j)$ by e_j where $|e_j| \leq d$ and each head $|E_j^i|, |F_j^i| \leq h$. If $w = \text{width}(h)$ and $t \geq f_2[d, h, w](w)$, then there is a successful final goal in the branch.*

Theorem 2 *(1) If $E \sim F$, then the decision procedure terminates with "successful" tableau. (2) If $|E|, |F| < n$, then the decision procedure with root $E \doteq F$ terminates within $f(n)$ steps.*

Proof: Assume that $E \sim F$. The tableau for $E \doteq F$ is built using the rules of the proof system. By completeness of rule application at each stage each frontier goal is true. Therefore, it is not possible to to reach an unsuccessful final goal. It is also not possible to become stuck because UNF can always apply. Hence the only issue is that the decision procedure doesn't terminate. Assume that there is an infinite branch of goals $g(0), \dots, g(n), \dots$ where $g(0)$ is the root goal $E \doteq F$ that does not contain a successful final goal. If there are only finitely many applications of BAL, then there is an infinite suffix of goals $g(i), \dots, g(n), \dots$ and there are no applications of BAL. However, by Lemma 1 there must be a successful final goal. Otherwise there are infinitely many applications of BAL. Let $F_0, F_1 \dots$ be the successive configurations used in applications of BAL. Start with F_0 and find the largest m_0 and the first interval $F_{i_1-1} F_{i_1}$ such that F_0 is m_0-low in this interval. By assumption, F_0 is not m'-low, for any $m' > m_0$ in any later interval. Next, consider F_{i_1}. Find the largest m_1 and the first interval $F_{i_2-1} F_{i_2}$, $i_2 > i_1$, such that F_1 is m_1-low in this interval. Using Proposition 2, the application of BAL using F_{i_1}, the goal $g(k_1)$, has a head/tail form $E_1^1 G_1^1 +$

$\ldots + E_{n_1}^1 G_{n_1}^1 \doteq F_1^1 G_1^1 + \ldots + F_{n_1}^1 G_{n_1}^1$ and the application of BAL using F_{i_2}, the goal $g(k_2)$, has head/tail form $E_1^2 G_1^2 + \ldots + E_{n_2}^2 G_{n_2}^2 \doteq F_1^2 G_1^2 + \ldots + F_{n_2}^2 G_{n_2}^2$ where $|E_j^i|, |F_j^i| \leq h = M^2 + 5M + 2$ and $g(k_2)$ extends $g(k_1)$ by e_1 where $|e_1| \leq d = M^2 + 2M$. The argument is now repeated giving an infinite subsequence of goals $g(k_1), g(k_2), \ldots, g(k_n), \ldots$ such that each $g(k_{i+1})$ is an extension of $g(k_i)$ by e_i with $|e_i| \leq d$, and the heads have bounded size no more than h. Therefore, by Lemma 2 there is a successful final goal within this subsequence. For part 2, the argument above is refined. Assume a branch $g(0), \ldots, g(t)$ where $g(0)$ is the root $E \doteq F$, and $|E|, |F| \leq n$ and $t \geq f(n)$. If there are no applications of BAL, then by Lemma 1 there is a successful final goal within $f_1(n)$ steps and $f_1(n) \leq f(n)$. Let $F_0, F_1, \ldots, F_l$ be successive configurations used in applications of BAL. Let d and h be as defined above and let $w = \text{width}(h)$ and let $b = f_2[d, h, w](w)$. Via Proposition 2, if no F_i has a low point that is greater than 0, then for $l \leq b$ the goals that are the result of these applications of BAL obey Lemma 2. By Proposition 1, $|F_{i+1}| \leq |F_i| + d$. Hence, it follows that the largest F_i is bounded by $u = n + bd$. Moreover, it follows that the maximum number of steps between F_i and F_{i+1} is at most $f_1(u)$ (because, otherwise there must be a successful final goal using Lemma 1). For the general case let $f'(b)$ be the number of successive configurations used in applications of BAL such that there is a subsequence of length b such that the goals that are the result of these applications of BAL obey Lemma 2. F_0 may have n low points. Therefore, $f'(b) = n \times f'(b-1)$. Subsequent F_i have only d low points, so $f'(b-1) = d \times f'(b-2)$. The overall bound is therefore $n \times d^b$. The maximum number of steps between F_i and F_{i+1} is again at most $f_1(u)$. Therefore, within $f(n)$ steps it follows that the decision procedure must terminate. □

References

1. Ginsberg, S., and Greibach, S. (1966). Deterministic context-free languages. *Information and Control*, 620-648.
2. Harrison, M. (1978). *Introduction to Formal Language Theory*, Addison-Wesley.
3. Harrison, M., and Havel, I. (1973). Strict deterministic grammars. *Journal of Computer and System Sciences*, **7**, 237-277.
4. Harrison, M., Havel, I., and Yehudai, A. (1979). On equivalence of grammars through transformation trees. *Theoretical Computer Science*, **9**, 173-205.
5. Korenjak, A and Hopcroft, J. (1966). Simple deterministic languages. *Procs. 7th Annual IEEE Symposium on Switching and Automata Theory*, 36-46.
6. Sénizergues, G. (1997). The equivalence problem for deterministic pushdown automata is decidable. *Lecture Notes in Computer Science*, **1256**, 671-681.
7. Sénizergues, G. (2001). L(A) = L(B)? decidability results from complete formal systems. *Theoretical Computer Science*, **251**, 1-166.
8. Sénizergues, G. (2001). L(A) = L(B)? a simplified decidability proof. pp.1-58. To appear in *Theoretical Computer Science*.
9. Stirling, C. (2001). Decidability of DPDA equivalence. *Theoretical Computer Science*, **255**, 1-31.
10. Stirling, C. (2001). An introduction to decidability of DPDA equivalence. *Lecture Notes in Computer Science*, **2245**, 42-56.

Two-Way Alternating Automata and Finite Models

Mikołaj Bojańczyk

Uniwersytet Warszawski, Wydział MIM, Banacha 2, Warszawa, Polska

Abstract. A graph extension of two-way alternating automata on trees is considered. The following problem: "does a given automaton accept any *finite* graph?" is proven EXPTIME complete. Using this result, the decidability of the finite model problem for the modal μ-calculus with backward modalities is shown.

1 Introduction

Alternating tree automata were introduced Muller and Schupp [12]. In terms of expressibility these automata define the same class of languages as the simpler nondeterministic tree automata on the one hand and as the complex monadic second order theory on trees (S2S) on the other. Nevertheless, the formalism of alternating automata offers a good balance between logical manageability and computational complexity. Testing emptiness for alternating tree automata is far easier than the non-elementary procedures in S2S; on the other hand, closure under conjunction and negation is trivial for alternating automata and very difficult for nondeterministic automata (cf. the famous "complementation lemma" [13]).

An important variant of alternating tree automata are *alternating tree automata with the parity condition*, introduced by Emmerson and Jutla in [3]. The parity condition assigns to each of the finite number of states of the automaton a natural number and those runs of the automaton are considered accepting where the least number occurring infinitely often is even. The parity condition is of growing importance in automata theory, particularly in connection with games [6,11].

Alternating automata are very closely connected to Kozen's modal μ-calculus [9]. Because of this close correspondence, the μ-calculus is a standard application for alternating automata [12,2]. In the same spirit, the satisfiability problem for the propositional μ-calculus with backward modalities is proved decidable by Vardi [16] via a reduction to two-way alternating automata.

The μ-calculus with *backwards modalities* augments the one-way μ-calculus with quantification over backward modalities, denoted by $\Diamond^-$ and $\Box^-$. A formula of the form $\Diamond^-\phi$ states that ϕ occurs in some predecessor of the current state, similarly for $\Box^-$. Analogously to the μ-calculus, a two-way automaton can, apart from the usual forward moves of one-way alternating automata, make backward moves [14].

P. Widmayer et al. (Eds.): ICALP 2002, LNCS 2380, pp. 833–844, 2002.
© Springer-Verlag Berlin Heidelberg 2002

Another application of the two-way automaton can be found in [5], where it is used to solve another satisfiability problem – this time for Guarded Fixed Point Logic, an extension of the Guarded Fragment [7] by fix-point operators.

There is an interesting common denominator to Guarded Fixed Point Logic, the μ-calculus with backward modalities and two-way alternating automata: none of them have a finite model property.

W say a logic has the *finite model* property if every satisfiable sentence is satisfiable in some finite structure. Modal logic and even the modal μ-calculus have the finite model property; the μ-calculus with backward modalities does not (consider the sentence $\nu X.(\Diamond X \wedge \mu Y.\Box^- Y)$). A similar situation occurs in the Guarded Fragment: the fix-point extension no longer has the finite model property, contrary to the "bare" Guarded Fragment [4] and some of its other extensions (most notably the Loosely Guarded Fragment [8]).

These observations give rise to the following decision problem: "Is a given sentence of the modal μ–calculus with backward modalities (or guarded fixed point logic) satisfiable in some *finite* model?" While tackling this problem, we took the automata approach. However, for reasons sketched out below, it turns out that we need a new definition of two-way alternating automata.

Most modal logics have a bisimulation-invariance property and the two-way μ-calculus is no exception. In particular, a sentence of the two-way μ-calculus cannot distinguish between a model and its tree unraveling (technically, its two-way tree unraveling) and so every satisfiable sentence is satisfiable in a tree-like structure. Thus for the purpose of deciding satisfiability, one can constrain attention to tree models. This was the approach taken by Vardi; in fact his automata were alternating two-way automata on *infinite trees*.

As much as the tree model property is helpful in researching the satisfiability problem, things get more complicated where the finite model problem is concerned. The reason is that, unfortunately, finite models rarely turn out to be trees. There are finitely satisfiable sentences that have no finite tree models, for instance $\nu X.\Diamond X$. For this reason, while investigating the finite model problem we will consider automata on arbitrary graphs, not on trees. These automata correspond very closely to the μ-calculus and consequently also lack a finite model property.

The suitable example, presented in more detail in Example 1, is as follows. An alternating automaton is constructed that accepts a graph if every reachable state has a successor and every backward path is finite. From this follows that every graph accepted by this automaton must be infinite. A typical graph accepted by $\mathcal{A}$ has natural numbers as vertices, with edges representing the successor relation.

The paper [1] presented an incomplete solution to the finite model problem. The automata considered used the *Büchi acceptance condition* or, equivalently, the parity condition for the colors $\{0,1\}$. The solution presented used a complicated technical pumping argument and the accompanying algorithm ran in doubly exponential time. Although an important step toward the solution of the general problem, the Büchi condition is insufficient for both the translations from μ-calculus and the guarded logics which use the full power of the parity condition. This old result is improved here in two ways: we solve the finite model problem for the general parity condition, and we do this in singly exponential time, which turns out to be optimal:

Theorem 1. *The finite model problem for alternating two-way automata with the full parity condition is EXPTIME complete*

Here we present an outline of the proof. First of all, we immediately get rid of graphs and start working on trees, since tree automata are much easier to work with. A finite graph is represented as an infinite tree – its two-way tree unraveling. Of course, not all infinite trees represent finite graphs – for this a special condition, called *bounded signature*, must be satisfied. This way, the finite model problem is reduced to the emptiness problem for the tree language consisting of trees with bounded signature. Although this language is not regular itself, its emptiness is equivalent to the emptiness of an appropriate (effectively found) regular language.

2   The Automaton

In this section we define our automata. Since our definition of acceptance uses games with the parity condition, we first briefly define them and recall some fundamental properties. A more detailed presentation can be found in [15].

2.1   Games with the Parity Condition

Games with the parity condition are powerful tool in the field of infinite tree automata. We take the game approach in defining the semantics of our two-way alternating automata.

Definition 2 (Parity condition game). *A game with the parity condition is a tuple* $G = \langle V_0, V_1, E, v_0, \Omega \rangle$, *where* V_0 *and* V_1 *are disjoint sets of positions, the function* $\Omega : V = V_0 \cup V_1 \to \{0, \ldots, N\}$ *is called the* coloring *function,* $E \subseteq V \times V$ *is the set of edges, and* $v_0 \in V$ *is some fixed starting position. We additionally assume that for every position* $v \in V$, *the set of outgoing edges* $(v, w) \in E$ *is finite.*

The game is played by two players, 0 and 1, and consists of moving a token from one position in V to another along the edges of the graph. The first position is v_0. If the token is in a position from V_0, the player 0 makes the move, otherwise the second player decides. If at some point one of the players cannot make a move, she loses. Otherwise, the winner depends on the infinite sequence $v_0, v_1, \ldots$ of vertices visited in the game. This infinite play is winning for player 0 if the sequence $\Omega(v_0), \Omega(v_1), \ldots$ satisfies the parity condition defined below, otherwise the play is winning for player 1.

Definition 3 (Parity condition). *An infinite sequence of natural numbers belonging to some finite set is said to satisfy the* parity condition *if the smallest number occurring infinitely often in the sequence is even.*

The case when the set of natural numbers in question is $\{0, 1\}$ is called the *Büchi condition.* The dual case of $\{1, 2\}$ is called the *co-Büchi condition.*

A strategy for the player $i \in \{0, 1\}$ is a mapping $s : V^* \times V_i \rightarrow V$ such that for each $v_0 v_1 \ldots v_j \in V^* V_i$, there is an edge in E from v_j to $s(v_0 v_1 \ldots v_j)$. We say a strategy is *memoryless* if $s(v_0 v_1 \ldots v_j)$ depends solely upon v_j. The concept of winning strategy is defined in the usual way. A very important theorem [3,11], which will enable us to consider only memoryless strategies, says:

Theorem 4 (Memoryless determinacy theorem). *Every game with the parity condition is* determined, *i. e. one of the players has a winning strategy. Moreover, the winner also has a memoryless winning strategy.*

2.2   Graphs and Trees

In this paper, when speaking of graphs, we mean *labeled graphs with a starting position.* Such a graph is a tuple $G = \langle V, E, \Sigma, e, v_0 \rangle$, where V is the set of *vertices*, $E \subseteq V \times V$ is the set of *edges*, the *labeling* is a function $e : V \rightarrow \Sigma$ and $v_0 \in V$ is the *starting position*. We assume that the set Σ of *labels* is finite. Given a graph G, its set of vertices V, or *domain*, is denoted as $\mathrm{dom}(G)$. For $W \subset V$, the *induced* subgraph of G is the graph of domain W obtained by restricting the edges and labeling in G to W.

Given $k \in \mathcal{N}$, a *k-ary tree* is a special kind of graph whose domain is the set $[k]^*$ of finite sequences over $[k]$ and the edge set is $\{(v, v \cdot i) : v \in [k]^*, i \in [k]\}$. The starting position is the empty sequence ε and there are no restrictions on the labeling.

2.3   Nondeterministic Automata

Definition 5 (Nondeterminitic parity automaton). *A* nondeterministic parity automaton on *k-ary trees is the tuple:*

$$\langle Q, q_0, \Sigma, \delta, \Omega \rangle$$

The finite set Σ *is called the* alphabet, Q *is a finite set of* states, $q_0 \in Q$ *is called the initial state and the function* $\Omega : Q \rightarrow \mathcal{N}$ *assigns to each state q its color* $\Omega(Q)$. *The transition function* δ *assigns to each pair* $(q, \sigma) \in Q \times \S$ *a set of transitions* $\delta(q, \sigma) \subseteq Q^k$.

Given a k-ary tree $\langle [k]^*, E, \Sigma, e, \epsilon \rangle$, a *run* of the automaton is a function $\rho : [k]^* \rightarrow Q$ such that for each vertex $v \in [k]^*$,

$$\langle \rho(v \cdot 1), \ldots, \rho(v \cdot k) \rangle \in \delta(\rho(v), e(v))$$

A run ρ is *accepting* if $\rho(\epsilon) = q_0$ and on each infinite path $v_0, v_1, \ldots$, the sequence $\Omega(\rho(v_0)), \Omega(\rho(v_1)) \ldots$ satisfies the parity condition. An automaton *accepts* a tree if there exists an accepting run.

Theorem 6 (Rabin). *Regular tree languages are closed under boolean operations and set quantification*

2.4 Alternating Automata

Two-way alternating automata on infinite trees were studied by Vardi in [16] as a tool for deciding the satisfiability problem of the modal μ-calculus with backward modalities. As opposed to "normal" alternating automata, two-way automata can travel backwards across edges. For reasons explained in the introduction, when dealing with the finite model problem, we consider a graph version of the automata.

Definition 7 (Two-way alternating automaton). *A* two-way alternating automaton *on Σ-labeled graphs is the tuple:*

$$\langle Q_\exists, Q_\forall, q_0, \Sigma, \delta, \Omega \rangle$$

$Q_\exists, Q_\forall$ are disjoint finite sets of states, *$q_0 \in Q = Q_\exists \cup Q_\forall$ is called the* starting state *and Ω is a function assigning to each state $q \in Q$ a natural number $\Omega(q)$ called the* color *of q. The* transition function *δ is of the form*

$$\delta : Q \times \Sigma \to P(\{0, +, -\} \times Q).$$

Intuitively, a run of $\mathcal{A}$ over some graph G is based on a game between two players: $\exists$ and $\forall$. The automaton starts in state q_0 in the starting position of G. Afterwards, $\mathcal{A}$ moves around the graph, the player deciding which move to make depending on whether the current state is in $Q_\exists$ or $Q_\forall$. The set of possible moves depends on the value assigned by δ to current state of $\mathcal{A}$ and the label of the current position in G. A choice of the form $(q, +)$ (respectively $(q, -)$) means the automaton changes the state to q and must move to some position along a forward (respectively backward) edge. Choosing $(q, 0)$ does not change the position, only the state of the automaton. The winner is determined as in the parity game.

A precise definition is as follows. Given a labeled graph $G = \langle V, E, \Sigma, e, v_0 \rangle$ and a two-way alternating automaton $\mathcal{A} = \langle Q_\exists, Q_\forall, q_0, \Sigma, \delta, \Omega \rangle$, we define the game $G(\mathcal{A}, G) = \langle V_0, V_1, E', v_0', \Omega' \rangle$:

- $V_0 = Q_\exists \times V$ and $V_1 = Q_\forall \times V$.
- $v_0' = (q_0, v_0)$.
- $((q, v)(q', v')) \in E'$ iff either:
  - $(0, q') \in \delta(q, e(v))$ and $v = v'$
  - $(-, q') \in \delta(q, e(v))$ and $(v', v) \in E$
  - $(+, q') \in \delta(q, e(v))$ and $(v, v') \in E$
- $\Omega'(q, v) = \Omega(q)$.

Definition 8 (Acceptance by the automaton). *We say that the automaton $\mathcal{A}$ accepts a graph G under the strategy s if s is a winning strategy for player $\exists$ in the game $G(\mathcal{A}, G)$. Such a strategy s is called* accepting. *We say $\mathcal{A}$ accepts graph G if there exists a strategy s such that $\mathcal{A}$ accepts G under s.*

Note that without loss of generality we assume that accepting strategies are memoryless. We will conclude this section with an example of an automaton that accepts a graph, yet no finite one.

Example 1. Consider the following two-way automaton

$$\mathcal{A} = \langle \{q_x\}, \{q_y, q_z\}, q_y, \{a, \delta, \Omega\rangle$$

Let Ω be $\{(q_x, 0), (q_y, 0), (q_z, 1)\}$ and the transition function δ be:

$$\delta(q_x, a) = \{(+, q_y)\} \qquad \delta(q_y, a) = \{(0, q_x), (-, q_z)\} \qquad \delta(q_z, a) = \{(-, q_z)\}$$

We will examine the game $G(\mathcal{A}, G)$, where $G = \langle \mathcal{N}, \{(n, n+1) : n \in \mathcal{N}\}, e, 0\}\rangle$, such that $e(n) = a$ for all $n \in \mathcal{N}$. Consider first the following example play. Since the starting state is q_y and the starting position is 0 , the play begins in $(q_y, 0)$. This is a game position for player $\forall$ – she has to choose a move from $\delta(q_y, a) = \{(0, q_x), (-, q_z)\}$, let's assume she picks $(0, q_x)$. This means we stay in the vertex 0 and now player $\exists$ has to choose from $\delta(q_x, a)$. There is only one possibility – namely $(+, q_y)$. This means $\exists$ has to choose a neighboring (in G) position along a forward edge. Again, $\exists$ has no choice, he has to choose 1; the game position is now $(q_y, 1)$. This goes on, through the game positions $(q_x, 1), (q_y, 2), (q_x, 2), \ldots$ until, say, we reach $(q_y, 10)$. Let's assume that this time player $\forall$ chooses $(-, q_z)$. Now it is her choice to choose a neighboring position in G, along a backward edge; she has to choose 9 – there is no other backward edge from 10. The play then goes on through game positions $(q_z, 9), (q_z, 8), \ldots, (q_z, 0)$ in which last position player $\forall$ loses for a lack of possible moves.

In the game $G(\mathcal{A}, G)$ there are essentially two kinds of play: a finite play like the one above, where player $\exists$ wins, or an infinite one where player $\forall$ always chooses $(+, q_x)$ from $\delta(q_x, a)$. The play goes through positions $(q_y, 0)$, $(q_x, 0)$, $(q_y, 1)$, $(q_x, 1)$, …. The only color appearing infinitely often in this play is 0, thus, again, player $\exists$ wins.

So we see that in the game $G(\mathcal{A}, G)$ player $\exists$ has a winning strategy, in other words, $\mathcal{A}$ accepts G. It can also be proven, that the automaton $\mathcal{A}$ accepts only graphs with an infinite forward path where no infinite backward path is ever reachable. In particular $\mathcal{A}$ accepts only infinite graphs.

3   The Finite Model Problem

The example above is a motivation for the following problem: "does a given alternating two-way automaton accept some finite graph?" This is the problem – called the *finite model problem* – tackled in this section.

Our plan is as follows. In the next subsection we prove an auxiliary decidability result about certain graphs with red and green edges. This is then used to prove the decidability of the finite model problem.

3.1 RG-Graphs

An RG-graph is a directed graph with two types of edges – red ones and green ones. More formally, it is a tuple $\langle V, E_r, E_g \rangle$ where $E_r, E_g \subseteq V \times V$. RG-graphs have no labels.

Definition 9. *Assuming that for $i \in \mathbb{Z}$, $(v_i, v_{i+1}) \in E_r \cup E_g$:*

- *A finite path is a sequence $v_0 v_1 \ldots v_n$.*
- *A backward infinite path is a sequence $\ldots v_{-1} v_0$.*
- *A forward infinite path is a sequence $v_0 v_1 \ldots$*
- *A path is any one of the above.*

Given a set X, an RG-graph over $\{\varepsilon, 0, \ldots, m\} \times X$ is (m, X)-*V-shaped* iff for all edges $((i, x), (j, y))$, either $i = \varepsilon$ or $j = \varepsilon$ or $i = j$. By VG_X^m we denote the set of (m, X)-V-shaped RG-graphs.

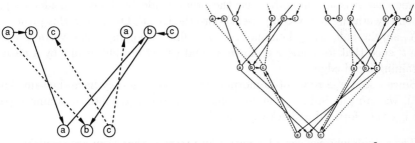

A 2-V-shaped graph over $\{a, b, c\}$ $\bar{t}$ for some tree over $\mathrm{VG}_{\{a,b,c\}}^2$

Given an m-ary tree t (which is not an RG-graph) whose vertices are labeled by elements of VG_X^m, the RG-graph $\bar{t}$ is defined as follows. The vertex set of $\bar{t}$ is $\mathrm{dom}(t) \times X$. If there is an edge from (i, x) to (j, y) in the label of v in t for $i, j \in \{\varepsilon, 0, \ldots, m\}$, then there is an edge from $(v \cdot i, x)$ to $(v \cdot j, y)$ of the same color in the graph $\bar{t}$. Note that $\bar{t}$ is not a tree.

The trees in the lemmas in this subsection are m-ary trees labeled by VG_X^m.

Definition 10. *A tree t is* path bounded *iff there is a finite bound on the number of red edges in paths in $\bar{t}$.*

Definition 11. *A tree t is* path finite *iff there is no (infinite) path with an infinite number of red edges in $\bar{t}$*

Given an $m \in \mathcal{N}$, a path is m-*acyclic* if it visits each vertex at most m times. Given $n \in \mathcal{N}$, we say a path π in a labeled tree is n-*upward* if there exist paths $\pi_1, \ldots, \pi_{n+1}$ and vertices $v_1 \leq \ldots \leq v_n$ such that:

- $\pi = \pi_1 v_1 \pi_2 v_2 \ldots \pi_n v_n \pi_{n+1}$
- All v_i have the same label.
- π contains a red edge for some $i \in \{2, \ldots, n\}$.

A path is n-vertical if it is n-upward or the reverse path is. The rather easy proof of the following lemma is omitted:

Lemma 1. *For every* $m, n \in \mathcal{N}$ *and every labeled tree* t, *there is a constant* $M_t^{m,n}$ *such that every* m-*acyclic path in* t *containing more than* $M_t^{m,n}$ *red edges is* n-*vertical.*

A labeled tree t is *regular* if it contains a finite number of nonisomorphic subtrees. The following observation is crucial in our proof:

Lemma 2. *A regular tree is path bounded iff it is path finite.*

Proof. The left to right part is obvious. For the other direction, assuming that a regular tree t is not path bounded we will prove it is not path finite.

Let $\tau_1, \ldots, \tau_n$ be the isomorphism types of t. Let t' be the $\{\tau_1, \ldots, \tau_n\} \times \mathrm{VG}_X^m$-labeled tree obtained from t by additionally labeling each vertex with its isomorphism type. Take some path π in $\bar{t}$ with more than $M_{t'}^{|X|, |X|}$ edges. We can assume that π is acyclic, otherwise the proof would be done. Consider its projection (by erasing the X component) onto the tree t'. Obviously, this projection is $|X|$-acyclic and, using Lemma 1, we can find two comparable vertices v and v' in t' such that for some $x \in X$, (v, x) and (v', x) are linked in $\bar{t}$ by a path π' containing a red edge.

Since v and v' a roots of isomorphic trees (because they have the same label in t'), we can pile infinitely many copies of π' on each other obtaining a path that is either forward or backward infinite.

The reason why we consider path finite trees is that they are regular:

Lemma 3. *There is a co-Büchi automaton recognizing the path finite trees.*

Proof. We will show a Büchi automaton recognizing the complement of this language. Our automaton needs to check if there is a path in $\bar{t}$ containing infinitely many red edges. The automaton accepts if either:

- There is a cycle $(v, x), \ldots, (v, x)$ containing a red edge.
- There exists an infinite acyclic path $v_0, v_1, \ldots$ in the tree t labeled by pairs $(x_0, x_0'), (x_1, x_1'), \ldots \in X \times X$ such that:
  - There is an edge in $\bar{t}$ between (v_i, x_i') and (v_{i+1}, x_{i+1}).
  - There is a path (perhaps of zero length) in $\bar{t}$ linking (v_i, x_i) with (v_i, x_i').
  - The edges and cycles mentioned above are directed accordingly to create either an infinite forward path or an infinite backward path.
  - This path has infinitely many red edges.

Both of these conditions can be checked by a nondeterministic Büchi automaton similar to the one constructed in [16].

The proof of the following can be found in [15]:

Theorem 12 (Rabin). *Every nonempty regular tree language contains a regular tree*

Theorem 13. *For a regular tree language L, the following conditions are equivalent:*

1. *L contains some path bounded tree.*
2. *L contains some regular path bounded tree.*
3. *L contains some regular path finite tree.*
4. *L contains some path finite tree.*

Proof. Since, by Lemma 3, the set of path finite trees in L is regular, 3 $\iff$ 4 follows from Theorem 12. On the other hand, the equivalence 2 $\iff$ 3 follows from Lemma 2.

The implication 2 $\Rightarrow$ 1 is obvious, for the other direction more care is needed because the set of path bounded trees is not regular. For a given M, however, it can be shown that the language L_M of trees where each path has at most M red vertices is regular. If L contains a path bounded tree, then L_M is nonempty for some M. Thus L_M contains a regular tree, which gives us 2.

Using Lemma 3 and Theorem 13 we obtain:

Lemma 4. *Given a regular language L over VG_X^m, it is decidable whether there is a path bounded tree in L.*

Assume the language L is recognizable by a Büchi nondeterministic automaton. Because emptiness for co-Büchi and Büchi automata is polynomial and the automaton in Lemma 3 is of exponential size with respect to m and $|X|$, the time taken by such a procedure is exponential in $|X| \cdot m$ and polynomial in the size of the automaton recognizing L.

3.2   Signatures and Induced RG-Graphs

We are now set to show the decidability of the finite model problem. We use Theorem 15, which represents finite graphs as trees satisfying a certain condition. This condition, in turn, can be expressed using bounded paths and, using the results from the previous section, the finite model problem is proved decidable.

Consider an alternating two-way automaton $\mathcal{A} = \langle Q_\exists, Q_\forall, q_0, \Sigma, \delta, \Omega \rangle$. Fix a graph G with vertices V and a strategy s for the player $\exists$ in the game $G(\mathcal{A}, G)$. For an odd color $i \in \Omega(Q)$ assumed by the acceptance condition we will define the relations $\rightarrow_i, \Rightarrow_i \subseteq (Q \times V)^2$ and the partial function $\text{Sig}_i : Q \times V \to \mathcal{N} \cup \{\infty\}$ as follows:

- $(q_1, v_1) \rightarrow_i (q_2, v_2)$ if $\Omega(q_1) \geq i$ and in the game $G(\mathcal{A}, G)$ there exists a play compliant with s with a move from (q_1, v_1) to (q_2, v_2). In other words, $\mathcal{A}$ can go from the position v_1 and state q_1 to the position v_2 and state q_2 in one move: either any move by $\forall$ or the single move by $\exists$ compliant with s. The transitive closure of this relation is denoted as $\rightarrow_i^*$. Note that $(q_1, v_1) \rightarrow_i (q_2, v_2)$ implies the position (q_1, v_1) is reachable in $G(\mathcal{A}, G)$ under s.
- $(q_1, v_1) \Rightarrow_i (q_2, v_2)$ if $(q_1, v_1) \rightarrow_i (q_2, v_2)$ and $\Omega(q_1) = i$.

- $\mathrm{Sig}_i(q, v)$ is defined only for states q such that $\Omega(q) \geq i$ and equals the upper bound on the $n \in \mathcal{N}$ such that for some $q_1, q_1', v_1, v_1' \ldots q_n, q_n', v_n, v_n'$:

$$(q, v) \rightarrow_i^* (q_1, v_1) \Rightarrow_i (q_1', v_1') \rightarrow_i^* (q_2, v_2) \Rightarrow_i \ldots \rightarrow_i^* (q_n, v_n) \Rightarrow_i (q_n', v_n')$$

Intuitively, given a strategy s for $\exists$, $\mathrm{Sig}_i(q, v)$ tells us how many "bad" states of odd color i can be visited by the automaton starting from (q, v) before they are either annulled by a state of color less than i or no more states of color i can be visited. This can be encoded by the very local relations $\rightarrow_i, \Rightarrow_i$ which say what are the possible moves between neighboring vertices of the graph.

We will illustrate the concept of signature using Example 1. Recall that we were dealing with the graph $G = \langle \mathcal{N}, \{(n, n+1) : n \in \mathcal{N}\}, e, 0 \rangle$. Consider the winning strategy for the player $\exists$ described in the example, call it s. We will define the signature for this strategy. There is only one odd color in the acceptance condition – so only the relations $\rightarrow_1, \Rightarrow_1$ and the function Sig_1 need be defined.

Consider first the state q_z. Since $\delta(q_z, a) = \{(-1, q_z)\}$, only backward moves to q_z are possible and we obtain that $(q_z, v) \rightarrow_1 (q, v')$ iff $q' = q_z$ and $v' = v - 1$. Since q_x, q_y are of color 0, $(q, v) \rightarrow_1 (q', v')$ does not hold for $q \in \{q_x, q_y\}$. Moreover, in this particular example, $\rightarrow_1$ is the same relation as $\Rightarrow_1$, since there are no states of color greater than 1.

Now we turn to the signature $\mathrm{Sig}_1(q, v)$, which is only defined if $q = q_z$. Given a vertex v, it is easy to see that at most v steps can be made using $\Rightarrow_1$:

$$(q_z, v) \Rightarrow_1 (q_z, v - 1) \Rightarrow_1, \ldots, \Rightarrow_1 (q_z, 0)$$

This shows that $\mathrm{Sig}_1(q_z, v)$ is v for all vertices $v \in \mathcal{N}$ of the example graph.

Definition 14. *A two-way k-ary tree is a graph whose domain is k^* and where all edges are exclusively either (v, vi) or (vi, v) for $v \in k^*$ and $i \leq k$.*

In particular, every k-ary tree is a two-way k-ary tree.

We say an automaton accepts a graph G with a *bounded signature* if there is some strategy s and bound $M \in \mathcal{N}$ such that $\mathrm{Sig}_i(q, v) \leq M$ for all q, v and odd i. The following was proved in [1] for automata with the Büchi acceptance condition, but the proof for the general parity condition is the same:

Theorem 15 (Bounded signature theorem). *An alternating two-way automaton accepts some finite graph iff it accepts some two-way $|Q|$-ary tree with a bounded signature.*

Given an automaton $\mathcal{A}$ with states Q and m colors in the acceptance condition, let $\mathrm{VG}(\mathcal{A})$ stand for $\mathrm{VG}_{Q \times [m]}^{|Q|}$. Note that $|\mathrm{VG}(\mathcal{A})|$ is exponential in $|Q|$. This alphabet will be used to encode the relations $\rightarrow_i$ and $\Rightarrow_i$ in the tree itself.

Definition 16. *Given a Σ-labeled two-way $|Q|$-ary tree t and a memoryless strategy s for the player $\exists$, the induced tree $\mathrm{T}_{t,s}$ is the unique $\mathrm{VG}(\mathcal{A})$-labeled tree such that $\overline{\mathrm{T}_{t,s}}$ has:*

- A *red edge* from (v, q, i) to (v', q', i) iff $(q, v) \Rightarrow_i (q', v')$.
- A *green edge* from (v, q, i) to (v', q', i) iff $(q, v) \to_i (q', v')$.

The construction of $\mathrm{T}_{t,s}$ yields:

Lemma 5. *The signature of t, s is bounded iff $\mathrm{T}_{t,s}$ is path bounded.*

To prove the next lemma, we need to encode strategies for $\exists$ in trees. Note that in a graph G, a memoryless strategy in the game $\mathrm{G}(\mathcal{A}, G)$ is a function $s : Q \times \mathrm{dom}(G) \to Q \times \mathrm{dom}(G)$ such that for $s(q, v) = (q', v')$, either $v = v'$ or $(v, v') \in E \cup E^{-1}$. If the graph in question is a two-way tree, we can use a different format:

$$\overline{s} : \mathrm{dom}(t) \to Q \to Q \times \{-1, \epsilon, 1, \ldots, k\}$$

Let us denote the finite set $Q \to Q \times \{-1, \epsilon, 1, \ldots, k\}$ by $\mathrm{S}_{\mathcal{A}}$. Given a strategy s over the tree t, the *encoding* of s is the relevant mapping $\overline{s} : \mathrm{dom}(t) \to \mathrm{S}_{\mathcal{A}}$. This allows us to encode strategies within a tree using an alphabet whose size is exponential in $|Q|$.

Lemma 6. *The set of all induced trees is a regular tree language*

Proof. It is easy to find a nondeterministic Büchi automaton that, given a tree t labeled by $\mathrm{VG}(\mathcal{A})$, guesses a labeling $\sigma : \mathrm{dom}(t) \to \Sigma$ and an encoding $\overline{s} : \mathrm{dom}(t) \to \mathrm{S}_{\mathcal{A}}$ and verifies that t is indeed the induced tree for σ and the strategy s. The state space of this automaton is exponential on $|Q|$.

Corollary 17. *The finite model problem is decidable in time exponential in $|Q|$.*

Proof. By Theorem 15, $\mathcal{A}$ accepts some finite graph iff $\mathcal{A}$ accepts some tree t with a bounded signature. This, by Lemma 5 is equivalent to the existence of an induced path bounded tree. Since the language of induced trees is regular by Lemma 6, we can use Lemma 4 to show the decidability.

We prove the lower bound by a reduction from the emptiness problem for *one-way automata*. Such an automaton is a two-way automaton where δ does not use the transitions $\{-\} \times Q$. This is the original alternating automaton [12,3]. In the one-way case, one can prove using Rabin's Theorem 12 that every nonempty one-way alternating automaton accepts some finite graph. This means that for one-way automata, the emptiness and finite model problems are equivalent. Since the emptiness for PDL, which is EXPTIME hard [10], can be reduced to the emptiness problem for alternating one-way automata, the finite model problem for one-way automata is also EXPTIME hard. The one-way automata being a special case of two-way automata, we obtain:

Theorem 18. *The finite model problem for alternating two-way automata is EXPTIME complete*

Without going into any details, we only state the application of our result to the μ-calculus. Readers interested in the translation should refer to [16,1].

Theorem 19. *The finite model problem for the propositional modal μ-calculus with backward modalities is decidable in EXPTIME.*

Before we conclude, we state one possible extension of this work. Two-way alternating automata are also used in the paper [5] to decide the satisfiability of formulas of the so-called Guarded Fragment with fixed points. It is possible that our result can be applied to solving the open problem of whether the finite model property for formulas of the Guarded Fragment with fixed points is decidable.

References

1. Mikołaj Bojańczyk. The finite graph problem for two-way alternating automata. In *FOSSACS 2001*, volume 2030 of *LNCS*, pages 88–103, 2001.
2. A. Saoudi D. E. Muller and P. E. Shupp. Weak alternating automata give a simple explanation why most temporal and dynamic logics are decidable in exponential time. In *Proceedings 3rd IEEE Symposium on Logic in Computer Science*, pages 422–427, 1988.
3. E. A. Emerson and C. Jutla. Tree automata, mu-calculus and determinacy. In *Proc. 32th IEEE Symposium on Foundations of Computer Science*, pages 368–377, 1991.
4. E. Grädel. On the restraining power of guards. *Journal of Symbolic Logic*, 1999.
5. E. Grädel and I. Walukiewicz. Guarded fixed point logic. In *Proceedings 14th IEEE Symp. on Logic in Computer Science*, pages 45–54, 1999.
6. Y. Gurevich and L. Harrington. Automata, trees and games. In *Proc. 14th. Ann. ACM Symp. on the Theory of Computing*, pages 60–65, 1982.
7. J. van Benthem H. Andreka and I. Nemeti. Modal logics and bounded fragments of predicate logic. *Journal of Philosophical Logic*, pages 217–274, 1998.
8. I. Hodkinson. Loosely guarded fragment has finite model property. *J. Symbolic Logic*, to appear.
9. D. Kozen. Results on the propositional μ-calculus. *Theoretical Computer Science*, 27:333–354, 1983.
10. R. Ladner M. Fischer. Propositional dynamic logic of regular programs. *Journal of Computer and System Sciences*, 18:194–211, 1979.
11. A. Mostowski. Games with forbidden positions. Technical report, University of Gdańsk, 1991.
12. D.E. Muller and P.E. Schupp. Alternating automata on infinite trees. *Theoretical Computer Science*, 54:267–276, 1987.
13. M. O. Rabin. Decidability of second-order theories and automata on infinite trees. *Trans. Amer. Math. Soc.*, 141, 1969.
14. G. Slutzki. Alternating tree automata. *Theoretical Computer Science*, 41:305–318, 1985.
15. Wolfgang Thomas. Languages, automata, and logic. In *Handbook of Formal Language Theory, III*, pages 389–455. Springer, 1997.
16. M. Vardi. Reasoning about the past with two-way automata. In *vol. 1443 LNCS*, pages 628–641, 1998.

Approximating Huffman Codes in Parallel

Piotr Berman[1,2*], Marek Karpinski[2**], and Yakov Nekrich[2***]

[1] Dept.of Computer Science and Engineering, The Pennsylvania State University
berman@cse.psu.edu
[2] Dept. of Computer Science, University of Bonn. {marek,yasha}@cs.uni-bonn.de.

Abstract. In this paper we present new results on the approximate parallel construction of Huffman codes. Our algorithm achieves linear work and logarithmic time, provided that the initial set of elements is sorted. This is the first parallel algorithm for that problem with the optimal time and work.

Combining our approach with the best known parallel sorting algorithms we can construct an almost optimal Huffman tree with optimal time and work. This also leads to the first parallel algorithm that constructs exact Huffman codes with maximum codeword length H in time $O(H)$ and with n processors. This represents a useful improvement since most practical situations satisfy $H = O(\log n)$.

Keywords: Parallel Algorithms, Approximation Algorithms, Huffman Codes

1 Introduction

A Huffman code for an alphabet $a_1, a_2, \ldots, a_n$ with weights $w_1, w_2, \ldots, w_n$ is a prefix code that minimizes the average codeword length, defined as $\sum_{i=1}^{n} w_i l_i$ so that $\sum 2^{-l_i} \leq 1$. The problem of construction of Huffman codes is closely related to the construction of Huffman trees (cf., e.g., [H51], [vL76]).

A problem of constructing a binary Huffman tree for a sequence $\bar{w} = w_1, w_2, \ldots, w_n$ consists in constructing a binary tree T with leaves, corresponding to the elements of the sequence, so that the *weighted path length* of T is *minimal*. The weighted path length of T, $wpl(T)$ is defined as follows:

$$wpl(T, \bar{w}) = \sum_{i=1}^{n} w_i l_i$$

where l_i is the depth of the leaf corresponding to the element w_i. A problem related to the problem discussed in this paper is the construction of optimal

* Research done in part while visiting Dept. of Computer Science, University of Bonn. Work partially supported by NSF grant CCR-9700053 and DFG grant Bo 56/157-1.
** Work partially supported by DFG grants, DIMACS, and IST grant 14036 (RAND-APX).
*** Work partially supported by IST grant 14036 (RAND-APX).

P. Widmayer et al. (Eds.): ICALP 2002, LNCS 2380, pp. 845–855, 2002.
© Springer-Verlag Berlin Heidelberg 2002

alphabetic codes. In case of the alpahbetic codes we have an additional limitation that $l_i \leq l_j$ for all $i < j$.

The classical sequential algorithm, described by Huffman ([H51]) can be implemented in $O(n \log n)$ time. Van Leeuwen has shown that if elements are sorted according to their weight, a Huffman code can be constructed in $O(n)$ time (see [vL76]). However, no optimal parallel algorithm is known. Teng [T87] has shown that construction of a Huffman code is in a class NC. His algorithm, uses the parallel dynamic programming method of Miller et al. [MR85] and works in $O(\log^2 n)$ time on n^6 processors. Attalah et al. have proposed an n^2 processor algorithm, working in $O(\log^2 n)$ time. This algorithm is based on the multiplication of concave matrices. The fastest n-processor algorithm is due to Larmore and Przytycka [LP95]. Their algorithm, based on reduction of Huffman tree construction problem to the *concave least weight subsequence* problem runs in $O(\sqrt{n} \log n)$ time.

Kirkpatrick and Przytycka [KP96] introduce an approximate problem of constructing, so called, almost optimal codes, i.e. the problem of finding a tree T' that is related to the Huffman tree T according to the formula $wpl(T') \leq wpl(T) + n^{-k}$ for a fixed error parameter k (assuming $\sum w_i = 1$). We call n^{-k} an error factor. In practical situations the nearly optimal codes, corresponding to nearly optimal trees, are as useful as the Huffman codes, because compressing a file of polynomial size with an approximate Huffman code leads to the compression losses limited only by a constant. Kirkpatrick and Przytycka [KP96] propose several algorithms for that problem. In particular, they present an algorithm that works in $O(k \log n \log^* n)$ time and with n processors on a CREW PRAM and an $O(k^2 \log n)$ time algorithm that works with n^2 processors on a CREW PRAM. In case of the optimal alphbetic codes the best known NC algorithm solves this problem in time $O(log^3 n)$ with $n^2 \log n$ processors (see [LPW93].

The problems considered in this paper were also partially motivated by a work of one of the authors on decoding the Huffman codes [N00b], [N00a].

In this paper we improve the before mentioned results by presenting an algorithm that works in $O(k \log n)$ time and with n processors. As we will see in the next section the crucial step in computing a nearly optimal tree is merging two sorted arrays and this operation is repeated $O(\log n^k)$ times. We have developed a method for performing such a merging in a constant time.

We also further improve this result and design an algorithm that constructs almost-optimal codes in time $O(\log n)$ and with $n/\log n$ processors, provided that elements are sorted. This results in an optimal speed-up of the algorithm of van Leeuwen [vL76]. Our algorithm works deterministically on a CREW PRAM and is the first parallel algorithm for that problem with the optimal time and work. Combining that algorithm with parallel radix sort algorithms we construct an optimal-work probabilistic algorithm that works in expected logarithmic time. We construct also a deterministic algorithm that works on a CRCW PRAM in $O(k \log n)$ time and with $n \log \log n / \log n$ processors.

The above described approach also leads to an algorithm for constructing exact Huffman trees that works in $O(H)$ time and with n processors, for H the

```
                    Algorithm Oblivious-Huffman
     1:   for i := m downto 1 do
     2:       if l(W_i) = 1)
     3:           W_{i-1} := merge(W_i, W_{i-1})
     4:       else
     5:           t := meld(W_i[1], W_i[2])
     6:           W_i := merge(t, W_i(3, l(W_i)))
     7:           a := l(W_i)
     8:           b := ⌊(a − 1)/2⌋
     9:           for i := 1 to b pardo
    10:               W_i[i] := meld(W_i[2i − 1], W_i[2i])
    11:           W_i := merge(W_i(1, b), W_i(2b + 1, a))
    12:           W_{i-1} := merge(W_{i-1}, W_i)
```

Fig. 1. Huffman tree construction scheme

height of Huffman tree. This is also an improvement of the algorithm of Larmore and Przytycka for the case when $H = o(\sqrt{n} \log n)$. We observe that in the most practical applications height of the Huffman tree is $O(\log n)$.

2   A Basic Construction Scheme

Our algorithm uses the following *tree* data structure. A single element is a tree, and if t_1 and t_2 are two trees, then $t = meld(t_1, t_2)$ is also a tree, so that $weight(t) = weight(t_1) + weight(t_2)$. Initial elements will be called leaves.

In a classical Huffman algorithm the set of trees is initialized with the set of weights. Then one melds consecutively two smallest elements in the set of trees until only one tree is left. This tree can be proven to be optimal.

Kirkpatrick and Przytycka [KP96] presented a scheme for parallelization of a Huffman algorithm. The set of element weights $p_1, p_2, \ldots, p_n$ is partitioned into sorted arrays $W_1, \ldots, W_m$, such that elements of array W_i satisfy the condition $1/2^i \leq p < 1/2^{i-1}$. In this paper we view (sorted) arrays as an abstract data type with the following operations: extracting of subarray $A[a, b]$, measuring the array length, $l(A)$, and merging two sorted arrays, $merge(A, B)$. The result of operation $merge(A, B)$ is a sorted array C which consists of elements of A and B. If we use n processors, then each entry of our sorted array has an associated processor.

Obviously elements of a class W_i can be melded in parallel before the elements of classes W_j, $j < i$ are processed. The scheme for the parallelization is shown on Figure 1, $W(a, b)$ denotes an array $W[a], W[a + 1], \ldots, W[b]$. We refer the reader to [KP96] for a more detailed description of this algorithm.

Because the total number of iterations of algorithm **Oblivious-Huffman** equals to the number of classes W_i and the number of classes is linear in the worst case, this approach does not lead to any improvements, if we want to construct an exact Huffman tree.

Kirkpatrick and Przytycka [KP96] also describe an approximation algorithm, based on **Oblivious-Huffman**. In this paper we convert **Oblivious-Huffman** into an approximation algorithm in a different way. We replace each weight w_i with $w_i^{new} = \lceil w_i n^k \rceil n^{-k}$. Let T^* denote an optimal tree for weights $w_1, \ldots, w_i$. Since $w_i^{new} < w_i + n^{-k}$,

$$\sum w_i^{new} l_i < \sum w_i l_i + \sum n^{-k} l_i < \sum w_i l_i + n^2 n^{-k}$$

because all l_i are smaller than n. Hence $wpl(T^*, \bar{p}_{new}) < wpl(T, \bar{w}) + n^{-k+2}$. Let T_A denote the (optimal) Huffman tree for weights w_i^{new}. Then

$$wpl(T_A, \bar{w}) \leq wpl(T_A, \bar{w}^{new}) \leq wpl(T^*, \bar{w}^{new}) < wpl(T^*, \bar{w}) + n^{-k+2}$$

Therefore we can construct an optimal tree for weights w^{new}, then replace w_i^{new} with w_i and the resulting tree will have an error of at most n^{-k+2}.

If we apply algorithm **Oblivious-Huffman** to the new set of weights, then the number of iterations of this algorithm will be $\lceil k \log_2 n \rceil$, since new elements will be divided into at most $\lceil k \log_2 n \rceil$ arrays. An additional benefit is that we will use registers with polynomially bounded values. Note that in [KP96] PRAM with an unbounded register capacity was used. That advantage of our algorithm will be further exploited in section 4.

3   An $O(k \log n)$ Time Algorithm

In this section we describe an $O(k \log n)$ time n-processor algorithm that works on CREW PRAM.

Algorithm **Oblivious-Huffman** performs $k \log n$ iterations and in each iteration only the merge operations are difficult to implement in a constant time. All other operations can be performed in a constant time. We will use the following simple fact, described in [V75]:

Proposition 1 *If array A has a constant number of elements and array B has at most n elements, then arrays A and B can be merged in a constant time and with n processors.*

Proof: Let $C = merge(A, B)$. We assign a processor to every possible pair $A[i], B[j]$, $i = 1, \ldots, c$ and $B = 1, \ldots, n$. If $A[i] < B[j] < A[i+1]$, then $B[j]$ will be the $i + j$-th element in array C. Also if $B[j] < A[i] < B[j+1]$, then $A[i]$ will be the $i + j$-th element in array C. $\square$

Proposition 1 allows to implement operation $merge(W_i(1, b), W_i[2b+1, a])$ (line 11 of Figure 1) in a constant time.

Operation $merge(W_{i-1}, W_i)$ is the slowest one, because array W_i can have linear size and merging two arrays of size n requires $\log \log n$ operations in general case (see [V75]). In this paper we propose a method, that allows us to perform every merge of **Oblivious-Huffman** in a constant time. The key to our method is that at the time of merging, all elements in both arrays know their predecessors in other array, and can thus compute their positions in a resulting array in a constant time. A merging operation itself is performed without comparisons. Comparisons will be used for the initial computation of predecessors and to update predecessors after each merge and meld operation.

We say that element e is of rank k, if $e \in W_k$. A relative weight $r(p)$ of an element p of rank k is $r(p) = p \cdot 2^k$. We will denote by $r(i, c)$ a relative weight of the c-th element in array W_i, $w[e]$ will denote the weight of element e, and $pos[e]$ will denote the position of an element e in its array W_i, so that $W_i[pos[e]] = e$. To make description more convenient we say that in every array W_k $W_k[0] = 0$ and $W_k[l(W_k) + 1] = +\infty$ At the beginning we construct a list R of all elements, sorted according to their relative weight. We observe that elements of the same class W_k will appear in R in a non-decreasing order of their weight. We assume that whenever $e \neq e'$, $r(e) \neq r(e')$.Besides that, if leaf e and tree t are of a rank k and t is the result of melding two elements t_1 and t_2 of rank $k + 1$, such that $r(t_1) > r(e)$ and $r(t_2) > r(e)$ ($r(t_1) < r(e)$ and $r(t_2) < r(e)$) then a weight of t is bigger (smaller) than a weight of e.

We also compute for every leaf e and every class i the value of $pred(e, i) = W_i[j]$, s.t. $r(i, j) < r(e) < r(i, j + 1)$. In other words, $pred(e, i)$ is the biggest element in class i, whose relative weight is smaller than or equal to $r(e)$. To find values of $pred(e, j)$ for some j we compute an array C^j with elements corresponding to all leaves, such that $C^j[i] = 1$ if $R[i] \in W_j$ and $C^j[i] = 0$ otherwise and compute prefix sums for elements of C^j. Since a prefix sum can be computed in a logarithmic time with $n/\log n$ processors we can compute $O(\log n)$ necessary prefix sums in a logarithmic time with n processors (see [B97]).

We use an algorithm from Figure 2 to update values of $pred(e, i)$ for all $e \in W_{i-1}, \ldots, W_1$ and values of $pred(e, t)$ for all $e \in W_i$ and $t = i - 1, \ldots, 1$ after melding of elements from W_i .

First we store the tentative new value of $pred(e, i)$ for all $e \in W_{i-1}, \ldots, W_1$ in array $temp$ (lines 1-3 of Figure 2). The values stored in $temp$ differ from the correct values by at most 1.

Next we meld the elements and change the values of $w[s]$ and $pos[s]$ for all $s \in W_i$ (lines 4-8 of Figure 2).

Finally we check whether the values of $pred(s, i)$ for $s \in W_1 \cup W_2 \cup \ldots \cup W_{i-1}$ are the correct ones. In order to achieve this we compare the relative weight of the tentative predecessor with the relative weight of s. If the relative weight of s is smaller, $pred(s, i)$ is assigned to the previous element of W_i. (lines 10-14 of Figure 2). In lines 15 and 16 we check whether the predecessors of elements in W_i have changed.

When the elements of W_i are melded and predecessor values $pred(e, i)$ are recomputed $pos[pred(W_i[j], i - 1)]$ equals to the number of elements in W_{i-1} that are smaller than or equal to $W_i[j]$. Analogically $pos[pred(W_{i-1}[j], i)]$ equals to the number of elements in W_i that are smaller than or equal to $W_{i-1}[j]$.

```
1:      for a < i, b ≤ l(W_a) pardo
2:          s := W_a[b]
3:          temp[s] := ⌈pos[pred(s,i)]/2⌉

4:      for c ≤ l(W_i)/2 pardo
5:          s := meld(W_i[2c − 1], W_i[2c])
6:          w[s] := w[W_i[2c − 1]] + w[W_i[2c]]
7:          pos[s] := c
8:          W_i[c] := s

9:      for a < i, b ≤ l(W_a) pardo
10:         s := W_a[b]
11:         c := temp[s]
12:         if r(i,c) > r(a,b)
13:             c := c − 1
14:             if r(a, b + 1) > r(i, c + 1)
15:                 pred(W_i[c + 1], a) := s
16:         pred(s, i) := W_i[c]
```

Fig. 2. Operation meld

Therefore indices of all elements in the merged array can be computed in a constant time.

After melding of elements from W_i every element of $W_{i-1} \cup W_{i-2} \cup \ldots \cup W_1$ has two candidates for a predecessor of rank $i-1$. We can find the new predecessor of element e by comparing $pred(e,i)$ and $pred(e, i-1)$. The pseudocode description of an operation $merge(W_{i-1}, W_i)$ (line 12 of Figure 1) is shown on Figure 3.

Since all operations of the algorithm **Oblivious-Huffman** can be implemented to work in a constant time, each iteration takes only a constant time. Therefore we have

Theorem 1. *An almost optimal tree with error factor $1/n^k$ can be constructed in $O(k \log n)$ time and with n processors on a CREW PRAM.*

The algorithm described in the previous section can also be applied to the case of exact Huffman trees. The difference is that in case of exact Huffman trees weights of elements are unbounded and number of classes W_i is $O(n)$ in the worst case. However, it is easy to see that number of classes W_i does not exceed $H + 2$ where H is the height of the resulting Huffman tree. We can sort elements and distribute them into classes in time $O(\log n)$ with n processors. We can then compute values of $pred$ for classes $H, H − 1, \ldots, H − \log n$ and perform first $\log n$ iterations of **Oblivious-Huffman** in time $O(\log n)$. Then, we compute values of $pred$ for the classes $H − \log n, H − \log n − 1, \ldots, H − 2 \log n$

> do simultaneously:
>
> 1: for $j \leq l(W_{i-1})$ pardo   for $j \leq l(W_i)$ pardo
> 2: $t := W_{i-1}[j]$ $t := W_i[j]$
> 3: $k := pos[pred(t, i)]$ $k := pos[pred(t, i-1)]$
> 4: $pos[t] := j + k$ $pos[t] := j + k$
> 5: $W_i[j + k] := t$ $W_i[j + k] := t$
>
> 6: for $a < i$, $b \leq l(W_a)$ pardo
> 7: $s := W_a[b]$
> 8:   if $(w[pred(s, i-1)] > w[pred(s, i)])$
> 9: $pred(s, i) := pred(s, i-1)$

Fig. 3. Operation $merge(W_i, W_{i-1})$

and perform the next $\log n$ iterations of the basic algorithm. Proceeding in the same manner we can perform H iterations in $O(H)$ time. Therefore we get

Theorem 2. *A Huffman tree can be constructed in time $O(H)$ with n processors, where H is the height of the Huffman tree.*

4   An $O(kn)$ Work Algorithm

In this section we describe a modification of the merging scheme, presented in the previous section. The modified algorithm works on a CREW PRAM in $O(\log n)$ time and with $n/\log n$ processors, provided that initial elements are sorted.

The main idea of our modified algorithm is that we do not use all values of $pred(e, i)$ at each iteration. In fact, if we know values of $pred(e, i - 1)$ for all $e \in W_i$ and values of $pred(e, i)$ for all $e \in W_{i-1}$ then merging can be performed in a constant time. Therefore, we will use function $\overline{pred}$ instead of $pred$ such that the necessary information is available at each iteration, but the total number of values in $\overline{pred}$ is limited by $O(n)$. We are also able to recompute values of $\overline{pred}$ in a constant time after each iteration. The technique that we use is reminiscent of the parallel merge sort algorithm of Cole ([C88]).

For an array R we denote by $sample_k(R)$ a subarray of R that consists of every 2^k-th element of R. We define $\overline{pred}(e, i)$ for $e \in W_l$, $l > i$ ($l < i$) as the biggest element $\tilde{e}$ in $sample_{l-i-1}(W_i)$ ($sample_{i-l-1}(W_i)$), such that $r(\tilde{e}) \leq r(e)$. Besides that we maintain the values of $\overline{pred}(e, i)$ only for $e \in sample_{l-i-1}(W_l)$. In other words for every 2^{l-i-1}-th element of W_l we know its predecessor in W_i with precision of up to 2^{l-i} elements. Obviously total number of values in $\overline{pred}$ is $O(n)$.

Now we will show how $\overline{pred}$ can be recomputed after elements in a class W_i are melded. Number of pairs (e, i) for which values $\overline{pred}(e, i)$ must be computed is $O(n)$, and we can assign one processor to every pair.

We denote by $sibling(e)$ an element with which e will be melded in **Oblivious-Huffman**. Consider an arbitrary pair (e, a), $e \in W_i$. First the value $\overline{pred}(e, a)$ is known, but the value of $\overline{pred}(s, i)$, where $s = sibling(e)$ may be unknown. We can set a tentative new value of $pred(e_m, a)$ where $e_m = meld(e, s)$ to $\overline{pred}(e, a)$.

Next we recompute the values of $\overline{pred}(s, i)$ for $s \in sample_{i-1}W_1 \cup sample_{i-2}W_2 \cup \ldots \cup sample_1 W_{i-1}$. Let $e_1 = pred(s, i)$, $e_2 = sibling(e_1)$ and $e = meld(e_1, e_2)$. The correct new values of $\overline{pred}(s, i)$ can be computed in a similar way as in section 3. If the relative weight of s is smaller than that of e, $\overline{pred}(s, i)$ is assigned to the element preceding e. Otherwise, we also compare the relative weight of s with the relative weight of the element following e. If the first one is bigger we set $\overline{pred}(s, i)$ to the element following e. We also can check whether the predecessors of elements in W_i are the correct ones at the same time. A pseudocode description of the parallel meld operation is shown on Figure 4.

When elements from W_i are melded the new elements will belong to W_{i-1}. Now we have to compute $\overline{pred}(e, a)$ in $sample_{i-a-2}(W_a)$ for every 2^{i-a-2}-th element of W_i. Suppose $\overline{pred}(e, a) = W_a[p \cdot 2^{i-a-1}]$. We can find the new "refined" value of $\overline{pred}(e, a)$ by comparing $r(e)$ with $r(W_l[p \cdot 2^{i-l-1} + 2^{i-l-2}])$. When the correct values of $\overline{pred}(e, i)$ $e \in sample_{l-i-1}(W_l)$ are known we can compute $\overline{pred}(e, i)$ for all e from $sample_{i-a-2}(W_a)$. Let e be a new element in $sample_{i-a-2}(W_a)$ and let e_p and e_n be the next and previous elements in $sample_{i-a-2}(W_a)$. Obviously e_n and e_p are in $sample_{i-a-1}(W_a)$ and $\overline{pred}(e, i)$ is between $pred(e_p, i)$ and $\overline{pred}(e_n, i)$. New correct values of $\overline{pred}(e, i)$ can be found in a constant time.

Using the values of $\overline{pred}$ we can merge W_{i-1} and the melded elements from W_i in a constant time in the same way as described in section 3. A detailed description of the meld and merge operations for the modified algorithm will be given in the full version of the paper.

Since all other operations can also be done in a constant time we can perform $\log n$ iterations of **Oblivious-Huffman** in a logarithmic time. Therefore we get

Theorem 3. *An almost optimal tree with error factor $1/n^k$ can be constructed in time $O(k \log n)$ and with $n/\log n$ processors, if elements are sorted according to their weight.*

We can combine the algorithm described above with different algorithms for the parallel bucket sort. Optimal time-processor product can be achieved under reasonable conditions.

Using a parallel bucket sort algorithm described in [H87] we can sort polynomially bounded integers in $O(\log n \log \log n)$ time and with $n/\log n$ processors on a priority CRCW PRAM. Using the algorithm described by Bhatt et al. [BDH$^+$91] we can also sort polynomially bounded integers in the same time and

```
1:      for a < i, b ≤ l(sample_{i-a-1}W_a) pardo
2:          s := W_a[b · 2^{i-a-1}]
3:          temp[s] := ⌈pos[pred(s,i)]/2⌉

4:      for c ≤ l(W_i)/2 pardo
5:          s := meld(W_i[2c − 1], W_i[2c])
6:          w[s] := w[W_i[2c − 1]] + w[W_i[2c]]
7:          pos[s] := c
8:          W_i[c] := s

9:      for a < i, b ≤ l(sample_{i-a-1}W_a) pardo
10:         d1 := 2^{i-a-1}
11:         d2 := 2^{i-a-2}
12:         s := W_a[b · d1]
13:         c := temp[s]
14:         if r(i, c · d2) > r(a, b · d1)
15:             c := c − 1
16:             if r(a, (b + 1) · d1) > r(i, (c + 1) · d2)
17:                 pred(W_i[(c + 1) · d2], a) := s
18:         else
19:             if r(i, (c + 1) · d2) < r(a, b · d1)
20:                 c := c + 1
21:             if r(a, (b − 1) · d1) < r(i, (c − 1) · d2)
22:                 pred(W_i[(c − 1) · d1], a) := W_a[(b − 1) · d2]
23:         pred(s, i) := W_i[c · d2]
```

Fig. 4. A melding operation for the improved algorithm

the processor bounds on arbitrary CRCW PRAM. Combining these results with our modified algorithm we get

Proposition 2 *An almost optimal tree with error $1/n^k$ can be constructed in $O(k \log n \log \log n)$ time and with $n/\log n$ processors on a priority CRCW PRAM or on an arbitrary CRCW PRAM.*

Applying an algorithm of Hagerup [H87] we get the following result

Proposition 3 *An almost optimal tree with error $1/n^k$ can be constructed for the set of n uniformly distributed random numbers with $n/\log n$ processors in time $O(k \log n)$ and with probability $1/C^{-\sqrt{n}}$ for any constant C .*

By using the results of Andersson, Hagerup, Nilsson and Raman [AHNR98], n integers in the range $0..n^k$ can be sorted in $O(\log n)$ time and with $n \log \log n/\log n$ processors on a unit-cost CRCW PRAM with machine word length $k \log n$. Finally [AHNR98] shows that n integers can be probabilistically sorted in an expected time $O(\log n)$ and expected work $O(n)$ on a unit-cost EREW PRAM with word length $O(\log^{2+\varepsilon} n)$.

Proposition 4 *An almost optimal tree with error* $1/n^k$ *can be constructed with expected time* $O(\log n)$ *and expected work* $O(n)$ *on a CREW PRAM with word size* $\log^{2+\varepsilon} n$.

The last statement shows that a Huffman tree can be probabilistically constructed on a CREW PRAM with polylogarithmic word length.

5   Conclusion

This paper describes the first optimal work approximate algorithms for constructing Huffman codes. The algorithms have polynomially bounded errors. We also show that a parallel construction of an almost optimal code for n elements is as fast as the best known deterministic and probabilistic methods for sorting n elements. In particular, we can deterministically construct an almost optimal code in logarithmic time and with linear number of processors on CREW PRAM or in $O(\log n)$ time and with $n \log \log n / \log n$ processors on CRCW PRAM. We can also probabilistically construct an almost-optimal tree with linear expected work in logarithmic expected time provided that the machine word size is $\log^{2+\varepsilon} n$. This is the first optimal work and the logarithmic time algorithm for that problem.

Our approach also leads to the improvement of the construction of Huffman trees for the case when $H = o(\sqrt{n} \log n)$, where H is the maximum codeword length. This gives the first parallel algorithm that works in $O(H)$ time and with n processors. In practical applications H is usually of order $O(\log n)$. The question of the existence of algorithms that deterministically sort polynomially bounded integers with linear time-processor product and achieve optimal speedup remains widely open. It will be also interesting to know, whether efficient construction of almost optimal trees is possible without sorting initial elements.

Acknowledgments. We thank Larry Larmore for stimulating comments and discussions.

References

[AHNR98] Andersson, A., Hagerup, T., Nilsson, S. Raman, R. *Sorting in Linear Time?*, Journal of Computer and System Sciences **57** (1998), pp. 74–93

[BDH+91] Bhatt, P., Diks, K., Hagerup, T, Prasad, V., T. Radzik, Saxena, S., *Improved deterministic parallel integer sorting*, Information and Computation **94** (1991), pp. 29–47.

[B97]   Blelloch, G. *Prefix Sums and Their Applications*, Reif, J., ed, Synthesis of Parallel Algorithms, pp. 35–60, 1997.

[C88]   Cole, R. *Parallel Merge Sort*, SIAM Journal on Computing **17** (1998), pp. 770–785.

[H87]   Hagerup, T., *Toward optimal parallel bucking sorting*, Information and
           Computation **75** (1987), pp. 39–51.

[H51]   Huffman, D. A. *A method for construction of minimum redundancy codes*,
           Proc. IRE, 40 (1951), pp. 1098–1101.

[KP96]   Kirkpatrick, D., Przytycka, T. *Parallel Construction of Binary Trees with
           Near Optimal Weighted Path Length*, Algorithmica (1996), pp. 172–192.

[LPW93]   Larmore, L.L., Przytycka. T., W. Rytter, *Parallel Construction of Optimal
           Alphabetic Trees*, Proc 5th ACM Symposium on Parallel Algorithms and
           Architectures (1993) pp. 214–223.

[LP95]   Larmore, L., Przytycka. T. *Constructing Huffman trees in parallel*, SIAM
           Journal on Computing, **24**(6) (1995) pp. 1163–1169.

[MR85]   Miller, G., Reif., J., *Parallel tree contraction and its applications*, Proc.
           26th Symposium on Foundations of Computer Science (1985), pp. 478–489.

[N00a]   Nekrich, Y., *Byte-oriented Decoding of Canonical Huffman Codes*, Pro-
           ceedings of the IEEE International Symposium on Information Theory
           2000, (2000), p. 371.

[N00b]   Nekrich, Y., *Decoding of Canonical Huffman Codes with Look-Up Tables*,
           Proceeding of the IEEE Data Compression Conference 2000, (2000) p. 342.

[T87]   Teng, S., *The construction of Huffman equivalent prefix code in NC*, ACM
           SIGACT **18** (1987), pp. 54–61.

[V75]   Valiant, L. *Parallelism in Comparison Problems*, SIAM Journal on Com-
           puting **4** (1975),pp. 348–355.

[vL76]   van Leeuwen, J. *On the construction of Huffman trees*, In 3rd Int. Collo-
           qium on Automata, Languages and Programming (1976), pp. 382–410.

Seamless Integration of Parallelism and Memory Hierarchy
(Extended Abstract)

Carlo Fantozzi, Andrea Pietracaprina, and Geppino Pucci

Università di Padova,
Dipartimento di Elettronica e Informatica
Via Gradenigo 6/A, I-35131, Padova, Italy
Fax: +39 049 8277826
{fantozzi,andrea,geppo}@artemide.dei.unipd.it

Abstract. We prove an analogue of Brent's lemma for BSP-like parallel machines featuring a hierarchical structure for both the interconnection and the memory. Specifically, for these machines we present a uniform scheme to simulate any computation designed for v processors on a v'-processor configuration with $v' \leq v$ and the same overall memory size. For a wide class of computations the simulation exhibits optimal $O(v/v')$ slowdown. The simulation strategy aims at translating communication locality into temporal locality. As an important special case ($v' = 1$), our simulation can be employed to obtain efficient hierarchy-conscious sequential algorithms from efficient fine-grained ones.

1 Introduction

Most modern, high-performance computer platforms consist of several processors, each endowed with its local memory, communicating through some interconnection. When looking at the overall aggregate memory available at the system, one observes a multi-level hierarchical structure, where lower (fast) levels are internal to the processing nodes while upper (slow) levels are implicitly imposed by the interconnection topology. This is the case, for example, of popular architectures such as clusters of workstations or UMA and NUMA multiprocessors. The performance of applications running on these platforms is determined by the maximum number of operations performed at a node (*computation*), and by the data movements required to bring the data close to the units that have to process them. These movements can either occur at individual nodes (*local accesses*) or involve several nodes (*communication*).

In the past, the development of fast algorithms has either focused on minimizing memory access time on sequential hierarchies, or on attaining optimal computation/communication tradeoffs on parallel architectures with flat local memories. Moreover, in the latter case, high parallelism has been often pursued under the assumption that performance could then be scaled down linearly with the number of processors. Such is the case, for instance, of PRAM algorithms designed for a large number of virtual processors (typically proportional to the input size), or algorithms for the more recent

P. Widmayer et al. (Eds.): ICALP 2002, LNCS 2380, pp. 856–867, 2002.

bulk-synchronous models such as BSP [1] and CGM [2], where virtual parallelism is employed to hide communication latency.

It is well known, however, that when implementing an algorithm designed for many (virtual) processors on a real machine with a fixed number of processors, in addition to the natural slowdown due to the loss of parallelism (Brent's lemma [3]), a further slowdown may be introduced if the (larger) subcomputations entrusted to each individual processor exhibit a low degree of locality of reference. The main contribution made by this paper is a rigorous framework integrating parallelism and memory hierarchy, and showing that in many cases this additional slowdown can be avoided by designing algorithms that expose parallelism in a structured fashion.

1.1 Previous Work

As mentioned before, a few works in the literature have dealt with both parallelism and memory hierarchy in an integrated fashion. The *Parallel Hierarchical Memory Model* (P-HMM), defined in [4], is a generalization of the sequential HMM model by [5] consisting of p HMM processors communicating through some powerful interconnection attached to the fastest levels of the local hierarchies. A similar model, also featuring block transfer, is the *Parallel Memory Hierarchy* (PMH) model of [6]. Unlike P-HMM, in PMH communication occurs via a shared memory module which is seen as connected to one of the slowest levels of each hierarchy. Both the P-HMM and the PMH models idealize communication either by assuming that its cost never dominates the time complexity (P-HMM) or by disregarding the impact of the interconnection topology (PMH). Another attempt at integrating parallelism and memory hierarchy is the H-PRAM [7], where processors can be partitioned into clusters operating as independent PRAM machines with smaller memories. However, in the H-PRAM access costs do not depend on the position of the data in memory, but solely on the size of the cluster, hence locality of reference is not captured.

The $M_d(n, p, m)$ model introduced by Bilardi and Preparata [8,9] features a better integration of parallelism and memory hierarchy and explicitly describes the network's topology. Specifically, an $M_d(n, p, m)$ is a d-dimensional mesh of p HMM nodes where the memory at each node has size nm/p, and access function $f(x) = \lceil (x+1)/m \rceil^{1/d}$. The cost for sending a constant-size message from a node to a neighbor is proportional to the cost of accessing the farthest cell in the node's local memory. Although the authors argue that the $M_d(n, p, m)$ is the only scalable architecture under speed of light limitations, its reliance on a specific interconnection may hamper the portability and generality of algorithm design based on this model.

In [8] it is shown that an $M_d(n, n, m)$ can be simulated by a $M_d(n, p, m)$, for $p < n$ and $d = 1, 2$, with slowdown $(n/p)\Lambda(n, p, m)$. This provides an analogue of Brent's lemma except for the factor $\Lambda(n, p, m)$, which represents an extra slowdown due to the interaction with the larger local memories. Such a slowdown, which can grow up to $(n/p)^{1/d}$, is proved to be unavoidable for certain computations [9]. Note that the $M_d(n, n, m)$ being simulated is fine-grained in the sense that each node has only a constant number of available memory cells (i.e., access time can be regarded as a constant).

The translation of parallelism into locality of reference is also studied in [10,11,12]. These works present strategies for efficiently simulating BSP-like computations designed for machine configurations with flat local memories on smaller configurations where a two-level (disk-RAM) hierarchy is provided at each node. The key of the simulation is a careful exploitation of the disks through striped access to the data.

1.2 Our Results

Continuing along the lines of [8,9,10,11], in this paper we show that for a wide class of computations, parallelism exposed in a structured fashion can be fully translated into locality of reference, affording the formulation of a Brent-like result proving that for such computations no slowdown due to the loss of locality is incurred when scaling down the number of processors.

Our results are based on the Decomposable BSP (D-BSP) model, a variant of BSP [1] introduced in [13] to account for submachine locality. Specifically, a v-processor D-BSP consists of collection of v processors communicating through a router. The processors are partitioned into several levels of independent submachines (clusters) induced by a binary decomposition tree for the machine. We slightly modify the original D-BSP definition by regarding each processor as an HMM and setting the cost of communication within a cluster to be proportional to the cost of accessing the farthest cell in a memory of size equal to the aggregate size of the cluster processors' memories. In this fashion, we integrate memory hierarchy and network by regarding the latter as a seamless continuation of the former.

Our main technical result is a uniform scheme that simulates any v-processor D-BSP computation on a v'-processor D-BSP, with $v' \leq v$ and the same aggregate memory size. For a large family of computations, including most prominent ones (e.g., sorting, FFT and matrix multiplication), our simulation exhibits an optimal $O(v/v')$ slowdown, thus providing an analogue of Brent's lemma. The simulation is based on a recursive strategy aimed at translating D-BSP submachine locality into temporal locality on the HMM. The strategy is similar in spirit to the one employed in [14] for porting DAG computations efficiently across sequential memory hierarchies while retaining temporal locality.

Our result complements those in [8,9] in the sense that it shows that the superlinear slowdown incurred by the simulation of fine-grained algorithms can only manifest itself for a restricted class of machine-dependent computations, namely those which take the fullest advantage of fine topological details. For these computations, some degree of network locality is bound to be lost when transformed into temporal locality.

Finally, as an important special case, by setting $v' = 1$ our simulation can be employed to obtain efficient hierarchy-conscious sequential algorithms from efficient fine-grained ones. In this fashion, a large body of algorithmic techniques exhibiting structured parallelism can be effortlessly transferred to the realm of sequential algorithms for memory hierarchies. We provide evidence of this fact by showing that for a number of prominent computations optimal sequential algorithms can be obtained in this fashion. In this respect, our work provides a generalization of the results by [10,11] to multi-level memory hierarchies.

We remark that in memory hierarchies another type of locality exists, namely *spatial locality*, exploitable through the block-transfer mechanism. A research avenue left open by this paper is to investigate whether spatial locality can be absorbed into the same framework.

The rest of the paper is organized as follows. Section 2 defines our reference models. Section 3 describes the simulation scheme first introducing its main ideas with two special cases (Subsections 3.1 and 3.2), and then deriving the general result (Subsection 3.3). Finally, in Section 4 we provide evidence of how our simulation yields efficient hierarchy-conscious sequential algorithms from efficient fine-grained ones.

2   Machine Models

HMM. The *Hierarchical Memory Model* (HMM) was introduced in [5] as a random access machine where access to memory location x requires time $f(x)$, for a given non-decreasing function $f(x)$. It is supposed that an n-ary operation (i.e., an operation involving memory cells $x_1, \ldots, x_n$) can be completed in time $1 + \sum_{i=1}^{n} f(x_i)$, regardless of the value of n. We refer to such a model as $f(x)$-HMM. As most works in the literature, we will focus our attention on *polynomially bounded* access functions, that is, functions $f(x)$ for which there exists a constant c such that $f(2x) \leq cf(x)$, for any x. Particularly interesting and widely studied special cases are the polynomial function $f(x) = x^\alpha$ and the logarithmic function $f(x) = \log x$. The following technical fact is proved in [5].

Fact 1. *If $f(x)$ is polynomially bounded, then the time to access the first n memory cells of an $f(x)$-HMM is $\Theta(nf(n))$.*

D-BSP. The *Decomposable Bulk Synchronous Parallel* (D-BSP) model was introduced in [13] for capturing submachine locality, and further investigated in [15,16,17]. In this paper, we employ a variant of the model equipped with hierarchical memory at the processors and whose cost function accounts for both memory accesses and communication costs in a uniform and reasonable fashion.

Let v be a power of two. A D-BSP $(v, \mu, f(x))$ is a collection of v processors $\{P_j : 0 \leq j < v\}$ communicating through a router, where each processor is an $f(x)$-HMM machine with memory size μ. In particular, a D-BSP $(1, \mu, f(x))$ coincides with an $f(x)$-HMM with memory size μ. For $0 \leq i \leq \log v$, the v processors are partitioned into 2^i fixed, disjoint *i-clusters* $C_0^{(i)}, C_1^{(i)}, \cdots, C_{2^i-1}^{(i)}$ of $v/2^i$ processors each, where the processors of a cluster are capable of communicating among themselves independently of the other clusters. The clusters form a hierarchical, binary decomposition tree of the D-BSP machine: specifically, $C_j^{\log v} = \{P_j\}$, for $0 \leq j < v$ and $C_j^{(i)} = C_{2j}^{(i+1)} \cup C_{2j+1}^{(i+1)}$, for $0 \leq i < \log v$ and $0 \leq j < 2^i$.

A D-BSP program consists of a sequence of *labelled supersteps*. In an *i-superstep*, $0 \leq i < \log v$, each processor executes internal computation on locally held data and sends messages exclusively to processors within its i-cluster (an output and an input pool for message exchange are provided in each processor's memory). The superstep is terminated by a barrier, which synchronizes processors within each i-cluster independently. It is assumed that messages sent in one superstep are available at the destinations only at the beginning of the next superstep. Also, it is reasonable to assume that any D-BSP computation ends with a global synchronization. If each processor spends at most τ units of time performing local computation during the superstep, and if the messages that are sent form an h-relation, $h > 0$, (i.e., each processor is the source or destination of at most h messages), then the cost of the i-superstep is upper bounded by $\tau + hf(\mu v/2^i)$. We call a program *full* if the communication required by every superstep is a $\Theta(\mu)$-relation, that is, each processor sends/receives an amount of data proportional to its local memory size. As we will see, several prominent problems can be efficiently solved by full programs.

Note that the communication costs in our variant of the model are as in a standard D-BSP where both bandwidth and latency parameters within i-clusters are set to $f(\mu v/2^i)$. Our choice for such parameters aims at creating a seamless hierarchy of memory access and communication costs. More specifically, the communication medium is regarded as an extension of the local memory hierarchies, with each message sent by a processor in an i-cluster C being charged with the cost of accessing the farthest memory cell in an $f(x)$-HMM with memory size equal to the aggregate memory size of C.

Although in this paper we will deal with arbitrary polynomially bounded access functions $f(x)$, we will support our findings by considering, as case studies, the aforementioned polynomial and logarithmic functions. For the polynomial function $f(x) = x^\alpha$, we observe that when $\alpha = 1/d$, the D-BSP (v, μ, x^α) can be regarded as an abstraction of the d-dimensional architecture proposed by Bilardi and Preparata in [8] as the only scalable machine in the limiting technology. In fact, this D-BSP can be simulated on such an architecture with constant slowdown.

3   The General Simulation Algorithm

In this section, we present a general scheme to simulate an arbitrary D-BSP $(v, \mu, f(x))$ program on a D-BSP $(v', \mu v/v', f(x))$, with $v' \leq v$. In order to introduce the main ideas underlying our simulation strategy, we first consider the simpler case $v' = 1$ and $f(x) = x^\alpha$, with $0 < \alpha < 1$ an arbitrary constant (Subsection 3.1). We then describe suitable modifications that are needed to extend the simulation to the case of arbitrary polynomially bounded functions (Subsections 3.2). Finally, in Subsection 3.3, we remove the assumption $v' = 1$.

3.1   Simulation of D-BSP (v, μ, x^α) on the x^α-HMM

Without loss of generality, we can restrict ourselves to consider D-BSP (v, μ, x^α) programs where the labels of consecutive supersteps differ by at most 1 (we refer to these programs as *smooth*). In a smooth program, an i-superstep can only by followed by a j-superstep, with $|j - i| \leq 1$. Indeed, for any non-smooth program there is an equivalent smooth program which exhibits the same asymptotic time complexity. This is easily seen as follows. When $f(x) = x^\alpha$, the cost of communication grows polynomially with the cluster size, hence one can insert, between any two consecutive supersteps with labels i and j, with $|j - i| > 1$, one dummy superstep for every distinct label between i and j. This transformation increases the time complexity of the program by at most a constant factor.

Consider now a smooth D-BSP (v, μ, x^α) program $\mathcal{P}$ to be simulated on an x^α-HMM. Let us regard the HMM memory as divided into v *blocks*, numbered from 0 to $v - 1$, of μ cells each, with block 0 at the top of memory (i.e., comprising memory cells $0, 1, \ldots, \mu - 1$), and block $v - 1$ at the bottom. During the course of the simulation every block will contain the *context* of a distinct D-BSP processor, that is, an image of the processor's local memory at a certain point of execution. At the beginning of the simulation, block i contains the context of processor P_i, $i = 0, 1, \ldots, v - 1$, but this association changes as the simulation proceeds. Additional constant-size space is needed for bookkeeping operations. For simplicity, we assume that $O(1)$ registers with unit-time access are employed for this purpose. Alternatively, we could use the memory cells at the top of memory and shift the processors' contexts down by a constant amount, thus paying a negligible time penalty.

The simulation of $\mathcal{P}$ on the HMM is organized in a number of *rounds*, where a round simulates the operations prescribed by a certain superstep of $\mathcal{P}$ for a certain cluster, and performs a number of context swaps to prepare the execution of the next round. Suppose that the supersteps of $\mathcal{P}$ are numbered consecutively and let i_s be the label of Superstep s, for $s \geq 0$ (i.e., Superstep s is executed independently within each i_s-cluster). The simulation algorithm is given by the following pseudo-code, where the loop iterations correspond to rounds.

while true **do**

1 $P \leftarrow$ processor whose context is on top of memory
   $s \leftarrow$ superstep number to be simulated next for P
   $i_s \leftarrow$ superstep index to be simulated next for P
   $C \leftarrow i_s$-cluster containing P
2 simulate Superstep s for cluster C
3 **if** P has finished its program **then break**
   {final superstep executed: simulation complete}
4 **if** $i_{s+1} = i_s - 1$ **then**
   swap the first $|C|$ blocks on top of memory
   with the next $|C|$ blocks

During the course of the simulation we say that a D-BSP processor P is *s-ready* if for all processors in the i_s-cluster of P (including P itself) Supersteps $0, 1, \ldots, s - 1$ have been simulated while Superstep s has not yet been simulated. As will be proved later, the following two invariants are maintained at the beginning of each round. Let s and C be defined as in Step 1 of the round.

INV1 The contexts of all processors in C are stored in the topmost $|C|$ blocks, sorted in increasing order by processor number. Moreover, for any other cluster C', the contexts of all processors in C' are stored in consecutive memory blocks (although not necessarily sorted).

INV2 All processors in C are s-ready.

If the two invariants hold, the cluster simulation in Step 2 can be correctly performed as follows.

for $j \leftarrow 0$ **to** $|C| - 1$ **do**

2.1   swap the contents of memory blocks 0 and j

2.2   simulate the local operations prescribed by Superstep s
       for the processor whose context is in block 0

2.3   swap the contents of memory blocks j and 0

2.4 simulate the message exchange prescribed by
       the superstep for cluster j

Note that the message exchange in Step 2.4 can be completed by scanning the output pools of the processor contexts sequentially and delivering each message to the input pool in the destination processor's context. Since by invariant **INV1** the contexts of the processors are sorted by processor number, the location of each input pool is easily determined based on the processor number.

Theorem 1. *The simulation algorithm is correct.*

Proof (sketch). The correctness of the entire simulation algorithm is immediately established once we show that **INV1** and **INV2** hold at the beginning of each round. This can be proved by induction on the number v of D-BSP processors. The claim is trivial for the basis $v = 2$, since in this case, the D-BSP program is simply a sequence of 0-supersteps, which are simulated in a straightforward fashion one after the other. Suppose that the claim is true for machines of up to v processors and consider the simulation of a program $\mathcal{P}$ of t supersteps for a D-BSP with $2v$ processors. First, consider the case that the t supersteps include a single 0-superstep (which by our former assumption must be the last superstep). If $t = 1$ then the claim trivially follows. Otherwise we can show that the algorithm will initially simulate the first $t - 1$ supersteps of $\mathcal{P}$ for cluster $C_0^{(1)}$ and its subclusters, as if it were simulating a program for a D-BSP with v processors. By the inductive hypothesis, the two invariants will hold for all the rounds performed in this initial phase, which ends with a round that simulates superstep $t - 1$ (a 1-superstep, by the smoothness of $\mathcal{P}$) for cluster $C_0^{(1)}$. At the end of such a round, the simulation algorithm swaps the contexts of the processors in $C_0^{(1)}$ with those in the sibling cluster $C_1^{(1)}$. Since this is the first time $C_1^{(1)}$ is brought to the top of the HMM memory and the first superstep of $\mathcal{P}$ has label strictly greater than 0, both invariants hold at the beginning of the next round. We can then reapply the inductive hypothesis for $C_1^{(1)}$ showing that the invariants will hold for all rounds up to and including the round that simulates superstep $t - 1$ for $C_1^{(1)}$. At the end of this latter round, $C_0^{(1)}$ and $C_1^{(1)}$ are swapped back and the next (final) round will simulate Superstep t, which involves the entire machine. Clearly, the invariants hold at the beginning of the final round. If $\mathcal{P}$ contains more than one 0-superstep, we can split the program into subprograms terminating at 0-superstep boundaries and iterate the above argument for each such subprogram.

Let us now evaluate the running time of a generic round of our algorithm, where a given superstep s for a given i_s-cluster C is simulated. Clearly, Steps 1 and 3 require constant time. Consider now Step 2, and note that, since each D-BSP processor is an x^α-HMM, the simulation of the local operations of each processor in C performed in Substep 2.2 incurs no slowdown, since the processor's context is brought at the top of the HMM memory. Note also that by virtue of invariant **INV1**, the message exchange in Substep 2.4 can be completed by accessing the first $\mu|C|$ HMM memory cells only a constant number of times. Substeps 2.1 and 2.3. also require accessing the first $\mu|C|$ HMM memory cells only a constant number of times. Hence, letting τ_s denote the maximum local computation time for a processor in Superstep s, the simulation of Step 2 is accomplished in time $O\left(|C|\left(\tau_s + \mu(\mu|C|)^\alpha\right)\right)$ because of Fact 1. Finally, observe that whenever a swap between $|C|$ blocks occurs in Step 4, we have just finished simulating the computation of a cluster of $|C|$ processors, hence the cost $\Theta\left((\mu|C|)^{\alpha+1}\right)$ required by such swap on a x^α-HMM is dominated by that of Step 2. The above argument suffices to prove the following theorem:

Theorem 2. *Consider a program $\mathcal{P}$ of a D-BSP (v, μ, x^α), where each processor performs local computation for $O(\tau)$ time, and there are λ_i i-supersteps, for $0 \le i < \log v$. Then, $\mathcal{P}$ can be simulated on a x^α-HMM in time $O\left(v\left(\tau + \mu \sum_{i=0}^{\log v - 1} \lambda_i(\mu v/2^i)^\alpha\right)\right)$.*

Recall that $\mathcal{P}$ is *full* if every superstep requires a $\Theta(\mu)$-relation. In this case, the simulation of a superstep for cluster C incurs a slowdown merely proportional to the cluster size, which is optimal due to the loss of parallelism. We have:

Corollary 1. *Any T-time full program $\mathcal{P}$ on a D-BSP (v, μ, x^α) can be simulated in optimal time $\Theta(Tv)$ on a x^α-HMM.*

3.2   General Simulation of D-BSP $(v, \mu, f(x))$ on the $f(x)$-HMM

The simulation presented in the preceding subsection crucially relies on the assumption that the program being simulated is smooth. Such an assumption was made without loss of generality since the access function $f(x) = x^\alpha$ is such that every D-BSP (v, μ, x^α) program admits an equivalent smooth program with the same asymptotic complexity. This is not true, however, if $f(x)$ is subpolynomial (although still polynomially bounded). In this case $f(x)$ grows so slowly that the introduction of dummy supersteps between adjacent supersteps with non-consecutive labels may increase the complexity of the transformed program by more than a constant factor.

We must adopt a different notion of smoothness and modify the simulation strategy accordingly. For a D-BSP $(v, \mu, f(x))$, let us define $\mathcal{G}(v, \mu, f(x))$ as the set of superstep labels $0 = j_0 < j_1 < \cdots < j_m < \log v$, with the property that for any $0 \le \ell < m$

$$\gamma_1 f(\mu v/2^{j_\ell}) \le f(\mu v/2^{j_{\ell+1}}) \le \gamma_2 f(\mu v/2^{j_\ell}),$$

where $0 < \gamma_1 < \gamma_2 < 1$ are suitable constants. We call a program $\mathcal{P}$ for a D-BSP $(v, \mu, f(x))$ $\mathcal{G}$-*smooth* if the label of every superstep of $\mathcal{P}$ belongs to $\mathcal{G}(v, \mu, f(x))$, and if for any pair of adjacent supersteps with labels j_ℓ and $j_{\ell'}$ we have $|\ell - \ell'| \le 1$. It is easy to see that for any arbitrary D-BSP $(v, \mu, f(x))$ program there exists a functionally equivalent $\mathcal{G}$-smooth program that exhibits the same asymptotic running time. Hence, without loss of generality, from now on we restrict ourselves to consider $\mathcal{G}$-smooth programs.

Note that under the new notion of $\mathcal{G}$-smoothness it is no longer true that adjacent supersteps involve clusters of size differing by at most a factor two, which is a property the simulation algorithm in the previous subsection relies upon when performing context swaps in Step 4. Consequently, we need to modify this step to make it work correctly under the new scenario. Consider a round that simulates a certain superstep s of a $\mathcal{G}$-smooth program $\mathcal{P}$ for a cluster C and suppose that the subsequent superstep $s+1$ in $\mathcal{P}$ is such that $i_{s+1} < i_s$. Let $\hat{C}$ be the i_{s+1}-cluster that contains C. Note that $\hat{C}$ is composed by $b = 2^{i_s - i_{s+1}} \geq 2$ i_s-clusters, including C, which we denote by $C_0, C_1, \ldots, C_{b-1}$. Suppose that $C = C_0$, that is, this is the first round executing Superstep s for processors in $\hat{C}$. The simulation will enforce the property that at the beginning of the round the contexts of all processors in $\hat{C}$ are at the top of memory sorted by processor number (i.e., the topmost contexts are those of the processors in C_0, followed by those of the processors in C_1, and so on). At this point the simulation enters a *cycle* consisting of b phases, each phase comprising one or more simulation rounds. In the k-th such phase, $0 \leq k < b$, the contexts of the processors in C_k are brought to the top of memory, then all supersteps up to Superstep s are simulated for these processors, and finally the contexts of C_k are put back to the positions occupied at the beginning of the cycle.

Since smaller cycles can open within a cycle, the simulation algorithm needs an auxiliary vector K to keep track of the nested phases. More precisely, during a cycle involving i_s-clusters $C_0, C_1, \ldots, C_{b-1}$, vector element $K[i_s]$ stores the index of the i_s-cluster which is currently being simulated. Vector K is initialized to 0 at the start of the whole simulation. For ease of presentation, we assume that K is stored in a separate memory with constant access time. However, the vector can be scattered in the HMM memory by storing element $K[i]$ immediately after the first $v/2^i$ memory blocks, without affecting the asymptotic performance of our simulation (more details will be provided in the full version of the paper).

As a whole, Step 4 of the simulation algorithm presented in the previous subsection must be replaced by the following piece of pseudo-code.

4   **if** $i_{s+1} < i_s$ **then do**
4.1 $b \leftarrow 2^{i_s - i_{s+1}}$
4.2   swap the $|C|$ blocks on top of memory
          with the $K[i_s]$-th group of $|C|$ blocks
          {empty step for $K[i_s] = 0$}
4.3 $K[i_s] = (K[i_s] + 1) \bmod b$
4.4   swap the $K[i_s]$-th group of $|C|$ blocks
          with the $|C|$ blocks on top of memory

The proof of the following theorem is along the lines of the one of Theorem 1 and is omitted for space limitations.

Theorem 3. *The generalized simulation algorithm is correct.*

Since the invariants **INV1** and **INV2** hold for all the relevant superstep indices, the simulation of a single i_s-cluster during Step 2 can still be performed as we described for the case $f(x) = x^\alpha$, hence Step 2 takes time $O\left(|C|(\tau_s + \mu f(\mu|C|))\right)$. The memory overhead connected with a cycle (Step 4 of the pseudo-code) needs more careful consideration. Memory blocks $0, 1, |C| - 1$ are accessed a constant number of times for each i_s-cluster that is being simulated, so this cost is dominated by the one for Step 2. Memory blocks $|C|, |C| + 1, \ldots, v/2^{i_{s+1}} - 1$ are accessed twice: the cost of these accesses is amortized by the cost of the future execution of Superstep $s + 1$ for the i_{s+1}-cluster. By combining these observations, we have:

Theorem 4. *Consider a program $\mathcal{P}$ of a D-BSP $(v, \mu, f(x))$ where each processor performs local computation for $O(\tau)$ time, and there are λ_i i-supersteps for $0 \leq i < \log v$. If $f(x)$ is polynomially bounded, then $\mathcal{P}$ can be simulated on a $f(x)$-HMM in time $O\left(v\left(\tau + \mu \sum_{i=0}^{\log v - 1} \lambda_i f(\mu v/2^i)\right)\right)$.*

Corollary 2. *If $f(x)$ is polynomially bounded then any T-time full program for a D-BSP $(v, \mu, f(x))$ can be simulated in optimal time $\Theta(Tv)$ on a $f(x)$-HMM.*

3.3   Analogue of Brent's Lemma

The following theorem generalizes the simulation results of the previous subsections providing an analogue of Brent's lemma [3] for parallel and hierarchical computations.

Theorem 5. *Consider a program $\mathcal{P}$ of a D-BSP $(v, \mu, f(x))$ where each processor performs local computation for $O(\tau)$ time, and there are λ_i i-supersteps, for $0 \leq i < \log v$. If $f(x)$ is polynomially bounded, then for any $1 \leq v' \leq v$, $\mathcal{P}$ can be simulated on a D-BSP $(v', \mu v/v', f(x))$ in time $O\left((v/v')\left(\tau + \mu \sum_{i=0}^{\log v - 1} \lambda_i f(\mu v/2^i)\right)\right)$.*

Proof (sketch). Let us refer to the D-BSP $(v, \mu, f(x))$ and the D-BSP $(v', \mu v/v', f(x))$ as guest and host machine, respectively. For every $0 \leq j < v'$, the processors in cluster $C_j^{(\log v')}$ of the guest machine are simulated by host processor P_j. For $i < \log v'$, each i-superstep of $\mathcal{P}$ is simulated by an i-superstep on the host. Suppose that the original superstep prescribes $O(\tau')$ local computation and the execution of an h-relation (note that $h \leq \mu$). It is easy to see that such a superstep can be simulated in time $O\left((v/v')(\tau' + \mu f(\mu v/2^i))\right)$, where the term $(v/v')\mu f(\mu v/2^i)$ accounts for both the execution of an (hv/v')-relation within i-clusters on the host and the data movements required by the simulation of the v/v' local computations at each host processor.

Instead, for $i \geq \log v'$, each i-superstep is simulated sequentially, using the strategy of Subsection 3.2: in fact, with a straightforward transformation any sequence of supersteps whose labels are equal or greater than $\log v'$ can be treated as a program of a D-BSP $(v/v', \mu, f(x))$. The slowdown incurred in the simulation of these supersteps can be obtained by applying Theorem 4.   ∎

Corollary 3. *If $f(x)$ is polynomially bounded then any T-time full program for a D-BSP $(v, \mu, f(x))$ can be simulated in optimal time $\Theta(Tv/v')$ on a D-BSP $(v, \frac{\mu v}{v'}, f(x))$, for any $1 \leq v' \leq v$.*

4   Application to Case Study Problems

In this section we show, on a number of prominent reference problems, how our simulation strategy can be employed to transform efficient D-BSP algorithms into optimal solutions for those problems on the HMM. This provides evidence that the structured parallelism exposed in D-BSP through submachine locality can be (automatically) transformed into temporal locality on a memory hierarchy. As a consequence, D-BSP can be profitably employed to develop efficient, portable algorithms for hierarchical architectures.

In order to emphasize the transformation of parallelism into temporal locality, we will consider *fine-grained* D-BSP algorithms where the size μ of each local memory is constant, that is, algorithms which are purely based on the exploitation of parallelism and need not bother with the local hierarchies. For concreteness, we will consider the access functions $f(x) = x^{\alpha}$, with $0 < \alpha < 1$, and $f(x) = \log x$. Under these functions, upper and lower bounds for our reference problems have been developed directly for the HMM in [5]. We will make use of these HMM results as a comparison stone for the results obtained through our simulation. Due to space limitations, all details are omitted and will appear in the full version of the paper.

Matrix multiplication. We call n-*MM* the problem of multiplying two $\sqrt{n} \times \sqrt{n}$ matrices on an n-processor D-BSP using only semiring operations. Both the input matrices and the output matrix are evenly and arbitrarily distributed among the D-BSP processors. We have

Proposition 1. *For the n-MM problem there is a D-BSP $(n, 1, x^{\alpha})$ algorithm that runs in time*

$$T_{\mathrm{MM}}(n) = \begin{cases} O(n^{\alpha}) & \text{for } 1/2 < \alpha < 1, \\ O(\sqrt{n} \log n) & \text{for } \alpha = 1/2, \\ O(\sqrt{n}) & \text{for } 0 < \alpha < 1/2, \end{cases}$$

and a D-BSP $(n, 1, \log x)$ algorithm that runs in time $T_{\mathrm{MM}}(n) = O(\sqrt{n})$. The simulation of these algorithms yields optimal performance on the x^{α}-HMM and the $\log x$-HMM, respectively.

Discrete Fourier Transform. We call n-*DFT* the problem of computing the Discrete Fourier Transform of an n-vector evenly and arbitrarily distributed among the n D-BSP processors. We have

Proposition 2. *For the n-DFT problem there is a D-BSP $(n, 1, x^{\alpha})$ algorithm that runs in time $T_{\mathrm{DFT}}^{\alpha}(n) = O(n^{\alpha})$, and a D-BSP $(n, 1, \log x)$ algorithm that runs in time $T_{\mathrm{DFT}}^{\log}(n) = O(\log n \log \log n)$. The simulation of these algorithms matches the best known bounds for the x^{α}-HMM and $\log x$-HMM, respectively.*

Sorting. We call n-*sorting* the problem in which n keys are initially evenly distributed among the n D-BSP processors and have to be redistributed so that the smallest key is held by processor P_0, the second smallest one by processor P_1, and so on. The following result is an immediate consequence of [16, Proposition 2], Corollary 1, and the lower bound proved in [5].

Proposition 3. *There is an n-sorting algorithm for D-BSP $(n, 1, x^{\alpha})$ that runs in time $T_{\mathrm{SORT}}^{\alpha}(n) = O(n^{\alpha})$. The simulation of this algorithm on the x^{α}-HMM exhibits optimal performance.*

We remark that all n-sorting strategies known in the literature for BSP-like models seem to yield $\Omega\left(\log^2 n\right)$-time algorithms when implemented as fine-grained on the D-BSP $(n, 1, \log x)$. By simulating one such algorithm on the $\log x$-HMM we get a running time of $\Omega\left(n \log^2 n\right)$, which is a $\log n / \log \log n$ factor away from optimal [5]. However, such non-optimality is due to the inefficiency of the D-BSP algorithm employed and not to a weakness of the simulation. In fact, our simulation implies a $\Omega\left(\log n \log \log n\right)$ lower bound for n-sorting on the D-BSP $(n, 1, \log x)$. No tighter lower bound or better algorithm are known so far.

References

1. Valiant, L.G.: A bridging model for parallel computation. Communications of the ACM **33** (1990) 103–111
2. Dehne, F., Fabri, A., Rau-Chaplin, A.: Scalable parallel geometric algorithms for coarse grained multicomputers. International Journal on Computational Geometry **6** (1996) 379–400
3. Brent, R.P.: The parallel evaluation of general arithmetic expressions. Journal of the ACM **21** (1974) 201–206
4. Vitter, J.S., Shriver, E.A.M.: Algorithms for parallel memory II: Hierarchical multilevel memories. Algorithmica **12** (1994) 148–169
5. Aggarwal, A., Alpern, B., Chandra, A.K., Snir, M.: A model for hierarchical memory. In: Proc. of the 19th ACM STOC. (1987) 305–314
6. Alpern, B., Carter, L., Ferrante, J.: Modeling parallel computers as memory hierarchies. In Programming Models for Massively Parallel Computers. IEEE Computer Society Press (1993) 116–123
7. Heywood, T., Ranka, S.: A practical hierarchical model of parallel computation. I. the model. Journal of Parallel and Distributed Computing **16** (1992) 212–232
8. Bilardi, G., Preparata, F.P.: Processor-time tradeoffs under bounded-speed message propagation: Part I, upper bounds. Theory of Computing Systems **30** (1997) 523–546
9. Bilardi, G., Preparata, F.P.: Processor-time tradeoffs under bounded-speed message propagation: Part II, lower bounds. Theory of Computing Systems **32** (1999) 531–559
10. Dehne, F., Dittrich, W., Hutchinson, D.: Efficient external memory algorithms by simulating coarse-grained parallel algorithms. In: Proc. of the 9th ACM SPAA. (1997) 106–115
11. Dehne, F., Dittrich, W., Hutchinson, D., Maheshwari, A.: Reducing I/O complexity by simulating coarse grained parallel algorithms. In: Proc. of the 13th IPPS. (1999) 14–20
12. Sibeyn, J., Kaufmann, M.: BSP-like external-memory computation. In: Proc. of the 3rd Italian Conference on Algorithms and Complexity. LNCS 1203 (1997) 229–240
13. De la Torre, P., Kruskal, C.P.: Submachine locality in the bulk synchronous setting. In: Proc. of EUROPAR 96. LNCS 1124 (1996) 352–358
14. Bilardi, G., Peserico, E.: A characterization of temporal locality and its portability across memory hierarchies. In: Proc. of ICALP 2001. LNCS 2076 (2001) 128–139
15. Bilardi, G., Pietracaprina, A., Pucci, G.: A quantitative measure of portability with application to bandwidth-latency models for parallel computing. In: Proc. of EUROPAR 99. LNCS 1685 (1999) 543–551
16. Fantozzi, C., Pietracaprina, A., Pucci, G.: Implementing shared memory on clustered machines. In: Proc. of IPDPS 2001. (2001)
17. Bilardi, G., Fantozzi, C., Pietracaprina, A., Pucci, G.: On the effectiveness of D-BSP as a bridging model of parallel computation. In: Proc. of ICCS 2001. LNCS 2074 (2001) 579–588

The Communication Complexity of Approximate Set Packing and Covering

Noam Nisan[*]

Institute of Computer Science,
The Hebrew University of Jerusalem
Israel
noam@cs.huji.ac.il

Abstract. We consider a setting where k players are each holding some collection of subsets of $\{1..n\}$. We consider the communication complexity of approximately solving two problems: The *cover number*: the minimal number of sets (in the union of their collections) whose union is $\{1...n\}$ and the *packing number*: the maximum number of sets (in the union of their collections) that are pair-wise disjoint. We prove that while computing a $(\ln n)$-approximation for the cover number and an $min(k, O(\sqrt{n}))$-approximation for the packing number can be done with polynomial (in n) amount of communication, getting a $(1/2 - \epsilon)\log n$ approximation for the cover number or a better than $min(k, n^{1/2-\epsilon})$-approximation for the packing number requires exponential communication complexity.

1   Introduction

The problems studied here were motivated by communication bottlenecks in combinatorial auctions. In a combinatorial auction n items are sold concurrently, and k bidders are bidding for them. The combinatorial nature of the auction comes from the fact that players value *bundles* of goods as opposed to individual goods. I.e. each player holds a valuation function v_i that assigns a non-negative value for each set of items $S \subset \{1...n\}$. The valuation function v is usually assumed to be monotone (items can be freely disposed). The problem that combinatorial auctions aim to solve is to find the optimal allocation between the different bidders, i.e. a partition $(S_1...S_k)$ of the items that maximizes $\sum_i v_i(S_i)$.

Even when each player is only interested in a single set of items (i.e. v_i is totally specified by a single set S_i and a price p_i), the allocation problem is computationally intractable even to approximate within a factor of $n^{1/2-\epsilon}$ (a simple reduction from approximate Clique [4]). However, it seems that in practice problems with hundreds of items can be solved [11,10], and those with thousands of items rather well approximated [13]. This paper explores another concern, that of actually transferring the required information about v_i – information that

---

[*] Supported by a grant from the Israeli Academy of Sciences.

P. Widmayer et al. (Eds.): ICALP 2002, LNCS 2380, pp. 868–875, 2002.

is exponentially larger than n. When each v_i takes only $0, 1$ values, each v_i is completely specified by a collection of "desired sets". An allocation in this case is exactly a set packing and we need to analyze the communication requirements of finding or approximating the optimum. (It turns out that the original, general, "weighted case" has essentially identical communication requirements.) From now on this paper deals with this non-weighted set-packing problem. More details regarding the applications to combinatorial auctions themselves may be obtained from a companion paper [8].

The set packing communication problem is as follows: k players are each holding a collection A_i of subsets of $\{1...n\}$, and the players are looking for the largest packing (i.e. collection of pair-wise disjoint sets) in the union of their collections. One may naturally also consider the related problem of set covering: each player holds a collection A_i of subsets of $\{1...n\}$, and the players are looking for the smallest covering (i.e. collection of sets whose union is $\{1...n\}$) in the union of their collections[1].

It is quite easy to prove an exponential lower bound on the communication needed to compute either the cover number or the packing number. What we are concerned here is approximating these numbers. Somewhat surprisingly, there has been very little work to date on approximation in the communication complexity model. We feel that studying approximation in a communication complexity setting is interesting for at least two reasons. First, it abstracts the communication requirements of approximation problems in exactly the same way as the exact version does for exact problems. Moreover, in many applications of communication complexity the input size is exponentially larger than the permissible amount of communication. In such cases one can rarely hope for exact solutions and thus approximation is the goal. Second, as the communication complexity model is mostly used for its lower bounds, relaxing the problem strengthens the lower bound, and hence may find further applications. E.g. the natural "number in the hand" model of k-player communication (where each player knows a single piece of the input) is only rarely considered (rather than the more powerful "number on the forehead" model) since increasing the number of players reduces the power of the model and hence is less interesting for lower bounds. However, when approximation enters the picture this may change since approximation factors may increase with the number of players. E.g. for the set packing problem, getting a k-approximation with k players is strictly harder than getting a $(k + 1)$-approximation with $k + 1$ players. Hence lower bounds for large numbers of players in the "number in the hand" model are of interest.

We are able to show tight bounds on the approximation factors obtainable with polynomial communication:

---

[1] The weighted version of the set covering problem also has a natural interpretation as "combinatorial procurement auctions" – the n items are now procured in an auction from bidders who are each bidding to supply subsets S for the price $v_i(S)$, and the auctioneers' goal is to minimize his total payment.

Theorem: The set cover problem may be approximated in polynomial communication to within a factor of $\ln n$. Any approximation to within a factor of $(1/2 - \epsilon) \log n$ requires exponential communication. The lower bound holds also for randomized or nondeterministic protocols.

Theorem: The set packing problem may be approximated in polynomial communication to within a factor of $min(k, \sqrt{n})$. Any approximation to within a factor of better than $min(k, n^{1/2-\epsilon})$ requires exponential communication. The lower bound holds also for randomized or nondeterministic protocols.

Both proofs rely on a single simple technique: probabilistic construction of a reduction. One is immediately struck with the similarity of these gaps to those in the computational setting. One may compute a $\ln n$-approximation for covering and a $O(\sqrt{n})$-approximation for packing in polynomial time, but it is NP-hard to approximate covering to within $(1/2 - \epsilon) \log n$ [7] and packing to within $n^{1/2-\epsilon}$ [4]. It seems that this similarity is rooted in the gap between the fractional solution and the integral one: The gap between the fractional cover number and the integral cover number is $\Theta(\log n)$ and the gap between the fractional packing number and the integral one is $\Theta(\sqrt{n})$ [6,1]. Moreover, for the case of the covering problem, the underlying combinatorial construction used here is indeed similar to that used in the NP-completeness result of [7].

It is interesting to observe that the fractional cover number and fractional packing number may be both computed exactly in a polynomial amount of communication. The linear programs describing the fractional cover number and the fractional packing number, each have a variable for each subset of the items, and a constraint for each item. The dual LP may thus be solved by a separation-based LP algorithm in polynomial time. Since each constraint of the dual LP corresponds to a subset that is held by one of the players, a separating hyperplane can be found by a single player, once the variable values are given. Thus each step of the algorithm may be implemented by the players using a polynomial amount of communication by broadcasting the variable values to all players, and one player responding with a separating hyperplane.

2   Set Packing

2.1   Upper Bound

We are in a situation that each of k players holds a collection $A_i \subseteq P(\{1...n\})$, and they want to approximately find a maximum packing.

Getting an approximation factor of k is easy by just picking the single player with largest packing in his collection. If $k > \sqrt{n}$ we can do better by using the following simple greedy protocol: at each stage each player announces the smallest set $S_i \in A_i$ that is disjoint from all previously chose sets (requiring n bits of communication from each of k players). The smallest such set is chosen to be in the packing. This is repeated until no more disjoint sets exist – after at most n rounds. It is easy to verify that this packing is at most a factor of

$\sqrt{n}$ smaller than the optimal packing. The total communication is bounded by $O(kn^2)$.

Note: The weighted case (where each player has a valuation function v_i : $P(\{1...n\}) \to R^+$ and the aim is to maximize $\sum_i v_i(S_i)$ over all packings $S_1...S_k$) can also be approximated this way to within the same factor, by choosing, at each stage the set S_i that maximizes $v_i(S_i)/\sqrt{|S_i|}$ [5].

2.2 Approximate Disjointness Lower Bound

Our lower bound for the set packing problem will use a reduction from the following "approximate-disjoitness" problem which is of independent interest, and which was studied in [2]:

The Approximate Disjointness Problem

k players are each holding an N-bit string. The string for player i specifies a subset $A_i \subseteq \{1...N\}$. They are required to distinguish between the following two extreme cases:

- Negative: $\cap_{i=1}^k A_i \neq \emptyset$
- Positive: for every $i \neq j$, $A_i \cap A_j = \emptyset$

Theorem 1. *[2] The approximate disjointness problem requires communication complexity of $\Omega(N/k^4)$. This lower bound applies also to randomized protocols with bounded 2-sided error, and also to nondeterministic protocols.*

The proof in [2] generalizes the probabilistic lower bound for disjointness in the two-player case of [9]. They only point out the randomized lower bound, but their proof gives a nondeterministic bound as well. We provide here a simple self-contained proof due to Jaikumar Radhakrishnan and Venkatesh Srinivasan of a somewhat stronger bound for the deterministic (and in fact also the nondeterministic) case.

Theorem 2. *(Radhakrishnan and Srinivasan) The approximate disjointness problem requires communication complexity of $\Omega(N/k)$.*

Proof. First note that any c-bit protocol for the approximate disjointness problem partitions the space of inputs into at most 2^c "boxes", where a box is a Cartesian product $S_1 \times S_2 \times ... \times S_k$, where for each i, $S_i \subseteq \{0,1\}^N$. Each box must be labeled with an answer, and thus the boxes must be "monochromatic" in the following sense: No box can contain both a positive instance and a negative instance. (There is no restriction on instances that are neither positive nor negative.)

We will show that there are exactly $(k+1)^N$ positive instances, but that any box that does not contain a negative instance can contain at most k^N positive instances. It follows that there must be at least $(1+1/k)^N$ boxes to cover all the positive instances and thus the number of bits communicated is at least the logarithm of this number, and the theorem follows.

To count the number of positive instances, note that any partition of the N items between the k players and "unallocated" provides a positive instance. The number of such partitions is exactly $(k+1)^N$.

Now consider a box $S_1 \times S_2 \times ... \times S_k$ that does not contain any negative instance. Note that for each item $i \in \{1...N\}$ there must exist a player $j = j_i$ such that for all $A \in S_j$, $i \notin A$. (Since otherwise we choose for all j, $A_j \in S_j$ with $i \in A_j$ and we have that $\cap_j A_j \neq \emptyset$ – a negative instance in the box.) We can obtain an upper bound to the number of positive instances in this box by noting that each such positive instance corresponds to a partition of the N items among the k players and "unallocated", but now with the added restriction that each item i can not be in the partition of j_i. Thus each item has only k possible locations for it and the number of such partitions is at most k^N.

2.3   Set Packing Lower Bound

We have k players, each holding a collection $A_i \subseteq P(\{1...n\})$. We will prove a lower bound on the communication complexity needed in order to distinguish between the case where the packing number is 1 and the case that it is k. I.e. to distinguish the case where there exists k disjoint sets $S_i \in A_i$ (a k-packing), and the case where any two sets $S_i \in A_i$ and $S_j \in A_j$ intersect (packing number is 1). We will reduce this problem from the approximate-intersection problem on vectors of size $t = e^{n/(2k^2)}/k$. The reduction uses a set of partitions $F = \{P^s | s = 1...t\}$, where each P^s is a partition $(P_1^s, ..., P_k^s)$ of $\{1...n\}$ into k subsets. This set of partitions will have the following property:

Definition 1. *A set of partitions* $F = \{P^s | s = 1...t\}$ *has the pair wise-intersection property if for every choice of* $1 \leq i \neq j \leq k$ *and every* $1 \leq s_i \neq s_j \leq t$ *we have that* $P_i^{s_i} \cap P_j^{s_j} \neq \emptyset$. *I.e. that any 2 parts from different partitions intersect.*

Lemma 1. *There exists a set* F *of partitions with the pair wise-intersection property of size* $|F| = t = e^{n/(2k^2)}/k$.

Proof. We will use the probabilistic method where each partition in the set will be chosen independently at random, and for each partition, each element is placed independently at random in one of the parts of the partition. Now fix $1 \leq i \neq j \leq k$ and two indices of partitions $1 \leq s_i \neq s_j \leq t$. The probability that they do not intersect can be calculated as:

$$Pr[P_i^{s_i} \cap P_j^{s_j} = \emptyset] = (1 - 1/k^2)^n \leq e^{-n/k^2}$$

Since there are at most $k^2 t^2$ such choices of indices, we get that as long as $k^2 t^2 < e^{n/k^2}$, the required set of partitions exists.

We can now specify the reduction of approximate disjointness on vectors of size t to the approximate set packing problem. Player i who gets as input the set $B_i \subset \{1...t\}$ will construct the collection $A_i = \{P_i^s | s \in B_i\}$. Now, if there exists $s \in \cap_i B_i$ then a k-packing exists: $P_1^s \in A_1...P_k^s \in A_k$. If, one the other hand, for all $i \neq j$, $B_i \cap B_j = \emptyset$, then for any two sets $P_i^{s_i} \in A_i$ and $P_j^{s_j} \in A_j$ we have $s_i \neq s_j$ and thus $P_i^{s_i} \cap P_j^{s_j} \neq \emptyset$.

From the $\Omega(N/k^4)$ lower bound for the approximate disjointness problem, it follows that the communication complexity of the approximate set packing problem is at least $\Omega((e^{n/(2k^2)}/k)/k^4) = \Omega(e^{n/(2k^2)-5\log k})$. We thus get:

Theorem 3. *Any k-player protocol for approximating the packing number to within a factor less than k requires $e^{n/(2k^2)-5\log k}$ communication. The lower bound holds also for randomized or nondeterministic protocols.*

In particular, as long as $k < n^{1/2-\epsilon}$, the communication complexity is exponential.

3 Set Cover

3.1 Upper Bound

We are in a situation that each of k players holds a collection $A_i \subseteq P(\{1...n\})$, and they want to approximately find a minimal cover. We use the well known greedy method: At each stage each player i announces a set $S_i \in A_i$ that covers the largest number of remaining elements (this takes n bits for each of the k players for each round), and the largest set among all players is chosen. The elements that are contained in this set are removed, and a new stage starts, now considering only the remaining elements. The protocol ends when all items have been covered – after at most n stages. Thus the total amount of communication is $O(kn^2)$. It is well known [6] that this procedure produces a $(\ln n + 1)$-approximation to the minimum cover.

Note: The weighted case (where each player has a valuation function v_i : $P(\{1...n\}) \rightarrow R^+$, where for all $S, T \subset \{1...n\}$ we have that $v(S \cup T) \leq v(S) + v(T)$, and the aim is to *minimize* $\sum_i v_i(S_i)$ over all covers $\{S_1...S_k\}$) can also be approximated this way to within the same factor, by choosing, at each stage the set S_i that minimizes $v_i(S_i)/|S_i|$ [3].

3.2 Lower Bound

Let Alice hold a collection $A \subseteq P(\{1...n\})$, and Bob hold a collection $B \subseteq P(\{1...n\})$. We will show that it requires exponential communication to separate between the case that for some $S \in A$ we also have $S^c \in B$ (in which case the cover number is 2), and the case where no r sets cover $\{1...n\}$ (for some r to be chosen).

Let us restrict both A to include only sets from a collection $C \subseteq P(\{1...n\})$ to be constructed below, and B to include only sets whose complement is in C. The class C will have the following property:

Definition 2. *A class C has the r-covering property if every collection of at most r sets who either are from C or whose complement is from C, does not cover all of $\{1...n\}$ as long as a set and its complement are not taken together. I.e. for all pair-wise different $S_1...S_r \in C$, and all $0 \leq j \leq r$ we have that $S_1 \cup ... \cup S_j \cup S^c_{j+1} \cup ... \cup S^c_r \neq \{1...n\}$.*

This property was also used in [7], but we use different parameters.

Lemma 2. *For any given $r \leq \log n - O(\log \log n)$, a set C of size t satisfying the r-covering property exists with $t = e^{n/(r2^r)}$.*

Proof. We will use the probabilistic method choosing sets $S_1...S_t$ to put in C at random, where each element $1 \leq i \leq n$ appears in each set S_j with probability $1/2$. For a random collection of r sets, and a single element i, we have that i is in their union with probability $1 - 2^{-r}$. The probability that the union of a given collection of r sets is all of $\{1...n\}$ is thus $(1 - 2^{-r})^n \leq e^{-n/2^r}$. The probability that some collection of r sets from among the t sets in C and the t set whose complement is in C is thus bounded from above by $\binom{2t}{r}e^{-n/2^r} \leq t^r e^{-n/2^r}$. Thus as long as $t < |C| = e^{n/r2^r}$ such a set exists.

At this point we can reduce the r-approximate covering problem from the disjointness problem on a vector of size t. Alice who is given a vector $x \in \{0,1\}^t$ will construct the collection $A = \{S_i | x_i = 1\}$. Bob who is given $y \in \{0,1\}^t$ will construct the collection $B = \{S^c_i | y_i = 1\}$. Now notice that if for some i, $x_i = 1 = y_i$ then the cover number is exactly 2 (given by S_i and S^c_i). Otherwise, the cover number will be at least $r+1$ since C has the r-covering property. Since the disjointness problem on vectors of length t has communication complexity $\Omega(t)$ [12], even for nondeterministic or randomized protocols [9] we get:

Theorem 4. *Any protocol that approximates the cover number to within a factor less than $r/2$ requires communication $\Omega(e^{n/(r2^r)})$. The lower bound holds also for randomized or nondeterministic protocols.*

In particular for any $r \leq (1/2 - \epsilon) \log n$, the communication complexity of r-approximation must be exponential.

Acknowledgements. The problems discussed here stem from my joint work with Ilya Segal on the communication compelxity of combinatorial auctions [8]. I thank Ziv Bar-Yosef for pointing my attention to [2], Jaikumar Radhakrishnan and Venkatesh Srinivasan for allowing me to include their proof of theorem 2, and Amir Ronen for comments on an earlier draft.

References

1. R. Aharoni, P. Erods, and N. Linial. Optima of dual integer linear programs. *Combinatorica*, 8, 1988.

2. N. Alon, Y. Matias, and M. Szegedy. The space complexity of approximating the frequency moments. *Journal of Computer and System Sciences*, 58(1):137–137, 1999.

3. V. Chvatal. A greedy heuristic for the set covering problem. *Math. Operations Reserach*, 4:233–235, 1979.

4. Johan Hastad. Clique is hard to approximate to within $n^{1-\epsilon}$. *Acta Mathematica*, 182, 1999.

5. Daniel Lehmann, Liadan Ita O'Callaghan, and Yoav Shoham. Truth revelation in rapid, approximately efficient combinatorial auctions. In *1st ACM conference on electronic commerce*, 1999.

6. L. Lov'asz. The ratio of optimal integral and fractional covers. *Discrete Mathematics*, 13, 1975.

7. Carsten Lund and Mihalis Yannakakis. On the hardness of approximating minimization problems. In *Proceedings of the Twenty-fifth Annual ACM Symposium on Theory of Computing*, pages 286–293, 1993.

8. Noam Nisan and Ilya Segal. The communication complexity of efficient allocation problems, 2001. Preliminary version available from http://www.cs.huji.ac.il/ noam/mkts.html.

9. A. A. Razborov. On the distributional complexity of disjointness. In *ICALP*, 1990.

10. T. Sandholm, S. Suri, A. Gilpin, and Levine D. Cabob: A fast optimal algorithm for combinatorial auctions. In *IJCAI*, 2001.

11. Rakesh Vohra and Sven de Vries. Combinatorial auctions: A survey, 2000. Availailabe from http://www.kellogg.nwu.edu/faculty/vohra/htm/res.htm.

12. Andrew Chi-Chih Yao. Some complexity questions related to distributive computing. In *ACM Symposium on Theory of Computing*, pages 209–213, 1979.

13. Edo Zurel and Noam Nisan. An efficient approximate allocation algorithm for combinatorial auctions. In *ACM conference onelectronic commerce*, 2001. Available from http://www.cs.huji.ac.il/ noam/mkts.html.

Antirandomizing the Wrong Game

Benjamin Doerr

Mathematisches Seminar II, Christian-Albrechts-Universität zu Kiel,
Ludewig-Meyn-Str. 4, D-24098 Kiel, Germany,
bed@numerik.uni-kiel.de,
http://www.numerik.uni-kiel.de/~bed/

Abstract. We study a variant of the tenure game introduced by
J. Spencer (Theor. Comput. Sci. 131 (1994), 415–429). In this version,
chips are not removed from the game, but moved down to the lowest
level instead. Though the rules of both versions differ only slightly, it
seems impossible to convert an upper bound strategy into a lower bound
one using the antirandomization approach of Spencer (which was very
effective for the original game and several others).

For the upper bound we give a potential function argument (both
randomized and derandomized). We manage to prove a nearly matching
lower bound using a strategy that can be interpreted as an antirandomization of Spencer's original game.

Keywords: Games, randomization, derandomization.

1 Introduction

Since the increasing interest in on-line problems at the latest, game theory has
gained attraction in theoretical computer science. In this paper we work on so-
called Pusher-Chooser games. These are two player perfect information games
where each round the player called 'Pusher' splits the position into two alternatives and 'Chooser' selects one thereof. Hence the theme of these games is
on-line balancing: Pusher has to find a balanced split (in the sense that neither
alternative is too favorable to Chooser), whereas Chooser tries to detect and
exploit such imbalances. Examples of such games are vector balancing games
([2,3,6,7,9,10]) and liar games. Concerning the latter, we refer to the survey [8]
its extensive bibliography of 120 references. Internet routing problems gave rise
to the related "guessing secrets" problem, that attracted attention recently ([1,
4,5]).

In his marvelous paper "Randomization, Derandomization and Antirandomization: Three Games", Spencer [11] shows a generic method to convert a random
strategy for Chooser in such a game into a deterministic algorithm. Moreover,
he also provides a method coined 'antirandomization' that produces a matching
lower bound constructively, i.e., including a strategy for Pusher.

The game Spencer demonstrated these methods most easily is the *Tenure
game*. We cite the rules from [11]:

P. Widmayer et al. (Eds.): ICALP 2002, LNCS 2380, pp. 876–887, 2002.

The tenure game is a perfect information game between two players, Paul — chairman of the department — and Carole — dean of the school. An initial position is given in which various faculty are at various pre-tenured positions. Paul will win if some faculty member receives tenure — Carole wins if no faculty member receives tenure. Each year Chair Paul creates a promotion list L of the faculty and gives it to Dean Carole who has two options: (1) Carole may promote all faculty on list L one rung and simultaneously fire all other faculty. (2) Carole may promote all faculty *not* on list L one rung and simultaneously fire all faculty on list L.

In this paper we study a slight variant of the Tenure game introduced by Spencer. We will assume that not-promoted faculty is not fired, but downgraded to the first rung instead. There are two reasons to investigate this game. In [6] we showed that good strategies for this game yield good strategies for the on-line vector balancing problem with aging, i.e., where decisions in the past become less important compared to the actual one. For reasons of space we just refer to the paper for this aspect.

Our main motivation is that this variant — though similar to the original game and clearly a Pusher-Chooser game — seemingly does not admit anti-randomization. Since the antirandomization method is a very powerful way to convert a randomized or derandomized strategy for Chooser into a matching strategy for Pusher, it is particularly interesting to investigate what happens if it cannot be applied. This work suggests two answers: First, things can become quite complicated without antirandomization, second, sometimes using the antirandomization of a different game can help.

Let us state the rules precisely. Whether Carole can win or not of course depends on the number of rungs we have. To remove this parameter without losing information about the game, we assume to have infinitely many rungs and play an optimization version of the game: The highest rung reached by some faculty member is called the pay-off for Paul. Naturally, he tries to maximize this pay-off, whereas Carole tries to minimize it. To further simplify the setting we assume that all faculty is on the first rung at the beginning of the game. Exchanging people by innocent chips and baptizing this version 'European Tenure Game', we have:

Rules of the European Tenure Game: The game is played with a fixed number d of chips which lie on levels numbered with the positive integers. At the start of the game, all d chips are on level one. The game is a two-player perfect information game. The first player, called 'Pusher', tries to get a chip to a possibly high level. The maximum level ever reached by a chip during the game is called pay-off to Pusher. Each round Pusher chooses a subset of chips he proposes to be promoted. If the second player ('Chooser') accepts, then these chips each move up one level, and the remaining chips are moved down to the first level. If Chooser rejects, the the remaining chips move up one level, and Chooser's choice is downgraded to level one. The game ends if pusher is satisfied with the position reached or a position reoccurs.

From the rules it is already clear that Pusher has some advantage in the European Tenure Game compared to Spencer's original game, which we shall call 'American Tenure Game'.

For the American Tenure Game, Spencer gave a complete solution even for arbitrary starting positions (which we do not regard in this paper). If the game is played with d chips, then the value of this game, i.e., the maximum level Pusher can reach, is $\lfloor \log d \rfloor + 1$, where $\log(\cdot)$ shall always denote the dyadic logarithm.

The European Tenure Game seems to be more difficult to analyze. Using straightforward reasoning, a bound of $\lfloor \log d \rfloor + \lfloor \log \log d \rfloor \leq v_d \leq \log d + \log \log d + 4$ for the value v_d of this game was shown in [6], where the game appeared first in a reduction of a vector balancing problem. In this paper we make some progress towards a full understanding of the game.

For the general case, we reduce the gap between lower and upper bound, so that there at most three possibilities for each d. For larger d, the gap reduces to at most two values, though we are able to determine a precise value for a set having positive lower density. To prove the lower bound we analyze a strategy that seems to be an antirandomization of the American Tenure Game strategy.

Theorem 1. *Let v_d denote the value of the European Tenure Game played with d chips. Then*

$$\lfloor \log d + \log \log d \rfloor \leq v_d \leq \lfloor \log d + \log \log d + 1.98 \rfloor$$

holds for all d. For d tending to infinity, these bounds improve to

$$\lfloor \log d + \log \log d + 1 + o(1) \rfloor \leq v_d \leq \lfloor \log d + \log \log d + 1.73 + o(1) \rfloor .$$

In particular, the set $S = \{d \in \mathbb{N} \,|\, v_d = \lfloor \log d + \log \log d + 1 \rfloor\}$ has lower density greater than $\frac{1}{5}$.[1]

2   Upper Bound: Chooser's Strategy

Let us assume $d \geq 3$ since smaller cases are trivial. We describe a position of the game by a function $P : \mathbb{N} \to \mathbb{N}_0$ such that $\sum_{i \in \mathbb{N}} P(i) = d$. Hence $P(i)$ denotes the number of chips on level i.

For the upper bound we are guided by [6]. Let $\lambda := \frac{2 \log d - 1}{\log d}$. Define a potential function v by $v(P) := \sum_{i \in \mathbb{N}} P(i) \lambda^{i-1}$ for all positions $P : \mathbb{N} \to \mathbb{N}_0$. We analyze the strategy for Chooser to choose that one of the alternatives which minimizes $v(P)$ for the resulting position. An easy induction shows that this ensures $v(P) \leq d \log d$ for all positions P occurring in the game, which in turn yields an upper bound for the value of the game.

---

[1] The lower density $\underline{d}(S)$ of a set $S \subseteq \mathbb{N}$ is $\underline{d}(S) := \liminf_{n \to \infty} \frac{1}{n} |\{s \in S \,|\, s \leq n\}|$. Roughly speaking, the last paragraph of the theorem states that we know the precise value of the game for more than a fifth of the values for d.

Lemma 1. *The value of the game played with d chips is at most*

$$\frac{\log\left(d\log d - d + 1\right)}{\log\left(2 - \frac{1}{\log d}\right)} + 1 \leq \log d + \log\log d + 1.73 + o(1).$$

Proof. Clearly we have $v(P) \leq d\log d$ for the starting position. Now let P be an arbitrary position of the game such that $v(P) \leq d\log d$. Denote by P_1, P_2 the two positions resulting from either accepting or rejecting Pusher's choice. Then

$$\min\{v(P_1), v(P_2)\} \leq \tfrac{1}{2}(v(P_1) + v(P_2))$$
$$= \tfrac{1}{2}(d + \lambda v(P))$$
$$\leq \tfrac{1}{2}(d + \tfrac{2\log d - 1}{\log d}d\log d)$$
$$= d\log d.$$

Hence Chooser's strategy of minimizing $v(P)$ ensures that $v(P) \leq d\log d$ holds throughout the game.

Let P be any position such that $v(P) \leq d\log d$. Let l denote the level of the highest-ranking chip. Since the remaining $d-1$ chips at least are on level one[2], we have

$$\lambda^{l-1} \leq d\log d - d + 1.$$

Hence

$$l \leq \log_\lambda(d\log d - d + 1) + 1 = \frac{\log\left(d\log d - d + 1\right)}{\log\left(2 - \frac{1}{\log d}\right)} + 1.$$

For $d \geq 3$, the latter term is strictly less than $\log d + \log\log d + 1.98$. For d tending to infinity, our bound becomes stronger and is optimal for quite a portion a values (as we will prove in the next section). We have

$$\frac{\log\left(d\log d - d + 1\right)}{\log\left(2 - \frac{1}{\log d}\right)} + 1 = \log d + \log\log d + 1 + \frac{1}{2\ln 2} + o(1).$$

For our purposes, the upper bound suffices.

Put $l = \log d$. Then $\log(2 - 1/l) = 1 + \log(l - 1/2) - \log l \geq 1 - \tfrac{1}{2}\frac{1}{(l-1/2)\ln 2}$, as the logarithm is concave. Thus

$$\frac{\log\left(d\log d - d + 1\right)}{\log\left(2 - \frac{1}{\log d}\right)} \leq \frac{l + \log l}{\log(2 - 1/l)}$$

$$\leq \frac{l + \log l}{1 - 1/(2(l - \tfrac{1}{2})\ln 2)}$$

$$= l + \log l + \frac{1}{2\ln 2} + \frac{\log l}{2l\ln 2 - 1 - \ln 2} + \frac{1 + \ln 2}{2\ln 2(2l\ln 2 - 1 - \ln 2)}.$$

$\square$

---

[2] This seems to be a negligible advantage. For large d in fact it is, but for smaller values this is enough to reduce the upper bound from $l + \log l + 4$ to $l + \log l + 1.98$.

Above we gave a deterministic strategy. Assume now that Chooser plays randomly, i.e., he flips a fair coin to decide which of the two alternatives to take. Then a similar argument as above shows that the expected v–value is bounded by $d \log d$, no matter what strategy Pusher is playing. Since the American Tenure Game is a finite perfect information game, we deduce that Chooser actually has a strategy ensuring $v(P) \leq d \log d$ throughout the game. Moreover, the one we proposed first is just the derandomization of this randomized proof.

What makes this game interesting, is that the corresponding antirandomization does not work.

3   Lower Bound: Pusher's Strategy

The antirandomization paradigm of Spencer's paper [11] would advise Pusher to match Chooser's strategy this way: "Play each round such that the outcome of both alternatives to Chooser has the same potential $v(\cdot)$. Thus Chooser can never gain an advantage."

For several reasons this does not work. Firstly, there is not analogue to the splitting lemma in [11]. Thus in general it is not possible to split the position into nearly equally valued (with respect to v) alternatives. Secondly, the starting position does not have a potential of $d \log d$, but only of d. Thus equally valued splits, even if they existed, would not be enough. A third point is that the potential function does not yield a good lower bound: A high potential $v(P)$ does not guarantee that there is a chip on a high level. To some extent it does, but these bounds are not strong enough. The reason is that the potential function v values chips on a lower level slightly higher than Spencer's potential function for the American Tenure Game. During play this is justified by the fact that these chips gain from the advantage of not being fired, but just downgraded. At the end of the game, this advantage does not exist anymore.

For these reasons an antirandomization argument corresponding to our upper bound strategies seems not to work. What does work, however, is the antirandomization of the American Tenure Game. The prize for using a non-corresponding antirandomization is that the proofs are more complicated.

3.1   Strategies for the American Tenure Game

At this point let us shortly review Chooser's derandomized strategy in the American Tenure Game. Recall that we described a position of the game by a function $P : \mathbb{N} \to \mathbb{N}_0$ such that $\sum_{i \in \mathbb{N}} P(i) = d$. Chooser can follow the strategy to choose that one of the alternatives that minimizes $f(P) = \sum_{i \in \mathbb{N}} P(i) 2^i$ for the resulting position. An easy calculation shows that this way the f–value (f–potential) of the position cannot increase during play. Thus it never exceeds the initial value of $2d$, and it is clear that no chip can reach a higher level than $\log d + 1$.

The corresponding antirandomization yields this strategy for Pusher: He always chooses a split which maximizes the minimum f–potential among the two alternatives to Chooser. Thus he maximizes the f–potential of the resulting position regardless of Chooser's move. For the American Tenure Game it can be shown that playing this way the f–potential can never drop below $2^{\lfloor \log d \rfloor}$, which matches the upper bound. Crucial for this result is the so-called splitting lemma.

In the following, we analyze this same strategy of maximizing the f–potential for Pusher in the European Tenure Game. The fact that this is not an antirandomization of our strategy for Chooser makes the proofs somewhat harder, but nevertheless we end up with a near-tight bound.

3.2 First Phase

The case $d = 2$ is solved by a moment's thought, so let us assume $d \geq 3$. We assume first that $d = 2^l$ is a power of 2 and deal with the general case at the end of this section. We shortly review the result in [6], as we use this as first part of our strategy.

It is clear that Pusher can change a position P such that all $P(i)$ are even, to the position P' defined by $P'(1) = \frac{1}{2}d$ and $P'(i+1) = \frac{1}{2}P(i)$ for all $i \in \mathbb{N}$. All pusher has to do is to select half of the chips of each level. Then, regardless of Chooser's choice, he ends up with position P'. We call this procedure an 'easy split'.

From the starting position with $d = 2^l$ chips on the first level, Pusher can do l easy splits and reach a position P with $P(i) = 2^{l-i}$ for all $i = 1, \ldots, l$, with $P(l+1) = 1$ and $P(i) = 0$ for $i \geq l+2$. Doing so was part of the strategy in [6], and will be part of ours as well. The interesting point is how to continue from this position. In [6] we gave an explicit strategy moving one chip up to level $l + \lfloor \log l \rfloor$. Spencer (private communication) noted that the position P has an f–potential of $(l+2)d$. Thus a pay-off of $l + \lfloor \log (l + 2) \rfloor$ can already be obtained in the American Tenure Game, which — as noted above — is less favorable for Pusher in the sense that he can get at most the same pay-off as in the European Tenure Game.

3.3 Second Phase

Guided by Spencer's observation, we now continue with a strategy that maximizes f. In the remainder of this paper, we will call $f(P)$ simply the potential of P omitting the f. Since in Phase 1 a greedy strategy of maximizing the function f was successful, one might be tempted to continue this. As each level has potential d (except level $l+1$, which has potential $2d$), it is not too difficult to split the levels into two parts having equal potential.[3] Thus the surviving part carries the whole potential (recall that moving up doubles the potential of a chip), and we gain a potential of 2 for each chip that is downgraded. We can continue this roughly $\log l$ times. If, while doing so, we partition the downgraded chips evenly, we can gain an extra potential of roughly $d \log l$. Since we needed roughly dl extra to prove our main result, we are not done yet.

---

[3] From the rules of the tenure games it is clear that it makes no difference whether Pusher proposes some set of chips or its complement. Therefore we may view any Pusher move simply as partition of the set of chips into two classes.

The problem is that having played this way, we might end up with one chip on level $l + \log l$ holding most of the potential of the whole position. Hence Chooser will downgrade this chip in his next move, and all our clever gains are gone.

The solution is modesty. Of course, we cannot prevent the chip on level $l + 1$ to move up to $l + \log l$ in $\log l - 1$ moves. Chooser can enforce this by simply downgrading that part of the chips that does not contain this highest-ranking one. Therefore, we partition the chips into classes having different potential: The one containing the highest-ranking chips has a that large potential, that we are immediately satisfied if it survives (ending with a potential of at least $2dl$). On the other hand, if the 'lower class' chips survive, we gain only little potential (an additional d), but end up with a flexible position (in particular having no too high-ranking chips, and allowing a similar step again). Here are the details:

We call the position

$$
P_k : \mathbb{N} \to \mathbb{N}_0; i \mapsto \begin{cases} 2^{l-i} & \text{if } i < k \\ 1 & \text{if } i = k \\ 2^{l+1-i} & \text{if } k < i \le l+1 \\ 0 & \text{otherwise.} \end{cases}
$$

a *logarithmic ladder with gap at level k*. Further on, we define for all $0 \le j < k$

$$
Q_{k,j} : \mathbb{N} \to \mathbb{N}_0; i \mapsto \begin{cases} d(1 - 2^{-j}) & \text{if } i = 1 \\ 2^{l+1-i} & \text{if } j+2 \le i \le k \\ 1 & \text{if } i = k+1 \\ 2^{l+2-i} & \text{if } k+2 \le i \le l+2 \\ 0 & \text{otherwise.} \end{cases}
$$

Lemma 2. *From a logarithmic ladder with gap P_k, Pusher can enforce for any $j < k$ one of the positions P_{l+1-j} and $Q_{k,j}$. In particular, he can advance from P_k to one of P_{k-1} and $Q_{k,l+2-k}$, if $k > l/2 + 1$.*

Proof. In position P_k, Pusher chooses those chips that have level at most j. If Chooser rejects, these $d(1 - 2^{-j})$ chips move down to level one, the remaining move up one level and position $Q_{k,j}$ is reached. Hence suppose that Chooser accepts. Then $d2^{-j} = 2^{l-j}$ chips move to level one, and the other chips move up one level. Now the number of chips on each level is a multiple of 2^{l-j}. Thus Pusher can play $l - j$ easy splits and reach position P_{l+1-j}. The second claim follows from the first by choosing $j = l + 2 - k$. □

Lemma 3. *For all $0 \le j < k \le l+1$, we have*

$$
f(P_k) = d(2l - k + 1) + 2^k,
$$
$$
f(Q_{k,j}) = d(4l - 2(k+j) + 4) + 2^{k+1} - 2^{l-j+1}).
$$

The proof is a simple calculation. Note that the levels of P_k below the gap (except the first one) each contribute d to the potential, whereas those above contribute $2d$.

From what we showed so far we already get a first lower bound:

Lemma 4. *For any $\lceil (l+1)/2 \rceil \leq s \leq l+1$, Pusher has a strategy enforcing one of the positions $Q_{k,l+2-k}$ for $k = s+1, \dots, l$, and P_s. For $s = \lceil (l+1)/2 \rceil$, this strategy yields a potential of at least $1.5d \log d$, and thus a lower bound for the value of the game of $\lfloor \log d + \log \log d + 0.58 \rfloor$.*

Proof. From the starting position with 2^l chips on level one, Pusher does l easy splits and reaches position $P_l = P_{l+1}$ (this is Phase 1). Once in Position P_i for some $l \geq i \geq s+1$, he applies Lemma 2 with $j = l+2-i$ and reaches $Q_{i,j}$ or P_{i-1}. This proves the statement concerning the possible positions. With $s = \lceil (l+1)/2 \rceil$, the bound for the value of the game follows directly from Lemma 3 and the discussion of the American Tenure Game.   □

Since the positions $Q_{k,l+2-k}$ all have a potential of more than $2dl$, the lower bounds of Lemma 4 just depend on the potential of $P_{\lceil (l+1)/2 \rceil}$ of about $\frac{3}{2}dl$. We therefore continue Pusher's strategy on this position.

3.4   Third Phase

The reason why we could not continue applying Lemma 2 is that the gap k and the position j where Pusher splits the levels would meet. Splitting the levels above the gap leads to slightly more complicated positions having two gaps. For $0 < r < s \leq l+2$, we define

$$
P_{r,s} : \mathbb{N} \to \mathbb{N}_0; i \mapsto
\begin{cases}
2^{l-i} & \text{if } i < r \\
1 & \text{if } i = r \\
2^{l+1-i} & \text{if } r < i < s \\
0 & \text{if } i = s \\
2^{l+2-i} & \text{if } s < i \leq l+2 \\
0 & \text{otherwise.}
\end{cases}
$$

We also need for $0 < r < j < s \leq l+2$

$$
Q_{r,s,j} : \mathbb{N} \to \mathbb{N}_0; i \mapsto
\begin{cases}
d(1 - 2^{-j+1}) & \text{if } i = 1 \\
1 & \text{if } i = r+1 \\
2^{l+2-i} & \text{if } j+2 \leq i < s+1 \\
0 & \text{if } i = s+1 \\
2^{l+3-i} & \text{if } s+1 < i \leq l+3 \\
0 & \text{otherwise.}
\end{cases}
$$

Again we compute their potentials:

Lemma 5.

$$
f(P_{r,s}) = d(4l - r - 2s + 5) + 2^r
$$
$$
f(Q_{r,s,j}) = d(8l - 4s - 4j + 14) + 2^{r+1} - 2^{l-j+2}.
$$

The following lemma shows that also logarithmic ladders with two gaps allow comprehensible strategies.

Lemma 6. *Let $0 < r < s \le l + 2$. For any j such that $r < j < s$, Pusher can advance position $P_{r,s}$ to one of $P_{l+2-j,l+r+2-j}$ and $Q_{r,s,j}$.*

Proof. Pusher chooses all chips on level at most j except the single chip on level r. If Chooser rejects, we are immediately in position $Q_{r,s,j}$. Otherwise, 2^{l-j+1} chips move to level one and Pusher's choice moves up one level. As all levels hold a multiple of 2^{l-j+1} chips, Pusher can play $l - j + 1$ easy splits and reach position $P_{l+2-j,l+r+2-j}$. □

Using Lemma 6, we apply a modesty strategy again: By Lemma 4 (and one extra step if l is odd), we reach $P_{(l+1)/2,l+1}$ or $P_{(l+2)/2,l+2}$. Once in position $P_{x,2x}$ for some $x \in [\lfloor (x + 7)/3 \rfloor, \lfloor (l + 2)/2 \rfloor]$, Pusher slowly increases the potential through the position $P_{x-1,2x-1}$ to $P_{x-1,2(x-1)}$. The first step increases the potential by roughly $3d$, the second by $2d$. If Chooser tries to foil this strategy, he immediately ends up with a Q–position having a potential of roughly $2dl$. Apart from a few small cases, this leads to a potential greater than $2dl$.

Lemma 7. *In the European Tenure Game played with $d = 2^l$ chips, Pusher can reach one of the positions*

- *$Q_{k,l+2-k}$ for $k = \lceil (l+1)/2 \rceil + 1, \ldots, l$,*
- *$Q_{x,2x,l+3-x}$ for $x = \lfloor (l + 7)/3 \rfloor, \ldots, \lfloor (l + 2)/2 \rfloor$,*
- *$Q_{x-1,2x-1,l+3-x}$ for $x = \lfloor (l + 7)/3 \rfloor, \ldots, \lfloor (l + 3)/2 \rfloor$,*
- *$P_{\lfloor (l+4)/3 \rfloor, 2\lfloor (l+4)/3 \rfloor}$.*

In consequence, Pusher can reach a potential of more than $2d(l - 1)$.

Proof. Applying Lemma 4 with $s = \lceil (l+1)/2 \rceil$, Pusher can get one of the positions $Q_{k,l+2-k}$ for $k = \lceil (l+3)/2 \rceil, \ldots, l$, or $P_{\lceil (l+1)/2 \rceil}$. Note that $P_{\lceil (l+1)/2 \rceil} = P_{\lceil (l+1)/2 \rceil, l+2}$.

If l is odd, we apply Lemma 6 with $j = \frac{l+1}{2} + 1$ and end up with either $Q_{(l+1)/2,l+2,(l+3)/2}$ (which is $Q_{x-1,2x-1,l+3-x}$ for $x = \lfloor (l + 3)/2 \rfloor$) or $P_{(l+1)/2,l+1}$. If l is even, our actual position is $P_{(l+2)/2,l+2}$.

The rest is an easy induction: Assume that we are in position $P_{x,2x}$ for some $x = \lfloor (l + 7)/3 \rfloor, \ldots, \lfloor (l + 2)/2 \rfloor$. Note that this implies $l \ge 4$. Applying Lemma 6 with $j = l + 3 - x$ on this position, we get $Q_{x,2x,l+3-x}$ or $P_{x-1,2x-1}$. Applying Lemma 6 on the latter with $j = l + 3 - x$ again, we reach position $Q_{x-1,2x-1,l+3-x}$ or $P_{x-1,2(x-1)}$. This proves the claim concerning the reachable positions.

For the potentials we look up in Lemma 3 and 5 and compute:

$$f(Q_{k,l+2-k}) = 2dl + 3 \cdot 2^{k-1},$$
$$f(Q_{x,2x,l+3-x}) = 4d(l - x) + 2d + 3 \cdot 2^{x-1},$$
$$f(Q_{x-1,2x-1,l+3-x}) = 4d(l - x) + 6d + 2^{x-1},$$
$$f(P_{\lfloor (l+4)/3 \rfloor, 2\lfloor (l+4)/3 \rfloor}) = d(4l - 5 \lfloor (l + 4)/3 \rfloor + 5) + 2^{\lfloor (l+4)/3 \rfloor}.$$

□

Note that all potentials above except the one of $Q_{x,2x,l+3-x}$ for $x = l/2 + 1$ and even $l \geq 4$ are at least $2dl$. We may remark that with some more effort one could avoid these exceptional cases and show a lower bound of $2dl$.

3.5   If d Is Not a Power of 2

So far we assumed that d is a power of two. Since we may always ignore some of the chips in our play, this immediately yields bounds for the general case as well. As we ignore less than half the chips, our loss is not very big. For the value of the game, we just lose an additive term of $1 + o(1)$. Unfortunately, our upper and lower bounds are already that close that such a loss is significant.

A first idea would be to partition the chips into subsets of cardinalities of powers of two, and then play the above strategies on each separately. It is a problem though to synchronize the strategies. It might happen (and Pusher cannot prevent this) that one subset already reached a Q–position ending the strategy, while another set is in the middle of a series of easy splits. To make this approach work, we would need a way to conserve the potential of a favorable position like a Q–position for several moves. This seems to be a difficult task.

Fortunately, an easy trick solves the problem and shows that the general case is not far away from the special case of powers of 2.

Lemma 8. *Let $i \in \mathbb{N}$ such that $2^i \leq d$. Then Pusher can earn a potential of $2 \cdot 2^i(i-1) \lfloor d/2^i \rfloor$. In consequence, Pusher has a strategy ensuring him a potential of at least $2d \log d(1 + o(1))$.*

Proof. Let $d_0 = \lfloor d/2^i \rfloor 2^i$, the largest multiple of 2^i not exceeding d. This is Pusher's strategy: He plays with d_0 chips only, ignoring the rest. The set of d_0 non-ignored chips is partitioned into $\lfloor d/2^i \rfloor$ groups of 2^i chips each. These groups will never be split in the course of the game, so we may assume these chips to be glued together forming 'big chips'. There are 2^i big chips, hence Pusher can follow his strategy for powers of 2 and ending up with a position of potential $2 \cdot 2^i(i - 1)$ in terms of big chips. Since each big chip consists of $\lfloor d/2^i \rfloor$ ordinary ones ('solving the glue again'), this position has a potential of $2 \cdot 2^i i \lfloor d/2^i \rfloor$.

Put $l := \log d$ and $i = \lfloor l - \log l \rfloor$. Then

$$2 \cdot 2^i(i - 1) \lfloor d/2^i \rfloor \geq 2(d - 2^i)(l - \log l - 2)$$
$$= 2dl(1 - 1/l)(1 - (\log l + 2)/l)$$
$$= 2dl(1 + o(1)).$$

$\square$

From Lemma 1 and 8 we deduce that we know the precise value of the game, namely $\lfloor \log d + \log \log d + 1 \rfloor$, for all sufficiently large d such that the fractional part of $\log d + \log \log d$ is contained in $[0, 0.27[$. Some elementary calculus leads to the conclusion that the set of all d such that we know the precise value of the game has lower density greater than $\frac{1}{5}$.

4   Remarks and Open Problems

An obvious problem left open in this paper is a precise determination of the value of the game for all or all but a few values of d. We only succeeded in doing so for a set of d having lower density $\frac{1}{5}$. For the remaining values apart from finitely many, two possibilities exist for the value of the game.

With quite some effort it is possible to continue Pusher's strategy from the Q–positions and thus gain a potential of γdl for some $\gamma > 2$. Unfortunately, these gains are not too big, in particular, they are not enough to determine the value of the game for asymptotically all numbers d.

More interesting than a slight increase of the set of numbers d such that the value of the game with d chips is determined might be the following: Assume that $d = 2^l$ is a power of two again. Then the proofs in Section 3 give a strategy for Pusher to obtain a potential of about $2dl$. A closer inspection of these proofs shows that Pusher might need more than l^2 moves to reach this aim. This is caused by the strategy which is quite unbalanced in the following sense: Whenever Chooser has two different alternatives, i.e., Pusher did not play an easy split, one of the alternatives immediately produces a potential of $2dl$, whereas the other only gains a modest additional potential of $\Theta(d)$ in up to $l - 1$ easy splits.

We do not know whether a more balanced strategy exists. If Pusher could produce two alternatives gaining an $\Omega(d)$ potential increase in one move (like the easy splits do), this would result in a strategy that needs $O(l)$ moves only.

Acknowledgments. I would like to thank Joel Spencer for several discussions on this topic. I am also grateful to the anonymous referees for their thorough reading.

References

1. N. Alon, V. Guruswam, T. Kaufman, and M. Sudan. Guessing secrets efficiently via list decoding. In *Proceedings of the 13th Annual ACM-SIAM Symposium on Discrete Algorithms*, pages 254–262, 2002.
2. I. Barany. On a class of balancing games. *J. Combin. Theory Ser. A*, 26:115–126, 1979.
3. I. Barany and V. S. Grunberg. On some combinatorial questions in finite dimensional spaces. *Linear Algebra Appl.*, 41:1–9, 1981.
4. F. Chung, R. Graham, and T. Leighton. Guessing secrets. *Electron. J. Comb.*, 8:Research Paper 13, approx. 25 pp. (electronic), 2001.
5. F. Chung, R. Graham, and L. Lu. Guessing secrets with inner product questions. In *Proceedings of the 13th Annual ACM-SIAM Symposium on Discrete Algorithms*, pages 247–253, 2002.
6. B. Doerr. Vector balancing games respecting aging. *J. Combin. Theory Ser. A*, 95:219–233, 2001.
7. J. Olson. A balancing strategy. *J. Combin. Theory Ser. A*, 40:175–178, 1985.
8. A. Pelc. Searching games with errors — fifty years of coping with liars. *Theor. Comput. Sci.*, 270:71–109, 2002.

9. H. Peng and C. H. Yan. Balancing game with a buffer. *Adv. Appl. Math.*, 21:193–204, 1998.
10. J. Spencer. Balancing games. *J. Combin. Theory Ser. B*, 23:68–74, 1977.
11. J. Spencer. Randomization, derandomization and antirandomization: Three games. *Theor. Comput. Sci.*, 131:415–429, 1994.

Fast Universalization of Investment Strategies with Provably Good Relative Returns

Karhan Akcoglu[1]*, Petros Drineas[1]**, and Ming-Yang Kao[2]***

[1] Department of Computer Science, Yale University, New Haven, CT 06520, USA.
{karhan.akcoglu,petros.drineas}@yale.edu
[2] Department of Computer Science, Northwestern University, Evanston, IL 60201,
USA. kao@cs.northwestern.edu

Abstract. A *universalization* of a parameterized investment strategy is an online algorithm whose average daily performance approaches that of the strategy operating with the optimal parameters determined offline in hindsight. We present a general framework for universalizing investment strategies and discuss conditions under which investment strategies are universalizable. We present examples of common investment strategies that fit into our framework. The examples include both trading strategies that decide positions in individual stocks, and portfolio strategies that allocate wealth among multiple stocks. This work extends Cover's universal portfolio work. We also discuss the runtime efficiency of universalization algorithms. While a straightforward implementation of our algorithms runs in time exponential in the number of parameters, we show that the efficient universal portfolio computation technique of Kalai and Vempala involving the sampling of log-concave functions can be generalized to other classes of investment strategies.

1 Introduction

An age-old question in finance deals with how to manage money on the stock market to obtain an "acceptable" return on investment. An *investment strategy* is an online algorithm that attempts to address this question by applying a given set of rules to determine how to invest capital. Typically, an investment strategy is parameterized by a vector $\mathbf{w} \in \mathbb{R}^* = \bigcup_{i=1}^{\infty} \mathbb{R}^i$ that dictates how the strategy operates. The optimal parameters that maximize the strategy's return are unknown when the algorithm is run and the parameters are usually chosen quite arbitrarily. A *universalization* of an investment strategy is an online algorithm based on the strategy whose average daily performance approaches that of the strategy operating with the optimal parameters determined offline in hindsight.

Consider the *constantly rebalanced portfolio* (CRP) investment strategy universalized by Cover [5] and the subject of several extensions and generalizations

* Supported in part by NSF Grant CCR-9988376.
** Supported in part by NSF Grant CCR-9896165.
*** Supported in part by NSF Grant CCR-9988376.

P. Widmayer et al. (Eds.): ICALP 2002, LNCS 2380, pp. 888–900, 2002.

[7,6,12,3,14]. The CRP strategy maintains a constant proportion of total wealth in each stock, where the proportions are dictated by the parameters given to the strategy. In a stock market with m stocks, the parameter space for the CRP strategy is $\mathcal{W}_m = \{\mathbf{w} \in [0,1]^m \mid \sum_{i=1}^{m} w_i = 1\}$, the set of vectors in $\mathbb{R}^m$ whose components are between 0 and 1 and add up to 1. Given a *portfolio vector* $\mathbf{w} = (w_1, \ldots, w_m) \in \mathcal{W}_m$, w_i tells us the proportion of wealth to invest in stock i, for $1 \leq i \leq m$. At the beginning of each day, the holdings are *rebalanced*, *i.e.*, money is taken out of some stocks and put into others, so that the desired proportions are maintained in each stock. As an example of the robustness of the CRP strategy, consider the following market with two stocks [6,12]. The price of one stock remains constant, while the other stock doubles and halves in price on alternate days. Investing in a single stock will at most double our money. With a CRP$(\frac{1}{2}, \frac{1}{2})$ strategy, our wealth will increase exponentially, by a factor of $(\frac{1}{2} \cdot 1 + \frac{1}{2} \cdot 2) \times (\frac{1}{2} \cdot 1 + \frac{1}{2} \cdot \frac{1}{2}) = \frac{3}{2} \times \frac{3}{4} = \frac{9}{8}$ every two days.

Cover developed an investment strategy that effectively distributes wealth uniformly over all portfolio vectors $\mathbf{w} \in \mathcal{W}_m$ on the first day and executes the CRP strategy with daily rebalancing according to each $\mathbf{w}$ on the (infinitesimally small) proportion of wealth initially allocated to each $\mathbf{w}$. Cover showed that the *average daily log-performance*[1] of such a strategy approaches that of the CRP strategy operating with the optimal, return-maximizing parameters chosen with hindsight.

This paper generalizes previous results and introduces a framework that allows universalizations of other parameterized investment strategies. As we see in Section 2, investment strategies fall under two categories; *trading strategies* operate on a single stock and dictate when to buy and short[2] the stock; *portfolio strategies*, such as CRP, operate on the stock market as a whole and dictate how to allocate wealth among multiple stocks. We present several examples of common trading and portfolio strategies that can be universalized in our framework. We discuss our universalization framework in Section 3. The proofs of our results are very general and, as with previous universal portfolio results, we make no assumptions on the underlying distribution of the stock prices; our results are applicable for all sequences of stock returns and market conditions. The running times of universalization algorithms are, in general, exponential in the number of parameters used by the underlying investment strategy. Kalai and Vempala [14] presented an efficient implementation of the CRP algorithm that runs in time polynomial in the number of parameters. In Section 4, we present general conditions on investment strategies under which the universalization algorithm can be efficiently implemented. We also give some investment strategies that satisfy these conditions.

---

[1] The average daily log-performance is the average of the logarithms of the factors by which our wealth changes on a daily basis. This notion is discussed further in Section 3.1.

[2] A short position in a stock, discussed in Section 2.1, allows us to earn a profit when the stock declines in value.

2   Types of Investment Strategies

Suppose we would like to distribute our wealth among m stocks[3]. *Investment strategies* are general classes of rules that dictate how to invest capital. At time $t > 0$, a strategy S takes as input an *environment vector* $\mathcal{E}_t$ and a *parameter vector* $\mathbf{w}$, and returns an *investment description* $S_t(\mathbf{w})$ specifying how to allocate our capital at time t. The environment vector $\mathcal{E}_t$ contains historic market information, including stock price history, trading volumes, *etc.*; the parameter vector $\mathbf{w}$ is independent of $\mathcal{E}_t$ and specifies exactly how the strategy S should operate; the investment description $S_t(\mathbf{w}) = (S_{t1}(\mathbf{w}), \dots, S_{tm}(\mathbf{w}))$ is a vector specifying the proportion of wealth to put in each stock, where we put a fraction $S_{ti}(\mathbf{w})$ of our holdings in stock i, for $1 \leq i \leq m$. For example, CRP is an investment strategy; coupled with a portfolio vector $\mathbf{w}$ it tells us to "rebalance our portfolio on a daily basis according to $\mathbf{w}$"; its investment description, $\mathrm{CRP}_t(\mathbf{w}) = \mathbf{w}$, is independent of the market environment $\mathcal{E}_t$.

There are two types of investment strategies. *Trading strategies* tell us whether we should take a *long* (bet that the stock price will rise) or a *short* (bet that the stock price will fall) position on a given stock. *Portfolio strategies* tell us how to distribute our wealth among various stocks. Trading strategies are denoted by T, and portfolio strategies are denoted by P. We use S to denote either kind of strategy. For $k \geq 2$, let

$$\mathcal{W}_k = \{\mathbf{w} = (w_1, \dots, w_k) \in [0,1]^k \mid \sum_{i=1}^{k} w_i = 1\}. \tag{1}$$

Remark 1. $\mathcal{W}_k$ is a $(k-1)$-dimensional simplex in $\mathbb{R}^k$. The investment strategies that we describe below are parameterized by vectors in $\mathcal{W}_k^\ell = \mathcal{W}_k \times \cdots \times \mathcal{W}_k$ (ℓ times) for some $k \geq 2$ and $\ell \geq 1$. We may write $\mathbf{w} \in \mathcal{W}_k^\ell$ in the form $\mathbf{w} = (\mathbf{w}_1, \dots, \mathbf{w}_\ell)$, where $\mathbf{w}_\iota = (w_{\iota 1}, \dots, w_{\iota k})$ for $1 \leq \iota \leq \ell$.

2.1   Trading Strategies

Suppose that our market contains a single stock. We have $m = 2$ potential investments: either a *long position* or *short position* in the stock. To take a *long position*, we buy shares in hopes that the share price will rise. We *close a long position* by selling the shares. The money we use to buy the shares is our *investment in the long position*; the *value* of the investment is the money we get when we close the position. If we let p_t denote the stock price at the beginning of day t, the value of our investment will change by a factor of $x_t = \frac{p_{t+1}}{p_t}$ from day t to $t+1$.

To take a *short position*, we borrow shares from our broker and sell them on the market in hopes that the share price will fall. We *close a short position* by

---

[3] We use the term "stocks" in order to keep our terminology consistent with previous work, but we actually mean a broader range of investment instruments, including both long and short positions in stocks.

buying the shares back and returning them to our broker. As collateral for the borrowed shares, our broker has a *margin requirement*: a fraction α of the value of the borrowed shares must be deposited in a *margin account*. Should the price of the security rise sufficiently, the collateral in our margin account will not be enough, and the broker will issue a *margin call*, requiring us to deposit more collateral. The margin requirement is our *investment in the short position*; the *value* of the investment is the money we get when we close the position.

Lemma 1. *Let the margin requirement for a short position be $\alpha \in (0, 1]$. Suppose that a short position is opened on day t and that the price of the underlying stock changes by a factor of $x_t = \frac{p_{t+1}}{p_t} < 1 + \alpha$ during the day. Then the value of our investment in the short position changes by a factor of $x'_t = 1 + \frac{1-x_t}{\alpha}$ during the day.*

Should the price of the underlying stock change by a factor greater than $1 + \alpha$, we will lose more money than we initially put in. We will assume that the margin requirement α is sufficiently large that the daily price change of the stock is always less than $1 + \alpha$.[4] If a short position is held for several days, assume that it is *rebalanced* at the beginning of each day: either part of the short is closed (if $x_t > 1$) or additional shares are shorted (if $x_t < 1$) so that the collateral in the margin account is exactly an α fraction of the value of the shorted shares. This ensures that the value of a short position changes by a factor, $x'_t = 1 + \frac{1-x_t}{\alpha}$, each day. Treating short positions in this way, they can simply be viewed as any other stock, so trading strategies are effectively investment strategies that decide between two potential investments: a long or a short position in a given stock. The investment description of a trading strategy T is $T_t = (T_{t1}, T_{t2})$, where T_{t1} and T_{t2} are the fraction of wealth to put in a long and short position respectively. Let $D = T_{t1} - T_{t2}/\alpha$ be the *net long position* of the investment description. In practice, if $D > 0$, investors should put a D fraction of their money in the long position and a $1 - D$ fraction in cash; if $D < 0$, investors should invest D in the short position and $1 - D$ in cash; if $D = 0$, investors should avoid the stock completely and keep all their money in cash. From a practical standpoint, it is desirable for the trading strategy to be *decisive*, *i.e.* $|D| = 1$, so that our allocation of money to the stock is always fully invested in the stock (either as a long or a short position).

We now describe some commonly used and researched trading strategies [17, 4,11,18] and show how they can be parameterized.

MA[k]: Moving Average Cross-over with k-day Memory. In traditional applications [11] of this rule, we compare the current stock price with the moving average over, say, the previous 200 days: if the price is above the moving average, we take a long position, otherwise we take a short position. Some general-

---

[4]   This assumption can be eliminated by purchasing a *call option* on the stock with some strike price $p < (1 + \alpha)p_t$. Should the stock price get too high, the call allows us to purchase the stock back for $p. Though its price detracts from the performance of our short trading strategy, the call protects us from potentially unlimited losses due to rising stock price.

izations of this rule have been made, where we compare a fast moving average (over, for example, the past five to 20 days) with a slow moving average (over the past 50 to 200 days). We generalize this rule further. Given day $t \geq 0$, let $\mathbf{v}_t = (v_{t1}, \ldots, v_{tk})$ be the *price-history vector* over the previous k days, where v_{tj} is the stock price on day $t - j$. Assume that the stock prices have been normalized such that $0 < v_{tj} \leq 1$. Let $(\mathbf{w}_F, \mathbf{w}_S) \in \mathcal{W}_k^2$ (where $\mathcal{W}_k$ is defined in (1)) be the weights to compute the fast moving and slow moving averages, so these averages on day t are given by $\mathbf{w}_F \cdot \mathbf{v}_t$ and $\mathbf{w}_S \cdot \mathbf{v}_t$ respectively. Since the prices have been normalized to the interval $(0, 1]$, $-1 \leq (\mathbf{w}_F - \mathbf{w}_S) \cdot \mathbf{v}_t \leq 1$. Let $g : [-1, 1] \to [0, 1]$ be the *long/short allocation function*. The idea is that $g((\mathbf{w}_F - \mathbf{w}_S) \cdot \mathbf{v}_t)$ represents the proportion of wealth that we invest in a long position. The full investment description for the MA $=$ MA[k] trading strategy is $\mathrm{MA}_t(\mathbf{w}_F, \mathbf{w}_S) = (g((\mathbf{w}_F - \mathbf{w}_S) \cdot \mathbf{v}_t), 1 - g((\mathbf{w}_F - \mathbf{w}_S) \cdot \mathbf{v}_t)$. Note that the dimension of the parameter space for MA[k] is $2(k - 1)$ since each of $\mathbf{w}_F$ and $\mathbf{w}_S$ are taken from $(k - 1)$-dimensional spaces. Possible functions for g include

$$g_s(x) = \begin{cases} 0 & \text{if } x < 0 \\ 1 & \text{otherwise} \end{cases} \qquad \text{(step function);} \qquad (2)$$

$$g_{(t)}(x) = \begin{cases} 0 & \text{if } x < -\frac{1}{t} \\ \frac{t}{2}(x + \frac{1}{t}) & \text{if } -\frac{1}{t} \leq x \leq \frac{1}{t} \\ 1 & \text{if } \frac{1}{t} < x \end{cases} \qquad \text{(linear step approximation);} \qquad (3)$$

and the line

$$g_\ell(x) = \frac{x + 1}{2} \qquad (4)$$

that intersects $g_s(x)$ at the extreme points $x = \pm 1$ of its domain. Note that $g_{(t)}(x)$ is parameterized by the day t during which it is called and that it converges to $g_s(x)$ on $[-1, 1] \setminus \{0\}$ as t increases.

Remark 2. The long/short allocation function used in traditional applications of this rule is the step function $g_s(\cdot)$. As we see in Section 3, in order for an investment strategy to be universalizable, its allocation function must be continuous, necessitating the continuous approximation $g_{(t)}(\cdot)$. The linear approximation $g_\ell(\cdot)$ can be used with the results of Section 4, to allow for efficient computation of the universalization algorithm.

In the full paper[5], we also describe a parameterization of another trading strategy, the support and resistance breakout, that has previously been discussed in the literature [18,4].

---

[5] Available at http://www.arxiv.org/abs/cs.CE/0204019, the full paper contains complete proofs of the results discussed in this extended abstract.

2.2   Portfolio Strategies

Portfolio strategies are investment strategies that distribute wealth among m stocks. The investment description of a portfolio strategy P is $P_t = (P_{t1}, \ldots, P_{tm})$, where $0 \leq P_{ti} \leq 1$ and $\sum_{i=1}^{m} P_{ti} = 1$. We put a fraction P_{ti} of our wealth in stock i at time t.

$\quad$ *CRP: Constantly Rebalanced Portfolio [5].* The parameter space for the CRP strategy is $\mathbb{W} = \mathcal{W}_m$. The investment description is $\mathrm{CRP}_t(\mathbf{w}) = \mathbf{w}$: at the beginning of each day, we invest a w_i proportion of our wealth in stock i.

$\quad$ In the full paper, we also describe several other portfolio strategies: a generalization of Cover and Ordentlich's constantly rebalanced portfolio with side information [7] and the indicator aggregation strategy.

3   Universalization of Investment Strategies

3.1   Universalization Defined

In a typical stock market, wealth grows geometrically. On day $t \geq 0$, let $\mathbf{x}_t$ be the *return vector* for day t, the vector of factors by which stock prices change on day t. The return vector corresponding to a trading strategy on a single stock is $(x_t, 1 + \frac{1-x_t}{\alpha})$, where x_t is the factor by which the price of the stock changes and $1 + \frac{1-x_t}{\alpha}$ is the factor by which our investment in a short position changes, as described in Lemma 1; the return vector corresponding to a portfolio strategy is $(x_{t1}, \ldots, x_{tm})$, where x_{ti} is the factor by which the price of stock i changes, where $1 \leq i \leq m$. Henceforth, we do not make a distinction between return vectors corresponding to trading and portfolio strategies; we assume that $\mathbf{x}_t$ is appropriately defined to correspond to the investment strategy in question. For an investment strategy S with parameter vector $\mathbf{w}$, the *return of $S(\mathbf{w})$ during the t-th day*—the factor by which our wealth changes on the t-th day when invested according to $S(\mathbf{w})$—is $S_t(\mathbf{w}) \cdot \mathbf{x}_t = \sum_{i=1}^{m} S_{ti}(\mathbf{w}) \cdot x_{ti}$ (recall that $S_t(\mathbf{w})$ is the investment description of $S(\mathbf{w})$ for day t, which is a vector specifying the proportion of wealth to put in each stock). Given time $n > 0$, let $\mathcal{R}_n(S(\mathbf{w})) = \prod_{t=0}^{n-1} S_t(\mathbf{w}) \cdot \mathbf{x}_t$ be the *cumulative return of $S(\mathbf{w})$ up to time n*; we may write $\mathcal{R}_n(\mathbf{w})$ in place of $\mathcal{R}_n(S(\mathbf{w}))$ if S is obvious from context. We analyze the performance of S in terms of the *normalized log-return* $\mathcal{L}_n(\mathbf{w}) = \mathcal{L}_n(S(\mathbf{w})) = \frac{1}{n} \log \mathcal{R}_n(\mathbf{w})$ of the wealth achieved.

$\quad$ For investment strategy S, let $\mathbf{w}_n^* = \arg\max_{\mathbf{w} \in \mathbb{R}^*} \mathcal{R}_n(S(\mathbf{w}))$ be the parameters that maximize the return of S up to day n.[6] An investment strategy U *universalizes* (or *is universal for*) S if[7] $\mathcal{L}_n(U) = \mathcal{L}_n(S(\mathbf{w}_n^*)) - o(1)$ for all environment vectors $\mathcal{E}_n$. That is, U is universal for S if the average daily log-return of U approaches the optimal average daily log-return of S as the length n of the time horizon grows, regardless of stock price sequences.

---

[6] As mentioned above, $\mathbf{w}_n^*$ can only be computed with hindsight.

[7] Unlike previously discussed investment strategies, the behavior of U is fully defined without an additional parameter vector $\mathbf{w}$.

3.2 General Techniques for Universalization

Given an investment strategy S, let $\mathbb{W}$ be the parameter space for S and let μ be the uniform measure over $\mathbb{W}$. Our universalization algorithm for S, $\mathcal{U}(S)$, is a generalization of Cover's original result [5]. The investment description $\mathcal{U}_t(S)$ for the universalization of S on day $t > 0$ is a weighted average of the $S_t(\mathbf{w})$ over $\mathbf{w} \in \mathbb{W}$, with greater weight given to parameters $\mathbf{w}$ that have performed better in the past (*i.e.* $\mathcal{R}_t(\mathbf{w})$ is larger). Formally, the investment description is

$$\mathcal{U}_t(S) = \frac{\int_{\mathbb{W}} S_t(\mathbf{w})\mathcal{R}_t(\mathbf{w})d\mu(\mathbf{w})}{\int_{\mathbb{W}} \mathcal{R}_t(\mathbf{w})d\mu(\mathbf{w})} = \frac{\int_{\mathbb{W}} S_t(\mathbf{w})\mathcal{R}_t(S(\mathbf{w}))d\mu(\mathbf{w})}{\int_{\mathbb{W}} \mathcal{R}_t(S(\mathbf{w}))d\mu(\mathbf{w})}, \tag{5}$$

where we take $\mathcal{R}_0(\mathbf{w}) = 1$ for all $\mathbf{w} \in \mathbb{W}$.[8] The definition of universalization can be expanded to include measures other than μ, but we consider only μ in our results.

Lemma 2 ([3,7]). *The cumulative n-day return of $\mathcal{U}(S)$ is $\mathcal{R}_n(\mathcal{U}(S)) = \int_{\mathbb{W}} \mathcal{R}_n(\mathbf{w})d\mu(\mathbf{w}) = \mathbb{E}(\mathcal{R}_n(\mathbf{w}))$, the μ-weighted average of the cumulative returns of the investment strategies $\{S(\mathbf{w}) \mid \mathbf{w} \in \mathbb{W}\}$.*

Rather than directly universalizing a given investment strategy S, we instead focus on a modified version of S that puts a nonzero fraction of wealth in each of the m stocks. Define the investment strategy $\bar{S}$ by $\bar{S}_t(\mathbf{w}) = (1 - \frac{\varepsilon}{2(t+1)^2})S_t(\mathbf{w}) + \frac{\varepsilon}{2m(t+1)^2}$ for $t \geq 0$ and some fixed $0 < \varepsilon < 1$. Rather than universalizing S, we instead universalize $\bar{S}$. Lemma 3 tells us that we do not lose much by doing this.

Lemma 3. *For all $n \geq 0$, (1) $\mathcal{R}_n(\mathcal{U}(\bar{S})) \geq (1-\varepsilon)\mathcal{R}_n(\mathcal{U}(S))$ and (2) $\mathcal{L}_n(\mathcal{U}(\bar{S})) = \mathcal{L}_n(\mathcal{U}(S)) - \frac{o(n)}{n}$. (3) If $\mathcal{U}(S)$ is a universalization of S, then $\mathcal{U}(\bar{S})$ is a universalization of S as well.*

Remark 3. Henceforth, we assume that suitable modifications have been made to S to ensure that $S_{ti}(\mathbf{w}) \geq \frac{\varepsilon}{2m(t+1)^2}$ for all $1 \leq i \leq m$ and $t \geq 0$.

Theorem 1. *Given an investment strategy S, let $\mathbb{W} = \mathcal{W}_k^{\ell}$ (for some $k \geq 2$ and $\ell \geq 1$) be its parameter space. For $1 \leq i \leq m$, $1 \leq \iota \leq \ell$ and $1 \leq j \leq k$, assume that there is a constant c such that $\left|\frac{\partial S_{ti}(\mathbf{w})}{\partial w_{\iota j}}\right| \leq c(t+1)$ for all $\mathbf{w} \in \mathbb{W}$. Then $\mathcal{U}(S)$ is a universalization of S.*

Proof. (Outline) From Lemma 2, the return of $\mathcal{U}(S)$ is the average of the cumulative returns of the investment strategies $\{S(\mathbf{w}) \mid \mathbf{w} \in \mathbb{W}\}$. Let $\mathbf{w}^* = \arg\max_{\mathbf{w} \in \mathbb{W}} \mathcal{R}_n(S(\mathbf{w}))$ be the parameters that maximize the return of S. We show that there is a set B of nonzero volume around $\mathbf{w}^*$ such that for $\mathbf{w} \in B$, the return $\mathcal{R}_n(\mathbf{w})$ is close to the optimal return $\mathcal{R}_n(\mathbf{w}^*)$. We then show that the contribution to the average return from B is sufficiently large to ensure

---

[8] Cover's algorithm is a special case of this, replacing $S_t(\mathbf{w})$ with $\mathbf{w}$.

universalizability. We begin by bounding the magnitude of the gradient vector $\nabla \mathcal{R}_n(\mathbf{w})$. From Remark 3 and our assumption in the statement of the theorem, for all $\mathbf{w}$, t, i, ι, and j $\frac{\left|\frac{\partial S_{ti}(\mathbf{w})}{\partial w_{\iota j}}\right|}{S_{ti}(\mathbf{w})} \leq c'm(t+1)^3$, where $c' = \frac{2c}{\varepsilon}$. The partial derivative of the return function $\mathcal{R}_n(\mathbf{w}) = \mathcal{R}_n(S(\mathbf{w})) = \prod_{t=0}^{n-1} r_t(S(\mathbf{w}))$ with respect to parameter $w_{\iota j}$ is $\left|\frac{\partial \mathcal{R}_n(\mathbf{w})}{\partial w_{\iota j}}\right| \leq \mathcal{R}_n(\mathbf{w}) \sum_{t=0}^{n-1} \frac{\left|\frac{\partial(S_t(\mathbf{w})\cdot\mathbf{x}_t)}{\partial w_{\iota j}}\right|}{S_t(\mathbf{w})\cdot\mathbf{x}_t} \leq c'\mathcal{R}_n(\mathbf{w})mn^4$ and

$$|\nabla \mathcal{R}_n(\mathbf{w})| \leq c'\mathcal{R}_n(\mathbf{w})mn^4\sqrt{k\ell}. \tag{6}$$

We would like to take our set B to be some d-dimensional ball around $\mathbf{w}^*$; unfortunately, if $\mathbf{w}^*$ is on (or close to) an edge of $\mathbb{W}$, the reasoning introduced at the beginning of this proof is not valid. In the full paper, we show how to perturb $\mathbf{w}^*$ to a point $\tilde{\mathbf{w}}$ that is at least $\rho = \frac{\gamma}{c'mn^4k^2\ell}$ away from all edges, where $0 < \gamma < 1$ is a constant, and such that $\mathcal{R}_n(\tilde{\mathbf{w}}) \geq \mathcal{R}_n(\mathbf{w}^*)(1-\gamma)$. For $0 \leq \iota \leq \ell$ let $C_\iota = \{\mathbf{w}_\iota \in \mathbb{R}^k \mid |\tilde{\mathbf{w}}_\iota - \mathbf{w}_\iota| \leq \rho\}$. From the construction of $\tilde{\mathbf{w}}$, $B_\iota = C_\iota \cap \mathbb{W}_k$ is a $(k-1)$-dimensional ball of radius ρ. We show in the full paper that for $\mathbf{w} \in B = B_1 \times \cdots \times B_\ell$, $\mathcal{R}_n(\mathbf{w}) \geq \mathcal{R}_n(\mathbf{w}^*)(1-2\gamma)$. By Lemma 2, $\mathcal{R}_n(\mathcal{U}(S)) \geq \int_B \mathcal{R}_n(\mathbf{w})d\mu(\mathbf{w}) \geq (1-2\gamma)\mathcal{R}_n(\mathbf{w}^*)\int_B d\mu(\mathbf{w}) \geq (1-2\gamma)\mathcal{R}_n(\mathbf{w}^*)\frac{\int_B d\mathbf{w}}{\int_\mathbb{W} d\mathbf{w}} = \mathcal{R}_n(\mathbf{w}^*)\Lambda(\gamma,m,k,\ell)n^{-4k\ell}$, where Λ is some constant depending on γ, m, k, and ℓ. Therefore, $\mathcal{L}_n(\mathbf{w}^*) - \mathcal{L}_n(\mathcal{U}(S)) \leq \frac{\log \Lambda(\gamma,m,k,\ell)}{n} + 4k\ell\frac{\log n}{n} = \frac{o(n)}{n}$, as claimed.

Remark 4. The techniques used in the proof of Theorem 1 can be generalized to other investment strategies with bounded parameter spaces $\mathbb{W}$ that are not necessarily of the form $\mathbb{W}_k^\ell$.

By proving an upper bound on $\left|\frac{\partial S_{ti}(\mathbf{w})}{\partial w_j}\right|$ for our investment strategies S, we show that they are universalizable.

Theorem 2. *The following investment strategies are universalizable:*

- *Moving average cross-over, MA[k], with long/short allocation functions $g_{(t)}(x)$ and $g_\ell(x)$ defined in (3) and (4) respectively.*
- *Constantly rebalanced portfolio, CRP.*

4   Fast Computation of Universal Investment Strategies

4.1   Approximation by Sampling

The running time of the universalization algorithm depends on the time to compute the integral in (5). A straightforward evaluation of it takes time exponential in the number of parameters. Following Kalai and Vempala [14], we propose to approximate it by sampling the parameters according to a biased distribution, giving greater weight to better performing parameters. Define the measure ζ_t on $\mathbb{W}$ by $d\zeta_t(\mathbf{w}) = \frac{\mathcal{R}_t(S(\mathbf{w}))}{\int_\mathbb{W} \mathcal{R}_t(S(\mathbf{w}))d\mu(\mathbf{w})}d\mu(\mathbf{w})$.

Lemma 4 ([14]). *The investment description $\mathcal{U}_t(S)$ for universalization is the average of $S_t(\mathbf{w})$ with respect to the ζ_t measure.*

In Section 4.2, we show that for certain strategies we can efficiently sample from a distribution $\bar{\zeta}_t$ that is "close" to ζ_t, *i.e.* given $\gamma_t > 0$, we generate samples from $\bar{\zeta}_t$ in $\mathcal{O}(\log \frac{1}{\gamma_t})$ time and such that

$$\int_{\mathbb{W}} |\zeta_t(\mathbf{w}) - \bar{\zeta}_t(\mathbf{w})| \, d\mu(\mathbf{w}) \leq \gamma_t. \tag{7}$$

Assume for now that we can sample from $\bar{\zeta}_t$, with $\gamma_t = \frac{\varepsilon^2}{4m(t+1)^4}$, where ε is the constant appearing in Remark 3. Let $\bar{\mathcal{U}}_t(S) = \int_{\mathbb{W}} S_t(\mathbf{w}) d\bar{\zeta}_t(\mathbf{w})$ be the corresponding approximation to $\mathcal{U}(S)$. Lemma 5 tells us that we do not lose much by sampling from $\bar{\zeta}_t$.

Lemma 5. *For all $n \geq 0$, (1) $\mathcal{R}_n(\bar{\mathcal{U}}(S)) \geq (1 - \varepsilon)\mathcal{R}_n(\mathcal{U}(S))$ and (2) if $\mathcal{U}(S)$ is a universalization of S, then $\bar{\mathcal{U}}(S)$ is a universalization of S as well.*

By sampling from $\bar{\zeta}_t$, we show in the full paper how to use a generalization of the Chernoff bound (due to Hoeffding [13]) to get an approximation $\tilde{\mathcal{U}}(S)$ to $\bar{\mathcal{U}}(S)$ such that with high probability $\tilde{\mathcal{U}}_{ti}(S) \geq (1 - \frac{\varepsilon}{2(t+1)^2})\bar{\mathcal{U}}_{ti}(S)$ for $0 \leq t < n$ and $1 \leq i \leq m$, so that $\tilde{\mathcal{U}}(S)$ is a universalization of S if $\bar{\mathcal{U}}(S)$ is a universalization of S.

4.2   Efficient Sampling

We now discuss how to sample from $\mathbb{W} = \mathcal{W}_k^\ell = \mathcal{W}_k \times \cdots \times \mathcal{W}_k$ according to distribution $\zeta_t(\cdot) \propto \mathcal{R}_t(\cdot) = \mathcal{R}_t(S(\cdot))$. $\mathbb{W}$ is a convex set of diameter $d = \sqrt{2\ell}$. We focus on a discretization of the sampling problem. Choose an orthogonal coordinate system on each $\mathcal{W}_k$ and partition it into hypercubes of side length δ_t, where δ_t is a constant chosen below. Let Ω be the set of centers of cubes that intersect $\mathbb{W}$ and choose the partition such that the coordinates of $\mathbf{w} \in \Omega$ are multiples of δ_t. For $\mathbf{w} \in \Omega$, let $C(\mathbf{w})$ be the cube with center $\mathbf{w}$. We show how to choose $\mathbf{w} \in \Omega$ with probability "close to" $\pi_t(\mathbf{w}) = \frac{\mathcal{R}_t(\mathbf{w})}{\sum_{\mathbf{w} \in \Omega} \mathcal{R}_t(\mathbf{w})}$. In particular, we sample from a distribution $\tilde{\pi}_t$ that satisfies

$$\sum_{\mathbf{w} \in \Omega} |\pi_t(\mathbf{w}) - \tilde{\pi}_t(\mathbf{w})| \leq \gamma_t = \frac{\varepsilon^2}{4m(t+1)^4}. \tag{8}$$

Note that this is a discretization of (7). We will also have that for each $\mathbf{w} \in \Omega$,

$$\frac{\tilde{\pi}_t(\mathbf{w})}{\pi_t(\mathbf{w})} \leq 2. \tag{9}$$

We would like to choose δ_t sufficiently small that $\mathcal{R}_t$ is "nearly constant" over $C(\mathbf{w})$ *i.e.* there is a small constant $\nu > 0$ such that

$$(1 + \nu)^{-1}\mathcal{R}_t(\mathbf{w}) \leq \mathcal{R}_t(\mathbf{w}') \leq (1 + \nu)\mathcal{R}_t(\mathbf{w}) \tag{10}$$

for all $\mathbf{w}' \in C(\mathbf{w})$. Such a δ_t can be chosen for investment strategies S that have bounded derivative, as we can see in Lemma 6, which we prove in the full paper.

Lemma 6. *Suppose that investment strategy S satisfies the condition for universalizability given in Theorem 1, i.e.* $\left| \frac{\partial S_{ti}(\mathbf{w})}{\partial w_j} \right| \le ct$. *Given $\nu > 0$, let $\delta_t = \delta_t(\nu) = \frac{\nu}{3c'mt^4k\ell}$, where c' is defined in the proof of Theorem 1. For $\mathbf{w}, \mathbf{w}' \in \mathbb{W}$ such that $|w_{ij} - w'_{ij}| \le \delta_t(\nu)$ for all $1 \le i \le \ell$ and $1 \le j \le k$, $(1+\nu)^{-1}\mathcal{R}_t(\mathbf{w}) \le \mathcal{R}_t(\mathbf{w}') \le (1+\nu)\mathcal{R}_t(\mathbf{w})$.*

We use a Metropolis algorithm [15] to sample from $\tilde{\pi}_t$. We generate a random walk on Ω according to a Markov chain whose stationary distribution is π_t. Begin by selecting a point $\mathbf{w}_0 \in \Omega$ according to $\tilde{\pi}_{t-1}$.[9] If $\mathbf{w}_\tau$ is the position of our random walk at time $\tau \ge 0$, we pick its position at time $\tau + 1$ as follows. Note that $\mathbf{w}_\tau$ has $2(k-1)\ell$ neighbors, two along each axis in the Cartesian product of ℓ $(k-1)$-dimensional spaces. Let $\mathbf{w}$ be a neighbor of $\mathbf{w}_\tau$, selected uniformly at random. If $\mathbf{w} \in \Omega$, set

$$\mathbf{w}_{\tau+1} = \begin{cases} \mathbf{w} & \text{with probability } p = \min(1, \frac{\mathcal{R}_t(\mathbf{w})}{\mathcal{R}_t(\mathbf{w}_\tau)}) \\ \mathbf{w}_\tau & \text{with probability } 1 - p. \end{cases}$$

If $\mathbf{w} \notin \Omega$, let $\mathbf{w}_{\tau+1} = \mathbf{w}_\tau$. It is well-known that the stationary distribution of this random walk is π_t. We must determine how many steps of the walk are necessary before the distribution has gotten sufficiently close to stationary. Let p_τ be the distribution attained after τ steps of the random walk. Applegate and Kannan [2] show that if the desired distribution π_t is proportional to a log-concave function F (*i.e.* $\log F$ is concave) the Markov chain is *rapidly mixing*, reaches its steady state in polynomial time. Frieze and Kannan [10] give an improved upper bound on the mixing time using Logarithmic Sobolev inequalities [8].

Theorem 3 (Theorem 1 of [10]). *Assume the diameter d of $\mathbb{W}$ satisfies $d \ge \delta_t\sqrt{k\ell}$ and that the target distribution π is proportional to a log-concave function. There is an absolute constant $\kappa > 0$ such that*

$$2\left(\sum_{\mathbf{w} \in \Omega} |\pi(\mathbf{w}) - p_\tau(\mathbf{w})| \right)^2 \le e^{-\frac{\kappa\tau\delta_t^2}{k\ell d^2}} \log \frac{1}{\pi_*} + \frac{M\pi_e k\ell d^2}{\kappa\delta_t^2}, \qquad (11)$$

where $\pi_* = \min_{\mathbf{w} \in \Omega} \pi(\mathbf{w})$, $M = \max_{\mathbf{w} \in \Omega} \frac{p_0(\mathbf{w})}{\pi(\mathbf{w})} \log \frac{p_0(\mathbf{w})}{\pi(\mathbf{w})}$, $p_0(\cdot)$ is the initial distribution on Ω, $\pi_e = \sum_{\mathbf{w} \in \Omega_e} \pi(\mathbf{w})$, and $\Omega_e = \{\mathbf{w} \in \Omega \,|\, \text{Vol}(C(\mathbf{w}) \cap \mathbb{W}) < \text{Vol}(C(\mathbf{w}))\}$ (the "e" in the subscripts of π_e and Ω_e stands for "edge").

---

[9] We can do this by "saving" our samples that were generated for the approximation of $\mathcal{U}_{t-1}$ at time $t-1$. For technical reasons, the number of samples generated at time $t-1$ may not be enough for our purposes so we may also use the samples generated at time $t-2$, so that the initial point $\mathbf{w}_0$ is actually chosen with distribution $\tilde{\pi}_{t-1}$ or $\tilde{\pi}_{t-2}$.

In the random walk described above, if $\mathbf{w}_\tau$ is on an edge of Ω, so it has many neighbors outside Ω, the walk may get "stuck" at $\mathbf{w}_\tau$ for a long time, as seen in the "π_e" term of Theorem 3. We must ensure that the random walk has low probability of reaching such edge points. We do this by applying a "damping function" to $\mathcal{R}_t$ that becomes exponentially small near the edges of $\mathbb{W}$. For $1 \le i \le \ell$, $1 \le j \le k$, and $\mathbf{w} = (\mathbf{w}_1, \ldots, \mathbf{w}_\ell) = ((w_{11}, \ldots, w_{1k}), \ldots, (w_{\ell 1}, \ldots, w_{\ell k})) \in \mathbb{W}$ let

$$f_{ij}(\mathbf{w}) = e^{\Gamma \min(-\sigma + w_{ij}, 0)}, \tag{12}$$

where $\sigma > 0$ and $\Gamma > 2$ are constants that we choose below, and let $F_t(\mathbf{w}) = \mathcal{R}_t(\mathbf{w}) \prod_{i=1}^{\ell} \prod_{j=1}^{k} f_{ij}(\mathbf{w})$. It is not difficult to prove the following lemma:

Lemma 7. F_t is log-concave if and only if $\mathcal{R}_t$ is log-concave.[10]

Choose $\sigma = \frac{1}{k}\delta_t(\frac{\gamma_t}{2})$, where $\delta_t(\cdot)$ is defined in Lemma 6 and γ_t is defined in (8). Let $\zeta_F \propto F_t$ be the probability measure proportional to F_t. We show in the full paper that for our purposes sampling from ζ_F is not much different than sampling from ζ_t. By Lemma 5, we can do this by proving the following lemma:

Lemma 8. $\int_{\mathbb{W}} |\zeta_t(\mathbf{w}) - \zeta_F(\mathbf{w})| d\mathbf{w} \le \gamma_t$.

Henceforth, we are concerned with sampling from $\mathbb{W}$ with probability proportional to $F_t(\cdot)$. We use the Metropolis algorithm described above, replacing $\mathcal{R}_t(\cdot)$ with $F_t(\cdot)$; we must refine our grid spacing δ_t so that (10) is satisfied by F_t; let δ_t' be the new grid spacing.

Lemma 9. *Suppose that the conditions of Lemma 6 are satisfied. Given $\nu > 0$, let $\delta_t'(\nu) = \delta_t' = \frac{\nu}{3\Gamma c' m t^4 k \ell} = \delta_t(\frac{\nu}{\Gamma})$, where Γ appears in (12). For $\mathbf{w}, \mathbf{w}' \in \mathbb{W}$ such that $|w_{ij} - w_{ij}'| \le \delta_t'(\nu)$ for all $1 \le i \le \ell$ and $1 \le j \le k$, $(1+\nu)^{-1} F_t(\mathbf{w}) \le F_t(\mathbf{w}') \le (1+\nu) F_t(\mathbf{w})$.*

We are now ready to use Theorem 3 to select τ so that the resulting distribution p_τ satisfies (8) (Theorem 4) and (9) (Theorem 5), with p_τ in place of $\tilde{\pi}_t$ and F_t in place of $\mathcal{R}_t$. We begin with some preliminary lemmas.

Lemma 10. *There is a constant $\beta > 0$ such that $\log \frac{1}{\pi_*} \le k \ell \Gamma \sigma + k \ell \log \frac{\beta}{\delta_t'} + t \log \frac{2mt^2}{\varepsilon}$, where ε is defined in Remark 3.*

Proof. This follows from the fact that for $\mathbf{w}_1, \mathbf{w}_2 \in \Omega$, the ratio of single-day returns on any day t' using $\mathbf{w}_1$ and $\mathbf{w}_2$ is $\frac{S_{t'}(\mathbf{w}_1) \cdot \mathbf{x}_{t'}}{S_{t'}(\mathbf{w}_2) \cdot \mathbf{x}_{t'}} \ge \frac{\varepsilon}{2m(t'+1)^2}$ (by Remark 3) and thus $\frac{\mathcal{R}_t(\mathbf{w}_1)}{\mathcal{R}_t(\mathbf{w}_2)} \ge \left(\frac{\varepsilon}{2mt^2}\right)^t$. The full proof is given in the full paper.

Lemma 11. $M \le 4 \left(\frac{2m(t+1)^2}{\varepsilon}\right)^2 \log \frac{2m(t+1)^2}{\varepsilon}$.

---

[10] We characterize investment strategies for which $\mathcal{R}_t$ is log-concave in Theorem 6

Proof. The initial distribution is either $p_0 = \tilde{\pi}_{t-1}$ or $\tilde{\pi}_{t-2}$. It turns out that the worst case happens when $p_0 = \tilde{\pi}_{t-2}$. For all $\mathbf{w} \in \Omega$, $\frac{\tilde{\pi}_{t-2}(\mathbf{w})}{\pi_{t-2}(\mathbf{w})} \leq 2$ by (9) and we show in the full paper that $\frac{\tilde{\pi}_{t-2}(\mathbf{w})}{\pi_t(\mathbf{w})} \leq (\frac{2m(t+1)^2}{\varepsilon})^2$. Together, these prove the result since $\frac{\tilde{\pi}_{t-2}(\mathbf{w})}{\pi_t(\mathbf{w})} = \frac{\tilde{\pi}_{t-2}(\mathbf{w})}{\pi_{t-2}(\mathbf{w})} \frac{\pi_{t-2}(\mathbf{w})}{\pi_t(\mathbf{w})}$.

Lemma 12. $\pi_e \leq (1+\nu)^4(1+\frac{\gamma_t}{2})e^{-\Gamma\sigma}$, *where ν appears in the definition of δ'_t in Lemma 9, γ_t appears in (8), and Γ and σ appear in (12).*

Proof. This is a technical result and its proof is deferred to the full paper. The basic idea is that we scale $\mathbb{W}$ outwards to a set $\mathbb{W}''$ that contains all edge cubes of $\mathbb{W}$ and inwards to a set $\mathbb{W}'$ that does not intersect any of the edge cubes and show that $\int_{\mathbb{W}''} F_t(\mathbf{w})d\mathbf{w}$ is not much bigger than $\int_{\mathbb{W}'} F_t(\mathbf{w})d\mathbf{w}$.[11]

Using the above results, we show in the full paper how to prove the following Corollary to Theorem 3.

Theorem 4. *Letting $\Gamma = \mathcal{O}^*(\frac{1}{\sigma}) = \mathcal{O}^*(\frac{m^2 t^8 k^2 \ell}{\varepsilon^2})$, the random walk reaches a distribution $\tilde{\pi}$ that satisfies (8) after $\tau = \mathcal{O}^*(\frac{k^7 \ell^6 m^6 t^{24}}{\kappa \nu^2 \varepsilon^4})$ steps.*[12]

Theorem 5. *Suppose that the distribution p_{τ_0} obtained after τ_0 steps satisfies $\sum_{\mathbf{w} \in \Omega} |\pi(\mathbf{w}) - p_{\tau_0}(\mathbf{w})| \leq \gamma_t$. After $\tau_0' \geq \frac{\tau_0}{\tau_0 - \log\frac{1}{\pi_*} - \log\frac{1}{\gamma_t}} \log\frac{1}{\pi_*} = \mathcal{O}^*(\tau_0(k\ell + t))$ steps, the resulting distribution $p_{\tau_0'}$ satisfies $\max_{\mathbf{w} \in \Omega} \frac{p_{\tau_0'}(\mathbf{w})}{\pi(\mathbf{w})} - 1 \leq 1$, which implies (9).*

Proof. This follows from results of Aldous and Fill [1, Equations (5) and (6)] and Sinclair [16, Proposition 1 (i)].

4.3 Application to Investment Strategies

The efficient sampling techniques of this section are applicable to investment strategies S whose return functions $\mathcal{R}_n(S(\cdot))$ are log-concave. Theorem 6 and Corollary 1 characterize such functions. They are proven in the full paper.

Theorem 6. *Given investment strategy S, suppose that for all parameters w_i and w_j, $\frac{\partial^2 S}{\partial w_i \partial w_j} = 0$. Then $\mathcal{R}_t(\mathbf{w}) = \mathcal{R}_t(S(\mathbf{w}))$ is log-concave.*

Corollary 1. *Universalizations of the following investment strategies can be computed using the sampling techniques of this section.*

1. *The trading strategy $MA[k]$ with long/short allocation function $g_\ell(x)$; and*
2. *The portfolio strategy CRP.*

---

[11] Note that we must also extend the definition of F_t to $\mathbb{W}''$, so that $\int_{\mathbb{W}''} F_t(\mathbf{w})d\mathbf{w}$ is defined.

[12] $\mathcal{O}^*(\cdot)$ notation ignores logarithmic and constant factors. For our purposes, $f(\cdot) = \mathcal{O}^*(g(\cdot))$ if there exists a constant $C \geq 0$ such that $f(\cdot) = \mathcal{O}(g(\cdot) \log^C(k\ell mt/\varepsilon))$.

References

1. D. Aldous and J. A. Fill. Advanced L^2 techniques for bounding mixing times. In *Reversible Markov Chains and Random Walks on Graphs.* 1999. Unpublished monograph. Available at http://stat-www.berkeley.edu/users/aldous/book.html.
2. D. Applegate and R. Kannan. Sampling and integration of near log-concave functions. In *Proceedings of the 23rd Annual ACM Symposium on Theory of Computing,* pages 156–163, 1991.
3. A. Blum and A. Kalai. Universal portfolios with and without transaction costs. *Machine Learning,* 35(3):193–205, 1999.
4. W. Brock, J. Lakonishok, and B. LeBaron. Simple technical trading rules and the stochastic properties of stock returns. *Journal of Finance,* 47(5):1731–1764, 1992.
5. T. M. Cover. Universal portfolios. *Mathematical Finance,* 1(1):1–29, January 1991.
6. T. M. Cover and E. Ordentlich. Online portfolio selection. In *Proceedings of the 9th Annual Conference on Computational Learning Theory,* 1996.
7. T. M. Cover and E. Ordentlich. Universal portfolios with side-information. *IEEE Transactions on Information Theory,* 42(2), March 1996.
8. P. Diaconis and L. Saloff-Coste. Logarithmic Sobolev inequalities for finite Markov chains. *Annals of Applied Probability,* 6:695–750, 1996.
9. G. B. Folland. *Real Analysis: Modern Techniques and their Applications.* John Wiley & Sons, New York, 1984.
10. A. Frieze and R. Kannan. Log-Sobolev inequalities and sampling from log-concave distributions. *Annals of Applied Probability,* 9:14–26, 1999.
11. H. M. Gartley. *Profits in the Stock Market.* Lambert Gann Publishing Company, Pomeroy, WA, 1935.
12. D. P. Helmbold, R. E. Schapire, Y. Singer, and M. K. Warmuth. On-line portfolio selection using multiplicative updates. *Mathematical Finance,* 8(4):325–347, 1998.
13. W. Hoeffding. Probability inequalities for sums of bounded random variables. *Journal of the American Statistical Association,* 58:13–30, March 1963.
14. A. Kalai and S. Vempala. Efficient algorithms for universal portfolios. In *Proceedings of the 41st Annual IEEE Symposium on Foundations of Computer Science,* pages 486–491, 2000.
15. N. Metropolis, A. W. Rosenbluth, M. N. Rosenbluth, A. H. Teller, and E. Teller. Equation of state calculation by fast computing machines. *Journal of Chemical Physics,* 21:1087–1092, 1953.
16. A. Sinclair. Improved bounds for mixing rates of Markov chains and multicommodity flow. *Combinatorics, Probability and Computing,* 1:351–370, 1992.
17. R. Sullivan, A. Timmermann, and H. White. Data-snooping, technical trading rules and the bootstrap. *Journal of Finance,* 54:1647–1692, 1999.
18. R. Wyckoff. *Studies in Tape Reading.* Fraser Publishing Company, Burlington, VT, 1910.

Randomized Pursuit-Evasion in Graphs

Micah Adler[1], Harald Räcke[2] *, Naveen Sivadasan[3] *, Christian Sohler[2] *, and
Berthold Vöcking[3] *

[1] Department of Computer Science University of Massachusetts, Amherst,
micah@cs.umass.edu
[2] Heinz Nixdorf Institute and Department of Mathematics and Computer Science, Paderborn
University, Germany, {harry,csohler}@upb.de
[3] Max-Planck-Institut für Informatik, Saarbrücken, Germany,
{ns,voecking}@mpi-sb.mpg.de

Abstract. We analyze a randomized pursuit-evasion game on graphs. This game
is played by two players, a *hunter* and a *rabbit*. Let G be any connected, undirected
graph with n nodes. The game is played in rounds and in each round both the hunter
and the rabbit are located at a node of the graph. Between rounds both the hunter
and the rabbit can stay at the current node or move to another node. The hunter
is assumed to be *restricted* to the graph G: in every round, the hunter can move
using at most one edge. For the rabbit we investigate two models: in one model
the rabbit is restricted to the same graph as the hunter, and in the other model the
rabbit is *unrestricted*, i.e., it can jump to an arbitrary node in every round.

We say that the rabbit is *caught* as soon as hunter and rabbit are located at the
same node in a round. The goal of the hunter is to catch the rabbit in as few
rounds as possible, whereas the rabbit aims to maximize the number of rounds
until it is caught. Given a randomized hunter strategy for G, the *escape length* for
that strategy is the worst case expected number of rounds it takes the hunter to
catch the rabbit, where the worst case is with regards to all (possibly randomized)
rabbit strategies. Our main result is a hunter strategy for general graphs with an
escape length of only $O(n \log(\mathrm{diam}(G)))$ against restricted as well as unrestricted
rabbits. This bound is close to optimal since $\Omega(n)$ is a trivial lower bound on the
escape length in both models. Furthermore, we prove that our upper bound is
optimal up to constant factors against unrestricted rabbits.

1   Introduction

In this paper we introduce a pursuit evasion game called the *Hunter vs. Rabbit* game. In
this round-based game, a pursuer (the *hunter*) tries to catch an evader (the *rabbit*) while
they both travel from vertex to vertex of a connected, undirected graph G. The hunter
catches the rabbit when in some round the hunter and the rabbit are both located on the
same vertex of the graph. We assume that both players know the graph in advance but they
cannot see each other until the rabbit gets caught. Both players may use a randomized
(also called *mixed*) strategy, where each player has a secure source of randomness which

---

* Partially supported by the IST Programme of the EU under contract number IST-1999-14186
(ALCOM-FT)

P. Widmayer et al. (Eds.): ICALP 2002, LNCS 2380, pp. 901–912, 2002.
© Springer-Verlag Berlin Heidelberg 2002

cannot be observed by the other player. In this setting we study upper bounds (i.e., good hunter strategies) as well as lower bounds (i.e., good rabbit strategies) on the expected number of rounds until the hunter catches the rabbit.

The problem we address is motivated by the question of how long it takes a single pursuer to find an evader on a given graph that, for example, corresponds to a computer network or to a map of a terrain in which the evader is hiding. A natural assumption is that both the pursuer and the evader have to follow the edges of the graph. In some cases however it might be that the evader has more advanced possibilities than the pursuer in the terrain where he is hiding. Therefore we additionally consider a stronger adversarial model in which the evader is allowed to jump arbitrarily between vertices of the graph. Such a jump between vertices corresponds to a short-cut between two places which is only known to the evader (like a rabbit using rabbit holes). Obviously, a strategy that is efficient against an evader that can jump is efficient as well against an evader who may only move along the edges of the graph.

One approach to use for a hunter strategy would be to perform a random walk on the graph G. Unfortunately, the hitting time of a random walk can be as large as $\Omega(n^3)$ with n denoting the number of nodes. Thus it would require at least $\Omega(n^3)$ rounds to find a rabbit even if the rabbit does not move at all. We show that one can do significantly better. In particular, we prove that for any graph G with n vertices there is a hunter strategy such that the expected number of rounds until a rabbit that is not necessarily restricted to the graph is caught is $\mathcal{O}(n \log n)$ rounds. Furthermore we show that this result cannot be improved in general as there is a graph with n nodes and an unrestricted rabbit strategy such that the expected number of rounds required to catch this rabbit is $\Omega(n \log n)$ for any hunter strategy.

1.1 Preliminaries

Definition of the game. In this section we introduce the basic notations and definitions used in the remainder of the paper. The Hunter vs. Rabbit game is a round-based game that is played on an undirected connected graph $G = (V, E)$ without self loops and multiple edges. In this game there are two players - the hunter and the rabbit - moving on the vertices of G. The hunter tries to catch the rabbit, i.e., he tries to move to the same vertex as the rabbit, and the rabbit tries not to be caught.

During the game both players cannot "see" each other, i.e., a player has no information about the movement decisions made by his opponent and thus does not know his position in the graph. The only interaction between both players occurs when the game ends because the hunter and the rabbit move to the same vertex in G and the rabbit is caught. Therefore the movement decisions of both players do not depend on each other. We want to find good strategies for both hunter and rabbit. We say that a hunter strategy has *expected (worst case) escape length* k, if for any rabbit strategy the expected number of rounds until the hunter catches the rabbit is k. Analogously a rabbit strategy is said to have *expected (worst case) escape length* k, if for any hunter strategy the expected number of rounds until the rabbit is caught is k. We can define a strategy for a player in the following way.

Definition 1. *A pure strategy for a player in the Hunter vs. Rabbit game on a graph $G = (V, E)$ is a sequence $S = S_0, S_1, S_2, \ldots$, where $S_t \in V$ denotes the position of* *the player in round $t \in \mathbb{N}_0$ of the game. A mixed strategy or strategy $\mathbb{S}$ for a player is a probability distribution over the set of pure strategies.*

Note that both players may use mixed strategies, i.e., we assume that they both have a source of random bits for randomizing their movements on the graph.

As mentioned in the previous section we assume that the hunter cannot change his position arbitrarily between two consecutive rounds but has to follow the edges of G. To model this we call a pure strategy S *restricted* (to G) if either $(S_t, S_{t+1}) \in E$ or $S_t = S_{t+1}$ holds for any $t \in \mathbb{N}_0$. A (mixed) strategy is called restricted if it is a probability distribution over the set of restricted pure strategies. For the analysis we will consider only restricted strategies for the hunter and both restricted as well as unrestricted strategies for the rabbit.

Notice that in our definition, the hunter may start his walk on the graph at an arbitrary vertex. However, we want to point out that defining a fixed starting position for the hunter would not asymptotically effect the results of the paper.

1.2 Previous Work

A first study of the Hunter vs. Rabbit game can be found in [1]. The presented hunter strategy is based on a random walk on the graph and it is shown that the hunter catches an unrestricted rabbit within $O(nm^2)$ rounds, where n and m denote the number of nodes and edges, respectively. In fact, the authors place some additional restrictions on the space requirements for the hunter strategy, which is an aspect that we do not consider in this paper.

In the area of mobile ad-hoc networks related models are used to design communication protocols (see e.g. [2,3]). In this scenario, some mobile users (the "hunters") aid in transmitting messages to the receivers (the "rabbits"). The expected number of rounds needed to catch the rabbit in our model corresponds directly to the expected time needed to deliver a message. We improve the deliver time of known protocols, which are based on random walks.

Deterministic pursuit-evasion games in graphs are well-studied. In the early work by Parsons [14,15] the graph was considered to be a system of tunnels in which the fugitive is hiding. Parsons introduced the concept of the *search number* of a graph which is, informally speaking, the minimum number of guards needed to capture a fugitive who can move with arbitrary speed. LaPaugh [9] showed that there is always a search strategy (a sequence of placing, removing, or moving a pebble along an edge) such that no edge that is cleared at a point of time can be recontaminated again, i.e., if the fugitive is known not to be in edge e then there is no chance for him to enter edge e again in the remainder of the search. Meggido et al. [11] proved that the computation of the search number of a graph is an NP-hard problem which implies its NP-completeness because of LaPaugh's result.

If an edge can be cleared without moving along it, but it suffices to 'look into' an edge from a vertex, then the minimum number of guards needed to catch the fugitive is called the *node search number* of a graph [8].

Pursuit evasion problems in the plane were introduced by Suzuki and Yamashita [16]. They gave necessary and sufficient condition for a simple polygon to be searchable

by a single pursuer. Later Guibas et al. [6] presented a complete algorithm and showed
that the problem of determining the minimal number of pursuers needed to clear a
polygonal region with holes is NP-hard. Recently, Park et al. [13] gave 3 necessary and
sufficient conditions for a polygon to be searchable and showed that there is an $\mathcal{O}(n^2)$
time algorithm for constructing a search path for an n-sided polygon.

Efrat et al. [4] gave a polynomial time algorithm for the problem of clearing a simple
polygon with a chain of k pursuers when the first and last pursuer have to move on the
boundary of the polygon.

1.3   New Results

We present a hunter strategy for general networks that improves significantly on the
results obtained by using random walks. Let $G = (V, E)$ denote a connected graph with
n vertices and diameter $\mathrm{diam}(G)$. Recall that $\Omega(n)$ is a trivial lower bound on the escape
length against restricted as well as against unrestricted rabbit strategies on every graph
with n vertices. Our hunter strategy achieves escape length close to this lower bound.
In particular, we present a hunter strategy that has an expected escape length of only
$\mathcal{O}(n \log(\mathrm{diam}(G)))$ against any unrestricted rabbit strategy. Clearly, an upper bound
on the escape length against unrestricted rabbit strategies implies the same upper bound
against restricted strategies.

Our general hunter strategy is based on a hunter strategy for cycles which is then
simulated on general graphs. In fact, the most interesting and original parts of our analysis
deal with hunter strategies for cycles. Observe that if hunter and rabbit are restricted
to a cycle, then there is a simple, efficient hunter strategy with escape length $\mathcal{O}(n)$.
(In every nth round, the hunter chooses a *direction* at random, either clockwise or
counterclockwise, and then it follows the cycle in this direction for the next n rounds.)
Against unrestricted rabbits, however, the problem of devising efficient hunter strategies
becomes much more challenging. (For example, for the hunter strategy given above, there
is a simple rabbit strategy that results in an escape length of $\Theta(n\sqrt{n})$.) For unrestricted
rabbits on cycles of length n, we present a hunter strategy with escape length $\mathcal{O}(n \log n)$.
Furthermore, we prove that this result is optimal by devising an unrestricted rabbit
strategy with escape length $\Omega(n \log n)$ against any hunter strategy on the cycle.

Generalizing the lower bound for cycles, we can show that our general hunter strategy
is optimal in the sense that for any positive integers n, d with $d < n$ there exists a graph
G with n nodes and diameter d such that any hunter strategy on G has escape length
$\Omega(n \cdot \log(\mathrm{diam}(G)))$. This gives rise to the question whether $n \cdot \log(\mathrm{diam}(G))$ is a
universal lower bound on the escape length in any graph. We can answer this question
negatively. In fact, in a full version of this paper, we present a hunter strategy with escape
length $\mathcal{O}(n)$ for complete binary trees against unrestricted rabbits.

Finally, we investigate the Hunter vs. Rabbit game on strongly connected directed
graphs. We show that there exists a directed graph for which every hunter needs $\Omega(n^2)$
rounds to catch a restricted rabbit. Furthermore, for every strongly connected directed
graph, there is a hunter strategy with escape length $\mathcal{O}(n^2)$ against unrestricted rabbits.
Due to space limitations, the analyses for directed graphs as well has to be moved to the
full version.

1.4 Basic Concepts

The strategies will be analyzed in phases. A phase consists of m consecutive rounds, where m will be defined depending on the context. Suppose that we are given an m-round hunter strategy $\mathbb{H}$ and an m-round rabbit strategy $\mathbb{R}$ for a phase. We want to determine the probability that the rabbit is caught during the phase. Therefore we introduce the indicator random variables $hit(t), 0 \le t < m$ for the event $\mathcal{H}_t = \mathcal{R}_t$ that the pure hunter strategy $\mathcal{H}$ and the pure rabbit strategy $\mathcal{R}$ chosen according to $\mathbb{H}$ and $\mathbb{R}$, respectively, meet in round t of the phase. Furthermore, we define indicator random variables $fhit(t), 0 \le t < n$ describing first hits, i.e., $fhit(t) = 1$ iff $hit(t) = 1$ and $hit(t') = 0$ for every $t' \in \{0, \ldots, t-1\}$. Finally we define $hits = \sum_{t=0}^{m-1} hit(t)$.

The goal of our analysis is to derive upper and lower bounds for $\mathbf{Pr}[hits \ge 1]$, the probability that the rabbit is caught in the phase. To analyze the quality of an m-round hunter strategy we fix a pure rabbit strategy $\mathcal{R}$ and derive a lower bound on the probability $\mathbf{Pr}[hits \ge 1]$. Similarly to analyze the quality of an m-round rabbit strategy we fix a pure hunter strategy and derive an upper bound on $\mathbf{Pr}[hits \ge 1]$. Then we apply Yao's min-max principle [12] to derive the bounds for the mixed strategies.

The following two non-standard but fundamental propositions are important tools in the analysis of the upper and lower bounds.

Proposition 1. *Let $\mathbb{R}$ be an m-round rabbit strategy and let $\mathcal{H}$ be a pure m-round hunter strategy. Then*

$$\mathbf{Pr}[hits \ge 1] = \frac{\mathbf{E}[hits]}{\mathbf{E}[hits \mid hits \ge 1]} .$$

Proposition 2. *Let $\mathbb{H}$ be an m-round hunter strategy and let $\mathcal{R}$ be a pure m-round rabbit strategy. Then*

$$\mathbf{Pr}[hits \ge 1] \ge \frac{\mathbf{E}[hits]^2}{\mathbf{E}[hits^2]} .$$

The proofs of these propositions are interesting exercises.

2 Efficient Hunter Strategies

In this section we prove that for a graph G with n nodes and diameter $\mathrm{diam}(G)$, there exists a hunter strategy with expected escape length $\mathcal{O}(n \cdot \log(\mathrm{diam}(G)))$. For this general strategy we cover G with a set of small cycles and then use a subroutine for searching these cycles. We first describe this subroutine: an efficient hunter strategy for catching the rabbit on a cycle. The general strategy is described in section 2.2.

2.1 Strategies for Cycles and Circles

We prove that there is an $\mathcal{O}(n)$-round hunter strategy on an n-node cycle that has a probability of catching the rabbit of at least $\frac{1}{2H_n+1} = \Omega(\frac{1}{\log(n)})$, where H_n is the nth

harmonic number which is defined as $\sum_{i=1}^{n} \frac{1}{i}$. Notice that by repeating this strategy until the rabbit is caught we get a hunter strategy with an expected escape length of $\mathcal{O}(n \cdot \log(n))$. In order to keep the description of the strategy as simple as possible, we introduce a continuous version of the Hunter vs. Rabbit game for cycles. In this version the hunter tries to catch the rabbit on the boundary of a circle with circumference n. The rules are as follows. In every round the hunter and the rabbit reside at arbitrary, i.e., continuously chosen points on the boundary of the circle. The rabbit is allowed to jump, i.e., it can change its position arbitrarily between two consecutive rounds whereas the hunter can cover at most a distance of one. For the notion of *catching*, we partition the boundary of the circle into n distinct half open intervals of length one. The hunter catches the rabbit if and only if there is a round in which both the hunter and the rabbit reside in the same interval. Since each interval of the boundary corresponds directly to a node of the cycle and vice versa we can make the following observation.

Observation 1 *Any hunter strategy for the Hunter vs. Rabbit game on the circle with circumference n can be simulated on the n-node cycle, achieving the same expected escape length.*

The $\mathcal{O}(n)$-round hunter strategy for catching the rabbit on the circle consists of two phases that work as follows. In an *initialization phase* that lasts for $\lceil n/2 \rceil$ rounds the hunter first selects a random position on the boundary as the *starting position* of the following *main phase*. Then the hunter goes to this position. Note that $\lceil n/2 \rceil$ rounds suffice for the hunter to reach any position on the circle boundary. We will not care whether the rabbit gets caught during the initialization phase. Therefore there is no need for specifying the exact route taken by the hunter to get to the starting position.

After the first $\lceil n/2 \rceil$ rounds the *main phase* lasting for n rounds starts. The hunter selects a velocity uniformly at random between 0 and 1 and proceeds in a clockwise direction according to this velocity. This means that a hunter with starting position $s \in [0, n)$ and velocity $v \in [0, 1]$ resides at position $(s + t \cdot v) \bmod n$ in the tth round of the main phase. Obviously this so called RANDOMSPEED-strategy is an $\mathcal{O}(n)$-round hunter strategy since at most $\lceil \frac{3}{2}n \rceil$ nodes are visited. The following analysis shows that it achieves the desired probability of catching the rabbit when simulated on the n-node cycle.

Theorem 2. *On an n-node cycle a hunter using the RANDOMSPEED-strategy catches the rabbit with probability at least $\frac{1}{2H_n+1} = \Omega(\frac{1}{\log(n)})$.*

Proof. We prove that the bound holds for the Hunter vs. Rabbit game on the circle. The theorem then follows from Observation 1.

Since the rabbit strategy is oblivious in the sense that it does not know the random choices made by the hunter we can assume that the rabbit strategy is fixed in the beginning before the hunter starts. Thus, let $\mathcal{R} = \mathcal{R}_0, \mathcal{R}_1, \dots, \mathcal{R}_{n-1}$ denote the rabbit strategy during the main phase, i.e., $\mathcal{R}_t$ is the interval containing the rabbit in round t of this phase.

For this rabbit strategy let *hits* denote a random variable counting how often the hunter catches the rabbit. This means *hits* is the number of rounds during the main phase in which the hunter and the rabbit reside in the same interval. The theorem follows by showing that for any rabbit strategy $\mathcal{R}$ the probability $\mathbf{Pr}\left[hits \geq 1\right] = \mathbf{Pr}\left[\text{hunter catches rabbit}\right]$ is larger than $\frac{1}{2H_n+1}$. For this we use Proposition 2 to estimate $\mathbf{E}\left[hits\right]$ and $\mathbf{E}\left[hits^2\right]$. Let $\Omega = [0,n) \times [0,1]$ denote the sample space of the random experiment performed by the hunter. Further let $S_i^t \subset \Omega$ denote the subset of random choices such that the hunter resides in the ith interval during the tth round of the main phase. The hunter catches the rabbit in round t iff his random choice $\omega \in \Omega$ is in the set $S_{\mathcal{R}_t}^t$. By identifying $S_{\mathcal{R}_t}^t$ with its indicator function we can write $hits(\omega) = \sum_{t=0}^{n-1} S_{\mathcal{R}_t}^t(\omega)$.

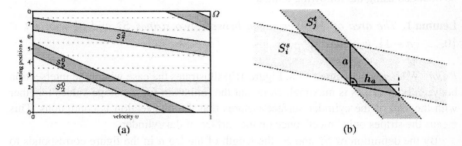

(a)   (b)

Fig. 1. (a) The sample space Ω of the RANDOMSPEED strategy can be viewed as the surface of a cylinder. The sets S_i^t correspond to stripes on this surface. (b) The intersection between two stripes of slope $-s$ and $-t$, respectively.

The following interpretation of the sets S_i^t will help derive bounds for $\mathbf{E}\left[hits\right]$ and $\mathbf{E}\left[hits^2\right]$. We represent Ω as the surface of a cylinder as shown in Figure 1(a). In this representation a set S_i^t corresponds to a stripe around the cylinder that has slope $\frac{ds}{dv} = -t$ and area 1. To see this recall that a point $\omega = (s,v)$ belongs to the set S_i^t iff the hunter position p_t in round t resulting from the random choice ω lies in the ith interval I_i. Since $p_t = (s + t \cdot v) \bmod n$ according to the RANDOMSPEED-strategy we can write S_i^t as $\{(s,v) \mid s = (p_t - t \cdot v) \bmod n \wedge p_t \in I_i\}$ which corresponds to a stripe of slope $-t$. For the area, observe that all n stripes S_i^t of a fixed slope t together cover the whole area of the cylinder which is n. Therefore each stripe has the same area of 1. This yields the following equation.

$$\mathbf{E}\left[hits\right] = \mathbf{E}\left[\sum_{t=0}^{n-1} S_{\mathcal{R}_t}^t\right] = \sum_{t=0}^{n-1} \mathbf{E}\left[S_{\mathcal{R}_t}^t\right] = \sum_{t=0}^{n-1} \int_\Omega \frac{1}{n} S_{\mathcal{R}_t}^t(\omega)\, d\omega = 1 \qquad (1)$$

Note that $\int_\Omega S_{\mathcal{R}_t}^t(\omega)\, d\omega$ is the area of a stripe and that $\frac{1}{n}$ is the density of the uniform distribution over Ω.

We now provide an upper bound on $\mathbf{E}\left[hits^2\right]$. By definition of $hits$ we have,

$$\mathbf{E}\left[hits^2\right] = \mathbf{E}\left[\left(\sum_{t=0}^{n-1} S^t_{\mathcal{R}_t}\right)^2\right] = \mathbf{E}\left[\sum_{s=0}^{n-1}\sum_{t=0}^{n-1} S^s_{\mathcal{R}_s} \cdot S^t_{\mathcal{R}_t}\right]$$

$$= \sum_{s=0}^{n-1}\sum_{t=0}^{n-1}\int_\Omega \frac{1}{n} S^s_{\mathcal{R}_s}(\omega) \cdot S^t_{\mathcal{R}_t}(\omega)\, d\omega \ . \tag{2}$$

$S^s_{\mathcal{R}_s}(\omega) \cdot S^t_{\mathcal{R}_t}(\omega)$ is the indicator function of the intersection between $S^s_{\mathcal{R}_s}$ and $S^t_{\mathcal{R}_t}$. Therefore $\int_\Omega S^s_{\mathcal{R}_s}(\omega) \cdot S^t_{\mathcal{R}_t}(\omega)\, d\omega$ is the area of the intersection of two stripes and can be bounded using the following lemma.

Lemma 1. *The area of the intersection between two stripes S^s_i and S^t_j with $s, t \in \{0, \dots, n-1\}$, is at most $\frac{1}{|t-s|}$.*

Proof. W.l.o.g. we assume $t > s$. Figure 1(b) illustrates the case where the intersection between both stripes is maximal. Note that the limitation for the slope values together with the size of the cylinder surface ensures that the intersection is contiguous. This means the stripes only "meet" once on the surface of the cylinder.

By the definition of S^s_i and S^t_j the length of the leg a in the figure corresponds to the length of an interval on the boundary of the circle. Thus $a = 1$. The length of h_a is $\frac{a}{t-s}$ and therefore the area of the intersection is $a \cdot h_a = \frac{a^2}{t-s} = \frac{1}{t-s}$. This yields the lemma. □

Using this Lemma we get

$$\sum_{t=0}^{n-1}\int_\Omega S^s_{\mathcal{R}_s}(\omega) \cdot S^t_{\mathcal{R}_t}(\omega)\, d\omega \le \sum_{t=0}^{s-1}\frac{1}{|t-s|} + \int_\Omega S^s_{\mathcal{R}_s}(\omega) \cdot S^s_{\mathcal{R}_s}(\omega)\, d\omega + \sum_{t=s+1}^{n-1}\frac{1}{|t-s|}$$

$$= \sum_{t=1}^{s}\frac{1}{t} + \int_\Omega S^s_{\mathcal{R}_s}(\omega)\, d\omega + \sum_{t=1}^{n-s-1}\frac{1}{t} \le 2H_n + 1 \ .$$

Plugging the above inequality into Equation 2 yields $\mathbf{E}\left[hits^2\right] \le 2H_n + 1$. Combining this with Proposition 2 and Equation 1 we get $\mathbf{Pr}\left[\text{hunter catches rabbit}\right] \ge \frac{1}{2H_n+1}$ which yields the theorem. □

2.2 Hunter Strategies for General Graphs

In this section we extend the upper bound of the previous section to general graphs.

Theorem 3. *Let $G = (V, E)$ denote a graph and let $\mathrm{diam}(G)$ denote the diameter of this graph. Then there exists a hunter strategy on G that has expected escape length $\mathcal{O}(|V| \cdot \log(\mathrm{diam}(G)))$.*

Proof. (Sketch) We *cover* the graph with $r = \Theta(n/d)$ cycles $C_1, \ldots, C_r$ of length d where $d = \Theta(diam(G))$, that is, each node is contained in at least one of these cycles (in order to obtain this covering, construct a tour of length $2n - 2$ along an arbitrary spanning tree, cut the tour into subpaths of length $d/2$ and then form a cycle of length d from each of these subpaths). From now on, if hunter or rabbit resides at a node of G corresponding to several cycle nodes, then we assume they *commit* to one of these virtual nodes and the hunter catches the rabbit only if they commit to the same node. This only slows down the hunter.

The hunter strategy now is to choose one of the r cycles uniformly at random and simulate the RANDOMSPEED-strategy on this cycle. Call this a *phase*. Observe that each phase takes only $\Theta(d)$ rounds. The hunter executes phase after phase, each time choosing a new random cycle, until the rabbit is caught. In the following we will show that the success probability within each phase is $\Omega(d/nH_d)$, which implies the theorem.

Let us focus on a particular phase. Suppose the hunter has chosen cycle C_i. Recall that the hunter chooses a random node v from C_i, walks to v and then starts traversing the cycle at a random speed. In the following consider only the hits in the *main phase*, i.e., those rounds after reaching v. The probability that hunter and rabbit reside in the same cycle at the beginning of the main phase is $1/r$. For simplicity, let us assume that the rabbit does not leave this cycle during the main phase. Under this simplifying assumption, Theorem 2 yields immediately that $\mathbf{Pr}\left[hits \geq 1\right] \geq \frac{1}{r(2H_d+1)} = \Omega(d/nH_d)$. In a full version we will prove this bound rigorously without this simplifying assumption. $\qquad\square$

3   Lower Bounds and Efficient Rabbit Strategies

We first prove that the hunter strategy for the cycle described in Section 2.1 is tight by giving an efficient rabbit strategy for the cycle. Then we provide lower bounds that match the upper bounds for general graphs given in Section 2.2.

3.1   An Optimal Rabbit Strategy for the Cycle

In this section we will prove a tight lower bound for any (mixed) hunter strategy on a cycle of length n. In particular, we describe a rabbit strategy such that every hunter needs $\Omega(n\log(n))$ expected time to catch the rabbit. We assume that the rabbit is unrestricted, i.e., can jump between arbitrary nodes, whereas the hunter is restricted to follow the edges of the cycle.

Theorem 4. *For the cycle of length n, there is a mixed, unrestricted rabbit strategy with escape length $\Omega(n\log(n))$ against any restricted hunter strategy.*

The rabbit strategy is based on a non-standard random walk. Observe that a standard random walk has the limitation that after n rounds, the rabbit is confined to a neighborhood of about $\sqrt{n}$ nodes around the starting position. Hence the rabbit is easily caught by a hunter that just sweeps across the ring (in one direction) in n steps. Also, the other extreme where the rabbit makes a jump to a node chosen uniformly at random in every round does not work, since in each round the rabbit is caught with probability exactly

$1/n$, giving an escape length of $O(n)$. But the following strategy will prove to be good for the rabbit. The rabbit will change to a randomly chosen position every n rounds and then, for the next $n - 1$ rounds, it performs a "heavy-tailed random walk". For this n-round strategy and any n-round hunter strategy, we will show that the hunter catches the rabbit with probability $\mathcal{O}(1/H_n)$. As a consequence, the expected escape length is $\Omega(n \log n)$, which gives the theorem.

A heavy-tailed random walk. We define a random walk on $\mathbb{Z}$ as follows. At time 0 a particle starts at position $X_0 = 0$. In a *step* $t \geq 1$, the particle makes a random jump $x_t \in \mathbb{Z}$ from position X_{t-1} to position $X_t = X_{t-1} + x_t$, where the jump length is determined by the following heavy-tailed probability distribution $\mathcal{P}$.

$$\mathbf{Pr}\,[x_t = k] \;=\; \mathbf{Pr}\,[x_t = -k] \;=\; \frac{1}{2(k+1)(k+2)} \;,$$

for every $k \geq 1$ and $\mathbf{Pr}\,[x_t = 0] = \frac{1}{2}$. Observe that $\mathbf{Pr}\,[|x_t| \geq k] = (k+1)^{-1}$, for every $k \geq 0$. The following lemma gives a property of this random walk that will be crucial for the proof of our lower bound. Due to space limitations, the proof for this lemma appears in the full version.

Lemma 2. *There is a constant $c_0 > 0$, such that, for every $t \geq 1$ and $\ell \in \{-t, \dots, t\}$, $\mathbf{Pr}\,[X_t = \ell] \geq c_0/t$.*

The rabbit strategy. Our n-round rabbit strategy starts at a random position on the cycle. Starting from this position, for the next $n-1$ rounds, the rabbit simulates the heavy-tailed random walk in a wrap-around fashion on the cycle. The following lemma immediately implies Theorem 4.

Lemma 3. *The probability that the hunter catches the rabbit within n rounds is $\mathcal{O}(1/H_n)$.*

Proof. Fix any n-round hunter strategy $\mathcal{H} = \mathcal{H}_0, \mathcal{H}_1, \dots, \mathcal{H}_{n-1}$. Because of Proposition 1 we only need to estimate $\mathbf{E}[hits]$ and $\mathbf{E}[hits \mid hits \geq 1]$. First, we observe that $\mathbf{E}[hits] = 1$. This is because the rabbit chooses its starting position uniformly at random so that $\mathbf{Pr}[hit(t) = 1] = 1/n$ for $0 \leq t < n$, and hence $\mathbf{E}[hit(t)] = \mathbf{Pr}[hit(t) = 1] = 1/n$. By linearity of expectation, we obtain $\mathbf{E}[hits] = \sum_{t=0}^{n-1} \mathbf{E}[hit(t)] = 1$. Thus, it remains only to show that $\mathbf{E}[hits \mid hits \geq 1] \geq c_0 H_n$. In fact, the idea behind the following proof is that we have chosen the rabbit strategy in such a way that when the rabbit is hit by the hunter in a round then it is likely that it will be hit additionally in several later rounds as well.

Claim. For every $\tau \in \{0, \dots, \frac{n}{2} - 1\}$, $\mathbf{E}[hits \mid fhit(\tau) = 1] \geq c_1 H_n$, for a suitable constant c_1.

Proof. Assume hunter and rabbit met at time τ for the first time, i.e., $fhit(\tau) = 1$. Observe that the hunter has to stay somewhere in interval $[\mathcal{H}_\tau - (t - \tau), \mathcal{H}_\tau + (t - \tau)]$ in round $t > \tau$ as he is restricted to the cycle. The heavy-tailed random walk will also have some tendency to stay in this interval. In particular, Lemma 2 implies, for every $t > \tau$, $\mathbf{Pr}[hit(t)] \geq c_0/(t-\tau)$. Consequently, $\mathbf{E}[hits \mid fhit(\tau) \geq 1] \geq 1 + \sum_{t=\tau+1}^{n-1} c_0/(t-\tau)$, which is $\Omega(H_n)$ since $\tau < n/2$. □

With this result at hand, we can now estimate the expected number of repeated hits as follows.

$$\mathbf{E}\left[hits \mid hits \geq 1\right] = \sum_{\tau=0}^{n-1} \mathbf{E}\left[hits \mid fhit(\tau) = 1\right] \cdot \mathbf{Pr}\left[fhit(\tau) = 1 \mid hits \geq 1\right]$$

$$\geq \sum_{\tau=0}^{n/2-1} \mathbf{E}\left[hits \mid fhit(\tau) = 1\right] \cdot \mathbf{Pr}\left[fhit(\tau) = 1 \mid hits \geq 1\right]$$

$$\geq c_1 H_n \sum_{\tau=0}^{n/2-1} \mathbf{Pr}\left[fhit(\tau) = 1 \mid hits \geq 1\right] \ .$$

Finally, observe that

$$\sum_{\tau=0}^{n/2-1} \mathbf{Pr}\left[fhit(\tau) = 1 \mid hits \geq 1\right] + \sum_{\tau=n/2}^{n-1} \mathbf{Pr}\left[fhit(\tau) = 1 \mid hits \geq 1\right] = 1 \ .$$

Thus, one of the two sums must be greater than or equal to $\frac{1}{2}$. If the first sum is at least $\frac{1}{2}$, then we directly obtain $\mathbf{E}\left[hits \mid hits \geq 1\right] \geq \frac{1}{2}c_1 H_n$. In the other case, one can prove the same lower bound by going backward instead of forward in time, that is, by summing over the last hits instead of the first hits. Hence Lemma 3 is shown. □

3.2 A Lower Bound in Terms of the Diameter

In this section, we show that the upper bound of Section 2.2 is asymptotically tight for the parameters n and $\operatorname{diam}(G)$. We will use the efficient rabbit strategy for cycles as a subroutine on graphs with arbitrary diameter.

Theorem 5. *For any positive integers n, d with $d < n$ there exists a graph G with n nodes and diameter d such that any hunter strategy on G has escape length $\Omega(n \cdot \log(d))$.*

Proof. For simplicity, we assume that n is odd, $d = 3d'$ and $N = (n-1)/2$ is a multiple of d'. The graph G consists of a *center* $s \in V$ and N/d' subgraphs called loops. Each *loop* consists of a cycle of length d' and a linear array of $d' + 2$ nodes such that the first node of the linear array is identified with one of the nodes on the cycle and the last node is identified with s. Thus, all loop subgraphs share the center s, otherwise the node sets are disjoint.

Every d' rounds the rabbit chooses uniformly at random one of the N/d' loops and performs the optimal d'-round cycle strategy from Section 3.1 on the cycle of this loop graph. Observe that the hunter cannot visit nodes in different cycles during a phase of length d'. Hence, the probability that the rabbit chooses a cycle visited by the hunter is at most d'/N. Provided that the rabbit chooses the cycle visited by the hunter the probability that it is caught during the next d' rounds is $\mathcal{O}(\frac{1}{H_{d'}})$ by Lemma 3. Consequently, the probability of being caught in one of the independent d'-round games is $\mathcal{O}(\frac{d'}{nH_{d'}})$. Thus, the escape length is $\Omega(nH_{d'})$ which is $\Omega(n \cdot \log(d))$. □

References

1. R. Aleliunas, R. M. Karp, R. J. Lipton, L. Lovasz, and C. Rackoff. Random walks, universal traversal sequences, and the complexity of maze problems. In *Proceedings of the 20th IEEE Symposium on Foundations of Computer Science (FOCS)*, pages 218–223, 1979.

2. I. Chatzigiannakis, S. Nikoletseas, N. Paspallis, P. Spirakis, and C. Zaroliagis. An experimental study of basic communication protocols in ad-hoc mobile networks. In *Proceedings of the 5thWorkshop on Algorithmic Engineering*, pages 159–171.

3. I. Chatzigiannakis, S. Nikoletseas, and P. Spirakis. Self-organizing ad-hoc mobile networks: The problem of end-to-end communication. In *Proceedings of the 20th ACM Symposium on Principles of Distributed Computing (PODC 2001)*, 2001.

4. A. Efrat, L. J. Guibas, S. Har-Peled, D. C. Lin, J. S. B. Mitchell, and T. M. Murali. Sweeping simple polygons with a chain of guards. In *Proceedings of the 11th ACM-SIAM Symposium on Discrete Algorithms (SODA)*, pages 927–936, 2000.

5. M. K. Franklin, Z. Galil, and M. Yung. Eavesdropping games: a graph-theoretic approach to privacy in distributed systems. *Journal of the ACM*, 47(2):225–243, 2000.

6. L. J. Guibas, J.-C. Latombe, S. M. LaValle, D. Lin, and R. Motwani. A visibility-based pursuit-evasion problem. *International Journal of Computational Geometry and Applications (IJCGA)*, 9(4):471–493, 1999.

7. K. Hatzis, G. Pentaris, P. Spirakis, and V. Tampakas. Implementation and testing eavesdropper protocols using the DSP tool. In *Proceedings of the 2nd Workshop on Algorithm Engineering (WAE)*, pages 74–85, 1998.

8. L. M. Kirousis and C. H. Papadimitriou. Searching and pebbling. *Theoretical Computer Science*, 47:205–218, 1986.

9. A. S. LaPaugh. Recontamination does not help to search a graph. *Journal of the ACM*, 40(2):224–245, 1993.

10. S. M. LaValle and J. Hinrichsen. Visibility-based pursuit-evasion: The case of curved environments. In *Proceedings of the IEEE International Conference on Robotics and Automation (ICRA)*, pages 1677–1682, 1999.

11. N. Megiddo, S. L. Hakimi, M. R. Garey, D. S. Johnson, and C. H. Papadimitriou. The complexity of searching a graph. *Journal of the ACM*, 35(1):18–44, 1988.

12. R. Motwani and P. Raghavan. *Randomized Algorithms*. Cambridge University Press, 1995.

13. S.-M. Park, J.-H. Lee, and K.-Y. Chwa. Visibility-based pursuit-evasion in a polgonal region by a searcher. In *Proceedings of the 28th International Colloquium on Automata, Languages and Programming (ICALP)*, pages 456–468, 2001.

14. T. D. Parsons. Pursuit-evasion in a graph. In Y. Alavi and D. Lick, editors, *Theory and Applications of Graphs*, Lecture Notes in Mathematics, pages 426–441. Springer, 1976.

15. T. D. Parsons. The search number of a connected graph. In *Proceedings of the 9th Southeastern Conference on Combinatorics, Graph Theory and Computing*, pages 549–554, 1978.

16. I. Suzuki and M. Yamashita. Searching for a mobile intruder in a polygonal region. *SIAM Journal on Computing*, 21(5):863–888, 1992.

The Essence of Principal Typings

J. B. Wells[1*]

Heriot-Watt University
http://www.cee.hw.ac.uk/~jbw/

Abstract. Let S be some type system. A *typing* in S for a typable term M is the collection of all of the information other than M which appears in the final judgement of a proof derivation showing that M is typable. For example, suppose there is a derivation in S ending with the judgement $A \vdash M : \tau$ meaning that M has result type τ when assuming the types of free variables are given by A. Then (A, τ) is a typing for M. A *principal typing* in S for a term M is a typing for M which somehow represents all other possible typings in S for M. It is important not to confuse this with a weaker notion in connection with the Hindley/Milner type system often called "principal types". Previous definitions of principal typings for specific type systems have involved various syntactic operations on typings such as *substitution* of types for type variables, *expansion, lifting*, etc.

This paper presents a new general definition of principal typings which does not depend on the details of any particular type system. This paper shows that the new general definition correctly generalizes previous system-dependent definitions. This paper explains why the new definition is the right one. Furthermore, the new definition is used to prove that certain polymorphic type systems using ∀-quantifiers, namely System F and the Hindley/Milner system, do not have principal typings. All proofs can be found in a longer version available at the author's home page.

1 Introduction

1.1 Background and Motivation

Why Principal Typings? A *term* represents a fragment of a program or other system represented in some calculus. In this paper, the examples will be drawn from the λ-calculus, but much of the discussion is independent of it. A typing t for a term in a specific type system for a calculus is *principal* if and only if all other typings for the same term can be derived from t by some set of semantically sensible operations. It is important not to confuse the *principal typings* property of a type system with the property of the Hindley/Milner system and the ML programming language often referred to (erroneously) as "principal types".

* This work was partly supported by EC FP5 grant IST-2001-33477, EPSRC grants GR/L 36963 and GR/R 41545, NATO grant CRG 971607, NSF grants 9806745, 9988529, and 0113193, and Sun Microsystems grant EDUD-7826-990410-US.

P. Widmayer et al. (Eds.): ICALP 2002, LNCS 2380, pp. 913–925, 2002.

Principal typings allow *compositional* type inference, where the procedure of finding types for a term uses only the analysis *results* for its immediate sub-fragments, which can be analyzed independently in any order. Compositional-ity helps with such things as performing separate analysis of program modules (and hence helps with separate compilation) and also helps in making a complete/terminating type inference algorithm. For a system lacking principal typ-ings, any type inference algorithm must either be incomplete (i.e., sometimes not finding a typing even though one exists), be noncompositional, or use some-thing other than typings of the system to represent intermediate results. An example of a noncompositional type inference algorithm is the $\mathcal{W}$ algorithm of Damas and Milner [6] for the Hindley/Milner (HM) type system which is used in programming languages like Haskell and Standard ML (SML). For an SML program fragment of the form (let val x = e_1 in e_2 end), the $\mathcal{W}$ algorithm first analyzes e_1 and then uses the result while analyzing e_2.

Why Automated Type Inference? Principal typings help with automated type inference in general. For higher-order languages, the necessary types can become quite complex and requiring all of the types to be supplied in advance is burdensome. It is desirable to have as much *implicit typing* as possible, where types are omitted by humans who write terms. With type inference, the compiler takes an untyped or partially typed term, and it either completes the typing of the term or reports an error if the term is untypable.

Algorithm $\mathcal{W}$ for HM is the most widely used type inference algorithm. HM supports polymorphism with quite restricted uses of $\forall$ quantifiers. In practice, HM's limitations on polymorphic types make some kinds of code reuse more difficult [19]. Programmers are sometimes forced into contortions to provide code for which the compiler can find typings. This has motivated a long search for more flexible type systems with good type inference algorithms. Candidates have included extensions of the HM system such as System F [7,25], $F_{\leq}$, F+η, or F_{ω}.

This search has yielded many negative results. For quite some time it seemed that HM was as good as a system could get and still have a complete/terminating type inference algorithm. Indeed, for many systems (F, F_{ω}, etc.), *typability* (whether an untyped program fragment can be given a type) has been proven undecidable, which means no type inference algorithm can be both complete and terminating for all program fragments. Wells proved this for System F [32,34], the finite-rank restrictions of F above 3 [16], and F+η [33]. Urzyczyn proved it for F_{ω} [29], Pottinger proved it for unrestricted intersection types [24], and Pierce proved that even the subtyping relation of $F_{\leq}$ is undecidable [22]. Even worse, for System F it seems hard to find an amount of type information less than total that is enough to obtain terminating type inference [28].

Along the way have been a few positive results, some extensions of the HM system, and some with restricted intersection types (cf. recent work on intersec-tion types of arbitrarily high finite ranks [17,15]).

A New Principal Typing Definition vs. $\forall$ Quantifiers. For many years it was not known whether some type systems with $\forall$ quantifiers could have principal typings. The first difficulty is simply in finding a sufficiently general system-independent definition of principal typings. The first such definition is given in this paper. A *typing* t is defined to be a pair $(A \vdash \tau)$ of a set A of *type assumptions* together with a *result type* τ. The meaning $\mathsf{Terms}(t)$ of a typing $t = (A \vdash \tau)$ in a particular type system is defined to be the set of all the program fragments M such that $A \vdash M : \tau$ is provable in the system (meaning "M is well typed with result type τ under the type assumptions A"). A typing t_1 is defined to be *stronger* than typing t_2 if and only if $\mathsf{Terms}(t_1) \subset \mathsf{Terms}(t_2)$. This is "stronger" because t_1 is a stronger predicate on terms than t_2 and provides more information about $M \in \mathsf{Terms}(t_1)$, allowing M to be used in more contexts. A typing t is defined to be *principal* in some system for program fragment M if and only if t is at least as strong as all other typings for M in that system.

Comparison with prior type-system-specific principal typing definitions shows the new definition either exactly matches the old definitions, or is slightly more liberal, admitting some additional principal typings. The new definition seems to be the best possible system-independent definition.

Importantly, the new definition can be used to show that various systems with $\forall$ quantifiers do not have principal typings. Using this definition, a this paper proves that HM and System F do not have principal typings. Because the definition used here is liberal, the failure of HM and System F to have principal typings by this definition can be taken to mean that there is no reasonable definition of "principal typings" such that these systems have them. The proof for System F can be adapted for related systems such as F+η, and System F's finite-rank restrictions Λ_k for $k \geq 3$.

If polymorphism is to be based only on $\forall$ quantifiers, it is not clear how to design a type system with the principal typings property. Even systems with extremely restricted uses of $\forall$ quantifiers such as the HM system do not have principal typings. The lack of principal typings has manifested itself as a difficulty in making type systems more flexible than the restrictive HM system that also have convenient type inference algorithms. But this difficulty appears to have been due to the use of the $\forall$ quantifier for supporting type polymorphism, because many type systems with intersection types have principal typings.

1.2 Summary of This Paper's Contributions

1. An explanation is given of the motivations behind the notion of "principal typing" and how this has affected type inference in widely used type systems such as the simply typed λ-calculus and HM.
2. A new, system-independent definition of principal typings is presented. It is shown that this definition correctly generalizes existing definitions.
3. The new definition is used to finally prove that HM and System F do *not* have principal typings. Proving this was impossible before.

2 Definitions

This paper restricts attention to the pure λ-calculus extended with one constant c of ground type. The type systems considered here derive judgements traditionally of the form $A \vdash M : \tau$. Many interesting type systems derive judgements with more information. To extend the machinery here to such type systems, the extra information should be considered part of the *typing* which is defined below.

Types. Each type system S will have a set of types $\mathsf{Types}(S)$. Let σ, τ, and ρ range over types. Included in the set of types are an infinite set of *type variables*, ranged over by α, β, and γ. There will also be one ground type, o, added to help illustrate certain typing issues. For a type system S considered in this paper, the set $\mathsf{Types}(S)$ will be some subset of the set Types of types given by the following pseudo-grammar:

$$\sigma, \tau, \rho ::= \alpha \mid \mathsf{o} \mid (\sigma \to \tau) \mid (\sigma \cap \tau) \mid (\forall \alpha\, \tau)$$

The *free type variables* $\mathsf{FTV}(\tau)$ of the type τ are those variables not bound by $\forall$. Types are identified which differ only by α-conversion. Let s range over *type substitutions*, finite maps from a set of type variables to types.

Typings. A pair $x{:}\sigma$ is a *type assumption*. A finite set of type assumptions is a *type environment*. Let A range over type environments. In this paper, type environments are required to mention each term variable at most once. Let $A(x)$ be undefined if x is not mentioned in A and otherwise be the unique type τ such that $(x{:}\tau) \in A$. Let $A_x = \{\, y{:}\tau \mid (y{:}\tau) \in A$ and $y \neq x \,\}$. Let $\mathsf{FTV}(A) = \bigcup_{(x:\tau) \in A} \mathsf{FTV}(\tau)$. Let $s(A) = \{\, (x{:}s(\tau)) \mid (x{:}\tau) \in A \,\}$.

Let a *typing judgement* be a triple of a type environment A, a λ-term M, and a result type τ with the meaning "M is well typed with result type τ under the type assumptions A". Rather than the traditional notation $A \vdash M : \tau$, this paper instead writes judgements in the form $M : (A \vdash \tau)$. Although this notation change simplifies the presentation, its real importance is that the new perspective will help dispel widespread misunderstanding of principal typings in the research community. The pair $A \vdash \tau$ is called a *typing*. Let t range over typings. If $t = (A \vdash \tau)$, let $\mathsf{FTV}(t) = \mathsf{FTV}(A) \cup \mathsf{FTV}(\tau)$ and let $s(t) = (s(A) \vdash s(\tau))$.

For a type system S, let the statement $S \rhd M : t$ hold iff the judgement $M : t$ is derivable using the typing rules of S. A type system S *assigns* typing t to term M iff $S \rhd M : t$. Let $\mathsf{Terms}_S(t) = \{\, M \mid S \rhd M : t \,\}$ and let $\mathsf{Typings}_S(M) = \{\, t \mid S \rhd M : t \,\}$.

Ordering Typings. This paper introduces a new ordering on typings in a type system S according to how much information they provide about terms to which they can be assigned, with typings that are lower in the order providing more information. Let $t_1 \leq_S t_2$ iff $\mathsf{Terms}_S(t_1) \subseteq \mathsf{Terms}_S(t_2)$.

Remark 1. Suppose $t_1 \leq_S t_2$. Then typing t_1 can be viewed as providing more information about any $M \in \mathsf{Terms}(t_1)$. If t_1 and t_2 are viewed as predicates on terms, then $t_1(M)$ implies $t_2(M)$ for any M. In practice, if all that is known about a term is whether t_1 or t_2 can be assigned to it, then knowing that $M \in \mathsf{Terms}(t_1)$ represents an increase in knowledge over knowing that $M \in \mathsf{Terms}(t_2)$. This increased knowledge enlarges the set of contexts in which it is *known* that it is safe to use M. □

Some Specific Type Systems. The formulation of the simply typed lambda calculus (STLC) presented here is in *Curry style*, meaning that type information is assigned to pure λ-terms [5]. The set Types(STLC) is the subset of Types containing all types which do not mention ∀ or ∩. The typing rules of STLC are CON, VAR, APP, and ABS:

CON		⇒	c : $(A \vdash o)$
VAR	$A(x) = \tau$	⇒	$x : (A \vdash \tau)$
APP	$(M : (A \vdash \sigma \to \tau)$ and $N : (A \vdash \sigma))$	⇒	$(MN) : (A \vdash \tau)$
ABS	$M : (A_x \cup \{x{:}\sigma\} \vdash \tau)$	⇒	$(\lambda x.M) : (A \vdash \sigma \to \tau)$

The Hindley/Milner type system (HM) is an extension of STLC which was introduced by Milner for use in the ML programming language [19,20]. The HM system introduces a syntactic form (let $x = M$ in N) for allowing definitions with polymorphic types. The set Types(HM) is the subset of Types containing all types which do not mention ∩ and which do not mention ∀ inside either argument of a function type constructor ("→"). In a typing $(A \vdash \tau)$, the type τ must not mention ∀. The typing rules of HM are CON, APP, and ABS from STLC and the new typing rules VAR$_{HM}$ and LET:

VAR$_{HM}$ $(A(x) = \forall \alpha.\tau$ and dom$(s) = \{\alpha\})$ ⇒ $x : (A \vdash s(\tau))$

LET $(M : (A \vdash \tau)$ and $\{\alpha\} = FTV(\tau)\backslash$FTV$(A)$ and $N : (A_x \cup \{x{:}\forall\alpha.\tau\} \vdash \sigma))$
⇒ (let $x = M$ in $N) : (A \vdash \sigma)$

Girard formulated System F [7] (independently invented by Reynolds [25]) in the *Church style*, with explicitly typed terms. The Curry style presentation of F which is given here was first published by Leivant [18]. The set Types(F) is the subset of Types containing all types which do not mention ∩. The typing rules of F are CON, VAR, APP, and ABS from STLC and the new typing rules INST and GEN:

INST $(M : (A \vdash \forall \alpha \sigma)$ and $s = \{\alpha \mapsto \tau\})$ ⇒ $M : (A \vdash s(\sigma))$
GEN $(M : (A \vdash \sigma)$ and $\alpha \notin FTV(A))$ ⇒ $M : (A \vdash \forall \alpha \sigma)$

3 History of Principal Typings

3.1 Basic Motivations and STLC

The notions of *principal type* and *principal typing* (which has also been called *principal pair*) first occurred in the context of type assignment systems for the λ-calculus or combinatory logic using simple types. The motivation was determining whether a term is typable and finding types for the term if it is typable. The key idea is to define the typing algorithm by structural recursion on terms. This means that in calculating types for a term M, the algorithm will invoke itself recursively on the immediate subterms of M. For this to work, the result returned by the recursive invocations must be sufficiently informative.

Example 1. This example illustrates the need for some sort of "most general" typing in the process of inferring type information. Consider these λ-terms:

$$M = \lambda z.NP \qquad N = \lambda w.w(wz) \qquad P = \lambda yx.x$$

A type inference algorithm Inf for STLC defined using structural recursion would generate a call tree like this:

$$\mathsf{Inf}(M) = \mathsf{Cmb}_\lambda(z, \mathsf{Inf}(NP)) = \mathsf{Cmb}_\lambda(z, \mathsf{Cmb}_@(\mathsf{Inf}(N), \mathsf{Inf}(P)))$$

Here the algorithm Inf uses two subalgorithms Cmb_λ and $\mathsf{Cmb}_@$ to combine the results from recursively processing the subterms.

Suppose the recursive call to $\mathsf{Inf}(P)$ were to return the typing $t_1 = (\emptyset \vdash \alpha \to \alpha \to \alpha)$, which is a derivable STLC typing for P. Unfortunately, there would be no way that $\mathsf{Cmb}_@$ could combine this result with any typing for N to yield a typing for NP. The application $(w(wz))$ inside N needs w to have a type of the shape $\sigma \to \sigma$ for some σ. This could be solved by using a typing like $t_2 = (\emptyset \vdash (\alpha \to \alpha) \to \alpha \to \alpha)$ for P.

However, the only thing that $\mathsf{Cmb}_@$ knows about the subterm P is the typing t_1, which does not imply the typing t_2. This can be seen from the following example of a term $P' \in \mathsf{Terms}(t_1) \setminus \mathsf{Terms}(t_2)$:

$$P' = \lambda xy.(\lambda z.x)(\lambda w.wx(wyx))$$

To see more precisely why $P' \in \mathsf{Terms}(t_1)$, here is a type-annotated version:

$$P' = \lambda x^\alpha.\lambda y^\alpha.(\lambda z^{(\alpha \to \alpha \to \alpha) \to \alpha}.x)(\lambda w^{\alpha \to \alpha \to \alpha}.wx(wyx))$$

It should also be clear why $P' \notin \mathsf{Terms}(t_2)$ — the types of x and y are forced to be the same by the applications (wx) and (wy), and this prevents P' from having a result type of the shape $\sigma \to \sigma$. Thus, $t_1 \nleq t_2$.

The problem here is that the result that $\mathsf{Inf}(P)$ returned is not the most general result. It would have been more useful to have returned the result $t_3 = (\emptyset \vdash \beta \to \alpha \to \alpha)$. In fact, t_3 is in a certain sense the *best* possible result, because it can be checked that $t_3 \leq t$ for every $t \in \mathsf{Typings}(P)$.   □

To avoid the kind of problems mentioned in example 1, type inference algorithms have been designed so that their intermediate results for subterms are, in some sense, most general. There have been several different ways of characterizing the needed notions of "most general". Hindley [9] gives the following definitions which were intended for use with STLC.

Definition 1 (Hindley's Principal Type). *A principal type in system S of a term M is a type τ such that*

1. *there exists a type environment A such that $S \rhd M : (A \vdash \tau)$, and*
2. *if $S \rhd M : (A' \vdash \tau')$, then there exists some s such that $\tau' = s(\tau)$.*   □

Notice that definition 1 completely ignores the type environments! Hindley used the name "principal pair" for what is called here "principal typing".

Definition 2 (Hindley's Principal Typing). *A principal typing in system S of a term M is a typing $t = (A \vdash \tau)$ such that*

1. *$S \rhd M : t$, and*
2. *if $S \rhd M : t'$ for some typing $t' = (A' \vdash \tau')$, then there exists some s such that $A' \supseteq s(A)$ and $\tau' = s(\tau)$.*   □

Clearly, if $t = (A \vdash \tau)$ is a principal typing for a term M, then the result type τ is a principal type. The key property satisfied by STLC w.r.t. these definitions is the following. (See [9] for the history of this discovery.)

Theorem 1 (Principality for STLC). *Every term typable in STLC has a principal typing and a principal type. Also, there is an algorithm that decides if a term is typable in STLC and if the answer is "yes" outputs the principal typing.*   □

Hindley gave a further definition of a *principal derivation* (called by Hindley "deduction") which is not needed for this discussion. These definitions of Hindley essentially represent the earlier approaches of Curry and Feys [5] and Morris [21]. There are two important aspects of this approach:

1. The notion of "more general" is tied to substitution and weakening. For STLC, this exactly captures what is needed, but this fails for more sophisticated type systems.
2. The literature using these definitions freely switches between "principal type" and "principal typing" (or "principal pair"). The algorithms for STLC which are described as having the goal of calculating the "principal type" in fact are designed to calculate principal typings. Because for STLC every term has both a principal typing and a principal type, many people did not pay much attention to the difference. For more sophisticated type systems the difference becomes important.

3.2   Type Polymorphism and HM

Although STLC is well behaved, in practice it is quite insufficient for programming languages. To overcome the limitations of STLC, various approaches to adding type polymorphism have been explored and for each approach efforts have been directed to the problem of type inference.

One approach to adding type polymorphism is System F, which was discovered around the beginning of the early 1970s. Towards the end of that decade people were thinking about Curry-style presentations of F and how to perform type inference for it [18]. In the mid-1990s, I proved that typability for F is undecidable and that therefore there is no complete and always terminating type inference algorithm for F [32,34]. Later in this paper, the further result that F does not have principal typings is proven.

So far, the most successful and widely used approach to adding type polymorphism is the Hindley/Milner (HM) system, an extension of STLC and also a restriction of F. The approach to type inference for HM differs from that for STLC, because Hindley's notion of principal typing (needed by the type inference algorithms used for STLC) quite clearly does not hold for HM.

Example 2. This example illustrates why definition 2 is not useful for HM. Consider the λ-terms $M = (\text{let } x = (\lambda y.y) \text{ in } N)$ and $N = (xx)$. The term M is typable in HM. For example, the judgement $M : t$ where $t = (\emptyset \vdash \alpha \to \alpha)$ can be derived in HM. A derivation of this typing might use as an intermediate step the assignment of the typing $t_1 = (\{x : \forall \beta.\beta \to \beta\} \vdash \alpha \to \alpha)$ to the subterm N. Given any σ, let σ^0 stand for σ and let σ^{i+1} stand for $\sigma^i \to \sigma^i$. Thus, using the new abbreviation, $t_1 = (\{x : \forall \beta.\beta^1\} \vdash \alpha^1)$. The subterm N can in fact be assigned for any $i \geq 0$ the typing $t_i = (\{x : \forall \beta.\beta^i\} \vdash \alpha^i)$. And for distinct $i, j \geq 0$, there is no substitution s such that $t_i = s(t_j)$. Note that the type $\forall \beta.\beta^i$ is closed and $s(\tau) = \tau$ for any closed type τ and substitution s. Furthermore, it is not hard to check for $i \geq 0$ that for any other typing t' assignable to N, that there is no substitution s such that $t_i = s(t')$. So N has no principal typing using definition 2. In contrast, the term M does have a principal typing by that definition, namely t. Although some HM-typable terms (e.g., N) have no principal typings by definition 2, it turns out that any HM-typable term with no free variables does. □

Until this paper, it was not known whether we were simply not clever enough to conceive of a set of operations which would yield all other HM typings from (hypothetical) HM principal typings. Later in this paper, it is shown that there is no reasonable replacement definition of principal typing that will work for HM.

Milner's cleverness was in finding a way around this problem. The key lies in in the following definition (a clear statement of which can be found in [8]).

Definition 3 (A-Typable and A-Principal).

1. *A term M is A-typable in HM iff there is some A' mentioning only monotypes (types without any occurrence of "$\forall$") and some type τ such that $M : (A \cup A' \vdash \tau)$ is derivable in HM.*
2. *A typing $(A \cup A' \vdash \tau)$ is A-principal for term M in HM iff*
   a) *A' mentions only monotypes,*
   b) *$M : (A \cup A' \vdash \tau)$ is derivable in HM, and*
   c) *whenever $M : ((s(A)) \cup A'' \vdash \tau')$ is derivable for A'' mentioning only monotypes, there is a substitution s' such that $A'' \supseteq (s'(A'))$, $\tau' = s'(\tau)$, and $s'(\alpha) = s(\alpha)$ for $\alpha \in \mathsf{FTV}(A)$.* □

The property that HM satisfies w.r.t. the above definition is the following, due to Damas and Milner [6].

Theorem 2 (Principality for HM). *If a term is A-typable in HM, then it has a A-principal typing in HM. Also, there is an algorithm that decides if a term is A-typable in HM and if the answer is "yes" outputs its A-principal typing.* □

It is not hard to see that a closed term (with no free variables) is A-typable iff it is $\emptyset$-typable. So theorem 2 implies that typability is decidable for closed programs and that closed programs have $\emptyset$-principal typings. This is good enough for use in a type inference algorithm for a programming language implementation and, indeed, the HM type system has been very widely used as a result. There are some drawbacks that should be noticed:

1. In order to take advantage of the notion of A-principality, any polymorphic types to be used in a term M must be determined before analyzing M. So in analyzing a subterm of the shape (let $x = N$ in M), the subterm N must be completely analyzed before M and the result of analyzing N is used in analyzing M. This is the behavior of Milner's algorithm $\mathcal{W}$ [19].

2. Because the only notion of principality requires fixing the part of the type environment containing polytypes, for an arbitrarily chosen HM-typable term there does not seem to be an HM typing which represents *all* possible typings for that term. Later in this paper it is shown in fact that this is not mere appearance — individual HM typings can not be used to represent all possible HM typings for a term. This makes it more difficult to use the HM type system for approaches that involve *incremental* or *separate* computation of type information. So HM may not be right for some applications.

3.3   Principal Typings with Intersection Types

At the present time, for a type system to support both type polymorphism and principal typings, the most promising approaches rely on the use of intersection types. There is not room in this paper to go into much detail about intersection types, so the discussion here gives only the highlights.

The first system of intersection types for which principal typings was proved was presented by Coppo, Dezani, and Venneri [2] (a later version is [3]). The same general approach has been followed by Ronchi Della Rocca and Venneri [27] and van Bakel [30] for other systems of intersection types. In this approach, finding a principal typing algorithm for a term M involves

- finding a normal form (or *approximate* normal form) M' for M,
- assigning a typing t to M',
- proving that any typing for the normal form M' is also a typing for the original term M, and
- proving that any other typing t' for the normal form M' can be obtained from t by a sequence of operations each of which is one of *expansion* (sometimes called *duplication*), *lifting* (implementing subtyping, sometimes called *rise*), or *substitution*.

This general approach is summarized nicely in [31, §5.3]. This is intrinsically an impractical method and hence is primarily of theoretical interest. The definitions of the operations on typings are sometimes quite complicated, so they will not be discussed in this paper.

The first unification-based approach to principal typing with intersection types is by Ronchi Della Rocca [26]. An always-terminating restriction is presented which bounds the height of types. Unfortunately, this approach uses a complicated approach to expansion and is thus quite difficult to understand.

The first potentially practical approaches to principal typing with intersection types were subsequent unification-based methods which focused on the rank-2 restriction of intersection types. Van Bakel presented a unification algorithm for principal typing for a rank-2 system [30]. Later independent work by Jim also attacks the same problem, but with more emphasis on handling practical programming language issues such as recursive definitions, separate compilation, and accurate error messages [14]. Successors to Jim's method include Banerjee's [1], which integrates flow analysis, and Jensen's [12], which integrates strictness analysis. Other approaches to principal typings and type inference with intersection types include [4] and [11].

The most recent development in this area is the introduction of the notion of *expansion variables* [17]. The key idea is that with expansion variables, the earlier notions of expansion and substitution can be integrated in a single notion of substitution called β-substitution. This results in a great simplification over earlier approaches beyond the rank-2 restriction. However, there are still many technical issues needing to be overcome before this is ready for general use.

3.4   An Observation about All Previous Definitions

The following holds for each system S with principal typings that I know about. In S, each typable term M can be assigned a typing t which is *principal for M* in the sense that for every other typing t' assignable to M, there exist operations $O_1, \ldots, O_n$ such that

- $t' = O_n(\cdots (O_1(t)) \cdots)$, and
- the operations are *sound* in the sense that for any term N, if $S \rhd N : t$, then $S \rhd N : t_i$ where $t_i = O_i(\cdots (O_1(t)) \cdots)$ for $1 \leq i \leq n$.

For some (but not all) systems a stronger statement about the soundness of the operations holds:

For $0 \leq i < n$ and any term N, if $S \rhd N : t'$, then $S \rhd N : O_i(t')$.

For STLC, these operations are substitution and weakening, which are sound in the stronger sense (provided weakening is defined to do nothing on typings already mentioning the term variable in question). For various systems with intersection types, these operations are expansion, lifting, rise, as well as substitution and weakening. In some systems with intersection types, these operations are sound in the stronger sense, and in others, they are sound in the weaker sense.

Observation 3 *In each such system S, if t is principal for M, then $t \leq_S t'$ for every $t' \in \mathsf{Typings}_S(M)$.* $\square$

4   A New Definition of Principal Typing

In designing a general definition of "principal typing", the important issue seems to be the following:

A *principal typing* in S for a term M should be a typing for M which somehow represents all other possible typings in S for M.

This paper has already introduced the new technical machinery necessary to capture this notion in the information order $\leq_S$ on typings. This suggests that the following new definition is the right one.

Definition 4 (Principal Typing). *A typing t is principal in system S for term M iff $S \rhd M : t$ and $S \rhd M : t'$ implies $t \leq_S t'$.* □

4.1   Positive Results

The first thing to check about the new definition is whether existing definitions for specific type systems are instances of the new definition. In all cases, every typing that is principal by one of the old definitions is principal by the new one. This is justified by observation 3.

It remains to be considered whether every typing in a system S that is principal by the new definition is also principal by the old one for S. For STLC, the new definition corresponds exactly to the old definition.

Theorem 4. *A typing t is principal in STLC for M according to definition 2 iff t is principal in STLC for M according to definition 4.* □

For some other type systems, the new definition will be slightly more liberal, accepting some additional typings as being principal. For example, consider the term $\omega = (\lambda x.xx)$. The usual principal typing of ω in a system with intersection types is $t_1 = (\emptyset \vdash ((\alpha \to \beta) \cap \alpha) \to \beta)$. With the new definition, $t_2 = (\emptyset \vdash ((\alpha \to \beta) \cap \alpha \cap \gamma) \to \beta)$ is also a principal typing, because in some systems with intersection types a term can be assigned t_1 iff it can be assigned t_2. This is merely a harmless quirk. The old definitions ruled out unneeded intersections with type variables because they were inconvenient.

4.2   Type Systems without Principal Typings

The new definition 4 of principal typings can be used to finally prove that certain type systems do *not* have principal typings. These results are significant, as clarified by this statement by Jim in [13] about the best previous knowledge:

> "This imprecision [in the definition of principal typings] makes it impossible for us to *prove* that a given type system lacks the principal type property."

Theorem 5. *The HM system does not have principal typings for all terms.* □

It is quite important that the research community is made aware of Theorem 5, because as Jim points out in [13], "a number of authors have published offhand claims that ML possesses the principal typings property".

Theorem 6. *System F does not have principal typings for all terms.* □

References

[1] A. Banerjee. A modular, polyvariant, and type-based closure analysis. In *Proc. 1997 Int'l Conf. Functional Programming*. ACM Press, 1997.

[2] M. Coppo, M. Dezani-Ciancaglini, B. Venneri. Principal type schemes and λ-calculus semantics. In Hindley and Seldin [10].

[3] M. Coppo, M. Dezani-Ciancaglini, B. Venneri. Functional characters of solvable terms. *Z. Math. Logik Grundlag. Math.*, 27(1), 1981.

[4] M. Coppo, P. Giannini. A complete type inference algorithm for simple intersection types. In *17th Colloq. Trees in Algebra and Programming*, vol. 581 of *LNCS*. Springer-Verlag, 1992.

[5] H. B. Curry, R. Feys. *Combinatory Logic I*. Studies in Logic and the Foundations of Mathematics. North-Holland, Amsterdam, 1958.

[6] L. Damas, R. Milner. Principal type schemes for functional programs. In *Conf. Rec. 9th Ann. ACM Symp. Princ. of Prog. Langs.*, 1982.

[7] J.-Y. Girard. *Interprétation Fonctionnelle et Elimination des Coupures de l'Arithmétique d'Ordre Supérieur*. Thèse d'Etat, Université de Paris VII, 1972.

[8] R. Harper, J. C. Mitchell. On the type structure of Standard ML. *ACM Trans. on Prog. Langs. & Systs.*, 15(2), 1993.

[9] J. R. Hindley. *Basic Simple Type Theory*, vol. 42 of *Cambridge Tracts in Theoretical Computer Science*. Cambridge University Press, 1997.

[10] J. R. Hindley, J. P. Seldin, eds. *To H. B. Curry: Essays on Combinatory Logic, Lambda Calculus, and Formalism*. Academic Press, 1980.

[11] B. Jacobs, I. Margaria, M. Zacchi. Filter models with polymorphic types. *Theoret. Comput. Sci.*, 95, 1992.

[12] T. Jensen. Inference of polymorphic and conditional strictness properties. In POPL '98 [23].

[13] T. Jim. What are principal typings and what are they good for? Tech. memo. MIT/LCS/TM-532, MIT, 1995.

[14] T. Jim. What are principal typings and what are they good for? In *Conf. Rec. POPL '96: 23rd ACM Symp. Princ. of Prog. Langs.*, 1996.

[15] A. J. Kfoury, H. G. Mairson, F. A. Turbak, J. B. Wells. Relating typability and expressibility in finite-rank intersection type systems. In *Proc. 1999 Int'l Conf. Functional Programming*. ACM Press, 1999.

[16] A. J. Kfoury, J. B. Wells. A direct algorithm for type inference in the rank-2 fragment of the second-order λ-calculus. In *Proc. 1994 ACM Conf. LISP Funct. Program.*, 1994.

[17] A. J. Kfoury, J. B. Wells. Principality and decidable type inference for finite-rank intersection types. In *Conf. Rec. POPL '99: 26th ACM Symp. Princ. of Prog. Langs.*, 1999.

[18] D. Leivant. Polymorphic type inference. In *Conf. Rec. 10th Ann. ACM Symp. Princ. of Prog. Langs.*, 1983.

[19] R. Milner. A theory of type polymorphism in programming. *J. Comput. System Sci.*, 17, 1978.

[20] R. Milner, M. Tofte, R. Harper, D. B. MacQueen. *The Definition of Standard ML (Revised)*. MIT Press, 1997.

[21] J. H. Morris. *Lambda-calculus Models of Programming Languages*. PhD thesis, Massachusetts Institute of Technology, Cambridge, Mass., U.S.A., 1968.

[22] B. Pierce. Bounded quantification is undecidable. *Inform. & Comput.*, 112, 1994.

[23] *Conf. Rec. POPL '98: 25th ACM Symp. Princ. of Prog. Langs.*, 1998.

[24] G. Pottinger. A type assignment for the strongly normalizable λ-terms. In Hindley and Seldin [10].

[25] J. C. Reynolds. Towards a theory of type structure. In *Colloque sur la Programmation*, vol. 19 of *LNCS*, Paris, France, 1974. Springer-Verlag.

[26] S. Ronchi Della Rocca. Principal type schemes and unification for intersection type discipline. *Theoret. Comput. Sci.*, 59(1–2), 1988.

[27] S. Ronchi Della Rocca, B. Venneri. Principal type schemes for an extended type theory. *Theoret. Comput. Sci.*, 28(1–2), 1984.

[28] A. Schubert. Second-order unification and type inference for Church-style polymorphism. In POPL '98 [23].

[29] P. Urzyczyn. Type reconstruction in $\mathbf{F}_\omega$. *Math. Structures Comput. Sci.*, 7(4), 1997.

[30] S. J. van Bakel. *Intersection Type Disciplines in Lambda Calculus and Applicative Term Rewriting Systems*. PhD thesis, Catholic University of Nijmegen, 1993.

[31] S. J. van Bakel. Intersection type assignment systems. *Theoret. Comput. Sci.*, 151(2), 1995.

[32] J. B. Wells. Typability and type checking in the second-order λ-calculus are equivalent and undecidable. In *Proc. 9th Ann. IEEE Symp. Logic in Comput. Sci.*, 1994. Superseded by [34].

[33] J. B. Wells. Typability is undecidable for F+eta. Tech. Rep. 96-022, Comp. Sci. Dept., Boston Univ., 1996.

[34] J. B. Wells. Typability and type checking in System F are equivalent and undecidable. *Ann. Pure Appl. Logic*, 98(1–3), 1999. Supersedes [32].

Complete and Tractable Local Linear Time Temporal Logics over Traces

Bharat Adsul* and Milind Sohoni

Department of Computer Science and Engineering,
Indian Institute of Technology, Mumbai 400 076, India
Fax: 91 22 572 0290,{abharat,sohoni}@cse.iitb.ac.in

Abstract. We present the first expressively complete and yet tractable temporal logics, PAST-TrLTL and TrLTL, to reason about distributed behaviours, modelled as Mazurkiewicz traces. Both logics admit singly exponential automata-theoretic decision procedures. General formulas of PAST-TrLTL are boolean combinations of local formulas which assert rich properties of local histories of the behaviours. PAST-TrLTL has the same expressive power as the first order theory of finite traces. TrLTL provides formulas to reason about recurring local PAST-TrLTL properties and equals the complete first order theory in expressivity. The expressive completeness criteria are based on new local normal forms for the first order logic. We illustrate the use of our logics for specification of global properties.

Keywords. Temporal Logics, Concurrency, Mazurkiewicz Traces, Limits

Introduction

Propositional Linear time Temporal Logic (LTL) has proved to be a useful formalism in the specification and analysis of reactive distributed systems [MP2]. While equalling the expressive power of the first order logic, LTL has the convenience of admitting an elementary decision procedure. However, since formulas of LTL are interpreted over sequences, important attributes of behaviours of distributed systems, such as causality and locality, are clumsily modelled. The correct structures for these behaviours are of course, Mazurkiewicz traces, or simply traces [DR].

It is argued in [MT] that a tractable and expressively complete logic over traces would be of considerable theoretical and practical use. It is possible to extend, naturally, LTL to traces. However this extension, although expressively complete [DG1,TW], is not tractable [Wal]. Many logics for traces have been proposed, some tractable and other expressive (see [MT] for a survey), but none

---

* The work of this author was supported by an INFOSYS FELLOWSHIP.

P. Widmayer et al. (Eds.): ICALP 2002, LNCS 2380, pp. 926–937, 2002.

–to date– both. A notable exception is [DG2] which describes a tractable and expressively complete logic for series-parallel graphs, a special class of traces.

We propose here new temporal logics over traces, PAST-TrLTL and TrLTL, which are expressively complete *and* tractable. In fact, both PAST-TrLTL and TrLTL, admit singly exponential decision procedures. Following [Thi], the decision procedures are based on asynchronous automata and hence local in nature. The core of PAST-TrLTL is a rich suite of local past operators and equals the first order theory of finite traces in expressive power. TrLTL mainly provides a local "infinitely often" operator to reason about recurring local PAST-TrLTL properties. TrLTL is shown to be expressively complete by proving a strengthening of McNaughton's theorem (see [Tho]). This states that global properties of infinite behaviours may also be described as boolean combinations of recurring properties of finite local histories of the behaviours.

The sections are organized as follows. In Section 1, we develop the notation for traces. Section 2 presents the syntax and semantics of PAST-TrLTL and TrLTL. A very brief outline of the elementary decision procedures is sketched in Section 3. Section 4 deals with the expressiveness issues. In Section 5, we present typical safety, liveness properties and program invariants.

1   Traces

Let $\mathcal{P} = \{1, 2, \ldots, K\}$ be a set of agents also called as processes. A distributed alphabet over $\mathcal{P}$ is a family $\widetilde{\Sigma} = (\Sigma_i)_{i \in \mathcal{P}}$. Let $\Sigma = \bigcup_{i \in \mathcal{P}} \Sigma_i$. For $a \in \Sigma$, we set $\mathrm{loc}(a) = \{i \in \mathcal{P} \mid a \in \Sigma_i\}$. By (Σ, I) we denote the corresponding trace alphabet, i.e., I is the *independence relation* $I = \{(a, b) \in \Sigma^2 \mid \mathrm{loc}(a) \cap \mathrm{loc}(b) = \emptyset\}$ induced by $\widetilde{\Sigma}$. The corresponding *dependence relation* $(\Sigma \times \Sigma) \setminus I$ is denoted by D.

A Σ-labelled poset is a structure $F = (E, \leq, \lambda)$, where E is a set, $\leq$ is a partial order on E and $\lambda : E \to \Sigma$ is a labelling function. For $e, e' \in E$, define $e \lessdot e'$ iff $e < e'$ and for each e'' with $e \leq e'' \leq e'$ either $e = e''$ or $e' = e''$. For $X \subseteq E$, let $\downarrow X = \{y \in E \mid y \leq x \text{ for some } x \in X\}$. For $e \in E$, we set $\downarrow e = \downarrow \{e\}$. A *trace* over $\widetilde{\Sigma}$ is a countable Σ-labelled poset $F = (E, \leq, \lambda)$ satisfying the following conditions.

- For each $e \in E$, $\downarrow e$ is a finite set.
- If $e, e' \in E$ with $e \lessdot e'$ then $(\lambda(e), \lambda(e')) \in D$.
- If $e, e' \in E$ with $(\lambda(e), \lambda(e')) \in D$ then $e \leq e'$ or $e' \leq e$.

Let $TR(\widetilde{\Sigma})$ denote the set of traces over $\widetilde{\Sigma}$. Henceforth, a trace means a trace over $\widetilde{\Sigma}$ unless specified otherwise. Fix $F = (E, \leq, \lambda) \in TR(\widetilde{\Sigma})$. We say F is a finite trace if E is finite, otherwise F is said to be an infinite trace. We denote sets of finite and infinite traces over $\widetilde{\Sigma}$ by $TR^*(\widetilde{\Sigma})$ and $TR^\omega(\widetilde{\Sigma})$ respectively. Often, we shall write TR, TR^* and TR^ω instead of $TR(\widetilde{\Sigma}), TR^*(\widetilde{\Sigma})$ and $TR^\omega(\widetilde{\Sigma})$ respectively. A trace language is just a subset of TR.

Let $F = (E, \leq, \lambda) \in TR$. E is the set of events in F and for an event e in F, $\mathrm{loc}(e)$ abbreviates $\mathrm{loc}(\lambda(e))$. Further, let $i \in \mathcal{P}$. The set of i-events in F is

$E_i = \{e \in E \mid i \in \mathrm{loc}(e)\}$. This is the set of events in which process i participates. It is clear that E_i is totally ordered with respect to $\leq$.

A subset $c \subseteq E$ is a *configuration* iff c is finite and $\downarrow c = c$. We let $\mathcal{C}_F$ denote the set of configurations of F. Notice that $\emptyset$, the empty set, is a configuration. More importantly, $\downarrow e$ is a configuration for every $e \in E$.

Let $c \in \mathcal{C}_F$ and $i \in \mathcal{P}$. Then $\downarrow^i(c)$ is the i-view of c and it is defined as: $\downarrow^i(c) = \downarrow(c \cap E_i)$. We note that $\downarrow^i(c)$ is also a configuration. It is the "best" configuration that the agent i is aware of at c. It is easy to see that if $\downarrow^i(c) \neq \emptyset$ then there exists $e \in E_i$ such that $\downarrow^i(c) = \downarrow e$. We say that $\downarrow^i(c)$ is an *i-local configuration*. Let $\mathcal{C}_F^i = \{\downarrow^i(c) \mid c \in \mathcal{C}_F\}$ be the set of i-local configurations. For $Q \subseteq \mathcal{P}$ and $c \in \mathcal{C}_F$, we let $\downarrow^Q(c)$ denote the set $\bigcup\{\downarrow^i(c) \mid i \in Q\}$. Once again, $\downarrow^Q(c)$ is a configuration. It represents the collective knowledge of the processes in Q about the configuration c.

2   New Linear Time Temporal Logics

We propose new temporal logics to reason about traces. We call them $\mathrm{TrLTL}(\widetilde{\Sigma})$ (Trace-based Linear time Temporal Logic) and $\mathrm{PAST\text{-}TrLTL}(\widetilde{\Sigma})$. Having fixed $\widetilde{\Sigma}$, we abbreviate $\mathrm{TrLTL}(\widetilde{\Sigma})$ and $\mathrm{PAST\text{-}TrLTL}(\widetilde{\Sigma})$ by TrLTL and $\mathrm{PAST\text{-}TrLTL}$ respectively. We start with the description of $\mathrm{PAST\text{-}TrLTL}(\widetilde{\Sigma})$.

2.1   PAST-TrLTL

Φ, the set of formulas of $\mathrm{PAST\text{-}TrLTL}$, is defined inductively via:

- $\top \in \Phi$.
- If α and β belong to Φ, so do $\neg\alpha$ and $\alpha \vee \beta$.
- For $i \in \mathcal{P}$ and $a \in \Sigma_i$, if $\alpha \in \Phi$ and $\beta \in \Phi$ then $\langle \bar{a} \rangle_i \{\alpha, \beta\} \in \Phi$.
- For $i \in \mathcal{P}$, if $\alpha, \beta \in \Phi$ then $\alpha \, \mathcal{S}_i \beta \in \Phi$.

We develop the necessary notation to present the semantics of $\mathrm{PAST\text{-}TrLTL}$. Let $F = (E, \leq, \lambda) \in TR$ and $c \in \mathcal{C}_F$. We denote by F_c the induced finite trace $F_c = (c, \leq', \lambda') \in TR^*$ where $\leq' = \leq \cap(c \times c)$ and $\lambda' : c \to \Sigma$ is λ restricted to c.

Let $i \in \mathcal{P}$. We define $\downarrow^i_{\mathrm{prev}}(c)$ to be the i-local configuration which corresponds to the knowledge of agent i before executing the last action. More formally, let e be the maximum i-event in c (assuming it exists, otherwise we set $\downarrow^i_{\mathrm{prev}}(c) = \emptyset$). We set $\downarrow^i_{\mathrm{prev}}(c) = \downarrow^i(c - \{e\})$.

We also set $\mathcal{N}(F_c, i)$ to be another finite trace $F' = (E', \leq', \lambda') \in TR^*$ where $E' = (\downarrow^i(c) - \downarrow^i_{\mathrm{prev}}(c) - \{e\}) \subseteq E$ and $\leq', \lambda'$ are $\leq, \lambda$ restricted to E' respectively (see Figure 1). The trace $\mathcal{N}(F_c, i)$ corresponds to the *additional* (new) knowledge gained by agent i by executing the last action e at the configuration c. We will abuse the notation to denote the "final" configuration E' of the trace $\mathcal{N}(F_c, i)$ by $\mathcal{N}(c, i)$. Note that, in general, $\mathcal{N}(c, i)$ (or for that matter, any other configuration of $\mathcal{N}(F_c, i)$) as a set, is merely a subset of E and is not a configuration of F.

An important observation is that with $c' = \downarrow^i(c)$, $\downarrow^i_{\mathrm{prev}}(c) = \downarrow^i_{\mathrm{prev}}(c')$ and $\mathcal{N}(c, i) = \mathcal{N}(c', i)$. In this respect, these operations are "local". Also, note that c' is the trace "concatenation" of $\downarrow^i_{\mathrm{prev}}(c')$, $\mathcal{N}(c', i)$ and e (see Figure 1).

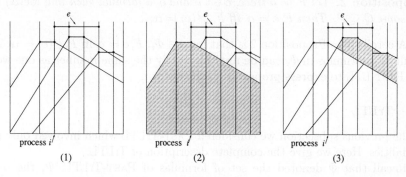

Fig. 1. (1) $c' = \downarrow e$ (2) $\downarrow^i_{\text{prev}}(c')$ (3) $\mathcal{N}(c', i)$

We can now present the semantics of PAST-TrLTL. A model is simply a trace $F = (E, \leq, \lambda) \in TR$. Let $c \in \mathcal{C}_F$ be a configuration and $\alpha \in \Phi$. Then $F, c \models \alpha$ denotes that α is satisfied at c in F and it is defined inductively. Note that for a finite trace $F = (E, \leq, \lambda)$, $E \in \mathcal{C}_F$. In this case, we abbreviate $F, E \models \alpha$ by $F \models \alpha$.

- $F, c \models \top$.
- $F, c \models \neg\alpha$ iff $F, c \not\models \alpha$.
- $F, c \models \alpha \vee \beta$ iff $F, c \models \alpha$ or $F, c \models \beta$
- $F, c \models \langle \bar{a} \rangle_i \{\alpha, \beta\}$ iff there exists $e \in E_i \cap c$ such that $\downarrow^i(c) = \downarrow e$, $\lambda(e) = a$, F, $\downarrow^i_{\text{prev}}(c) \models \alpha$ and $\mathcal{N}(F_c, i) \models \beta$ (recall that $\mathcal{N}(F_c, i) \models \beta$ stands for $\mathcal{N}(F_c, i), \mathcal{N}(c, i) \models \beta$).
- $F, c \models \alpha \, \mathcal{S}_i \beta$ iff there exists $c' \in \mathcal{C}_F$ such that $c' \subseteq c$ and $F, \downarrow^i(c') \models \beta$. Moreover, for every $c'' \in \mathcal{C}_F$, if $\downarrow^i(c') \subsetneq \downarrow^i(c'') \subseteq \downarrow^i(c)$ then $F, \downarrow^i(c'') \models \alpha$.

Thus PAST-TrLTL is an action based agent-wise generalization of PAST-LTL. Indeed both in terms of its syntax and semantics, PAST-LTL corresponds to the case where there is only one agent. With $\mathcal{P} = \{1\}$ and $\widetilde{\Sigma} = \{\Sigma_1\}$ one then writes a instead of $\langle \bar{a} \rangle_1 \{\top, \top\}$, $\ominus\alpha$ instead of $\bigvee_{a \in \Sigma_1} \langle \bar{a} \rangle_1 \{\alpha, \top\}$ and $\alpha \, \mathcal{S}\beta$ instead of $\alpha \, \mathcal{S}_1 \beta$. Note that here actions in Σ_1 are treated as atomic propositions. The semantics of PAST-TrLTL when specialized down to this case yields the usual PAST-LTL semantics.

We let α, β range over Φ. Abusing notation, we use loc to denote the map which associates a set of locations with each formula.

$$\text{loc}(\top) = \emptyset, \ \text{loc}(\langle \bar{a} \rangle_i \{\alpha, \beta\}) = \text{loc}(\alpha \, \mathcal{S}_i \beta) = \{i\},$$
$$\text{loc}(\neg\alpha) = \text{loc}(\alpha), \ \text{loc}(\alpha \vee \beta) = \text{loc}(\alpha) \cup \text{loc}(\beta).$$

In what follows, $\Phi_i = \{\alpha \mid \text{loc}(\alpha) = \{i\}\}$ is the set of i-type formulas.

Basic important observations concerning the syntax and semantics of PAST-TrLTL can be phrased as follows:

Proposition 1. *Every formula in Φ is a boolean combination of formulas taken from* $\bigcup_{i \in \mathcal{P}} \Phi_i$.

Proposition 2. *Let F be a trace, $c \in C_F$ and α a formula such that $\mathrm{loc}(\alpha) \subseteq Q$ for some $Q \subseteq \mathcal{P}$. Then $F, c \models \alpha$ iff $F, \downarrow^Q(c) \models \alpha$.*

A corollary to Proposition 2 is that for $\alpha \in \Phi_i$, $F, c \models \alpha$ iff $F, \downarrow^i(c) \models \alpha$. As a result, the formulas in Φ_i can be used in exactly the same manner as one would use PAST-LTL to express properties of the agent i.

2.2   TrLTL

In the previous subsection we considered PAST-TrLTL which involved only past modalities. Here we give the complete description of TrLTL.

Recall that Φ denoted the set of formulas of PAST-TrLTL. Ψ, the set of formulas of TrLTL, is defined inductively via:

- If α and β belong to Ψ, so do $\neg\alpha$ and $\alpha \vee \beta$.
- For $i \in \mathcal{P}$, if $\alpha \in \Phi$ then $\boxtimes_i^* \alpha, \boxtimes_i^\omega \alpha \in \Psi$.

Let $F = (E, \leq, \lambda) \in TR$. Recall that, for $c \in C_F$ and $\alpha \in \Phi$, $F, c \models \alpha$ denoted that the PAST-TrLTL formula α is satisfied at c in F.

We now present the semantics of TrLTL. A model is simply an infinite trace $F = (E, \leq, \lambda) \in TR^\omega$. By a slight abuse of notation, for $\alpha \in \Psi$, $F \models \alpha$ denotes that the TrLTL formula α holds in F and is defined (making use of the models relation of PAST-TrLTL) inductively as follows:

- $F \models \neg\alpha$ iff $F \not\models \alpha$.
- $F \models \alpha \vee \beta$ iff $F \models \alpha$ or $F \models \beta$
- $F \models \boxtimes_i^* \alpha$ iff E_i is finite and $F, \downarrow E_i \models \alpha$.
- $F \models \boxtimes_i^\omega \alpha$ iff E_i is infinite and for infinitely many $e \in E_i$, $F, \downarrow e \models \alpha$.

We need two notions of $\boxtimes_i$ to distinguish whether or not process i is making progress. The finite notion $\boxtimes_i^* \alpha$ asserts that process i participates in a finite number of events. Moreover, at the end, the PAST-TrLTL formula α holds in it's view. On the other hand, $\boxtimes_i^\omega \alpha$ says that infinitely many local views of i assert the PAST-TrLTL formula α. Note that $\boxtimes_i^\omega$ is a local version of the derived modality $\square\lozenge$ ("infinitely often") of LTL.

3   Elementary Decision Procedures

In this section we briefly sketch the decision procedures to solve the satisfiability problems for PAST-TrLTL and TrLTL. See [AS] for complete details.

A formula α of PAST-TrLTL is said to be satisfiable iff there exists a *finite* trace F such that $F \models \alpha$ or *equivalently*, iff there exists a trace F and $c \in C_F$ such that $F, c \models \alpha$.

Our strategy is to effectively associate an asynchronous automaton (see [DR]) $\mathcal{A}_\alpha$ with α which exactly recognizes the models of α. Since the emptiness problem for asynchronous automata is decidable, this yields the desired decision procedure. Here we just remark that, as in [Thi], the *gossip automaton* [MS] plays a very crucial role in the construction of $\mathcal{A}_\alpha$.

Theorem 1. *The satisfiability problem for* PAST-TrLTL *can be solved in time* $2^{O(n2^K K^3 \log K)}$ *where n is the size of the input formula.*

Now we turn to the satisfiability problem for TrLTL. A formula α of TrLTL is said to be satisfiable iff there exists an infinite trace F such that $F \models \alpha$. As before we are interested in deciding if a given formula of TrLTL is satisfiable.

Let α be a formula of TrLTL of size n. It is easy to construct a deterministic asynchronous Muller automaton with $2^{O(n2^K K^3 \log K)}$ global states recognizing precisely the models of α using the asynchronous automata corresponding to the PAST-TrLTL subformulas of α. Instead, it is also possible to use ideas in [MT] to design an asynchronous Büchi automaton of the same size as the Muller automaton. The emptiness problem for asynchronous Büchi automaton is easy and leads to a decision procedure which is singly exponential in the size of the input formula.

Theorem 2. *The satisfiability problem for complete* TrLTL *can be solved in time* $2^{O(n2^K K^3 \log K)}$ *where n is the size of the input formula.*

4   Expressiveness Issues

Our main aim here is to show that TrLTL is equal in expressive power with the first order theory of infinite traces.

4.1   First-Order Theory

We now recall $\mathrm{FO}(\widetilde{\Sigma})$, the first order theory of traces over $\widetilde{\Sigma}$. One starts with a countable set of individual variables $V = \{x_0, x_1, \ldots\}$ with x, y, z with or without subscripts ranging over V. For each $a \in \Sigma$ there is a unary predicate symbol R_a. There is also a binary predicate symbol $\leq$.

$R_a(x)$ and $x \leq y$ are atomic formulas. If φ and φ' are formulas, so are $\neg\varphi$, $\varphi \vee \varphi'$ and $(\exists x)\varphi$. The structures for this first order theory are elements of TR. Let $F \in TR$ with $F = (E, \leq, \lambda)$ and let $\mathcal{I} : V \to E$ be an interpretation. Then $F \models_{\mathcal{I}} R_a(x)$ iff $\lambda(\mathcal{I}(x)) = a$ and $F \models_{\mathcal{I}} x \leq y$ iff $\mathcal{I}(x) \leq \mathcal{I}(y)$. The remaining semantic definitions go along the expected lines.

For a sentence φ of $\mathrm{FO}(\widetilde{\Sigma})$, we write $F \models \varphi$ with the intended meaning that F is a model of φ. Each sentence φ defines the trace languages $L_\varphi^\omega = \{F \in TR^\omega \mid F \models \varphi\}$ and $L_\varphi^* = \{F \in TR^* \mid F \models \varphi\}$. We say that $L \subseteq TR^\omega$ (TR^*) is FO-definable iff there exists a sentence φ in $\mathrm{FO}(\widetilde{\Sigma})$ such that $L = L_\varphi^\omega$ $(L = L_\varphi^*)$.

4.2   Main Result

We start with a description of the trace languages defined by formulas of PAST-TrLTL and TrLTL. L_α, the trace language defined by a formula α of TrLTL is, $L_\alpha = \{F \in TR^\omega \mid F \models \alpha\}$. Similarly, a formula α of PAST-TrLTL defines the trace language $L_\alpha = \{F \in TR^* \mid F \models \alpha\}$. We say that $L \subseteq TR^\omega$ (TR^*) is TrLTL-definable (PAST-TrLTL-definable) iff there exists a formula α of TrLTL (PAST-TrLTL) such that $L = L_\alpha$.

The theorem that we wish to prove is:

Theorem 3. $L \subseteq TR^\omega$ *is* TrLTL-*definable iff it is also* FO-*definable.*

We describe the main ingredients required in showing that TrLTL is expressible within the first order theory of traces. The key observation (see [MT]) is that configurations of a trace can be described in $FO(\widetilde{\Sigma})$ using predicates of *bounded* dimension. Similarly, other constructs like $\downarrow_{\text{prev}}^{-}(\_)$, $\mathcal{N}(\_,\_)$ can also be described using predicates of bounded dimension. Using these constructs and essentially following ideas in [MT], for each formula α of TrLTL, we can define a sentence $\text{SAT}(\alpha)$ in $FO(\widetilde{\Sigma})$ through induction such that for every $F \in TR^\omega$, $F \models \alpha$ iff $F \models \text{SAT}(\alpha)$. This immediately leads to one direction of Theorem 3. We remark that the construction of $\text{SAT}(\_)$ also shows that PAST-TrLTL is expressible within the first order theory of traces.

Now we concentrate on proving the other direction of Theorem 3 which says that every FO-definable trace language is also TrLTL-definable.

4.3   Local Normal Forms

As a first step we state two normal forms for FO-definable trace languages.

We introduce some notation before going further. Let $F = (E, \leq, \lambda) \in TR$ and $i \in \mathcal{P}$. Recall that $E_i = \{e \in E \mid \lambda(e) \in \Sigma_i\}$ is the set of i-events and $\mathcal{C}_F^i$ is the set of i-local configurations, i.e., configurations of the form $\downarrow e$ for some $e \in E_i$ (except of the empty configuration). Also, recall that, for $c \in \mathcal{C}_F$, F_c denotes the induced finite trace $(c, \leq', \lambda')$ where $\leq', \lambda'$ are $\leq, \lambda$ restricted to c respectively. Along the same lines, we define $F_i = (E', \leq', \lambda') \in TR$ where $E' = \downarrow E_i$ and $\leq', \lambda'$ are $\leq, \lambda$ restricted to E' respectively. The trace F_i represents the knowledge of process i about the trace F.

Fix $i \in \mathcal{P}$. Let $L \subseteq TR^*(TR^\omega)$ be an FO-definable trace language. We say that L is an i-local FO-definable trace language if, for each $F \in TR^*(TR^\omega)$, $F \in L$ iff $F_i \in L$. Further, L is said to be a local FO-definable trace language if it is i-local FO-definable for some $i \in \mathcal{P}$.

Let $L \subseteq TR^*$ be an i-local FO-definable trace language for some $i \in \mathcal{P}$. We define $\lim_i^*(L) = \{F \in TR^\omega \mid E_i$ is finite and $F_i \in L\}$ and $\lim_i^\omega(L) = \{F \in TR^\omega \mid$ for infinitely many $c \in \mathcal{C}_F^i, F_c \in L\}$. It is not very difficult to see that $\lim_i^*(L)$ and $\lim_i^\omega(L)$ are also FO-definable. These languages are referred to as local limit FO-definable trace languages.

Now we are in a position to describe the new normal forms for FO-definable trace languages. See [AS] for complete proofs.

Theorem 4 (First Normal Form). *The class of* FO-*definable trace languages equals the boolean closure of the family of local* FO-*definable trace languages.*

Theorem 5 (Second Normal Form). *The class of* FO-*definable trace languages of* TR^ω *equals the boolean closure of the family of local limit* FO-*definable trace languages.*

A basic step in the proof of the first normal form is Proposition 3 which yields a decomposition of a global first order sentence into a boolean combination of local first order sentences. The first normal form follows immediately from this decomposition. The second normal form follows from an adaptation of the techniques in [DM] applied to the first normal form.

We develop further notation to state Proposition 3. For a sentence φ of $FO(\widetilde{\Sigma})$, let $qd(\varphi)$ denote the quantifier depth of φ. It is defined to be the maximum number of nested quantifiers in φ. Let n be a non-negative integer. For $F \in TR$, we denote the n-theory of F by $Th_n(F)$ where $Th_n(F) = \{\varphi \mid \varphi$ is a sentence of $FO(\widetilde{\Sigma})$ with $qd(\varphi) \leq n$ and $F \models \varphi\}$.

Proposition 3. *There exists a positive integer κ depending only on $\widetilde{\Sigma}$ such that, for all $n \geq 0$ and $F \in TR$, the n-theory $Th_n(F)$ is determined by the collection $\{Th_{n+\kappa}(F_i)\}_{i \in \mathcal{P}}$ of $n+\kappa$-theories.*

The proof of this proposition involves construction of a winning strategy for a global Ehrenfeucht-Fraïssé game (see [Tho]) from such strategies for process-wise local games. See [AS] for complete details.

4.4   Expressive Completeness of TrLTL

The first normal form enables to prove the next theorem which shows that PAST-TrLTL equals the first order theory of finite traces in expressive power.

Theorem 6. $L \subseteq TR^*$ *is* PAST-TrLTL-*definable iff it is* FO-*definable.*

We assume Theorem 6 and invoke the second normal form to complete the proof of Theorem 3.

Let $L \subseteq TR^*$ be an i-local FO-definable trace language for some $i \in \mathcal{P}$. It follows from Theorem 6 that there exists a formula α of PAST-TrLTL such that $L_\alpha = L$. Now it is completely straightforward to check that $\lim_i^*(L) = L_{\boxtimes_i^* \alpha}$ and $\lim_i^\omega(L) = L_{\boxtimes_i^\omega \alpha}$. This shows that local limit FO-definable trace languages are TrLTL-definable. By Theorem 5, we can conclude then that every FO-definable trace language of TR^ω is also TrLTL-definable.

4.5   Expressive Completeness of PAST-TrLTL

Now we turn our attention to the proof of Theorem 6. As observed earlier, PAST-TrLTL is expressible within the first order theory of finite traces. It remains to show the converse.

We first recall the notion of a linearization of a finite trace. Let $F \in TR^*$ with $F = (E, \leq, \lambda)$. A linearization of F is a total ordering $\rho = e_0 e_1 \ldots$ of E such that whenever $e, e' \in E$ and $e < e'$, e appears before e' in ρ. Let $\rho = e_0 e_1 \ldots$ be a linearization of F. Then $\lambda(\rho)$ denotes the corresponding word $\lambda(e_0)\lambda(e_1)\ldots$ in Σ^*. Abusing notation, we also call $\lambda(\rho)$ as a linearization of F. Let $L \subseteq TR^*$ and $L' = \{w \in \Sigma^* \mid w$ is a linearization of some $F \in L\} \subseteq \Sigma^*$. In this notation:

Proposition 4 ([EM]). *L is* FO-*definable iff L' is definable in the first order theory of sequences.*

We also recall the following automata-theoretic characterization of first order definable word languages.

Proposition 5 ([McNP]). *Let $M \subseteq \Sigma^*$. Then M is definable in the first order theory of sequences iff the minimal deterministic automaton of M is counter-free.*

Now we show that every FO-definable trace language of TR^* is also PAST-TrLTL-definable. The proof involves an induction on K (recall that K is the number of processes). As pointed out earlier, PAST-TrLTL when specialized to $K = 1$ corresponds to PAST-LTL. As PAST-LTL is equal in expressive power with the first order theory of finite sequences [MP1], the base case with $K = 1$ is indeed true. What we are really after is a generalization of this result in the trace setting. Let $K > 1$. We assume by induction that the result is true for $1, 2, \ldots, K-1$.

In view of Theorem 4, it suffices to prove the result for local FO-definable trace languages. Fix $i \in \mathcal{P}$ and $L \subseteq TR^*$, an i-local FO-definable trace language. Consider the word language $L' = \{w \in \Sigma^* \mid w$ is a linearization of some $F \in L\} \subseteq \Sigma^*$. Proposition 4 implies that L' is definable in the first order theory of finite sequences and hence, by Proposition 5, the minimal deterministic automaton $\mathcal{A}$ of L' is counter-free.

We write $\mathcal{A} = (S, \{\delta_a\}_{a \in \Sigma}, s_{in}, S_{fin})$ where

- S is a finite non-empty set of states with $s_{in} \in S$ and $S_{fin} \subseteq S$.
- For $a \in \Sigma$, $\delta_a : S \to S$ is a transition function.

For $w \in \Sigma^*$, let $\delta_w : S \to S$ denote the obvious induced transition function. As $\mathcal{A}$ is counter-free, for every non-empty word $w \in \Sigma^*$ and for every $R \subseteq S$ of size atleast two, δ_w does not induce a non-trivial permutation on R.

It follows easily from the partial commutativity properties of the syntactic monoid of L' that $\mathcal{A}$ is I-*consistent*, that is, for $(a, b) \in I$, $\delta_a \circ \delta_b = \delta_b \circ \delta_a$. Therefore we can talk of a run of $\mathcal{A}$ over a finite trace $F \in TR^*$ starting at some state $s \in S$. Since $\mathcal{A}$ is deterministic, there is a unique state $s' \in S$ reached at the end of this run. We denote this by $s \xrightarrow{F}_A s'$.

With $s, s' \in S$, we will associate a formula $\alpha_{s \to_A s'}$ of PAST-TrLTL such that

$$L_{\alpha_{s \to_A s'}} = \{F = (E, \leq, \lambda) \in TR^* \mid s \xrightarrow{F_i}_A s'\}$$

It is easy to see that L can be expressed as a boolean combination of languages taken from the collection $\{L_{\alpha_{s \to_A s'}}\}_{s, s' \in S}$. This in turn implies that L is PAST-TrLTL-definable.

We associate two deterministic sequential transition systems with i. The first one is easy to define. Let $\mathcal{A}-i = (S, \{\delta_b\}_{b \in \Sigma - \Sigma_i})$. Thus $\mathcal{A}-i$ is the same $\mathcal{A}$ restricted to $\Sigma - \Sigma_i$. Clearly agent i does not participate in $\mathcal{A}-i$. An important observation is that $\mathcal{A}-i$ is an I-consistent (with respect to actions in $\Sigma - \Sigma_i$) counter-free deterministic sequential automaton.

The second transition system, associated with i, is $\mathcal{A}_i = (S, \{\delta_\pi\}_{\Pi_i})$ where

$$\Pi_i = \{a \in \Sigma_i \mid \mathrm{loc}(a) = \{i\}\} \cup \left\{\langle a, s, s' \rangle \, \middle| \, \begin{array}{l} \mathrm{loc}(a) \supsetneq \{i\} \text{ and } s, s' \in S \text{ such} \\ \text{that } s' \text{ is reachable from } s \text{ in } \mathcal{A}-i \end{array} \right\}$$

and for $\pi \in \Pi_i$, δ_π is defined as: if $\pi = a$ then $\delta_\pi = \delta_a$, otherwise, $\pi = \langle a, s, s' \rangle$ and $\delta_\pi(s) = s''$ where s'' is the unique state in S such that $\delta_a(s') = s''$. Note that, in general, δ_π is a partial function. We can add a dummy loop-state and extend δ_π uniquely to a complete function. To keep the notation clean, we will not mention this dummy state. This will not affect our presentation. It is easy to see that $\mathcal{A}_i$ is a counter-free deterministic sequential automaton over the alphabet Π_i.

For $s, s' \in S$ and $P \subseteq \mathcal{P} - \{i\}$, it is easy to see that the language

$$M = \{F = (E, \leq, \lambda) \in TR^* \mid \downarrow^P E = E \text{ and } s \xrightarrow{F}_{\mathcal{A}-i} s'\}$$

is FO-definable. It essentially follows from the facts that $\mathcal{A}-i$ is counter-free and the condition $\downarrow^P E = E$ is FO-definable. As $\mathcal{A}-i$ does not involve actions of the agent i, no trace in this language involves an i-action. Therefore by induction, there exists a formula $\beta^P_{s \to_{\mathcal{A}-i} s'}$ of PAST-TrLTL such that $L_{\beta^P_{s \to_{\mathcal{A}-i} s'}} = M$.

Now we consider $\mathcal{A}_i$ operating on the alphabet Π_i. As in the case when $K = 1$, for each $s, s' \in S$, we can find formulas $\gamma_{s \to_{\mathcal{A}_i} s'}$ of PAST-LTL such that

$$L_{\gamma_{s \to_{\mathcal{A}_i} s'}} = \{\sigma \in \Pi_i^* \mid s \xrightarrow{\sigma}_{\mathcal{A}_i} s'\}$$

Note that the PAST-LTL formulas $\{\gamma_{s \to_{\mathcal{A}_i} s'}\}_{s,s' \in S}$ are defined over the alphabet Π_i. We let PAST-LTL(Π_i) denote PAST-LTL over the alphabet Π_i. Recall that formulas of PAST-LTL(Π_i) are defined via: $\pi \in \Pi_i \mid \neg\gamma \mid \gamma_1 \vee \gamma_1 \mid \ominus\gamma \mid \gamma_1 \mathcal{S}\gamma_2$.

Each $\pi \in \Pi_i$ is either of the form a or $\langle a, s, s' \rangle$ for some $a \in \Sigma_i$ and $s, s' \in S$. We can inductively translate a formula γ of PAST-LTL(Π_i) into a formula $\chi(\gamma)$ of PAST-TrLTL formula as:

$$\chi(a) = \langle \bar{a} \rangle_i \{\top, \top\}, \quad \chi(\langle a, s, s' \rangle) = \langle \bar{a} \rangle_i \{\top, \beta^{loc(a)-\{i\}}_{s \to_{\mathcal{A}-i} s'}\},$$
$$\chi(\neg\gamma) = \neg\chi(\gamma), \quad \chi(\gamma_1 \vee \gamma_2) = \chi(\gamma_1) \vee \chi(\gamma_2),$$
$$\chi(\ominus\gamma) = \ominus_i\chi(\gamma), \quad \chi(\gamma_1 \mathcal{S}\gamma_2) = \chi(\gamma_1) \mathcal{S}_i \chi(\gamma_2).$$

We claim that, for $s, s' \in S$, we can take $\alpha_{s \to_{\mathcal{A}} s'}$ to be $\chi(\gamma_{s \to_{\mathcal{A}_i} s'})$. This can be seen from the special *factorization* that we discuss next.

Let $F = (E, \leq, \lambda) \in TR^*$ such that $E = \downarrow^i E$. Suppose $E_i = \{e_1, e_2, \ldots, e_n\}$ with $e_1 \leq e_2 \ldots \leq e_n$. Note that $E = \downarrow e_n$. For each m with $1 \leq m \leq n$, let $F_m = \mathcal{N}(\downarrow e_m, i) = (E_m, \leq_m, \lambda_m) \in TR^*$ and $F'_m = (\{e_m\}, \emptyset, \lambda'_m) \in TR^*$ with $\lambda'_m(e_m) = \lambda(e_m)$. Observe that $E_m = \downarrow^{loc(e_m)-\{i\}} E_m$ for $1 \leq m \leq n$. It is easy to see that F admits the following factorization: $F = F_1 F'_1 F_2 F'_2 \cdots F_n F'_n$.

Fix $s_0 \in S$. Let $s'_0, s_1, s'_1, \ldots, s_n$ be in S such that for each m with $1 \leq m \leq n$, $s_{m-1} \xrightarrow{F_m}_{\mathcal{A}} s'_{m-1}$ and $s'_{m-1} \xrightarrow{F'_m}_{\mathcal{A}} s_m$. Note that this implies that $s_0 \xrightarrow{\downarrow e_m}_{\mathcal{A}} s_m$ for each $1 \leq m \leq n$.

By the definition of $\mathcal{A}-i$ and F_m, it follows that $s_{m-1} \xrightarrow{F_m}_{\mathcal{A}-i} s'_{m-1}$ for each $1 \leq m \leq n$. Also, it follows from the definition of $\mathcal{A}_i$ and F'_m that $s_{m-1} \xrightarrow{\pi_m}_{\mathcal{A}_i} s_m$, where $\pi_m = \lambda(e_m)$ if $loc(e_m) = \{i\}$ and otherwise $\pi_m = \langle \lambda(e_m), s_{m-1}, s'_{m-1} \rangle$. This implies that $s_0 \xrightarrow{\sigma}_{\mathcal{A}_i} s_n$ where $\sigma = \pi_1 \pi_2 \ldots \pi_n$.

Thus a run of $\mathcal{A}$ over F corresponds to several compatible runs of $\mathcal{A}{-}i$ starting at different states, one for each F_m and a single run of $\mathcal{A}_i$ over σ where $\sigma = \pi_1 \pi_2 \ldots \pi_n$ as above. Conversely, it is easy to see that given a run of $\mathcal{A}_i$ and *compatible* runs of $\mathcal{A}{-}i$, we can synthesize a run of $\mathcal{A}$. Using this correspondence between runs of $\mathcal{A}$, $\mathcal{A}{-}i$ and $\mathcal{A}_i$, it is straightforward but tedious to see that for $s, s' \in S$

$$L_{\chi(\gamma_{s \to \mathcal{A}_i s'})} = \{F = (E, \leq, \lambda) \in TR^* \mid s \xrightarrow{F_i}_{\mathcal{A}} s'\}$$

Thus we may take $\alpha_{s \to \mathcal{A} s'}$ to be $\chi(\gamma_{s \to \mathcal{A}_i s'})$. As pointed out earlier, this finishes the proof of Theorem 6.

5   Discussion

In this section, we examine the practicality of our logics for specifying typical global safety and liveness properties of distributed systems.

TrLTL is useful in expressing program or protocol invariants. For example, in a typical "bakery" implementation of mutual exclusion, the safety condition translates to a pre-condition which ensures that process i enters $\mathtt{crit}_i$ only when i sees that all processes j which raised their $\mathtt{flag}_j$ (causally) before $\mathtt{flag}_i$ have already entered the critical section. With this notation, for $j \in \mathcal{P}$, there is a formula $\mathtt{jBeforEi}$ which asserts that j has raised $\mathtt{flag}_j$ before i and has not yet entered $\mathtt{crit}_j$. Thus a typical *necessary* safety condition which process i ensures before executing $\mathtt{crit}_i$ is $\phi_i \equiv \neg(\vee_{j \in \mathcal{P}} \mathtt{jBeforEi})$.

Typical liveness properties are easier to specify, for example, $(\boxtimes_i^\omega (\mathtt{flag}_i) \Rightarrow \boxtimes_i^\omega (\mathtt{crit}_i))$ says that process i eventually enters $\mathtt{crit}_i$ after every occurrence of $\mathtt{flag}_i$ (assuming i does so infinitely often).

The specification of global safety is a delicate matter since violations of safety must be noticed in the past of certain key future events (in contrast to its "implementation" ϕ_i). In order to make this precise, we introduce some notation.

Latest Information. Fix $F = (E, \leq, \lambda) \in TR$ and $c \in \mathcal{C}_F$. Let $i, j, k \in \mathcal{P}$. We define $\mathsf{latest}_{i \to j}(c) = e$ where e is the maximum j-event in $\downarrow^i(c)$. If e does not exist then $\mathsf{latest}_{i \to j}(c)$ is set to a fictitious event $\Im$ which is deemed to be below every event. Further, we define $\mathsf{latest}_{i \to j \to k}(c)$ to be the maximum k-event in $\downarrow \mathsf{latest}_{i \to j}(c)$. Comparisons of such events are easily expressed in PAST-TrLTL.

An important use of the latest information is in encoding global safety properties. The typical requirement is that in the trace F, there should be no *concurrent* events $e_i \in E_i$ and $e_j \in E_j$ such that $\lambda(e_i) = \mathtt{crit}_i$ and $\lambda(e_j) = \mathtt{crit}_j$.

Proposition 6. *If $e_i \in E_i$ and $e_j \in E_j$ are two concurrent events with another event e such that $e_i \leq e$ and $e_j \leq e$, then there is a k and an $f \in E_k$ such that $\mathsf{latest}_{k \to i}(\downarrow f) \geq e_i \geq \mathsf{latest}_{k \to j \to i}(\downarrow f)$ and $\mathsf{latest}_{k \to j}(\downarrow f) \geq e_j \geq \mathsf{latest}_{k \to i \to j}(\downarrow f)$.*

Thus the safety requirement on the events e_i and e_j is enforceable at a suitable common future f as above. The above proposition extends to say, *three* concurrent events; but then we require the operators $\mathsf{latest}_{i \to j \to k \to l}$ with *three* hops. In essence, the latest operators are effective tools to navigate the past and locate finite sets of events in significant relationships.

Acknowledgement. We have benefitted greatly from discussions with K. Narayan Kumar and Madhavan Mukund. The first author also thanks Prof. Wolfgang Thomas and Christof Löding for their warm hospitality and many insights into the connections between logics, automata and languages.

References

[AS]   B. ADSUL AND M. SOHONI: First-Order Logic over Traces, Technical Report, Dept. of Computer Science & Engineering, I. I. T. Mumbai, INDIA (2002). Also see http://www.cse.iitb.ac.in/~abharat/logics.html

[DG1]   V. DIEKERT AND P. GASTIN: LTL is Expressively Complete for Mazurkiewicz Traces, *Proc. ICALP '00, LNCS* **1853** (2000) 211–222.

[DG2]   V. DIEKERT AND P. GASTIN: Local Temporal Logic is Expressively Complete for Cograph Dependence Alphabets, *Proc. LPAR '01, LNAI* **2250** (2000) 55–69.

[DM]   V. DIEKERT AND A. MUSCHOLL: Deterministic Asynchronous Automata for Infinite Traces, *Acta Inf.*, **31** (1993) 379–397.

[DR]   V. DIEKERT AND G. ROZENBERG (Eds.): *The Book of Traces*, World Scientific, Singapore (1995).

[EM]   W. EBINGER AND A. MUSCHOLL: Logical Definability on Infinite Traces, *Proc. ICALP '93, LNCS* **700** (1993) 335–346.

[MP1]   O. MALER AND A. PNUELI: Tight bounds on the Complexity of Cascaded Decomposition of Automata, *Proc. 31st IEEE FOCS '90* (1990) 672–682.

[MP2]   Z. MANNA AND A. PNUELI: *The Temporal Logic of Reactive and Concurrent Systems (Specification)*, Springer-Verlag, Berlin (1991).

[McNP]   R. MCNAUGHTON AND S. PAPERT: *Counter-free Automata*, MIT Press, Cambridge (1971).

[MS]   M. MUKUND AND M. SOHONI: Keeping Track of the Latest Gossip in a Distributed System, *Distributed Computing*, **10**(3) (1997) 137–148.

[MT]   M. MUKUND AND P.S. THIAGARAJAN: Linear Time Temporal Logics over Mazurkiewicz Traces, *Proc. MFCS'96, LNCS* **1113** (1996) 62–92.

[Thi]   P.S. THIAGARAJAN: A Trace Based Extension of Linear Time Temporal Logic, *Proc. 9th IEEE LICS* (1994) 438–447.

[Tho]   W. THOMAS: Languages, Automata and Logic, in: *Handbook of Formal Languages Vol. 3 – Beyond Words*, Springer-Verlag, New York (1997) 389-456.

[TW]   P.S. THIAGARAJAN AND I. WALUKIEWICZ: An Expressively Complete Linear Time Temporal Logic for Mazurkiewicz Traces, *Proc. 12th IEEE LICS* (1997) 183–194.

[Wal]   I. WALUKIEWICZ: Difficult Configurations - On the Complexity of LTrL, *Proc. ICALP '98, LNCS* **1443** (1998) 140–151.

An Elementary Expressively Complete Temporal Logic for Mazurkiewicz Traces*

Paul Gastin[1] and Madhavan Mukund[2]

[1] LIAFA, Université Paris 7, 2, place Jussieu, F-75251 Paris Cedex 05
Fax: 33 1 44 27 68 49 — Paul.Gastin@liafa.jussieu.fr
[2] Chennai Mathematical Institute, 92 G N Chetty Road, Chennai 600 017, India
Fax: 91 44 823 3671 — madhavan@cmi.ac.in

Abstract. In contrast to the classical setting of sequences, no temporal logic has yet been identified over Mazurkiewicz traces that is equivalent to first-order logic over traces and yet admits an elementary decision procedure. In this paper, we describe a local temporal logic over traces that is expressively complete and whose satisfiability problem is in PSPACE. Contrary to the situation for sequences, past modalities are essential for such a logic. A somewhat unexpected corollary is that first-order logic with three variables is expressively complete for traces.

Keywords. Temporal logics, Mazurkiewicz traces, concurrency

1 Introduction

Linear-time temporal logic (LTL) [15] has established itself as a useful formalism for specifying the interleaved behaviour of reactive systems. To combat the combinatorial blow-up involved in describing computations of concurrent systems in terms of interleavings, there has been a lot of interest in using temporal logic more directly on labelled partial orders.

Mazurkiewicz traces [12] are labelled partial orders generated by dependence alphabets of the form (Σ, D), where D is a *dependence* relation over Σ. If $(a, b) \notin D$, a and b are deemed to be independent actions that may occur concurrently. Traces are a natural formalism for describing the behaviour of static networks of communicating finite-state agents [26].

LTL over Σ-labelled sequences is equivalent to $FO_\Sigma(<)$, the first-order logic over Σ-labelled linear orders [11] and thus defines the class of aperiodic languages over Σ. Though $FO_\Sigma(<)$ permits assertions about both the past and the future, future modalities suffice for establishing the expressive completeness of LTL with respect to $FO_\Sigma(<)$ [9]. From a practical point of view, a finite-state program may be checked against an LTL specification relatively efficiently [2,19,20].

Though a number of LTL-like temporal logics have been proposed for reasoning directly about Mazurkiewicz traces, it has proved elusive to define a tractable

---

* Work done while the second author was visiting LIAFA, Université Paris 7. Partial support of CEFIPRA-IFCPAR Project 2102-1 (ACSMV) is gratefully acknowledged.

P. Widmayer et al. (Eds.): ICALP 2002, LNCS 2380, pp. 938–949, 2002.

logic that is expressively complete with respect to $\mathrm{FO}_\Sigma(<)$, the first order logic over traces. The first expressively complete temporal logic over traces was described in [17]. The result was refined in [3,5] to show expressive completeness without past modalities, using an extension of the proof technique developed for LTL in [24,25]. Formulae in both these logics are defined at *global configurations* (maximal antichains). Unfortunately, reasoning at the level of global configurations makes the complexity of deciding satisfiability non-elementary [21].

Computational tractability seems to require interpreting formulae at *local states*—effectively at individual events. This approach was followed (sometimes indirectly) in early temporal logics over traces [1,13,16,18]. While model-checking is relatively efficient for these logics, they are not known to be expressively complete. In fact, the logic TLC described in [1] is not even first-order definable.

Recently, an expressively complete local temporal logic has been defined for the subclass of Mazurkiewicz traces where the dependence graph of the underlying alphabet has a series-parallel structure [4]. Unfortunately, this logic *cannot* be expressively complete over all traces—the logic uses only future modalities and, unlike LTL over sequences, there exist two first-order inequivalent traces that cannot be distinguished using only future modalities [23].

In this paper, we define a new local temporal logic over traces that is expressively complete and whose satisfiability problem is in PSPACE. The observation in [23] about the shortcoming of future modalities necessitates the introduction of past modalities in our logic. As in [3,4], we show expressive completeness using an extension to traces of the proof technique introduced in [24,25] for LTL over sequences. However, the earlier proofs exploit the fact that the past cannot affect the truth of a formula. The introduction of past modalities in our logic requires the use of new techniques for "relativizing" formulae placed in new past contexts. The satisfiability problem for our logic is settled using two-way alternating automata, extending techniques from [1,4].

A corollary of our main result is that $\mathrm{FO}^3_{(\Sigma,D)}(<)$, the subclass of first-order logic formulae with at most three variables, is expressively complete over traces, extending the same result for sequences. This is somewhat surprising—intuition suggests that we require variables proportional to the width of a trace. (This result has been independently established by Walukiewicz [22].)

The paper is organized as follows. We begin with some preliminaries about traces. In Section 3 we define our new temporal logic. Section 4 describes a syntactic partition of traces that is used in Section 5 to establish expressive completeness. The decision procedure for satisfiability is sketched in Section 6. We conclude with a discussion of open questions. Many proofs have had to be omitted in this extended abstract.

2   Preliminaries

We briefly recall some notions about Mazurkiewicz traces (see [6] for background). A *dependence alphabet* is a pair (Σ, D) where the alphabet Σ is a finite set of actions and the *dependence relation* $D \subseteq \Sigma \times \Sigma$ is reflexive and symmetric.

The *independence relation* I is the complement of D. For $A \subseteq \Sigma$, the set of letters independent of A is denoted by $I(A) = \{b \in \Sigma \mid (a,b) \in I$ for all $a \in A\}$ and the set of letters depending on (some action in) A is denoted by $D(A) = \Sigma \setminus I(A)$.

A *Mazurkiewicz trace* is a labelled partial order $t = [V, \leq, \lambda]$ where V is a set of vertices labelled by $\lambda : V \to \Sigma$ and $\leq$ is a partial order over V satisfying the following conditions: For all $x \in V$, the downward set $\downarrow x = \{y \in V \mid y \leq x\}$ is finite, $(\lambda(x), \lambda(y)) \in D$ implies $x \leq y$ or $y \leq x$, and $x \lessdot y$ implies $(\lambda(x), \lambda(y)) \in D$, where $\lessdot = < \setminus <^2$ is the immediate successor relation in t.

The *alphabet* of a trace t is the set $\mathrm{alph}(t) = \lambda(V) \subseteq \Sigma$ and its *alphabet at infinity*, $\mathrm{alphinf}(t)$, is the set of letters occurring infinitely often in t. The set of all traces is denoted by $\mathbb{R}(\Sigma, D)$ or simply by $\mathbb{R}$. A trace t is called *finite* if V is finite. For $t = [V, \leq, \lambda] \in \mathbb{R}$, we define $\min(t) \subseteq V$ as the set of all minimal vertices of t. We can also read $\min(t) \subseteq \Sigma$ as the set of labels of the minimal vertices of t. It will be clear from the context what we actually mean.

Let $t_i = [V_i, \leq_i, \lambda_i]$, $i \in \{1, 2\}$, be a pair of traces such that $V_1 \cap V_2 = \emptyset$ and $\mathrm{alphinf}(t_1) \times \mathrm{alph}(t_2) \subseteq I$. We define the concatenation of t_1 and t_2 to be $t_1 \cdot t_2 = [V, \leq, \lambda]$ where $V = V_1 \cup V_2$, $\lambda = \lambda_1 \cup \lambda_2$ and $\leq$ is the transitive closure of $\leq_1 \cup \leq_2 \cup (V_1 \times V_2 \cap \lambda^{-1}(D))$. The set of finite traces is then a monoid, denoted $\mathbb{M}(\Sigma, D)$ or simply $\mathbb{M}$, with the empty trace $1 = (\emptyset, \emptyset, \emptyset)$ as unit.

Here is some useful notation for subclasses of traces. For $C \subseteq \Sigma$, let $\mathbb{R}_C = \{t \in \mathbb{R} \mid \mathrm{alph}(t) \subseteq C\}$ and $\mathbb{M}_C = \mathbb{M} \cap \mathbb{R}_C$. Also, $(\mathrm{alphinf} = C) = \{t \in \mathbb{R} \mid \mathrm{alphinf}(t) = C\}$ and $(\min = C) = \{t \in \mathbb{R} \mid \min(t) = C\}$. For $A, C \subseteq \Sigma$, we set $\mathbb{R}_C^A = \mathbb{R}_C \cap (\mathrm{alphinf} = A)$. Observe that $\mathbb{M}_C = \mathbb{R}_C^\emptyset$.

The first order logic over traces $\mathrm{FO}_\Sigma(<)$ is given by the syntax:

$$\varphi ::= P_a(x) \mid x < y \mid \neg\varphi \mid \varphi \vee \varphi \mid \exists x \varphi,$$

where $a \in \Sigma$ and $x, y \in \mathrm{Var}$ are first order variables. For a trace $t = [V, \leq, \lambda]$ and a valuation $\sigma : \mathrm{Var} \to V$, $t, \sigma \models \varphi$ denotes that t satisfies φ under σ. We interpret each predicate P_a by the set $\{x \in V \mid \lambda(x) = a\}$ and the relation $<$ as the strict partial order relation of t. The semantics then lifts to all formulas as usual. Since the meaning of a closed formula (sentence) φ is independent of the valuation σ, we can associate with each sentence φ the language $\mathcal{L}(\varphi) = \{t \in \mathbb{R} \mid t \models \varphi\}$. A trace language $L \subseteq \mathbb{R}$ is said to be expressible in $\mathrm{FO}_\Sigma(<)$ if there is a sentence $\varphi \in \mathrm{FO}_\Sigma(<)$ such that $L = \mathcal{L}(\varphi)$. For $n > 0$, $\mathrm{FO}_\Sigma^n(<)$ denotes the set of formulae with at most n distinct variables (possibly bound and reused several times).

We use the algebraic notion of recognizability (see e.g. [14]). Let $h : \mathbb{M} \to S$ be a morphism to a finite monoid S. For $t, u \in \mathbb{R}$, we say that t and u are h-similar, denoted $t \sim_h u$, if either $t, u \in \mathbb{M}$ and $h(t) = h(u)$ or t and u have infinite factorizations in non-empty finite traces $t = t_1 t_2 \cdots$, $u = u_1 u_2 \cdots$ with $h(t_i) = h(u_i)$ for all i. The transitive closure $\approx_h$ of $\sim_h$ is an equivalence relation. Since S is finite, this equivalence relation is of finite index with at most $|S|^2 + |S|$ equivalence classes. A trace language $L \subseteq \mathbb{R}$ is *recognized* by h if it is saturated by $\approx_h$ (or equivalently by $\sim_h$), i.e., $t \in L$ implies $[t]_{\approx_h} \subseteq L$ for all $t \in \mathbb{R}$.

Let $L \subseteq \mathbb{R}$ be recognized by a morphism $h : \mathbb{M} \to S$. For $B \subseteq \Sigma$, $L \cap \mathbb{M}_B$ and $L \cap \mathbb{R}_B$ are recognized by $h \lceil_{\mathbb{M}_B}$ the restriction of h to $\mathbb{M}_B$.

A finite monoid S is *aperiodic* if there is an $n \geq 0$ such that $s^n = s^{n+1}$ for all $s \in S$. A trace language $L \subseteq \mathbb{R}$ is *aperiodic* if it is recognized by some morphism to a finite and aperiodic monoid.

Theorem 1 ([8,7]). *A language $L \subseteq \mathbb{R}(\Sigma, D)$ is expressible in* $\mathrm{FO}_\Sigma(<)$ *if and only if it is aperiodic.*

3 Local Temporal Logic

Our local temporal logic over traces consists of two types of formulae: *internal formulae*, evaluated at arbitrary events within a trace, and *initial formulae*, asserting properties of minimal events of the trace. Internal formulae only talk about the structure within a connected portion of a trace. Initial formulae permit us to combine local assertions across disjoint components and form "global" assertions about the structure of the trace.

The set LocTL^i_Σ of internal formulae over the alphabet Σ is defined as follows:

$$\varphi ::= a \in \Sigma \mid \neg\varphi \mid \varphi \vee \varphi \mid \mathsf{EX}\,\varphi \mid \varphi\,\mathcal{U}_C\,\varphi, C \subseteq \Sigma \mid \mathsf{EY}\,\varphi \mid \varphi\,\mathcal{S}_C\,\varphi, C \subseteq \Sigma$$

Let $t = [V, \leq, \lambda] \in \mathbb{R}$ be a finite or infinite trace and let $x \in V$ be some vertex of t. We write $t, x \models \varphi$ to denote that trace t at node x satisfies the formula $\varphi \in \mathrm{LocTL}^i_\Sigma$. This is defined inductively as follows:

$$
\begin{array}{lll}
t, x \models a & \text{if} & \lambda(x) = a \\
t, x \models \neg\varphi & \text{if} & t, x \not\models \varphi \\
t, x \models \varphi \vee \psi & \text{if} & t, x \models \varphi \text{ or } t, x \models \psi \\
t, x \models \mathsf{EX}\,\varphi & \text{if} & \exists y.\ x \lessdot y \text{ and } t, y \models \varphi \\
t, x \models \mathsf{EY}\,\varphi & \text{if} & \exists y.\ y \lessdot x \text{ and } t, y \models \varphi \\
t, x \models \varphi\,\mathcal{U}_C\,\psi & \text{if} & \exists z \geq x.\ [t, z \models \psi \text{ and } \forall y.\ (x \leq y < z) \Rightarrow t, y \models \varphi \text{ and} \\
& & \quad\quad \forall y.\ (\lambda(y) \in C \text{ and } y \leq z) \Rightarrow y \leq x] \\
t, x \models \varphi\,\mathcal{S}_C\,\psi & \text{if} & \exists z \leq x.\ [t, z \models \psi \text{ and } \forall y.\ (z < y \leq x) \Rightarrow t, y \models \varphi \text{ and} \\
& & \quad\quad \forall y.\ (\lambda(y) \in C \text{ and } y \leq x) \Rightarrow y \leq z]
\end{array}
$$

The modalities EX and EY are existential extensions of the next state and previous state operators X and Y of LTL. The modality $\mathcal{U}_C$ extends the "universal" until operator $\mathcal{U}$ defined in [4] with an alphabetic *filter* that restricts the actions performed while fulfilling an until requirement. If $t, x \models \varphi\,\mathcal{U}_C\,\psi$ then no C-actions are performed when going from x to the node satisfying ψ. Since the filter also applies to events that are concurrent to x, the modality $\mathcal{U}_C$ no longer looks purely at the future of an event. Similarly, $\mathcal{S}_C$ is a filtered extension of the LTL since operator. The intuitive meanings of $\mathcal{U}_C$ and $\mathcal{S}_C$ are depicted below.

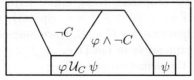 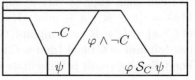

Past modalities are essential, as indicated by the following example from [23], where the dependence relation is $a - b - c - d$. These two traces are not first-order equivalent but are bisimilar at the level of events and thus cannot be distinguished by pure future modalities. However, the two traces below can be distinguished using the filtered until, which is not pure future.

$$a \to b \to c \to b \to c \cdots \qquad\qquad d \to c \to b \to c \to b \cdots$$
$$\uparrow \qquad\qquad\qquad\qquad\qquad\qquad \uparrow$$
$$d \to c \qquad\qquad\qquad\qquad\qquad\qquad a \to b$$

For the same alphabet, one can also show that the language $ad(bc)^* \subseteq \mathbb{M}$ of finite traces is first order but cannot be expressed by pure future modalities.

As usual, we can derive useful operators. Using unfiltered versions of until and since, $\varphi \mathcal{U} \psi = \varphi \mathcal{U}_\emptyset \psi$ and $\varphi \mathcal{S} \psi = \varphi \mathcal{S}_\emptyset \psi$, we recover the classical modalities *eventually in the future* $\mathsf{F} \varphi = \top \mathcal{U} \varphi$, *eventually in the past* $\mathsf{F}^{-1} \varphi = \top \mathcal{S} \varphi$, *always in the future* $\mathsf{G} \varphi = \neg \mathsf{F} \neg \varphi$, and *always in the past* $\mathsf{G}^{-1} \varphi = \neg \mathsf{F}^{-1} \neg \varphi$. We also use some filtered versions of these modalities: $\mathsf{F}_b \varphi = \top \mathcal{U}_b \varphi$ (eventually in the future but above the same b's), $\mathsf{G}_b \varphi = \neg \mathsf{F}_b \neg \varphi$ (for all future vertices that are above the same b's). The existence of a successor (resp. a predecessor) above the same b's can be written as follows:

$$\mathsf{EX}_b \varphi = \bigvee_{d \neq b} \mathsf{EX}(d \wedge \varphi) \wedge \top \mathcal{U}_b d = \bigvee_c c \wedge \mathsf{EX}(\neg b \wedge \varphi \wedge \top \mathcal{S}_b c),$$
$$\mathsf{EY}_b \varphi = \bigvee_c \mathsf{EY}(c \wedge \varphi) \wedge \top \mathcal{S}_b c = \bigvee_{d \neq b} d \wedge \mathsf{EY}(\varphi \wedge \top \mathcal{U}_b d).$$

Finally, $\mathsf{F}^\infty a = \mathsf{F} a \wedge \mathsf{G}(a \Rightarrow \mathsf{EX} \mathsf{F} a)$ expresses the existence of infinitely many vertices labelled with a above the current vertex. The filtered version of this formula, $\mathsf{F}_b^\infty a = \mathsf{F}_b a \wedge \mathsf{G}_b(a \Rightarrow \mathsf{EX}_b \mathsf{F}_b a)$, requires in addition that the vertices labelled with a are above the same b's.

For $A \subseteq \Sigma$, let A denote the internal formula $\bigvee_{a \in A} a$. An *alphabetic since* is a formula of the form $A \mathcal{S}_C a$ for $A, C \subseteq \Sigma$ and $a \in \Sigma$.

We turn now to initial formulae. These are boolean combinations of formulae $\mathsf{EM} \varphi$ asserting that some minimal vertex in the trace satisfies the internal formula φ. Formally, the set LocTL_Σ of initial formulae over Σ is given by:

$$\alpha ::= \bot \mid \mathsf{EM} \varphi, \varphi \in \mathrm{LocTL}_\Sigma^i \mid \neg\alpha \mid \alpha \vee \alpha$$

The semantics is defined as follows:

$$t \models \mathsf{EM} \varphi \quad \text{if} \quad \exists x. \, (x \in \min(t) \text{ and } t, x \models \varphi)$$
$$t \models \neg\alpha \quad \text{if} \quad t \not\models \alpha$$
$$t \models \alpha \vee \beta \quad \text{if} \quad t \models \alpha \text{ or } t \models \beta$$

An initial formula $\alpha \in \mathrm{LocTL}_\Sigma$ defines the trace language $\mathcal{L}(\alpha) = \{t \in \mathbb{R} \mid t \models \alpha\}$. We can then express various alphabetic properties using initial formulae: $\mathcal{L}(\mathsf{EM} a) = \{t \in \mathbb{R} \mid a \in \min(t)\}$, $\mathcal{L}(\mathsf{EM} \mathsf{F} a) = \{t \in \mathbb{R} \mid a \in \mathrm{alph}(t)\}$, and $\mathcal{L}(\mathsf{EM} \mathsf{F}^\infty a) = \{t \in \mathbb{R} \mid a \in \mathrm{alphinf}(t)\}$. Therefore, for $C \subseteq \Sigma$, trace languages such as $(\mathrm{alphinf} = C)$ and $(\min = C)$ are expressible in LocTL_Σ.

The following result is immediate from the definition of LocTL_Σ.

Proposition 2. *If a trace language is expressible in* LocTL_Σ, *then it is expressible in* $\mathrm{FO}^3_\Sigma(<)$.

4   Decomposition of Traces

The proof of our main result is a case analysis based on partitioning the set of traces according to the structure of the trace. Fix a letter $b \in \Sigma$ and set $B = \Sigma \setminus \{b\}$. Using the notation introduced in the middle of Section 2, let $\Gamma^A = \{t \in \mathbb{R}^A_B \mid \min(t) \subseteq D(b)\}$, $\Gamma = \Gamma^\emptyset$, and $\Omega_A = \{t \in \mathbb{R}_{I(A)} \mid \min(t) \subseteq \{b\}\}$.

Each trace $t \in \mathbb{R}$ has a unique factorization $t = t_0 b t_1 b t_2 \cdots$ with $t_0 \in \mathbb{R}_B$ and $t_i \in \mathbb{R}_B \cap (\min \subseteq D(b))$ for all $i > 0$. We obtain the following partition of $\mathbb{R}$.

$$\mathbb{R} = \mathbb{M}_B(b\Gamma)^\infty \uplus \left(\bigcup_{A \neq \emptyset} \mathbb{R}^A_B \Omega_A \right) \uplus \left(\bigcup_{A \neq \emptyset} \mathbb{M}_B(b\Gamma)^* b\Gamma^A \Omega_A \right)$$

If all t_i are finite, then we are in the first block. If $t_0 \notin \mathbb{M}$, then we are in the second block. Otherwise, we are in the last block. Note that $\mathbb{M}_B(b\Gamma)^* = \mathbb{M}$. The three possibilities are depicted below.

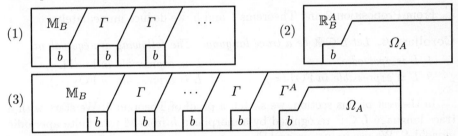

The following two results will allow us to use this decomposition effectively in proving the expressive completeness of our logic.

Lemma 3. *All these sets are expressible in* LocTL_Σ. *More precisely,*

1. *The sets* $\mathbb{M}_B(b\Gamma)^*$ *and* $\mathbb{M}_B(b\Gamma)^\omega$ *are expressible in* LocTL_Σ.
2. $\mathbb{R}^A_B \cdot \Omega_A = \mathcal{L}(\alpha)$ *where*

$$\alpha = \left(\bigwedge_{a \in A} \mathsf{EM}(\neg b \wedge \mathsf{F}^\infty_b a) \right) \wedge \left(\bigwedge_{a \notin A} \neg \mathsf{EM}(\neg b \wedge \mathsf{F}^\infty_b a) \right).$$

3. $\mathbb{M}_B(b\Gamma)^* b\Gamma^A \Omega_A = \mathbb{M}b\Gamma^A \Omega_A = \mathcal{L}(\beta)$ *where*

$$\beta = \bigvee_{C \subseteq \Sigma} \left((\mathrm{alphinf} = C) \wedge \mathsf{EM}\,\mathsf{F}\left(b \wedge \bigwedge_{c \in C} \mathsf{F}^\infty c \wedge \bigwedge_{a \in A} \mathsf{F}^\infty_b a \wedge \bigwedge_{a \notin A} \neg \mathsf{F}^\infty_b a \right) \right).$$

Note that "the" b in $\mathbb{M}b\Gamma^A \Omega_A$ *is characterized by the formula* $b \wedge \mathsf{F}^\infty_b a$, *where a is any letter in A.*

Lemma 4. *Let $A \subseteq \Sigma$ and let $L \subseteq \mathbb{R}$ be a trace language recognized by a morphism h from $\mathbb{M}$ into a finite monoid S. Then,*

(1) $L \cap \mathbb{M}_B(b\Gamma)^\infty = \underset{finite}{\bigcup} (L_1 \cap \mathbb{M}_B) \cdot (L_2 \cap (b\Gamma)^\infty)$

(2) $L \cap \mathbb{R}_B^A \Omega_A = \underset{finite}{\bigcup} (L_1 \cap \mathbb{R}_B^A) \cdot (L_2 \cap \Omega_A)$

(3) $L \cap \mathbb{M} b\Gamma^A \Omega_A = \underset{finite}{\bigcup} (L_1 \cap \mathbb{M}_B)(L_2 \cap (b\Gamma)^*)b(L_3 \cap \Gamma^A)(L_4 \cap \Omega_A)$

where the trace languages $L_i \subseteq \mathbb{R}$ are recognized by h.

5 Expressive Completeness

We now show that our temporal logic for traces is expressively complete with respect to $\mathrm{FO}_\Sigma(<)$. In light of Theorem 1, it suffices to establish the following.

Theorem 5. *Every aperiodic trace language $L \subseteq \mathbb{R}$ is expressible in LocTL_Σ.*

From Proposition 2 and Theorems 1 and 5, we deduce immediately

Corollary 6. *Let $L \subseteq \mathbb{R}$ be a trace language. The following are equivalent*

1. L is aperiodic,
3. L is expressible in $\mathrm{FO}_\Sigma^3(<)$,

2. L is expressible in LocTL_Σ,
4. L is expressible in $\mathrm{FO}_\Sigma(<)$.

In the rest of this section, we sketch a proof of Theorem 5. We start with a trace language $L \subseteq \mathbb{R}$ recognized by a morphism h from $\mathbb{M}$ to a finite aperiodic monoid S. We use induction on $|\Sigma|$.

The base case is when $\Sigma = \{a\}$, a singleton. Then L is either a finite set or the union of a finite set and a set of the form $a^n a^*$, $n \geq 0$. In both cases, L is expressible in LocTL_Σ. For instance, $\{a^\omega\}$ and $\{a^2\}$ correspond to the formulae $\mathsf{EM\,G\,EX\,\top}$ and $\mathsf{EM\,EX\,\neg\,EX\,\top}$. Also, $a^3 a^* \cup \{a^\omega\}$ is expressed by $\mathsf{EM\,EX\,EX\,\top}$.

For the induction step, with $|\Sigma| > 1$, we fix a letter $b \in \Sigma$ and use the decomposition introduced in Section 4. Therefore, we have

$$L = \left(L \cap \mathbb{M}_B(b\Gamma)^\infty \right) \uplus \left(L \cap \bigcup_{A \neq \emptyset} \mathbb{R}_B^A \Omega_A \right) \uplus \left(L \cap \bigcup_{A \neq \emptyset} \mathbb{M}_B(b\Gamma)^* b\Gamma^A \Omega_A \right).$$

We show separately that the three languages above are expressible in LocTL_Σ. From Lemma 4, it suffices to establish the following.

Proposition 7. *Recall that $B = \Sigma \setminus \{b\}$. Let $A \subseteq \Sigma$ be non-empty and let $L \subseteq \mathbb{R}$ be a trace language recognized by h. Then,*

(1) $(L \cap \mathbb{R}_B) \cdot (\min \subseteq \{b\})$ is expressible in LocTL_Σ.
(2) $\mathbb{M}_B \cdot (L \cap (b\Gamma)^\infty)$ is expressible in LocTL_Σ.
(3) $L \cap \mathbb{M}_B(b\Gamma)^\infty$ is expressible in LocTL_Σ.

(4) $\mathbb{R}_B^A \cdot (L \cap \Omega_A)$ *is expressible in* LocTL$_\Sigma$.
(5) $L \cap \mathbb{R}_B^A \Omega_A$ *is expressible in* LocTL$_\Sigma$.
(6) $M_B(L \cap (b\Gamma)^*) b\Gamma^A \Omega_A$ *is expressible in* LocTL$_\Sigma$.
(7) $Mb(L \cap \Gamma^A)\Omega_A$ *is expressible in* LocTL$_\Sigma$.
(8) $Mb\Gamma^A(L \cap \Omega_A)$ *is expressible in* LocTL$_\Sigma$.
(9) $L \cap Mb\Gamma^A \Omega_A$ *is expressible in* LocTL$_\Sigma$.

In all cases, we use the induction hypothesis to derive a formula for the factor over a smaller alphabet and then *relativize* this formula to the full class. As a representative example, we sketch the proof of Proposition 7(1,2,3).

Proof of Proposition 7(1). The language $L \cap \mathbb{R}_B$ is recognized by $h \restriction_{M_B}$ (Section 2). By the induction hypothesis, it is expressible in LocTL$_B$: $L \cap \mathbb{R}_B = \mathcal{L}_B(\alpha) = \{t \in \mathbb{R}_B \mid t \models \alpha\}$ for some $\alpha \in$ LocTL$_B$. Let $\tilde{\alpha} \in$ LocTL$_\Sigma$ be the formula given by Lemma 8 below. We have $(L \cap \mathbb{R}_B) \cdot (\min \subseteq \{b\}) = \mathcal{L}_\Sigma(\tilde{\alpha})$. □

Lemma 8. *Let* $\alpha \in$ LocTL$_B$. *There exists a formula* $\tilde{\alpha} \in$ LocTL$_\Sigma$ *such that for all* $t = t_1 t_2$ *with* $t_1 \in \mathbb{R}_B$ *and* $\min(t_2) \subseteq \{b\}$, $t_1 \models \alpha$ *iff* $t \models \tilde{\alpha}$.

Proof. We show simultaneously a similar result for internal formulae: for all $\varphi \in$ LocTL$_B^i$, there exists a formula $\tilde{\varphi} \in$ LocTL$_\Sigma^i$ such that for all $t = t_1 t_2$ with $t_1 \in \mathbb{R}_B$ and $\min(t_2) \subseteq \{b\}$, and for all $x \in t_1$, $t_1, x \models \varphi$ iff $t, x \models \tilde{\varphi}$. We use structural induction on α and φ. We have $\widetilde{\alpha \vee \beta} = \tilde{\alpha} \vee \tilde{\beta}$, $\widetilde{\neg \alpha} = \neg \tilde{\alpha}$, $\widetilde{\mathsf{EM}\,\varphi} = \mathsf{EM}(\neg b \wedge \tilde{\varphi})$, $\widetilde{\varphi \vee \psi} = \tilde{\varphi} \vee \tilde{\psi}$, $\widetilde{\neg \varphi} = \neg \tilde{\varphi}$, $\tilde{a} = a$, $\widetilde{\mathsf{EX}\,\varphi} = \mathsf{EX}_b\,\tilde{\varphi}$, $\widetilde{\mathsf{EY}\,\varphi} = \mathsf{EY}\,\tilde{\varphi}$, $\widetilde{\varphi\,\mathcal{U}_C\,\psi} = \tilde{\varphi}\,\mathcal{U}_{C \cup \{b\}}\,\tilde{\psi}$, and $\widetilde{\varphi\,\mathcal{S}_C\,\psi} = \tilde{\varphi}\,\mathcal{S}_C\,\tilde{\psi}$. □

To prove Proposition 7(2), we use a reduction to the word case. Let LTL$_T$(XU) denote LTL over T-labelled sequences with the strong until modality XU—for $t = t_1 t_2 \cdots \in T^\infty$, $t \models f$ XU g if there exists $j > 1$ such that $t_j t_{j+1} \cdots \models g$ and $t_k t_{k+1} \cdots \models f$ for all $1 < k < j$. From the theory of LTL over sequences, we know that every aperiodic word language $K \in T^\infty$ is expressible in LTL$_T$(XU).

We fix $T = h(b\Gamma)$ and define $\sigma : (b\Gamma)^\infty \to T^\infty$ by $\sigma(x) = h(bx_1)h(bx_2)\cdots$ if $x = bx_1 bx_2 \cdots$ with $x_i \in \Gamma$ for $i \geq 1$. The map σ is well-defined since each trace $x \in (b\Gamma)^\infty$ has a unique factorization $x = bx_1 bx_2 \cdots$ with $x_i \in \Gamma$ for $i \geq 1$.

We begin with the following result.

Lemma 9. *Let* $L \subseteq \mathbb{R}$ *be recognized by* h. *Then,* $L \cap (b\Gamma)^\infty = \sigma^{-1}(K)$ *for some* K *expressible in* LTL$_T$(XU).

Proof. Let $K = \sigma(L \cap (b\Gamma)^\infty)$. We first show that $L \cap (b\Gamma)^\infty = \sigma^{-1}(K)$. The inclusion $\subseteq$ is clear. Conversely, let $x = bx_1 bx_2 \cdots \in L \cap (b\Gamma)^\infty$ and $y \in \sigma^{-1}(\sigma(x))$. Then, $y = by_1 by_2 \cdots \in (b\Gamma)^\infty$ with $h(bx_i) = h(by_i)$ for all i. Therefore, $y \sim_h x$ and $x \in L$ implies $y \in L$.

Next, we show that K is recognized by the evaluation morphism $e : T^* \to S$ defined by $e(u) = u$ for all $u \in T$.

Let $s, t \in T^\infty$ with $s \sim_e t$ and $s \in K$. Write $s = s_1 s_2 \cdots \in T^\infty$ and $t = t_1 t_2 \cdots \in T^\infty$. Since $T = h(b\Gamma)$ we find $x = bx_1 bx_2 \cdots \in L$ with $x_i \in \Gamma$ and

$h(bx_i) = s_i$ for all $i > 1$; and $y = by_1by_2 \cdots$ with $y_i \in \Gamma$ and $h(by_i) = t_i$ for all $i > 1$. From $s \sim_e t$ we deduce that $x \sim_h y$. Since $x \in L$, $y \in L$ and $t = \sigma(y) \in K$.

Since S is aperiodic, $K \subseteq T^\infty$ is an aperiodic language over T and is thus expressible in $\mathrm{LTL}_T(\mathsf{XU})$—this is the only place where we use S aperiodic. □

The next step is to show that if $K \subseteq T^\infty$ is expressible in $\mathrm{LTL}_T(\mathsf{XU})$ then $\sigma^{-1}(K)$ is expressible in LocTL_Σ.

Lemma 10. *Let* $f \in \mathrm{LTL}_T(\mathsf{XU})$. *There exists* $\tilde{f} \in \mathrm{LocTL}^i_\Sigma$ *such that for all* $t = t_1 t'$ *with* $t_1 \in \mathbb{M}$ *and* $t' \in (b\Gamma)^\infty \setminus \{1\}$, *we have* $\sigma(t') \models f$ *iff* $t, \min(t') \models \tilde{f}$.

Proof Sketch. We use a structural induction on f. Clearly, $\widetilde{\neg f} = \neg \tilde{f}$ and $\widetilde{f \vee g} = \tilde{f} \vee \tilde{g}$. We claim that $\widetilde{f \mathsf{XU} g} = \mathsf{EX}((\neg b \vee \tilde{f}) \mathcal{U} (b \wedge \tilde{g}))$.

Let $t' = t_2 t_3 \cdots$ with $t_i \in b\Gamma$ for all $i > 1$ and let $x = \min(t')$.

Assume that $\sigma(t') \models f \mathsf{XU} g$ and let $j > 2$ be such that $\sigma(t_j t_{j+1} \cdots) \models g$ and $\sigma(t_k t_{k+1} \cdots) \models f$ for $2 < k < j$. Let $x' \in t$ with $x \lessdot x'$ and $z = \min(t_j)$. We have $x' \leq z$, $\lambda(z) = b$ and $t, z \models \tilde{g}$, by induction. Suppose $y \in t$ with $x' \leq y < z$. Either $\lambda(y) \neq b$ and $t, y \models \neg b$, or there exists $2 < k < j$ with $y = \min(t_k)$ and, by induction, $t, y \models \tilde{f}$. Therefore, $t, x' \models (\neg b \vee \tilde{f}) \mathcal{U} (b \wedge \tilde{g})$ and $t, x \models \mathsf{EX}((\neg b \vee \tilde{f}) \mathcal{U} (b \wedge \tilde{g}))$.

Conversely, assume that $t, x \models \mathsf{EX}((\neg b \vee \tilde{f}) \mathcal{U} (b \wedge \tilde{g}))$. Let $x', z \in t$ with $x \lessdot x' \leq z$, $t, z \models b \wedge \tilde{g}$ and $t, y \models \neg b \vee \tilde{f}$ for all $x' \leq y < z$. Since $\lambda(z) = b$, there exists $j > 2$ with $z = \min(t_j)$ and by induction, we get $\sigma(t_j t_{j+1} \cdots) \models g$. Now, let $2 < k < j$ and $y = \min(t_k)$. We have $x' \leq y < z$ and since $\lambda(y) = b$ we get $t, y \models \tilde{f}$ and $\sigma(t_k t_{k+1} \cdots) \models f$ by induction. Therefore, $\sigma(t') \models f \mathsf{XU} g$.

It remains to deal with a formula of the form $f = s \in T$. For all $r \in S$, the trace language $h_B^{-1}(r) \subseteq \mathbb{M}_B$ is aperiodic and, by induction on $|\Sigma|$, $h_B^{-1}(r) = \mathcal{L}_B(\alpha_r)$ for some formula $\alpha_r \in \mathrm{LocTL}_B$. Let $\widetilde{\alpha_r} \in \mathrm{LocTL}^i_\Sigma$ be the formula obtained using Lemma 11. We claim that $\tilde{s} = \bigvee_{h(b) \cdot r = s} \widetilde{\alpha_r}$.

We write $t' = bt_2 t_3$ with $t_2 \in \Gamma$ and $t_3 \in (b\Gamma)^\infty$. Let $r = h(t_2)$. Then, $t_2 \models \alpha_r$ and by Lemma 11 we obtain $t, \min(t') \models \widetilde{\alpha_r}$. Now, if $\sigma(t') \models s$ then $s = h(bt_2) = h(b)r$ and we get $t, \min(t') \models \tilde{s}$.

Conversely, assume that $t, \min(t') \models \widetilde{\alpha_r}$ for some r with $s = h(b)r$. By Lemma 11 we get $t_2 \models \alpha_r$ and $h(t_2) = r$. Hence, $h(bt_2) = s$ and $\sigma(t') \models s$. □

Lemma 11. *For all* $\alpha \in \mathrm{LocTL}_B$, *there is a formula* $\tilde{\alpha} \in \mathrm{LocTL}^i_\Sigma$ *such that for all* $t = t_1 bt_2 t_3 \in \mathbb{R}$ *with* $t_1 \in \mathbb{M}$, $t_2 \in \mathbb{R}_B$, $\min(t_2) \subseteq D(b)$ *and* $\min(t_3) \subseteq \{b\}$ *we have* $t_2 \models \alpha$ *iff* $t, \min(bt_2 t_3) \models \tilde{\alpha}$.

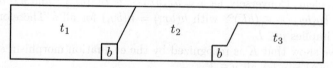

Proof. We simultaneously establish an equivalent result for internal formulae: for all $\varphi \in \mathrm{LocTL}^i_B$, there exists a formula $\tilde{\varphi} \in \mathrm{LocTL}^i_\Sigma$ such that for all $t =$

$t_1 b t_2 t_3 \in \mathbb{R}$ with $t_1 \in \mathbb{M}$, $t_2 \in \mathbb{R}_B$, $\min(t_2) \subseteq D(b)$ and $\min(t_3) \subseteq \{b\}$ and for all $x \in t_2$, we have $t_2, x \models \varphi$ iff $t, x \models \widetilde{\varphi}$.

We proceed by structural induction on α and φ. As always, we have $\widetilde{\neg \alpha} = \neg \widetilde{\alpha}$, $\widetilde{\alpha \vee \beta} = \widetilde{\alpha} \vee \widetilde{\beta}$, $\widetilde{\neg \varphi} = \neg \widetilde{\varphi}$, and $\widetilde{\varphi \vee \psi} = \widetilde{\varphi} \vee \widetilde{\psi}$. Now, $\widetilde{\mathsf{EM}\,\varphi} = \mathsf{EX}(\neg b \wedge \widetilde{\varphi})$, $\widetilde{a} = a$, $\widetilde{\mathsf{EX}\,\varphi} = \mathsf{EX}_b\, \widetilde{\varphi}$, $\widetilde{\mathsf{EY}\,\varphi} = \mathsf{EY}_b(\neg b \wedge \widetilde{\varphi})$,

$$\widetilde{\varphi\, \mathcal{U}_C\, \psi} \;=\; \bigvee_{E \subseteq \Sigma} \widetilde{\varphi}\, \mathcal{U}_{(C \cap E) \cup \{b\}}\, (\widetilde{\psi} \wedge E\,\mathcal{S}\, b)$$

$$\widetilde{\varphi\, \mathcal{S}_C\, \psi} \;=\; \bigvee_{E \subseteq \Sigma} (E\,\mathcal{S}\, b) \wedge (\widetilde{\varphi}\, \mathcal{S}_{(C \cap E) \cup \{b\}}\, (\widetilde{\psi} \wedge \neg b)).$$

We prove the formula for $\mathcal{U}_C$. Let $x \in t_2$ with $t_2, x \models \varphi\, \mathcal{U}_C\, \psi$. Let $z \in t_2$ be such that $x \leq z$, $t_2, z \models \psi$, $t_2, y \models \varphi$ for all $x \leq y < z$, and for all $y \in t_2$, $y \leq z$ and $\lambda(y) \in C$ implies $y \leq x$. By induction, $t, z \models \widetilde{\psi}$ and $t, y \models \widetilde{\varphi}$ for all $x \leq y < z$.

Let $E = \mathrm{alph}(\downarrow_{t_2} z) = \mathrm{alph}(\{y \in t_2 \mid y \leq z\})$. Clearly, $t, z \models E\,\mathcal{S}\, b$. Let $y \in t$ with $y \leq z$. If $\lambda(y) = b$, then $y \leq \min(b t_2 t_3) \leq x$. If $\lambda(y) \in C \cap E$, then there exists $y' \in t_2$ such that $\lambda(y') = \lambda(y) \in C$ and $y \leq y'$. From $y' \leq z$, we deduce $y' \leq x$ and then $y \leq y' \leq x$.

Conversely, let $x \in t_2$ and $E \subseteq \Sigma$ with $t, x \models \widetilde{\varphi}\, \mathcal{U}_{(C \cap E) \cup \{b\}}\, (\widetilde{\psi} \wedge E\,\mathcal{S}\, b)$. Let $z \in t$ with $x \leq z$, $t, z \models \widetilde{\psi} \wedge E\,\mathcal{S}\, b$, $t, y \models \widetilde{\varphi}$ for all $x \leq y < z$, and for all $y \leq z$, $\lambda(y) \in (C \cap E) \cup \{b\}$ implies $y \leq x$. If $z \in t_3$ then $y = \min(t_3)$ satisfies $\lambda(y) = b$, $y \leq z$ and $y \not\leq x$, a contradiction. Therefore, $z \in t_2$ and by induction, we get $t_2, z \models \psi$ and $t_2, y \models \varphi$ for all $x \leq y < z$. Now, let $y \in t_2$ with $y \leq z$ and $\lambda(y) \in C$. Since $t, z \models E\,\mathcal{S}\, b$ we deduce that $\mathrm{alph}(\downarrow_{t_2} z) \subseteq E$. Therefore, $\lambda(y) \in C \cap E$ and we obtain $y \leq x$.   □

Proof of Proposition 7(2). By Lemma 9, $L \cap (b\Gamma)^\infty = \sigma^{-1}(\mathcal{L}_T(f))$ for some $f \in \mathrm{LTL}_T(\mathsf{XU})$. Let $\widetilde{f} \in \mathrm{LocTL}_\Sigma^i$ be given by Lemma 10 and define the formula $\gamma = \mathsf{EM}(b \wedge \widetilde{f}) \vee \mathsf{EM}(\neg b \wedge \mathsf{F}_b \mathsf{EX}(b \wedge \widetilde{f}))$. We claim that

$$\mathbb{M}_B \cdot (L \cap (b\Gamma)^\infty \setminus \{1\}) = \mathbb{M}_B \cdot (b\Gamma)^\infty \cap \mathcal{L}(\gamma).$$

Indeed, let $t = t_1 t'$ with $t_1 \in \mathbb{M}_B$, $t' \in L \cap (b\Gamma)^\infty \setminus \{1\}$ and let $x = \min(t')$. We have $\sigma(t') \models f$ and by Lemma 10 we get $t, x \models \widetilde{f}$. Now, if $x \in \min(t)$ we have $t \models \mathsf{EM}(b \wedge \widetilde{f})$ and otherwise we have $t \models \mathsf{EM}(\neg b \wedge \mathsf{F}_b \mathsf{EX}(b \wedge \widetilde{f}))$.

Conversely, let $t = t_1 t'$ with $t_1 \in \mathbb{M}_B$, $t' \in (b\Gamma)^\infty$ and assume that $t \models \gamma$. Necessarily, $b \in \mathrm{alph}(t)$ and therefore $t' \neq 1$. From $t \models \gamma$ we deduce that $t, x \models b \wedge \widetilde{f}$ with $x = \min(t')$. By Lemma 10 we get $\sigma(t') \models f$. Therefore, $t' \in \sigma^{-1}(\mathcal{L}_T(f)) \subseteq L$ and t belongs to the left-hand side.

Since $\mathbb{M}_B \cdot (b\Gamma)^\infty$ and $\mathbb{M}_B$ are both expressible in LocTL_Σ, we deduce that $\mathbb{M}_B \cdot (L \cap (b\Gamma)^\infty)$ is expressible in LocTL_Σ.   □

Proof of Proposition 7(3). From Lemma 4, $L \cap \mathbb{M}_B(b\Gamma)^\infty$ is a finite union of languages of the form $(L_1 \cap \mathbb{M}_B) \cdot (L_2 \cap (b\Gamma)^\infty)$ where the trace languages $L_i \subseteq \mathbb{R}$ are recognized by h. Now, since the product $\mathbb{M}_B(b\Gamma)^\infty$ is unambiguous, we have

$$(L_1 \cap \mathbb{M}_B) \cdot (L_2 \cap (b\Gamma)^\infty) = (L_1 \cap \mathbb{M}_B) \cdot (b\Gamma)^\infty \cap \mathbb{M}_B \cdot (L_2 \cap (b\Gamma)^\infty)$$
$$(L_1 \cap \mathbb{M}_B) \cdot (b\Gamma)^\infty = (\mathbb{M}_B \cdot (b\Gamma)^\infty) \cap ((L_1 \cap \mathbb{R}_B) \cdot (\min \subseteq \{b\})).$$

We conclude using Lemma 3 and Proposition 7(1,2).   □

We can prove Proposition 7(1-3) using no EY and only sinces of the form $A\,\mathcal{S}_C\,a$ ("alphabetic" sinces). If $L \subseteq \mathbb{M}$ then $L = L \cap \mathbb{M}_B(b\Gamma)^*$, so we only need to consider the first case in the decomposition. From this, we have the following.

Theorem 12. *Let $L \subseteq \mathbb{M}$ be a trace language recognized by h. Then, L is expressible in* LocTL$_\Sigma$ *without using* EY *and with alphabetic sinces only.*

6 Satisfiability

To decide satisfiability, we work with alternating automata over sequentializations of traces. The states of the alternating automaton consist of subformulas of the main formula together with bookkeeping information about the letters traversed while satisfying until and since requirements. The basic technique we use was introduced in [1] and refined in [4]. The additional complication for our logic is that we have to deal with past requirements, both explicitly in the operators EY and $\mathcal{S}_C$, and implicitly, because of the alphabetic filters in $\mathcal{U}_C$ and $\mathcal{S}_C$. We thus need two-way alternating automata. Nevertheless, checking emptiness for this class remains within PSPACE [10] for each fixed dependence alphabet (Σ, D). Due to space constraints, we omit all details of the construction.

7 Open Problems

As we have seen in Section 3, the introduction of past modalities is unavoidable when constructing an expressively complete local temporal logic over traces. A natural question is whether unfiltered versions of $\mathcal{U}_C$ and $\mathcal{S}_C$ are sufficient for expressive completeness. We also do not know whether the temporal logic based only on EX and $\mathcal{U}_C$ is expressively complete. Another alternative is to keep the unfiltered until and to use a concurrency modality asserting the existence of a concurrent vertex satisfying some formula. Also, we have seen in Theorem 12 that no EY are required for finite traces and that alphabetic sinces are sufficient. We do not know whether the same restriction applies to infinite traces.

References

1. R. Alur, D. Peled, and W. Penczek. Model-checking of causality properties. In *Proceedings of LICS'95*, pages 90–100, 1995.
2. C. Courcoubetis, M.Y. Vardi, P. Wolper, and M. Yannakakis. Memory efficient algorithms for the verification of temporal properties. *Formal Methods in System Design*, 1:275–288, 1992.

3. V. Diekert and P. Gastin. LTL is expressively complete for Mazurkiewicz traces. In *Proceedings of ICALP 2000*, number 1853 in LNCS, pages 211–222. Springer Verlag, 2000.

4. V. Diekert and P. Gastin. Local temporal logic is expressively complete for cograph dependence alphabets. In *Proceedings of LPAR'01*, number 2250 in LNAI, pages 55–69. Springer Verlag, 2001.

5. V. Diekert and P. Gastin. LTL is expressively complete for Mazurkiewicz traces. *Journal of Computer and System Sciences*, 2002. To appear.

6. V. Diekert and G. Rozenberg, editors. *The Book of Traces*. World Scientific, Singapore, 1995.

7. W. Ebinger and A. Muscholl. Logical definability on infinite traces. *Theoretical Computer Science*, 154:67–84, 1996.

8. W. Ebinger. *Charakterisierung von Sprachklassen unendlicher Spuren durch Logiken*. Dissertation, Institut für Informatik, Universität Stuttgart, 1994.

9. D. Gabbay, A. Pnueli, S. Shelah, and J. Stavi. On the temporal analysis of fairness. In *Proceedings of PoPL'80*, pages 163–173, Las Vegas, Nev., 1980.

10. T. Jiang and B. Ravikumar. A note on the space complexity of some decision problems for finite automata. *Information Processing Letters*, 40:25–31, 1991.

11. J.A.W. Kamp. *Tense Logic and the Theory of Linear Order*. PhD thesis, University of California, Los Angeles, California, 1968.

12. A. Mazurkiewicz. Concurrent program schemes and their interpretations. DAIMI Rep. PB 78, Aarhus University, Aarhus, 1977.

13. P. Niebert. A ν-calculus with local views for sequential agents. In *Proceedings of MFCS'95*, number 969 in LNCS, pages 563–573. Springer Verlag, 1995.

14. D. Perrin and J.-E. Pin. Infinite words. Technical report, LITP, Avril 1997.

15. A. Pnueli. The temporal logics of programs. In *Proceedings of the 18th IEEE FOCS, 1977*, pages 46–57, 1977.

16. R. Ramanujam. Locally linear time temporal logic. In *Proceedings of LICS'96*, pages 118–128, 1996.

17. P.S. Thiagarajan and I. Walukiewicz. An expressively complete linear time temporal logic for Mazurkiewicz traces. In *Proceedings of LICS'97*, pages 183–194, 1997.

18. P.S. Thiagarajan. A trace based extension of linear time temporal logic. In *Proceedings of LICS'94*, pages 438–447, 1994.

19. M.Y. Vardi and P. Wolper. An automata-theoretic approach to automatic program verification. In *Proceedings of LICS'86*, pages 322–331, 1986.

20. M.Y. Vardi. An automata-theoretic approach to linear temporal logic. In *Logics for Concurrency: Structure versus Automata*, number 1043 in LNCS, pages 238–266. Springer Verlag, 1996.

21. I. Walukiewicz. Difficult configurations – on the complexity of LTrL. In *Proceedings of ICALP'98*, number 1443 in LNCS, pages 140–151. Springer Verlag, 1998.

22. I. Walukiewicz. Private communication, 2001.

23. I. Walukiewicz. Local logics for traces. *Journal of Automata, Languages and Combinatorics*, 2002. To appear.

24. Th. Wilke. Classifying discrete temporal properties. Habilitationsschrift (post-doctoral thesis), April 1998.

25. Th. Wilke. Classifying discrete temporal properties. In *Proceedings of STACS'99*, number 1563 in LNCS, pages 32–46. Springer Verlag, 1999.

26. W. Zielonka. Notes on finite asynchronous automata. *R.A.I.R.O. — Informatique Théorique et Applications*, 21:99–135, 1987.

Random Numbers and an Incomplete Immune Recursive Set

Vasco Brattka[*]

Theoretische Informatik I, Informatikzentrum
FernUniversität, 58084 Hagen, Germany
vasco.brattka@fernuni-hagen.de

Abstract. Generalizing the notion of a recursively enumerable (r.e.) set to sets of real numbers and other metric spaces is an important topic in computable analysis (which is the Turing machine based theory of computable real number functions). A closed subset of a computable metric space is called *r.e. closed*, if all open rational balls which intersect the set can be effectively enumerated and it is called *effectively separable*, if it contains a dense computable sequence. Both notions are closely related and in case of Euclidean space (and complete computable metric spaces in general) they actually coincide. Especially, both notions are generalizations of the classical notion of an r.e. subset of natural numbers. However, in case of incomplete metric spaces these notions are distinct. We use the immune set of random natural numbers to construct a recursive immune "tree" which shows that there exists an r.e. closed subset of some incomplete subspace of Cantor space which is not effectively separable. Finally, we transfer this example to the incomplete space of rational numbers (considered as a subspace of Euclidean space).

Keywords: Computable analysis, r.e. closed sets, random numbers, Kolmogorov complexity, immune sets.

1 Introduction

In classical recursion theory the notions of a recursively enumerable (r.e.) and a recursive set play an essential role [6]. One interesting topic of research in computable analysis (which is the theory of real number functions which can be computed by Turing machines) is concerned with the generalization of these notions to Euclidean space and other types of spaces [9]. In the case of computable metric spaces some central notions of effectivity for closed subsets and their mutual relationship can be visualized as shown in Figure 1. The displayed results have been obtained in [3] from a very uniform point of view and each arrow in the diagram does not only indicate an implication but an effective reducibility. Below, we will precisely define the notions which are relevant to the present paper. In case of Euclidean space (and other computable metric spaces

---

[*] Work supported by DFG Grant BR 1807/4-1

P. Widmayer et al. (Eds.): ICALP 2002, LNCS 2380, pp. 950–961, 2002.

which roughly speaking have to be "effectively locally compact") the vertical arrows can be reversed and thus the three horizontal layers of the diagram collapse [4]. However, a number of examples have been presented in [3] which prove that some of these notions have to be distinguished in the general case of computable metric spaces.

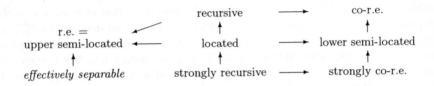

Fig. 1. Notions of effectivity for closed subsets of computable metric spaces

In this paper we will deal with another example of this type which has already been announced but not proved in [3] and which shows that there are naturally defined computable metric spaces and examples of closed subsets which fulfill all displayed effectivity notions besides effective separability. The results of [3] imply that such spaces and sets necessarily have to be incomplete (since recursive enumerability implies effective separability in case of complete spaces or sets). Even the space of rational numbers (endowed with subtopology of the Euclidean metric) admits subsets of the mentioned type. A glance at Figure 1 shows that it suffices to show that there exists a strongly recursive closed subset which is not effectively separable to prove the claim.

In the following Section 2 we define some basic concepts from computable analysis and we introduce the relevant notions of effectivity for closed subsets. In Section 3 we use the set of natural random numbers to construct a recursive subset of Cantor space. This "tree" is our basic example of a strongly recursive closed subset (of some incomplete subspace of Cantor space) which is not effectively separable. Actually, we can prove even a stronger result which shows that the constructed tree is immune in the sense that it does not include an infinite computable sequence. In Section 4 we will transfer our example to the space of rational numbers.

2 Preliminaries from Computable Analysis

In this section we introduce some basic concepts from computable analysis; for details we refer the reader to [9]. We start with computable metric spaces.

Definition 1 (Computable metric space). We will call a triple (X, d, α) a *computable metric space*, if

1. $d : X \times X \to \mathbb{R}$ is a metric on X,
2. $\alpha : \mathbb{N} \to X$ is dense in X, i.e. the closure of range(α) is equal to X,
3. $d \circ (\alpha \times \alpha) : \mathbb{N}^2 \to \mathbb{R}$ is a computable (double) sequence of real numbers.

Here we assume that the reader is familiar with the notion of a computable sequence of real numbers in sense of computable analysis (cf. for instance [7,5,10, 9]). We mention some standard examples of computable metric spaces which we will use in the following. Therefore, we first introduce some technical notations. By $\{0,1\}^*$ we denote the set of *finite words* over the alphabet $\{0,1\}$ and by $\mathcal{C} := \{0,1\}^\omega$ we denote the set of infinite binary sequences (and occasionally we assume $\{0,1\}^\omega \subseteq \mathbb{N}^\mathbb{N}$). By wp we denote the concatenation of a word w with a sequence (or a finite word) p, and the *prefix order* will be denoted by "$\sqsubseteq$", i.e. $w \sqsubseteq p$ holds, whenever w is a prefix of p. By 0^ω we denote the zero sequence and by $\mathcal{Z} := \{0,1\}^*0^\omega$ the set of all binary sequences which are eventually zero. Similarly, we denote by $w\{0,1\}^\omega$ the set of sequences with prefix w. By $\langle n,k \rangle := 1/2(n+k)(n+k+1)+k$ we denote the *Cantor pairing* of $n,k \in \mathbb{N}$ which can be extended inductively to arbitrary finite tuples and we define $\widehat{0} := 0$ and $\widehat{n+1} := 1$ for all $n \in \mathbb{N}$.

Example 2 (Computable metric spaces).

1. $(\mathbb{R}, d_\mathbb{R}, \alpha_\mathbb{R})$ with the *Euclidean metric*

$$d_\mathbb{R}(x,y) := |x-y|$$

   and some standard enumeration $\alpha_\mathbb{R}\langle i,j,k \rangle := \frac{i-j}{k+1}$ of the set $\mathbb{Q}$ of rational numbers, is a computable metric space.
2. $(\mathbb{Q}, d_\mathbb{Q}, \alpha_\mathbb{Q})$ with the restriction $d_\mathbb{Q}$ of $d_\mathbb{R}$ to $\mathbb{Q} \times \mathbb{Q}$ in the source and the restriction $\alpha_\mathbb{Q}$ of $\alpha_\mathbb{R}$ to $\mathbb{Q}$ in the target, is a computable metric space.
3. $(\{0,1\}^\omega, d_\mathcal{C}, \alpha_\mathcal{C})$ with the *Cantor metric*

$$d_\mathcal{C}(p,q) := \begin{cases} 2^{-\min\{i \in \mathbb{N} \,:\, p(i) \neq q(i)\}} & \text{if } p \neq q \\ 0 & \text{else} \end{cases}$$

   and some enumeration $\alpha_\mathcal{C}\langle k, \langle n_1, ..., n_k \rangle \rangle := \widehat{n_1}...\widehat{n_k}0^\omega$ of $\mathcal{Z} := \{0,1\}^*0^\omega$ is a computable metric space.
4. $(\mathcal{Z}, d_\mathcal{Z}, \alpha_\mathcal{Z})$ with the restriction $d_\mathcal{Z}$ of $d_\mathcal{C}$ to $\mathcal{Z} \times \mathcal{Z}$ in the source and the restriction $\alpha_\mathcal{Z}$ of $\alpha_\mathcal{C}$ to $\mathcal{Z}$ in the target, is a computable metric space.

The proofs that these spaces are actually computable metric spaces are straightforward (cf. [2,8,9]). In the following we will refer to these spaces simply by writing $\mathbb{R}, \mathbb{Q}, \mathcal{C}$, and $\mathcal{Z}$, respectively. The spaces $\mathbb{Q}$ and $\mathcal{Z}$ are typical examples of *incomplete* computable metric spaces and $\mathbb{R}$ and $\mathcal{C}$ are their natural completions, respectively.

In computable analysis a function $f :\subseteq X \to Y$ is called *computable*, if there exists a Turing machine which transfers each infinite sequence $p \in \Sigma^\omega$ (over some alphabet Σ) that represents some input $x \in X$ into some sequence $q \in \Sigma^\omega$ which represents the result $f(x)$. Of course, such a Turing machine has to compute infinitely long, but in the long run each infinite input sequence is transformed into an appropriate output sequence. Here and in the following, the inclusion symbol "$\subseteq$" is used to denote functions which are possibly partial. It is a reasonable restriction that only Turing machines with one-way output tape are allowed (because otherwise the output after some finite time would be useless, since it could be changed by the machine later on). More formally, a *representation* of a set X is a surjective mapping $\delta :\subseteq \Sigma^\omega \to X$. Using this notion we can define computable functions precisely.

Definition 3 (Computable functions). Let δ and δ' be representations of X and Y, respectively. A function $f :\subseteq X \to Y$ is called (δ, δ')–*computable*, if there exists a Turing machine M such that $f\delta(p) = \delta' F_M(p)$ for all $p \in \mathrm{dom}(f\delta)$.

Here, $F_M :\subseteq \Sigma^\omega \to \Sigma^\omega$ denotes the function computed by Turing machine M. The diagram in Figure 2 illustrates the situation.

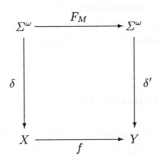

Fig. 2. Computability with respect to representations

With each computable metric space (X, d, α) we can canonically associate its *Cauchy representation* δ_X, where $\delta_X(01^{n_0+1}01^{n_1+1}01^{n_2+1}...) = \lim_{i\to\infty} \alpha(n_i)$ for all $n_i \in \mathbb{N}$ such that $d(\alpha(n_i), \alpha(n_j)) \le 2^{-j}$ for all $i > j$. Roughly speaking, $\delta_X(p) = x$, if p encodes a Cauchy sequence in $\mathrm{range}(\alpha)$ which rapidly converges to x. Occasionally, we will also use some standard representation $\delta_\mathbb{N}$ of the natural numbers $\mathbb{N} := \{0, 1, 2, ...\}$. In the following we will say for short that a function on $\mathbb{N}, \mathbb{Q}, \mathbb{R}, \mathcal{Z}$ and $\mathcal{C}$ is *computable*, if it is computable with respect to the corresponding representation $\delta_\mathbb{N}, \delta_\mathbb{Q}, \delta_\mathbb{R}, \delta_\mathcal{Z}$ and $\delta_\mathcal{C}$, respectively. The class of functions $f : \Sigma^\omega \to \Sigma^\omega$ over the alphabet $\Sigma := \{0, 1\}$ which are computable by Turing machines coincides with the class of computable functions with respect to $\delta_\mathcal{C}$. Thus, the introduced notions are consistent.

A point $x \in X$ of a recursive metric space (X, d, α) with Cauchy representation δ_X is called *computable*, if there exists some computable p such that $\delta_X(p) = x$. A sequence $(x_n)_{n\in\mathbb{N}}$ in X is called *computable*, if the corresponding function $f : \mathbb{N} \to X$ with $f(n) := x_n$ is $(\delta_\mathbb{N}, \delta_X)$–computable.

We close this section with a definition of those effectivity notions for subsets which will be used throughout this paper. We will use the notation $B(x, \varepsilon) := \{y \in X : d(x, y) < \varepsilon\}$ for the *open ball* with center x and radius ε and analogously we denote by $\overline{B}(x, \varepsilon) := \{y \in X : d(x, y) \le \varepsilon\}$ the corresponding *closed ball*. In general, $\overline{A}$ denotes the *topological closure* of a subset $A \subseteq X$. Moreover, we use the abbreviation $\overline{n} := \alpha_\mathbb{R}(n)$ for the rational numbered by $n \in \mathbb{N}$.

Definition 4 (Recursively enumerable closed subsets). Let (X, d, α) be a computable metric space and let $A \subseteq X$ be a closed subset.

1. A is called *r.e. closed*, if the set $\{\langle n, k \rangle \in \mathbb{N} : A \cap B(\alpha(n), \overline{k}) \neq \emptyset\}$ is r.e.
2. A is called *strongly co-r.e. closed*, if the set $\{\langle n, k \rangle \in \mathbb{N} : A \cap \overline{B}(\alpha(n), \overline{k}) = \emptyset\}$ is r.e.
3. A is called *strongly recursive closed*, if A is r.e., as well as strongly co-r.e. closed.

Our aim is to construct a strongly recursive closed subset which is not effectively separable. Therefore we define effective separability.

Definition 5 (Effective separability). Let (X, d, α) be a computable metric space and let $A \subseteq X$ be a subset. Then A is called *effectively separable*, if there exists a computable sequence $f : \mathbb{N} \to X$ such that $\text{range}(f) = \{f(n) : n \in \mathbb{N}\}$ is dense in A.

3   An Immune Recursive Tree

In this section we construct a subset $T \subseteq \{0, 1\}^\omega$ with some interesting properties. Using a non-standard but intuitive terminology, we will call such subsets *trees* for short (even if T is not closed). The tree T that we will construct is recursive in the sense that we can effectively decide which nodes belong to a path of the tree. Especially, T is a strongly recursive closed subset of $\mathcal{Z} = \{0, 1\}^* 0^\omega$. Since $T \subseteq \mathcal{Z}$, all paths of T are computable, but T has the property that there is no algorithm which, given a node of T as input, finds some path in T which goes through the given node. This implies that T is not effectively separable. And more than this, we will even prove that T does not contain an infinite computable sequence.

For the construction of T we will use some notions from recursion theory [6]. Let $\varphi : \mathbb{N} \to P$ denote some *admissible Gödel numbering* of the set of partial computable functions $P := \{f :\subseteq \mathbb{N} \to \mathbb{N} : f \text{ computable}\}$. Then the *Kolmogorov complexity* of a number $n \in \mathbb{N}$ is defined by $K(n) := \min\{i \in \mathbb{N} : \varphi_i(0) = n\}$, which is the "shortest program" that can produce n. Without loss of generality, we can assume that $K(\varphi_i \circ \varphi_j(n))$ is bounded by a term linear in i, j, n, i.e. we can assume that there exists some constant $c \in \mathbb{N}$ such that for all $i, j, n \in \mathbb{N}$ with $n \in \text{dom}(\varphi_i \circ \varphi_j)$

$$K(\varphi_i \circ \varphi_j(n)) \leq c(i+1)(j+1)(n+1) + c.$$

By $L := \{n \in \mathbb{N} : K(n) < n\}$ we denote the set of *nonrandom* or *lawful numbers*. Intuitively, a number $n \in \mathbb{N}$ is *random*, i.e. $n \notin L$, if n is its own shortest description (thus there is no algorithm i, smaller than n itself, which produces $\varphi_i(0) = n$). Without loss of generality, we assume that $0, 1 \in \mathbb{N} \setminus L$ are random numbers. It is known that the set of nonrandom numbers L is *simple*, i.e. it is r.e. and its complement is infinite but does not contain any infinite r.e. subset. Thus, the set of random numbers $\mathbb{N} \setminus L$ is *immune*: no algorithm can produce more than finitely many random numbers (cf. [6] for the definition and properties of random numbers). Now we use the set L of nonrandom numbers to construct a tree T.

Definition 6 (The tree T). Let $s : \mathbb{N} \to \mathbb{N}$ be some computable function which enumerates the set of nonrandom numbers, i.e. range$(s) = L$ and let $t :\subseteq \mathbb{N} \to \mathbb{N}$ be defined by $t(k) := \min\{m \in \mathbb{N} : s(m) = k\}$. Now let $T \subseteq \mathcal{Z} = \{0,1\}^* 0^\omega$ be the set which contains all sequences

$$p = 1^{n_0+1} 0^{t(k_0)+1} 1^{n_1+1} 0^{t(k_1)+1} 1^{n_2+1} 0^{t(k_2)+1} ... 1^{n_{j-1}+1} 0^{t(k_{j-1})+1} 1^{n_j+1} 0^\omega$$

such that $j \in \mathbb{N}$, $n_0, ..., n_j \in \mathbb{N}$, $k_0, ..., k_{j-1} \in L$ and $k_j \in \mathbb{N} \setminus L$ and the equations

$$\begin{cases} k_0 &= n_0 \\ k_{i+1} &= k_i^2 + n_{i+1} \end{cases}$$

hold for all $i = 0, ..., j - 1$.

By construction T consists only of sequences which are eventually zero. More than this, T is even closed in $\mathcal{Z}$ but it is not complete and hence not closed in $\{0,1\}^\omega$. The closure $\overline{T}$ of T contains sequences with infinitely many ones.

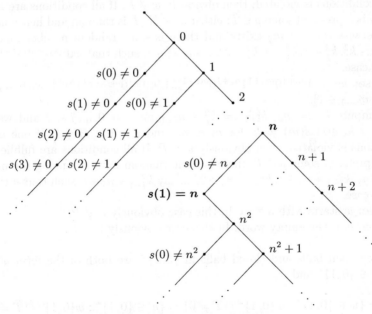

Fig. 3. A part of tree T with a nonrandom number $n = s(1) \in L$

The intuition of the construction of T is as follows (cf. Figure 3). Let us consider T as a subset of the full binary tree $\{0,1\}^\omega$ and consider the rightmost path of ones in this tree: we enumerate consecutive nodes on this path with natural numbers $0, 1, 2, ...$ and so on. The path p through node n which continues to the left with zeroes, i.e. $p = 1^{n+1}0^\omega$, belongs to T, if and only if n is a random number, i.e. $n \in \mathbb{N} \setminus L$. With any node i of path p we could associate a test which tries to prove that n is nonrandom, i.e. it is checked whether $n = s(i)$? If the test fails, then we proceed within the next step and check $n = s(i+1)$? If n is random, we follow the path p in this way until infinity. If n is nonrandom, then some test $n = s(i)$ will succeed and the consecutive sequence of zeroes is stopped at this node $i = t(n)$. In this case the path p is continued to the right, i.e. it has $1^{n+1}0^{t(n)+1}1$ as prefix. Again we associate numbers $n^2, n^2+1, n^2+2, ...$ and so on with the nodes of the following rightmost path and we continue in this subtree as described before. Actually, this intuitive description of the construction of T leads to the following result.

Proposition 7. *The set $P := \{w \in \{0,1\}^* : (\exists p \in T) \, w \sqsubseteq p\}$ is recursive.*

Proof. Let $w \in \{0,1\}^*$ be some word. In order to describe how one can decide whether $w \in P$ holds, we distinguish four cases.

1. Case: $w = 1^{n_0+1}0^{m_0+1}1^{n_1+1}0^{m_1+1}...1^{n_j+1}0^{m_j+1}$, $n_0, ..., n_j, m_0, ..., m_j \in \mathbb{N}$.
We compute $k_0 := n_0$, $k_{i+1} := k_i^2 + n_{i+1}$ for $i = 0, ..., j-1$ and we verify $s(m_i) = k_i$ for $i = 0, ..., j-1$ and $s(m) \neq k_i$ for $m < m_i$ and $i = 0, ..., j$. If one of these conditions is violated, then obviously $w \notin P$. If all conditions are fulfilled, then w is a prefix of some $p \in T$: either $k_j \in \mathbb{N} \setminus L$ is random and hence $w0^\omega \in T$ or otherwise $t(k_j) \geq m_j$ exists and there is some random number among the numbers $k_j^2, k_j^2+1, k_j^2+2, k_j^2+3, ...$, say k_j^2+l, such that $w0^{t(k_j)-m_j}1^{l+1}0^\omega \in T$ in this case.

2. Case: $w = 1^{n_0+1}0^{m_0+1}1^{n_1+1}0^{m_1+1}...1^{n_{j-1}+1}0^{m_{j-1}+1}1^{n_j+1}$ with $n_0, ..., n_j$, $m_0, ..., m_{j-1} \in \mathbb{N}$.
We compute $k_0 := n_0$, $k_{i+1} := k_i^2 + n_{i+1}$ for $i = 0, ..., j-2$ and we verify $s(m_i) = k_i$ and $s(m) \neq k_i$ for $m < m_i$ and $i = 0, ..., j-1$. If one of these conditions is violated, then obviously $w \notin P$. If all conditions are fulfilled, then w is a prefix of some $p \in T$: there is some random number among the numbers $k_{j-1}^2+n_j, k_{j-1}^2+n_j+1, k_{j-1}^2+n_j+2, ...$, say $k_{j-1}^2+n_j+l$, such that $w1^l0^\omega \in T$ in this case.

3. Case: w starts with a zero. In this case obviously $w \notin P$.
4. Case: w is the empty word. In this case obviously $w \in P$.   □

Since open balls and closed balls in $\{0,1\}^\omega$ are both of the form $w\{0,1\}^\omega$ with $w \in \{0,1\}^*$ and

$$P = \{w \in \{0,1\}^* : w\{0,1\}^\omega \cap T \neq \emptyset\} = \{w \in \{0,1\}^* : w\{0,1\}^\omega \cap \overline{T} \neq \emptyset\},$$

we obtain the following corollary.

Corollary 8. *T is a strongly recursive closed subset of $\mathcal{Z}$ and its closure $\overline{T}$ is a strongly recursive closed subset of $\{0,1\}^\omega$.*

Although the tree T is recursive there is no computable strategy which determines a path through a given node. In order to prove this result we start with a lemma which shows how the square terms in the definition of the tree T can be used.

Lemma 9. *Let $f :\subseteq \mathbb{N} \to \mathbb{N}$ be a computable function with $f\langle n, 0\rangle \geq n$ for all $n \in \mathbb{N}$ and let $g :\subseteq \mathbb{N} \to \mathbb{N}$ be defined by*

$$\begin{cases} g\langle n, 0\rangle & := f\langle n, 0\rangle \\ g\langle n, k+1\rangle := (g\langle n, k\rangle)^2 + f\langle n, k+1\rangle \end{cases}$$

Then g is computable and the sequence $g\langle n, 1\rangle, g\langle n, 2\rangle, g\langle n, 3\rangle \ldots$ is either finite or all values are defined. In both cases the sequence does contain no random number for almost all fixed values $n \in \mathbb{N}$.

Proof. Obviously, g is a computable function since f is computable. Let $i \in \mathbb{N}$ be such that $\varphi_i = g$. Then there is some constant $c \in \mathbb{N}$ such that

$$K(g\langle n, k\rangle) \leq c(i+1)(n+1)(k+1) + c$$

holds for all $n, k \in \mathbb{N}$ such that $g\langle n, k\rangle$ exists. On the other hand, $g\langle n, k\rangle \geq n^{(2^k)}$ for all values $k \in \mathbb{N}$ such that $g\langle n, k\rangle$ exists. Thus, if n is sufficiently large and itself nonrandom, then $g\langle n, k\rangle > K(g\langle n, k\rangle)$ follows for all $k > 0$ such that $g\langle n, k\rangle$ exists. Thus, for almost all $n \in \mathbb{N}$ the sequence $g\langle n, 1\rangle, g\langle n, 2\rangle, g\langle n, 3\rangle \ldots$ does not contain a random value. $\square$

With the help of this lemma we can prove the main result of this section.

Theorem 10. *T is a strongly recursive closed subset of $\mathcal{Z}$ which is not effectively separable.*

Proof. It has already been proved in Corollary 8 that T is strongly recursive closed. It remains to show that T is not effectively separable. Let us assume that there is a computable sequence $(p_i)_{i \in \mathbb{N}}$ in $\{0, 1\}^* 0^\omega$ such that $\{p_i : i \in \mathbb{N}\}$ is dense in T. Then there exists a computable function $h : \mathbb{N} \to \mathbb{N}$ such that $1^{n+1} 0 \sqsubseteq p_{h(n)}$, i.e. h finds a path associated with node n. Let us define a function $f :\subseteq \mathbb{N} \to \mathbb{N}$ by $f\langle n, 0\rangle := n$ and $f\langle n, j+1\rangle = k$, if and only if

$$(\exists n_1, \ldots, n_j, m_0, \ldots, m_j \in \mathbb{N}) \ 1^{n+1} 0^{m_0+1} 1^{n_1+1} 0^{m_1+1} \ldots 1^{n_j+1} 0^{m_j+1} 1^{k+1} 0 \sqsubseteq p_{h(n)}$$

for all $n, j, k \in \mathbb{N}$. Then f is computable too. We mention that $f\langle n, j\rangle$ is undefined for all random numbers $n \in \mathbb{N} \setminus L$ and $j > 0$ and it is defined for all nonrandom numbers $n \in L$ and a certain initial segment $j = 0, \ldots, \iota$. Then the function $g :\subseteq \mathbb{N} \to \mathbb{N}$, defined according to Lemma 9, is computable as well and there exists some sufficiently large nonrandom $n \in L$ such that the sequence $g\langle n, 0\rangle, g\langle n, 1\rangle, g\langle n, 2\rangle \ldots$ does not contain a random number. For this fixed n there exists some ι such that

$$p_{h(n)} = 1^{f\langle n, 0\rangle+1} 0^{tg\langle n, 0\rangle+1} \ldots 1^{f\langle n, \iota-1\rangle+1} 0^{tg\langle n, \iota-1\rangle+1} 1^{f\langle n, \iota\rangle+1} 0^\omega \in T$$

and thus $g\langle n, \iota\rangle \in \mathbb{N} \setminus L$ is random by definition of T. But this is a contradiction to the choice of n! $\square$

Now we can use the same method to prove even a stronger result which shows
that there is no infinite computable sequence in T at all. In the following we will
call a closed subset $A \subseteq X$ of a computable metric space X *immune*, if there
exists no computable sequence $f : \mathbb{N} \to X$ such that range$(f) = \{f(n) : n \in \mathbb{N}\}$
is infinite and included in A. The proof of the following result is a refined version
of the previous proof and it is also based on Lemma 9. In the previous proof
density allowed to construct a function f starting from the rightmost path of T.
Now we will construct a similar function f but starting deeper within the tree.
We will use the fact that $\mathbb{N} \setminus L$ is immune in order to prove that any infinite
sequence in the tree goes arbitrary deep into the tree.

Theorem 11. T *is an immune set.*

Proof. For each

$$p = 1^{n_0+1}0^{t(k_0)+1}1^{n_1+1}0^{t(k_1)+1}1^{n_2+1}0^{t(k_2)+1}...1^{n_{j-1}+1}0^{t(k_{j-1})+1}1^{n_j+1}0^\omega \in T$$

let $\zeta(p) := k_j := k_{j-1}^2 + n_j$ and $\mu(p) := j$. Thus, $\zeta(p)$ is "the random number"
encoded by p and $\mu(p) + 1$ is "the depth" of p, i.e. the number of occurrences
of "10" in p. Let us consider $\zeta :\subseteq \{0,1\}^\omega \to \mathbb{N}$ as function. Obviously, ζ with
dom$(\zeta) := \{p : \mu(p) = n\}$ is computable for any fixed $n \in \mathbb{N}$.

Let us assume that $(p_i)_{i \in \mathbb{N}}$ is a computable sequence in $\{0,1\}^\omega$ such that the
range $\{p_i : i \in \mathbb{N}\}$ is infinite and included in T.

We prove by induction on n that for all $n \in \mathbb{N}$ there are infinitely many
different sequences p_i with $\mu(p_i) \geq n$. This is obvious for $n = 0$. Let us assume
that we have proved the claim for n. We will show that the assumption that
there are only finitely many p_i with $\mu(p_i) \geq n + 1$ leads to a contradiction.
By induction hypothesis and this assumption there are infinitely many p_i with
$\mu(p_i) = n$. Since $\{p_i : \mu(p_i) \geq n + 1\}$ is finite, it follows that the set of indices
$\{i \in \mathbb{N} : \mu(p_i) < n+1\}$ is r.e. On the other hand, $\{i \in \mathbb{N} : \mu(p_i) \geq n\}$ is r.e. as well
and thus $M := \{i \in \mathbb{N} : \mu(p_i) = n\}$ is r.e. too. Moreover, $i \mapsto \zeta(p_i)$ is computable
on M and thus $Z := \{\zeta(p_i) : i \in M\}$ is r.e. too. Since $\{p_i : \mu(p_i) = n\}$ is infinite,
it follows that the set of numbers $k \in \mathbb{N}$ such that 1^k is a subword of some p_i
with $\mu(p_i) = n$, has to be infinite too and since $\zeta(p_i) \geq k - 1$ in this case, it
follows that Z is infinite as well. But this is a contradiction since $Z \subseteq \mathbb{N} \setminus L$ and
$\mathbb{N} \setminus L$ is immune. Thus, there have to be infinitely many p_i with $\mu(p_i) \geq n + 1$.
This finishes the induction.

Since $0, 1 \in \mathbb{N} \setminus L$ are random, we obtain $\mu(p) \geq n \implies \zeta(p) \geq 2^{(2^n)}$ for
all $n \geq 1$. Thus, as a consequence of the previous claim, for each $n \in \mathbb{N}$ there
are infinitely many p_i with $\zeta(p_i) \geq n$. We can conclude that there exist total
computable functions $\gamma, \sigma, h : \mathbb{N} \to \mathbb{N}$ such that for all $n \in \mathbb{N}$ and $i = \gamma(n)$ there
exist $n_0, ..., n_i, m_0, ..., m_{i-1} \in \mathbb{N}$ such that

$$\sigma(n) = s(m_{i-1})^2 + n_i > n \text{ and } 1^{n_0+1}0^{m_0+1}1^{n_1+1}...0^{m_{i-1}+1}1^{n_i+1}0 \sqsubseteq p_{h(n)}.$$

Thus $h(n)$ finds a path $p_{h(n)}$ which on depth $\gamma(n) + 1$ "tries to prove" that the
value $\sigma(n) > n$ is nonrandom. Now let us define a function $f :\subseteq \mathbb{N} \to \mathbb{N}$ by

$f\langle n, 0\rangle := \sigma(n)$ and $f\langle n, j+1\rangle = k$, if and only if there exist $n_0, ..., n_{\gamma(n)+j}$, $m_0, ..., m_{\gamma(n)+j} \in \mathbb{N}$ such that

$$1^{n_0+1}0^{m_0+1}...1^{n_{\gamma(n)}+1}0^{m_{\gamma(n)}+1}...1^{n_{\gamma(n)+j}+1}0^{m_{\gamma(n)+j}+1}1^{k+1}0 \sqsubseteq p_{h(n)}$$

for all $n, j, k \in \mathbb{N}$. Then f is computable too. We mention that $f\langle n, j\rangle$ is undefined for all $n \in \mathbb{N}$ such that $\sigma(n) \in \mathbb{N} \setminus L$ is random and $j > 0$ and it is defined for all $n \in \mathbb{N}$ such that $\sigma(n) \in L$ is nonrandom and a certain initial segment $j = 0, ..., \iota$. Since the set $\mathbb{N} \setminus L$ is immune, it follows that $\sigma(n) \in L$ for almost all $n \in \mathbb{N}$. Altogether, the function $g :\subseteq \mathbb{N} \to \mathbb{N}$, defined according to Lemma 9, is computable and there exists some sufficiently large $n \in \mathbb{N}$ such that $\sigma(n) \in L$ and such that the sequence $g\langle n, 0\rangle, g\langle n, 1\rangle, g\langle n, 2\rangle...$ does not contain a random number. For this fixed n there exists some $\iota > 0$ and $n_0, ..., n_{\gamma(n)}, m_0, ..., m_{\gamma(n)-1}$ such that

$$p_{h(n)} = 1^{n_0+1}0^{m_0+1}...1^{n_{\gamma(n)}+1}0^{tg\langle n,0\rangle+1}1^{f\langle n,1\rangle+1}0^{tg\langle n,1\rangle+1}...$$
$$1^{f\langle n,\iota-1\rangle+1}0^{tg\langle n,\iota-1\rangle+1}1^{f\langle n,\iota\rangle+1}0^{\omega} \in T$$

and thus $g\langle n, \iota\rangle \in \mathbb{N} \setminus L$ is random by definition of T. But this is a contradiction to the choice of n! $\qquad\square$

If we generalize the notion of constructively immune sets [6] to subsets of computable metric spaces appropriately, then Theorem 11 together with Proposition 7 imply that T is even a constructively immune subset of $\mathcal{Z}$.

4   The Rational Case

The aim of this section is to transfer the tree T and its properties from Cantor space to Euclidean space. The main idea is to use the transformation which is given in the following lemma.

Lemma 12. *The function* $\Gamma : \{0,1\}^{\omega} \to \mathbb{R}$, $p \mapsto \sum_{i=0}^{\infty} 2^{-2^i} p(i)$ *has the following properties:*

1. *Γ is a computable embedding, i.e. Γ is injective and Γ as well as its partial inverse Γ^{-1} are computable.*
2. *Γ preserves "the dense subset", i.e. $\Gamma(\mathcal{Z}) \subseteq \mathbb{Q}$ and $\Gamma(\{0,1\}^{\omega} \setminus \mathcal{Z}) \subseteq \mathbb{R} \setminus \mathbb{Q}$.*

The stated properties can easily be verified. Now we formulate a proposition which shows that Γ also preserves several effectivity properties of sets.

Proposition 13. *Let $A \subseteq \{0,1\}^{\omega}$ be a subset. Then the following holds:*

1. *A immune $\iff$ $\Gamma(A)$ immune.*
2. *A effectively separable in $\{0,1\}^{\omega}$ $\iff$ $\Gamma(A)$ effectively separable in $\mathbb{R}$.*
3. *A r.e. closed in $\{0,1\}^{\omega}$ $\implies$ $\Gamma(A)$ r.e. closed in $\mathbb{R}$.*
4. *A strongly co-r.e. closed in $\{0,1\}^{\omega}$ $\implies$ $\Gamma(A)$ strongly co-r.e. closed in $\mathbb{R}$.*
5. *A strongly rec. closed in $\{0,1\}^{\omega}$ $\implies$ $\Gamma(A)$ strongly rec. closed in $\mathbb{R}$.*

Proof. 1. and 2. immediately follow from the fact that Γ is a computable embedding.

3. Let A be an r.e. closed subset of $\{0,1\}^\omega$. Then the set

$$P := \{w \in \{0,1\}^* : w\{0,1\}^\omega \cap A \neq \emptyset\}$$

is r.e. For all $a, b \in \mathbb{Q}$ we denote by (a, b) the corresponding open interval and we obtain

$$(a, b) \cap \Gamma(A) \neq \emptyset \iff \Gamma^{-1}(a, b) \cap A \neq \emptyset$$
$$\iff (\exists w \in P) \, w\{0,1\}^\omega \subseteq \Gamma^{-1}(a, b)$$

Using a theorem on effective continuity (e.g. Theorem 6.2 in [1]) we can conclude that $\Gamma(A)$ is r.e. closed in $\mathbb{R}$.

4. Let A be a strongly co-r.e. closed subset of $\{0,1\}^\omega$. Then $\{0,1\}^* \setminus P$, with P defined as in 3., is r.e. For all $a, b \in \mathbb{Q}$ we denote by $[a, b]$ the corresponding closed interval and we obtain

$$[a, b] \cap \Gamma(A) = \emptyset \iff \Gamma^{-1}[a, b] \cap A = \emptyset$$
$$\iff (\exists \text{ open } U \subseteq \{0,1\}^\omega) \, \Gamma^{-1}[a, b] \subseteq U \text{ and } U \cap A = \emptyset$$
$$\iff (\exists w_1, ..., w_n \in \{0,1\}^* \setminus P) \, \Gamma^{-1}[a, b] \subseteq \bigcup_{i=1}^{n} w_i\{0,1\}^\omega.$$

Here, the last equivalence holds since $\Gamma^{-1}[a, b]$ is compact (as a closed subset of the compact Cantor space $\{0,1\}^\omega$). Using a theorem on effective continuity (Theorem 6.2 in [1]) and the effective Heine-Borel Theorem (Theorem 4.10 (2) in [3]), we can conclude that $\Gamma(A)$ is strongly co-r.e. closed in $\mathbb{R}$.

5. is a direct consequence of 3. and 4. □

Now we formulate a lemma which allows to transfer effectivity properties to subspaces and back.

Lemma 14. *Let $A \subseteq \mathbb{Q}$ be a subset which is closed in $\mathbb{Q}$ and let $\overline{A}$ denote the closure of A in $\mathbb{R}$. Then*

1. *$\overline{A}$ is r.e. closed in $\mathbb{R}$ $\iff$ A is r.e. closed in $\mathbb{Q}$.*
2. *$\overline{A}$ is strongly co-r.e. closed in $\mathbb{R}$ $\iff$ A is strongly co-r.e. closed in $\mathbb{Q}$.*
3. *$\overline{A}$ is strongly recursive closed in $\mathbb{R}$ $\iff$ A is strongly recursive closed in $\mathbb{Q}$.*
4. *A is effectively separable in $\mathbb{R}$ $\iff$ A is effectively separable in $\mathbb{Q}$.*

Proof. 1. For all $a, b \in \mathbb{Q}$ we obtain $(a, b) \cap A \neq \emptyset \iff (a, b) \cap \overline{A} \neq \emptyset$, since (a, b) is open.

2. For all $a, b \in \mathbb{Q}$ we obtain $[a, b] \cap A = \emptyset \iff [a, b] \cap \overline{A} = \emptyset$. Here, on the one hand, "$\Longleftarrow$" follows since $A \subseteq \overline{A}$. And, on the other hand, "$\Longrightarrow$" follows since $[a, b] \cap A = \emptyset$ implies $[a, b] \cap \overline{A} \subseteq \partial[a, b] = \{a, b\} \subseteq \mathbb{Q}$ and thus $[a, b] \cap \overline{A} = \emptyset$ since A is closed in $\mathbb{Q}$.

3. Is a direct consequence of 1. and 2.

4. A sequence $f : \mathbb{N} \to \mathbb{Q}$ is dense in A and computable with respect to $\mathbb{Q}$, if and only if it is dense in A and computable with respect to $\mathbb{R}$ (considered as a sequence $f : \mathbb{N} \to \mathbb{R}$). □

It should be mentioned that our results show that direction "$\Longrightarrow$" of 4. does not hold if A is replaced by its closure $\overline{A}$ on the left-hand side. Finally, we are prepared to prove our main result on $\Gamma(T)$ as a subset of $\mathbb{Q}$.

Theorem 15. $\Gamma(T)$ *is a strongly recursive closed subset of* $\mathbb{Q}$*, which is immune and thus not effectively separable.*

Proof. On the one hand, $\overline{T}$ is strongly recursive closed in $\{0,1\}^{\omega}$ by Corollary 8 and thus $\overline{\Gamma(T)} = \Gamma(\overline{T})$ is strongly recursive closed in $\mathbb{R}$ by Proposition 13.5. Hence $\Gamma(T)$ is strongly recursive closed in $\mathbb{Q}$ by Lemma 14.3. since $\Gamma(T)$ is closed in $\mathbb{Q}$. On the other hand, T is immune by Theorem 11 and thus $\Gamma(T)$ is immune by Proposition 13.1. $\qquad\square$

One could also directly use Theorem 10, Proposition 13.2. and Lemma 14.4. to conclude that $\Gamma(T)$ is not effectively separable in $\mathbb{Q}$ (without using Theorem 11). The reader should notice that immunity in Theorem 15 means that there is no infinite computable sequence of reals contained in $\Gamma(T)$. This implies the weaker statement that there is no infinite sequence of rationals included in $\Gamma(T)$, which is computable in the classical discrete sense.

Acknowledgement. The author would like to thank the anonymous referees for several useful remarks and suggestions.

References

1. Vasco Brattka. Computability of Banach space principles. Informatik Berichte 286, FernUniversität Hagen, Fachbereich Informatik, Hagen, June 2001.
2. Vasco Brattka. Computability over topological structures. In S. Barry Cooper and Sergey Goncharov, editors, *Computability and Models*. Kluwer Academic Publishers, Dordrecht, (in preparation).
3. Vasco Brattka and Gero Presser. Computability on subsets of metric spaces. *Theoretical Computer Science*, accepted for publication.
4. Vasco Brattka and Klaus Weihrauch. Computability on subsets of Euclidean space I: Closed and compact subsets. *Theoretical Computer Science*, 219:65–93, 1999.
5. Ker-I Ko. *Complexity Theory of Real Functions*. Progress in Theoretical Computer Science. Birkhäuser, Boston, 1991.
6. Piergiorgio Odifreddi. *Classical Recursion Theory*, volume 125 of *Studies in Logic and the Foundations of Mathematics*. North-Holland, Amsterdam, 1989.
7. Marian B. Pour-El and J. Ian Richards. *Computability in Analysis and Physics*. Perspectives in Mathematical Logic. Springer, Berlin, 1989.
8. Klaus Weihrauch. Computability on computable metric spaces. *Theoretical Computer Science*, 113:191–210, 1993. Fundamental Study.
9. Klaus Weihrauch. *Computable Analysis*. Springer, Berlin, 2000.
10. Mariko Yasugi, Takakazu Mori, and Yoshiki Tsujii. Effective properties of sets and functions in metric spaces with computability structure. *Theoretical Computer Science*, 219:467–486, 1999.

A Banach–Mazur Computable But Not Markov Computable Function on the Computable Real Numbers

(Extended Abstract)

Peter Hertling

Lehrgebiet Theoretische Informatik I, Informatikzentrum, Fernuniversität Hagen,
58084 Hagen, Germany,
peter.hertling@fernuni-hagen.de

Abstract. We consider two classical computability notions for functions mapping all computable real numbers to computable real numbers. It is clear that any function that is computable in the sense of Markov, i.e., computable with respect to a standard Gödel numbering of the computable real numbers, is computable in the sense of Banach and Mazur, i.e., it maps any computable sequence of real numbers to a computable sequence of real numbers. We show that the converse is not true. This solves a long–standing open problem; see Kushner [9].

Keywords: Computable real numbers; computable sequences of real numbers; Banach–Mazur computable real function; Markov computable real function; effective continuity.

1 Introduction

While there is a widely accepted notion of computability for real numbers, the situation is different for functions mapping real numbers to real numbers. For a discussion of various approaches we refer the reader to the recent monograph by Weihrauch [19]. In this paper we are concerned with two classical computability notions dating back to the early days of computable analysis. Both apply to functions mapping computable real numbers to computable real numbers.

The first of them is due to Banach and Mazur [2,12]. They considered functions that map computable sequences of real numbers to computable sequences of real numbers.

The second notion is due to Markov [11], who founded the Russian school of constructive analysis. For an overview of this constructive direction of mathematics we refer the reader to Kushner [8]. While this direction of mathematics is based on a constructive, hence, nonclassical logic, the notions in this field which are relevant for us have natural computability theoretic counterparts. We shall be concerned only with those. We say that a function f mapping computable real numbers to computable real numbers is computable in the sense of Markov if it is effective with respect to a standard Gödel numbering of the computable

P. Widmayer et al. (Eds.): ICALP 2002, LNCS 2380, pp. 962–972, 2002.

real numbers: that means that there is an algorithm which, given a Gödel number of a computable real number x in the domain of definition of f, computes a Gödel number of the computable real number $f(x)$. For an exposition of computable analysis based on this computability notion and on recursion theory see Aberth [1].

It is clear that any function computable in the sense of Markov is Banach–Mazur computable. Is the converse true? It is easy to see that the answer is No if one considers partial functions (see Section 6). Is the converse true at least for total functions, i.e., for functions defined on all computable real numbers? Kushner [9, p. 278] writes (using the symbol $\mathbb{D}$ for the constructive counterpart of the set of computable real numbers, and "c.f." for "constructive function"):

> "It is evident that every c.f. is computable by Banach–Mazur. As for the inverse statement it is still an open problem, though an old (and sophisticated) counterexample of Friedberg for the similar problem in the Baire space suggests that there are functions that are computable by Banach–Mazur, but not by Markov. It is easy to see that if a Banach–Mazur computable function is computably continuous on $\mathbb{D}$ than it can be represented as a c.f. Since on the other hand a Banach–Mazur computable function cannot have a constructive discontinuity, the required counterexample cannot be a simple one."

In this paper we construct a counterexample: a Banach–Mazur computable but not Markov computable function mapping all computable real numbers to computable real numbers. The proof borrows ideas from a related result by Friedberg [3], mentioned above by Kushner. We mention again that we work in the framework of classical mathematics, using the language and the concepts of recursion theory.

In the following two sections we introduce some basic notation and basic computability notions from recursion theory. Then we define computable sequences, Banach–Mazur computable functions, and Markov computable functions for arbitrary numbered sets. In Section 5 we give precise definitions of computable real numbers and the required computability notions on computable real numbers. Then we formulate the main result. In Section 7 we discuss a notion of effective continuity for real number functions and its relation to the other two computability notions. Note that this section is also important for the proof of the main result. In Section 8 we discuss Friedberg's result [3]. Finally, in Section 9 we give a sketch of the proof of the main result.

2   Basic Notation

By $\mathbb{N}$ we denote the set of natural numbers, i.e., nonnegative integers, and by $\mathbb{R}$ the set of real numbers. For two sets X and Y, by $f :\subseteq X \to Y$ we denote a possibly partial function whose domain of definition is a subset of X, and whose range is a subset of Y. We denote the domain of definition of f by dom f and the range of f by range f. If dom $f = X$, we call the function f *total* and may indicate this by writing $f : X \to Y$ instead of $f :\subseteq X \to Y$. A sequence $x_0, x_1, x_2, \ldots$ over a set X is nothing but a function $x : \mathbb{N} \to X$ and will often be denoted by $(x_n)_n$ or $(x_i)_i$, etc. We use the standard bijection $\langle \cdot, \cdot \rangle : \mathbb{N}^2 \to \mathbb{N}$ defined by $\langle i, j \rangle = (i+j)(i+j+1)/2 + j$, for all $i, j \in \mathbb{N}$. Inductively, we define $\langle i_1, \ldots, i_k, i_{k+1} \rangle := \langle \langle i_1, \ldots, i_k \rangle, i_{k+1} \rangle$, for $k \geq 2$.

3 Computability on the Natural Numbers

For an integer $k \geq 1$, we denote by $P^{(k)}$ the set of all computable —in the sense of recursion theory— functions $f :\subseteq \mathbb{N}^k \to \mathbb{N}$. Note that a function $f :\subseteq \mathbb{N}^k \to \mathbb{N}$ is computable if and only if the function $g :\subseteq \mathbb{N} \to \mathbb{N}$ defined by $g(\langle x_1, \ldots, x_k \rangle) := f(x_1, \ldots, x_k)$ is computable. As usual, a set $A \subseteq \mathbb{N}^k$ is called *computably enumerable*, if and only if there is a computable function $f :\subseteq \mathbb{N}^k \to \mathbb{N}$ with dom $f = A$. It is *decidable*, if and only if both it and its complement are computably enumerable.

We fix a total standard numbering φ of all computable natural number functions, i.e., a total surjective function $\varphi : \mathbb{N} \to P^{(1)}$ satisfying the following two conditions: (1) (universality) the function $u_\varphi :\subseteq \mathbb{N}^2 \to \mathbb{N}$ defined by $u_\varphi(i, j) := \varphi(i)(j)$, for all $i, j \in \mathbb{N}$, is computable; (2) (smn–property) for any computable function $f :\subseteq \mathbb{N}^2 \to \mathbb{N}$ there exists a total computable function $r : \mathbb{N} \to \mathbb{N}$ with $f(i, j) = \varphi(r(i))(j)$, for all $i, j \in \mathbb{N}$. Often we write φ_i instead of $\varphi(i)$, and $\varphi_i(j)$ instead of $\varphi(i)(j)$. By setting $W_i := \text{dom}\,\varphi_i$, we obtain a total numbering of the set of all computably enumerable sets of natural numbers.

4 Computability on Numbered Sets

Using a suitable numbering, one can transfer computability notions from the natural numbers to other countable sets. A *numbering* of a countable set X is a surjective function $\nu :\subseteq \mathbb{N} \to X$. For example, the identity function $\text{id}_\mathbb{N} : \mathbb{N} \to \mathbb{N}$ is a numbering of $\mathbb{N}$, even a total numbering.

Definition 1. Let ν be a numbering of a set X, and ν' be a numbering of a set X'.

1. A function $f :\subseteq X \to X'$ is called (ν, ν')–*computable*, if and only if there is a computable function $F :\subseteq \mathbb{N} \to \mathbb{N}$ with $f\nu(i) = \nu'F(i)$, for all $i \in \text{dom}(f\nu)$.
2. A sequence $(x_n)_n$ in X is called ν–*computable*, if and only if the function $n \mapsto x_n$ is $(\text{id}_\mathbb{N}, \nu)$–computable.
3. A function $f :\subseteq X \to X'$ is called *Banach–Mazur* (ν, ν')–*computable*, if and only if for any ν–computable sequence $(x_n)_n$ with $x_n \in \text{dom}\,f$, for all n, the sequence $(f(x_n))_n$ is ν'–computable.

What is the relation between these two computability notions for functions? The following lemma is straightforward to check.

Lemma 2. *Let (X, ν) and (X', ν') be numbered sets. Any (ν, ν')–computable function $f :\subseteq X \to X'$ is Banach–Mazur (ν, ν')–computable.*

When is the converse true? It is easy to see that a Banach–Mazur (ν, ν')–computable function $f :\subseteq X \to X'$ is (ν, ν')–computable if dom $f\nu$ is computably enumerable; cf. Hertling [5]. But in general the converse is not true. In this paper we shall show that the converse to Lemma 2 is not true for total functions mapping computable real numbers to computable real numbers, with respect to a standard numbering of the set of computable real numbers.

We conclude this section by mentioning that the computability notions introduced in Definition 1 do not depend on insignificant details in the definitions of the numberings ν and ν'. Assume that both ν and $\widetilde{\nu}$ are numberings of a set X, and both ν' and $\widetilde{\nu}'$ are numberings of a set X'. The numbering ν is called *reducible* to $\widetilde{\nu}$ if there is a computable function $F :\subseteq \mathbb{N} \to \mathbb{N}$ with $\nu(i) = \widetilde{\nu}(F(i))$, for all $i \in$ dom ν. If ν is reducible to $\widetilde{\nu}$, and $\widetilde{\nu}$ is reducible to ν, then ν and $\widetilde{\nu}$ are called *equivalent*. Assume that ν and $\widetilde{\nu}$ are equivalent, and that ν' and $\widetilde{\nu}'$ are equivalent. Then, clearly, any sequence in X is ν–computable if and only if it is $\widetilde{\nu}$–computable, any function $f :\subseteq X \to X'$ is (ν, ν')–computable if and only if it is $(\widetilde{\nu}, \widetilde{\nu}')$–computable, and it is Banach–Mazur (ν, ν')–computable if and only if it is Banach–Mazur $(\widetilde{\nu}, \widetilde{\nu}')$–computable.

5 Computability on the Computable Real Numbers

A real number should be called computable if one can compute rational approximations to it with any desired precision. This can be made precise for example as follows. We define a total numbering I of the set of all non–degenerate, i.e., nonempty, open intervals with rational endpoints by

$$I\langle i, j, k, l, m\rangle := \left(\frac{i-j}{k+1} - \frac{l+1}{m+1}, \frac{i-j}{k+1} + \frac{l+1}{m+1}\right) .$$

and a numbering $\overline{I}$ of all non–degenerate closed intervals with rational endpoints by $\overline{I}(n) := \text{closure}(I(n))$. We call a sequence $(J_n)_n$ of closed intervals a *name* for a real number x if it satisfies

$$x \in J_n \quad \text{and} \quad \text{length}(J_n) \leq \frac{1}{n+1}$$

for all $n \in \mathbb{N}$. A real number x is called *computable* if there is an $\overline{I}$–computable sequence of intervals which is a name for x. The set of all computable real numbers is denoted by $\mathbb{R}_c$. It is a countable set. We define a numbering $\nu_{\mathbb{R}_c} :\subseteq \mathbb{N} \to \mathbb{R}_c$ by

$$\nu_{\mathbb{R}_c}(i) = x \iff \text{dom } \varphi_i = \mathbb{N} \text{ and } \left(\overline{I}(\varphi_i(n))\right)_n \text{ is a name for } x.$$

A sequence $(x_n)_n$ of real numbers is called *computable* if its members x_n are computable real numbers and the sequence $(x_n)_n$ is $\nu_{\mathbb{R}_c}$–computable. Clearly, this amounts to demanding that one can compute any member x_n of the sequence with any desired precision, uniformly in n. We call a function $f :\subseteq \mathbb{R}_c \to \mathbb{R}_c$

Banach–Mazur computable if it is Banach–Mazur $(\nu_{\mathbb{R}_c}, \nu_{\mathbb{R}_c})$–computable. We call a function $f : \subseteq \mathbb{R}_c \to \mathbb{R}_c$ *Markov computable* if it is $(\nu_{\mathbb{R}_c}, \nu_{\mathbb{R}_c})$–computable.

In view of the comment after Lemma 2 it is interesting to note that the domain of definition of $\nu_{\mathbb{R}_c}$ is not computably enumerable. The same is true for any numbering of $\mathbb{R}_c$ that is equivalent to $\nu_{\mathbb{R}_c}$. In fact, by a simple diagonalisation one can show the stronger statement that there does not exist any computable sequence of real numbers that contains all computable real numbers; see e.g. Weihrauch [19].

6   Banach–Mazur Computable Functions versus Markov Computable Functions

In this section we discuss the relation between Banach–Mazur computable functions and Markov computable functions on the computable real numbers and formulate the main result of the paper.

As we have seen in Lemma 2, any Markov computable function $f : \subseteq \mathbb{R}_c \to \mathbb{R}_c$ is Banach–Mazur computable. Is the converse true? That the answer is No if one considers partial functions can easily be seen by adapting an argument by Pour–El [13]. Let $A \subseteq \mathbb{N}$ be an immune set, i.e., an infinite set of natural numbers not containing any infinite computably enumerable subset. Then A may also be considered to be a subset of $\mathbb{R}_c$. There are uncountably many functions $f : \subseteq \mathbb{R}_c \to \mathbb{R}_c$ with dom $f = A$. All of them are Banach–Mazur computable because any computable sequence $(x_n)_n$ of real numbers with $\{x_n \mid n \in \mathbb{N}\} \subseteq A$ can take only finitely many different values, i.e., the set $\{x_n \mid n \in \mathbb{N}\}$ must be finite. But not all such functions can be Markov computable because there are only countably many Markov computable functions $g : \subseteq \mathbb{R}_c \to \mathbb{R}_c$ with dom $g = A$.

Is the converse at least true if one restricts oneself to functions with a very simple domain of definition, for example to total functions $f : \mathbb{R}_c \to \mathbb{R}_c$? This is a long–standing open question, see Kushner [9, p. 278] resp. the quotation in the introduction. We shall show that the answer is No.

Theorem 3. *There exists a total function $f : \mathbb{R}_c \to \mathbb{R}_c$ which is Banach–Mazur computable but not Markov computable.*

A sketch of the proof will be given in Section 9. In Section 8 we discuss a related result by Friedberg [3]. In Section 7 we introduce another important computability notion for functions on real numbers and discuss Theorem 3 in view of this notion. Section 7 is also important for the proof.

The proof will actually show a bit more. We will see that there is a bounded, continuous, total, linear spline function $F : \mathbb{R} \to \mathbb{R}$ with finitely many rational breakpoints in any compact interval such that the restriction $f := F|_{\mathbb{R}_c}$ of F to $\mathbb{R}_c$ is Banach–Mazur computable but not Markov computable. Note that a linear spline function with finitely many rational breakpoints in any compact interval maps any computable real number to a computable real number. One can even construct a total, Banach–Mazur computable but not Markov computable function which is uniformly continuous and vanishes outside the open unit interval

$(0, 1)$. Instead of a linear spline function one can also construct a differentiable function.

Finally, we mention that, for arbitrary $k \geq 1$, one can define the set $\mathbb{R}_c^k$ of computable vectors of real numbers with k components and a standard numbering of $\mathbb{R}_c^k$ in the same way as we have defined $\mathbb{R}_c$ and $\nu_{\mathbb{R}_c}$. Theorem 3 is true also for multivariate functions $f : \mathbb{R}_c^k \to \mathbb{R}_c^l$, $k, l \geq 1$.

7   Effective Continuity

One of the most interesting properties obtained by Mazur [12] about Banach–Mazur computable functions is that every Banach–Mazur computable function $f :\subseteq \mathbb{R}_c \to \mathbb{R}_c$ whose domain is of the form $J \cap \mathbb{R}_c$, where J is an arbitrary interval, is continuous. This can easily be generalizedto functions with a computably separable domain and to the setting of computable metric spaces; see Hertling [5]. A subset $D \subseteq \mathbb{R}_c$ is called *computably separable* if there is a computably enumerable set $A \subseteq \text{dom}\,\nu_{\mathbb{R}_c}$ such that $\nu_{\mathbb{R}_c}(A)$ is a dense subset of D. For example, the set $\mathbb{R}_c$ itself is computably separable. Markov computable functions with a computably separable domain are not only continuous but even effectively continuous. Since effective continuity is an important notion which will also be important in the proof, we define it in detail.

Definition 4.   1. A set P of pairs of open intervals *describes* a function $h :\subseteq \mathbb{R} \to \mathbb{R}$ if

$$(\forall (K, L) \in P)\ h(K \cap \text{dom}\,h) \subseteq L, \quad \text{and}$$
$$(\forall x \in \text{dom}\,h)\,(\forall \varepsilon > 0)\,(\exists (K, L) \in P)\,(x \in K\ \&\ \text{length}(L) \leq \varepsilon\}.$$

2. A function $f :\subseteq \mathbb{R} \to \mathbb{R}$ is called *effectively continuous* if there is a computably enumerable set $A \subseteq \mathbb{N}$ such that the set

$$\text{Pairs}(A) := \{(I(k), I(l)) \mid \langle k, l \rangle \in A\}$$

describes f.

Note that a set P of pairs of open intervals can describe many different functions. If P describes a function $h :\subseteq \mathbb{R} \to \mathbb{R}$, then it describes also every restriction of h to a subset of $\text{dom}\,h$. Of course, every effectively continuous function is continuous. This computability notion for functions mapping real numbers —arbitrary real number, not necessarily computable ones!— to real numbers goes back to Grzegorczyk [4] and Lacombe [10]. It is the computability notion treated in the monographs by Pour–El, Richards [14], Ko [6], Weihrauch [19].

How is effective continuity connected to the computability notions for functions on computable real numbers which we considered so far in this paper? The following lemma is well known and easy to verify. We omit its proof.

Lemma 5. *Any effectively continuous function $f :\subseteq \mathbb{R} \to \mathbb{R}$ maps any computable real number in $\text{dom}\,f$ to a computable real number. The restriction $f|_{\mathbb{R}_c}$ of f to the computable real numbers is a Markov computable function.*

The inverse statement is not true in general. There exists even a discontinuous, Markov computable function $f :\subseteq \mathbb{R}_c \to \mathbb{R}_c$; compare Slisenko [15] or Weihrauch [19, p. 259]. But for functions with sufficiently simple domains the inverse statement is true. The following theorem is a special case of a result by Tseitin [17,18] which he stated and proved in the context of computable metric spaces.

Theorem 6. *Every Markov computable function $f :\subseteq \mathbb{R}_c \to \mathbb{R}_c$ with computably separable domain is effectively continuous.*

Thus, a total function $f : \mathbb{R}_c \to \mathbb{R}_c$ is computable if and only if it is effectively continuous. We shall need this result in the proof of Theorem 3. Indeed, in the proof we shall construct a total Banach–Mazur computable function and we will make sure that it is not effectively continuous.

We end this section with a caveat concerning effectively continuous functions on $\mathbb{R}_c$ and on $\mathbb{R}$. The following result is a slight strengthening of a result by Pour–El and Richards about the difference between Banach–Mazur computability and effective continuity [14, Theorem 6, p. 67].

Theorem 7. *There exists a total, uniformly continuous, and differentiable, but not effectively continuous function $f : \mathbb{R} \to \mathbb{R}$ with $f(x) = 0$ for all $x \notin [0,1]$, with $f(x) \in [0,1]$ for all x, and such that f maps computable real numbers to computable real numbers, and its restriction $f|_{\mathbb{R}_c}$ to the set of computable real numbers is Markov computable.*

This implies that the function $f|_{\mathbb{R}_c} : \mathbb{R}_c \to \mathbb{R}_c$ is effectively continuous but cannot be extended to an effectively continuous function defined on all real numbers.

In fact, the function f constructed by Pour–El and Richards in the proof of [14, Theorem 6, p. 67] already has the properties stated in the theorem, although Pour–El and Richards state only Banach–Mazur computability of $f|_{\mathbb{R}_c}$ instead of Markov computability. In order to see that $f|_{\mathbb{R}_c}$ is Markov computable, one has to observe that the sequence of rational numbers constructed in the proof of their Effective Modulus Lemma [14, Theorem 5, p. 65], on which the construction of the function f is based, also has slightly stronger properties than those formulated in the Effective Modulus Lemma. Essentially the same construction is given and analyzed by Aberth in the proof of [1, Theorem 5.4] and used for the construction of a similar function in [1, Theorem 7.3].

8   Friedberg's Result

In this section we discuss a related result by Friedberg [3]. We consider only total functions $f : R^{(1)} \to \mathbb{N}$ where $R^{(1)} := \{\varphi \in P^{(1)} \mid \operatorname{dom} \varphi_i = \mathbb{N}\}$ is the set of all total, computable natural number functions. For such functions f one can consider three computability notions in analogy to the notions considered above for functions on computable real numbers. Such functions f can be:

1. uniformly partial recursive, as Friedberg [3] calls it. For completeness sake we define this property for total functions from $R^{(1)}$ to $\mathbb{N}$. First, we define a total numbering $\widetilde{I}$ of a class of subsets of $R^{(1)}$ (this class is actually a base of the natural topology on $R^{(1)}$) by $\widetilde{I}(0) = R^{(1)}$ and

$$\widetilde{I}(1 + \langle k, \langle n_0, \dots, n_k \rangle \rangle) := \{f \in R^{(1)} \mid f(i) = n_i \ \text{for} \ i \le k\}.$$

A total function $f : R^{(1)} \to \mathbb{N}$ is *uniformly partial recursive* if there is a computably enumerable set $A \subseteq \mathbb{N}$ with the following two properties:
   a) $f(\widetilde{I}(i)) = \{k\}$ for all $\langle i, k \rangle \in A$.
   b) For every $p \in R^{(1)}$ there exists some $\langle i, k \rangle \in A$ such that $p \in \widetilde{I}(i)$.
This notion is an analogy to effective continuity.
2. $(\varphi', \mathrm{id}_\mathbb{N})$–computable, where $\varphi' :\subseteq \mathbb{N} \to R^{(1)}$ is the partial numbering of $R^{(1)}$ defined by $\mathrm{dom}\,\varphi' := \varphi^{-1}(R^{(1)})$, and $\varphi'(i) := \varphi(i)$ for all $i \in \varphi^{-1}(R^{(1)})$.
3. Banach–Mazur $(\varphi', \mathrm{id}_\mathbb{N})$–computable.

It is easy to see that the first property implies the second, and we have seen in Lemma 2 that the second implies the third. That also the second implies the first —this is the analogous statement to Theorem 6— was shown by Kreisel, Lacombe, and Shoenfield [7] and by Tseitin [16,17,18]. Thus, the first two properties are equivalent. Friedberg [3] showed that the third property does not imply the first two properties, i.e., that there is a Banach–Mazur $(\varphi', \mathrm{id}_\mathbb{N})$–computable total function $f : R^{(1)} \to \mathbb{N}$ which is not uniformly partial recursive respectively $(\varphi', \mathrm{id}_\mathbb{N})$–computable. This result is an analogy to Theorem 3.

We sketch how Friedberg [3] proceeds. First he defines a numbering of a class of continuous functions $f : R^{(1)} \to \mathbb{N}$ which contains all Banach–Mazur $(\varphi', \mathrm{id}_\mathbb{N})$–computable functions. Then he shows that with respect to this numbering the index set of the set of all Banach–Mazur $(\varphi', \mathrm{id}_\mathbb{N})$–computable functions is a Π_4–set, while the index set of the set of all uniformly partial recursive functions is a Σ_4–complete set (this is the difficult part of the proof). Since a Π_4–set and a Σ_4–complete set cannot be identical and since the first set contains the second, the first set must be strictly larger than the second.

In the proof of Theorem 3 we shall not classify index sets of effectively continuous functions or of Banach–Mazur computable functions. Instead, we will explicitly construct a total function $f : \mathbb{R}_c \to \mathbb{R}_c$ which is Banach–Mazur computable but not effectively continuous, hence, according to Theorem 6, not Markov computable. Nevertheless, our construction still uses ideas used also in the part of Friedberg's proof where he shows that the index set of the uniformly partial recursive functions $f : R^{(1)} \to \mathbb{N}$ is Σ_4–complete. On the other hand, it is not difficult to transfer our construction to the case of functions $f : R^{(1)} \to \mathbb{N}$ and to construct directly a Banach–Mazur $(\varphi', \mathrm{id}_\mathbb{N})$–computable function which is not uniformly partial recursive, hence, not $(\varphi', \mathrm{id}_\mathbb{N})$–computable.

9 Sketch of the Proof

In this section we describe the strategy of the proof of Theorem 3.

We shall construct a total, continuous, linear spline function $F : \mathbb{R} \to \mathbb{R}$ with finitely many rational breakpoints in any compact interval and with range$(F) \subseteq [0, 2]$. In fact, the function F will be zero everywhere except that for some numbers $i \in \mathbb{N}$, the interval $\left[i + \frac{1}{4}, i + \frac{3}{4}\right]$ will contain a rational subinterval on which F will have a triangular shape. Clearly, any such function maps all computable real numbers to computable real numbers. We wish to make sure that the restriction $f := F|_{\mathbb{R}_c}$ of F to the computable real numbers is Banach–Mazur computable but not Markov computable. Thus, f should satisfy the following two conditions.

- For every Markov computable function $g : \mathbb{R}_c \to \mathbb{R}_c$ we have $f \neq g$.
- For every computable sequence $(x_n)_n$ of real numbers, the sequence $(f(x_n))_n$ is also a computable sequence of real numbers.

The strategy for satisfying the first condition is as follows. For every i we shall define a computable real number r_i such that for every function $h :\subseteq \mathbb{R} \to \mathbb{R}$ described by Pairs(W_i), we have either $r_i \notin \mathrm{dom}\, h$ or $f(r_i) \neq h(r_i)$. Since f will be a total function on $\mathbb{R}_c$, in any case we have $f \neq h$. Since every effectively continuous function is described by Pairs(W_i) for some i, this ensures that f cannot be effectively continuous. Since every total, Markov computable function $h : \mathbb{R}_c \to \mathbb{R}_c$ is effectively continuous due to Theorem 6, f cannot be Markov computable.

The number r_i will lie in the interval $[i + \frac{1}{4}, i + \frac{3}{4}]$. We shall proceed in stages, and try to list more and more elements of W_i, hence, of interval pairs in Pairs(W_i). At the same time we keep a list of rational intervals which are candidates for containing r_i. Only when by enumerating W_i we have found information which tells us the approximate values that any function h described by Pairs(W_i) can take in a possible candidate interval for r_i, then we fix r_i and the value $F(r_i)$. In fact, then we define r_i to be the midpoint of this candidate interval and define F to have triangular shape on this rational interval, making sure that $F(r_i)$ is different from any possible value of $h(r_i)$. Furthermore, we set $F(x) := 0$ for all numbers $x \in [i, i + 1]$ that lie outside this rational interval. If we never find suitable information, then F will take the value zero on the whole interval. In order to obtain a computable real number r_i in that case as well, we will make sure that in the list of possible intervals for r_i there is at least one which gets smaller and smaller and, finally, converges to r_i. In that case, none of the functions described by W_i is defined at r_i, but f is. Hence, in any case, f is not described by W_i.

On the other hand, we have to make sure that f satisfies the second condition. Let $(x_n)_n$ be a computable sequence of real numbers. The first idea for ensuring that $(f(x_n))_n$ is a computable sequence of real numbers is simply to define $f(x_n) := 0$ for all n. Of course, this cannot work in general since the sequence $(x_n)_n$ might for example be a dense sequence, and then fixing the values $f(x_n)$ for all n in one step does not leave us any freedom to give f perhaps a value different from zero at r_i at some later stage. Remember that any total Banach–Mazur computable function is continuous. Hence, its values at a dense sequence determine its values everywhere. Therefore, our strategy will be to fix the value

$f(x_n)$ only once we have been able to compute the first $n+1$ numbers $x_0, \ldots, x_n$ with some sufficiently high precision, so that we still have the freedom to choose a suitable point r_i sufficiently far away from $x_0, \ldots, x_n$ where we can set $f(r_i)$ to a value different from zero, if necessary. The only possibility for $f(x_n)$ to be assigned a nonzero value at some stage is that we have fixed r_i and the value of F at r_i and in the interval $[i, i+1]$ before we were able to compute x_n with sufficient precision. Then, when it turns out later that x_n lies in the interval $[i, i+1]$, the value $f(x_n)$ is already fixed.

We can do this for all computable sequences because, as for effectively continuous functions, we have a natural partial numbering of all computable sequences of real numbers. For every computable sequence $(x_n)_n$ of real numbers, there exists a $j \in \mathbb{N}$ such that for every n the sequence $(\overline{I}(\varphi_j \langle n, k \rangle))_k$ of intervals is a name for x_n. Then we might say that j is an index for the sequence $(x_n)_n$. While we are constructing F on the interval $[i, i+1]$ and enumerating W_i, we also enumerate more and more elements in W_j for $j < i$, in order to compute more and more elements of the sequences with index $j < i$ with higher and higher precision.

Note that during the construction of F on $[i, i+1]$ we take into account only sequences with index j smaller than i. This is due to the fact that we also have to make sure that the number r_i is computable if the construction of F on $[i, i+1]$ does not stop. Therefore, we have to make sure that in the list of the intervals which we keep in mind for r_i there is at least one interval which is "free" infinitely often and becomes smaller and smaller. Here by "free" we mean that it has empty intersection with the intervals which we have computed as approximations for those elements of the sequences with index $j < i$ which we have taken into account so far. In the construction step above where we fixed F on the interval $[i, i+1]$ and gave it a triangular shape on some candidate interval for r_i, of course this candidate interval had to be free. When we compute more elements of the sequences with index $j < i$, it may happen that a currently free candidate interval for r_i is covered ("blocked") by intervals for elements of the sequences. If all of the candidate intervals for r_i are blocked, we have to add a new, free one to the list. On the other hand, it can also happen that some subinterval I of a currently blocked candidate J for r_i becomes free, namely when we manage to compute those elements of the sequences with higher precision which were responsible for blocking J. In that case, we replace J in the list of candidates for r_i by I. Furthermore, during the construction we also have to take care that the free candidate intervals for r_i become smaller and smaller.

On the other hand, it is not a tragedy that during the construction for $[i, i+1]$ we take into account only the sequences with index $j < i$. This still means that a sequence $(x_n)_n$ with index j is taken into account in the constructions for all intervals $[i, i+1]$ for $i > j$. And the behavior of the constructed function F on $(-\infty, j+1]$ can be described by finite information. On negative real numbers F will be equal to zero anyway. Thus, we can compute the values of f at the elements x_m of the computable sequence $(x_m)_m$ by first computing some integer i with $x_m \in \left(i - \frac{1}{4}, i + \frac{5}{4}\right)$ and then distinguishing whether $i \le j$ or $i > j$. In the first case we can use the finite information describing F on $[-\infty, j+1]$, and in the second case we follow the construction of F on $[i, i+1]$.

This ends the sketch of the proof.

References

1. O. Aberth. *Computable Analysis*. McGraw-Hill, New York, 1980.
2. S. Banach and S. Mazur. Sur les fonctions calculables. *Ann. Soc. Pol. de Math.*, 16:223, 1937.
3. R. M. Friedberg. 4-quantifier completeness: A Banach–Mazur functional not uniformly partial recursive. *Bulletin de l'Academie Polonaise des Sciences, Série des sci. math., astr. et phys.*, 6(1):1–5, 1958.
4. A. Grzegorczyk. Computable functionals. *Fundamenta Mathematicae*, 42:168–202, 1955.
5. P. Hertling. Banach-Mazur computable functions on metric spaces. In J. Blanck, V. Brattka, and P. Hertling, editors, *Computability and Complexity in Analysis*, volume 2064 of *Lecture Notes in Computer Science*, pages 69–81, Berlin, 2001. Springer. 4th International Workshop, CCA 2000, Swansea, UK, September 2000.
6. K.-I. Ko. *Complexity Theory of Real Functions*. Progress in Theoretical Computer Science. Birkhäuser, Boston, 1991.
7. G. Kreisel, D. Lacombe, and J. Shoenfield. Partial recursive functionals and effective operations. In A. Heyting, editor, *Constructivity in Mathematics*, Studies in Logic and the Foundations of Mathematics, pages 290–297, Amsterdam, 1959. North-Holland. Proc. Colloq., Amsterdam, Aug. 26–31, 1957.
8. B. A. Kušner. *Lectures on Constructive Mathematical Analysis*, volume 60. American Mathematical Society, Providence, 1984.
9. B. A. Kušner. Markov's constructive analysis; a participant's view. *Theoretical Computer Science*, 219:267–285, 1999.
10. D. Lacombe. Classes récursivement fermés et fonctions majorantes. *Comptes Rendus Académie des Sciences Paris*, 240:716–718, June 1955. Théorie des fonctions.
11. A. A. Markov. On the continuity of constructive functions (Russian). *Uspekhi Mat. Nauk (N.S.)*, 9:226–230, 1954.
12. S. Mazur. *Computable Analysis*, volume 33. Razprawy Matematyczne, Warsaw, 1963.
13. M. B. Pour-El. A comparison of five "computable" operators. *Zeitschrift für Mathematische Logik und Grundlagen der Mathematik*, 6:325–340, 1960.
14. M. B. Pour-El and J. I. Richards. *Computability in Analysis and Physics*. Perspectives in Mathematical Logic. Springer, Berlin, 1989.
15. A. O. Slisenko. Examples of a nondiscontinuous but not continuous constructive operator in a metric space. *Tr. Mat. Inst. Steklov*, 72:524–532, 1964. (in Russian, English trans. in AMS Trans. 100, 1972).
16. G. S. Tseitin. Uniform recursiveness of algorithmic operators on general recursive functions and a canonical representation for constructive functions of a real argument. In *Proc. Third All–Union Math. Congr., Moscow 1956*, volume 1, pages 188–189, Moscow, 1956. Izdat. Akad. Nauk SSSR. (in Russian).
17. G. S. Tseitin. Algorithmic operators in constructive complete separable metric spaces. *Doklady Akad. Nauk*, 128:49–52, 1959. (in Russian).
18. G. S. Tseitin. Algorithmic operators in constructive metric spaces. *Tr. Mat. Inst. Steklov*, 67:295–361, 1962. (in Russian, English trans. in AMS Trans. 64, 1967).
19. K. Weihrauch. *Computable Analysis*. Springer, Berlin, 2000.

Polynomial-Time Approximation Schemes for the Euclidean Survivable Network Design Problem*

Artur Czumaj[1], Andrzej Lingas[2], and Hairong Zhao[1]

[1] Department of Computer Science, New Jersey Institute of Technology, Newark, NJ
07102-1982, USA, {czumaj,hairong}@cis.njit.edu
[2] Department of Computer Science, Lund University, Box 118, S-22100 Lund, Sweden,
Andrzej.Lingas@cs.lth.se

Abstract. The survivable network design problem is a classical problem in combinatorial optimization of constructing a minimum-cost subgraph satisfying predetermined connectivity requirements. In this paper we consider its geometric version in which the input is a complete Euclidean graph. We assume that each vertex v has been assigned a connectivity requirement r_v. The output subgraph is supposed to have the vertex- (or edge-, respectively) connectivity of at least $\min\{r_v, r_u\}$ for any pair of vertices v, u.

We present the first polynomial-time approximation schemes (PTAS) for basic variants of the survivable network design problem in Euclidean graphs. We first show a PTAS for the Steiner tree problem, which is the survivable network design problem with $r_v \in \{0, 1\}$ for any vertex v. Then, we extend it to include the most widely applied case where $r_v \in \{0, 1, 2\}$ for any vertex v. Our polynomial-time approximation schemes work for both vertex- and edge-connectivity requirements in time $\mathcal{O}(n \log n)$, where the constants depend on the dimension and the accuracy of approximation. Finally, we observe that our techniques yield also a PTAS for the multigraph variant of the problem where the edge-connectivity requirements satisfy $r_v \in \{0, 1, \ldots, k\}$ and $k = \mathcal{O}(1)$.

1 Introduction

In this paper we consider a geometric version of the *survivable network design problem*. The survivable network design problem is a classical problem in combinatorial optimization because of its evident applications in telecommunication, communication network design, VLSI design, etc. In its most general formulation, there is given an undirected graph $G = (V, E)$, a cost function $c : E \to \mathbb{R}$, and a connectivity requirement function r mapping any pair of vertices to $\mathbb{N}$. The task is to find a minimum-cost subgraph of G such that for any pair of vertices $v, u \in V$, the subgraph has $r_{v,u}$ internally vertex-disjoint (or edge-disjoint, respectively) paths between v and u. Often, the output is allowed to be a multigraph [23].

In many applications of this problem, often regarded as the most interesting ones [8, 13], the connectivity requirement function is specified with the help of a one-argument function which assigns to each vertex v its connectivity type $r_v \in \mathbb{N}$. Then, for any pair of

* Research supported in part by NSF grant CCR-0105701, SBR grant No. 421090, and TFR grant 221-99-344.

vertices $v, u \in V$, the connectivity requirement $r_{u,v}$ is simply given as $\min\{r_u, r_v\}$ [10, 11,12,13,20,23]. Following the literature, we assume this standard simplification of the connectivity requirements function in this paper. Notice that, in particular, this includes the *Steiner tree problem* [21], in which $r_v \in \{0, 1\}$ for any vertex $v \in V$. It also includes the most widely applied variant of the survivability problem in which $r_v \in \{0, 1, 2\}$ for any vertex $v \in V$ (see, e.g., [13,20,23]).

In the geometric version of the survivable network design problem, the input vertices are points in $\mathbb{R}^d$ and the cost of each link is equal to the Euclidean distance between its endpoints (which is a good approximation in many applications, since often the "installation" and the "service" cost is roughly proportional to the length of the link [20]). We focus on two most basic variants of the geometric survivable network design problem when the connectivity requirements satisfy $r_v \in \{0, 1\}$ or $r_v \in \{0, 1, 2\}$ for all $v \in V$. The arguments provided by Grötschel *et. al.* [13] (see also [20,23]), suggest that the second special case of the survivability problem models well many applications, e.g., the problem of designing survivable fiber telephone networks [20,23]. In the case of fiber communication networks for telephone companies, network topologies with connectivity requirements in $\{0, 1, 2\}$ provide an adequate level of survivability for the distinguished central nodes of connectivity type 2. Simply, most failures usually can be repaired relatively quickly and, as statistical studies have revealed, it is unlikely that a second failure will occur for their duration.

1.1   New Contributions

We design the *first polynomial-time approximation schemes* (PTASs) for the two afore-mentioned basic variants of the geometric survivable network design problem.

First, we consider the simplest case in which $r_v \in \{0, 1\}$ for any vertex $v \in V$, that is, the Steiner tree problem. We design an algorithm that, for any constant d and any constant ε, returns a Steiner tree whose cost is at most $(1 + \varepsilon)$ times larger than the minimum. The algorithm runs in time $\mathcal{O}(n \log n)$. For general d and ε, its running time is $\mathcal{O}(n \log n \, (d/\varepsilon)^{\mathcal{O}(d)}) + \mathcal{O}(n \, (d/\varepsilon)^{(d/\varepsilon)^{\mathcal{O}(d^2)}})$.

Next, we consider the case when $r_v \in \{0, 1, 2\}$ for any vertex $v \in V$; this is the classical problem investigated thoroughly by Grötschel and Monma *et. al.* [10,11,12,13, 20,23]. We extend the algorithm for the Steiner tree problem to design an algorithm that, for any constant d and any constant ε, returns a graph satisfying all the vertex (or edge, respectively) connectivity requirements and having the cost at most $(1 + \varepsilon)$ times larger than the minimum. The algorithm runs in time $\mathcal{O}(n \log n)$. When d and ε are allowed to vary arbitrarily, its running time is $\mathcal{O}(n \log n \, (d/\varepsilon)^{\mathcal{O}(d)}) + \mathcal{O}(n \, (d/\varepsilon)^{(d/\varepsilon)^{\mathcal{O}(d^2)}})$.

Finally, we observe that our techniques yield also a PTAS for the multigraph variant where the edge-connectivity requirements satisfy $r_v \in \{0, 1, \dots, k\}$ and $k = \mathcal{O}(1)$.

Our polynomial-time approximation schemes follow an approach similar to those used in the recent PTASs for finding TSP, (complete) Steiner trees, and minimum-cost biconnected spanning subgraph in Euclidean graphs, see [1,4,22]. However, there are many important differences that make the new results significantly more complicated. First of all, we have to deal with the restriction of the Steiner points to the set given *a priori* (unlike in the minimum-cost Euclidean (complete) Steiner trees problem, in which

Steiner points are allowed to be any points in $\mathbb{R}^d$). Furthermore, we have to deal with non-uniform connectivity requirements. The substantial differences and complications occur in the so called filtering phase and searching phase (dynamic programming).

As presented in the paper, all our PTASs are randomized and achieve the promised approximation guarantees and running time on the average. However, all our algorithms can be *derandomized* in a way similar to that used by Rao and Smith in [22]. The derandomization preserves the running time of $\mathcal{O}(n \log n)$ for constant d and ε. Because of space limitations we defer this discussion to the full version of this paper.

1.2 Related Works

There has been a lot of research on the survivable network design problem. Typically, the research addresses either practical heuristics and algorithms (see, e.g., [2,10], [11,12, 13,20,23]) or the general problem for arbitrary networks (see, e.g., [5,6,8,17,24]), or the problem restricted to very specific networks. In particular, the celebrated result due to Jain [17] gives a polynomial-time 2-approximation algorithm for the edge-connected survivable network design problem (for *arbitrary connectivity requirements*). Also, polynomial-time 2-approximation algorithms for arbitrary networks in the case $r_{v,u} \in \{0, 1, 2\}$ for every v, u have been recently presented [5,6]. We are not aware of any other good polynomial-time approximation algorithm specialized for the geometric version of the survivability problem except the case when $r_v \in \{0, 1\}$ for every v [21]. If $r_v \in \{0, 1\}$ for every v and $U = \{v \in V : r_v = 1\}$, then one can easily show that a minimum spanning tree of U guarantees the approximation ratio of 2 in any metric space (and thus, in particular, in any Euclidean space $\mathbb{R}^d$). Importantly, in this case the geometric survivability problem is a generalization of the classical *Euclidean (complete) Steiner tree problem* (see, e.g., [9,15,16,21,25]). The Euclidean (complete) Steiner tree problem for a finite set of points S in $\mathbb{R}^d$ is to construct a minimum length tree whose vertex set consists of all points in S and possibly some other points in $\mathbb{R}^d$. Thus, the latter problem can be regarded as a survivable network design problem on an infinite vertex domain, i.e., $V = \mathbb{R}^d$ and $r_v = 1$ for any $v \in S$, and $r_v = 0$ otherwise. By the celebrated results due to Arora [1] and Mitchell [19] (see also [22]), the Euclidean (complete) Steiner tree problem admits a PTAS for any constant d.

Organization of the paper. Section 2 provides basic terminology used in our approximation schemes. In Section 3 we outline our new PTAS for the Steiner minimum tree problem. Section 4 outlines a generalization of the PTAS for Steiner minimum tree problem to include the survivability problem with the connectivity requirements in $\{0, 1, 2\}$. Finally, in Section 5 we briefly discuss possible further extensions of our PTASs to include the survivability problem in multigraphs with the edge-connectivity requirements in $\{0, 1, \ldots, k\}$, as well as other ℓ_p^d metrics. Due to space limitations most of our technical claims and their proofs are deferred to the full version of this paper.

2 Definitions

We introduce more specific notation on Euclidean (called also geometrical) graphs. An *Euclidean* graph, is a pair $G = (P, E)$, where P is a set of points in an Euclidean space

$\mathbb{R}^d$ and E is a subset of the pairs of points in P. Every Euclidean graph is weighted and the *cost* of edge (x, y) is equal to the Euclidean distance between points x and y. The *cost of the graph* is the sum of the costs of its edges. Additionally, we shall also consider Euclidean multigraphs, which are as Euclidean graphs but may contain parallel edges. Consistently with our definition, the edges of an Euclidean graph or multigraph $G = (P, E)$ are in one-to-one correspondence with the straight-line segments (in $\mathbb{R}^d$) connecting the incident vertices. (Such graphs are frequently called straight-line graphs in the literature.) Sometimes, for technical reasons, we shall also allow to *bend* some edges. A bent edge between a pair of points in P will be identified with a *straight-line path* (a path consisting of straight-line segments connecting the points).

Classes of Euclidean graphs. We shall discuss various classes of Euclidean graphs. Let P be a set of points in $\mathbb{R}^d$. A graph G on P is called a *t-spanner* of P, $t \geq 1$, if for any pair of points $p, q \in P$ there exists a path in G from p to q of length at most t times the Euclidean distance between p and q. Gudmundsson *et. al.* [14] gave a very efficient construction of t-spanners of small maximum degree and small cost.

Let P_0 and P_1 be sets of points in $\mathbb{R}^d$. An Euclidean tree is called a *Steiner tree* of P_1 if its vertex set includes all the points P_1. All the vertices of a Steiner tree of P_1 outside P_1 are called its *Steiner points*. If the Steiner points are restricted to a point set P_0, the tree is called a *Steiner tree of P_1 with respect to P_0* and the points in P_0 are called *Steiner point candidates*. The *Euclidean (complete) Steiner minimal tree* of P_1 is a Steiner tree of P_1 having the minimum cost. A Steiner tree of P_1 with respect to P_0 having the minimum cost will be called a *Steiner minimum tree (SMT)* of P_1 with respect to P_0. Observe the difference between our definition of Euclidean (complete) Steiner tree and Steiner minimum tree; in this paper the abbreviation SMT is used only to denote a Steiner minimum tree.

Definition 1. (Survivable network design problem) *Let P be a set of n points in $\mathbb{R}^d$. Suppose that for any point $p \in P$ there is associated a value $r_p \in \{0, \ldots, n-1\}$. The* survivable network design *problem is to find a minimum-cost Euclidean graph on P in which for any pair of points $v, u \in P$ there are at least $\min\{r_v, r_u\}$ disjoint paths between v and u. In the* vertex-connected *version of the problem the paths must be internally vertex-disjoint and in the* edge-connected *version of the problem the paths must be edge-disjoint.*

In this paper we focus mostly on the variant of the problem when $r_p \in \{0, 1, 2\}$ for any $p \in P$. We shall call this variant of the problem the $\{0, 1, 2\}$-*vertex- or - edge-connectivity* problem, depending on whether the vertex-connected or the edge-connected version of the problem is considered. (Notice that the $\{0, 1, 2\}$-vertex- and -edge-connectivity problem includes the SMT problem in which $r_p \in \{0, 1\}$ for any $p \in P$.) Furthermore, we can repeat the arguments used in [4] (which were also used earlier in [7, Section 3]) to show that in metric spaces $\{0, 1, 2\}$-vertex-connectivity and $\{0, 1, 2\}$-edge-connectivity are essentially equivalent (the arguments in [4] and [7] were given only for biconnectivity vs. two-edge-connectivity).

Lemma 2. *Let P_0, P_1, P_2 be any three sets of points in a metric space. Let H be a multigraph with the vertex set $P_0 \cup P_1 \cup P_2$ such that for any pair of vertices $u \in P_i$ and*

$v \in P_j$ there are at least $\min\{i, j\}, 0 \leq i, j \leq 2$, edge-disjoint paths from u to v in H. Then, in linear time, one can transform H into a graph G without increasing the total cost such that for any pair of vertices $u \in P_i$ and $v \in P_j$ there are at least $\min\{i, j\}$ internally vertex-disjoint paths form u to v in G. □

This lemma allows us to concentrate only on the $\{0, 1, 2\}$-edge-connectivity problem, and to allow the output to be given in a form of a multigraph.

Partitioning the space. An important component of our approximation algorithms is a partitioning scheme introduced by Arora in [1] and later extended in [3,4].

Definition 3. (Dissection, 2^d-ary tree) *Given a set S of points in $\mathbb{R}^d$, a bounding box of S is the smallest d-dimensional axis-parallel cube L^d containing the points in S. A (2^d-ary) dissection [1] of S is the recursive partitioning of the cube into smaller sub-cubes, called* regions. *Each region U^d of volume > 1 is recursively partitioned into 2^d regions $(U/2)^d$. A 2^d-ary tree (for a given 2^d-ary dissection) is a tree whose root corresponds to L^d, and whose other non-leaf nodes correspond to the regions containing at least two points from S. For a non-leaf node v of the tree corresponding to a region R, its children in the tree correspond to the 2^d regions that partition R in the dissection.*

For any d-vector $\mathbf{a} = (a_1, \ldots, a_d)$, where all a_i are integers $0 \leq a_i \leq L$, the $\mathbf{a}$-shifted dissection [1,3] of a set X of points in the cube L^d in $\mathbb{R}^d$ is the dissection of the set X^* in the cube $(2L)^d$ in $\mathbb{R}^d$ obtained from X by transforming each point $\mathbf{x} \in X$ to $\mathbf{x} + \mathbf{a}$. A random shifted dissection of a set of points X in a cube L^d in $\mathbb{R}^d$ is an $\mathbf{a}$-shifted dissection of X with $\mathbf{a} = (a_1, \ldots, a_d)$ and the elements $a_1, \ldots, a_d$ chosen independently and uniformly at random from $\{0, 1, \ldots, L\}$.

In the paper we shall also study a special class of geometric graphs with respect to a given dissection [4].

Definition 4. (r-locally-light graphs) *A graph is r-locally-light [4] with respect to a shifted dissection if for each region in the dissection there are at most r edges having exactly one endpoint in the region.*

The main reason of introducing this class of graphs is that while it is $\mathcal{NP}$-hard to solve the $\{0, 1, 2\}$-vertex- or -edge-connectivity problem (or even the minimum-cost Steiner tree problem) for arbitrary Euclidean graphs, we can show how to solve this problem efficiently for r-locally-light graphs in Section 4.

3 Steiner Minimum Tree Problem

In our attempt to provide an efficient approximation scheme for the $\{0, 1, 2\}$-connectivity problem in Euclidean graphs, we consider the SMT problem first. We apply a method that can be seen as a combination of the approach of Arora [1] and Rao and Smith [22] developed to design a PTAS for the TSP problem with the approach of Czumaj and Lingas [4] developed to design a PTAS for k-connectivity problems. Our algorithms is based on efficient implementations of the following three procedures.

Filtering: *Let P_0 and P_1 be sets of points in $\mathbb{R}^d$ and let t be any positive real number. Find a subset X of P_0 that satisfies the following two properties:*
- *The cost of the SMT of P_1 with respect to X is at most $1 + t$ times the cost of the SMT of P_1 with respect to P_0.*
- *The cost of the minimum spanning tree of $X \cup P_1$ is upper bounded by $\lambda_{d,t}$ times the cost of the minimum spanning tree of P_1, where $\lambda_{d,t}$ is a function of d and t only ($\lambda_{d,t}$ will be set to $2^{\mathcal{O}(d^4)}/t^{\mathcal{O}(d)}$).*

Lightening: *Let P_0 and P_1 be sets of points in $\mathbb{R}^d$ and let t be any positive number. Let G be any $(1 + t)$-spanner of $P_0 \cup P_1$ satisfying the so called $(t', 1 + t)$-leapfrog property [14] for $1 < t' < 1 + t$. Modify G to obtain an r-locally-light graph with the vertex set $P_0 \cup P_1$ that has as its subgraph a Steiner tree of P_1 with respect to P_0 whose cost is at most $(1 + 2t)$ times the cost of the SMT of P_1 with respect to P_0.*

Searching: *Let P_0 and P_1 be sets of points in $\mathbb{R}^d$ and let r be any positive integer. Let G be any r-locally-light graph on $P_0 \cup P_1$. Find a minimum-cost Steiner tree of P_1 with respect to P_0 that is a subgraph of G.*

3.1 Filtering for SMT

In this section we show how to perform the filtering phase efficiently. We first prove that the following algorithm finds a subset X of P_0 that satisfies the required filtering property with $t = \frac{3}{2}\epsilon$.

1. Build a $(1+\epsilon)$-spanner S on P_1 with $n \cdot \xi_{d,\epsilon}$ edges whose total cost is upper bounded by $\xi_{d,\epsilon}$ times the cost of the minimum spanning tree of P_1, where $\xi_{d,\epsilon} = (d/\epsilon)^{\mathcal{O}(d)}$, [14].
2. For each edge e of the spanner whose cost exceeds the $|P_1|^{-4}$ fraction of the cost of minimum spanning tree of P_1, circumscribe a d-dimensional ball $B\langle e \rangle$ with the center at the middle of e and of radius $R\langle e \rangle$ equal to ρ_d/ϵ times the length of the edge, where ρ_d is a function depending only on d, $\rho_d = 2^{\mathcal{O}(d^3)}$.
3. Let Y be a subset of P_0 that includes all points contained in the constructed balls and possibly some other points in P_0 at distance at most $4R\langle e \rangle$ from the center of such a ball $B\langle e \rangle$.
4. For each ball $B\langle e \rangle$ define a (rectilinear) d-dimensional cube $C\langle e \rangle$ of side length $8R\langle e \rangle$ that is co-centric with the ball $B\langle e \rangle$. Within each cube $C\langle e \rangle$ introduce a grid with interspacing $|e| \, \epsilon^2/(8 \Delta \rho_d \sqrt{d})$, where Δ is the bound of maximum degree of any MST of n points in dimension d. Let set X be initially empty. Repeatedly, in the increasing length order of the edges e of S, assign each point $p \in Y$ associated with ball $B\langle e \rangle$ to the closest point of the grid $C\langle e \rangle$. For each grid point in $C\langle e \rangle$ if there is at least one point $p \in Y$ assigned to it, add one such a point to X.

Lemma 5. (SMT Filtering) *For any point sets P_0 and P_1 in $\mathbb{R}^d$ and any positive real number ϵ, the subset X of P_0 satisfies the filtering property.* $\qquad\square$

The proof of the lemma above is highly non-trivial and very involved. It is based on a very subtle analysis of properties of spanners and optimal Steiner trees and some ideas from [22]. Because of space limitations we defer the proof to the full version of the paper.

The construction of the set X described above has been specially tuned to enable an efficient implementation.

Lemma 6. *The algorithm above can be implemented to determine the set X in time $(d/\epsilon)^{\mathcal{O}(d)} \cdot n \log n$, where $n = |P_0 \cup P_1|$.*

Proof. By [14], Step 1 can be implemented in time $(d/\epsilon)^{\mathcal{O}(d)} \cdot n + \mathcal{O}(d \cdot n \cdot \log n)$.

To implement Steps 2, 3, i.e., to determine the set Y, we use a core data structure for *approximate point location in equal balls* due to Indyk and Motwani [18]. For a given approximation factor $c \geq 1$ and a given real r, they designed a static data structure for a set of points $S \subseteq \mathbb{R}^d$, called (c, r)-PLEB, such that for any point $q \in \mathbb{R}^d$, if S contains a point within distance r from q then (c, r)-PLEB outputs a point $q \in S$ that is promised to be within a distance at most cr from q. Indyk and Motwani [18, Theorem 3] show that there is an algorithm for $(3/2, r)$-PLEB that after an $\mathcal{O}(|S| 2^{\mathcal{O}(d)})$-time preprocessing achieves $\mathcal{O}(d)$ query time. Note that in our algorithm above, the costs of the spanner edges from which the balls originate fall into a logarithmic number of intervals of the form $[2^i l_0, 2^{i+1} l_0)$, where l_0 is their minimum cost (which is lower bounded by the cost of the minimum spanning tree of P_1 over $|P_1|^4$). To determine Y, for $i = 0, \ldots, \mathcal{O}(\log n)$, we build the $(3/2, 2^{i+1} l_0 \, \rho_d/\epsilon)$-PLEB data structure for all the center points of spanner edges having cost in the interval $[2^i l_0, 2^{i+1} l_0)$, and then query it with all the points in P_0. The time needed for the construction of the logarithmic number of PLEB data structures is $\mathcal{O}(\log n)$ times the number of spanner edges (which is $n \, \xi_{d,\epsilon} = n \, (d/\epsilon)^{\mathcal{O}(d)}$) and the time needed for the $|P_0|$ queries is $\mathcal{O}(\log n) \times |P_0| \times \mathcal{O}(d)$ [18]. Hence, the total time required in Steps 2 and 3 is $\mathcal{O}(n \, (d/\epsilon)^{\mathcal{O}(d)} \log n)$.

Observe that during the construction of Y, we can also insert each point accounted to Y into a list corresponding to the smallest ball it belongs to (approximately) by processing the $\mathcal{O}(\log n)$ PLEB queries without increasing the asymptotic time performance. These lists are useful in the implementation of Step 4. Simply, for each of the balls $B\langle e \rangle$, and for each point p in the corresponding list $L\langle e \rangle$, we check whether or not the point p after rounding off to the nearest point on the grid $C\langle e \rangle$ is already marked. If not, we mark the grid point and add p to X. It is easy to see that in this way Step 4 can be implemented in time $\mathcal{O}((d/\epsilon)^{\mathcal{O}(d)} \cdot n)$. □

3.2 Lightening for SMT

For the Lightening phase we use a framework developed in [4] to transform spanners into r-locally-light graphs maintaining connectivity properties. Using this framework we can prove the following lemma.

Lemma 7. *Let P_0 and P_1 be sets of points in $\mathbb{R}^d$ and let $\epsilon > 0$. Let $r = (d/\epsilon)^{\mathcal{O}(d^2)}$. Let G be any $(1 + \frac{1}{4}\epsilon)$-spanner of $P_0 \cup P_1$ that has $n \, (d/\epsilon)^{\mathcal{O}(d)}$ edges, whose total cost is upper bounded by $(d/\epsilon)^{\mathcal{O}(d)}$ times the cost of the minimum spanning tree of P_1 and that satisfies the $(t, 1 + \frac{1}{4}\epsilon)$-leapfrog property, where $1 < t < 1 + \frac{1}{4}\epsilon$. Then, one can transform G to obtain a graph H with vertex set $P_0 \cup P_1$ (i) that is r-locally-light and*

(ii) that contains as its subgraph a Steiner tree of P_1 with respect to P_0 whose cost is at most $(1 + \frac{1}{2}\epsilon)$ times the cost of the SMT of P_1 with respect to P_0. Moreover, this transformation can be performed in time $\mathcal{O}(d^{3/2} n \log n) + n (2^{d^{\mathcal{O}(d)}} + (d/\epsilon)^{\mathcal{O}(d)})$. □

Informally, the $(t, 1 + \frac{1}{4}\epsilon)$-leapfrog property means that if there is an edge between u and v in the $(1 + \frac{1}{4}\epsilon)$-spanner then any path in the spanner between u and v that does not use the edge (u, v) has cost greater than t times the cost of (u, v). For a formal definition see Gudmundsson *et. al.* [14] where also an efficient construction of the spanner used in Lemma 7 is given.

Remark 8. The key property of the construction in Lemma 7 is that the obtained graph H has $P_0 \cup P_1$ as its vertex set. This distinguishes it from previous constructions, e.g., [1,22], where a related graph H was allowed to use arbitrary points outside $P_0 \cup P_1$. This property is critical in the approximation of SMT and other connectivity problems (in contrast to, e.g., TSP approximation). Indeed, suppose that H has as a vertex a point $q \notin P_0 \cup P_1$ which is a cut-vertex in H and that H contains some tree T to be pruned to a Steiner tree for P (or its subset). Then, if the degree of q in T is very high, it might be impossible to remove q from T to obtain a tree of the cost as good (or almost as good) as the cost of T (in contrast, TSP on a superset of P can be easily modified to obtain a TSP on P without any cost increase). Observe that this construction allows to bend the edges, see our discussion at the beginning of Section 2. ♠

3.3   Searching for SMT

Our approach in the Searching phase is to apply dynamic programming to find an optimal Steiner tree in a r-locally-light graph. Using dynamic programming approach essentially the same as the one used in PTAS algorithms for TSP (and the Euclidean complete Steiner tree problems) due to Arora [1,22], we can prove the following result.

Lemma 9. *Let P_0 and P_1 be sets of jointly n points in $\mathbb{R}^d$ and let r be any integer. Let G be an r-locally-light (with respect to a certain given shifted dissection) graph on $P_0 \cup P_1$. Then, in time $\mathcal{O}(n \cdot r^{\mathcal{O}(2^d r)})$ one can find a minimum-cost Steiner tree of P_1 with respect to P_0 that is a subgraph of G.* □

3.4   Polynomial-Time Approximation Scheme for SMT

Now, we show how to combine all our arguments from Sections 3.1–3.3 to obtain a PTAS for the Euclidean SMT problem. The input to our problem consists of two sets P_0 and P_1 of points in $\mathbb{R}^d$ of total size $|P_0| + |P_1| = n$. Our goal is to find a Steiner tree of P_1 with respect to P_0 whose cost is less than or equal to $1 + \varepsilon$ times the minimum.

We first apply Lemma 5 to find a subset X of P_0 having the promised properties (with $t = \frac{1}{4}\varepsilon$). Then, we take a $(1 + \frac{1}{4}\varepsilon)$-spanner G for $X \cup P_1$ and apply Lemma 7 to modify G in order to obtain an r-locally-light graph H that has as a subgraph a Steiner tree of P_1 with respect to X whose cost is upper bounded by $(1 + \frac{1}{2}\varepsilon)$ times the cost of the SMT of P_1 with respect to X. Finally, we apply Lemma 9 to find a minimum-cost Steiner tree of P_1 with respect to X that is a subgraph of H and output it. This leads to the following theorem.

Theorem 10. *There exists a polynomial-time approximation scheme for the Steiner Minimum Problem in Euclidean space $\mathbb{R}^d$. In particular, for any sets of points P_0 and P_1 in $\mathbb{R}^d$ of total size n, in time $\mathcal{O}(n \log n \, (d/\varepsilon)^{\mathcal{O}(d)}) + \mathcal{O}(n \, (d/\varepsilon)^{(d/\varepsilon)^{\mathcal{O}(d^2)}})$ one can find a Steiner tree of P_1 with respect to P_0 whose cost is at most $1 + \varepsilon$ times the minimum.*

For constant d and ε, the running time of this algorithm is $\mathcal{O}(n \log n)$.   □

4   {0, 1, 2}-Connectivity Problem

One can extend the algorithm from the previous section to obtain a polynomial-time approximation scheme for the $\{0, 1, 2\}$-Connectivity Problem in Euclidean graphs. Actually, we have paid special attention to present the algorithm for the SMT problem in a form extendable to include the $\{0, 1, 2\}$-Connectivity Problem.

Due to Lemma 2, we may consider only the $\{0, 1, 2\}$-edge-connectivity problem and allow the output to be given in a form of a multigraph. Our algorithm uses similar three phases as the algorithm for the SMT problem. The Filtering phase is essentially the same as for the SMT problem except that we work on the spanner of $P_1 \cup P_2$ in order to find X and Y. For the Lighting phase, we argue analogously as for the SMT.

Lemma 11. *Let P_0, P_1 and P_2 be sets of points in $\mathbb{R}^d$. Let $\epsilon > 0$ and $r = (d/\epsilon)^{\mathcal{O}(d^2)}$. Let G be any $(1 + \frac{1}{4}\epsilon)$-spanner of $P_0 \cup P_1 \cup P_2$ that has $n \, (d/\epsilon)^{\mathcal{O}(d)}$ edges, whose total cost is upper bounded by $(d/\epsilon)^{\mathcal{O}(d)}$ times the cost of the minimum spanning tree of $P_1 \cup P_2$, and that satisfies the $(t, 1 + \frac{1}{4}\epsilon)$-leapfrog property, where $1 < t < 1 + \frac{1}{4}\epsilon$. Then, in time $\mathcal{O}(d^{3/2} n \log n) + n \, (2^{d^{\mathcal{O}(d)}} + (d/\epsilon)^{\mathcal{O}(d)})$, one can transform G to obtain a graph H with vertex set $P_0 \cup P_1 \cup P_2$ that is r-locally-light and for which there is a sub-multigraph M (i) whose induced graph[1] is H, (ii) that satisfies the connectivity requirement of every vertex in M, and (iii) whose cost is at most $(1 + \frac{1}{2}\epsilon)$ times the cost of the minimum-cost multigraph having the same connectivity property.*   □

Remark 12. One can modify this lemma to hold for arbitrary $\{0, 1, \ldots, k\}$-edge-connectivity in multigraphs with the same time complexity, provided that k is a constant.   ♠

Searching phase is the only phase that is completely different and rather tricky.

4.1   Dynamic Programming for {0, 1, 2}-Edge-Connectivity

Consider a point set $P = P_0 \cup P_1 \cup P_2$ in $\mathbb{R}^d$, where P_i is the set of points whose connectivity requirement is i. For an arbitrary shifted dissection, let G be an Euclidean graph on P that is r-locally-light with respect to the this dissection. We apply dynamic programming approach to find in time $\mathcal{O}(n \cdot r^{\mathcal{O}(2^d r)})$ the minimum-cost $\{0, 1, 2\}$-edge-connected *multigraph* H on P whose induced graph[1] is a subgraph of G.

The main idea of our approach is similar to the method used by Arora in [1] and in the follow-up papers (see, e.g., [3,4,22]). We apply dynamic programming to find an optimal solution by finding optimal solutions to subproblems induced by all regions in

---

[1] The *graph induced* by a multigraph M is the graph obtained by reducing the multiplicity of each edge of M to one.

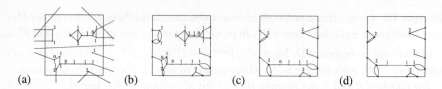

Fig. 1. (a) A part of graph G inside a region R, (b) a multigraph M_R within R, (c) contraction of two-edge-connected components in M_R, and (d) contraction of paths in M_R.

the shifted dissection, in a bottom-up manner. That is, for each region in the dissection we find in particular all solutions in a form of a "sub"-multigraph on the edges (possible duplicated) of the input graph within the region that may be extended to a minimum-cost $\{0, 1, 2\}$-edge-connected multigraph on P. For a non-leaf region, we do it by combing solutions for the 2^d child subregions in the dissection. To reduce the number of solutions considered, we introduce various "connectivity types" for each region, and produce solely a single optimal multigraph for each of the types.

For any region R of the dissection, after duplicating some of the edges of G, a *multigraph within R* is the part of the multigraph contained within R resulting from the removal of all edges that cross R and have no endpoint inside R (see, Figure 1 (a–b)). The multigraph within R has two types of points, those from P, which we call the *input points*, and those defined by the crossings of the edges with the border of P, which we shall call the *border points*.

The aforementioned *connectivity type* of a multigraph within any region R is obtained by (see also, Figure 1 (c–d)):

- Contracting each maximal two-edge-connected component in the multigraph induced by its input points into a single vertex, and associating with the new vertex the connectivity requirement equal to the maximum connectivity requirement among all input points of the component.
- Contracting each maximal path composed of input and/or contracted points of degree two into the single edge with the endpoints being the first and the last vertex at the path, and associating with each endpoint of the new edge the connectivity requirement equal to the maximum connectivity requirement among all the points on the path.

Two multigraphs H_1 and H_2 within a region R are said to have the same connectivity type if their connectivity types admit isomorphism preserving the border points.

To apply dynamic programming we need the following lemma which ensures that among all multigraphs within any region in the dissection it is sufficient to consider only those having minimum cost with respect to some connectivity type.

Lemma 13. *If H is the minimum-cost $\{0, 1, 2\}$-edge-connected multigraph on P whose induced graph is a subgraph of G, then for any region R in the shifted dissection the sub-multigraph H_R of H within R has the minimum-cost over all multigraphs within R that have the same connectivity type as H_R.* □

If one ignores parallel edges having one endpoint as a border point and another as an input point, any connectivity type is a forest with no path having three consecutive vertices of degree 2. Furthermore, since G is r-locally light, any region R in the dissection is crossed by at most r edges with one endpoint in R. Therefore, the number of connectivity types within R is upper bounded by $r^{\mathcal{O}(r)}$.

Our dynamic programming procedure determines for each region and for each possible connectivity type, the minimum-cost multigraph of this type within the region in a bottom-up fashion, in a similar way as it is done in [1,3,4,22] and in Section 3.3. We can compute the optimal solution for each connectivity type in a region containing only a single point from P in $r^{\mathcal{O}(r)}$ time. Then, for each region containing more than a single point we compute the optimal solution for each connectivity type in the region by considering all sub-regions in total time $\mathcal{O}(2^d r) \cdot \left(r^{\mathcal{O}(r)} \right)^{2d} = r^{\mathcal{O}(2^d r)}$. Summarizing, the total time required by the dynamic programming is $\mathcal{O}(n \cdot r^{\mathcal{O}(2^d r)})$.

Hence, we obtain the following theorem.

Theorem 14. *There exists a polynomial-time approximation scheme for the* $\{0, 1, 2\}$-*vertex/edge-connectivity problem in Euclidean space* $\mathbb{R}^d$. *For* $\varepsilon > 0$ *and any sets of points* P_0, P_1, P_2 *in* $\mathbb{R}^d$ *of total size* n, *the algorithm in time* $n \log n \, (d/\epsilon)^{\mathcal{O}(d)} + n \, (d/\epsilon)^{(d/\epsilon)^{\mathcal{O}(d^2)}}$ *finds a graph on* $P_1 \cup P_2$ *with possible Steiner points in* P_0 *that satisfies the connectivity requirement for any pair of points and whose cost is at most* $1 + \varepsilon$ *times the minimum.*

For constant d *and* ε, *the running time of this algorithm is* $\mathcal{O}(n \log n)$.   □

5   Extensions

1. All the arguments from Section 4 except the dynamic programming part can be easily generalized to include $\{0, 1, \ldots, k\}$-*edge-connectivity for multigraphs*. Regarding the dynamic programming, we can combine the dynamic programming for the k-edge-connectivity from [3] with our method of dealing with non-uniform connectivity requirements. Thus, the variant of the geometric survivability problem, where the edge-connectivity requirements satisfy $r_v \in \{0, 1, \ldots, k\}$ and $k = \mathcal{O}(1)$, admits also a PTAS. The running time is however, significantly greater than $\mathcal{O}(n \log n)$, though it is still polynomial for constant d and ε.
2. Using the same arguments as in [1], we can extend our PTASs to include other ℓ_p^d metrics as well.
3. Our PTASs could be also extended to an infinite domain for Steiner candidate points (e.g., if $P_0 = \mathbb{R}^d$ we have the Euclidean complete Steiner problem) provided we can determine the set Y (or its good approximation) as fast as claimed in Lemma 6.

References

1. S. Arora. Polynomial time approximation schemes for Euclidean traveling salesman and other geometric problems. *J. Assoc. Comput. Mach.*, 45(5):753–782, 1998.
2. S. Chopra and C.-Y. Tsai. A branch-and-cut approach for minimum cost multi-level network design. *Discrete Math.*, 242:65–92, 2002.
3. A. Czumaj and A. Lingas. On approximability of the minimum-cost k-connected spanning subgraph problem. In *Proc. 10th ACM-SIAM SODA*, pp. 281–290, 1999.

4. A. Czumaj and A. Lingas. Fast approximation schemes for Euclidean multi-connectivity problems. In *Proc. 27th ICALP*, pp. 856–868, 2000.
5. L. Fleischer. A 2-approximation for minimum cost $\{0, 1, 2\}$ vertex connectivity. In *Proc. 8th IPCO*, pp. 115–129, 2001.
6. L. Fleischer, K. Jain, and D. P. Williamson. An iterative rounding 2-approximation algorithm for the element connectivity problem. In *Proc. 42nd FOCS*, pp. 339–347, 2001.
7. G. N. Frederickson and J. JáJá. On the relationship between the biconnectivity augmentation and Traveling Salesman Problem. *Theoret. Comput. Sci.*, 19(2):189–201, 1982.
8. H. N. Gabow, M. X. Goemans, and D. P. Williamson. An efficient approximation algorithm for the survivable network design problem. *Math. ProgrammingB*, 82:13–40, 1998.
9. E. N. Gilbert and H. O. Pollak. Steiner minimal trees. *SIAM J. Appl. Math.*, 16(1):1–29, 1968.
10. M. Grötschel and C. L. Monma. Integer polyhedra arising from certain network design problems with connectivity constraints. *SIAM J. Discr. Math.*, 3(4):502–523, 1990.
11. M. Grötschel, C. L. Monma, and M. Stoer. Computational results with a cutting plane algorithm for designing communication networks with low-connectivity constraints. *Operations Research*, 40(2):309–330, 1992.
12. M. Grötschel, C. L. Monma, and M. Stoer. Polyhedral and computational investigations for designing communication networks with high survivability requirements. *Operations Research*, 43:1012–1024, 1995.
13. M. Grötschel, C. L. Monma, and M. Stoer. Design of survivable networks. In M. O. Ball, T. L. Magnanti, C. L. Monma, and G. L. Nemhauser, editors, *Handbooks in Operations Research and Management Science*, volume 7: Network Models, chapter 10, pp. 617–672. North-Holland, Amsterdam, 1995.
14. J. Gudmundsson, C. Levcopoulos, and G. Narasimhan. Improved greedy algorithms for constructing sparse geometric spanners. In *Proc. 7th SWAT*, pp. 314–327, 2000.
15. F. K. Hwang and D. S. Richards. Steiner tree problems. *Networks*, 22:55–89, 1991.
16. F. K. Hwang, D. S. Richards, and P. Winter. *The Steiner Tree Problem*. North-Holland, Amsterdam, 1992.
17. K. Jain. A factor 2 approximation algorithm for the generalized Steiner network problem. *Combinatorica*, 21(1):39–60, 2001.
18. P. Indyk and R. Motwani. Approximate nearest neighbors: Towards removing the curse of dimensionality. In *Proc. 30th ACM STOC*, pp. 604–613, 1998.
19. J. S. B. Mitchell. Guillotine subdivisions approximate polygonal subdivisions: A simple polynomial-time approximation scheme for geometric TSP, k-MST, and related problems. *SIAM J. Comput.*, 28(4):1298–1309, August 1999.
20. C. L. Monma and D. F. Shallcross. Methods for designing communications networks with certain two-connected survivability constraints. *Operations Research*, 37(4):531–541, July 1989.
21. H. J. Prömel and A. Steger. *The Steiner Tree Problem. A Tour Through Graphs, Algorithms and Complexity*. Vieweg Verlag, Wiesbaden, 2002.
22. S. B. Rao and W. D. Smith. Approximating geometrical graphs via "spanners" and "banyans." In *Proc. 30th ACM STOC*, pp. 540–550, 1998.
23. M. Stoer. *Design of Survivable Networks*, volume 1531 of *Lecture Notes in Mathematics*. Springer-Verlag, Berlin, 1992.
24. D. P. Willimason, M. X. Goemans, M. Mihail, and V. V. Vazirani. A primal-dual approximation algorithm for generalized Steiner network problem. *Combinatorica*, 15:435–454, 1995.
25. P. Winter. Steiner problem in networks: A survey. *Networks*, 17:129–167, 1987.

Finding a Path of Superlogarithmic Length

A. Björklund and T. Husfeldt

Department of Computer Science, Lund University Box 118, 221 00 Lund, Sweden.
thore@cs.lth.se.

Abstract. We consider the problem of finding a long, simple path in an undirected graph. We present a polynomial-time algorithm that finds a path of length $\Omega\big((\log L/\log\log L)^2\big)$, where L denotes the length of the longest simple path in the graph. This establishes the performance ratio $O\big(|V|(\log\log|V|/\log|V|)^2\big)$ for the Longest Path problem, where V denotes the graph's vertices.

1 Introduction

Given an unweighted, undirected graph $G = (V, E)$ the *longest path* problem is to find the longest sequence of distinct vertices $v_1 \cdots v_k$ such that $v_i v_{i+1} \in E$. This is a classical NP-hard problem (number ND29 in Garey and Johnson [5]) with a considerable body of research devoted to it, yet its approximability remains elusive:

> "For most canonical NP-hard problems, either dramatically improved approximation algorithms have been devised, or strong negative results have been established, leading to a substantially improved understanding of the approximability of these problems. However, there is one problem which has resisted all attempts at devising either positive or negative results — longest paths and cycles in undirected graphs. Essentially, there is no known algorithm which guarantees approximation ratio better than $|V|/\mathrm{polylog}|V|$ and there are no hardness of approximation results that explain this situation." [4]

Indeed, the quoted ratio has been obtained only for special classes of graphs (for example, Hamiltonian graphs), while in the general case the best known ratio prior to the present paper was of order $|V|/\log|V|$.

We present a polynomial-time algorithm for the general case that finds a path of length $\Omega\big((\log L/\log\log L)^2\big)$ in a graph with longest path length L; the best previous bound was $\Omega(\log L)$. This corresponds to a performance ratio of order

$$O\left(\frac{|V|\big(\log\log|V|\big)^2}{\log^2|V|}\right). \tag{1}$$

For bounded degree graphs we improve the ratio to $O\big(|V|\log\log|V|/\log^2|V|\big)$. For three-connected graphs we establish the perormance ratio (1) for the *longest cycle* problem.

P. Widmayer et al. (Eds.): ICALP 2002, LNCS 2380, pp. 985–992, 2002.

Previous Work

The first approximation algorithms for longest path are due to Monien [7] and Bodlaender [2], both finding a path of length $\Omega(\log L/\log\log L)$. Neither of these algorithms can be used to find a $\log|V|$ path if it exists, but Papadimitriou and Yannakakis conjectured that such a polynomial-time algorithm exists [8]. This was confirmed by Alon, Yuster, and Zwick [1], introducing the important method of *colour-coding*. Especially, this algorithm finds an $\Omega(\log L)$-path and corresponds to a performance ratio of

$$O\left(\frac{|V|}{\log|V|}\right),$$

which is the best ratio known prior to the present paper.

The problem has received additional study for restricted classes of graphs, where the "$\log|V|$-barrier" has been broken by Vishwanathan [9]. His algorithm achieves the same performance ratio (1) as ours, but works only for Hamiltonian graphs. In *sparse* Hamiltonian graphs, Feder, Motwani, and Subi [4] find even longer paths.

The hardness results for this problem are mainly due to Karger, Motwani, and Ramkumar [6]: The longest path problem does not belong to APX and cannot be approximated within $2^{\log^{1-\epsilon}|V|}$ unless $\mathrm{NP}\subseteq\mathrm{DTIME}\big(2^{O(\log^{1/\epsilon}n)}\big)$ for any $\epsilon>0$.

2   Preliminaries

In the remainder, we consider a connected graph $G=(V,E)$ with $n=|V|$ vertices and $e=|E|$ edges. We write $G[W]$ for the graph induced by the vertex set W.

Paths and Cycles

The *length* of a path and a cycle is its number of edges. The length of a cycle C is denoted $l(C)$. A k-cycle is a cycle of length k, a k^+-cycle is a cycle of length k or larger. A k-path and k^+-path is defined similarly. For vertices x and y, an xy-path is a (simple) path from x to y, and if P is a path containing u and v we write $P[u,v]$ for the subpath from u to v. We let $L_G(v)$ denote the length of the longest path from a vertex v in the graph G, and sometimes abbreviate $L_W(v)=L_{G[W]}(v)$. The *path length* of G is $\max_{v\in V}L_G(v)$.

We need the following result, Theorem 5.3(i) of [2]:

Theorem 1 (Bodlaender) *Given a graph, two of its vertices s,t, and an integer k, one can find a k^+-path from s to t (if it exists) in time $O\big((2k)!2^{2k}n+e\big)$.*

Corollary 1 *A k^+-cycle through a given vertex can be found in time $t(k)=O\big(((2k)!2^{2k}n+e)n\big)$, if it exists.*

Proof. Let s be the given vertex. For all neighbours t of s apply the Theorem on the graph with the edge st removed.   □

We also need the following easy lemma.

Lemma 1 *If a connected graph contains a path of length r then every vertex is an endpoint of a path of length at least $\frac{1}{2}r$.*

Proof. Given vertices $u, v \in V$ let $d(u, v)$ denote the length of the shortest path between u and v.

Let $P = p_0 \cdots p_r$ be a path and let v be a vertex. Find i minimising $d(p_i, v)$. By minimality there is a path Q from v to p_i that contains no other vertices from P. Now either $QP[p_i, p_r]$ or $QP[p_i, p_0]$ has length at least $\frac{1}{2}r$.   □

The next lemma is central to our construction: Assume that a vertex v originates a long path P and v lies on a cycle C; then the removal of C decomposes G into connected components, one of which must contain a large part of P.

Lemma 2 *Assume that a connected graph G contains a simple path P of length $L_G(v) > 1$ originating in vertex v. There exists a connected component $G[W]$ of $G[V - v]$ such that the following holds.*

1. *If $G[W + v]$ contains no k^+-cycle through v then every neighbour $u \in W$ of v is the endpoint of a path of length*

$$L_W(u) \geq L_G(v) - k.$$

2. *If C is a cycle in $G[W + v]$ through v of length $l(C) < L_{G[W+v]}(v)$ then there exists a connected component H of $G[W - C]$ that contains a neighbour u of $C - v$ in $G[W + v]$. Moreover, every such neighbour u is the endpoint of a path in H of length*

$$L_H(u) \geq \frac{L_G(v)}{2l(C)} - 1.$$

Proof. Let $r = L_G(v)$ and $P = p_0 \cdots p_r$, where $p_0 = v$. Note that $P[p_1, p_r]$ lies entirely in one of the components $G[W]$ of $G[V - v]$.

First consider statement 1. Let $u \in W$ be a neighbour of v. Since $G[W]$ is connected, there exists a path Q from u to some vertex of P. Consider such a path. The first vertex p_i of P encountered on Q must have $i < k$ since otherwise the three paths vu, $Q[u, p_i]$ and $P[p_0, p_i]$ form a k^+-cycle. Thus the path $Q[u, p_i]P[p_i, p_r]$ has length at least $r - k + 1 > r - k$.

We proceed to statement 2. Consider any cycle C in $G[W + v]$ through v.

Case 1. First assume that $P \cap C = v$, so that one component H of $G[W - C]$ contains all of P except v. Let N be the set of neighbours of $C - v$ in H. First note that N is nonempty, since $G[W]$ is connected. Furthermore, the path length of H is at least $r - 1$, so Lemma 1 gives $L_H(u) \geq (r - 1)/2$ for every $u \in N$.

Case 2. Assume instead that $|P \cap C| = s > 1$. Enumerate the vertices on P from 0 to r and let $i_1, \ldots, i_s$ denote the indices of vertices in $P \cap C$, in particular

$i_1 = 0$. Let $i_{s+1} = r$. An averaging argument shows that there exists j such that $i_{j+1} - i_j \geq r/s$. Consequently there exists a connected component H of $G(W - C)$ containing a simple path of length $r/s - 2$. At least one of the i_jth or i_{j+1}th vertices of P must belong to $C - v$, so the set of neighbours N of $C - v$ in H must be nonempty. As before, Lemma 1 ensures $L_H(u) \geq r/2s - 1$ for every $u \in N$, which establishes the bound after noting that $s \leq l(C)$.   □

3   Result and Algorithm

The construction in this section and its analysis establishes the following theorem, accounting for the performance ratio (1) claimed in the introduction in the worst case.

Theorem 2 *If a graph contains a simple path of length L then we can find a simple path of length*

$$\Omega\left(\left(\frac{\log L}{\log \log L}\right)^2\right)$$

in polynomial time.

3.1   Construction of the Cycle Decomposition Tree

Given a vertex v in G, our algorithm constructs a rooted node-weighted tree $T_k = T_k(G, v)$, the cycle decomposition tree. Every node of T_k is either a *singleton* or a *cycle* node: A singleton node corresponds to a single vertex $u \in G$ and is denoted $\langle u \rangle$, a cycle node corresponds to a cycle C with a specified vertex $u \in C$ and is denoted $\langle C, u \rangle$. Every singleton node has unit weight and every cycle node $\langle C, u \rangle$ has weight $\frac{1}{2}l(C)$.

The tree is constructed as follows. Initially T_k contains a singleton node $\langle v \rangle$, and a call is made to the following procedure with arguments G and v.

1. For every maximal connected component $G[W]$ of $G[V - v]$, execute step 2.
2. Search for a k^+-cycle through v in $G[W + v]$ using Theorem 1. If such a cycle C is found then execute step 3. Otherwise pick an arbitrary neighbour $u \in G[W + v]$ of v, insert the node $\langle u \rangle$ and the tree edge $\langle v \rangle\langle u \rangle$, and recursively compute $T_k(G[W], u)$.
3. Insert the cycle node $\langle C, v \rangle$ and the tree edge $\langle v \rangle\langle C, v \rangle$. For every connected component H of $G[W - C]$ choose an arbitrary neighbour $u \in H$ of $C - v$, and insert the singleton node $\langle u \rangle$ and the tree edge $\langle C, v \rangle\langle u \rangle$. Then, recursively compute $T_k(H, u)$.

Note that each recursive step constructs a tree that is connected to other trees by a single edge, so T_k is indeed a tree. Also note that the ancestor of every cycle node must be a singleton node. The root of T_k is $\langle v \rangle$.

3.2   Paths in the Cycle Decomposition Tree

The algorithm finds a path of greatest weight in T_k. This can be done in linear time by depth first search. The path found in T_k represents a path in G, if we interpret paths through cycle vertices as follows. Consider a path in T_k through a cycle vertex $\langle C, u \rangle$. Both neighbours are singleton nodes, so we consider the subpath $\langle u \rangle \langle C, u \rangle \langle v \rangle$. By construction, v is connected to some vertex $w \in C$ with $w \neq u$. One of the two paths from u to w in C must have length at least half the length of C, call it P. We will interpret the path $\langle u \rangle \langle C, u \rangle \langle v \rangle$ in T_k as a path uPv in G. If a path ends in a cycle node $\langle C, u \rangle$, we may associate it with a path of length $l(C) - 1$, by moving along C from u in any of its two directions. Thus a path of weight m in T_k from the root to a leaf identifies a path of length at least m in G.

We need to show that T_k for some small k has a path of sufficient length:[1]

Lemma 3 *If G contains a path of length $r \geq 2^8$ starting in v then $T_k = T_k(G, v)$ for*

$$k = \left\lceil \frac{2 \log r}{\log \log r} \right\rceil$$

contains a weighted path of length at least $\frac{1}{8}k^2 - \frac{1}{4}k - 1$.

Proof. We follow the construction of T_k in §3.1.

We need some additional notation. For a node $x = \langle w \rangle$ or $x = \langle C, w \rangle$ in T_k we let $L(x)$ denote the length of the longest path from w in the component $G[X]$ corresponding to the subtree rooted at x. More precisely, for every successor y of x (including $y = x$), the set X contains the corresponding vertices w' (if $y = \langle w' \rangle$ is a singleton node) or C' (if $y = \langle w', C' \rangle$ is a cycle node).

Furthermore, let $\mathbf{S}(n)$ denote the singleton node children of a node n and let $\mathbf{C}(n)$ denote its cycle node children. Consider any singleton node $\langle v \rangle$.

Lemma 2 asserts that

$$L(v) \leq \max \left\{ \max_{w \in \mathbf{S}\langle v \rangle} L(w) + k, \max_{\substack{\langle C, v \rangle \in \mathbf{C}\langle v \rangle \\ w \in \mathbf{S}\langle C, v \rangle}} (2L(w) + 2) l(C) \right\}. \tag{2}$$

Define $n(v) = w$ if $\langle w \rangle$ maximises the right hand side of the inequality (2) and consider a path $Q = \langle x_0 \rangle \cdots \langle x_t \rangle$ from $\langle v \rangle = \langle x_0 \rangle$ described by these heavy nodes. To be precise we have either $n(x_i) = x_{i+1}$ or $n(x_i) = x_{i+2}$, in the latter case the predecessor of $\langle x_{i+2} \rangle$ is a cycle node.

We will argue that the gaps in the sequence

$$L(x_0) \geq L(x_1) \geq \cdots \geq L(x_t).$$

---

[1] All logarithms are to the base 2 and the constants involved have been chosen aiming for simplicity of the proof, rather than optimality.

cannot be too large due to the inequality above and the fact that $L(x_t)$ must be small (otherwise we are done), and therefore Q contains a lot of cycle nodes or even more singleton nodes.

Let s denote the number of cycle nodes on Q. Since every cycle node has weight at least $\frac{1}{2}k$ the total weight of Q is at least $\frac{1}{2}sk + (t-s) = s(\frac{1}{2}k - 1) + t$.

Consider a singleton node that is followed by a cycle node. There are s such nodes, we will call them *cycle parents*. Assume $\langle x_j \rangle$ is the first cycle parent node. Thus according to the first part of Lemma 2 its predecessors $\langle x_0 \rangle, \ldots, \langle x_j \rangle$ satisfy the relation $L(x_{i+1}) \geq L(x_i) - k$, so

$$L(x_j) \geq r - jk \geq r - \tfrac{1}{8}k^3 \geq \tfrac{7}{8}r,$$

since $j \leq t \leq \frac{1}{8}k^2$ (otherwise we are finished) and $r \geq k^3$.

From the second part of Lemma 2 we have

$$L(x_{j+2}) \geq \frac{7r}{16l(C)} - 1 \geq \frac{r}{k^2}.$$

where we have used $l(C) \leq \frac{1}{4}k^2$ (otherwise we are finished) and $r \geq \frac{4}{3}k^2$.

This analysis may be repeated for the subsequent cycle parents as long as their remaining length after each cycle node passage is at least k^3. Note that Q must pass through as many as $s' \geq \lceil \frac{1}{4}k - 1 \rceil$ cycle nodes before

$$\frac{r}{k^{2s'}} < k^3,$$

at which point the remaining path may be shorter than k^3. Thus we either have visited $s \geq s'$ cycle nodes, amounting to a weighted path Q of length at least

$$s(\tfrac{1}{2}k + 1) \geq \tfrac{1}{8}k^2 - \tfrac{1}{4}k - 1$$

(remembering that any two consecutive cycle nodes must have a singleton node in-between), or there are at most $s < s'$ cycle nodes on Q. In that case there is a tail of singleton nodes starting with some $L(x) \geq k^3$. Since $L(x_j) \leq L(x_{j+1}) + k$ for the nodes on the tail, the length of the tail (and thus the weight of Q) is at least k^2. □

3.3 Summary

Our algorithm divides the input graph into its connected components and performs the following steps for each. It picks a vertex v in the component and constructs cycle decomposition trees T_k for all $k = 6, \ldots, \lceil 2 \log n / \log \log n \rceil$. Corollary 1 tells us that this is indeed a polynomial time task. Moreover, Lemma 1 ensures that v originates a path of at least half the length of the longest path in the component. The algorithm then finds paths in G identified by the longest weighted paths in T_k in linear time. Finally, Lemma 3 establishes the desired approximation ratio.

4   Extensions

4.1   Bounded Degree Graphs

As in [9], the class of graphs with their maximum degree bounded by a constant admits a relative $\log \log n$-improvement over the performance ratio shown in this paper. All paths of length $\log n$ can be enumerated in polynomial time for these graphs. Consequently, we can replace the algorithm from Theorem 1 by an algorithm that efficiently finds cycles of logarithmic length or larger through any given vertex if they exist.

Proposition 1 *If a constant degree graph contains a simple path of length L then we can find a simple path of length*

$$\Omega\left(\frac{\log^2 L}{\log \log L}\right)$$

in polynomial time.

This gives the performance ratio $O\big(|V| \log \log |V| / \log^2 |V|\big)$ for the longest path problem in constant degree graphs.

4.2   Three-Connected Graphs

Bondy and Locke [3] have shown that every 3-connected graph with path length l must contain a cycle of length at least $2l/5$. Moreover, their construction is easily seen to be algorithmic and efficient. This implies the following result on the longest cycle problem:

Proposition 2 *If a 3-connected graph contains a simple cycle of length L then we can find a simple cycle of length*

$$\Omega\left(\left(\frac{\log L}{\log \log L}\right)^2\right)$$

in polynomial time.

This gives the performance ratio $O\big(|V| (\log \log |V| / \log |V|)^2\big)$ for the longest cycle problem in 3-connected graphs. Note that for 3-connected cubic graphs, [4] show a considerably better bound.

Acknowledgement. We thank Andrzej Lingas for bringing [9] to our attention, and Gerth Stølting Brodal for commenting on a previous version of this paper.

References

1. N. Alon, R. Yuster, and U. Zwick. Color-coding. *Journal of the ACM*, 42(2):844–856, 1995.
2. H. L. Bodlaender. On linear time minor tests with depth-first search. *Journal of Algorithms*, 14(1):1–23, 1993.
3. J. A. Bondy and S. C. Locke. Relative length of paths and cycles in 3-connected graphs. *Discrete Mathematics*, 33:111–122, 1981.
4. T. Feder, R. Motwani, and C. S. Subi. Finding long paths and cycles in sparse Hamiltonian graphs. In *Proc. 32th STOC*, pages 524–529. ACM, 2000.
5. M. Garey and D. Johnson. *Computers and intractability: A guide to the theory of NP-completeness*. W. H. Freeman, San Francisco, 1979.
6. D. Karger, R. Motwani, and G.D.S. Ramkumar. On approximating the longest path in a graph. *Algorithmica*, 18(1):82–98, 1997.
7. B. Monien. How to find long paths efficiently. *Annals of Discrete Mathematics*, 25:239–254, 1985.
8. C. H. Papadimitriou and M. Yannakakis. On limited nondeterminism and the complexity of the V–C dimension. *Journal of Computer and Systems Sciences*, 53(2):161–170, 1996.
9. S. Vishwanathan. An approximation algorithm for finding a long path in Hamiltonian graphs. In *Proc. 11th SODA*, pages 680–685, 2000.

Linear Time Algorithms on Chordal Bipartite and Strongly Chordal Graphs

Ryuhei Uehara[**]

Natural Science Faculty, Komazawa University
uehara@komazawa-u.ac.jp

Abstract. Chordal bipartite graphs are introduced to analyze non-symmetric matrices, and form a large class of perfect graphs. There are several problems, which can be solved efficiently on the class using the characterization by the doubly lexical ordering of the bipartite adjacency matrix. However, the best known algorithm for computing the ordering runs in $O(\min\{m \log n, n^2\})$, which is the bottleneck of the problems. We show a linear time algorithm that computes the ordering of a given chordal bipartite graph. The result improves the upper bounds of several problems, including recognition problem, from $O(\min\{m \log n, n^2\})$ to linear time. Strongly chordal graphs are well-studied subclass of chordal graphs, and that have similar characterization. The upper bounds of several problems on a given strongly chordal graph are also improved from $O(\min\{m \log n, n^2\})$ to linear time.

Keywords: Chordal bipartite graphs, design and analysis of algorithms, lexicographic breadth first search, vertex elimination ordering, strongly chordal graphs.

1 Introduction

Chordal bipartite graphs are bipartite graphs such that every cycle of length greater than 4 has a chord. The graph class is introduced by Golumbic and Goss [11], and it has applications to nonsymmetric matrices (see [10]) including Gaussian elimination [1], integer programming [13], and matrix analysis [15,14]. The class also forms a large class of perfect graphs (see [20]). On the class, several problems can be solved efficiently. Most of the algorithms are based on the following characterization; a bipartite graph G is chordal bipartite if and only if the doubly lexical ordering of the bipartite adjacency matrix of G is Γ-free. However, the best known algorithm for computing the doubly lexical ordering runs in $O(\min\{m \log n, n^2\})$ time, and this upper bound is the bottleneck of the efficiency of many problems. Typically, the best known upper bound for recognition of the class is $O(\min\{m \log n, n^2\})$ [13,19,21,27]. All these recognition algorithms use the same underlying idea: First compute a doubly lexical ordering of the bipartite adjacency matrix of a given bipartite graph and then check whether it is Γ-free.

[**] Work done while visiting University of Waterloo.

P. Widmayer et al. (Eds.): ICALP 2002, LNCS 2380, pp. 993–1004, 2002.

Our main result is a linear time algorithm for the bottleneck. We first introduce a vertex elimination ordering, called *weak elimination ordering* (WEO), which characterizes a chordal bipartite graph. Next we construct a linear time algorithm that generates a WEO for given chordal bipartite graph. Using the WEO, we can improve the upper bounds of the following problems on a chordal bipartite graph from $O(\min\{m \log n, n^2\})$ to linear time: (1) the recognition problem [13,19,21,27], and constructing Γ-free bipartite adjacency matrix, which means that we can perform Gaussian elimination efficiently [10,1], and (2) computing a perfect edge without vertex elimination ordering and a list of all maximal complete bipartite subgraphs [17], and (3) computing a strong T-elimination ordering and finding a maximum matching [4]. We also show that a maximum vertex-weighted matching can be found in linear time.

Next we consider strongly chordal graphs, which are originally introduced by Farber as a subclass of chordal graphs for which the domination problem can be solved efficiently [8,9]. This class has similar characterization to the class of chordal bipartite graphs, and has the same bottleneck. We also improve the upper bounds of the following problems on a strongly chordal graph from $O(\min\{m \log n, n^2\})$ to linear time: (1) the recognition problem and computing a strong elimination ordering [13,19,21,27] (see also [24]), and (2) finding a maximum (vertex-weighted) matching [3,7], an independent dominating set and a minimum weighted dominating set [9], a minimum dominating clique [18], and a domatic partition [23].

From the viewpoint of algorithmic graph theory, it is worth remarking that our algorithm for chordal bipartite graph recognition uses the lexicographic breadth first search (LexBFS). For recognition of a chordal graph, two linear time algorithms are well known; the lexicographic breadth first search (LexBFS) [25], and the maximum cardinality search (MCS) [28]. The LexBFS has been investigated deeply [26,22], and recently, it is used for recognition of not only chordal graphs, but also the other graph classes including interval graphs [5], distance hereditary graphs, etc (see [2, Chapter 5] for references). Our results extend the list of the graph classes, which can be recognized by LexBFS.

2   Preliminaries

Given a graph $G = (V, E)$ and a subset $U \subseteq V$, the *subgraph of G induced by U* is the graph (U, F), where $F = \{\{u, v\} | \{u, v\} \in E$, and $u, v \in U\}$, and denoted by $G[U]$. A vertex set I is an *independent set* if $G[I]$ contains no edges. The *(open) neighborhood* of a vertex v in G is the set $N_G(v) = \{u \in V | \{u, v\} \in E\}$, and the *closed neighborhood* is $N_G[v] = N_G(v) \cup \{v\}$. We will omit the index G if no confusion can arise. The *degree* of a vertex v is $|N(v)|$ and denoted by $d(v)$. For a vertex set U, $N(U)$ denotes the set $\cup_{u \in U}\{v | v \in N(u)\}$. A sequence of the vertices $v_1, \cdots, v_l$ is a *path*, denoted by $(v_1, \cdots, v_l)$, if $\{v_j, v_{j+1}\} \in E$ for each $1 \leq j \leq l - 1$. The *length* of a path is the number of edges in the path. A *cycle* is a path beginning and ending with the same vertex. An edge which joins two vertices of a cycle but is not itself an edge of the cycle is a *chord* of that cycle. A graph $G = (V, E)$ is *bipartite* if V can be divided into two sets X and Y with $X \cup Y = V$ and $X \cap Y = \emptyset$ such that every edge joins a vertex in X and another one in Y. A graph G is bipartite if and only if G contains no cycle of odd length [16]. Throughout the paper, we denote the number of vertices and edges of the input graph $G = (V, E)$ by n and m, respectively. An *ordering* α of $G = (V, E)$ is a bijection $\alpha : V \to \{1, \cdots, n\}$. Often it will be convenient to denote an ordering by using it to index the vertex set, so that $\alpha(v_i) = i$ for $1 \le i \le n$ where i will be referred to as the *index* of v_i. Let $v_1, \cdots, v_n$ be an ordering of V. For $1 \le i \le n$, we define $\mathcal{L}_i$ to be the set of vertices with indices greater than $i - 1$: $\mathcal{L}_i = \{v_i, \cdots, v_n\}$. For given graph and ordering, we denote by G_i the graph $G[\mathcal{L}_i]$. We also denote by G_v, for vertex v, the graph $G[\mathcal{L}_{\alpha(v)}]$. For given vertices u, v we denote by $u < v$ if $\alpha(u) < \alpha(v)$, and $N_{>v}(u)$ denotes the set of neighbors of u greater than v.

Given bipartite graph $G = (V_1, V_2, E)$ with $V_1 = \{v_1, \cdots, v_{n_1}\}$ and $V_2 = \{u_1, \cdots, u_{n_2}\}$, we define *bipartite adjacency matrix* by $n_1 \times n_2$ $(0, 1)$-matrix $A = (a_{ij})$ with $a_{ij} = 1$ if and only if $\{v_i, u_j\} \in E$. Given graph $G = (V, E)$ with $V = \{v_1, \cdots, v_n\}$, its *adjacency matrix* is defined by $n \times n$ $(0, 1)$-matrix $A = (a_{ij})$ with $a_{ij} = 1$ if and only if either $i = j$ or $\{v_i, v_j\} \in E$. An ordered $(0, 1)$-matrix is Γ-*free* if it has no ordered submatrix which is a $\begin{pmatrix} 1 & 1 \\ 1 & 0 \end{pmatrix}$.

Lemma 1 (see [13,19] for the details). *(1) A bipartite graph is chordal bipartite if and only if there is an ordering of vertices such that its bipartite adjacency matrix is Γ-free. (2) A graph is strongly chordal if and only if there is an ordering of vertices such that its adjacency matrix is Γ-free.*

3   Characterizations of Chordal Bipartite Graph

A *chordal bipartite graph* is the bipartite graph such that each cycle of length greater than 4 has a chord. In this section, we show two characterizations of a chordal bipartite graph. The first characterization is based on the following folklore; a bipartite graph $G = (V, E)$ is chordal bipartite if and only if G has an ordering $v_1, v_2, \cdots, v_n$ such that for each $\{x_1, y_2\}, \{x_2, y_1\}, \{x_1, x_2\} \in E$ with $x_1 < y_1$ and $x_2 < y_2$, we have $\{y_1, y_2\} \in E$. (Although we omit the proof of the folklore since it appears in literature, e.g. [6], it can be proved using Lemma 1.) A vertex v is *weak-simplicial* if (1) $N(v)$ is an independent set, and (2) for each $x, y \in N(v)$, either $N(x) \subseteq N(y)$ or $N(y) \subseteq N(x)$. A vertex ordering $v_1, v_2, \cdots, v_n$ is a *weak elimination ordering* (WEO) if each vertex v_i is weak-simplicial on G_i, and, for each $x, y \in N_{>v_i}(v_i)$, $x < y$ implies $N_{>v_i}(x) \subseteq N_{>v_i}(y)$.

Theorem 1. *A graph is chordal bipartite if and only if it admits a WEO.*

Proof. ($\Rightarrow$) Let G be a chordal bipartite graph. Then we have an ordering $v_1, v_2, \cdots, v_n$ such that for each $\{x_1, y_2\}, \{x_2, y_1\}, \{x_1, x_2\} \in E$ with $x_1 < y_1$ and $x_2 < y_2$, we have $\{y_1, y_2\} \in E$. We show the ordering is a WEO. Since

G is bipartite, $N(v)$ is an independent set for each $v \in V$. Let x, y be vertices that $x, y \in N(v)$ and $v < x < y$. Let x' be any vertex in $N_{>v}(x)$. Then $\{v, y\}, \{x, x'\}, \{v, x\} \in E$, and $x < y$. Thus, by the folklore, $v < x'$ implies $\{x', y\} \in E$. Therefore, any vertex x' in $N_{>v}(x)$ is adjacent to y, consequently, $N_{>v}(x) \subseteq N_{>v}(y)$. ($\Leftarrow$) Assume that an ordering $v_1, v_2, \cdots, v_n$ is a WEO of G. We first show that each cycle of length greater than 4 has a chord. We suppose that $(u_1, u_2, u_3, \cdots, u_4, u_5, u_1)$ be a chordless cycle of length greater than 4. Without loss of generality, we assume that u_1 has the smallest index on the cycle. That is, G_{u_1} contains the cycle. Then, since u_1 is weak-simplicial on G_{u_1}, either $N(u_2) \subseteq N(u_5)$ or $N(u_5) \subseteq N(u_2)$. Thus we have $\{u_3, u_5\} \in E$ or $\{u_2, u_4\} \in E$, which contradicts that the cycle is chordless. Thus every cycle of length greater than 4 has a chord. We next show that G is bipartite. We assume that G has a cycle of odd length. Since every cycle of G of length greater than 4 has a chord, G contains at least one cycle of length 3. Let w, x, y be the vertices on the cycle of length 3 with $w < x < y$. Then, $N(w)$ is not independent set on G_w, which is a contradiction. Thus G is bipartite. □

We next show the second characterization, which is independent of the ordering of vertices. In a graph, a pair of two vertices x and y is said to be *two-pair* if each chordless path between x and y has exactly two edges. (The notion two-pair is originally introduced by Hayward, Hoàng, and Maffray to characterize weakly chordal graphs [12].) We here introduce two notions of *semi-clique* and *semi-simplicial*. We say a vertex set S is *semi-clique* if every pair of two vertices x and y in S is a two-pair in $G[(V - S) \cup \{x, y\}]$. Intuitively, there is no path joining $N(x) - N(y)$ and $N(y) - N(x)$ not through $N(x) \cap N(y)$. We say a vertex v is *semi-simplicial* if $N(v)$ is a semi-clique.

Theorem 2. *A graph is chordal bipartite if and only if every vertex is semi-simplicial.*

Proof. ($\Rightarrow$) Suppose that G is chordal bipartite and G contains a vertex s that is not semi-simplicial. By definition, $N(s)$ contains x and y such that they are not two-pair on $G[(V - N(s)) \cup \{x, y\}]$. Then there is a chordless path P joining x and y of length not equal to 2. If the length of P is 1, (x, s, y) and P induce a cycle of length 3, which contradicts that G is bipartite. On the other hand, if the length of P is greater than 2, (x, s, y) and P induce a chordless cycle of length greater than 4, which contradicts that G is chordal bipartite. (Note: There is no chord joining s and a vertex on P since x and y are the only neighbors of s in $G[(V - N(s)) \cup \{x, y\}]$.)

($\Leftarrow$) We first show that G has no chordless cycle of length greater than 4. Suppose that every vertex is semi-simplicial and G contains a chordless cycle of length greater than 4. We let the chordless cycle be $C = (v_1, v_2, v_3, \cdots, v_4, v_5, v_1)$. Then $\{v_2, v_5\} \notin E$, $\{v_2, v_4\} \notin E$, and $\{v_3, v_5\} \notin E$. Hence, since v_1 is semi-simplicial, v_2 and v_5 have a common neighbor on the path joining v_3 and v_4. However, this contradicts that C is chordless. Thus G has no chordless cycle of length greater than 4. Next we show that G is bipartite. We assume that G contains a cycle of odd length. Then, since G has no chordless cycle of length greater than 4, G has at least one cycle of length 3. Let x, y, s be the vertices on the cycle. However, this contradicts that s is semi-simplicial since $\{x, y\} \in E$. Thus G is chordal bipartite. □

4   Linear Time Algorithm for Generating WEO

In this section we first construct an algorithm GEN that generates a WEO for a given chordal bipartite graph. We next simplify the algorithm for implementation. For lack of space, linear time implementation is omitted (see [29] for details). The LexBFS contains two phases [25]: (1) generate an ordering α of G, and (2) compute a fill-in F and check if $F = \emptyset$. GEN uses the first phase of the LexBFS. It always generates an ordering α, and α is a WEO if given graph is chordal bipartite.

Algorithm description

To describe GEN, we introduce some notions. Let α be an ordering, and $A = \{a_1, \cdots, a_k\}$ and $B = \{b_1, \cdots, b_h\}$ be two sets of vertices with $a_1 > \cdots > a_k$ and $b_1 > \cdots > b_h$. Then we say A is *lexicographically larger* (lex-larger) than B if there is an index i such that $a_j = b_j$ for all $j < i$, and either $a_i > b_i$ or $h = i - 1$. (For example, we have $\{5, 4, 3\} > \{5, 4, 2, 1\} > \{5, 4, 2\} > \{5, 4\} > \{4, 3\}$.) We denote by $N^+(u)$ ($N^-(u)$) the sets of numbered (unnumbered, resp.) neighbors of u. $N^*(u)$ denotes the set of vertices defined by $N^*(u) = \cup_{u' \in N^-(u)} \{w | w \in N^+(u')\}$. GEN computes the ordering from n down to 1 as follows:

A0. the first vertex v_n is arbitrary chosen;

A1. let v be the largest numbered vertex that has unnumbered neighbors; let L be the set of unnumbered vertices in $N(v)$;

A2. number all vertices in L as follows:

A2.1. pick up the vertex u such that u is not numbered, and $N^+(u)$ is lex-larger than any other $N^+(u')$ with unnumbered u' in L;

A2.2. number u; set $U := \{u\}$;

A2.3. set $W := \{w \mid w \in L$ is not numbered, and $N^-(w) \cap N^-(u') \neq \emptyset$ for some $u' \in U\}$;

A2.4. if $W = \emptyset$ and L contains the vertices that are not numbered, go to step A2.1;

A2.5. while W contains unnumbered vertex, pick up the vertex w such that $N^*(w)$ is lex-larger than any other $N^*(w')$ with unnumbered w' in W, ties are broken by $|N^-(w)|$ (that is, two vertices w and w' with $N^*(w) = N^*(w')$ are $w > w'$ if $|N^-(w)| > |N^-(w')|$);

A2.6. set $U := W$ and go to step A2.3;

A3. if unnumbered vertices remain then go to step A1.

We note that GEN is the LexBFS; the ordinary LexBFS in [25] decides that the vertices in L are tie, which can be broken arbitrary, and GEN checks and sorts them in step A2 and its substeps.

In GEN, for each vertex u in $V - \{v_n\}$, there exists a unique vertex v such that v is picked up in step either A1, and u is numbered as a neighbor of v in step A2. We say v is the *parent* of u, and denote by $p(u) = v$. Here we observe that $p(v_1) < p(v_2)$ implies $v_1 < v_2$ by step A1.

For a parent v, we have a set L of unnumbered neighbors. The main loop of A2 divides the vertices in L into some *groups*. The first group is constructed as follows; GEN first picks up a vertex u in step A2.1, constructs W in step A2.3, numbers all vertices in W in step A2.5, updates U by W in step A2.6, updates W in step A2.3 for new U, and the process is repeated until $W = \emptyset$. Then, if L contains some vertices not numbered, GEN picks up the next u in step A2.1 again, constructs the second group, and so on. We note that each group has unique vertex u picked up in step A2.1.

Correctness of GEN

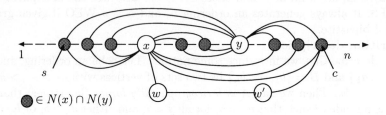

Fig. 1. The vertices not linearly ordered

Lemma 2. *For a given chordal bipartite graph,* GEN *generates a WEO.*

Proof. We assume that the ordering generated by GEN is not a WEO, and derive contradictions. By assumption, there are vertices $s, x, y,$ and w such that $s < x < y$, $s < w$, $s \in N(x) \cap N(y)$, $\{x, w\} \in E$, and $\{y, w\} \notin E$ (Figure 1). We assume that x is greater than any other x' that has the same properties.

Since $\{s, y\} \in E$, we have $s' = p(s) \geq y$. On the other hand, $w > s$ and $w \notin N(y)$. Thus $p(w) > s' \geq y$. Let $w' = p(w)$. Then, since $w' > y$, $p(w') > w' > y$. We here show that $p(w') \in N(x) \cap N(y)$. Suppose that $p(w') \notin N(x) \cap N(y)$. Then, since $p(w')$ is in G_y, there is a (shortest) path P between y and $p(w')$ on G_y. By Theorem 2, s is semi-simplicial. Thus there is a vertex c on P in $N(x) \cap N(y)$. Since $c \neq p(w')$, the path $(c, x, w, w', p(w'))$ and the part of P joining $p(w')$ and c induce a chordless cycle of length greater than 4. Thus we have $p(w') = c \in N(x) \cap N(y)$. That is, we have $w' = p(w) > s' \geq y$, and $p(w') = c \in N(x) \cap N(y)$ (see Figure 1 again).

Since $c \in N(x) \cap N(y)$, we have $c \leq p(x)$ and $c \leq p(y)$. (We remind that if x is unnumbered and c is chosen in step A1, then x will be numbered in step A2 as a neighbor of c, consequently, c will be the parent of x.) If $p(y) > c = p(w')$, we have $y > w'$, which is a contradiction. Thus $c = p(y)$. On the other hand, $c < p(x)$ implies that $p(x) > c = p(y)$, which contradicts that $x < y$. Hence we have $c = p(x) = p(y)$. Since $y \leq s' < w'$, we also have $c = p(s')$. That is, we have $c = p(s') = p(x) = p(y) = p(w')$.

We now suppose that c is chosen in step A1, and $x, y, w',$ and s' are in L, and GEN performs step A2. We first assume that y has another numbered neighbor c'.

Since c is the largest numbered vertex that has unnumbered neighbors, $c' < c$. Since $y < w'$, w' also has numbered neighbor c'' with $c' \leq c'' < c$. We have a path P joining c' and c'' on $G_{c'}$. Thus, $(c', y, s, x, w, w', c'')$ and P induce a cycle of length at least 6. Here, $\{y, w\} \notin E$ and $\{s, w'\} \notin E$ since $s' = p(s) < w'$. Thus $\{c', x\}, \{c'', x\} \in E$. If $c' \neq c''$, x is lex-larger than y and we have $y < x$, which is a contradiction. Thus we have $c' = c''$. Thus, x, y, and w' are lexicographically tie. The ordering will satisfy $x < y < w'$. Thus w' will be chosen before x and y, and it will be numbered in step A2.1 or A2.5. When w' is numbered, it will be put into U, and x will be put into W, and numbered (or x may be numbered in the same W of w'). Thus, since $x < y < w'$, they are numbered in the same group. Let u be the first vertex in the group. The vertices in the group will be numbered step by step. During ith step, we let the set U_i defines the set W_i; that is, $U_0 = \{u\}$, W_0 is the set of unnumbered vertices that share unnumbered neighbor with u, $U_1 = W_0$, W_1 is the set of unnumbered vertices that share unnumbered neighbor with u' for some $u' \in U_1$, $U_2 = W_1$, and so on. We assume that w' in U_i for some $i \geq 0$. Then x is either in W_i or U_i. We have two cases.

(1) x is in U_i. Since $x < y < w'$, we have $y \in U_i$. We consider the set W_{i-1}, which is the set of unnumbered vertices, including x, y, w'. Since y is in W_{i-1}, we have at least one unnumbered vertex y' in $N^-(y) \cap N(U_{i-1})$. y' has a neighbor in U_{i-1}, and so does w'. Thus, there is a path P between y' and w' not through any vertices in $\{x, y, s, w\}$. Hence (y', y, s, x, w, w') and P induce a cycle of length at least 6. Here, $\{w, y\} \notin E$, and $\{s, w'\} \notin E$. Thus $\{x, y'\}$ is in E. That is $N^-(y) \cap N(U_{i-1}) \subseteq N^-(x) \cap N(U_{i-1})$, and $N^*(y) \subseteq N^*(x)$ even if any other vertices in W are numbered. Thus, we have $N^*(y) = N^*(x)$ since $x < y$. Thus, x and y are tie in step A2.5, and when w' is numbered, the tie is broken by $|N^-(x)| > |N^-(y)|$ since $w \in N^-(x) - N^-(y)$. We finally have $x > y$, which is a contradiction.

(2) x is in W_i. Using the same argument in (1), any unnumbered neighbor y' of y is also incident to x. Thus, by the definiton of W in step A2.3, y is also in W_i (not in U_i). Thus, using the same argument in (1), $N^*(y) \subseteq N^*(x)$. Hence, when w' is numbered, the ties are broken by the vertex $w \in N^-(x) - N^-(y)$, which contradicts that $x < y$.

Thus GEN generates a WEO.   □

Simplified GEN

For any given bipartite graph $G = (X, Y, E)$, we can obtain another bipartite graph $G' = (X \cup \{x\}, Y \cup \{y\}, E')$ with $E' = E \cup \{\{x, y'\} | y' \in Y\} \cup \{\{x', y\} | x' \in X\} \cup \{x, y\}$. The vertices x and y are called *universal vertices*. Then G is chordal bipartite iff G' is chordal bipartite. For the graph G', GEN can generate a WEO such that the largest two vertices are the universal vertices. Then we can obtain a WEO of G by removing the last x and y. The WEO for G satisfies that all vertices in X are less than each vertex in Y. That is, any given chordal bipartite graph $G = (X, Y, E)$, there is a WEO such that $\max\{x \in X\} < \min\{y \in Y\}$. (In fact, the assumption has no effect on the proof of Theorem 1.) This fact simplifies the algorithm GEN. The modified algorithm GEN' runs on G and generates the

WEO of G, which is obtained from the WEO of G' generated by GEN running on G' by removing the last universal vertices:

A'1. number all vertices in Y as follows:
    A'1.1. pick up any vertex y_n;
    A'1.2. number y; set $U := \{y\}$;
    A'1.3. set $W := \{w \mid w \in Y$ is not numbered, and $N(w) \cap N(U) \neq \emptyset\}$;
    A'1.4. while W contains unnumbered vertex, pick up the vertex w such that $N^*(w)$ is lex-larger than any other $N^*(w')$ with unnumbered w' in W, ties are broken by $|N(w)|$;
    A'1.5. set $U := W$ and go to step A'1.3;
A'2. number all vertices in X such that $x > x'$ if and only if $N(x)$ is lex-larger than $N(x')$;

Linear time implementation of GEN' is omitted for lack of space, and can be found in [29].

5   Applications on Chordal Bipartite Graph

Recognition of chordal bipartite graph
We assume that given graph is bipartite, which can be determined in $O(n + m)$ time by the depth first search. Let $G = (V_1, V_2, E)$ be a given bipartite graph, and suppose that $V = V_1 \cup V_2$ is ordered by GEN.

Lemma 3. *Given bipartite graph is chordal bipartite iff the bipartite adjacency matrix A ordered by GEN is Γ-free.*

Proof. By Lemma 1(1), "only if" part is trivial. Let v_i, v_j, u_k, and u_l be the vertices with $v_i < v_j$, $u_k < u_l$, and $a_{ik} = a_{jk} = a_{il} = 1$. We show that $a_{jl} = 1$. Without loss of generality, we assume that $v_i < u_k$. We have two cases; $v_i < u_k < v_j < u_l$ and $v_i < v_j < u_k < u_l$. In both cases, since $\{v_i, u_k\}, \{v_i, u_l\} \in E$ and $u_k < u_l$, we have $N_{>v_i}(u_k) \subseteq N_{>v_i}(u_l)$ by the definition of WEO. Thus $v_j \in N_{>v_i}(u_k) \subseteq N_{>v_i}(u_l)$, or $a_{jl} = 1$. □

Corollary 1. *Given graph $G = (V, E)$, the ordering $v_1, \cdots, v_n$ is WEO iff $i < j$, $k < l$, $v_k, v_l \in N(v_i)$, and $v_k \in N(v_j)$ imply $v_l \in N(v_j)$ for each i, j, k and l.*

Theorem 3. *The recognition problem for chordal bipartite graph can be solved in $O(n + m)$ time.*

Proof. (Sketch) Using Lubiw's algorithm shown in [19, Section 5], we can check whether the ordered matrix by GEN is Γ-free in $O(n + m)$ time. □

Corollary 2. *Any given $n_1 \times n_2$ $(0,1)$-matrix A with m 1's, we can compute the doubly lexical ordering of A in $O(n_1 + n_2 + m)$ time if A is Γ-free.*

Other characterizations

Let $G = (V_1, V_2, E)$ be a bipartite graph. Then $\{x, y\} \in E$ is called a *bisimplicial edge* if $G[N(x) \cup N(y)]$ is a complete bipartite graph. Let $e_1, e_2, \cdots, e_m$ be an ordering of the edges of G. For $i = 0, 1, \cdots, m$ define the subgraph $G^i = (V_1, V_2, E_i)$ by $E_0 = E$, and $E_j = E_{j-1} - \{e_j\}$ for $j \geq 1$. The ordering $e_1, e_2, \cdots, e_m$ is a *perfect edge without vertex elimination ordering* (pewveo) for G if each edge e_i is bisimplicial in G^{i-1}. A bipartite graph is chordal bipartite iff it has a pewveo (see [2, Remark 5.9.1]).

Given bipartite graph $G = (S, T, E)$ with $S = \{s_1, s_2, \cdots, s_p\}$ and $T = \{t_1, t_2, \cdots, t_q\}$, a pair of two orderings $s_1 < s_2 < \cdots < s_p$ and $t_1 < t_2 < \cdots < t_q$ is a *strong T-elimination ordering* if, for each $1 \leq i \leq q$ and $1 \leq j < k \leq p$ where $s_j, s_k \in N(t_i)$, we have that $N(s_j) \subseteq N(s_k)$ on the graph $G[S \cup \{t_h \in T | h \geq i\}]$. Chang shows that a bipartite graph is chordal bipartite iff it has a strong T-elimination ordering [4].

Theorem 4. *Given chordal bipartite graph, we can compute a pewveo and a strong T-elimination ordering in $O(n + m)$ time.*

Proof. (Sketch) Let $v_1, v_2, \cdots, v_n$ be the WEO of the chordal bipartite graph G. Then the following algorithm produces the pewveo:

For each $v = v_1, v_2, \cdots, v_n$, let $N(v) = \{u_1, u_2, \cdots, u_{d(v_i)}\}$ with $u_1 < u_2 < \cdots < u_{d(v)}$; for each $u = u_1, u_2, \cdots, u_{d(v)}$, output $\{v, u\}$, and delete the edge from G.

The correctness of the algorithm follows from the results by Kloks and Kratsch [17] with Lemma 3. The algorithm runs in $O(n + m)$ time. The WEO directly gives a strong T-elimination ordering. $\square$

Other problems

A *matching M* is a subset of E such that no two edges in M are incident to a common vertex. A *maximum matching* is a matching of maximum cardinality.

Theorem 5. *The following problems on a chordal bipartite graph can be solved in linear time: (1) Computing the list of all maximal complete bipartite subgraphs, and (2) finding a maximum matching.*

Proof. (Sketch) Using Theorem 4, the results by Kloks and Kratsch in [17], and the results by Chang in [4], we have the theorem. $\square$

Any given matching M, if an edge $\{u, v\}$ belongs to M, we say that M *covers* the vertices u and v. Given a real weight for each vertex of G, the *vertex-weight of a matching M* is defined to be the sum of the weights of the vertices covered by M. A *maximum vertex-weighted matching* is one which has the maximum vertex-weight among all possible matchings of G.

Theorem 6. *A maximum vertex-weighted matching on a chordal bipartite graph can be found in linear time.*

Proof. Campêlo and Klein presented a linear time algorithm for a maximum vertex-weighted matching in given strongly chordal graph with its strong elimination ordering [3]. Their algorithm uses the property "for each i, j, k and l, if $i < j, k < l$, $v_k, v_l \in N[v_i]$, and $v_k \in N[v_j]$, then $v_l \in N[v_j]$". However, it is sufficient for their algorithm to satisfy the property "for each i, j, k and l, if $i < j, k < l$, $v_k, v_l \in N(v_i)$, and $v_k \in N(v_j)$, then $v_l \in N(v_j)$". Thus, by Corollary 1, the algorithm also finds a maximum vertex-weighted matching of a chordal bipartite graph with the WEO in linear time.   □

It is worth mentioning that Chang's algorithm for a maximum matching is very simple, while the algorithm for a maximum vertex-weighted matching due to Campêlo and Klein uses an integer linear programming.

6   Applications on Strongly Chordal Graph

We now turn to strongly chordal graphs, which are introduced by Farber as follows[8]: a *strong elimination ordering* of a graph $G = (V, E)$ is an ordering $v_1, v_2, \cdots, v_n$ of V such that $i < j$, $k < l$, $v_k, v_l \in N[v_i]$, and $v_k \in N[v_j]$ imply $v_l \in N[v_j]$ for each i, j, k and l. A graph is *strongly chordal* if it admits a strong elimination ordering.

Recognition of strongly chordal graph and generating strong elimination ordering

Theorem 7. *The recognition problem for strongly chordal graph can be solved in $O(n+m)$ time. Moreover, we can compute the strong elimination ordering in linear time.*

Proof. (Sketch) Given graph $G = (V, E)$ with $V = \{v_1, v_2, \cdots, v_n\}$ and $|E| = m$, we define $G' = (V \cup U, E_1 \cup E_2)$ as follows: $U = \{u_1, u_2, \cdots, u_n\}$, E_1 contains $\{v_i, u_j\}$ and $\{v_j, u_i\}$ for each $\{v_i, v_j\} \in E$, and E_2 contains $\{u_i, v_i\}$ for each $1 \leq i \leq n$. Since $|V \cup U| = 2n$ and $|E_1 \cup E_2| = 2m+n$, the construction of G' can be done in linear time. We show that G is strongly chordal iff G' is chordal bipartite. By Lemma 1, if G is strongly chordal, G' is chordal bipartite. Thus we assume that G' is chordal bipartite, and show that G is strongly chordal. Since G' is chordal bipartite, G' has a WEO over $V \cup U$. We show the WEO over V is the strong elimination ordering of G. Since the ordering is the WEO of G', for each $i < j$ and $k < l$, $\{v_i, u_k\} \in E_1$, $\{v_i, u_l\} \in E_1$ and $\{v_j, u_k\} \in E_1$ imply $\{u_l, v_j\} \in E_1$. Then, it is sufficient to show that, on G, for each $i < j$ and $k < l$, $v_k, v_l \in N[v_i]$ and $v_k \in N[v_j]$ imply $v_l \in N[v_j]$. Since $\{v_i, v_j\} \in E$ on G iff $\{v_i, u_j\}, \{v_j, u_i\} \in E_1$ on G', we have nothing to do when i, j, k and l are all different. Thus we have three cases to consider: $k = i$, $l = i$, and $k = j$. For lack of space, we only consider the first case here. Assume $k = i$. On G', we have $\{v_i, u_l\} \in E_1$ and $\{v_j, u_i\} \in E_1$. Since $i < j$, $i < l$, $\{v_i, u_j\} \in E_1$, and the ordering is WEO, $N(u_j) \subseteq N(u_l)$ or $N(u_l) \subseteq N(u_j)$. If $N(u_j) \subseteq N(u_l)$, $v_j \in N(u_j) \subseteq N(u_l)$. Thus $\{v_j, u_l\}, \{v_l, u_j\} \in E_1$, thus $\{v_j, v_l\} \in E$. Hence

$v_l \in N[v_j]$ on G. The case $N(u_l) \subseteq N(u_j)$ is symmetrical. The second and third cases are similar, and omitted here.

Therefore to determine whether G is strongly chordal, we check whether G' is chordal bipartite, which can be done in linear time by Theorem 3. The strong elimination ordering of G can be obtained from the WEO of G'. □

Other problems on strongly chordal graph
For a given graph $G = (V, E)$, $U \subseteq V$ is a *dominating set* if $N[U] = V$. The *domatic number* of G, denoted by $dn(G)$, is the maximum number of pairwise disjoint dominating sets in G. The *domatic partition* is a partition V into $dn(G)$ disjoint dominating sets.

If a strongly chordal graph is given with its strong elimination ordering, several problems can be solved in linear time. Thus we immediately have the following by Theorem 7.

Theorem 8. *Given strongly chordal graph, we can find a maximum (vertex-weighted) matching, an independent dominating set, a minimum weighted dominating set, a minimum dominating clique, and a domatic partition in linear time.*

Proof. A maximum matching is due to Dahlhaus and Karpinski [7], a weighted maximum matching is due to Campêlo and Klein [3], an independent dominating set and a minimum weighted dominating set are due to Farber [9], a minimum dominating clique is due to Kratsch [18], and a domatic partition is due to Peng and Chang [23]. □

Acknowledgments. The author is grateful to Jeremy Spinrad and Feodor Dragan for many useful discussions. They pointed out the errors of GEN in an early version. The idea of universal vertices, which made the algorithm and its proof much simpler, is due to Jeremy Spinrad. The author is also grateful to Haiko Müller for helpful comments. He introduces a notion of two-pair, which simplify the definitions of semi-clique and semi-simplicial.

References

1. M. Bakonyi and A. Bono. Several Results on Chordal Bipartite Graphs. *Czechoslovak Math. J.*, 47:577–583, 1997.
2. A. Brandstädt, V.B. Le, and J.P. Spinrad. *Graph Classes: A Survey.* SIAM, 1999.
3. M.B. Campêlo and S. Klein. Maximum Vertex-Weighted Matching in Strongly Chordal Graphs. *Disc. Appl. Math.*, 84:71–77, 1998.
4. M.-S. Chang. Algorithms for Maximum Matching and Minimum Fill-in on Chordal Bipartite Graphs. In *7th ISAAC*, pp. 146–155. LNCS Vol. 1178, Springer-Verlag, 1996.
5. D.G. Corneil, S. Olariu, and L. Stewart. The Ultimate Interval Graph Recognition Algorithm? In *9th SODA*, pp. 175–180. ACM, 1998.
6. E. Dahlhaus. Generalized Strongly Chordal Graphs. Technical Report TR93-458, University of Sydney, 1993.

7. E. Dahlhaus and M. Karpinski. Matching and Multidimensional Matching in Chordal and Strongly Chordal Graphs. *Disc. Appl. Math.*, 84:79–91, 1998.

8. M. Farber. Characterization of Strongly Chordal Graphs. *Disc. Math.*, 43:173–189, 1983.

9. M. Farber. Domination, Independent Domination, and Duality in Strongly Chordal Graphs. *Disc. Appl. Math.*, 7:115–130, 1984.

10. M.C. Golumbic. *Algorithmic Graph Theory and Perfect Graphs*. Academic Press, 1980.

11. M.C. Golumbic and C.F. Goss. Perfect Elimination and Chordal Bipartite Graphs. *J. of Graph Theory*, 2:155–163, 1978.

12. R. Hayward, C. Hoàng, and F. Maffray. Optimizing Weakly Triangulated Graphs. *Graphs and Combinatorics*, 5(4):339–349, 1989. [Erratum, *ibid.*, 6(1):33–35, 1990.].

13. A.J. Hoffman, A.W.J. Kolen, and M. Sakarovitch. Totally-Balanced and Greedy Matrices. *SIAM J. Alg. Disc. Meth.*, 6(4):721–730, 1985.

14. C.R. Johnson and J. Miller. Rank Decomposition Under Combinatorial Constraints. *Linear Algebra and Its Applications*, 251:97–104, 1997.

15. C.R. Johnson and G.T. Whitney. Minimum Rank Completions. *Linear and Multilinear Algebra*, 28:271–273, 1991.

16. D. König. *Theorie der endlichen und unendlichen Graphen (in German)*. Akademische Verlagsgesellschaft, 1936.

17. T. Kloks and D. Kratsch. Computing a Perfect Edge Without Vertex Elimination Ordering of a Chordal Bipartite Graph. *IPL*, 55:11–16, 1995.

18. D. Kratsch. Finding Dominating Cliques Efficiently, in Strongly Chordal Graphs and Undirected Path Graphs. *Disc. Math.*, 86:225–238, 1990.

19. A. Lubiw. Doubly Lexical Orderings of Matrices. *SIAM J. on Computing*, 16(5):854–879, 1987.

20. T.A. McKee and F.R. McMorris. *Topics in Intersection Graph Theory*. SIAM, 1999.

21. R. Paige and R.E. Tarjan. Three Partition Refinement Algorithms. *SIAM J. on Computing*, 16(6):973–989, 1987.

22. B.S. Panda. New Linear Time Algorithms for Generating Perfect Elimination Ordering of Chordal Graphs. *IPL*, 58:111–115, 1996.

23. S.-L. Peng and M.-S. Chang. A Simple Linear Time Algorithm for the Domatic Partition Problem on Strongly Chordal Graphs. *IPL*, 43:297–300, 1992.

24. N.K.R. Prasad and P.S. Kumar. On Generating Strong Elimination Orderings of Strongly Chordal Graphs. In *FST&TCS*, pp. 221–232. LNCS Vol. 1530, Springer-Verlag, 1998.

25. D.J. Rose, R.E. Tarjan, and G.S. Lueker. Algorithmic Aspects of Vertex Elimination on Graphs. *SIAM J. on Computing*, 5(2):266–283, 1976.

26. K. Simon. A Note on Lexicographic Breadth First Search for Chordal Graphs. *IPL*, 54:249–251, 1995.

27. J.P. Spinrad. Doubly Lexical Ordering of Dense 0-1 matrices. *IPL*, 45:229–235, 1993.

28. R.E. Tarjan and M. Yannakakis. Simple Linear-Time Algorithms to Test Chordality of Graphs, Test Acyclicity of Hypergraphs, and Selectively Reduce Acyclic Hypergraphs. *SIAM J. on Computing*, 13(3):566–579, 1984.

29. R. Uehara. Linear Time Algorithms on Chordal Bipartite and Strongly Chordal Graphs.
http://www.komazawa-u.ac.jp/~uehara/ps/chordal-linear-full.ps.gz, 2002.

Improved Inapproximability Results for Vertex Cover on k-Uniform Hypergraphs

Jonas Holmerin

Department of Numerical Analysis and Computer Science, Royal Institute of
Technology, SE-100 44 Stockholm, Sweden. joho@kth.se

Abstract. We prove that Minimum Vertex Cover on k-uniform hypergraphs is **NP**-hard to approximate within $\Omega(k^{1-\epsilon})$. The result follows by a new reduction and a PCP characterization of **NP** by Håstad and Khot [11]. We also give an alternate construction of a PCP with the required properties. We also show that Minimum Vertex Cover on 3-uniform hypergraphs is **NP**-hard to approximate within $3/2 - \epsilon$. Of independent interest may be a 3 query PCP for **NP** with perfect completeness where answers are from a domain of size d and where the soundness is $2/d$.

1 Introduction

In this paper we study lower bounds on the approximability of Minimum Vertex Cover generalized to k-uniform hypergraphs (or, equivalently, Minimum Hitting Set where all sets have size exactly k). A simple k-approximation algorithm is well-known, and the best known approximation algorithm approximates the problem to within $k - o(1)$ [9]. Turning to lower bounds, the case of graphs (i.e, $k = 2$) has been well studied; Dinur and Safra [4] recently proved that this problem is **NP**-hard to approximate to within $10\sqrt{5} - 21 - \epsilon \approx 1.36$, improving on the previous best lower bound of $7/6 - \epsilon$ by Håstad [10].

Also the case when the size of the edges is unbounded has been well studied; this problem is equivalent to Minimum Set Cover for which Feige [7] proved that it is "almost" **NP**-hard to approximate within a factor $(1 - \epsilon)\ln n$ for any $\epsilon > 0$. This result is essentially tight since there is an $1 + \ln n$-approximation algorithm [14].

For the case when k is a constant greater than or equal to 3, up to recently this problem seems not to have been studied at all, except for a folk-lore approximation preserving reduction which makes any lower bound for Minimum Vertex Cover on graphs also a lower bound for Minimum Vertex Cover on k-uniform hypergraphs for $k \geq 3$.

However, recently Trevisan [18] proved that asymptotically, the problem is **NP**-hard to approximate to within $\Omega(k^{1/19})$, and Holmerin [13] proved that for 4-uniform hypergraphs, the problem is **NP**-hard to approximate to within $2 - \epsilon$. Also, Goldreich [8], using the FGLSS reduction, gave simple proofs of inapproximability results which, while inferior to the results mentioned above, inspired the current paper.

P. Widmayer et al. (Eds.): ICALP 2002, LNCS 2380, pp. 1005–1016, 2002.

1.1 Our Contributions

In this paper we improve upon the result of Trevisan and prove

Theorem 1. *For any $\epsilon > 0$, Minimum Vertex Cover on k-uniform hypergraphs is **NP**-hard to approximate within $\Omega(k^{1-\epsilon})$.*

For 3-uniform hypergraphs, we prove the following:

Theorem 2. *For any $\epsilon > 0$, Minimum Vertex Cover on 3-uniform hypergraphs is **NP**-hard to approximate within $3/2 - \epsilon$.*

Previously the only known lower bounds for the case of 3-uniform hypergraphs have been the ones implied by the results on Minimum Vertex Cover for graphs.

There are two sources of our improvement over Trevisan's result. Firstly, we give an improved reduction from a PCP system to Minimum Vertex Cover on hypergraphs. By this reduction, we have that if **NP** can be characterized by a q query PCP with answers from a domain of size d, with perfect completeness and soundness $s(d)$ (i.e, correct proofs are accepted with probability 1 and incorrect proofs are accepted with probability at most $s(d)$), then Minimum Vertex Cover on $q(d-1)$-uniform hypergraphs is **NP**-hard to approximate to within $s(d)^{1/q}$.

Secondly, we use a different PCP characterization **NP**. Trevisan constructs a 3 query PCP with perfect completeness, answer domain of size $d = 2^l$ and soundness $(3/4)^l = d^{\log(3/4)} \approx d^{-0.415}$. We instead use a PCP with properties similar to the query efficient PCP of Håstad and Khot [11]. Our PCP has perfect completeness, makes $2t+t^2$ queries from a domain of size $d \geq 3$ and has soundness $(d/2)^{-t^2}$. For the result for 3-uniform hypergraphs, we use the same reduction as for general k, applying it to a variant of the PCP used by Håstad [10] to prove inapproximability of Maximum E3-Sat on satisfiable instances. But instead of treating the PCP as a black box in the analysis, we are more careful and use properties of the PCP to get the lower bound of $3/2 - \epsilon$.

We also note that the special case $t = 1$ for the PCP mentioned above gives the best currently known characterization of **NP** by a 3 query PCP with perfect completeness and answers from a domain of size d for $d \geq 3$. We get soundness $2/d$ which improves on the PCP by Trevisan mentioned above. This result may be of independent interest.

2 Preliminaries

Definition 1. *A k-uniform hypergraph $H = (V, E)$ consists of a set of vertices, V and a set of edges E. An edge $e \in E$ is a subset of the vertices of size k.*

A vertex cover of H is a subset S of the vertices such that for all edges $e \in E$, $e \cap S \neq \emptyset$.

For any subset S of the vertices, we say that an edge e is covered by S if $S \cap e \neq \emptyset$.

2.1   PCPs and the Basic Two Prover Protocol

Recall that in the PCP model membership in a language is checked by a probabilistic verifier which is given oracle access to the proof and which is allowed to accept incorrect proofs with some probability. We define the class $\textbf{naPCP}_{c,s}[r, q, d]$ to consist of languages L that can be checked by a non-adaptive verifier (i.e, the queries depends only the input x and the random string ρ, not on the answers to previous queries) which uses $r(n)$ random bits, makes $q(n)$ queries from an oracle with answer domain of size $d(n)$. The verifier has the property that if $x \in L$, then there is a proof which the verifier accepts with probability at least c, and if $x \notin L$, no proof is accepted with probability more than s. We call c and s the completeness and the soundness of the verifier, respectively. When $c = 1$, we say that the verifier has *perfect completeness*. In terms defined above, the remarkable PCP theorem of Arora et al [1] says that $\textbf{NP} = \textbf{naPCP}_{1,1/2}[O(\log n), O(1), 2]$.

The starting point for our PCPs will be the standard two-prover one-round protocol for $\textbf{NP}$ which we will now describe. There is an approximation preserving reduction from general E3-Sat formulas to formulas where each variable occurs in exactly five clauses [15]. From the PCP theorem and this reduction, it follows that it is possible to reduce any problem in $\textbf{NP}$ to the problem of distinguishing between the two cases that a E3-Sat formula ϕ is satisfiable, or that at most a fraction G of the clauses are satisfiable. The formula has the property that each variable occurs in exactly 5 clauses. The protocol consists of two provers, P_1 and P_2, and one verifier. Given an instance, i.e., an E3-Sat formula ϕ, the verifier picks u clauses $(C_1, \dots, C_u)$ each uniformly at random from the instance. For each C_i, it also picks a variable x_i from C_i uniformly at random. The verifier then sends $(x_1, \dots, x_u)$ to P_1 and the clauses $(C_1, \dots, C_u)$ to P_2. It receives an assignment to $(x_1, \dots, x_u)$ from P_1 and an assignment to the variables in $(C_1, \dots, C_u)$ from P_2, and accepts if these assignments are consistent and satisfy $C_1 \wedge \dots \wedge C_u$. The completeness of this proof system is 1, and it follows by a general result by Raz [16] that the soundness is at most c_G^u, where $c_G < 1$ is some constant depending on G but not on u or the size of the instance.

2.2   The Long Code over $\mathbb{Z}_d$ and the Standard Written Proof

A PCP verifier will be given access to what is supposed to be encodings of the answers from the provers in the protocol of the previous sections. The code which we will use is a the *long d-code* of [10], which is a generalization of the long code of Bellare et al [2].

Let $d \geq 3$ be an integer and let $\omega = e^{2\pi i/d}$. We consider $\mathbb{Z}_d$ to consist of the dth roots of unity $1, \omega, \dots, \omega^{d-1}$. The group operation is thus multiplication. For a set U of variables let $\{0, 1\}^U$ denote all possible assignments to variables in U. Consider a set W of variables and a subset U of W. Then for any assignment y to the variables in W, write $y|_U$ for the restriction of y to U. Let $\mathcal{F}_d(U) = \{f : \{0, 1\}^U \to \mathbb{Z}_d\}$. For a string $x \in \{0, 1\}^U$, we define *the long d-code of x*, $\mathcal{LC}^d(x)$, to be the function $\mathcal{LC}^d(x) : \mathcal{F}_d(U) \to \mathbb{Z}_d$ given by $\mathcal{LC}^d(x)(f) = f(x)$.

Definition 2. *A* Standard Written *d*-Proof *consists of, for each set U of u variables a table* $A_U : \mathcal{F}_d(U) \to \mathbb{Z}_d$ *and for each set W of u clauses a table* $A_W : \mathcal{F}_d(W) \to \mathbb{Z}_d$.

The tables A_U and A_W are supposed to be encoding of answers from the provers P_1 and P_2 on query U and W respectively. We say that Standard Written *d*-Proof is a *correct proof* for formula ϕ if x is a satisfying assignment to ϕ and $A_U = \mathcal{LC}^d(x|_U)$ and $A_W = \mathcal{LC}^d(x|_W)$.

A verifier will thus have access to tables A_W and A_U which are purported long *d*-codes. It will be useful for us to be able to ensure that a purported long *d* code A has certain properties. For a function $A : \mathcal{F}_d(U) \to \mathbb{C}$, we define $A_{\mathbb{Z}_d}$, A *folded over* $\mathbb{Z}_d$ by partitioning $\mathcal{F}_d(U)$ into sets of functions $\{f, \omega f, \ldots, \omega^{d-1} f\}$, and for each part picking one representative f (by e.g, letting f be the function with a 1 in the first non-zero coordinate). Then we let $A_{\mathbb{Z}_d}(\gamma f) = \gamma A(f)$.

We also say that any table A which has the property that $A(\gamma f) = \gamma A(f)$ is folded over $\mathbb{Z}_d$. In the case when $d = 2$ we use the terminology *folded over* **true** instead of *folded over* $\mathbb{Z}_2$, and the notation $A_{\textbf{true}}$ for $A_{\mathbb{Z}_2}$. For functions $h : \{0,1\}^U \to \{0,1\}$, and $F \in \mathcal{F}_d(U)$ we define

$$(f \wedge h)(x) = \begin{cases} f(x) \text{ if } h(x) = 1 \\ 1 \quad \text{if } h(x) = 0 \end{cases}$$

It will be useful for us to be able to ensure that a A has the property that $A(f) = A(f \wedge h)$. Since we want to be able to fold over $\mathbb{Z}_d$ at the same time, we have to be a bit careful. We omit the details from this extended abstract. We use the notation $A_{h,\mathbb{Z}_d}$ for A conditioned on h and folded over $\mathbb{Z}_d$. We also say that any function A which has the property that $A(f) = A(f \wedge h)$ is conditioned upon h.

2.3   The Fourier Transform over $\mathbb{Z}_2$

To use the analysis methods of [10], we need the Fourier inversion theorem on a function $A : \mathcal{F}_2(U) \to \mathbb{C}$.

Let $\alpha \subseteq \{1,-1\}^U$, $\chi_\alpha(f) = \prod_{x \in \alpha} f(x)$ and $\hat{A}_\alpha = 2^{-2^{|U|}} \sum_{f \in \mathcal{F}_2(U)} A(f)\chi_\alpha(f)$. Then we have that a A can be written as

$$A(f) = \sum_{\alpha \subseteq \{1,-1\}^U} \hat{A}_\alpha \chi_\alpha(f) \tag{1}$$

This is an extremely useful tool since the basis functions $\chi_\alpha(f)$ are just products of long codes for different assignments to the variables in U. Thus a coefficient $\hat{A}_\alpha$ is the correlation of A with a certain product of long codes. We also need Parseval's equality: let $A : \mathcal{F}_2(U) \to \mathbb{C}$. Then

$$|\mathcal{F}_2(U)|^{-1} \sum_{f \in \mathcal{F}_2(U)} |A(f)|^2 = \sum_{\alpha \subseteq \{-1,1\}^U} |\hat{A}_\alpha|^2$$

Folding over **true** and conditioning has consequences for the distribution of the Fourier coefficient, as stated in the following lemmas (for the straightforward proofs, see [10]):

Lemma 1. *If $A : \mathcal{F}_d(U) \to \mathbb{C}$ has the property that $A(-f) = -A(f)$ then $\hat{A}_\alpha = 0$ unless $|\alpha|$ is odd. In particular $\hat{A}_\emptyset = 0$.*

Lemma 2. *If $A : \mathcal{F}_2(U) \to \mathbb{C}$ has the property that $A(f) = A(f \wedge h)$, then for all α such that for some x, $h(x) = 0$ and $x \in \alpha$, we have $\hat{A}_\alpha = 0$.*

Let W be a set of variables and let $U \subset W$. For $\beta \subset \{0,1\}^W$, *the projection of β on U, $\pi^U(\beta)$ is defined as $x \in \pi^U(\beta)$ if there is an y in β such that $y|_U = x$.*

3   The Reduction

In this section we prove that if $\mathbf{NP} \subseteq \mathbf{naPCP}_{1,s(d)}[O(\log n), q, d]$, then, for $k = q(d-1)$, Minimum Vertex Cover on k-uniform hypergraphs is $\mathbf{NP}$-hard to approximate within $s(d)^{-1/q}$. Our reduction is inspired by Goldreich [8], and its analysis uses techniques similar to those used by Trevisan [18].

Assume that $\mathbf{NP} \subseteq \mathbf{naPCP}_{1,s(d)}[O(\log n), q, d]$ and for some $\mathbf{NP}$-complete language L let V be the verifier which thus exists. First we modify the verifier to make sure that each of the q queries are uniformly distributed in the proof. We do this by first making all positions equally likely by if necessary duplicating some positions in the proof and when querying a duplicated position choosing one of the duplicates randomly. If some positions are never queried then we simply omit them from the proof. We also randomly permute the order of the queries. This is possible since the verifier is non-adaptive.

We now describe the reduction to Minimum Vertex Cover. On input x we construct the hypergraph H. The vertices in H are pairs (p, a) for all positions p in the proofs and all the possible answers a. Hence if P is the number of positions in the proof, the graph H consists of P *layers* S_p, each layer of size d.

For each layer p we add an edge S_p containing the nodes in layer p. We call these edges *layer-edges*. The purpose of the layer-edges is to force a vertex cover to contain at least one answer for each position.

Now we add edges for all the random strings σ. Suppose the verifier queries the positions $p_1, \ldots, p_q$ on random string σ. Consider the corresponding q layers $S_{p_1}, \ldots, S_{p_q}$. We want to ensure that for any vertex cover of H, there is at least one way of picking $(p_1, a_2), \ldots, (p_q, a_q)$ from the cover such that $a_1, \ldots a_q$ is accepted by the verifier. We can achieve this by for each set for which this is not the case, simply adding, as an edge in H, the complement in the subgraph induced by the layers $S_{p_1}, \ldots, S_{p_q}$

More formally, for all non-empty subsets $S_1 \subseteq S_{p_1}, \ldots, S_q \subseteq S_{p_q}$ such that there is no way of choosing $(p_1, a_1) \in S_1, \ldots, (p_q, a_q) \in S_q$, such that $a_1, \ldots, a_q$ is accepted by the verifier on random string σ, we add $(S_{p_1} \setminus S_1) \cup \ldots \cup (S_{p_q} \setminus S_q)$ as an edge in H.

We call these edges *non layer-edges*.

Lemma 3. *If $x \in L$ then there is a vertex cover of size P in H.*

Proof. If $x \in L$ then there is a proof π which makes the verifier accept with probability 1. Construct S by for each layer p putting $(p, \pi(p))$ in S. The layer-edges S_p are clearly covered. Suppose we have an uncovered non layer-edge $e = (S_{p_1} \setminus S_1) \cup \ldots \cup (S_{p_q} \setminus S_q)$. Then we must have $(p_1, \pi(p_1)) \in S_1, \ldots, (p_q, \pi(p_q)) \in S_q$. But then $\pi(p_1), \ldots, \pi(p_q)$ must be rejected by the verifier on random string σ, a contradiction. Thus all edges must be covered.

Lemma 4. *If $x \notin L$, then no set of size less than $s(d)^{-1/q}P$ is a vertex cover of H.*

Proof. Suppose there is a cover S of size tP. To begin with, all layer-edges S_p must be covered, so each layer contains at least one vertex from the cover.

For layer p let $L(p) = S_p \cap S$. Now define a random proof π by for each position p picking uniformly at random a such that $(p, a) \in L(p)$ and letting $\pi(p) = a$. Consider a random string σ, and note that since S was a cover, there must be at least one way of choosing $(p_1, a_1) \in L(p_1), \ldots, (p_q, a_q) \in L(p_q)$ such that the verifier accepts on random string σ when seeing the answers $a_1, \ldots, a_q$, since otherwise $(S_{p_1} \setminus L(p_1)) \cup \ldots \cup (S_{p_q} \setminus L(p_q))$ would be an edge in H which is not covered by S. The probability that we picked these answers when constructing π is $\prod_{i=1}^{q} |L(p_i)|^{-1}$. Also, $E[|L(p_i)|] = E[|L(p_1)|] = t$, since all the queries are uniformly distributed in the proof. Hence the expected number of σ for which the verifier accepts on π is at least

$$\mathop{E}_{\sigma}\left[\prod_{i=1}^{q}|L(p_i)|^{-1}\right] \geq e^{-E_\sigma[\sum_{i=1}^{q}\ln|L(p_i)|]} = e^{-qE_\sigma[\ln|L(p_1)|]} \geq$$

$$e^{-q\ln E[|L(p_1)|]} = e^{-q\ln t} = t^{-q}$$

where we used that $E[e^X] \geq e^{E[X]}$ and $E[-\ln X] \geq -\ln E[X]$.

If follows that there must be some π which makes the verifier accept with at least probability t^{-q}

However, we know that no proof makes the verifier accept with probability greater than $s(d)$, so we must have $t \geq s(d)^{-1/q}$

We combine the two previous lemmas into the following theorem:

Theorem 3. *Suppose that* **NP** $\subseteq$ **naPCP**$_{1,s(d)}[O(\log n), q, d]$. *Then Minimum Vertex Cover on $q(d-1)$-uniform hypergraphs is* **NP**-*hard to approximate within $s(d)^{-1/q}$.*

4   A Query Efficient PCP for NP and the Result for k-Uniform Hypergraphs

In this section we give an simplified construction as compared to that of Håstad and Khot [11] of a query efficient non-adaptive PCP with domain d. We then use it to prove the $\Omega(k^{1-\epsilon})$ lower bound for Minimum Vertex Cover on k-uniform hypergraphs.

The test is based on simple 3-query PCP with perfect completeness and domain d. Similarly to Samorodnitsky and Trevisan [17] this test is then iterated in a query-efficient way. The resulting test is similar to the test of Samorodnitsky and Trevisan, and in particular to the generalization of this test to larger domains by Engebretsen [5]. The test is analyzed using techniques from Håstad and Widgerson's [12] simplified analysis of the PCP of Samorodnitsky and Trevisan. The basic test works as follows: The input to the verifier is a formula ϕ and a Standard Written d-Proof. The verifier V picks U and W as in the two-prover one-round game. Then V picks random f and g in $\mathcal{F}_d(U)$ and $\mathcal{F}_d(W)$ respectively, and picks h in $\mathcal{F}_d(W)$ by for each y choosing $h(y) \in \{1, \omega\}$ uniformly. The verifier accepts if

$$A_{U,\mathbb{Z}_d}(f)A_{W,\mathbb{Z}_d,\phi_W}(g)A_{W_j,\mathbb{Z}_d,\phi_W}((fg)^{-1}h) \in \{1, \omega\}.$$

We note at this point that this test is similar to the test used by Engebretsen and Guruswami [6] to prove inapproximability of linear inequations over $\mathbb{Z}_d$. In their test $h(y)$ is chosen uniformly in $\mathbb{Z}_d \setminus \{1\}$ and the test accepts if $A_U(f)A_W(g)A_W((fg)^{-1}h) \neq 1$.

We give the iterated verifier below.

IterTest. Input: A formula $\phi = C_i \wedge \ldots \wedge C_m$ with n variables and m clauses. Oracle access to a Standard Written d-Proof with parameter u, with each table A_W conditioned upon $\phi_W = \bigwedge_{C_i \in W} C_i$, and all the tables folded over $\mathbb{Z}_d$.

1. Select uniformly at random a set $U = \{x_1, \ldots, x_u\}$ of u variables.
2. Select t sets $W_1, \ldots W_t$ by independently for each j selecting W_j by for each $x_l \in U$ picking uniformly at random a clause C_l in which x_l occurs, and letting $W_j = \{C_1, \ldots, C_u\}$.
3. For $1 \le i \le t$ select uniformly at random $f_i \in \mathcal{F}_d(U)$
4. For $1 \le j \le t$ select uniformly at random $g_j \in \mathcal{F}_d(W_j)$.
5. For $1 \le j \le t$ Select $h_{ij} \in \mathcal{F}_d(W)$ by choosing $h_{ij}(y) \in \{1, \omega\}$ uniformly and independently.
6. Accept if, for all $1 \le i, j \le t$,

$$A_{U,\mathbb{Z}_d}(f_i)A_{W,\mathbb{Z}_d,\phi_W}(g_j)A_{W_j,\mathbb{Z}_d,\phi_W}((f_ig_j)^{-1}h_{ij}) \in \{1, \omega\}.$$

The completeness is straightforward, and we omit the proof of the following lemma:

Lemma 5. *If ϕ is satisfiable, a correct proof is accepted by* **IterTest** *with probability* 1.

The analysis of the soundness is also relatively straightforward, given the techniques of [12]. Due to space limitations we omit the proof of the following lemma from this extended abstract:

Lemma 6. *If IterTest accepts with probability* $2^{t^2}d^{-t^2}+\delta$, *then there is a strat-egy for the two-prover one-round game with success probability at least* $d^2\delta^2$.

Combining the two previous lemmas, we get:

Theorem 4. *For any* $t \geq 1$, $d \geq 3$, *for any* $\delta > 0$,

$$NP = \mathbf{naPCP}_{1,2^{t^2}d^{-t^2}+\delta}[O(\log n), 2t + t^2, d]$$

This can be compared with the result of [11] that, for primes p,

$$\mathbf{NP} = \mathbf{naPCP}_{1,p^{-t^2}+\delta}[O(\log n), 4t + t^2, p].$$

Combining Theorem 4 and Theorem 3, we get Theorem 1.

5   Minimum Vertex Cover on 3-Uniform Hypergraphs

In this section we will use the same reduction as in Section 3, but instead of using the PCP from Section 4 we will use the non-adaptive 3-query boolean PCP with perfect completeness and soundness $3/4 + \epsilon$ from Håstad [10]. When analyzing the reduction we will not use the PCP as a black box, instead we will do a careful analysis of the number of covered non layer-edges.

We need to modify the proof system as described by Håstad somewhat, since we have a bipartite situation where the verifier queries tables of two classes (corresponding to sets U and sets W), it is important that the total number of positions in the two classes are equal. To this end, let $D(n, u)$ be the number of copies we need of each position corresponding to an U to make the classes the same size. This is a polynomial in n whose degree depend on u. We make the classes the same size by for $1 \leq \kappa \leq D(n, u)$ for each U having alternate tables $A_{U,\kappa}$ and picking one at random when making a query. Again, positions which are never queried are omitted from the proof. We give the verifier below.

Input: A formula $\phi = C_i \wedge \ldots \wedge C_m$ with n variables and m clauses and oracle access to a Standard Written Proof with parameter u, with each table A_W conditioned upon $\phi_W = \bigwedge_{C_i \in W} C_i$, and all the tables folded over true.

1. Set $t = \lceil \delta^{-1} \rceil$, let $\epsilon_1 = \delta$ and $\epsilon_i = \delta^{1+2/c} 2^{-1/c} \epsilon_{i-1}$ for $i \in \{2, \ldots, t\}$ where $c = 1/35$ is a constant from [10].
2. Choose $i \in \{1, \ldots, t\}$ with uniform probability.
3. Select uniformly at random a set $U = \{x_1, \ldots, x_u\}$ of u variables.
4. Select uniformly at random $\kappa \in \{1, \ldots, D(u, n)\}$.
5. Select W by for each $x_k \in U$ picking uniformly at random a clause C_k in which x_k occurs. Let $W = \{C_1, \ldots, C_u\}$.
6. Select uniformly at random $f : \{0,1\}^U \to \{-1,1\}$ and $g_1 : \{0,1\}^W \to \{-1,1\}$.
7. Select $g_2 : \{0,1\}^W \to \{-1,1\}$ by: if $f(y|_U) = 1$, let $g_2(y) = -g_1(y)$. If $f(y|_U) = -1$, with probability $1 - \epsilon_i$ let $g_2(y) = g_1(y)$, and with probability ϵ_i choose $g_2(y)$ uniformly.

8. Accept if $(1 + A_{U,\kappa,\textbf{true}}(f))(1 + A_{W,\phi_W,\textbf{true}}(g_1)A_{W,\phi_W,\textbf{true}}(g_2)) = 0$, else reject.

Next we apply the reduction described in Section 3 to this PCP. The vertices of the hypergraph consists of pairs (p, a) where p is a position in the proof and $a \in \{-1, 1\}$.

Remember that because of conditioning and folding, not all f and g are queried by the verifier. Denote by $\tilde{f}$ and $\tilde{g}$ the functions actually queried when the verifier chooses f and g respectively.

Note that, since we omit positions in the proofs which are never queried, a position in the proof is given either by the triple $(U, \kappa, \tilde{f})$ or by the pair $(W, \tilde{g})$.

Thus a layer looks either as $S_{(U,\kappa,\tilde{f})} = \{((U, \kappa, \tilde{f}), -1), ((U, \kappa, \tilde{f}), 1)\}$, or as $S_{(W,\tilde{g})} = \{((W, \tilde{g}), -1), ((W, \tilde{g}), 1)\}$.

For a random string σ for which the verifier makes queries $(U, \kappa, \tilde{f})$ $(W, \tilde{g}_1)$ and $(W, \tilde{g}_2)$ the non-layer edges are the complements of the rejecting views. For example, in the case when $\tilde{f} = f$, $\tilde{g}_1 = g_1$ and $\tilde{g}_2 = g_2$, the rejecting views are $(1, 1, 1)$ and $(1, -1, -1)$, so we add the edges corresponding to the answers $(-1, -1, -1)$ and $(-1, 1, 1)$. Hence the size of the edges will be at most 3.

Let P be the total number of positions in the proof. By Lemma 3 it follows that if ϕ is satisfiable, then there is a cover of size P.

Let Q_1 be the set of all triples $(U, \kappa, \tilde{f})$ and Q_2 be the set of all pairs (W, g). By construction, $P/2 = |Q_1| = |Q_2|$.

Suppose there is a cover S of size tP in H. Now since S is a cover, all layer-edges must be covered. We may thus view S as a function $S : Q_1 \cup Q_2 \to \{\{-1\}, \{1\}, \{1, -1\}\}$ defined by the intersection of S and S_p for a position p.

Define tables $B_{U,\kappa}$ and B_W by (for $f = \tilde{f}$ i.e, f which are actually in the proof)

$$B_{U,\kappa}(\tilde{f}) = \begin{cases} -1 \text{ if } S(U, \kappa, \tilde{f}) = \{-1\} \\ 1 \ \text{ if } S(U, \kappa, \tilde{f}) = \{1\} \\ 0 \ \text{ if } S(U, \kappa, \tilde{f}) = \{1, -1\} \end{cases}$$

For the f which are not in the proof $-f$ must be in the proof so we define $B_{U,\kappa}(f) = -B_{U,\kappa}(-f)$ (i.e, folding over **true**). Define $B_W(\tilde{g})$ in the same way and extend the table to all g by folding over **true** and conditioning upon ϕ_W. We have the following lemma, the proof of which we omit from this extended abstract:

Lemma 7. *The fraction of non-layer edges which is covered is*

$$1 - \frac{\mathbb{E}_{U,\kappa,W,f,g_1,g_2} \left[B_{U,\kappa}(f)^2 B_W(g_1)^2 B_W(g_2)^2 (1 + B_{U,\kappa}(f))(1 + B_W(g_1)B_W(g_2)) \right]}{8}$$

$$(2)$$

To analyze (2) we make use of the following technical lemma, the proof of which is again omitted:

Lemma 8. *Let $C_{U,\kappa} : \mathcal{F}_2(U) \to [-1,1]$, and let $C_W : \mathcal{F}_2(W) \to [-1,1]$, where the C_W are conditioned upon ϕ_W. Let $U, \kappa, W, f, g_1, g_2, i$ be chosen as in Test NAPC3-δ. Suppose that*

$$\mathop{\mathrm{E}}_{U,\kappa,W,f,g_1,g_2,i} [C_{U,\kappa}(f) C_W(g_1) C_W(g_2)]$$

differ more than 5δ from

$$\mathop{\mathrm{E}}_{U,\kappa,W,i} \left[\sum_{\substack{\beta \\ \text{all } s_x \text{ even}}} \hat{C}_{U,\kappa,\emptyset} \hat{C}^2_{W,\beta} \prod_{x \in \{-1,1\}^U} \frac{1}{2}((-1)^{s_x} + (1 - \epsilon_i)^{s_x}) \right] \quad (3)$$

where s_x is the number of $y \in \beta$ such that $y|_U = x$. Then there is a strategy for the two-prover one-round game with success probability $\delta^{O(\delta^{-1})}$.

Lemma 9. *Let $t \leq 3/2 - \epsilon$. Then there is a choice of the parameters u and δ such that if ϕ is not satisfiable, then for any set of size tP which covers all layer-edges, the fraction of non-layer edges which is covered is at $1 - \frac{\epsilon^2}{2}$.*

Proof. Choose δ such that $5\delta \leq \frac{\epsilon^2}{2}$. Choose u such that the soundness of the two-prover one-round game c_G^u is less than the success probability of Lemma 8.

Let p_1 be the fraction of (U, κ, f) for which $B_{U,\kappa}(f)$ is 0, and let p_2 be the fraction of (W, g) for which $B_W(g)$ is 0. Then $tP = (p_1 + 1)|Q_1| + (p_2 + 1)|Q_2| = ((p_1 + p_2)/2 + 1)P$. So $p_1 + p_2 \leq 1 - 2\epsilon$. Let $D_{U,\kappa}(f) = B^2_{U,\kappa}(f)$, and let $D_W(g) = B^2_W(g)$.

By construction, the tables $B_{U,\kappa}$ and B_W are folded over **true**, and the tables B_W and D_W are conditioned upon ϕ_W. We also have that $\hat{D}_{U,\kappa,\emptyset} = \mathrm{Pr}_f[B_{U,\kappa}(f) \neq 0]$ (and similarly for B_W), so $\mathrm{E}_{U,\kappa}[\hat{D}_{U,\kappa,\emptyset}] = 1 - p_1$, and $\mathrm{E}_W[\hat{D}_{W,\emptyset}] = 1 - p_2$. Now consider the terms in (2). Since $B_{U,\kappa}(f)^3 = B_{U,\kappa}(f)$ (and of course similarly for B_W), these are on the form $\mathrm{E}[B_{U,\kappa}(f)^{e_1} B_W(g_1)^{e_2} B_W(g_2)^{e_2}]$, where $e_1, e_2 \in \{1,2\}$. Since B_W is folded over **true**, $\hat{B}_{U,\kappa,\emptyset} = 0$ and thus by the choice of u and Lemma 8 with $C_{U,\kappa} = B_{U,\kappa}$, $D_W = B_W$, we have that each term where $e_1 = 1$ is at least -5δ. Similarly for the term where $e_1 = 2$ and $e_2 = 1$, we note that the fact that all s_x are even in (3) implies that β is even, and hence $\hat{B}_{W,\beta} = 0$, and thus this terms also is at least -5δ. Finally we have the term where $e_1 = e_2 = 2$, i.e,

$$\mathop{\mathrm{E}}_{U,\kappa,W,i,f,g_1,g_2} [B_{U,\kappa}(f)^2 B_W(g_1)^2 B_W(g_2)^2] \quad (4)$$

Since $\hat{D}_{U,\kappa,\emptyset} = \mathrm{Pr}_f[B_{U,\kappa}(f) \neq 0] \geq 0$, $\hat{D}^2_{W,\beta} \geq 0$, and

$$\prod_{x \in \{-1,1\}^U} \frac{1}{2}((-1)^{s_x} + (1 - \epsilon_i)^{s_x}) \geq 0$$

when all s_x are even, we have that (4) is at least $-5\delta + E_{U,\kappa,W}[\hat{D}_{U,\kappa,\emptyset}\hat{D}^2_{W,\emptyset}]$. Let $S_{U,\kappa} = E_W[\hat{D}_{W,\emptyset}|U]$. Then $E_{U,\kappa}[S_{U,\kappa}] = 1 - p_2$, and $E_W[\hat{D}^2_{W,\emptyset}|U] \geq S^2_{U,\kappa}$.
Furthermore we have that

$$\mathop{E}_{U,\kappa}[\hat{D}_{U,\kappa,\emptyset} + S_{U,\kappa}] = (1 - p_1) + (1 - p_2) \geq 1 + 2\epsilon.$$

Note that, when X, Y are random variables taking values in $[0, 1]$, we have that $(X - 1)(Y - 1) \geq 0$, and thus $XY \geq X + Y - 1$. Taking the expectation, $E[XY] \geq E[X] + E[Y] - 1$. Also, $E[XY^2] \geq E[X^2Y^2] \geq E[XY]^2$. Combining these two inequalities, we get that $E[XY^2] \geq (E[X + Y] - 1)^2$. Hence

$$\mathop{E}_{U,\kappa,W}[\hat{D}_{U,\kappa,\emptyset}\hat{D}^2_{W,\emptyset}] \geq \mathop{E}_{U,\kappa}[\hat{D}_{U,\kappa,\emptyset}S^2_{U,\kappa}]$$

$$\geq (\mathop{E}_{U,\kappa}[\hat{D}_{U,\kappa,\emptyset} + S_{U,\kappa}] - 1)^2$$

$$\geq 4\epsilon^2$$

To sum up, we have that the fraction of covered non-layer edges is at most

$$1 - \frac{4\epsilon^2 - 20\delta}{4} = 1 - \epsilon^2 + 5\delta \leq 1 - \frac{\epsilon^2}{2}.$$

Now we are ready to prove the inapproximability result for 3-uniform hypergraphs:

Proof (Proof of Theorem 2). Choose the parameters u and δ as in Lemma 9. Construct the graph H as described above. Then if ϕ is satisfiable, there is a vertex cover of size P of H. If ϕ is not satisfiable, for each set S of size at most $(3/2 - \epsilon)P$ there must be some uncovered edge, since either some layer-edge is uncovered, or some fraction of non layer-edges is uncovered. Hence if we could distinguish between these two cases, we could decide whether ϕ is satisfiable.

6   Subsequent Results

The $\Omega(k^{1-\epsilon})$ bound for has subsequently been improved to $k/3$ by Dinur [3]. It has been communicated to me by Dinur, Guruswami, Khot and Regev that they have since improved this to $k - 1$, thus also improving our $3/2 - \epsilon$ bound for the 3-regular case.

Acknowledgments. This work was inspired by Oded Goldreich's very nice observation that the FGLSS reduction can be used to give simple proofs for hardness of approximation of Minimum Vertex Cover in hypergraphs [8]. It was also inspired by Johan Håstad who suggested applying a similar reduction to two-prover games to get $k^{\Omega(1)}$ inapproximability for Minimum Vertex Cover in k-uniform hypergraphs. Thanks also to Johan Håstad for helpful discussions, and to Oded Goldreich for comments on an earlier version of this paper.

References

1. Sanjeev Arora, Carsten Lund, Rajeev Motwani, Madhu Sudan, and Márió Szegedy. Proof verification and the hardness of approximation problems. *Journal of the ACM*, 45(3):501–555, May 1998.

2. Mihir Bellare, Oded Goldreich, and Madhu Sudan. Free bits, PCPs and non-approximability—towards tight results. *SIAM Journal on Computing*, 27(3):804–915, June 1998.

3. Irit Dinur. Vertex-cover on k-uniform hypergraphs is NP-hard to approximate to within $\Omega(k)$. Manuscript, January 2002.

4. Irit Dinur and Shmuel Safra. The importance of being biased. In *Proceedings of the Thirty-fourth Annual ACM Symposium on Theory of Computing*, Montreal, Canada, 19–21 May 2002.

5. Lars Engebretsen. The non-approximability of non-Boolean predicates. Technical Report TR00-042, revision 1, ECCC, August 2001.

6. Lars Engebretsen and Venkatesan Guruswami. Is constraint satisfaction over two variables always easy? Manuscript, November 2001.

7. Uriel Feige. A threshold of $\ln n$ for approximating set cover. *Journal of the ACM*, 45(4):634–652, July 1998.

8. Oded Goldreich. Using the FGLSS-reduction to prove inapproximability results for minimum vertex cover in hypergraphs. Technical Report TR01-102, ECCC, December 2001.

9. Eran Halperin. Improved approximation algorithms for the vertex cover problem in graphs and hypergraphs. In *Proceedings of the Eleventh Annual ACM-SIAM Symposium on Discrete Algorithms*, pages 329–337, San Francisco, California, 9–11 January 2000.

10. Johan Håstad. Some optimal inapproximability results. *Journal of the ACM*, 48(4):798–859, July 2001.

11. Johan Håstad and Subhash Khot. Query efficient pcps with perfect completeness. In *42nd Annual Symposium on Foundations of Computer Science*, pages 610–619, Las Vegas, Nevada, 14-17 October 2001.

12. Johan Håstad and Avi Widgerson. Simple analysis of graph tests for linearity and PCP. In *Proc. of Conference on Computational Complexity*, pages 244–255, Chicago, June 2001.

13. Jonas Holmerin. Vertex cover on 4-regular hyper-graphs is hard to approximate within $2 - \epsilon$. In *Proceedings of the Thirty-fourth Annual ACM Symposium on Theory of Computing*, Montreal, Canada, 19–21 May 2002.

14. David S. Johnson. Approximation algorithms for combinatorial problems. *Journal of Computer and System Sciences*, 9:256–278, December 1974.

15. Christos H. Papadimitriou and Mihalis Yannakakis. Optimization, approximation, and complexity classes. *Journal of Computer and System Sciences*, 43(3):425–440, December 1991.

16. Ran Raz. A parallel repetition theorem. *SIAM Journal on Computing*, 27(3):763–803, June 1998.

17. Alex Samorodnitsky and Luca Trevisan. A PCP characterization of NP with optimal amortized query complexity. In *Proceedings of the Thirty-second Annual ACM Symposium on Theory of Computing*, pages 191–199, Portland, Oregon, 21–23 May 2000.

18. Luca Trevisan. Non-approximability results for optimization problems on bounded degree instances. In *Proceedings of the Thirty-third Annual ACM Symposium on Theory of Computing*, pages 453–461, Hersonissos, Crete, 6–8 July 2001.

Efficient Testing of Hypergraphs

(Extended Abstract)

Yoshiharu Kohayakawa[1*], Brendan Nagle[2**], and Vojtěch Rödl[3***]

[1] Instituto de Matemática e Estatística, Universidade de São Paulo, Rua do Matão 1010, 05508–090 São Paulo, Brazil. yoshi@ime.usp.br
[2] School of Mathematics, Georgia Institute of Technology, Atlanta, GA, 30332, USA, and Department of Mathematics, University of Nevada, Reno, Nevada, 89557, USA. nagle@math.gatech.edu
[3] Department of Mathematics and Computer Science, Emory University, Atlanta, GA, 30322, USA. rodl@mathcs.emory.edu

Abstract. We investigate a basic problem in combinatorial property testing, in the sense of Goldreich, Goldwasser, and Ron [9,10], in the context of 3-uniform hypergraphs, or 3-graphs for short. As customary, a 3-graph F is simply a collection of 3-element sets. Let $\mathrm{Forb}_{\mathrm{ind}}(n, F)$ be the family of all 3-graphs on n vertices that contain no copy of F as an induced subhypergraph. We show that the property "$H \in \mathrm{Forb}_{\mathrm{ind}}(n, F)$" is testable, for any 3-graph F. In fact, this is a consequence of a new, basic combinatorial lemma, which extends to 3-graphs a result for graphs due to Alon, Fischer, Krivelevich, and Szegedy [2,3].

Indeed, we prove that if more than ζn^3 ($\zeta > 0$) triples must be added or deleted from a 3-graph H on n vertices to destroy all induced copies of F, then H must contain $\geq cn^{|V(F)|}$ induced copies of F, as long as $n \geq n_0(\zeta, F)$. Our approach is inspired in [2,3], but the main ingredients are recent hypergraph regularity lemmas and counting lemmas for 3-graphs.

1 Introduction

We consider combinatorial property testing, in the sense of Goldreich, Goldwasser, and Ron [9,10] (see also Ron [16] for a recent survery). We address the problem of testing hypergraph properties; we in fact focus on testing induced subhypergraphs in 3-uniform hypergraphs. (For a recent result on hypergraph property testing, see Czumaj and Sohler [6], where k-colourability is proved to be testable.)

The main ingredients in our methods are hypergraph regularity lemmas and counting lemmas for 3-graphs, which are the first elements of a novel circle of results that, once complete, should allow one to tackle problems for hypergraphs that currently seem unapproachable with elementary techniques.

* Partially supported by MCT/CNPq through ProNEx Programme (Proc. CNPq 664107/1997–4), by CNPq (Proc. 300334/93–1, 910064/99–7, and 468516/2000–0).
** Partially supported by NSF Grant INT–0072064.
*** Partially supported by NSF Grants 0071261 and INT–0072064.

P. Widmayer et al. (Eds.): ICALP 2002, LNCS 2380, pp. 1017–1028, 2002.

We start with some basic definitions. A k-*uniform hypergraph* H, k-*graph* for short, is a family of k-element sets. When $k = 2$, we speak of *graphs*, and when $k = 3$, we have *triple systems*. A k-*graph property* $\mathcal{P}$ is an infinite class of k-graphs closed under isomorphism. A k-graph H *satisfies* property $\mathcal{P}$ if $H \in \mathcal{P}$. A k-graph H is said to be ζ-*far* from property $\mathcal{P}$ if every k-graph $\tilde{H} \in \mathcal{P}$ with $V(\tilde{H}) = V(H)$ differs from H in at least $\zeta|V(H)|^k$ k-tuples of vertices (i.e., the symmetric difference $H \bigtriangleup \tilde{H}$ has size at least $\zeta|V(H)|^k$).

A ζ-*test* for property $\mathcal{P}$ is a randomized algorithm which, given as input a k-graph H on $n = |\bigcup H|$ vertices, is allowed to make queries whether any given k-tuple of vertices belongs to H or not, and distinguishes, with high probability, between the case that H satisfies $\mathcal{P}$ and the case that H is ζ-far from $\mathcal{P}$. A property $\mathcal{P}$ is said to be *testable* if, for every $\zeta > 0$, there exists a function $f(\zeta)$ and an ζ-test for $\mathcal{P}$ which makes a total of $f(\zeta)$ queries for any input k-graph. Note that, in particular, the number of queries does not depend on the order of the input k-graph. A ζ-test is said to be a *one-sided* test if when H satisfies property $\mathcal{P}$, the test determines that this is the case with probability 1. A property $\mathcal{P}$ is said to be *strongly-testable* if for every $\zeta > 0$, there exists a one sided ζ-test for $\mathcal{P}$.

For a k-graph F, we let $\mathrm{Forb}_{\mathrm{ind}}(n, F)$ be the property of all k-graphs on n vertices not containing a copy of F as an induced subhypergraph. Recently, Alon, Fischer, Krivelevich and Szegedy [2,3] studied the graph properties $\mathrm{Forb}_{\mathrm{ind}}(n, F)$ for arbitrary graphs F. Among other results, they proved that all graph properties of the type $\mathrm{Forb}_{\mathrm{ind}}(n, F)$ are strongly-testable. A statement central to their main result was the following combinatorial theorem.

Theorem 1. *For every $\zeta > 0$ and every graph F, there exists $c > 0$ so that if a graph G on $n > n_0(\zeta, F)$ vertices is ζ-far from $\mathrm{Forb}_{\mathrm{ind}}(n, F)$, then G must contain at least $cn^{|V(F)|}$ copies of F as an induced subgraph.*

Observe that Theorem 1 implies that $\mathrm{Forb}_{\mathrm{ind}}(n, F)$ is a strongly-testable property; if $f = |V(F)|$, then the maximum number of queries required is $O(f^2/c)$: we simply sample α/c random vertices from G, where $\alpha > 0$ is some large enough constant, and check $O(f^2/c)$ suitable adjacencies. This is a strong ζ-test: suppose a given graph G on n vertices contains no copy of F as an induced subgraph. Then, our test will certainly find no copy of F in G as an induced subgraph. Consequently, the test correctly decides with probability 1 that $G \in \mathrm{Forb}_{\mathrm{ind}}(n, F)$. On the other hand, if G is ζ-far from $\mathrm{Forb}_{\mathrm{ind}}(n, F)$, then, by Theorem 1, the graph G contains cn^f copies of F as an induced subgraph. This means that our randomized algorithm (which makes $O(f^2/c)$ queries) is able to locate a copy of F with high probability.

The goal of this paper is to extend Theorem 1 to 3-graphs. Since the connection between testability and the combinatorial property illustrated in Theorem 1 remains unchanged from graphs to hypergraphs, we choose to present our work in a purely combinatorial language.

Theorem 2. *For every* $\zeta > 0$ *and 3-graph* F, *there exists* $c > 0$ *so that if a 3-graph* G *on* $n > n_0(\zeta, F)$ *vertices is* ζ-*far from* $\text{Forb}_{\text{ind}}(n, F)$, *then* G *contains at least* $cn^{|V(F)|}$ *copies of* F *as an induced subhypergraph.*

We conjecture that Theorems 1 and 2 are true for general k-graphs.

Conjecture 3. *For every* $\zeta > 0$ *and* k-*graph* F, *there exists* $c > 0$ *so that if a* k-*graph* G *on* $n > n_0(\zeta, F)$ *vertices is* ζ-*far from* $\text{Forb}_{\text{ind}}(n, F)$, *then* G *contains at least* $cn^{|V(F)|}$ *copies of* F *as an induced subhypergraph.*

In Sections 1 and 1 below, we discuss Conjecture 3 and some related problems. Our discussion below may be thought of as an explanation of why the generalization from graphs to triple systems (which we are able to achieve here) and then from triple systems to general k-graphs should be hard.

On Conjecture 3: The Special Case $F^{(k)} = K_{k+1}^{(k)}$.

The validity of Conjecture 3 in the special case $F^{(k)} = K_{k+1}^{(k)}$, the complete k-graph on $k+1$ vertices, has an interesting connection to the following well known and deep problem concerning arithmetic progressions.

Let $r_k(n)$ be the maximum cardinality of a set of integers $A \subset \{1, \ldots, n\}$ which contains no arithmetic progressions of length k. A conjecture of Erdős and Turán from 1936 stated that $r_k(n) = o(n)$. Roth [17] proved $r_3(n) = o(n)$ in 1953. Szemerédi [20] was able to introduce some genuinely new ideas 16 years later, to prove that $r_4(n) = o(n)$. Roth [18] later incorporated some of Szemerédi's ideas into his analytical approach to give an alternative proof for $r_4(n) = o(n)$. In 1975, Szemerédi [21] finally proved his celebrated result that $r_k(n) = o(n)$, confirming the Erdős–Turán conjecture in full. Furstenberg [8] and, more recently, Gowers [12] have given alternative proofs (see [11] for Gowers's proof of $r_4(n) = o(n)$). The sharpest result to date for $k = 3$ is due to Bourgain [5].

In [19], Ruzsa and Szemerédi solved an extremal combinatorial problem yielding alternative proof to Roth's theorem [17]. In [7], the following extremal problem was considered. Let $F_1^{(k+1)}$ be the $(k + 1)$-graph consisting of two edges intersecting in k points. Let $F_2^{(k+1)}$ be the $(k + 1)$-graph with $2k + 2$ vertices $\{a_1, \ldots, a_{k+1}, b_1, \ldots, b_{k+1}\}$ and all $(k + 1)$-tuples of the form $\{a_1, \ldots, a_{k+1}, b_i\} \setminus \{a_i\}$, $1 \leq i \leq k + 1$. Let $\text{ex}(n, \{F_1^{(k+1)}, F_2^{(k+1)}\})$ denote the maximum number of edges of any $(k + 1)$-graph $H^{(k+1)}$ not containing $F_1^{(k+1)}$ or $F_2^{(k+1)}$ as a (not necessarily induced) subhypergraph. In [7], a constructive argument was given showing that

$$\text{ex}(n, \{F_1^{(k+1)}, F_2^{(k+1)}\}) \geq r_{k+1}(n)n^{k-1}. \tag{1}$$

To verify Conjecture 3 for general k, even in the special case when $F^{(k)} = K_{k+1}^{(k)}$, will not be easy. Indeed, if Conjecture 3 is proved for general k and $F^{(k)} = K_{k+1}^{(k)}$, then one may quickly deduce Szemerédi's theorem, that is, that $r_{k+1}(n) = o(n)$. We state this assertion as a formal claim.

Claim 4. *Conjecture 3 for general k and $F^{(k)} = K_{k+1}^{(k)}$ implies $r_{k+1}(n) = o(n)$.*

The reader is referred to Frankl and Rödl [7] for the proof of Claim 4.

On the Non-induced Case.   Note that Conjecture 3 is formulated for classes of k-graphs not containing F as an *induced* subhypergraph (a feature of no importance when F is a clique). One may also consider the class $\text{Forb}(n, F)$ of all k-graphs on n vertices which do not contain a copy of F as a (not necessarily induced) subhypergraph. We state the following analogue to Conjecture 3.

Conjecture 5. *For every $\zeta > 0$ and k-graph F, there exists $c > 0$ so that if a k-graph G on $n > n_0(\zeta, F)$ vertices is ζ-far from $\text{Forb}(n, F)$, then G contains at least $cn^{|V(F)|}$ copies of F as a subhypergraph.*

For $k = 2$ and $k = 3$, Conjecture 5 is true. For $k = 2$, Conjecture 5 follows by a standard application of Szemerédi's regularity lemma. For $k = 3$, the result follows from results of [7] and [14]. Recently, Conjecture 5 was proved for $k = 4$ and $F = K_5^{(4)}$ (see [15]).

For general k and k-graph F, let us define a function $C(\zeta, F) = C_k(\zeta, F)$ as follows. The quantification of Conjecture 5 is of the form "$(\forall \zeta, F)(\exists c, n_0)$". For given F and ζ, let $C(\zeta, F)$ be the supremum of all c for which the implication in Conjecture 5 holds (for the given value of ζ and the given k-graph F) for some large enough n_0. As Conjecture 5 is true for $k = 2, 3$ and F arbitrary and $k = 4$ and $F = K_5^{(4)}$, the quantities $C_2(\zeta, F)$, $C_3(\zeta, F)$ and $C_4(\zeta, K_5^{(4)})$ are finite.

The question of how $C(\zeta, F)$ behaves for a fixed F as a function of ζ has been addressed recently by Alon [4], who proved the following theorem.

Theorem 6. *For a fixed graph F, the function $C(\zeta, F)$ is polynomial in $1/\zeta$ if and only if F is bipartite.*

Preliminary results suggest that Theorem 6 extends to k-graphs. Indeed, if F is a k-partite k-graph with partition classes of size $t_1, \ldots, t_k$, then, for a constant $c_1 = c_1(F) > 0$ depending only on F, we have $\log C(\zeta, F) \le c_1 t_1 \ldots t_k \log(k/\zeta)$. On the other hand, there exist non k-partite F (e.g., $F = K_{k+1}^{(k)}$), for which there is a constant $c_2 = c_2(F) > 0$ depending only on F, so that $\log C(\zeta, F) \ge c_2 (\log 1/\zeta)^2$.

2   Preliminary Results

Graph Concepts.   We begin with some basic notation. As is customary, if X is a set, we write $[X]^k$ for the set of k-element subsets of X. For a graph P and two disjoint sets $X, Y \subset V(P)$, we set $P[X, Y] = \{\{x, y\} \in P : x \in X, \, y \in Y\}$. We define the *density* $d_P(X, Y)$ of P with respect to X and $Y \ne \emptyset$ by $d_P(X, Y) = |P[X, Y]|/|X||Y|$.

For a graph P, we let $\mathcal{K}_3(P)$ be the set of vertex sets of triangles in P. Thus, $\{x, y, z\} \in \mathcal{K}_3(P)$ if and only if x, y, and z are mutually adjacent in P.

A graph P with a fixed k-partition $V_1 \cup \cdots \cup V_k$ is referred to as a *k-partite cylinder*. We write $P = \bigcup_{1 \le i < j \le k} P^{ij}$, where $P^{ij} = P[V_i, V_j]$, $1 \le i < j \le k$. For $B \in [k]^3$, we sometimes write $P(B)$ to denote the subgraph of P induced on the vertex set $\bigcup_{i \in B} V_i$. When $k = 3$, we call P a *triad*.

We proceed with the following definitions.

Definition 7 ((α, ε)-regularity). *For α and $\varepsilon > 0$ reals, we say that a bipartite graph P with vertex bipartition $X \cup Y$ is (α, ε)-regular if $\alpha(1-\varepsilon) < d_P(X_0, Y_0) < \alpha(1 + \varepsilon)$ for every pair of subsets $X_0 \subseteq X$ and $Y_0 \subseteq Y$ with $|X_0| > \varepsilon|X|$ and $|Y_0| > \varepsilon|Y|$.*

Definition 8 ((ℓ, ε, k)-cylinder). *For an integer ℓ and a real $\varepsilon > 0$, we call a k-partite cylinder $P = \bigcup_{1 \le i < j \le k} P^{ij}$ an (ℓ, ε, k)-cylinder if each bipartite graph P^{ij}, $1 \le i < j \le k$, is $(1/\ell, \varepsilon)$-regular.*

Hypergraph Concepts. We refer to any k-partite 3-uniform hypergraph $\mathcal{H}$ with a fixed k-partition $V_1 \cup \cdots \cup V_k$ as a *k-partite 3-cylinder*. For $B \in [k]^3$, we set $\mathcal{H}(B)$ to be the set of triples of $\mathcal{H}$ induced on the vertex set $\bigcup_{i \in B} V_i$.

Let a 3-uniform hypergraph $\mathcal{H}$ and a graph P be given so that $V(\mathcal{H}) = V(P)$. We say that P *underlies* $\mathcal{H}$ if $\mathcal{H} \subseteq \mathcal{K}_3(P)$. In the remainder of this paper, we only consider hypergraphs $\mathcal{H}$ together with graphs P that underlie them. We continue with the following technical definitions.

Definition 9 (Density of $\vec{Q}$; density of a triad). *Let $\mathcal{H}$ be a 3-partite 3-cylinder with underlying 3-partite cylinder $P = P^{12} \cup P^{23} \cup P^{13}$. Let $\vec{Q} = (Q(1), \ldots, Q(r))$ be an r-tuple of 3-partite cylinders $Q(s) = Q^{12}(s) \cup Q^{23}(s) \cup Q^{13}(s)$ satisfying that, for every $s \in \{1, 2, \ldots, r\}$ and for each $\{i, j\}$, $1 \le i < j \le 3$, we have $Q^{ij}(s) \subseteq P^{ij}$. Let $\mathcal{K}_3(\vec{Q}) = \bigcup_{s=1}^r \mathcal{K}_3(Q(s))$. We define the density $d_{\mathcal{H}}(\vec{Q})$ of $\vec{Q}$ as*

$$
d_{\mathcal{H}}(\vec{Q}) = \begin{cases} \dfrac{|\mathcal{H} \cap \mathcal{K}_3(\vec{Q})|}{|\mathcal{K}_3(\vec{Q})|} & \text{if } |\mathcal{K}_3(\vec{Q})| > 0, \\ 0 & \text{otherwise.} \end{cases}
$$

If $r = 1$, we have the notion of the density $d_{\mathcal{H}}(P)$ of a (single) triad P with respect to $\mathcal{H}$.

Definition 10 ((α, δ, r)-regularity). *Let $\mathcal{H}$ be a 3-partite 3-cylinder with underlying 3-partite cylinder $P = P^{12} \cup P^{23} \cup P^{13}$. Let a positive integer r and a real $\delta > 0$ be given. We say that the 3-cylinder $\mathcal{H}$ is (α, δ, r)-regular with respect to P if for any r-tuple of 3-partite cylinders $\vec{Q} = (Q(1), \ldots, Q(r))$ as above, if $|\mathcal{K}_3(\vec{Q})| = \left| \bigcup_{s=1}^r \mathcal{K}_3(Q(s)) \right| > \delta|\mathcal{K}_3(P)|$, then $|d_{\mathcal{H}}(\vec{Q}) - \alpha| < \delta$.*

We say $\mathcal{H}$ is (δ, r)-regular with respect to P if it is (α, δ, r)-regular for some α. If the regularity condition fails to be satisfied for every α, we say that $\mathcal{H}$ is (δ, r)-irregular with respect to P.

A Hypergraph Regularity Lemma.   In this section, we state the a Hypergraph Regularity Lemma, due to Frankl and Rödl [7]. This important theorem essentially states that every (large enough) hypergraph $\mathcal{H}$ may be decomposed into a constant number of "random-like" blocks. These so-called random-like blocks are 3-partite 3-cylinders which observe the regularity property in Definition 10.

Before stating the Hypergraph Regularity Lemma, we first state a number of supporting definitions.

Definition 11 (($\ell, t, \gamma, \varepsilon$)-**partition**).   *Let V be a set of cardinlaity $|V| = N$. An $(\ell, t, \gamma, \varepsilon)$-partition $\mathcal{P}$ of $[V]^2$ is an (auxiliary) partition $V = V_1 \cup \cdots \cup V_t$ of V, together with a system of edge-disjoint bipartite graphs $\{P_\alpha^{ij} : 1 \leq i < j \leq t, \, 0 \leq \alpha \leq \ell\}$, such that*

(i) $V = \bigcup_{1 \leq i \leq t} V_i$ *is a t-equitable partition, i.e., we have $\lfloor N/t \rfloor \leq |V_i| \leq \lceil N/t \rceil$ for all $1 \leq i \leq t$,*

(ii) $\bigcup_{\alpha=0}^{\ell} P_\alpha^{ij} = K(V_i, V_j)$ *for all i, j, $1 \leq i < j \leq t$, where $K(V_i, V_j)$ denotes the complete bipartite graph with vertex bipartition $V_i \cup V_j$,*

(iii)   *for all but $\gamma \binom{t}{2} \ell$ indices $1 \leq i < j \leq t$, $1 \leq \alpha \leq \ell$, the graph P_α^{ij} is (ℓ^{-1}, ε)-regular.*

If we do not have or do not care about (iii), and $P_\alpha^{ij} = \emptyset$ for all $1 \leq i < j \leq t$, we say that $\mathcal{P}$ above is an (ℓ, t)-partition of $[V]^2$. For an (ℓ, t)-partition $\mathcal{P} = \{P_\alpha^{ij} : 1 \leq i < j \leq t, \, 0 \leq \alpha \leq \ell\}$ of $[V]^2$, the set of triads generated by $\mathcal{P}$ is

$$\mathrm{Triad}(\mathcal{P}) = \left\{ P = P_\alpha^{ij} \cup P_\beta^{jk} \cup P_\gamma^{ik} : 1 \leq i < j < k \leq t, \, 0 \leq \alpha, \beta, \gamma \leq \ell \right\}.$$

Definition 12 ((δ, r)-**regular partition**).   *Let $\mathcal{H}$ be a 3-uniform hypergraph with vertex set V where $|V| = N$. We say that an $(\ell, t, \gamma, \varepsilon)$-partition $\mathcal{P}$ of $[V]^2$ is (δ, r)-regular for $\mathcal{H}$ if*

$$\sum \left\{ |\mathcal{K}_3(P)| : P \in \mathrm{Triad}(\mathcal{P}), \, \mathcal{H} \text{ is } (\delta, r)\text{-irregular with respect to } P \right\} < \delta N^3.$$

Theorem 13 (Hypergraph Regularity Lemma [7]).   *For every δ and γ with $0 < \gamma \leq 2\delta^4$, for all integers t_0 and ℓ_0 and for all integer-valued functions $r(t, \ell)$ and all functions $\varepsilon(\ell) > 0$, there exist T_0, L_0, and N_0 such that any 3-uniform hypergraph $\mathcal{H} \subseteq [N]^3$, $N \geq N_0$, admits a $(\delta, r(t, \ell))$-regular $(\ell, t, \gamma, \varepsilon(\ell))$-partition for some t and ℓ satisfying $t_0 \leq t \leq T_0$ and $\ell_0 \leq \ell \leq L_0$.*

The Counting Lemma.   In this section, we present the Counting Lemma, Lemma 15 below. We begin by describing the context in which this lemma applies. We note that this context is the same rendered by an appropriate application of the Hypergraph Regularity Lemma.

Setup 14.   *Fix integers f, ℓ, and r. Let δ and $\varepsilon > 0$ be given, together with an indexed family $\{\alpha_B : B \in [f]^3\}$ of positive reals. We consider the following conditions on a given hypergraph $\mathcal{H}$ and underlying graph P.*

(i) $\mathcal{H}$ is an f-partite 3-cylinder with f-partition $V_1 \cup \cdots \cup V_f$, where $|V_1| = \cdots = |V_f| = m$.

(ii) $P = \bigcup_{1 \leq i < j \leq f} P^{ij}$ is an underlying (ℓ, ε, f)-cylinder of $\mathcal{H}$.

(iii) For all $B \in [f]^3$, the 3-partite 3-cylinder $\mathcal{H}(B)$ is (α_B, δ, r)-regular with respect to the triad $P(B)$ (cf. Definition 10).

Finally, suppose $\mathcal{F}$ is a 3-uniform hypergraph with vertex set $V(\mathcal{F}) = [f]$. For $B \in [f]^3$, we let $\rho_B = \alpha_B$ if $B \in \mathcal{F}$, and we let $\rho_B = 1 - \alpha_B$ if $B \notin \mathcal{F}$. In the main lemma of this section, we are concerned with the number of induced, *transversal* copies of $\mathcal{F}$ in $\mathcal{H}$, by which we mean the number of functions $\iota \colon V(\mathcal{F}) = [f] \hookrightarrow V(\mathcal{H})$ with $\iota(i) \in V_i$ for all $1 \leq i \leq f$ that induces an isomorphism of $\mathcal{F}$ onto the image $\mathcal{H}[\mathrm{im}\,\iota] = \mathcal{H} \cap [\mathrm{im}\,\iota]^3$ of ι. We denote the set of such ι by $\mathcal{F}_{\mathrm{ind}}(\mathcal{H}; V_1, \ldots, V_f)$.

Lemma 15 (Counting Lemma [14]). *Let $f \geq 3$ be a fixed integer. For all constants α and $\beta > 0$, there exists $\delta > 0$ for which the following holds. For all integers ℓ, there exist an integer r and a real $\varepsilon > 0$ so that whenever an f-partite 3-cylinder $\mathcal{H}$ with an underlying cylinder P satisfy the conditions of Setup 14 with constants f, δ, ℓ, r and ε and an indexed family $\{\alpha_B \colon B \in [f]^3\}$ of reals, with $\alpha \leq \alpha_B \leq 1 - \alpha$ for all $B \in [f]^3$, we have*

$$\frac{\Pi_{B \in [f]^3} \rho_B}{\ell^{\binom{f}{2}}} m^f (1 - \beta) \leq |\mathcal{F}_{\mathrm{ind}}(\mathcal{H}; V_1, \ldots, V_f)| \leq \frac{\Pi_{B \in [f]^3} \rho_B}{\ell^{\binom{f}{2}}} m^f (1 + \beta)$$

for any triple system $\mathcal{F}$ on $[f]$.

The Counting Lemma is a companion statement to the Hypergraph Regularity Lemma. The Counting Lemma roughly states that with hypergraph $\mathcal{H}$ and graph P satisfying the hypothesis of Setup 14 (with appropriate parameters), the number of induced transversal copies of a fixed hypergraph $\mathcal{F}$ in $\mathcal{H}$ is essentially the same as would be expected in the corresponding random environment. Indeed, suppose each 3-partite 3-cylinder $\mathcal{H}(B)$, $B \in [f]^3$, of Condition (iii) in Setup 14 is replaced by the random hypergraph $\mathcal{H}_{\mathrm{rand}}(B)$ consisting of all triples of $\mathcal{K}_3(P(B))$ independently included with probability α_B. Let $\mathcal{H}_{\mathrm{rand}}$ be the union of $\mathcal{H}_{\mathrm{rand}}(B)$ over all $B \in [f]^3$. Then, with ρ_B $(B \in [f]^3)$ given above, we have

$$\mathbb{E}\left[|\mathcal{F}_{\mathrm{ind}}(\mathcal{H}_{\mathrm{rand}}; V_1, \ldots, V_f)|\right] \sim \frac{\Pi_{B \in [f]^3} \rho_B}{\ell^{\binom{f}{2}}} m^f.$$

3   The Perfect Reduction Lemma

It is well-known that in the regularity lemma of Szemerédi, one cannot avoid the existence of irregular pairs [1]. It proved to be vital for the proof of Theorem 1 to develop a version of the regularity lemma that, within large subsets of a Szemerédi partition, admits no irregular pairs. Here, we need a similar version of Theorem 13 that admits no irregular triads.

In this section, we consider the following setup.

Setup 16. *Let $\mathcal{H}$ be a 3-uniform hypergraph on n vertices. Let $V(\mathcal{H}) = V_1 \cup \cdots \cup V_t$ be a t-equitable partition, that is, for all $1 \leq i \leq t$, suppose we have $\lfloor n/t \rfloor \leq |V_i| \leq \lceil n/t \rceil$. Let $\mathcal{P} = \{ P_\alpha^{ij} : 1 \leq i < j \leq t, \, 1 \leq \alpha \leq \ell \}$ be an (ℓ, t)-partition of $[V(\mathcal{H})]^2$.*

Observe that $\alpha \geq 1$ in Setup 16, since in (ℓ, t)-partitions we have $P_0^{ij} = \emptyset$.

Definition 17 $((\delta_0, \tau)$-**reduction**)**.** *Let $\mathcal{H}$ and $\mathcal{P}$ as in Setup 16 be given. For δ_0, $\tau > 0$, we say the graph Q is a (δ_0, τ)-reduction of $\mathcal{P}$ if the following hold:*

1. *Q has a t-partition $V(Q) = \bigcup_{1 \leq i \leq t} W_i$, where $W_i \subseteq V_i$ for all $1 \leq i \leq t$, and $|W_1| = \cdots = |W_t| \geq \tau n$.*
2. *For all but $\delta_0 \binom{t}{3} \ell^3$ indices $1 \leq i < j < k \leq t$, $1 \leq \alpha, \beta, \gamma \leq \ell$, the triads $P = P_\alpha^{ij} \cup P_\beta^{jk} \cup P_\gamma^{ik}$ and $Q_P = Q \cap P$ satisfy $|d_{\mathcal{H}}(Q_P) - d_{\mathcal{H}}(P)| < \delta_0$.*

Definition 18 $((\delta, r, \varepsilon, \ell')$-**perfect reduction**)**.** *Let $\mathcal{H}$ and $\mathcal{P}$ as in Setup 16 be given. Let a graph Q be a (δ_0, τ)-reduction of $\mathcal{P}$. For reals δ and $\varepsilon > 0$ and for integers r and ℓ', we say that Q is a $(\delta, r, \varepsilon, \ell')$-perfect reduction of $\mathcal{P}$ if Q satisfies the following conditions:*

1. *For each $1 \leq i < j \leq t$, $1 \leq \alpha \leq \ell$, the graph $Q \cap P_\alpha^{ij}$ is $((\ell'\ell)^{-1}, \varepsilon)$-regular.*
2. *For each $1 \leq i < j < k \leq t$, $1 \leq \alpha, \beta, \gamma \leq \ell$, $\mathcal{H}$ is (δ, r)-regular with respect to the triad $Q \cap \left(P_\alpha^{ij} \cup P_\beta^{jk} \cup P_\gamma^{ik} \right)$.*

We are now able to state the main lemma in this section.

Lemma 19 (Perfect Reduction Lemma). *For all integers t_0, for all $\delta_0 > 0$, for all functions $\delta(\ell)$, $\varepsilon(\ell)$, and $r(\ell)$, there exist $\tau > 0$ and integers T_0, L_0, L_0' and N_0 for which the following holds: for any $\mathcal{H}$ on $n \geq N_0$ vertices, there exist integers $t_0 \leq t \leq T_0$, $1 \leq \ell \leq L_0$, $1 \leq \ell' \leq L_0'$, such that*

(a) *there exists an (ℓ, t)-partition $\mathcal{P} = \{ P_\alpha^{ij} : 1 \leq i < j \leq t, \, 1 \leq \alpha \leq \ell \}$, as in Setup 16,*
(b) *there exists a (δ_0, τ)-reduction Q of $\mathcal{P}$ that is a $(\delta(\ell), r(\ell'\ell), \varepsilon(\ell'\ell), \ell')$-perfect reduction of $\mathcal{P}$ (see Definitions 17 and 18).*

4   Sketch of the Proof of Theorem 2

The proof of Theorem 2 follows the same basic strategy as the proof of Theorem 1. For a hypergraph $\mathcal{H}$ as in the statement of the theorem, we first apply Lemma 19 to obtain a partition $\mathcal{P}$ and perfect reduction Q of $\mathcal{P}$. Unfortunately, this application does not allow us to control triples of $\mathcal{H}$ which are not transversal with respect to the vertex partition of Q. To remedy this situation, we apply Theorem 13 to further subdivide the partition Q. Finally, using the hypothesis that $\mathcal{H}$ is ζ-far from $\mathrm{Forb}_{\mathrm{ind}}(n, \mathcal{F})$, we find within the subdivision of Q an environment satisfying Setup 14. Consequently, we may apply Lemma 15 to this environment to count induced transversal copies of $\mathcal{F}$ in $\mathcal{H}$.

While the proof of Theorem 2 is not difficult given the machinery developed so far, it is rather technical. We now provide a sketch of the argument, omitting technical details. For the complete proof, the reader is encouraged to see [13].

Theorem 2 asserts that for any triple system $\mathcal{F}$ and any $\zeta > 0$, there is a constant $c > 0$ for which we have $|\mathcal{F}_{\mathrm{ind}}(\mathcal{H})| \geq cn^{|V(\mathcal{F})|}$, whenever $\mathcal{H}$ is ζ-far from $\mathrm{Forb}_{\mathrm{ind}}(n, \mathcal{F})$. Thus, suppose we are given $\mathcal{F}$ and $\zeta > 0$. We let $f = |V(\mathcal{F})|$.

The proof scheme described above leads naturally to the definition of several constants and functions for our lemmas to apply one after the other in a compatible manner. Because of space considerations, we do not give these definitions, but only mention that one obtains the constants α_0, δ_0, δ_{CL}, τ, τ', t_0, T_0, L_0, L_0', K_1, and K_2 and functions $\varepsilon(\ell)$ and $r(\ell)$. Finally, we define the constant $c = c(f, \zeta) > 0$ by putting

$$c = \frac{\alpha_0^{\binom{f}{3}} (\tau')^f}{2(L_0 L_0' K_2)^{\binom{f}{2}}}.$$

Claim 20. *The constant $c = c(f, \zeta) > 0$ defined above will do in Theorem 2.*

Proof of Claim 20.   We split the proof into Steps I to IV. Suppose we are given a hypergraph $\mathcal{H}$ on $V = \bigcup \mathcal{H}$ as in the statement of the theorem.

Step I.   We first apply Lemma 19, the Perfect Reduction Lemma, to obtain

(a) an (ℓ, t)-partition $\mathcal{P}$ of $[V]^2$ on $V = \bigcup_{1 \leq i \leq t} V_i$,
(b) a $(\delta_{\mathrm{CL}}, r(\ell\ell'), \varepsilon(\ell\ell'), \ell')$-perfect (δ_0, τ)-reduction Q of $\mathcal{P}$,

where $t_0 \leq t \leq T_0$, $1 \leq \ell \leq L_0$, and $1 \leq \ell' \leq L_0'$.

Step II.   Applying a variant of Theorem 13 (see Nagle and Rödl [14]) and several further combinatorial arguments (Turán type arguments and Ramsey type arguments), we obtain the setup we shall now describe. The objects that constitute our setup are described in (S1)–(S3) below.

(S1) *For every $1 \leq i \leq t$, we have a partition $W_i' = \bigcup_{1 \leq i' \leq f} A_{ii'}$ of a subset W_i' of W_i (see Definition 17). Moreover, we suppose that the $V_i \supset W_i'$ are totally ordered in such a way that $A_{i1} < \cdots < A_{if}$ for all i.*

(S2) *For every $1 \leq i < j \leq t$, $1 \leq i', j' \leq f$, and $1 \leq \alpha \leq \ell$, we have a subgraph $*Q_\alpha^{ij}(i', j')$ of the bipartite graph $Q_\alpha^{ij}(i', j') = Q_\alpha^{ij} \cap K(A_{ii'}, A_{jj'}) = Q \cap P_\alpha^{ij} \cap K(A_{ii'}, A_{jj'})$. Moreover, for every $1 \leq i \leq t$ and $1 \leq i', i'' \leq f$ with $i' \neq i''$, we have a subgraph $Q(i; i', i'')$ of $K(A_{ii'}, A_{ii''})$.*

(S3) *We have functions χ_1^*, χ_2^*, and χ_3^* with domains and codomains as follows:*

$$\chi_1^* \colon [t] \to \{\pm 1\}, \qquad \chi_2^* \colon [t] \times [t] \setminus \Delta \to \{\pm 1\}^{I_2}, \tag{2}$$

where $\Delta = \{(i, i) \colon 1 \leq i \leq t\}$ and $I_2 = [\ell] \times [\ell]$, and

$$\chi_3^* \colon [t]^3 \to \{0, \pm 1\}^{I_3}, \tag{3}$$

where $I_3 = [\ell] \times [\ell] \times [\ell]$.

The various objects in (S1)–(S3) satisfy the properties (P4)–(P7) given below.

(P4) We have $|A_{ii'}| = m$ for all $1 \leq i \leq t$, $1 \leq i' \leq f$, where $m \geq \tau'n$.

(P5) All ${}^*Q^{ij}_\alpha(i',j')$ $(1 \leq i < j \leq t,\ 1 \leq i',j' \leq f,\ 1 \leq \alpha \leq \ell)$ and all $Q(i;i',i'')$ $(1 \leq i \leq t,\ 1 \leq i',i'' \leq f,\ i' \neq i'')$ are $(1/\ell\ell'k_2, \varepsilon(\ell\ell'k_2))$-regular for some $k_2 \leq K_2 = K_2(f,\zeta)$.

(P6) We have the following regularity conditions for the triads defined by the bipartite graphs in (S2).

(i)   For any $1 \leq i \leq t$ and $1 \leq i',i'',i''' \leq f$, set $Q^i(i',i'',i''') = Q(i;i',i'') \cup Q(i;i'',i''') \cup Q(i;i',i''')$. Then all the triads $Q^i(i',i'',i''')$ are $(\delta_{\mathrm{CL}}, r_{\mathrm{CL}}(\ell\ell'k_2))$-regular.

(ii) For any $1 \leq i < j \leq t$, $1 \leq i',i'',j' \leq f$ with $i' \neq i''$, and any $1 \leq \alpha,\beta \leq \ell$, set $Q^{ij}_{\alpha\beta}(i',i'',j') = {}^*Q^{ij}_\alpha(i',j') \cup {}^*Q^{ij}_\beta(i'',j') \cup Q(i;i',i'')$. Then all the triads $Q^{ij}_{\alpha\beta}(i',i'',j')$ are $(\delta_{\mathrm{CL}}, r_{\mathrm{CL}}(\ell\ell'k_2))$-regular.

(iii) For any $1 \leq i < j < k \leq t$, $1 \leq i',j',k' \leq f$, and any $1 \leq \alpha,\beta,\gamma \leq \ell$, set $Q^{ijk}_{\alpha\beta\gamma}(i',j',k') = {}^*Q^{ij}_\alpha(i',j') \cup {}^*Q^{jk}_\beta(j',k') \cup {}^*Q^{ik}_\gamma(i',k')$. Then all the triads $Q^{ijk}_{\alpha\beta\gamma}(i',j',k')$ are $(\delta_{\mathrm{CL}}, r_{\mathrm{CL}}(\ell\ell'k_2))$-regular.

(P7) The triads defined in (P6) satisfy the 'density-coherence' properties given below.

(i)   For any $1 \leq i \leq t$, the following holds.
   (a) If $\chi_1^*(i) = -1$, then $d_\mathcal{H}(Q^i(i',i'',i''')) < 1/2$ for all $1 \leq i' < i'' < i''' \leq f$.
   (b) If $\chi_1^*(i) = 1$, then $d_\mathcal{H}(Q^i(i',i'',i''')) \geq 1/2$ for all $1 \leq i' < i'' < i''' \leq f$.

(ii) For any $1 \leq i,j \leq t$ with $i \neq j$ and any $1 \leq \alpha,\beta \leq \ell$, the following holds.
   (a) If $(\chi_2^*(i,j))(\alpha,\beta) = -1$, then $d_\mathcal{H}(Q^{ij}_{\alpha\beta}(i',i'',j')) < 1/2$ for all $1 \leq i' < i'' \leq f$ and for all $1 \leq j' \leq f$.
   (b) If $(\chi_2^*(i,j))(\alpha,\beta) = 1$, then $d_\mathcal{H}(Q^{ij}_{\alpha\beta}(i',i'',j')) \geq 1/2$ for all $1 \leq i' < i'' \leq f$ and for all $1 \leq j' \leq f$.

(iii) For any $1 \leq i < j < k \leq t$ and any $1 \leq \alpha,\beta,\gamma \leq \ell$, the following holds.
   (a) If $(\chi_3^*(\{i,j,k\}))(\alpha,\beta,\gamma) = -1$, then $d_\mathcal{H}(Q^{ijk}_{\alpha\beta\gamma}(i',j',k')) < \alpha_0$ for all $1 \leq i',j',k' \leq f$.
   (b) If $(\chi_3^*(\{i,j,k\}))(\alpha,\beta,\gamma) = 0$, then $\alpha_0 \leq d_\mathcal{H}(Q^{ijk}_{\alpha\beta\gamma}(i',j',k')) \leq 1 - \alpha_0$ for all $1 \leq i',j',k' \leq f$.
   (c) If $(\chi_3^*(\{i,j,k\}))(\alpha,\beta,\gamma) = 1$, then $d_\mathcal{H}(Q^{ijk}_{\alpha\beta\gamma}(i',j',k')) > 1 - \alpha_0$ for all $1 \leq i',j',k' \leq f$.

Step III.   Based on the functions χ_1^*, χ_2^*, and χ_3^* from (S3), we define a 'perturbation' of $\mathcal{H}$. Using the main hypothesis on $\mathcal{H}$, namely, that any small perturbation of $\mathcal{H}$ does not destroy all the induced copies of $\mathcal{F}$ present in $\mathcal{H}$, we deduce that $\mathcal{H}'$ contains at least one induced copy of $\mathcal{F}$.

Let us define the hypergraph $\mathcal{H}'$. We shall have $\mathcal{H}' \subset [V]^3$, where $V = \bigcup_{1 \leq i \leq t} V_i$. Let

$$\mathcal{H}'_{\mathrm{tr}} = \{E \in \mathcal{H}' : |E \cap V_i| \leq 1\ (1 \leq i \leq t)\} \tag{4}$$

be the subhypergraph of $\mathcal{H}'$ formed by the 'transversal' triples in $\mathcal{H}'$. Recall that we have the (ℓ,t)-partition $\mathcal{P}$ of $[V]^2$. We thus have the family of triads $\mathrm{Triad}(\mathcal{P})$

given by $\mathcal{P}$. Now, the family of triple systems $\mathcal{K}_3(P)$ $(P \in \mathrm{Triad}(\mathcal{P}))$ partitions the family $\{E \in [V]^3 \colon |E \cap V_i| \leq 1 \ (1 \leq i \leq t)\}$ of the transversal triples in $[V]^3$. Therefore, to define $\mathcal{H}'_{\mathrm{tr}}$, we may define $\mathcal{H}'_{\mathrm{tr}} \cap \mathcal{K}_3(P)$ independently for all $P \in \mathrm{Triad}(\mathcal{P})$. For each $P \in \mathrm{Triad}(\mathcal{P})$, if $P = P^{ijk}_{\alpha\beta\gamma} = P^{ij}_{\alpha} \cup P^{jk}_{\beta} \cup P^{ik}_{\gamma}$, we let

$$
\mathcal{H}'_{\mathrm{tr}} \cap \mathcal{K}(P) = \begin{cases} \emptyset & \text{if } (\chi^*_3(\{i,j,k\}))(\alpha,\beta,\gamma) = -1 \\ \mathcal{H} \cap \mathcal{K}_3(P) & \text{if } (\chi^*_3(\{i,j,k\}))(\alpha,\beta,\gamma) = 0 \\ \mathcal{K}_3(P) & \text{if } (\chi^*_3(\{i,j,k\}))(\alpha,\beta,\gamma) = 1. \end{cases} \tag{5}
$$

We now proceed to define $\mathcal{H}' \setminus \mathcal{H}'_{\mathrm{tr}}$. We have the following natural partition of the non-transversal triples in $[V]^3$:

$$
\{E \in [V]^3 \colon |E \cap V_i| \geq 2 \text{ some } i\} = \bigcup_{1 \leq i \leq t} [V_i]^3 \cup \bigcup_{i,j,\alpha,\beta} \mathcal{K}_3(i,j;\alpha,\beta),
$$

where the last union is over all $1 \leq i, j \leq t$ with $i \neq j$ and $1 \leq \alpha, \beta \leq \ell$, and $\mathcal{K}_3(i,j;\alpha,\beta)$ is the set of triples $\{u,v,w\} \in [V_i \cup V_j]^3$ with $u,v \in V_i$, $u < v$, $w \in V_j$, $\{u,w\} \in P^{ij}_{\alpha}$, and $\{v,w\} \in P^{ij}_{\beta}$. To define $\mathcal{H}' \setminus \mathcal{H}'_{\mathrm{tr}}$, we may therefore define $(\mathcal{H}' \setminus \mathcal{H}'_{\mathrm{tr}}) \cap [V_i]^3$ $(1 \leq i \leq t)$ and $(\mathcal{H}' \setminus \mathcal{H}'_{\mathrm{tr}}) \cap \mathcal{K}_3(i,j;\alpha,\beta)$ $(1 \leq i,j \leq t,$ $i \neq j, 1 \leq \alpha, \beta \leq \ell)$ independently. Fix $1 \leq i \leq t$. We let $(\mathcal{H}' \setminus \mathcal{H}'_{\mathrm{tr}}) \cap [V_i]^3 = \emptyset$ if $\chi^*(i) = -1$ and we let $(\mathcal{H}' \setminus \mathcal{H}'_{\mathrm{tr}}) \cap [V_i]^3 = [V_i]^3$ if $\chi^*(i) = 1$. Now fix $1 \leq i, j \leq t$ with $i \neq j$ and $1 \leq \alpha, \beta \leq \ell$. We let $(\mathcal{H}' \setminus \mathcal{H}'_{\mathrm{tr}}) \cap \mathcal{K}_3(i,j;\alpha,\beta) = \emptyset$ if $(\chi^*_2(i,j))(\alpha,\beta) = -1$ and we let $(\mathcal{H}' \setminus \mathcal{H}'_{\mathrm{tr}}) \cap \mathcal{K}_3(i,j;\alpha,\beta) = \mathcal{K}_3(i,j;\alpha,\beta)$ if $(\chi^*_2(i,j))(\alpha,\beta) = 1$.

A crucial property about $\mathcal{H}'$ defined above is given in Claim 21 below.

Claim 21. *The hypergraphs $\mathcal{H}'$ and $\mathcal{H}$ are ζ-close. Therefore, $\mathcal{H}'$ contains an induced copy of $\mathcal{F}$.*

Step IV. We now fix an induced embedding $\iota \colon \bigcup \mathcal{F} \hookrightarrow \bigcup \mathcal{H}$ of $\mathcal{F}$ into $\mathcal{H}$. Using this embedding, one may obtain a k-partite 3-cylinder $\mathcal{H}^* \subset \mathcal{H}$ with underlying 2-cylinder P to which one may apply Lemma 15, the Counting Lemma.

To simplify the notation, we may and shall assume that the image $\mathrm{im}\,\iota$ of ι meets only $V_1, \ldots, V_g$. Let $h_i = |\mathrm{im}\,\iota \cap V_i|$ $(1 \leq i \leq g)$. One may then check that $\mathcal{H}^*$ and P may be obtained as described in Claim 22 below.

Claim 22. *There is a choice of indices $1 \leq \alpha(i,i',j,j') \leq \ell$ $(1 \leq i < j \leq g, 1 \leq i', j' \leq f)$ for which the following holds. Consider the vertex sets $A_{ii'}$ $(1 \leq i \leq g, 1 \leq i' \leq h_i)$, and consider the k-partite cylinder P on the partition $\bigcup_{i,i'} A_{ii'}$, where the union is over all $1 \leq i \leq g$ and $1 \leq i' \leq h_i$, given by the following bipartite graphs: (i) for each $1 \leq i \leq g$ and each $1 \leq i' < i'' \leq h_i$, take $Q(i;i',i'')$ and (ii) for each $1 \leq i < j \leq g$ and for each $1 \leq i' \leq h_i, 1 \leq j' \leq h_j$, take ${}^*Q^{ij}_{\alpha(i,i',j,j')}(i',j')$. Let $\mathcal{H}^* \subset \mathcal{H}$ be the subhypergraph of $\mathcal{H}$ formed by the triples in $\mathcal{H} \cap \mathcal{K}_3(P)$. Then the hypotheses in Lemma 15 apply, and, consequently, $|\mathcal{F}_{\mathrm{ind}}(\mathcal{H})| \geq cn^k$.*

References

1. N. Alon, R. A. Duke, H. Lefmann, V. Rödl, and R. Yuster, *The algorithmic aspects of the regularity lemma*, J. Algorithms **16** (1994), no. 1, 80–109.
2. N. Alon, E. Fischer, M. Krivelevich, and M. Szegedy, *Efficient testing of large graphs (extended abstract)*, 40th Annual Symposium on Foundations of Computer Science (New York City, NY), IEEE Comput. Soc. Press, 1999, pp. 656–666.
3. _____, *Efficient testing of large graphs*, Combinatorica **20** (2000), no. 4, 451–476.
4. Noga Alon, *Testing subgraphs in large graphs*, Proceedings of the 42nd IEEE Annual Symposium on Foundations of Computer Science (FOCS 2001), 2001, pp. 434–439.
5. J. Bourgain, *On triples in arithmetic progression*, Geom. Funct. Anal. **9** (1999), no. 5, 968–984.
6. A. Czumaj and C. Sohler, *Testing hypergraph coloring*, Proc. of ICALP, 2001, pp. 493–505.
7. Peter Frankl and Vojtěch Rödl, *Extremal problems on set systems*, Random Structures and Algorithms **20** (2002), no. 2, 131–164.
8. Harry Furstenberg, *Ergodic behavior of diagonal measures and a theorem of Szemerédi on arithmetic progressions*, J. Analyse Math. **31** (1977), 204–256.
9. Oded Goldreich, Shafi Goldwasser, and Dana Ron, *Property testing and its connection to learning and approximation*, 37th Annual Symposium on Foundations of Computer Science (Burlington, VT, 1996), IEEE Comput. Soc. Press, Los Alamitos, CA, 1996, pp. 339–348.
10. _____, *Property testing and its connection to learning and approximation*, Journal of the Association for Computing Machinery **45** (1998), no. 4, 653–750.
11. W. T. Gowers, *A new proof of Szemerédi's theorem for arithmetic progressions of length four*, Geom. Funct. Anal. **8** (1998), no. 3, 529–551.
12. _____, *A new proof of Szemerédi's theorem*, Geom. Funct. Anal. **11** (2001), no. 3, 465–588.
13. Y. Kohayakawa, B. Nagle, and V. Rödl, *Testing hypergraphs*, In preparation, 2002.
14. Brendan Nagle and Vojtěch Rödl, *Regularity properties for triple systems*, Random Structures & Algorithms (2002), to appear.
15. Vojtěch Rödl and Jozef Skokan, *Uniformity of set systems*, manuscript, 2001.
16. Dana Ron, *Property testing*, Handbook of randomized algorithms (P. M. Pardalos, S. Rajasekaran, J. Reif, and J. D. P. Rolim, eds.), Kluwer Academic Publishers, 2001, to appear.
17. K. F. Roth, *On certain sets of integers*, Journal of the London Mathematical Society **28** (1953), 104–109.
18. _____, *Irregularities of sequences relative to arithmetic progressions. IV*, Period. Math. Hungar. **2** (1972), 301–326, Collection of articles dedicated to the memory of Alfréd Rényi, I.
19. I. Z. Ruzsa and E. Szemerédi, *Triple systems with no six points carrying three triangles*, Combinatorics (Proceedings of the Fifth Hungarian Colloquium, Keszthely, 1976), Vol. II (Amsterdam–New York), Colloq. Math. Soc. János Bolyai, no. 18, North-Holland, 1978, pp. 939–945.
20. E. Szemerédi, *On sets of integers containing no four elements in arithmetic progression*, Acta Mathematica Academiae Scientiarum Hungaricae **20** (1969), 89–104.
21. _____, *On sets of integers containing no k elements in arithmetic progression*, Acta Arithmetica **27** (1975), 199–245, collection of articles in memory of Juriĭ Vladimirovič Linnik.

Optimal Net Surface Problems with Applications[*]

Xiaodong Wu[**] and Danny Z. Chen

Department of Computer Science and Engineering
University of Notre Dame
Notre Dame, IN 46556, USA
FAX: 219-631-9260
{xwu,chen}@cse.nd.edu

Abstract. In this paper, we study an interesting geometric graph called *multi-column graph* in the d-D space ($d \geq 3$), and formulate two combinatorial optimization problems called the *optimal net surface problems* on such graphs. Our formulations capture a number of important problems such as surface reconstruction with a given topology, medical image segmentation, and metric labeling. We prove that the optimal net surface problems on general d-D multi-column graphs ($d \geq 3$) are NP-hard. For two useful special cases of these d-D ($d \geq 3$) optimal net surface problems (on the so-called *proper ordered multi-column graphs*) that often arise in applications, we present polynomial time algorithms. We further apply our algorithms to some surface reconstruction problems in 3-D and 4-D, and some medical image segmentation problems in 3-D and 4-D, obtaining polynomial time solutions for these problems. The previously best known algorithms for some of these applied problems, even for relatively simple cases, take at least exponential time. Our approaches for these applied problems can be extended to higher dimensions.

Keywords: Geometric Graphs, Algorithms, NP-hardness, Surface Reconstructions, 3-D Image Segmentations

1 Introduction

We study an interesting geometric graph $G = (V, E)$ in the d-D space ($d \geq 3$), defined as follows. Given any undirected graph $B = (V_B, E_B)$ embedded in $(d - 1)$-D (called the *base graph*) and an integer $K > 0$, $G = (V, E)$ is an undirected graph in d-D *generated* by B and K. For each vertex $i = (x_0, x_1, \ldots, x_{d-2}) \in V_B$, there is a set V_i of K vertices in G corresponding to i; $V_i = \{(x_0, x_1, \ldots, x_{d-2}, k) : k = 0, 1, \ldots, K - 1\}$, called the *i-column* of G. We

---

[*] This research was supported in part by the National Science Foundation under Grant CCR-9988468 and the 21st Century Research and Technology Fund from the State of Indiana.

[**] The work of this author was supported in part by a fellowship from the Center for Applied Mathematics, University of Notre Dame, Notre Dame, Indiana, USA.

denote the vertex $(x_0, x_1, \ldots, x_{d-2}, k)$ of V_i by i_k. If an edge $(i, j) \in E_B$, we say that the i-column and j-column are *adjacent* to each other in G. The edges in G can only connect pairs of vertices in adjacent columns. We call G thus defined a d-D *multi-column graph* generated by its $(d-1)$-D base graph B with a *height* K. A *net surface* in G (briefly called a *net*) is a subgraph of G defined by a function $\mathcal{N}: V_B \to \{0, 1, \ldots, K-1\}$, such that for each edge $(i, j) \in E_B$, $(i_{\mathcal{N}(i)}, j_{\mathcal{N}(j)})$ is an edge in E. For simplicity, we denote a net by its function $\mathcal{N}$. Intuitively, a net $\mathcal{N}$ in G is a special mapping of the $(d-1)$-D base graph B to the d-D space, such that $\mathcal{N}$ "intersects" each i-column at exactly one vertex and $\mathcal{N}$ preserves the topology of B. Let $V(H)$ and $E(H)$ denote the vertices and edges of a graph H. We consider two optimal net surface problems.

Optimal V-weight net surface problem: Given a d-D multi-column graph $G = (V, E)$, each vertex $v \in V$ having a real-valued weight $w(v)$, find a net $\mathcal{N}$ in G such that the weight $\alpha(\mathcal{N})$ of $\mathcal{N}$, with $\alpha(\mathcal{N}) = \sum_{v \in V(\mathcal{N})} w(v)$, is minimized.

Optimal VE-weight net surface problem: Given a d-D multi-column graph $G = (V, E)$, each vertex $v \in V$ having a real-valued weight $w(v)$ and each edge $e \in E$ having a real-valued cost $c(e)$, find a net $\mathcal{N}$ in G such that the cost $\beta(\mathcal{N})$ of $\mathcal{N}$, with $\beta(\mathcal{N}) = \sum_{v \in V(\mathcal{N})} w(v) + \sum_{e \in E(\mathcal{N})} c(e)$, is minimized.

These optimal net surface problems find applications in several areas such as medical image analysis, computational geometry, computer vision, and data mining [34]. For example, we model a 3-D/4-D image segmentation problem as computing an optimal V-weight net surface, and a surface reconstruction problem in $\Re^d$ with a given underlying topology as an optimal VE-weight net problem, motivated by the deformable model in image analysis [25,32]. See Section 4 for more details of these problems.

We are particularly interested in d-D multi-column graphs ($d \geq 3$) with a special property called *proper ordering*. Let $V_B(i)$ denote the set of vertices adjacent to a vertex i in the base graph B. A multi-column graph G is said to be *proper ordered* if the following two conditions hold on every i-column V_i of G. (1) For each vertex $i_k \in V_i$, i_k is connected to a non-empty sequence of consecutive vertices in every adjacent j-column V_j of V_i (i.e., $j \in V_B(i)$), say $j_{k'}, j_{k'+1}, \ldots, j_{k'+s}$ ($s \geq 0$); we call $(j_{k'}, j_{k'+1}, \ldots, j_{k'+s})$, in this order, the *edge interval* of i_k on V_j, denoted by $I(i_k, j)$. (2) For any two consecutive vertices i_k and i_{k+1} in V_i, $First(I(i_k, j)) \leq First(I(i_{k+1}, j))$ and $Last(I(i_k, j)) \leq Last(I(i_{k+1}, j))$ for each $j \in V_B(i)$, where $First(I)$ (resp., $Last(I)$) denotes the d-th coordinate of the first (resp., last) vertex in an edge interval I. We call such a graph G a *proper ordered multi-column graph* (briefly, a proper ordered graph). Many multi-column graphs for applied problems are proper ordered (e.g., medical image segmentation and surface reconstruction in 3-D and 4-D, as shown in Section 4).

We are also interested in a useful special case of the optimal VE-weight net surface problem, called the **optimal VCE-weight net surface problem.** The optimal VCE-weight net surface problem is defined on a d-D proper ordered graph $G = (V, E)$ such that the cost of each edge $(i_k, j_{k'}) \in E$ is $f_{ij}(|k - k'|)$, herein $f_{ij}(\cdot)$ is a convex non-decreasing function associated with each edge $(i, j) \in E_B$.

These d-D optimal net surface problems ($d \geq 3$), even on proper ordered graphs, might appear to be computationally intractable at first sight. In fact, even for a very restricted case of the optimal V-weight net problem on a 3-D proper ordered graph arising in 3-D image segmentation, the previously best known exact algorithms [33,15] take at least exponential time. (The 2-D case of this image segmentation problem can be solved in polynomial time by computing shortest paths [13,33].) Interestingly, we are able to develop polynomial time algorithms for the optimal V-weight net problem and optimal VCE-weight net problem on proper ordered graphs. Let $n_B = |V_B|$, $m_B = |E_B|$, $n = |V|$, $m = |E|$, and K be the height of G. Denote by $T(n', m')$ the time for finding a minimum s-t cut in an edge-weighted directed graph with $O(n')$ vertices and $O(m')$ edges [3]. Our main results are summarized as follows.

- Proving that in d-D ($d \geq 3$), the optimal V-weight net problem on a general multi-column graph is NP-hard , and the optimal VE-weight net problem is NP-hard even on a proper ordered graph.
- A $T(n, m_B K)$ time algorithm for the optimal V-weight net problem on a proper ordered graph G.
- A $T(n, \kappa m_B K)$ time algorithm for the optimal VCE-weight net problem, where $\kappa = \max_{(i_k, j_{k'}) \in E} |k - k'|$.
- Modeling a surface reconstruction problem in $\Re^d$ ($d = 3, 4$) with a given topology of the underlying surface as a geometric optimization problem, and solving it as an optimal VCE-weight net problem in polynomial time (depending on a tolerance error $\epsilon > 0$).
- Modeling 3-D and 4-D medical image segmentation of smooth objects as an optimal V-weight net problem on a proper ordered graph, and solving it in $T(n, n)$ time, where n is the number of input image voxels. Our solutions for these problems can be extended to higher dimensions.

In fact, our work on optimal nets can be viewed as a rather general theoretical framework, which is likely to be interesting in its own right. This framework captures a number of other problems, such as metric labeling problems [26,17,12] and a class of integer optimization problems [22,19,21]. All those problems can be transformed to a V-weight or VE-weight net problem on a special proper ordered graph. Our polynomial time algorithms are inspired by Hochbaum's solutions for the minimum closure problem [30,20] and s-excess problem [21], and use a technique for integer optimization over some special monotone constraints [19]. Hochbaum [19] defined a class of integer programs with some special monotone constraints and cast that problem as a minimum cut problem on graphs. We generalize the graph construction in [19] to proper ordered graphs through a judicious characterization of the structures of the proper ordering.

Kleinberg and Tardos [26], Gupta and Tardos [17], and Chekuri et al. [12] studied the metric labeling problem, formulating the problem as optimizing a combinatorial function consisting of assignment costs and separation costs. We are able to reduce this metric labeling problem to the optimal VE-weight net problem on a special proper ordered graph. The metric labeling problem is known to be NP-hard when the separation costs are non-convex [26]. Boykov et al. [11]

and Ishikawa and Geiger [24] showed a direct reduction to the minimum s-t cuts for a special case of metric labeling when the label set is $L = \{0, 1, \ldots, l\}$ and the separation costs are linear; but, their algorithms cannot handle real-valued vertex weights.

We omit the proofs of the lemmas due to the space limit.

2   Hardness of the Net Surface Problems on Multi-column Graphs

This section presents some hardness results for the two optimal net problems. We show that the optimal V-weight net problem and optimal VE-weight net problem on a general d-D multi-column graph are both NP-hard ($d \geq 3$). Actually, the optimal VE-weight net problem is NP-hard even on a 3-D proper ordered graph.

Section 3.1 shows that finding an optimal V-weight net in a d-D proper ordered graph is polynomially solvable ($d \geq 3$). The proper ordering property is crucial to our polynomial time algorithms. Here, we prove that without the proper ordering, the optimal V-weight net problem on a d-D multi-column graph is NP-hard ($d \geq 3$), by a sophisticated reduction from the 3SAT problem [16]. In fact, even finding any net in such a graph in 3-D is NP-complete.

Lemma 1. *Deciding whether there exists a net in a 3-D multi-column graph (EXNET) is NP-complete.*

The NP-hardness of the optimal V-weight net problem thus follows from Lemma 1.

Theorem 1. *The optimal V-weight net problem on a d-D multi-column graph is NP-hard ($d \geq 3$).*

Now, we show the NP-hardness of the optimal VE-weight net problem on a proper ordered graph (whose edge costs need not form convex non-decreasing functions). Kleinberg and Tardos studied the metric labeling problem and pointed out its NP-completeness [26]. We can transform the metric labeling problem to an instance of the optimal VE-weight net problem on a special proper ordered graph with any two adjacent columns forming a complete bipartite graph.

Theorem 2. *The optimal VE-weight net problem on a proper ordered graph is NP-hard.*

3   Algorithms for Net Surface Problems on Proper Ordered Graphs

This section presents our polynomial time algorithms for computing an optimal V-weight net and an optimal VCE-weight net on any proper ordered graph G.

We use a unified approach which formulates both the optimal V-weight and VCE-weight net problems as computing a minimum s-t cut [3] in another graph transformed from G. Our approach is inspired by Hochbaum's solutions for the minimum closure problem [30,20] and minimum s-excess problem [21]. Note that the minimum s-excess problem is an extension of the minimum closure problem.

Without loss of generality (WLOG), we assume that the base graph B is connected. Otherwise, we can solve these optimal net problems with respect to each connected component of B separately.

3.1   The Optimal V-Weight Net Surface Problem

Consider an n_B-vertex and m_B-edge base graph $B = (V_B, E_B)$ and an arbitrarily generated proper ordered graph $G = (V, E)$ of height K, each vertex v of G having a real-valued weight $w(v)$. Let $n = |V| = n_B * K$ and $m = |E|$. This subsection presents a $T(n, m_B K)$-time algorithm for finding an optimal V-weight net in G, where $T(n', m')$ denotes the time bound for finding a minimum s-t cut in an edge-weighted directed graph with $O(n')$ vertices and $O(m')$ edges [3]. Interestingly, the time bound of our algorithm is independent of the number of edges in G.

We begin with reviewing the minimum closure problem [30,20]. A *closed set* $\mathcal{C}$ in a directed graph is a set of vertices such that all successors of any vertex in $\mathcal{C}$ are also contained in $\mathcal{C}$. Given a directed graph $G' = (V', E')$ with real-valued vertex weights, the minimum closure problem seeks a closed set in G' with the minimum total vertex weight.

We now discuss our construction. First, we build a directed graph $\tilde{G} = (\tilde{V}, \tilde{E})$ from G as follows. $\tilde{V} = \{\tilde{i}_k : i_k \in V\}$. The weight $\tilde{w}(\tilde{i}_k)$ of each vertex $\tilde{i}_k$ in $\tilde{G}$ is assigned in the following way: For each $i \in V_B$, the weight of vertex $\tilde{i}_0$ is set to $w(i_0)$, and for every $k = 1, 2, \ldots, K-1$, $\tilde{w}(\tilde{i}_k) = w(i_k) - w(i_{k-1})$. Next, we define the edges for $\tilde{G}$. For each $i \in V_B$ and every $k = 1, 2, \ldots, K - 1$, vertex $\tilde{i}_k$ has a directed edge $(\tilde{i}_k, \tilde{i}_{k-1})$ to vertex $\tilde{i}_{k-1}$ with a cost $+\infty$ (note that the edge costs of $\tilde{G}$ are for computing a minimum closed set later). We call the directed path $\tilde{i}_{K-1} \to \tilde{i}_{K-2} \to \cdots \to \tilde{i}_0$ in $\tilde{G}$ the *i-chain*. For any ordered pair of adjacent columns in G, say (i-column, j-column), and each $k \in \{0, 1, \ldots, K - 1\}$, let the edge interval of the vertex i_k on V_j be $I(i_k, j) = (j_p, j_{p+1}, \ldots, j_{p+s})$, with $p \geq 0$, $s \geq 0$, and $p + s < K$. Then, let a directed edge with a cost $+\infty$ go from vertex $\tilde{i}_k$ to vertex $\tilde{j}_p$ in $\tilde{G}$. Note that the same construction is also done for the ordered pair (j-column, i-column). Figure 1(a) shows part of the graph G that is associated with an edge $(i, j) \in E_B$, and Figure 1(b) illustrates the corresponding constructions in $\tilde{G}$. Let $\tilde{V}_0 = \{\tilde{i}_0 : i \in V_B\}$, and $\tilde{G}_0 \subseteq \tilde{G}$ be the induced subgraph of the vertex set $\tilde{V}_0$ in $\tilde{G}$. The following lemma characterizes several properties of $\tilde{G}_0$, which are useful to our construction.

Lemma 2. *(1) $\tilde{G}_0$ is a strongly connected component of $\tilde{G}$. (2) $\tilde{V}_0$ is a closed set in $\tilde{G}$. (3) For any non-empty closed set S in $\tilde{G}$, $\tilde{V}_0 \subseteq S$. (4) $\tilde{G}_0$ corresponds to a net in G.*

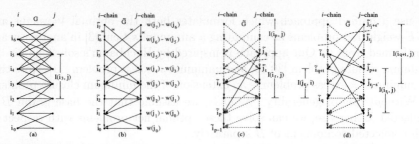

Fig. 1. (a) Illustrating the proper ordering of G. (b) Constructing graph $\tilde{G}$ from G. (c) Illustrating the proof of Lemma 3. (d) Illustrating the proof of Lemma 4.

The next two lemmas show the relations between a V-weight net in G and a closed set in $\tilde{G}$.

Lemma 3. *A V-weight net $\mathcal{N}$ in G corresponds to a non-empty closed set S in $\tilde{G}$ with the same weight.*

Lemma 4. *A non-empty closed set S in $\tilde{G}$ corresponds to a V-weight net $\mathcal{N}$ in G with the same weight.*

Based on Lemmas 3 and 4, we have the following lemma.

Lemma 5. *The net $\mathcal{N}^*$ corresponding to a non-empty minimum closed set S^* in $\tilde{G}$ is an optimal V-weight net in G.*

Note that the characterizations in the above lemmas are all concerned with non-empty closed sets in $\tilde{G}$. However, the minimum closed set S^* in $\tilde{G}$ can be empty (with a weight zero), and when this is the case, $S^* = \emptyset$ gives little useful information on $\tilde{G}$. Fortunately, our careful construction of $\tilde{G}$ still enables us to overcome this difficulty. If the minimum closed set in $\tilde{G}$ is empty, then it implies that the weight of every non-empty closed set in $\tilde{G}$ is non-negative. To obtain a minimum *non-empty* closed set in $\tilde{G}$, we do the following: Let M be the total weight of vertices in $\tilde{V}_0$; pick an arbitrary vertex $\tilde{i}_0 \in \tilde{V}_0$ and assign a new weight $\tilde{w}(\tilde{i}_0) - M - 1$ to $\tilde{i}_0$. We call this a *translation operation* on $\tilde{G}$. From Lemma 2, $\tilde{V}_0 \neq \emptyset$ is a closed set in $\tilde{G}$ and is a subset of any non-empty closed set in $\tilde{G}$. Further, observe that the total weight of vertices in the closed set $\tilde{V}_0$ (after a translation operation on $\tilde{G}$) is negative. This implies that any minimum closed set in $\tilde{G}$ (after a translation operation on $\tilde{G}$) cannot be empty. Also based on Lemma 2, we have the following lemma.

Lemma 6. *For a non-empty closed set S in $\tilde{G}$, let $\tilde{w}(S)$ denote the total weight of S before any translation operation on $\tilde{G}$. Then after a translation operation, the weight of S is $\tilde{w}(S) - M - 1$.*

Now, we can simply find a minimum closed set S^* in $\tilde{G}$ after performing a translation operation on $\tilde{G}$. Based on Lemma 6, S^* is a minimum *non-empty* closed set in $\tilde{G}$ before the translation.

As in [30,20], we compute a minimum non-empty closed set in $\tilde{G}$, as follows. Let $\tilde{V}^+$ and $\tilde{V}^-$ denote the set of vertices in $\tilde{G}$ with non-negative and negative weights, respectively. Define a new directed graph $\tilde{G}_{st} = (\tilde{V} \cup \{s,t\}, \tilde{E} \cup E_{st})$. The vertex set of $\tilde{G}_{st}$ is the vertex set $\tilde{V}$ of $\tilde{G}$ plus a source s and a sink t. The edge set of $\tilde{G}_{st}$ is the edge set $\tilde{E}$ of $\tilde{G}$ plus a new edge set E_{st}. E_{st} consists of the following edges: The source s is connected to each vertex $v \in \tilde{V}^-$ by a directed edge of cost $-\tilde{w}(v)$; every vertex $v \in \tilde{V}^+$ is connected to the sink t by a directed edge of cost $\tilde{w}(v)$. Let $(\mathcal{S}, \bar{\mathcal{S}})$ denote a finite-cost s-t cut in $\tilde{G}_{st}$ with $s \in \mathcal{S}$ and $t \in \bar{\mathcal{S}}$, and $C(\mathcal{S}, \bar{\mathcal{S}})$ denote the total cost of the cut. Note that the directed edges in the cut $(\mathcal{S}, \bar{\mathcal{S}})$ are either in $(\mathcal{S} \cap \tilde{V}^+, t)$ or in $(s, \bar{\mathcal{S}} \cap \tilde{V}^-)$. Let $\tilde{w}(V')$ denote the total weight of vertices in a subset $V' \subseteq \tilde{V}$. Then, we have $C(\mathcal{S}, \bar{\mathcal{S}}) = -\tilde{w}(\tilde{V}^-) + \sum_{v \in \mathcal{S}} \tilde{w}(v)$.

Note that the term $-\tilde{w}(\tilde{V}^-)$ is fixed and is the sum over all vertices with negative weights in $\tilde{G}$. The term $\sum_{v \in \mathcal{S}} \tilde{w}(v)$ is the total weight of all vertices in the source set $\mathcal{S}$ of $(\mathcal{S}, \bar{\mathcal{S}})$. But, $\mathcal{S} - \{s\}$ is a closed set in $\tilde{G}$ [30,20]. Thus, the cost of a cut $(\mathcal{S}, \bar{\mathcal{S}})$ in $\tilde{G}_{st}$ and the weight of the corresponding closed set in $\tilde{G}$ differ by a constant, and the source set of a minimum cut in $\tilde{G}_{st}$ corresponds to a minimum closed set in $\tilde{G}$. Since the graph $\tilde{G}$ has n vertices and $O(m_B K)$ edges, we have the following result.

Theorem 3. *Given an n_B-vertex, m_B-edge base graph B and a generated proper ordered vertex-weight graph $G = (V, E)$ with n vertices, m edges, and a height K, an optimal V-weight net in G can be computed in $T(n, m_B K)$ time.*

3.2   The Optimal VCE-Weight Net Surface Problem

In this section, we study the optimal VCE-weight net problem. Besides each vertex v in the proper ordered graph $G = (V, E)$ having a real-valued weight $w(v)$ as for the optimal V-weight net problem, each edge $e = (i_k, j_{k'}) \in E$ has a cost $c(e) = f_{ij}(|k - k'|)$, where $f_{ij}(\cdot)$ is a convex and non-decreasing function associated with the edge $(i, j) \in E_B$. We give a $T(n, \kappa m_B K)$-time algorithm for the optimal VCE-weight net problem, where $\kappa = \max_{(i_k, j_{k'}) \in E} |k - k'|$, and $T(n', m')$ is the time bound for finding a minimum s-t cut in an edge-weighted directed graph with $O(n')$ vertices and $O(m')$ edges [3].

Hochbaum [18,21] studied the minimum s-excess problem, which is a relaxation of the minimum closure problem [30,20]. Given a directed graph $G' = (V', E')$, each vertex $v' \in V''$ having an arbitrary weight $w'(v')$ and each edge $e' \in E'$ having a cost $c'(e') \geq 0$, the problem seeks a vertex subset $\mathcal{S}' \subseteq V'$ such that the cost of $\mathcal{S}'$, $\gamma(\mathcal{S}') = \sum_{v' \in \mathcal{S}'} w'(v') + \sum_{\substack{(u',v') \in E' \\ u' \in \mathcal{S}', v' \in \bar{\mathcal{S}}'}} c'(u', v')$, is minimized, where $\bar{\mathcal{S}}' = V' - \mathcal{S}'$. Instead of forcing all successors of each vertex in $\mathcal{S}'$ to be in $\mathcal{S}'$, the s-excess problem charges a penalty onto the edges leading to such immediate successors that are not included in $\mathcal{S}'$.

Below we construct a directed graph $\tilde{G} = (\tilde{V}, \tilde{E})$ from the graph G, and argue the equivalence between the optimal VCE-weight net problem on G and the minimum s-excess problem on $\tilde{G}$.

As in Section 3.1, $\tilde{V} = \{\tilde{i}_k : i_k \in V\}$. The graph $\tilde{G}$ includes the same construction of the i-chains for all $i \in V_B$ and the same weight assignment $\tilde{w}(\cdot)$ for each vertex in $\tilde{G}$. The directed edges of cost $+\infty$ associated with the edge intervals in G, $I(i_k, j)$ and $I(j_k, i)$ for every $(i, j) \in E_B$ and $0 \leq k < K$, are put into $\tilde{G}$ in the same way as well. The main difficulty here is how to enforce the edge penalty in $\tilde{G}$ such that when an edge is on a net $\mathcal{N}$ in G, its cost is charged appropriately to the corresponding cut in $\tilde{G}$. We overcome this difficulty by using a novel edge cost penalty embedding scheme below.

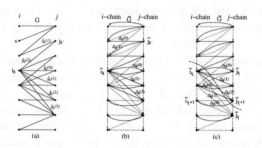

Fig. 2. (a) Part of a proper ordered graph G (associated with an edge $(i, j) \in E_B$) with weighted edges. (b) Constructing the graph $\tilde{G}$ from G (the dashed arrows are for edges with cost $+\infty$, and the solid arrows enforce the edge penalty for charging the costs of the corresponding edges that are on an arbitrary net $\mathcal{N}$ in G). (c) Illustrating the proof of Lemma 9.

For every ordered pair of adjacent columns in G, say (i-column, j-column), consider each edge interval $I(i_k, j) = (j_p, j_{p+1}, \ldots, j_{p+s})$ of $i_k \in V_i$ on V_j, with $0 \leq k < K$, $p \geq 0$, $p + s < K$, and $s \geq 0$. If $k > p$, then for every $k' = 0, 1, \ldots, k - p - 1$, there is a directed edge in $\tilde{G}$ from $\tilde{i}_k$ to $\tilde{j}_{k-k'}$, called a *cost-penalty edge*, whose cost is $\tilde{c}(\tilde{i}_k, \tilde{j}_{k-k'}) = \Delta_{ij}(k')$ (see Figure 2). Herein, the functions $\Delta_{ij}(\cdot)$ are defined as follows.

$$\Delta_{ij}(0) = f_{ij}(1) - f_{ij}(0)$$
$$\Delta_{ij}(x + 1) = [f_{ij}(x + 2) - f_{ij}(x + 1)] - [f_{ij}(x + 1) - f_{ij}(x)], \quad x = 0, 1, \ldots$$

Lemma 7. *Every function $\Delta_{ij}(\cdot)$ is non-negative.*

WLOG, we assume that each $f_{ij}(0) = 0$ (otherwise, we can subtract $f_{ij}(0)$ from the cost of each edge between every two adjacent i-column and j-column in G, without affecting essentially the costs of the VCE-weight nets in G). Let $\tilde{V}_0 = \{\tilde{i}_0 : i \in V_B\}$, and $\tilde{G}_0 \subseteq \tilde{G}$ be the induced subgraph of the vertex set $\tilde{V}_0$ in $\tilde{G}$. Then $\tilde{G}_0$ has several nice properties.

Lemma 8. *(1) $\tilde{G}_0$ is a strongly connected component of $\tilde{G}$. (2) $\tilde{V}_0$ is an s-excess set in $\tilde{G}$ with a finite cost. (3) For any non-empty s-excess set S in $\tilde{G}$ with a finite cost $\gamma(S)$, $\tilde{V}_0 \subseteq S$. (4) $\tilde{G}_0$ corresponds to a net in G.*

Next, we demonstrate that computing an optimal VCE-weight net $\mathcal{N}^*$ in G is equivalent to finding a minimum non-empty s-excess set $\mathcal{S}^*$ in $\tilde{G}$.

Lemma 9. *A VCE-weight net $\mathcal{N}$ in G corresponds to an s-excess set $\mathcal{S} \neq \emptyset$ in $\tilde{G}$ with the same cost.*

Lemma 10. *A non-empty s-excess set $\mathcal{S}$ with a finite cost $\gamma(\mathcal{S})$ in $\tilde{G}$ defines a VCE-weight net $\mathcal{N}$ in G with $\beta(\mathcal{N}) = \gamma(\mathcal{S})$.*

Following Lemmas 9 and 10, a minimum non-empty s-excess set $\mathcal{S}^*$ in $\tilde{G}$ corresponds to an optimal VCE-weight net $\mathcal{N}^*$ in G, as stated in Lemma 11.

Lemma 11. *A minimum s-excess set $\mathcal{S}^* \neq \emptyset$ in $\tilde{G}$ defines an optimal VCE-weight net $\mathcal{N}^*$ in G.*

In the case when the minimum s-excess set in $\tilde{G}$ is empty, we apply a similar translation operation onto $\tilde{G}$ as in Section 3.1, in which $M = \tilde{w}(\tilde{V}_0)$. Thus, from Lemma 8, we obtain the lemma below.

Lemma 12. *A minimum s-excess set $\mathcal{S}^* \neq \emptyset$ in $\tilde{G}$ can be found after a translation operation on $\tilde{G}$.*

A minimum s-excess set in $\tilde{G}$ can be computed by using a minimum s-t cut algorithm, as in [21]. If the minimum s-excess set in $\tilde{G}$ thus obtained is empty, then by Lemma 12, we can first perform a translation operation on $\tilde{G}$, and then apply the s-t cut based algorithm to obtain a minimum *non-empty* s-excess set in $\tilde{G}$. As in Section 3.1, we define a directed graph $\tilde{G}_{st}$ from $\tilde{G}$. Note that the costs of the directed edges from the source s in $\tilde{G}_{st}$ are the negatives of the weights of the corresponding vertices. Let $C(\mathcal{A}, \mathcal{B})$ denote the sum of costs of all directed edges in a cut $(\mathcal{A}, \mathcal{B})$ in $\tilde{G}_{st}$ with their heads in $\mathcal{A}$ and tails in $\mathcal{B}$. Let $\mathcal{S}$ be an s-excess set in $\tilde{G}$ with a finite cost $\gamma(\mathcal{S})$. Then, we have
$\gamma(\mathcal{S}) = -\tilde{w}(\tilde{V}^-) + C(\{s\} \cup \mathcal{S}, \{t\} \cup \bar{\mathcal{S}})$.

Note that the term $-\tilde{w}(\tilde{V}^-)$ is fixed. Hence, the set $\mathcal{S} \subseteq \tilde{V}$ is a set with the minimum s-excess cost in the graph $\tilde{G}$ if and only if $\mathcal{S} \cup \{s\}$ is the source set of a minimum cut in $\tilde{G}_{st}$. Since the graph $\tilde{G}_{st}$ has $O(n)$ vertices and $O(\kappa m_B K)$ edges, where $\kappa = \max_{(i_k, j_{k'}) \in E} |k - k'|$, we have the following result.

Theorem 4. *An optimal VCE-weight net in a proper ordered graph can be found in $T(n, \kappa m_B K)$ time.*

4   Applications

This section discusses two application problems for our optimal net results on proper ordered graphs: surface reconstruction and medical image segmentation. For $d = 3, 4$, we present efficient algorithms for a surface reconstruction problem in $\Re^d$ with a given topology and for d-D medical image segmentation. Our algorithms for these problems can also be extended to higher dimensions.

4.1   Surface Reconstruction in $\Re^d$ with a Given Topology

Surface reconstruction is an important problem in many applications, such as CAD, computer graphics, computer vision, and mathematical modeling. It often involves computing a piecewise linear approximation to a desired surface from a set of sample points.

Surface reconstruction problems in $\Re^3$ have been intensively studied in computer graphics and computer vision (e.g., see [8,23,27]), and computational geometry (e.g., see [5][4][9][14]). In this paper, we study a surface reconstruction problem in $\Re^d$ ($d = 3, 4$) with a given topology of the underlying surface. Considerable work has been done on surface reconstruction with some given topological constraints or structures (e.g., see [10][7][29][6]).

Our focus is on computing a piecewise linear ϵ-approximation to a "smooth" surface from sample data. Agarwal and Desikan [1] and Agarwal and Suri [2] studied the problem of computing a piecewise linear function with the minimum complexity to approximate a 3-D xy-monotone surface within a tolerance error $\epsilon > 0$. However, our criterion for measuring the quality of the output ϵ-approximate surface is different from those in [1,2].

Let a set of n sample points in $\Re^d$ ($d = 3, 4$), $\mathcal{P} = \{(x, \mathcal{P}(x)) : x \in \mathcal{D} \subset \Re^{d-1}, \mathcal{P}(x) \in \Re\}$, of some underlying smooth surface $\mathcal{S}$ be given, where $\mathcal{D}$ is an n-point set in $\Re^{d-1}$. The topology of $\mathcal{S}$ in $\Re^d$ is completely specified by a neighborhood system $\mathcal{H}$ on $\mathcal{D}$ (i.e., $\mathcal{H}$ is the edge set of a graph defined on the vertex set $\mathcal{D}$). For example, $\mathcal{H}$ is for a triangulated planar graph embedded in the plane $z = 0$ for $\mathcal{S}$ in $\Re^3$ (e.g., a Delaunay triangulation of n planar points). For two points $x, y \in \mathcal{D}$, if $(x, y) \in \mathcal{H}$, then we say the points $(x, \mathcal{P}(x))$ and $(y, \mathcal{P}(y))$ in $\Re^d$ are *adjacent* on $\mathcal{S}$. We seek a mapping for an approximation surface (whose topology has already been specified by $\mathcal{H}$ on $\mathcal{D}$), $\mathcal{F} : \mathcal{D} \to \Re$, that satisfies the following conditions. (For simplicity, we let $\mathcal{F}$ denote the approximation surface.)

- $\mathcal{F}$ is ϵ-*approximate*: Given a tolerance error $\epsilon > 0$, for any $x \in \mathcal{D}$, $|\mathcal{F}(x) - \mathcal{P}(x)| \le \epsilon$. This means that the approximation surface $\mathcal{F}$ should not deviate too much from the sample data.
- $\mathcal{F}$ is δ-*smooth*: Given a smoothness system $\delta = \{\delta_{ij} \ge 0 : (x_i, x_j) \in \mathcal{H}\}$ for all pairs of adjacent points in $\mathcal{P}$, $|\mathcal{F}(x_i) - \mathcal{F}(x_j)| \le \delta_{ij}$ holds for every $(x_i, x_j) \in \mathcal{H}$.

Such a surface is said to be ϵ-*approximate and δ-smooth*. Our goal is to compute a piecewise linear ϵ-approximate and δ-smooth surface $\mathcal{F}$ (if it exists) such that the *energy cost* $\mathcal{E}(\mathcal{F})$ of the surface, with

$$\mathcal{E}(\mathcal{F}) = \sum_{x \in \mathcal{D}} g(\mathcal{F}(x), \mathcal{P}(x)) + \sum_{(x_i, x_j) \in \mathcal{H}} f((x_i, \mathcal{F}(x_i)), (x_j, \mathcal{F}(x_j))),$$

is minimized, where $g(\cdot, \cdot)$ is a *data force* function that attracts $\mathcal{F}$ to the sought surface $\mathcal{S}$, and $f(\cdot, \cdot)$ represents the *internal tension* of the surface $\mathcal{F}$. In fact, the "smoother" $\mathcal{F}$ is, the smaller its internal tension becomes. We use the Euclidean distance as the internal tension function (of course, other convex functions are applicable). This model, which captures both the local and global information in determining a surface based on sample points, is inspired by the deformable model in [25][32] for image analysis. We also assume an accuracy factor τ for $\mathcal{F}$ on all points $x_i \in \mathcal{D}$, with $0 < \tau < \epsilon$.

We solve this surface reconstruction problem as an optimal VCE-weight net problem. Our algorithm for this problem has some details different from those in Section 3.2, which we leave to the full version due to the space limit. Let $K = 2\lceil \frac{\epsilon}{\tau} \rceil + 1$ and $\kappa = \max_{\delta_{ij} \in \delta} \frac{\delta_{ij}}{\tau}$. We have the theorem below.

Theorem 5. *Given an n-point set $\mathcal{P} = \{(x, \mathcal{P}(x)) : x \in \mathcal{D} \subset \Re^{d-1}, \mathcal{P}(x) \in \Re\}$ in $\Re^d$ ($d = 3, 4$), a neighborhood system $\mathcal{H}$ defined on $\mathcal{D}$, a tolerance $\epsilon > 0$, and a smoothness system δ, we can compute an optimal ϵ-approximate, δ-smooth surface for $\mathcal{P}$ within an accuracy of τ in $T(nK, \kappa|\mathcal{H}|K)$ time.*

4.2   Medical Image Segmentation in 3-D and 4-D

Segmentation is one of the most challenging problems in image analysis. Medical image segmentation in 3-D and 4-D is essentially useful [28]. However, little work has been done on 4-D medical image segmentation. In this section, we study some 3-D and 4-D medical image segmentation problems and give efficient algorithms based on our optimal V-weight net solution.

Segmenting a 3-D volumetric image is to identify 3-D surfaces representing object boundaries in the 3-D space [31]. Let $\mathcal{I}(X, Y, Z) = \{(x, y, z) : 0 \leq x < X, 0 \leq y < Y, 0 \leq z < Z\}$ be a 3-D image of size $X \times Y \times Z$, with each voxel having a density value. Here, Y is the number of 2-D slices in the image, X is the number of columns in each slice, and Z is the number of voxels in each column. Note that an image $\mathcal{I}(X, Y, Z)$ can be viewed as representing a cylindrical object such as a vessel, or a spherical object such as a left ventricle [33]. 3-D images for such objects can be transformed such that the sought surface $\mathcal{S}$ in $\mathcal{I}(X, Y, Z)$ for the object boundary contains exactly one voxel in each column of every xz-slice [33]. Further, since many anatomical structures (e.g., human brains, iliac arteries, and ventricles) are smooth, one may desire the sought surface $\mathcal{S}$ to be sufficiently "smooth". Precisely, given two integers M_1 and M_2, for any (x_1, y_1, z_1) and (x_2, y_2, z_2) on $\mathcal{S}$, if $|x_1 - x_2| = 1$ and $y_1 = y_2$, then $|z_1 - z_2| \leq M_1$, and if $x_1 = x_2$ and $|y_1 - y_2| = 1$, then $|z_1 - z_2| \leq M_2$. We call such a surface in $\mathcal{I}(X, Y, Z)$ an (M_1, M_2)-*smooth surface*. Note that M_1 characterizes the smoothness within each xz-slice, and M_2 specifies the smoothness across the neighboring xz-slices. A weight $w(x, y, z)$ is assigned to each voxel (x, y, z) such that the weight is inversely related to the likelihood in that the voxel may appear at the desired object boundary, which is usually determined by using simple low-level image features [33]. Thus, a segmentation problem on $\mathcal{I}(X, Y, Z)$ is to obtain an (M_1, M_2)-smooth surface $\mathcal{S}$ such that the total weight $W(\mathcal{S})$ of $\mathcal{S}$, $W(\mathcal{S}) = \sum_{(x,y,z) \in \mathcal{S}} w(x, y, z)$, is minimized. The previously best known exact algorithms for this problem take at least exponential time [15,31,33].

We solve the problem of segmenting an optimal (M_1, M_2)-smooth surface from $\mathcal{I}(X, Y, Z)$ by transforming it to the optimal V-weight net problem on a proper ordered graph.

Theorem 6. *Given a 3-D image $\mathcal{I}(X, Y, Z)$ and two integers $M_1 \geq 0$ and $M_2 \geq 0$, an optimal (M_1, M_2)-smooth surface in $\mathcal{I}(X, Y, Z)$ can be computed in $T(n, n)$ time, with $n = X \times Y \times Z$.*

Next, we consider segmenting 4-D medical images, which are crucial for dynamic anatomical structure studies and pathology monitoring through time [28].

Let $\mathcal{I}$ denote a 4-D image representing a sequence of T 3-D images $(\mathcal{I}_0, \ldots, \mathcal{I}_{T-1})$ acquired at T different time points $t_0 < \cdots < t_{T-1}$ with each $\mathcal{I}_i = \mathcal{I}_i(X, Y, Z)$ being (M_1, M_2)-smooth. In addition, along the time dimension, the movement of the 3-D object is assumed to be continuous and smooth, which is specified by the third smoothness parameter M_3. A sequence of 3-D surfaces in $\mathcal{I}$, $\mathcal{S} = (\mathcal{S}_0, \ldots, \mathcal{S}_{T-1})$, that satisfies all three smoothness constraints is called an (M_1, M_2, M_3)-smooth surface in 4-D. Our goal is to find an (M_1, M_2, M_3)-smooth surface $\mathcal{S} = (\mathcal{S}_0, \ldots, \mathcal{S}_{T-1})$ whose total weight $\mathcal{W}(\mathcal{S}) = \sum_{t=0}^{T-1} \sum_{(x,y,z,t) \in \mathcal{S}_t} w(x, y, z, t)$ is minimized, where $w(x, y, z, t)$ is the weight of a voxel $(x, y, z, t) \in \mathcal{I}$, assigned in a similar way as in the 3-D case. We are not aware of any previous exact algorithm for this 4-D segmentation problem.

We reduce this 4-D segmentation problem to the optimal V-weight net problem on a 4-D proper ordered graph. The base graph is a 3-D $X \times Y \times T$ grid.

Theorem 7. *Given an n-voxel 4-D image $\mathcal{I} = \{\mathcal{I}_0, \mathcal{I}_1, \ldots, \mathcal{I}_{T-1}\}$, with each $\mathcal{I}_t$ being a 3-D image at a time point t, and smoothness parameters M_1, M_2 and M_3, an (M_1, M_2, M_3)-smooth surface $\mathcal{S} = \{\mathcal{S}_0, \mathcal{S}_1, \ldots, \mathcal{S}_{T-1}\}$ in $\mathcal{I}$ whose total weight $\mathcal{W}(\mathcal{S})$ is minimized can be computed in $T(n, n)$ time.*

Acknowledgments. The authors are very grateful to Shuang Luan for helpful discussions on the hardness of the optimal net surface problems.

References

1. P.K. Agarwal and P. Desikan, An Efficient Algorithm for Terrain Simplification, *Proc. 8th ACM-SIAM SODA*, 1997, 139-147.
2. P.K. Agarwal and S. Suri, Surface Approximation and Geometric Partitions, *SIAM J. Comput.*, 19(1998), 1016-1035.
3. R.K. Ahuja, T.L. Magnanti, and J.B. Orlin, *Network Flows: Theory, Algorithms, and Applications*, Prentice Hall, Inc., 1993.
4. N. Amenta, S. Choi, T.K. Dey, and N. Leekha, A Simple Algorithm for Homeomorphic Surface Reconstruction, *Proc. 16th ACM Symp. Comp. Geom.*, 2000, 213-222. Also *International J. of Computational Geometry and Applications*, to appear.
5. N. Amenta, S. Choi, and R.K. Kolluri, The Power Crust, Unions of Balls, and the Medial Axis Transform, *International J. of Computational Geometry and Applications*, to appear.
6. C. Bajaj and G. Xu, Modeling, and Visualization of C^1 and C^2 Scattered Function Data on Curved Surfaces, *Proc. of 2nd Pacific Conference on Computer Graphics and Applications*, 1997, 19-29.
7. R.E. Barnhill, K. Opitz, and H. Pottman, Fat Surfaces: A Trivariate Approach to Triangle-based Interpolation on Surface, *CAGD*, 9(5)(1992), 365-378.
8. F. Bernardini, C. Bajaj, J. Chen, and D. Schikore, Automatic Reconstruction of 3-D CAD Models from Digital Scans, *International J. of Computational Geometry and Applications*, 9(4-5)(1999), 327-361.

9. J.D. Boissonnat and F. Cazals, Smooth Surface Reconstruction via Natural Neighbor Interpolation of Disance Functions, *Proc. 16th ACM Symp. Comput. Geom.*, 2000, 223-232.

10. R.M. Bolle and B.C. Vemuri, On Three-dimensional Surface Reconstruction Methods, *IEEE Trans. Pat. Anal. Mach. Intell.*, 13(1)(1991), 1-13.

11. Y. Boykov, O. Veksler, and R. Zabih, Markov Random Fields with Efficient Approximations, *Proc. IEEE Conf. on Computer Vision and Pattern Recognition*, 1998.

12. C. Chekuri, A. Khanna, J. Naor, and L. Zosin, Approximation Algorithms for the Metric Labeling Problem via a New Linear Programming Formulation, *Proc. 12th ACM-SIAM SODA*, 2001, 109-118.

13. D.Z. Chen, J. Wang, and X. Wu, Image Segmentation with Monotonicity and Smoothness Constraints, *Proc. 12th Annual International Symp. on Algorithms and Computation*, New Zealand, December 2001, 467-479.

14. T.K. Dey and J. Giesen, Detecting Undersampling in Surface Reconstruction, *Proc. 17th Sympos. Comput. Geom.*, 2001, 257–263.

15. R.J. Frank, D.D. McPherson, K.B. Chandran, and E.L. Dove, Optimal Surface Detection in Intravascular Ultrasound Using Multi-dimensional Graph Search, *Computers in Cardiology, IEEE*, Los Alamitos, CA, 1996, 45-48.

16. M.R. Garey and D.S. Johnson, *Computers and Intractability: A Guide to the Theory of NP-Completeness*, W.H. Freeman, New York, NY, 1979.

17. A. Gupta and E. Tardos, Constant Factor Approximation Algorithms for a Class of Classification Problem, *Proc. 32nd ACM STOC*, 2000, 652-658.

18. D.S. Hochbaum, The Pseudoflow Algorithm and the Pseudoflow-Based Simplex for the Maximum Flow Problem, Lecture Notes in Computer Science, Vol. 1412, *Proc. 6th International IPCO Conf.*, 1998, 325-337.

19. D.S. Hochbaum, Instant Recognition of Half Integrality and 2-Approximations, Lecture Notes in Computer Science, Vol. 1444, *Proc. APPROX'98*, 1998, 99-110.

20. D.S. Hochbaum, A New-old Algorithm for Minimum Cut in Closure Graphs, *Networks*, 37(4)(2001), 171-193.

21. D.S. Hochbaum, An Efficient Algorithm for Image Segmentation, Markov Random Fields and Related Problems, *J. of the ACM*, 48(2001), 686-701.

22. D.S. Hochbaum and J.G. Shanthikumar, Convex Separable Optimization Is Not Much Harder Than Linear Optimization, *J. of the ACM*, 37(4)(1990), 843-862.

23. H. Hoppe, T. DeRose, T. Duchamp, J. McDonald and W. Stuetzle, Surface Reconstruction from Unorganized Points, *SIGGRAPH'92*, 1992, 71-78.

24. H. Ishikawa and D. Geiger, Segmentation by Grouping Junctions, *IEEE Conf. on Computer Vision and Pattern Recognition*, 1998, 125-131.

25. M. Kass, A. Witkin, and D. Terzopoulos, Snakes: Active Contour Models, *Int. J. Comput. Vision*, 1(4)(1988), 321-331.

26. J. Kleinberg and E. Tardos, Approximation Algorithms for Classification Problems with Pairwise Relationships: Metric Labeling and Markov Random Fields, *Proc. 40th IEEE Symp. on Foundations of Computer Science*, 1999, 14-23.

27. R. Mencl and H. Muller, Interpolation and Approximation of Surfaces from 3-D Scattered Data Points, Research Report No. 662, Fachbereich Informatik, Lehrstuhl VII, University of Dortmund, Germany, 1997.

28. J. Montagnat and H. Delingette, Space and Time Shape Constrained Deformable Surfaces for 4D Medical Image Segmentation, Lecture Notes in Computer Science, Vol. 1935, *Proc. Medical Image Computing and Computer-Assisted Intervention*, 2000, 196-205.

29. G.M. Nielson, T.A. Foley, B. Hamann, and D. Lane, Visualizing and Modeling Scattered Multivariate Data, *IEEE CG&A*, 11(3)(1991), 47-55.
30. J.C. Picard, Maximal Closure of a Graph and Applications to Combinatorial Problems, *Management Science*, 22(1976), 1268-1272.
31. M. Sonka, V. Hlavac, and R. Boyle, *Image Processing, Analysis, and Machine Vision*, 2nd edition, Brooks/Cole Publishing Company, Pacific Grove, CA, 199-205.
32. D. Terzopoulos, A. Witkin, and M. Kass, Constraints on Deformable Models: Recovering 3-D Shape and Nonrigid Motion, *Artif. Intelligence*, 36(1)(1988), 91-123.
33. D.R. Thedens, D.J. Skorton, and S.R. Fleagle, Methods of Graph Searching for Border Detection in Image Sequences with Applications to Cardiac Magnetic Resonance Imaging, *IEEE Trans. on Medical Imaging*, 14(1)(1995), 42-55.
34. T. Tokuyama, personal communication, June 2001.

Wagner's Theorem on Realizers

Nicolas Bonichon, Bertrand Le Saëc, and Mohamed Mosbah

LaBRI-Université Bordeaux 1, 351 Cours de la Libération,
33405 Talence, France
{bonichon, lesaec, mosbah}@labri.fr

Abstract. A realizer of a maximal plane graph is a set of three particular spanning trees. It has been used in several graph algorithms and particularly in graph drawing algorithms. We propose colored flips on realizers to generalize Wagner's theorem on maximal planar graphs to realizers. From this result, it is proved that $\xi_0 + \xi_1 + \xi_2 - \Delta = n - 1$ where ξ_i is the number of inner nodes in the tree T_i, Δ is the number of three colored faces in the realizer and n is the number of vertices. As an application of this formula, we show that orderly spanning trees with at most $\lfloor \frac{2n+1-\Delta}{3} \rfloor$ leaves can be computed in linear time.

1 Introduction

Schnyder showed that every maximal plane graph admits a special decomposition of its interior edges into three trees, called realizer [13,14]. Such a decomposition can be constructed in linear time [14]. Using realizers, it has been proved in [14] that every plane graph with $n \geq 3$ vertices has a planar straight-line drawing in a rectangular grid area $(n - 2) \times (n - 2)$. The existence of straight line embeddings for planar graphs was independently proven by Wagner [16], Fary [8] and Stein [15]. However, the question whether a grid of polynomial size for such an embedding exists was raised by Rosenstiehl and Tarjan [12].

Realizers are useful for many graph algorithms, of course for graph drawing [14,2] but also for graph encoding [3]. They are strongly connected with canonical ordering (or shelling order) [9,11], with 3-orientations [4], and with orderly spanning trees [2]. They can also be used to characterize planar graphs in terms of the order of their incidence, i.e., a graph G is planar iff the dimension of the incidence order of vertices and edges is at most 3 [13]. Realizers of the same graph have already been investigated. Suitable operations transforming a realizer of a graph to another realizer of the same graph have been investigated in [4,1]. A particular normal form is also characterized. Moreover, the structure of the set of realizers of a given graph turns out to be a distributive lattice [4].

In this paper, we deal with realizers of size n, i.e. realizers of maximal plane graphs of size n. Wagner [16] showed that any two maximal planar graphs having the same number of vertices are equivalent under diagonal flip transformations. A diagonal flip is a graph operation on maximal planar graphs which consists of removing the diagonal (u_1, u_3) of a quadrilateral (u_1, u_2, u_3, u_4), and inserting the opposite diagonal (u_2, u_4) (see Fig. 4). Hence, one can obtain all maximal

P. Widmayer et al. (Eds.): ICALP 2002, LNCS 2380, pp. 1043–1053, 2002.

plane graphs of size n using diagonal flips. An extension of Wagner's result to torus graphs has been proved in [5]. We generalize Wagner's theorem to realizers. Moreover, we show that the set of all realizers is a poset, i.e. a partially ordered set, with some appropriate relationship. Instead of the flip operation, we introduce two new operations, that we will call colored or oriented flips which will be used to get a realizer from another one of the same size.

Flip operations are also related to the four color theorem. In [6], a signed version of the diagonal flip has been used to define transformations between signed triangulations of a polygon. The existence of a sequence of signed flips between two given triangulations of a polygon turns out to be equivalent to the four color theorem [6,7,10]. Our approach involves an "oriented" version of diagonal flips.

As an application of our main result, we characterize the number of inner nodes of realizers. Precisely, we prove that $\xi_0 + \xi_1 + \xi_2 - \Delta = n - 1$ where ξ_i is the number of inner nodes in the tree T_i and Δ is the number of three colored faces in the realizer. As an application of this result, we prove that an orderly spanning tree with at most $\lfloor \frac{2n+1-\Delta}{3} \rfloor$ leaves can be computed in linear time for a maximal plane graphs. This bound is a precise formulation of that given in [2].

The rest of this paper is organized as follows. In Section 2, we present realizers and we give some basic properties. An extension of Wagner's theorem on realizers and the structure of the set of realizers are investigated in Section 3. We prove in Section 4 that $\xi_0 + \xi_1 + \xi_2 - \Delta = n - 1$, and we apply it to orderly spanning trees. Section 5 concludes the paper.

2   Preliminaries

2.1   Definitions

We assume that the reader is familiar with graph theory. In this paper we deal with simple graphs. A drawing of a graph is a mapping of each vertex to a point of the plane and of each edge to the continuous curve joining the two ends of this edge. A planar drawing or *plane graph* is a drawing without crossing edges except, possibly, on a common extremity. A graph that has a planar drawing is a planar graph. A plane graph splits the plane into topologically connected regions, called *face regions*. A *face* is the counterclockwise walk of the boundary of a face region. One of the regions is unbounded and its associated face is named the *external face* of the plane graph. The vertices and edges of this face are also qualified as *external*, the other vertices are called *inner* ones. A *cycle* is an eulerian connected partial subgraph (i.e. with vertices of even degree only). A cycle is *elementary* if all its vertices are of degree 2. A *circuit* is a cycle where each vertex has an out-degree equal to its in-degree. A *k-cycle* is a cycle of k edges. Since we deal with plane graphs, a cycle defines an interior region. For simplicity, we say the *region C*, for the region delimited by the cycle C.

A planar graph G is *maximal* (or *triangulated*) if all the other graphs with the same number of vertices that contain G, are not planar. The faces of a maximal plane graph are triangular. In this case, we denote v_0, v_1, v_2 the three vertices of the external face of this plane graph.

Definition 1 *[13] A realizer of a maximal plane graph G is a partition of the interior edges of G in three sets T_0, T_1, T_2 of directed edges such that for each interior vertex u it holds thats*

1. *u has out-degree exactly one in each of T_0, T_1, T_2.*
2. *The counterclockwise order of the edges incident with u is: leaving in T_0, entering in T_2, leaving in T_1, entering in T_0, leaving in T_2 and entering in T_1 (see Fig. 1).*

So a realizer is a set of three rooted trees where their edges are oriented to their roots, which are the external vertices v_0, v_1, v_2. For simplicity, we write $i + 1$ as

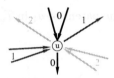

Fig. 1. Edge coloration and orientation around a vertex

a shorthand for $(i + 1) \mod 3$ and $i - 1$ as a shorthand for $(i + 2) \mod 3$. In the rest of the article, we assume that the edges of the tree T_i are colored with color i, where $i \in \{0, 1, 2\}$ and that the external edges (v_i, v_{i+1}) are of the color $i + 1$. In fact, we consider a slightly different definition of realizers by coloring the external edges in order to reduce the number of particular cases.

An example of a graph, and a realizer of this graph are given in Fig. 2.

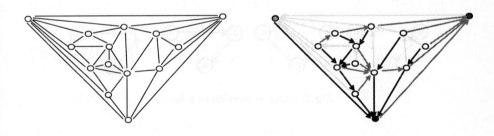

Fig. 2. An example of a realizer (a graph on the left side, and one of its realizer on the right side)

We denote by $deg_i(v)$ the number of ingoing edges of v in T_i. $u_1 \xrightarrow{i} u_2$ denotes the path colored i from u_1 to u_2. We write $u_1 >^i_{ccw} u_2$ (resp. $u_1 >^i_{cw} u_2$) if u_1 is after u_2 in the *counterclockwise preordering* (resp. *clockwise preordering*) of the tree T_i. Counterclockwise (resp. clockwise) preordering of a tree means visiting the root, then recursively the subtrees in the counterclockwise (resp. clockwise) order.

2.2 Properties of Realizers

Let $F = (e_0, e_1, e_2)$ be a face of G with $e_j = \{u_j, u_{j+1}\}$.

Property 1 *If u_1 is the parent of u_2 in T_i then $u_1 >_{cw}^{i+1} u_2$ and $u_2 >_{cw}^{i-1} u_1$.*

The proof of this property is based on the following facts.

Fact 1 *Let e_0 be colored i.*
If e_0 and e_1 are oriented towards u_1 then e_1 is colored i.

Proof. If e_1 is not colored i, the parent of u_1 in T_{i+1} would be inside the face F. This is not possible.

Fact 2 *Let e_0 be colored i.*
If e_0 and e_1 are respectively oriented towards u_1 and u_2 then e_1 is colored $i + 1$. Similarly, if e_0 and e_2 are respectively oriented towards u_0 and u_1 then e_2 is colored $i - 1$.

Proof. Assume that e_0 and e_1 are respectively oriented towards u_1 and u_2. If e_1 is not colored $i + 1$, the parent of u_1 in T_{i+1} would be inside the face F. This is not possible. A similar argument can be used to prove the second part of the fact.

As a consequence of the above facts, there are four possible colorations of a face which are given in Fig. 3. Notice that the last two colorations use the three colors. We will say the faces are *three-colored* ones.

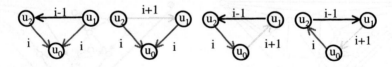

Fig. 3. Edge coloration of a face

2.3 Realizer Distributive Lattice

In this section, we recall results on the set of the realizers of a given graph. In particular, its structure is a distributive lattice.

Since we work with embedded graphs, a directed cycle is *oriented* either clockwise or counterclockwise, also noted as *cw-cycle* (resp. *ccw-cycle*).

Lemma 1 *[1] Let R be a realizer of a maximal plane graph G. Let C be a ccw-cycle (resp. cw-cycle) in R. Then we obtain a new realizer R' of G by*

1. *reversing the direction of the edges of C*
2. *setting for each edge of C the new color to be the color succeeding (resp. preceding) its original color,*
3. *setting for each edge inside C the new color to be the color preceding (resp. succeeding) its original color,*
4. *leaving the color of any other edge unchanged.*

Definition 2 *[1]*

- *Let G be a maximal plane graph. Define $\mathcal{R}(G) := \{R: R \text{ is a realizer of } G\}$.*
- *For two realizers R_1 and R_2 of G, $R_1 \preceq R_2$ iff R_1 can be obtained from R_2 by re-coloring some cw-cycles, i.e. there exists a re-coloration sequence which transforms R_1 into R_2.*
- *Let L_G (resp. R_G) be the realizer of G without any cw-cycle, (resp. ccw-cycle).*

Theorem 1 *[1] Let G be a maximal plane graph. $(\mathcal{R}(G), \preceq)$ is a distributive lattice.*

In the following section, we will deal with the set of all realizers of size n.

3   Diagonal Flips on Realizers

3.1   Diagonal Flip

In [16], Wagner proved that it is possible to obtain all maximal planar graphs of size n using a graph rewriting rule called diagonal flip. In this section, we extend this result to realizers using colored flips.

Definition 3 *Let G be an embedded graph. Let u_2, u_1, u_4 and u_3, u_4, u_1 be two adjacent faces where u_2 is not neighbor of u_3. A diagonal flip consists of removing the edge (u_1, u_4) and inserting the edge (u_2, u_3) (see Fig. 4).*

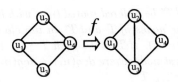

Fig. 4. Diagonal flip operation

Theorem 2 *[16] Let G_1 and G_2 be two maximal planar graphs with n vertices. There exists a sequence of diagonal flips which transforms G_1 into G_2.*

3.2 Generalization to Realizers

As shown in Fig. 5, we propose colored diagonal flips for realizers using two kinds of flips: f_1^i and f_2^i. It is easy to see that the application of a diagonal flip f_1^i or f_2^i on a realizer gives another realizer.

The choice between f_1^i and f_2^i depends of the quadrilateral configuration. Note that if the edge (u_2, u_1) is colored $i - 1$ and oriented towards u_1, and if the edge (u_3, u_1) is colored $i + 1$ and oriented towards u_1, then f_1^i or f_2^i can be applied.

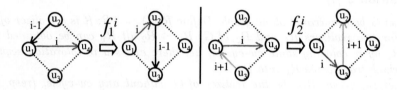

Fig. 5. Flips on realizer

Unfortunately, it is not always possible to apply one of the two operations. This occurs for the configuration of the quadrilateral of Fig. 6.

Fig. 6. Configuration for which colored flip cannot be directly applied

Now, we define two orders on trees, $\leq_{cw}$ and $\leq_{ccw}$, that are useful to express some flip properties.

Definition 4 *Let T and T' be ordered rooted trees with k nodes. Let $n_1, n_2, ..., n_k$ and $m_1, m_2, ..., m_k$ be the nodes of T and T' in the clockwise preordering (resp. counterclockwise preordering). If $T = T'$ then $T \leq_{cw} T'$ and $T \leq_{ccw} T'$. Else, let i be the rank of the first node where $deg(n_i) \neq deg(m_i)$. If $deg(n_i) < deg(m_i)$ then $T \leq_{cw} T'$ (resp. $T \leq_{ccw} T'$).*

Naturally, $T \geq_{cw} T'$ means that $T' \leq_{cw} T$. Also, $T <_{cw} T'$ means that $T \leq'_{cw} T$ and $T \neq T'$. If we consider the example of Fig. 7, we can see that the two grey nodes are the first nodes in T and T' with different degrees, with respect to the clockwise preordering. As the grey node in T has more children than in T', $T >_{cw} T'$. Similarly, the two black nodes are the first which have different

$$T \qquad\qquad T'$$

Fig. 7. Illustration of order relation between trees

degrees in T and T', in the counterclockwise preordering. Since the black node in T' has more children than that in T, $T <_{ccw} T'$.

Property 2 *Let $R = (T_0, T_1, T_2)$ be a realizer. Let $R' = (T'_0, T'_1, T'_2)$ be a realizer obtained from R by applying a flip f_1^i (resp. f_2^i). We have the following properties: $T'_i <_{cw} T_i$, $T'_{i-1} >_{ccw} T_{i-1}$ (resp. $T'_i <_{ccw} T_i$, $T'_{i+1} >_{cw} T_{i+1}$)*

Proof. Let us consider the flip f_1^i of Fig. 5. The edge (u_1, u_3) can be colored $i-1$ and oriented towards u_3 or colored $i+1$ and oriented towards u_1. In both cases $u_1 <_{ccw}^{i-1} u_3$. As the number of children of u_1 is greater in T'_{i-1} than in T_{i-1}, $T'_{i-1} >_{ccw} T_{i-1}$. The edge (u_2, u_4) can be colored i and oriented towards u_4 or colored $i+1$ and oriented towards u_2. In both cases, $u_4 <_{cw}^i u_2$. Since the number of children of u_4 is lesser in T'_i than in T_i, $T'_i <_{cw} T_i$.

3.3   Structure of $\mathcal{R}_n$ and Wagner's Theorem

Let $\mathcal{R}_n$ be the set of realizers of graphs of size n. The set of all realizers of size n can be represented by an oriented colored graph where vertices stand for realizers, and an edge colored i between two vertices R and R' represents the flip f_1^i transforming R into R'. Fig. 8 shows the graph of realizers of size 6. Each vertex represents a realizer. There is a directed edge colored i from a realizer R to another one R' if R' can be obtained from R by a flip f_1^i. The right part of the figure displays the transformation of the realizer 6 into the realizer 5 by a flip f_1^0.

We write $R(f_1^i | f_1^{i+1})^* R'$ if R can be transformed into R' by a sequence of flips f_1^i and f_1^{i+1}. Let $(\mathcal{R}_n, f_1^i | f_1^{i+1})$ be the set of realizers of size n, equipped with the relation $(f_1^i | f_1^{i+1})^*$.

Property 3 *$(\mathcal{R}_n, f_1^i | f_1^{i+1})$ is a poset, i.e. a partially ordered set.*

Proof. When applying f_1^i, $T'_i <_{cw} T_i$ and $T'_{i-1} >_{ccw} T_{i-1}$. When applying f_1^{i+1}, $T'_{i+1} <_{cw} T_{i+1}$ and $T'_i >_{ccw} T_i$. So, using the flips f_1^i or f_1^{i+1}, T_{i-1} is strictly decreasing considering the order $>_{ccw}$ and T_{i+1} is strictly increasing considering the order $>_{cw}$. Hence $(f_1^i | f_1^{i+1})^*$ is antisymmetric. An empty sequence transforms R into R, and thus the relation is reflexive. The relation is also transitive since flip sequences can be concatenated.

Let D_n^i be the realizer of size n where all the inner vertices are on the same branch of T_i and T_{i+1} and T_{i-1} are trees of depth 1 (see Fig. 9).

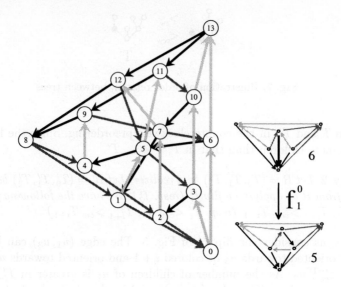

Fig. 8. $\mathcal{R}_6$ with f_1 operations.

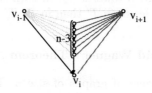

Fig. 9. Realizer D_n^i

Property 4 D_n^{i-1} *is the upper bound of* $(\mathcal{R}_n, f_1^i | f_1^{i+1})$ *and* D_n^{i+1} *is the lower bound.* D_n^{i-1} *is the lower bound of* $(\mathcal{R}_n, f_2^{i-1} | f_2^i)$ *and* D_n^{i+1} *is the upper bound.*

Proof. Let $R = (T_0, T_1, T_2)$ be a realizer of size n. If $R \neq D_n^{i+1}$ then there is an inner node in T_i or in T_{i-1}. If u is an inner node of T_i, a flip f_1^{i+1} can be applied on its outgoing edge colored $i+1$. Similarly, if u is an inner node of T_{i-1}, a flip f_1^i can be applied on its outgoing edge colored i. Hence a flip f_1^i or a flip f_1^{i+1} can be applied. In the same way, we show that D_n^{i-1} is the upper bound of $(\mathcal{R}_n, f_1^i | f_1^{i+1})$. The second part of the property comes directly from the fact that f_2^{i-1} is the inverse of f_1^i and f_2^i is the inverse of f_1^{i+1}.

Theorem 3 *There exists a sequence of colored flips that transforms any realizer R with n vertices into any other realizer R' with n vertices.*

Proof. Let R and R' be two realizers with n vertices. As D_n^{i-1} is the unique maximal element of $(\mathcal{R}_n, f_1^i | f_1^{i+1})$, there is a flip sequence that transforms R into D_n^{i-1}. There is also a flip sequence that transforms D_n^{i-1} into R'. Hence, the concatenation of the two previous sequences transforms R into R'.

In the previous theorem, flips f_1 and f_2 are allowed to transform a realizer to another one. It is possible to use only flips f_1 as stated in the following corollary.

Corollary 1 *There exists a sequence of colored flips $(f_1^0|f_1^1|f_1^2)^*$ that transforms any realizer R with n vertices into any other realizer R' with n vertices.*

Proof. Since D_n^0 is the upper bound of $(\mathcal{R}_n, f_1^1|f_1^2)$ there exists a flip sequence $(f_1^1|f_1^2)^*$ that transforms R into D_n^0. Since D_n^0 is the upper bound of $(\mathcal{R}_n, f_2^1|f_2^2)$, there exists a sequence $(f_2^1|f_2^2)^*$ that transforms R' into D_n^0. The inverse flip sequence transforms D_n^0 into R' and is composed of flips f_1^0 and flips f_1^1. Hence, the concatenation of the two appropriated sequences gives a flip sequence $(f_1^0|f_1^1|f_1^2)^*$ that transforms R into R'.

4   Three-Colored Faces and Number of Inner Nodes in Realizers

Let Δ be the number of three-colored faces of a realizer R. Let ξ_i be the number of inner nodes of the tree T_i.

Lemma 2 *Let R be a realizer. Let R' be a realizer obtained from R with a flip f_1^i. The sum $\xi_0 + \xi_1 + \xi_2 - \Delta$ is the same for R and R'.*

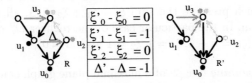

Fig. 10. Example of flip configuration

Proof. To prove the lemma, we need to check the modifications induced by the flip f_1^i on the quadrilateral and the adjacent face of the edge (u_1, u_3). If we consider the configuration of Fig. 10, we can see that u_2 is an inner node of T_1 but not of T_1'. Moreover, R has a three-colored face whereas R' does not. So the lemma holds for this configuration.

More generally, there are 32 possible configurations of the quadrilateral and the adjacent face of the edge (u_1, u_3) for the flip f_1^i. Fig. 11 shows the 32 possible configurations. For all these configurations, the lemma is verified.

Theorem 4 *For any realizer R of a maximal plane graph G, we have $\xi_0 + \xi_1 + \xi_2 - \Delta = n - 1$.*

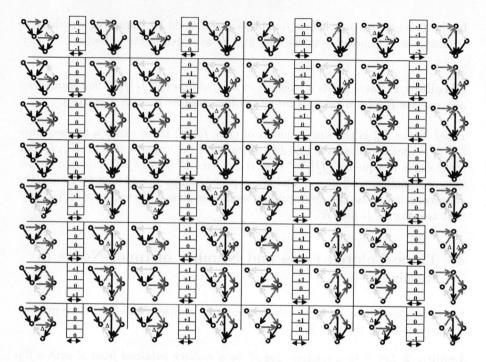

Fig. 11. 32 configurations of flips f_1

Proof. Since all realizers of size n can be obtained from one another by applying operations f_1 which preserve the sum $\xi_0 + \xi_1 + \xi_2 - \Delta$, all realizers of size n have the same sum. In the particular case of the realizer D_n^0 of Fig. 9, this sum is equal to $n - 1$ ($\xi_0 = n - 3, \xi_1 = 1, \xi_2 = 1$ and $\Delta = 0$).

The *orderly spanning tree* [2] of a maximal plane graph can be obtained from a Schnyder tree T_i by adding the edge (v_{i+1}, v_i) to T_i.

Corollary 2 *Let R be a realizer of a plane graph G. An orderly spanning tree can be obtained in linear time from R with at most $\lfloor \frac{2n+1-\Delta}{3} \rfloor$ leaves.*

Proof. This corollary follows from the fact that the number of leaves in a Schnyder tree T_i is $n - 1 - \xi_i$ and that the orderly spanning tree obtained from T_i has one more leaf, the node v_{i+1}.

5   Conclusion and Remarks

In this paper, we proved Wagner's theorem for realizers of maximal plane graphs of size n. The bound for the number of leaves of an orderly spanning tree, given in Corollary 2, improves that given in [2]. This result can be useful particularly for classes of graphs where $\Delta > 0$, such as 4-connected graphs. As an application of this bound, there exists an $n \times \left(\frac{2n+1-\Delta}{3}\right)$ grid for drawing a plane graph of size n.

Besides, we have shown that $(\mathcal{R}_n, f_1^i | f_1^{i+1} | f_2^{i+1})$ has a structure of a bounded poset. We have also looked for richer structure such as lattices. Unfortunately, neither $(\mathcal{R}_6, f_1^i | f_1^{i+1} | f_2^{i+1})$, $(\mathcal{R}_6, f_1^i | f_1^{i+1})$ nor $(\mathcal{R}_6, f_1^i | f_2^{i+1})$ has a structure of a lattice. We conjecture that $(\mathcal{R}_n, f_1^i | f_1^{i+1} | f_2^{i+1})$ is a poset. We have shown that this conjecture is true for $n \leq 6$.

References

1. E. Brehm. *3-orientations and Schnyder 3-tree-decompositions*. PhD thesis, FB Mathematik und Informatik, Freie Universität Berlin, 2000.
2. Yi-Ting Chiang Ching-Chi and Hsueh-I Lu. Orderly spanning trees with applications to graph encoding and graph drawing. In *Proc. 12th Symp. Discrete Algorithms*, pages 506–515. ACM and SIAM, 2001.
3. Richie Chih-Nan Chuang, Ashim Garg, Xin He, Ming-Yang Kao, and Hsueh-I Lu. Compact encodings of planar graphs via canonical ordering and multiple parentheses. In *Proc. 25th International Colloquium on Automata, Languages, and Programming (ICALP'98)*, volume 1443, pages 118–129, 1998.
4. P. Ossona de Mendez. *Orientations bipolaires*. PhD thesis, Ecole des Hautes Etudes en Sciences Sociales, 1994.
5. A. K. Dewdney. Wagner's theorem for torus graphs. *Discrete Math.*, 4:139–149, 1973.
6. S. Eliahou. Signed diagonal flips and the four color theorem. *Europ. J. Combinatorics*, 20:641–646, 1999.
7. S. Eliahou, S. Gravier, and C. Payan. Three moves on signed surface triangulations. *Les cahiers du laboratoire leibniz*, 2000.
8. I. Fary. On straight lines representation of planar graphs. *Acta Sci. Math. Szeged*, 11:229–233, 1948.
9. H. De Frayseix, J. Pach, and J. Pollack. How to draw a planar graph on a grid. *Combinatorica*, 10:41–51, 1990.
10. S. Gravier and C. Payan. Flips signés et triangulations d'un polygone. *Les cahiers du laboratoire leibniz*, 2000.
11. G. Kant. Drawing planar graphs using the canonical ordering. *Algorithmica*, 16:4–32, 1996.
12. P. Rosenstiehl and R.E. Tarjan. Rectilinear planar layouts and bipolar orientations of planar graphs. *Discrete Comput. Geom.*, 1:343–353, 1986.
13. W. Schnyder. Planar graphs and poset dimension. *Order*, 5:323–343, 1989.
14. W. Schnyder. Embedding planar graphs on the grid. *Proc. 1st ACM-SIAM Symp. Discrete Algorithms*, pages 138–148, 1990.
15. S.K. Stein. Convex maps. In *Proc. Amer. Math. Soc.*, volume 2, pages 464–466, 1951.
16. K. Wagner. Bemerkungen zum Vierfarbenproblem. In *Jahresber. Deutsche Math.-Verein.*, volume 46, pages 26–32, 1936.

Appendix

A colored version of the configurations is available at the following URL:
`http://dept-info.labri.fr/~bonichon/Realizer`.

Circular Arrangements

Vincenzo Liberatore*

Electrical Engineering and Computer Science Department
Case Western Reserve University
vxl11@po.cwru.edu

Abstract. Motivated by a scheduling problem in multicast environments, we consider the problem of arranging a weighted graph around a circle so as to minimize the total weighted arc length. We describe the first polynomial-time approximation algorithms for this problem, and specifically an $O(\log n)$-approximation algorithm for undirected circular arrangements and a $\widetilde{O}(\sqrt{n})$-approximation algorithm for directed circular arrangements. We will show that a simplification of the latter algorithm has better performance than previous heuristics on graphs obtained from a busy Web server log.

Keywords: Approximation algorithms, combinatorial optimization, scheduling, broadcast disks, multicast.

1 Introduction

A fundamental issue in server design is to ensure the server ability to support an arbitrarily large number of requests (*scalability*) [23]. A scalable server has the tremendous advantage of providing constant service time to clients even during request bursts and peak times. Server scalability can be ensured by *multicast*, whereby a single data unit is duplicated several times within the network infrastructure [25]. Consequently, a single multicast server needs only to send a data unit once to reach an arbitrarily large number of clients, which guarantees the server scalability. Multicast methods have originated research (e.g., [20,19, 9]) and commercial companies [1,2,3]. Applications range from heavily loaded Web servers [5] to high-throughput database systems [17], and we believe that multicast applications will significantly impact the performance of most Web transfers [11].

A fundamental question is the order in which the server should multicast data (*scheduling*). The scheduling problem must take into account that client access patterns often show dependencies between consecutive requests, so that the request for a data unit will make it more likely or less likely that certain other data will be requested next. In this scenario, we have modeled the server data set as a weighted directed graph where nodes represent server data units and arc weights represent the strength of the dependency. The scheduling problem becomes the following question in combinatorial optimization: given a weighted

* 10900 Euclid Avenue, Cleveland, Ohio 44106-7221, USA. Ph: (216) 368 4088, Fax: (216) 368 6039. This work has been supported in part under NSF grant ANI-0123929.

P. Widmayer et al. (Eds.): ICALP 2002, LNCS 2380, pp. 1054–1065, 2002.

directed graph $G = (N, A)$, arrange the nodes N around a circle so as to minimize the total weighted arc length [22]. We call such question the *directed circular arrangement problem* and show that it is NP-hard. The main objectives of this paper are to

- Present the first polynomial-time approximation algorithms for the circular arrangement problem, and
- Measure algorithm performance on graphs obtained from the workload of a heavily loaded Web site.

Arrangements: Circular and Linear. A problem related to circular arrangements is that of finding *linear arrangements*, where the graph is to be arranged along a line (rather than a circle) so as to minimize the total weighted length of the arrangement. If the graph is directed, it is further required that the graph be acyclic and the linear arrangement be a topological ordering. The linear arrangement problem is NP-hard [15] and can be approximated to within an $O(\log n)$ factor [24]. Furthermore, the minimum linear arrangement problem admits a polynomial-time approximation scheme if the graph is dense and weights are uniform [7]. The linear arrangement problem naturally suggests its analogous on a circle, but, in spite of superficial similarities, the two problems are unrelated as far as the approximation ratio is concerned. Specifically, an optimum linear arrangement can cost $\Omega(n)$ times as much as a circular arrangement of the same directed graph, while an $O(n)$-approximation algorithm for circular arrangements is trivial [22]. Consequently, the approximation of the two problems is in general unrelated. Indeed, the circular arrangement problem poses a specific technical issue that we discuss next.

Technical Issues. The circular arrangement problem presents two main technical obstacles. The first obstacle is the weakness of the linear relaxation of the problem. Such hurdle is common to other sequencing problems such as linear arrangements [8]. The second difficulty is specific to circular arrangements and is due to the weakness of the divide-and-conquer approach (e.g., [14,24]). Indeed, we will argue that the "divide" cost does not depend only on the problem partition, but also on how each of the individual subproblems is solved. In particular, an arc will "cross" to the other graph component depending not only on how the top level separator is chosen but also on how the nodes are arranged within one of the two components. Due to this issue, we will not be able to use the known divide-and-conquer paradigm directly and we will provide a different approach to partition the original instance into subinstances.

Our Results. In order to attack the directed circular arrangement problem, we will first consider a version of the problem on undirected graphs. As it turns out, the undirected version of the problem is simpler in that it can be reduced to linear arrangements. More to the point, the reduction highlights certain structural properties that will be used for the directed version of the problem. Our result on undirected graphs is that

Theorem 1. *There is a polynomial-time $O(\log n)$-approximation algorithm for the optimum undirected circular arrangement problem.*

The directed version of the problem is the one that is directly applicable to our multicast application. The directed problem will not be reduced to linear arrangements, and, actually, the directed circular arrangement problem does not use in any way results for the linear arrangement problem. However, the algorithm will use certain structural properties that we originally proved for the undirected case. We obtain that

Theorem 2. *There is a polynomial-time $\widetilde{O}(\sqrt{n})$-approximation algorithm for the optimum directed circular arrangement problem.*

Finally, we will show that a simplification of the algorithm in Theorem 2 has better performance than previous heuristics on graphs obtained from a busy Web server log.

Contents. The paper is organized as follows. Section 2 contains our results on undirected graphs and section 3 contains our results on directed graphs. Section 4 reports on our algorithm engineering and on our experimental results in the context of multicast-based data dissemination.

2   Undirected Graphs

We first consider the decision version of the optimum undirected circular arrangement problem, which we call the *Undirected Circular Arrangement Problem* (CA).

Instance: An undirected graph $G = (V, E)$, positive arc weights $w(e) \in \mathbb{N}$ for each $e \in A$, and a positive integer K.
Question: Is there a one-to-one function $f : V \to \{0, 1, \ldots, n-1\}$ such that $\sum_{e \in E} w(e)h(e) \leq K$, where $n = |V|$ and $h(\{u, v\}) = \min\{(f(v) - f(u)) \bmod n, (f(u) - f(v)) \bmod n\}$?

Proposition 1. *The Undirected Circular Arrangement Problem is NP-complete.*

Proof (Sketch). The proof is a reduction from the optimal linear arrangement problem (GT42) [15]. $\square$

We consider the minimum circular arrangement problem where we seek a solution that minimizes K. We begin with some definitions. Henceforth, $m = |E|$. If $X \subseteq E$ is a set of edges, then we define $w(X) = \sum_{e \in X} w(e)$. We will fix the *canonical orientation* of $\{u, v\} \in E$ in the circular arrangement f to be (u, v) if $(f(v) - f(u)) \bmod n < (f(u) - f(v)) \bmod n$. If $(f(v) - f(u)) \bmod n = (f(u) - f(v)) \bmod n$, we fix arc orientation arbitrarily. By definition, $h((u, v)) = (f(v) - f(u)) \bmod n$. We will say that an edge (u, v) *crosses* $x \in V$ in an arrangement f if $(f(x) - f(u)) \bmod n < h(e)$. Hence, $e = (u, v)$ crosses exactly the $h(e)$ vertices in $C((u, v)) = \{x : (f(x) - f(u)) \bmod n < h(e)\}$.

Definition 1. *The nodes $v_1, v_2, \ldots, v_l$ are placed next to each other in a cirular arrangement f if $f(v_{i+1}) = (f(v_i) + 1) \bmod n$ $(1 \le i < l)$.*

Definition 2. *Let $G = (V, E)$ be a graph, $U \subseteq V$, and f a circular arrangement of G. A vertex $u \in U$ is the U-successor of a vertex v in the arrangement f if and only if $(f(u) - f(v)) \bmod n \le (f(x) - f(v)) \bmod n$ for all $x \in U$.*

The U-successor of node v in f will be denoted by $s_{f,U}(v)$, or simply as $s(v)$ when f and U are clear from the context. It is immediate to see that $s(v)$ is unique and belongs to U.

Definition 3. *Let $G = (V, E)$ be a graph, $U \subseteq V$, and f a circular arrangement of G. A vertex $u \in U$ is the U-predecessor of a vertex v in the arrangement f if and only if $v = s_{f,U}^{-1}(s_{f,U}(v))$.*

The following set $I(v)$ basically represents all vertices between v and its successor $s(v)$.

Definition 4. *Let $G = (V, E)$ and $v \in V$. Let $d_{f,U}(v) = (f(s(v)) - f(v)) \bmod n$ and $I_{f,U}(v) = \{u : 0 < (f(u) - f(v)) \bmod n < d(v)\}$.*

Subscripts will be omitted when f and U are clear from the context. The following properties are easily proven:

Lemma 1. *The following properties hold on $d(v)$ and $I(v)$ for any f and U:*

1. *$d(v) = |I(v)| + 1$*
2. *$I(v) \cap U = \emptyset$*
3. *For any $U \subseteq V$, $\sum_{v \in U} d(v) = n$.*
4. *$I(v) \cap I(u) = \emptyset$ for all $u, v \in U$, $u \ne v$.*

Definition 5. *A graph $G = (V, E)$ is said to be the unbiased union of graphs $G_1 = (V_1, E_1), G_2 = (V_2, E_2), \ldots, G_k = (V_k, E_k)$ if and only if G is the union of $G_1, G_2, \ldots, G_k$, $V_i \cap V_j = \emptyset$, and $|V_i| \le |V|/2$ $(1 \le i, j \le k, i \ne j)$.*

Lemma 2. *Let $G = (V, E)$ be the unbiased union of $G_1 = (V_1, E_1), G_2 = (V_2, E_2), \ldots, G_k = (V_k, E_k)$ and f any circular arrangement of G. If an edge $e \in E_j$ crosses in f a vertex $v \in V_j$, then e crosses all vertices in $I_{f,V_j}(v)$.*

Lemma 3. *Let $G = (V, E)$ be the unbiased union of $G_1 = (V_1, E_1), G_2 = (V_2, E_2), \ldots, G_k = (V_k, E_k)$. Let $C_j(e) = C(e) \cap V_j$. Let $\rho_v^j = \sum_{e \in E_j : v \in C_j(e)} w(e)$ be the total weight of edges in E_j that cross $v \in V_j$. Then, the cost of the circular arrangement is at least $\sum_{j=1}^k \sum_{v \in V_j} \rho_v^j d(v)$.*

Proof (Sketch). The proof applies Lemma 1 and 2. □

The following lemma is critical in order to compare algorithm and optimum costs.

Lemma 4. *Let $G = (V, E)$ be the unbiased union of $G_1 = (V_1, E_1), G_2 = (V_2, E_2), \ldots, G_k = (V_k, E_k)$. Then, any minimum circular arrangement places the vertices of V_j next to each other $(1 \le j \le k)$.*

We define a β-separator of $G = (V, E)$ as a partition of V into V_1 and V_2 with the property that $\min\{|V_1|, |V_2|\} \geq \beta n$. A merely technical assumption that we will use later on is that since $\lfloor n/2 \rfloor \geq \min\{|V_1|, |V_2|\} \geq \lceil \beta n \rceil$, we can assume that $\lfloor n/2 \rfloor \geq \lceil \beta n \rceil$. For example, if $\beta = 2/5$, then $n \geq 4$. The cost of the separator is defined as the cost of the edges that cross from V_1 to V_2. An optimum β-separator has minimum cost among all β-separators. The problem of finding a β-separator of small cost is NP-hard even when all edges have the same weight [21].

Theorem 3 ([21]). *There exists a polynomial-time algorithm that, given a weighted, undirected graph G, finds a $(1/3)$-separator whose cost is $\alpha = O(\log n)$ times the cost of an optimum $(2/5)$-separator.*

Lemma 5. *Let G be a graph with more than $n = 4$ vertices. Let s^* be the cost of an optimum $(2/5)$-separator and a^* the minimum cost of a circular arrangement. Then, $s^* \leq (2a^*)/n$.*

A component of our algorithm will be an approximate solution for the *minimum linear arrangement* problem: given an undirected graph $G = (V, E)$ with positive integer edge weights $w(e)$ for all $e \in E$, find a one-to-one correspondence $f : V \to \{1, 2, \ldots, |V|\}$ that minimizes $\sum_{e=\{u,v\}\in E} w(e)|f(v) - f(u)|$.

We can now state our approximation algorithm. First, we find a $(1/3)$-separator as in Theorem 3, and denote by H_1 the smallest of the two sets of vertices. We remove H_1 and all edges incident on H_1, and find a $(1/3)$-separator for the rest of the graph. We call the smallest vertex set of this second separator H_2 and let $H_3 = V - H_1 - H_2$. The decomposition is illustrated in Fig. 1.

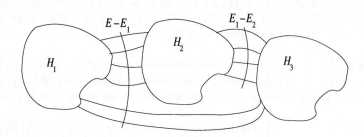

Fig. 1. Decomposition of a graph into the components H_1, H_2, and H_3.

Observe that $|H_3|/n \leq (2/3)^2 < 1/2$. We now run the $O(\log n)$-approximation algorithm for minimum linear arrangement independently on H_1, H_2, H_3 to obtain linear orderings f_1, f_2, f_3. We finally patch the three linear orderings together:

$$f(v) = \begin{cases} f_1(v) - 1 & \text{if } v \in H_1 \\ f_2(v) + |H_1| - 1 & \text{if } v \in H_2 \\ f_3(v) + |H_1| + |H_2| - 1 & \text{otherwise} \end{cases}.$$

The analysis of the algorithm is based on the following definitions and general considerations.

Definition 6. *If $X \subseteq A$ is a set of arcs and f a circular arrangement, then the contribution of X to the cost of f is $c_f(X) = \sum_{e \in X} w(e)h(e)$.*

The subscript f will be omitted when the circular arrangement is clear from the context. The main idea of the analysis is as follows. Consider a subgraph $G' = (V, E')$ of $G = (V, E)$ on the same vertex set V. Then, the minimum cicular arrangement of G' does not cost more than the cost a^* of the minimum circular arrangement of G. Our algorithm considers a sequence of graphs $G_0, G_1 = (V, E_1), \ldots, G_3 = (V, E_3)$. We define $G_0 = G$, $G_1 = (V, E_1)$ as the graph obtained from G by removing edges that are incident on both H_1 and $V - H_1$, $G_2 = (V, E_2)$ as the graph obtained by removing from E_1 edges that are incident on both H_2 and H_3, and $G_3 = (V, \emptyset)$ (see Fig. 1). The decomposition has the properties that $E_i \subset E_{i+1}$ ($0 \le i < k$). We will show that $c_f(E_i - E_{i+1}) \le \alpha a^*$ for some $\alpha = O(\log n)$. As a result, the algorithm achieves an $O(\log n)$ approximation factor.

Lemma 6. *The cost a^* of the optimum circular arrangement is at least equal to the sum of the costs of the optimum linear arrangements of H_1, H_2, H_3 (i.e., $c(E_2 - E_3) \le a^*$).*

Lemma 7. *For the edge sets E_0 and E_1 defined above, $c(E_0 - E_1) \le \alpha a^*$ for some $\alpha = O(\log n)$.*

Lemma 8. *For the edge sets E_1 and E_2 defined above, $c(E_1 - E_2) \le \alpha a^*$ for some $\alpha = O(\log n)$.*

We can now prove the main result of this section:

Proof (Theorem 1). By summing up the identities in the three previous lemmata, we obtain that $c(E) = c(E - E_1) + c(E_1 - E_2) + c(E_2) = \alpha a^*$, for some $\alpha = O(\log n)$, which proves the theorem. □

3   Directed Graphs

We consider the following decision version of the optimum circular arrangement problem, which we call the *Directed Circular Arrangement Problem* (DCA):

Instance: A directed graph $G = (N, A)$, non-negative arc weights $w(e) \in \mathbf{N}$ for each $e \in A$, and a positive integer K.

Question: Is there a one-to-one function $f : N \to \{0, 1, \ldots, |N| - 1\}$ such that, for $n = |N|$, $\sum_{e=(u,v) \in A} w(e)h(e) \le K$, where $n = |N|$ and $h((u,v)) = (f(v) - f(u)) \bmod n$?

We consider the optimum circular arrangement problem where we seek a solution that minimizes K. In order to simplify our notation, we make the following assumptions without loss of generality. We assume that G contains no loops as they do not contribute to the solution cost, and that $m = |A| = n(n-1)$. If $X \subseteq A$ is a set of arcs, then we define $w(X) = \sum_{e \in X} w(e)$. We will say that an arc (u, v) crosses node $x \in N$ if $(f(x) - f(u)) \bmod n < (f(v) - f(u)) \bmod n$. Hence, arc $e = (u, v)$ crosses exactly the $h(e)$ vertices in $C((u, v)) = \{x : (f(x) - f(u)) \bmod n < (f(v) - f(u)) \bmod n\}$.

As discussed in the introduction, the divide-and-conquer approach (e.g., [14, 24]) fails in that the "divide" cost depends on how the individual subproblems are solved. Specifically, suppose that the directed circular arrangement problem on a graph G is decomposed to the subgraphs $H_1, H_2, \ldots, H_k$. In the divide-and-conquer approach, we would ordinarily divide the cost into arcs that cross between two H_i's, and arcs that have both endpoints within a single H_i. Consider an arc (u, v) with u, v belonging to the same subgraph H_i. If we arrange $f(v) = (f(u) + 1) \bmod n$, then indeed arc (u, v) crosses only nodes in H_i. However, if $f(u) = (f(v) + 1) \bmod n$, then (u, v) can cross nodes that are not in H_i. As a result, the contribution of an arc depends not only on the node set partition, but also on how nodes are arranged within a single partition. Thus, this fact is a major obstacle to a divide-and-conquer approach. Our solution will balance partitions so as to sidestep this problem.

Definition 7 ([4]). *Let $G = (N, A)$ be a directed graph. The* head *of an arc (u, v) is node v and the* tail *is node u.*

Lemma 9. *Suppose that e crosses u, and let u be the node that e does not cross and that minimizes the quantity $(f(u) - f(v)) \bmod n$. Then, v is e's head.*

Lemma 10. *The cost of a circular arrangement f is equal to $\sum_{v \in V} \rho_v$, where $\rho_v = \sum_{e : v \in C(e)} w(e)$ is the total weight of arcs crossing v.*

Lemma 11. *Let $G = (N, A)$ be a directed graph, and assume that the node set N can be partitioned into $N_1, N_2, \ldots, N_k$ with the property that for $e \in A$ there is a unique N_j $(1 \leq j \leq k)$ on which e is incident. Then, there is a minimum circular arrangement that places the vertices of N_j next to each other $(1 \leq j \leq k)$.*

The major differences between Lemma 4 and Lemma 11 are that $|N_j|$ can be larger than $n/2$, and that a property of one optimum arrangement is described, rather than a property of all optimum arrangements.

Proof (Sketch). The proof is similar to that of Lemma 4, with the following difference. The arcs that cross v_j do not cross only nodes of N_j in g, but have to circumnavigate around the nodes in $N - N_j$. □

A first component of our algorithm is a polynomial-time approximation algorithm for the *minimum feedback arc set* problem: given a directed graph $G = (N, A)$ with non-negative integer arc weights $w(e)$ for all $e \in A$, find an arc set of minimum total weight that intersects every directed cycle in the graph G. The minimum feedback arc set problem is APX-complete [18] and has a polynomial-time algorithm that approximates the optimum solution to within an $O(\log n \log \log n)$ factor [13]. Another component of our algorithm is expressed by the following lemma, which will give the termination condition of a sequence of separator computations.

Lemma 12. *Let* $G = (N, A)$ *be a directed acyclic graph and assume that the node set* N *can be partitioned into* $N_1, N_2, \ldots, N_k$ *with the property that for* $e \in A$ *there is a unique* N_j *(* $1 \leq j \leq k$ *) on which* e *is incident. Let* A_j *be the set of arcs incidents on* N_j *and consider the cost* ℓ_j *of any linear arrangement of the graph* (N_j, A_j)*. Then,* $\ell_j \leq (|N_j| - 1)c_{f^*}(A_j)$*.*

The main ideas of the proof are as follows. Consider a subgraph $G' = (N, A')$ of $G = (N, A)$ on the same node set N. Then, the minimum circular arrangement of G' costs no more than the minimum circular arrangement of G. Our algorithm will then consider a sequence of graphs $G_0, G_1 = (N, A_1), \ldots, G_k = (N, A_k)$ with the properties that $G_0 = G$, that $A_{i+1} \subset A_i$ $(0 \leq i < k)$, that $A_k = \emptyset$, and that $c_f(A_i - A_{i+1}) \leq \beta a^*$ for some $\beta \in \widetilde{O}(\sqrt{n})$ and where a^* is the minimum cost of a circular arrangement of G. As a result, the algorithm achieves an $\widetilde{O}(k\sqrt{n})$ approximation factor. Unlike the undirected case, we cannot show a decomposition with $k = O(1)$. Instead, we will remove a feedback arc set to make G acyclic, then we will keep partitioning the graph until the size of each connected components is so small that we can arrange those components with any topological ordering. Such decomposition guarantees a value of $k = O(\log n)$, so that the approximation factor will be $\widetilde{O}(\sqrt{n})$. The graph $G_i = (N, A_i)$ will be partitioned into components $X_1, X_2, \ldots, X_p \subseteq N$ with the property that no arc is incident on more than one X_j. In other words, the X_j's contain weakly connected components of G_i. The cost of a circular arrangement of G_i is the sum of the costs due to the X_j's: $c_f(A_i) = \sum_{j=1}^{p} \sum_{e=(x,y):x,y \in X_j} w(e)h(e)$. For the purpose of the analysis, we will sometimes consider a subgraph G'' obtained by removing nodes from a G_i along with all incident arcs and observe that a minimum circular arrangement of G'' costs no more than that of G_i. We will also need the following definition.

Definition 8. *Given a complete weighted directed graph* $G = (N, A)$*, the graph obtained by removing arc orientation from* G *is the complete weighted undirected graph* $G' = (N, E)$ *with weights* $w(\{u, v\}) = w((u, v)) + w((v, u))$ *(* $u, v \in N$*,* $u \neq v$*).*

We can now state our approximation algorithm. In the first step, we remove arc orientation from G. Then, we compute H_1, H_2, H_3 exactly as in the algorithm for undirected graph. At this point, $|H_1|, |H_2|, |H_3| < n/2$. Let G_1 be the directed graph obtained by removing from G arcs incident on both H_1 and $N - H_1$ and G_2 be the directed graph obtained by removing from G_1 arc incident on both H_2 and $N - H_2$. Observe that H_1, H_2, H_3 have the property that they contain weakly connected components of G_2. Our algorithm then computes a feedback arc set for G_2 with an $O(\log n \log \log n)$-approximation algorithm [13], removes the feedback arcs, and obtains a graph G_3. At this point, G_3 is acyclic and we start a procedure to be described below that finds $(1/3)$-separators for the H_j's while component size is bigger than a parameter $\theta = \sqrt{n \log n}$. At any step of this recursive procedure, the graph $G_i = (N, A_i)$ has the properties that the node set N can be partitioned into $X_1, X_2, \ldots, X_p$ and that arcs are incident on a unique X_j. In other words, the X_j's have the property that they contain the weakly connected components of G_i. Initially, G_3 has $X_j = H_j$ $(j = 1, 2, 3)$. To obtain G_{i+1} from G_i, we remove arc orientation from G_i, and for each X_j, if $|X_j| \geq \theta$, we find a $(1/3)$-separator of the undirected subgraph induced by X_j, and correspondingly break X_j into X_{j1} and X_{j2}. Thus, the previous X_j's contain the new X_j's. The procedure continues until $|X_j| < \theta$ $(1 \leq j \leq p)$.

Lemma 13. *Let $G = (N, A)$ be a directed acyclic graph and assume that the node set N can be partitioned into $N_1, N_2, \ldots, N_k$ with the property that for $e \in A$ there is a unique N_j $(1 \leq j \leq k)$ on which e is incident. Let A_j be the set of arcs incident of N_j. Then, $c_f(A) = \sum_{j=1}^{k} c_f(A_j)$.*

Lemma 14. *The total contribution to the circular arrangement cost of the arcs that have endpoints in different H_j's is no more than an $O(\log n)$ factor away from the cost a^* of an optimum circular arrangement (i.e., $c_f(A - A_2) \leq \alpha a^*$ for some $\alpha = O(\log n)$).*

Proof (Sketch). Similar to the proof of Theorem 1, except that $h(e)$ can be as high as $n - 1$. However, this modification only doubles the constant factor.   □

Lemma 15. *The total contributions to the circular arrangement cost of the arcs in the feedback arc set is no more than an $O(\log n \log \log n)$-factor away from the cost a^* of an optimum circular arrangement (i.e., $c_f(A_2 - A_3) \leq \gamma a^*$ for some $\gamma = O(\log n \log \log n)$).*

We now start with the DAG G_2 and, for each of components, H_1, H_2, H_3, we compute an approximate $(1/3)$-separator. We continue breaking the components with approximate $(1/3)$-separators until component size is no more than a threshold $\theta = \sqrt{n \log n}$, at which point we arrange nodes within each component in topological order.

Lemma 16. *Let G be a directed graph with more than $n = 4$ vertices and G' be undirected graph obtained from G by removing arc orientation. Let s^* be the cost of an optimum $(2/5)$-separator of G' and a^* the minimum cost of a circular arrangement of G. Then, $s^* \leq (2a^*)/n$.*

Corollary 1. *Let $G = (N, A)$ be a weighted directed graph with a node subset $H \subseteq N$ such that $|H| \geq 4$ and there is no arc that is incident on both H and $N - H$. Let A' be the set of arcs incident on H. Let $G' = (H, E')$ be the undirected subgraph obtained by removing arc orientation from the subgraph of G induced by H. Let s^* be the cost of an optimum $(2/5)$-separator of G'. Then, $s^* \leq 2c_{f^*}(A')/|H|$.*

Lemma 17. *The arcs eliminated during each stage of the procedure above contribute to the arrangement cost no more than $O(\sqrt{n \log n})$ times the optimum cost a^* of a circular arrangement (i.e., $c_f(A_i - A_{i+1}) \leq \sqrt{n \log n}$).*

Proof (Theorem 2). The discussion above and Lemmata 14, 15, and 17 yield a $\tilde{O}(k\sqrt{n})$-aproximation algorithm. Since $(1/3)$-separators are guaranteed to split node subsets into subset containing at least $1/3$ of the original number of nodes, the splitting procedure terminates after $k = O(\log \theta) = O(\log n)$ steps, thus yielding the theorem.   □

4   Algorithm Engineering

We have implemented a fast version of our algorithm for directed graphs as well as three previous algorithms for the same problem. The approximation algorithm in section 3 requires the computation of feedback arc sets and balanced separators, which in turn require the solution of multicommodity flow problems and the computation of spreading metrics. Furthermore, spreading metrics require the solution of several linear programs with an exponential number of constrains by means of the ellipsoid method [13,16]. Thus, we do not believe that the algorithm can be practical in its original form and we made the following simplifications. First, we do not directly compute balanced separators. Instead, we find all the minimum $s - t$ cuts in the graph, which results in a separator tree [10]. Then, we start from the largest edge in the separator tree and we repeatedly add the largest incident edge to the current component until we isolate a set of vertices of the desired size. By construction, the resulting cut is a balanced separator and the cut value should not be too large as the separator edges have greedily joined the chosen component. Thus, we believe that the proposed heuristic is a reasonable way to compute a separator. The second simplification is in the calculation of the feedback arc set, where we replace an approximation algorithm with a greedy heuristic [12]. The resulting algorithm is denoted as *Bsep*. In addition to the Bsep algorithm, we experimented with the following three algorithms. The first algorithm (*Rnd*) is a random ordering of the graph nodes. The second algorithm is based on maximum spanning trees (*MST*) to cluster nodes close to each other [22]. The final algorithm originates from the observation that the heuristic for feedback arc set computes a linear ordering of the underlying graph and then discards all arcs that violate the topological ordering, that is, all arcs that link higher-numbered vertices to lower-numbered vertices [12]. Thus, such heuristic can also be used to obtain directly a circular arrangement of the original graph, and it is the last algorithm that we consider in this paper (*Fesh*). We also considered the application of integer linear programming to solve the circular arrangement problem exactly. However, after two weeks of computation on a Sun Ultra60, Cplex had not found any integer solution other than the one that we manually inserted as a starting point, even when clique cuts and ordered sets were added. We explain such behavior by noticing that the linear relaxation of the integer formulation places an equal fraction of every node in each position of the circular arrangement, and that branch-and-cut algorithms suffer when the relaxation does not effectively direct the selection process.

We have tested the algorithms in the context of multicast-based data dissemination on directed graphs obtained from the access pattern to the Web server of the World Cup 98. The World Cup trace includes more than one billion requests over a period of 1 1/2 month and is one of the largest trace analyzed to date [6]. Furthermore, the World Cup servers received up to 10 million requests per hour. As a result, the World Cup site is one of the most busy recorded so far, which makes it an ideal testbed for multicast data dissemination. Additional information on this workload can be found in [6,22].

The four algorithms are compared in table 1. The Rnd cost is roughly $n/2$,

Table 1. Cost of the circular arrangement returned by the four algorithms. Additionally, the worst-case cost on these graphs is $n-1$, where n is the number of nodes in the graph. Thirty random arrangements were tried, and we report their cost in the form of average $\pm$ standard deviation.

	Graph 1	Graph 2	Graph 3	Graph 4
n	225	174	206	200
Rnd	112.2294 ± 0.7440	87.0159 ± 1.2416	102.7072 ± 0.5072	100.0331 ± 1.5135
MST	107.26	79.6681	97.6602	103.693
Fesh	99.0344	88.6276	98.052	103.619
Bsep	97.1283	82.2139	89.5094	95.4161

and in most cases the other algorithms improve over it. The MST and Fesh algorithm had comparable cost on graph 3 and 4, but MST is better than Fesh on graph 1 and Fesh is better than MST on graph 2. Thus, we believe that in general the performance of MST and Fesh is roughly comparable. Finally, the Bsep algorithm had the best performance in three graphs out of four, and on the other graph, its performance trails the best algorithm by 3%. Thus, we believe that Bsep is overall the best algorithm in terms of circular arrangement cost. However, Bsep is slower than the other algorithms even with all the simplifications above. For example, Bsep took 17.3s on an unloaded Ultra60 for graph 1 while Fesh took only 1s on the same machine and with the same compiler (g++ -O).

References

1. http://www.digitalfountain.com/.
2. http://www.hns.com/.
3. http://www.panamsat.com/.
4. Ravindra K. Ahuja, Thomas L. Magnanti, and James B. Orlin. *Network flows.* Prentice Hall Inc., Englewood Cliffs, NJ, 1993. Theory, algorithms, and applications.
5. K. C. Almeroth, M. H. Ammar, and Z. Fei. Scalable delivery of Web pages using cyclic best-effort (UDP) multicast. In *Proceedings of the Seventeenth Annual Joint Conference of the IEEE Computer and Communications Societies (INFO-COM 1998)*, 1998.

6. Martin Arlitt and Tai Jin. Workload characterization of the 1998 World Cup web site. Technical Report HPL-1999-35R1, HP Labs, 1999.

7. Sanjeev Arora, Alan Frieze, and Haim Kaplan. A new rounding procedure for the assignment problem with applications to dense graph arrangement problems. In *37th Annual Symposium on Foundations of Computer Science (Burlington, VT, 1996)*, pages 21–30. IEEE Comput. Sec. Press, Los Alamitos, CA, 1996.

8. Kenneth R. Baker. *Introduction to sequencing and scheduling*. Wiley, New York, 1974.

9. John W. Byers, Michael Luby, Michael Mitzenmacher, and Ashutosh Rege. A digital fountain approach to reliable distribution of bulk data. In *Proc. Sigcomm*, 1998.

10. C. K. Cheng and T. C. Hu. Ancestor tree for arbitrary multi-terminal cut functions. *Ann. Oper. Res.*, 33(1–4):199–213, 1991. Topological network design (Copenhagen, 1989).

11. Panos K. Chrysanthis, Vincenzo Liberatore, and Kirk Pruhs. Middleware support for multicast-based data dissemination: A working reality. White Paper, 2001.

12. William W. Cohen, Robert E. Schapire, and Yoram Singer. Learning to Order things. *Journal of Artificial Intelligence Research*, 10:243–270, 1999.

13. G. Even, J. Naor, B. Schieber, and M. Sudan. Approximating minimum feedback sets and multicuts in directed graphs. *Algorithmica*, 20(2):151–174, 1998.

14. Guy Even, Joseph (Seffi) Naor, Satish Rao, and Baruch Schieber. Divide-and-conquer approximation algorithms via spreading metrics. In *Proceedings of the 36th Annual Symposium on Foundations of Computer Science*, pages 62–71, October 1995.

15. Michael R. Garey and David S. Johnson. *Computers and intractability*. W. H. Freeman and Co., San Francisco, Calif., 1979. A guide to the theory of NP-completeness, A Series of Books in the Mathematical Sciences.

16. L. G. Hačijan. A polynomial algorithm in linear programming. *Dokl. Akad. Nauk SSSR*, 244(5):1093–1096, 1979.

17. Gary Herman, Gita Gopal, K. C. Lee, and Abel Weinrib. The datacycle architecture for very high throughput database systems. In *Proceedings of the 1987 ACM SIGMOD Conference International Conference on Management of Data*, pages 97–103, 1987.

18. V. Kann. *On the Approximability of NP-complete Optimization Problems*. PhD thesis, Royal Institute of Technology, Stockholm, 1992.

19. Claire Kenyon, Nicolas Schabanel, and Neal Young. Polynomial-time approximation scheme for data broadcast. In *Proceedings of the Thirtisecond ACM Symposium on the Theory of Cornputing*, 2000.

20. Sanjeev Khanna and Vincenzo Liberatore. On broadcast disk paging. *SIAM Journal on Computing*, 29(5):1683–1702, 2000.

21. Tom Leighton and Satish Rao. Multicommodity max-flow min-cut theorems and their use in designing approximation algorithms. *J. ACM*, 46(6):787–832, 1999.

22. Vincenzo Liberatore. Multicast scheduling for list requests. In *21st Annual Joint Conference of the IEEE Computer und Communications Societies (INFOCOM 2002)*, 2002. To appear.

23. Larry L. Peterson and Bruce S. Davie. *Computer Networks*. Morgan Kaufmann, 2000.

24. Satish Rao and Andréa W. Richa. New approximation techniques for some ordering Problems. In *Proceedings of the Ninth Annual ACM-SIAM Symposium on Discrete Algorithms (San Francisco, CA, 1998)*, pages 211–218, New York, 1998. ACM.

25. W. Richard Stevens. *Unix Network Programming*. PTR PH, 1998.

Author Index

Lecture Notes in Computer Science

For information about Vols. 1–2293
please contact your bookseller or Springer-Verlag

Vol. 2330: P.M.A. Sloot, C.J.K. Tan, J.J. Dongarra, A.G. Hoekstra (Eds.), Computational Science – ICCS 2002. Proceedings, Part II. XLI, 1115 pages. 2002.

Vol. 2331: P.M.A. Sloot, C.J.K. Tan, J.J. Dongarra, A.G. Hoekstra (Eds.), Computational Science – ICCS 2002. Proceedings, Part III. XLI, 1227 pages. 2002.

Vol. 2332: L. Knudsen (Ed.), Advances in Cryptology – EUROCRYPT 2002. Proceedings, 2002. XII, 547 pages. 2002.

Vol. 2333: J.-J.C. Meyer, M. Tambe (Eds.), Intelligent Agents VIII. Revised Papers, 2001. XI, 461 pages. 2001. (Subseries LNAI).

Vol. 2334: G. Carle, M. Zitterbart (Eds.), Protocols for High Speed Networks. Proceedings, 2002. X, 267 pages. 2002.

Vol. 2335: M. Butler, L. Petre, K. Sere (Eds.), Integrated Formal Methods. Proceedings, 2002. X, 401 pages. 2002.

Vol. 2336: M.-S. Chen, P.S. Yu, B. Liu (Eds.), Advances in Knowledge Discovery and Data Mining. Proceedings, 2002. XIII, 568 pages. 2002. (Subseries LNAI).

Vol. 2337: W.J. Cook, A.S. Schulz (Eds.), Integer Programming and Combinatorial Optimization. Proceedings, 2002. XI, 487 pages. 2002.

Vol. 2338: R. Cohen, B. Spencer (Eds.), Advances in Artificial Intelligence. Proceedings, 2002. X, 197 pages. 2002. (Subseries LNAI).

Vol. 2340: N. Jonoska, N.C. Seeman (Eds.), DNA Computing. Proceedings, 2001. XI, 392 pages. 2002.

Vol. 2342: I. Horrocks, J. Hendler (Eds.), The Semantic Web – ISCW 2002. Proceedings, 2002. XVI, 476 pages. 2002.

Vol. 2345: E. Gregori, M. Conti, A.T. Campbell, G. Omidyar, M. Zukerman (Eds.), NETWORKING 2002. Proceedings, 2002. XXVI, 1256 pages. 2002.

Vol. 2346: H. Unger, T. Böhme, A. Mikler (Eds.), Innovative Internet Computing Systems. Proceedings, 2002. VIII, 251 pages. 2002.

Vol. 2347: P. De Bra, P. Brusilovsky, R. Conejo (Eds.), Adaptive Hypermedia and Adaptive Web-Based Systems. Proceedings, 2002. XV, 615 pages. 2002.

Vol. 2348: A. Banks Pidduck, J. Mylopoulos, C.C. Woo, M. Tamer Ozsu (Eds.), Advanced Information Systems Engineering. Proceedings, 2002. XIV, 799 pages. 2002.

Vol. 2349: J. Kontio, R. Conradi (Eds.), Software Quality – ECSQ 2002. Proceedings, 2002. XIV, 363 pages. 2002.

Vol. 2350: A. Heyden, G. Sparr, M. Nielsen, P. Johansen (Eds.), Computer Vision – ECCV 2002. Proceedings, Part I. XXVIII, 817 pages. 2002.

Vol. 2351: A. Heyden, G. Sparr, M. Nielsen, P. Johansen (Eds.), Computer Vision – ECCV 2002. Proceedings, Part II. XXVIII, 903 pages. 2002.

Vol. 2352: A. Heyden, G. Sparr, M. Nielsen, P. Johansen (Eds.), Computer Vision – ECCV 2002. Proceedings, Part III. XXVIII, 919 pages. 2002.

Vol. 2353: A. Heyden, G. Sparr, M. Nielsen, P. Johansen (Eds.), Computer Vision – ECCV 2002. Proceedings, Part IV. XXVIII, 841 pages. 2002.

Vol. 2355: M. Matsui (Ed.), Fast Software Encryption. Proceedings, 2001. VIII, 169 pages. 2001.

Vol. 2358: T. Hendtlass, M. Ali (Eds.), Developments in Applied Artificial Intelligence. Proceedings, 2002 XIII, 833 pages. 2002. (Subseries LNAI).

Vol. 2359: M. Tistarelli, J. Bigun, A.K. Jain (Eds.), Biometric Authentication. Proceedings, 2002. XII, 373 pages. 2002.

Vol. 2360: J. Esparza, C. Lakos (Eds.), Application and Theory of Petri Nets 2002. Proceedings, 2002. X, 445 pages. 2002.

Vol. 2361: J. Blieberger, A. Strohmeier (Eds.), Reliable Software Technologies – Ada-Europe 2002. Proceedings, 2002 XIII, 367 pages. 2002.

Vol. 2363: S.A. Cerri, G. Gouardères, F. Paraguaçu (Eds.), Intelligent Tutoring Systems. Proceedings, 2002. XXVIII, 1016 pages. 2002.

Vol. 2364: F. Roli, J. Kittler (Eds.), Multiple Classifier Systems. Proceedings, 2002. XI, 337 pages. 2002.

Vol. 2366: M.-S. Hacid, Z.W. Raś, D.A. Zighed, Y. Kodratoff (Eds.), Foundations of Intelligent Systems. Proceedings, 2002. XII, 614 pages. 2002. (Subseries LNAI).

Vol. 2367: J. Fagerholm, J. Haataja, J. Järvinen, M. Lyly. P. Råback, V. Savolainen (Eds.), Applied Parallel Computing. Proceedings, 2002. XIV, 612 pages. 2002.

Vol. 2368: M. Penttonen, E. Meineche Schmidt (Eds.), Algorithm Theory – SWAT 2002. Proceedings, 2002. XIV, 450 pages. 2002.

Vol. 2369: C. Fieker, D.R. Kohel (Eds.), Algebraic Number Theory. Proceedings, 2002. IX, 517 pages. 2002.

Vol. 2370: J. Bishop (Ed.), Component Deployment. Proceedings, 2002. XII, 269 pages. 2002.

Vol. 2373: A. Apostolico, M. Takeda (Eds.), Combinatorial Pattern Matching. Proceedings, 2002. VIII, 289 pages. 2002.

Vol. 2374: B. Magnusson (Ed.), ECOOP 2002 – Object-Oriented Programming. XI, 637 pages. 2002.

Vol. 2375: J. Kivinen, R.H. Sloan (Eds.), Computational Learning Theory. Proceedings, 2002. XI, 397 pages. 2002. (Subseries LNAI).

Vol. 2380: P. Widmayer, F. Triguero, R. Morales, M. Hennessy, S. Eidenbenz, R. Conejo (Eds.), Automata, Languages and Programming. Proceedings, 2002. XXI, 1069 pages. 2002.

Vol. 2382: A. Halevy, A. Gal (Eds.), Next Generation Information Technologies and Systems. Proceedings, 2002. VIII, 169 pages. 2002.

Vol. 2384: L. Batten, J. Seberry (Eds.), Information Security and Privacy. Proceedings, 2002. XII, 514 pages. 2002.

Vol. 2385: J. Calmet, B. Benhamou, O. Caprotti, L. Henocque, V. Sorge (Eds.), Artificial Intelligence, Automated Reasoning, and Symbolic Computation. Proceedings, 2002. XI, 343 pages. 2002. (Subseries LNAI).

Vol. 2386: E.A. Boiten, B. Möller (Eds.), Mathematics of Program Construction. Proceedings, 2002. X, 263 pages. 2002.

Vol. 2389: E. Ranchhod, N.J. Mamede (Eds.), Advances in Natural Language Processing. Proceedings, 2002. XII, 275 pages. 2002. (Subseries LNAI).